Lecture Notes in Computer Science 12654

Hisao Ishibuchi · Qingfu Zhang ·
Ran Cheng · Ke Li · Hui Li ·
Handing Wang · Aimin Zhou (Eds.)

Evolutionary Multi-Criterion Optimization

11th International Conference, EMO 2021
Shenzhen, China, March 28–31, 2021
Proceedings

Editors
Hisao Ishibuchi
Southern University of Science and Technology
Shenzhen, China

Ran Cheng
Southern University of Science and Technology
Shenzhen, China

Hui Li
Xi'an Jiaotong University
Xi'an, China

Aimin Zhou
East China Normal University
Shanghai, China

Qingfu Zhang
City University of Hong Kong
Kowloon Tong, China

Ke Li
University of Exeter
Exeter, UK

Handing Wang
Xidian University
Xi'an, China

ISSN 0302-9743 ISSN 1611-3349 (electronic)
Lecture Notes in Computer Science
ISBN 978-3-030-72061-2 ISBN 978-3-030-72062-9 (eBook)
https://doi.org/10.1007/978-3-030-72062-9

LNCS Sublibrary: SL1 – Theoretical Computer Science and General Issues

This Springer imprint is published by the registered company Springer Nature Switzerland AG
The registered company address is: Gewerbestrasse 11, 6330 Cham, Switzerland

Preface

The International Conference on Evolutionary Multi-Criterion Optimization (EMO) is a well-established, widely recognized, bi-annual international conference series devoted to both theoretical foundations and practical applications of the theme area: evolutionary multi-criterion optimization. The first EMO conference was born in Zürich (Switzerland) in 2001. Afterwards, it took place on different continents every two years, i.e., in Faro (Portugal), Guanajuato (Mexico), Matsushima (Japan), Nantes (France), Ouro Preto (Brazil), Sheffield (UK), Guimarães (Portugal) and East Lansing (USA).

The 11th International Conference on Evolutionary Multi-Criterion Optimization (EMO 2021) took place in Shenzhen, China, during March 28–31, 2021. It has been 14 years since EMO came to Asia and it is the first time that EMO has been located in China. Nowadays, China has the largest number of students, researchers and practitioners working at EMO. Shenzhen occupies an important position in China's high-tech industries, financial services, foreign trade exports, marine transportation, creative culture and other aspects. It shoulders the important mission of experiments and demonstration in China's institutional innovation and opening-up. Born in Shenzhen, SUSTech is an innovative university and has been recognized as a trailblazer in the Chinese higher education system. It aims to become an international high-level research university rapidly. SUSTech has the main disciplines of science, engineering and medicine, concurrently, some disciplines in the humanities and social sciences, all of which make it the perfect place to host EMO 2021 in its modernized building.

EMO 2021 received 120 submissions and a total of 61 papers were accepted for presentation and publication in this volume (acceptance rate around 51%), all of whose authors were invited to give an oral presentation. Given the COVID-19 pandemic, this EMO was special and maybe unique in the conference's history in that it took place in a mixed onsite and online format.

The conference profited from the presentations of four keynote speakers from both academia and high-profile industry: Akira Oyama (JAXA and University of Tokyo, Japan), Tapabrata Ray (University of New South Wales Canberra, Australia), Lixin Tang (Northeastern University, China) and Mark Harman (FACEBOOK and University of College London, UK). In addition, two tutorials were delivered about past, current and future developments of EMO and multi-criteria decision making from two pioneers in the field: Kalyanmoy Deb (Michigan State University, USA) and Kaisa Miettinen (University of Jyväskylä, Finland).

We express our gratitude to all the authors, reviewers, and organizers who made the conference a success. We are grateful to all the authors for submitting their best and latest work, to all the reviewers for the generous way they spent their time and provided their valuable expertise in preparing their reviews, to the program and publication chairs for their hard work in compiling an ambitious scientific program, to the keynote and tutorial speakers for delivering impressive talks, to the competition chairs for creating the framework for the HUAWEI Logistics Challenge, to the publication and publicity chairs for taking care of the EMO proceedings and website, and to the local organizers who helped to make EMO 2021 happen. Last but not least, we would like take this opportunity to appreciate the generous support from HUAWEI Noah's Ark Lab for sponsoring the Logistics Challenge.

March 2021

Hisao Ishibuchi
Qingfu Zhang

Organization

EMO 2021 was organized by the Department of Computer Science, Southern University of Science and Technology, Shenzhen, China.

General Chairs

Hisao Ishibuchi	Southern University of Science and Technology, China
Qingfu Zhang	City University of Hong Kong, Hong Kong SAR, China

Program Chairs

Hui Li	Xi'an Jiaotong University, China
Handing Wang	Xidian University, China

Publication Chairs

Ke Li	University of Exeter, UK
Aimin Zhou	East China Normal University, China

Organizing Chair

Ran Cheng	Southern University of Science and Technology, China

MCDM Chair

Kaisa Miettinen	University of Jyväskylä, Finland

Industrial Sessions Chair

Mingxuan Yuan	Noah's Ark Lab, Huawei, China

Publicity Chairs

Tea Tušar	Jožef Stefan Institute, Slovenia
Gregorio Toscano Pulido	Cinvestav Tamaulipas, Mexico
Hemant Kumar Singh	The University of New South Wales, Australia

Competition Chair

Zhenkun Wang	Southern University of Science and Technology, China

CoI Chair

Xingyi Zhang	Anhui University, China

Online Platform Co-chairs

Cheng He	Southern University of Science and Technology, China
Zhichao Lu	Southern University of Science and Technology, China

Web Chair

Hui Bai	Southern University of Science and Technology, China

Program Committee

Janos Abonyi	University of Pannonia, Hungary
Mohamed Abouhawwash	Mansoura University, Egypt
Shaukat Ali	Simula Research Laboratory, Norway
Richard Allmendinger	The University of Manchester, UK
Carlos Antunes	University of Coimbra, Portugal
Sunith Bandaru	University of Skövde, Sweden
Helio Barbosa	Laboratório Nacional de Computação Científica, Brazil
Slim Bechikh	University of Carthage, Tunisia
Julian Blank	Michigan State University, USA
Juergen Branke	The University of Warwick, UK
Dimo Brockhoff	Inria, France
Xinye Cai	Nanjing University of Aeronautics and Astronautics, China
Lei Chen	Guangdong University of Technology, China
Cheng He	Southern University of Science and Technology, China
Ran Cheng	Southern University of Science and Technology, China
Sung-Bae Cho	Yonsei University, Korea
João Clímaco	Universidade de Coimbra, Portugal
Carlos Coello Coello	CINVESTAV-IPN, Mexico
Lino Costa	University of Minho, Portugal
Kalyanmoy Deb	Michigan State University, USA
Alexandre Delbem	University of São Paulo, Brazil
Bilel Derbel	University of Lille, France
Clarisse Dhaenens	University of Lille, France
Michael Doumpos	Technical University of Crete, Greek
Matthias Ehrgott	Lancaster University, UK
Saber Elsayed	UNSW at Canberra, Australia
Michael Emmerich	Leiden University, Netherlands
Wang Fan	Dalian University of Technology, China
Jonathan Fieldsend	University of Exeter, UK
Guangtao Fu	University of Exeter, UK

Hongwei Ge	Dalian University of Technology, China
Dunwei Gong	China University of Mining and Technology, China
Wenyin Gong	China University of Geosciences, China
Erik Goodman	Michigan State University, USA
Salvatore Greco	University of Catania, Italy
Crina Grosan	Brunel University London, UK
Abhishek Gupta	Nanyang Technological University, Singapore
Julia Handl	University of Manchester, UK
Jin-Kao Hao	University of Angers, France
Wenjing Hong	University of Science and Technology of China, China
Min Jiang	Xiamen University, China
Shouyong Jiang	Newcastle University, UK
Yi Jun	Chongqing University of Science and Technology, China
Sandeep Kulkarni	Michigan State University, USA
Sam Kwong	City University of Hong Kong, China
Hao Li	Xidian University, China
Ke Li	University of Exeter, UK
Miqing Li	University of Birmingham, UK
Arnaud Liefooghe	University of Lille, France
Qiuzhen Lin	Shenzhen University, China
Jia Liu	Xidian University, China
Manuel López-Ibáñez	University of Manchester, UK
Mariano Luque	Universidad de Málaga, Spain
Yi Mei	Victoria University of Wellington, New Zealand
Efrén Mezura-Montes	University of Veracruz, Mexico
Martin Middendorf	University of Leipzig, Germany
Julian Molina	University of Málaga, Spain
Sanaz Mostaghim	Universität Magdeburg, Germany
Boris Naujoks	Cologne University of Applied Sciences, Germany
Antonio J. Nebro	University of Málaga, Spain
Frank Neumann	The University of Adelaide, Australia
Amos Ng	University of Skövde, Sweden
Bach Nguyen	Victoria University of Wellington, New Zealand
Ender Özcan	University of Nottingham, UK
Trinadh Reddy Pamulapati	Kyungpook National University, Korea
Linqiang Pan	Huazhong University of Science and Technology, China
Luís Paquete	University of Coimbra, Portugal
Jun Peng	Chongqing University of Science and Technology, China
Pasqualina Potena	RISE Research Institutes of Sweden, Sweden
Robin Purshouse	University of Sheffield, UK
Zhao Qi	Beijing University of Technology, China
Tapabrata Ray	University of New South Wales Canberra, Australia
Proteek Roy	KLA, USA

Günter Rudolph	TU Dortmund, Germany
Francisco Ruiz	University of Málaga, Spain
Sergio Santander-Jiménez	Universidade de Lisboa, Portugal
Marc Schoenauer	Inria Saclay Île-de-France, France
Oliver Schütze	CINVESTAV-IPN, Mexico
Haitham Seada	AltairPD/Ford Motor Company
Marc Sevaux	Université de Bretagne Sud, France
Deepak Sharma	Indian Institute of Technology Guwahati, India
Pradyumn Kumar Shukla	Karlsruhe Institute of Technology, Germany
Patrick Siarry	Université Paris-Est Créteil, France
Johannes Siebert	University of Bayreuth, Germany
Hemant Singh	The University of New South Wales Canberra, Australia
Michael Stiglmayr	University of Wuppertal, Germany
Chaoli Sun	Taiyuan University of Science and Technology, China
Liang Sun	Dalian University of Technology, China
Ricardo H. C. Takahashi	Universidade Federal de Minas Gerais, Brazil
Jürgen Teich	University of Erlangen-Nuremberg, Germany
Gregorio Toscano-Pulido	CINVESTAV-IPN, Mexico
Heike Trautmann	University of Münster, Germany
Alexis Tsoukiàs	CNRS – LAMSADE, France
Miguel A. Vega-Rodríguez	University of Extremadura, Spain
Jiahai Wang	Sun Yat-sen University, China
Rui Wang	National University of Defense Technology, China
Wenjun Wang	Heriot-Watt University, UK
Yuping Wang	Xidian University, China
Elizabeth Wanner	Aston University, UK
Ka-Chun Wong	City University of Hong Kong, China
Bai Yan	Beijing University of Technology, China
Shengxiang Yang	De Montfort University, UK
Kai Ye	Xi'an Jiaotong University, China
Gary Yen	Oklahoma State University, USA
Caitong Yue	Zhengzhou University, China
Hu Zhang	Harbin Institute of Technology, China
Mengjie Zhang	Victoria University of Wellington, New Zealand
Tao Zhang	National University of Defense Technology, China
Xingyi Zhang	Anhui University, China
Aimin Zhou	East China Normal University, China
Heiner Zille	Otto von Guericke University Magdeburg, Germany

Contents

Theory

Algorithms

Theory

It Is Hard to Distinguish Between Dominance Resistant Solutions and Extremely Convex Pareto Optimal Solutions

Qite Yang[1], Zhenkun Wang[1,2(✉)], and Hisao Ishibuchi[2]

[1] School of System Design and Intelligent Manufacturing, Southern University of Science and Technology, Shenzhen, China
yangqt@mail.sustech.edu.cn

[2] Department of Computer Science and Engineering, Southern University of Science and Technology, Shenzhen, China
{wangzk3,hisao}@sustech.edu.cn

Abstract. It has been acknowledged that dominance resistant solutions (DRSs) often exist in the feasible region of multi-objective optimization problems. DRSs can severely degrade the performance of many multi-objective evolutionary algorithms (MOEAs). In previous work, some coping strategies (e.g., the ϵ-dominance and the modified objective calculation) have been demonstrated to be effective in eliminating DRSs. However, these strategies may in turn cause algorithm inefficiency in other aspects. We argue that these coping strategies prevent the algorithm from obtaining extremely convex Pareto optimal solutions (ECPOSs), which are located around the boundary of the convex Pareto front (PF). That is, there is a dilemma between eliminating DRSs and preserving ECPOSs. To illustrate such a dilemma, we propose a new multi-objective optimization test problem with the extremely convex PF as well as the hardly dominated boundaries. Using this test problem, we investigate the performance of six representative MOEAs in terms of ECPOS preservation and DRS elimination. The results indicate that it is indeed challenging to distinguish between ECPOSs and DRSs.

Keywords: Multi-objective optimization · Dominance resistant solution · Hardly dominated boundary · Extremely convex Pareto optimal solution

1 Introduction

In dealing with multi-objective optimization problems (MOPs) [1–3], the Pareto dominance criterion is one of the most commonly used methods to evaluate the merits of solutions. Many multi-objective evolutionary algorithms (MOEAs) are developed based on the Pareto dominance criterion, such as NSGA-II [4], SPEA2

H. Ishibuchi et al. (Eds.): EMO 2021, LNCS 12654, pp. 3–14, 2021.
https://doi.org/10.1007/978-3-030-72062-9_1

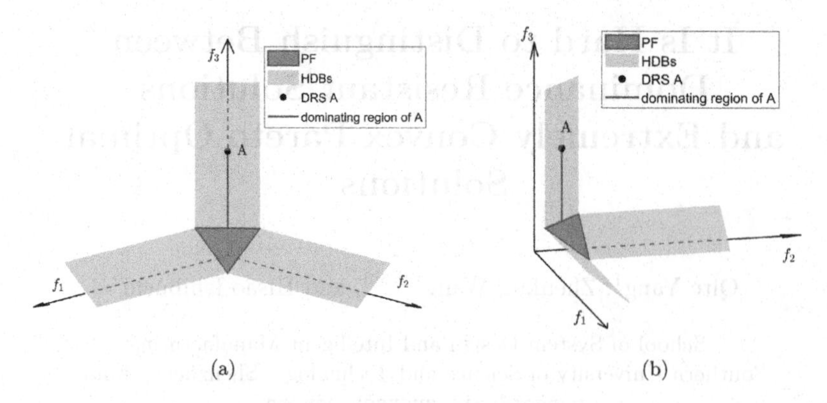

Fig. 1. Illustration of HDBs and a DRS. The same figure is shown from different angles in (a) and (b). (Color figure online)

[5] and PAES [6]. However, recent studies reveal that these Pareto dominance-based MOEAs may be hindered by dominance resistant solutions (DRSs) [7–10].

DRSs are the solutions located on the hardly dominated boundaries (HDBs). These solutions are hardly dominated by but obviously worse than the solutions on the Pareto front (PF). As shown in Fig. 1, solution A is a DRS on the HDB that parallels to the axis of f_3. An obvious feature of A is that it is extremely inferior to other solutions on the PF in f_3 but it has optimal or near-optimal values in f_1 and f_2 [11]. Thus, A can only be dominated by the solutions in the dominating region (red line), whereas the probability of finding those solutions is very small. Therefore, A is almost always non-dominated in the current population in Pareto dominance-based MOEAs. Since the preservation of A is beneficial for the population diversity, it can almost always remain during the population update. Once too many DRSs exist in the current population, the performance of these MOEAs degrades severely. In view of the fact that HDB is an often encountered MOP property, enabling MOEAs to eliminate DRSs from the current population tends to be imperative [7].

In recent years, some coping strategies (i.e., countermeasures against DRSs) have been demonstrated to be effective in eliminating DRSs. Among them, the ϵ-dominance criterion [12] is one of the most utilized methods. The ϵ-dominance is a relaxation form of the Pareto dominance criterion. It enables each solution to have a larger dominating region (see in Fig. 2). Therefore, DRSs are more likely to be dominated by other solutions. Except for the ϵ-dominance, a modified NSGA-II (denoted as mNSGA-II in this paper) is proposed in [13]. In mNSGA-II, the objective functions are calculated in a modified manner so as to improve the convergence ability. It is also shown in [7,13] that some decomposition based MOEAs (e.g., MOEA/D-PBI [14] and MOEA/D-Gen [15]) are capable of getting rid of DRSs. For instance, by using a small θ in MOEA/D-PBI, the optimization of the subproblems will no longer be trapped in DRSs.

In this paper, we argue that these coping strategies may in turn prevent the algorithm from obtaining solutions located around the boundary of the convex

PF (e.g., around the two extreme points of a two-objective PF, around the boundary of a three-objective PF). We refer to such solutions as Extremely Convex Pareto Optimal Solutions (ECPOSs). The preservation of ECPOSs has been shown to be very important in many real-world applications (e.g., the agile satellite mission planning [16] and the search based software engineering [17]). This is because ECPOSs may be the preferred solutions when the decision-maker is biased towards one or multiple objectives. One of the most effective methods to obtain ECPOSs is to transform MOP into several single objective problems by using extreme bias weights or weighted min-max scores [18], and optimize them then. Nevertheless, the preservation of ECPOSs is sensitive to the domination criterion. Figure 2 gives an illustration of ECPOS on the two-objective convex PF. Solution A and B are Pareto optimal solutions, and A is an ECPOS. The black dash line and the blue dot-dash line represent the boundaries of the regions Pareto dominated and ϵ-dominated by B, respectively. It is clear that A is not Pareto dominated by B, but A is ϵ-dominated by B. Once B is achieved, A can no longer survive in an ϵ-dominance based MOEA. Likewise, ECPOS C in Fig. 2 is hard to survive too.

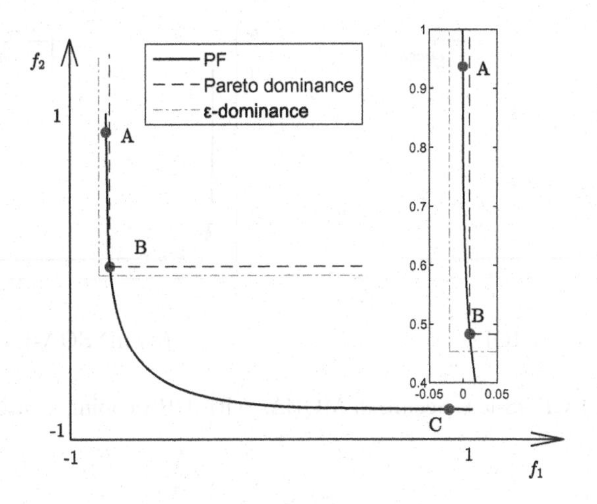

Fig. 2. Illustration of ECPOS.

To demonstrate that the above-mentioned coping strategies can result in ECPOS loss, we first formulate a two-objective test problem with a convex PF. Then we apply MOEAs with the DRS coping strategies to the test problem. Experimental results demonstrate the existence of a dilemma between eliminating DRSs and preserving ECPOSs. To further examine such a dilemma, we propose a scalable MOP with a convex PF and HDBs. Using this test problem, we investigate the performance of six representative MOEAs in terms of ECPOS preservation and DRS elimination. Our experimental results show that it is difficult for these MOEAs to distinguish between DRSs and ECPOSs.

The remainder of this paper is organized as below. Section 2 illustrates the ECPOS loss using a two-objective test problem. In Sect. 3, we propose a scalable test problem and explain it in detail. Then Sect. 4 shows the experimental results of the six representative MOEAs on the proposed test problem. Finally, we conclude this paper in Sect. 5.

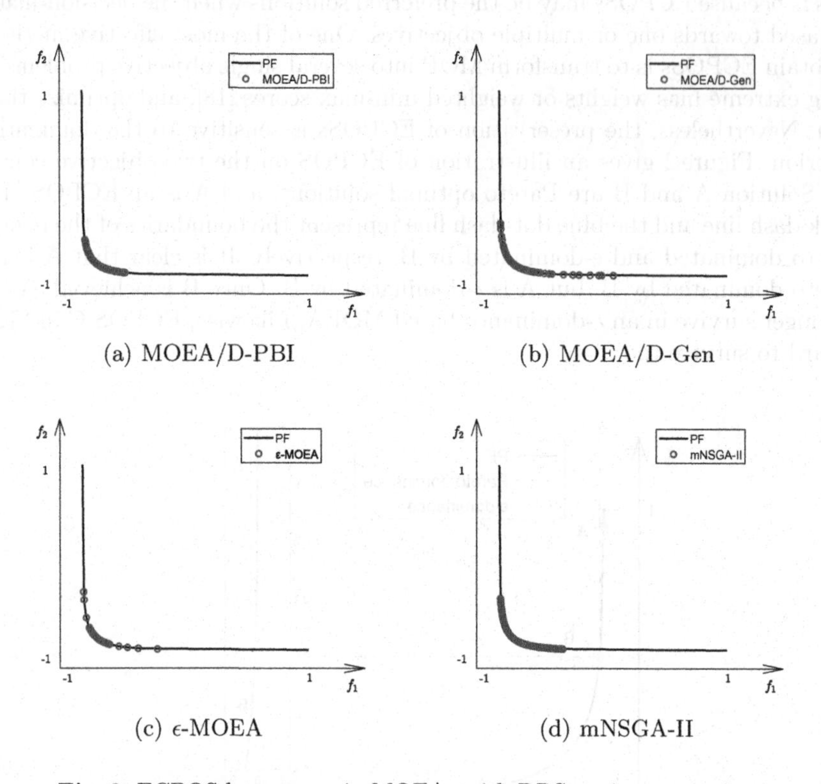

(a) MOEA/D-PBI (b) MOEA/D-Gen

(c) ε-MOEA (d) mNSGA-II

Fig. 3. ECPOS loss occurs in MOEAs with DRS coping strategies.

2 ECPOS Loss Caused by the DRS Coping Strategies

In this section, we illustrate that MOEAs with DRS coping strategies can result in the ECPOS loss. For this purpose, a two-objective test problem with the convex PF is first formulated as:

$$\text{minimize} \quad \begin{cases} f_1(\mathbf{x}) = (0.5 - x_1 + x_2)^4, \\ f_2(\mathbf{x}) = (0.5 + x_1 - x_2)^4, \end{cases} \tag{1}$$

where $\mathbf{x} = (x_1, x_2)^\top \in [0, 0.5]^2$. The PF of this problem is $f_1^{\frac{1}{4}}(\mathbf{x}) + f_2^{\frac{1}{4}}(\mathbf{x}) = 1$, where $0 \leq f_i(\mathbf{x}) \leq 1$, $i = 1, 2$. MOEA/D-PBI, MOEA/D-Gen, ε-MOEA [19] and mNSGA-II are applied to this problem. The setting of the relevant parameters in these algorithms is the same as in Sect. 4.

Figure 3 shows the obtained solutions by a single run of each algorithm. The experimental results in Fig. 3(a)–(d) are very similar. It can be seen that all the obtained solutions are only around the center of the PF, although their distribution is different in each figure. Besides, the areas around the extreme points of the PF are not covered by any solutions. That is, ECPOSs are not obtained by any algorithm in Fig. 3. We only give the result of one run here, but similar results always appear in multiple independent experiments of each algorithm. These results indicate that ECPOSs are easily missed by these MOEAs with DRS coping strategies.

3 Scalable MOP with a Convex PF and HDBs

As we mentioned above, many coping strategies can be used for eliminating DRSs. However, MOEAs with these strategies face the challenge of preserving ECPOSs. This suggests that there exists a dilemma between DRS elimination and ECPOS preservation. To verify it experimentally, we propose a scalable MOP with a convex PF and HDBs (denoted as MOP-CH) in this section.

In MOP-CH, the decision vector $\mathbf{x} = (x_1, \cdots, x_n)^\top \in [0,1]^n$ is divided into two independent parts: $\mathbf{x_I} = (x_1, \cdots, x_{m-1})^\top$ and $\mathbf{x_{II}} = (x_m, \cdots, x_n)^\top$, where $\mathbf{x_I}$ are position variables and $\mathbf{x_{II}}$ are distance variables. Then an m-objective MOP-CH can be written as:

$$\text{minimize} \quad f_k(\mathbf{x}) = (1 - h_k^p(\mathbf{x_I}))(1 + g_k(\mathbf{x_{II}})), \quad \text{for } k = 1, \cdots, m; \tag{2}$$

where p $(0 < p < 1)$ is a parameter determining the curvature of the convex PF. The smaller p is, the more convex of the PF. In this paper, the p is set to 0.25.

The position functions in MOP-CH are designed as:

$$h_k(\mathbf{x_I}) = \begin{cases} 1 - x_1, & k = 1; \\ (1 - x_k)\prod_{i=1}^{k-1} x_i, & k = 2, \cdots, m-1; \\ \prod_{i=1}^{m-1} x_i, & k = m. \end{cases} \tag{3}$$

The distance functions in MOP-CH are designed as:

$$g_k(\mathbf{x_{II}}) = 100\Big(|\phi_k| + \sum_{i \in \phi_k} \left((x_i - 0.5)^2\right) - cos\big(20\pi(x_i - 0.5)\big)\Big), \tag{4}$$

where $\phi_k = \{m - 1 + k, 2m - 1 + k, \cdots, \lfloor \frac{n+1-k}{m} \rfloor m - 1 + k\}$ for $k = 1, \cdots, m$.

Let $\mathbf{y}^* = (y_1^*, \cdots, y_m^*)$ be a Pareto optimal solution in the objective space. Then the PF of MOP-CH can be written as:

$$\sum_{i=1}^{m} (y_i^* - 1)^4 = 1, \tag{5}$$

where $0 \leq y_i^* \leq 1$ for $i = 1, \cdots, m$. The PF and HDBs of the three-objective MOP-CH are shown in Fig. 4.

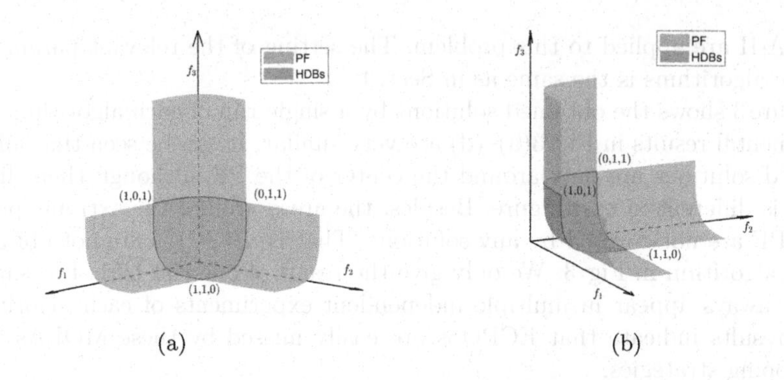

Fig. 4. The PF and HDBs of the three-objective MOP-CH. The same figure is shown from different angles in (a) and (b).

4 Experiments on Three-Objective MOP-CH

4.1 Performance Metrics

In this paper, the IGD [20] and HV [21] metrics are used to evaluate the performance of the obtained solution sets. A smaller IGD value indicates a better performance of a solution set. On the contrary, the smaller the HV value is, the worse the performance of a solution set is.

Considering P^* is a set of uniformly distributed solutions along the PF, and P is an approximation set of the PF in the objective space, then the IGD is calculated as:

$$IGD(P^*, P) = \frac{\sum_{\mathbf{v} \in P^*} d(\mathbf{v}, P)}{|P^*|}, \tag{6}$$

where $d(\mathbf{v}, P)$ is the Euclidean distance between the point $\mathbf{v}$ and the nearest member of P.

For a solution set S, the HV metric is defined as:

$$\mathrm{HV}(S) = \mathrm{VOL}\Big(\underset{\mathbf{x} \in S}{\cup} [f_1(\mathbf{x}), z_1^r] \times \cdots \times [f_m(\mathbf{x}), z_m^r]\Big), \tag{7}$$

where $\mathbf{z}^r = (z_1^r, \cdots, z_m^r)$ is a reference point, and VOL is the Lebesgue measure. In this paper, $\mathbf{z}^r$ is set to $(1.2, \cdots, 1.2)$.

4.2 Experimental Results and Analyses

To illustrate the dilemma between DRS elimination and ECPOS preservation, six typical MOEAs: NSGA-II, MOEA/D-PBI, MOEA/D-Gen, ϵ-MOEA, mNSGA-II and SMS-EMOA [22] are applied to the three-objective MOP-CH. The simulated binary crossover [23] and polynomial mutation operators are used with a distribution index of 20. The population size for general MOEAs is set to 200. Note that the population size in decomposition based MOEAs is the same as the number of weight vectors. So the population size in MOEA/D-PBI and

Table 1. HV and IGD values of the six algorithms

Metric		NSGA-II	MOEA/D-PBI	MOEA/D-Gen	ϵ-MOEA	mNSGA-II	SMS-EMOA
HV	Mean	4.53E−02	1.41E+00	1.04E+00	1.38E+00	1.45E+00	**1.46E+00**
	Std	4.52E−02	5.93E−02	5.45E−02	1.62E−02	2.99E−03	**2.40E−04**
IGD	Mean	1.69E+00	1.21E−01	2.14E−01	1.34E−01	7.28E−02	**5.55E−02**
	Std	1.35E+00	4.99E−02	6.99E−03	2.14E−02	8.35E−03	**4.65E−04**

MOEA/D-Gen is 190. Each algorithm runs 31 times independently. The termination condition of each algorithm is specified by 100,000 solution evaluations. According to the relevant references [7,13,14], θ in MOEA/D-PBI is set to 5, ρ in MOEA/D-Gen is set to 0.01, ϵ in ϵ-MOEA is set to 0.06, and α in mNSGA-II is set to 0.1. In this paper, the reference point in SMS-EMOA for hypervolume calculation is set to $\mathbf{f}^{max} + 5$, where $\mathbf{f}^{max} = (f_1^{max}, \cdots, f_m^{max})$, and f_i^{max} is the maximum value of the i-th objective in the current population.

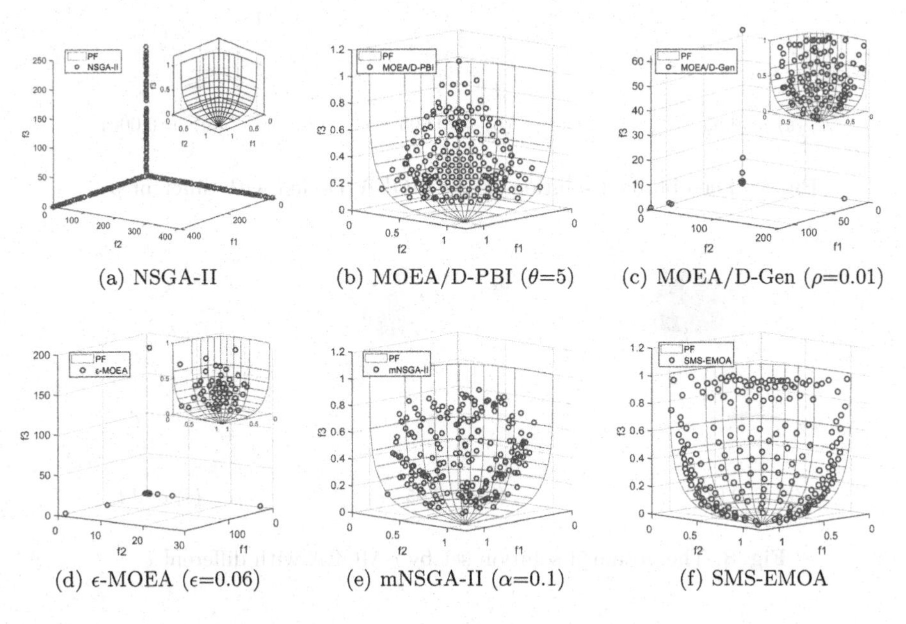

(a) NSGA-II (b) MOEA/D-PBI (θ=5) (c) MOEA/D-Gen (ρ=0.01)

(d) ϵ-MOEA (ϵ=0.06) (e) mNSGA-II (α=0.1) (f) SMS-EMOA

Fig. 5. The obtained solution set with the median HV value in 31 runs of each algorithm on MOP-CH

Experimental results are shown in Table 1 where the best results are highlighted in bold. From these results, it can be seen that NSGA-II performs worst since it has the lowest HV and highest IGD mean values among the six algorithms. The decomposition based MOEAs (MOEA/D-PBI and MOEA/D-Gen) and the modified Pareto dominance-based MOEAs (ϵ-MOEA and mNSGA-II) perform better than NSGA-II. However, none of them performs as well as the hypervolume-based MOEA (SMS-EMOA).

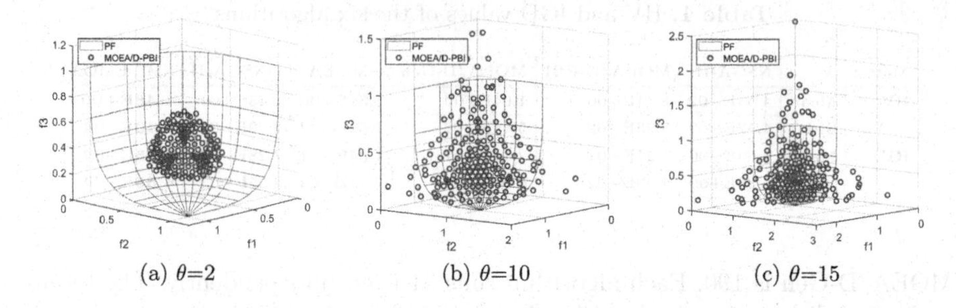

(a) θ=2 (b) θ=10 (c) θ=15

Fig. 6. The obtained solution set by MOEA/D-PBI with different θ

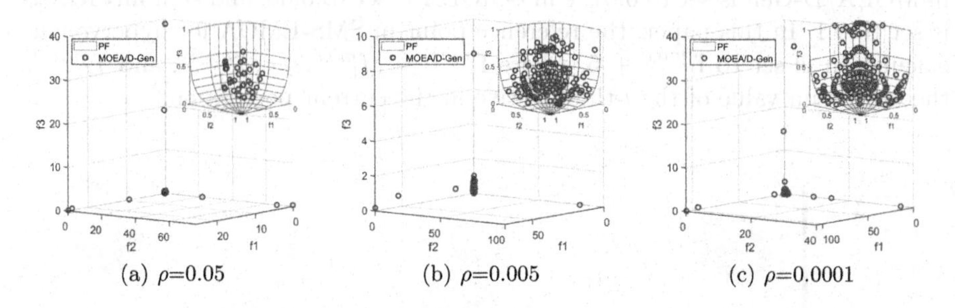

(a) ρ=0.05 (b) ρ=0.005 (c) ρ=0.0001

Fig. 7. The obtained solution set by MOEA/D-Gen with different ρ

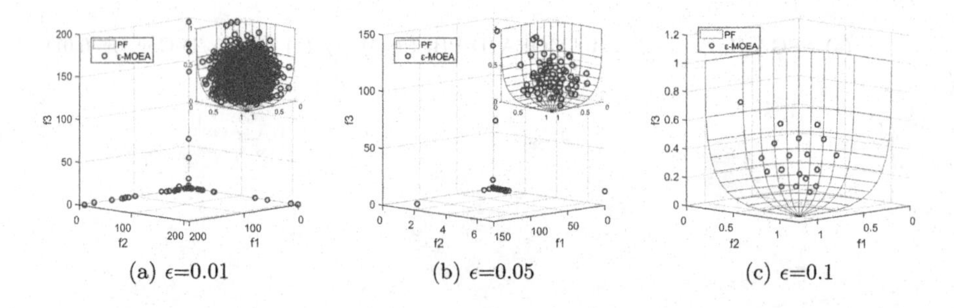

(a) ϵ=0.01 (b) ϵ=0.05 (c) ϵ=0.1

Fig. 8. The obtained solution set by ϵ-MOEA with different ϵ

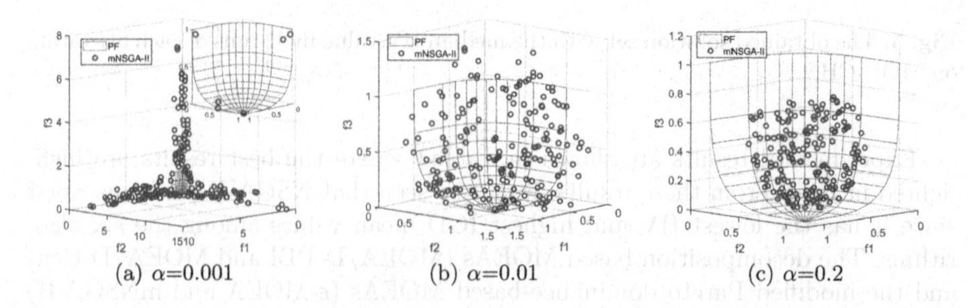

(a) α=0.001 (b) α=0.01 (c) α=0.2

Fig. 9. The obtained solution set by mNSGA-II with different α

Figure 5 presents the obtained solution set with the median HV value in the 31 runs of each algorithm. It can be seen that the solutions obtained by NSGA-II suffer from DRSs and fail to approximate the PF. The solutions by MOEA/D-PBI can avoid the DRSs and approximate the PF well. However, at the same time, the ECPOSs located around the three corners of the PF are missed. As for MOEA/D-Gen, most of the obtained solutions approximate the PF well, whereas some DRSs still exist. In Fig. 5(d), DRSs are not fully eliminated from the obtained solution set by ϵ-MOEA. Besides, the distribution of the solutions is not good. Compared with ϵ-MOEA, mNSGA-II in Fig. 5(e) performs better. All DRSs are eliminated from the obtained solution set by mNSGA-II, although some ECPOSs around the boundary of the PF are missing. The best approximation set is obtained in Fig. 5(f) by SMS-EMOA with respect to both the convergence and the diversity. This result suggests that it is hard for MOEAs with DRS coping strategies to distinguish between DRSs and ECPOSs. Some MOEAs (e.g., MOEA/D-Gen and ϵ-MOEA) cannot eliminate DRSs. Other MOEAs (MOEA/D-PBI and mNSGA-II) cannot find ECPOSs. The hypervolume-based MOEA, SMS-MOEA, is the best algorithms in avoiding this dilemma. This is because SMS-EMOA does not use dominance relationship but hypervolume contribution as the criterion to judge the merits and demerits between solutions, which means it cannot be disturbed by DRSs. In addition, most of ECPOSs on SMS-EMOA can be preserved since hypervolume contributions on them are very large [24]. However, it is not easy to apply SMS-EMOA to many-objective problems (e.g., with ten objectives).

According to [13,25], the performance of the MOEAs with DRS coping strategies may be affected by their parameters mentioned above. Therefore, experimental studies with different parameter values are conducted on MOEA/D-PBI, MOEA/D-Gen, ϵ-MOEA and mNSGA-II. Figure 6, 7, 8 and 9 show the solutions obtained by these algorithms, respectively.

Figure 6 shows the solutions obtained by MOEA/D-PBI with $\theta = 2$, 10 and 15. With the increase of θ, the distribution of the obtained solutions becomes wider, and more ECPOSs are preserved. Nevertheless, the number of DRSs increases at the same time. In Fig. 6(b) and (c), the ECPOSs located around the three corners of the PF are missing whereas many DRSs are preserved. This indicates that MOEA/D-PBI with a large value of θ may misjudge the merits of DRSs and ECPOSs, and preserve DRSs instead of ECPOSs.

Figure 7 presents the solutions obtained by MOEA/D-Gen with $\rho = 0.05$, 0.005 and 0.0001. By decreasing the value of ρ, MOEA/D-Gen can get more widely distributed solutions. However, the entire PF is not approximated well even when a very small value is used (i.e., there are no solutions around the three corners of the PF in the three figures in Fig. 7). Besides, it is noteworthy that DRSs have never been eliminated completely. This indicates that MOEA/D-Gen is also caught in the dilemma between DRS elimination and ECPOS preservation.

Figure 8 shows the solutions obtained by ϵ-MOEA with $\epsilon = 0.01$, 0.05 and 0.1, respectively. The number of the obtained solutions by ϵ-MOEA is related to the value of ϵ. The larger the value of ϵ, the less the number of the obtained solutions. It can be seen that the DRSs can be eliminated from the solution set by increasing the value of ϵ, which leads to the severe decrease in the number of obtained solutions as shown in Fig. 8(c). Moreover, almost all of these solutions are located around the center of the PF. ECPOSs are missing whereas DRSs are eliminated in Fig. 8(c).

Figure 9 presents the solutions obtained by mNSGA-II with $\alpha = 0.001$, 0.01 and 0.2, respectively. Similar to ϵ-MOEA in Fig. 8, when α is very small, the convergence of mNSGA-II is severely deteriorated by DRSs. With the increase of α, DRSs in the obtained solution sets become less, and more solutions approximate the PF. However, when the value of α is too large, a lot of ECPOSs are missing. Figure 9 shows that it may be difficult for mNSGA-II to set an appropriate α for getting rid of the dilemma between DRS elimination and ECPOS preservation.

Based on the above-mentioned results, it can be concluded that these MOEAs are very sensitive to their parameters. All the algorithms except for SMS-EMOA are trapped in the dilemma between DRS elimination and ECPOS preservation. The above experimental results show that it is indeed hard for these MOEAs with DRS coping strategies to distinguish between DRSs and ECPOSs.

5 Conclusion

In this paper, we discussed the difficulty of distinguishing between DRSs and ECPOSs, and illustrated the dilemma between DRS elimination and ECPOS preservation. We first used a simple two-objective example to demonstrate that ECPOSs are easy to be lost by MOEAs with DRS coping strategies. Then, we proposed a new MOP with a convex PF and HDBs (called MOP-CH). Six representative MOEAs were applied to the proposed problem to examine their performance in terms of eliminating DRSs and preserving ECPOSs. Our experimental results showed that NSGA-II suffers from DRSs and fails in approximating the PF. The examined MOEAs with DRS coping strategies were easily caught in the dilemma between DRS elimination and ECPOS preservation. Nevertheless, the results by SMS-EMOA showed that the hypervolume-based MOEA is effective in getting rid of such a dilemma. However, the diversity of the solutions obtained by SMS-EMOA is not perfect. This is because the optimal distribution of solutions for hypervolume maximization is not uniform when the PF is nonlinear. In the future, we will explore the approach to distinguish between DRSs and ECPOSs, and try to find an ideal balance between DRS elimination and ECPOS preservation. Since the proposed test problem is scalable, we will discuss these future research topics not only for the three-objective problem for visual comparisons but also for many-objective problems for scalability examination.

References

1. Zitzler, E., Deb, K., Thiele, L.: Comparison of multiobjective evolutionary algorithms: Empirical results. Evol. Comput. 8(2), 173–195 (2000)

2. Deb, K., Thiele, L., Laumanns, M., Zitzler, E.: Scalable test problems for evolutionary multiobjective optimization. In: Abraham, A., Jain, L., Goldberg, R. (eds.) Evolutionary Multiobjective Optimization. Advanced Information and Knowledge Processing., pp. 105–145. Springer, London (2005). https://doi.org/10.1007/1-84628-137-7_6
3. Huband, S., Barone, L., While, L., Hingston, P.: A scalable multi-objective test problem toolkit. In: Coello Coello, C.A., Hernández Aguirre, A., Zitzler, E. (eds.) EMO 2005. LNCS, vol. 3410, pp. 280–295. Springer, Heidelberg (2005). https://doi.org/10.1007/978-3-540-31880-4_20
4. Deb, K., Pratap, A., Agarwal, S., Meyarivan, T.: A fast and elitist multiobjective genetic algorithm: NSGA-II. IEEE Trans. Evol. Comput. **6**(2), 182–197 (2002)
5. Zitzler, E., Laumanns, M., Thiele, L.: SPEA2: Improving the strength pareto evolutionary algorithm. TIK-report, vol. 103 (2001)
6. Knowles, J., Corne, D.: The Pareto archived evolution strategy: a new baseline algorithm for Pareto multiobjective optimisation. In: Proceedings of the Congress on Evolutionary Computation, vol. 1, pp. 98–105. IEEE (1999)
7. Wang, Z., Ong, Y.-S., Ishibuchi, H.: On scalable multiobjective test problems with hardly dominated boundaries. IEEE Trans. Evol. Comput. **23**(2), 217–231 (2018)
8. Batista, L.S., Campelo, F., Guimarães, F.G., Ramírez, J.A.: A comparison of dominance criteria in many-objective optimization problems. In: Proceedings of the Congress on Evolutionary Computation, pp. 2359–2366. IEEE (2011)
9. Yuan, Y., Xu, H., Wang, B., Yao, X.: A new dominance relation-based evolutionary algorithm for many-objective optimization. IEEE Trans. Evol. Comput. **20**(1), 16–37 (2015)
10. Le, K., Landa-Silva, D.: Obtaining better non-dominated sets using volume dominance. In: Proceedings of the Congress on Evolutionary Computation, pp. 3119–3126. IEEE (2007)
11. Ikeda, K., Kita, H., Kobayashi, S.: Failure of Pareto-based MOEAs: does non-dominated really mean near to optimal? In: Proceedings of the Congress on Evolutionary Computation, vol. 2, pp. 957–962. IEEE (2001)
12. Laumanns, M., Thiele, L., Deb, K., Zitzler, E.: Combining convergence and diversity in evolutionary multiobjective optimization. Evol. Comput. **10**(3), 263–282 (2002)
13. Ishibuchi, H., Matsumoto, T., Masuyama, N., Nojima, Y.: Effects of dominance resistant solutions on the performance of evolutionary multi-objective and many-objective algorithms. In: Proceedings of the Genetic and Evolutionary Computation Conference, pp. 507–515. Springer (2020). https://doi.org/10.1145/3377930.3390166
14. Zhang, Q., Li, H.: MOEA/D: a multiobjective evolutionary algorithm based on decomposition. IEEE Trans. Evol. Comput. **11**(6), 712–731 (2007)
15. Giagkiozis, I., Purshouse, R.C., Fleming, P.J.: Generalized decomposition. In: Purshouse, R.C., Fleming, P.J., Fonseca, C.M., Greco, S., Shaw, J. (eds.) EMO 2013. LNCS, vol. 7811, pp. 428–442. Springer, Heidelberg (2013). https://doi.org/10.1007/978-3-642-37140-0_33
16. Li, L., Chen, H., Li, J., Jing, N., Emmerich, M.: Preference-based evolutionary many-objective optimization for agile satellite mission planning. IEEE Access **6**, 40963–40978 (2018)
17. Zhang, Y., Finkelstein, A., Harman, M.: Search based requirements optimisation: existing work and challenges. In: Paech, B., Rolland, C. (eds.) REFSQ 2008. LNCS, vol. 5025, pp. 88–94. Springer, Heidelberg (2008). https://doi.org/10.1007/978-3-540-69062-7_8

18. Hughes, E.J.: Multiple single objective Pareto sampling. In: Proceedings of the Congress on Evolutionary Computation, vol. 4, pp. 2678–2684. IEEE (2003)
19. Deb, K., Mohan, M., Mishra, S.: Towards a quick computation of well-spread pareto-optimal solutions. In: Fonseca, C.M., Fleming, P.J., Zitzler, E., Thiele, L., Deb, K. (eds.) EMO 2003. LNCS, vol. 2632, pp. 222–236. Springer, Heidelberg (2003). https://doi.org/10.1007/3-540-36970-8_16
20. Zitzler, E., Thiele, L., Laumanns, M., Fonseca, C.M., Da Fonseca, V.G.: Performance assessment of multiobjective optimizers: an analysis and review. IEEE Trans. Evol. Comput. **7**(2), 117–132 (2003)
21. Zitzler, E., Thiele, L.: Multiobjective evolutionary algorithms: a comparative case study and the strength pareto approach. IEEE Trans. Evol. Comput. **3**(4), 257–271 (1999)
22. Beume, N., Naujoks, B., Emmerich, M.: SMS-EMOA: multiobjective selection based on dominated hypervolume. Eur. J. Oper. Res. **181**(3), 1653–1669 (2007)
23. Deb, K., Agrawal, R.B., et al.: Simulated binary crossover for continuous search space. Complex Syst. **9**(2), 115–148 (1995)
24. Ishibuchi, R.I., Masuyama, N., Nojima, Y.: Dynamic specification of a reference point for hypervolume calculation in SMS-EMOA. In: Proceedings of the Congress on Evolutionary Computation, pp. 1–8. IEEE (2018)
25. Hernandez-Diaz, A.G., Santana-Quintero, L.V., Coello Coello, C.A., Molina, J.: Pareto-adaptive ε-dominance. Evol. Comput. **15**(4), 493–517 (2007)

On Analysis of Irregular Pareto Front Shapes

Shouyong Jiang[1(✉)], Jinglei Guo[2], Bashar Alhnaity[1], and Qingyang Zhang[3(✉)]

[1] School of Computer Science, University of Lincoln, Lincoln LN6 7TS, UK
[2] School of Computer Science, Central China Normal University, Wuhan, China
guojinglei@mail.ccnu.edu.cn
[3] School of Computer Science and Technology, Jiangsu Normal University, Xuzhou, China

Abstract. After decades of effort, evolutionary algorithms have been able to solve a variety of multiobjective optimisation problems with diverse characteristics. However, the presence of irregularity in the Pareto-optimal front is increasingly recognised as a big challenge to some well-established algorithms. In order to further our understanding of this irregularity and its effect on algorithms, we develop a generic framework of constructing irregular Pareto-optimal front shapes, and use it as a tool to examine the performance of some well-known algorithms. Experimental results reveal that conventional algorithms are not always inferior to the state of the arts, and all the algorithms considered in this paper face some unexpected challenges when dealing with irregularity of Pareto-optimal front. The findings suggest that a systematic evaluation and analysis is needed for any newly-developed algorithms to avoid biases.

Keywords: Multi-objective optimization · Evolutionary algorithm · Irregular Pareto front · Test problem

1 Introduction

Problems with multiple objectives to optimise simultaneously are called multi-objective optimisation problems (MOPs). A mathematical representation of MOPs can be described as follows:

$$\begin{aligned} &min\ \mathbf{F}(x) = (f_1(x), ..., f_M(x))^T \\ &s.t. \begin{cases} h_i(x) = 0, & i = 1, ..., n_h \\ g_i(x) \geq 0, & i = 1, ..., n_g \\ x \in \Omega_x \end{cases} \end{aligned} \tag{1}$$

where M is the number of conflicting objectives, n_h and n_g are the number of equity and inequity constraints, respectively. $\Omega_x \subseteq R^n$ is the decision space. $F(x)$: $\Omega_x \rightarrow R^M$ is the objective function vector that evaluates the solution x.

The multi-objectivity in (1) results in a set of solutions trade-offing objectives for an MOP. This set is called the Pareto-optimal set (PS) in decision space and

H. Ishibuchi et al. (Eds.): EMO 2021, LNCS 12654, pp. 15–25, 2021.
https://doi.org/10.1007/978-3-030-72062-9_2

its image in objective space is called the Pareo-optimal front (PF). The PF can have various characteristics to which algorithms are sensitive [1,2]. In this paper we focus on MOPs with *irregular PF* defined below, although 'irregularity' may refer to different things in other studies [1].

Definition 1 (Irregular PF). *The PF of an M-objective problem is considered irregular if the projection of the PF surface, after min-max normalisation, onto the hyperplane* $\sum_{i=1}^{M} f_i = 1$ *is not the standard M-simplex.*

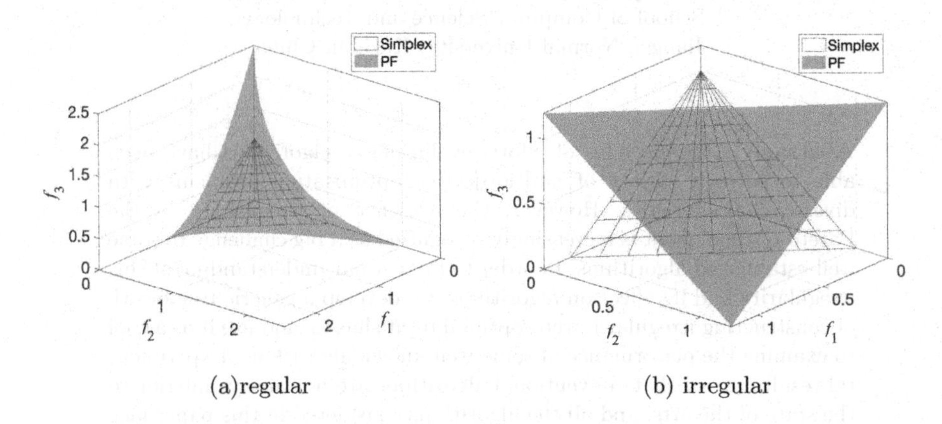

Fig. 1. Illustration of regular and irregular PF with 3 objectives.

Accompanied with this definition are the two examples provided in Fig. 1, where the inverted triangular PF is considered irregular since its projection on the simplex is only a part of the simplex. Most of existing test problems, including the DTLZ [3], WFG [4] and CEC09 test suites [5], have a regular PF shape and have been extensively studied. However, PF irregularity has only recently been found to present new challenges to some evolutionary algorithms (EAs) that work well for regular PF shapes [6].

Loosely speaking, PF irregularity may influence EAs in the following aspects.

1. Ineffective population normalisation: EAs relying on extreme points for population normalisation can be affected. An example is NSGA-III [7] which employs an achievement scalarising function to identify an extreme point for each objective axis. This approach works well when true extreme points on the PF are on or near objective axes. However, this may not be the case for irregular PF shapes, where true extreme points are far from objective axes. Incorrect identification of true extreme points may lead to ineffective population normalisation, which in turn affects population evolution.
2. Sub-optimal environmental selection: Due to PF irregularity, it can be difficult to select a population of N well distributed solutions across the PF for the next cycle of evolution. In particular, EAs relying on extreme solutions for

environmental selection are most influenced. For example, a number of EAs employing the quickselect approach [8,9] start environmental selection with 'corner' points on the PF. The selected solution set may not be well diversified if 'corner' points are incorrectly identified.

3. Computational waste: EAs with the aid of reference points or weight vectors may waste computational resources during the search. This has been observed especially in decomposition-based algorithms [10] where some weight vectors have no intersection with the PF. So computational resources allocated to such weight vectors are wasted.
4. Increased search difficulty: The search for an irregular PF encounters more challenges than for a regular PF. An example is an irregular PF consisting of multiple disconnected segments. EAs need to figure out the balance between these segments in computational resource allocation in order to find all the segments effectively. In contrast, a regular PF just has one segment and is therefore relatively easier to be searched.

The above challenges suggest there is a great need of research on investigating irregular PF shapes. Despite that disconnected PF shapes have been seen in numerous popular test suites, other types of irregular PF shapes are not much studied. Some recent studies have suggested a few irregular PF shapes resulting from an adaptation of existing test problems [6]. These PF shapes are like inverted simplexes when they are projected onto the standard simplex, but they have also been found very challenging for some popular algorithms in spite of simple PF geometries. Further to these studies, we would like to investigate how EAs are affected by other types of irregular PF shapes. To do so, we first develop a problem generator for irregular PF shapes, and then we test a variety of algorithms on different PF shapes instantiated from this generator. Experimental results show that these irregular PF shapes impact significantly the performance of many algorithms.

The remainder of this paper is organized as follows. Section II presents the proposed framework for creating irregular PF shapes. Section III describes experimental settings and discussion of results. Section IV concludes the paper.

2 LinPF: A Framework for Constructing Irregular PF Shapes

We suggest for constructing irregular PF shapes the following simple framework:

$$\mathbf{LinPF}: \begin{cases} f_1 = g(\mathbf{x})\,(c_1 + 1 - x_1\cos(\pi x_2)) \\ f_i = g(\mathbf{x})\left(c_i + i + x_1\Big(\cos(\pi x_2) - \sum\limits_{j=2, j\leq i-1}^{i-1} \sin(\pi x_j) + \sin(\pi x_i)\Big)\right) \\ f_M = g(\mathbf{x})\left(c_M + M + x_1\Big(\cos(\pi x_2) - \sum\limits_{j=2, j\leq M-1}^{M-1} \sin(\pi x_j)\Big)\right) \end{cases} \tag{2}$$

where $\mathbf{x} = (x_1, ...x_M, ..x_n)$ is the decision variable vector and the g function, which has a minimum value of one, is related to the distance of a solution x to

Table 1. Test problems generated by Eq.(2) with search space $[0,1]^n$, where $n = 12$. For notation convenience, the first $M-1$ decision variables $x_i (1 \le i \le M-1)$ used in Eq.(2) are working variables transformed from the original search space via a linear mapping function $x_i \to ax_i + b$ (see this mapping for each instance for detail).

Prob	Definition	Main characteristics of PF
LinPF1	$x_i \leftarrow 2x_i, i=2,\ldots M\text{–}1$ $g = 1 + (1+\sin(0.5\pi x_1)) \sum_{i=M}^{n} (x_i - 0.5)^2$	Hyper-eclipse
LinPF2	$x_1 \leftarrow 0.2x_1 + 0.8$ $x_i \leftarrow 2x_i, i=2,\ldots M\text{–}1$ $g = 1 + (1+\sin(0.5\pi x_1)) \sum_{i=M}^{n} (x_i - 0.5)^2$	Hyper-ring
LinPF3	$x_1 \leftarrow 0.5x_1 + 0.5$ $x_i \leftarrow 0.5x_i + 2, i=2,\ldots M\text{–}1$ $g = 1 + (1+\sin(0.5\pi x_1)) \sum_{i=M}^{n} (x_i - 0.5)^2$	A hyper-ring segment
LinPF4	$x_1 \leftarrow 0.2x_1 + 0.8$ $x_i \leftarrow 0.5x_i + 0.25, i=2,\ldots M\text{–}1$ $\sin(t) \leftarrow \sin(t^2)$ $g = 1 + (1+\sin(0.5\pi x_1)) \sum_{i=M}^{n} (x_i - 0.5)^2$	S-shape manifold
LinPF5	$x_i \leftarrow 2x_i, i=2,\ldots M\text{–}1$ $g = 1 + (1+\sin(0.5\pi x_1)) \sum_{i=M}^{n} (x_i - 0.5)^2$ $+\sin(0.5\pi \lceil 2x_i \rfloor)$	A hyper-eclipse and a hyper-ring
LinPF6	$x_i \leftarrow 2x_i, i=2,\ldots M\text{–}1$ $g = 1 + (1+\sin(0.5\pi x_1))$ $\sum_{i=M}^{n} (x_i - 0.5)^2 + \sin(0.5\pi \lfloor 2x_i + 10^{-5} \rfloor)$	A hyper-eclipse and a circle

$\lceil a \rfloor$ rounds a to its nearest integer.
$\lfloor a \rfloor$ rounds a to the nearest integer less than or equal to a.

the PF . This framework is called LinPF because the resulting PF is on a linear (hyper)surface. Indeed, the PF of LinPF is part of an $(M-1)$-dimensional hyper-eclipse centred at $(c_1, \ldots, c_M)$, and the hyper-eclipse is located on the hyperplane:

$$\sum_{i=1}^{M-1} (M-i)(f_i - c_i) + (f_M - c_M) = 0 \quad (3)$$

Through the LinPF framework, we generate a set of six MOPs whose definition is detailed in Table 1. The PF of these problems on a linear surface is shown in Fig. 2. It is worth noting that variable ranges can be adjusted by the approach in [4]. PF shapes on nonlinear surfaces can be also generated easily from this framework. For example, a nonlinear mapping $f_i \to f_i^k$ ($k > 0$ and $k \neq 1$) in objective functions can lead to the PF front being on a nonlinear surface. However, we have no intention to nonlinearity, and we focus mainly on linear cases to show EAs can be affected irregular PF shapes.

3 Experimental Results

In this section, we first describe experimental settings, followed by discussion of results. We chose for experimentation a total of six algorithms belonging to three groups. The first group consists of three decomposition-based algorithms, i.e., RVEA [11], MOEA/D [10] and SPEA/R [12], which have shown good performance for general PF shapes. The second group comprises MOEA/D-AWA [2]

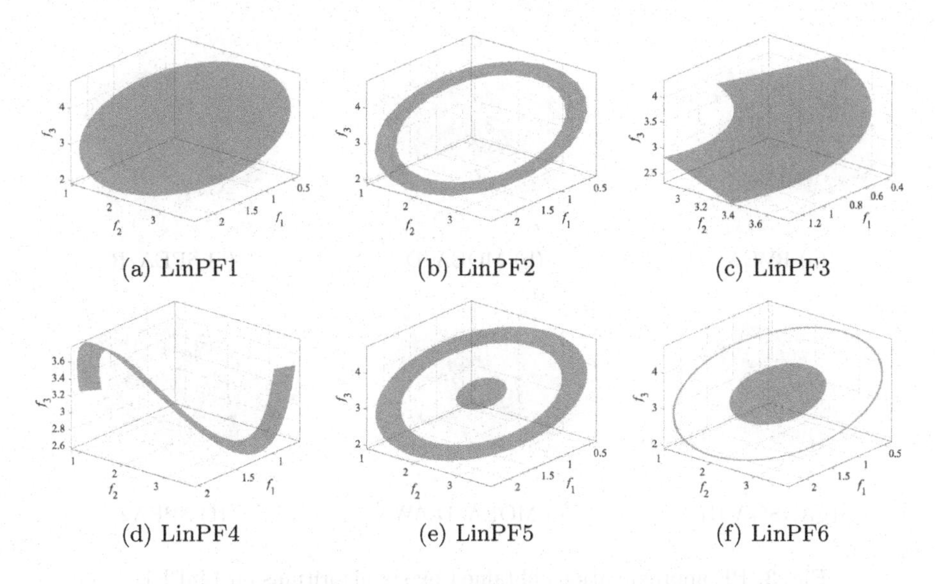

(a) LinPF1 (b) LinPF2 (c) LinPF3

(d) LinPF4 (e) LinPF5 (f) LinPF6

Fig. 2. Illustration of the PF of 6 LinPF instances with 3 objectives.

and ANSGA-III [7], and each of them has a population adaptation approach to addressing irregular PF shapes. Additionally, SPEA2 [13] as a conventional algorithm was chosen for testing, since it does not require reference points or weight vectors and has demonstrated great effectiveness in multi-objective optimisation. For ease of visualisation, we focused only on the 3-objective case of LinPF problems, although the proposed LinPF can be easily evaluated in many-objective cases. All the algorithms performed a maximum 300 generational cycles with a population size of 105. Other parameters in these algorithms remain the same as in their original papers.

3.1 Performance Measure

In this work, we used the following measure to quantify the performance of algorithms.

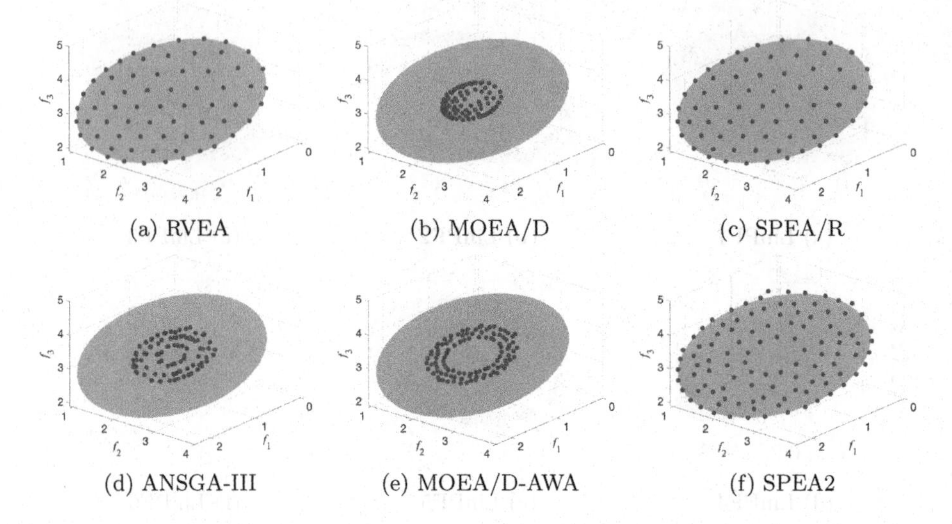

Fig. 3. PF approximations obtained by six algorithms on LinPF1.

Hypervolume. The hypervolume (HV) [12] metric measures the size of the objective space dominated by the approximated solution set S and bounded by a reference point $R = (R_1, \ldots, R_M)^T$ that is dominated by all points on the PF, and is computed by:

$$HV(S) = Leb(\bigcup_{x \in S} [f_1(x), R_1] \times \cdots \times [f_M(x), R_M]), \tag{4}$$

where $Leb(A)$ is the Lebesgue measure of a set A. Here, $R = (1.1, \ldots, 1.1)^T$ was used after min-max normalisation of objective values in a PF approximation PF^*.

3.2 Discussion of Results

Figure 3 shows the PF approximations obtained by six algorithms for LinPF1. It is observed that RVEA, SPEA/R and SPEA2 approximate the PF with a set of uniformly distributed solutions. However, the former two algorithms have much fewer solutions than SPEA2, and SPEA2 misses solutions near the centre of the PF. MOEA/D, ANSGA-III and MOEA/D-AWA favour the inner area of the PF and do not find any solutions on the outer area.

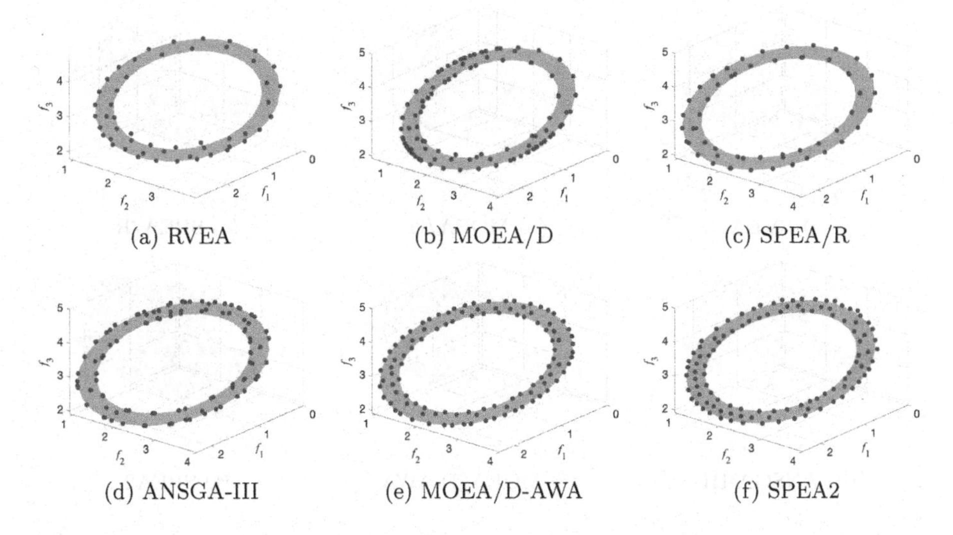

(a) RVEA (b) MOEA/D (c) SPEA/R

(d) ANSGA-III (e) MOEA/D-AWA (f) SPEA2

Fig. 4. PF approximations obtained by six algorithms on LinPF2.

All the six algorithms have no difficulty in identifying the PF of LinPF2, as indicated by Fig. 4. However, they differ in solution distribution. Again, RVEA and SPEA/R have more spare solutions on the PF, despite excellent diversification. The solutions of MOEA/D and ANSGA-III are not well diversified. SPEA2 achieves the best PF approximation for LinPF2.

Figure 5 shows PF approximations for LinPF3. Most of the algorithms except MOEA/D and ANSGA-III perform well on this problem. The solutions of MOEA/D are not well distributed, and those of ANSGA-III cover only a small part of the PF. The poor performance of ANSGA-III may be due to that the extreme points of LinPF3 for population normalisation cannot be effectively identified.

For LinPF4, the PF approximations of six algorithms are illustrated in Fig. 6. It is seen that MOEA/D-AWA and SPEA2 solve this problem very well. The other algorithms, however, are unable to provide a good approximation to the PF, largely due to poor solution distribution. In addition, RVEA seems to be not well converged for this problem.

The disconnectivity of LinPF Fig. 7 poses a big challenge to most of the algorithms except SPEA2. Despite great proximity to the PF, the solutions obtained by most of the algorithms are either too sparse or poorly distributed. MOEA/D-AWA approximates very well the outer segment of the PF but has no solution to the inner island.

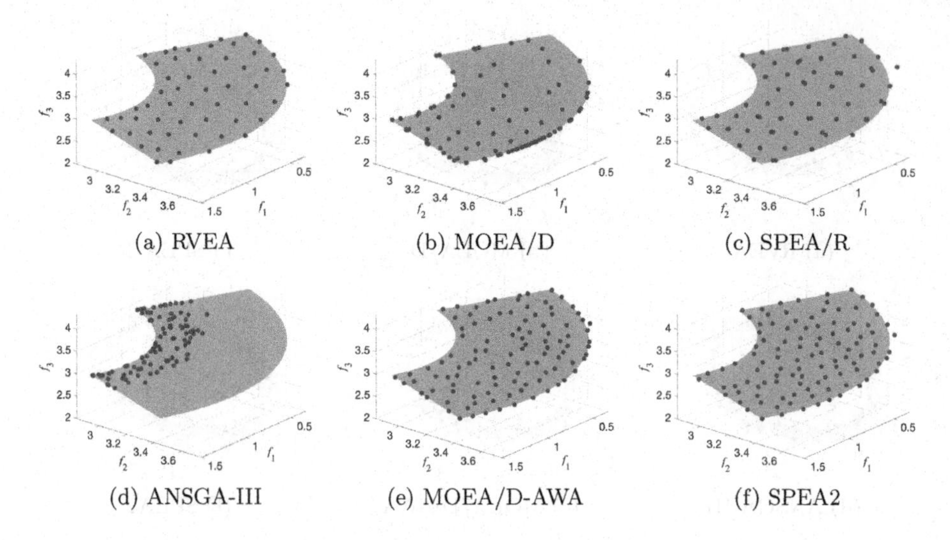

Fig. 5. PF approximations obtained by six algorithms on LinPF3.

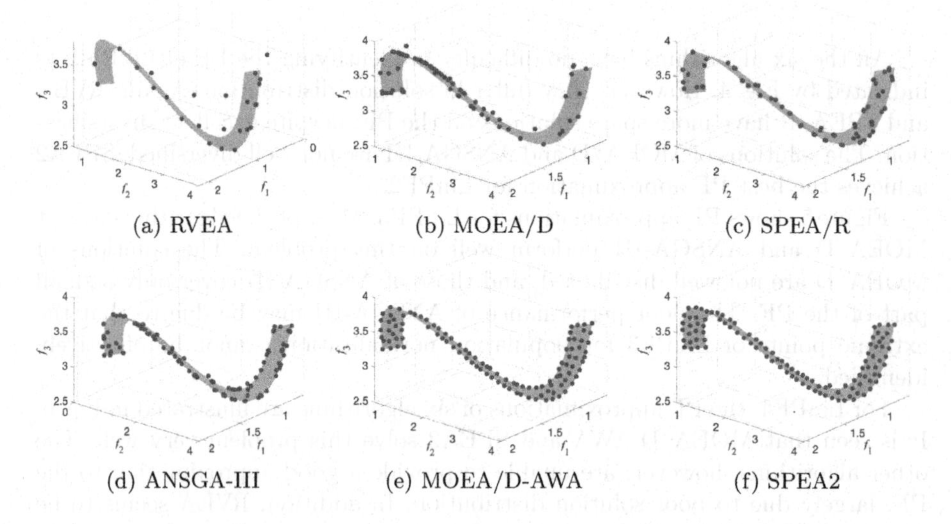

Fig. 6. PF approximations obtained by six algorithms on LinPF4.

Figure 8 shows that LinPF6 poses more challenges to the algorithms. Specifically, only RVEA and SPEA/R can approximate the complete PF of LinPF6, and SPEA/R seems to have a better approximation due to the good convergence of solutions. The rest of algorithms all encounter difficulties in finding the curve as part of the PF. Additionally, three algorithms (ANSGA-III, MOEA/D-AWA, and SPEA2) miss solutions near the centre the PF.

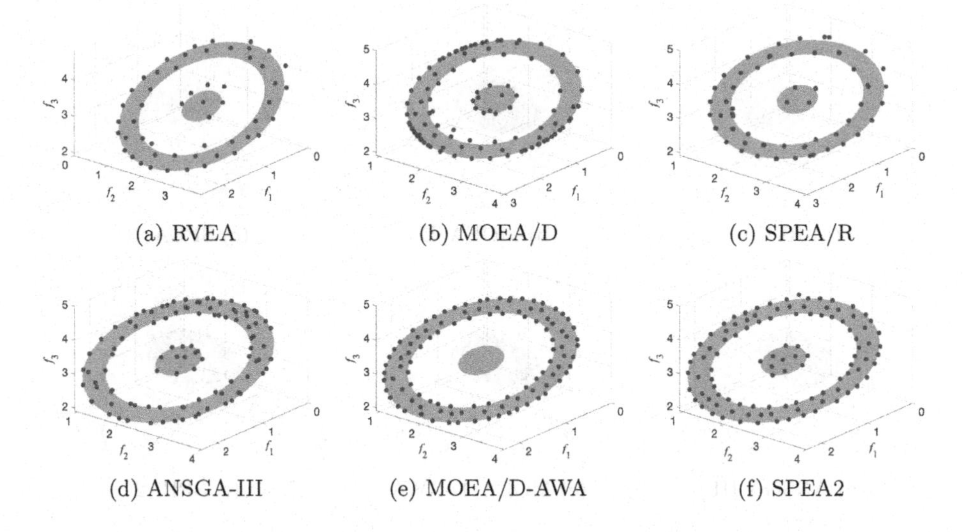

Fig. 7. PF approximations obtained by six algorithms on LinPF5.

Table 2 shows the normalised HV values for all the algorithms. According to their overall performance indicated by total ranks, the top-performing algorithms are SPEA2, MOEA/D-AWA and SPEA/R, and the others are significantly inferior. The results suggest that 1) the state of the arts are not always better than conventional algorithms like SPEA2; 2) reference adaptation (i.e., MOEA/D-AWA) improves decomposition-based algorithms; and 3) algorithms (i.e., ANSGA-III) relying on extreme points for population normalisation are vulnerable to irregular PF shapes in spite of the use of reference adaptation.

Table 2. Average HV values obtained by six algorithms and best values are highlighted in grey. For each algorithm, individual rank resulting from rank-sum test is shown in parentheses. The last row are the sum of individual ranks (and ordered rank)

Problem	RVEA	MOEA/D	SPEA/R	ANSGA-III	MOEA/D-AWA	SPEA2
LinPF1	2.25e−1 (2)	1.40e−1 (6)	2.27e−1 (1)	1.48e−1 (5)	1.80e−1 (4)	2.05e−1 (3)
LinPF2	1.98e−1 (6)	2.16e−1 (2)	2.08e−1 (5)	2.15e−1 (3)	2.21e−1 (1)	2.14e−1 (4)
LinPF3	6.28e−2 (5)	6.73e−2 (3)	6.78e−2 (2)	6.12e−2 (6)	6.63e−2 (4)	6.98e−2 (1)
LinPF4	8.34e−2 (6)	8.93e−2 (5)	8.96e−2 (4)	9.37e−2 (3)	9.78e−2 (1)	9.65e−2 (2)
LinPF5	2.07e−1 (5)	2.06e−1 (6)	2.15e−1 (4)	2.28e− 1(2)	2.27e−1 (3)	2.29e−1 (1)
LinPF6	2.16e−1 (2)	1.33e−1 (6)	2.22e−1 (1)	1.35e−1 (5)	1.56e−1 (3)	1.47e−1 (4)
rank	26 (5)	28 (6)	17(3)	24 (4)	16 (2)	15 (1)

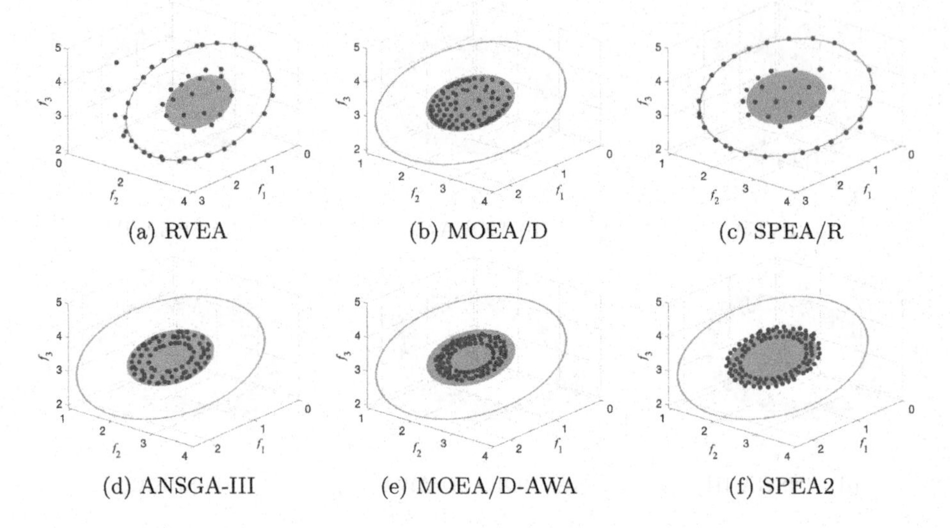

(a) RVEA (b) MOEA/D (c) SPEA/R

(d) ANSGA-III (e) MOEA/D-AWA (f) SPEA2

Fig. 8. PF approximations obtained by six algorithms on LinPF6.

4 Conclusions

In this work, we have developed a generic framework for constructing irregular PF shapes, and we have investigated the effect of PF irregularity on EAs using the developed framework. Our key findings are listed as follows.

1. SPEA2 is not inferior to the state-of-the-art algorithms considered in this work in dealing with PF irregularity, and it even outperforms some of these algorithms in many cases.
2. Reference and decomposition-based algorithms are generally disadvantageous for irregular PF shapes, but improved performance can be observed when reference adaptation approaches are used in such algorithms. However, reference adaptation seems to have side-effects; for example, the coverage of population on the PF decreases to some extent, as seen in MOEA/D-AWA for LinPF6-7.
3. MOEA/D does not provide a uniform distribution of solutions on the PF as expected. Instead, its solutions are dense on some parts of the PF and sparse on other parts.
4. Overall, there is no clear winner, among the considered algorithms in this paper, for the LinPF test instances. Indeed, each algorithm succeeding in one LinPF test problem encounters some difficulties in another.

These findings suggest that the proposed LinPF framework is an important tool for critically evaluating multiobjective optimisers, providing us with a greater understanding of irregular PF shapes and their impact on evolutionary multi-objective optimisation.

Our future work will be performing experimental analysis on irregular PF shapes in many-objective scenarios. Also, we will develop algorithms to overcome the challenges posed by PF irregularity.

Acknowledgements. This work was supported by National Natural Science Foundation of China (Grant No. 62006103).

References

1. Jiang, S., Yang, S.: An improved multiobjective optimization evolutionary algorithm based on decomposition for complex pareto fronts. IEEE Trans. Cybern. **46**(2), 421–437 (2016)
2. Qi, Y., Ma, X., Liu, F., Jiao, L., Sun, J., Wu, J.: MOEA/D with adaptive weight adjustment. Evol. Comput. **22**(2), 231–264 (2014)
3. Deb, K., Thiele, L., Laumanns, M., Zitzler, E.: Scalable test problems for evolutionary multiobjective optimization. In: Abraham, A., Jain, L., Goldberg, R. (eds.) Evolutionary Multiobjective Optimization. Advanced Information and Knowledge Processing. Springer, London (2005). https://doi.org/10.1007/1-84628-137-7_6
4. Huband, S., Hingston, P., Barone, L., While, L.: A review of multiobjective test problems and a scalable test problem toolkit. IEEE Trans. Evol. Comput. **10**(5), 477–506 (2006)
5. Zhang, Q., Zhou, A., Zhao, S., Suganthan, P.N., Liu, W., Tiwari, S.: Multiobjective optimization test instances for the CEC 2009 special session and competition (2009)
6. Ishibuchi, H., Setoguchi, Y., Masuda, H., Nojima, Y.: Performance of decomposition-based many-objective algorithms strongly depends on pareto front shapes. IEEE Trans. Evol. Comput. **21**(2), 169–190 (2017)
7. Jain, H., Deb, K.: An evolutionary many-objective optimization algorithm using reference-point based nondominated sorting approach, part II: handling constraints and extending to an adaptive approach. IEEE Trans. Evol. Comput. **18**(4), 602–622 (2013)
8. Tseng, L.Y., Chen, C.: Multiple trajectory search for unconstrained/constrained multi-objective optimization. In: 2009 IEEE Congress on Evolutionary Computation, pp. 1951–1958. IEEE (2009)
9. He, X., Zhou, Y., Chen, Z., Zhang, Q.: Evolutionary many-objective optimization based on dynamical decomposition. IEEE Trans. Evol. Comput. **23**(3), 361–375 (2018)
10. Zhang, Q., Li, H.: MOEA/D: a multiobjective evolutionary algorithm based on decomposition. IEEE Trans. Evol. Comput. **11**(6), 712–731 (2007)
11. Cheng, R., Jin, Y., Olhofer, M., Sendhoff, B.: A reference vector guided evolutionary algorithm for many-objective optimization. IEEE Trans. Evol. Comput. **20**(5), 773–791 (2016)
12. Jiang, S., Yang, S.: A strength Pareto evolutionary algorithm based on reference direction for multiobjective and many-objective optimization. IEEE Trans. Evol. Comput. **21**(3), 329–346 (2017)
13. Zitzler, E., Laumanns, M., Thiele, L.: SPEA 2: improving the strength pareto evolutionary algorithm. TIK-report 103 (2001)

On Statistical Analysis of MOEAs with Multiple Performance Indicators

Hao Wang[1], Carlos Igncio Hernández Castellanos[2], and Tome Eftimov[3](✉)

[1] Leiden University, Leiden, The Netherlands
h.wang@liacs.leidenuniv.nl
[2] Cinvestav-IPN, Mexico City, Mexico
chernandez@computacion.cs.cinvestav.mx
[3] Computer Systems Department, Jožef Stefan Institute, Ljubljana, Slovenia
tome.eftimov@ijs.si

Abstract. Assessing the empirical performance of Multi-Objective Evolutionary Algorithms (MOEAs) is vital when we extensively test a set of MOEAs and aim to determine a proper ranking thereof. Multiple performance indicators, e.g., the generational distance and the hypervolume, are frequently applied when reporting the experimental data, where typically the data on each indicator is analyzed independently from other indicators. Such a treatment brings conceptual difficulties in aggregating the result on all performance indicators, and it might fail to discover significant differences among algorithms if the marginal distributions of the performance indicator overlap. Therefore, in this paper, we propose to conduct a multivariate $\mathcal{E}$-test on the joint empirical distribution of performance indicators to detect the potential difference in the data, followed by a post-hoc procedure that utilizes the linear discriminative analysis to determine the superiority between algorithms. This performance analysis's effectiveness is supported by an experimentation conducted on four algorithms, 16 problems, and 6 different numbers of objectives.

Keywords: Many-objective optimization · Benchmarking · Performance analysis · Performance indicators · Hypothesis testing

1 Introduction

A variety of Multi-Objective Evolutionary Algorithms (MOEAs) have been proposed to tackle the multi-objective optimization problem (MOP), e.g., NSGA-II [7] and SMS-EMOA [1], which aim to optimize several objective functions simultaneously. In contrast to the single-objective optimization problems (SOP), for which the optimal solution is a single point in the search space, in MOPs, we seek the optimal trade-off between several objective functions, which is to approximate the Pareto front. Therefore, the difficulty of comparing MOEAs lies in quantifying approximation sets' quality to the Pareto front. Many performance indicators have been proposed and widely applied in many benchmarking and competition scenarios to determine a ranking among many tested

H. Ishibuchi et al. (Eds.): EMO 2021, LNCS 12654, pp. 26–37, 2021.
https://doi.org/10.1007/978-3-030-72062-9_3

algorithms, e.g., the hypervolume indicator [25] and the inverse generational distance (IGD) [4]. Typically, in the performance analysis, we consider a set of performance indicators and report the empirical performance of MOEAs on each indicator separately. This treatment, however, makes it difficult to draw an overall conclusion on which algorithm is generally the best, since each performance indicator emphasizes on different aspects of the empirical performance, e.g., GD measures the distance of the approximation set to the Pareto front while the hypervolume indicator also considers the spread of approximation sets. Naturally, the performance indicators induce a multi-criteria decision-making problem which would pick the algorithm showing the best trade-off between performance indicators based on some decision criteria.

The performance analysis becomes more complicated when it comes to the statistical significance: we usually apply a univariate non-parametric test (e.g., the Mann–Whitney U test) on each performance indicators independently for generating a ranking of algorithms, and aggregate such ranks over performance indicators when we would like to obtain an overview of the results [10]. This procedure would be sub-optimal and potentially fail to detect the significant difference between algorithms when the joint empirical distribution of performance indicators is multi-modal (each mode belongs to one algorithm) and, at the same time, its marginal distributions for each algorithm are overlapping. Therefore, we argue for the necessity of applying the multivariate testing procedure to the joint empirical distribution of performance data. This paper shall justify this proposal on the experimental data obtained on the combination of four algorithms, 16 problems, and six different choices for objective dimensionality (see Sect. 3.2 for the experimental setup).

This paper is organized as follows. In the next section, we will provide the background and related approaches for assessing the empirical performance of MOEAs. In Sect. 3, the proposed performance analysis procedure is discussed in detail. In Sect. 4, we demonstrate the outcome of our approach using the experimental data, followed by concluding remarks and future directions.

2 Background

Here, we consider continuous multi-objective optimization problems of the form $\min_{x \in Q} F(x)$, where $Q \subseteq \mathbb{R}^n$ and F is defined as the vector of the objective functions $F : Q \to \mathbb{R}^k, F(x) = (f_1(x), \ldots, f_k(x))^\top$, and where each objective $f_i : Q \to \mathbb{R}$ is continuous. In this paper, we assume the Pareto order/dominance in the objective space $\mathbb{R}^k$ and denote the set of (Pareto) efficient points and the corresponding Pareto front by P_Q and $F[P_Q]$, respectively. When $k > 4$, it is called a many-objective evolutionary problem.

2.1 Evolutionary Multi-objective Optimization

Multi-objective evolutionary algorithms [5] are among the most widely-applied algorithms for solving MOPs: 1) *Dominance-based methods* guide the search points using the dominance relationship directly. In this class, one of the most popular methods is the non-dominated sorting genetic algorithm II/III (NSGA-II/III) [7,8], which is underpinned by the non-dominated sorting procedure for ranking the current solutions with respect to the Pareto dominance, and by measuring the crowding distance for diversifying the solutions. In this work, we will also test the SPEA2+SDE (Shift-Based Density Estimation for Pareto-Based Algorithms) algorithm which has been found to have good results in many-objective optimization problems [18]. 2) *Decomposition-based methods* [19,24] are built upon scalarization techniques, which can generate a good distribution of the Pareto front given a set of well-distributed scalarization problems. The most prominent algorithm in this class is MOEA/D (MOEA based on decomposition) [24] that decomposes the original MOP into multiple different SOPs, each of which is tackled with the information from the neighboring SOP. The decomposition is usually realized via scalarizations, such as, the weighted sum, weighted Tchebycheff, or Penalty-based boundary intersection. 3) *Indicator-based methods* [21] rely on performance indicators to assign the contribution to individual points. SMS-EMOA [1] has received the most attention in this class.

2.2 Performance Assessment for Multi-objective Optimization

To assess the performance of multi-objective optimization, which is not a single solution but rather an approximation set, it can be analyzed with different criteria related to convergence and diversity. As we have previously mentioned, to do this, performance indicators are usually employed. These indicators are mathematical functions that map an approximation set to a real number [20]. The most commonly used performance indicators are the hypervolume [25], the generational distance [4], inverse generational distance [4], epsilon [16], spread [9], and generalized spread [9]. The selection of the performance indicator is a user preference and allows the user to select the performance indicator for which the newly developed algorithm performs the best, so in most cases such kind of comparisons can be biased. The influence of the selection of different performance indicators in comparisons studied has been investigated in [17].

In some cases, one could be interested in using several performance indicators to assess the performance or to guide search of a MOEA. This could be helpful to improve the overall performance of an algorithm and help adapt it to different preferences from the user. For instance, in [15], the authors proposed an ensemble approach which learns the best weights for a combination of five performance indicators. Further, advanced usage of the hypervolume indicators [2] over the space of indicators for automatic design of MOEAs.

Recently, a Deep Statistical Comparison (DSC) approach has been presented for performing more robust statistical analysis for single-objective optimization that avoids the influence of outliers and small differences that exist in the data values [12]. The main benefit of this approach is its ranking scheme, which is based on comparing distributions and not using only one descriptive statistic (either mean or median) to perform the analysis. The benefit of using the DSC approach for analyzing single performance indicator data has been shown in [11]. Further, different ensembles have been proposed that combine the DSC results obtained for a set of performance indicators: an average ensemble [10], a hierarchical majority vote [10], and data-driven ensemble [13]. The average ensemble performs average of the DSC rankings obtained for different performance indicators for each algorithm on a specific problem. The hierarchical majority vote ensemble checks which algorithm wins (i.e. with regard to the DSC ranking) in the most performance indicators on each benchmark problem separately. The data-driven ensemble uses the preference of each performance indicator, which is estimated by its entropy. The PROMETHEE method [3] is then used to determine the rankings.

3 Methodology

We typically consider multiple performance indicators at the same time to evaluate the performance of multi-objective optimization algorithms. When it comes to applying the statistical procedure to compare them, commonly, univariate hypothesis testing procedures are applied to each performance indicator, and the results are either presented separately or aggregated using simple methods. The potential risk of such approaches is that if the performance indicators are correlated and hence employing univariate tests on each indicator might fail to report significant differences among algorithms, which could only be discovered when we consider the joint distribution of performance indicators (see Fig. 3a and its explanation in Sect. 4.2 for an example). This consideration leads to our proposal that is to apply multivariate tests on the joint distribution of performance data.[1]

3.1 Our Approach

All ranking schemes of combining different performance indicators first compared the algorithms for each performance indicator separately and then combined the results from different comparisons using different heuristics. To avoid this, we extended the DSC ranking scheme, where the information obtained from several performance indicators is not analyzed separately but their joint distribution is compared. So instead of comparing one-dimensional data which is in the case

[1] The data files and the source code of this study can be accessed here: https://github.com/wangronin/EMO21.

of the DSC, we compared the high-dimensional joint distribution of the performance indicators. For this reason, we used the *multivariate $\mathcal{E}$-test* [22], which may be one of the most powerful tests available for high-dimensional data.

Further, we briefly reintroduce the DSC ranking scheme. Let us assume that m algorithms are involved in the comparison. Let α be the significance level used for the *multivariate $\mathcal{E}$-test* statistical test, where all $m \cdot (m-1)/2$ pairwise comparisons between the algorithms are performed and the p-values are organized in a $m \times m$ matrix, N_i, where i is the benchmark problem:

$$N_i[l_1, l_2] = \begin{cases} p_{\text{value}}, & l_1 \neq l_2 \\ 1, & l_1 = l_2 \end{cases}, \tag{1}$$

where l_1 and l_2 are different algorithms and $l_1, l_2 = 1, \ldots, m$.

Since multiple pairwise comparisons are made, the family-wise error (FWER) is controled by applying the *Bonferroni correction*. Looking at the matrix N_i, we can say that is always reflexive, and symmetric, but its transitivity should be checked. To do this, the matrix $N_i^{'}$ is introduced using the following equation:

$$N_i^{'}[l_1, l_2] = \begin{cases} 1, & N_i[l_1, l_2] \geq \alpha_{\text{X}}/C_m^2 \\ 0, & N_i[l_1, l_2] < \alpha_{\text{X}}/C_m^2 \end{cases}. \tag{2}$$

Having the matrix $N_i^{'}$, if transitivity is satisfied, we can split the algorithms into disjoint sets of algorithms such that the algorithms that belong to the same set have the same joint distribution of the quality indicators and should be ranked as the same.

Formally, given a set of m algorithms $\mathcal{A} = \{A_1, A_2, \ldots, A_m\}$ and a set of performance indicators $\mathcal{I} = \{I_1, I_2, \ldots, I_p\}$ (all subject to maximization, w.l.o.g.), this test returns a family of disjoint sets of algorithms that are significantly different from each other with respect to $\mathcal{I}$, namely $\mathcal{G} = \{G_1, G_2, \ldots\}$, where $\forall G \neq G' \in \mathcal{G}, G, G' \subseteq \mathcal{A}, G \cap G' = \emptyset$, and $\cup_{G \in \mathcal{G}} G = \mathcal{A}$. Based on this testing result, we proceed to determine which group is superior using the widely-applied *Fisher's linear discriminative analysis* (LDA), which aims to find the best linear scalarization of performance indicators that maximizes the separation of the algorithm groups when scalarizing the points in each group. The rationale behind this approach is: having known that two groups are significantly different, we seek a subspace (representing the linear scalarization) onto which the projection of grouped performance points would respect this statistical significance to the maximal degree, such that in this subspace the decision of superiority for algorithms, even via simple descriptive statistics, e.g., the projected group mean, would not be affected by the randomness with high probability. We shall denote by $\mathbf{w}$ the linear weights and use $\{\boldsymbol{\mu}_i\}_{i=1}^{|\mathcal{G}|}$ and $\boldsymbol{\mu}$ for the group means and overall mean of the performance values, respectively. The LDA procedure is to maximize the following separation measure: $S = \mathbf{w}^\top \boldsymbol{\Sigma} \mathbf{w} / \mathbf{w}^\top \mathbf{C} \mathbf{w}$, subject to $\mathbf{w} \in \mathbb{R}^p_{\geq 0}$, where $\boldsymbol{\Sigma} = \sum_{i=1}^{|\mathcal{G}|} (\boldsymbol{\mu}_i - \boldsymbol{\mu})(\boldsymbol{\mu}_i - \boldsymbol{\mu})^\top / |\mathcal{G}|$ is the sample covariance of the group means and $\mathbf{C}$ is the sample covariance of all performance points.

Note that, it is necessary to enforce all components of $\mathbf{w}$ to be non-negative since we will loose the interpretability of the scalarization if performance indicators (subject to maximization) were combined using weights having different signs. Also, when this procedure fails to deliver a feasible $\mathbf{w}$ in analyzing our experimental data, we simply take the equal weights for all performance indicators. Having obtained the weights $\mathbf{w}$ from LDA, we could calculate the scalarized performance values, i.e., $\sum_{k=1}^{p} w_i I_k$, which will be referred as *LD values* in this paper. To determine a ranking of algorithm groups, we propose to take the mean of scalarized points for each group and simply sort the resulting mean value.

3.2 Experimental Setup

We elaborate the experimental setting as follows. Algorithms: NSGA-II, NSGA-III, MOEA/D, and SPEA-SDE; Problems: DTLZ 1–7 and WFG 1–9, which are denoted as F1-F7 and F8-F16 in this paper, respectively; Number of objectives: 3, 5, 7, 10, and 15; Performance indicators: Δ_2 and the hypervolume HV; Function evaluation budget: 125,000; Independent runs: 20; We analyze the final population for each algorithm. The experiment is implemented using **Platemo** [23].

4 Performance Analysis

Firstly, the empirical performance of the tested algorithms are plotted in Fig. 1, where the performance values from 20 repetitions are scatted in the performance space spanned by HV and Δ_2. We re-scale the HV and Δ_2 values to the unit interval for each pair of objective dimension and problem, and additionally transform the normalized Δ_2 value to $1 - \Delta_2$, such that the resulting values are subject to maximization. From the figure, it is obvious to see that variability of both performance indicators can change drastically across different algorithms, dimensions, and problems, for instance, both HV and Δ_2 are very stable for each algorithm on problem F2 across all dimensions while HV exhibits a relatively huge dispersion on problems F8, F9, and F10 on dimensions 7, 10, and 15.

4.1 Deep Statistical Comparison

Here, we applied the state-of-the-art statistical ranking approaches from the deep statistical comparison methodology[2], where the well-known two-sample Anderson-Darling (AD) [6] test is taken to generate the ranks. In details, four statistical procedures are conducted: 1) the standard DSC ranking scheme on the HV samples, 2) the standard DSC ranking scheme on the Δ_2 samples, 3) the so-called DSC hierarchical ranking on both HV and Δ_2, which determines the superiority of an algorithm by the majority vote from a set of performance indicators, and 4) a data-driven DSC ranking scheme on both HV and Δ_2, in

[2] For applying the DSC ranking scheme, we used the **DSCTool**, which is developed as RESTful web services to the DSC ranking functionalities [14].

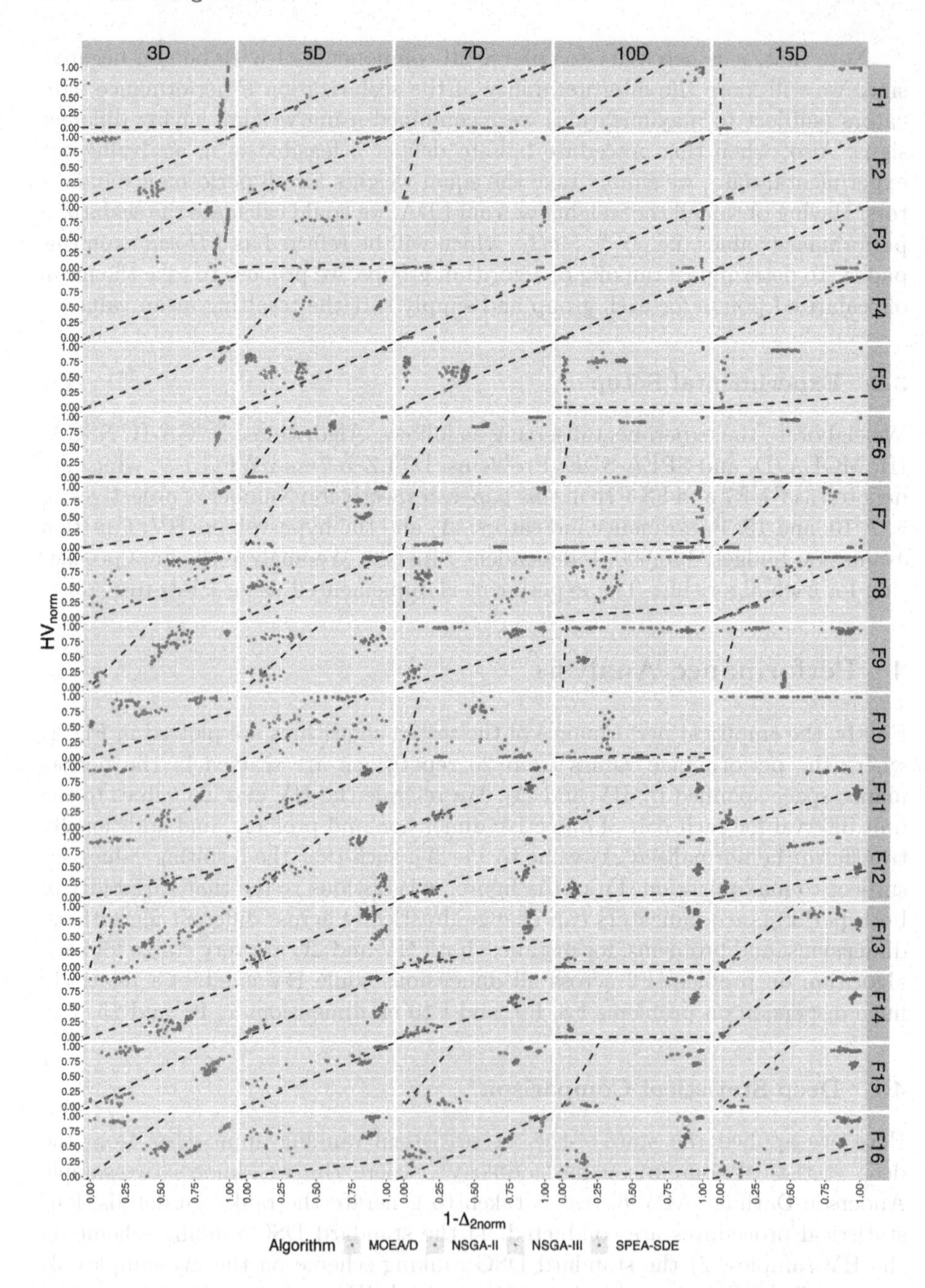

Fig. 1. The normalized samples of two performance indicators, the hypervolume (HV) and Δ_2, are shown for each pair of objective dimension and test function for four MOO algorithms, where weights $\mathbf{w}$ of the best linear discriminative function is indicated by the dash line. The best discriminative function is determined by maximizing the separation between algorithm groups that are statistically significant (see Table 1), after projecting them onto span$\{\mathbf{w}\}$.

which we compute the entropy carried by each performance indicator and impose an entropy-based weight on rankings from each performance indicator when aggregating them. We demonstrate the resultant DSC rankings as heatmaps in Fig. 2, where ranks obtained on each problem and objective dimension are averaged over 16 problems for a high-level comparison. When comparing the rankings from HV to that from Δ_2, we observed some disagreements of those two indicators in deciding the ordering of algorithm, e.g., judging from both HV, NSGA-III performs about the same as SPEA-SDE except that it shows better performance on 15D, while SPEA-SDE takes the lead when the dimensionality gets higher in terms of Δ_2. Such disagreements are intrinsic to the performance indicators since HV prefers solutions around the knee of the Pareto front while Δ_2 prefers solutions close to the approximation set, which is chosen to be well-distributed along the Pareto front. Also, the hierarchical and data-driven DSC ranking schemes aggregate the rankings from each performance indicator to yield an overview of the empirical performance. Those two DSC ranking approaches exhibit, on our experimental data, quite consistent results across all dimensions of the objective space. It is worth noting that we always employ a univariate test to determine the rankings, even if aggregations are applied in those four comparison approaches.

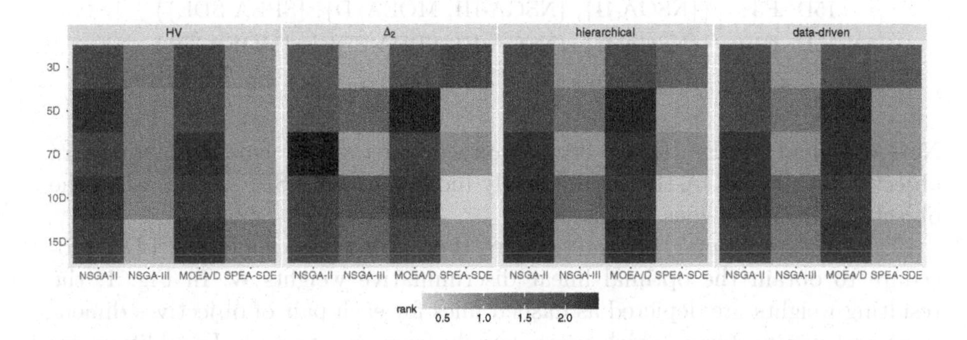

Fig. 2. Averaged deep statistical comparison rankings of four algorithms over 16 test problems with four different ranking schemes applied (from left to right): the DSC ranking on HV, the DSC ranking on Δ_2, the DSC hierarchical ranking, and the data-driven DSC ranking.

4.2 Multi-indicator Analysis

As proposed previously, we firstly applied the multivariate $\mathcal{E}$-test to compare the algorithms in a pairwise manner (with Bonferroni correction at a significance level of 5%), which takes the 2D sample points in the performance space spanned by HV and Δ_2 and results in groups of algorithms that are significantly different from each other. We show, in Table 1, the cases where there is at least one non-singleton set in the grouping. From the table, it is seen than algorithms

Table 1. The result of applying the multivariate $\mathcal{E}$-test on the bi-variate sample points (HV, Δ_2) with 5% statistical significance and the Bonferroni correction. For each objective dimension and test problem, the groups of algorithms are significantly different from each other and we only show the cases where there is at least one non-singleton group in $\mathcal{G}$.

k	$F(x)$	$\mathcal{G}$
3D	F1	{{NSGA-II}, {NSGA-III, MOEA/D}, {SPEA-SDE}}
3D	F3	{{NSGA-II}, {NSGA-III, MOEA/D}, {SPEA-SDE}}
3D	F4	{{NSGA-II}, {NSGA-III, MOEA/D}, {SPEA-SDE}}
3D	F8	{{NSGA-II}, {NSGA-III, SPEA-SDE}, {MOEA/D}}
5D	F1	{{NSGA-II}, {NSGA-III, MOEA/D}, {SPEA-SDE}}
5D	F2	{{NSGA-II}, {NSGA-III, MOEA/D}, {SPEA-SDE}}
5D	F3	{{NSGA-II}, {NSGA-III, MOEA/D}, {SPEA-SDE}}
5D	F4	{{NSGA-II}, {NSGA-III, MOEA/D}, {SPEA-SDE}}
7D	F3	{{NSGA-II}, {NSGA-III, MOEA/D}, {SPEA-SDE}}
7D	F8	{{NSGA-II, MOEA/D}, {NSGA-III}, {SPEA-SDE}}
10D	F2	{{NSGA-II}, {NSGA-III, MOEA/D}, {SPEA-SDE}}
10D	F8	{{NSGA-II}, {NSGA-III, SPEA-SDE}, {MOEA/D}}
15D	F3	{{NSGA-II}, {NSGA-III, MOEA/D}, {SPEA-SDE}}
15D	F10	{{NSGA-II}, {NSGA-III, SPEA-SDE}, {MOEA/D}}

NSGA-III and MOEA/D are often grouped together for smaller numbers of objectives while NSGA-III is statistically indifferent from SPEA-SDE when the objective dimension goes larger.

Taking the grouping information, we then proceed to apply the LDA procedure to obtain the optimal linear discriminative weights $\mathbf{w}$. In Fig. 1, the resulting weights are depicted as dashed lines for each pair of objectives dimensions and test problem, which varies largely from case to case. In addition, we zoom into the scenario of the five-dimensional F16 problem (Fig. 3a) and also render the distributions of the LD values for each algorithm (Fig. 3b. Note that all groups are singletons in this case). The effectiveness of the proposed methodology lies in the fact that a univariate test applied on the marginal distributions of performance indicators would not lead to a significant discovery because the marginal distributions are largely overlapping for all algorithms (see the top and right sub-plots in Fig. 3a, where the marginals are estimated using the kernel density estimation), as opposed to the outcome of the multivariate $\mathcal{E}$-test that all algorithms are significant different in this two-dimensional performance space. The LDA procedure, which could be considered as a post-hoc analysis to the multivariate $\mathcal{E}$-test, visualizes this statistical difference on the 1D subspace $\mathrm{span}\{\mathbf{w}\}$ (the dashed line in Fig. 3a): the 2D performance values, after projected onto $\mathrm{span}\{\mathbf{w}\}$, are more distinguishable than the two marginals of performance values (see the histogram/density in Fig. 3b). Finally, to decide the ranking of algorithm groups, we could take the mean LD values since we have already gained the confidence on the group differences from the multivariate $\mathcal{E}$-test.

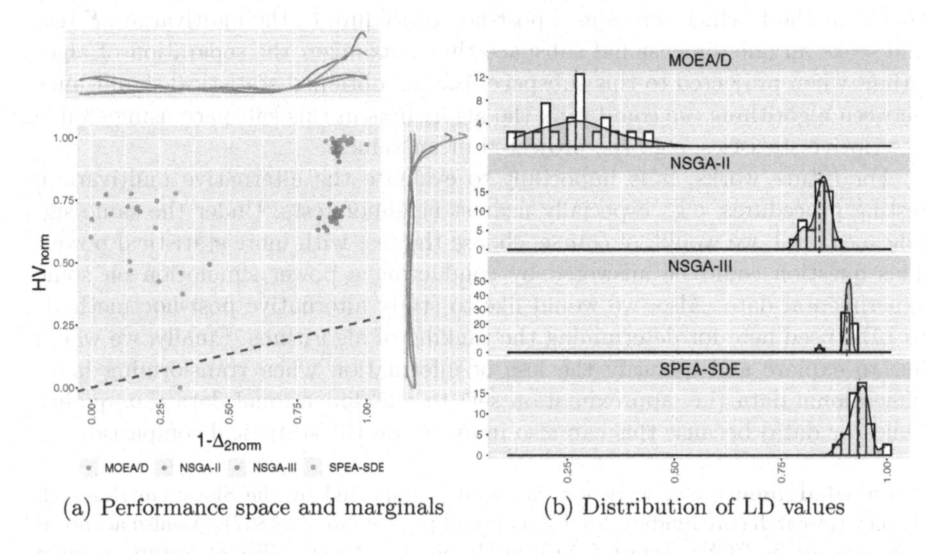

(a) Performance space and marginals (b) Distribution of LD values

Fig. 3. On the five-dimensional F16 problem, the performance points are shown in the scatter plot on the left, where the linear discriminative weights are indicated by the dashed line. Accompanying the scatter plot, the marginal distributions of HV are largely overlapping for all algorithms (the same for Δ_2), in contrast to the clear difference of the LD values across algorithms (on the right).

5 Conclusions and Future Work

In this paper, we delve into the question on how to utilize multiple performance indicators together in a statistical testing procedure to compare multi-objective optimization (MOO) algorithms, and how to best visualize and rank the performance data from multiple performance indicators when observing a significant difference between algorithms. For this purpose, we have chosen four evolutionary MOO algorithms, NSGA-II, NSGA-II, MOEA/D, and SPEA-SDE, which are independently executed for 20 times on DTLZ 1–7 and WFG 1–9 problems with the number of objectives $k \in \{3, 5, 7, 10, 15\}$. Subsequently, we compare those algorithms using the state-of-the-art Deep Statistical Comparison (DSC) method. Although it is easy to understand the resulting rankings, those are still computed using univariate testing procedures, which might fail to discover significant difference among algorithm, if the performance indicators exhibit a correlation structure. This comes from the fact that the selection of the performance indicators in DSC is still done by the user. Thus, we propose to directly apply a multivariate test ($\mathcal{E}$-test in this paper) in the multi-dimensional performance space spanned by those indicators. The result shows that this test could identify significant difference which would not be detected if we only apply univariate tests on the marginal distribution. Furthermore, we propose to visualize and rank the multivariate performance data by the linear discriminative analysis (LDA) method, which serves as a post-hoc procedure to the multivariate $\mathcal{E}$-test and seeks an one-dimensional subspace that maximizes the separation of algorithms when projected to this subspace. Having obtained statistical significance between algorithms, we could rank the algorithms in this subspace using simple descriptive statistics, e.g., the projected mean value.

For future works, it is important to evaluate the alternative multivariate testing procedures, e.g., especially nearest neighbor tests. Under the same significance level, we would, of course, choose the test with more statistical power. This question could be answered by conducting a power simulation on some experimental data. Also, we would like to study alternative post-hoc methods to LDA used here for determining the ranking of algorithms. Finally, we would like to explore and quantify the loss of information when transforming high-dimensional data (i.e. approximation set) to one-dimensional data (i.e. quality indicator data) because this can also influence on the statistical comparison.

Acknowledgment. Our work was financially supported by the Slovenian Research Agency (research core funding No. P2-0098 and project No. Z2-1867). We also acknowledge support by COST Action CA15140 "Improving Applicability of Nature-Inspired Optimisation by Joining Theory and Practice (ImAppNIO)".

References

1. Beume, N., Naujoks, B., Emmerich, M.: SMS-EMOA: multiobjective selection based on dominated hypervolume. Eur. J. Oper. Res. **181**(3), 1653–1669 (2007)
2. Bezerra, L.C.T., López-Ibáñez, M., Stützle, T.: Automatic component-wise design of multiobjective evolutionary algorithms. IEEE Trans. Evol. Comput. **20**(3), 403–417 (2016)
3. Brans, J.-P., Mareschal, B.: Promethee methods. Multiple Criteria Decision Analysis: State of the Art Surveys. ISORMS, vol. 78, pp. 163–186. Springer, New York (2005). https://doi.org/10.1007/0-387-23081-5_5
4. Coello, C.A.C., Cortés, N.C.: Solving multiobjective optimization problems using an artificial immune system. Genet. Program. Evolvable Mach. **6**(2), 163–190 (2005)
5. Coello, C.A.C., van Veldhuizen, D.A., Lamont, G.B.: Evolutionary algorithms for solving multi-objective problems, Genetic algorithms and evolutionary computation. Kluwer, vol. 5 (2002)
6. D'Agostino, R.B.: Goodness-of-Fit-Techniques, vol. 68. CRC Press, United States (1986)
7. Deb, K., Pratap, A., Agarwal, S., Meyarivan, T.: A fast and elitist multiobjective genetic algorithm: NSGA-II. IEEE Trans. Evol. Comput. **6**(2), 182–197 (2002)
8. Deb, K., Jain, H.: An evolutionary many-objective optimization algorithm using reference-point-based nondominated sorting approach. IEEE Trans. Evol. Comput. **18**(4), 577–601 (2014)
9. Deb, K., Sindhya, K., Hakanen, J.: Multi-objective optimization. In: Decision Sciences: Theory and Practice, pp. 145–184. CRC Press (2016)

10. Eftimov, T., Korošec, P., Seljak, B.K.: Comparing multi-objective optimization algorithms using an ensemble of quality indicators with deep statistical comparison approach. In: 2017 IEEE Symposium Series on Computational Intelligence (SSCI), pp. 1–8. IEEE (2017)
11. Eftimov, T., Korošec, P., Koroušić Seljak, B.: Deep statistical comparison applied on quality indicators to compare multi-objective stochastic optimization algorithms. In: Nicosia, G., Pardalos, P., Giuffrida, G., Umeton, R. (eds.) MOD 2017. LNCS, vol. 10710, pp. 76–87. Springer, Cham (2018). https://doi.org/10.1007/978-3-319-72926-8_7
12. Eftimov, T., Korošec, P., Seljak, B.K.: A novel approach to statistical comparison of meta-heuristic stochastic optimization algorithms using deep statistics. Inf. Sci. **417**, 186–215 (2017)
13. Alves, F., Pereira, A.I., Fernandes, A., Leitão, P.: Optimization of home care visits schedule by genetic algorithm. In: Korošec, P., Melab, N., Talbi, E.-G. (eds.) BIOMA 2018. LNCS, vol. 10835, pp. 1–12. Springer, Cham (2018). https://doi.org/10.1007/978-3-319-91641-5_1
14. Eftimov, T., Petelin, G., Korošec, P.: DSCTool: a web-service-based framework for statistical comparison of stochastic optimization algorithms. Appl. Soft Comput. **87**, 105977 (2020)
15. Falcón-Cardona, J.G., Liefooghe, A., Coello Coello, C.A.: An ensemble indicator-based density estimator for evolutionary multi-objective optimization. In: Bäck, T., et al. (eds.) PPSN 2020. LNCS, vol. 12270, pp. 201–214. Springer, Cham (2020). https://doi.org/10.1007/978-3-030-58115-2_14
16. Knowles, J., Thiele, L., Zitzler, E.: A tutorial on the performance assessment of stochastic multiobjective optimizers. Tik Rep. **214**, 327–332 (2006)
17. Korošec, P., Eftimov, T.: Multi-objective optimization benchmarking using DSCTool. Mathematics **8**(5), 839 (2020)
18. Li, M., Yang, S., Liu, X.: Shift-based density estimation for pareto-based algorithms in many-objective optimization. IEEE Trans. Evol. Comput. **18**(3), 348–365 (2014)
19. Moubayed, N.A., Petrovski, A., McCall, J.A.W.: D^2MOPSO: MOPSO based on decomposition and dominance with archiving using crowding distance in objective and solution spaces. Evol. Comput. **22**(1), 47–77 (2014)
20. Riquelme, N., Von Lücken, C., Baran, B.: Performance metrics in multi-objective optimization. In: 2015 Latin American Computing Conference (CLEI), pp. 1–11. IEEE (2015)
21. Schütze, O.: A scalar optimization approach for averaged Hausdor approximations of the Pareto front. Eng. Optim. **48**(9), 1593–1617 (2019)
22. Székely, G.J., Rizzo, M.L.: Testing for equal distributions in high dimension. InterStat **5**, 1–6 (2004)
23. Tian, Y., Cheng, R., Zhang, X., Jin, Y.: PlatEMO: a MATLAB platform for evolutionary multi-objective optimization [educational forum]. IEEE Comput. Intell. Mag. **12**(4), 73–87 (2017)
24. Zhang, Q., Li, H.: MOEA/D: a multi-objective evolutionary algorithm based on decomposition. IEEE Trans. Evol. Comput. **11**(6), 712–731 (2007)
25. Zitzler, E., Thiele, L.: Multiobjective evolutionary algorithms: a comparative case study and the strength pareto approach. IEEE Trans. Evol. Comput. **3**(4), 257–271 (1999)

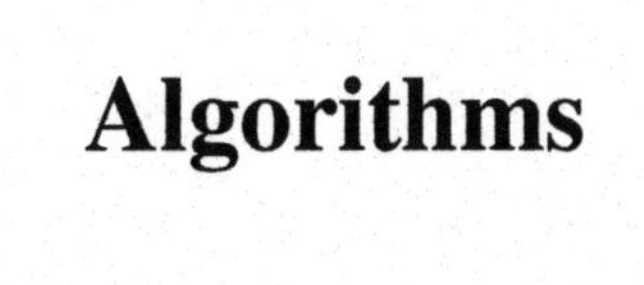

Algorithms

Population Sizing of Evolutionary Large-Scale Multiobjective Optimization

Cheng He and Ran Cheng(✉)

Guangdong Provincial Key Laboratory of Brain-Inspired Intelligent Computation, Department of Computer Science and Engineering, Southern University of Science and Technology, Shenzhen 518055, China

Abstract. Large-scale multiobjective optimization problems (LSMOPs) are emerging and widely existed in real-world applications, which involve a large number of decision variables and multiple conflicting objectives. Evolutionary algorithms (EAs) are naturally suitable for multiobjective optimization due to their population-based property, allowing the search of optima simultaneously. Nevertheless, LSMOPs are challenging for conventional EAs, mainly due to the huge volume of search space in LSMOPs. Thus, it is important to explore the impact of the population sizing on the performance of conventional multiobjective EAs (MOEAs) in solving LSMOPs. In this work, we compare several representative MOEAs with different settings of population sizes on some transformer ratio error estimation (TREE) problems in the power system. These test cases are defined on combinations of three population sizes, three TREE problems, and five MOEAs. Our results indicate that the performances of conventional MOEAs with different population sizes in solving LSMOPs are different. The impact of population sizing is most significant for differential evolution based and particle swarm based MOEAs.

Keywords: Large-scale optimization · Multiobjective optimization · Population size · Transformer ratio error estimation

1 Introduction

In real-world applications, to the satisfaction of different decision makers, the optimization of a problem may involve multiple conflicting criterion. That is, an optimization problem can involve multiple objectives, known as the multiobjective optimization problem [12,37,41]. Without loss of generality, a minimization MOP can be mathematically formulated as

$$\begin{aligned} &\text{minimize } \mathbf{f}(\mathbf{x}) = (f_1(\mathbf{x}), \ldots, f_M(\mathbf{x}))^T \\ &\text{subject to } \mathbf{x} \in \mathbb{R}^D, \end{aligned} \tag{1}$$

where $\mathbf{x}$ denotes a solution in the D dimensional decision space $\mathbb{R}^D$ and M is the number of objectives [1]. Typically, the Pareto dominance relationship is

H. Ishibuchi et al. (Eds.): EMO 2021, LNCS 12654, pp. 41–52, 2021.
https://doi.org/10.1007/978-3-030-72062-9_4

adopted to distinguish the qualities of two candidate solutions $\mathbf{x}_1$ and $\mathbf{x}_2$ in an MOP. For $i, j \in \{1, 2, \ldots, M\}$, if for all objectives $f_i(\mathbf{x}_1) \leqslant f_i(\mathbf{x}_2)$ and there is at least one index j that $f_j(\mathbf{x}_1) < f_j(\mathbf{x}_2)$, $\mathbf{x}_1$ Pareto dominates $\mathbf{x}_2$ (denoted as $\mathbf{x}_1 \prec \mathbf{x}_2$). To solve MOPs, a variety of multiobjective evolutionary algorithms (MOEAs) have been proposed and can be categorized into three classes [19].

Dominance based MOEAs. These MOEAs compare pairwise candidate solutions according to a specific dominance relationship, e.g., Pareto dominance [24], fuzzy dominance [33], corner dominance [34], or improved strength Pareto dominance [42]. Typical algorithms include the fast and elitist multiobjective genetic algorithm (NSGA-II) [12] and the region-based MOEA (PESA-II) [9].

Decomposition based MOEAs. By decomposing an MOP into a series of subproblems and optimizing them simultaneously, MOEAs of this class have shown good scalability to the number of objectives, such as MOEA/D [37] and MOEA/D-M2M [23].

Indicator based MOEAs. An indicator is designed to assess the quality of a candidate solution considering both convergence and diversity [41]. The indicator-based MOEAs adopt the indicator to assess the contribution of a candidate solution to the entire population, and candidate solutions with better indicator values are selected to survive during the evolution (e.g., dominated hypervolume based MOEA (SMS-EMOA) [3]).

In addition to these three classes of MOEAs, some other MOEAs use the general framework of these classical MOEAs but adopt different offspring generation strategies, including the multiobjective particle swarm optimizer (MOPSO) [22] and the third evolution step of generalized differential evolution (GDE3) [21].

Most existing MOEAs have shown promising performance in solving MOPs with less than 30 decision variables but may fail to solve MOPs with more than 100 decision variables efficiently (*a.k.a.* large-scale MOPs (LSMOPs) [6,8]). Their failure in solving LSMOPs is mainly attributed to the huge volume of search space, known as "the curse of dimensionality" [40]. Without specific guidance in searching through the huge search space, existing population-based MOEAs may suffer from slow convergence rate and ineffective global search. Generally, four main types of large-scale MOEAs are proposed in the recent two decades.

Cooperative Coevolution Based MOEAs. Under the framework of cooperative coevolution (CC), the decision variables are first divided into different groups for problem decomposition without prior knowledge. Then the problem is solved in a divide-and-conquer manner [36]. Typical algorithms include the CC based GDE3 (CCGDE3) [2] and the CC based MOEA with improved resource allocation (CCMOGA-ICRA) [30].

Decision Variable Analysis Based MOEAs. Similar to the CC-based large-scale MOEAs, algorithms of this type divide the decision variables into different groups (i.e., convergence-related variable, diversity-related variable, or mixed variable). Then, the algorithm optimizes different types of variables independently. Since the variable grouping is on the basis of variable analysis, this method can also be regarded as MOEAs with posteriori knowledge. For instances, the decision variable analysis based MOEA (MOEA/DVA) [25], the decision variable cluster based MOEA (LMEA) [38], the MOEA based on covariance matrix adaptation evolution strategy with scalable small subpopulations (S3CMA-ES) [4], and the recently proposed universal strengthened searching module based on MOEA (MOEA-SS) [26] have used decision variable analysis techniques to handle LSMOPs.

Problem Reformulation Based MOEAs. These approaches aim to transfer the original LSMOP into some simpler MOP(s). Algorithms in this type can be further divided into two different categories, i.e., reformulations with and without feedback from the LSMOP. Taking the weight optimization based framework (WOF) as an example. It first divides the decision variables into different groups by using some grouping techniques and then assigns each group a weight to be optimized via some transfer function. In WOF, neither prior nor posteriori knowledge is used during the optimization, and the problem transformation is independent with the original problem [40]. On the contrary, the problem reformulation based framework (LSMOF) uses some promising candidate solutions during the problem reformulation without grouping [5,18]. Moreover, the problem reformulation and the problem optimization are processed in an interactive way, during which the knowledge of the population and the original problem is involved.

Offspring Generation Based MOEAs. Different from the aforementioned MOEAs that solve LSMOPs by manipulating the decision space directly, those offspring generation based MOEAs aim to use some effective and efficient offspring generation strategies for accelerating the optimization. For instances, the competitive swarm optimizer (CSO) [7] used in large-scale single-objective optimization is adopted and tailored for large-scale multiobjective optimization. Two representative algorithms are the CSO based MOEA (CMOPSO) [39] and the large-scale CSO (LMCSO) [32]. Recently, a direction guided MOEA (DGEA) inspired by problem reformulation is also proposed to solve LSMOPs via effective offspring generation [15].

Most existing large-scale have shown promising performance in solving some artificial benchmark problems, e.g., DTLZ [13] and WFG [29] test suites. Nevertheless, their effectiveness and efficiency are still not satisfactory in solving some complex large-scale optimization problems (e.g., LSMOP test suite [8]) or real-world applications (e.g., TREE test suite [17]). The development of large-scale multiobjective optimization is still in the infancy. On the other side, the potential of conventional MOEAs in solving LSMOPs remains to be seen since most large-scale MOEAs are developed on the basis of conventional MOEAs. It is an intuitive idea to use large population size in conventional MOEAs to conquer the vast search space in large-scale optimization. Specifically, the use of large population size is expected to enhance the effectiveness of conventional MOEAs in solving LSMOPs. Thus, it is essential to investigate the impact of population sizing on the performance of conventional MOEA in solving LSMOPs.

There are some studies about the impact of population sizing on the performance of MOEAs in multiobjective optimization. For instances, Kenneth *et al.* analyzed the interacting roles of population sizing and crossover and genetic algorithm in 1990. It was concluded that a larger population resulted in better solutions in single-objective optimization [10]. Later, instead of addressing the decision variable related difficulty, Hisao *et al.* explored the impact of population sizing for solving MOPs with more than three objectives, and large population size was also suggested for many-objective optimization [20]. In 2015, Olympia *et al.* tested the impact of population sizing on both genetic and ant algorithms performance in case of cultivation process modelling. It was summarized that the increment in population size increases the accuracy of the solution in a genetic algorithm. However, the use of larger populations does not improve the solution accuracy and only increase the needed computational resources in the ant colony optimization [28]. Very recently, Geoffrey *et al.* investigated the combined impact of population sizing and sub-problem selection in MOEA/D, suggesting that a larger population performs better as the problem difficulty increases [27].

Most existing studies recommend a large population for genetic algorithms in solving complex MOPs, which supports our assumption about the impact of population sizing on the performance of MOEAs. Nevertheless, there is little research on the effect of population sizing when facing LSMOPs. Motivated by this, we explore the impact of the population sizing on the performance of conventional MOEAs in large-scale multiobjective optimization. All in all, five representative MOEAs are tested on three representative TREE problems, where different population sizes are adopted for solving TREE problems with up to 12,000 decision variables.

The remainder of this paper is organized as follows. We present the basic introduction to the five involved MOEAs and three TREE problems in Sect. 2. The experimental setup is given in Sect. 3 while the results are present in Sect. 4. Finally, we conclude this work in Sect. 5.

2 Background

In this part, we first give a brief introduction to five representative MOEAs, i.e., NSGA-II [12], MOEA/D [37], IBEA [41], MOPSO [22], and GDE3 [21]. Then the details of TREE1, TREE4, and TREE6 are demonstrated.

2.1 Representative MOEAs

As one of the most classical MOEAs in the area of multiobjective optimization, NSGA-II has been popular for decades due to its simplicity and robustness in solving conventional MOPs. It mainly includes two components for environmental selection, i.e., non-dominated sorting and crowding distance measurement, for balancing the convergence enhancement and diversity maintenance. The basic idea of non-dominated sorting is simple using the Pareto dominance, where those candidate solutions that are non-dominated by each other are grouped into the same front. Afterwards, the candidate solutions in those well-converged fronts are selected, and some remaining ones will be selected from the same front based on the crowding distance.

MOEA/D adopts a totally different mechanism. It first decomposes an MOP into a set of single-objective subproblem using some aggression method and then optimizes them simultaneously. The Tchebycheff aggression is widely used, which can be formulated as (2).

$$\text{minimize } g(\mathbf{x}|\gamma, \mathbf{z}^*) = \max_{1 \leq i \leq M} \{\gamma_i |f_i(\mathbf{x}) - z_i^*|\}, \tag{2}$$

where $\gamma = (\gamma_1, \ldots, \gamma_M)$ is a weight vector (M is the number of objectives), f_i is the ith objective value of solution $\mathbf{x}$, and $\mathbf{z}^* = (z_1^*, \ldots, z_M^*)$ is the reference point with $z_i^* = \min\{f_i(\mathbf{x})|\mathbf{x} \in \Omega\}$ (Ω is the decision space) [16].

Different with above two MOEAs, IBEA iteratively removes the worst individual from the population and updating the fitness values of the remaining individuals. The adopted indicator for assessing the fitness of a candidate is

$$F_0(\mathbf{x_i}) = \sum -\exp^{I(\mathbf{x}_j, \mathbf{x}_i)/\kappa} \text{ with } \mathbf{x}_j \in P \backslash \{\mathbf{x}_i\}, \tag{3}$$

where P is the current population, κ is a parameter, and function $I(\mathbf{x}_i, \mathbf{x}_j)$ is a binary quality indicator reflecting the differences between two solutions.

GDE3 is an extension of differential evolution (DE) for multiobjective optimization considering the constraints [21], which extended the DE/rand/1/bin method and considered the crowding distance for obtaining better-distributed solutions. Similarly, MOPSO is an extension of single-objective particle swarm optimizer. In MOPSO, the Pareto dominance is applied to determine the flight direction of a particle, and the previously found non-dominated vectors are maintained for guiding other particles later. The update of velocity $\mathbf{v}_i$ for the ith particle $\mathbf{x}_i$ follows Eq. (4).

$$\mathbf{v}_i = \omega \cdot \mathbf{v}_i + r_1 \cdot (\mathbf{p_{best}}_i - \mathbf{x}_i) + r_2 \cdot (\mathbf{rep}_h - \mathbf{x}_i), \tag{4}$$

where ω is an inertia weight, r_1 and r_2 are two random values selected from [0, 1], $\mathbf{p_{best}}_i$ is the best position that the ith particle has had, $\mathbf{rep}_h$ is a value that is taken from a repository with h being selecting in a specific way. Readers can find the implementation in [31].

Since different strategies have been used for environmental selection and offspring generation in the aforementioned five MOEAs, and the impact of population sizing on those algorithms could be different.

2.2 TREE Problems

TREE problems are real-world optimization problems extracted from the power delivery systems, whose goal is to estimate the real value of the measured voltages for calibrating the voltage transformers (VTs). Generally, there are three types of TREE problems according to the topologies of the VTs, including substations with primary side VTs (e.g., TREE1), substations with primary side and secondary side VTs (e.g., TREE4), and substations with both voltage and phase angle values (e.g., TREE6). In TREE1 and TREE4, the two objective functions are the total time-varying ratio errors and the ratio error variation over time. The detailed formulations of the first objectives are

$$f_1(\mathbf{x}) = \sum_{j=1}^{T}\sum_{k=1}^{K}\sum_{i=1}^{P} e_{k,j}^{i}, \tag{5}$$

$$f_2(\mathbf{x}) = \sum_{i=1}^{P} \sqrt{std((\Delta e_{1,1}^{i}, \cdots, \Delta e_{L,T}^{i})}), \tag{6}$$

where e denotes the ratio error and Δe is the variation of ratio errors. Specifically, TREE6 has the third objective, i.e., the variance of the phase angle ratio error variations for different VTs. The detailed formulation of the third objective is

$$f_3(\mathbf{x}) = \sum_{i=1}^{P} \sqrt{std((\Delta e_{\lfloor K/2 \rfloor,1}^{i}, \cdots, \Delta e_{K,T}^{i})}), \tag{7}$$

where K, T are problem related parameters. Note that, the original TREE problems can reflect different characters of real-world applications, e.g., variable interaction and constraints.

Since this work mainly focuses on large-scale multiobjective optimization, the constraints are ignored in the rest of this paper. Readers can refer to [17] for more details of these problems.

3 Experimental Setup

In this work, we test NSGA-II, MOEA/D, IBEA, GDE3, and MOPSO on TREE1, TREE4, and TREE6 with different settings of population size. Specifically, each test instance is examined for 20 independent runs to obtain a statistical result for reflecting the general performance of each algorithm. Experiments are conducted in the PlatEMO [31] on a server with Intel Xeon Gold 6130 CPU, 128GB RAM, and Windows 10 operating system. In the following, we briefly introduce the adopted performance indicator and the parameter settings in both the tested algorithms and problems.

1) Performance Indicator. Since the Pareto optimal fronts of those TREE problems are unknown, we adopt the normalized hypervolume (NHV) indicator to assess the performance of MOEAs. Note that, NHV takes both the convergence and uniformity into consideration, and a larger NHV value indicate the better performance [35].

2) Parameter Settings in TREE. In TREE1, TREE4, and TREE6, parameter K is set to 3, 6, and 12 respectively, and parameter T is set to 1,000. Consequently, the numbers of decision variables are 3,000, 6,000, and 12,000 in the three problems, respectively. Besides, both TREE1 and TREE4 are bi-objective LSMOPs while TREE6 is a tri-objective LSMOP as introduced in Sect. 2.

3) Parameter Settings in MOEAs. There is no specific parameter in NSGA-II and GDE3. In MOEA/D, the penalty based aggression method (as given in (2)) is adopted, and the neighborhood size T is set to $\lfloor N/10 \rfloor$ with N being the population size. In MOPSO, the number of divisions *div* in each objective is set to ten, and the fitness scaling factor κ is set to 0.05 in IBEA.

3) Other Settings. The maximum number of function evaluations (FEs) is used as the terminate condition for the compared algorithms. A total number of 3,000$\times D$ FEs are used for a test instance with D decision variables. For NSGA-II, IBEA, GDE3, and MOPSO, the population size N is set to 100, 300, and 1000 for each test instance. Since the population size N in MOEA/D should be equal to the number of uniform weight vectors, we set N to 91, 300, and 990 for each test instance by using the uniform weight sampling method in [37].

4 Empirical Analysis

In this section, we first give the general performance of the five compared MOEAs on three TREE problems. Then the convergence profiles of each tested MOEA are analyzed.

Table 1. The statics of NHV values obtained by five compared MOEAs with different population sizes. the best result in each row is highlighted.

Problem	N	NSGA-II	MOEA/D	IBEA	GDE3	MOPSO
TREE1	100	**8.63E-1(1.71E-5)**	**8.63E-1(8.58E-6)**	**8.63E-1(2.39E-5)**	**8.60E-1(2.04E-4)**	8.11E-1(3.03E-3)
	300	8.62E-1(3.73E-5)	8.63E-1(7.72E-6)	8.62E-1(4.18E-5)	8.53E-1(2.12E-4)	8.16E-1(1.81E-3)
	1000	8.61E-1(9.70E-5)	8.63E-1(1.50E-5)	8.61E-1(6.78E-5)	8.36E-1(6.59E-4)	**8.21E-1(1.70E-3)**
TREE4	100	**9.78E-1(1.75E-4)**	**9.79E-1(8.90E-5)**	**9.79E-1(1.71E-4)**	**9.76E-1(1.97E-4)**	9.07E-1(4.73E-3)
	300	9.64E-1(1.01E-3)	9.77E-1(1.30E-4)	9.66E-1(7.74E-4)	9.70E-1(2.71E-4)	9.17E-1(3.73E-3)
	1000	9.58E-1(1.00E-3)	9.73E-1(2.64E-4)	9.60E-1(7.97E-4)	9.56E-1(9.65E-4)	**9.24E-1(2.40E-3)**
TREE6	100	**8.64E-1(2.50E-4)**	**8.78E-1(6.27E-4)**	**8.70E-1(2.67E-4)**	8.54E-1(2.39E-3)	8.34E-1(3.14E-3)
	300	8.62E-1(4.10E-4)	8.77E-1(1.69E-4)	8.66E-1(3.66E-4)	8.62E-1(3.28E-3)	8.43E-1(1.89E-3)
	1000	8.57E-1(7.49E-4)	8.76E-1(2.75E-4)	8.63E-1(3.68E-4)	**8.63E-1(2.90E-3)**	**8.49E-1(1.25E-3)**

Table 1 presents the statistical NHV results achieved by each MOEA with different population sizes. On the one hand, NSGA-II, MOEA/D, and IBEA perform the best with a small population size on three TREE problems; by contrast, MOPSO has performed the best with large population size. As for GDE3, it has performed the best on TREE1 and TREE4 with a small population size but the best on TREE6 with large population size. On the other hand, MOEA/D with small population size has achieved the best result among all the test instances.

In spite of the different performances among MOEAs with different population sizes, the achieved NHV values are not significantly distinguishable. Thus, it is essential to observe the convergence profiles of the algorithms. For TREE1 problem with 3000 decision variables, the convergence profiles of the five compared MOEAs are displayed in Fig. 1(a). It can be observed that NSGA-II, MOEA/D, and IBEA have performed quite similar, which is attributed to the use

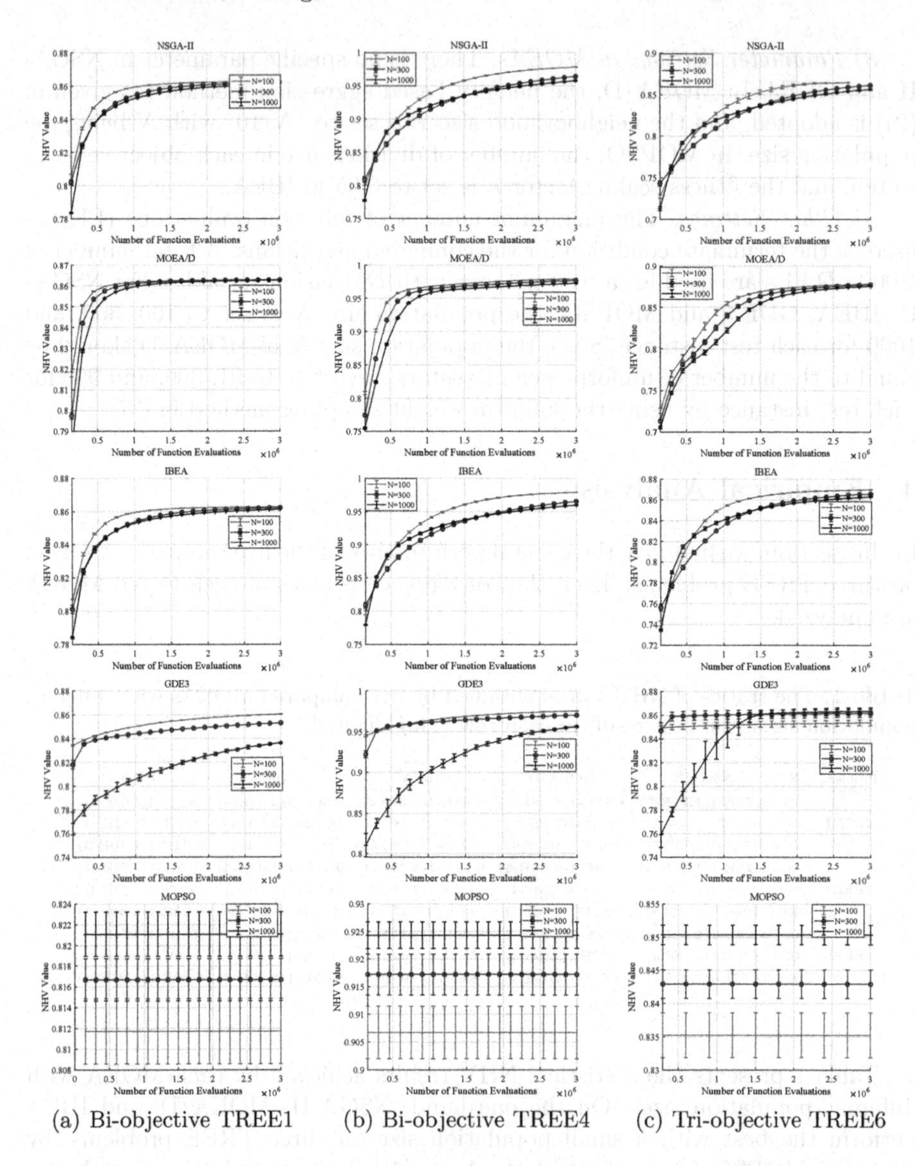

(a) Bi-objective TREE1 (b) Bi-objective TREE4 (c) Tri-objective TREE6

Fig. 1. The convergence profiles of the five compared MOEAs with different population sizes on three TREE problems, respectively.

of the same offspring reproduction strategy, i.e., simulated binary crossover [11] and polynomial crossover [14]. Despite that these three MOEAs with large population size have achieved worse at the early stage of the evolution, MOEAs with larger population size seems to converge faster. In GDE3, the differential evolution operator is adopted for offspring generation with strong randomness. Thus, a smaller population size will allow the algorithm to evolve for more generations, leading to significantly better performance. As for MOPSO, the particle swarm based update strategy is adopted for offspring generation, but it fails to converge after the first several generations, and the variation of NHV value is too small to be observed in this plot. That is, particle swarm based strategy may fail to solve LSMOPs, and it may suffer from premature even at the very early stage of evolution.

For TREE4 problem with 6000 decision variables, the convergence profiles of the five compared MOEAs are displayed in Fig. 1(b). Note that TREE4 is similar to TREE1 but with more decision variables. Consequently, the five compared algorithms with different settings of population sizes have performed similarly to their performance on TREE1. Nevertheless, the impact of population sizing on the five compared MOEAs is more significant. That is, NSGA-II, IBEA, and GDE3 encourages the use of small population size at the early stage; by contrast, MOPSO performs better with larger population size.

In comparison with TREE1 and TREE4 problems, TREE6 problem is more challenging due to its triple objectives and a larger number of decision variables. The convergence profiles of the five compared MOEAs on this problem are presented in Fig. 1(c). NSGA-II, MOEA/D, and IBEA have performed similarly to their performance on TREE1 and TREE4 problems. It is highlighted that GDE3 tends to perform better with large population size for TREE6 problem.

All the experimental results indicate that the population sizing could impact the performance of conventional MOEAs in large-scale multiobjective optimization. Generally, for those genetic operator based MOEAs, e.g., NSGA-II, MOEA/D, and IBEA, large population size can lead to a fast convergence rate at an early stage but a poor final result. For DE based MOEAs, it is good to use a small population size to encourage longer evolution process. Nevertheless, if the number of decision variables is quite large, the large population size should be used to achieve better final results. As for the particle swarm based MOEAs, they can use a large population size to enhance the quality of the global best particle(s) for better global optimization.

5 Conclusions

In this work, we have investigated the impact of population sizing on the performance of conventional MOEAs in solving LSMOPs. Five representative MOEAs, i.e., NSGA-II, MOEA/D, IBEA, GDE3, and MOPSO, are examined on three TREE problems. The number of objectives in these problems ranges from two to three, and the number of decision variables ranges in $\{3000, 6000, 12000\}$. Both the general performance and convergence profiles of the five compared MOEAs are assessed via the normalized hypervolume indicator. Generally, MOEA/D has achieved the best over nine test instances, followed by NSGA-II, IBEA, and GDE3. To be more specific, MOEA/D has shown to be robust to the population sizing; NSGA-II and IBEA with small population size have performed the best; the performance of GDE3 is relatively sensitive to population sizing; by contrast, MOPSO has stagnated at the early stage, and MOPSO with large population size has shown better performance. It can be concluded that the population sizing could affect the performance of conventional MOEAs in solving LSMOPs, especially for those DE and particle swarm based ones. Generally, a relatively small population size, e.g., 100, is suggested for conventional crossover based MOEAs for large-scale multiobjective optimization. On the contrary, a relatively large population size, e.g., 300 to 1000, is suggested for DE or PSO based MOEAs. Future work to design adaptive population sizing strategies is desirable for handling LSMOPs.

Acknowledgment. This work was supported in part by the National Natural Science Foundation of China under Grant 61903178 and Grant 61906081; in part by the Program for Guangdong Introducing Innovative and Entrepreneurial Teams under Grant 2017ZT07X386; in part by the Shenzhen Peacock Plan under Grant KQTD2016112514355531; and in part by the Program for University Key Laboratory of Guangdong Province under Grant 2017KSYS008.

References

1. Afshari, H., Hare, W., Tesfamariam, S.: Constrained multi-objective optimization algorithms: review and comparison with application in reinforced concrete structures. Appl. Soft Comput. **83**, 105631 (2019)
2. Antonio, L.M., Coello Coello, C.: Use of cooperative coevolution for solving large scale multiobjective optimization problems. In: Proceedings of 2013 IEEE Congress on Evolutionary Computation, pp. 2758–2765 (2013)
3. Beume, N., Naujoks, B., Emmerich, M.: SMS-EMOA: multiobjective selection based on dominated hypervolume. Eur. J. Oper. Res. **181**(3), 1653–1669 (2007)
4. Chen, H., Cheng, R., Wen, J., Li, H., Weng, J.: Solving large-scale many-objective optimization problems by covariance matrix adaptation evolution strategy with scalable small subpopulations. Inf. Sci. **509**, 457–469 (2020)
5. Cheng, H., Ran, C., Ye, T., Xingyi, Z.: Evolutionary many-objective optimization by NSGA-II and MOEA/D with large populations. In: IEEE Congress on Evolutionary Computation (CEC 2020). IEEE (2020)
6. Cheng, R.: Nature inspired optimization of large problems. Ph.D. thesis, University of Surrey (2016)
7. Cheng, R., Jin, Y.: A competitive swarm optimizer for large scale optimization. IEEE Trans. Cybern. **45**(2), 191–204 (2015)
8. Cheng, R., Jin, Y., Olhofer, M., Sendhoff, B.: Test problems for large-scale multiobjective and many-objective optimization. IEEE Trans. Cybern. **47**, 4108–4121 (2017)
9. Corne, D.W., Jerram, N.R., Knowles, J.D., Oates, M.J.: PESA-II: region-based selection in evolutionary multi-objective optimization. In: Proceedings of the Genetic and Evolutionary Computation Conference (GECCO2001), pp. 283–290. Citeseer (2001)
10. De Jong, K.A., Spears, W.M.: An analysis of the interacting roles of population size and crossover in genetic algorithms. In: Schwefel, H.-P., Männer, R. (eds.) PPSN 1990. LNCS, vol. 496, pp. 38–47. Springer, Heidelberg (1991). https://doi.org/10.1007/BFb0029729
11. Deb, K., Agrawal, R.B.: Simulated binary crossover for continuous search space. Complex Syst. **9**(4), 115–148 (1995)

12. Deb, K., Pratap, A., Agarwal, S., Meyarivan, T.: A fast and elitist multi-objective genetic algorithm: NSGA-II. IEEE Trans. Evol. Comput. **6**, 182–197 (2002)
13. Deb, K., Thiele, L., Laumanns, M., Zitzler, E.: Scalable test problems for evolutionary multiobjective optimization. In: Advanced Information and Knowledge Processing. Springer, London (2005)
14. Deb, K., Goyal, M.: A combined genetic adaptive search (GeneAS) for engineering design. Comput. Sci. Inf. **26**, 30–45 (1996)
15. He, C., Cheng, R., Danial, Y.: Adaptive offspring generation for evolutionary large-scale multiobjective optimization. IEEE Trans. Syst. Man Cybern. Syst. (2020). https://doi.org/10.1109/TSMC.2020.3003926
16. He, C., Cheng, R., Tian, Y., Zhang, X.: Iterated problem reformulation for evolutionary large-scale multiobjective optimization. In: 2020 IEEE Congress on Evolutionary Computation (CEC), pp. 1–8. IEEE (2020)
17. He, C., Cheng, R., Zhang, C., Tian, Y., Chen, Q., Yao, X.: Evolutionary large-scale multiobjective optimization for ratio error estimation of voltage transformers. IEEE Trans. Evol. Comput. **24**(5), 868–881 (2020)
18. He, C., et al.: Accelerating large-scale multiobjective optimization via problem reformulation. IEEE Trans. Evol. Comput. **23**(6), 949–961 (2019)
19. He, C., Tian, Y., Jin, Y., Zhang, X., Pan, L.: A radial space division based many-objective optimization evolutionary algorithm. Appl. Soft Comput. **61**, 603–621 (2017)
20. Ishibuchi, H., Sakane, Y., Tsukamoto, N., Nojima, Y.: Evolutionary many-objective optimization by NSGA-II and MOEA/D with large populations. In: 2009 IEEE International Conference on Systems, Man and Cybernetics, pp. 1758–1763. IEEE (2009)
21. Kukkonen, S., Lampinen, J.: GDE3: the third evolution step of generalized differential evolution. In: 2005 IEEE Congress on Evolutionary Computation, vol. 1, pp. 443–450. IEEE (2005)
22. Lechuga, M., Coello, E.: MOPSO: a proposal for multiple objective particle swarm optimization. In: Proceedings of the 2002 Congress on Evolutionary Computation, Part of the 2002 IEEE World Congress on Computational Intelligence, pp. 2051–11056 (2002)
23. Liu, H.L., Gu, F., Zhang, Q.: Decomposition of a multiobjective optimization problem into a number of simple multiobjective subproblems. IEEE Trans. Evol. Comput. **18**, 450–455 (2014)
24. Loshchilov, I., Schoenauer, M., Sebag, M.: Dominance-based Pareto-surrogate for multi-objective optimization. In: Deb, K., et al. (eds.) SEAL 2010. LNCS, vol. 6457, pp. 230–239. Springer, Heidelberg (2010). https://doi.org/10.1007/978-3-642-17298-4_24
25. Ma, X., et al.: A multiobjective evolutionary algorithm based on decision variable analyses for multi-objective optimization problems with large scale variables. IEEE Trans. Evol. Comput. **20**, 275–298 (2016)
26. Pan, A., Wang, L., Guo, W.: A universal strengthened searching module for multi-objective optimization based on variable properties. Appl. Soft Comput. **91**, 106199 (2020)
27. Pruvost, G., Derbel, B., Liefooghe, A., Li, K., Zhang, Q.: On the combined impact of population size and sub-problem selection in MOEA/D. In: Paquete, L., Zarges, C. (eds.) EvoCOP 2020. LNCS, vol. 12102, pp. 131–147. Springer, Cham (2020). https://doi.org/10.1007/978-3-030-43680-3_9

28. Roeva, O., Fidanova, S., Paprzycki, M.: Population size influence on the genetic and ant algorithms performance in case of cultivation process modeling. In: Fidanova, S. (ed.) Recent Advances in Computational Optimization. SCI, vol. 580, pp. 107–120. Springer, Cham (2015). https://doi.org/10.1007/978-3-319-12631-9_7
29. Huband, S., Hingston, P., Barone, L., While, L.: A review of multiobjective test problems and a scalable test problem toolkit. IEEE Trans. Evol. Comput. **10**, 477–506 (2006)
30. Shen, X., Guo, Y., Li, A.: Cooperative coevolution with an improved resource allocation for large-scale multi-objective software project scheduling. Appl. Soft Comput. **88**, 106059 (2020)
31. Tian, Y., Cheng, R., Zhang, X., Jin, Y.: PlatEMO: a MATLAB platform for evolutionary multi-objective optimization [educational forum]. IEEE Comput. Intell. Mag. **12**, 73–87 (2017)
32. Tian, Y., Zhang, X., Wang, C., Jin, Y.: An evolutionary algorithm for large-scale sparse multiobjective optimization problems. IEEE Trans. Evol. Comput. **24**(2), 380–393 (2019)
33. Wang, G., Jiang, H.: Fuzzy-dominance and its application in evolutionary many objective optimization. In: Proceedings of the 2007 International Conference on Computational Intelligence and Security Workshops, pp. 195–198. IEEE (2007)
34. Wang, H., Yao, X.: Corner sort for Pareto-based many-objective optimization. IEEE Trans. Cybern. **44**(1), 92–102 (2014)
35. While, L., Hingston, P., Barone, L., Huband, S.: A faster algorithm for calculating hypervolume. IEEE Trans. Evol. Comput. **10**, 29–38 (2006)
36. Yang, Z., Tang, K., Yao, X.: Large scale evolutionary optimization using cooperative coevolution. Inf. Sci. **178**(15), 2985–2999 (2008)
37. Zhang, Q., Li, H.: MOEA/D: a multiobjective evolutionary algorithm based on decomposition. IEEE Trans. Evol. Comput. **11**, 712–731 (2007)
38. Zhang, X., Tian, Y., Jin, Y., Cheng, R.: A decision variable clustering-based evolutionary algorithm for large-scale many-objective optimization. IEEE Trans. Evol. Comput. **22**, 97–112 (2016)
39. Zhang, X., Zheng, X., Cheng, R., Qiu, J., Jin, Y.: A competitive mechanism based multi-objective particle swarm optimizer with fast convergence. Inf. Sci. **427**, 63–76 (2018)
40. Zille, H., Ishibuchi, H., Mostaghim, S., Nojima, Y.: A framework for large-scale multi-objective optimization based on problem transformation. IEEE Trans. Evol. Comput. **22**, 260–275 (2018)
41. Zitzler, E., Künzli, S.: Indicator-based selection in multiobjective search. In: Yao, X., et al. (eds.) PPSN 2004. LNCS, vol. 3242, pp. 832–842. Springer, Heidelberg (2004). https://doi.org/10.1007/978-3-540-30217-9_84
42. Zitzler, E., Laumanns, M., Thiele, L.: SPEA2: improving the strength Pareto evolutionary algorithm. TIK-Rep. **103** (2001)

Kernel Density Estimation for Reliable Biobjective Solution of Stochastic Problems

Marius Bommert(✉) and Günter Rudolph

TU Dortmund University, 44221 Dortmund, Germany
marius.bommert@tu-dortmund.de

Abstract. Stochastic objective functions can be optimized by finding values in decision space for which the expected output is optimal and the uncertainty is minimal. We investigate the optimization of expensive stochastic black box functions $f : \mathbb{R}^a \times \mathbb{R}^b \to \mathbb{R}$ with controllable parameter $x \in \mathbb{R}^a$ and b-dimensional random variable C. Our original approach, MIOS-PDE, presented in previous work, requires that the distribution class of C is known. We extend our approach such that it is applicable in situations where the distribution class is unknown. We achieve this with kernel density estimation. With the extended approach, MIOS-KDE, expectation $\mathrm{E}(f(x, C))$ and standard deviation $\mathrm{S}(f(x, C))$ are estimated with a high quality and good approximations of the true Pareto frontiers are obtained. No quality is lost with MIOS-KDE compared to MIOS-PDE. Additionally, MIOS-KDE is robust with respect to outliers, while MIOS-PDE is not.

Keywords: Biobjective optimization · Stochastic black box function · Metamodel · Kernel density estimation

1 Introduction

Many optimization problems are influenced by random effects. The objective functions of such problems are stochastic. Stochastic functions can be denoted as $f(x, C)$ where x is a controllable parameter vector and C a random variable. An example is a production process where some real-valued quality measure of the product is assessed and some parameters that influence the production process (e.g. force, speed) are controllable while others (e.g. air temperature, humidity) are not. In previous work [2,3], we have proposed an approach for optimizing expensive stochastic black box functions. This original approach requires that C is observable and that its distribution class is known. In this work, we consider the setting where the distribution class of C is unknown, which is more common in practice. Therefore, we extend our previous work using kernel density estimation.

The evaluation of a stochastic function f in any point x does not yield a deterministic output. Instead, $f(x, C)$ is a random variable whose distribution

H. Ishibuchi et al. (Eds.): EMO 2021, LNCS 12654, pp. 53–64, 2021.
https://doi.org/10.1007/978-3-030-72062-9_5

depends on x. In our original approach [2,3], we optimize f by simultaneously optimizing the expectation $\mathrm{E}(f(x,C))$ and the standard deviation $\mathrm{S}(f(x,C))$. Depending on the context, $\mathrm{E}(f(x,C))$ is minimized or maximized. $\mathrm{S}(f(x,C))$ is always minimized to achieve small uncertainty. So, in order to optimize the single-objective stochastic function f, we optimize the biobjective deterministic function $(\mathrm{E}(f(x,C)), \mathrm{S}(f(x,C)))$:

$$\mathrm{E}(f(x,C)) \rightarrow \text{min/max!} \quad \text{and} \quad \mathrm{S}(f(x,C)) \rightarrow \text{min!}.$$

The expectation $\mathrm{E}(f(x,C))$ provides information about the central location of the distribution when f is evaluated in x. It describes the expected output of $f(x,C)$. If f is evaluated in x many times, the mean output value will be close to $\mathrm{E}(f(x,C))$ (law of large numbers). A single evaluation of f in x does not necessarily yield a value close to $\mathrm{E}(f(x,C))$. The variance $\mathrm{V}(f(x,C))$ gives the expected quadratic deviation of an evaluation of $f(x,C)$ from $\mathrm{E}(f(x,C))$. It is a measure of the spread of the distribution and can be used to assess the uncertainty about how far $f(x,C)$ will likely deviate from $\mathrm{E}(f(x,C))$. As the variance uses a quadratic scale, the standard deviation $\mathrm{S}(f(x,C)) = \sqrt{\mathrm{V}(f(x,C))}$ is often used instead. In our previous work [2,3], we have shown that $\mathrm{E}(f(x,C))$ and $\mathrm{S}(f(x,C))$ can be estimated using a metamodel for f and numerical integration. This approach for estimating $\mathrm{E}(f(x,C))$ and $\mathrm{S}(f(x,C))$ is able to outperform the standard approach: moment estimation of $\mathrm{E}(f(x,C))$ and $\mathrm{S}(f(x,C))$. With our approach, more accurate estimates are obtained. Additionally, with our approach, it is possible to estimate $\mathrm{E}(f(x,C))$ and $\mathrm{S}(f(x,C))$ for any value $x \in \mathbb{R}^a$, while with the standard approach, estimates are only possible if f has been evaluated at least twice in the x-value of interest.

Optimizing a stochastic black box function by maximizing the expectation and minimizing a risk measure is a common strategy in portfolio optimization. The risk is often assessed by the variance of f or other domain specific risk measures like the value at risk. The mean-variance model introduced by Markowitz [14] is a very popular approach in this field. For this model, expectation and variance are scalarized into a single objective function and then optimized. Portfolio optimization has also been analyzed as a multi-objective problem, see for example [15]. An overview of methods for optimizing stochastic functions is given in [10]. One approach for the optimization of stochastic functions presented in [10] also includes the simultaneous optimization of the expectation and a risk measure like the standard deviation. Methods for estimating expectation and standard deviation are not suggested by the authors.

Another field of research which deals with the optimization of stochastic functions is stochastic programming [12]. In this field, stochastic functions f are optimized by optimizing the expectation $\mathrm{E}(f(x,C))$. One approach for estimating $\mathrm{E}(f(x,C))$ is sample average approximation [12]. In contrast to our setting, f usually is known analytically and not an expensive function in this field. In stochastic programming, the optimization of a measure of uncertainty like the standard deviation $\mathrm{S}(f(x,C))$ is not considered.

In the field of optimizing stochastic black box functions f, many contributions exist. But, to the best of our knowledge, none of them optimize $\mathrm{E}(f(x,C))$ and

$S(f(x, C))$ simultaneously. Huang et al. [11] optimize the expectation $E(f(x, C))$ of a stochastic black box function f, using sequential Kriging optimization. Swamy [19] optimizes the expectation $E(f(x, C))$ of a stochastic black box function f while controlling a risk measure using a probability constraint. In contrast to our work, both [11] and [19] do not optimize the standard deviation $S(f(x, C))$. In [11], the randomness in f is assumed to be additive noise, so $S(f(x, C))$ is assumed to be constant with respect to x. In [19], the risk measure is used for constraining the set of x-values for the optimization of $E(f(x, C))$.

In this paper, we extend our approach of employing Metamodels and numerical Integration for Optimizing Stochastic black box functions (MIOS). The extended approach, MIOS-KDE, uses kernel density estimation while the original approach, MIOS-PDE, employs a parametric density estimation, that is, the parameters of a known distribution class are estimated. Non-parametric density estimation is more difficult than estimating parameters, but does not require the assumption of a known distribution class. When the distribution class is known, both approaches are applicable but MIOS-KDE does not use the information about the distribution class. In the setting of an unknown distribution class, only MIOS-KDE can be used. We compare MIOS-PDE and MIOS-KDE with respect to the quality of the obtained results. We analyze scenarios in which the true distribution class is known for investigating whether MIOS-KDE can achieve the same quality of results as MIOS-PDE without having the advantage of knowing the true distribution class. We also consider a setting with outliers, that is, a setting in which the true distribution class slightly differs from the one assumed by MIOS-PDE to analyze the robustness of the two approaches.

The remainder of this article is organized as follows: In Sect. 2, basic concepts and methods are described. In Sect. 3, the estimation of $E(f(x, C))$ and $S(f(x, C))$ using a metamodel and numerical integration is explained. The design of our comparative experiments is presented in Sect. 4. In Sect. 5, MIOS-PDE and MIOS-KDE are compared with respect to the quality of the estimation of $E(f(x, C))$ and $S(f(x, C))$ and the approximation quality of the Pareto frontiers. Concluding remarks are presented in Sect. 6.

2 Concepts and Methods

In the following, Kriging, Pareto optimality, kernel density estimation and the Δ_p-indicator are explained.

2.1 Kriging

Let $x_1, \ldots, x_n$ denote n points which are evaluated with a deterministic function f. Kriging is an interpolating method to build a metamodel $\hat{f}$ of f using the n given points. For that purpose, it is assumed that $f(x_1), \ldots, f(x_n)$ are realizations of a Gaussian random field. Numerical optimization is performed to fit the model to the data. For more information on Kriging see [17].

2.2 Pareto Set and Pareto Frontier

Let $f : X \to \mathbb{R}^m, f(x) = (f_1(x), \ldots, f_m(x))$ denote an objective function where all components should be minimized. $x \in X$ weakly dominates $y \in X$ (notation: $x \preceq y$ or $f(x) \preceq f(y)$) iff $\forall i \in \{1, \ldots, m\} : f_i(x) \leq f_i(y)$. The point x dominates the point y iff it weakly dominates it and $\exists i \in \{1, \ldots, m\} : f_i(x) < f_i(y)$. $x^* \in X$ is Pareto optimal if there is no $x \in X$ which dominates it. The set of all Pareto optimal points is called Pareto set and the corresponding image is the Pareto frontier. For more information on Pareto frontiers and Pareto sets see [5].

2.3 Kernel Density Estimation

Kernel density estimation [7] can be used to estimate a probability density function p_C. Given r samples $\tilde{c}_i \in \mathbb{R}^b, i = 1, \ldots, r$, from a b-dimensional random variable C, the estimate $\hat{p}_C(c)$ of $p_C(c)$ for a value $c \in \mathbb{R}^b$ is given as

$$\hat{p}_C(c) = \frac{1}{r} \sum_{i=1}^{r} K_H(c - \tilde{c}_i).$$

K_H denotes a kernel function. The kernel is used for weighting the samples $\tilde{c}_i$ depending on their distance to the observation c. The kernel function depends on a bandwidth matrix H which determines the shape of the kernel. K_H is commonly chosen as a unimodal probability density function that is symmetric around zero. The kernel density estimate $\hat{p}_C(c)$ for c is a weighted average based on the values $\tilde{c}_i, i = 1, \ldots, r$. The weight of sample $\tilde{c}_i$ decreases for increasing distance between c and $\tilde{c}_i$. This distance is measured with respect to the contour lines of K_H. For this paper, K_H is chosen as

$$K_H(c - \tilde{c}_i) = \frac{1}{\sqrt{(2\pi)^b \det(H)}} \exp\left(-\frac{1}{2}(c - \tilde{c}_i)^\top H^{-1}(c - \tilde{c}_i)\right),$$

the probability density function of the multivariate normal distribution with expectation $\tilde{c}_i$ and covariance matrix H. The bandwidth matrix H has to be estimated using the samples $\tilde{c}_1, \ldots, \tilde{c}_r$. In the software implementation used for this paper [8], the bandwidth matrix H is calculated according to [20] in the one-dimensional case and according to [9] in the multidimensional case.

2.4 Δ_p-indicator

Let $X = \{x_1, \ldots, x_N\}$ and $Y = \{y_1, \ldots, y_M\}$ denote two non empty sets. The Δ_p-indicator [18] can be used to measure the distance between the sets X and Y. The generational distance GD_p and the inverted generational distance IGD_p between X and Y are defined as

$$\text{GD}_p(X, Y) = \left(\frac{1}{N} \sum_{i=1}^{N} d(x_i, Y)^p\right)^{\frac{1}{p}} \text{ and } \text{IGD}_p(X, Y) = \left(\frac{1}{M} \sum_{i=1}^{M} d(y_i, X)^p\right)^{\frac{1}{p}}$$

with $d(x_i, Y) = \min_{y \in Y} ||x_i - y||_p$, $d(y_i, X) = \min_{x \in X} ||y_i - x||_p$ and $p \geq 1$. Then the Δ_p-indicator is defined as

$$\Delta_p(X, Y) = \max\{\mathrm{GD}_p(X, Y), \mathrm{IGD}_p(X, Y)\}.$$

3 Approach for Estimating Expectation and Standard Deviation of *f*

3.1 General Idea

Let $f(x, C)$ be a stochastic function where $x = (x_1, \ldots, x_a) \in \mathbb{R}^a$ is a controllable parameter and C is a b-dimensional random variable. Our aim is optimizing $f(x, C)$ by simultaneous minimization of the expectation

$$\mathrm{E}(f(x, C)) = \int_{-\infty}^{\infty} \cdots \int_{-\infty}^{\infty} f(x, c_1, \ldots, c_b) \cdot p_C(c_1, \ldots, c_b) dc_1 \ldots dc_b$$

and the standard deviation $\mathrm{S}(f(x, C)) = \sqrt{\mathrm{V}(f(x, C))}$ with

$$\mathrm{V}(f(x, C)) = \int_{-\infty}^{\infty} \cdots \int_{-\infty}^{\infty} (f(x, c_1, \ldots, c_b) - \mathrm{E}(f(x, C)))^2 \cdot p_C(c_1, \ldots c_b) dc_1 \ldots dc_b.$$

$p_C(c_1, \ldots, c_b)$ denotes the probability density function of C at the point $c = (c_1, \ldots, c_b)$. We assume that the distribution of C is continuous. For a discrete distribution, a summation over the support of C is needed instead of the integration. If the expectation should be maximized, this maximization problem can be transformed into a minimization problem by multiplication with -1.

3.2 Estimation of Expectation and Standard Deviation

For the estimation of $\mathrm{E}(f(x, C))$ and $\mathrm{S}(f(x, C))$, the integrals presented in Subsect. 3.1 are estimated with empirical quantities and numerical integration. A metamodel $\hat{f}$ is used instead of the function f and the estimated probability density function $\hat{p}_C$ replaces p_C. For the estimation of $\mathrm{V}(f(x, C))$, the estimated expectation $\hat{\mathrm{E}}(f(x, C))$ is used instead of $\mathrm{E}(f(x, C))$. The integration limits $-\infty$ and ∞ are replaced by finite integration limits chosen based on the realizations of C. For more details, see [2,3]. For MIOS-PDE, $\hat{p}_C$ is obtained by estimating the parameters of the known distribution class of C. For MIOS-KDE, $\hat{p}_C$ is estimated non-parametrically with kernel density estimation.

4 Experiments

We describe the design of our experiments for analyzing and comparing MIOS-PDE and MIOS-KDE and explain some computational aspects.

4.1 Design of Experiments

In this analysis, we consider inexpensive objective functions in order to be able to perform a large number of experiments. But, we limit the budget of function evaluations like it is common for expensive black box optimization. We analyze the optimization of two objective functions. The first objective function is

$$f_1(x, C) = x\cos(C - x^2)$$

with $x \in [-\frac{\pi}{2}, \frac{\pi}{2}]$. For the random variable C we consider three different scenarios: (a) $C \sim \mathcal{N}(5, 1)$, (b) $C \sim \mathcal{N}(1, 1)$ and (c) $C \sim 0.95 \times \mathcal{N}(1, 1) + 0.05 \times \mathcal{N}(10, 1)$. In setting (a) and (b), C is normally distributed with different expectations which leads to different shapes of the Pareto frontier as displayed in Fig. 1. Setting (c) is a perturbed version of setting (b). When drawing samples from this Gaussian mixture model, with probability 0.95, an observation is drawn from $\mathcal{N}(1, 1)$, and with probability 0.05 from $\mathcal{N}(10, 1)$. As second objective function we consider

$$f_2(x, C) = f(x_1, x_2, C_1, C_2) = 5x_1\cos(C_1 - x_1^2) + 20\sin(x_2 - C_2) + x_1x_2$$

with $x = (x_1, x_2), x_1, x_2 \in [-\pi, \pi]$ and $C = (C_1, C_2)$. For the random variable C, a normal distribution $C \sim \mathcal{N}\left(\begin{pmatrix}50\\70\end{pmatrix}, \begin{pmatrix}9 & -6\\-6 & 9\end{pmatrix}\right)$ is chosen. For both objective functions, the expectation is maximized and the uncertainty is minimized. Therefore, $-\mathrm{E}(f(x, C))$ and $\mathrm{S}(f(x, C))$ are minimized.

To build the metamodel $\hat{f}_i, i \in \{1, 2\}$, a data set with n values for x, c and $f_i(x, c)$ is needed. For f_1, we choose n x-values equidistantly in the whole interval $[-\frac{\pi}{2}, \frac{\pi}{2}]$. For f_2, n x-values in $[-\pi, \pi] \times [-\pi, \pi]$ are chosen using an improved latin hypercube design. For the function f_1, $n \in \{10, 20, 50, 100, 200, 500\}$ x-values are used. f_2 is evaluated in $n \in \{50, 100, 200, 500, 1\,000, 2\,000, 5\,000\}$ x-values. For each x-value, a realization c of the respective random variable C is drawn. The objective function f_i is evaluated in the n pairs of x-values and corresponding c-values. A metamodel $\hat{f}_i$ for f_i is built based on the resulting data set.

The probability density function p_C is estimated based on the n samples of C. For MIOS-PDE, we always assume that C is normally distributed. This is correct for f_1 (a), f_1 (b) and f_2. For f_1 (c), this assumption is slightly wrong because the normal distribution is perturbed.

For evaluating the quality of the results obtained with both approaches, a ground truth has to be assessed. For all settings, we calculate $\mathrm{E}(f(x, C))$ and $\mathrm{S}(f(x, C))$ for a certain set of x-values. The Pareto frontiers obtained based on these sets are called the "true" Pareto frontiers in this paper. For f_1, $\mathrm{E}(f(x, C))$ and $\mathrm{S}(f(x, C))$ are evaluated in 101 equidistant x-values in the whole interval $[-\frac{\pi}{2}, \frac{\pi}{2}]$. This set of x-values is denoted as $\mathcal{X}_1$. For f_2, an improved latin hypercube design [1] with 20 000 x-values in $[-\pi, \pi] \times [-\pi, \pi]$ is used. This set is denoted as $\mathcal{X}_2$.

Figure 1 displays the true values $-\mathrm{E}(f(x, C))$ and $\mathrm{S}(f(x, C))$ for the considered objective functions evaluated in the sets $\mathcal{X}_1$ for f_1 and $\mathcal{X}_2$ for f_2. The red

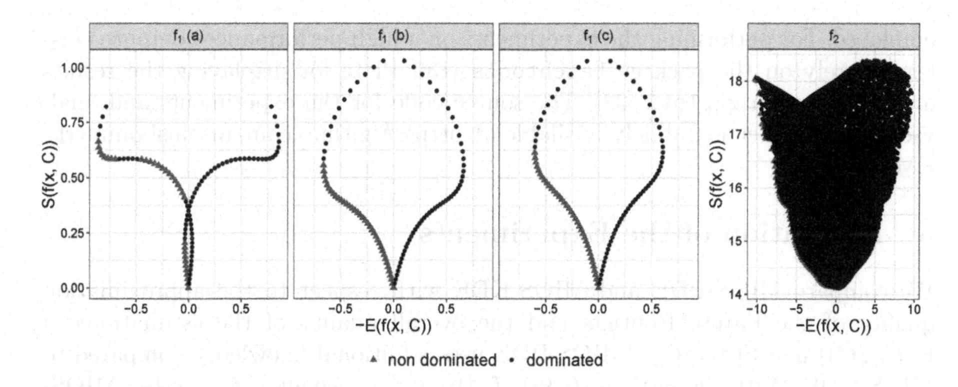

Fig. 1. Plot of $-\mathrm{E}(f(x,C))$ and $\mathrm{S}(f(x,C))$ for the considered objective functions evaluated in $\mathcal{X}_1$ for f_1 and $\mathcal{X}_2$ for f_2. The Pareto frontiers are displayed in red. (Color figure online)

triangles in the plots are the true Pareto frontiers based on $\mathcal{X}_1$ or $\mathcal{X}_2$. It can be seen that there are convex and concave parts of the Pareto frontiers. Our aim is achieving good approximations of the Pareto frontiers as well as a high overall quality of the estimations of $\mathrm{E}(f(x,C))$ and $\mathrm{S}(f(x,C))$ for all x-values in $\mathcal{X}_1$ or $\mathcal{X}_2$.

To quantify the quality of the results of both approaches, two different measures are used. For calculating these measures, $\hat{\mathrm{E}}(f(x,C))$ and $\hat{\mathrm{S}}(f(x,C))$ are evaluated in all x-values in $\mathcal{X}_1$ for f_1 and $\mathcal{X}_2$ for f_2. To measure the overall quality of the estimation of $\mathrm{E}(f(x,C))$ and $\mathrm{S}(f(x,C))$, we calculate the mean Euclidean distance between the points $(\mathrm{E}(f(x,C)), \mathrm{S}(f(x,C)))$ and $(\hat{\mathrm{E}}(f(x,C)), \hat{\mathrm{S}}(f(x,C)))$ for the k x-values in $\mathcal{X}_1$ for f_1 and $\mathcal{X}_2$ for f_2:

$$\mathrm{MED} = \frac{1}{k}\sum_{i=1}^{k}\left\| \begin{pmatrix}\mathrm{E}(f(x_i,C))\\ \mathrm{S}(f(x_i,C))\end{pmatrix} - \begin{pmatrix}\hat{\mathrm{E}}(f(x_i,C))\\ \hat{\mathrm{S}}(f(x_i,C))\end{pmatrix}\right\|_2 .$$

To compare the quality of the approximations of the true Pareto frontiers, the distance between each Pareto frontier estimate and the true Pareto frontier is measured with the Δ_p-indicator with $p = 2$. For both quality measures, small values are desirable. Because the results of the experiments are influenced by random effects, we repeat the experiments for each scenario 100 times.

4.2 Computational Aspects

For our experiments, we use the software `R` [16]. For generating a latin hypercube sampling the package `lhs` [6] is employed. As metamodel we apply Kriging which is implemented in the package `DiceKriging` [17]. Kernel density estimation is implemented in the package `ks` [8]. For numerical integration the package `mvQuad` [21] is used. To determine the Pareto frontiers, the package `ecr` [4] is employed. For performing the experiments on a high performance compute cluster, we rely on the package `batchtools` [13]. Plots for displaying the results are created with `ggplot2` [22]. The source code for the experiments and analyses of this paper is publicly available at https://github.com/mariusbommert/EMO2021.

5 Evaluation of the Experiments

We compare MIOS-PDE and MIOS-KDE with respect to the approximation quality of the Pareto frontiers and the overall quality of the estimations of $\mathrm{E}(f(x, C))$ and $\mathrm{S}(f(x, C))$. MIOS-PDE uses additional knowledge compared to MIOS-KDE. With the settings f_1 (a), f_1 (b) and f_2, we analyze whether MIOS-KDE can achieve the same quality of results as MIOS-PDE without having the advantage of knowing the true distribution class. Setting f_1 (c) allows robustness analyses. In this scenario, the true distribution class slightly differs from the one assumed by MIOS-PDE. With this scenario, we analyze whether MIOS-KDE outperforms MIOS-PDE in situations where the class assumption is slightly wrong.

First, we analyze the effect of the number of observations n for building the metamodel on the quality of the results. Figure 2 displays the Δ_p-indicator and the mean Euclidean distance for the four considered scenarios. Looking at the Δ_p-measure, it can be observed that the quality of the approximations of the true Pareto frontiers is improved for increasing value of n for most scenarios. For f_2, there is one exception for $n = 1\,000$ which is due to random effects. The overall estimation quality, measured by the mean Euclidean distance, is improved for increasing value of n for most scenarios. For the three parameter settings of f_1, it seems to be sufficient to use $n = 100$ observations for building the metamodel because there is only little improvement in both quality measures for larger numbers of observations. For f_2, it seems to be necessary to use $n = 2\,000$ to $n = 5\,000$ observations for building the metamodel.

In setting f_1 (c) for MIOS-PDE, the values of both quality measures are improved up to $n = 50$ and worsen again for larger values of n. For explaining this behavior, boxplots of the mean Euclidean distance between p_C and $\hat{p}_C$ in setting f_1 (c) are displayed in Fig. 3. For MIOS-PDE, especially for values up to $n = 50$, a decrease in variation of the estimation quality of p_C can be observed. When estimating the parameter of a normal density based on data drawn from a Gaussian mixture model that models data with outliers, the quality of the estimate depends on the number of outliers in the data set. The smaller the value of n, the more frequently extreme numbers of outliers - none or very few, or very many compared to the total number n - are included in the data sets. This means that, the smaller the value of n, the more frequently extremely good or extremely bad estimates of p_C are observed. For values larger than $n = 50$, the convergence of $\hat{p}_C$ to a function different from p_C causes a decrease in quality of the results obtained with MIOS-PDE. With MIOS-KDE, there is no problem with estimating p_C. The mean Euclidean distance between p_C and $\hat{p}_C$ decreases for increasing value of n.

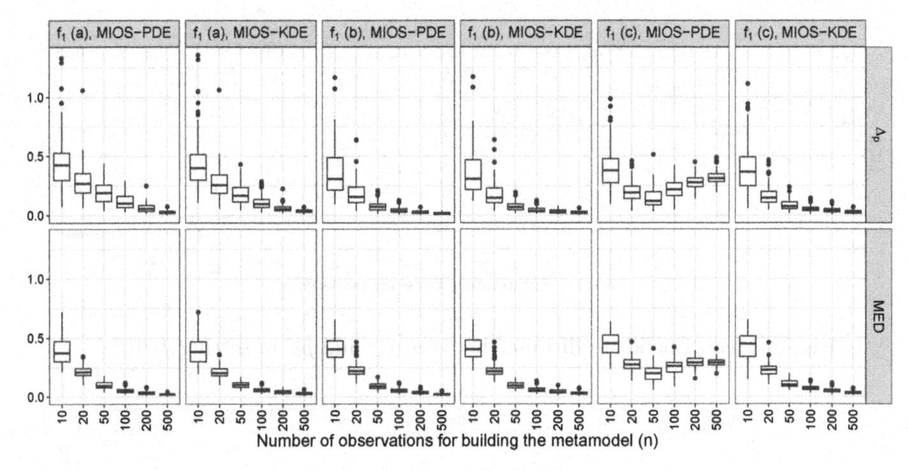

(a) Function f_1

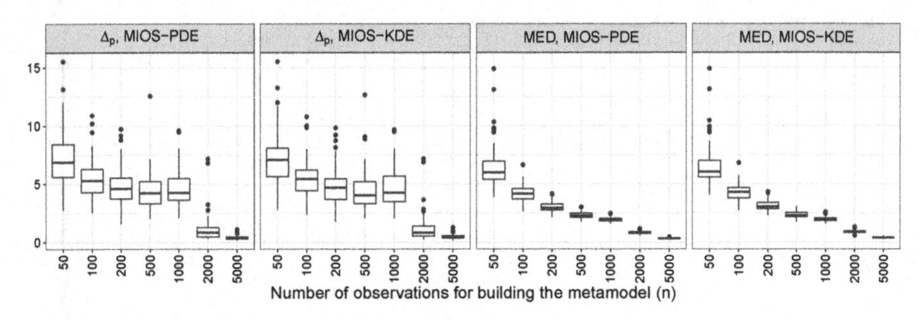

(b) Function f_2

Fig. 2. Quality of the two approaches: Δ_p and mean Euclidean distance.

Now, we analyze the difference in quality between MIOS-PDE and MIOS-KDE when applied to the same Kriging model. The differences displayed in Figure 4 are only due to knowing or not knowing the distribution class of C. For all scenarios except f_1 (c), it can be seen that the median values of differences in quality are always very close to zero. For all settings except f_1 (c), the boxes are small which means that the variation between repetitions is small. The boxplots show that for all settings in which MIOS-PDE knows the true distribution class, results of almost identical quality are obtained with both approaches. This holds independent of the number of observations used for building the metamodel. Sometimes MIOS-PDE is slightly better and sometimes MIOS-KDE. In setting f_1 (c), MIOS-KDE achieves a much higher quality than MIOS-PDE which confirms the results gained from Fig. 2.

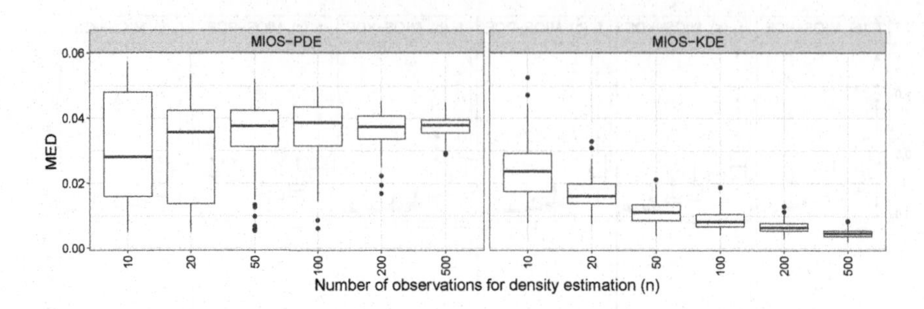

Fig. 3. Mean Euclidean distance between p_C and $\hat{p}_C$ in setting f_1 (c).

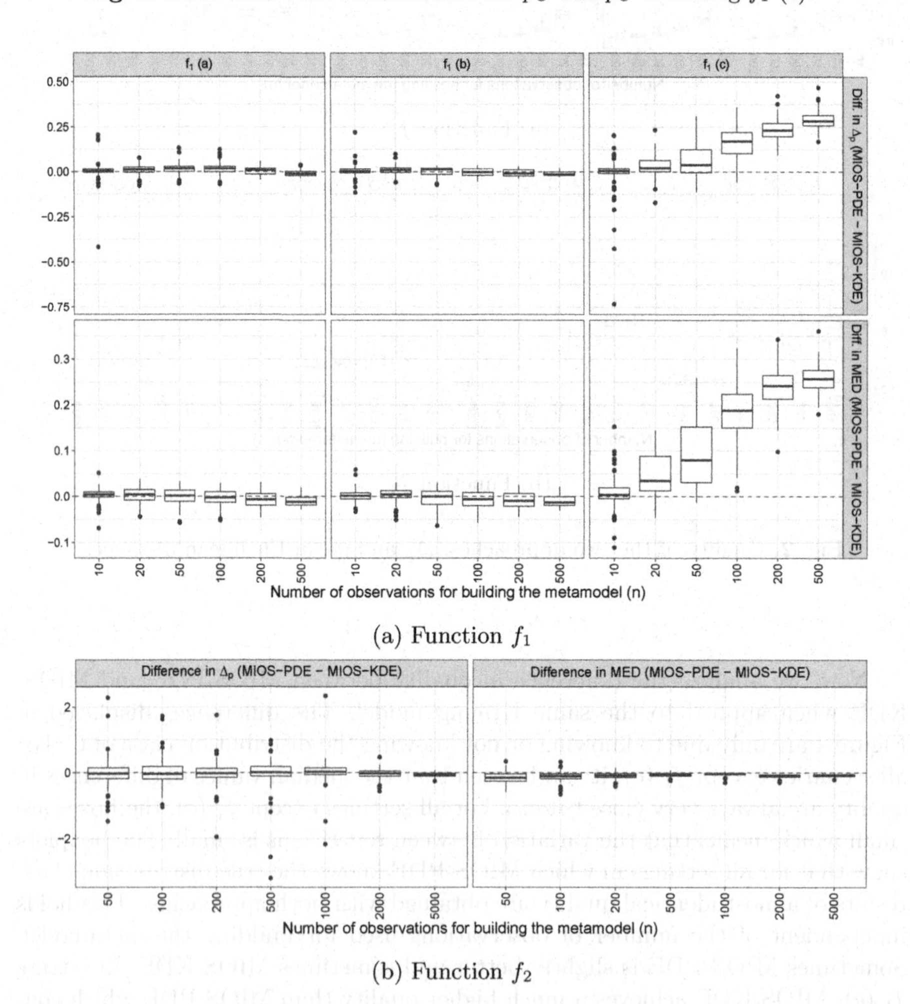

(a) Function f_1

(b) Function f_2

Fig. 4. Difference in Δ_p and mean Euclidean distance for both objective functions. A negative (positive) value means that the quality for MIOS-PDE (MIOS-KDE) is better than for MIOS-KDE (MIOS-PDE). (Color figure online)

6 Conclusion

We considered the optimization of expensive stochastic black box functions $f(x,C)$ with $x \in \mathbb{R}^a$ and b-dimensional random variable C. In our previous work [2,3], we transformed the problem of optimizing the stochastic function f into the biobjective optimization of the deterministic expectation $\mathrm{E}(f(x,C))$ and standard deviation $\mathrm{S}(f(x,C))$ and we proposed estimating $\mathrm{E}(f(x,C))$ and $\mathrm{S}(f(x,C))$ by building a metamodel for f and using numerical integration. Compared to the standard approach of moment estimation of $\mathrm{E}(f(x,C))$ and $\mathrm{S}(f(x,C))$, the approach provides the advantage that $\mathrm{E}(f(x,C))$ and $\mathrm{S}(f(x,C))$ can be estimated in any x-value, regardless if f has been evaluated in x or not. The approach allows estimating $\mathrm{E}(f(x,C))$ and $\mathrm{S}(f(x,C))$ with a high quality and yields a good approximation of the true Pareto frontier at the cost of requiring that C is observable.

In our previous work [2,3], we assumed that the distribution class of C is known. In this paper, we extended our approach to make it applicable in scenarios where the distribution class of C is unknown. To achieve this, we used kernel density estimation. We compared the original approach, MIOS-PDE, and the extended approach, MIOS-KDE, with respect to the quality of estimating $\mathrm{E}(f(x,C))$ and $\mathrm{S}(f(x,C))$ and of approximating the Pareto frontier. We conclude that with MIOS-KDE, no quality is lost compared to MIOS-PDE. For MIOS-PDE, it is very important that the correct distribution class of C is known. A small deviation from the assumed distribution can lead to comparably bad results. MIOS-KDE is robust in settings with outliers and achieves high quality results. So, MIOS-KDE achieves the same or better quality of results compared to MIOS-PDE and is applicable to a lot more real world problems. Now we are in the position to apply our approach in subsequent work for approximating the Pareto frontier.

Acknowledgments. The authors gratefully acknowledge the computing time provided on the Linux HPC cluster at TU Dortmund University (LiDO3), partially funded in the course of the Large-Scale Equipment Initiative by the German Research Foundation (DFG) as project 271512359.

References

1. Beachkofski, B.K., Grandhi, R.V.: Improved distributed hypercube sampling. In: 43rd AIAA/ASME/ASCE/AHS/ASC Structures, Structural Dynamics, and Materials Conference, p. 1274 (2002)
2. Bommert, M., Rudolph, G.: Reliable biobjective solution of stochastic problems using metamodels. In: Deb, K., et al. (eds.) EMO 2019. LNCS, vol. 11411, pp. 581–592. Springer, Cham (2019). https://doi.org/10.1007/978-3-030-12598-1_46
3. Bommert, M., Rudolph, G.: Reliable solution of multidimensional stochastic problems using metamodels. In: Nicosia, G., et al. (eds.) LOD 2020. LNCS, vol. 12565, pp. 215–226. Springer, Cham (2020). https://doi.org/10.1007/978-3-030-64583-0_20

4. Bossek, J.: ecr 2.0: A modular framework for evolutionary computation in R. In: Proceedings of the Genetic and Evolutionary Computation Conference Companion, pp. 1187–1193. ACM (2017)
5. Miettinen, K., Ruiz, F., Wierzbicki, A.P.: Introduction to multiobjective optimization: interactive approaches. In: Branke, J., Deb, K., Miettinen, K., Słowiński, R. (eds.) Multiobjective Optimization. LNCS, vol. 5252, pp. 27–57. Springer, Heidelberg (2008). https://doi.org/10.1007/978-3-540-88908-3
6. Carnell, R.: lhs: Latin Hypercube Samples, R package version 1.0.1 (2019)
7. Chacón, J.E., Duong, T.: Multivariate Kernel Smoothing and Its Applications. CRC Press, United States (2018)
8. Duong, T.: ks: Kernel Smoothing, R package version 1.11.7 (2020)
9. Duong, T., Hazelton, M.: Plug-in bandwidth matrices for bivariate kernel density estimation. J. Nonparametric Stat. **15**(1), 17–30 (2003)
10. Gutjahr, W.J., Pichler, A.: Stochastic multi-objective optimization: a survey on non-scalarizing methods. Ann. Oper. Res. **236**(2), 475–499 (2016)
11. Huang, D., Allen, T.T., Notz, W.I., Zeng, N.: Global optimization of stochastic black-box systems via sequential kriging meta-models. J. Global Optim. **34**(3), 441–466 (2006)
12. Kleywegt, A.J., Shapiro, A., Homem-de-Mello, T.: The sample average approximation method for stochastic discrete optimization. SIAM J. Optim. **12**(2), 479–502 (2002)
13. Lang, M., Bischl, B., Surmann, D.: batchtools: Tools for R to work on batch systems. J. Open Source Softw. **2**(10), 135 (2017)
14. Markowitz, H.: Portfolio selection. J. Finance **7**(1), 77–91 (1952)
15. Ponsich, A., Jaimes, A.L., Coello Coello, C.A.: A survey on multiobjective evolutionary algorithms for the solution of the portfolio optimization problem and other finance and economics applications. IEEE Trans. Evol. Comput. **17**(3), 321–344 (2013)
16. R Core Team: R: A Language and Environment for Statistical Computing. R Foundation for Statistical Computing, Vienna, Austria (2020). https://www.R-project.org/
17. Roustant, O., Ginsbourger, D., Deville, Y.: DiceKriging, DiceOptim: Two R packages for the analysis of computer experiments by kriging-based metamodeling and optimization. J. Stat. Softw. **51**(1), 1–55 (2012)
18. Schütze, O., Esquivel, X., Lara, A., Coello Coello, C.A.: Using the averaged hausdorff distance as a performance measure in evolutionary multiobjective optimization. IEEE Trans. Evol. Comput. **16**(4), 504–522 (2012)
19. Swamy, C.: Risk-averse stochastic optimization: probabilistically-constrained models and algorithms for black-box distributions. In: Proceedings of the 22nd annual ACM-SIAM symposium on Discrete Algorithms, pp. 1627–1646. SIAM (2011)
20. Wand, M., Jones, C.: Multivariate plug-in bandwidth selection. Comput. Stat. **9**(2), 97–116 (1994)
21. Weiser, C.: mvQuad: Methods for Multivariate Quadrature (2016)
22. Wickham, H.: ggplot2: Elegant Graphics for Data Analysis. Springer-Verlag, New York (2016)

Approximating Pareto Fronts in Evolutionary Multiobjective Optimization with Large Population Size

Hui Li[1](✉), Yuxiang Shui[1], Jianyong Sun[1], and Qingfu Zhang[2]

[1] Xi'an Jiaotong University, 710049 Xi'an, Shaanxi, China
{lihui10,j.sun}@xjtu.edu.cn
[2] City University of Hong Kong, Kowloon Tong, Hong Kong

Abstract. Approximating the Pareto fronts (PFs) of multiobjective optimization problems (MOPs) with a population of nondominated solutions is a common strategy in evolutionary multiobjective optimization (EMO). In the case of two or three objectives, the PFs of MOPs can be well approximated by the populations including several dozens or hundreds of nondominated solutions. However, this is not the case when approximating the PFs of many-objective optimization problems (MaOPs). Due to the high dimensionality in the objective space, almost all EMO algorithms with Pareto dominance encounter the difficulty in converging towards the PFs of MaOPs. In contrast, most of efficient EMO algorithms for many-objective optimization use the idea of decomposition in fitness assignment. It should be pointed out that small population size is often used in these many-objective optimization algorithms, which focus on the approximation of PFs along some specific search directions. In this paper, we studied the extensions of two well-known algorithms (i.e., NSGA-II and MOEA/D) with the ability to find a large population of nondominated solutions with good spread. A region-based archiving method is also suggested to reduce the computational complexity of updating external population. Our experimental results showed that these two extensions have good potential to find the PFs of MaOPs.

Keywords: Multiobjective optimization · NSGA-II/LP · MOEA/D-LP · Large population · Relaxed epsilon dominance

1 Introduction

Many real-world optimization problems involve multiple objectives. The optimality of these problems is often defined in terms of Pareto dominance. A solution is called Pareto-optimal if it is not dominated by any other solutions in the search space. Under this definition, there is no solution optimal to all objectives. The set of all Pareto-optimal solutions is called Pareto front in the objective space. According to the relationship between search and decision, multiobjective optimization methods can be divided into three classes [1]: the prior methods, the

H. Ishibuchi et al. (Eds.): EMO 2021, LNCS 12654, pp. 65–76, 2021.
https://doi.org/10.1007/978-3-030-72062-9_6

posterior methods, and the interactive methods. In traditional multiobjective optimization, an MOP is often converted into a single objective optimization problem by aggregating all objectives into a single objective with prior weight vectors. This is also called decomposition or scalarization. The most widely-used traditional methods are the weighted sum method and the weighted Tchebycheff approach. In both methods, a prior weight vector should be first specified by decision makers, and an aggregation function is optimized for finding only one Pareto-optimal solution.

Over the past three decades, the research on the development of EMO algorithms has attracted a lot of interests in the community of evolutionary computation [2,3]. Unlike the prior traditional multiobjective methods, most EMO algorithms attempt to find a population of nondominated solutions with good distribution. Among all EMO algorithms, NSGA-II [4] and MOEA/D [5] are two representative algorithms based on Pareto dominance and decomposition respectively. In NSGA-II, the union of parent and offspring populations are grouped into a number of fronts, and the diversity of fronts is maintained by using crowding distance. In MOEA/D, an MOP is decomposed into a number of single objective optimization subproblems, which can be optimized by any existing single objective optimizers. Both algorithms have been widely used in the comparison studies on the performance of EMO algorithms on many multiobjective benchmark test problems.

The research on the EMO algorithms on many-objective optimization has become very popular during the past ten years. Note that the populations of EMO algorithms based on Pareto dominance fail to converge towards the true PFs of MaOPs due to the weak selection pressure. In contrast, the EMO algorithms based on decomposition have no difficulty in converging towards the true PFs of MaOPs. In the well-known many-objective EMO algorithm NSGA-III [6], both nondominated sorting and decomposition are simultaneously addressed for balancing the convergence and the diversity of population. It should be noted that the use of small population size is considered in the majority of research work on many-objective EMO algorithms. The goal of these algorithms is to find a set of Pareto-optimal solutions along a number of preferred search directions.

In fact, some EMO algorithms for multi-objective combinatorial optimization problems aim at finding all possible nondominated solutions, which are often stored in an external population [7,8]. However, when solving continuous MaOPs, only a fixed number of nondominated solutions are obtained by evolving a population in many-objective EMO algorithms. In the case of no preferred search directions, the EMO algorithms with small population sizes could fail to find a good approximation of the PFs of MaOPs. Instead, the EMO algorithms with large population sizes should be a better choice when approximating the PFs of MaOPs. But the use of large population sizes in EMO algorithms will cause high computational cost in fitness computation and maintenance of external population. In this paper, we propose the extensions of two representative EMO algorithms (NSGA-II and MOEA/D), where fitness computation is still performed regarding a population with small and fixed size, and all nondominated solutions found in the history search is stored in an external population with large and dynamic size. Some computational experiments are conducted to study the performance of these two extensive EMO algorithms.

The remainder of this paper is organized as follows. Section 2 gives a brief review on the EMO algorithms and the strategies of population sizes in these algorithms. In Sect. 3, the extensions of NSGA-II and MOEA/D are illustrated in detail. Section 4 provides the analysis and the discussions of some experimental results. The final section concludes the paper.

2 Brief Review on EMO Population Size

In the area of multiobjective optimization, many EMO algorithms have been studied due to their ability to find multiple Pareto solutions in a single run. Convergence and diversity are two major issues in the EMO research. To ensure the convergence of population, three strategies, i.e., Pareto dominance, decomposition, and performance indicator, are used to define the fitness values of individuals. The common schemes for maintaining the diversity of nondominated solutions include crowding distance, nearest neighbor method, grid-based density estimation, uniform decomposition, and so on. The combination of EMO algorithms with existing single objective optimizers [9], such as differential evolution [10], particle swarm optimization [11], ant colony optimization [12], covariance matrix adaptation evolution strategy [13], has also gained some attention. Apart from the above issues, the setting of population sizes has a great influence on the performance of EMO algorithms. However, this issue has not yet been studied too much in the community of EMO research.

The use of a fixed population size is a common strategy in the development of EMO algorithms. For example, the union of current population and offspring population in NSGA-II is always reduced to a new current population with a fixed size. In the first version of MOEA/D, the number of subproblems is also fixed during the evolution of population. It should be noted that the performance of EMO algorithms could be poor if the population size is set to be either too small or too large.

- As analyzed in [5], the population of NSGA-II cannot converge towards the PFs of ZDT test problems when the population size is less than 20. The reason of poor convergence is due to the weak selection pressure. That is, all solutions in the current population are very likely to be nondominated to each other although they are not close to PFs. In contrast, the performance of MOEA/D is more robust than that of NSGA-II in convergence when a small population size is considered. However, the diversity of population in MOEA/D with small population size should be carefully preserved.
- In [14], the performance of EMO algorithms with large population sizes was investigated when solving multiobjective knapsack problems. The major problems in these EMO algorithms lie in the high computational cost of fitness assignment or diversity computation. If the number of function evaluations is limited, the population with large size could not be well converged. On

the other hand, the storage of a large-sized external population containing all possible nondominated solutions also involves high computational cost.

When the goal of EMO algorithms is the approximation of the whole PFs, the use of dynamic population size [7,15] or large population size should be a better choice than the use of fixed population size. In [7], a hybrid EMO algorithm, called multi-objective genetic local search (MOGLS), maintains a dynamic population, of which the maximal size is twenty times of the initial size. In each iteration of MOGLS, a temporary mating population is determined regarding an aggregation function with a random weight vector. MOGLS also employs an external population to save all nondominated solutions found during the history search. It should be pointed out that the maintenance of a large population will cause very high computational cost in the selection of mating parents in MOGLS.

To find all possible nondominated solutions in the PFs, MOEA/D with small population size can also work well. In [16], the combination of MOEA/D with simulated annealing, denoted by EMOSA, considers the use of small population size. But the weight vectors of subproblems are dynamically adjusted according to the diversity of population in the objective space and the diversity of weight vectors. When the current solutions or the weight vectors of two neighboring subproblems are too close, the corresponding weight vectors should be replaced in order to increase the distance between them. In EMOSA, an external population is maintained in terms of the relaxed epsilon dominance. In this way, two nondominated solutions very close to each other cannot be saved in the external population at the same time.

Note that the study of EMO algorithms with large population sizes mainly focuses on multi-objective combinatorial optimization problems. Very little work has been done on the EMO algorithms with large population sizes when dealing with continuous MOPs.

3 The Extensions of NSGA-II and MOEA/D

In this section, the motivations of studying the EMO algorithms with large population size on continuous MOPs are first discussed. Then, the extensions of NSGA-II and MOEA/D with large population sizes, denoted by NSGA-II/LP and MOEA/D-LP, are illustrated separately.

3.1 Motivations

Over the past ten years, two well-known EMO frameworks, i.e., NSGA-II and MOEA/D, have attracted a lot of research interests in the community of EMO research. When solving continuous MOPs, both algorithms often use a fixed population size, and each of them has its own advantages. NSGA-II has less sensitivity on the shapes of PFs while MOEA/D has more flexibility to integrate existing optimizers. When solving ZDT and DTLZ test problems, both algorithms are able to find many nondominated solutions along the PFs. In MOEA/D, the distribution of these solutions highly depends on how the subproblems are defined. In most of cases, both NSGA-II and MOEA/D only need to find a population of nondominated solutions with fixed but not large size for approximating the PFs of continuous MOPs.

In fact, finding a large number of nondominated solutions should be necessary in continuous many-objective optimization when the preferred search directions are not available in EMO algorithms. Since the dimensionality of many-objective PFs is high, a good approximation of many-objective PFs requires a huge number of nondominated solutions. However, the use of current population with huge size in both NSGA-II and MOEA/D is of low efficiency in fitness assignment and diversity estimation. In this paper, we study the idea of using two populations in both NSGA-II and MOEA/D, where the size of one population is small but fixed and that of the other population can be very large. The following two issues must be further addressed:

- Since the number of Pareto solutions of a continuous MOPs is often infinite, it is impractical to find all possible nondominated solutions. If some nondominated solutions are very close to each other, only one representative solution should survive. Therefore, the update of external population should be done in terms of relaxed epsilon dominance.
- Both NSGA-II and MOEA/D perform their genetic operators on the current population with fixed size as usual. The current population in these two algorithms could contain the members of external population. This should be helpful in diversity when aiming at find a large number of nondominated solutions with good distribution along PFs.

Based on the above two issues, both NSGA-II and MOEA/D have ability to find a large number of nondominated solutions within an acceptable computational time. Moreover, the management of a large population of nondominated solutions is also feasible since the computers of nowadays have very powerful computational ability.

3.2 NSGA-II/LP

The extension of NSGA-II proposed in this paper, denoted by NSGA-II/LP, works very similarly as NSGA-II. The following three major changes are addressed.

- Two populations are maintained. The current population $\mathcal{P}_c$ with the size *pop* is operated as usual. The external population $\mathcal{P}_e$ is used to store all possible nondominated solutions.
- The update of $\mathcal{P}_e$ is conducted in terms of relaxed epsilon dominance. If two solutions dominate each other, the one closer to the reference point wins the selection.
- When the size of $\mathcal{P}_e$ exceeds $2 \times pop$, i.e., the double of the size of $\mathcal{P}_c$, the union population is replaced by $2 \times pop$ solutions randomly selected from $\mathcal{P}_e$.

Algorithm 1: NSGA-II/LP

Input: The population size: *pop*.
Output: A set of nondominated solutions: $\mathcal{P}_e$.

1 Initialize $\mathcal{P}_c$ with *pop* solutions generated randomly, and set $\mathcal{P}_e = \emptyset$;
2 **while** the stopping condition is not met **do**
3 Select mating parents from $\mathcal{P}_c$ via **tournament selection** and produce an offspring population $\mathcal{P}_o$ with *pop* solutions via **genetic operators**;
4 Update $\mathcal{P}_e$ with the offspring solutions in $\mathcal{P}_o$ in terms of **relaxed epsilon dominance**;
5 **if** $|\mathcal{P}_e| \leq 2 \times pop$ **then**
6 $\mathcal{Q} = \mathcal{P}_c \cup \mathcal{P}_o$;
7 Divide the members of $\mathcal{Q}$ into a number of fronts via **nondominated sorting**.
8 **else**
9 Select $2 \times pop$ distinct solutions from $\mathcal{P}_e$ and save them into $\mathcal{Q}$.
10 **end**
11 Determine the best *pop* solutions in $\mathcal{Q}$ via **crowding distance**, and save them into $\mathcal{P}_c$.
12 **end**
13 Output $\mathcal{P}_e$.

The detailed illustration of NSGA-II/LP is shown in Algorithm 1. The major operators in NSGA-II/LP, such as tournament selection and genetic operators in line 3, nondominated sorting in line 7, and crowding distance in line 11, are the same as those in NSGA-II.

3.3 MOEA/D-LP

The extension of MOEA/D with large population size, denoted by MOEA/D-LP, also needs to decompose an MOP into a number of subproblems. When dealing with continuous MOPs, the commonly-used decomposition methods in MOEA/D include the weighted Tchebycheff methods and the penalty-based intersection method. In these methods, a number of weight vectors, or search directions, or reference points, should be specified in advance. Again, MOEA/D-LP works very similarly as MOEA/D. Take the weighted Tchebycheff approach [1] as a decomposition example, the subproblem in MOEA/D-LP is defined by:

$$\text{minimize}_{x \in \Omega} \; g^{tch}(x|\lambda, z) = \max_{i \in \{1,\dots,m\}} \lambda_i |f_i(x) - z_i| \tag{1}$$

where

- $\Omega \subseteq \mathbb{R}^n$ is the feasible region in the decision space;
- $\lambda = (\lambda_1, \dots, \lambda_m)^T$ is a weight vector with $\lambda_i \geq 0, i = 1, \dots, m$ and $\sum_{i=1}^m \lambda_i = 1$;
- $z = (z_1, \dots, z_m)^T$ is a reference point satisfying:

$$z_i = \min_{x \in \Omega} f_i(x), i = 1, \dots, m.$$

Algorithm 2: MOEA/D-LP

Input: The population size: *pop*; A set of weight vectors: $\{\lambda^1, \lambda^2, \ldots, \lambda^{pop}\}$
Output: A set of nondominated solutions: $\mathcal{P}_e$.

1 Initialize $\mathcal{P}_c = \{x^1, \ldots, x^{pop}\}$ randomly, set $\mathcal{P}_e$ as the set of nondominated solutions in $\mathcal{P}_c$, and initialize $z_j = \min_{x \in \mathcal{P}_c} f_j(x), j = 1, \ldots, m$;
2 Initialize the neighborhoods of all subproblems by setting $B(i)$ as the set of indexes of T nearest neighbors of subproblem i for all $i \in \{1, \ldots, pop\}$.
3 **while** the stopping condition is not met **do**
4   **foreach** $i \in \{1, \ldots, pop\}$ **do**
5   **Select** mating parents from $B(i)$, and replace the duplicated parents with distinct solutions chosen from $\mathcal{P}_e$ randomly if necessary;
6   **Produce** an offspring solution x_o via genetic operators on the selected mating parents and update the reference point z by: $z_j = f_j(x_o)$ if $f_j(x_o) < z_j, j = 1, \ldots, m$;
7   **Update** $\mathcal{P}_e$ with x_o in terms of relaxed epsilon dominance;
8   **Replace** any neighboring solution x^r with x_o if:
$$g^{tch}(x_o|\lambda^r, z) \le g^{tch}(x^r|\lambda^r, z), r \in B(i).$$
9   **end**
10 **end**
11 Output $\mathcal{P}_e$.

The detailed illustration of MOEA/D-LP is shown in **Algorithm** 2.

The first two changes in NSGA-II/LP, i.e., the use of two populations ($\mathcal{P}_c$ and $\mathcal{P}_e$), as well as the update of $\mathcal{P}_e$ in terms of relaxed epsilon dominance, are also considered in MOEA/D-LP. The selection of mating parent and the update of subproblems are modified in the following ways.

- Two mating parents are selected from the neighborhood of current subproblem. If both parents are identical, then one of them is replaced by one solution randomly selected from $\mathcal{P}_e$.
- If the size of $\mathcal{P}_e$ exceeds $2 \times pop$, the selection of mating parents from $\mathcal{P}_e$ is performed with the probability $\frac{|\mathcal{P}_e| - pop}{|\mathcal{P}_e|}$.

3.4 The Update of External Population

As studied in [17], the diversity of external population can be maintained by the use of relaxed epsilon dominance. To avoid the storage of two very similar solutions in $\mathcal{P}_e$, the following rules are suggested for the update of $\mathcal{P}_e$ when an offspring solution is compared with the members of $\mathcal{P}_e$.

- Case 1: If the offspring solution is dominated by any member of $\mathcal{P}_e$, then it loses the selection.
- Case 2: If the offspring solution and one member of $\mathcal{P}_e$ epsilon-dominate each other, the one closer to the reference point wins the selection.
- Case 3: If the offspring solution doesn't lose the selection in the above two cases, then it is added into $\mathcal{P}_e$.

4 Computational Experiments

4.1 Experimental Settings

Table 1. The Settings of Parameters used in NSGA-II/LP and MOEA/D-LP

Test instance		Population size	# of FEs	ϵ
ZDT1		100	50000	0.001
ZDT2		100	50000	0.001
DTLZ1	m = 3	300	150000	0.01
	m = 5	495	500000	0.03
	m = 8	792	800000	0.05
DTLZ2	m = 3	300	150000	0.01
	m = 5	495	500000	0.03
	m = 8	792	800000	0.05

In this section, the performance of both NSGA-II/LP and MOEA/D-LP were verified on a set of benchmark test problems with regular PF shapes, i.e., two bi-objective test problems (ZDT1, ZDT2), two scalable test problems (DTLZ1 and DTLZ2) with 3, 5, and 8 objectives. The settings of population size, computational budgets, and the epsilon value of relaxed dominance are listed in Table 1. Since all test instances have no interacting variables, the simulated binary crossover (SBX) and polynomial mutation are applied to produce offspring solutions in both algorithms as in [4].

The IGD metric is not used in this work since the number of reference points required in the computation of IGD values might be huge since the final size of $\mathcal{P}_e$ might be very large. Instead, the set coverage [18] is used to compare the performance of NSGA-II/LP and MOEA/D-LP in and convergence and diversity. Both algorithms were implemented in Matlab and executed in the computer (Intel i5 CPU (@1.6 GHZ, 1.8 GHZ) and 8G memory) running Windows 10 operating system. Each instance was tested by both algorithm for ten runs.

4.2 Experimental Results

Figure 1 shows that the nondominated solutions in the final populations found by both NSGA-II/LP and MOEA/D-LP on the test instances with 2 or 3 objectives, i.e., ZDT1, ZDT2, DTLZ1, and DTLZ2. It can be observed from this figure that both algorithms are able to approximate the PFs of these test instances very well. As reported in [5], the performance of NSGA-II in diversity on DTLZ1 and DTLZ2 is not good. Here, our results showed that the use of external population can improve the performance of NSGA-II on three-objective test instances.

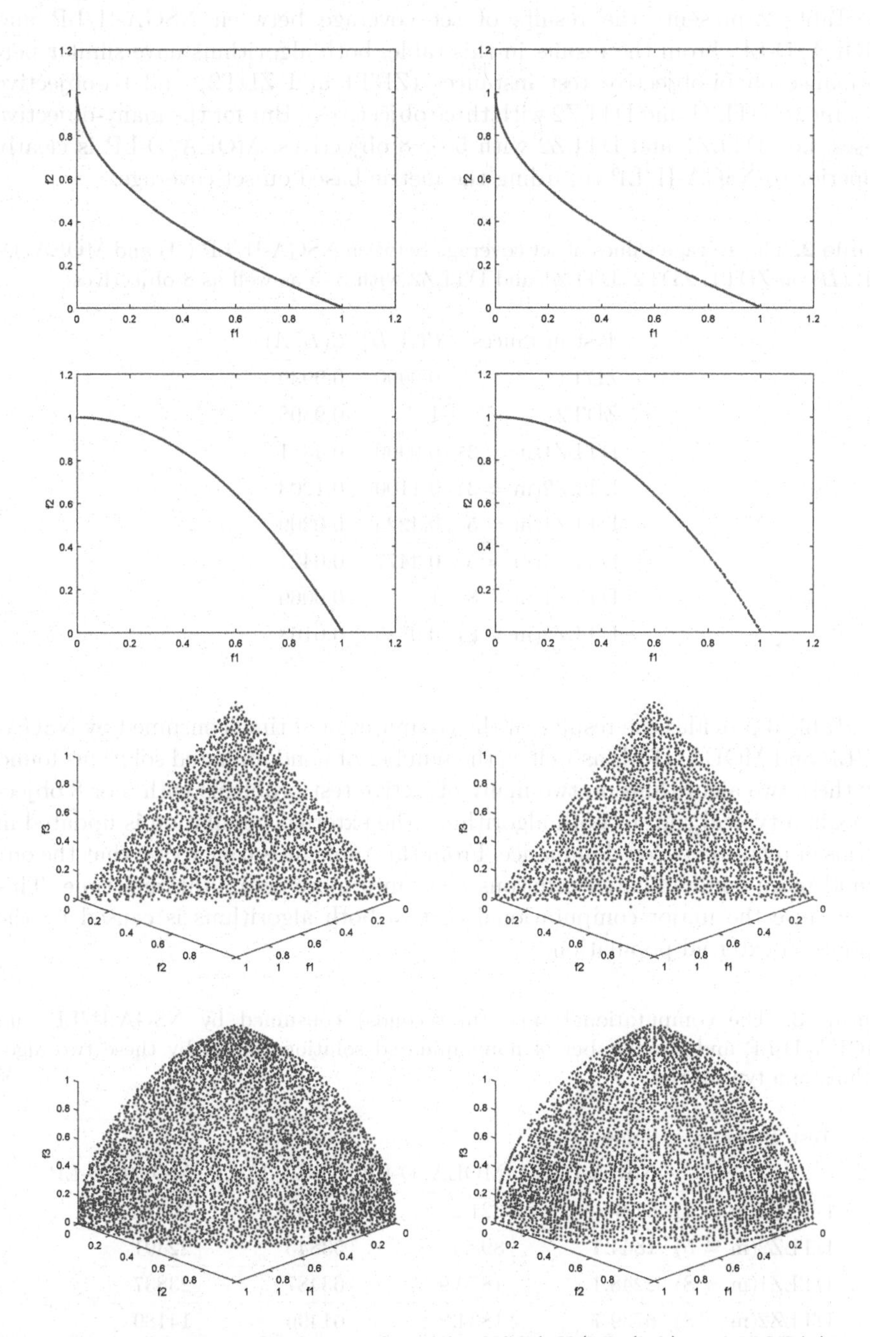

Fig. 1. The nondominated solutions found by NSGA-II/LP (left) and MOEA/D-LP (right) on ZDT1, ZDT2, DTLZ1 and DTLZ2 positoned in order from the top to the bottom.

Table 2 presents the results of set coverage between NSGA-II/LP and MOEA/D-LP. From the results in this table, both algorithms have similar performance on bi-objective test instances (ZDT1 and ZDT2) and tri-objective instances (DTLZ1 and DTLZ2 with three objectives). But for the many-objective cases, i.e., DTLZ1 and DTLZ2 with 5 or 8 objectives, MOEA/D-LP is clearly superior to NSGA-II/LP regarding the metric based on set coverage.

Table 2. The average values of set coverage between NSGA-II/LP (A) and MOEA/D-LP (B) on ZDT1, ZDT2, DTLZ1 and DTLZ2 with 3, 5 as well as 8 objectives.

Test instances	$\mathcal{C}(A, B)$	$\mathcal{C}(B, A)$
ZDT1	0.9990	0.9939
ZDT2	1	0.9898
DTLZ1(m = 3)	0.0960	0.4331
DTLZ2(m = 3)	0.1100	0.1203
DTLZ1(m = 5)	0.4225	0.0209
DTLZ2(m = 5)	0.4477	0.0421
DTLZ1(m = 8)	1	0.0069
DTLZ2(m = 8)	0.9785	0.0168

Table 3 provides the results on the computational time consumed by NSGA-II/LP and MOEA/D-LP as well as the number of nondominated solutions found by these two algorithms on two many-objective test problems with 5 or 8 objectives in a typical run. In both algorithms, the external population is updated in terms of relaxed epsilon dominance. From this table, it is easy to see that the one found more nondominated solutions consumes more computational time. This is because the major computational cost in both algorithms is caused by the update of external population.

Table 3. The computational time (in seconds) consumed by NSGA-II/LP and MOEA/D-LP and the number of nondominated solutions found by these two algorithms in a typical run.

Test instance	Time		Number	
	NSGA-II/LP	MOEA/D-LP	NSGA-II/LP	MOEA/D-LP
DTLZ1(m = 5)	61.6	321.2	5986	8051
DTLZ2(m = 5)	1844.1	898.1	53845	22491
DTLZ1(m = 8)	2248.1	1870.9	63487	23837
DTLZ2(m = 8)	6739.7	1834.2	61496	14189

To reduce the cost for updating the external population, we suggest a region-based archiving strategy in this paper. That is, a number of directions are pre-specified to divide the objective space into multiple subregions. The offspring solutions generated by genetic operators only compare with those nondominated solutions in the subregion, where they are located. Table 4 reports the average computational time spent by NSGA-II/LP and MOEA/D-LP on all test instances. These results clearly show that the computational time used by both algorithms with the region-based archiving strategy is reduced dramatically in comparison with the results in Table 3.

Table 4. The average computational time (in seconds) consumed by NSGA-II/LP and MOEA/D-LP with the region-based archiving strategy.

Test instance	NSGA-II/LP	MOEA/D-LP
ZDT1	2.5	2.5
ZDT2	2.4	2.3
DTLZ1(m = 3)	10.4	16.0
DTLZ2(m = 3)	15.8	20.5
DTLZ1(m = 5)	44.35	76.7
DTLZ2(m = 5)	176.1	121.9
DTLZ1(m = 8)	205.6	144.8
DTLZ2(m = 8)	256.5	213.8

5 Conclusions

In this paper, we studied the extensions of two popular EMO algorithms, i.e., NSGA-II and MOEA/D, where an external population with a very large size could be maintained. In both NSGA-II/LP and MOEA/D-LP, the current population always has a fixed size and is operated as usual. Therefore, the conventional operators in both NSGA-II/LP and MOEA/D-LP don't involve high computational cost. The proposed two extensive EMO algorithms have been tested on a number of multiobjective test instances with 2, 3, 5, 8 objectives. The reported experimental results show that (1) NSGA-II/LP has the potential to solve many-objective test problems if an external population with a large size is used; (2) MOEA/D-LP performs clearly better than NSGA-II/LP in convergence and diversity on many-objective test instances. Moreover, the computational time on the update of external population can be significantly reduced by the region-based archiving strategy. In our future work, the performance of NSGA-II/LP and MOEA/D-LP on the multiobjective test instances with more difficult features will be studied.

Acknowledgment. The authors would like to thank the anonymous reviewers for their insightful comments. This work was supported by National Natural Science Foundation of China (NSFC) grants 11991023, 62072364, and 62076197.

References

1. Miettinen, K.: Nonlinear Multiobjective Optimization. Kluwer Academic Publishers, Amsterdam (1999)
2. Deb, K.: Multi-Objective Optimization using Evolutionary Algorithms. Wiley, Hoboken (2001)
3. Zhou, A.M., et al.: Multiobjective evolutionary algorithms: a survey of the state of the art. Swarm Evol. Comput. **1**(1), 32–49 (2011)
4. Deb, K., Agrawal, S., Pratap, A., Meyarivan, T.: A fast and elitist multiobjective genetic algorithm: NSGA-II. IEEE Trans. Evol. Comput. **6**(2), 182–197 (2002)
5. Zhang, Q., Li, H.: MOEA/D: a multiobjective evolutionary algorithm based on decomposition. IEEE Trans. Evol. Comput. **11**(6), 712–731 (2007)
6. Deb, K., Jain, H.: An evolutionary many-objective optimization algorithm using reference-point-based nondominated sorting approach, part I: Solving problems with box constraints. IEEE Trans. Evol. Comput. **18**(4), 577–601 (2014)
7. Jaszkiewicz, A.: On the performance of multiple-objective genetic local search on the 0/1 knapsack problem - a comparative experiment. IEEE Trans. Evol. Comput. **6**(4), 402–412 (2002)
8. Ishibuchi, H., Yoshida, T., Murata, T.: Balance between genetic search and local search in memetic algorithms for multiobjective permutation flowshop scheduling. IEEE Trans. Evol. Comput. **7**(2), 204–223 (2003)
9. Fred, G., Gary, K.: Handbook of Metaheuristics. Kluwer Academic Publishers, Boston (2003)
10. Li, H., Zhang, Q.: Multiobjective optimization problems with complicated pareto sets, MOEA/D and NSGA-II. IEEE Trans. Evol. Comput. **13**(2), 284–302 (2009)
11. Dai, C., Wang, Y.P., Ye, M.: A new multi-objective particle swarm optimization algorithm based on decomposition. Inf. Sci. **325**, 541–557 (2015)
12. Ke, L.J., Zhang, Q., Battiti, R.: MOEA/D-ACO: a multiobjective evolutionary algorithm using decomposition and antcolony. IEEE Trans. Cybern. **43**, 1845–1859 (2013)
13. Li, H., Zhang, Q., Deng, J.: Biased multiobjective optimization and decomposition algorithm. IEEE Trans. Cybern. **47**(1), 52–66 (2017)
14. Ishibuchi, H., Sakane, Y., Tsukamoto, N., Nojima, Y.: Evolutionary many-objective optimization by NSGA-II and MOEA/D with large populations. In: 2009 IEEE International Conference on Systems, Man and Cybernetics, pp. 1758–1763 (2009)
15. Tan, K.C., Lee, T.H., Khor, E.F.: Evolutionary algorithms with dynamic population size and local exploration for multiobjective optimization. IEEE Trans. Evol. Comput. **5**(6), 565–588 (2001)
16. Li, H., Landa-Silva, D.: An adaptive evolutionary multi-objective approach based on simulated annealing. Evol. Comput. **19**(4), 561–595 (2011)
17. Li, H., Deng, J.D., Zhang, Q., Sun, J.Y.: Adaptive epsilon dominance in decomposition-based multiobjective evolutionary algorithm. Swarm Evol. Comput. **45**, 52–67 (2019)
18. Zitzler, E., Deb, K., Thiele, L.: Comparison of multiobjective evolutionary algorithms: empirical results. Evol. Comput. **8**(2), 173–195 (2000)

Multitask Feature Selection for Objective Reduction

Genghui Li[1,2(✉)] and Qingfu Zhang[1,2]

[1] Department of Computer Science, City University of Hong Kong, Hong Kong, China
genghuili2-c@my.cityu.edu.hk, qingfu.zhang@cityu.edu.hk

[2] The City University of Hong Kong Shenzhen Research Institute, Shenzhen, China

Abstract. Objective reduction has been regarded as a major tool for solving many-objective optimization problems (MaOPs). This paper proposes a multitask feature selection method for objective reduction. In our proposed method, each objective is formulated as a positive linear combination of a small number of essential objectives, and sparse regularization is employed to identify redundant objectives. Our numerical experiment shows the effectiveness and robustness of the proposed method by comparing it with some state-of-the-art objective reduction methods.

Keywords: Many-objective optimization · Multitask feature selection · Objective reduction

1 Introduction

A multi-objective optimization problem (MOP) can be formulated as

$$\begin{aligned} &\text{minimize} \quad F(\mathbf{x}) = (f_1(\mathbf{x}), \cdots, f_M(\mathbf{x})) \\ &\text{subject to} \quad \mathbf{x} \in \boldsymbol{\Omega} \end{aligned}, \tag{1}$$

where $\mathbf{x} = (x_1, \cdots, x_d)^T \in \boldsymbol{\Omega}$ is a solution, and d is the number of decision variables; M is the number of objectives. If $M > 3$, it is called a many-objective optimization problem (MaOP). Some basic definitions of multi-objective optimization used in this paper are:

Pareto Dominance: A solution $\mathbf{x}_1$ dominates solution $\mathbf{x}_2$, denoted by $\mathbf{x}_1 \preccurlyeq \mathbf{x}_2$, iff $\forall i \in \{1, \cdots, M\}$, $f_i(\mathbf{x}_1) \leq f_i(\mathbf{x}_2)$, and $\exists i \in \{1, \cdots, M\}$, $f_i(\mathbf{x}_1) < f_i(\mathbf{x}_2)$.

Pareto Solution: A solution $\mathbf{x}^*$ is called a Pareto optimal solution if there is no solution $\mathbf{x} \in \boldsymbol{\Omega}$ such that $\mathbf{x} \preccurlyeq \mathbf{x}^*$.

This work was supported by the National Natural Science Foundation of China (Grant No: 61876163) and ANR/RGC Joint Research Scheme sponsored by the Research Grants Council of the Hong Kong Special Administrative Region, China and France National Research Agency (Project No: A-CityU101/16).

H. Ishibuchi et al. (Eds.): EMO 2021, LNCS 12654, pp. 77–88, 2021.
https://doi.org/10.1007/978-3-030-72062-9_7

Pareto Set and Pareto Front: The Pareto set (PS) is the set of all Pareto optimal solutions $\mathbf{x}^* \in \boldsymbol{\Omega}$. The Pareto front PF $= \{F(\mathbf{x})|\mathbf{x} \in \text{PS}\}$ is the image of the PS in the objective space.

Multiobjective evolutionary algorithms (MOEAs) are to find a set of solutions that should be as close as possible to the true PF and distribute uniformly. Many MOEAs have been developed over the past two decades [8,25]. However, their search ability degrades considerably as the number of objective increases [16]. Dominance-based MOEAs [8] lack selection pressure for many objective optimization, some new dominance relations [22,24] have been developed to overcome this shortcoming. Indicator-based MOEAs [3] suffer from high complexity when the the number of objectives increases, high-efficiency approximation of hypervolume [2,10] and other computationally efficient performance indicators [11] have been proposed. Decomposition-based MOEAs [25] need a set of well distributed weight vectors for dealing with many objective optimization, some enhanced generation of well-distributed weight vector methods [21] have been studied.

Along with the above efforts, some attempts have been made to identify the redundant objectives in MaOP. Objective reduction methods can be roughly classified into the following two categories [23]:

1) *Objective selection*: Objective selection is to select a small set of objectives to approximate the original MaOP as much as possible. There are two objective selection methods: correlation-based and dominance-based objective selection methods.

Correlation-based objective selection methods select the essential objectives by exploiting the correlation information among all objectives. For instance, the principal component analysis has been used to select the essential objectives in [7,19]. An unsupervised feature selection in [17] is used to identify essential objectives, in which the similarity is measured based on the correlation among objectives. Objective selection was formulated as a bi-objective optimization problem in [23], in which the first objective is the number of the selected objectives, and the second objective is a criterion that measures the degree of the correlation structure change between the selected objectives and the original objectives.

Dominance-based objective selection methods select essential objectives by keeping the dominance structure as much as possible. By analyzing how adding and removing an objective affects the dominance structure, δ-MOSS and k-EMOSS were proposed in [4]. δ-MOSS selects a minimal-objective subset for a given error δ, while k-EMOSS is to select k objectives with minimal error. δ-MOSS and k-EMOSS were combined as a bi-objective optimization problem in [23]. [20] claimed that the corner points can represent the main features of the PF. It uses only the corner points to decide whether an objective is essential or redundant via detecting the change of the ratio of the non-dominated solution after removing this objective. However, [13] argued that the corner points maybe not enough to represent the overall features of the PF, and it used the number of conflicting pairs to identify the essential objectives on a set of well-distributed representative non-dominated solutions instead of corner points.

Unlike correlation-based and dominance-based objective selection methods, a hyperplane-approximation-based method is proposed in [18] for objective selection, in which the objective with the non-zero coefficient of the hyperplane is identified as an essential objective.

2) *Objective extraction (or objective transformation)*: Objective extraction is to transform the objectives of the original MOP into a small set of other objectives. [5,6] formulated the essential objective as a positive linear combination of the original objectives, and the combination weights were determined by minimizing the correlation of each pair of the essential objectives under a sparse constraint. [27] followed [5,6] and obtained the combination weights by the fuzzy clustering method. Besides, [14] mapped solutions from a high-dimensional objective space into a 2-D polar coordinate plot. [26] sequentially decomposed redundant objectives which can be linearly represented by their positive correlated objectives until the expected number of preserved objectives is reached.

As claimed in [23], the commonly used correlation-based and dominance-based objective selection methods suffer from some limitations. To be specific, the correlation-based methods cannot identify the essential objective set exactly on some problems. The dominance-based methods are sensitive to the noise (also called misdirection) and suffer from high complexity. Moreover, the number of extracted objectives should be prescribed in advance for the objective extraction method, it is not a trivial task.

Unlike other existing methods, this paper formulates the objective selection as a multitask feature selection problem, each task is to construct a regression model for one objective, and the combination weights are obtained by minimizing the approximation error subject to a sparse constraint. The optimal combination weight matrix can be used to distinguish essential objectives and redundant ones effectively.

The remainder of this paper is organized as follows. Section 2 details the proposed method. Section 3 provides the numerical experimental results and corresponding analysis. Section 4 concludes this paper and discusses some future works.

2 Multitask Feature Selection for Objective Reduction (MTFSOR)

Before detailing the proposed algorithm, we introduce some necessary definitions [23]. To simplify the notation, let the original objective set of a given MOP be $\mathcal{F}_0 = \{f_1, \ldots, f_M\}$, and the corresponding PF be PF($\mathcal{F}_0$). Similarly, for a subset $\mathcal{F}' \subset \mathcal{F}_0$, we denote its PF as PF($\mathcal{F}'$). The notation $\mathbf{u}^{\mathcal{F}}$ is a subvector of $\mathbf{u}$ by the given nonempty subset $\mathcal{F}$. For example, if $\mathbf{u} = (f_1, f_2, f_3)^T$ and $\mathcal{F} = \{f_2, f_3\}$, $\mathbf{u}^{\mathcal{F}} = (f_2, f_3)^T$.

- **Definition 1**: Let $\mathcal{F} = \mathcal{F}_0 \setminus \mathcal{F}'$, the objective subset $\mathcal{F}' \subset \mathcal{F}_0$ is redundant iff $\forall \mathbf{u}^{\mathcal{F}_0} \in$ PF($\mathcal{F}_0$), we have $\mathbf{u}^{\mathcal{F}} \in$ PF($\mathcal{F}$), and $\forall \mathbf{u}^{\mathcal{F}_0} \notin$ PF($\mathcal{F}_0$), we have $\mathbf{u}^{\mathcal{F}} \notin$ PF($\mathcal{F}$).

– **Definition 2**: The objective set $\mathcal{F}'$ is an essential objective set iff $\mathcal{F}_0 - \mathcal{F}'$ is a redundant objective subset with largest cardinality.

2.1 Method

Let $\mathcal{X} = \{\mathbf{x}_1, \cdots, \mathbf{x}_N\}$ be a set of non-dominated solutions, and N be the number of solutions in $\mathcal{X}$. The objective vector of $\mathcal{X}$ is denoted by $\mathbf{F} = [\mathbf{f}_1, \cdots, \mathbf{f}_M]$, where $\mathbf{f}_k = [f_{1,k}, \cdots, f_{N,k}]^T, 1 \leq k \leq M$ and $f_{i,k} = f_k(\mathbf{x}_i), 1 \leq i \leq N$. The task corresponding to the k-th objective is formulated as

$$\hat{f}_k(\mathbf{x}) = \sum_{i=1}^{M} w_{i,k} f_i(\mathbf{x}), 1 \leq k \leq M, \tag{2}$$

where $w_k = [w_{1,k}, \cdots, w_{M,k}]^T \geq \mathbf{0}$ is the combination weight vector of the k-th task.

We assume that some objectives are redundant, which means that not all objectives are necessary in (2). In other words, the matrix $W = [w_1, \cdots, w_M]$ may has some zero rows. To obtain W, we define the following loss function [1]:

$$\mathcal{L}(W) = \frac{1}{N} \sum_{k=1}^{M} \sum_{i=1}^{N} (f_{i,k} - \hat{f}_{i,k})^2 + \lambda \|W\|_{2,1}, \tag{3}$$

where λ is a regularization parameter, which is used to adjust the row sparse degree of W; $\|W\|_{2,1} = \sum_{i=1}^{M} \|w^i\|_2$ is the (2,1)-norm of matrix W, and w^i is the i-th row of W. Our approach for objective selection is to solve the following optimization problem

$$\begin{aligned} \text{minimize} \quad & \mathcal{L}(W) = \frac{1}{N} \sum_{k=1}^{M} \sum_{i=1}^{N} (f_{i,k} - \hat{f}_{i,k})^2 + \lambda \|W\|_{2,1} \\ \text{subject to} \quad & W \geq \mathbf{0}. \end{aligned} \tag{4}$$

Although $\|W\|_{2,1}$ is not smooth, it is convex. Therefore, (4) is a convex optimization problem [1]. Here, the Matlab solver CVX [12], a package is specifically designed for dealing with convex problems, is employed to solve (4). After obtaining the optimal solution W^* of (4), we can select the objectives based on their importance, which is defined as

$$I_k = \|w^{*k}\|_2, 1 \leq k \leq M, \tag{5}$$

where I_k is the importance of the k-th objective, and w^{*k} is the k-th row of W^*. Finally, we can select the objectives based on the following rules:

– If the number (e.g., k) of the selected objectives is given in advance, the k objectives with the largest importance are chosen.

- If the number of the selected objectives is unknown, the objective will be selected based on its importance ratio, which is defined as

$$IR_k = \frac{I_k}{max\{I_i, i = 1, \cdots, M\}}, \quad 1 \leq k \leq M. \tag{6}$$

The k-th objective will be selected if $IR_k > \epsilon$, where ϵ is a small positive constant, and it is set to 0.001 in this paper.

Clearly, if $\mathbf{w}^{*k} = \mathbf{0}$, $1 \leq k \leq M$, the k-th objective will be treated as a redundant objective based on above rules due to $I_k = 0$ and $IR_k = 0$. The reason why any objective corresponding to a zero row in W^* is a redundant objective is explained in the following theorem.

Theorem 1. *For a given MOP with* $\mathcal{F}_0 = \{f_1, \cdots, f_M\}$ *and* $\mathrm{PF}(\mathcal{F}_0)$, W^* *is the optimal solution of (4). If* $w^{*k} = \mathbf{0}$, *and the error* $\sum_{k=1}^{M}\sum_{i=1}^{|\mathrm{PF}(\mathcal{F}_0)|}(f_{k,i} - \hat{f}_{k,i})^2 = 0$, *then* f_k *is a redundant objective.*

Proof. We first prove that $\forall \mathbf{u}^{\mathcal{F}_0} \in \mathrm{PF}(\mathcal{F}_0)$, $\mathbf{u}^{\mathcal{F}} \in \mathrm{PF}(\mathcal{F})$. Let $\mathcal{F} = \mathcal{F}_0 - \{f_k\}$, for $\forall \mathbf{u}^{\mathcal{F}_0} \in \mathrm{PF}(\mathcal{F}_0)$, assuming $\mathbf{u}^{\mathcal{F}} \notin \mathrm{PF}(\mathcal{F})$, then there exists $\mathbf{v}^{\mathcal{F}} \in \mathrm{PF}(\mathcal{F})$ and $\mathbf{v}^{\mathcal{F}} \preccurlyeq \mathbf{u}^{\mathcal{F}}$. Thus, for $\forall f_i \in \mathcal{F}, \mathbf{v}^{\{f_i\}} \leq \mathbf{u}^{\{f_i\}}$, and $\exists f_i \in \mathcal{F}, \mathbf{v}^{\{f_i\}} < \mathbf{u}^{\{f_i\}}$. Since the approximation error is 0 and $w^{*k} = \mathbf{0}$, we have $f_k = \sum_{i=1,i\neq k}^{M} w_{i,k}^* f_i$. Therefore $\mathbf{v}^{\{f_k\}} = \sum_{i=1,i\neq k}^{M} w_{i,k}^* \mathbf{v}^{\{f_i\}}$ and $\mathbf{u}^{\{f_k\}} = \sum_{i=1,i\neq k}^{M} w_{i,k}^* \mathbf{u}^{\{f_i\}}$. Since $w_{i,k}^* \geq 0$, we have $\mathbf{v}^{\{f_k\}} \leq \mathbf{u}^{\{f_k\}}$. Therefore, $\mathbf{v}^{\mathcal{F}+\{f_k\}} \preccurlyeq \mathbf{u}^{\mathcal{F}+\{f_k\}}$, which means $\mathbf{v}^{\mathcal{F}_0} \preccurlyeq \mathbf{u}^{\mathcal{F}_0}$. It contradicts $\mathbf{u}^{\mathcal{F}_0} \in \mathrm{PF}(\mathcal{F}_0)$. Therefore, for $\forall \mathbf{u}^{\mathcal{F}_0} \in \mathrm{PF}(\mathcal{F}_0)$, we have $\mathbf{u}^{\mathcal{F}} \in \mathrm{PF}(\mathcal{F})$.

Now we prove that $\forall \mathbf{u}^{\mathcal{F}_0} \notin \mathrm{PF}(\mathcal{F}_0)$, $\mathbf{u}^{\mathcal{F}} \notin \mathrm{PF}(\mathcal{F})$. Let $\mathcal{F} = \mathcal{F}_0 - \{f_k\}$, for $\forall \mathbf{u}^{\mathcal{F}_0} \notin \mathrm{PF}(\mathcal{F}_0)$, there exists $\mathbf{v}^{\mathcal{F}_0} \in \mathrm{PF}(\mathcal{F}_0)$ such that $\mathbf{v}^{\mathcal{F}_0} \preccurlyeq \mathbf{u}^{\mathcal{F}_0}$. Thus, for $\forall f_i \in \mathcal{F}_0, \mathbf{v}^{\{f_i\}} \leq \mathbf{u}^{\{f_i\}}$ and $\exists f_i \in \mathcal{F}_0, \mathbf{v}^{\{f_i\}} < \mathbf{u}^{\{f_i\}}$. Assuming $\forall f_i \in \mathcal{F}, \mathbf{v}^{\{f_i\}} = \mathbf{u}^{\{f_i\}}$. Since $f_k = \sum_{i=1,i\neq k}^{M} w_{i,k}^* f_i$ and $w_{i,k}^* \geq 0$, $\mathbf{v}^{\{f_k\}} = \mathbf{u}^{\{f_k\}}$, then $\mathbf{v}^{\mathcal{F}_0} = \mathbf{u}^{\mathcal{F}_0}$. It contradicts $\mathbf{v}^{\mathcal{F}_0} \preccurlyeq \mathbf{u}^{\mathcal{F}_0}$. Therefore, the assumption $\forall f_i \in \mathcal{F}, \mathbf{v}^{\{f_i\}} = \mathbf{u}^{\{f_i\}}$ is not true, then $\exists f_i \in \mathcal{F}, \mathbf{v}^{\{f_i\}} < \mathbf{u}^{\{f_i\}}$ and $\mathbf{v}^{\mathcal{F}} \preccurlyeq \mathbf{u}^{\mathcal{F}}$, therefore $\mathbf{u}^{\mathcal{F}} \notin \mathrm{PF}(\mathcal{F})$. Therefore, for $\forall \mathbf{u}^{\mathcal{F}_0} \notin \mathrm{PF}(\mathcal{F}_0)$, we have $\mathbf{u}^{\mathcal{F}} \notin \mathrm{PF}(\mathcal{F})$.

Therefore, $\forall \mathbf{u}^{\mathcal{F}_0} \in \mathrm{PF}(\mathcal{F}_0)$, $\mathbf{u}^{\mathcal{F}} \in \mathrm{PF}(\mathcal{F})$, and $\forall \mathbf{u}^{\mathcal{F}_0} \notin \mathrm{PF}(\mathcal{F}_0)$, $\mathbf{u}^{\mathcal{F}} \notin \mathrm{PF}(\mathcal{F})$. It follows that f_k is a redundant objective. This completes the proof.

It should be pointed out that **Theorem 1** is based on the assumption $\frac{1}{|\mathrm{PF}(\mathcal{F}_0)|}\sum_{k=1}^{M}\sum_{i=1}^{|\mathrm{PF}(\mathcal{F}_0)|}(f_{k,i} - \hat{f}_{k,i})^2 = 0$. It is not often the case. However, our MTFSOR does not directly select the objective based on whether its coefficient vector is zero or not. Instead, it uses the importance (or importance rate) to distinguish essential objectives and redundant ones.

2.2 An Example of MTFSOR

To explain how MTFSOR works, we show the detailed procedures of MTFSOR on a modified DTLZ2(3) [9], denoted by MDTLZ2(3+2), in which the first three objectives are the original three objectives of DTLZ2(3), and the 4-th and 5-th objectives are defined as $f_4 = 0.5f_1 + 0.2f_2 + 0.3f_3$ and $f_5 = 1/3\sqrt{f_1} + 1/3f_2 + 1/3f_3^2$, respectively. In MDTLZ2(3+2), the essential objective set is $\{f_1, f_2, f_3\}$, and the redundant objectives are $\{f_4, f_5\}$. 300 points are randomly sampled from its PF, and λ is set to 0.05. The corresponding W^* by solving (4) is

$$W^* = \begin{bmatrix} 8.1e{-}01 & 8.2e{-}02 & 3.9e{-}02 & 1.0e{-}01 & 4.0e{-}01 \\ 7.6e{-}02 & 8.3e{-}01 & 3.0e{-}02 & 1.8e{-}01 & 3.8e{-}02 \\ 4.1e{-}02 & 3.4e{-}02 & 9.2e{-}01 & 2.8e{-}01 & 2.0e{-}02 \\ \mathbf{4.6e{-}10} & \mathbf{5.3e{-}10} & \mathbf{6.0e{-}10} & \mathbf{4.8e{-}10} & \mathbf{4.2e{-}10} \\ \mathbf{7.7e{-}10} & \mathbf{4.2e{-}10} & \mathbf{4.1e{-}10} & \mathbf{4.4e{-}10} & \mathbf{5.6e{-}10} \end{bmatrix}, \tag{7}$$

and the corresponding importance and importance ratio are $I = \{9.2e{-}01, 8.5e{-}01, 9.6e{-}01, 1.1e{-}09, 1.2e{-}09\}$ and $IR = \{9.5e{-}01, 8.8e{-}01, 1.0e+00, 1.1e{-}09, 1.2e-09\}$, respectively. If the number of the essential objectives is known (i.e., $k = 3$), MTFSOR chooses $\{f_1, f_2, f_3\}$ as the essential objective set based on the importance. Moreover, even if the number of the essential objectives is unknown, MTFSOR still selects $\{f_1, f_2, f_3\}$ as the essential objective set based on the importance ratio. It is worth pointing out that although f_5 is a nonlinear combination of $\{f_1, f_2, f_3\}$, MTFSOR can identify it as a redundant objective correctly, this means that our MTFSOR is able to deal with nonlinearly degenerate PF.

2.3 Difference Between Our MTFSOR and NLHA

Our MTFSOR shares similarity with NLHA [18]. However, they have essential differences. NLHA uses a hyperplane to approximate the PF, and the objectives with non-zero coefficients are treated as essential objectives. A hyperplane can be described as

$$\sum_{i=1}^{M} w_i f_i = 1. \tag{8}$$

For a given training set with objective vectors $\mathbf{F} = [\mathbf{f}_1, \cdots, \mathbf{f}_M]$, the objective reduction model is formulated as

$$\begin{aligned} &\text{minimize} \quad \|\mathbf{F}\mathbf{w} - \mathbf{1}\|_2^2 + \lambda\|w\|_1 \\ &\text{subject to} \quad \mathbf{w} \geq \mathbf{0}. \end{aligned} \tag{9}$$

In theory, NLHA only works well when the structure of $\mathbf{F}$ can be represented by one equation. However, if the dimension of $\mathbf{F}$ is smaller than $M - 1$, only using one equation to approximate $\mathbf{F}$ is wrong. Our MTFSOR can overcome this shortcoming since it uses M equations to extract the structure of $\mathbf{F}$.

A MOP with six objectives $\{f_1(\mathbf{x}), \cdots, f_6(\mathbf{x})\}$ is defined as

$$\begin{cases} f_1(\mathbf{x}) = x_1 + \sum_{j=3}^{10}(x_j - 0.5)^2 \\ f_2(\mathbf{x}) = 1 - x_1 + \sum_{j=3}^{10}(x_j - 0.5)^2 \\ f_3(\mathbf{x}) = 0.2f_1(\mathbf{x}) + 0.8f_2(\mathbf{x}) \\ f_4(\mathbf{x}) = x_2 + \sum_{j=11}^{20}(x_j - 0.5)^2 \\ f_5(\mathbf{x}) = \sqrt{(1 - x_2^2)} + \sum_{i=11}^{20}(x_j - 0.5)^2 \\ f_6(\mathbf{x}) = 0.8f_4(\mathbf{x}) + 0.2f_5(\mathbf{x}) \end{cases}, \tag{10}$$

where $0 \leq x_i \leq 1, i = 1, \cdots, 20$. The PF satisfies $f_1 + f_2 = 1$ and $f_4^2 + f_5^2 = 1$. Obviously, it needs two equations to describe the PF precisely and its essential objectives are $\{f_1, f_2, f_4, f_5\}$.

We use 300 points that are randomly sampled from the PF to analyze. The optimal $\mathbf{w}^*$ obtained by NLHA is $\mathbf{w}^* = (4.2e-01, 8.5e-01, 4.2e-01, 8.9e-02, 1.2e-01, \mathbf{5.5e-09})$, therefore NLHA treats $\{f_1, f_2, f_3, f_4, f_5\}$ as an essential objective set. While the optimal W^* obtained by our MTFSOR is

$$W^* = \begin{bmatrix} 7.7e-01 & 1.4e-10 & 4.9e-01 & 9.7e-02 & 2.3e-01 & 1.8e-01 \\ 1.6e-10 & 7.3e-01 & 8.2e-10 & 9.8e-02 & 2.2e-01 & 1.8e-01 \\ \mathbf{1.2e-09} & \mathbf{2.1e-10} & \mathbf{1.0e-09} & \mathbf{5.8e-10} & \mathbf{7.8e-10} & \mathbf{7.1e-10} \\ 6.6e-02 & 8.3e-02 & 1.5e-01 & 7.8e-01 & 2.6e-10 & 5.9e-01 \\ 9.5e-02 & 1.0e-01 & 2.0e-01 & 3.6e-10 & 7.0e-01 & 7.9e-02 \\ \mathbf{6.3e-10} & \mathbf{6.6e-10} & \mathbf{8.0e-10} & \mathbf{1.6e-09} & \mathbf{3.7e-10} & \mathbf{1.4e-09} \end{bmatrix}. \tag{11}$$

Finally, MTFSOR identifies $\{f_1, f_2, f_4, f_5\}$ as the essential objectives.

Therefore, for a MaOP with M objectives, if the dimension of the PF is less than $M - 1$. Our MTFSOR can do better than NLHA.

3 Experimental Design and Results

3.1 Experimental Setting

We test MTFSOR on DTLZ5(M,m) [15] and MAOP(M,m) [6] problems, where M and m are the number of all objectives and essential objectives, respectively. Their essential objective set $\mathcal{F}_e$ of DTLZ5(M,m) and MAOP(M,m) are, respectively,

- DTLZ5(M,m): $\mathcal{F}_e = \{f_k, f_{M-m+2}, \cdots, f_M\}, k \in \{1, \cdots, M - m + 1\}$.
- MAOP(M,m): $\mathcal{F}_e = \{f_1, f_2, \cdots, f_m\}$.

We use the δ-metric [6] to evaluate the performance of the objective reduction methods, which is defined as

$$\delta = \frac{|E(\mathcal{F}', \mathcal{X}) \cap E(\mathcal{F}_0, \mathcal{X})|}{|E(\mathcal{F}_0, \mathcal{X})|}, \tag{12}$$

where $\mathcal{X}$ is a set of non-dominated solutions; $\mathcal{F}_0 = \{f_1, \cdots, f_M\}$ is the original objective set, and $\mathcal{F}'$ is the objective set obtained by the algorithm; $E(\mathcal{F}, \mathcal{X})$ is the set of non-dominated solutions in $\mathcal{X}$ regarding objective set $\mathcal{F}$. Obviously, $\delta \in [0,1]$, and $\delta = 1$ means that the non-dominated solutions with regard to $\mathcal{F}'$ are equal to those of the original objective set $\mathcal{F}_0$.

It should be pointed out that when δ-metric is employed to evaluate the performance of objective reduction methods, the number of the reduced objectives should be kept the same for all compared algorithms. Otherwise, this metric is inappropriate since it does not consider the cardinality of $\mathcal{F}'$. Therefore, to evaluate the performance of objective reduction methods that can keep different numbers of objectives, we propose a δ'-metric as

$$\delta' = (1 - \frac{|\mathcal{F}'|}{M}) \cdot \frac{|E(\mathcal{F}', \mathcal{X}) \cap E(\mathcal{F}_0, \mathcal{X})|}{|E(\mathcal{F}_0, \mathcal{X})|}. \tag{13}$$

Intuitively, δ' is larger if the objective set has the smaller cardinality and higher δ-metric value. The training set $\mathcal{X}$ of 300 points randomly sampled from the true PF is used in the experiments.

Five state-of-the-art objective reduction methods are used to compare. They are OEM [6], NLHA [18], NSGA-II-η [23], FCM [27], and ORV [26]. Among them, OEM, FCM, and ORV are objective extraction methods, and they need to prescribe the number of extracted objectives in advance. While NLHA, NSGA-II-η [23], and our MTFSOR are objective selection methods, and they do not need to specify the number of selected objective in advance. Therefore, we first compare all algorithms based on the assumption that the number of the essential objective is available. In this comparison, the number of the selected (or extracted) objectives of all algorithms is set to the number of essential objectives, and the δ-metric is employed to evaluate the performance. Then, we compare NLHA, NSGA-II-η, and our MTSFOR when the number of the essential objectives is unavailable, and the δ'-metric is applied to measure the performance. For all compared algorithms, their parameters are kept the same as those of their original papers. For MTFSOR, λ is set to 0.05 and 0.3 on DTLZ5(M,m) problems and MAOP(M,m) problems, respectively.

3.2 Experimental Results

1) When the number of the essential objectives is available

On DTLZ5(M,m) problems, the average δ-metric of 20 independent runs of all algorithms is reported in Table 1. As shown in Table 1, all objective reduction methods, excluding OEM, can obtain an optimal (or nearly optimal) δ value, which means that these objective reduction methods (excluding OEM) can find or extract the essential objectives of DTLZ5 peoblem if the training set is a good approximation of the real PF and the number of the essential objectives is available. Obviously, our MTFSOR keeps competitive with these compared algorithms on DTLZ5(M,m) problems.

Table 1. The average δ-metric of 20 independent runs of all compared algorithms on DTLZ5 problems.

Problem	OEM	NLHA	NSGA-II-η	FCM	ORV	MTFSOR
DTLZ5(5, 2)	**1**	**1**	**1**	**1**	**1**	**1**
DTLZ5(20, 2)	0.998	**1**	**1**	**1**	**1**	**1**
DTLZ5(50, 2)	0.990	**1**	**1**	**1**	**1**	**1**
DTLZ5(5, 3)	0.831	**1**	**1**	**1**	**1**	**1**
DTLZ5(20, 3)	0.906	**1**	**1**	**1**	**1**	**1**
DTLZ5(10, 5)	0.891	**1**	**1**	**1**	**1**	**1**
DTLZ5(20, 5)	0.616	**1**	**1**	0.998	**1**	**1**
DTLZ5(10, 7)	0.964	**1**	**1**	0.999	**1**	**1**
DTLZ5(20, 7)	0.727	**1**	**1**	**1**	**1**	**1**

Table 2. The average δ-metric of 20 independent runs of all compared algorithms on MAOP problems.

Problem	OEM	NLHA	NSGA-II-η	FCM	ORV	MTFSOR
MAOP1(5, 3)	**1**	**1**	0.673	0.790	**1**	**1**
MAOP2(5, 3)	**1**	**1**	0.670	0.336	0.516	**1**
MAOP3(10, 4)	0.603	**1**	0.389	0.493	0.960	**1**
MAOP4(10, 4)	0.983	**1**	0.476	0.436	0.983	**1**
MAOP5(10, 6)	0.867	**1**	0.929	0.817	0.937	**1**
MAOP6(10, 6)	0.997	**1**	0.932	0.965	0.997	**1**
MAOP7(15, 5)	0.777	**1**	0.486	0.476	0.720	**1**
MAOP8(15, 5)	0.490	**1**	0.607	0.577	0.873	**1**
MAOP9(15, 7)	0.657	**1**	0.725	0.685	0.757	**1**
MAOP10(15, 7)	0.613	**1**	0.741	0.694	0.790	**1**
Mean	0.799	**1**	0.663	0.627	0.853	**1**

On MAOP(M,m) problems, the average δ-metric of all algorithms is provided in Table 2. From Table 2, we can find that MTFSOR and NLHA obtain the optimal δ value on all MAOP problems, OEM gets the optimal δ value only on MAOP1 and MAOP2, ORV obtains the optimal δ value only on MAOP1, and NSGA-II-η and FCM fail to get the optimal δ value on any MAOP problems. The performance of NSGA-II-η, FCM, and ORV on MAOP(M,m) problems degrade significantly compared with that on DTLZ5(M,m) problems. Clearly, in this comparison, MTFSOR and NLHA perform similarly, and they outperform other algorithms on MAOP(M,m).

2) When the number of the essential objectives is unavailable.

The objective extraction methods (e.g., OEM, FCM, and ORV) become infeasible. Thus, we only compare the performance of NLHA, NSGA-II-η, and our MTFSOR by using δ'-metric. In this paper, the final selected objective set of NLHA and NSGA-II-η is set as:

- NLHA [18]: $\mathcal{F}' = \{f_k | w_k > 0.1\max(\mathbf{w}), 1 \leq k \leq M\}$, where $\mathbf{w} = (w_1, \cdots, w_M)$ is the optimal coefficient vector of the hyperplane. If $|\mathcal{F}'| = 1$, the one in $\mathcal{F}_0 - \mathcal{F}'$ has the most negative correlation with the objective in $\mathcal{F}'$ will also be selected.
- NSGA-II-η: The solution that is closest to the ideal point is selected as the final solution.

The average δ'-metric of NLHA, NSGA-II-η and MTFSOR is reported in Table 3. On DTLZ5(M,m) problems, MTFSOR obtains the best δ' value on 8 out of 9 problems, and NLHA performs best on DTLZ5(20,7) problem. NSGA-II-η is able to get the best δ' value on DTLZ5 problems that have 2 essential objectives, while it performs badly on other DTLZ5 problems. On MAOP(M,m) problems, NLHA and MTFSOR perform equally, and they get better δ' value than NSGA-II-η on all MAOP(M,m) problems. In summary, when the number of the essential objectives is unavailable, our MTFSOR performs more robust than NLHA and NSGA-II-η on DTLZ5(M,m) problems and MAOP(M,m) problems.

Table 3. The average δ'-metric of 20 independent runs of NLHA, NSGA-II-η, and MTFSOR on DTLZ5 and MAOP problems.

Problem	NLHA	NSGA-II-η	MTFSOR	Problem	NLHA	NSGA-II-η	MTFSOR
DTLZ5(5, 2)	0.400	**0.600**	**0.600**	MAOP1(5,3)	**0.400**	0.270	**0.400**
DTLZ5(20, 2)	0.790	**0.900**	**0.900**	MAOP2(5,3)	**0.400**	0.255	**0.400**
DTLZ5(50, 2)	0.923	**0.960**	**0.960**	MAOP3(10,4)	**0.600**	0.081	**0.600**
DTLZ5(5, 3)	0.340	0.016	**0.400**	MAOP4(10,4)	**0.600**	0.280	**0.600**
DTLZ5(20, 3)	0.775	0.025	**0.850**	MAOP5(10,6)	**0.400**	0.223	**0.400**
DTLZ5(10, 5)	0.460	0.059	**0.478**	MAOP6(10,6)	**0.400**	0.454	**0.400**
DTLZ5(20, 5)	0.714	0.081	**0.729**	MAOP7(15,5)	**0.666**	0.279	**0.666**
DTLZ5(10, 7)	0.371	0.119	**0.379**	MAOP8(15,5)	**0.666**	0.267	**0.666**
DTLZ5(20, 7)	**0.686**	0.277	0.602	MAOP9(15,7)	**0.533**	0.268	**0.533**
–	–	–	–	MAOP10(15, 7)	**0.533**	0.203	**0.533**
Mean	0.607	0.337	**0.655**	Mean	**0.520**	0.258	**0.520**

4 Conclusion

In this paper, we have proposed a multitask feature selection method for objective reduction (MTFSOR), in which each objective can be represented by some essential objectives. The objective with zero coefficient in all the tasks can be treated as the redundant objective. The multitask feature selection method formulates the objective reduction problem as a convex optimization problem. It can be easily solved by some well-developed convex optimization software package. The performance of MTFSOR has been demonstrated by comparing it with several state-of-the-art objective reduction methods on DTLZ5(M,m) and MAOP(M,m) problems.

In MTFSOR, each objective is formulated a linear combination of the essential objectives. It may be unable to deal with various non-linearly degenerate PF. In future, we will develop a nonlinear multitask feature learning method for objective reduction. The proposed objective reduction method will be used in MOEAs for solving MaOPs.

References

1. Argyriou, A., Evgeniou, T., Pontil, M.: Convex multi-task feature learning. Mach. Learn. **73**, 243–272 (2008)
2. Bader, J., Zitzler, E.: HypE: an algorithm for fast hypervolume-based many-objective optimization. Evol. Comput. **19**(1), 45–76 (2011)
3. Brockhoff, D., Friedrich, T., Neumann, F.: Analyzing hypervolume indicator based algorithms. In: Rudolph, G., Jansen, T., Beume, N., Lucas, S., Poloni, C. (eds.) PPSN 2008. LNCS, vol. 5199, pp. 651–660. Springer, Heidelberg (2008). https://doi.org/10.1007/978-3-540-87700-4_65
4. Brockhoff, D., Zitzler, E.: Objective reduction in evolutionary multiobjective optimization: theory and applications. Evol. Comput. **17**(2), 135–166 (2009)
5. Cheung, Y.M., Gu, F.: Online objective reduction for many-objective optimization problems. In: 2014 IEEE Congress on on Evolutionary Computation, pp. 1166–1171 (2014)
6. Cheung, Y.M., Gu, F., Liu, H.L.: Objective extraction for many-objective optimization problems: algorithm and test problems. IEEE Trans. Evol. Comput. **20**(5), 755–772 (2016)
7. Deb, K., Saxena, D.: Searching for pareto-optimal solutions through dimensionality reduction for certain large-dimensional multi-objective optimization problems. Technical Report. Indian Institute of Technology, Kanpur
8. Deb, K., Pratap, A., Agarwal, S., Meyarivan, T.: A fast and elitist multiobjective genetic algorithm: NSGA-II. IEEE Trans. Evol. Comput. **6**(2), 182–197 (2002)
9. Deb, K., Thiele, L., Laumanns, M., Zitzler, E.: Scalable test problems for evolutionary multiobjective optimization (2006)
10. Deng, J., Zhang, Q.: Combining simple and adaptive Monte Carlo methods for approximating hypervolume. IEEE Trans. Evol. Comput. **24**(5), 896–907 (2020)
11. Gómez, R.H., Coello, C.A.C.: Mombi: a new metaheuristic for many-objective optimization based on the R2 indicator. In: 2013 IEEE Congress on Evolutionary Computation, pp. 2488–2495. IEEE (2013)
12. Grant, M.: CVX : Matlab software for disciplined convex programming, version 1.21 (2011). http://cvxr.com/cvx
13. Guo, X., Wang, Y., Wang, X.: An objective reduction algorithm using representative Pareto solution search for many-objective optimization problems. Soft. Comput. **20**, 4881–4895 (2016)
14. He, Z., Yen, G.G.: Visualization and performance metric in many-objective optimization. IEEE Trans. Evol. Comput. **20**(3), 386–402 (2016)

15. Ishibuchi, H., Masuda, H., Nojima, Y.: Pareto fronts of many-objective degenerate test problems. IEEE Trans. Evol. Comput. **20**(5), 807–813 (2016)
16. Ishibuchi, H., Tsukamoto, N., Nojima, Y.: Evolutionary many-objective optimization: a short review. In: 2008 IEEE Congress on Evolutionary Computation (IEEE World Congress on Computational Intelligence), pp. 2419–2426. IEEE (2008)
17. López Jaimes, A., Coello, C.A.C., Urías Barrientos, J.E.: Online objective reduction to deal with many-objective problems. In: Ehrgott, M., Fonseca, C.M., Gandibleux, X., Hao, J.-K., Sevaux, M. (eds.) EMO 2009. LNCS, vol. 5467, pp. 423–437. Springer, Heidelberg (2009). https://doi.org/10.1007/978-3-642-01020-0_34
18. Li, Y., Liu, H.L., Goodman, E.D.: Hyperplane-approximation-based method for many-objective optimization problems with redundant objectives. Evol. Comput. **27**(2), 313–344 (2019)
19. Saxena, D.K., Duro, J.A., Tiwari, A., Deb, K., Zhang, Q.: Objective reduction in many-objective optimization: linear and nonlinear algorithms. IEEE Trans. Evol. Comput. **17**(1), 77–99 (2013)
20. Singh, H.K., Isaacs, A., Ray, T.: A Pareto corner search evolutionary algorithm and dimensionality reduction in many-objective optimization problems. IEEE Trans. Evol. Comput. **15**(4), 539–556 (2011)
21. Wang, R., Purshouse, R.C., Fleming, P.J.: Preference-inspired co-evolutionary algorithms using weight vectors. Eur. J. Oper. Res. **243**(2), 423–441 (2015)
22. Yang, S., Li, M., Liu, X., Zheng, J.: A grid-based evolutionary algorithm for many-objective optimization. IEEE Trans. Evol. Comput. **17**(5), 721–736 (2013)
23. Yuan, Y., Ong, Y.S., Gupta, A., Xu, H.: Objective reduction in many-objective optimization: evolutionary multiobjective approaches and comprehensive analysis. IEEE Trans. Evol. Comput. **22**(2), 189–210 (2018)
24. Yuan, Y., Xu, H., Wang, B., Yao, X.: A new dominance relation-based evolutionary algorithm for many-objective optimization. IEEE Trans. Evol. Comput. **20**(1), 16–37 (2015)
25. Zhang, Q., Li, H.: MOEA/D: a multiobjective evolutionary algorithm based on decomposition. IEEE Trans. Evol. Comput. **11**(6), 712–731 (2007)
26. Zhen, L., Li, M., Peng, D., Yao, X.: Objective reduction for visualising many-objective solution sets. Inf. Sci. **512**, 278–294 (2020)
27. Zhou, A., Wang, Y., Zhang, J.: Objective extraction via fuzzy clustering in evolutionary many-objective optimization. Inf. Sci. **509**, 343–355 (2020)

Embedding a Repair Operator in Evolutionary Single and Multi-objective Algorithms - An Exploitation-Exploration Perspective

Kalyanmoy Deb[1(✉)], Sukrit Mittal[2], Dhish Kumar Saxena[2], and Erik D. Goodman[1]

[1] BEACON NSF Center and Michigan State University, East Lansing, USA
{kdeb,goodman}@egr.msu.edu
[2] Indian Institute of Technology Roorkee, Roorkee, India
{smittal1,dhish.saxena}@me.iitr.ac.in

Abstract. Evolutionary algorithms (EAs) are population-based search and optimization methods whose efficacy strongly depends on a fine balance between *exploitation* caused mainly by its selection operators and *exploration* introduced mainly by its variation (crossover and mutation) operators. An attempt to improve an EA's performance by simply adding a new and apparently promising operator may turn counter-productive, as it may trigger an imbalance between the exploitation-exploration trade-off. This necessitates a proper understanding of mechanisms to restore the balance while accommodating a new operator. In this paper, we introduce a new *repair* operator based on an AI-based mapping between past and current good population members to improve an EA's convergence properties. This operator is applied to problems with different characteristics, including single-objective (with single and multiple global optima) and multi-objective problems. The focus in this paper is to highlight the importance of restoring the exploitation-exploration balance, when a new operator is introduced. We show how different combinations of problems and EA characteristics pose different opportunities for restoration of this balance, enabling the *repair*-based EAs/EMOs to outperform the original EAs/EMOs in most cases.

Keywords: Evolutionary algorithms · Exploration · Exploitation · Convergence · Repair operator · *Innovization*

1 Introduction

Evolutionary algorithms (EAs) are population-based search and optimization algorithms that start with a set of random solutions to achieve a single best

This research work has been supported by Government of India under SPARC project P66.

H. Ishibuchi et al. (Eds.): EMO 2021, LNCS 12654, pp. 89–101, 2021.
https://doi.org/10.1007/978-3-030-72062-9_8

solution or multiple equally good (non-dominated) solutions. Different EAs use various operators, like selection, crossover, mutation, etc., to generate offspring solutions at each iteration. While deploying these operators, there are two key characteristics, namely, *exploration* and *exploitation*. Exploitation refers to the extent of emphasis provided to the current population-best (or above-average population members) in mating- or survival-selection. In contrast, exploration refers to the extent of search introduced by the variation operators, such as crossover and mutation operators, in creating an offspring population. Any working EA maintains a balance between these two by using a combination of evolutionary operators; however, failure to do so may cause poor or premature convergence. This issue is discussed further in Sect. 2.

Even though learning during the search process is not new [1], recent studies [7,8,13,14] have focused on *online innovization*, which attempts to extract variable patterns from certain good solutions and utilize those through an additional *repair* operator (apart from crossover and mutation) in subsequent generations. This operator has been discussed in Sect. 3. Here, the repair operator's disruption to the otherwise existing exploitation-exploration balance and the necessary changes to restore this balance have also been discussed. Subsequently, this repair operator is applied in single-objective, multi-modal, and multi-objective problems. Sections 4, 5 and 6 demonstrate the application and efficacy of this operator from the exploitation-exploration perspective and discuss some ways to restore the balance that existed before introducing this new *repair* operator. Finally, Sect. 7 presents the conclusions of this study.

2 Exploitation-Exploration Perspective

In a canonical EA, the exploitation is established in a population by the mating selection and the survival selection operators, and exploration refers to the extent of *search* by introducing variation using crossover and mutation operators. Goldberg and Deb [9] showed that if the exploitation-exploration balance is not established, the resulting Genetic Algorithm (GA) fails to solve even a simple linear and unconstrained "onemax" problem.

While it is crucial to make a suitable balance between exploitation and exploration within an EA's population by selecting and parameterizing its operators, their quantification is not very clear. Goldberg and Deb [9] used the selection pressure (number of copies of the population-best solution in mating pool) as an indicator of exploitation and the crossover probability (p_c) for exploration in an EA (mutation was not used). The higher the value of these parameters, the greater is the extent of each issue. It is a matter for future research to devise more comprehensive indicators for exploitation and exploration in a population, considering every operation implemented in an EA. But here, we discuss how the balance is affected by introducing a new (and notably, promising) operator.

The majority of EA applications use tournament selection for creating a mating pool and real-parameter crossover and mutation operators for creating an offspring population. The best half of a combined parent and offspring population is often used as the survival selection operator. In some applications, the top $\kappa\%$ of parent population members are included in the next population to preserve elites. EA application then requires fine-tuning of the associated hyper-parameters (such as tournament size, crossover and mutation probabilities, and κ) to develop an efficient EA for a specific problem. When an EA with standard operators is used with their optimized hyper-parameters, an important question arises: if a novel operator is to be introduced—how shall the imbalance between exploitation and exploration caused by the new operator be restored to make the overall algorithm efficient again? We address these issues in context of introducing a new repair operator, first for single-objective optimization to search for single and multiple solutions, and then for multi-objective optimization.

3 An 'Innovized' Repair Operator

Since the beginning, EAs have been treated as Markov chains, in which the new population at generation $t+1$ is created entirely from the population at generation t, with some exceptions such as particle swarm optimization (PSO) [12]. But a recent focus in the optimization literature, particularly in population-based methods, has been on *learning* a mapping of current solutions with relevant solutions from the past generations and then utilizing the learned mapping to *repair* newly created offspring in the hope of improving them further. Recent advances in machine learning (ML) methods allow us to consider such a repair operator as a new additional operator for creating better offspring solutions. Learning from past generations for knowledge acquisition is commonly called *innovization* in context of evolutionary multi-objective optimization (EMO) [5].

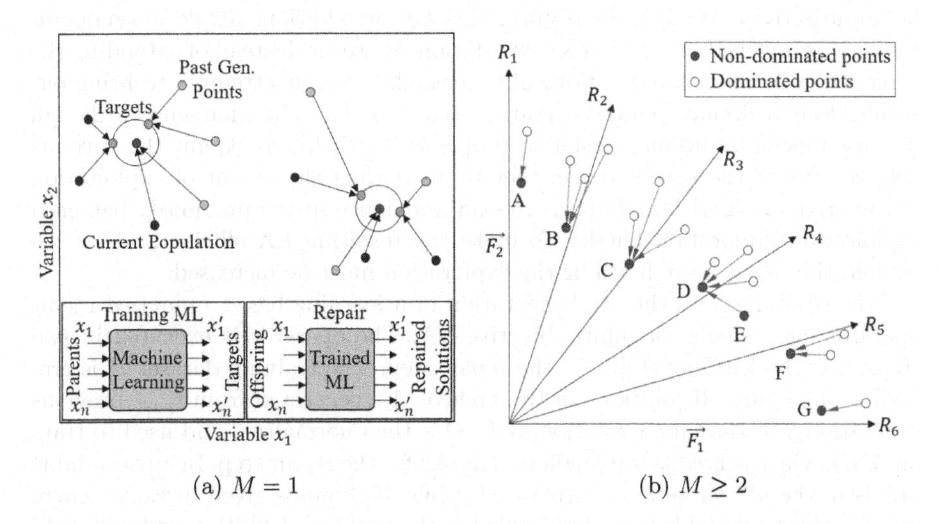

Fig. 1. The mapping process for ML training in the *innovized repair* operator. (Color figure online)

Figure 1a illustrates the principle of *innovized* repair (IR) operator in the context of a generic multi-modal optimization problem, seeking to find multiple optimal solutions. The red points are population-best solutions with high fitness. They and their neighboring high-fitness solutions are target points. Past and current population members are mapped to one of the target points based on distance in variable space. In the training phase, an ML method is used to learn the mapping between a past or current population member with a single target solution in the variable space, having high fitness value. Once the mapping is learned, an offspring can be repaired using the learned (or trained) ML model to create a new and hopefully better solution.

Offspring created using crossover and mutation operators applied on the current parent population are expected to possess good properties of parents, but the success of this process depends on the crossover and mutation operators used. However, if the mapping of past populations to the current-best target points is captured, a not-so-good offspring can be repaired before it is evaluated in the hope of making the solution better. This is the basic motivation for the proposed IR operator. A comparison with the other learning-based methods like *surrogate-assisted* EAs and crossovers like *Differential Evolution* has been drawn in [14], and an extensive performance analysis of IR operator is presented in [15]. The scope of this paper is limited to discussing the disruption of exploration-exploitation balance in context of IR operator and restoring that balance.

4 Search for a Single Optimal Solution

With inclusion of the IR operator for a single-objective optimization task (number of objectives, $M = 1$), the resulting EA has an additional exploration operator, as it is involved in creating a new solution. However, instead of extending the exploration to unexplored regions of the search space, it attempts to bring offspring close to already found good target solutions. Thus, in some sense, although it is used as an additional exploration operator, it helps to exploit the current-best regions of the search space, thereby increasing the extent of exploitation of the original algorithm. This creates an imbalance in the previously-balanced exploitation-exploration battle. To make the resulting EA efficient, either the exploitation must be reduced or the exploration must be increased.

Algorithm 1 shows the implementation of a learning-based *innovized* repair operator for a single- or multi-objective EA. The operator is executed in two steps, i.e., the learning step and the repair step. The training dataset D is created by mapping all solutions in the archive A_t (generated from t_{past} previous generations) to the target solution(s) T. D is then normalized and used to train an ANN, which serves as the prediction model for the repair step. In an n-variable problem, the kth variable is normalized using the bounds given in Eq. 1, where $[x_k^l, x_k^u]$ are the variable bounds specified in the problem definition and $[x_k^{l,t}, x_k^{u,t}]$ are the minimum and maximum values of the kth variable in the archive A_t [13]:

$$x_k^{\min} = 0.5(x_k^{l,t} + x_k^l), \quad x_k^{\max} = 0.5(x_k^{u,t} + x_k^u). \tag{1}$$

Algorithm 1. Generation t of EA with IR operator

Require: Parent population P_t, Archive A_t.
1: $A_t \leftarrow A_T \bigcup P_t \backslash P_{t-t_{past}}$ `% Archive Update`
2: $P_{mating} \leftarrow$ Tournament Selection on P_t
3: $Q_t \leftarrow$ Crossover and mutation on P_{mating}
4: $T \leftarrow$ Variable vector(s) of target solution(s) identified from P_t
5: $X \leftarrow$ All variable vectors in A_t
6: $D \leftarrow$ Map the solutions in X to T
7: $D \leftarrow$ Dynamic normalization of D `%using equation 1`
8: $Model \leftarrow$ Train the ANN using D
9: $Q_t \leftarrow$ Repair randomly selected 50% offsprings Q_t using $Model$
10: Evaluate Q_t
11: $P_{t+1} \leftarrow$ Survival Selection on $P_t \bigcup Q_t$
12: **return** Next Parent Population P_{t+1}

The ANN used here has a pre-determined architecture with two hidden layers of 30 neurons each. The input and output layers have neurons equal to the number of decision variables n in the optimization problem. The loss function used is *Mean Squared Error* (MSE), activation function is *sigmoid* and the optimizer is *adam*[1]. The training is terminated if the MSE is stagnant for 50 epochs. The maximum epoch limit is set to 2,500. For the repair step, each offspring is first normalized using the already acquired bounds from Eq. 1, repaired using the trained ANN and then denormalized to the original scale. In this process, it is possible that some variable may go out of the bounds $[x_k^l, x_k^u]$ as defined in the problem. To address this, another repair is done using an Inverse Parabolic Spread Distribution [16], individually, for the variables lying out of bounds.

There are different ways in which one ($N_T = 1$) or multiple ($N_T > 1$) target solutions can be selected. We start with the obvious choice of one target solution—i.e., the best fitness solution as discussed in the Subsect. 4.1. Learning to map all existing solutions to a single target solution would emphasize the exploitation aspect of this IR operator. In order to reduce that exploitation, multiple diverse target solutions can be selected from P_t. Even though some targets may have worse fitness values, selecting a target that is dissimilar to the best solution can help. The selection of multiple target solutions and how it practically differs from using IR with a single-target is discussed in Subsect. 4.2.

4.1 Repairing a Single-Target Solution

Here, all solutions in A_t are mapped to the current best-fitness solution T. For this study, four 20-variable single-objective test problems are chosen: Ackley, Different Powers Function (DPF) [6], Rastrigin, and Zakharov.

A comparison is done between an EA with and one without the IR operator. The EA with repair is referred to as EA-IR. The EA (here, GA) is used with population size $N = 100$, binary tournament selection, SBX crossover ($\eta_c = 3$,

[1] The parameters are taken from https://keras.io/api/optimizers/adam/.

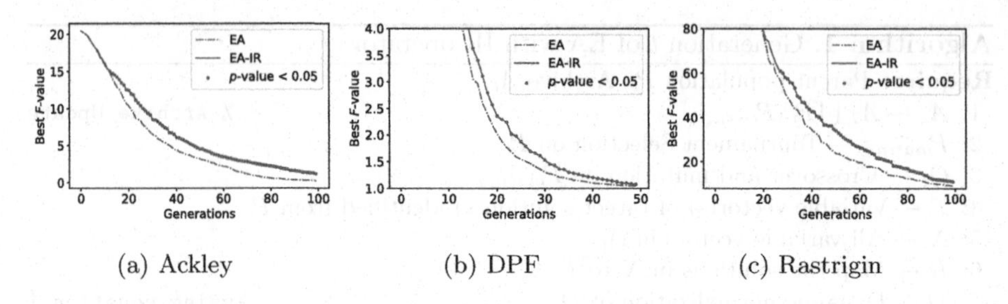

(a) Ackley (b) DPF (c) Rastrigin

Fig. 2. Median fitness plots with EA using IR with a single target.

$p_c = 0.9$), polynomial mutation ($\eta_m = 20$, $p_m = 1/n$), a survival selection operator where the best half of the $P_t \bigcup Q_t$ are selected and $t_{past} = 5$ [13]. For each test instance, 31 independent seed runs are done and the median of the best fitness F-value at each generation is shown in Fig. 2. The p-value is calculated using the Wilcoxon Ranksum Test. It is evident from the figure that EA-IR applied in this way performs worse in all cases than the EA without repair. The IR operator, focusing on a single solution, increases the exploitation factor, thereby disrupting the original exploration-exploitation balance of the EA.

Table 1 shows various EA operators and their effects on the exploration or exploration aspects. Thus, when a new operator to be implemented within an EA is expected to disrupt the balance between exploration and exploitation, one of the operators can be modified as depicted in the table to try to restore the balance.

Table 1. Effect of operators and their parameters on exploration-exploitation aspects.

Operator	Modification	Effect
Initialization	Biased ↑	Exploration ↓
Crossover	η_c ↓ or p_c ↑	Exploration ↑
Mutation	η_m ↓ or p_m ↑	Exploration ↑
IR	t_{freq} ↑ or N_T ↑	Exploitation ↓
Mating selection	Tournament size ↑	Exploitation ↑
Survival selection	Niche-preservation	Exploitation ↓
Elites	% Population preserved ↑	Exploitation ↑
Child acceptance	Steady-state	Exploitation ↑

4.2 Repairing to Multiple Target Solutions

The increase in exploitation (discussed in Subsect. 4.1) can be reduced if a distributed set of N_T neighboring solutions to the current best-fitness solution are used as targets. This requires the following two considerations:

1. A clearing strategy is used to choose N_T good and diverse solutions. First, the population is sorted from best to worst according to fitness. Then, the first solution on the list is chosen as the first target solution. Thereafter, all members within ϵ Euclidean distance in the variable space from the chosen solution are cleared and the next un-cleared solution is chosen as the second target solution. Then, all subsequent solutions within ϵ Euclidean distance from chosen two target solutions are cleared and the next solution in the list

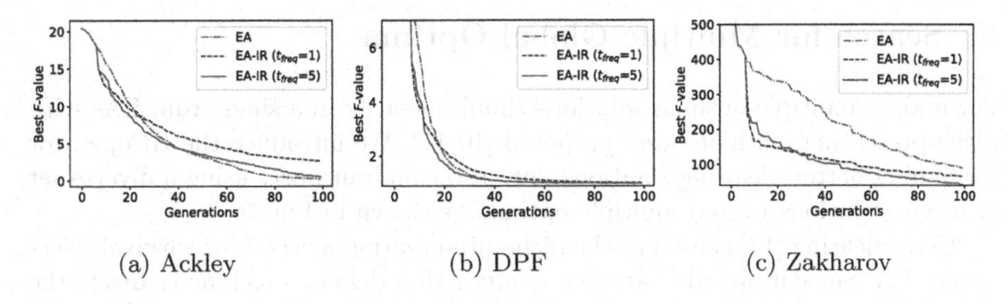

(a) Ackley (b) DPF (c) Zakharov

Fig. 3. Median fitness plots with EA using IR with multiple diverse targets.

is accepted as the next target solution. This is continued until N_T target solutions are picked. Thus, the target solutions are guaranteed to have a minimum Euclidean distance of ϵ from each other.

2. For training phase, every past solution is associated with the nearest target solution in variable space. Since only N_T targets are chosen from P_t, the other $(N - N_T)$ members remain part of the archive A_t. To avoid mapping of an archive solution to a target having a worse fitness value, the algorithm ensures that the target must have a better fitness than the associated archive solution. Otherwise, the pairing is excluded from training dataset D.

Keeping all other parameters similar to the EA-IR used in Subsect. 4.1, experiments are repeated using $N_T = 5$. The parameter ϵ is adaptively fixed at every generation by evaluating the average Euclidean distance of the kth (=5 used here) nearest neighbor of members of P_t. Besides selecting multiple targets for learning, another way to reduce the exploitation of the IR operator is to use it after every t_{freq} (> 1) generations than $t_{freq} = 1$ (see Table 1). In this subsection, we choose $t_{freq} = 1$ and 5. The median of best-fitness value is plotted in Fig. 3, suggesting that $t_{freq} = 5$ performs better than $t_{freq} = 1$.

There is a significant improvement in fitness values on the Zakharov problem, while for the remaining problems, the fitness improves in the beginning, but degrades slightly in the later generations. This can be explained as follows. Ensuring a diverse set of target solutions (by using ϵ) improves the search in initial generations by increasing the exploration to make a balance with the increased exploitation caused in the IR operator. However, the ϵ gets smaller with generation, as population members come closer to the optimal solution. With a reduced diversity in target solutions, the exploration gets reduced, thereby disrupting the exploitation-exploration balance. However, the exploitation effect of IR operator gets reduced when t_{freq} is increased to 5 restoring the balance and the performance of EA-IR gets better.

As discussed in Table 1, a niche-preserving operation can also reduce the effect of increased exploitation due to the IR operator. Thus, it is motivating for us to introduce the IR operator in a multi-modal EA to find multiple optimal solutions with an increased efficiency. The implementation and results are discussed in the next section.

5 Search for Multiple Global Optima

For finding multiple optimal solutions simultaneously in a single run, EAs with niche-preserving methods were proposed [10,17]. We introduce the IR operator for finding better offspring solutions, but also simultaneously using a diverse set of target solutions to find multiple optima, as shown in Fig. 1a.

Here, clearing [17] is used as the niche-preservation method for survival selection in EA. Since it already has a parameter ϵ that defines a niching radius in the variable space, the same ϵ is used to identify the target solutions for ANN training dataset D. The rest of the IR operator is the same as used in Subsect. 4.1 for selecting multiple target solutions. For the experiments, we use the Himmelblau function with $n = 2$ and four optimal solutions and the MMP problem [4] with $n = 15$ having eight optimal solutions. The EA is used with binary tournament selection, SBX crossover ($\eta_c = 3$, $p_c = 0.9$), and polynomial mutation ($\eta_m = 10$, $p_m = 1/n$). The niching parameter $\epsilon = 0.1$ and 0.2, and population size $N = 40$ and 100 are used for the Himmelblau and MMP problems, respectively. Two values of target solutions N_T (5 and 10) are tried for the IR operator. Since there are multiple optimal solutions, the best fitness value is not an appropriate measure. Hence, IGDX (IGD computed in the variable space with known $\mathbf{x}^*$ set) is used as the performance indicator, as shown in Fig. 4.

Since in this problem, multiple optima in the variable space are naturally expected to be distant from each other, and a training of the ANN is likely to learn mapping of solutions to multiple distant target solutions simultaneously, an increase in exploration caused by artificially spreading solutions near diverse targets gets somewhat balanced by the increase in exploitation caused by convergence of solutions to each target by the IR operator. The figure shows that using $t_{freq} = 1$ (applying IR at every generation) performs better than either the original niching EA, or the IR-based EA with $t_{freq} = 5$. Pushing offspring solutions near multiple target solution makes a good balance of exploitation and exploration, and applying it at every generation provides maximum advantage.

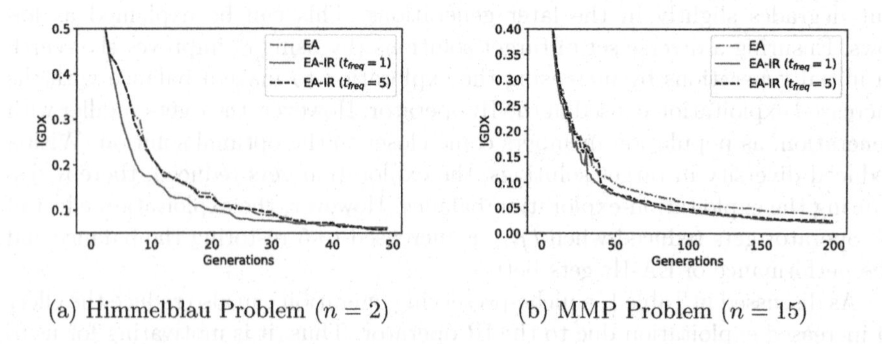

(a) Himmelblau Problem ($n = 2$) (b) MMP Problem ($n = 15$)

Fig. 4. Median IGDX comparison for multi-modal problems with clearing-based multi-modal EA (with and without IR operator).

If the frequency of applying the IR operator is reduced (i.e., t_{freq} is increased to 5), the fullest advantage is lost and the performance of the IR-based EA drops.

Following the above, it is clear that if implemented well for an exploitation-exploration perspective, the IR operator should demonstrate a performance enhancement in multi-objective optimization as well, where the goal is to find a well-distributed set of Pareto-optimal (PO) solutions. This is presented next.

6 Search for Multi-Objective Solutions

With proper implementation, the increased focus on multiple non-dominated but distant variable space solutions may prove to be beneficial in an evolutionary multi-objective optimization (EMO) procedure. A preliminary study [13] motivated us to launch this extensive study. In EMO, all non-dominated solutions at generation t are equally good and, hence, can serve as the target solutions. Their count N_T may range from 1 to N. The archive solutions are mapped to the non-dominated solutions in P_t that act as the target solutions T. The main question that now remains is how an archive solution can be mapped to a respective target solution, particularly in the context of multiple objectives.

Figure 1b shows the schematic of associating archive solutions in A_t to the target solution set T. The idea is motivated from the goal of achieving better solutions, improving orthogonally to the PO front. As in Fig. 1b, N equidistant reference vectors are generated (R1–R6) and the non-dominated solutions (A–G) are assigned to the reference vectors using the Achievement Scalarizing Function (ASF) [18]. As evident, even though E is a non-dominated solution, it is not selected as a target since choosing D for R4 could provide a more diverse target set. The remaining solutions in A_t are then attached to their nearest reference vectors using ASF, and consequently are associated to the target solutions assigned to those particular reference vectors.

This modified IR operator for the multi-objective framework is used with NSGA-II and NSGA-III, and its performance is evaluated on a number of test functions from ZDT [19] and WFG [11] test suites. The ZDT problems are

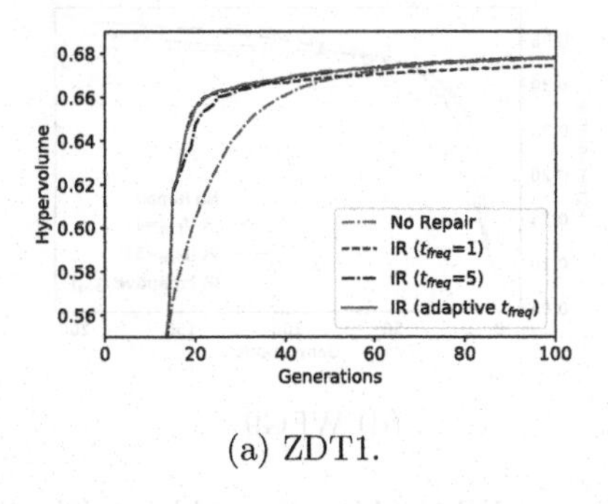

(a) ZDT1.

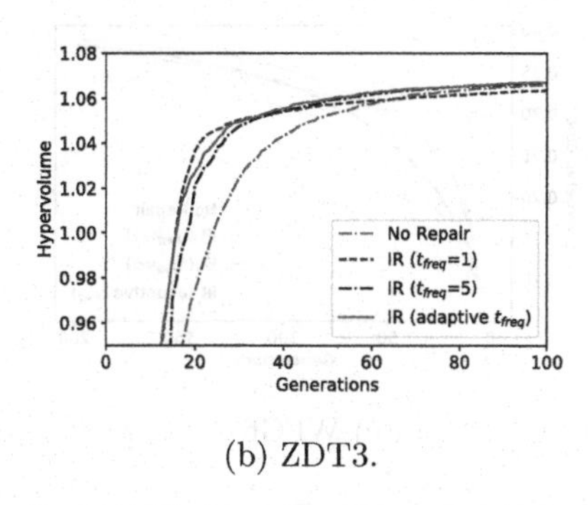

(b) ZDT3.

Fig. 5. Median HV plots for two-objective ZDT problems using NSGA-II (with and without IR operator).

modified to shift the optima at $x_k = 0.5$ for $k = 2, \ldots, 30$, in order to create non-boundary optimal solutions. NSGA-II uses a population size $N = 100$, SBX crossover ($\eta_c = 10$, $p_c = 0.9$) and polynomial mutation ($\eta_m = 20, p_m = 1/n$). N reference-vectors are generated using the Das and Dennis method [2] for mapping in the training dataset. To make the learning from some non-dominated solutions effective, we wait to apply the IR operator until at least 50% of P_t lies on the non-dominated front. For NSGA-III, the perpendicular distance matrix (PDM) [3] is used for reference-vector mapping instead of ASF (as was used for NSGA-II). This is to ensure that the targets of the survival selection and the IR operator are aligned. We use $N = 105$ for three-objective WFG problems.

The results on two-objective ZDT problems and three-objective WFG problems are shown in Figs. 5 and 6, respectively. We explore the effect with $t_{freq} = 1$ and 5. The plots represent the generation-wise median hypervolume (HV) from 31 independent runs. With the exception of ZDT6, all other plots reveal that IR with $t_{freq} = 1$ (more exploitation) performs better in earlier generations, while $t_{freq} = 5$ (less exploitation) works well in later generations. This is consistent with our previous argument in Subsect. 4.2 about the aggressive use the IR operator in the beginning when a natural diversity exists in the population. This suggests that adjusting the exploitation of the IR operator with generations by self-adapting t_{freq} parameter can achieve a better overall performance (by maintaining the exploration-exploitation balance from start the end).

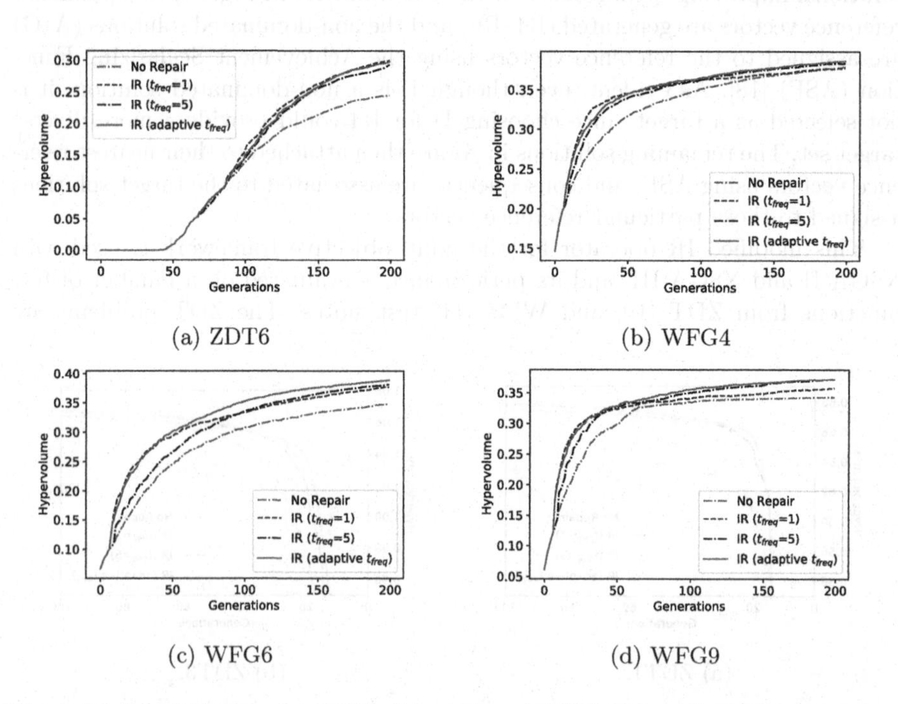

Fig. 6. Median HV plots for ZDT6 ($M = 2$) and WFG ($M = 3$) problems. Three-objective problems are solved using NSGA-III (with and without IR operator).

Towards this, t_{freq} is initialized at two. Then, the count of total survived offspring (SO) is recorded at every generation. At any generation t when the offspring are repaired using the IR operator, the count of total surviving offspring SO_t and the count of offspring that survived in the previous generation SO_{t-1} are compared. If $SO_t \geq SO_{t-1}$, then t_{freq} is reduced by one, else t_{freq} is increased by one.

Figures 5 and 6 show the results for the IR operator with adaptive t_{freq}, which works better than fixing $t_{freq} = 1$ or 5 at all generations. It performs better by having higher exploitation at earlier generations and relaxing it in later generations, but importantly by making a balance with the exploration effect caused by diversity preservation among ND solutions, crossover and mutation operators. This study also shows the importance of balancing the exploitation-exploration issue adaptively, if possible, within an algorithm.

7 Conclusions

In this paper, a learning-based innovized repair (IR) operator for single-objective (including multi-modal) and multi-objective optimization problems has been introduced, and analyzed from the exploitation-exploration perspective. The repair operator learns the mapping of past not-so-good (dominated) solutions with relevant current good (non-dominated) solutions and uses the learnt mapping to repair created offspring solutions in the hope of pushing them close to the better solutions. While this seems to be a promising operator for any EA or EMO algorithm, its standalone addition to a standard single-objective EA has not shown to improve EA's performance. It has been argued that the introduced repair operator puts an over-emphasis on the exploitation aspect and disrupts its balance with exploration. Other adjustments (mainly to reduce the extent of exploitation, in this case) were needed to restore the balance so that the resulting modified algorithm starts to work better than the original algorithm.

Next, we have executed insightful moderations of the extent of exploitation (through mechanisms such as the use of multiple targets, niching, adaptive frequency of repair, etc.), in an attempt to restore the balance in multi-modal problems. The modified EA has been shown to produce comparable or better performance than its original version, highlighting that the crux of a good population-based optimization algorithm design is intricately related to maintaining a balance between exploitation and exploration aspects of its operators.

In multi-objective problems, EMO algorithms distribute the effect of exploitation to a number of well-distributed multiple optimal solutions in the search space. While the increase in exploitation by the introduction of the innovized repair operator is not as severe as in unimodal EAs, its aggressive use in the beginning and less-frequent use at the end of a run has been found to maintain the exploitation-exploration balance. This concept has us led to develop a self-adaptive EMO procedure in which the frequency of use of the IR operator is adapted based on the success of repaired offspring solutions, resulting in better performance of the original EMO algorithm.

While the modified EA and EMO algorithms embedded with the IR operator have been found to perform better than their original versions, the main message of this paper stays in demonstrating the importance of maintaining a balance of two important factors in a population-based optimization algorithm: exploitation caused by convergence-enhancing operators and exploration caused by diversity-enhancing operators. By keeping the balance in mind, EA and EMO researchers will have a better understanding of the working of the algorithms, will better able to modify and introduce new operators for performance enhancement, and importantly, will be able to apply them with confidence.

References

1. Battiti, R., Brunato, M.: Reactive search optimization: learning while optimizing. In: Gendreau, M., Potvin, J.Y. (eds.) Handbook of Metaheuristics. ISOR, vol. 146, pp. 543–571. Springer, Boston (2010). https://doi.org/10.1007/978-1-4419-1665-5_18
2. Das, I., Dennis, J.: Normal-boundary intersection: a new method for generating the pareto surface in nonlinear multicriteria optimization problems. SIAM J. Optim. **8**(3), 631–657 (1998)
3. Deb, K., Jain, H.: An evolutionary many-objective optimization algorithm using reference-point based non-dominated sorting approach, part I: solving problems with box constraints. IEEE Trans. Evol. Comput. **18**(4), 577–601 (2014)
4. Deb, K., Saha, A.: Multimodal optimization using a bi-objective evolutionary algorithms. Evol. Comput. J. **20**(1), 27–62 (2012)
5. Deb, K., Srinivasan, A.: Innovization: innovating design principles through optimization. In: Proceedings of the Genetic and Evolutionary Computation Conference (GECCO-2006), pp. 1629–1636. ACM, New York (2006)
6. Finck, S., Hansen, N., Ros, R., Auger, A.: Real-parameter black-box optimization benchmarking 2010: presentation of the noiseless functions. Technical report 2009/20, Research Center PPE (2010)
7. Gaur, A., Deb, K.: Effect of size and order of variables in rules for multi-objective repair-based innovization procedure. In: Proceedings of Congress on Evolutionary Computation (CEC-2017) Conference. IEEE Press, Piscatway (2017)
8. Ghosh, A., Goodman, E., Deb, K., Averill, R., Diaz, A.: A large-scale bi-objective optimization of solid rocket motors using innovization. In: 2020 IEEE Congress on Evolutionary Computation (CEC), pp. 1–8 (2020)
9. Goldberg, D.E., Deb, K.: A comparison of selection schemes used in genetic algorithms. In: Foundations of Genetic Algorithms 1 (FOGA-1), pp. 69–93 (1991)
10. Goldberg, D.E., Richardson, J.: Genetic algorithms with sharing for multimodal function optimization. In: Proceedings of the First International Conference on Genetic Algorithms and Their Applications, pp. 41–49 (1987)
11. Huband, S., Hingston, P., Barone, L., While, L.: A review of multiobjective test problems and a scalable test problem toolkit. IEEE Trans. Evol. Comput. **10**(5), 477–506 (2006)
12. Kennedy, J., Eberhart, R.: Particle swarm optimization. In: Proceedings of IEEE International Conference on Neural Networks, pp. 1942–1948. IEEE Press (1995)
13. Mittal, S., Saxena, D.K., Deb, K.: Learning-based multi-objective optimization through ANN-assisted online innovization. In: Proceedings of the 2020 Genetic and Evolutionary Computation Conference Companion, GECCO 2020, pp. 171–172. Association for Computing Machinery, New York (2020)

14. Mittal, S., Saxena, D.K., Deb, K., Goodman, E.: An ANN-assisted repair operator for evolutionary multiobjective optimization. Technical report COIN report no. 2020005, Michigan State University, East Lansing, USA (2020). https://www.egr.msu.edu/~kdeb/papers/c2020005.pdf
15. Mittal, S., Saxena, D.K., Deb, K., Goodman, E.: Enhanced innovized repair operator for evolutionary multi- and many-objective optimization. Technical report COIN report no. 2020020, Michigan State University, East Lansing, USA (2020). https://www.egr.msu.edu/~kdeb/papers/c2020020.pdf
16. Padhye, N., Deb, K., Mittal, P.: Boundary handling approaches in particle swarm optimization. In: Bansal, J., Singh, P., Deep, K., Pant, M., Nagar, A. (eds.) BIC-TA 2012. AISC, vol. 201, pp. 287–298. Springer, Cham (2013). https://doi.org/10.1007/978-81-322-1038-2_25
17. Pétrowski, A.: A clearing procedure as a niching method for genetic algorithms. In: IEEE 3rd International Conference on Evolutionary Computation (ICEC 1996), pp. 798–803 (1996)
18. Wierzbicki, A.P.: The use of reference objectives in multiobjective optimization. In: Fandel, G., Gal, T. (eds.) Multiple Criteria Decision Making Theory and Applications. LNE, vol. 177, pp. 468–486. Springer, Heidelberg (1980). https://doi.org/10.1007/978-3-642-48782-8_32
19. Zitzler, E., Deb, K., Thiele, L.: Comparison of multiobjective evolutionary algorithms: empirical results. Evol. Comput. **8**(2), 173–195 (2000)

Combining User Knowledge and Online *Innovization* for Faster Solution to Multi-objective Design Optimization Problems

Abhiroop Ghosh([✉]), Kalyanmoy Deb, Ronald Averill, and Erik Goodman

Michigan State University, East Lansing, MI 48824, USA
ghoshab1@msu.edu, {kdeb,averillr,goodman}@egr.msu.edu
https://www.coin-lab.org/

Abstract. Real-world optimization problems often come with additional user knowledge that, when accommodated within an algorithm, may produce a faster convergence to acceptable solutions. Modern population-based optimization algorithms, aided by recent advances in AI and machine learning, can also learn and utilize patterns of variables from past iterations to improve convergence in subsequent iterations – an approach termed *innovization*. In this paper, we discuss ways to combine user-supplied heuristics and machine-learnable patterns and rule sets in developing efficient multi-objective optimization algorithms. Two practical large-scale design problems are chosen to demonstrate the usefulness of integrating human-machine information within a multi-objective optimization in finding similar quality solutions as that obtained by the original algorithm with less computational time.

Keywords: Multi-objective optimization · *Innovization* · Collaborative optimization · Knowledge-driven design optimization

1 Introduction

Multi-Objective Optimization (MOO) algorithms are widely used for solving design optimization problems. A set of trade-off solutions between multiple conflicting objectives, known as Pareto-Optimal (PO) solutions, are the final outcome. This then requires that the decision-makers perform a post-optimal decision-making task to choose a single preferred solution for implementation.

MOO algorithms, on their own and without any additional problem knowledge, can solve real-world design problems involving various practicalities, such as large-dimensional search and objective spaces, nonlinearities, multimodalities, isolation of optima, etc., but the time taken to achieve a reasonably good solution set may not make the application practical. On the other hand, in most cases, the engineers or domain experts, have critical insights about certain

H. Ishibuchi et al. (Eds.): EMO 2021, LNCS 12654, pp. 102–114, 2021.
https://doi.org/10.1007/978-3-030-72062-9_9

regularity and relationships among variables and objectives, which can be used as key knowledge (or heuristic) in the optimization process. This may result in improved performance and help arrive at an acceptable solution quickly.

Heuristics have been applied across a range of optimization problems covering diverse fields. In combinatorial optimization problems, like the traveling salesman problem, specific studies exist [11]. An algorithm inspired by the knapsack problem has been used in [7] for solving a pharmaceutical clinical trial planning problem. A meta-heuristic optimization algorithm has been used in [2] to achieve collaboration between an engineer and an architect for designing pre-fabricated wall-floor building systems. A two-step heuristic-based algorithm is presented in [12] for performing network topology design.

Innovization [9] is a process of finding hidden common characteristics among the decision variables of current non-dominated solutions which can be retained as important knowledge derived from the optimization procedure, ready to be applied to hasten the convergence properties of an optimization algorithm. In early studies [3,9], innovization was presented as a post-optimization step to extract key knowledge for future analysis. Further works [10] demonstrated the benefits of applying the derived innovization principles even within the same optimization run in an *online* manner, as heuristics or as repair operators. Here, we combine the two sources of knowledge—human-supplied and machine-derived—to develop an efficient optimization procedure.

The rest of the paper is organized as follows. Section 2 introduces the knowledge-driven design optimization framework. Section 3 introduces our proposed online innovization approaches. Sections 4 and 5 introduce the truss design and Demand Side Management optimization problems and the results obtained with our proposed approaches. Section 6 presents the conclusions of our study and possible future extensions.

2 Knowledge-Driven Design Optimization

The proposed knowledge-driven design optimization process is shown in Fig. 1. There are two entities involved in this process: (i) the user and (ii) the optimization algorithm developer. The user can be an engineer or domain expert who describes the problem objectives, constraints, and design parameters (Problem Design Block). The user also often has additional knowledge about good solutions that can be prescribed as additional constraints, but often such information is not entirely quantitative and must be used more as a recommendation, rather than supplied rigid constraints. The algorithm developer is responsible for collecting the problem formulation and any additional problem heuristics from the user (User Knowledge Block) and designing a suitable optimization algorithm (Optimization Process Design Block). Encouraging collaboration between these two entities is the main goal of the Knowledge-driven Design Optimization framework and this paper makes a step towards this goal. Two types of knowledge are used during the optimization: a priori knowledge (U) provided by the user (Past Knowledge Block), and knowledge learned during the optimization

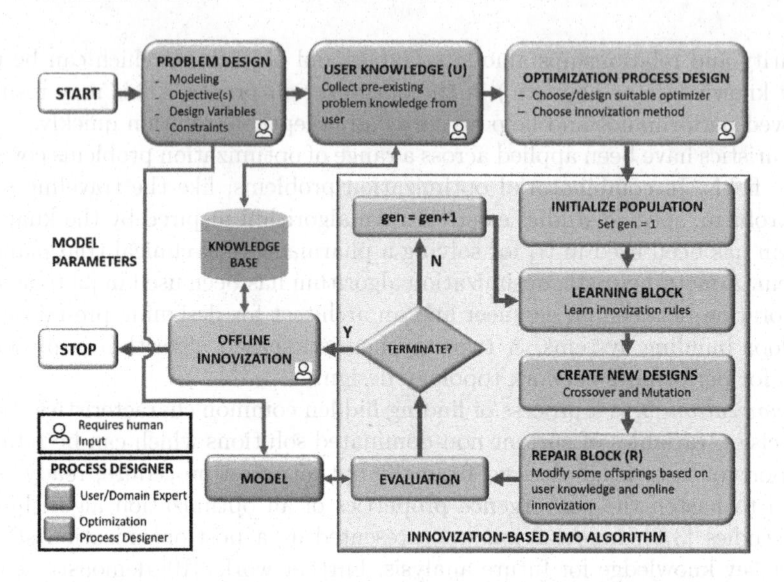

Fig. 1. Knowledge-driven design optimization procedure.

(Learning Block). A repair operator (Repair Block) ensures that both types of knowledge are used while generating new solutions (offspring). Both Learning and Repair Blocks constitute the online innovization process. As is evident in Fig. 1, any Evolutionary Multi-Objective Optimization (EMO) or Evolutionary Many-Objective Optimization (EMaO) algorithm can be used along with a pattern learning method.

3 Innovization-Based Design Optimization Procedure

In a population-based optimization algorithm, non-dominated (ND) solutions represent best solutions of the current population. Hence, they are expected to share some common properties related to the design variables that determine them. For example, coordinates of adjacent nodes in a truss may follow some monotonic property or cross-sectional sizes may follow some symmetric relationships. If such properties are extracted from ND solutions, they can be considered as good derived heuristics, which may or may not have been known and supplied by the designer. If an EMO algorithm makes a correct convergence towards Pareto-optimal solutions, such properties of ND solutions should also firm up and stabilize with generations. Thus, extracting common properties from ND solutions and using them to "repair" future offspring solutions seems to be a useful strategy. Since such a task does not use any additional knowledge from the user, we associate this approach with user knowledge 'None'. With no prior knowledge supplied by the user, the task of identifying such common properties, especially in a space of many design variables, demands a large amount of data, in order to be able to extract a reliable relationship by any machine learning method. However, if the designer provides some indication of likely relationships (such as, symmetry or monotonicity) among a cluster of variables, the learning algorithms are likely to work better. Such a collaboration between a designer with expert knowledge and a machine's ability to extract details should turn out to be a winning strategy, deserving of significant attention.

There are various ways a priori knowledge can be provided to an optimization algorithm. Here, we first discuss the simplest form of providing such knowledge. Initially, the user puts variables into several clusters and for each of them (say, the k-th one C_k), the variables ($\mathbf{x}_k = \{x_i\}$, for $i \in C_k$) within the cluster are expected to possess some unknown relationships for the overall solution to be good or near-optimal. Such unspecified information can also be exploited by a learning algorithm. But if more specific ordering information $I_k(\mathbf{x}_k)$ of variable values $x_i \in C_k$ are available, such as $x_i \leq x_j$, they can also be supplied. More specific relationships, such as $x_i = c_i$, $x_i = 2x_j$ or $\phi(x_i, x_j) = c_{ij}$ (where ϕ and c denote a function and a constant, respectively), can also be supplied, if available.

An algorithm can utilize the user-supplied knowledge to any desired extent, from 'lightly' to 'extensively', within an algorithm. An extensive use of supplied knowledge within an optimization algorithm may allow fast convergence, but possibly to a sub-optimal solution, if the supplied knowledge for some reason was not congruent to the true optimal solution set. On the other hand, light use of the supplied knowledge may allow the algorithm to recover from sub-optimal convergence, but may not produce efficient use of the knowledge, if the supplied information was actually correct. In this paper, we propose three repair approaches implementing different extents (low, intermediate, and high) of the use of supplied user knowledge, and investigate their performance.

The proposed repair operations implemented in this study follow a specific procedure. Repair operators are activated once 80% of the population constitutes the non-dominated (ND) set, to prevent any knowledge extraction from low quality solutions. Then, the average ($x_{i_{avg}}$) and standard deviation (σ_i) of the i-th variable values in the current ND set are first computed. If monotonicity or any other specific relationships (e.g., ordering $I_k(\mathbf{x}_k)$ of variable values $x_i \in C_k$) was also supplied, they are noted. Then, at every generation, the offspring population members are repaired using the above information, described next.

3.1 Low-Knowledge Repair Operator

Here we discuss the procedure when no ordering or more specific information is provided. First, we observe whether any pair of variables from the same designated cluster follows the same pattern as they did in the average values derived from the ND set. If variable values of an offspring x_i and x_j ($i, j \in C_k$) have the same relationship as that between $x_{i_{avg}}$ and $x_{j_{avg}}$ of the ND set, no further repair is made. Otherwise, x_i and x_j are repaired as follows:

$$(x_{i_{rep}}, x_{j_{rep}}) = \begin{cases} \left(\frac{x_i + x_{i_{avg}}}{2}, \frac{x_j + x_{j_{avg}}}{2}\right), & \text{for } (x_{i_{avg}} - x_{j_{avg}})(x_i - x_j) < 0, \\ (x_i, x_j), & \text{otherwise.} \end{cases} \quad (1)$$

The variables x_i and x_j are pushed closer to the average $x_{i_{avg}}$. The same operation is also performed on all other consecutive pairs of variables within the cluster and on other clusters.

Now, we discuss the repair procedure if ordering between x_i and x_j was supplied. If the relationship between x_i^u and x_j^u (for an offspring) and between $x_{i_{avg}}$ and $x_{j_{avg}}$ (the average of ND solutions) are in agreement, a repair operation will be performed using Eq. 1; otherwise, Eq. 2 will be used.

$$\left(x_{i_{rep}}, x_{j_{rep}}\right) = \begin{cases} \left(x_{ij_{avg}} + \frac{x_{ij_{avg}} - x_{i_{avg}}}{2}, x_{ij_{avg}} - \frac{x_{ij_{avg}} - x_{j_{avg}}}{2}\right), & \text{if } x_i^u \geq x_j^u, \\ \left(x_{ij_{avg}} - \frac{x_{ij_{avg}} - x_{i_{avg}}}{2}, x_{ij_{avg}} + \frac{x_{ij_{avg}} - x_{j_{avg}}}{2}\right), & \text{if } x_i^u < x_j^u, \end{cases} \tag{2}$$

where $x_{ij_{avg}} = 0.5\left(x_{i_{avg}} + x_{j_{avg}}\right)$.

3.2 Intermediate-Knowledge Repair Operator

This repair operator follows a similar update of offspring variables, but uses a closer relationship of the variable pairs obtained from the history of the optimization run. In this case, the repair operator for two variables x_i and x_j is similar to that shown in Eq. 1, but $x_{i_{avg}}$ is replaced by $x_{i_{ref}}$, calculated by adding a momentum parameter (γ) as shown in Eq. 3.

$$x_{i_{ref}}(t) = x_{i_{avg}}(t) + \gamma v(t), \tag{3}$$

where $v(t) = x_{i_{avg}}(t) - x_{i_{avg}}(t-1)$. If the ordering knowledge between x_i and x_j is available, the same procedure is followed as in the low-knowledge usage operator, but $\mathbf{x}_{avg}$ is replaced with $\mathbf{x}_{ref}$.

3.3 High-Knowledge Repair Operator

This repair operator updates offspring variable vectors more closely to the history of their evolution statistically by staying within one standard deviation (σ_i) of the reference variable value $x_{i_{ref}}$: $[x_{i_{ref}} - \sigma_i, x_{i_{ref}} + \sigma_i]$, where σ_i is the standard deviation of x_i in the ND set:

$$\left(x_{i_{rep}}, x_{j_{rep}}\right) = \left(\mathcal{U}(x_{i_{ref}} - \sigma_i, x_{i_{ref}} + \sigma_i), \mathcal{U}(x_{j_{ref}} - \sigma_j, x_{j_{ref}} + \sigma_j)\right), \tag{4}$$

where $\mathcal{U}(a, b)$ is a uniformly distributed number between a and b. If the ordering knowledge between x_i and x_j is available, we use the same process as in the low-knowledge repair operator, but using $x_{i_{ref}}$, rather than $x_{i_{avg}}$.

We now present the results of an EMO algorithm with the above three repair operators on two engineering design problems.

4 Truss Design Problem

The large-scale truss design problem used in this article is a modified version of the one given in the Bright Optimizer ISCSO 2019 competition [1].

The truss with 260 members and 76 nodes (shown in Fig. 2) is simply-supported at two ends and vertical loads of 5 kN are applied to all top nodes in the negative z-direction. The truss length and width are fixed at 72 m and 4 m respectively. The length of the vertical members can vary from 0.5 to 30 m. A good design is determined by two criteria: low weight and low compliance. There are two sets of design variables: member sizes ($\mathbf{A}$), and coordinates of bottom nodes ($\mathbf{z}$). Finite element analysis (FEA) is performed to compute the weight (f_1), displacement vector ($\mathbf{U}$) and compliance (f_2). The optimization problem formulation is given below:

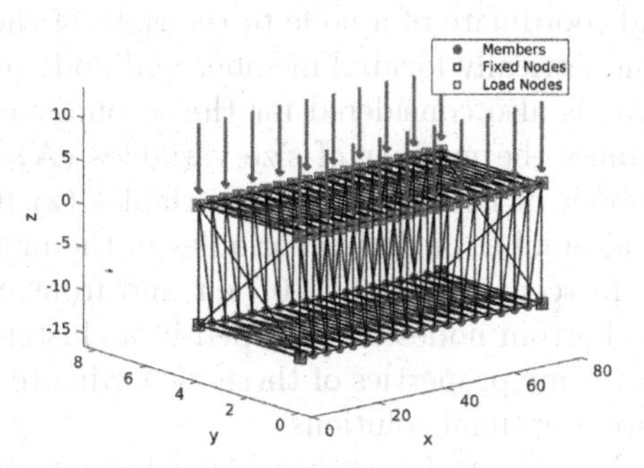

Fig. 2. Truss and its loading.

$$\text{Minimize} \quad (f_1(\mathbf{A},\mathbf{z}),\ f_2(\mathbf{A},\mathbf{z})) = \left(\rho \sum_{k=1}^{N_m} A_k \ell_k(\mathbf{z}),\ \sum_{k=1}^{N_n} \mathbf{F}_k \cdot \mathbf{U}_k(\mathbf{A},\mathbf{z})\right), \quad (5)$$

$$\text{Subject to} \quad \sigma_{max}(\mathbf{A},\mathbf{z}) \leq 248.2 \text{ MPa}, \quad (6)$$

$$|U_k(\mathbf{A},\mathbf{z})| \leq 25 \text{ mm}, \quad \text{for } k = 1,\ldots,N_n, \quad (7)$$

where ρ is the material density, $\ell_k(\mathbf{z})$ is the length of the kth member, which depends on the coordinates of the two end nodes connecting the member, N_m is the total number of members, N_n is the total number of nodes, F_k is the force developed at each node, and σ_{max} is the maximum stress developed in the truss.

4.1 User Knowledge

For the truss shown in Fig. 2, any experienced structural engineer, when asked to provide any expert knowledge on properties of good solutions, will think of a few:

1. Symmetry of the truss around a vertical plane at the mid-point in the x-direction and around another vertical plane at the mid-point in the y-direction can be assumed, due to the symmetry of support and loading conditions.
2. Vertical members on the left half should follow a monotonic property to have a smoothly changing shape from the support towards the middle, due to a smooth transfer of load from the middle of the truss toward the end supports.

To inject the above a priori user knowledge into the optimization process, we consider four scenarios as supplied heuristics U_{t1} to U_{t4}. In all scenarios, symmetry of members on left and right of the y-z plane at $x = 36$ (at the mid-point along the x-axis) is provided. This makes the cross-section size of a member and coordinate of a node to the right of the symmetry plane be identical with a symmetrically located member and node to the left of the symmetry plane. The same is also considered for the second symmetry plane mentioned above. This reduces the number of size variables ($\mathbf{A}$) from the original 260 to 86 and the number of node coordinate variables ($\mathbf{z}$) from the original 38 to 10. Thus, the total number of design variables in the modified problem is 96.

In scenario U_{t1}, all 10 (left and front of symmetry planes) z-coordinates of the bottom nodes are grouped in a cluster ($C_1 = \{i\}$ for $i \in [1, 10]$) to indicate that some properties of these z-coordinate values are expected to appear on the Pareto-optimal solutions.

In scenario U_{t2}, further knowledge is provided for the C_1 variable group by stating that the 10 z-coordinates of the bottom nodes are likely to exhibit a monotonic property: the vertical members will monotonically increase in length from the support toward the middle of the truss. Thus, if x_1 is the bottom z-coordinate of the first vertical member and x_2 is the same for the second vertical member and so on, then $x_1 \leq x_2 \leq \ldots$ is additionally assumed.

Table 1. Knowledge scenarios U_{t3} (with no ordering) and U_{t4} (with ordering).

Cluster	Variable type	Variable indices	Ordering (U_{t4})
C_1	z_i of bottom nodes	[1–10]	Increasing
C_2	A_i of top Longitudinal members	[11–19]	Increasing
C_3	A_i of bottom Longitudinal members	[20–28]	Increasing
C_4	A_i of vertical members	[29–38]	Increasing

In scenario U_{t3}, further knowledge can be simulated by providing additional clustering of variables in three more clusters (see Table 1). Scenario U_{t4} uses U_{t3} but provides ordering of variables in each cluster as mentioned the final column of the table. The cluster C_1 of variables with a monotonic increase in z-coordinates as in U_{t3} is imposed. Additionally, clusters C_2 and C_3 contain the cross-section size variables of the 10 top longitudinal members (along x) and 10 bottom longitudinal members (along x), respectively, which are assumed to increase monotonically from the support toward the middle. Finally, cluster C_3 contains the cross-section size variable of 10 vertical members, and a similar monotonic increase in their values is assumed. The reason for such an increase in cross-section may come from the fact that the bending moment induced in a simply-supported beam increases from the support toward the middle of the beam and hence a monotonically larger cross-sectional size is needed to withstand a larger bending moment.

4.2 Results and Discussion

NSGA-II [8] is used as the optimization algorithm. Population size is set to 200 and a maximum 2,500 generations are used with a maximum number of function evaluations (FEs) of 500k. The hypervolume (HV) metric is used to measure the performance. The mean final HV over 10 runs obtained by NSGA-II with no knowledge usage is set as the target performance ($\text{HV}^T = 0.98$). For the HV calculation, both objectives are normalized and the reference point is set as [1.1, 1.1]. An algorithm is terminated if the specified HV^T is reached or the maximum allowed FEs of 500k have elapsed.

Comparison of the optimization performances is shown in Table 2. Each entry represents the mean and standard deviation of the number of function evaluations taken to reach HV^T. For cases where the algorithm failed to achieve HV^T within 500k FEs, the final HV obtained is additionally mentioned within braces. Each row represents a particular scenario of user-provided knowledge and each column denotes the specific repair operators (described in Sect. 3) used. In the first row, all columns except the first are marked with 'N/A' since the repair operators cannot operate without any initial user knowledge. From the table, it can be inferred that, for this problem, for every scenario of user knowledge integration, the intermediate-knowledge operator is able to achieve the target HV with the fewest evaluations. This establishes the fact that an extensive follow-up of the average variable profile (in high-knowledge repair) reduces the diversity of the variable vectors among ND solutions in an evolving population. Also, a low-knowledge-based repair does not exploit the supplied information well. Keeping the intermediate-knowledge repair strategy within NSGA-II, the table also shows that scenario U_{t3} provides the right amount of prior information for NSGA-II with the innovized repair operator to work the best. U_{t4} provides the maximum a priori knowledge to the optimization. However, none of the repair operators was able to achieve HV^T. This shows that too much a priori knowledge constrains the optimization and lowers the variable vector diversity among ND solutions in an evolving population. As in the case of knowledge usage for the repair operators, there exists an optimal level of user-provided knowledge beyond which performance starts to degrade.

Table 2. FEs required to reach target $\text{HV}^T = 0.98$.

User knowledge	Repair operator used			
	None	Low-knowledge	Intermediate-knowledge	High-knowledge
None	500k	N/A	N/A	N/A
U_{t1}	500k	433k ± 17k	**323k** ± 6k	328k ± 2k
U_{t2}	500k	400k ± 8k	**302k** ± 12k	302k ± 12k
U_{t3}	500k	350k ± 10k	**167k** ± 16k	230k ± 12k
U_{t4}	**500k**	500k (0.73)	500k (0.74)	500k (0.74)

Figure 3a shows the ND solutions obtained with U_{t3}. It shows that the intermediate-knowledge operator is able to provide the best quality solutions, dominating those given by the other scenarios. Figure 3b shows the HV evolution with the FEs for U_{t3} with the repair operations being started after 80k FEs. Figure 3c shows the minimum weight and minimum compliance trusses for the intermediate-knowledge repair operator with user knowledge U_{t3}.

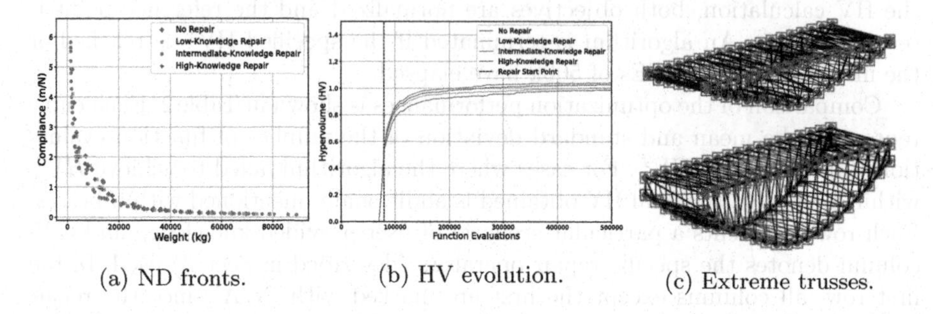

(a) ND fronts. (b) HV evolution. (c) Extreme trusses.

Fig. 3. Performance comparison for user knowledge U_{t3}. In (c), top is the minimum weight truss and bottom is the minimum compliance truss.

5 Demand Side Management (DSM) Problem

Rise in electric power demand in metropolitan areas implies that existing supply capabilities may struggle to handle peak demands [5]. A battery storage-based DSM method [14] can be a viable technology. In this paper, a simplified multiobjective DSM optimization problem has been created from existing literature [13–15], which will serve as a test problem to demonstrate the benefit of knowledge-driven optimization approaches. An artificial hourly electricity demand (in kWh) dataset over a period of two weeks has been used, as shown in Fig. 4. The daily peak loads are marked in red. A custom version of a real-world tariff model [6] has been used. A battery model based on the one given in [14] has been used.

5.1 Optimization Problem Formulation

A multiobjective optimization problem has been formulated which aims to optimize two conflicting objectives: (i) electricity tariff (C_t), and (ii) battery degradation (L_{deg}).

$$\text{Minimize} \quad C_t(\boldsymbol{P_b}, \boldsymbol{E_d}) = C_e(\boldsymbol{P_b}, \boldsymbol{E_d}) + C_p(\boldsymbol{P_b}, \boldsymbol{E_d}), \tag{8}$$

$$\text{Minimize} \quad L_{deg} = \sum_t \max\{0, L(D(t+1)) - L(D(t))\}, \tag{9}$$

$$\text{subject to} \quad P_{min} <= P_b(t) <= P_{max}, \; D_{min} <= D(t) <= D_{max}, \tag{10}$$

$$D(t+1) = D(t) - \eta \times \frac{P_b(t)}{S_b}, \tag{11}$$

where C_t is the total electricity bill, C_e is the energy delivery charges, C_p is the daily peak demand charges, L_{deg} is the total battery degradation, L is the battery degradation loss function, η is the battery charge/discharge efficiency, S_b is the total battery size (kWh), and $E_d(t)$ is the energy demand at time t assumed to be known beforehand. Depth of discharge (D) is defined as the proportion of total battery capacity S_b that has been consumed so far. After a certain number of charge-discharge cycles, the battery capacity degrades to a certain percentage of its initial capacity which makes it unusable. A loss function given in [14] determines the battery degradation for a depth of discharge transition $D(t)$ to $D(t+1)$. $P_b(t)$ is the power supplied by the battery at time t and constitutes the decision variable set.

5.2 User Knowledge

In a DSM problem, an intuitive optimal battery operation is to discharge the battery during the peak daily loads. Since the peak demand charges are high [4, 6], the maximum savings can be made with minimal battery degradation if we operate the battery during the peaks. This knowledge (U_{d1}) can be expressed as shown below: $P_b(t_{peak}(i)) = \max_{t \in [24(i-1)+1,\ldots,24i]} P_b(t)$, where $t_{peak}(i) = \arg\max_{t \in [24(i-1)+1,24i]} E_d(t)$ for the ith day. As a part of U_{d1}, multiple clusters are also specified, with each cluster consisting of 24 variables determining the hourly power outputs. For example, if we are performing the optimization over the load data of 1 week, we will have 7 clusters C_1 to C_7 with each cluster $C_i = \{k\}$ for $i \in [1,7]$ and $k \in [24(i-1)+1, 24i]$.

Scenario U_{d2} is an extension of U_{d1} with the additional knowledge that the battery is charged during the period of lowest demand. This knowledge (U_{d2}) can be expressed as: U_{d1} and $P_b(t_{low}) = \min_{t \in [1,24]} P_b(t)$, where $t_{low} = \arg\min_{t \in [1,\ldots,24]} E_d(t)$. Since there are max and min terms in the above scenario equations, they can also be decomposed into a set of inequality relations as shown below:

$$P_b(t_{peak}) > P_b(t \neq t_{peak}), \quad \forall t \in [1,\ldots,24], \tag{12}$$

$$P_b(t_{low}) < P_b(t \neq t_{low}), \quad \forall t \in [1,\ldots,24]. \tag{13}$$

In order to illustrate the effects of adding more information to the optimization, we create a new scenario U_{d3}. This is an extension of U_{d2} where variable ordering relations are imposed between the individual daily peak loads. For every week, during the daily peak demand hours, the power output can be set to be proportional to the demand. For example, if, for 3 days, the peak loads $E_d(t_{peak}(i))$ for $i \in [1,3]$ in descending order are $\{E_p(3), E_p(1), E_p(2)\}$, the relationship enforced is expressed as $P_b(t_{peak}(3)) \geq P_b(t_{peak}(1)) \geq P_b(t_{peak}(2))$.

5.3 Results and Discussion

This section presents the experimental results on the DSM problem for a two-week period (thus having $24 \times 14 = 336$ variables). NSGA-II [8] has been used as the optimization algorithm to solve this problem. For comparing the performances with different types of user knowledge and repair operators, 10 runs were performed with different random seeds. Maximum FEs was set to be 500k. Identical NSGA-II parameters are used as in the truss problem. HV has been used to measure the performance, and the mean final HV obtained by NSGA-II with no user knowledge is set as the target performance. We set $HV^T = 0.60$. Performance comparison in terms of FEs required to reach HV^T is made in Table 3.

Table 3. FEs required to reach base NSGA-II performance (336-variable DSM problem, $HV^T = 0.60$).

User knowledge	Repair operator used			
	None	Low-knowledge	Intermediate-knowledge	High-knowledge
None	500k	N/A	N/A	N/A
U_{d1}	500k	412k ± 8.0k	**393k** ± 9.0k	406k ± 6.0k
U_{d2}	500k	366k ± 7.5k	**310k** ± 6.0k	382k ± 12.0k
U_{d3}	**500k**	500k (0.52)	500k (0.51)	500k (0.51)

From the table, it is seen that the intermediate-knowledge repair operator is able to reach the target HV in the least number of function evaluations. This supports the conclusions reached for the truss problem that following the average values of the clustered variables too closely constrains the optimization and reduces the diversity of solutions among ND solutions in an evolving population. On the other hand, a low level of knowledge usage does not exploit the available knowledge well enough. Scenario U_{d2} provides the right amount of prior information for the best optimization performance for this problem. U_{d3} provides the most information, but none of the repair operators were able to achieve the target HV. This shows that incorporating more information than necessary can affect the optimization negatively.

Figures 4a show the Pareto-front and HV evolution for U_{d2} with the repair operations being started after 120k FEs. It is seen that the intermediate-knowledge operator is able to generate better quality solutions compared to the other algorithms. The intermediate-knowledge operator is able to achieve a higher HV, faster, which is an indicator of its superior performance. Figure 4b shows the energy supplied by the grid and battery for one of the Pareto-optimal solutions. The figure shows that battery output is utilized, as desired by the user, during most of the peak demands.

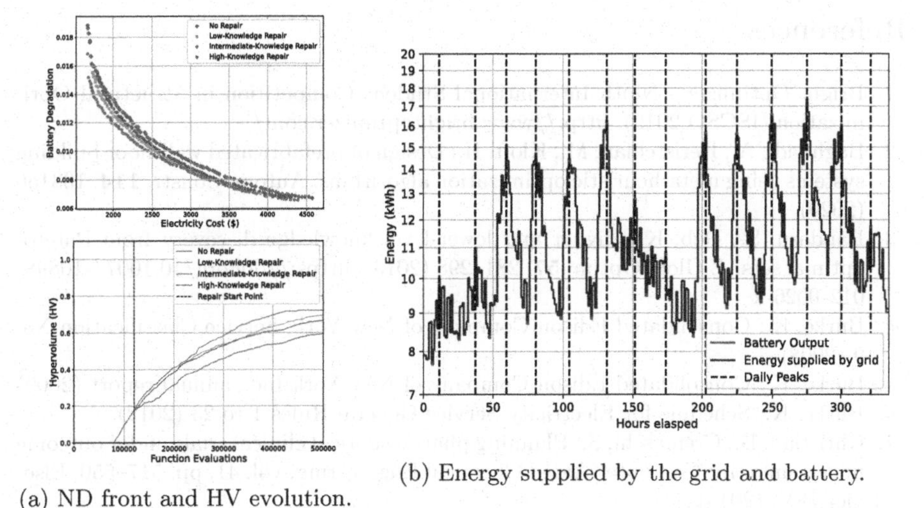

(a) ND front and HV evolution.

(b) Energy supplied by the grid and battery.

Fig. 4. Performance comparison for user knowledge U_{d2}. (Color figure online)

6 Conclusions and Future Work

The study has aimed to demonstrate the usefulness of incorporating user knowledge into an optimization algorithm and has studied the effect of different extents of knowledge usage within an optimization process. Three online innovization-based repair operators have been proposed to exploit any patterns found during the optimization. The proposed methods have been tested on two large-scale practical optimization problems: a 260-member truss optimization problem with 96 variables, and a battery-based DSM problem with 168 variables. The results show the existence of an optimum level of knowledge usage for best performance. If too much or too little knowledge is used, performance is observed to drop. Although an adequate extent of knowledge usage is problem dependent, this study has clearly shown the importance of using the right amount of problem knowledge for a computationally fast optimization process.

Our future work will focus on integrating interactive optimization into the framework such that the user can provide useful feedback during the optimization process. This will further enhance the human-machine collaboration in the knowledge-driven design optimization process. The three separate repair operators can also be unified under an ensemble method which will remove the need to perform separate experiments. Studies on the generalizability of the proposed methods to other types of practical design problems need to be performed.

Acknowledgment. This research was developed with funding from the Defense Advanced Research Projects Agency (DARPA). The views, opinions and/or findings expressed are those of the authors and should not be interpreted as representing the official views or policies of the Department of Defense or the U.S. Government. We thank Qiren Gao and Ming Liu for their valuable inputs.

References

1. Bright Optimizer - Ninth International Student Competition in Structural Optimization (ISCSO 2019). http://www.brightoptimizer.com/
2. Baghdadi, A., Heristchian, M., Kloft, H.: Design of prefabricated wall-floor building systems using meta-heuristic optimization algorithms. Autom. Constr. **114**, 103156 (2020)
3. Bandaru, S., Deb, K.: Higher and lower-level knowledge discovery from Pareto-optimal sets. J. Glob. Optim. **57**, 281–298 (2013). https://doi.org/10.1007/s10898-012-0026-x
4. Burke, K.: Consolidated Edison Company of New York, Service Classification No. 9 (2008)
5. Burke, K.: Consolidated Edison Company of New York Inc., annual report (2009)
6. Burke, K.: Schedule for Electricity Service General Rules 1 to 23 (2012)
7. Christian, B., Cremaschi, S.: Planning pharmaceutical clinical trials under outcome uncertainty. In: Computer Aided Chemical Engineering, vol. 41, pp. 517–550. Elsevier B.V. (2018)
8. Deb, K., Pratap, A., Agarwal, S., Meyarivan, T.: A fast and elitist multiobjective genetic algorithm: NSGA-II. IEEE Trans. Evol. Comput. **6**(2), 182–197 (2002)
9. Deb, K., Srinivasan, A.: Innovization: innovating design principles through optimization. In: GECCO 2006 - Genetic and Evolutionary Computation Conference, vol. 2, pp. 1629–1636 (2006)
10. Gaur, A., Deb, K.: Effect of size and order of variables in rules for multi-objective repair-based innovization procedure. In: Proceedings of the 2017 IEEE Congress on Evolutionary Computation, CEC 2017, pp. 2177–2184. Institute of Electrical and Electronics Engineers Inc. (2017)
11. Jaszkiewicz, A., Zielniewicz, P.: Pareto memetic algorithm with path relinking for bi-objective traveling salesperson problem. Eur. J. Oper. Res. **193**(3), 885–890 (2009)
12. Li, H.: Network topology design. In: Communications for Control in Cyber Physical Systems, pp. 149–180. Elsevier (2016)
13. Tooryan, F., Ghosh, A., Wang, Y., Srivastava, S., Arvanitis, E., Angelis, V.D.: Microgrid energy storage design for reliability and cost performances. In: Proceedings of the 2020 IEEE PES General Meeting (2020)
14. Wang, Y., Song, Z., De Angelis, V., Srivastava, S.: Battery life-cycle optimization and runtime control for commercial buildings demand side management: a New York City case study. Energy **165**, 782–791 (2018)
15. Yadav, G.G., et al.: Regenerable Cu-intercalated MnO_2 layered cathode for highly cyclable energy dense batteries. Nat. Commun. **8**(1), 1–9 (2017)

Improving the Efficiency of R2HCA-EMOA

Ke Shang, Hisao Ishibuchi(✉), Longcan Chen, Weiyu Chen, and Lie Meng Pang

Guangdong Provincial Key Laboratory of Brain-Inspired Intelligent Computation, Department of Computer Science and Engineering, Southern University of Science and Technology, Shenzhen 518055, China
kshang@foxmail.com, {hisao,panglm}@sustech.edu.cn, {11813009,11711904}@mail.sustech.edu.cn

Abstract. R2HCA-EMOA is a recently proposed hypervolume-based evolutionary multi-objective optimization (EMO) algorithm. It uses an R2 indicator variant to approximate the hypervolume contribution of each solution. Meanwhile, it uses a utility tensor structure to facilitate the calculation of the R2 indicator variant. This makes it very efficient for solving many-objective optimization problems. Compared with HypE, another hypervolume-based EMO algorithm designed for many-objective problems, R2HCA-EMOA runs faster and at the same time achieves better performance. Thus, R2HCA-EMOA is more attractive for practical use. In this paper, we further improve the efficiency of R2HCA-EMOA without sacrificing its performance. We propose two strategies for the efficiency improvement. One is to simplify the environmental selection, and the other is to change the number of direction vectors depending on the state of evolution. Numerical experiments clearly show that the efficiency of R2HCA-EMOA is significantly improved without deteriorating its performance.

Keywords: Hypervolume indicator · R2HCA-EMOA · Evolutionary multi-objective optimization

1 Introduction

In the field of Evolutionary Multi-objective Optimization (EMO), new algorithms are proposed every year. In general, these algorithms are classified into three categories: Pareto dominance-based (e.g., NSGA-II [5]), decomposition-based (e.g., MOEA/D [18]), and indicator-based (e.g., SMS-EMOA [2,7]). Among these three categories, indicator-based EMO algorithms transform a multi-objective optimization problem into a single-objective optimization problem where the single-objective is to optimize the value of an indicator (e.g., GD [16], IGD [3], Hypervolume [19], R2 [8]). The performance of EMO algorithms is usually evaluated by indicators. Each indicator-based EMO algorithm is likely to achieve a good performance with respect to the indicator used in

H. Ishibuchi et al. (Eds.): EMO 2021, LNCS 12654, pp. 115–125, 2021.
https://doi.org/10.1007/978-3-030-72062-9_10

the algorithm. That is, an IGD-based algorithm is likely to achieve a good IGD performance, and a hypervolume-based algorithm is likely to achieve a good hypervolume performance. This is the main advantage of the indicator-based algorithms.

Among different indicators, the hypervolume indicator is the most popular one in the EMO community since it has solid theoretical foundations and is comprehensively investigated [14]. The hypervolume-based EMO algorithms use the hypervolume indicator as the optimization criterion, which have been successfully applied to solve multi- and many-objective optimization problems. Some representative hypervolume-based algorithms are SMS-EMOA [2,7], FV-MOEA [11], HypE [1], and R2HCA-EMOA [12]. SMS-EMOA and FV-MOEA use the exact hypervolume calculation. The advantage of these two algorithms is that a good hypervolume performance can be achieved. The main disadvantage of these two algorithms is that the exact hypervolume calculation is very time-consuming in high-dimensional objective spaces. Thus, their efficiency is very low when solving many-objective problems.

In order to solve the above issue, HypE and R2HCA-EMOA are proposed. These two algorithms use hypervolume approximation methods in the algorithm implementations, and thus they can solve many-objective problems efficiently. Since the hypervolume is approximated in these two algorithms, their hypervolume performance is worse than SMS-EMOA and FV-MOEA. As reported in [12], R2HCA-EMOA is able to run faster and at the same time achieve better hypervolume performance than HypE. Thus, R2HCA-EMOA is more efficient and effective than HypE for solving many-objective problems.

However, compared with other popular EMO algorithms (e.g., NSGA-II, MOEA/D, NSGA-III [4]), R2HCA-EMOA is still not very efficient (i.e., slower than those algorithms). Thus, the question is whether we can further improve the efficiency of R2HCA-EMOA without sacrificing its high performance. In this paper, we investigate this issue and propose two strategies to further improve the efficiency of R2HCA-EMOA. The improved version of R2HCA-EMOA is more efficient and its performance is not deteriorated. That is, the improved version is more attractive for practical use.

The rest of the paper is organized as follows. R2HCA-EMOA is briefly introduced in Sect. 2. Two efficiency improvement strategies are presented in Sect. 3. Numerical experiments are conducted in Sect. 4. The conclusions are drawn in Sect. 5.

2 R2HCA-EMOA

R2HCA-EMOA follows the framework of SMS-EMOA. In each generation, one offspring is generated and added into the population. Then one individual with the least hypervolume contribution in the last front is removed from the population. The main difference between R2HCA-EMOA and SMS-EMOA is that the hypervolume contribution is approximated in R2HCA-EMOA instead of exactly calculated. The pseudocode of R2HCA-EMOA is shown in Algorithm 1.

Algorithm 1. R2HCA-EMOA

Input: Population Size N, Maximum Function Evaluations $\text{FEs}^{\max}$.
Output: Final Population P.

1: Initialize Population P, Direction Vector Set Λ, Utility Tensor $\mathbf{T}$, reference point $\mathbf{r}$, $\text{FEs} = N$;
2: **while** $\text{FEs} \leq \text{FEs}^{\max}$ **do**
3: $\quad \mathbf{q} = \texttt{GenerateOffspring}(P)$;
4: $\quad P = P \cup \{\mathbf{q}\}$;
5: $\quad P' = \texttt{Normalize}(P)$;
6: $\quad \mathbf{T} = \texttt{UpdateUtilityTensor}(\mathbf{T}, P', \Lambda)$;
7: $\quad \{F_1, F_2, ..., F_l\} = \texttt{NondominatedSort}(P')$;
8: $\quad$ **if** $|F_l| == 1$ **then**
9: $\quad\quad R_2^{\text{HCA}}(\mathbf{s}) = 0, \forall \mathbf{s} \in F_l$;
10: $\quad$ **else if** $|F_l| == N + 1$ **then**
11: $\quad\quad R_2^{\text{HCA}}(\mathbf{s}) = \texttt{CalculateR2HCA}(\mathbf{T}), \forall \mathbf{s} \in F_l$;
12: $\quad$ **else**
13: $\quad\quad \mathbf{T}' = \texttt{ExtractUtilityTensor}(\mathbf{T}, F_l)$;
14: $\quad\quad R_2^{\text{HCA}}(\mathbf{s}) = \texttt{CalculateR2HCA}(\mathbf{T}'), \forall \mathbf{s} \in F_l$;
15: $\quad$ **end if**
16: $\quad \mathbf{s}_{\text{worst}} = \arg\min_{\mathbf{s} \in F_l} R_2^{\text{HCA}}(\mathbf{s})$;
17: $\quad P = P \setminus \{\mathbf{s}_{\text{worst}}\}$;
18: $\quad \text{FEs} = \text{FEs} + 1$;
19: **end while**

In R2HCA-EMOA, the hypervolume contribution is approximated by an R2 indicator variant, and a utility tensor structure is used to facilitate the calculation of the R2 indicator variant. Next, the R2 indicator variant and the utility tensor structure are briefly explained.

2.1 R2-Based Hypervolume Contribution Approximation

In R2HCA-EMOA, an R2 indicator variant is used to approximate the hypervolume contribution of a solution $\mathbf{s}$ to a solution set A in an m-dimensional objective space[1]:

$$R_2^{\text{HCA}}(\mathbf{s}) = \frac{1}{|\Lambda|} \sum_{\boldsymbol{\lambda} \in \Lambda} \min\left\{ \min_{\mathbf{a} \in A \setminus \{\mathbf{s}\}} \left\{ g^{*2\text{tch}}(\mathbf{a}|\boldsymbol{\lambda}, \mathbf{s}) \right\}, g^{\text{mtch}}(\mathbf{r}|\boldsymbol{\lambda}, \mathbf{s}) \right\}^m, \tag{1}$$

where A is a non-dominated solution set with $\mathbf{s} \in A$, Λ is a direction vector set, each direction vector $\boldsymbol{\lambda} = (\lambda_1, \lambda_2, ..., \lambda_m) \in \Lambda$ satisfies $\|\boldsymbol{\lambda}\|_2 = 1$ and $\lambda_i \geq 0$ for $i = 1, ..., m$, and $\mathbf{r}$ is the reference point.

The $g^{*2\text{tch}}(\mathbf{a}|\boldsymbol{\lambda}, \mathbf{s})$ function is defined for minimization problems as

$$g^{*2\text{tch}}(\mathbf{a}|\boldsymbol{\lambda}, \mathbf{s}) = \max_{j \in \{1,...,m\}} \left\{ \frac{a_j - s_j}{\lambda_j} \right\}. \tag{2}$$

[1] In this paper, the solutions (individuals) are assumed to be points in the objective space, i.e., a solution $\mathbf{s}$ denotes an objective vector.

For maximization problems, it is defined as

$$g^{*2tch}(\mathbf{a}|\boldsymbol{\lambda}, \mathbf{s}) = \max_{j\in\{1,\ldots,m\}} \left\{\frac{s_j - a_j}{\lambda_j}\right\}. \quad (3)$$

The g^{mtch} function is defined for both minimization and maximization problems as

$$g^{mtch}(\mathbf{r}|\boldsymbol{\lambda}, \mathbf{s}) = \min_{j\in\{1,\ldots,m\}} \left\{\frac{|s_j - r_j|}{\lambda_j}\right\}. \quad (4)$$

The geometric meaning of the R2 indicator variant is illustrated in Fig. 1. Given a non-dominated solution set $\{\mathbf{a}^1, \mathbf{a}^2, \mathbf{a}^3\}$ and a set of line segments with different directions in the hypervolume contribution region of $\mathbf{a}^2$, the R2 indicator variant in (1) is calculated as $R_2^{\text{HCA}}(\mathbf{a}^2) = \frac{1}{|\Lambda|}\sum_{i=1}^{|\Lambda|} l_i^m$ where l_i is the length of the ith line segment. For more detailed explanations of the R2 indicator variant, please refer to [13].

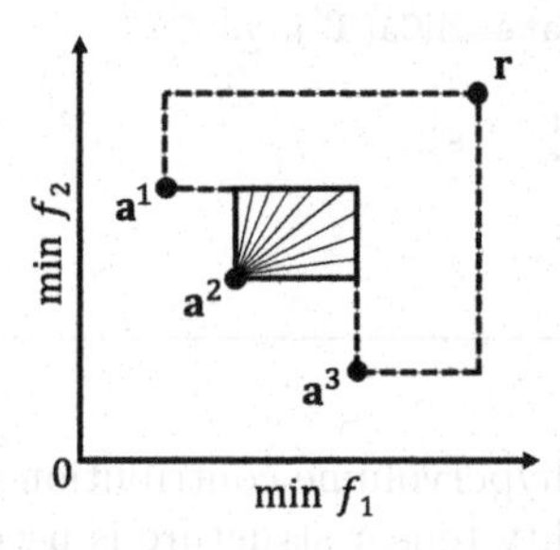

Fig. 1. An illustration of the geometric meaning of R_2^{HCA} for the hypervolume contribution approximation.

2.2 Utility Tensor Structure

In R2HCA-EMOA, a utility tensor structure is introduced to facilitate the calculation of R_2^{HCA}. From the definition of R_2^{HCA} in (1), we can see that in order to compute the R_2^{HCA} value of a solution $\mathbf{s} \in A$, we need to compute $g^{*2tch}(\mathbf{a}|\boldsymbol{\lambda}, \mathbf{s})$ for all $\mathbf{a} \in A\backslash\{\mathbf{s}\}$ and all $\boldsymbol{\lambda} \in \Lambda$, and $g^{mtch}(\mathbf{r}|\boldsymbol{\lambda}, \mathbf{s})$ for all $\boldsymbol{\lambda} \in \Lambda$. Thus, we can store g^{*2tch} and g^{mtch} values in a matrix $\mathbf{M}$ as follows:

$$\mathbf{M} = \begin{bmatrix} g^{*2tch}(\mathbf{a}_1|\boldsymbol{\lambda}_1, \mathbf{s}) & g^{*2tch}(\mathbf{a}_1|\boldsymbol{\lambda}_2, \mathbf{s}) & \cdots & g^{*2tch}(\mathbf{a}_1|\boldsymbol{\lambda}_{|\Lambda|}, \mathbf{s}) \\ \vdots & \vdots & \ddots & \vdots \\ g^{*2tch}(\mathbf{a}_{|A|-1}|\boldsymbol{\lambda}_1, \mathbf{s}) & g^{*2tch}(\mathbf{a}_{|A|-1}|\boldsymbol{\lambda}_2, \mathbf{s}) & \cdots & g^{*2tch}(\mathbf{a}_{|A|-1}|\boldsymbol{\lambda}_{|\Lambda|}, \mathbf{s}) \\ g^{mtch}(\mathbf{r}|\boldsymbol{\lambda}_1, \mathbf{s}) & g^{mtch}(\mathbf{r}|\boldsymbol{\lambda}_2, \mathbf{s}) & \cdots & g^{mtch}(\mathbf{r}|\boldsymbol{\lambda}_{|\Lambda|}, \mathbf{s}) \end{bmatrix}_{|A|\times|\Lambda|}$$

After obtaining the above matrix, we can calculate $R_2^{\text{HCA}}(\mathbf{s})$ as follows:

$$R_2^{\text{HCA}}(\mathbf{s}) = \frac{1}{|\Lambda|}\sum_{j=1}^{|\Lambda|}\left(\min_{i=\{1,2,\ldots,|A|\}} M_{ij}\right)^m. \quad (5)$$

Since we need to calculate R_2^{HCA} for each $\mathbf{s} \in A$, we need to have $|A|$ matrices for all solutions in A. These $|A|$ matrices form a tensor

$$\mathbf{T} = \left[\mathbf{M}_1, \mathbf{M}_2, ..., \mathbf{M}_{|A|}\right],$$

where $\mathbf{M}_k$ is the matrix for $\mathbf{a}_k \in A$.

We use tensor $\mathbf{T}$ as a memory to store useful information for the calculation of R_2^{HCA}. As shown in Algorithm 1, since only one offspring is generated and one individual is removed in each generation, we do not need to recalculate tensor $\mathbf{T}$ in each generation. Instead, we only update $\mathbf{T}$ with respect to the changed elements to maximize the efficiency. Furthermore, if all the solutions in the current population are non-dominated, we can use (5) to calculate R_2^{HCA}. However, if there are some dominated solutions, we need to extract useful information from $\mathbf{T}$ to calculate R_2^{HCA} for the last front. For more detailed explanation of tensor $\mathbf{T}$ and its operations, please refer to [12].

3 Efficiency Improvement Strategies

In this section, we propose two strategies to further improve the efficiency of R2HCA-EMOA. The first strategy is to simplify the environmental selection, and the second strategy is to change the number of direction vectors depending on the state of evolution.

3.1 Simplify the Environmental Selection

The first strategy is to simplify the environmental selection in R2HCA-EMOA. As explained in the previous section, we use tensor $\mathbf{T}$ to calculate R_2^{HCA} values, i.e., the approximated hypervolume contributions. If there are dominated solutions, we cannot directly use tensor $\mathbf{T}$ to calculate R_2^{HCA} for the last front. Thus, a tensor extraction is performed to get a subtensor $\mathbf{T}'$. This will lower the efficiency of R2HCA-EMOA.

The idea is to simplify the environmental selection when there are dominated solutions. Instead of extracting a subtensor $\mathbf{T}'$ and calculating R_2^{HCA} for the last front, we simply randomly remove one solution from the last front. In this way, we can avoid the tensor extraction procedure and the algorithm efficiency can be improved. To be more specific, lines 8–15 in Algorithm 1 are replaced by the following lines:

1: **if** $|F_l| == N + 1$ **then**
2: $\quad R_2^{\mathrm{HCA}}(\mathbf{s}) = \mathtt{CalculateR2HCA}(\mathbf{T})$, $\forall \mathbf{s} \in F_l$;
3: **else**
4: $\quad R_2^{\mathrm{HCA}}(\mathbf{s}) = \mathtt{rand}(0, 1)^2$, $\forall \mathbf{s} \in F_l$;
5: **end if**

The question is whether this simple strategy can deteriorate the performance of R2HCA-EMOA. The answer is NO (which will be verified in the experiments).

[2] rand(0, 1) means a random number between 0 and 1.

The reason can be explained as follows. In many-objective optimization, all or almost all solutions in the current population are non-dominated. In this situation, it is not likely that dominated solutions play an important role in multi-objective evolution. That is, it is likely that all dominated solutions in the current population are almost equally useless (if compared with the non-dominated solutions). As a result, it is not needed to carefully evaluate each dominated solution to identify the worst one with the least contribution.

3.2 Change the Number of Direction Vectors

The second strategy is to dynamically change the number of direction vectors used in R2HCA-EMOA. As discussed in [12], the worst-case time complexity of one generation of R2HCA-EMOA is $O(N^2|\Lambda|)$. We can see that the time complexity is related to the number of direction vectors $|\Lambda|$. We can improve the efficiency of R2HCA-EMOA by decreasing the value of $|\Lambda|$. However, a smaller value of $|\Lambda|$ means a lower approximation quality of R_2^{HCA}. Thus, if we simply set $|\Lambda|$ as a small value, the performance of R_2^{HCA} may be deteriorated.

The idea is to use a small number of $|\Lambda|$ in the early stage of the evolution and use a large number of $|\Lambda|$ in the late stage of the evolution. In the early stage of the evolution, we do not need a very high approximation quality of R_2^{HCA} since the main focus is the convergence of the population. Also, as we discussed in the first strategy, the solutions in the last front are randomly removed. In this case, we do not need to calculate R_2^{HCA}. In the late stage of the evolution, we use a large number of $|\Lambda|$ to achieve a good hypervolume performance.

The question is the timing to change the number of direction vectors. Ideally, the timing for the change is when the population has converged to the Pareto front. However, the Pareto front is usually unknown in practice. Thus, we need to have a mechanism to decide the timing for the change. In this paper, we use the following mechanism:

$$|\Lambda| = \begin{cases} \text{a small number, if FEs} \leq \theta \text{FEs}^{\max} \\ \text{a large number, } \quad\quad \text{otherwise} \end{cases} \tag{6}$$

where $\theta \in [0, 1]$ is a timing control parameter. The small number and the large number are set to 10 and 100, respectively, in our experiments.

We need to note that when $|\Lambda|$ is changed, the direction vector set Λ is regenerated, and tensor $\mathbf{T}$ needs to be reinitialized since the size of $\mathbf{T}$ is also changed.

4 Experiments

In this section, we examine the effect of the proposed two strategies on the performance of R2HCA-EMOA through computational experiments.

4.1 Experimental Settings

1. **Platforms.** Our experiments are performed on PlatEMO [15], which is a MATLAB-based open source platform for evolutionary multi-objective optimization. The source code of R2HCA-EMOA is available at https://github.com/HisaoLabSUSTC/R2HCA-EMOA. The hardware platform is a PC equipped with Intel Core i7-8700K CPU@3.70 GHz, 16 GB RAM.
2. **Test Problems.** DTLZ1-4 [6], WFG1-9 [9], and their minus versions (i.e., MinusDTLZ1-4 and MinusWFG1-9) [10] are selected as test problems in our experiments. The number of objectives is set to $m = 5$. The number of decision variables is set to $m + 4$ for DTLZ1 and MinusDTLZ1, and $m + 9$ for the other DTLZ and MinusDTLZ problems. For WFG and MinusWFG problems, the distance and position decision variables are set to 24 and $m - 1$, respectively.
3. **Parameter Settings.** The population size N is set to 100. For DTLZ1, DTLZ3, WFG1 and their minus versions, the maximum function evaluations FEs^{max} is set to 100,000, while FEs^{max} is set to 30,000 for the other test problems. The simulated binary crossover (SBX) and the polynomial mutation are adopted in the algorithms where the distribution index is specified as 20 in both operators. The crossover and mutation rates are set to 1.0 and $1/n$ respectively, where n is the number of decision variables. Each algorithm is run 20 time independently on each test problem.
4. **Performance Metrics.** The hypervolume indicator is used to evaluate the performance of the algorithms, and the runtime is recorded to evaluate the efficiency of the algorithms. We employ the WFG algorithm [17] to calculate the exact hypervolume of the obtained solution sets. The hypervolume is calculated as follows. First, using the true ideal point $\mathbf{z}^*$ and the true nadir point $\mathbf{z}^{nad}$ calculated from the true PF, the objective space is normalized, i.e., the two points $\mathbf{z}^*$ and $\mathbf{z}^{nad}$ are $(0, ..., 0)$ and $(1, ..., 1)$ after the normalization. Then, the reference point $\mathbf{r}$ is specified as $(1.1, ..., 1.1)$ to calculate the hypervolume of the obtained solution set in the normalized objective space.
5. **Algorithm Configurations.** The compared algorithms are R2HCA-EMOA and its improved version (denoted as R2HCA-EMOA-II) equipped with the two strategies. For R2HCA-EMOA, the number of direction vectors $|\Lambda|$ is set to 100. For R2HCA-EMOA-II, $|\Lambda|$ is initially set to 10, and finally changed to 100 according to (6). The timing control parameter θ is set to $\{0.0, 0.3, 0.5, 0.8\}$, i.e., we examine the effect of different θ values. All the results are analyzed by the Wilcoxon rank sum test with a significance level of 0.05. '+', '−' and '≈' indicate that R2HCA-EMOA-II is 'significantly better than', 'significantly worse than' and 'statistically similar to' the original R2HCA-EMOA, respectively.

4.2 Results

The mean hypervolume results and the mean runtime results over 20 runs are shown in Table 1 and Table 2, respectively.

Table 1. Statistical results of hypervolume values obtained by R2HCA-EMOA and R2HCA-EMOA-II. The best result for each test problem is shaded.

Problem	R2HCA-EMOA	R2HCA-EMOA-II			
		$\theta = 0.0$	$\theta = 0.3$	$\theta = 0.5$	$\theta = 0.8$
DTLZ1	1.5657e+0	1.5656e+0 ≈	1.5647e+0 ≈	1.5636e+0 ≈	1.5651e+0 ≈
DTLZ2	1.2852e+0	1.2861e+0 ≈	1.2844e+0 ≈	1.2852e+0 ≈	1.2850e+0 ≈
DTLZ3	1.2826e+0	1.2823e+0 ≈	1.2825e+0 ≈	1.2826e+0 ≈	1.2797e+0 −
DTLZ4	1.1013e+0	1.1123e+0 ≈	1.1372e+0 ≈	1.1152e+0 ≈	1.1387e+0 ≈
MinusDTLZ1	1.6317e−2	1.6310e−2 ≈	1.6233e−2 ≈	1.6327e−2 ≈	1.6294e−2 ≈
MinusDTLZ2	2.0310e−1	2.0346e−1 ≈	2.0264e−1 ≈	2.0282e−1 ≈	2.0270e−1 ≈
MinusDTLZ3	2.0302e−1	2.0287e−1 ≈	2.0283e−1 ≈	2.0298e−1 ≈	2.0280e−1 ≈
MinusDTLZ4	2.0184e−1	2.0203e−1 ≈	2.0190e−1 ≈	2.0208e−1 ≈	2.0185e−1 ≈
WFG1	1.6027e+0	1.6005e+0 ≈	1.5981e+0 ≈	1.5945e+0 ≈	1.5888e+0 ≈
WFG2	1.5915e+0	1.5921e+0 ≈	1.5913e+0 ≈	1.5902e+0 ≈	1.5872e+0 −
WFG3	3.1829e−1	2.7968e−1 ≈	2.9585e−1 ≈	2.6692e−1 −	2.6957e−1 −
WFG4	1.2607e+0	1.2636e+0 ≈	1.2593e+0 ≈	1.2585e+0 ≈	1.2546e+0 −
WFG5	1.1983e+0	1.1988e+0 ≈	1.1987e+0 ≈	1.1979e+0 ≈	1.1972e+0 ≈
WFG6	1.2002e+0	1.2066e+0 ≈	1.2061e+0 +	1.1985e+0 ≈	1.2016e+0 ≈
WFG7	1.2767e+0	1.2774e+0 ≈	1.2768e+0 ≈	1.2760e+0 ≈	1.2739e+0 −
WFG8	1.1532e+0	1.1527e+0 ≈	1.1462e+0 −	1.1434e+0 −	1.1419e+0 −
WFG9	1.2124e+0	1.2186e+0 +	1.2149e+0 ≈	1.2145e+0 ≈	1.1968e+0 −
MinusWFG1	4.1141e−3	4.1346e−3 ≈	4.1124e−3 ≈	4.0299e−3 −	4.0278e−3 −
MinusWFG2	1.3725e−2	1.3776e−2 ≈	1.3681e−2 ≈	1.3743e−2 ≈	1.3651e−2 ≈
MinusWFG3	1.6000e−2	1.5931e−2 ≈	1.5991e−2 ≈	1.5972e−2 ≈	1.5887e−2 ≈
MinusWFG4	2.0259e−1	2.0243e−1 ≈	2.0226e−1 ≈	2.0247e−1 ≈	2.0230e−1 ≈
MinusWFG5	1.9853e−1	1.9895e−1 ≈	1.9904e−1 ≈	1.9814e−1 ≈	1.9717e−1 −
MinusWFG6	2.0067e−1	2.0059e−1 ≈	2.0092e−1 ≈	2.0039e−1 ≈	2.0101e−1 ≈
MinusWFG7	2.0072e−1	2.0093e−1 ≈	2.0006e−1 ≈	2.0056e−1 ≈	2.0029e−1 ≈
MinusWFG8	2.0230e−1	2.0211e−1 ≈	2.0221e−1 ≈	2.0231e−1 ≈	2.0270e−1 ≈
MinusWFG9	1.9961e−1	1.9952e−1 ≈	1.9956e−1 ≈	1.9945e−1 ≈	1.9858e−1 −
+/−/≈	-	1/0/25	1/1/24	0/3/23	0/10/16

From Table 1, we can see that R2HCA-EMOA-II achieves a comparable hypervolume performance to R2HCA-EMOA when $\theta = 0.0, 0.3$. When $\theta = 0.0$ (i.e., we only consider the first strategy), R2HCA-EMOA-II is significantly better on WFG9. When $\theta = 0.3$, it is significantly better on WFG6 and significantly worse on WFG8, and has no significant difference on all the other test problems. It should be noted that the difference in the average hypervolume value between the two algorithms is very small even when there exists a statistically significant difference (see the results on WFG6, WFG8 and WFG9). When $\theta = 0.5$, R2HCA-EMOA-II is slightly worse than R2HCA-EMOA. It is significantly worse on 3 out of the 26 test problems, and has no significant difference on all the other test problems. When $\theta = 0.8$, the performance of R2HCA-EMOA-II is clearly deteriorated. It is significantly worse on 10 out of the 26 test problems. From these results, we can see that the first strategy does not degrade the performance of R2HCA-EMOA-II (i.e., when $\theta = 0.0$). We can also see that the second strategy has a large effect on the performance of R2HCA-EMOA-II depending on

Table 2. Statistical results of runtime (in seconds) of R2HCA-EMOA and R2HCA-EMOA-II. The best result for each test problem is shaded.

Problem	R2HCA-EMOA	R2HCA-EMOA-II			
		$\theta = 0.0$	$\theta = 0.3$	$\theta = 0.5$	$\theta = 0.8$
DTLZ1	4.5574e+2	4.5307e+2 ≈	3.5623e+2 +	2.7796e+2 +	1.6856e+2 +
DTLZ2	1.6576e+2	1.6766e+2 ≈	1.3070e+2 +	1.0260e+2 +	6.1985e+1 +
DTLZ3	4.0426e+2	3.8419e+2 +	3.2252e+2 +	2.5434e+2 +	1.5958e+2 +
DTLZ4	1.6334e+2	1.6439e+2 ≈	1.3528e+2 +	1.0462e+2 +	6.2214e+1 +
MinusDTLZ1	5.3837e+2	5.3888e+2 ≈	4.3997e+2 +	3.4244e+2 +	2.0172e+2 +
MinusDTLZ2	1.6480e+2	1.7038e+2 ≈	1.3480e+2 +	1.0591e+2 +	6.1981e+1 +
MinusDTLZ3	5.2682e+2	5.2273e+2 ≈	4.9828e+2 +	3.3254e+2 +	1.9823e+2 +
MinusDTLZ4	1.8391e+2	1.7922e+2 +	1.4052e+2 +	1.1022e+2 +	6.5740e+1 +
WFG1	4.6995e+2	4.7799e+2 ≈	3.9469e+2 +	3.1191e+2 +	1.8881e+2 +
WFG2	1.5686e+2	1.5491e+2 ≈	1.3662e+2 +	1.0389e+2 +	6.2514e+1 +
WFG3	1.7011e+2	1.6653e+2 ≈	1.3885e+2 +	1.0886e+2 +	6.4457e+1 +
WFG4	1.5867e+2	1.5813e+2 ≈	1.3048e+2 +	1.0211e+2 +	6.2384e+1 +
WFG5	1.6327e+2	1.8309e+2 ≈	1.3231e+2 +	1.0401e+2 +	6.6410e+1 +
WFG6	1.4974e+2	1.5208e+2 ≈	1.2421e+2 +	9.5330e+1 +	6.2164e+1 +
WFG7	1.7443e+2	1.7106e+2 ≈	1.3610e+2 +	1.0749e+2 +	6.4252e+1 +
WFG8	1.3313e+2	1.3389e+2 ≈	1.1160e+2 +	8.9537e+1 +	5.6189e+1 +
WFG9	1.7798e+2	1.7765e+2 ≈	1.4325e+2 +	1.1197e+2 +	6.6564e+1 +
MinusWFG1	4.6240e+2	4.6057e+2 ≈	3.6451e+2 +	2.9485e+2 +	1.8118e+2 +
MinusWFG2	1.5324e+2	1.5700e+2 ≈	1.2389e+2 +	9.8582e+1 +	6.0794e+1 +
MinusWFG3	1.7885e+2	1.7575e+2 ≈	1.4634e+2 +	1.1536e+2 +	6.9738e+1 +
MinusWFG4	1.7316e+2	1.7272e+2 ≈	1.4242e+2 +	1.1180e+2 +	6.7144e+1 +
MinusWFG5	1.6915e+2	1.6888e+2 ≈	1.3963e+2 +	1.0950e+2 +	6.5649e+1 +
MinusWFG6	1.6638e+2	1.6679e+2 ≈	1.3882e+2 +	1.0950e+2 +	6.6004e+1 +
MinusWFG7	1.8269e+2	1.8287e+2 ≈	1.4910e+2 +	1.1705e+2 +	7.8022e+1 +
MinusWFG8	1.7359e+2	1.7494e+2 ≈	1.4802e+2 +	1.1267e+2 +	6.7384e+1 +
MinusWFG9	1.8258e+2	1.8270e+2 ≈	1.4986e+2 +	1.1747e+2 +	7.0116e+1 +
+/ − / ≈		2/0/24	26/0/0	26/0/0	26/0/0

the value of θ. A too large θ value can deteriorate the performance of R2HCA-EMOA-II. In our results, $\theta = 0.5$ is an acceptable value since the performance of R2HCA-EMOA-II is almost the same as that of R2HCA-EMOA.

From Table 2, we can see that the efficiency of R2HCA-EMOA-II is slightly improved by the first strategy when $\theta = 0.0$. The runtime of R2HCA-EMOA-II is significantly better on 2 out of the 26 test problems. This is because the first strategy only works when there exist dominated solutions. If all the solutions are non-dominated, the two algorithms are exactly the same. It is very likely that the first strategy only works for a small number of generations. For the second strategy, the efficiency of R2HCA-EMOA-II is significantly improved compared with R2HCA-EMOA. We can also see that R2HCA-EMOA-II runs faster as θ increases. Take DTLZ1 as an example, when $\theta = 0.3$, the runtime of R2HCA-EMOA-II is about 0.78 times that of R2HCA-EMOA. When $\theta = 0.5$, the runtime of R2HCA-EMOA-II is about 0.61 times that of R2HCA-EMOA. When $\theta = 0.8$, this number is about 0.37. Similar results are obtained for other test problems.

Based on Table 1 and Table 2, we can see that if θ is properly specified (e.g., $\theta = 0.3, 0.5$), we can improve the efficiency of R2HCA-EMOA without sacrificing its performance, which suggests the usefulness of the proposed two strategies.

5 Conclusions

In this paper, we proposed two strategies to further improve the efficiency of R2HCA-EMOA. The first strategy is to randomly remove one individual in the last front when there are dominated solutions. The second strategy is to use a small number of direction vectors in the early stage and a large number of direction vectors in the late stage of the evolution. The experimental results verified the effectiveness of the proposed two strategies. The efficiency of R2HCA-EMOA was improved without sacrificing its hypervolume performance. The results in this paper make R2HCA-EMOA more attractive for practical use.

In the future, the improved R2HCA-EMOA can be tested on more test problems and real-world problems. It is also interesting to propose other strategies to further improve the efficiency of R2HCA-EMOA. For example, removing the non-dominated sorting procedure and using the R2 indicator variant (i.e., R_2^{HCA}) to evaluate all the solutions (not only the solutions in the last front) is an interesting way to go.

Acknowledgments. This work was supported by National Natural Science Foundation of China (Grant No. 62002152, 61876075), Guangdong Provincial Key Laboratory (Grant No. 2020B121201001), the Program for Guangdong Introducing Innovative and Entrepreneurial Teams (Grant No. 2017ZT07X386), Shenzhen Science and Technology Program (Grant No. KQTD2016112514355531), the Program for University Key Laboratory of Guangdong Province (Grant No. 2017KSYS008).

References

1. Bader, J., Zitzler, E.: HypE: an algorithm for fast hypervolume-based many-objective optimization. Evol. Comput. **19**(1), 45–76 (2011)
2. Beume, N., Naujoks, B., Emmerich, M.: SMS-EMOA: multiobjective selection based on dominated hypervolume. Eur. J. Oper. Res. **181**(3), 1653–1669 (2007)
3. Coello Coello, C.A., Reyes Sierra, M.: A study of the parallelization of a coevolutionary multi-objective evolutionary algorithm. In: Monroy, R., Arroyo-Figueroa, G., Sucar, L.E., Sossa, H. (eds.) MICAI 2004. LNCS (LNAI), vol. 2972, pp. 688–697. Springer, Heidelberg (2004). https://doi.org/10.1007/978-3-540-24694-7_71
4. Deb, K., Jain, H.: An evolutionary many-objective optimization algorithm using reference-point-based nondominated sorting approach, part I: solving problems with box constraints. IEEE Trans. Evol. Comput. **18**(4), 577–601 (2013)
5. Deb, K., Pratap, A., Agarwal, S., Meyarivan, T.: A fast and elitist multiobjective genetic algorithm: NSGA-II. IEEE Trans. Evol. Comput. **6**(2), 182–197 (2002)
6. Deb, K., Thiele, L., Laumanns, M., Zitzler, E.: Scalable test problems for evolutionary multiobjective optimization. In: Abraham, A., Jain, L., Goldberg, R. (eds.) Evolutionary Multiobjective Optimization. AI&KP, pp. 105–145. Springer, London (2005). https://doi.org/10.1007/1-84628-137-7_6

7. Emmerich, M., Beume, N., Naujoks, B.: An EMO algorithm using the hypervolume measure as selection criterion. In: Coello Coello, C.A., Hernández Aguirre, A., Zitzler, E. (eds.) EMO 2005. LNCS, vol. 3410, pp. 62–76. Springer, Heidelberg (2005). https://doi.org/10.1007/978-3-540-31880-4_5
8. Hansen, M.P., Jaszkiewicz, A.: Evaluating the quality of approximations to the non-dominated set. IMM, Department of Mathematical Modelling, Technical University of Denmark (1998)
9. Huband, S., Hingston, P., Barone, L., While, L.: A review of multiobjective test problems and a scalable test problem toolkit. IEEE Trans. Evol. Comput. **10**(5), 477–506 (2006)
10. Ishibuchi, H., Setoguchi, Y., Masuda, H., Nojima, Y.: Performance of decomposition-based many-objective algorithms strongly depends on pareto front shapes. IEEE Trans. Evol. Comput. **21**(2), 169–190 (2017)
11. Jiang, S., Zhang, J., Ong, Y.S., Zhang, A.N., Tan, P.S.: A simple and fast hypervolume indicator-based multiobjective evolutionary algorithm. IEEE Trans. Cybern. **45**(10), 2202–2213 (2015)
12. Shang, K., Ishibuchi, H.: A new hypervolume-based evolutionary algorithm for many-objective optimization. IEEE Trans. Evol. Comput. **24**(5), 839–852 (2020)
13. Shang, K., Ishibuchi, H., Ni, X.: R2-based hypervolume contribution approximation. IEEE Trans. Evol. Comput. **24**(1), 185–192 (2020)
14. Shang, K., Ishibuchi, H., He, L., Pang, L.M.: A survey on the hypervolume indicator in evolutionary multi-objective optimization. IEEE Trans. Evol. Comput. **25**(1), 1–20 (2021)
15. Tian, Y., Cheng, R., Zhang, X., Jin, Y.: PlatEMO: a MATLAB platform for evolutionary multi-objective optimization [educational forum]. IEEE Comput. Intell. Mag. **12**(4), 73–87 (2017)
16. Van Veldhuizen, D.A.: Multiobjective evolutionary algorithms: classifications, analyses, and new innovations. Ph.D. thesis, Department of Electrical and Computer Engineering, Graduate School of Engineering, Air Force Institute of Technology, Wright-Patterson AFB, Ohio (1999)
17. While, L., Bradstreet, L., Barone, L.: A fast way of calculating exact hypervolumes. IEEE Trans. Evol. Comput. **16**(1), 86–95 (2012)
18. Zhang, Q., Li, H.: MOEA/D: a multiobjective evolutionary algorithm based on decomposition. IEEE Trans. Evol. Comput. **11**(6), 712–731 (2007)
19. Zitzler, E., Thiele, L., Laumanns, M., Fonseca, C., da Fonseca, V.: Performance assessment of multiobjective optimizers: an analysis and review. IEEE Trans. Evol. Comput. **7**(2), 117–132 (2003)

Pareto Front Estimation Using Unit Hyperplane

Tomoaki Takagi(✉), Keiki Takadama(✉), and Hiroyuki Sato(✉)

The University of Electro-Communications, 1-5-1 Chofugaoka, Chofu, Tokyo 182-8585, Japan
{tomtkg,h.sato}@uec.ac.jp, keiki@inf.uec.ac.jp

Abstract. This work proposes a method to estimate the Pareto front even in areas without objective vectors in the objective space. For the Pareto front approximation, we use a set of non-dominated points, objective vectors, in the objective space. To finely approximate the Pareto front, we need to increase the number of objective vectors. It is worth to estimate the Pareto front with a limited number of objective vectors. The proposed method uses the Kriging approximation and estimates the Pareto front using the unit hyperplane in the objective space. In the experiment using representative simple and complicated Pareto fronts derived from the DTLZ family, we visually show the estimation quality of the proposed method. Also, we show that the shape of the Pareto front and the distribution of sample objective vectors affect the estimation quality.

Keywords: Multi-objective optimization · Pareto front estimation · Kriging

1 Introduction

An advantage of the evolutionary multi-objective optimization is to acquire a solution set approximating the Pareto front in a single run [1]. In two-objective optimization, we approximate a line Pareto front with a set of points, which are objective vectors in the objective space. In three-objective optimization, we approximate a surface Pareto front with a set of points in the objective space. Thus, we need to approximate any-dimensional Pareto front with a set of points. To finely approximate the Pareto front, we need to increase the number of solutions, and it should be further boosted as the number of objectives increases. Also, in optimization problems with computationally expensive solution evaluation, it is not easy to improve the approximation quality of the Pareto front by generating and evaluating a number of solutions. Against this issue, it is worth to estimate the Pareto front even in areas where objective vectors do not exist in the objective space. When a rough shape of the Pareto front could be obtained, we can provide the decision-maker the objective trade-off that cannot be clarified only with a small number of objective vectors. Also, in reference-based search

H. Ishibuchi et al. (Eds.): EMO 2021, LNCS 12654, pp. 126–138, 2021.
https://doi.org/10.1007/978-3-030-72062-9_11

algorithms such as R-NSGA-II [2] and iMOEA/D [3], a rough shape of the Pareto front helps to set an accurate reference point in the objective space. In the algorithms searching the entire Pareto front, the decomposition approach using a set of weight vectors has been promising recently [4,5]. The appropriate distribution of weight vectors depends on the shape of the Pareto front. When a rough shape of the Pareto front could be obtained, it helps to arrange appropriate weight vectors in the objective space.

This work aims to estimate the Pareto front even in areas without objective vectors in the objective space. The proposed method estimates the Pareto front using the unit hyperplane in the objective space. We use the Kriging approximation with the Design and Analysis of Computer Experiments (DACE) model [6]. The input of the model is a point on the unit hyperplane in the objective space. The output of the model is the distance from the origin point on the line passing through the origin point and the input point in the objective space. The Kriging interpolation based on the constructed model represents the estimated Pareto front. Also, the normally distributed confidence interval represents the estimation range at any point on the estimated Pareto front.

In the research area of evolutionary optimization, the Kriging approximation is frequently employed, especially for optimization problems with computationally expensive solution evaluation [7]. In this case, the input is a solution, the output is the objective function value, and the objective function is estimated. On the other hand, in this paper, we only consider the objective space without considering the decision variable space. In the proposed method, the input is a point on the unit hyperplane in the objective space, and the output is the distance from the origin point on the line passing through the origin point and the point on the unit hyperplane, and the Pareto front is estimated as the distance from the origin point for the line. In the experiment, we estimate representative simple and complicated Pareto fronts in two- and three-objective cases.

2 Kriging Approximation with DACE Model

The Kriging approximation with the Design and Analysis of Computer Experiments (DACE) model approximates responses to input vectors based on the Bayesian statistics [6]. The response, the output, $y(\boldsymbol{x})$ of an input vector $\boldsymbol{x} = (x_1, x_2, \ldots, x_n)$ is estimated by

$$\hat{y}(\boldsymbol{x}) = \mu(\boldsymbol{x}) + \epsilon(\boldsymbol{x}), \tag{1}$$

where, $\mu(\boldsymbol{x})$ is the mean value of the responses of the given samples, $\epsilon(\boldsymbol{x})$ is the probabilistic deviation from $\mu(\boldsymbol{x})$ based on the standard normal distribution. The accuracy of the response estimation $\hat{y}(\boldsymbol{x})$ on the DACE model depends on the distance from the given samples. In this paper, we use the Gaussian correlation function to calculate the correlations among elements $x_1, x_2, \ldots, x_n$ of input vector $\boldsymbol{x}$. The Gaussian correlation function is calculated with weight coefficients $\theta_1, \theta_2, \ldots, \theta_n$ for elements $x_1, x_2, \ldots, x_n$ in input vector $\boldsymbol{x}$. Characteristic of the DACE model is that estimation values and their estimation errors can be obtained at the same time.

3 Proposal: Pareto Front Estimation Using Unit Hyperplane

We propose a method to estimate the Pareto front by the Kriging approximation with the DACE model. The core idea is to estimate the Pareto front using the unit hyperplane in the objective space.

3.1 Method

This method uses objective vector $\boldsymbol{f} = (f_1, f_2, \ldots, f_m)$, which is a point in the m-dimensional objective space. For the Kriging-based Pareto front estimation, we use the L^1 unit vector of objective vector $\boldsymbol{f}$ as input and the L^1 norm of the objective vector $\boldsymbol{f}$ as output. The L^1 norm $||\boldsymbol{f}||$ of objective vector $\boldsymbol{f}$ is given by

$$||\boldsymbol{f}|| = \sum_{i=1}^{m} f_i. \tag{2}$$

The L^1 unit vector $\boldsymbol{e}$ of the objective vector $\boldsymbol{f}$ is given by

$$\boldsymbol{e} = \frac{\boldsymbol{f}}{||\boldsymbol{f}||}. \tag{3}$$

Note that the output is scalar $||\boldsymbol{f}||$, and the input is vector $\boldsymbol{e}_{1:m-1}$ involving $m-1$ elements $e_1, e_1, \ldots, e_{m-1}$ of $\boldsymbol{e}$. For a given set of sample objective vectors $\mathcal{F} = \{\boldsymbol{f}^1, \boldsymbol{f}^2, \ldots, \boldsymbol{f}^N\}$, we construct the DACE model. That is, Eq. (1) mentioned in the previous section is treated as

$$||\hat{\boldsymbol{f}}||(\boldsymbol{e}_{1:m-1}) = \mu(\boldsymbol{e}_{1:m-1}) + \epsilon(\boldsymbol{e}_{1:m-1}). \tag{4}$$

The sum of all the elements of L^1 unit vector $\boldsymbol{e}$ is always 1. The objective vector $\boldsymbol{f}$ is obtained from the L^1 norm $||\boldsymbol{f}||$ and the L^1 unit vector $\boldsymbol{e}$ as

$$\boldsymbol{f} = ||\boldsymbol{f}|| \cdot \boldsymbol{e}. \tag{5}$$

For the constructed DACE model, we can obtain the estimated objective vectors $\hat{\boldsymbol{f}}$ by specifying unit vectors $\boldsymbol{e}$, which is directions in the objective space.

3.2 Example

Figure 1 show an example of the proposed method on a $m = 2$ objective case.

Figure 1(a) shows the $m = 2$ dimensional objective space. The red filled circles are $N = 5$ sample objective vectors $\mathcal{F} = \{\boldsymbol{f}^1, \boldsymbol{f}^2, \ldots, \boldsymbol{f}^5\}$. We first calculate their L^1 norms $||\boldsymbol{f}^1||, ||\boldsymbol{f}^2||, \ldots, ||\boldsymbol{f}^5||$ and their L^1 unit vectors $\boldsymbol{e}^1, \boldsymbol{e}^2, \ldots, \boldsymbol{e}^5$. As mentioned above, since the input vector $\boldsymbol{e}_{1:m-1}$ involves $m-1$ elements of unit vector $\boldsymbol{e}$, the inputs becomes $e_1^1, e_1^2, \ldots, e_1^5$ in the $m = 2$ dimensional case.

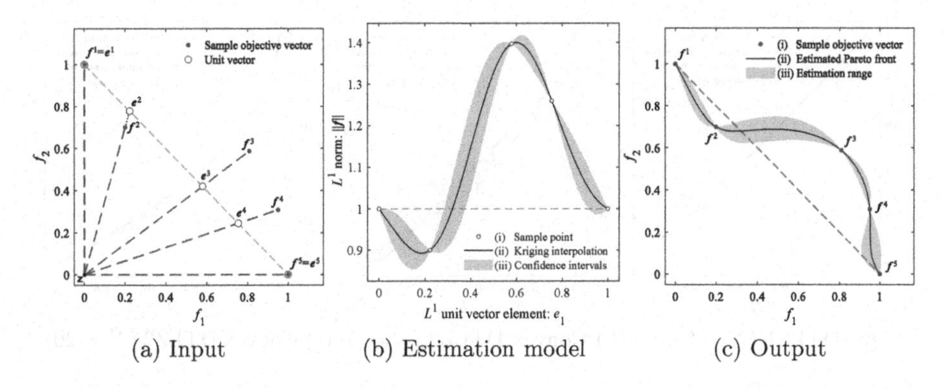

Fig. 1. Example of the proposed Pareto front estimation procedure (Color figure online)

Then, we construct a DACE model for the Kriging approximation of the Pareto front with these first elements of the L^1 unit vectors as inputs and L^1 norms as their outputs.

Figure 1(b) shows the relationship between the input unit element e_1 and the output $||\boldsymbol{f}||$ obtained by the constructed DACE model. In this figure, the red circles are sample points calculated from the sample objective vectors $\mathcal{F}$, the black line is the Kriging interpolation, which is the estimated L^1 norms by the constructed DACE model. The gray areas are the normal distributed confidence interval, which is the estimated error range of L^1 norms. This can be mapped to the $m = 2$ dimensional objective space as shown in Fig. 1(c). The black lines in Fig. 1(b) and (c) are corresponded, and the gray areas in Fig. 1(b) and (c) are corresponded. In Fig. 1(c), the black line is called the estimated Pareto front, and the gray area is called the estimation range in this paper.

3.3 Discussion

The proposed method is able to suggest the estimated areas of the Pareto front with a limited number of objective vectors in the objective space. However, note that the estimated areas are not Pareto dominance compatible. That is, the estimated areas may involve the dominating areas of the Pareto optimal points specified as the sample objective vectors $\mathcal{F}$. Also, although the proposed method assumes that the Pareto front continuously exists in the objective space, the actual Pareto front may be disconnected in the objective space. The proposed method provides an estimated objective point on every unit vector. However, there may be no actual objective points on a subset of unit vectors when the actual Pareto front is discontinuous.

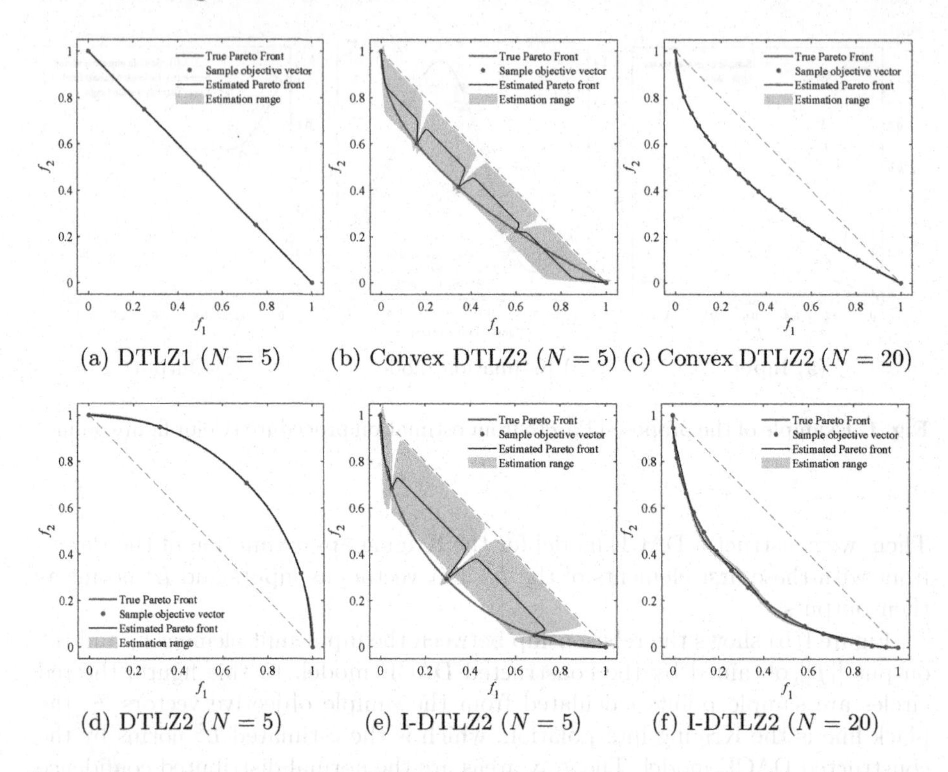

Fig. 2. Pareto front estimation of simple Pareto fronts in $m = 2$ objective case (Color figure online)

4 Experimental Results and Discussion

4.1 Experimental Settings

We apply the proposed method to several Pareto fronts with different shapes. As simple Pareto fronts, we use plane ones from DTLZ1 [8], convex ones from the convex DTLZ2 [5] and the inverted DTLZ2 (I-DTLZ2) [9], and concave ones from DTLZ2 [8]. As complicated Pareto fronts, we use disconnected one from DTLZ7 [8] and mixed shaped ones from C3-DTLZ1 and C3-DTLZ4 [5].

In the Kriging approximation with the DACE model, we use the zero-order polynomial regression function and the Gaussian correlation function to build the DACE model. The $m-1$ dimensional weight coefficients $\boldsymbol{\theta} = (\theta_1, \theta_2, \ldots, \theta_{m-1})$ is obtained by the constrained non-linear least-squares fitting in the range $[10^{-3}, 10^{+3}]^{m-1}$ with the initial $[10^0]^{m-1}$.

4.2 Simple Pareto Fronts in Two-Objective Case

Figure 2 shows the estimated Pareto fronts with simple shapes on $m = 2$ objective case. Each figure shows the objective space, the red points are the sample objective vectors $\mathcal{F}$ used for the Pareto front estimation, and the blue line is

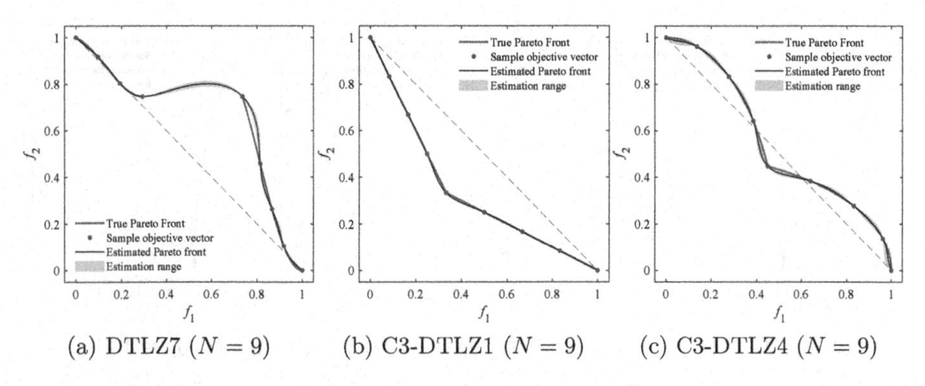

(a) DTLZ7 ($N = 9$) (b) C3-DTLZ1 ($N = 9$) (c) C3-DTLZ4 ($N = 9$)

Fig. 3. Pareto front estimation of complicated Pareto fronts in $m = 2$ objective case (Color figure online)

the Pareto front, which should be estimated. In this experiment, we picked the red points, the sample objective vectors $\mathcal{F}$, from the blue Pareto front. Therefore, the red points are on the blue Pareto front. The black line is the estimated Pareto front. The blue Pareto front and the back estimated Pareto front should be overlapped. Also, the gray areas are the estimation ranges. The gray areas should be small since the smaller gray areas, the lower the uncertainty of the Pareto front estimation.

Figure 2(a) is the result on the plane Pareto front derived from DTLZ1. This case has $N = 5$ sample objective vectors shown as the red points. In this case, we see that the black estimated Pareto front and the blue Pareto front are overlapped visually. Also, the gray estimation ranges are small. This result reveals that the proposed Pareto front estimation works well on this case with $N = 5$ sample objective vectors. We see a similar tendency in Fig. 2(d), which is the result of the concave Pareto front derived from DTLZ2.

Figure 2(b) and (e) are the results with $N = 5$ sample objective vectors on convex Pareto fronts derived from the convex DTLZ2 [5] and I-DTLZ2 [9], respectively. From these figures, we see that the black estimated Pareto front is drastically changed as the distance from the red sample point increases. Figure 2(c) and (f) are the results with $N = 20$ sample objective vectors on the same convex Pareto fronts of the convex DTLZ2 and I-DTLZ2, respectively. From these results, we see that the drastic change of the black estimated Pareto front is suppressed and the estimation ranges are contracted compared with the case with $N = 5$ sample objective vectors shown in Fig. 2(b) and (e).

Thus, the estimation accuracy of the Pareto front is affected by the shape of the Pareto front. However, the gray estimation ranges can show the uncertainty of the Pareto front estimation. The additional sample objective vectors to these areas can reduce the uncertainty of the Pareto front estimation.

4.3 Complicated Pareto Fronts in Two-Objective Case

Figure 3 shows the estimated Pareto fronts with complicated shapes on $m = 2$ objective case. Since the proposed method with $N = 5$ sample objective vectors

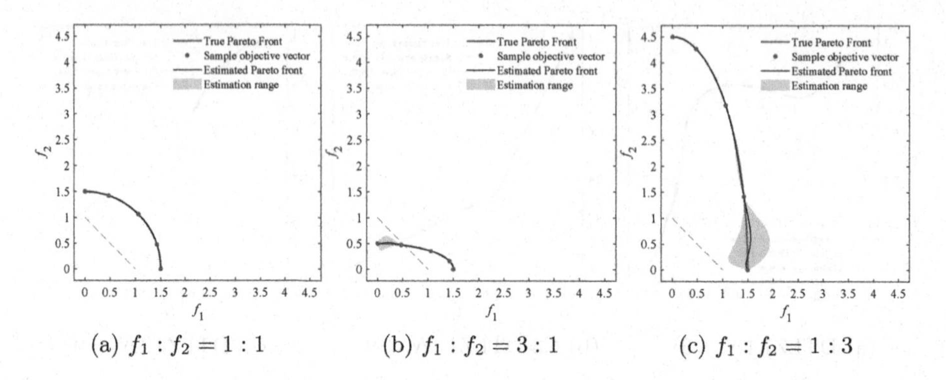

Fig. 4. Pareto front estimation of Pareto fronts with different objective value ranges (Color figure online)

are not enough for the Pareto front estimation in these cases, Fig. 3 shows the results with $N = 9$ sample objective vectors.

Figure 3(a) is the result of a disconnected Pareto front derived from DTLZ7. We see that the blue Pareto front is disconnected around the line $f_2 = 0.748$. Since the proposed method continuously estimates the Pareto front in the objective space, we see that the black estimated Pareto front exists in the gap area of the blue Pareto front. However, we see that a red sample objective vector dominates the estimated Pareto front in the gap area.

Figure 3(b) and (c) are results of the Pareto fronts derived from C3-DTLZ1 and C3-DTLZ4. We see that each shape of the blue Pareto fronts drastically changes at a knee point in the center of the Pareto front. Since the proposed method estimates the Pareto front as a smooth line, the estimation errors around the center of the Pareto front are larger than other areas. We see the importance of having red sample objective vectors around such the knee point to improve the Pareto front estimation accuracy.

4.4 Pareto Front with Different Objective Value Ranges

Figure 4 shows the estimated Pareto fronts with different objective value ranges on $m = 2$ objective case. In this section, we focus on the concave Pareto front of DTLZ2 previously shown in Fig. 2(d) as the baseline.

Figure 4(a) is the case that both objective values of the baseline Pareto front shown in Fig. 2(d) are multiplied by 1.5. That is, objective values of the Pareto front in Fig. 4(a) is evenly increased from the baseline Pareto front in Fig. 2(d), and the ratio of f_1 to f_2 is 1:1. From Fig. 4(a), we see that the black estimated Pareto front and the blue Pareto front are overlapped visually, and the estimation ranges are narrow. That is, such an even scale change of objective values does not affect the estimation accuracy.

Figure 4(b) is the case that f_2 of the Pareto front in Fig. 4(a) is divided by 3. That is, the ratio of f_1 to f_2 is 3:1. In the left part of the Pareto front of Fig. 4(b), we see that the black estimated Pareto front differs from the blue

Pareto front, and the estimation range is large compared with other part of the Pareto front. Thus, the estimation accuracy of the Pareto front is biased when the objective value ranges of f_1 and f_2 are different. Figure 4(c) is the case that f_2 of the Pareto front in Fig. 4(a) is multiplied by 3. That is, the ratio of f_1 to f_2 is 1:3. In the right part of the Pareto front of Fig. 4(c), the estimation errors are emphasized.

In this way, since the difference of objective value ranges affects the estimation accuracy, it is desirable to normalize the objective values before performing the proposed Pareto front estimation.

4.5 Impacts of Objective Vector Sampling

We compare three sampling methods of objective vectors. They are the simplex-lattice design (SLD) [10], the incremental lattice design (ILD) [11], and the Hammersley method [12].

Figure 5 shows $m = 3$ dimensional objective space. Each figure involves $N = 10$ red points, which are sample objective vectors. Since each sampler generates points on the unit hyperplane, the generated points are transformed onto the unit sphere of the concave DTLZ2. Each sampler has characteristics with its point distribution. SLD regularly arranges points in the objective space, and the nine points out of the total ten points are on the edge of the objective space. ILD also regularly arranges points in the objective space. However, its number of points on the edge of the objective space is smaller than SLD. The Hammersley method irregularly arranges points in the objective space and does not provide any points on the edge of the objective space. Also, in Fig. 5, the surface is the estimated Pareto front, and the estimation error is represented as the heatmap. In the case of the Hammersley method, we see that the estimation accuracy of the Pareto front in the extreme areas around the f_1, f_2, and f_3 axes is lower than the other two samplers. SLD and ILD involve three unit vectors $(1, 0, 0)$, $(0, 1, 0)$, and $(0, 0, 1)$ as sample objective vectors, and the Hammersley method does not. This difference affects the estimation quality of the extreme area of the Pareto front.

Figure 6 shows the estimated L^1 norm $||\hat{\boldsymbol{f}}||$ in the model space, $e_1 - e_2$ space. Figure 7 shows the estimated error in $e_1 - e_2$ space. Each figure involves the transformed points of the sample objective vectors shown in Fig. 5 as the red points. Estimated $||\hat{\boldsymbol{f}}||$ and errors are respectively represented as the heatmap. Since input vector $\boldsymbol{e}_{1:m-1} = (e_1, e_2)$ are always on the $m = 3$ dimensional unit hyperplane in the objective space, the plots are limited to the half lower left of the $e_1 - e_2$ space. From these results, we see that the differences among objective vector sets sampled by the three methods, the estimated $||\hat{\boldsymbol{f}}||$, and the estimated errors visually.

Table 1 shows the quantitative results of the three samplers. Table 1 involves θ_1, θ_2, and $\mu(e_1, e_2)$ of the constructed DACE model. We see that $\theta_1 = \theta_2$ in ILD but $\theta_1 \neq \theta_2$ in SLD and the Hammersley method. In this example, the Pareto front has the axisymmetric characteristic that the L^1 norm does not change when e_1 and e_2 are switched. That is, the correlation weight coefficients θ_1 and

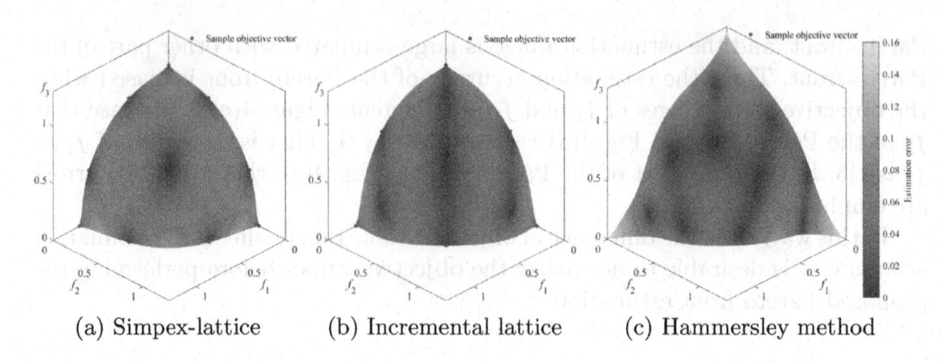

(a) Simpex-lattice (b) Incremental lattice (c) Hammersley method

Fig. 5. Pareto front estimation with three objective vector samplers (objective space) (Color figure online)

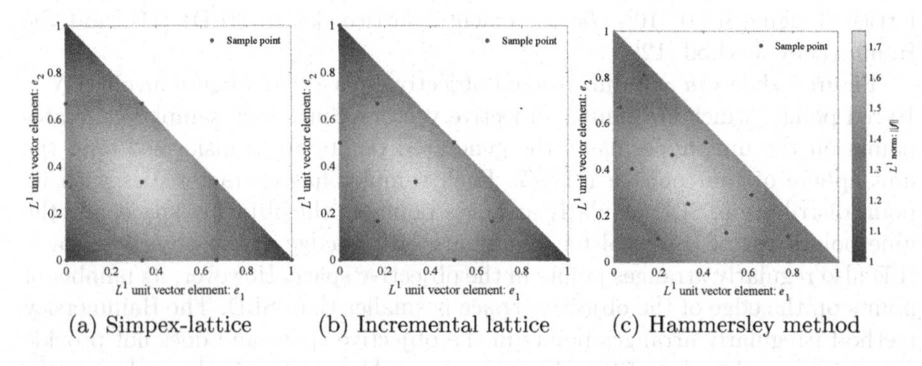

(a) Simpex-lattice (b) Incremental lattice (c) Hammersley method

Fig. 6. Estimated L^1 norm $||\hat{\boldsymbol{f}}||$ with three objective vector samplers (model space) (Color figure online)

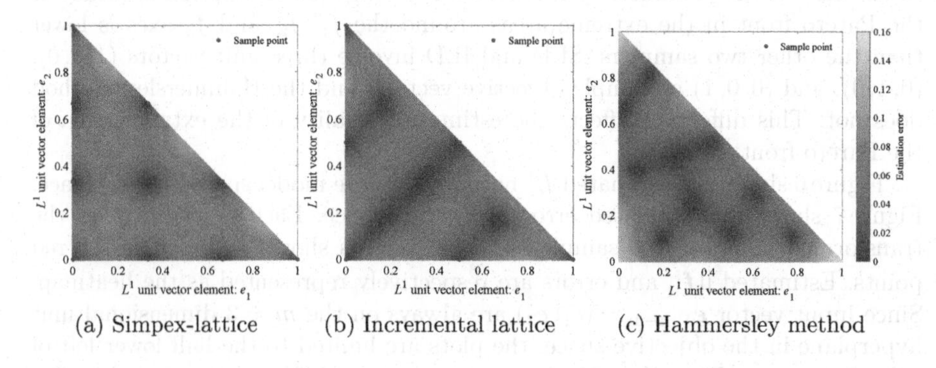

(a) Simpex-lattice (b) Incremental lattice (c) Hammersley method

Fig. 7. Estimated error with three objective vector samplers (model space) (Color figure online)

θ_2 should be the same. From this viewpoint, the ILD sampler is better than the other two samplers in this example. $\mu(e_1, e_2)$ is the mean L^1 norm of the sample objective vectors. In this example, $\mu(e_1, e_2)$ increases as the number of sample

Table 1. Results on different samplers of objective vectors

Objective value sampler	Constructed model			L^1 norm $\|\|\hat{f}\|\|$		Estimation error	
	θ_1	θ_2	$\mu(e_1, e_2)$	Min	Max	Ave	Max
SLD [10]	0.87055	0.93303	1.27820	1.00000	1.73244	0.05728	0.10303
ILD [11]	0.70711	0.70711	1.32173	1.00000	1.73213	0.04442	0.09415
Hammersley [12]	0.84090	0.70711	1.45998	1.18931	1.73207	0.04636	0.15510

objective vectors located inside of the Pareto front increases since the distance between the unit hyperplane and the unit sphere increases toward the center of the Pareto front. The sample objective vectors of SLD tend to be distributed outside of the Pareto front, and its $\mu(e_1, e_2)$ becomes the lowest among the three samplers. The Hammersley method does not have any sample objective vectors on the edge of the Pareto front, and its $\mu(e_1, e_2)$ becomes the highest among three samplers. The objective vectors sampled by ILD are balanced between the inside and the outside of the Pareto front, and its $\mu(e_1, e_2)$ becomes the middle among the three samplers.

Table 1 also involves the minimum and the maximum values of the estimated L^1 norm $||\hat{f}||$. The true minimum value is 1 on the three corners of the Pareto front, and the true maximum value is $\sqrt{m} = \sqrt{3} = 1.73205$ at the center of the concave Pareto front on the unit sphere. The minimum estimated $||\hat{f}||$ values of SLD and ILD are 1 since they have unit vectors $(1, 0, 0)$, $(0, 1, 0)$, and $(0, 0, 1)$ as sample objective vectors at the extreme three corners of the Pareto front. The maximum estimated $||\hat{f}||$ value of SLD is the highest among three objective vector samplers and higher than the true maximum value $\sqrt{m} = \sqrt{3} = 1.73205$. That is, the estimation accuracy of SLD in the central part of the Pareto front is lower than ILD. In the case of the Hammersley method, the minimum estimated $||\hat{f}||$ is higher than 1 since sample objective vectors on the edge of the Pareto front are not included. Also, from Fig. 6(c), the color gradation of the estimated $||\hat{f}||$ is not axisymmetric with respect to the line $e_1 = e_2$ since the sample objective vectors irregularly arranged. Table 1 also involves the estimated errors. We see the ILD achieves the best in the viewpoints of the average and the maximum estimated errors.

4.6 Simple Pareto Fronts in Three-Objective Case

Figure 8 shows the estimated Pareto fronts with simple shapes in $m = 3$ objective case. ILD generates $N = 64$ sample objective vectors on the unit hyperplane and transforms them matched to each Pareto front shape. The color gradation surface is the estimated Pareto front, and the color gradation represents the estimation errors.

From the results, we see that the plane Pareto front of DTLZ1 and the concave one of DTLZ2 can be visually observed with relatively small prediction errors. Although the convex DTLZ2 also has a smooth Pareto front, the estimated Pareto front cannot be smooth, and the estimation errors are larger than DTLZ1 and DTLZ2 cases. Thus, the estimation accuracy is influenced by the

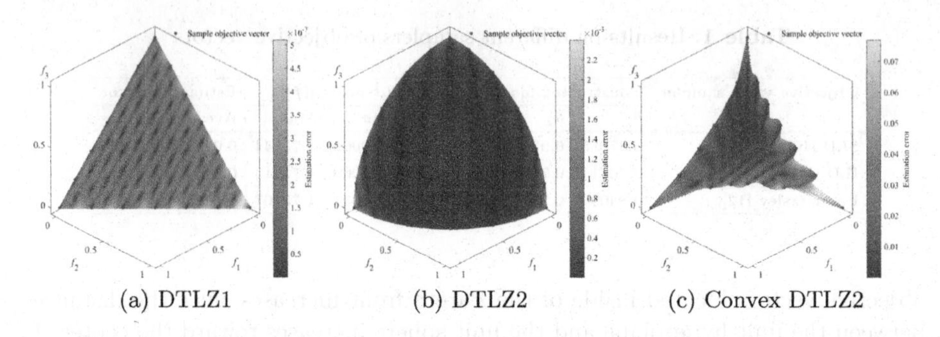

(a) DTLZ1 (b) DTLZ2 (c) Convex DTLZ2

Fig. 8. Pareto front estimation of simple Pareto fronts in $m = 3$ objective case

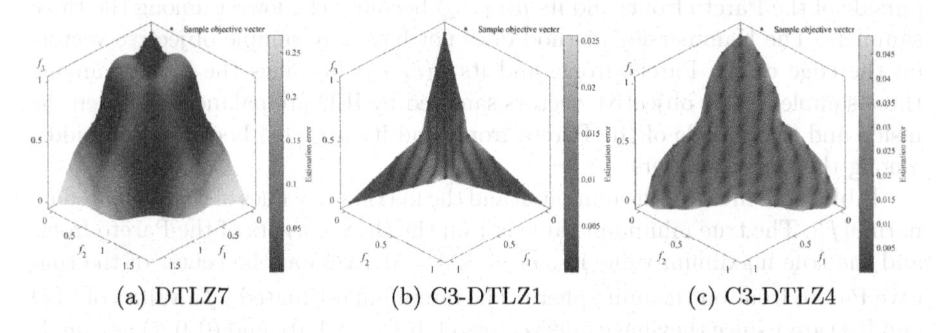

(a) DTLZ7 (b) C3-DTLZ1 (c) C3-DTLZ4

Fig. 9. Pareto front estimation of complicated Pareto fronts in $m = 3$ objective case

shape of the Pareto front. However, areas with a low estimation accuracy can be seen by the prediction errors with the color gradation, and it suggests adding new sample objective vectors to these areas.

4.7 Complicated Pareto Fronts in Three-Objective Case

Figure 9 shows the estimated Pareto fronts with complicated shapes in $m = 3$ objective case. Each figure involves $N = 64$ sample objective vectors picked from its Pareto front.

DTLZ7 has the disconnected Pareto front. Since the proposed method continuously estimates the Pareto front, the disconnected parts cannot be found clearly. However, areas with high estimation errors suggest the disconnected parts. In this case, areas with estimation errors greater than about 0.01 will be disconnected parts. The Pareto fronts of C3-DTLZ1 and C3-DTLZ4 have areas that objective values are drastically changed. In these areas, the prediction error increases compared with areas gradually changing objective values. However, the entire pictures of the Pareto fronts can be observed by the proposed method with $N = 64$ sample objective vectors.

5 Conclusions

To estimate the Pareto front even in areas without objective vectors, we proposed a method to estimate the Pareto front using the unit hyperplane in the objective space. The proposed method is based on the Kriging approximation with the DACE model. The proposed method represents not only the estimated Pareto front but also the estimation confidence interval at any point on the estimated Pareto front. The experimental results using simple and complicated Pareto fronts derived from the DTLZ problem family visually showed the estimation quality of the proposed method. The experimental results also showed that the shape of the Pareto front and the distribution of the sample objective vectors affected the estimation quality.

As future works, we are planning to conduct experiments on problems with many objectives. Also, we will build a visualization method of a high dimensional Pareto front to show the estimated Pareto front and the estimated range simultaneously.

References

1. Deb, K.: Multi-Objective Optimization Using Evolutionary Algorithms. Wiley, Hoboken (2001)
2. Deb, K., Sundar, J.: Reference point based multi-objective optimization using evolutionary algorithms. In: 2006 Genetic and Evolutionary Computation (GECCO 2006), pp. 635–642 (2006)
3. Gong, M., Liu, F., Zhang, W., Jiao, L., Zhang, Q.: Interactive MOEA/D for multi-objective decision making. In: 2011 Genetic and Evolutionary Computation (GECCO 2011), pp. 721–728 (2011)
4. Zhang, Q., Li, H.: MOEA/D: a multiobjective evolutionary algorithm based on decomposition. IEEE Trans. Evol. Comput. **11**(6), 7120–731 (2007)
5. Deb, K., Jain, H.: An evolutionary many-objective optimization algorithm using reference-point based nondominated sorting approach, part I: solving problems with box constraints. IEEE Trans. Evol. Comput. **18**(4), 577–601 (2014)
6. Sacks, J., Welch, W.J., Mitchell, T.J., Wynn, H.P.: Design and analysis of computer experiments. Stat. Sci. **4**(4), 409–435 (1989)
7. Jin, Y.: Surrogate-assisted evolutionary computation: recent advances and future challenges. Swarm Evol. Comput. **1**(2), 61–70 (2011)
8. Deb, K., Thiele, L., Laumanns, M., Zitzler, E.: Scalable multi-objective optimization test problems. In: 2002 IEEE Congress on Evolutionary Computation (CEC 2002), pp. 825–830 (2002)
9. Jain, H., Deb, K.: An evolutionary many-objective optimization algorithm using reference-point based nondominated sorting approach, part II: handling constraints and extending to an adaptive approach. IEEE Trans. Evol. Comput. **18**(4), 602–622 (2014)
10. Das, I., Dennis, J.E.: Normal-boundary intersection: a new method for generating the Pareto surface in nonlinear multicriteria optimization problems. SIAM J. Optim. **8**(3), 631–657 (1998)

11. Takagi, T., Takadama, K., Sato, H.: Incremental lattice design of weight vector set. In: 2020 Genetic and Evolutionary Computation Conference Companion (GECCO 2020), pp. 1486–1494 (2020)
12. Molinet Berenguer, J.A., Coello Coello, C.A.: Evolutionary many-objective optimization based on Kuhn-Munkres' algorithm. In: Gaspar-Cunha, A., Henggeler Antunes, C., Coello, C.C. (eds.) EMO 2015. LNCS, vol. 9019, pp. 3–17. Springer, Cham (2015). https://doi.org/10.1007/978-3-319-15892-1_1

Towards Multi-objective Co-evolutionary Problem Solving

Anirudh Suresh(✉), Kalyanmoy Deb, and Vishnu Naresh Boddeti

Michigan State University, East Lansing, MI, USA
{suresha2,kdeb,vishnu}@msu.edu

Abstract. Co-evolutionary algorithms deal with two co-evolving populations, each having its own objectives. The linking of the two populations comes from the fact that the evaluation of the objective of a member of the first population requires a companion member from the second population, and vice versa. These algorithms are of great interest in cooperative and competing games and search tasks in which multiple agents having different interests are in play. While there have been significant studies devoted to single-objective co-evolutionary optimization algorithms and applications, they have not been paid much attention when each population must be evolved for more than one conflicting objectives. In this paper, we propose a multi-objective co-evolutionary (MOCoEv) algorithm and present its working on two interesting problems. This proof-of-principle study suggests that the presence of multiple Pareto solutions for each population and the ensuing multi-criterion decision-making complexities make the MOCoEv research and application be challenging. This paper should spur immediate further attention to multi-objective co-evolutionary problem solving studies.

Keywords: Co-evolutionary algorithm · Multi-payoff game · Multi-objective optimization · Multi-criterion decision making

1 Introduction

Many problems in practice, particularly in secure and trustworthy system design, must be considered from the point of view of two independent agents who have either altruistic or adversarial interests [10]. Recent studies in generative adversarial networks (GANs) [9] in computer vision literature in arriving at deep neural networks (DNNs) for generating realistic fake images, or the dynamics of adversarial attacks [13] and defenses of DNNs are some examples. Designing power systems, for example, from possible adversarial attacks require a model of generating plausible attacks so that a better secured system can be developed. Often, such a consideration leads to an 'arms race' in evolving defender-and-attacker systems simultaneously. Another wide application of such co-evolutionary system design is in multi-agent games [11].

These problems give rise to two inter-connected sub-problems, each having its own decision variables (or strategies) which must be searched for a better

H. Ishibuchi et al. (Eds.): EMO 2021, LNCS 12654, pp. 139–151, 2021.
https://doi.org/10.1007/978-3-030-72062-9_12

combination to maximize a goal. A difficulty arises due to the fact that the evaluation of the goal (or objective) requires a combination of both sub-problem's decision variables. Unfortunately, the second sub-problem's decision variables cannot be controlled by the first sub-problem and vice versa. Thus, these problems are usually posed as co-evolutionary optimization problems and two populations (one dealing with each sub-problem) are evolved with a close interaction with each other in an evolutionary framework [14]. Competitive co-evolutionary methods have been used to solve single objective min-max problems [1]. Some recent works have focused on finding the objective trade-offs for multi-objective games (MOGs) and have proposed a new solution concept based on rational strategies from game theory [6–8,11]. Eisenstadt et al. [6] used co-evolution to compute multiple rational strategies in an MOG. No efficient aggregate performance is computed to reduce the cardinality of the final solution set, but an elaborate multi-objective decision making (MCDM) approach [5] is used to choose the desired solutions. Żychowski et al. [15] used memetic co-evolution to solve MOGs using the nadir point (worst case) as the representative vector of rational fronts.

Despite all these studies, co-evolutionary frameworks have not been studied enough for optimizing multiple conflicting objectives efficiently at each sub-problem. Considering multiple conflicting objectives in a co-evolutionary framework introduces a number of challenges. First, each sub-problem must find a set of Pareto-optimal (PO) solutions optimizing its own objectives well. This requires a widely diverse population to be maintained at each sub-problem throughout the optimization process. This calls for developing efficient methodologies for computing aggregate fitness vectors that would emphasize survival of a diverse population. Second, the final PO solutions from each sub-population must be analyzed using a multi-criterion decision-making approach and decision-makers' preferences to pick a single preferred solution. We address all the above concerns in this paper.

In the remainder, we define a multi-objective co-evolutionary (MOCoEv) problem in Sect. 2 by highlighting the challenges they introduce to any algorithm. Then, in Sect. 3, we present our proposed MOCoEv algorithm in detail. Two multi-criterion decision-making (MCDM) approaches are discussed in this section. Results on two problems are presented in Sect. 4. Finally, conclusions are discussed in Sect. 5.

2 Multi-objective Co-evolutionary Problems

Without loss of generality, we consider two co-evolving problems in this study. Problems P_1 and P_2 deal with variable vectors $\mathbf{x}$ and $\mathbf{y}$ of sizes n_1 and n_2, respectively. The objective and constraint functions for P_i is given as $\mathbf{f}^{(i)}(\mathbf{x}, \mathbf{y})$ of size M_i and $\mathbf{g}^{(i)}(\mathbf{x}, \mathbf{y}) \leq \mathbf{0}$ of size J_i. The optimization problem formulation is given below (for simplicity, we do not consider equality constraints here):

Problem P_1	Problem P_2
$\min_{\mathbf{x}} \left(f_1^{(1)}(\mathbf{x},\mathbf{y}),\ldots,f_{M_1}^{(1)}(\mathbf{x},\mathbf{y})\right),$ $s.t.\ g_j^{(1)}(\mathbf{x},\mathbf{y}) \le 0,\ j=1,\ldots,J_1,\quad (1)$ $x_i^{(L)} \le x_i \le x_i^{(U)},\ i=1,\ldots,n_1.$	$\min_{\mathbf{y}} \left(f_1^{(2)}(\mathbf{x},\mathbf{y}),\ldots,f_{M_2}^{(2)}(\mathbf{x},\mathbf{y})\right),$ $s.t.\ g_j^{(2)}(\mathbf{x},\mathbf{y}) \le 0,\ j=1,\ldots,J_2,\quad (2)$ $y_j^{(L)} \le y_j \le y_j^{(U)},\ j=1,\ldots,n_2.$

It is expected that the objectives of each problem are in conflict with each other, thereby resulting in its own Pareto-optimal (PO) solutions. However, since each objective function of a Problem depends on a variable vector of the other Problem, each multi-objective optimization problem cannot be optimized independently. In other words, if we ignore P_2, and optimize P_1 alone by including $\mathbf{y}$ in its variable set, the resulting PO front is expected to be better than the PO front of the above MOCoEv problem P_1. The same is true for the independently optimized P_2 as well. The solution consists of PO front for each problem, with choices and objective preferences of the other problem unknown. This necessitates the use of a co-evolutionary search and optimization algorithm to solve such inter-linked problems and find a compromised PO front for each problem simultaneously, which is the focus of this paper.

2.1 Cooperative and Competing Problems

Let us first consider that there is a single objective function for each problem. Thus, Problem 1 will correspond to a single optimal solution $\mathbf{z}^{(1),*} = (\mathbf{x}^{(1),*}, \mathbf{y}^{(1),*})$ and Problem 2 will correspond to another optimal solution $\mathbf{z}^{(2),*} = (\mathbf{x}^{(2),*}, \mathbf{y}^{(2),*})$. If these two solutions are exactly the same (or close to each other in their respective variable spaces), a co-evolutionary framework will work in a cooperative manner, thereby making the problem relatively easier to solve.

On the other hand, if these two optimal solutions are distant in their respective variable spaces, a standard co-evolutionary framework (which emphasizes its own variables only based on its performance on its own objective and constraint functions) must preserve population members close to each of these optima. Population 1 has no motivation to preserve points close to $\mathbf{z}^{(2),*}$, since its goal is to minimize $\mathbf{f}_1^{(1)}$ with its constraints, and Population 2 has no motivation for saving solutions close to $\mathbf{z}^{(1),*}$. It is clear that to solve competing problems, the survival of solutions in each population must also consider the importance of its own variables in the other population.

For multi-objective co-evolutionary problems, Population 1 will correspond to a set of PO solutions ($\mathbf{Z}^{(1)} = (\mathbf{z}^{(1),i},\ i=1,\ldots,N_1)$). Similarly, Population 2 will correspond to another set of PO solutions ($\mathbf{Z}^{(2)} = (\mathbf{z}^{(2),i},\ i=1,\ldots,N_2)$). In a cooperative MOCoEv problem, these two sets are expected to be close to each other in their respective variable spaces and in a completely competing MOCoEv problem, the two sets are expected to be far away from each other. However, in a generic case, it is expected that the two sets will have some common solutions and some other solutions that are different from each other. To find and maintain two sets of solutions, each population must, not only emphasize its own best solutions, but also must emphasize its own solutions corresponding to good solutions of the other population. This co-evolutionary problem is different from a traditional many-objective problem with objectives from both populations. In the many objective problem, the solution would involve explicit trade-off between objectives of each population. However, in the co-evolutionary problem, the objectives of the two populations implicitly reach a compromise that helps in achieving a selfish trade-off amongst each population's own objectives. Next, we present a new multi-objective co-evolutionary algorithm based on a well-known EMO algorithm – NSGA-II [4].

3 Proposed Multi-objective Co-evolutionary (MOCoEv) Algorithm

In an MOCoEv algorithm, there are two interacting populations each containing its own decision variables. Like in other evolutionary algorithms, initial populations $Pop_1^{(0)}$ and $Pop_2^{(0)}$ of two problems can contain random solutions $\mathbf{x}$ and $\mathbf{y}$ of sizes N_1 and N_2, respectively. At iteration t, genetic operations can be performed on $Pop_1^{(t-1)}$ as usual using its own objectives and constraints and a new population $Pop_1^{(t)}$ can be obtained. Here, we follow the NSGA-II's operations for this purpose. A total τ_1 iterations can be continued as above before the next population is updated for a consecutive τ_2 iterations. Then, again Population 1 can be updated for another τ_1 iterations. This process can continue until a termination condition is satisfied. For simplicity, we ignore constraints in presenting our proposed algorithm. A pseudo-algorithm of the MOCoEv procedure is presented in Algorithm 1.

However, there is a complication in the evaluation process which we discuss next. For evaluating a single population member $\mathbf{x}^{(1),i}$ of $P_1^{(t)}$, we need a specific variable vector $\mathbf{y}$ from the second population. This is where a number of strategies can be adopted in an MOCoEv algorithm. We consider three ways to evaluate a solution:

1. **Mean Aggregate Fitness:** Population member $\mathbf{x}^{(1),k}$ is paired with every member $\mathbf{y}^{(2),l}$ of the second population one at a time and objectives are evaluated. Then, a mean fitness value of i-th objective function of Population 1 is computed as follows:

$$F_i(\mathbf{x}^{(1),k}) = \frac{1}{N_2}\sum_{l=1}^{N_2} f_i(\mathbf{x}^{(1),k}, \mathbf{y}^{(2),l}). \tag{3}$$

Figure 1 illustrates the procedure for a specific solution in the first population for the entire population members of the second population. Similarly, the fitness of each member of the second population can also be computed by averaging the respective objective values over all Population 1 members. Note that instead of considering all N_2 population members as above, the equation can be used for non-dominated solutions of the second population only. This would be particularly useful for cooperative MOCoEv problems.

Algorithm 1: Proposed multi-objective co-evolutionary algorithm.

Input: Population sizes N_1 and N_2 for Pop_1 and Pop_2, number of generations T, number of generations of each population τ_1 and τ_2
Output: Final populations $Pop_1^{(T)}$ and $Pop_2^{(T)}$
Initialize populations $Pop_1^{(0)}$ and $Pop_2^{(0)}$
for *gen = 0 to T or another termination condition not satisfied* **do**
  $Pop_1^{(0)} = Pop_1^{(gen)}$, $Pop_2^{(0)} = Pop_2^{(gen)}$
  for $t = 1$ *to* τ_1 **do**
    Generate offspring $Q_1^{(t)}$ for $Pop_1^{(t-1)}$ using selection and variation operators
    Merge and redefine $Pop_1^{(t)} = Pop_1^{(t-1)} \cup Q_1^{(t)}$
    Evaluate $Pop_1^{(t)}$ with all members of $Pop_2^{(gen)}$
    Survive elite population of $Pop_1^{(t)}$ of size N_1
  end
  $Pop_1^{(gen)} = Pop_1^{(t)}$
  for $t = 1$ *to* τ_2 **do**
    Generate offsprings $Q_2^{(t)}$ for $Pop_2^{(t-1)}$ using selection and variation operators
    Merge and redefine $Pop_2^{(t)} = Pop_2^{(t-1)} \cup Q_2^{(t)}$
    Evaluate $Pop_2^{(t)}$ with all members of $Pop_1^{(gen)}$
    Survive elite population of $Pop_2^{(t)}$ of size N_2
  end
  $Pop_2^{(gen)} = Pop_2^{(t)}$
end

2. **Best Aggregate Fitness:** Instead of the mean value, the best objective value can be assigned as a fitness: $F_i(\mathbf{x}^{(1),k}) = \min_{l=1}^{N_2} f_i(\mathbf{x}^{(1),k}, \mathbf{y}^{(2),l})$. This is useful for a greedy evaluation of a solution, if desired in a problem. Figure 1 shows the best aggregate fitness value.
3. **Worst Aggregate Fitness:** The worst objective value can be assigned as a fitness: $F_i(\mathbf{x}^{(1),k}) = \max_{l=1}^{N_2} f_i(\mathbf{x}^{(1),k}, \mathbf{y}^{(2),l})$. Figure 1 shows the worst aggregate fitness value. This will be particularly suitable for a pessimistic (conservative) evaluation of a solution, if suitable for a problem. For example, in min-max problems, this can be an effective aggregation function.

After the aggregate fitness function for each objective is computed by using one of the above strategies, they can be used for domination check and other niche-preserving operator of the EMO algorithm. Consider Fig. 2, in which the non-domination evaluation procedure using the mean aggregate fitness function is illustrated for Population 1. Every member is evaluated for both F_1 and F_2 using member of the second population. Then, the mean fitness vector is computed (shown with a star) for each member and is used for domination check. For the shown example, yellow and blue colored points of Population 1 belong to the first non-dominated (ND) front, followed by the brown point in the second ND

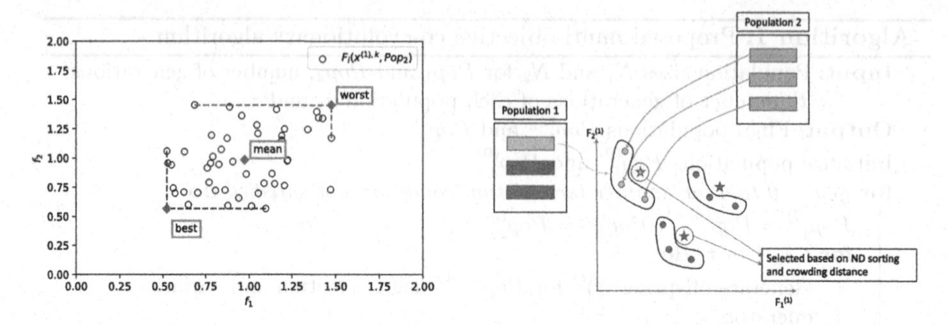

Fig. 1. Three aggregation procedures.

Fig. 2. Non-dominated sorting using mean aggregate fitness values. (Color figure online)

front. The crowding distance values for each member is also computed using the aggregate fitness values. Thus, the final trade-off set of solutions of each population ($\mathbf{Z}^{(1)}$ and $\mathbf{Z}^{(2)}$) will correspond to non-domination principle of the chosen aggregate fitness functions.

3.1 Multi-criterion Decision Making in MOCoEv Problems

The next step in a multi-objective optimization problem solving task is to choose a single preferred solution based on decision-maker's preferences. There exist a number of methods in the multi-criterion decision-making (MCDM) literature [12] for this purpose. Here, we discuss a few MCDM methods, specifically adapted for solving MOCoEv problems. Note that a decision-making process for choosing a preferred solution for Population 1 must now involve likely decision-making in the second population and vice versa.

Pseudo-weight Approach: From the obtained ND set of i-th population, a pseudo-weight vector [2] can be computed for k-th member ($\mathbf{x}^{(i),k}$) using the aggregate fitness function considered during the optimization process:

$$w_m^{(i),k} = \frac{\left(F_m^{(i),\max} - F_m(\mathbf{x}^{(i),k})\right) / \left(F_m^{(i),\max} - F_m^{(i),\min}\right)}{\sum_{m=1}^{M_i} \left(F_m^{(i),\max} - F_m(\mathbf{x}^{(i),k})\right) / \left(F_m^{(i),\max} - F_m^{(i),\min}\right)}, \quad \forall m = 1, \ldots, M_i, \quad (4)$$

where $F_m^{(i),\min}$ and $F_m^{(i),\max}$ are the minimum and maximum fitness values of the final ND members of the i-th population, respectively. The pseudo-weight vector $(w_1^{(i),k}, \ldots, w_{M_i}^{(i),k})$ for k-th member of i-th population indicates the relative importance of each objective in the span of the entire ND set of the population. Thereafter, the member ($\mathbf{x}^{(i),T}$) which has the pseudo-weight vector closest to a desired preference vector $\mathbf{d} = (f_1^{(i),T}, \ldots, f_{M_i}^{(i),T})$ is chosen for the i-th population. For two-population coevolution, the interaction between two populations in arriving at the final preferred solutions ($\mathbf{x}^{(1),T}$ and $\mathbf{y}^{(2),T}$) come from the aggregation used during the optimization process. The same can also be applied to Population 2 members.

Minimum Sensitivity Approach: Another pragmatic decision-making (DM) approach is proposed here. When a specific ND solution is chosen for Population 1 for implementation, there is a distribution of the fitness vector caused by the presence of different ND solutions at the final generation of Population 2. Thus, one DM idea would be to pick that solution from Population 1 which makes the tighter distribution in the resulting fitness vectors. One way to compute the extent of distribution would be to compute the normalized average distance of every objective vector from the mean objective vector:

$$\mathbf{x}^{(1),*} = \text{argmin}_{k=1}^{N_1} \frac{1}{N_2} \sum_{l=1}^{N_2} \|\mathbf{F}^{(1)}_{\text{mean}}(\mathbf{x}^{(1),k}), \mathbf{F}(\mathbf{x}^{(1),k}, \mathbf{y}^{(2),l})\|_2, \quad (5)$$

where $\|\cdot\|_2$ is the normalized Euclidean distance measure and $\mathbf{F}^{(1)}_{\text{mean}}(\mathbf{x}^{(1),k})$ is the mean fitness vector of ND solution $\mathbf{x}^{(1),k}$. There are other well-known DM approaches [12], such as, the trade-off method, compromise programming approach, which can also be used.

The above approaches can be made more pragmatic by assuming scenarios in which the DM information for the second population (say, the hacker's problem) is guessed from their past practices and are available. We can simulate this scenario by choosing a specific set of neighboring ND solutions ($\mathbf{Z}^{(2),S}$) from Population 2's final ND set. The defender's (Population 1) DM task will then be to choose a single preferred solution that is most appropriate to the likely chosen solutions of Population 2. One of the above procedures can be repeated using the neighboring ND solutions of Population 2, instead of the entire population. An advantage of first finding multiple Pareto strategies is that an iterative decision-making procedure can choose the most appropriate strategy from the obtained Pareto sets, thereby allowing an online and computationally fast procedure for choosing multi-objective co-evolutionary results.

4 Results

We consider two example problems to demonstrate the effectiveness of the proposed algorithm. We first consider an existing problem – Tug of War – as a proof-of-concept result and thereafter apply our MOCoEv algorithm and decision-making approaches to a numerical problem.

4.1 Tug-of-War Problem

Tug-of-war is a differential game proposed in [6]. It is a two player, non-cooperative, imperfect information game with one decision variable for each player. The game consists of a mass m placed on a frictionless horizontal 2-dimensional plane, shown in Fig. 3. Player 1 applies a force F_1 at an angle $x = \theta_1$ and Player 2 applied F_2 at an angle $y = \theta_2$ aiming to move the mass. The players aim to maximize or minimize the resulting Cartesian coordinates of the mass, m_x and m_y.

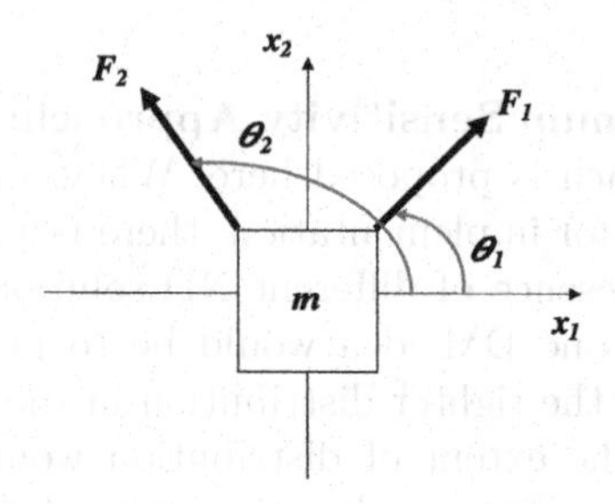

Fig. 3. Tug of war problem.

Forces and acceleration are considered constants. Acceleration along axes are: $\ddot{x_1} = F_1 \cos(\theta_1) + F_2 \cos(\theta_2)$ and $\ddot{x_2} = F_1 \sin(\theta_1) + F_2 \sin(\theta_2)$. Final position of the mass are: $x_1(t) = x_1(0) + \dot{x}_1(0)t + \frac{1}{2}\ddot{x}_1 t^2$ and $x_2(t) = x_2(0) + \dot{x}_2(0)t + \frac{1}{2}\ddot{x}_2 t^2$. To simplify the problem, initial conditions are assumed as $x_1(0) = x_2(0) = 0$ and $\dot{x}_1(0) = \dot{x}_2(0) = 0$. $F_1 = F_2 = 1N$ and $m = 1\,\text{kg}$ are assumed. The position after $t_f = \sqrt{2}\,\text{s}$ is used to construct the objective functions with $x = \theta_1$ and $y = \theta_2$, both in the range $[0, 2\pi]$: $m_x(x, y) = \cos(x) + \cos(y)$ and $m_y(x, y) = \sin(x) + \sin(y)$. Thus, according to our MOCoEv problem formulation, the minimization objectives are given as follows:

$$
\begin{aligned}
f_1^{(1)}(x, y) &= m_x(x, y), & f_2^{(1)}(x, y) &= m_y(x, y), \\
f_1^{(2)}(x, y) &= -m_x(x, y), & f_2^{(2)}(x, y) &= -m_y(x, y).
\end{aligned}
$$

It is clear that $f_i^{(1)}$ and $f_i^{(2)}$ are in conflict with each other. To achieve minimum coordinate values in Population 1, variable θ_1 is expected to lie within $[\pi, 1.5\pi]$, ideally with θ_2 values also in the same range. However, variable θ_2 lying in $[0, \pi/2]$ is expected to be optimal for Population 2, ideally with θ_1 values also in the same range. In these independent cases, $\theta_1 = \theta_2$ for all PO solutions of each population and the respective PO front will lie on the third quadrant of a circle of radius 2.0 for Population 1 and on the first quadrant of the same circle for Population 2. However, since neither of these ranges is optimal for the other population, the PO front for each population of the MOCoEv problem will be compromised from these idealized fronts, as we shall demonstrate with our results.

In our simulations, we use a population size of $N_1 = N_2 = 50$, SBX crossover operator [3] with $\eta_c = 20$ and $p_c = 0.5$, and polynomial mutation with $\eta_m = 50$, $p_m = 0.5$ and $T = 50$ generations for each population. Frequency of each iteration $\tau_1 = \tau_2 = 1$ is used here. We use the mean aggregation fitness function for both populations. The gray points in Fig. 4a show all final x-y positions of mass for each solution of $\mathbf{Z}^{(1)}$ evaluated with every solution from $\mathbf{Z}^{(2)}$. The distribution of the mass position for a specific solution ($\theta_1 = 225.71°$), lying in the middle of the PO front of Population 1, occurs due to a spread in the second population θ_2 solutions, as shown by the dark colored circles. For this specific θ_1 in the third quadrant, several θ_2 values in the first quadrant creates a range of PO solutions. Similarly, in Fig. 4b, we show distributions of each θ_2 from Population 2 for every θ_1 from Population 1. For a specific $\theta_2 = 46.05°$, the respective PO front is shown in dark color. These two figures make it clear that by keeping a good distribution of θ_1 values in the third quadrant and θ_2 values in the first quadrant, both populations can achieve a compromise PO front for each player.

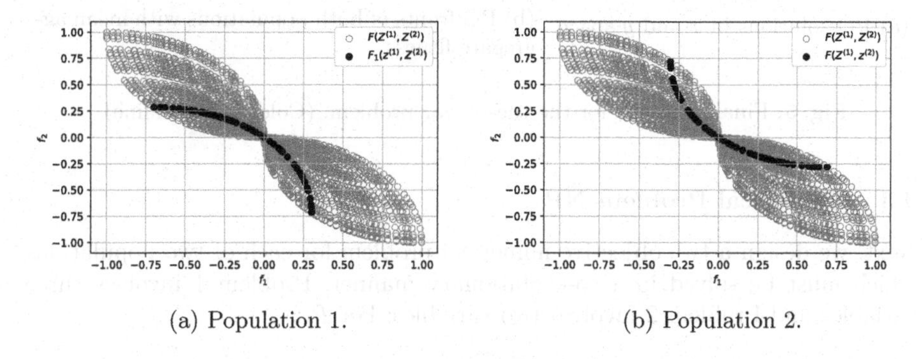

(a) Population 1. (b) Population 2.

Fig. 4. Final positions of mass for PO solutions of two populations.

Figure 5a shows the θ_1 and θ_2 of the final populations. It can be seen that they lie in third and first quadrant, respectively, with a good uniform spread. Figure 5b shows the PO front with the mean fitness values for each population. The compromise of these two fronts from their idealized independent fronts (shown in red and blue colors) to achieve each population's effect on each other is very clear from this figure. Population 1 mean fitness values are worse than the independent P_1 PO front, as a co-evolutionary P_2 optimization does not allow independent P_1 front to survive. These results are similar to that in [6].

4.2 Decision-Making for Tug-of-War Problem

Figure 5b represents the PO front with respect to the mean performance. These points can be used to choose a preferred point when the opponent's choice is unknown. If we desire to choose a point from the PO front that has pseudo-weight (according to Eq. 4) close to (0.5, 0.5) (having equal importance to both objectives) for both the populations, points marked with a star are points in Populations 1 and 2, respectively: $\theta_1 = 225.71°$, $\theta_2 = 46.05°$. They seem to lie in the middle of each PO front, ensuring an equal importance to each objective. The dark colored points used in Fig. 4 used these points as well.

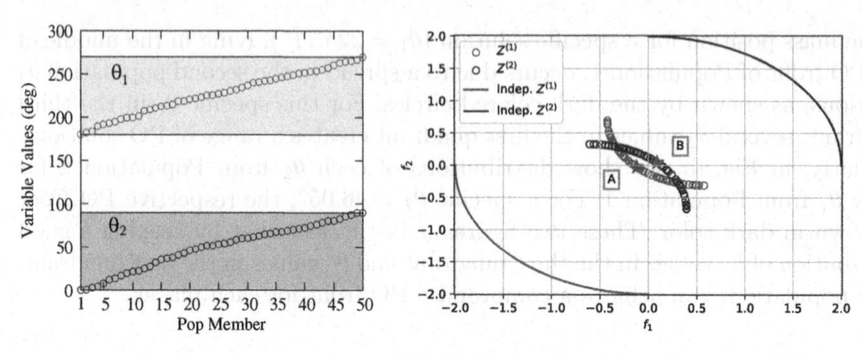

(a) PO solutions in x and y space.

(b) PO fronts of both populations with mean aggregate fitness.

Fig. 5. Final solutions for the tug-of-war problem. (Color figure online)

4.3 Numerical Problem NP

Next, we design a two-objective numerical problem for each of two populations which must be solved in a co-evolutionary manner. Problem 1 involves three variables and Problem 2 involves two variables: For P_1:

$$f_1^{(1)}(\mathbf{x},\mathbf{y}) = x_1^2 + x_2^2 + x_3^2 + y_1^2, \quad f_2^{(1)}(\mathbf{x},\mathbf{y}) = -(x_1 + x_2)y_1 + 10y_2 + x_3^2,$$
$$f_1^{(2)}(\mathbf{x},\mathbf{y}) = x_1^2 + x_2^2 - x_3^2 + y_1^2, \quad f_2^{(2)}(\mathbf{x},\mathbf{y}) = -(x_1 + x_2)y_1 + 10y_2 - x_3^2,$$

where, $-20 \leq (x_1, x_2, x_3) \leq 20$; $0.5 \leq (y_1, y_2) \leq 4$. The objectives of P_1 needs to be minimized and P_2 needs to be maximized. Identical parameter setting to that in the tug-of-war problem is used here. One aspect is clear: since x_3 comes as a positive quadratic form in P_1 and in negative quadratic form in P_2, it is intuitive that $x_3 = 0$ will for all PO solutions. With $x_3 = 0$, $f_i^{(1)} = f_i^{(2)}$ (for $i = 1, 2$), but since P_1 and P_2 objectives are to minimized and maximized, respectively, they will provide a competing scenario within the MOCoEv optimization. Moreover, smaller absolute values of x_1 and x_2 will be better for $f_1^{(1)}$, but worse for $f_2^{(1)}$, thereby indicating that each problem will constitute a PO front.

First, the mean fitness aggregation is used for both populations. Figure 6a displays mean fitness values of each population computed with all members of the other population. The distribution of mean fitness values of a specific $\mathbf{x}^{(1)} = (12.6201, 12.2150, 0.0230)$ due to variation in $\mathbf{Z}^{(2)}$ is marked with dark colored circles. A similar plot is shown in Fig. 6b with $\mathbf{y}^{(2)} = (2.5714, 3.9999)$. In Fig. 6a, every near-vertical line describes evaluation of a specific member of Population 1 and all members of Population 2. It is observed that all PO solutions of both populations, $y_2 = 4$ (upper bound). This variable is directly controlled by P_2. Since a higher value of y_2 makes $f_2^{(2)}$ better and y_2 does not appear in $f_1^{(2)}$, our MOCoEv decides to fix y_2 at its maximum allowable value. We also observe that y_1 values are distributed among population members within its entire range [0.5, 4.0]. Since y_1 makes a trade-off between $f_1^{(2)}$ and $f_2^{(2)}$ as shown in their

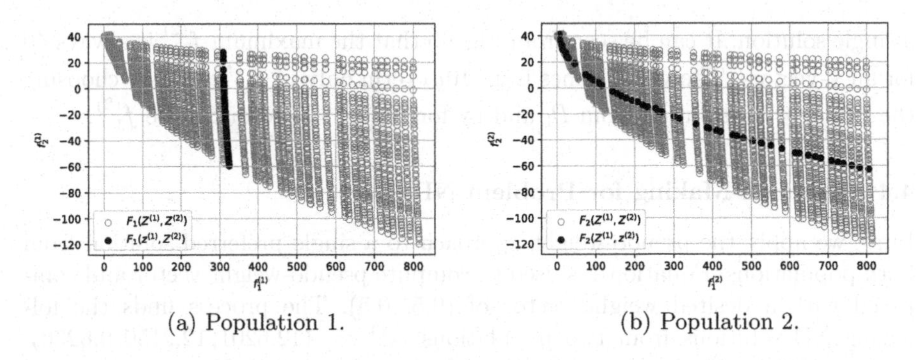

(a) Population 1. (b) Population 2.

Fig. 6. Evaluation score of middle point of Pareto front and all evaluations of Generation 50. Optimized using mean performance.

expressions, it is also expected, provided the term $(x_1 + x_2)$ is non-negative. Since P_1 controls the **x**-vector, it is clear that a positive value of $(x_1 + x_2)$ will dominate a negative value. we observe that both x_1 and x_2 is well distributed within $[0, 20]$. With the above details, it is clear that the limits of $f_1^{(1)}$ on the PO front will be $[0.5^2, 20^2 + 20^2 + 4^2]$ or $[0.25, 816]$, which matches with the figure. On a similar effort, $f_2^{(1)}$ lies in $[-120, 40]$.

The PO front with the mean aggregation fitness functions is shown in Fig. 7a. The respective PO front for Population 2 is shown in Fig. 7b with black circles. Despite a near-vertical nature of the fitness values, there exists a trade-off in these points.

To compare the PO fronts with the best and worst fitness aggregation functions, we rerun MOCoEv algorithm with identical parameter setting and present the final points in Fig. 7. It is clear that the mean aggregation fitness vectors are sandwiched in between the best and worst aggregation fitness vectors. It is interesting that when best case aggregation fitness is used, Population 2 converges to

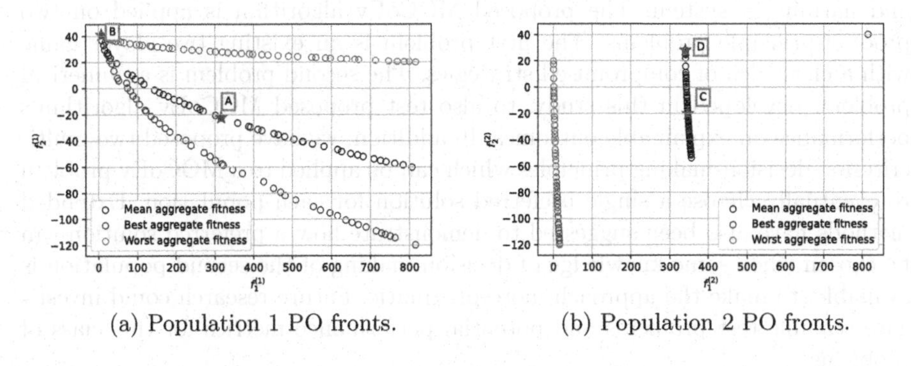

(a) Population 1 PO fronts. (b) Population 2 PO fronts.

Fig. 7. PO fronts for the numerical problem NP.

a single solution. It can be seen from Fig. 6b that the maximum $f_2^{(2)}$ is always 40 for all **y**-vectors. The value comes from $10y_2$ term only in $f_2^{(2)}$, thereby choosing the $x_1 = x_2 = 0$ solution from P_1 and by forcing $y_1 = 4$ to maximize $f_1^{(2)}$.

4.4 Decision-Making for Problem NP

First, we apply the pseudo-weight approach to a single preferred solution from both populations. Equation 4 is used to compute pseudo-weight vector and compared with a desired weight vector of (0.5, 0.5). The process finds the following ND solutions from two populations: $\mathbf{x}^{(1)} = (12.6201, 12.2150, 0.0230)$, $\mathbf{y}^{(2)} = (2.5714, 3.9999)$. This point is marked as A and C on the mean aggregation fitness plots in Fig. 7.

Next, we follow the minimum sensitivity approach and finds solutions B and D as the most preferred ones for two problems. The solutions are given as: $\mathbf{x}^{(1)} = (-0.2155, -0.1699, 0.0231)$, $\mathbf{y}^{(2)} = (0.5003, 3.9999)$. and also marked in mean aggregate fitness plot in Fig. 6a. It is clear that this happens for the minimum $f_1^{(1)}$ solution.

The above shows how a systematic approach of starting with finding a number of PO solutions using the proposed MOCoEv algorithm and then using a decision-making strategy to choose a single preferred solution can be obtained in a multi-objective co-evolutionary problem.

5 Conclusions

This paper has dealt with developing a multi-objective co-evolutionary (MOCoEv) optimization algorithm involving two interacting populations co-evolving for more than one conflicting objectives of their own. The motivation for pursuing this study has stemmed from recent surge in interests in solving adversarial network design problems and 'arms-race' problems which involve a defender and an attacker who clearly have more than one goals of protecting and harming a system. The proposed MOCoEv algorithm is applied on two proof-of-principle problems. The first problem is an existing two-player game with a clear idea of compromised strategies. The second problem is a numerical problem, developed in this study, to also test proposed MOCoEv algorithm's performance on explainable outcomes. In addition, we have proposed two multi-criterion decision-making principles which can be applied to a MOCoEv problem to eventually choose a single preferred solution for each population. Extended methods have also been suggested to demonstrate how a preferred solution can be chosen when some knowledge of decision-making of the second population is available, to make the approach more pragmatic. Future research could investigate scalable test problems and potential performance metrics for this class of problems.

To the best of our knowledge, this study stays as one of the few studies on simultaneous multi-objective solution of co-evolutionary problems. Our results in this paper have shown that MOCoEv problem solving is quite a challenging task, but extensions of evolutionary multi-objective optimization methods may make MOCoEv problem solving tractable and attractive. The proposed algorithm can now be applied to GAN, MOGs, and other co-evolutionary problems.

Acknowledgments. Authors acknowledge the Facebook Research Award for conducting this study.

References

1. Barbosa, H.J.C.: A genetic algorithm for min-max problems. In: Proceedings of the First International Conference on Evolutionary Computation and Its Application (EvCA 1996), pp. 99–109 (1996)
2. Deb, K.: Multi-Objective Optimization Using Evolutionary Algorithms. Wiley, Chichester (2001)
3. Deb, K., Agrawal, R.B.: Simulated binary crossover for continuous search space. Complex Syst. **9**(2), 115–148 (1995)
4. Deb, K., Agrawal, S., Pratap, A., Meyarivan, T.: A fast and elitist multi-objective genetic algorithm: NSGA-II. IEEE Trans. Evol. Comput. **6**(2), 182–197 (2002)
5. Eisenstadt, E., Moshaiov, A.: Decision-making in non-cooperative games with conflicting self-objectives. J. Multi-Criteria Decis. Anal. **25**(5–6), 130–141 (2018)
6. Eisenstadt, E., Moshaiov, A., Avigad, G.: Co-evolution of strategies for multi-objective games under postponed objective preferences. In: 2015 IEEE Conference on Computational Intelligence and Games (CIG), pp. 461–468. IEEE (2015)
7. Eisenstadt-Matalon, E., Avigad, G., Moshaiov, A., Branke, J.: Rationalizable strategies in multi-objective games under undecided objective preferences (2016)
8. Eisenstadt-Matalon, E., Moshaiov, A.: Mutual rationalizability in vector-payoff games. In: Deb, K., et al. (eds.) EMO 2019. LNCS, vol. 11411, pp. 593–604. Springer, Cham (2019). https://doi.org/10.1007/978-3-030-12598-1_47
9. Goodfellow, I., et al.: Generative adversarial nets. In: Ghahramani, Z., Welling, M., Cortes, C., Lawrence, N.D., Weinberger, K.Q. (eds.) Advances in Neural Information Processing Systems, vol. 27, pp. 2672–2680. Curran Associates (2014)
10. Li, X., Yao, X.: Cooperatively coevolving particle swarms for large scale optimization. IEEE Trans. Evol. Comput. **16**(2), 210–224 (2012)
11. Matalon-Eisenstadt, E., Moshaiov, A., Avigad, G.: The competing travelling salespersons problem under multi-criteria. In: Handl, J., Hart, E., Lewis, P.R., López-Ibáñez, M., Ochoa, G., Paechter, B. (eds.) PPSN 2016. LNCS, vol. 9921, pp. 463–472. Springer, Cham (2016). https://doi.org/10.1007/978-3-319-45823-6_43
12. Miettinen, K.: Nonlinear Multiobjective Optimization. Kluwer, Boston (1999)
13. Nguyen, A., Yosinski, J., Clune, J.: Deep neural networks are easily fooled: high confidence predictions for unrecognizable images. In: IEEE Conference on Computer Vision and Pattern Recognition (2015)
14. Paredis, J.: Coevolutionary constraint satisfaction. In: Parallel Problem Solving from Nature III (PPSN-III), pp. 46–55 (1994)
15. Żychowski, A., Gupta, A., Mańdziuk, J., Ong, Y.S.: Addressing expensive multi-objective games with postponed preference articulation via memetic co-evolution. Knowl.-Based Syst. **154**, 17–31 (2018)

MOEA/D for Multiple Multi-objective Optimization

Jingjing Chen[1(✉)], Qingfu Zhang[1,2(✉)], and Genghui Li[1,2(✉)]

[1] Department of Computer Science, City University of Hong Kong, Hong Kong, China
{jingjche2-c,genghuili2-c}@my.cityu.edu.hk, qingfu.zhang@cityu.edu.hk
[2] The City University of Hong Kong Shenzhen Research Institute, Shenzhen, China

Abstract. This paper defines a new multi-objective optimization problem, called multiple multi-objective optimization problem (MMOP). An MMOP is composed of several multi-objective optimization problems (MOPs) with different decision spaces and the same objective space, and its optimal solutions are non-dominated solutions among Pareto optimal solutions of all the individual MOPs. We construct a set of benchmark test instances with different characteristics. We propose a decomposition-based multi-objective evolutionary algorithm for solving MMOP (MOEA/D-MM). Experimental results on benchmarks show that MOEA/D-MM is more effective than some well-known traditional multi-objective evolutionary algorithms on MMOP.

Keywords: Multi-objective optimization · Multiple multi-objective optimization · MOEA/D

1 Introduction

A multi-objective optimization problem (MOP) can be formulated mathematically as follows:

$$\begin{aligned} \text{minimize} \quad & F(\mathbf{x}) = (f_1(\mathbf{x}), \ldots, f_m(\mathbf{x}))^T \\ \text{subject to} \quad & \mathbf{x} \in \Omega, \end{aligned} \tag{1}$$

where Ω is the decision space and $\mathbf{x} \in \Omega \subset R^n$ is the decision vector. $F : \Omega \rightarrow R^m$ consists of m real-valued objective functions, and R^m is the objective space. Since an MOP usually involves multiple conflicting objectives, no solution in Ω minimizes all objectives simultaneously. Therefore, decision makers use *Pareto optimality* to obtain the best trade-offs among the objectives.

Let $\mathbf{x}, \mathbf{y} \in \Omega$, $\mathbf{x}$ is said to *dominate* $\mathbf{y}$, denoted by $\mathbf{x} \prec \mathbf{y}$, if and only if $f_i(\mathbf{x}) \leq f_i(\mathbf{y})$ for all $i \in \{1, \ldots, m\}$, and $f_j(\mathbf{x}) < f_j(\mathbf{y})$ for at least one

This work was supported by the National Key Research and Development Project, Ministry of Science and Technology, China (Grant No. 2018AAA0101301) and the National Natural Science Foundation of China (Grant No. 61876163).

H. Ishibuchi et al. (Eds.): EMO 2021, LNCS 12654, pp. 152–163, 2021.
https://doi.org/10.1007/978-3-030-72062-9_13

index $j \in \{1, \ldots, m\}$. A decision vector $\mathbf{x}^*$ is called *Pareto optimal* if there does not exist another decision vector $\mathbf{x}$ such that $\mathbf{x}$ dominates $\mathbf{x}^*$. The set of Pareto optimal decision vectors is called the *Pareto set* (PS). Correspondingly, an objective vector is Pareto optimal if the corresponding decision vector is Pareto optimal and the set of Pareto optimal objective vectors is the *Pareto front* (PF) [10].

Given K MOPs defined as Eq. (1):

$$\begin{aligned} \text{minimize} \quad & F_k(\mathbf{x}_k) = (f_{k,1}(\mathbf{x}_k), \ldots, f_{k,m}(\mathbf{x}_k))^T \\ \text{subject to} \quad & \mathbf{x}_k \in \Omega_k \subset R^{n_k}, \\ & k = 1, \ldots, K, \end{aligned} \tag{2}$$

where $f_{k,j}(x_k)$ denotes the j-th objective function in the k-th MOP. The solutions of different MOPs can be compared since all MOPs, as defined above, share the same objective space. Let PS_k and PF_k be the PS and PF of the k-th MOP, respectively. In this paper, we introduce a new optimization problem, named as the multiple multi-objective optimization problem (MMOP). The PS of the MMOP is defined as

$$\text{PS} = \left\{\mathbf{x} \in \cup_{k=1}^{K} \text{PS}_k | \neg\exists \mathbf{x}' \in \cup_{k=1}^{K} \text{PS}_k, \mathbf{x}' \prec \mathbf{x}\right\}. \tag{3}$$

Clearly, the Pareto optimal solutions of the MMOP are the non-dominated solutions among Pareto optimal solutions of all MOPs. Correspondingly, the PF of the MMOP is composed of the non-dominated objective vectors in the union of all the individual PFs (i.e., $\cup_{k=1}^{K} \text{PF}_k$).

Some real-world applications can be modeled as an MMOP [6,12,13]. For example, the sensor coverage problem [13] has two objectives: coverage and cost. There are several different types of sensors for consideration. Different sensor types will lead to different decision variables. Each fixed sensor type corresponds to one corresponding traditional MOP:

$$\begin{aligned} \text{minimize} \quad & (f_1, f_2)^T \\ \text{subject to} \quad & f_1 = 1 - \text{coverage}(T_k), \\ & f_2 = \text{cost}(T_k), \end{aligned} \tag{4}$$

where T_k is a set of decision variables determined by the k-th sensor type. Thus, this problem can be naturally modelled as an MMOP. A direct approach for solving MMOP is to first solve all the individual MOPs using multi-objective evolutionary algorithms (MOEAs) [3,15,21], and then find the PS and the PF of the MMOP [2,18]. However, this approach may not be very efficient [11] since there is no need to obtain the complete PF of each MOP. For example, if any Pareto optimal solution of one MOP is dominated by at least one solution of other MOPs, there is no need to allocate the computational resource [9] to it.

In this paper, we first construct a set of MMOP benchmark test problems. To solve MMOP more efficiently, we propose a decomposition-based multi-objective evolutionary algorithm for solving MMOP (MOEA/D-MM), in which the computational resources are adaptively assigned to subproblems.

The rest of this paper is organized as follows. Section 2 constructs several MMOP benchmark test problems. Section 3 gives the details of MOEA/D-MM. Section 4 presents numerical experiments. Finally, Sect. 5 concludes this paper.

2 Benchmark Test Problems

Widely used MOP benchmark problems [4,17,20] can be directly used as individual problems to construct MMOP benchmark test problems. In this paper, we use UF problems [17] as individual problems. For the details of UF problems, please refer to [17].

Table 1. MMOP benchmark test problems.

Test problem	Objective function	Decision space
MMOP1	$f_1 = f_1^{\text{UF1}}$ $f_2 = f_2^{\text{UF1}}$	$[0,1] \times [-1,1]^{19}$
	$f_1 = (f_1^{\text{UF1}})^2$ $f_2 = \sqrt{f_2^{\text{UF1}}}$	$[0,1] \times [-1,1]^{29}$
MMOP2	$f_1 = f_1^{\text{UF3}}$ $f_2 = f_2^{\text{UF3}}$	$[0,1]^{20}$
	$f_1 = f_1^{\text{UF2}}$ $f_2 = \frac{f_2^{\text{UF2}}}{2} + 0.2$	$[0,1] \times [-1,1]^{29}$
MMOP3	$f_1 = f_1^{\text{UF4}}$ $f_2 = f_2^{\text{UF4}}$	$[0,1] \times [-2,2]^{19}$
	$f_1 = f_1^{\text{UF1}}$ $f_2 = \frac{f_2^{\text{UF1}}}{2} + 0.2$	$[0,1] \times [-1,1]^{29}$
MMOP4	$f_1 = f_1^{\text{UF4}}$ $f_2 = f_2^{\text{UF4}}$	$[0,1] \times [-2,2]^{19}$
	$f_1 = f_1^{\text{UF7}}$ $f_2 = \sqrt{f_2^{\text{UF7}}}$	$[0,1] \times [-1,1]^{29}$
MMOP5	$f_1 = f_1^{\text{UF7}}$ $f_2 = f_2^{\text{UF7}}$	$[0,1] \times [-1,1]^{19}$
	$f_1 = f_1^{\text{UF4}}$ $f_2 = f_2^{\text{UF4}}$	$[0,1] \times [-2,2]^{29}$
MMOP6	$f_1 = f_1^{\text{UF7}}$ $f_2 = f_2^{\text{UF7}}$	$[0,1] \times [-1,1]^{19}$
	$f_1 = f_1^{\text{UF3}}$ $f_2 = f_2^{\text{UF3}}$	$[0,1]^{30}$

As shown in Table 1, six MMOP benchmark problems (MMOP1–MMOP6) are constructed, and each of them consists of two MOPs. Each MOP has two objective functions. Figure 1 plots the PFs of these six MMOP benchmark test problems. Based on the true PF, these six test problems can be classified into two categories:

1) MMOP1–MMOP4: the MMOP's PF is a combination of the PF segments of two individual MOPs.
2) MMOP5–MMOP6: the MMOP's PF is the PF of one individual MOP.

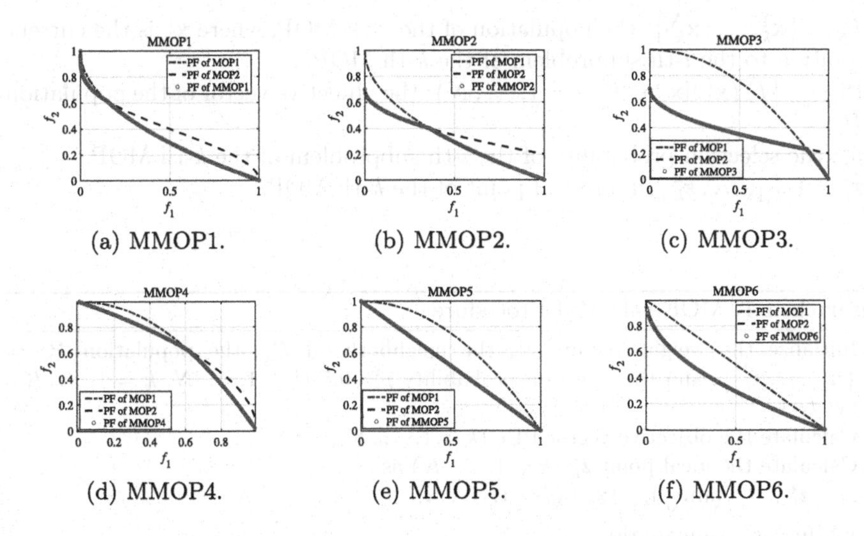

(a) MMOP1. (b) MMOP2. (c) MMOP3.

(d) MMOP4. (e) MMOP5. (f) MMOP6.

Fig. 1. The PFs of MMOP1–MMOP6.

3 Proposed Algorithm: MOEA/D-MM

3.1 General Framework

As discussed earlier, the PF of an MMOP can be the same as the PF of one individual MOP or consists of PF segments of all the individual MOPs. In the former case, an algorithm only needs to identify the most promising MOP and then allocates the computational resources to it. In the latter case, an algorithm needs to optimize all the individual MOPs. However, searching for all the non-dominated solutions of all MOPs is not necessary since the PF segment of a certain MOP may be dominated by a PF segment of another MOP.

To make reasonable computational resources allocation, the decomposition framework [14,15] for multi-objective optimization is adopted. Under the decomposition framework, each MOP can be decomposed into N ($N > 1$) subproblems, and then they will be allocated a certain amount of computational resources according to their utility function. In the design of a utility function, we should consider that optimal solutions of some subproblems are not Pareto optimal. This is different from some existing resource allocation strategies, such as MOEA/D-DRA [16] and MOEA/D-GRA [19].

The proposed MOEA/D-MM is shown in Algorithm 1. In each generation, MOEA/D-MM maintains and updates the following data ($k = 1, \ldots, K$).

- $P_k = \{\mathbf{x}_k^1, \ldots, \mathbf{x}_k^N\}$: the population of the k-th MOP, where $\mathbf{x}_k^i$ is the current solution to the i-th subproblem of the k-th MOP.
- $\mathrm{PF}_k = \{F_k(\mathbf{x}_k^i)|\mathbf{x}_k^i \in P_k, i = 1, \ldots, N\}$: the objective vector of the population P_k.
- p_k^i: the selection probability of the i-th subproblem of the k-th MOP.
- $\mathbf{z}_k^* = (z_{k,1}^*, \ldots, z_{k,m}^*)$: an ideal point of the k-th MOP.

Algorithm 1. MOEA/D-MM Procedure

1: Initialize the weight vectors λ^i, the neighborhood B_k^i, the population $P_k = \{\mathbf{x}_k^1, \ldots, \mathbf{x}_k^N\}$, and the selection probability p_k^i=0.5 ($i = 1, \ldots, N$, $k = 1, \ldots, K$). Set t=1.
2: Calculate the objective vector PF_k ($k = 1, \ldots, K$).
3: Calculate the ideal point $\mathbf{z}_k^*$ ($k = 1, \ldots, K$) as
$$z_{k,j}^* = \min_{i=1,\ldots,N} f_{k,j}\left(\mathbf{x}_k^i\right) \; (j = 1, \ldots, m).$$
4: **while** not terminate **do**
5: **for** $k = 1$ to K **do**
6: **for** $i = 1$ to N **do**
7: **if** *rand*() $\leq p_k^i$ **then**
8: Generate a trial solution $\mathbf{y}$ for subproblem i of the k-th MOP.
9: Update the ideal point $\mathbf{z}_k^*$ by
$$z_{k,j}^* = \min\left\{z_{k,j}^*, f_{k,j}\left(\mathbf{x}_k^i\right)\right\} \; (j = 1, \ldots, m).$$
10: Update the population P_k and objective vector PF_k by $\mathbf{y}$ and $F_k(\mathbf{y})$.
11: **end if**
12: **end for**
13: **end for**
14: Update the selection probability p_k^i ($i = 1, \ldots, N$, $k = 1, \ldots, K$).
15: t=t+1.
16: **end while**
17: **Output:** The non-dominated solutions of $\cup_{k=1}^K P_k$.

3.2 Implementation

1) Subproblem Definition. Based on the Tchebycheff method [10], the i-th subproblem of the k-th MOP is defined as follows:

$$g_k^i\left(\mathbf{x} \mid \lambda^i, \mathbf{z}^*\right) = \max_{1 \leq j \leq m} \left\{\lambda_j^i \left|f_{k,j}(\mathbf{x}) - z_{k,j}^*\right|\right\}, \tag{5}$$

where $\lambda^i = (\lambda_1^i, \ldots, \lambda_m^i)^T (i = 1, \ldots, N)$ is a set of uniformly distributed weight vectors that satisfy $\lambda_j^i \geq 0$ and $\sum_{j=1}^m \lambda_j^i = 1$. $\mathbf{z}_k^* = \left(z_{k,1}^*, \ldots, z_{k,m}^*\right)^T$ is an ideal point of the k-th MOP.

2) Offspring Reproduction. To generate a new solution $\mathbf{y}$ for the i-th subproblem of the k-MOP (line 8 in Algorithm 1), Differential Evolution (DE) operator (same as reproduction in MOEA/D-DE [8]) is used. The details of offspring reproduction can be found in [8].

3) Population Replacement. MOEA/D-MM updates the population by a candidate solution $\mathbf{y}$ (Line 10 in Algorithm 1) in the same way as MOEA/D [15]. Refer to [15] for details on population replacement for multi-objective optimization.

4) Selection Probability Update. The selection probability p_k^i $(i = 1, \ldots, N, k = 1, \ldots, K)$ is used for adaptively allocating computational resources (Line 7 in Algorithm 1). It is updated every ΔT generations. In each generation of MOEA/D-MM, the selection probability is determined by two factors: the relative improvement and the non-dominant rank. The pseudo-code of the selection probability update is shown in Algorithm 2.

Algorithm 2. Selection Probability Update

1: **Input:** The population P_k and objective vector PF_k, $k = 1, \ldots, K$.
2: **if** mod(t, ΔT)==0 **then**
3:   **for** $k = 1$ to K **do**
4:     **for** $i = 1$ to N **do**
5:       Calculate the relative improvement μ_k^i as Eq. (6).
6:       Calculate the probability of improvement PoI_k^i as Eq. (7).
7:     **end for**
8:   **end for**
9:   Apply fast non-dominate sorting for $\cup_{k=1}^K \mathrm{PF}_k$ to obtain the non-dominated rank matrix r and calculate the dominance utility u as Eq. (8).
10:   Calculate the probability of dominance PoD as Eq. (9).
11:   Update the selection probability p as Eq. (10).
12: **end if**
13: **Output:** The selection probability p.

Firstly, to allocate more computational resources to the subproblems that are easier to be improved, Like MOEA/D-GRA [19], we employ the relative improvement as the utility function to differentiate subproblems. The relative improvement of the i-th subproblem of the k-th MOP in the last ΔT generations is defined as follows:

$$\mu_k^i = \frac{g_k^i(\mathbf{x}_{k,t-\Delta T}^i) - g_k^i(\mathbf{x}_{k,t}^i)}{g_k^i(\mathbf{x}_{k,t-\Delta T}^i) + \epsilon}, \tag{6}$$

where $\mathbf{x}_{k,t}^i$ and $\mathbf{x}_{k,t-\Delta T}^i$ denote solutions of the i-th subproblem of the k-th MOP in the current generation t and the $t - \Delta T$ generation, respectively. ϵ is a small positive constant which is set to 10^{-50} in this paper to avoid division by zero. Based on this utility function, the Probability of Improvement (PoI) of each subproblem is defined as:

$$PoI_k^i = \frac{\mu_k^i + \epsilon}{\max\limits_{i=1,\ldots,N} \mu_k^i + \epsilon}. \tag{7}$$

As defined in Eq. (7), the subproblem that has a higher PoI will receive more computational resources.

On the other hand, the selection probability should also consider the dominance relationship of solutions of all subproblems. To this end, MOEA/D-MM first conducts fast non-dominated sorting [3] on all objective vectors ($\cup_{k=1}^{K} \mathrm{PF}_k$). To be more specific, let r_k^i be the non-dominated rank of the i-th individual of PF_k. The dominance utility of the i-th subproblem of the k-th MOP, based on the non-dominance rank, is denoted by u_k^i, where

$$u_k^i = \max_{\substack{n=1,\ldots,N,\\ m=1,\ldots,K}} r_m^n - r_k^i + 1. \tag{8}$$

Clearly, the lower the rank, the higher the dominance utility is.

Analogously, based on the dominance utility, one can define the Probability of Dominance (PoD) as follows:

$$PoD_k^i = \frac{u_k^i}{\max\limits_{i=1,\ldots,N} u_k^i}. \tag{9}$$

The subproblem that has a higher PoD will have more chance to be selected.

Finally, considering both the PoI and the PoD, the selection probability of the i-th subproblem of the k-th MOP is defined as

$$p_k^i = \alpha PoI_k^i + (1 - \alpha) PoD_k^i, \tag{10}$$

where α is a control parameter. It is used to balance the PoI and the PoD.

4 Experiments

We evaluate the performance of the proposed MOEA/D-MM by comparing it with NSGA-II, MOEA/D, and MOEA/D-GRA on the MMOP benchmark problems. For NSGA-II, MOEA/D, and MOEA/D-GRA, each MOP of MMOP is solved independently using $\frac{maxFES}{K}$ function evaluations ($maxFES$ is the maximum number of function evaluations), and the final solutions are the non-dominated solutions in the union of obtained non-dominated solutions of all the individual MOPs. We assume that the computational resources of one function evaluation for different MOPs are the same. The experimental setup, parameter settings for all compared algorithms, performance metrics, as well as the experimental results and analysis are presented as follows.

4.1 Experimental Setup

- Stopping criterion: The maximum number of function evaluations ($maxFES$) is 300,000 for each MMOP.
- Total runs: 20. Each algorithm conducts 20 independent runs on each MMOP.

4.2 Parameter Settings

- NSGA-II: The population size $N = 300$, the crossover probability $p_c = 0.8$ and the mutation probability $p_m = 1/n$.
- MOEA/D: The population size $N = 300$, the neighbourhood size $T = 20$, $\delta = 0.9$, $p_m = 1/n$, $CR = 1.0$, $F = 0.5$, $\eta = 20$, and $nr = 2$.
- MOEA/D-GRA: The population size $N = 300$, the neighbourhood size $T = 20$, $\delta = 0.9$, $p_m = 1/n$, $CR = 1.0$, $F = 0.5$, $\eta = 20$, $nr = 2$, and $\Delta T = 20$.
- MOEA/D-MM: The population size $N = 300$, the neighbourhood size $T = 20$, $\delta = 0.9$, $p_m = 1/n$, $CR = 1.0$, $F = 0.5$, $\eta = 20$, $nr = 2$, $\Delta T = 20$, and $\alpha = 0.5$.

4.3 Performance Metrics

Two metrics are employed to evaluate the performance of algorithms: the inverted generational distance (IGD) [22] and hypervolume difference (I_H^-) [7]. Let PF^* be the true PF, F^* be a nadir point, and PF be the approximated PF obtained by the algorithm. The IGD and I_H^- are defined as follows:

$$\mathrm{IGD}(PF^*, PF) = \frac{1}{|PF^*|} \sum_{\mathbf{x} \in PF^*} \min_{\mathbf{y} \in PF} \|\mathbf{x} - \mathbf{y}\|_2, \tag{11}$$

$$I_H^-(PF, PF^*, F^*) = I_H(PF^*, F^*) - I_H(PF, F^*), \tag{12}$$

where $I_H(PF, F^*)$ is the hypervolume [1,5] between the PF and the nadir point F^*.

4.4 Experimental Results and Analysis

The mean and standard deviation of the IGD and I_H^- of all compared algorithms are presented in Table 2 and Table 3, respectively. As shown in Table 2, the IGD of MOEA/D-MM is the best on all benchmark problems except for MMOP4. On MMOP4, NSGA-II obtains the best IGD. Similarly, as shown in Table 3, based on the I_H^- metric, MOEA/D-MM performs better than NSGA-II, MOEA/D, and MOEA/D-GRA on MMOP1–MMOP3 and MMOP5–MMOP6, and NSGA-II performs best on MMOP4. Obviously, on 5 out of 6 benchmark problems, MOEA/D-MM gets a better approximation to the true PF than that of NSGA-II, MOEA/D, and MOEA/D-GRA. The reason is that MOEA/D-MM can adaptively allocate more computational resources to the subproblems that correspond to the Pareto optimal solutions of MMOP.

Table 2. IGD of NSGA-II, MOEA/D, MOEA/D-GRA, and MOEA/D-MM.

Test problem	NSGA-II	MOEA/D	MOEA/D-GRA	MOEA/D-MM
MMOP1	$0.0687_{0.0142}$	$0.0064_{0.0012}$	$0.0061_{0.0010}$	$\mathbf{0.0059_{0.0012}}$
MMOP2	$0.0802_{0.0088}$	$0.0396_{0.0143}$	$0.0312_{0.0179}$	$\mathbf{0.0291_{0.0157}}$
MMOP3	$0.0769_{0.0128}$	$0.0132_{0.0060}$	$0.0114_{0.0020}$	$\mathbf{0.0101_{0.0012}}$
MMOP4	$\mathbf{0.0637_{0.0022}}$	$0.0639_{0.0014}$	$0.0655_{0.0020}$	$0.0648_{0.0018}$
MMOP5	$0.0885_{0.0369}$	$0.0796_{0.0522}$	$0.0612_{0.0525}$	$\mathbf{0.0594_{0.0559}}$
MMOP6	$0.1011_{0.0115}$	$0.0380_{0.0144}$	$0.0334_{0.0153}$	$\mathbf{0.0290_{0.0134}}$

Table 3. I_H^- of NSGA-II, MOEA/D, MOEA/D-GRA, and MOEA/D-MM.

Test problem	NSGA-II	MOEA/D	MOEA/D-GRA	MOEA/D-MM
MMOP1	$0.0812_{0.0188}$	$0.0096_{0.0041}$	$0.0093_{0.0029}$	$\mathbf{0.0087_{0.0025}}$
MMOP2	$0.0952_{0.0036}$	$0.0456_{0.0145}$	$0.0548_{0.0262}$	$\mathbf{0.0399_{0.0158}}$
MMOP3	$0.0875_{0.0288}$	$0.0286_{0.0196}$	$0.0224_{0.0056}$	$\mathbf{0.0208_{0.0047}}$
MMOP4	$\mathbf{0.0908_{0.0006}}$	$0.1011_{0.0026}$	$0.1040_{0.0027}$	$0.1023_{0.0028}$
MMOP5	$0.1239_{0.0135}$	$0.1316_{0.0850}$	$0.1035_{0.0872}$	$\mathbf{0.0983_{0.0912}}$
MMOP6	$0.2316_{0.0139}$	$0.0694_{0.0238}$	$0.0618_{0.0239}$	$\mathbf{0.0539_{0.0218}}$

It should be highlighted that the better performance of NSGA-II on MMOP4 may benefit from its dominance-based search mechanism; however, it does not mean that our resources allocation mechanism is ineffective on MMOP4. The reason is that MOEA/D-MM considers the dominance relationship and performs better than MOEA/D-GRA on MMOP4.

The obtained best PF in 20 independent runs of all compared algorithms is illustrated in Fig. 2. As shown in Fig. 2, MOEA/D-MM can obtain a better approximation to the true PF on most MMOP benchmark problems.

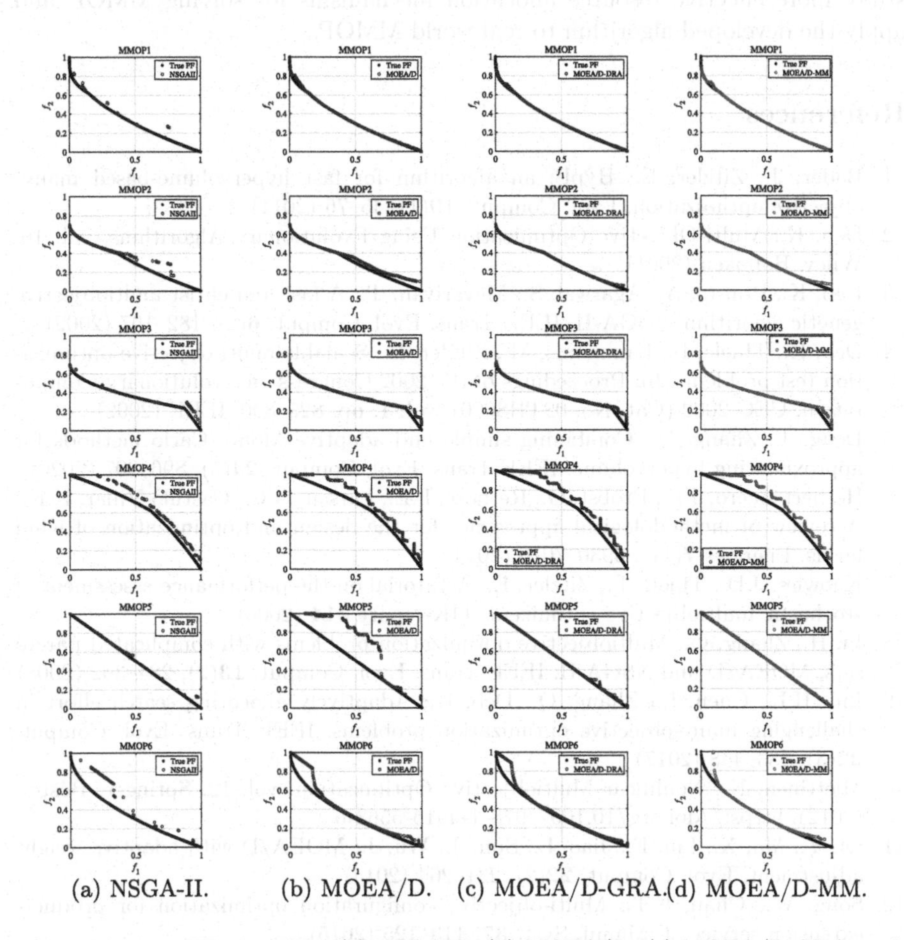

(a) NSGA-II. (b) MOEA/D. (c) MOEA/D-GRA. (d) MOEA/D-MM.

Fig. 2. The best PFs found by (a) NSGA-II, (b) MOEA/D, (c) MOEA/D-GRA, and (d) MOEA/D-MM on MMOP1–MMOP6.

5 Conclusion

In this paper, we have introduced a new multi-objective optimization problem, namely, multiple multi-objective optimization problem (MMOP). Its Pareto optimal solutions are the non-dominated solutions of the Pareto optimal solutions of all individual MOPs. We have constructed a set of MMOP benchmark problems for performance evaluation. We have designed a decomposition-based multi-objective evolutionary algorithm to solve the MMOP (MOEA/D-MM).

Our proposed algorithm can adaptively allocate computational resources to subproblems based on their relative improvement and non-dominance rank. The experimental results have demonstrated its effectiveness. In the future, we will study more effective resource allocation mechanisms for solving MMOP and apply the developed algorithm to real-world MMOP.

References

1. Bader, J., Zitzler, E.: HypE: an algorithm for fast hypervolume-based many-objective optimization. Evol. Comput. **19**(1), 45–76 (2011)
2. Deb, K.: Multi-Objective Optimization Using Evolutionary Algorithms, vol. 16. Wiley, Hoboken (2001)
3. Deb, K., Pratap, A., Agarwal, S., Meyarivan, T.: A fast and elitist multiobjective genetic algorithm: NSGA-II. IEEE Trans. Evol. Comput. **6**(2), 182–197 (2002)
4. Deb, K., Thiele, L., Laumanns, M., Zitzler, E.: Scalable multi-objective optimization test problems. In: Proceedings of the 2002 Congress on Evolutionary Computation, CEC 2002 (Cat. No. 02TH8600), vol. 1, pp. 825–830. IEEE (2002)
5. Deng, J., Zhang, Q.: Combining simple and adaptive Monte Carlo methods for approximating hypervolume. IEEE Trans. Evol. Comput. **24**(5), 896–907 (2020)
6. Herbert-Acero, J.F., Probst, O., Réthoré, P.E., Larsen, G.C., Castillo-Villar, K.K.: A review of methodological approaches for the design and optimization of wind farms. Energies **7**(11), 6930–7016 (2014)
7. Knowles, J.D., Thiele, L., Zitzler, E.: A tutorial on the performance assessment of stochastic multiobjective optimizers. TIK-report **214** (2006)
8. Li, H., Zhang, Q.: Multiobjective optimization problems with complicated pareto sets, MOEA/D and NSGA-II. IEEE Trans. Evol. Comput. **13**(2), 284–302 (2008)
9. Liu, H.L., Chen, L., Zhang, Q., Deb, K.: Adaptively allocating search effort in challenging many-objective optimization problems. IEEE Trans. Evol. Comput. **22**(3), 433–448 (2017)
10. Miettinen, K.: Nonlinear Multiobjective Optimization, vol. 12. Springer, Boston (2012). https://doi.org/10.1007/978-1-4615-5563-6
11. Qi, Y., Ma, X., Liu, F., Jiao, L., Sun, J., Wu, J.: MOEA/D with adaptive weight adjustment. Evol. Comput. **22**(2), 231–264 (2014)
12. Song, W., Chan, F.T.: Multi-objective configuration optimization for product-extension service. J. Manuf. Syst. **37**, 113–125 (2015)
13. Ting, C.K., Lee, C.N., Chang, H.C., Wu, J.S.: Wireless heterogeneous transmitter placement using multiobjective variable-length genetic algorithm. IEEE Trans. Syst. Man Cybern. Part B (Cybern.) **39**(4), 945–958 (2009)
14. Trivedi, A., Srinivasan, D., Sanyal, K., Ghosh, A.: A survey of multiobjective evolutionary algorithms based on decomposition. IEEE Trans. Evol. Comput. **21**(3), 440–462 (2016)
15. Zhang, Q., Li, H.: MOEA/D: a multiobjective evolutionary algorithm based on decomposition. IEEE Trans. Evol. Comput. **11**(6), 712–731 (2007)
16. Zhang, Q., Liu, W., Li, H.: The performance of a new version of MOEA/D on CEC09 unconstrained MOP test instances. In: 2009 IEEE Congress on Evolutionary Computation, pp. 203–208. IEEE (2009)
17. Zhang, Q., Zhou, A., Zhao, S., Suganthan, P.N., Liu, W., Tiwari, S.: Multiobjective optimization test instances for the CEC 2009 special session and competition (2008)

18. Zhou, A., Qu, B.Y., Li, H., Zhao, S.Z., Suganthan, P.N., Zhang, Q.: Multiobjective evolutionary algorithms: a survey of the state of the art. Swarm Evol. Comput. **1**(1), 32–49 (2011)
19. Zhou, A., Zhang, Q.: Are all the subproblems equally important? Resource allocation in decomposition-based multiobjective evolutionary algorithms. IEEE Trans. Evol. Comput. **20**(1), 52–64 (2015)
20. Zitzler, E., Deb, K., Thiele, L.: Comparison of multiobjective evolutionary algorithms: empirical results. Evol. Comput. **8**(2), 173–195 (2000)
21. Zitzler, E., Laumanns, M., Thiele, L.: SPEA2: improving the strength pareto evolutionary algorithm. TIK-report **103** (2001)
22. Zitzler, E., Thiele, L., Laumanns, M., Fonseca, C.M., Da Fonseca, V.G.: Performance assessment of multiobjective optimizers: an analysis and review. IEEE Trans. Evol. Comput. **7**(2), 117–132 (2003)

Using a Genetic Algorithm-Based Hyper-Heuristic to Tune MOEA/D for a Set of Benchmark Test Problems

Lie Meng Pang, Hisao Ishibuchi(✉), and Ke Shang

Guangdong Provincial Key Laboratory of Brain-Inspired Intelligent Computation, Department of Computer Science and Engineering, Southern University of Science and Technology, Shenzhen 518055, China
{panglm,hisao}@sustech.edu.cn, kshang@foxmail.com

Abstract. The multi-objective evolutionary algorithm based on decomposition (MOEA/D) is one of the most popular algorithms in the EMO community. In the last decade, the high performance of MOEA/D has been reported in many studies. In general, MOEA/D needs a different implementation for a different type of problems with respect to its components such as a scalarizing function, a neighborhood structure, a normalization mechanism and genetic operators. For MOEA/D users who do not have the in-depth knowledge about the algorithm, it is not easy to implement an appropriate algorithm that is suitable for their problems at hand. In our previous studies, we have suggested an offline genetic algorithm-based hyper-heuristic method to tune MOEA/D for a single problem. However, in real-world situations, users may want to use an algorithm with robust performance over many problems. In this paper, we improve the offline genetic algorithm-based hyper-heuristic method for tuning a set of problems. The offline hyper-heuristic procedure is applied to 26 benchmark test problems. The obtained MOEA/D implementations are compared with six decomposition-based EMO algorithms. The experimental results show that the tuned MOEA/D outperformed the compared algorithms on many test problems. The tuned MOEA/D also shows good (and stable) performance over a set of test problems.

Keywords: Decomposition-based algorithms · Hyper-heuristics · MOEA/D · Parameter tuning

1 Introduction

Evolutionary multi-objective optimization (EMO) algorithms have shown to be a useful tool in solving multi-objective optimization problems. Since Schaffer introduced the first EMO algorithm in 1985 [1], many EMO algorithms have been proposed in the literature. As the number of EMO algorithms is increasing every year, it is not easy for a user (who often does not have the in-depth knowledge about EMO algorithms) to choose a suitable EMO algorithm to solve his/her problem(s). It is very often (and natural) that a user tends to choose a popular algorithm in the field.

H. Ishibuchi et al. (Eds.): EMO 2021, LNCS 12654, pp. 164–177, 2021.
https://doi.org/10.1007/978-3-030-72062-9_14

Among many EMO algorithms in the literature, the multi-objective evolutionary algorithm based on decomposition (MOEA/D) is currently very popular in the EMO community [2]. It decomposes a multi-objective problem into a set of single-objective subproblems using a scalarizing function and solves the subproblems collaboratively by exploiting neighborhood information. Since its proposal in 2007, the high performance of MOEA/D has been reported in many studies [3, 4]. In fact, many recently proposed decomposition-based EMO algorithms are designed or improved based on the MOEA/D framework [28].

In general, MOEA/D needs a different implementation for a different type of problems with respect to its components such as a scalarizing function, a decomposition strategy, a neighborhood structure, a normalization mechanism and genetic operators. However, appropriate specifications of the components and the related parameter values in MOEA/D are not always easy since there exist a wide variety of choices and many characteristics of a real-world problem are usually unknown. As a result, users usually tend to use the default setting in the original paper or a frequently-used setting in the literature. However, the default setting has often been specified based on limited benchmark test problems. These test problems may not have strong relations to real-world problems to be solved [5]. Thus, it is very likely that the performance of MOEA/D is not always high if we simply use the default setting to solve real-world problems.

A hyper-heuristic method can be useful to determine the appropriate specification of the MOEA/D components and the related parameter values. In our previous studies [6, 7], a genetic algorithm-based hyper-heuristic method was used to offline tune MOEA/D for a single problem. It was shown that the performance of the properly tuned MOEA/D outperformed the MOEA/D with default settings. However, in real-world situations, users may want to solve not only a single problem but a set of similar problems. So, one question is how MOEA/D can be tuned to solve a set of problems. In this paper, the focus is on using an offline genetic algorithm-based hyper-heuristic method to optimize the implementations of MOEA/D for a set of problems.

This paper is organized as follows. First, we briefly introduce multi-objective optimization problems, the MOEA/D algorithm and its important components and parameters in Sect. 2. Next, we introduce the genetic algorithm-based hyper-heuristic method for tuning MOEA/D for a set of problems in Sect. 3. Then, in Sect. 4, we report experimental results on the three-objective benchmark test problems. Finally, we conclude this paper in Sect. 5.

2 Background

2.1 Multi-objective Optimization Problems

A multi-objective optimization problem can be mathematically expressed as follows:

$$\text{Minimize} \boldsymbol{f}(\boldsymbol{x}) = (f_1(\boldsymbol{x}), \cdots, f_M(\boldsymbol{x})) \text{ subject to } \boldsymbol{x} \in \mathbf{X}, \tag{1}$$

where $\boldsymbol{x} = (x_1, \cdots, x_D)$ is a D-dimensional vector for decision variables, $\mathbf{X}$ is the feasible space for $\boldsymbol{x}$, and $f_i(\boldsymbol{x})$ is the i-th objective to be minimized ($i = 1, 2, \cdots, M$).

Given two solution vectors x^1 and x^2, x^1 is dominated by x^2 if and only if $f_i(x^2) \leq f_i(x^1)$ for all $i \in \{1, 2, \cdots, M\}$ and $f_j(x^2) < f_j(x^1)$ for at least one index j ($j \in \{1, 2, \cdots, M\}$). A solution x^* is a Pareto optimal solution if and only if there do not exist any other solutions that dominate x^*. All Pareto optimal solutions form the Pareto set. The Pareto front is the mapping of the Pareto set onto the objective space.

2.2 MOEA/D

The general idea of MOEA/D is to decompose a multi-objective problem into a number of single-objective subproblems using a set of weight vectors together with a scalarizing function. Then, the subproblems are solved in a cooperative manner. The weight vectors are generated with the condition of satisfying the following relation [2]:

$$\sum_{i=1}^{M} w_i = 1, \tag{2}$$

where $0 \leq w_i \leq 1$ for $i = 1, 2, \cdots, M$. The Das and Dennis method [8] is the most frequently-used weight vector generation method in the standard implementation of MOEA/D. It generates a set of uniformly distributed weight vectors, in which each individual weight w_i of a weight vector $w = (w_1, w_2, \ldots, w_M)$ gets a value from

$$\left\{\frac{0}{H}, \frac{1}{H}, \frac{2}{H}, \ldots, \frac{H}{H}\right\}, \tag{3}$$

where H (H can be any positive integer) is used to specify the number of the generated weight vectors. Apart from the commonly-used uniform decomposition strategy, we also consider a random strategy in this paper. That is, a set of weight vectors that fulfils (2) is randomly generated before the execution of MOEA/D. Each subproblem has a different weight vector. Thus the number of subproblems equals to the number of generated weight vectors. In MOEA/D, the population size also equals to the number of generated weight vectors (i.e., the number of subproblems) because only a single best solution is stored by each subproblem.

Scalarizing Functions. One of the essential components in MOEA/D is a scalarizing function. Since different scalarizing functions exhibit different search behavior [9], the choice of an appropriate scalarizing function strongly depends on the characteristics of the problem at hand. Whereas there are quite a number of scalarizing functions available in the literature, we consider five commonly-used scalarizing functions in this paper, as listed in Table 1. Let $w = (w_1, w_2, \ldots, w_M)$ be a weight vector, then the formulation of each scalarizing function is also given in Table 1.

In Table 1, $z^* = (z_1^*, z_2^*, \ldots, z_M^*)$ is the estimated ideal point (reference point). The minimum value of each objective $f_i(x)$ over all solutions examined so far is used to specify each element of z^*. $z^N = (z_1^N, z_2^N, \ldots, z_M^N)$ is the estimated nadir point. Each element z_i^N of z^N is specified by the maximum value of each objective $f_i(x)$ in the current population. For the MTCH function, w_i is set to 10^{-6} if $w_i = 0$. θ is the user-definable penalty parameter (which can be any non-negative real value) for the PBI and IPBI functions.

Neighborhood Structures. The neighborhood structure is a feature of MOEA/D. Each weight vector (i.e., subproblem) has its own neighborhood. The neighborhood is defined by the Euclidean distance between weight vectors. Each weight vector has a pre-defined number of neighbors, where the neighborhood size is defined by the user. For each subproblem, a pair of parents is randomly selected from its neighborhood to generate an offspring (i.e., new solution) using genetic operators. This process is known as the local mating since the mating process takes place in a local neighborhood (known as the mating neighborhood).

The newly generated offspring is compared with its current solution as well as all the solutions in its neighborhood. The new solution will replace its own current solution and its neighbor solutions if it has a better scalarizing function value. This process is known as the local replacement process. The same neighborhood is used for both mating and replacement in the original MOEA/D. That is, the neighborhood size specifications are the same for the mating and replacement neighborhoods. Since the use of two different neighborhood structures (i.e., one for mating and the other for replacement) could improve the performance of MOEA/D as reported in [12], two neighborhood structures are examined in this study.

Table 1. Scalarizing functions for MOEA/D.

Scalarizing function	Formulation	
Weighted Sum (WS) [2]	Minimize $g^{\mathrm{WS}}(\boldsymbol{x}\mid\boldsymbol{w}) = w_1 f_1(\boldsymbol{x}) + \ldots + w_M f_M(\boldsymbol{x})$.	(4)
Tchebycheff (TCH) [2]	Minimize $g^{\mathrm{TCH}}(\boldsymbol{x}\mid\boldsymbol{w}, \boldsymbol{z}^*) = \max_{i=1,2,\ldots,M} \{w_i \cdot \lvert z_i^* - f_i(\boldsymbol{x})\rvert\}$.	(5)
Modified TCH (MTCH) [11]	Minimize $g^{\mathrm{MTCH}}(\boldsymbol{x}\mid\boldsymbol{w}, \boldsymbol{z}^*) = \max_{i=1,2,\ldots,M} \left\{\frac{1}{w_i} \cdot \lvert z_i^* - f_i(\boldsymbol{x})\rvert\right\}$.	(6)
Penalty-based Boundary Intersection (PBI) [2]	Minimize $g^{\mathrm{PBI}}(\boldsymbol{x}\mid\boldsymbol{w}, \boldsymbol{z}^*) = d_1 + \theta d_2$,	(7)
	$d_1 = \left\lvert (f(\boldsymbol{x}) - \boldsymbol{z}^*)^T \boldsymbol{w}\right\rvert / \lVert\boldsymbol{w}\rVert$,	(8)
	$d_2 = \lVert f(\boldsymbol{x}) - \boldsymbol{z}^* - d_1(\boldsymbol{w}/\lVert\boldsymbol{w}\rVert)\rVert$.	(9)
Inverted PBI (IPBI) [10]	Maximize $g^{\mathrm{IPBI}}\left(\boldsymbol{x}\mid\boldsymbol{w}, \boldsymbol{z}^N\right) = d_1 - \theta d_2$,	(10)
	$d_1 = \left\lvert \left(\boldsymbol{z}^N - f(\boldsymbol{x})\right)^T \boldsymbol{w}\right\rvert / \lVert\boldsymbol{w}\rVert$,	(11)
	$d_2 = \lVert \boldsymbol{z}^N - f(\boldsymbol{x}) - d_1(\boldsymbol{w}/\lVert\boldsymbol{w}\rVert)\rVert$.	(12)

Normalization Mechanism. A normalization mechanism is needed for problems with differently scaled objectives. In this paper, we use a simple normalization mechanism as proposed in [13]. The objective value z_i is normalized by the estimated ideal point $\boldsymbol{z}^*$ and nadir point $\boldsymbol{z}^N$, as follows:

$$z_i := \frac{z_i - z_i^*}{z_i^N - z_i^* + \varepsilon}, \tag{13}$$

where ε is known as the normalization parameter. The normalization parameter can be any positive number. The purpose of the normalization parameter is to prevent division by zero in the case of $z_i^N = z_i^*$. It was shown in [13] that the value of ε have some effects on the performance of the normalization mechanism.

Genetic Operators. Crossover and mutation are the essential components in evolutionary algorithms. The SBX crossover [14] and the polynomial mutation [15] are commonly-used genetic operators by many EMO algorithms in the literature. In this paper, we also examine the use of some other operators, i.e., the local arithmetic crossover (LAX) [16] and the random mutation [17] in MOEA/D.

3 Genetic Algorithm-Based Hyper-Heuristics Method

3.1 Algorithm Design Space

In general, an appropriately defined algorithm design space is important before applying any hyper-heuristics to generate an evolutionary algorithm. Table 2 shows the algorithm design space defined for MOEA/D in our experiments, which includes all MOEA/D components and parameters that have been described in Sect. 2. The column "Domain" in Table 2 lists the candidate values for each component or parameter. For example, one of {10%, 20%, 30%, 40%, 50%, 60%, 70%, 80%, 90%, 100%} is selected as a neighborhood size. In this paper, the MOEA/D algorithm implementations means the selection of a possible value in the domain for each component/parameter in Table 2.

Table 2. Algorithm design space for MOEA/D

Component/Parameter	Domain
Decomposition strategy	{Uniform, Random}
g (Scalarizing function)	{WS, TCH, MTCH, PBI, IPBI}
θ (Penalty parameter for PBI or IPBI)	[0:0.5:10]
% of mating neighborhood size (with respect to the population size)	{10%, 20%, 30%, 40%, 50%, 60%, 70%, 80%, 90%, 100%}
% of replacement neighborhood size (with respect to the population size)	{10%, 20%, 30%, 40%, 50%, 60%, 70%, 80%, 90%, 100%}

(*continued*)

Table 2. (*continued*)

Component/Parameter	Domain
Normalization	{With normalization, No normalization}
ε (Normalization parameter)	{0.000001, 0.00001, 0.0001, 0.001, 0.01, 0.1, 1, 5, 10, 15, 20, 25}
Crossover	{SBX, LAX}
Distribution index (for SBX)	{10, 20, 30, 40, 50, 60, 70, 80, 90, 100}
P_c (Crossover rate)	{0, 0.1, 0.2, 0.3, 0.4, 0.5, 0.6, 0.7, 0.8, 0.9, 1}
Mutation	{Polynomial, Random}
Distribution index (for polynomial mutation)	{10, 20, 30, 40, 50, 60, 70, 80, 90, 100}
P_m (Mutation rate)	{0, 0.1, 0.2, 0.3, 0.4, 0.5, 0.6, 0.7, 0.8, 0.9, 1}

3.2 Genetic Algorithm-Based Hyper-Heuristic Procedure

The process of finding a good parameter vector for an evolutionary algorithm can be regarded as a highly non-linear and complex optimization problem due to the existence of interactive variables, multiple local optima and noise [18]. Since evolutionary algorithms are very good at solving this kind of optimization problems, it is a natural idea to use an evolutionary approach to obtain an optimized configuration vector for an evolutionary algorithm [18]. Hence, in this paper, a genetic algorithm is used as a tuner to optimize the configuration vector of MOEA/D. This is because genetic algorithms have shown to be a useful hyper-heuristic and have been widely used [19].

The following specifications are used for the genetic algorithm implementation in our computational experiments:

Coding: Integer encoding (13 integers),
Population size μ: 50,
Crossover: Uniform crossover with probability 1,
Mutation: Random mutation with probability 1/13,
Generation update: $(\mu + \mu)$ strategy,
Selection: Tournament selection with tournament size 3,
Fitness evaluation: Average rank,
Termination condition: 30 generations.

Since the domain of each component or parameter can be discretized into a categorical variable as shown in Table 2, the integer encoding is used in our genetic algorithm implementation. The search space size (i.e., the total number of possible integer strings) is $2 \times 5 \times 21 \times 10 \times 10 \times 2 \times 12 \times 2 \times 10 \times 11 \times 2 \times 10 \times 11 = 2.4 \times 10^{10}$. A relatively fewer tuning resources (i.e., only 30 generations) are used in our study in order to simulate the real-world condition (in which the available tuning resources are often very limited). We will briefly explain the procedure for using the genetic algorithm to optimize the configuration vector of MOEA/D for a set of problems in the next few paragraphs.

First, a population of 50 individuals is randomly generated at the initialization stage. Each individual (i.e., each integer string) is a configuration vector for MOEA/D. That is, there are 50 MOEA/D implementations at the initialization stage. The performance of each MOEA/D implementation is evaluated over a set of problems by applying it to each problem with the termination condition of 10,000 solution evaluations. Five independent runs of each MOEA/D implementation are performed for each problem in order to deal with the random nature of evolutionary algorithms. The hypervolume indicator is used to evaluate the performance of each MOEA/D implementation for each problem. During the tuning process, it is assumed that the true Pareto fronts of each test problem is unknown. Thus, all solution sets obtained from five runs of all MOEA/D implementations in each generation for a test problem are combined, and only non-dominated solutions among them are selected to approximate the Pareto front. The estimated ideal point and nadir point are calculated from the obtained non-dominated solution set. Then, the objective space is normalized using the estimated ideal point and nadir point. The hypervolume value is calculated by specifying the reference point $\boldsymbol{r} = (r, r, \ldots, r)$ with $r = 1.1$, that is, r is a value slightly worse than the estimated nadir point in the normalized objective space.

Based on the average hypervolume value obtained for each problem from five runs of each MOEA/D implementation, a rank is assigned to each MOEA/D implementation. A smaller rank indicates better performance. Then, each MOEA/D implementation is assigned with an average rank by averaging the ranks it receives over all problems. The average rank of each MOEA/D implementation is used as the fitness indicator in the genetic algorithm tuner.

At every generation, 50 offspring are produced by genetic operators. The $(\mu + \mu)$ generation update strategy is used. That is, 50 parents and 50 offspring are merged into a combined population and evaluated. Since all the non-dominated solutions obtained from the parents and the offspring (i.e., 100 MOEA/D implementations) at the current generation are used to approximate the Pareto front, the reference point for HV calculation for each problem is updated at every generation. For fair comparison, the parents are evaluated again using the current reference point. That is, the solution sets obtained by the parents in the previous generations are used for the new evaluations. Then, the best 50 individuals among the 100 individuals are selected. The selected 50 individuals will become the parents of the next generation. The same process is iterated until the termination condition is met.

4 Experimental Results

The genetic algorithm-based hyper-heuristic procedure is used to optimize the implementations of MOEA/D for the three-objective DTLZ1-4 [20], WFG1-9 [21], and their minus versions [22]. Based on the Pareto front shape of each test problem, we roughly categorize them into six set of test problems, i.e., DTLZ1-4 (Set 1), WFG1-3 (Set 2), WFG4-9 (Set 3), Minus-DTLZ1-4 (Set 4), Minus-WFG1-3 (Set 5), and Minus-WFG4-9 (Set 6). In our experiments, we fix the population size in MOEA/D to 91. The number of decision variables (i.e., D) is set as 7 for DTLZ1 and Minus-DTLZ1, and 12 for the other test problems. The obtained MOEA/D implementation for each set of test problems

Table 3. The obtained MOEA/D implementations for each set of test problems.

Component/Parameter	DTLZ1-4 (Set 1)	WFG1-3 (Set 2)	WFG4-9 (Set 3)
Decomposition strategy	Uniform	Uniform	Uniform
g (Scalarizing function)	MTCH	MTCH	TCH
θ (Penalty parameter for PBI or IPBI)	–	–	–
Mating neighborhood size	90%	100%	30%
Replacement neighborhood size	100%	90%	80%
Normalization	No	Yes	Yes
ε (Normalization parameter)	–	10^{-2}	10^{-6}
Crossover	SBX	SBX	SBX
Distribution index (for SBX)	50	10	60
P_c (Crossover rate)	1	0.7	0.8
Mutation	Random	Polynomial	Polynomial
Distribution index (for Polynomial mutation)	–	30	20
P_m (Mutation rate)	0.1	0.2	0.2

Table 4. The obtained MOEA/D implementations for each set of test problems.

Component/Parameter	Minus-DTLZ1-4 (Set 4)	Minus-WFG1-3 (Set 5)	Minus-WFG4-9 (Set 6)
Decomposition strategy	Random	Random	Uniform
g (Scalarizing function)	MTCH	MTCH	WS
θ (Penalty parameter for PBI or IPBI)	–	–	–
Mating neighborhood size	40%	30%	40%
Replacement neighborhood size	100%	40%	90%
Normalization	Yes	Yes	Yes
ε (Normalization parameter)	10^{-2}	10^{-3}	10^{-4}
Crossover	SBX	SBX	SBX
Distribution index (for SBX)	50	70	10
P_c (Crossover rate)	1	0.9	1
Mutation	Random	Polynomial	Polynomial
Distribution index (for Polynomial mutation)	–	10	10
P_m (Mutation rate)	0.1	0.2	0.1

by the hyper-heuristic procedure is presented in Table 3 (DTLZ1-4 and WFG4-9) and Table 4 (their minus versions).

We can see that the genetic algorithm-based tuner is able to suggest a promising (reasonable) MOEA/D implementation for each set of test problems. For example, the tuner suggests that a normalization mechanism is not needed for DTLZ1-4 whereas it is needed for WFG1-9. These suggestions are consistent with our knowledge on the DTLZ and WFG test problems. That is, DTLZ test problems do not have differently scaled objective space, whereas the WFG test problems have differently scaled objective space. Another interesting observation from Table 4 is that a random decomposition strategy is suggested for the Minus-DTLZ1-4 and Minus-WFG1-3 test problems. This observation also in-line with our knowledge that a uniformly generated weight vectors with a triangular shape do not work well on the test problems with inverted triangular Pareto fronts (i.e., the Minus-DTLZ and Minus-WFG test problems). Next, we validate the obtained MOEA/D implementations in Tables 3 and 4 with a series of experiments. The obtained MOEA/D implementations (denoted as MOEA/D-Auto in the subsequent tables) for each set of test problems are compared with MOEA/D-TCH [2], MOEA/D-PBI [2], MOEA/DD [23], MOEA/D-DE [24], MOEA/D-DRA [25], MOEA/D-DU [27]. Our experiments are performed on the PlatEMO [26] platform. For comparison purposes, the following default settings are used in each compared algorithm:

Population size: 91,
Crossover: SBX with probability 1,
Mutation: Polynomial mutation with probability $1/D$,
Distribution index for the SBX and Polynomial mutation: 20,
Termination condition: 10,000 solution evaluations.

It should be noted that some compared algorithms (i.e., MOEA/D-PBI, MOEA/D-DU, and MOEA/DD) required extra parameter settings. In our study, we do not tune their settings and simply use the default settings provided by PlatEMO. For each test problem, each algorithm is run 31 times independently. The hypervolume indicator is used to evaluate the performance of each algorithms. The larger hypervolume value indicates better performance. It should be noted that the hypervolume value is normalized using the true Pareto front of each test problem. That is, the true ideal point and the true nadir point are used to normalize the hypervolume value for performance evaluation. The reference point $\boldsymbol{r} = (1.1, 1.1, 1.1)$ is used for calculating the hypervolume value.

The experimental results for each test problem set are summarized in Tables 5, 6, 7, 8, 9, and 10. The Wilcoxon's rank sum test at a significant level of 5% is used to evaluate whether there is a statistically difference between MOEA/D-Auto and each of the compared algorithms. The sign "+", "−", and "=" are used to indicate that the compared algorithm is significantly better than, worse than or equivalent to MOEA/D-Auto.

Tables 5, 6 and 7 show the experimental results on the DTLZ and WFG test problems. MOEA/D-Auto shows good and stable performance over a set of test problems. Even though it does not show the best performance on some test problems, its performance is comparable to the best performance obtained by the other compared algorithms.

Table 5. The mean hypervolume value and the standard deviation over 31 runs of each algorithm on each three-objective DTLZ1-4 test problems using 10,000 solution evaluations. The best result is highlighted by bold and yellow color.

Problem	MOEA/D-TCH	MOEA/D-PBI	MOEA/DD	MOEA/D-DE	MOEA/D-DRA	MOEA/D-DU	MOEA/D-Auto
DTLZ1	**5.9309e-1 (3.11e-1) =**	5.4770e-1 (3.25e-1) =	2.4455e-1 (2.52e-1) -	1.1477e-1 (2.33e-1) -	8.7018e-3 (4.51e-2) -	1.6665e-1 (2.21e-1) -	5.3551e-1 (2.98e-1)
DTLZ2	5.2931e-1 (1.79e-3) -	5.5533e-1 (7.30e-4) =	5.5446e-1 (1.13e-3) =	5.1389e-1 (4.28e-3) -	4.5829e-1 (2.41e-2) -	**5.5731e-1 (5.18e-4) +**	5.5446e-1 (3.04e-3)
DTLZ3	0.0000e+0 (0.00e+0)=	0.0000e+0 (0.00e+0)=	0.0000e+0 (0.00e+0)=	**2.5850e-2 (9.80e-2) =**	2.2037e-2 (6.98e-2) =	0.0000e+0 (0.00e+0)=	4.2393e-3 (2.27e-2)
DTLZ4	3.1471e-1 (1.86e-1) -	3.6877e-1 (1.69e-1) -	5.5444e-1 (9.31e-4) -	4.9253e-1 (3.64e-2) -	3.8322e-1 (8.19e-2) -	5.3985e-1 (6.75e-2) -	**5.5672e-1 (1.02e-3)**
+/-/=	0/2/2	0/1/3	0/2/2	0/3/1	0/3/1	1/2/1	

Table 6. The mean hypervolume value and the standard deviation over 31 runs of each algorithm on each three-objective WFG1-3 test problems using 10,000 solution evaluations. The best result is highlighted by bold and yellow color.

Problem	MOEA/D-TCH	MOEA/D-PBI	MOEA/DD	MOEA/D-DE	MOEA/D-DRA	MOEA/D-DU	MOEA/D-Auto
WFG1	**5.4403e-1 (5.49e-2) =**	3.9960e-1 (4.69e-2) -	2.8435e-1 (1.40e-2) -	2.6389e-1 (1.59e-2) -	2.0182e-1 (2.68e-2) -	2.5278e-1 (8.21e-2) -	5.2564e-1 (5.03e-2)
WFG2	8.6396e-1 (1.57e-2) -	8.2546e-1 (3.17e-2) -	8.9491e-1 (1.06e-2) =	8.5842e-1 (1.24e-2) -	8.6436e-1 (1.42e-2) -	**9.0392e-1 (7.89e-3) +**	8.7952e-1 (5.96e-2)
WFG3	3.5340e-1 (1.38e-2) -	2.8154e-1 (3.36e-2) -	2.6865e-1 (4.58e-2) -	2.8868e-1 (2.67e-2) -	2.6077e-1 (2.67e-2) -	2.7973e-1 (1.58e-2) -	**3.9041e-1 (9.97e-3)**
+/-/=	0/2/1	0/3/0	0/2/1	0/3/0	0/3/0	1/2/0	

Table 7. The mean hypervolume value and the standard deviation over 31 runs of each algorithm on each three-objective WFG4-9 test problems using 10,000 solution evaluations. The best result is highlighted by bold and yellow color.

Problem	MOEA/D-TCH	MOEA/D-PBI	MOEA/DD	MOEA/D-DE	MOEA/D-DRA	MOEA/D-DU	MOEA/D-Auto
WFG4	5.0499e-1 (5.68e-3) -	4.9019e-1 (9.91e-3) -	5.2029e-1 (3.52e-3) -	4.4397e-1 (9.14e-3) -	4.4472e-1 (1.24e-2) -	5.2115e-1 (3.59e-3) -	**5.2784e-1 (4.22e-3)**
WFG5	4.6017e-1 (3.46e-3) -	4.8043e-1 (4.97e-3) -	**4.9577e-1 (4.23e-3) +**	4.5801e-1 (4.79e-3) -	4.6166e-1 (6.01e-3) -	4.9019e-1 (4.31e-3) +	4.8651e-1 (5.73e-3)
WFG6	4.4357e-1 (1.84e-2) -	4.3628e-1 (1.74e-2) -	4.4944e-1 (3.57e-2) -	3.8485e-1 (2.66e-2) -	3.7707e-1 (1.87e-2) -	4.5934e-1 (1.38e-2) -	**4.7197e-1 (2.21e-2)**
WFG7	5.0636e-1 (5.70e-3) -	4.3521e-1 (2.64e-2) -	5.0452e-1 (1.64e-2) -	4.5923e-1 (8.52e-3) -	4.4672e-1 (1.66e-2) -	5.1657e-1 (5.13e-3) -	**5.2735e-1 (5.62e-3)**
WFG8	4.2188e-1 (7.32e-3) -	4.0602e-1 (2.66e-2) -	4.3174e-1 (7.72e-3) -	3.4381e-1 (1.49e-2) -	3.0352e-1 (4.24e-2) -	4.3762e-1 (4.14e-3) =	**4.3847e-1 (5.50e-3)**
WFG9	4.5398e-1 (3.85e-2) =	4.1271e-1 (4.05e-2) -	4.8287e-1 (2.20e-2) +	4.5661e-1 (2.41e-2) =	4.3784e-1 (3.24e-2) -	**4.9402e-1 (2.10e-2) +**	4.4841e-1 (5.45e-2)
+/-/=	0/5/1	0/6/0	2/4/0	0/5/1	0/6/0	2/3/1	

Tables 8, 9 and 10 show the experimental results for the Minus-DTLZ and Minus-WFG test problems. Recently, it has been pointed out that test problems with inverted triangular Pareto fronts are more closely related to the real-world problems (e.g., the Pareto fronts of three-objective knapsack problems) [27]. The test problems in Tables 8, 9 and 10 have inverted triangular Pareto fronts. In Tables 8, 9 and 10, we can observe that MOEA/D-Auto shows the best performance on almost all test problems.

As shown in Tables 5, 6, 7, 8, 9 and 10, good results are obtained by the tuned MOEA/D implementations. More specifically, the best results are obtained by the tuned MOEA/D implementations on 17 out of the 26 test problems in Tables 5, 6, 7, 8, 9 and 10. The tuned MOEA/D implementations outperforms the compared algorithms on many test problems (and is outperformed by each algorithm on only a few test problems). These observations clearly show that a good MOEA/D implementation is obtained for a set of test problems by the proposed hyper-heuristic approach with a relatively small computation load (30 generations with the population size 50 where 1500 implementations are examined). These observations also suggest that the careful tuning of MOEA/D for each test problem is not necessarily needed (i.e., good results can be obtained by its tuning for a set of test problems).

Table 8. The mean hypervolume value and the standard deviation over 31 runs of each algorithm on each three-objective Minus-DTLZ1-4 test problems using 10,000 solution evaluations. The best result is highlighted by bold and yellow color.

Problem	MOEA/D-TCH	MOEA/D-PBI	MOEA/DD	MOEA/D-DE	MOEA/D-DRA	MOEA/D-DU	MOEA/D-Auto
Minus-DTLZ1	1.9375e-1 (5.75e-3) -	1.8007e-1 (6.45e-3) -	7.9551e-2 (1.43e-2) -	1.5041e-1 (6.51e-3) -	1.3346e-1 (5.77e-3) -	9.9425e-2 (3.26e-3) -	**2.0605e-1 (2.68e-3)**
Minus-DTLZ2	5.1568e-1 (1.39e-3) +	**5.1915e-1 (2.27e-3) +**	3.0245e-1 (1.83e-2) -	5.0865e-1 (3.12e-3) -	5.0116e-1 (8.79e-3) -	2.1869e-1 (1.56e-2) -	5.1281e-1 (4.45e-3)
Minus-DTLZ3	4.8028e-1 (1.09e-2) -	4.6069e-1 (1.34e-2) -	3.1319e-1 (2.06e-2) -	3.5004e-1 (1.59e-2) -	2.7478e-1 (2.17e-2) -	2.1637e-1 (1.53e-2) -	**5.0674e-1 (5.20e-3)**
Minus-DTLZ4	4.7164e-1 (1.33e-1) -	3.9421e-1 (1.93e-1) =	2.9861e-1 (2.22e-2) -	4.9378e-1 (1.08e-2) -	4.9060e-1 (1.14e-2) -	1.3882e-1 (1.69e-2) -	**5.0559e-1 (4.49e-3)**
+/-/=	1/3/0	1/2/1	0/4/0	0/4/0	0/4/0	0/4/0	

Table 9. The mean hypervolume value and the standard deviation over 31 runs of each algorithm on each three-objective Minus-WFG1-3 test problems using 10,000 solution evaluations. The best result is highlighted by bold and yellow color.

Problem	MOEA/D-TCH	MOEA/D-PBI	MOEA/DD	MOEA/D-DE	MOEA/D-DRA	MOEA/D-DU	MOEA/D-Auto
Minus-WFG1	8.0444e-2 (2.45e-2) -	5.3607e-2 (2.10e-2) -	3.0949e-2 (1.10e-2) -	8.9528e-2 (3.08e-3) -	8.2859e-2 (2.30e-3) -	2.7998e-2 (1.33e-2) -	**1.0366e-1 (1.53e-3)**
Minus-WFG2	**2.8267e-1 (9.13e-4) +**	2.6777e-1 (9.47e-3) -	1.4965e-1 (1.14e-3) -	2.5704e-1 (5.41e-3) -	2.3937e-1 (1.63e-2) -	1.4218e-1 (2.61e-2) -	2.7537e-1 (4.11e-3)
Minus-WFG3	1.8768e-1 (1.48e-3) -	1.8446e-1 (4.23e-3) -	5.6298e-2 (7.85e-3) -	1.4796e-1 (6.15e-3) -	1.4041e-1 (7.31e-3) -	9.8652e-2 (6.29e-3) -	**2.0272e-1 (1.65e-3)**
+/-/=	1/2/0	0/3/0	0/3/0	0/3/0	0/3/0	0/3/0	

Table 10. The mean hypervolume value and the standard deviation over 31 runs of each algorithm on each three-objective Minus-WFG4-9 test problems using 10,000 solution evaluations. The best result is highlighted by bold and yellow color.

Problem	MOEA/D-TCH	MOEA/D-PBI	MOEA/DD	MOEA/D-DE	MOEA/D-DRA	MOEA/D-DU	MOEA/D-Auto
Minus-WFG4	5.0008e-1 (2.98e-3) -	5.0629e-1 (4.76e-3) -	2.8599e-1 (3.59e-2) -	4.9321e-1 (4.13e-3) -	4.8655e-1 (6.53e-3) -	2.0823e-1 (1.69e-2) -	**5.3047e-1 (1.87e-4)**
Minus-WFG5	5.0078e-1 (1.44e-3) -	5.0519e-1 (3.53e-3) -	2.8204e-1 (2.68e-2) -	4.5640e-1 (8.33e-3) -	4.5013e-1 (1.01e-2) -	2.9090e-1 (7.63e-3) -	**5.2789e-1 (6.49e-4)**
Minus-WFG6	5.0185e-1 (1.19e-3) -	5.0258e-1 (5.03e-3) -	2.8030e-1 (2.42e-2) -	4.4806e-1 (5.80e-3) -	4.3053e-1 (8.54e-3) -	3.3534e-1 (9.93e-3) -	**5.2956e-1 (2.69e-4)**
Minus-WFG7	4.9279e-1 (5.80e-3) -	4.9529e-1 (7.93e-3) -	2.6619e-1 (2.32e-2) -	4.8884e-1 (5.57e-3) -	4.6812e-1 (1.77e-2) -	2.4085e-1 (1.52e-2) -	**5.3016e-1 (2.57e-4)**
Minus-WFG8	5.0225e-1 (9.64e-4) -	5.1105e-1 (2.45e-3) -	2.8047e-1 (2.60e-2) -	4.9709e-1 (2.46e-3) -	4.9175e-1 (4.70e-3) -	3.8009e-1 (9.61e-3) -	**5.3034e-1 (2.86e-4)**
Mnus-WFG9	4.8394e-1 (6.14e-3) -	4.8979e-1 (5.90e-3) -	2.5767e-1 (1.52e-2) -	4.8032e-1 (5.38e-3) -	4.7359e-1 (6.49e-3) -	3.7077e-1 (1.25e-2) -	**5.2629e-1 (2.57e-3)**
+/-/=	0/6/0	0/6/0	0/6/0	0/6/0	0/6/0	0/6/0	

5 Conclusions

In this paper, we presented a hyper-heuristic method based on a genetic algorithm to tune MOEA/D for a set of problems. The tuning procedure was applied to six sets of test problems. The experimental results showed that a hyper-heuristic tuning procedure is beneficial (and necessary) for MOEA/D.

In this paper, MOEA/D was tuned for a relatively small set of test problems. It will be interesting to examine the usefulness of the proposed approach on a much larger set of test problems such as a set of all test problems examined in this paper (i.e., DTLZ, WFG and their minus versions). It will be also interesting to examine the performance of the tuned MOEA/D implementation on test problems which are not used for tuning. For example, the performance of MOEA/D tuned on WFG1–WFG6 can be examined on WFG7–WFG9.

As other future studies, the tuning procedure can be extended to other EMO algorithms. It is important to identify the dominant components and parameters of EMO algorithms, which is needed for specifying an appropriate algorithm design space. Another important research direction is to improve the efficiency of the hyper-heuristic procedure. Applications of the automatic tuning procedure to real-world problems is also an important research direction.

Acknowledgments. This work was supported by National Natural Science Foundation of China (Grant No. 62002152, 61876075), Guangdong Provincial Key Laboratory (Grant No. 2020B121201001), the Program for Guangdong Introducing Innovative and Enterpreneurial Teams (Grant No. 2017ZT07X386), Shenzhen Science and Technology Program (Grant No. KQTD2016112514355531), the Program for University Key Laboratory of Guangdong Province (Grant No. 2017KSYS008).

References

1. Schaffer, J.D.: Multiple objective optimization with vector evaluated genetic algorithms. In: Proceedings of 1st ICGA, pp. 93–100 (1985)
2. Zhang, Q., Li, H.: MOEA/D: a multiobjective evolutionary algorithm based on decomposition. IEEE Trans. Evol. Comput. **11**(6), 712–731 (2007)
3. Ishibuchi, H., Sakane, Y., Tsukamoto, N., Nojima, Y.: Evolutionary many-objective optimization by NSGA-II and MOEA/D with large populations. In: Proceedings of IEEE International Conference on System, Man, and Cybernetics, San Antonio, USA, pp. 1758–1763 (2009)
4. Ishibuchi, H., Akedo, N., Nojima, Y.: Behavior of multi-objective evolutionary algorithms on many-objective knapsack problems. IEEE Trans. Evol. Comput. **19**(2), 264–283 (2015)
5. Wessing, S., Beume, N., Rudolph, G., Naujoks, B.: Parameter tuning boosts performance of variation operators in multiobjective optimization. In: Schaefer, R., Cotta, C., Kołodziej, J., Rudolph, G. (eds.) PPSN 2010. LNCS, vol. 6238, pp. 728–737. Springer, Heidelberg (2010). https://doi.org/10.1007/978-3-642-15844-5_73
6. Pang, L.M., Ishibuchi, H., Shang, K.: Offline automatic parameter tuning of MOEA/D using genetic algorithms. In: 2019 IEEE SSCI, Xiamen, China, pp. 1889–1897 (2019)
7. Pang, L.M., Ishibuchi, H., Shang, K.: Decomposition-based multi-objective evolutionary algorithm design under two algorithm frameworks. IEEE Access **8**, 163197–163208 (2020)
8. Das, I., Dennis, J.E.: Normal-boundary intersection: a new method for generating the Pareto surface in nonlinear multicriteria optimization problems. SIAM J. Optim. **8**(3), 631–657 (1998)
9. Ishibuchi, H., Akedo, N., Nojima, Y.: A study on the specification of a scalarizing function in MOEA/D for many-objective knapsack problems. In: Nicosia, G., Pardalos, P. (eds.) LION 2013. LNCS, vol. 7997, pp. 231–246. Springer, Heidelberg (2013). https://doi.org/10.1007/978-3-642-44973-4_24
10. Sato, H.: Inverted PBI in MOEA/D and its impact on the search performance on multi and many-objective optimization. In: Proceedings of the 2014 Annual Conference on Genetic and Evolutionary Computation (GECCO 2014), Vancouver, Canada, pp. 645–652 (2014)
11. Yuan, Y., Xu, H., Wang, B., Zhang, B., Yao, X.: Balancing convergence and diversity in decomposition-based many-objective optimizers. IEEE Trans. Evol. Comput. **20**(2), 180–198 (2016)
12. Ishibuchi, H., Akedo, N., Nojima, Y.: Relation between neighborhood size and MOEA/D performance on many-objective problems. In: Purshouse, R.C., Fleming, P.J., Fonseca, C.M., Greco, S., Shaw, J. (eds.) EMO 2013. LNCS, vol. 7811, pp. 459–474. Springer, Heidelberg (2013). https://doi.org/10.1007/978-3-642-37140-0_35
13. Ishibuchi, H., Doi, K., Nojima, Y.: On the effect of normalization in MOEA/D for multiobjective and many-objective optimization. Complex Intell. Syst. **3**(4), 279–294 (2017)
14. Deb, K., Argawal, R.B.: Simulated binary crossover for continuous search space. Complex Syst. **9**, 115–148 (1995)
15. Deb, K., Deb, D.: Analysing mutation schemes for real-parameter genetic algorithms. Int. J. Artif. Intell. Soft Comput. **4**(1), 1–28 (2014)
16. Yu, X., Gen, M.: Introduction to Evolutionary Algorithms. Springer, London (2010). https://doi.org/10.1007/978-1-84996-129-5
17. Deb, K.: Multi-Objective Optimization Using Evolutionary Algorithms. Wiley, Chichester (2001)
18. Eiben, A.E., Smit, S.K.: Parameter tuning for configuring and analyzing evolutionary algorithms. Swarm Evol. Comput. **1**(1), 19–31 (2011)
19. Burke, E.K., Gendreau, M., Hyde, M., Kendall, G., Ochoa, G., Özcan, E., Qu, R.: Hyper-heuristics: a survey of the state of the art. J. Oper. Res. Soc. **64**, 1695–1724 (2013)

20. Deb, K., Thiele, L., Laumanns, M., Zitzler, E.: Scalable multi-objective optimization test problems. In: Proceedings of IEEE Congress on Evolutionary Computation (CEC), Honolulu, USA, pp. 825–830 (2002)
21. Huband, S., Hingston, P., Barone, L., While, L.: A review of multiobjective test problems and a scalable test problem toolkit. IEEE Trans. Evol. Comput. **10**(5), 477–506 (2006)
22. Ishibuchi, H., Setoguchi, Y., Masuda, H., Nojima, Y.: Performance of decomposition-based many-objective algorithms strongly depends on pareto front shapes. IEEE Trans. Evol. Comput. **21**(2), 169–190 (2017)
23. Li, K., Deb, K., Zhang, Q., Kwong, S.: An evolutionary many-objective optimization algorithm based on dominance and decomposition. IEEE Trans. Evol. Comput. **19**(5), 694–716 (2015)
24. Li, H., Zhang, Q.: Multiobjective Optimization Problems with Complicated Pareto Sets, MOEA/D and NSGA-II. IEEE Trans. Evol. Comput. **13**(2), 284–302 (2009)
25. Zhang, Q., Liu, W., Li, H.: The performance of a new version of MOEA/D on CEC 2009 unconstrained MOP test instances. In: Proceedings of the IEEE Congress on Evolutionary Computation (CEC), Trondheim, Norway, pp. 1–6 (2009)
26. Tian, Y., Cheng, R., Zhang, X., Jin, Y.: PlatEMO: A MATLAB platform for evolutionary multi-objective optimization [Educational Forum]. IEEE Comput. Intell. Mag. **12**(4), 73–87 (2017)
27. Ishibuchi, H., He, L., Shang, K.: Regular pareto front shape is not realistic. In: Proceedings of IEEE Congress on Evolutionary Computation (CEC), Wellington, New Zealand, pp. 2034–2041 (2019)
28. Xu, Q., Xu, Z., Ma, T.: A survey of multiobjective evolutionary algorithms based on decomposition: variants, challenges, and future directions. IEEE Access **8**, 41588–41614 (2020)

Diversity-Driven Selection Operator for Combinatorial Optimization

Eduardo G. Carrano[1], Felipe Campelo[1,2], and Ricardo H. C. Takahashi[3(✉)]

[1] Department of Electrical Engineering, Universidade Federal de Minas Gerais, Belo Horizonte, Brazil
egcarrano@ufmg.br

[2] School of Engineering and Applied Science, Aston University, Birmingham, UK
f.campelo@aston.ac.uk

[3] Department of Mathematics, Universidade Federal de Minas Gerais, Belo Horizonte, Brazil
taka@mat.ufmg.br

Abstract. A new selection operator for genetic algorithms dedicated to combinatorial optimization, the *Diversity Driven* selection operator, is proposed. The proposed operator treats the population diversity as a second objective, in a multiobjectivization framework. The Diversity Driven operator is parameterless, and features low computational complexity. Numerical experiments were performed considering four different algorithms in 24 instances of seven combinatorial optimization problems, showing that it outperforms five classical selection schemes with regard to solution quality and convergence speed. Besides, the Diversity Driven selection operator delivers good and considerably different solutions in the final population, which can be useful as design alternatives.

Keywords: Multiobjectivization · Combinatorial optimization · Genetic algorithms · Selection operator · Diversity preservation

1 Introduction

Genetic algorithms (GAs) are known to face the exploitation-exploration dilemma: too much focus on exploitation can result in premature convergence, whereas too much effort in exploration can negatively affect both convergence speed and solution quality. Different approaches have been proposed to address this issue, such as: changes in the population structure [1]; adaptation of crossover or mutation operators [18] or of operator parameters [18]; ranking-based selection schemes [2]; and niche-sharing/crowding-based selection schemes [14]. In those procedures, the algorithm can be adjusted in order to become either more exploitation-oriented or more exploration-oriented, and the choice of a specific setting should be performed specifically for each class of problem.

This work was supported by the Brazilian agencies CNPq, CAPES and FAPEMIG.

H. Ishibuchi et al. (Eds.): EMO 2021, LNCS 12654, pp. 178–190, 2021.
https://doi.org/10.1007/978-3-030-72062-9_15

An important drawback of those schemes is that the issue of diversity has been often considered as a merit factor by itself. However, diversity by itself does not provide a considerable contribution for the quality of the final solutions delivered; instead, diversity is valuable when the diverse solutions cover promising areas of search space [18,31]. Those references have not employed a search mechanism based on multiobjective optimization principles for dealing with those two objectives.

The formal setting of *multiobjective optimization* for handling the objectives of fitness and diversity was firstly introduced by [36]. Within this setting, it becomes possible to properly define which solutions should be found and stored, and to establish suitable search strategies for finding them. Other works followed the same approach of defining a diversity measure as an auxiliary objective in a multiobjective setting within evolutionary algorithms [8,17,23,32,37], but many rely on diversity metrics that depend on the computation of the distance from all individuals to all other ones, which causes a computational complexity of $\mathcal{O}(n^2)$ [17,23,32,36,37]. More recent works have focused on the use of multiobjective indicators to optimize diversity [28] or in developing frameworks for the development of algorithms aimed at optimising in the quality-diversity landscape [15], which indicates an ongoing interest into the development of evolutionary methods capable of generating both diverse and high-quality final populations.

This paper proposes a further development on the use of multiobjective optimization principles for achieving *useful diversity* in single-objective combinatorial optimization. A selection operator based on the selection mechanism of NSGA-II [16] is proposed: the *Diversity-Driven* (DiD) selection operator. The DiD operator is parameterless, and can be used as an off-the-shelf operator for building problem-specific GAs, keeping compatibility with almost any crossover, mutation, and local search operators. As a main feature, instead of using a raw distance measure as a new objective, the DiD operator builds an auxiliary objective function that combines the solution ordering induced by the original objective function with the distance from solution to solution, in the genotype space, between solutions that are consecutive in this ordering. To define distance metrics that adhere to combinatorial problems, the DiD operator uses geometric descriptions of combinatorial spaces [10,12,25,27].

The remainder of this paper is structured as follows: Sect. 2 presents some formal definitions and states the diversity-related objective function used for problem multiobjectivization. Section 3 describes the proposed diversity-driven selection operator (DiD operator). Distance metrics in discrete variable spaces are discussed in section 3.1. The test framework, including problems, instances, algorithms and comparison procedures is presented in Sect. 4. The results achieved by the DiD operator and the other five ones employed for benchmark are discussed in Sect. 5. Some concluding remarks are drawn in Sect. 6.

2 Multiobjective Selection Preference

The optimization problem of interest is defined as stated in (1), where $\mathcal{X}$ represents the space of discrete decision variables, $\mathbf{x} \in \mathcal{X}$ is a point in this space denoted as a *candidate solution* (an *individual*); and $f(\cdot) : \mathcal{X} \mapsto \mathbb{R}$ is an objective function to be minimized.

$$\begin{aligned} &\text{Find:} && \mathbf{x}^* = \arg\min_{\mathbf{x}} f(\mathbf{x}) \\ &\text{Subject to:} && \mathbf{x} \in \mathcal{X} \end{aligned} \tag{1}$$

Let $\mathcal{P} = \{\mathbf{x}_1, \ldots, \mathbf{x}_p\}$ denote a *population* of current candidate solutions. Assume that a *total order* $[\mathbf{x}_1 \leftarrow \mathbf{x}_2 \leftarrow \ldots \leftarrow \mathbf{x}_p]$ is induced on this set by the comparison operator $\leq$ over the objective function values, such that the set is ordered in increasing order of objective function values ($f(\mathbf{x}_i) \leq f(\mathbf{x}_{i+1}) \forall i \leq p-1$). Suppose, *w.l.g.*, that when there are replicas of the same point $\mathbf{x}_i$, those replicas are always ordered with consecutive indices. Although the aim of a single-objective optimization algorithm is to find the minimal element of this order, $\mathbf{x}^*$, it should be noticed that the direct usage of the raw ordering $\mathbf{x}_1 \leftarrow \mathbf{x}_2 \leftarrow \ldots \leftarrow \mathbf{x}_p$ inside the search engine may cause the loss of diversity because nothing prevents the selection of very similar individuals, or even several replicas of the same individual.

The formal statement of orderings that are oriented by more than one objective should consider the dominance relation operator ($\prec$):

$$\mathbf{a} \prec \mathbf{b} \Leftrightarrow \{a_i \leq b_i \ \forall i = 1, \ldots, m \text{ and } \exists j \in \{1, \ldots, m\} \text{ such that } a_j < b_j\} \tag{2}$$

It is possible that neither $\mathbf{a} \prec \mathbf{b}$ nor $\mathbf{b} \prec \mathbf{a}$ is true; in such a case $\mathbf{a}$ and $\mathbf{b}$ are said to be *non-comparable* vectors. Considering a set $\mathcal{S}$ of vectors, in which the $\prec$ operation can be applied, a *minimal element* $\mathbf{a}^*$ within $\mathcal{S}$ is an element for which the following relation holds:

$$\nexists\, \mathbf{a} \in \mathcal{S} \text{ such that } \mathbf{a} \prec \mathbf{a}^* \tag{3}$$

As a consequence of the existence of non-comparable solutions, there may exist several different minimal elements in $\mathcal{S}$. The solution of a multiobjective optimization problem is defined as the set of minimal elements of this partial order in the problem domain, the *Pareto-optimal set* of the problem.

The idea in this paper is to employ, in the selection operator, the ordering established by the dominance relation operator ($\prec$) on the population, considering the objective function and an auxiliary objective stated as a particular distance to a neighbor, as follows. Let the *left neighbor* of an element $\mathbf{x}_i$ be the element $\mathbf{x}_{i-1}$. A function f_α is defined as:

$$f_\alpha(\mathbf{x}_i) = -\|(\mathbf{x}_i) - (\mathbf{x}_{i-1})\| \tag{4}$$

in which $\|\cdot\|$ denotes a norm defined in $\mathcal{X}$, and $f_\alpha(\mathbf{x}_1) = -\infty$ by convention. In situations where multiple copies of the same individual are present in $\mathcal{P}$, the relation $\mathbf{x}_i = \mathbf{x}_{i+1}$ will hold for some pairs of individuals. In this case, $f(\mathbf{x}_i) = f(\mathbf{x}_{i+1})$ and $f_\alpha(\mathbf{x}_{i+1}) = 0$ hold. Now, define f_d as:

$$f_d(\mathbf{x}_i) = \begin{cases} f_\alpha(\mathbf{x}_i), & \text{if } \ f_\alpha(\mathbf{x}_i) < 0 \text{ or } f_\alpha(\mathbf{x}_{i-1}) < 0 \\ f_d(\mathbf{x}_{i-1}) + 1, & \text{otherwise} \end{cases} \tag{5}$$

Function f_d will return smaller values for individuals at a greater distance from their left neighbors, which can be interpreted as a measure of diversity. The following properties of f_d are of interest:

(i) The best individual, $\mathbf{x}_1$, has its smallest possible value: $f_d(\mathbf{x}_1) = -\infty$;

(ii) For any two individuals $\mathbf{x}_i \neq \mathbf{x}_{i+1}$ the condition $f_d(\mathbf{x}_{i+1}) < 0$ holds, even if $f(\mathbf{x}_i) = f(\mathbf{x}_{i+1})$;

(iii) For any set of $j+1$ copies of the same individual, $\mathbf{x}_i = \mathbf{x}_{i+1} = \ldots = \mathbf{x}_{i+j}$ the conditions $f_d(\mathbf{x}_i) < 0$, $f_d(\mathbf{x}_{i+1}) = 0$ and $f_d(\mathbf{x}_{k+i+1}) = f_d(\mathbf{x}_{k+i}) + 1$ for $k = 1, \ldots, j-1$ hold.

The idea of the proposed selection method (the DiD operator) is to apply a non-dominated sorting (NDS) selection procedure [16] to $\mathcal{P}$, using functions $f(\mathbf{x})$ and $f_d(\mathbf{x})$ as the selection criteria.

3 Diversity-Driven Selection Operator

The non-dominated sorting procedure [16] relies on the assignment of a front number to each individual in $\mathcal{P}$, such that the non-dominated individuals are put in front $\mathcal{F}_0$, the individuals that are dominated only by individuals in the front $\mathcal{F}_0$ are put in $\mathcal{F}_1$, the individuals that are dominated only by individuals in the fronts $\mathcal{F}_0$ and $\mathcal{F}_1$ are placed in $\mathcal{F}_2$, and so forth.

By proceeding in this way with functions $f(\mathbf{x})$ and $f_d(\mathbf{x})$, the resulting fronts $\mathcal{F}_i$ are such that the best individual is always the only one in the front $\mathcal{F}_0$, since it dominates all other individuals in $\mathcal{P}$. This ensures that the best individual is always selected. The second front, $\mathcal{F}_1$, will contain the second copy of the best individual (if any), the first copy of the second best individual, and also the first copy of some other individuals (only one copy per individual). The following front $\mathcal{F}_2$ will contain the third copy of the best individual (if any), the second copy of the second best individual (if any), and other individuals. This pattern is repeated in the other fronts until all individuals are placed in some front.

Consider two individuals $\mathbf{x}_b$ and $\mathbf{x}_a$ such that $f(\mathbf{x}_b) > f(\mathbf{x}_a)$. These individuals will be placed in the same front only if $f_d(\mathbf{x}_b) < f_d(\mathbf{x}_a)$, which may happen if: (i) $\mathbf{x}_b$ is more distant from its left neighbor than $\mathbf{x}_a$, or (ii) the previous fronts already selected include more copies of $\mathbf{x}_a$ than copies of $\mathbf{x}_b$. In both cases, the situation can be interpreted as $\mathbf{x}_b$ being more important for diversity than $\mathbf{x}_a$. Otherwise, $\mathbf{x}_b$ will be placed in a front with higher index than $\mathbf{x}_a$.

After finding the k fronts $\{\mathcal{F}_0, \ldots, \mathcal{F}_{k-1}\}$, the individuals within each front $\mathcal{F}_i$ are ordered according to the one-point contribution metric [7]. For two-objective problems, such as the one considered here, the one-point contribution of a given point $\mathbf{x}_a$ is defined as the area A shown in Fig. 1.

The set $\mathcal{S}$ of the selected individuals resulting from DiD selection operator is filled with the first n_s individuals, according to the total ordering that is established considering first the increasing ordering of fronts, breaking the ties by the decreasing ordering of one-point contributions within any front. Both

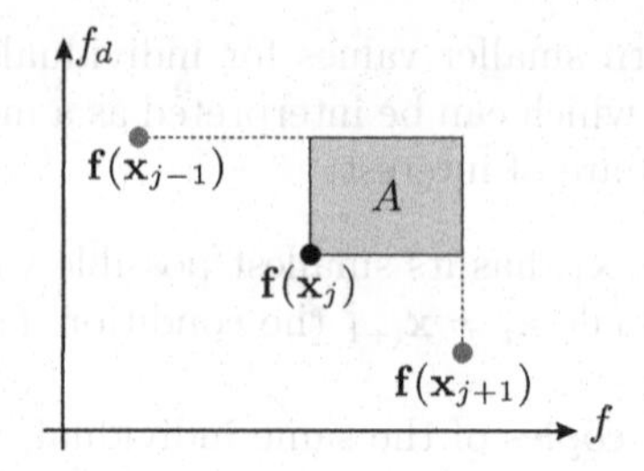

Fig. 1. Evaluation of the one-point contribution for the individual $\mathbf{x}_j$.

population diversity and objective function values are considered in the selection by the fitness function $f(\cdot)$ and the diversity function $f_d(\cdot)$.

The computational complexity of the DiD operator is found by analyzing the operations within it. Population sorting can be accomplished in $\mathcal{O}(n \log n)$ time. Evaluation of f_d requires $\mathcal{O}(n)$ distance calculations. Efficient implementations of NDS [38] have complexity of $\mathcal{O}(m \cdot n \log n)$, which reduces to $\mathcal{O}(n \log n)$ since $m = 2$. One-point contribution evaluation requires, in the worst case, $n-2$ multiplication operations. The complexity of the DiD selection procedure is therefore determined by population sorting and NDS: $\mathcal{O}\left(n \log n\right)$ in the worst case.

3.1 Distance Metrics in Combinatorial Vector Spaces

The proposed selection operator relies on the assumption of availability of a distance between the individuals in the combinatorial problem. A brief explanation of the formalism for defining distances in combinatorial spaces is presented next.

The geometric structure of the space of a combinatorial problem may be stated using the *edit move*, defined as the smallest change of variables that transforms one solution into another one [27]. A *path* $\mathbf{P}$ from $\mathbf{x}^a$ to $\mathbf{x}^b$ is a sequence of solutions that start in $\mathbf{x}^a$ and reaches $\mathbf{x}^b$ by edit move steps, and the length of a path is the number of edit moves that generate all solutions on the path. The *distance* from $\mathbf{x}^a$ to $\mathbf{x}^b$ is the length of the path of minimum length.

A more finely grained definition of a norm can be stated, the *weighted norm*, as proposed in [12] for tree networks. Such a definition starts with the definition of weights for the several components of the decision vector. Each component of each vector receives its weight, such that $x_i^a \mapsto w_i^a$. Those weights are defined such that the components whose change have a greater impact in the individual phenotype receive greater weight. The possibility of defining such weights depends on the problem. Then, the lengths of the paths between individuals become weighted. Let $\ell_w(\cdot)$ denote a weighted length of a path, and let $\mathbf{x}^b$ be generated from $\mathbf{x}^a$ by an edit move that changes only the components i and j. The weighted length of this edit move is defined as:

$$\ell_w(\mathbf{x}^a - \mathbf{x}^b) = w_i^a + w_i^b + w_j^a + w_j^b \quad (6)$$

The weighted length of a path $\mathbf{P} = \{\mathbf{p}^1, \mathbf{p}^2, \ldots, \mathbf{p}^m\}$ in which $\mathbf{p}^1 = \mathbf{x}^a$ and $\mathbf{p}^m = \mathbf{x}^b$ becomes defined by the sum of the weighted lengths of the edit moves that compose the path:

$$\ell_w(\mathbf{x}^a - \mathbf{x}^b) = \ell_w(\mathbf{p}^m - \mathbf{p}^{m-1}) + \ldots + \ell_w(\mathbf{p}^2 - \mathbf{p}^1) \tag{7}$$

From this definition of weighted path length, comes the weighted distance between individuals.

4 Test Framework

4.1 Test Problems

Seven combinatorial problems are considered: (i) degree-constrained minimum spanning tree; (ii) quadratic minimum spanning tree; (iii) optimal communication spanning tree; (iv) linear ordering; (v) common due-date single machine scheduling; (vi) makespan in unrelated parallel machine scheduling, and; (vii) generalized assignment problem. These problems can be divided in three classes: network problems, permutation problems, and assignment problems.

Degree-constrained minimum spanning tree problem (DCMST) is defined as the search for the minimum cost spanning tree with constraints on the degrees of its vertices. In [22], the authors propose a class of instances known as SHRD (structured hard), which are harder than the Euclidean ones. This class is used in the experiments here. Two instances, with 30 (DCMST30) and 50 vertices (DCMST50), have been generated, following the procedure described in [22].

Quadratic minimum spanning tree problem (QMST) involves searching for a spanning tree which minimizes a quadratic cost function that depends on the cost of edges and on the interaction between each pair of edges [33]. Instances with 25 (QMST25) and 50 nodes (QMST50) [12] are considered in this study.

Optimal communication spanning tree problem (OCST) which aims at obtaining the minimum cost spanning tree, with the cost considering the communication requirements between each pair of nodes in the system [33]. Two instances, OCST25 and OCST50 [12], are used in this paper.

Linear ordering (LinOrder) problem [21], in which there is a set of n objects that are to be ordered in a linear sequence. For every pair (i, j) of objects there are coefficients $c_{i,j}$ that define the preference for having i before j. The goal is to find a linear sequence that maximizes the sum of the coefficients that are compatible with the preferences. Six instances from the LOLIB-library [30] are used in this study: be75eec, be75np, be75oi, stabu1, stabu2 and stabu3. The first three have 50 objects and the others have 60 objects.

Common due date single machine scheduling (Scheduling) which consists of the search for the optimal processing sequence for a given set of tasks, in such a way that the weighted sum of earliness and tardiness is minimized. Instances sch50

(50 tasks, $K = 1$ and $h = 0.60$), sch100 (100 tasks, $K = 5$ and $h = 0.60$) and sch200 (200 tasks, $K = 1$ and $h = 0.60$) of the OR-Library [5] are used.

Makespan in unrelated parallel machine scheduling (Makespan) consists in assigning set of tasks to a set of machines, while guaranteeing that each task is assigned to a single machine [29]. In this work the *makespan* is the function to be minimized. Three random instances have been generated: $Mk25 \times 5$ (25 tasks, 5 machines), $Mk50 \times 10$ (50 tasks, 10 machines), $Mk100 \times 10$ (100 tasks, 10 machines). The processing times have been generated as integer numbers in the interval $1 \leq p_{i,j} \leq 10$, following a uniform distribution.

Generalized assignment problem (GAP), which consists in assigning n tasks to m agents. The objective function is the total cost required for performing the tasks [4]. Six instances from the OR-Library [5] are considered in this study, and coded as "*Am* × *n*": *A*5 × 100 (5 agents, 100 tasks), *A*5 × 200, *A*10 × 100, *A*10 × 200, *A*20 × 100 and *A*20 × 200.

4.2 Algorithms

Four different genetic algorithms are used for solving the problems described above. These algorithms vary on the crossover, mutation, and local search operators employed, which are chosen according to the specific problem under consideration, but have exactly the same generational structure. The operators employed by each algorithm are briefly described below:

VETreeOpt. (DCMST, QMST and OCST) is an algorithm employed for network problems. It has been built with the same crossover (binary) and mutation (unary) operators as proposed in [12]. These operators have been chosen because they have achieved very good performance on this class of problems.

PermutationGA. (LinOrder and Scheduling) is an algorithm used for permutation problems, in which each individual is encoded as a permutation. Partially Matched Crossover (PMX) [19] and Swap Mutation [19] are employed to perform crossover and mutation. In the local search operator, all first order swap operations are tested, one by one, until the first improvement is reached.

AssignmentGA. (Makespan) encodes the individuals as chains of integer numbers, in which the integer lying in position i (x_i) identifies the machine that should perform task i. Crossover and mutation are performed using Uniform Crossover [35] and Flip Mutation [13]. The local search operator evaluates the tasks (one by one) on every machine, until it detects the first improvement.

GAPGA. (GAP) is very similar to the AssignmentGA, since Makespan and GAP have similar structures. The only difference of GAPGA is the mutation operation, which is performed using the technique proposed in [34]. This operator has been designed specifically for GAP, and cannot be employed on unrelated parallel machine scheduling.

Each one of those algorithms is tested with six selection operators:

- DiD: the Diversity-Driven operator proposed in this paper;
- SW: stochastic wheel without ranking [20];
- SWLR: stochastic wheel with linear ranking [2];
- SUS: stochastic universal sampling without ranking [24];
- SUSLR: stochastic universal sampling with linear ranking [3];
- ST: stochastic binary tournament [20,24].

Therefore, six different versions of each algorithm (one for each selection operator) are employed for solving each instance. All algorithms are executed considering the same parameters: population size of 50 individuals; stop criterion: $1000n$ function evaluations (FEs)[1]; crossover: 0.80 per pair; mutation: 0.40 per individual; interval between local searches: 50 generations; maximum number of FEs used in local search: 500; algorithm runs per instance: 50. These settings are in agreement with what is commonly employed in the literature. In addition, preliminary testing of the algorithms showed that they are generally robust to small variations on these parameters.

4.3 Performance Assessment

Two merit criteria are employed for comparing the selection operators:

Convergence Quality Criterion (CQC): the quality of the outcomes delivered by the algorithm is estimated using the objective function value of the best individual obtained at the end of each algorithm run. For each algorithm and for each run, the value of the objective function of the final best individual is stored. After r runs, the CQC performance of an algorithm on a given instance is determined from the r objective function values stored.

Convergence Speed Criterion (CSC), which estimates how fast an algorithm converges to a reasonable solution. For each algorithm and for each run, the best individual at the end of each generation and the number of function evaluations required for reaching that solution are stored. The worst final solution amongst all runs of all algorithms is found. Then, for each algorithm and for each run, the number of function evaluations that has been required to reach the first solution that is better than or equal to that worst one is found. The CSC metric is determined from this information.

The CQC and CSC values achieved by the algorithms in each individual instance are analyzed using the procedure described below:

1. For CQC and CSC criteria:
   (a) For each Instance k:
      i. For each pair of operators Op_i, Op_j, with $i < j$, estimate the p-value for the comparison of means of the algorithm equipped with operators Op_i and Op_j ($pv_{i,j}$) based on Welch's t-test [26].

[1] n is the problem size: *number of vertices* for DCMST, QMST and OCST; *number of objects* for LinOrder. *number of tasks* for Scheduling, Makespan and GAP.

ii. Using the p-values obtained on step 1(a)i, build a partial order for the methods using the Benjamini-Hochberg Multiple Comparison procedure [6], considering a global significance $q^* = 0.05$.

(b) Based on the partial orders achieved for CQC and CSC on step 1(a)ii, use the Pareto dominance principle to identify the set of efficient algorithms.

The output of this procedure is a set of algorithms which are efficient with regard to CQC and CSC. An algorithm i is efficient if no other algorithm j is better than i in one criterion without being worse in the other one. This comparison approach is similar to the one proposed in [11]. The only difference of the newer procedure is the replacement of the ANOVA and Tukey tests by the Welch's t-test and the Benjamini-Hochberg Multiple Comparison procedure, which makes the process more robust to heteroscedasticity [9].

This procedure makes it possible to establish a ranking of the algorithms: (*i*) algorithms that are not significantly outperformed by any others are ranked in the first position; (*ii*) the algorithms that lose only to rank 1 methods are placed on rank 2, and so forth. In the specific case of this work, the statistical comparisons generate two orders for the algorithms, one for CQC and other for CSC. These orders are considered in a Pareto dominance analysis, in order to find the set of efficient operators in each instance.

5 Results

A summary of the results achieved in all instances is shown in Table 1. The DiD operator was not outperformed by any other selection operator, staying in the first position for all instances in both criteria, CQC and CSC. In addition, it was the only operator that was efficient in all 24 instances. Amongst the other five operators, SUSLR was efficient in 8 instances, SWLR and ST were efficient in 7, and SW and SUS were Pareto-optimal in only a single instance each.

The genetic algorithm using the DiD operator was also able to determine new "best known solutions" for two scheduling instances, namely sch50 and sch100, when compared to the current best reported in the OR-Library (see http://people.brunel.ac.uk/~mastjjb/jeb/orlib/schinfo.html):

- **sch50** (k = 1, h = 0.60), former best: 17990, new best: **17976**
- **sch100** (k = 5, h = 0.60), former best: 55291, new best: **55286**

The genetic algorithms using other selection operators were not able to achieve those new best objective function values.

Table 1. Operator positions in each criterion for each instance

Instance	CQC						CSC					
	DiD	ST	SUS	SUSLR	SW	SWLR	DiD	ST	SUS	SUSLR	SW	SWLR
DCMST30	1	2	3	2	3	2	1	2	3	2	3	2
DCMST50	1	2	3	2	3	2	1	2	3	2	3	2
QMST25	1	1	1	1	1	1	1	1	2	1	2	1
QMST50	1	1	1	1	1	1	1	2	4	2	4	3
OCST25	1	1	2	1	2	1	1	1	1	1	1	1
OCST50	1	1	1	1	1	1	1	2	2	2	2	2
be75eec	1	1	1	1	1	1	1	1	2	1	2	1
be75np	1	2	2	2	2	2	1	1	2	1	2	1
be75oi	1	2	2	2	2	2	1	1	2	1	2	1
stabu1	1	1	1	1	1	1	1	1	2	1	3	1
stabu2	1	1	1	1	1	1	1	1	2	1	3	1
stabu3	1	2	2	2	2	2	1	1	2	1	3	1
sch50	1	1	1	1	1	1	1	2	3	1	3	2
sch100	1	1	2	1	2	2	1	2	4	2	4	3
sch200	1	1	2	1	2	1	1	1	2	1	2	1
Mk25 × 5	1	2	1	2	2	2	1	2	2	2	2	2
Mk50 × 10	1	4	2	6	3	5	1	2	2	2	2	2
Mk100 × 10	1	3	2	5	2	4	1	1	1	1	1	1
A5 × 100	1	1	1	1	1	1	1	2	2	2	2	2
A10 × 100	1	1	1	1	1	1	1	3	2	3	3	3
A20 × 100	1	1	1	1	1	1	1	1	1	1	1	1
A5 × 200	1	1	1	1	1	1	1	3	2	3	3	3
A10 × 200	1	2	2	2	2	2	1	3	2	2	3	3
A20 × 200	1	2	2	2	2	2	1	2	2	2	2	2

6 Conclusions

This paper proposed a multiobjectivization approach for dealing with the exploration/exploitation dilemma in genetic algorithms for combinatorial problems. The trade-off between population fitness and diversity is processed inside a single parameter-less operator, the *Diversity-Driven (DiD) selection operator*, which selects the individuals according to the objective function and a function that is built over a specific measure of distance to one neighbor. The specific neighborhood structure induced by the solution ordering scheme employed within DiD allows the computation of distances with $\mathcal{O}(n \log n)$ worst-case complexity. The DiD operator was conceived as an off-the-shelf operator that can be used within problem-specific GA's while keeping compatibility with almost any crossover, mutation and local search operators.

Numerical experiments were performed on 24 instances of seven different combinatorial problems, for which some specific GA's were built and equipped with either the DiD operator or several other existing selection methods. The comparisons were performed with regards to convergence speed and solution quality. The results obtained indicate that the proposed DiD operator was able to significantly and consistently outperform the competing selection approaches in both merit criteria.

Additional studies are needed in order to assess the behavior of DiD operator on combinatorial problems with different structures, for instance in the case of satisfiability problems (maximum satisfiability, constraint satisfaction). It is expected that the development of distance metrics specially adapted for different problem structures will allow the application of the DiD operator in a greater variety of situations.

Other extensions of the DiD operators may be interesting to explore. These include the incorporation of preference information into the operator, to allow differential weighting of the dual objectives of solution quality or diversity; and the exploration of this operator in other problem contexts, such as continuous optimization.

References

1. Bäck, T., Fogel, D.B., Michalewicz, Z. (eds.): Handbook on Evolutionary Computation. Oxford University Press, Oxford (1997)
2. Baker, J.E.: Adaptive selection methods for genetic algorithms. In: Proceedings of the 1st International Conference on Genetic Algorithms, pp. 101–111 (1985)
3. Baker, J.E.: Reducing bias and inefficiency in the selection algorithm. In: Proceedings of the 2nd International Conference on Genetic Algorithms and their Application, pp. 14–21 (1987)
4. Balachandran, V.: An integer generalized transportation model for optimal job assignment in computer networks. Oper. Res. **24**, 742–759 (1976)
5. Beasley, J.E.: OR-Library. http://people.brunel.ac.uk/mastjjb/jeb/orlib/schinfo.html. Accessed 20 Sep 2020
6. Benjamini, Y., Hochberg, Y.: Controlling the false discovery rate: a practical and powerful approach to multiple testing. J. Royal Stat. Soc. Ser. B **57**(1), 289–300 (1995)
7. Brandstreet, L., While, L., Barone, L.: A fast incremental hypervolume algorithm. IEEE Trans. Evol. Comp. **12**(6), 714–723 (2008)
8. Bui, L.T., Abbass, H.A., Branke, J.: Multiobjective optimization for dynamic environments. In: Proceedings on IEEE Congress on Evolutionary Computation (CEC 2005), Edinburgh, UK, vol. 3, pp. 2349–2356 (2005)
9. Campelo, F., Takahashi, F.: Sample size estimation for power and accuracy in the experimental comparison of algorithms. J. Heuristics **25**(2), 305–338 (2018). https://doi.org/10.1007/s10732-018-9396-7
10. Carrano, E.G., Ribeiro, G., Cardoso, E., Takahashi, R.H.C.: Subpermutation based evolutionary multiobjective algorithm for load restoration in power distribution networks. IEEE Trans. Evol. Comp. **20**, 546–562 (2016)

11. Carrano, E.G., Wanner, E.F., Takahashi, R.H.C.: A multicriteria statistical based comparison methodology for evaluating evolutionary algorithms. IEEE Trans. Evol. Comp. **15**(6), 848–870 (2011)
12. Carrano, E.G., Takahashi, R.H.C., Fonseca, C.M., Neto, O.M.: Nonlinear network optimization - an embedding vector space approach. IEEE Trans. Evol. Comp. **14**(2), 206–226 (2010)
13. Chicano, F., Alba, E.: Exact computation of the expectation curves of the bit-flip mutation using landscapes theory. In: Proceedings of the Genetic and Evolutionary Computation Conference (GECCO 2011), Dublin, Ireland, pp. 2027–2034 (2011)
14. Cioppa, A., De Stefano, C., Marcelli, A.: Where are the niches? Dynamic fitness sharing. IEEE Trans. Evol. Comp. **11**(4), 453–465 (2007)
15. Cully, A., Demiris, Y.: Quality and diversity optimization: a unifying modular framework. IEEE Trans. Evol. Comput. **22**(2), 245–259 (2018)
16. Deb, K., Pratap, A., Agarwal, S., Meyarivan, T.: A fast and elitist multiobjective genetic algorithm: NSGA-II. IEEE Trans. Evol. Comp. **6**, 182–197 (2002)
17. Garcia-Najera, A., Bullinaria, J.A.: An improved multi-objective evolutionary algorithm for the vehicle routing problem with time windows. Comp. Oper. Res. **38**, 287–300 (2011)
18. Ginley, B.M., Maher, J., O'Riordan', C., Morgan, F.: Maintaining healthy population diversity using adaptive crossover, mutation, and selection. IEEE Trans. Evol. Comp. **15**(5), 692–714 (2011)
19. Goldberg, D.E., Lingle, R.: Alleles, loci, and the traveling salesman problem. In: Proceedings of the Conference on Genetic Algorithms, pp. 154–159 (1985)
20. Goldberg, D.: Genetic Algorithms in Search, Optimization and Machine Learning, 1st edn. Addison-Wesley, Boston (1989)
21. Grötschel, M., Jünger, M., Reinelt, G.: Optimal triangulation of large real world input-output matrices. Statistische Hefte **25**, 261–295 (1984)
22. Krishnamoorthy, M., Ernst, A.T., Sharaiha, Y.M.: Comparison of algorithms for the degree constrained minimum spanning tree. J. Heuristics **7**, 587–611 (2001)
23. Landa Silva, J.D., Burke, E.K.: Using diversity to guide the search in multi-objective optimization, chapter 30, pp. 727–751. World Scientific (2004)
24. Luke, S.: Essentials of Metaheuristics. Lulu (2016). https://cs.gmu.edu/sean/book/metaheuristics/Essentials.pdf
25. Martins, F.V.C., Carrano, E.G., Wanner, E.F., Takahashi, R.H.C., Mateus, G.R., Nakamura, F.G.: On a vector space representation in genetic algorithms for sensor scheduling in wireless sensor networks. Evol. Comp. **22**, 361–403 (2014)
26. Montgomery, D., Runger, G.: Applied Statistics and Probability for Engineers. Wiley, Newyork (2003)
27. Moraglio, A.: Towards a geometric unification of evolutionary algorithms. Ph.D. thesis, University of Essex (2007)
28. Neumann, A., Gao, W., Wagner, M., Neumann, F.: Evolutionary diversity optimization using multi-objective indicators. In: Proceedings of the Genetic and Evolutionary Computation Conference. ACM (2019)
29. Pfund, M., Fowler, J.W., Gupta, J.N.D.: A survey of algorithms for single and multi-objective unrelated parallel-machine deterministic scheduling problems. J. Chin. Inst. Ind. Eng. **21**, 230–241 (2004)
30. Reinelt, G.: LOLIB. http://comopt.ifi.uni-heidelberg.de/software/LOLIB/. Accessed 20 Sep 2020
31. Segura, C., Coello Coello, C.A., Miranda, G., León, C.: Using multi-objective evolutionary algorithms for single-objective optimization. Ann. Oper. Res. **11**(3), 201–228 (2013). https://doi.org/10.1007/s10288-013-0248-x

32. Segura, C., Coello Coello, C.A., Segredo, E., Miranda, G., Leon, C.: Improving the diversity preservation of multi-objective approaches used for single-objective optimization. In: Proceedings of the IEEE Congress on Evolutionary Computation (CEC 2013), Cancun, Mexico, pp. 3198–3205 (2013)
33. Soak, S., Corne, D.W., Ahn, B.: The edge-window-decoder representation for tree-based problems. IEEE Trans. Evol. Comp. **10**, 124–144 (2006)
34. Subtil, R.F., Carrano, E.G., Souza, M.J., Takahashi, R.H.C.: Using an enhanced integer NSGA-II for solving the multiobjective generalized assignment problem. In: Proceedings of the IEEE World Congress on Computational Intelligence, Barcelona, Spain (2010)
35. Syswerda, G.: Uniform crossover in genetic algorithms. In: Proceedings of the 3rd International Conference on Genetic Algorithms, pp. 2–9 (1989)
36. Toffolo, A., Benini, E.: Genetic diversity as an objective in multi-objective evolutionary algorithms. Evol. Comp. **11**(2), 151–167 (2003)
37. Wessing, S., Preuss, M., Rudolph, G.: Niching by multiobjectivization with neighbor information: trade-offs and benefits. In: Proceedings of the IEEE Congress on Evolutionary Computation (CEC 2013), Cancun, Mexico, pp. 103–110 (2013)
38. Zhang, X., Tian, Y., Cheng, R., Jin, Y.: An efficient approach to nondominated sorting for evolutionary multiobjective optimization. IEEE Trans. Evol. Comput. **19**(2), 201–213 (2015)

Dynamic Multi-objective Optimization

An Online Machine Learning-Based Prediction Strategy for Dynamic Evolutionary Multi-objective Optimization

Min Liu[1,2](✉), Diankun Chen[1], Qiongbing Zhang[1], and Lei Jiang[1,2]

[1] School of Computer Science and Engineering, Hunan University of Science and Technology, Xiangtan 4111201, China
liumin_xt@163.com, 2059426443@qq.com, mrtly2@whu.edu.cn, jleihn@hotmail.com

[2] Key Laboratory of Knowledge Processing and Networked Manufacturing, College of Hunan Province, Xiangtan 4111201, China

Abstract. Due to the impact of environmental changes, dynamic evolutionary multi-objective optimization algorithms need to track the time-varying Pareto optimal solution set of dynamic multi-objective optimization problems (DMOPs) as soon as possible by effectively mining historical data. Since online machine learning can help algorithms dynamically adapt to new patterns in the data in machine learning community, this paper introduces Passive-Aggressive Regression (PAR, a common online learning technology) into dynamic evolutionary multi-objective optimization research area. Specifically, a PAR-based prediction strategy is proposed to predict the new Pareto optimal solution set of the next environment. Furthermore, we integrate the proposed prediction strategy into the multi-objective evolutionary algorithm based on decomposition with a differential evolution operator (MOEA/D-DE) to handle DMOPs. Finally, the proposed prediction strategy is compared with three state-of-the-art prediction strategies under the same dynamic MOEA/D-DE framework on CEC2018 dynamic optimization competition problems. The experimental results indicate that the PAR-based prediction strategy is promising for dealing with DMOPs.

Keywords: Evolutionary multi-objective optimization · Dynamic environment · Prediction strategy · Online machine learning

1 Introduction

In the real world, there exist a lot of dynamic multi-objective optimization problems (DMOPs) which usually involve several conflicting objective functions and may be subject to a number of constraints. Furthermore, the objective functions, constraints and/or relative parameters of DMOPs may change over time [1].

H. Ishibuchi et al. (Eds.): EMO 2021, LNCS 12654, pp. 193–204, 2021.
https://doi.org/10.1007/978-3-030-72062-9_16

In this work, we consider the following continuous DMOPs:

$$\begin{cases} \min\limits_{v \in \Omega} F(v,t) = \left(f_1(v,t), f_2(v,t), ..., f_m(v,t)\right)^{\mathrm{T}} \\ s.t. \quad h_i(v,t) \leq 0, i = 1,2,...,p; \quad l_j(v,t) = 0, j = 1,2,...,r \end{cases} \tag{1}$$

where v is the decision vector defined in the decision space Ω, $F(v,t)$ is the objective vector that calculates a numerical value for each objective of the solution v at time t, m is the number of objectives. The functions h_i and l_j are the inequality and equality constraints which change over t, respectively. As a result, the Pareto optimal solution set (POS) in the decision space and/or Pareto optimal front (POF) in the objective space of DMOPs may also change over time, which imposes a big challenge to evolutionary multi-objective optimization (EMO) researches.

Over the past decade or so, there are increasing interests in designing dynamic evolutionary multi-objective optimization (DEMO) algorithms to handle DMOPs. The task of a DEMO is to trace the movement of the POS and/or POF with reasonable computational costs. How to take action to respond to a new change is the key issue of DEMO, since an effective change reaction method can help the DEMO algorithm adapt to the new environment quickly. A lot of change reaction methods have been proposed to handle changes. These methods include diversity introduction after a change occurs [2,3] or maintain diversity throughout the run [4], multi-population approaches [5], memory schemes [6,7], and prediction strategies [8–14].

In recent years, prediction strategies have gained much attention. This kind of methods predicts POS of the new environment in advance by exploiting different machine learning techniques. According to the amount of historical information used for prediction, the prediction strategies can be roughly categorized into two types: local information-based prediction (LIP) and time series-based prediction (TSP). LIP approaches usually only use information obtained from several recent environments to make predictions, so that the history data has not been fully mined. Some representative LIP methods are simple linear model [9], Kalman filter [11], differential model [12], and transfer learning [13]. In contrast, TSP approaches make use of much more historical environmental information than LIP approaches. TSP approaches first collect approximate optimal solutions from past environments as training data, and then feed them to some machining learning model to predict the location of the new optimal solutions. Therefore, TSP approaches generally have higher training costs than LIP approaches. Autoregressive model [8,10] and support vector regression [14] are some well-known models used in TSP approaches.

The Motivation of This Work. Dealing with a dynamic and uncertain environment is not a unique challenge to evolutionary optimization. There are also active research activities within online machine learning community that try to tackle similar challenges from changing data streams [15]. Online learning is used to update the prediction model for future data at each time step, as opposed to batch learning techniques which generate the best predictor by learning on the entire training data set at once [16]. Online learning is also used in situations where it is necessary for the algorithm to dynamically adapt to new patterns in the data [17]. In addition, online learning algorithms are typically easy to implement and generally have low computational cost [18], which is particularly suitable for dynamic optimization. Inspired by the aforementioned advantages of online learning, we introduce online learning into DEMO in order to help dynamic optimization algorithms better adapt to changing environments.

The Contribution of This Work. We first introduce Passive-Aggressive Regression (PAR, a common online learning technique) into DEMO, and then propose a PAR-based prediction strategy (PARS) to react the new change of DMOPs. Furthermore, we incorporate this prediction strategy into MOEA/D-DE [19] to deal with DMOPs. Finally, in order to fairly evaluate the performance of the proposed prediction strategy, it is compared with three state-of-the-art prediction methods under the same dynamic MOEA/D-DE framework.

The rest of this paper is organized as follows. Section 2 provides a brief survey on the prediction strategies for DEMO, and introduces the basic knowledge of PAR. In Sect. 3, PARS is proposed, and then it is incorporated into MOEA/D-DE to handle DMOPs. In the following section, some experimental results are reported to show the effectiveness of our proposed strategy. The final section concludes this paper.

2 Related Works

2.1 Prediction Strategies for DEMO Algorithm

Local Information-Based Prediction Strategies: Zhou et al. [9] proposed a prediction strategy (PRE) in 2007. In PRE, a simple linear prediction model with Gaussian noise is used to reinitialize population in new environment by utilizing only optimal individuals of last two environments. The idea of this simple linear prediction model was employed by other algorithms which focused predicted optimal individuals on some special points [7,20] or multiple directions [21]. Instead of making predictions based on data from the previous two environments, Cao et al. [12] proposed a differential model for predicting the movement of the centroid by its locations in three former environments. Muruganantham et al. [11] proposed a Kalman filter (KF) prediction model to guide the search toward a new POS. This KF model was essentially implemented as first- or second-order linear prediction model [14]. Recently, Jiang et al. [13] only utilized the approximate optimal solutions under two sequential environments to construct a prediction model by adopting transfer learning.

Time Series-Based Prediction Strategies: Zhou et al. [10] proposed a population-based prediction strategy (PPS). The movement of centers of the population is learnt by a univariate autoregressive model. The predicted center point and estimated manifold are utilized to generate a new population for the next environment. Cao et al. [14] proposed a support vector regression (SVR) predictor for better solving DMOPS with nonlinear correlation. Whenever there is a new change taken place, the historical approximate optimal solutions in time series are first used to train a group of SVR models. Then the POS of the next environment can be predicted by using the trained SVR models.

2.2 Online Passive-Aggressive Regression

In 2006, Crammer et al. [18] proposed a Passive-Aggressive Regression (PAR) which has become a common online machine learning method. To deal with noises of the samples, two PAR variants (i.e., PAR-I and PAR-II) were also proposed in [18]. In this paper, we only focus on PAR-II just for simplicity. On each time step, the PAR-II algorithm receives an instance $\boldsymbol{x}^t \in R^n$ and predicts a target value $\hat{y}^t \in R$ using its regression function, that is, $\hat{y}^t = \boldsymbol{w}^t \cdot \boldsymbol{x}^t$ where $\boldsymbol{w}^t$ is the incrementally learned vector. After prediction, the algorithm obtains the true target value y^t and suffers an instantaneous loss. PAR-II uses the ε-insensitive hinge loss function:

$$\ell_\varepsilon(\boldsymbol{w}^t; (\boldsymbol{x}^t, y^t)) = \begin{cases} 0, & |\boldsymbol{w}^t \cdot \boldsymbol{x}^t - y^t| \le \varepsilon \\ |\boldsymbol{w}^t \cdot \boldsymbol{x}^t - y^t| - \varepsilon, & otherwise \end{cases} \tag{2}$$

where ε is a positive parameter which controls the sensitivity to prediction errors. At the end of each time step, the algorithm uses $\boldsymbol{w}^t$ and the example $(\boldsymbol{x}^t, y^t)$ to generate a new weight vector $\boldsymbol{w}^{t+1}$, which will be utilized to continue making a prediction on the next time step. Firstly, $\boldsymbol{w}^1$ is initialized to (0, . . . ,0). On each time step, the PAR-II algorithm updates the new weight vector to be,

$$\boldsymbol{w}^{t+1} = \arg \min_{\boldsymbol{w} \in R^n} \frac{1}{2}||\boldsymbol{w} - \boldsymbol{w}^t||^2 + C\xi^2 \quad s.t. \quad \ell_\varepsilon(\boldsymbol{w}; (\boldsymbol{x}^t, y^t)) \le \xi \tag{3}$$

where ξ is a non-negative slack variable and C is a positive parameter which controls the influence of the slack term on the objective function. Specifically, larger values of C imply a more aggressive update step and therefore C is referred to as the aggressiveness parameter of the algorithm. Using the shorthand $\ell^t = \ell_\varepsilon(\boldsymbol{w}^t; (\boldsymbol{x}^t, y^t))$, the update given in Eq. (3) has a closed form solution as follows,

$$\boldsymbol{w}^{t+1} = \boldsymbol{w}^t + sign(y^t - \hat{y}^t)\tau^t \boldsymbol{x}^t \quad where \quad \tau^t = \frac{\ell^t}{||\boldsymbol{x}^t||^2 + \frac{1}{2C}} \tag{4}$$

This update keeps a balance between the amount of progress made on each time step and the amount of information retained from previous time steps [18]. On one hand, this update requires $\boldsymbol{w}^{t+1}$ to correctly predict the current example with a sufficiently high accuracy and thus progress is made. On the other hand, $\boldsymbol{w}^{t+1}$ must be as close to $\boldsymbol{w}^t$ as possible in order to retain the information learned from previous time steps. After given the loss function Eq. (2) and the update rule Eq. (4), the pseudo-code of PAR-II is presented in Algorithm 1.

Algorithm 1. Passive-Aggressive Regression II (PAR-II) [18]

Input: aggressiveness parameter $C > 0$;
1: Initialize $\boldsymbol{w}^1 = (0, 0, \ldots, 0)$;
2: **for** t=1, 2, ... **do**
3: receive instance $\boldsymbol{x}^t$;
4: receive label y^t;
5: predict: $\hat{y}^t = \boldsymbol{w}^t \cdot \boldsymbol{x}^t$;
6: suffer loss: $\ell^t = \ell_\varepsilon(\boldsymbol{w}^t; (\boldsymbol{x}^t, y^t))$; /*according to Eq. 2*/
7: update: /* according to Eq. 4*/
8: $\tau^t = \ell^t/(||\boldsymbol{x}^t||^2 + 1/(2C))$;
9: $\boldsymbol{w}^{t+1} = \boldsymbol{w}^t + sign(y^t - \hat{y}^t)\tau^t \boldsymbol{x}^t$
10: **end for**

3 Proposed PAR-Based Prediction Strategy

Whenever there is a new change taken place, DEMO algorithm need make some operation to response the new change so as to adapt to new environment as quickly as possible. In the following, we will propose a new prediction strategy to adapt to the new change by using PAR-II. Algorithm 2 presents our PAR-based strategy (PARS) for predicting the new POS of the next environment. We assume that there are N individuals in the population and each individual has d decision variables. Because we perform regression prediction on each dimension variable separately, there are a total of $N * d$ PAR-II regression models in this algorithm. In addition, we assume that the dimension of the weight vector $\boldsymbol{w}$ of PAR-II is q. Firstly line 3 is to generate a time series s from the stored populations. Then, M samples are gotten from s in line 4. In order to reduce

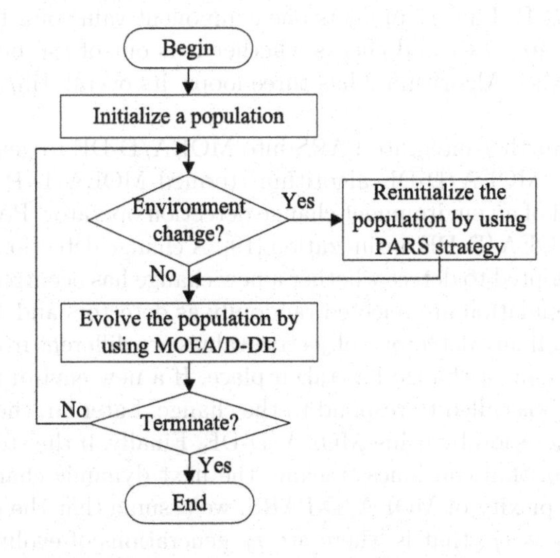

Fig. 1. MOEA/D-PARS algorithm framework

Algorithm 2. Passive-Aggressive Regression-based Strategy (PARS)

Input: the stored populations of the preceding time steps $\boldsymbol{P}^1$,$\boldsymbol{P}^2$,...,$\boldsymbol{P}^t$, where $\boldsymbol{P}^k = \{\boldsymbol{v}_{ij}^k | i = 1, 2, \ldots, N; j = 1, 2, \ldots d; k = 1, 2, \ldots, t\}$
Output: a predicted population of the next time step P^{t+1}.
1: **for** i=1, 2, ..., N **do**
2: **for** j=1,2,...,d **do**
3: Generate a time series $s = (\boldsymbol{v}_{ij}^1, \boldsymbol{v}_{ij}^2, ..., \boldsymbol{v}_{ij}^t)$ from the stored populations;
4: Get M training samples from s;
5: Initialize weight vector of PAR-II model: $\boldsymbol{w}^1 = (0, 0, ..., 0)$;
6: **for** k=1,2,...,M **do** /*Train the model*/
7: receive instance $\boldsymbol{x}^k = (\boldsymbol{v}_{ij}^{t-q-k+1}, ..., \boldsymbol{v}_{ij}^{t-k})$ from the time series s;
8: receive label $y^k = \boldsymbol{v}_{ij}^{t-k+1}$ from time series s;
9: predict: $\hat{y}^k = \boldsymbol{w}^k \cdot \boldsymbol{x}^k$;
10: suffer loss: $\ell^k = \ell_\varepsilon(\boldsymbol{w}^k; (\boldsymbol{x}^k, y^k))$; /*according to Eq. 2*/
11: update: /* according to Eq. 4*/
12: $\tau^k = \ell^k/(||\boldsymbol{x}^k||^2 + 1/(2C))$;
13: $\boldsymbol{w}^{k+1} = \boldsymbol{w}^k + sign(y^k - \hat{y}^k)\tau^k \boldsymbol{x}^k$
14: **end for**/*end of the training procedure*/
15: set $\boldsymbol{w}^{t+1} = \boldsymbol{w}^{M+1}$;
16: set test instance $\boldsymbol{x}^{t+1} = (\boldsymbol{v}_{ij}^{t-q+1}, ..., \boldsymbol{v}_{ij}^t)$;
17: predict: $\boldsymbol{v}_{ij}^{t+1} = \boldsymbol{w}^{t+1} \cdot \boldsymbol{x}^{t+1}$, and repair $\boldsymbol{v}_{ij}^{t+1}$ if it is out of the boundary [10];
18: **end for**
19: **end for**
20: **return** $\boldsymbol{P}^{t+1} = (\boldsymbol{v}_1^{t+1}, ..., \boldsymbol{v}_N^{t+1})$.

the training time, the size of training set is limited to no more than 20, so the size M is set to $min(20, t-q)$. Referring to Algorithm 1, line 5 to 15 is used to implement a PAR-II. Line 17 predicts one component value of a Pareto optimal solution on time step t+1 and checks whether it is out of the boundary of the decision space. Since Algorithm 2 has three loops, its overall time complexity is $O(N * d * M)$.

In Fig. 1, we further integrate PARS into MOEA/D-DE to generate a PARS assisted dynamic MOEA/D-DE algorithm (termed MOEA/D-PARS) which is mainly composed of an environment change detection operator, PARS prediction strategy and a MOEA/D-DE optimization [19]. A change detection operator suggested in [2] is adopted to detect whether a new change has occurred. 10% of individuals in the population are selected randomly as detectors and re-evaluated in every generation. If any detector's objective values are different from its previous ones, then we assume a change has taken place. If a new environmental change is detected, PARS is called to respond to the change. Later on, the population is evolved for a generation by using MOEA/D-DE. Finally, if the stop condition is not met, the algorithm continues tracking the next dynamic change. Considering the time complexity of MOEA/D-PARS, we assume that the environmental change frequency is τ_T, that is, there are τ_T generations of evolutionary multi-objective optimization in an environmental change. The time complexities of the change detect operator, PARS, and MOEA/D-DE are $O(mN)$, $O(N^*d^*M)$, and $O(mNT)$ [19] respectively, where T is the number of weight vectors, and m is the number of objective functions. Therefore, the time complexity of MOEA/D-PARS during an environmental change is $O(\tau_T mNT)$.

4 Computational Experiments

4.1 Experimental Settings

Benchmark Problems. Fourteen different benchmark problems for CEC2018 competition on dynamic multi-objective optimization [22] are chosen here to compare the performance of DEMO algorithms.

Compared Prediction Strategies and DEMO Algorithms. Three well-known prediction strategies, i.e., PRE [9], PPS [10], and KF [11], are chosen here to validate the performance of the proposed PARS. For fairly comparison, we choose MOEA/D-DE as their basic multi-objective optimizer. In addition to the proposed MOEA/D-PARS algorithm, we integrate the aforementioned PRE, PPS, and KF into MOEA/D-DE, and then we get other three compared dynamic MOEA/D-DE algorithms which are termed MOEA/D-PRE, MOEA/D-PPS, and MOEA/D-KF respectively. All the algorithms are implemented by using Python programming.

Performance Metrics. Mean Inverted Generational Distance (MIGD) [11,14] is selected as performance metrics. It provides a quantitative measurement for both the proximity and diversity goals of DEMO algorithms. A lower value of MIGD implies that the algorithm has better optimization performance. In addition, the IGD(t) [23] is used to illustrated the evolutionary tracking curves of the compared DEMO algorithms.

Parameter Setting. The parameter settings of all the DEMO algorithms tested in this paper were almost the same as those of their original papers. Some key parameters in these algorithms were set as follows:

1) Population size (N) and the number of decision variables (d): They were set to be 100 and 10 respectively for all benchmark problems.
2) Common control parameters in MOEA/D-DE: $CR = 0.5$, $F = 0.5$ in the differential evolution operator. Distribution $\eta_m = 20$, mutation probability $p_m = 1/d$ in the polynomial mutation operator. In addition, neighborhood size $T = 20$ and neighborhood selection probability $\delta = 0.8$.
3) Parameters setting for PARS strategy: $q = 4$, $C = 1$, $\varepsilon = 0.05$.

Table 1. Mean and standard deviation values of MIGD obtained by four algorithms on the DF1–DF7 problems

Problem	(τ_T, n_T)	MOEA/D-PRE	MOEA/D-PPS	MOEA/D-KF	MOEA/D-PARS
DF1	(5, 5)	4.53e−2(3.10e−3)	5.86e−2(1.09e−2)	5.48e−2(8.83e−3)	**4.17e−2(9.36e−3)**
	(5, 10)	**1.88e−2(1.54e−4)**	3.52e−2(7.16e−3)	2.10e−2(2.00e−3)	2.75e−2(2.20e−3)
	(10, 5)	1.42e−2(3.26e−3)	1.33e−2(2.08e−3)	1.33e−2(1.90e−3)	**1.14e−2(4.11e−4)**
	(10, 10)	8.40e−3(2.87e−4)	1.24e−2(2.41e−3)	**8.19e−3(3.31e−4)**	1.00e−2(4.23e−4)
DF2	(5, 5)	1.07e−1(5.37e−3)	6.81e−2(5.86e−3)	1.40e−1(9.48e−3)	**4.73e−2(3.69e−3)**
	(5, 10)	6.90e−2(5.62e−3)	5.38e−2(6.50e−3)	7.84e−2(5.06e−3)	**4.30e−2(3.88e−3)**
	(10, 5)	2.08e−2(1.98e−3)	1.76e−2(9.86e−4)	2.50e−2(1.87e−3)	**1.29e−2(8.66e−4)**
	(10, 10)	1.52e−2(1.39e−3)	1.56e−2(1.21e−3)	1.66e−2(2.64e−3)	**1.26e−2(9.10e−4)**
DF3	(5, 5)	3.38e−1(3.12e−3)	4.74e−1(1.32e−2)	3.22e−1(2.49e−2)	**2.89e−1(3.36e−2)**
	(5, 10)	**2.34e−1(2.13e−2)**	4.79e−1(8.50e−3)	3.02e−1(2.99e−2)	3.31e−1(2.92e−2)
	(10, 5)	**2.16e−1(2.40e−2)**	3.92e−1(1.68e−2)	2.43e−1(2.64e−2)	2.26e−1(4.55e−2)
	(10, 10)	**1.54e−1(2.55e−3)**	3.91e−1(2.53e−2)	1.99e−1(2.48e−2)	2.40e−1(1.67e−2)
DF4	(5, 5)	7.84e−2(1.07e−3)	9.20e−2(1.95e−3)	7.99e−2(2.53e−3)	**7.64e−2(5.35e−3)**
	(5, 10)	**8.11e−2(1.97e−3)**	9.81e−2(3.16e−3)	8.23e−2(2.12e−3)	8.39e−2(4.81e−3)
	(10, 5)	6.36e−2(3.97e−4)	6.39e−2(5.47e−4)	6.42e−2(7.85e−3)	**6.25e−2(8.87e−4)**
	(10, 10)	**7.17e−2(6.42e−4)**	7.20e−2(1.07e−3)	7.23e−2(6.64e−4)	7.19e−2(9.22e−4)
DF5	(5, 5)	1.56e−2(4.27e−4)	2.15e−2(3.98e−3)	**1.26e−2(2.31e−4)**	1.26e−2(7.38e−4)
	(5, 10)	1.26e−2(3.26e−4)	1.96e−2(1.80e−3)	**1.14e−2(2.31e−4)**	1.82e−2(4.49e−4)
	(10, 5)	8.18e−3(9.03e−5)	8.77e−3(8.97e−5)	7.37e−3(3.91e−5)	**6.63e−3(8.36e−5)**
	(10, 10)	7.27e−3(4.95e−5)	8.75e−3(1.26e−4)	**6.69e−3(5.15e−5)**	8.74e−3(1.22e−4)
DF6	(5, 5)	2.18e+0(4.60e−1)	1.38e+0(3.52e−1)	1.73e+0(6.02e−1)	**9.86e−1(3.37e−1)**
	(5, 10)	2.01e+0(7.23e−1)	1.66e+0(4.47e−1)	2.81e+0(1.24e+0)	**8.42e−1(3.31e−1)**
	(10, 5)	1.05e+0(4.71e−1)	**9.14e−1(2.43e−1)**	1.43e+0(2.92e−1)	1.04e+0(5.01e−1)
	(10, 10)	1.25e+0(5.92e−1)	1.13e+0(4.01e−1)	2.89e+0(1.27e+0)	**5.99e−1(2.23e−1)**
DF7	(5, 5)	1.53e+0(1.09e−1)	1.81e+0 (2.35e−2)	**1.43e+0(5.00e−2)**	1.51e+0(1.51e−1)
	(5, 10)	5.73e−1(1.37e−1)	8.58e−1(4.28e−2)	**5.64e−1(1.56e−1)**	7.41e−1(7.56e−2)
	(10, 5)	1.44e+0(1.19e−1)	1.69e+0(4.19e−2)	**1.28e+0(3.12e−2)**	1.71e+0(2.94e−2)
	(10, 10)	4.78e−1(1.19e−1)	8.19e−1(3.24e−2)	**2.91e−1(9.20e−2)**	6.39e−1(1.18e−1)

4) Number of runs and stopping condition: For each tested algorithm, it was executed 20 runs on each test instance, and the experimental results were recorded to obtain the statistical information. In each run, each algorithm stopped at pre-specified 100 environmental changes. During each environmental change, each algorithm has the same number of function evaluations.

4.2 Experimental Results

The experiments were conducted with different combinations of change severity levels (n_T) and frequencies (τ_T) in order to study the impact of dynamics in changing environments. They were set to 5 and 10, respectively. Comparison results of four DEMO algorithms (i.e., MOEA/D-PRE, MOEA/D-PPS, MOEA/D-KF, and MOEA/D-PARS) are reported to study the influence of different prediction strategies on DEMO algorithms. Table 1 and Table 2 detail MIGD values of these algorithms on DF benchmark problems. These two tables show that MOEA/D-PARS obtains competitive results on the majority of the

Table 2. Mean and standard deviation values of MIGD obtained by four algorithms on the DF8–DF14 problems

Problem	(τ_T, n_T)	MOEA/D-PRE	MOEA/D-PPS	MOEA/D-KF	MOEA/D-PARS
DF8	(5, 5)	8.70e−2(2.55e−3)	1.04e−1(4.42e−3)	8.66e−2(7.66e−4)	**8.45e−2(1.43e−3)**
	(5, 10)	**8.31e−2(2.47e−3)**	1.25e−1(2.75e−2)	9.33e−2(5.41e−3)	8.55e−2(8.97e−3)
	(10, 5)	7.65e−2(1.06e−3)	8.72e−2(1.51e−3)	**7.55e−2(9.02e−4)**	7.90e−2(4.68e−4)
	(10, 10)	7.21e−2(1.35e−3)	9.31e−2(1.34e−2)	**7.23e−2(1.47e−3)**	7.77e−2(1.18e−1)
DF9	(5, 5)	2.75e−1(9.68e−2)	2.00e−1(1.90e−2)	4.56e−1(4.22e−2)	**1.41e−1(1.13e−2)**
	(5, 10)	**1.17e−1(1.81e−2)**	1.81e−1(1.93e−2)	1.65e−1(1.68e−2)	1.51e−1(1.61e−2)
	(10, 5)	1.23e−2(2.62e−2)	1.04e−1(7.33e−3)	2.16e−1(2.95e−2)	**7.96e−2(8.62e−3)**
	(10, 10)	7.83e−2(1.12e−2)	9.50e−2(1.14e−2)	**7.64e−2(6.40e−3)**	7.83e−2(1.46e−2)
DF10	(5, 5)	1.74e−1(3.16e−3)	2.35e−1(1.05e−2)	2.18e−1(1.33e−2)	**1.50e−1(3.78e−3)**
	(5, 10)	1.75e−1(1.16e−2)	2.37e−1(9.47e−3)	1.99e−1(1.48e−2)	**1.49e−1(5.59e−3)**
	(10, 5)	1.39e−1(1.86e−3)	1.75e−1(2.28e−2)	1.53e−1(1.27e−2)	**1.31e−1(1.79e−3)**
	(10, 10)	1.40e−1(2.58e−3)	1.65e−1(4.46e−3)	1.45e−1(3.35e−3)	**1.30e−1(2.40e−3)**
DF11	(5, 5)	7.24e−2(1.12e−4)	9.50e−2(6.58e−3)	7.22e−2(5.28e−4)	**6.51e−2(6.98e−4)**
	(5, 10)	6.83e−2(5.66e−4)	1.01e−1(7.12e−3)	6.92e−2(9.85e−4)	**6.47e−2(7.43e−4)**
	(10, 5)	5.80e−2(2.88e−4)	6.66e−2(2.75e−3)	5.84e−2(1.58e−4)	**5.55e−2(3.98e−4)**
	(10, 10)	5.69e−2(4.11e−4)	6.90e−2(7.34e−3)	5.69e−2(2.75e−4)	**5.52e−2(2.02e−4)**
DF12	(5, 5)	**2.92e−1(1.15e−2)**	3.91e−1(1.02e−2)	2.94e−1(2.34e−2)	3.00e−1(1.14e−2)
	(5, 10)	2.91e−1(1.01e−2)	3.88e−1(1.70e−2)	2.88e−1(1.87e−2)	**2.81e−1(6.32e−3)**
	(10, 5)	**2.58e−1(1.15e−2)**	3.05e−1(1.14e−2)	2.80e−1(1.62e−2)	2.74e−1(5.61e−3)
	(10, 10)	2.61e−1(1.15e−2)	3.07e−1(5.25e−3)	2.68e−1(1.43e−2)	**2.59e−1(6.21e−3)**
DF13	(5, 5)	2.25e−1(2.17e−4)	2.31e−1(6.13e−3)	**2.23e−1(1.97e−3)**	2.25e−1(2.84e−3)
	(5, 10)	2.22e−1(1.48e−3)	2.41e−1(2.63e−2)	**2.18e−1(2.67e−3)**	2.28e−1(1.57e−3)
	(10, 5)	2.26e−1(1.92e−3)	**2.23e−1(2.33e−3)**	2.27e−1(1.64e−3)	2.30e−1(2.35e−3)
	(10, 10)	2.23e−1(2.12e−3)	**2.20e−1(1.90e−3)**	2.24e−1(2.66e−3)	2.24e−1(1.96e−3)
DF14	(5, 5)	6.68e−2(1.79e−3)	7.71e−2(8.41e−3)	**6.28e−2(1.09e−3)**	6.86e−2(6.23e−4)
	(5, 10)	6.42e−2(4.86e−4)	7.84e−2(1.09e−2)	**6.38e−2(1.19e−3)**	6.86e−2(6.23e−4)
	(10, 5)	5.67e−2(2.27e−4)	6.01e−2(2.64e−3)	**5.56e−2(6.18e−4)**	5.58e−2(1.02e−3)
	(10, 10)	5.70e−2(1.77e−4)	6.01e−2(3.66e−4)	**5.64e−2(2.35e−4)**	5.85e−2(1.97e−4)

DF instances, which implies that MOEA/D-PARS has good performance regarding convergence and distribution. Specifically, MOEA/D-PARS obtains the best results on DF2, DF10, and DF11 instances, whereas MOEA/D-KF performs the best on DF5, DF7 and DF14 instances. MOEA/D-PARS also achieves the best results on DF1, DF3, DF4, DF6, and DF9 under some certain combinations of severity level and frequency. Besides, MOEA/D-PRE achieves the best results in a few cases. In most cases, the results obtained by MOEA/D-PPS are relatively poor.

Apart from presenting the data in the above tables, we also provide the average tracking performance of these algorithms over 20 runs in Fig. 2, where the average IGD values are plotted versus the time. For the sake of limitation of paper length, only results of four problems with $\tau_T = 10$ and $n_T = 5$ are presented here. It is clear that the tracking performance of MOEA/D-PARS is better than the other three algorithms on DF2 and DF10 problems. On DF4 problem, the tracking performance of the four algorithms is very similar. On DF6 problem, although MOEA/D-PARS keeps fluctuating in IGD values in the first 50 time steps like the other three algorithms, it clearly outperformed the other competitors in the remaining time.

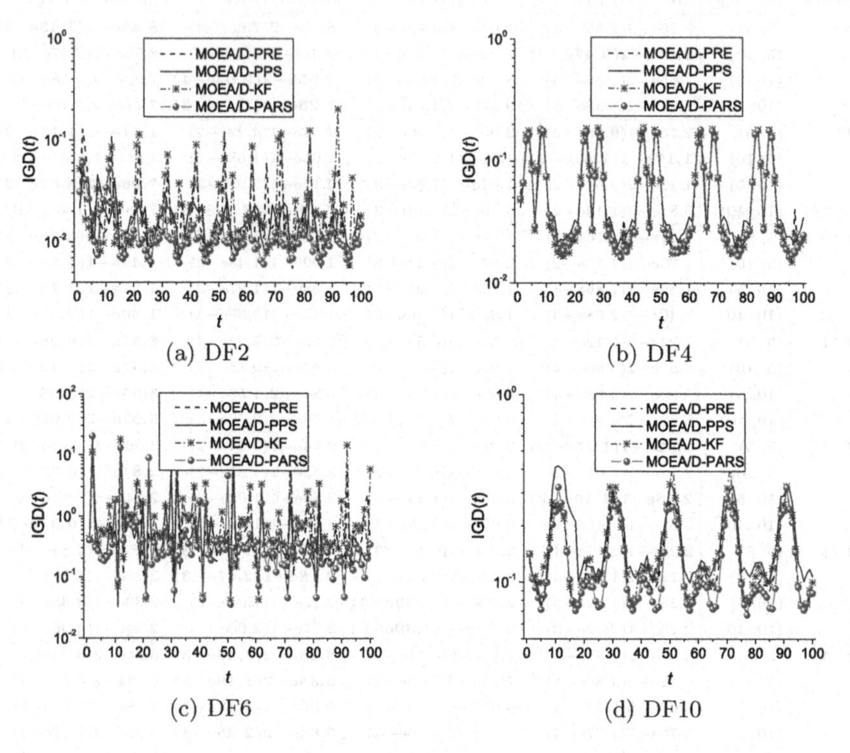

Fig. 2. Evolutionary average IGD(t) curves obtained by four algorithms over 20 runs on the DF2, DF4, DF6, and DF10 problems with $\tau_T = 10$ and $n_T = 5$

5 Conclusions

Both evolutionary dynamic optimization and online machine learning communities face the challenge of dynamic and uncertain environment change problems. Hence, there may be an excellent opportunity for cross fertilization between them [15]. This paper first introduces PAR, a common online learning technique, into DEMO, and then proposes a PAR-based prediction strategy (PARS) to react the new environmental change of DMOPs. Furthermore, the proposed prediction strategy is integrated into MOEA/D-DE to deal with DMOPs. Finally, the proposed PARS is compared with three state-of-the-art prediction strategies under the same dynamic MOEA/D-DE framework. The experimental results show that MOEA/D-PARS is promising for handling DMOPs, which demonstrates that PARS is a competitive prediction strategy for DEMO.

The work presented here is preliminary, and there are some possible directions for future work. We would like to investigate the influence of parameter settings of PAR-II and other PAR variants on algorithm performance. In addition, we will try to design other prediction strategy based on other kind of online machine learning techniques for better dealing with DMOPs.

Acknowledgment. This work was supported by China Scholarship Council (Grant No. 201609480012) and a Project Supported by Scientific Research Fund of Hunan Provincial Education Department(Grant Nos. 18K060, 18C0331, 18B199).

References

1. Farina, M., Deb, K., Amato, P.: Dynamic multiobjective optimization problems: test cases, approximations, and applications. IEEE Trans. Evol. Comput. **8**(5), 425–442 (2004)
2. Deb, K., Rao N., U.B., Karthik, S.: Dynamic multi-objective optimization and decision-making using modified NSGA-II: a case study on hydro-thermal power scheduling. In: Obayashi, S., Deb, K., Poloni, C., Hiroyasu, T., Murata, T. (eds.) EMO 2007. LNCS, vol. 4403, pp. 803–817. Springer, Heidelberg (2007). https://doi.org/10.1007/978-3-540-70928-2_60
3. Liu, M., Zheng, J.H., Wang, J.N., Liu, Y.Z., Jiang, L.: An adaptive diversity introduction method for dynamic evolutionary multiobjective optimization. In: Proceedings of 2014 IEEE Congress on Evolutionary Computation, CEC2014, Beijing, China, pp. 3160–3167. IEEE (2014)
4. Ruan, G., Yu, G., Zheng, J.H., Zou, J., Yang, S.X.: The effect of diversity maintenance on prediction in dynamic multi-objective optimization. Appl. Soft Comput. **58**, 631–647 (2017)
5. Goh, C.K., Tan, K.C.: A competitive-cooperative coevolutionary paradigm for dynamic multiobjective optimization. IEEE Trans. Evol. Comput. **13**(1), 103–127 (2009)
6. Liu, M., Zeng, W.H.: Memory enhanced dynamic multi-objective evolutionary algorithm based on decomposition. Ruan Jian Xue Bao/J. Softw. **24**(7), 1571–1588 (2013)
7. Peng, Z., Zheng, J., Zou, J., Liu, M.: Novel prediction and memory strategies for dynamic multiobjective optimization. Soft Comput. **19**(9), 2633–2653 (2014). https://doi.org/10.1007/s00500-014-1433-3
8. Hatzakis, I., Wallace, D.: Dynamic multi-objective optimization with evolutionary algorithms: a forward-looking approach. In: Proceedings of Genetic and Evolutionary Computation Conference, GECCO 2006, Seattle, USA, pp. 1201–1208. ACM (2006)
9. Zhou, A., Jin, Y., Zhang, Q., Sendhoff, B., Tsang, E.: Prediction-based population re-initialization for evolutionary dynamic multi-objective optimization. In: Obayashi, S., Deb, K., Poloni, C., Hiroyasu, T., Murata, T. (eds.) EMO 2007. LNCS, vol. 4403, pp. 832–846. Springer, Heidelberg (2007). https://doi.org/10.1007/978-3-540-70928-2_62
10. Zhou, A.M., Jin, Y.C., Zhang, Q.F.: A population prediction strategy for evolutionary dynamic multiobjective optimization. IEEE Trans. Cybern. **44**(1), 40–53 (2014)

11. Muruganantham, A., Tan, K.C., Vadakkepat, P.: Evolutionary dynamic multiobjective optimization via Kalman filter prediction. IEEE Trans. Cybern. **46**(12), 2862–2873 (2016)
12. Cao, L.L., Xu, L., Goodman, E.D., Li, H.: Decomposition-based evolutionary dynamic multiobjective optimization using a difference model. Appl. Soft Comput. **76**, 473–490 (2019)
13. Jiang, M., Huang, Z.Q., Qiu, L.M., Huang, W.Z., Yan, G.G.: Transfer learning-based dynamic multiobjective optimization algorithms. IEEE Trans. Evol. Comput. **22**(4), 501–514 (2018)
14. Cao, L.L., Xu, L., Goodman, E.D., Bao, C., Zhu, S.: Evolutionary dynamic multiobjective optimization assisted by a support vector regression predictor. IEEE Trans. Evol. Comput. **24**(2), 305–319 (2020)
15. Yao, X.: Challenges and opportunities in dynamic optimisation. In: Proceedings of the 15th Annual Conference Companion on Genetic and Evolutionary Computation, Amsterdam, The Netherlands, pp. 1761–1762. ACM (2013)
16. Sun, J.Y., Zhang, H., Zhou, A.M., Zhang, Q.F.: Learning from a stream of nonstationary and dependent data in multiobjective evolutionary optimization. IEEE Trans. Evol. Comput. **23**(4), 541–555 (2019)
17. Minku, L.L., Xin, Y.: DDD: a new ensemble approach for dealing with concept drift. IEEE Trans. Knowl. Data Eng. **24**(4), 619–633 (2012)
18. Crammer, K., Dekel, O., Keshet, J., Shalev-Shwartz, S., Singer, Y.: Online passive-aggressive algorithms. J. Mach. Learn. Res. **7**, 551–585 (2006)
19. Li, H., Zhang, Q.F.: Multiobjective optimization problems with complicated pareto sets, MOEA/D and NSGA-II. IEEE Trans. Evol. Comput. **13**(2), 284–302 (2009)
20. Zou, J., Li, Q.Y., Yang, S.X., Bai, H., Zheng, J.H.: A prediction strategy based on center points and knee points for evolutionary dynamic multi-objective optimization. Appl. Soft Comput. **61**, 806–818 (2017)
21. Rong, M., Gong, D., Zhang, Y., Jin, Y.C., Pedrycz, W.: Multidirectional prediction approach for dynamic multiobjective optimization problems. IEEE Trans. Cybern. **49**(9), 3362–3374 (2019)
22. Jiang, S., Yang, S., Yao, X., Tan, K.C., Kaiser, M., Krasnogor, N.: Benchmark problems for CEC 2018 competition on dynamic multiobjective optimisation. In: CEC 2018 Competition (2018)
23. Li, X.D., Branke, J., Kirley, M.: On performance metrics and particle swarm methods for dynamic multiobjective optimization problems. In: proceedings of 2007 IEEE Congress on Evolutionary Computation, Singapore, pp. 576–583, IEEE (2007)

Generalized Test Suite for Continuous Dynamic Multi-objective Optimization

Chang Shao[1,2], Qi Zhao[1,3], Yuhui Shi[1(✉)], and Jing Jiang[2(✉)]

[1] Department of Computer Science and Engineering, Southern University of Science and Technology, Shenzhen 518055, China
shaoc@mail.sustech.edu.cn, {zhaoq,shiyh}@sustech.edu.cn

[2] The Australian Artificial Intelligence Institute (AAII), Faculty of Engineering and Information Technology, University of Technology Sydney, Ultimo, Australia
jing.jiang@uts.edu.au

[3] School of Computer Science and Technology, University of Science and Technology of China, Hefei 230027, China

Abstract. Dynamic multi-objective optimization (DMO) has recently attracted increasing interest. Suitable benchmark problems are crucial for evaluating the performance of DMO solvers. However, most of the existing DMO benchmarks mainly focus on Pareto-optimal solutions (PS) varying on the hyperplane, which may produce some unexpected bias for algorithmic analysis. Furthermore, they do not comprehensively consider the general time-linkage property, yet which is commonly observed in real-world applications. To alleviate these two issues, we designed a generalized test suite (GTS) for DMO with the following two advantages over previous existing benchmarks: 1) the PS can change on the hypersurface over time, to better compare the tracking ability of different DMO solvers; 2) the general time-linkage feature is included to systemically investigate the algorithmic robustness in the dynamic environment. Experimental results on five representative DMO algorithms demonstrated the proposed GTS can efficiently discriminate the performance of DMO algorithms and is more general than existing benchmarks.

Keywords: Dynamic multi-objective optimization · Time-linkage · Benchmark problems · Performance metric

1 Introduction

Many practical problems can be modeled as continuous dynamic multi-objective optimization problems (DMOPs), such as, scheduling [17], process control [21], industrial design [5], and resource allocation [20]. These problems have multiple conflicting objectives, and the objective functions or constraints are time-dependent.

C. Shao and Q. Zhao—Contribute equally.

The source code of GTS is available at https://dynamicoptimization.github.io.

H. Ishibuchi et al. (Eds.): EMO 2021, LNCS 12654, pp. 205–217, 2021.
https://doi.org/10.1007/978-3-030-72062-9_17

Suitable benchmark problems are crucial for evaluating the performance of DMOP solvers. Studies on dynamic multi-objective optimization (DMO) benchmarks can be divided into three categories. The earlier category introduces dynamics on the basis of some well-known static multi-objective optimization problems. For example, Farina et al. developed the FDA test suite [6] based on the static ZDT [28] and DTLZ [3] problems. Jin and Sendhoff [15] aggregated multiple static objectives and simulated dynamics by changing the weight of each objective. The ZJZ [27] benchmarks improved the FDA by considering nonlinear correlation between decision variables and Mehnen [16] proposed a new generic scheme DTF, which is a generalized FDA function.

To diversify problem properties, subsequent studies consider special and difficult test scenarios. For instance, in [11], the cases with time-varying numbers of decision variables and objectives were investigated. The test suites in [9,10,26] and [1] simulated various complicated Pareto-optimal Solutions (PS) and disconnected Pareto-optimal Front (PF). A common feature of these benchmarks is each dimension of the PS holding a separate geometry. This kind of PS geometry is not intuitive and makes the static problem difficult to solve. As a result, it would be unclear whether algorithms' performance on these benchmarks attributes to the dynamics or the static properties of problems.

To emphasize the effect of dynamics on DMO benchmarks, the recent test suites simplify the static form of problems and focus on emulating different PF features. Gee et al. [7] suggested two forms of DMO benchmarks, i.e., the additive form and multiplicative form. PFs of their problems have diverse characteristics, e.g., modality, connectedness, and degeneracy. Jiang et al. proposed the JY [13], DF [14], and SDP [12] test suites by combining the merits of previous benchmarks. The proposed test suites mainly focus on PF properties, such as connectivity, degeneration, PF shrinkage/expansion, search favorability, etc.

The DF [14] and SDP [12] benchmarks described various PS/PF characteristics and have gained great popularity in recent studies [2,22–24]. However, there are still some limitations from the perspective of dynamics. Firstly, to reduce the difficulty of static problem-solving, DF and SDP benchmarks set the dynamics of PS to be performed over hyperplanes, which is obviously not always the case in practice, such as, scheduling [17] and process control [21] . Secondly, in addition to the dynamics that depend on historical problems, there are also DMOPs with dynamics based on historical solutions, i.e., the so-called time-linkage problems [19]. Although time-linkage has been considered in [11] and [12], the PS of these problems is either a straight line segment or a rectangular plane segment. This kind of PS is a very special case which is far from sufficient to represent the time-linkage property.

To address the above limitations, in this paper, we proposed an generalized test suite with innovations in emulating more general dynamics and time-linkage properties for DMO. Our main work is summarized as follows.

1) We proposed a simple method to construct diverse PS dynamics that perform on hypersurfaces. This kind of dynamics is more general in practice than these on hyperplanes designed in recent benchmarks and therefore can effectively

and efficiently test the tracking ability of DMO solvers in the dynamic environment.
2) We mathematically formularized the time-linkage property in DMOPs to investigate the robustness of DMO solvers. Our proposed formulation, for the first time, emulates the time-deception and error accumulation characteristics of time-linkage DMOPs.
3) According to the proposed PS construction method and the new mathematical time-linkage formulation, we presented the GTS to facilitate the development of practically more powerful DMO algorithms. Experimental results demonstrated the effectiveness and efficiency of these new GTS functions.

The remainder of the paper is organized as follows. Section 2 details the proposed test suite. Section 3 reports and analyzes the experimental results. Finally, Sect. 4 concludes the paper.

2 Generalized Test Suite for Dynamic Multi-objective

2.1 Basic Framework

The following DMOP is considered [6]

$$\min\ \mathbf{f}(\mathbf{x},t) = (f_1(\mathbf{x},t), f_2(\mathbf{x},t), ..., f_M(\mathbf{x},t))^T, \\ \text{s.t. } \mathbf{x} \in \Omega\ , \tag{1}$$

with

$$f_{i=1:M}(\mathbf{x},t) = g(\mathbf{x},t)\mu_i(\mathbf{x},t). \tag{2}$$

In (1), $\mathbf{x} = (x_1, x_2, ..., x_D)^T$ is a candidate solution, $\mathbf{f} : \Omega \rightarrow \mathbb{R}^M$ contains M objective functions to be minimized, $\Omega \subseteq \mathbb{R}^D$ is the decision space, $\mathbb{R}^M$ is the objective space, and t is a discrete time parameter controlling dynamics. In (2), commonly $g(\mathbf{x},t) > 0$, the solution to minimizing $g(\mathbf{x},t)$, $(\mathbf{x}^*,t)$ characterizes PS properties, e.g., line, curve and surface; the PF is obtained though $\mu(\mathbf{x}^*,t) = (\mu_1(\mathbf{x}^*,t), \cdots, \mu_M(\mathbf{x}^*,t))$, and $\mu(\mathbf{x}^*,t)$ describes PF geometry shapes, i.e., convexity and concavity. $\mathbf{x} = (x_1, x_2, ..., x_D)^T$ is usually divided into two sub-vectors, i.e., $\mathbf{x}_I = (x_1, \cdots, x_m)^T$, $M-1 \leq m < D$, for $\mu(\mathbf{x},t)$ and $\mathbf{x}_{II} = (x_{m+1}, \cdots, x_D)^T$ for $g(\mathbf{x},t)$. These sub-vectors are used for separately designing the PS and PF, respectively.

Following [6], t controls the dynamics via

$$t = \frac{1}{n_t}\left\lfloor \frac{\tau}{\tau_t} \right\rfloor, \tag{3}$$

where n_t governs the severity of change, τ is the generation counter, and τ_t is the number of generations available at each time step.

2.2 Components with General Dynamics

The $g(\mathbf{x}, t)$ function in (2) is the key component for constructing the PS/PF shapes and understanding the PS/PF changing patterns. To emulate general PS changing patterns, we divide $\mathbf{x}_{II}$ into P parts

$$\mathbf{x}_{II,p}, p = \{1, 2, ..., P\},$$

where $1 \leq P \leq D - M$.

Obviously, $P = 1$ and $P = D - M$ represent two extreme cases. The former has been widely used in the FDA[6], ZJZ[27], dMOPs [8], DF [14], JY [13] and SDP[12] [7] test suites. In such cases, all x_j in $\mathbf{x}_{II}$ are the same, resulting in PS varying on a static hyperplane, which is not always the case in practice. The latter has been considered in [9–11,26] and [1]. In such cases, each dimension of the PS holds a separate geometry, which makes the static problem being difficult to solve. As a result, it would be unclear whether algorithms' performance on these benchmarks attributes to the dynamics or the static properties of problems.

To address the above drawbacks and to make a balance between the generality of dynamics and the complexity of static problem-solving, we propose an generalized $g(\mathbf{x}, t)$ function with $1 < P < D - M$. For visualization, we present the $g(\mathbf{x}, t)$ with $P = 2$ in this paper to construct the new benchmarks, as shown in Eq. (4). The benchmarks can be easily extended to any $1 < P < D - M$ cases following our proposal.

$$g(\mathbf{x}, t) = 1 + \sum_{x_i \in \mathbf{x}_{II,1}} \theta_i \left|x_i - h_1(\mathbf{x}_I, t)\right|^k + \sum_{x_j \in \mathbf{x}_{II,2}} \delta_j \left|x_j - h_2(\mathbf{x}_I, t)\right|^k \quad (4)$$

where $k = 1$ or 2, θ_i, δ_j and $h_1(\mathbf{x}_I, t)$, $h_2(\mathbf{x}_I, t)$ are provided in Tables 1 and 2.

The proposed component function (4) emulates the dynamics of PS over hypersurfaces. These dynamics are more general cases with respect to the dynamics on hyperplanes in other test suites. Figures 1 and 2 show two representative PS dynamics in our proposed benchmarks. Furthermore, variables in $\mathbf{x}_{II,1}$ and $\mathbf{x}_{II,2}$ are linear with respect to each other, respectively, which reduces the difficulty of static problem-solving. Therefore, the proposed benchmarks can focus on evaluating the performance of algorithms on dynamic-handling.

2.3 Components with Time-Linkage

Many practical DMOPs have time-linkage [18,19], i.e., historical solutions determine the behavior of future problems. An improper decision may cause model errors in future environments; thus, these problems can be time-deceptive, and the error will accumulate.

Time-linkage has been considered in [11] and [12]. However, the PS of the problems in [11] and [12] is either a straight line segment or a rectangular plane segment, which is a very special case and far from sufficient to represent the time-linkage property. Here we present a generalized component function to construct time-linkage DMOPs by adding a parameter $\phi(t)$ to the $g(\mathbf{x}, t)$ in Eq. (4)

$$g(\mathbf{x},t) = 1 + \sum_{x_i \in \mathbf{x}_{II,1}} \theta_i \left|x_i - \phi(t)h_1(\mathbf{x}_I,t)\right|^k + \sum_{x_j \in \mathbf{x}_{II,2}} \delta_j \left|x_j - \phi(t)h_2(\mathbf{x}_I,t)\right|^k \quad (5)$$

with

$$\phi(t) = \begin{cases} 1, & \text{if} t = 1, \\ 1 + \left\| f(\mathbf{x}_{knee}, t-1) - f(\hat{\mathbf{x}_{knee}}, t-1) \right\|, & \text{otherwise}, \end{cases} \quad (6)$$

where $f(\mathbf{x}_{knee}, t-1)$ is the objective value of the knee point of the true PF at time $t-1$, and $f(\hat{\mathbf{x}_{knee}}, t-1)$ is the objective value of the knee point obtained by algorithms at time $t-1$. Here we choose the knee point representing the decision made at time $t-1$.

According to Eq. (6), if the obtained solutions approximate to the true PF, $\phi(t)$ will approach 1. Subsequently the problem the algorithm faces at time t will be quite similar to the predefined problem, i.e., the problem with $\phi(t) = 1$. On the contrast, if the algorithm fails to converge at time $t-1$, $\phi(t)$ will be larger than 1. The larger the $\phi(t)$ value, the greater the model error at time t, and the error will accumulate over time. The above settings well emulate the time-deception and error accumulation characteristics of time-linkage. Furthermore, the predefined problem at each time step, i.e., the problem with $\phi(t) = 1$, is employed as the baseline for comparison among different algorithms; thus, a fair experimental comparison can be guaranteed.

2.4 Generalized Test Suite

We present the GTS by incorporating the proposed component functions with the basic framework. The GTS contains 11 benchmark problems, in which GTS1-5 are two-objective problems with dynamics performing over hypersurfaces, GTS6-8 are problems with time-linkage, and GTS9-11 are three-objective problems. Symbols used in the benchmarks are provided in Table 1, the benchmarks are summarized in Table 2, and the PS and PF properties are described in Table 3.

Table 1. Symbols used in the GTS

$G(t) = \sin(0.5\pi t)$	$\delta_j \equiv j \pmod 3 + 1$	$\theta_i \equiv i \pmod 3 + 1$	$\omega_t = \lfloor 10G(t) \rfloor$
$\beta_t = 0.2 + 2.8\left\|G(t)\right\|$	$p_t = \lfloor 6G(t) \rfloor$	$a = \sin(0.5\pi t)$	$\alpha_t = 5\cos(0.5\pi t)$
$b = 1 + \left\|\cos(0.5\pi t)\right\|$	$y = 0.5 + G(t)(x_1 - 0.5)$	$k_t = 10\sin(\pi t)$	$H(t) = 1.5 + G(t)$
$\gamma_t = 1$, when $0.5 - \frac{1}{16}\left\|\sin(0.3\pi t)\right\| \le x_1 \le 0.5 + \frac{1}{16}\left\|\sin(0.3\pi t)\right\|$, $\gamma_t = 0$, otherwise			

3 Experiments

3.1 Experimental Setup

We only test 10 dimension scenario and four state-of-the-art algorithms, i.e., MDP [23], PBDMO [24], MOEA/D-SVR [2], MOEA/D-MoE [22] and the classical DNSGAII-A [4] were employed for experimental study. Among them, MDP,

Table 2. Definitions of the GTS

Problem	Definition
GTS1	$f_1(\mathbf{x},t)=x_1^{\gamma_t}$, $f_2(\mathbf{x},t)=(g(\mathbf{x},t)(1-(\frac{x_1}{g(\mathbf{x},t)})^{H(t)}))^{\gamma_t}$ $g(\mathbf{x},t)=1+\sum_{i=2}^{\lfloor \frac{D}{2}\rfloor}\theta_i \lvert x_i-\cos(0.5\pi t)\rvert+\sum_{j=\lfloor \frac{D}{2}\rfloor+1}^{D}\delta_j \left\lvert x_j-(G(t)+x_1^{H(t)})\right\rvert$ search space: $[0,1]\times[-1,1]^{\lfloor \frac{D}{2}\rfloor-1}\times[-1,2]^{\lceil \frac{D}{2}\rceil}$ PS(t): $0\le x_1\le 1, x_i=\cos(0.5\pi t), i=2,\cdots,\lfloor \frac{D}{2}\rfloor, x_j=G(t)+x_1^{H(t)}, j=\lfloor \frac{D}{2}\rfloor+1,\cdots,D$ PF(t): a part of $f_2=1-f_1^{H(t)}, 0\le f_1\le 1$
GTS2	$f_1(\mathbf{x},t)=0.5x_1+x_2$, $f_2(\mathbf{x},t)=g(\mathbf{x},t)(2.8-(\frac{0.5x_1+x_2}{g(\mathbf{x},t)})^{H(t)})$ $g(\mathbf{x},t)=1+\sum_{i=3}^{\lfloor \frac{D}{2}\rfloor+1}\theta_i \left\lvert x_i-\frac{1}{\pi}\left\lvert\arctan(\cot(3\pi t^2))\right\rvert\right\rvert+\sum_{j=\lfloor \frac{D}{2}\rfloor+2}^{D}\delta_j \left\lvert x_j-(G(t)+x_1^{H(t)})\right\rvert$ search space: $[0,1]^2\times[0,1]^{\lfloor \frac{D}{2}\rfloor-1}\times[-1,2]^{\lceil \frac{D}{2}\rceil-1}$ PS(t): $0\le x_{1,2}\le 1, x_i=\frac{1}{\pi}\left\lvert\arctan(\cot(3\pi t^2))\right\rvert, i=3,\cdots,\lfloor \frac{D}{2}\rfloor+1, x_j=G(t)+x_1^{H(t)}, j=\lfloor \frac{D}{2}\rfloor+2,\cdots,D$ PF(t): $f_2=1-f_1^{H(t)}, 0\le f_1\le 1.5$
GTS3	$f_1(\mathbf{x},t)=g(x)(x_1+0.1\sin(3\pi x_1))^{\beta_t}$, $f_2(\mathbf{x},t)=g(\mathbf{x},t)(1-x_1+0.1\sin(3\pi x_1))^{\beta_t}$ $g(\mathbf{x},t)=1+\sum_{i=2}^{\lfloor \frac{D}{2}\rfloor}\theta_i(x_i-\frac{G(t)\sin(4\pi x_1)}{1+\lvert G(t)\rvert})^2+\sum_{j=\lfloor \frac{D}{2}\rfloor+1}^{D}\delta_j(x_j-(G(t)+x_1^{H(t)}))^2$ search space: $[0,1]\times[-1,1]^{\lfloor \frac{D}{2}\rfloor-1}\times[-1,1]^{\lceil \frac{D}{2}\rceil}$ PS(t): $0\le x_1\le 1, x_i=\frac{G(t)\sin(4\pi x_1)}{1+\lvert G(t)\rvert}, i=2,\cdots,\lfloor \frac{D}{2}\rfloor, x_j=G(t)+x_1^{H(t)}, j=\lfloor \frac{D}{2}\rfloor+1,\cdots,D$ PF(t): $f_1^{\frac{1}{\beta_t}}+f_2^{\frac{1}{\beta_t}}=1+0.2\sin(3\pi\frac{f_1^{\frac{1}{\beta_t}}-f_2^{\frac{1}{\beta_t}}+1}{2}), 0\le f_1\le 1$
GTS4	$f_1(\mathbf{x},t)=g(\mathbf{x},t)\frac{1+t}{x_1+3}$, $f_2(\mathbf{x},t)=g(\mathbf{x},t)\frac{x_1+3}{1+t}$ $g(\mathbf{x},t)=1+\sum_{i=2}^{\lfloor \frac{D}{2}\rfloor}\theta_i(x_i-\lvert G(t)\rvert)^2+\sum_{j=\lfloor \frac{D}{2}\rfloor+1}^{D}\delta_j(x_j-\frac{G(t)\sin(4\pi x_1)}{1+\lvert G(t)\rvert})^2-0.5+0.25\sin(0.3\pi t)$ search space: $[0,1]\times[0,1]^{\lfloor \frac{D}{2}\rfloor-1}\times[-1,1]^{\lceil \frac{D}{2}\rceil}$ PS(t):$0\le x_1\le 1, x_i=\lvert G(t)\rvert, i=2,\cdots,\lfloor \frac{D}{2}\rfloor, x_j=\frac{G(t)\sin(4\pi x_1)}{1+\lvert G(t)\rvert}, j=\lfloor \frac{D}{2}\rfloor+1,\cdots,D$ PF(t): $f_2=\frac{1}{f_1}, \frac{1+t}{16}\le f_1\le\frac{1+t}{4}$
GTS5	$f_1(\mathbf{x},t)=g(\mathbf{x},t)((0.5x_1+x_2))+0.02\sin(\omega_t\pi(0.5x_1+x_2)))$, $f_2(\mathbf{x},t)=g(\mathbf{x},t)(1.6-(0.5x_1+x_2)+0.02\sin(\omega_t\pi(0.5x_1+x_2)))$ $g(\mathbf{x},t)=1+\sum_{i=3}^{\lfloor \frac{D}{2}\rfloor+1}\theta_i(x_i-\cos(0.5\pi t)^2+\sum_{j=\lfloor \frac{D}{2}\rfloor+2}^{D}\delta_j(x_j-(G(t)+x_1^{H(t)}))^2+0.5+0.5*G(t)$ search space: $[0,1]^2\times[-1,1]^{\lfloor \frac{D}{2}\rfloor-1}\times[-1,2]^{\lceil \frac{D}{2}\rceil-1}$ PS(t): $0\le x_{1,2}\le 1, x_i=\cos(0.5\pi t), i=3,\cdots,\lfloor \frac{D}{2}\rfloor+1, x_j=G(t)+x_1^{H(t)}, j=\lfloor \frac{D}{2}\rfloor+2,\cdots,D$ PF(t): $(f_1+f_2)=(1.6+0.5G(t))(1+0.04\sin(\omega_t\pi\frac{\frac{1}{1.6+0.5G(t)}(f_1-f_2)+1.6}{2})), 0\le f_1\le 3$
GTS6	$f_1(\mathbf{x},t)=x_1$, $f_2(\mathbf{x},t)=g(\mathbf{x},t)(1-(\frac{x_1}{g(\mathbf{x},t)})^{H(t)})$ $g(\mathbf{x},t)=1+\sum_{i=2}^{\lfloor \frac{D}{2}\rfloor}\theta_i\lvert x_i-\phi(t)\cos(0.5\pi t)\rvert+\sum_{j=\lfloor \frac{D}{2}\rfloor+1}^{D}\delta_j\left\lvert x_j-\phi(t)(G(t)+x_1^{H(t)})\right\rvert$ search space: $[0,1]\times[-1,1]^{\lfloor \frac{D}{2}\rfloor-1}\times[-1,2]^{\lceil \frac{D}{2}\rceil}$ PS(t): $0\le x_1\le 1, x_i=\cos(0.5\pi t), i=2,\cdots,\lfloor \frac{D}{2}\rfloor, x_j=G(t)+x_1^{H(t)}, j=\lfloor \frac{D}{2}\rfloor+1,\cdots,D$ PF(t): $f_2=1-f_1^{H(t)}, 0\le f_1\le 1$
GTS7	$f_1(\mathbf{x},t)=g(\mathbf{x},t)\lvert x_1-a\rvert^{H(t)}$, $f_2(\mathbf{x},t)=g(\mathbf{x},t)\lvert x_1-a-b\rvert^{H(t)}$ $g(\mathbf{x},t)=1+\sum_{i=2}^{\lfloor \frac{D}{2}\rfloor}\theta_i\lvert x_i-\phi(t)\cos(0.5\pi t)\rvert+\sum_{j=\lfloor \frac{D}{2}\rfloor+1}^{D}\delta_j\left\lvert x_j-\phi(t)\frac{1}{1+e^{\alpha_t(x_1-0.5)}}\right\rvert$ search space: $[-2,2]\times[-1,1]^{\lfloor \frac{D}{2}\rfloor-1}\times[0,1]^{\lceil \frac{D}{2}\rceil}$ PS(t): $a\le x_1\le b, x_i=\cos(0.5\pi t), i=2,\cdots,\lfloor \frac{D}{2}\rfloor, x_j=\frac{1}{1+e^{\alpha_t(x_1-0.5)}}, j=\lfloor \frac{D}{2}\rfloor+1,\cdots,D$ PF(t): $f_2=(b-f_1^{\frac{1}{H(t)}})^{H(t)}), 0\le f_1\le b^{H(t)}$
GTS8	$f_1(\mathbf{x},t)=(0.5x_1+x_2)$, $f_2(\mathbf{x},t)=g(\mathbf{x},t)(2.8-(\frac{(0.5x_1+x_2)}{g(\mathbf{x},t)})^{H(t)})$ $g(\mathbf{x},t)=1+\sum_{i=3}^{\lfloor \frac{D}{2}\rfloor+1}\theta_i\left\lvert x_i-\phi(t)\frac{1}{1+e^{\alpha_t(x_1-0.5)}}\right\rvert+\sum_{j=\lfloor \frac{D}{2}\rfloor+2}^{D}\delta_j\left\lvert x_j-\phi(t)(G(t)+x_1^{H(t)})\right\rvert+0.25*\lvert\cos(0.3\pi t)\rvert$ search space: $[0,1]^2\times[0,1]^{\lfloor \frac{D}{2}\rfloor-1}\times[-1,2]^{\lceil \frac{D}{2}\rceil-1}$ PS(t): $0\le x_{1,2}\le 1, x_i=\frac{1}{1+e^{\alpha_t(x_1-0.5)}}, i=3,\cdots,\lfloor \frac{D}{2}\rfloor+1, x_j=G(t)+x_1^{H(t)}, j=\lfloor \frac{D}{2}\rfloor+2,\cdots,D$ PS(t): $\frac{f_2}{1+0.25\lvert\cos(0.3\pi t)\rvert}=1-(\frac{f_1}{1+0.25\lvert\cos(0.3\pi t)\rvert}))^{H(t)}, 0\le f_1\le 1.5$
GTS9	$f_1(\mathbf{x},t)=g(\mathbf{x},t)\cos(0.5\pi x_1)\cos(0.5\pi x_2)$, $f_2(\mathbf{x},t)=g(\mathbf{x},t)\cos(0.5\pi x_1)\sin(0.5\pi x_2)$, $f_3(\mathbf{x},t)=g(\mathbf{x},t)\sin(0.5\pi x_1)$ $g(\mathbf{x},t)=1+\sum_{i=3}^{\lfloor \frac{D}{2}\rfloor+1}\theta_i\left\lvert x_i-\frac{1}{e^{\alpha_t(x_1-0,5)}}\right\rvert+\sum_{j=\lfloor \frac{D}{2}\rfloor+2}^{D}\delta_j\lvert x_j-\sin(tx_1)\rvert+\left\lvert\prod_{j=1}^{2}\sin(\lfloor k_t(2x_j-1)\rfloor\frac{\pi}{2})\right\rvert$ search space: $[0,1]^2\times[0,1]^{\lfloor \frac{D}{2}\rfloor-1}\times[-1,1]^{\lceil \frac{D}{2}\rceil-1}$ PS(t): $0\le x_{1,2}\le 1, x_i=\frac{1}{e^{\alpha_t(x_1-0,5)}}, i=3,\cdots,\lfloor \frac{D}{2}\rfloor+1, x_j=\sin(tx_1), j=\lfloor \frac{D}{2}\rfloor+2,\cdots,D$ PF(t): a part of $\sum_{i=1}^{3}f_i^2=(1+\left\lvert\prod_{j=1}^{2}\sin(\lfloor k_t(2x_j-1)\rfloor\frac{\pi}{2})\right\rvert)^2, 0\le f_{i=1,2,3}\le 2$
GTS10	$f_1(\mathbf{x},t)=g(\mathbf{x},t)\cos^2(0.5\pi x_1)$, $f_2(\mathbf{x},t)=g(\mathbf{x},t)\cos^2(0.5\pi x_2)$, $f_3(\mathbf{x},t)=g(\mathbf{x},t)\sum_{j=1}^{2}(\sin^2(0.5\pi x_j)+\sin(0.5\pi x_j)\cos^2(p_t\pi x_j))$ $g(\mathbf{x},t)=1+\sum_{i=3}^{\lfloor \frac{D}{2}\rfloor+1}\theta_i\lvert x_i-\lvert G(t)\rvert\rvert+\sum_{j=\lfloor \frac{D}{2}\rfloor+2}^{D}\delta_j\left\lvert x_j-(-0.5+\frac{\lvert G(t)\sin(4\pi x_1)\rvert}{0.5(1+\lvert G(t)\rvert)})\right\rvert$ search space: $[0,1]^2\times[0,1]^{\lfloor \frac{D}{2}\rfloor-1}\times[-1,1]^{\lceil \frac{D}{2}\rceil-1}$ PS(t): $0\le x_{1,2}\le 1, x_i=\lvert G(t)\rvert, i=3,\cdots,\lfloor \frac{D}{2}\rfloor+1, x_j=(-0.5+\frac{\lvert G(t)\sin(4\pi x_1)\rvert}{0.5(1+\lvert G(t)\rvert)}), j=\lfloor \frac{D}{2}\rfloor+2,\cdots,D$
GTS11	$f_1(\mathbf{x},t)=g(\mathbf{x},t)(1.05-y+0.05\sin(6\pi y))$, $f_2(\mathbf{x},t)=g(\mathbf{x},t)(1.05-x_2+0.05\sin(6\pi x_2))(y+0.05\sin(6\pi y))$, $f_3(\mathbf{x},t)=g(\mathbf{x},t)(x_2+0.05\sin(6\pi x_2))(y+0.05\sin(6\pi y))$ $g(\mathbf{x},t)=1+\sum_{i=3}^{\lfloor \frac{D}{2}\rfloor+1}\theta_i\lvert x_i-\lvert G(t)\rvert\rvert+\sum_{j=\lfloor \frac{D}{2}\rfloor+2}^{D}\delta_j\left\lvert x_j-(G(t)+x_1^{H(t)})\right\rvert$ search space: $[0,1]^2\times[0,1]^{\lfloor \frac{D}{2}\rfloor-1}\times[-1,2]^{\lceil \frac{D}{2}\rceil-1}$ PS(t): $0\le x_{1,2}\le 1, x_i=\lvert G(t)\rvert, i=3,\cdots,\lfloor \frac{D}{2}\rfloor+1, x_j=G(t)+x_1^{H(t)}, j=\lfloor \frac{D}{2}\rfloor+2,\cdots,D$

Table 3. Characteristics of the GTS

Problem	M	PS	PF
GTS1	2	Disconnected PS, PS varing on the hypersurface, simple variable-linkage	Mixed convexity-concavity, disconnected PF
GTS2	2	PS moving on the hypersurface, simple variable-linkage	Mixed convexity-concavity, fixed intersection point,
GTS3	2	PS changing in decision space, simple variable-linkage, the length of PS changing	Dynamic PF, knee point
GTS4	2	Time-varing PS, simple variable-linkage, degeneration	Dynamic PF, PF range changing
GTS5	2	Dynamic PS, simple variable-linkage	Dynamic PF, knee point, degeneration
GTS6	2	Time-linkage PS, PS moving on the hypersurface, simple variable-linkage	Mixed convexity-concavity
GTS7	2	Time-linkage PS, PS changing in the decision space, the length of PS changing, simple variable-linkage	Mixed convexity-concavity, the length and curvature of PF changing
GTS8	2	Time-linkage PS, PS varing in the decision space, simple variable-linkage	Mixed convexity-concavity
GTS9	3	PS moving in the space, time-varing geometry shape of PS, simple variable-linkage	Disconnection
GTS10	3	PS varing on the hypersurface, simple variable-linkage, degeneration	Disconnected segments
GTS11	3	PS varing on the hypersurface, simple variable-linkage	Degeneration

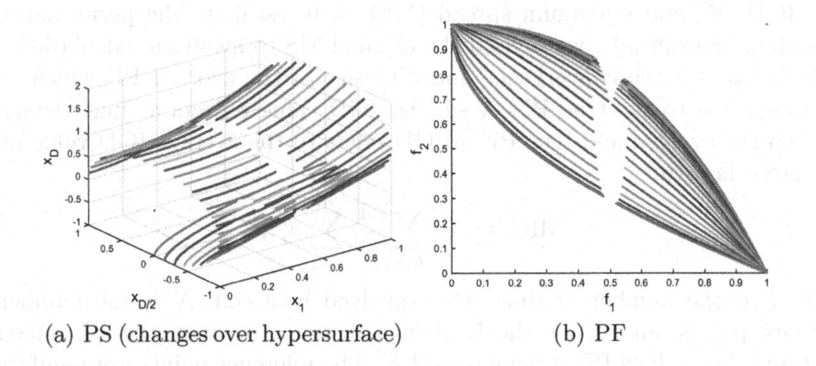

(a) PS (changes over hypersurface) (b) PF

Fig. 1. Illustration of the true PS and PF of GTS1.

PBDMO and MOEA/D-SVR were three novel prediction-based approaches, and MOEA/D-MoE integrated four predictors in a single run. Parameters of each algorithm were set the same as in the original papers [2,22–24] and [4], respectively.

Dynamics in the benchmark problems were controlled by (3), where the severity of change was set to $n_t = 10$ and the number of generations available in each time step was $\tau_t = \{10, 30\}$. $\tau_t = 10$ represented fast change scenarios, while $\tau_t = 30$ indicated a general changing speed. $T = 30$ represented time steps involved in each run. 50 generations were executed for each algorithm before the first change to minimize the effect of static optimization [14]. Algorithms

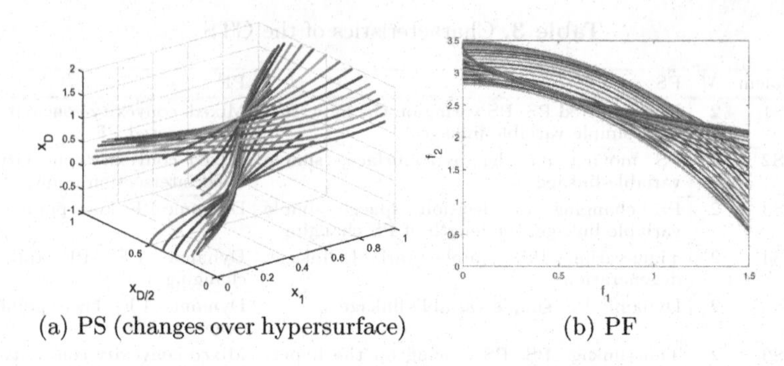

(a) PS (changes over hypersurface) (b) PF

Fig. 2. Illustration of the true PS and PF of GTS8.

terminated after $N * \tau_t * T$ fitness evaluations, where N is the population size. Each algorithm was run 31 times on each test instance.

3.2 Performance Metrics

We employed the mean inverted generational distance (MIGD) [26] adapted from IGD [25] and maximum spread (MS) [8] to estimate the performance of algorithms. Specifically, we revised the original MS to avoid miscalculation.

IGD assesses the convergence and diversity of the evolved PF which is an approximation to the true PF. A smaller IGD value reflects a small deviation between the evolved and true PF. MIGD refers to the average IGD value of all time steps in a run

$$\text{MIGD} = \frac{1}{T}\sum_{K=1}^{T}\frac{1}{N}\sum_{i=1}^{N} d_i^K, \tag{7}$$

where T is the number of time steps involved in a run, N is the number of reference points, and d_i^K is the Euclidean distance between the ith reference point and the evolved PF at time period K. The reference points were uniformly sampled from the true PF at each time step, where $N = 1500$ for the two-objective GTS1-8 and $N = 2500$ for the three-objective GTS9-11 problems.

The original MS indicator misjudges when all the obtained solutions are out of the bounds of the true PF. To avoid this unexpected situation, we proposed a revised version called MS2 which measures the distribution of the obtained solutions, i.e., the coverage of the obtained solutions over the true PF. A larger MS2 result refers to a better solution distribution. Similar to the MIGD metric, we calculate the average MS2 value of all the T time periods in a run

$$\text{MS2} = \frac{1}{T}\sum_{K=1}^{T} \text{MS2}^K, \tag{8}$$

where

$$\mathrm{MS2}^K = \prod_{j=1}^{M} \Theta(\overline{\mathrm{PF}_j^K} - \mathbf{P}_{j,\tau_t}^K) \sqrt{\frac{1}{M} \sum_{i=1}^{M} \left[\frac{\min(\overline{\mathbf{P}_{j,\tau_t}^K}, \overline{\mathrm{PF}_j^K}) - \max(\underline{\mathbf{P}_{j,\tau_t}^K}, \underline{\mathrm{PF}_j^K})}{\overline{\mathrm{PF}_j^K} - \underline{\mathrm{PF}_j^K}} \right]^2} \tag{9}$$

In Eq. (9), $\Theta(x) = 1$, when $x \geq 0$, $\Theta(x) = 0$, when $x < 0$. $\overline{\mathbf{P}_{j,\tau_t}^K}$ and $\underline{\mathbf{P}_{j,\tau_t}^K}$ are the maximum and minimum values of the jth objective in the final population $\mathbf{P}_{\tau_t}^K$ at time period K, respectively; while $\overline{\mathrm{PF}_j^K}$ and $\underline{\mathrm{PF}_j^K}$ are the maximum and minimum values of the jth objective in the ture PF^K.

3.3 Experimental Results and Analysis

The IGD and MS2 performance of the referred algorithms on the proposed GTS is given in Table 4 and 5, respectively.

Table 4. Mean and standard deviation of MIGD results

	τ_t, n_t	DNSGAII	MDP	PBDMO	MOEA/D-SVR	MOEA/D-MoE
GTS1	10,10	0.7147±2.20E-02	1.2851±6.07E-01	0.6868±4.16E-03	0.6586±2.66E-02	0.6502±1.32E-02
	30,10	0.7114±2.54E-02	1.6245±6.19E-01	0.6670±1.28E-02	0.4131±1.72E-02	0.4113±2.37E-02
GTS2	10,10	2.2159±1.83E-01	2.4533±4.99E-01	3.2366±2.22E-01	0.9675±7.64E-02	1.1027±1.19E-01
	30,10	1.9389±1.55E-01	2.3082±5.74E-01	0.9456±1.64E-01	0.2418±1.59E-02	0.2545±2.07E-02
GTS3	10,10	0.1484±2.24E-02	0.5640±2.07E-01	0.1253±9.83E-03	0.1257±1.55E-02	0.1634±1.56E-02
	30,10	0.1197±1.58E-02	0.4666±1.67E-01	0.1091±1.22E-02	0.0943±1.00E-03	0.1170±1.60E-02
GTS4	10,10	0.1250±2.03E-02	0.6183±1.57E-01	0.2751±2.80E-02	0.1005±1.56E-02	0.0969±1.13E-02
	30,10	0.0784±9.82E-03	0.6618±2.24E-01	0.0416±9.58E-03	0.0290±1.60E-02	0.0479±2.07E-02
GTS5	10,10	0.2048±1.76E-02	2.1074±5.75E-01	0.0271±3.12E-03	0.0940±1.14E-02	0.0478±1.26E-02
	30,10	0.1258±9.34E-03	2.1015±5.97E-01	0.0135±9.60E-04	0.0159±3.30E-04	0.0139±5.56E-04
GTS6	10,10	1.0916±9.48E-02	4.3034±9.60E-01	1.1740±1.47E-01	0.6073±4.24E-02	0.4218±6.65E-02
	30,10	0.9570±6.59E-02	4.8204±1.19E+00	0.2254±5.86E-02	0.2287±8.24E-02	0.2043±6.84E-02
GTS7	10,10	0.4675±7.76E-02	1.8074±3.56E-01	0.4900±8.61E-02	0.5873±7.94E-02	0.2690±5.63E-02
	30,10	0.3860±3.39E-02	1.7899±4.77E-01	0.2517±6.00E-02	0.1435±3.96E-03	0.1231±5.89E-03
GTS8	10,10	2.3968±1.56E-01	3.1737±5.23E-01	2.2291±1.87E-01	0.5481±4.45E-02	0.4180±4.91E-02
	30,10	1.9583±1.18E-01	2.9801±4.91E-01	0.3986±1.16E-01	0.1661±6.71E-03	0.1419±1.90E-02
GTS9	10,10	0.7298±4.87E-02	1.5894±3.61E-01	0.3919±1.67E-02	0.4477±3.55E-02	0.5260±2.78E-02
	30,10	0.5697±2.37E-02	2.0431±5.88E-01	0.2694±1.24E-02	0.2656±1.71E-02	0.2935±1.78E-02
GTS10	10,10	1.0827±7.61E-02	1.3851±1.52E-01	1.0759±6.67E-02	0.9197±1.43E-01	1.0831±8.41E-02
	30,10	0.8368±5.26E-02	1.4818±2.83E-01	0.6492±2.32E-02	0.4437±4.16E-02	0.5598±5.02E-02
GTS11	10,10	0.5458±5.53E-02	1.1469±2.01E-01	0.5537±1.03E-01	0.5029±5.12E-02	0.3193±3.61E-02
	30,10	0.3968±3.14E-02	1.2807±4.46E-01	0.3015±6.53E-02	0.1730±1.01E-02	0.1518±1.24E-02

>: A smaller MIGD value indicates a better convergence and diversity of solutions found by an algorithm and the best is with gray backgound

GTS with General Dynamics: GTS1-5 hold various PS and PF geometrical shapes, such as scaling, shifting, reversing, and degeneration. More importantly, the PS dynamics perform over hypersurfaces, rather than hyperplanes in other test suites. These dynamics pose great challenges to algorithms' PS tracking ability. According to Tables 4 and 5, it is not surprising that there was no algorithm constantly better than the others. This observation illustrates that the prediction methods employed in different algorithms are only competitive in specific problems, and there is still no PS tracking scheme qualified for various changing patterns. Furthermore, DNSGAII-A and MDP showed ineligible performance because the random reinitialization and linear prediction are not sufficient for PS moving on hypersurfaces, especially for high-dimensional decision spaces.

Table 5. Mean and standard deviation of MS2 results

	τ_t, n_t	DNSGAII	MDP	PBDMO	MOEA/D-SVR	MOEA/D-MoE
GTS1	10,10	0.6721±3.16E-02	0.3809±3.01E-01	0.7085±2.47E-03	0.6818±6.32E-02	0.6913±1.25E-02
	30,10	0.6696±3.14E-02	0.2258±2.98E-01	0.7134±4.68E-03	0.5996±5.15E-02	0.5928±4.83E-02
GTS2	10,10	0.0000±0.00E+00	0.0000±0.00E+00	0.0179±1.46E-02	0.3783±8.46E-02	0.3233±6.94E-02
	30,10	0.0010±3.76E-03	0.0000±0.00E+00	0.3158±9.16E-02	0.8000±4.24E-02	0.7676±6.33E-02
GTS3	10,10	0.9905±2.03E-02	0.4922±3.70E-01	1.0000±9.02E-08	0.9923±4.31E-02	0.8245±1.14E-01
	30,10	0.9954±1.24E-02	0.5513±3.37E-01	0.9938±1.80E-02	1.0000±1.18E-07	0.8952±6.58E-02
GTS4	10,10	0.5847±6.98E-02	0.0131±2.38E-02	0.0872±3.99E-02	0.2446±1.69E-01	0.2046±8.69E-02
	30,10	0.7107±3.91E-02	0.0158±3.32E-02	0.7999±6.15E-02	0.8918±1.45E-01	0.6567±1.99E-01
GTS5	10,10	0.9746±1.25E-02	0.7187±1.18E-01	0.9991±3.34E-04	0.9935±1.59E-03	0.9968±1.08E-03
	30,10	0.9765±1.19E-02	0.6943±1.86E-01	0.9998±1.01E-04	0.9995±7.24E-05	0.9995±8.13E-05
GTS6	10,10	0.0007±3.86E-03	0.0000±0.00E+00	0.1739±6.77E-02	0.3937±1.37E-01	0.4318±1.38E-01
	30,10	0.0089±2.69E-02	0.0000±0.00E+00	0.7259±1.14E-01	0.6694±2.40E-01	0.6581±1.79E-01
GTS7	10,10	0.9910±9.30E-03	0.9989±3.02E-03	0.9475±1.03E-02	0.9476±2.89E-03	0.9351±6.84E-03
	30,10	0.9612±8.04E-03	0.9964±6.33E-03	0.9298±1.03E-02	0.9120±1.45E-03	0.9100±2.26E-03
GTS8	10,10	0.0000±0.00E+00	0.0000±0.00E+00	0.0254±3.84E-02	0.5911±4.82E-02	0.6674±6.45E-02
	30,10	0.0002±1.07E-03	0.0000±0.00E+00	0.4761±9.48E-02	0.8037±1.97E-02	0.8112±4.33E-02
GTS9	10,10	0.9942±6.67E-03	0.9967±8.63E-03	1.0000±0.00E+00	0.9906±2.50E-03	0.9922±1.83E-03
	30,10	0.9841±8.37E-03	0.9507±5.65E-02	1.0000±0.00E+00	0.9747±3.10E-03	0.9738±3.44E-03
GTS10	10,10	0.9999±3.15E-04	0.9836±1.18E-02	1.0000±7.22E-05	1.0000±1.31E-09	0.9998±1.83E-04
	30,10	1.0000±1.75E-05	0.9734±1.50E-02	1.0000±0.00E+00	1.0000±2.44E-09	1.0000±6.52E-10
GTS11	10,10	0.8754±2.80E-02	0.3123±3.22E-01	0.6954±4.06E-02	0.7870±3.70E-02	0.8980±3.63E-02
	30,10	0.8870±2.80E-02	0.2827±2.46E-01	0.8554±2.78E-02	0.9453±3.52E-02	0.9537±2.70E-02

>: A larger MS2 result refers to a better distribution of solutions found by an algorithm and the best is with gray backgound

GTS with General Dynamics and Time-Linkage: GTS6-8 are benchmarks with time-linkage, i.e., future dynamics depending on historical solutions. They face challenging problems: for algorithms with less convergence at the initial time step, future problems they face deviate from the predefined ones. Thus, the convergence loss accumulates. Compared with the other four algorithms, the performance of MoE was relatively stable and promising, since MoE ensembles multiple predictors in a single run, and different predictors may be eligible at different time periods. Therefore, the convergence of the algorithm could be maintained as time goes by.

GTS with General Dynamics and Scalability: GTS9-11 are three-objective problems. MOEA/D-MoE and MOEA/D-SVR were in leading positions, which was similar to the performance on GTS1-5. These results show the scalability of our design of dynamics on problems with three objectives.

3.4 Advantage of the Generalized Test Suite

To demonstrate the reasonability, computationally efficiency, and easily tunability of our GTS functions, we employed five state-of-the-art and representative DMO algorithms for empirical study. According to the results of Tables 4 and 5, we see that GTS can efficiently discriminate different algorithms in terms of convergence, diversity, and solutions distribution. This discrimination can be attributed to the dynamics performing over hypersurfaces, which is more general and challenging than the dynamic property in other test suites.

In addition, we emulated time-linkage features in the GTS functions. These features can efficiently test the robustness of algorithms in cases of previous decision-marking deciding future optimization environment.

Overall, from the perspective of PS dynamics, the proposed GTS has a great advantage in discriminating algorithms and understanding the sensitivity of algorithms in more general and diverse optimization scenarios.

4 Conclusion

Having recognized the limitations of PS dynamics in existing DMO benchmarks, in this paper, we proposed a simple method to construct diverse PS dynamics. The constructed PS dynamics change on hypersurfaces over time, which are more general than the dynamics on hyperplanes in recent benchmarks, and can fairly compare the tracking ability of DMO algorithms. In addition, this method gives a reasonable trade-off between dynamic and static properties of benchmarks, to efficiently test the convergence, diversity, and solution distribution of DMO solvers. Furthermore, we designed a kind of new formulations for modeling the time-linkage property in DMOPs to systematically study the robustness of DMO algorithms. The proposed formulation, for the first time, emulates the time-deception and error accumulation characteristics abstracted from real-world applications. Experimental results demonstrated the effectiveness and efficiency of our proposed benchmarks. In the future, we plan to include more diverse PF dynamics.

Acknowledgement. This work is supported by the National Science Foundation of China under the Grant No. 61761136008, the Science and Technology Innovation Committee Foundation of Shenzhen under the Grant No. JCYJ20200109141235597, the Shenzhen Peacock Plan under the Grant No. KQTD2016112514355531, and the Program for Guangdong Introducing Innovative and Entrepreneurial Teams under the Grant No. 2017ZT07X386.

References

1. Biswas, S., Das, S., Suganthan, P.N., Coello, C.A.C.: Evolutionary multiobjective optimization in dynamic environments: a set of novel benchmark functions. In: 2014 IEEE Congress on Evolutionary Computation (CEC), pp. 3192–3199. IEEE (2014)
2. Cao, L., Xu, L., Goodman, E.D., Bao, C., Zhu, S.: Evolutionary dynamic multiobjective optimization assisted by a support vector regression predictor. IEEE Trans. Evol. Comput. **24**(2), 305–319 (2020)
3. Deb, K., Thiele, L., Laumanns, M., Zitzler, E.: Scalable multi-objective optimization test problems. In: 2002 IEEE Congress on Evolutionary Computation (CEC), vol. 1, pp. 825–830. IEEE (2002)
4. Deb, K., Rao N., U.B., Karthik, S.: Dynamic multi-objective optimization and decision-making using modified NSGA-II: a case study on hydro-thermal power scheduling. In: Obayashi, S., Deb, K., Poloni, C., Hiroyasu, T., Murata, T. (eds.) EMO 2007. LNCS, vol. 4403, pp. 803–817. Springer, Heidelberg (2007). https://doi.org/10.1007/978-3-540-70928-2_60
5. Di Barba, P.: Dynamic multiobjective optimization: a way to the shape design with transient magnetic fields. IEEE Trans. Magn. **44**(6), 962–965 (2008)
6. Farina, M., Deb, K., Amato, P.: Dynamic multiobjective optimization problems: test cases, approximations, and applications. IEEE Trans. Evol. Comput. **8**(5), 425–442 (2004)
7. Gee, S.B., Tan, K.C., Abbass, H.A.: A benchmark test suite for dynamic evolutionary multiobjective optimization. IEEE Trans. Cybern. **47**(2), 461–472 (2017)

8. Goh, C., Tan, K.C.: A competitive-cooperative coevolutionary paradigm for dynamic multiobjective optimization. IEEE Trans. Evol. Comput. **13**(1), 103–127 (2009)
9. Helbig, M., Engelbrecht, A.P.: Benchmarks for dynamic multi-objective optimisation. In: 2013 IEEE Symposium on Computational Intelligence in Dynamic and Uncertain Environments (CIDUE), pp. 84–91. IEEE (2013)
10. Helbig, M., Engelbrecht, A.P.: Benchmarks for dynamic multi-objective optimisation algorithms. ACM Comput. Surv. **46**(3), 37:1–37:39 (2014). https://doi.org/10.1145/2517649
11. Huang, L., Suh, I.H., Abraham, A.: Dynamic multi-objective optimization based on membrane computing for control of time-varying unstable plants. Informat. Sci. **181**(11), 2370–2391 (2011)
12. Jiang, S., Kaiser, M., Yang, S., Kollias, S., Krasnogor, N.: A scalable test suite for continuous dynamic multiobjective optimization. IEEE Trans. Cybern. **50**(6), 2814–2826 (2020)
13. Jiang, S., Yang, S.: Evolutionary dynamic multiobjective optimization: benchmarks and algorithm comparisons. IEEE Trans. Cybern. **47**(1), 198–211 (2017)
14. Jiang, S., Yang, S., Yao, X., Tan, K.C., Kaiser, M., Krasnogor, N.: Benchmark Problems for CEC2018 Competition on Dynamic Multiobjective Optimisation, pp. 1–8 (2018)
15. Jin, Y., Sendhoff, B.: Constructing dynamic optimization test problems using the multi-objective optimization concept. In: Raidl, G.R., et al. (eds.) EvoWorkshops 2004. LNCS, vol. 3005, pp. 525–536. Springer, Heidelberg (2004). https://doi.org/10.1007/978-3-540-24653-4_53
16. Mehnen, J., Rudolph, G., Wagner, T.: Evolutionary optimization of dynamic multiobjective functions. Technical report, Technische Universität Dortmund, Dortmund, Germany (2006). https://doi.org/10.17877/DE290R-631
17. Nguyen, S., Zhang, M., Johnston, M., Tan, K.C.: Automatic design of scheduling policies for dynamic multi-objective job shop scheduling via cooperative coevolution genetic programming. IEEE Trans. Evol. Comput. **18**(2), 193–208 (2014)
18. Nguyen, T.T., Yang, Z., Bonsall, S.: Dynamic time-linkage problems - the challenges. In: 2012 IEEE RIVF International Conference on Computing & Communication Technologies, Research, Innovation, and Vision for the Future, pp. 1–6. IEEE (2012)
19. Nguyen, T.T., Yao, X.: Dynamic time-linkage problems revisited. In: Giacobini, M., et al. (eds.) EvoWorkshops 2009. LNCS, vol. 5484, pp. 735–744. Springer, Heidelberg (2009). https://doi.org/10.1007/978-3-642-01129-0_83
20. Palaniappan, S., Zein-Sabatto, S., Sekmen, A.: Dynamic multiobjective optimization of war resource allocation using adaptive genetic algorithms. In: Proceedings of the IEEE SoutheastCon 2001 (Cat. No. 01CH37208), pp. 160–165. IEEE (2001)
21. Qiao, J., Zhang, W.: Dynamic multi-objective optimization control for wastewater treatment process. Neural Comput. Appl. **29**(11), 1261–1271 (2016). https://doi.org/10.1007/s00521-016-2642-8
22. Rambabu, R., Vadakkepat, P., Tan, K.C., Jiang, M.: A mixture-of-experts prediction framework for evolutionary dynamic multiobjective optimization. IEEE Trans. Cybern. 1–14 (2019, in press). https://doi.org/10.1109/TCYB.2019.2909806
23. Rong, M., Gong, D., Pedrycz, W., Wang, L.: A multimodel prediction method for dynamic multiobjective evolutionary optimization. IEEE Trans. Evol. Comput. **24**(2), 290–304 (2020)

24. Zhang, Q., Yang, S., Jiang, S., Wang, R., Li, X.: Novel prediction strategies for dynamic multiobjective optimization. IEEE Trans. Evol. Comput. **24**(2), 260–274 (2020)
25. Zhang, Q., Zhou, A., Jin, Y.: RM-MEDA: a regularity model-based multiobjective estimation of distribution algorithm. IEEE Trans. Evol. Comput. **12**(1), 41–63 (2008)
26. Zhou, A., Jin, Y., Zhang, Q.: A population prediction strategy for evolutionary dynamic multiobjective optimization. IEEE Trans. Cybern. **44**(1), 40–53 (2014)
27. Zhou, A., Jin, Y., Zhang, Q., Sendhoff, B., Tsang, E.: Prediction-based population re-initialization for evolutionary dynamic multi-objective optimization. In: Obayashi, S., Deb, K., Poloni, C., Hiroyasu, T., Murata, T. (eds.) EMO 2007. LNCS, vol. 4403, pp. 832–846. Springer, Heidelberg (2007). https://doi.org/10.1007/978-3-540-70928-2_62
28. Zitzler, E., Deb, K., Thiele, L.: Comparison of multiobjective evolutionary algorithms: empirical results. Evol. Comput. **8**(2), 173–195 (2000)

A Special Point and Transfer Component Analysis Based Dynamic Multi-objective Optimization Algorithm

Ruochen Liu(✉), Nanxi Li, Luyao Peng, and Kai Wu

Key Lab of Intelligent Perception and Image Understanding of Ministry of Education, International Center of Intelligent Perception and Computation, Xidian University, Xi'an 710071, China
ruochenliu@xidian.edu.cn

Abstract. To solve a dynamic multi-objective optimization problem better, algorithms need to quickly adapt to environmental changes and track its changing Pareto fronts fast. In this paper, an algorithm (SPTr-RMMEDA) based on special point and transfer component analysis is presented, in which, by using some special points and their neighborhoods solutions, the prediction strategy are used to generated the transfer learning prediction model, together with part new initial solutions to generate next population when change occurs. In order to better adapt to environmental changes, in addition to reusing some important historical information by transfer component analysis, the algorithm also introduces the adaptive diversity introduction strategy. The algorithm performs on 12 groups test problems. And the experimental results show that the proposed algorithm is superior to other four algorithms, such as Tr-NSGA-II, Tr-RMMEDA, DNSGA-II-A and DNSGA-II-B on most test problems in term of convergence and computation complexity.

Keywords: Special point · Transfer component analysis · Dynamic multi-objective optimization

1 Introduction

Dynamic multi-objective optimization problem (DMOP) refers to the optimization of a multi-objective problem in time-varying environment. Affected by the degree and frequency of environmental changes, the objective function, constraint function and parameters of DMOPs will change [4]. A good algorithm for solving DMOPs not only needs a faster convergence rate, but also exists various populations after each change to cope with [6], which is still a challenge in DMOP field. At present, evolutionary algorithm is the most common method to solve DMOPs [2,11].

The existing algorithms can be divided into four categories: diversity-based methods, memory-based methods, multi-population methods and prediction-based methods [1]. In 2007, Deb proposed a dynamic NSGA-II algorithm on

H. Ishibuchi et al. (Eds.): EMO 2021, LNCS 12654, pp. 218–231, 2021.
https://doi.org/10.1007/978-3-030-72062-9_18

the basis of static NSGA-II [5]. Goh and Tan [6] used a temporary memory method in their proposed co-evolutionary multi-objective algorithm to process past solutions in archival sets. The self-organizing scouts' method is proposed by Branke et al. [3]. In 2006, Hatzakis and Wallace [7] adopted a prediction method based on the forward-looking model, but determining the correlation of historical solutions does not guarantee correctness. Jiang et al. [10] took the lead in proposing to apply transfer learning into DMOPs. However, the method does not extract effective historical information, the diversity of the population is poor, and the time complexity is greatly increased.

To overcome the shortcomings mentioned above, this paper proposes a new algorithm based on special point and transfer component analysis. Its achievements are in two aspects: One is to use the prediction model, constructed by the transfer component analysis to predict a new population, in which the transfer model is developed by using selecting special points and their neighborhoods; another is to add adaptive diversity introduction strategy to improve the performance of the population.

The rest of this paper is organized as follows. Section 2 will introduce the related work of the proposed algorithm. Section 3 elaborates the algorithm's idea and frame in detail. Section 4 shows the experimental settings and comparative experimental results and their analysis. Section 5 summarizes the thesis and discusses future research work.

2 Related Work

2.1 Dynamic Multi-objective Optimization Problem

The dynamic multi-objective optimization problem (DMOP) can be defined as follows:

$$\begin{cases} \min \boldsymbol{f}(\boldsymbol{x},t) = \{f_1(\boldsymbol{x},t), f_2(\boldsymbol{x},t), \ldots, f_M(\boldsymbol{x},t)\} \\ \text{s.t. } \boldsymbol{g}(\boldsymbol{x},t) \leq 0 \\ \quad\quad \boldsymbol{h}(\boldsymbol{x},t) = 0 \\ \quad\quad\quad \boldsymbol{x} \in D \end{cases} \tag{1}$$

where t is the time variable, $\boldsymbol{x} = (x_1, x_2, \ldots, x_n)^T$ are n decision variables, D is the search space of decision variable, $\boldsymbol{g}(\boldsymbol{x},t)$ and $\boldsymbol{h}(\boldsymbol{x},t)$ are equality and inequality constraints, respectively. $\boldsymbol{f}(\boldsymbol{x})$ defines the mapping of D to the objective space, and is an M dimensional objective vector. The optimal solution obtained by optimization is called Pareto front (PF) in the objective space and Pareto solution (PS) in the decision space.

2.2 Transfer Component Analysis

Transfer component analysis (TCA) [9] reuses the knowledge obtained from the source domain and performs tasks in the target domain, which is related to but different from the source domain. When the domains are in different distributions, the data will be mapped together into a reproducing kernel Hilbert Space (RKHS) $\mathcal{H}$ [12], where the distance between the source and target is minimized and their internal attributes are retained to the maximum extent. TCA makes this method of calculating distances universal and simple.

TCA [9] assumes that the marginal distribution of the source domain and the target domain is different, i.e. $P(X_s) \neq P(Y_t)$, where X_s and Y_t represent the source domain and the target domain, respectively. It has a feature mapping ϕ, so that $P(\phi(X_s)) \approx P(\phi(Y_t))$. Maximum mean discrepancy (MMD) is used to calculate the distance, and is given as follows:

$$dist\left(X_s^{'}, Y_t^{'}\right) = \left\| \frac{1}{n_1} \sum\nolimits_{i=1}^{n_1} \phi(x_i) - \frac{1}{n_2} \sum\nolimits_{i=1}^{n_2} \phi(y_i) \right\|_{\mathcal{H}}^2 \tag{2}$$

where $X_s = \{x_1, x_2, \ldots, x_{n_1}\}$, and $Y_t = \{y_1, y_2, \ldots, y_{n_2}\}$. $X_s^{'}$ and $Y_t^{'}$ represent $\phi(X_s)$ and $\phi(Y_t)$, respectively. $\phi(\cdot): \mathcal{X} \rightarrow \mathcal{H}$ is the domain mapped in RKHS, to facilitate the calculation, TCA introduces a kernel matrix K and a coefficient matrix L as follows:

$$K = \begin{bmatrix} K_{s,s} & K_{s,t} \\ K_{t,s} & K_{t,t} \end{bmatrix} \in \mathbb{R}^{(n_1+n_2)\times(n_1+n_2)} \quad L_{ij} = \begin{cases} \frac{1}{n_1^2} & x_i, x_j \in X_s \\ \frac{1}{n_2^2} & y_i, y_j \in Y_t \\ -\frac{1}{n_1 n_2} & otherwise \end{cases} \tag{3}$$

where $K_{u,v}$ is a kernel matrix. For instance in $K_{s,t}$, each element $k_{i,j} = k(x_i, y_j) = \phi(x_i)^T \phi(y_j)$.

Thus, (2) above can be changed into the following form:

$$dist\left(X_s^{'}, Y_t^{'}\right) = tr(KL) \tag{4}$$

where $tr(\cdot)$ is the trace of matrix. Here (2) can be translated into a semi-definite programming problem. Let $\varphi(x) = W^T k_x \in \mathbb{R}^d$, and the MMD can be calculated by a dimensional reduction way, where W is an $(n_1+n_2)\times d$ $(d \ll n_1+n_2)$ weight matrix of samples, $k_x = [k(x_1, x), \ldots, k(x_{n_1}, x), k(y_1, x), \ldots, k(y_{n_2}, x)]^T$, and the kernel function K can be converted as the following formula:

$$\begin{aligned} \widetilde{K} &= [\varphi(x_1), \ldots, \varphi(x_{n_1}), \varphi(y_1), \ldots, \varphi(y_{n_2})]^T [\varphi(x_1), \ldots, \varphi(x_{n_1}), \varphi(y_1), \ldots, \varphi(y_{n_2})] \\ &= \left(KK^{-\frac{1}{2}}\widetilde{W}\right)\left(\widetilde{W}K^{-\frac{1}{2}}K\right) = KWW^TK \end{aligned} \tag{5}$$

where $\widetilde{W}$ is a dimensional reduction matrix, which transform the kernel map into a d-dimensional space, and K is a symmetric matrix.

After sorting out the above formulas, TCA can be expressed as:

$$\begin{aligned} &\arg\min_{w}\ tr\left(W^T KLKW\right) + \mu\, tr\left(W^T W\right) \\ &\text{s.t.}\quad W^T KHKW = I \end{aligned} \tag{6}$$

where $H = I_{n_1+n_2} - \frac{1}{n_1+n_2}\mathbf{1}\mathbf{1}^T$ is a central matrix added in order to maintain the respective data characteristics, I is the identity matrix and $\mathbf{1}$ is an all-ones matrix. For solving the above optimization problem, the second regularization term of (6) is added by the Lagrange duality to get W.

3 The Proposed Algorithm

This section introduces the proposed algorithm (SPTr-RMMEDA). In SPTr-RMMEDA, when changes were detected, an adaptive mechanism would predict the population under new environment, by using TCA and an adaptive diversity introduction, then predicted initial population was optimized by static algorithm, namely, regularity model based multi-objective estimation of distribution algorithm (RMMEDA) which is proposed by Zhang in 2008. Compared with classical NSGA-II, RMMEDA showed more advantages with TCA in experiments [10]. The detail of RMMEDA can be found in [13]. The flow of SPTr-RMMEDA is shown in Algorithm 1.

Algorithm 1. SPTr-RMMEDA

Input:

A DMOP; The termination condition, *Ter*

N: the population size and a *kernel function.*

Output:

Approximation of PS: X; Approximation of PF: $F(X)$.

Initializion: The number of time change; Initializes the parent population Pop_0.

Step1: Randomly select 5% individuals to revaluated, if environment changes, go to Step 2. Otherwise, go to Step 5.

Step2: Predict a new population *nPop* by the prediction strategy in Algorithm 2.

Step3: Calculate the number of diversity introduction $N_{div} = N - N_{Pop_{non}} - N_{nPop}$, and introduce individuals.

Step4: Get Pop_{t+1} by Step 2 and 3 and the non-dominant solutions Pop_{non} in t time.

Step5: Optimize Pop_{t+1} with the static RMMEDA. $t = t + 1$.

Step6: If *Ter* is met, stop. Otherwise go to the Step 2.

3.1 Special Point

Special points can help outline the PF for most multi-state testing problems. Five kinds of special points are introduced in this paper.

The boundary point [14] refers to the point of minimum value of each target in the target space in minimization problem. Its mathematical expression is given as follows:

$$B = \{f(f_{1min}(X), \ldots, f_M(X)), \ldots, f(f_1(X), \ldots, f_{Mmin}(X))\} \tag{7}$$

where M is the number of objectives.

The knee point [14] has the maximum marginal regression rate in the objective space. It can be defined as the point farthest from the vertical distance of the line L or plane S composed of boundary points. The formula is shown as follows:

$$D(K, L) = \frac{|ax_K + by_K + c|}{\sqrt{a^2 + b^2}} \tag{8}$$

where K is the knee point, D is the distance between knee point and the line L, a and b are the coefficients of L.

PCTB is the closest point to the boundary points that can be understood as the point with the smallest distance to the sum of all boundary points. The formula is presented in the following:

$$\min_{1 \leq i \leq k} \sum\nolimits_{j=1}^{M} \|X_i - K_j\| \tag{9}$$

where K_j is the boundary point, and k is the number of non-dominated points X_i.

The ideal point is the point composed of the minimum values in each dimension in the target space. In general, the ideal point does not exist in the non-dominated set, and is a point of construction. E is the ideal point, which is given in the following:

$$E = \min\{f(P^1), f(P^2), \ldots, f(P^M)\} \tag{10}$$

The center point usually refers to the point at the center of all points in the non-dominated set. The calculation formula of the center point is shown as follows:

$$C_i = \frac{1}{|P|} \sum_{x \in P} x_i, \quad i = 1, 2 \ldots, n \tag{11}$$

where n is the dimension of the decision space. P is the non-dominated set.

3.2 Change Adaptation Mechanism

When the environmental change is detected, the prediction part is mainly based on the special point in PF at previous moment. Each special point and its neighborhood will be partly sampled, and then the set sampled will be predicted through TCA method to get the predicted population at next time. In TCA prediction method, the individuals at t time as the source domain, objective values of time $t + 1$ as the target domain. When establishing the mapping relationship, the individuals in new environment can be predicted by the transfer learning of the sampled special point population. The prediction strategy is given in detail in Algorithm 2, where l represents the individuals from special points $SPop$.

Because the number of boundary points is related to the objective dimension, the number of special points in two objectives and three objectives is 6 and 7

Algorithm 2. Prediction Strategy based on TCA

Input:

The objective function $F(\cdot)$; the PF at time t; $k(\cdot\,,\cdot)$: a *kernel function*; the sampling rate, γ_s; the size of population, N.

Output:

An initial population $nPop$ at time $t+1$.

Step1: Calculate the special points, then partial sampling.

Step2: Randomly generate two sets of initial solutions X_s and Y_t, then calculate objective function values $F_t(X_s)$ and $F_{t+1}(Y_t)$ at time t and $t+1$, respectively.

Step3: Use TCA on $F_t(X_s)$ and $F_{t+1}(Y_t)$ to determine the mapping relationship and obtain the matrix W by (6).

Step4: Map the sampled population $SPop$ at time t by $k(\cdot\,,\cdot)$.

Step5: Use $\varphi(x) = W^T k_x$ to calculate the characteristic mapping of $SPop$ in high-dimension space, $\varphi(SPop) = W^T k_{Spop}$.

Step6: $\varphi(SPop)$ will search one by one for mapping individuals at $t+1$, $(F_{t+1}(x))$ closest to it. $x \leftarrow \arg\min_x \|\varphi(F_{t+1}(x)) - \varphi(F_{SPop}(l))\|$.

Step7: The solution obtained in Step 6 is the predicted initial population $nPop$.

respectively, so the size of partial sampling is $6\times\gamma_s\times N$ and $7\times\gamma_s\times N$ respectively. The number of sampling $\gamma_s \times N$. The adaptive diversity introduction strategy means that after the special point prediction, if the predicted population $nPop$ and the non-dominated set Pop_{non} at the previous moment $(N_{nPop} + N_{Pop_{non}})$ is less than the population size N, the remain $N-(N_{nPop}+N_{Pop_{non}})$ individuals are generated randomly. Otherwise, N initial solutions will be selected by non-dominated sort selection method to form the initial population Pop_{t+1}.

4 Experimental Results

In this section, experiments are performed to compare the performance of the proposed SPTr-RMMEDA with four other dynamic multi-objective optimization evolutionary algorithms, including Tr-NSGA-II [10], Tr-RMMEDA [10], DNSGA-II-A [5], DNSGA-II-B [5].

4.1 Performance Metrics

MIGD [15] is to calculate the average inverted generational distance (IGD) value, and it is adopted to evaluate the performance in this paper. The definition is shown as follows:

$$MIGD = \frac{1}{|T|} \sum_{t\in T} \frac{\sum_{v\in PF_t} d(v, P_t)}{|PF_t|} \tag{12}$$

where T means a set of discrete time points in a run. PF_t is Pareto optimal points set distributed uniformly in PF, P_t represents the approximation PF obtained by algorithms, $d(v, P_t) = min_{u\in p_t}\|F(v) - F(u)\|$ directs the distance between v and P_t.

The running time of the algorithm is also an important factor to be considered in practical application. The running time represents the cost. Therefore, in this paper, the CPU running time of each test function **Time** will also be taken as an important performance evaluation metrics.

Table 1. Functions settings

Function	Type	Dim	Objectives	Feature
DMOP3	I	10	2	Convex
DIMP2	I	10	2	Continuous
FDA4	I	12	3	Non-convex
DMOP2	II	10	2	Convex to concave
DMOP2$_{iso}$	II	10	2	Isolated
DMOP2$_{dec}$	II	10	2	Deceptive
FDA5	II	12	3	Non-convex
FDA5$_{iso}$	II	12	3	Isolated
FDA5$_{dec}$	II	12	3	Deceptive
HE2	III	30	2	Discontinuous
HE7	III	10	2	Complex PSs
HE9	III	10	2	Complex PSs

Table 2. Functions parameters settings

	C1	C2	C3	C4	C5	C6	C7	C8
n_t	10	10	10	10	1	1	20	20
τ_T	100	200	500	1000	200	1000	200	1000
τ_t	5	10	25	50	10	50	10	50

4.2 Experimental Settings

In order to test the performance and universal applicability of the algorithm, three types of DMOP problems [8] were selected as shown in Table 1. It is worth noting that the isolated and deceptive problems' coefficients A, B and C are $G(t)$, 0.001, and 0.05 respectively. What's more, in terms of influencing factors, different parameters for different environments $C1$-$C8$ are set as shown in Table 2, where n_t, τ_T and τ_t are the severity of change, the maximum number of iterations, and frequency of change, respectively. The parameters of SPTr-RMMEDA, SPTr-NSGA-II, Tr-RMMEDA and Tr-NSGA-II: the population size is 200, the sample rate γ_s is 0.05, the kernel function is a Gaussian kernel function, and the expected dimension value m was set as 20, and the tradeoff parameter μ is set at 0.5. Our algorithm is performed on 12 groups of test questions and 8 groups of changes as shown in Table 2.

4.3 Results

We compared the Tr-RMMEDA and Tr-NSGA-II proposed by Jiang [10] with the SPTr-RMMEDA and SPTr-NSGA-II in which the prediction strategy is proposed in this paper. Experiments explain two parts of data, including MIGD metric in Table 3 and the CPU running time in Table 4.

continued

In Table 3, for the analysis of 96 experimental results, it can be seen that the partial sampling prediction framework we proposed has greatly improved the performance compared with the whole population prediction, among which a total of 64 experimental results are superior to Tr-RMMEDA and Tr-NSGA-II in diversity and convergence. Problems with isolated PF, in particular, showed better performance on 8 variations. This is because the characteristics of PF of these problems are significant, and our algorithm can accurately find the special points, thus improving the convergence rate of the algorithm. For SPTr-RMMEDA, compared with Tr-RMMEDA and Tr-NSGA-II, DMOP3 and DMOP2, the type I and the type II functions, showed significant improvement in convergence. There is not doubt that RMMEDA is a good optimizer on the contribution of convergence and diversity. But, it still can be seen that these five kinds of special points improve the efficiency of Tr-RMMEDA.

What's more, in terms of experimental cost, CPU running time metric is adopted. Here we have adopted the average CPU time for the 8 changes for each test function. The results are shown in Table 4. The proposed algorithm SPTr-RMMEDA and SPTr-NSGA-II have less time complexity and faster operation speed, which greatly saves the calculation cost. Almost all functions run half as fast, especially for DMOP2 and FDA5, the running time is shortened by more than 70%.

In order to prove the universal applicability of SPTr-RMMEDA proposed in this paper, we also compared it with the famous D-NSGA-II-A and D-NSGA-II-B. The test functions are consistent with Table 4, and the experimental results are shown in Table 5, here we put the average MIGD of each function. Note that the ++ represents that SPTr-RMMEDA performs better than other algorithms except the boldface one. As can be seen from the table, the experimental results of 9 groups of functions show that SPTr-RMMEDA has better performance than D-NSGA-II-A and D-NSGA-II-B. And for type I and type II problems, our algorithm is superior to D-NSGA-II-A and D-NSGA-II-B for most functions, especially for a series of DMOP. For type III problems, the complex HE7 and HE9, although our algorithm is better than D-NSGA-II-A and D-NSGA-II-B, it has poor convergence compared with Tr-RMMEDA.

Table 3. The MIGD of 12 test functions compared with Tr-RMMEDA and Tr-NSGA-II

		Tr-RMMEDA	Tr-NSGA-II	SPTr-NSGA-II	SPTr-RMMEDA
DMOP3	C1	1.2863E+00	1.0754E+00	4.3357E−01	**3.3944E−01**
	C2	3.5463E−01	7.4905E−01	2.7393E−01	**6.7255E−02**
	C3	3.6507E−02	1.9665E−01	4.1606E−02	**4.3151E−03**
	C4	1.7887E−03	4.1599E−02	4.8844E−03	**1.3911E−03**
	C5	2.0006E+01	3.1536E+01	**8.9771E+00**	2.3269E+01
	C6	1.3151E+01	1.9600E+01	**1.2878E−01**	1.9108E+01
	C7	3.3009E−01	6.4625E−01	1.0384E−01	**5.5857E−02**
	C8	2.2889E−03	2.9116E−02	5.6663E−03	**1.5273E−03**
DIMP2	C1	1.3994E+01	1.6513E+01	1.6402E+01	**1.1329E+01**
	C2	1.1777E+01	1.1729E+01	1.0791E+01	**8.9868E+00**
	C3	6.1650E+00	6.6555E+00	6.2163E+00	**5.5050E+00**
	C4	4.3800E+00	4.5751E+00	**3.8872E+00**	4.4165E+00
	C5	1.1400E+01	1.1692E+01	**6.4065E+00**	9.0067E+00
	C6	4.5364E+00	5.0571E+00	**2.2073E+00**	4.6903E+00
	C7	1.2146E+01	1.1775E+01	**6.9015E+00**	9.2480E+00
	C8	4.5156E+00	4.5501E+00	**3.0262E+00**	4.6632E+00
FDA4	C1	**8.5935E−02**	6.2513E−02	1.0185E−01	9.5432E−02
	C2	5.9661E−02	**5.0425E−02**	8.8898E−02	6.3630E−02
	C3	**3.4118E−02**	6.5065E−02	9.1288E−02	3.8320E−02
	C4	**2.8592E−02**	5.1857E−02	6.3435E−02	3.1015E−02
	C5	**5.9974E−02**	6.8167E−02	6.8813E−02	2.2795E−01
	C6	**2.2941E−02**	3.6584E−02	5.0706E−02	4.3543E−02
	C7	5.6338E−02	**4.6327E−02**	8.3578E−02	6.7860E−02
	C8	**2.9195E−02**	5.3299E−02	7.2096E−02	3.0138E−02
DMOP2	C1	3.5707E+00	1.2645E+00	1.1525E+00	**4.5787E−01**
	C2	9.3208E−01	9.5185E−01	6.6927E−01	**1.1683E−01**
	C3	5.6712E−02	2.2691E−01	5.8186E−02	**8.5256E−03**
	C4	2.4209E−03	3.2114E−02	4.2297E−02	**1.8152E−03**
	C5	2.9607E+01	3.3273E+01	**1.0092E+01**	2.3767E+01
	C6	1.9701E+01	2.0066E+01	**1.1911E−01**	1.9113E+01
	C7	9.4594E−01	9.7600E−01	4.0502E−01	**8.1779E−02**
	C8	6.6918E−03	2.8489E−02	4.0326E−02	**1.8401E−03**
DMOP2_{iso}	C1	**3.7482E−04**	4.8529E−04	4.9446E−02	3.7893E−04
	C2	3.7430E−04	4.7425E−04	4.8700E−02	**3.7397E−04**
	C3	3.8167E−04	5.0120E−04	4.8666E−02	**3.7660E−04**
	C4	3.7531E−04	5.0526E−04	4.8586E−02	**3.7515E−04**
	C5	**1.2711E−01**	1.2900E−01	2.3627E−01	1.2761E−01
	C6	1.2723E−01	1.2866E−01	2.3673E−01	**1.2703E−01**
	C7	3.7736E−04	4.7151E−04	4.6265E−02	**3.7687E−04**
	C8	3.7787E−04	4.9554E−04	4.6610E−02	**3.7492E−04**
DMOP2_{dec}	C1	2.9048E+00	1.5265E+00	1.6849E+00	**9.8003E−01**
	C2	1.6965E+00	1.1236E+00	7.8413E−01	**4.4401E−01**
	C3	2.6503E−01	5.0484E−01	1.5520E−01	**1.2488E−01**
	C4	5.7880E−02	1.4913E−01	6.1786E−02	**5.7634E−02**
	C5	6.3960E+00	6.4900E+00	4.5222E+00	**2.6376E+00**
	C6	2.5014E−01	1.0417E+00	8.8060E−01	**2.2967E−01**
	C7	1.8043E+00	1.0513E+00	6.0663E−01	**3.1359E−01**
	C8	7.1776E−02	1.4461E−01	7.2294E−02	**3.9884E−02**

Table 3. continued

		Tr-RMMEDA	Tr-NSGA-II	SPTr-NSGA-II	SPTr-RMMEDA
FDA5	C1	9.4148E−01	1.1005E+00	1.0247E+00	**7.2009E−01**
	C2	8.0605E−01	9.5679E−01	7.8639E−01	**6.5414E−01**
	C3	6.0611E−01	1.1574E+00	6.5425E−01	**6.0141E−01**
	C4	6.3127E−01	9.7688E−01	7.3302E−01	**6.3126E−01**
	C5	9.8630E−01	1.1558E+00	**6.2167E−01**	1.2127E+00
	C6	**4.8942E−01**	1.1341E+00	6.4161E−01	5.1395E−01
	C7	7.8530E−01	9.5554E−01	7.0544E−01	**6.2135E−01**
	C8	6.0839E−01	8.5646E−01	7.0436E−01	**6.0710E−01**
FDA5$_{iso}$	C1	9.3516E−01	9.7656E−01	7.7709E−01	**6.0538E−01**
	C2	8.2748E−01	8.7581E−01	7.5873E−01	**6.2419E−01**
	C3	6.4934E−01	6.5635E−01	7.3241E−01	**5.9193E−01**
	C4	6.2490E−01	6.2631E−01	7.1902E−01	**6.2485E−01**
	C5	6.8010E−01	**5.7857E−01**	5.9942E−01	7.2335E−01
	C6	4.7161E−01	**4.5308E−01**	5.8836E−01	4.7091E−01
	C7	8.0447E−01	7.6986E−01	7.2956E−01	**6.0168E−01**
	C8	6.0205E−01	**5.7622E−01**	7.1850E−01	6.0213E−01
FDA5$_{dec}$	C1	1.9532E+00	6.2846E+00	**1.2769E+00**	1.3032E+00
	C2	1.6195E+00	4.8859E+00	**8.0435E−01**	1.1515E+00
	C3	1.3194E+00	2.8816E+00	**6.4712E−01**	8.2091E−01
	C4	1.0118E+00	9.7583E−01	**6.2837E−01**	7.6108E−01
	C5	5.5564E+00	8.0650E+01	**1.1117E+00**	1.7701E+00
	C6	**5.3990E−01**	5.0291E+00	**7.8663E−01**	6.0937E−01
	C7	1.6717E+00	1.8825E+00	**7.3973E−01**	1.0383E+00
	C8	1.1122E+00	9.1008E−01	**6.9787E−01**	8.2960E−01
HE2	C1	1.24E+00	1.08E+00	7.48E−01	**5.18E−01**
	C2	8.19E−01	2.06E−01	**9.71E−02**	4.05E−01
	C3	3.60E−01	8.54E−02	**6.31E−02**	1.46E−01
	C4	9.41E−02	7.81E−02	6.19E−02	**4.41E−02**
	C5	7.20E−01	**5.95E−02**	1.74E−01	2.28E−01
	C6	6.25E−02	**4.73E−02**	1.47E−01	4.99E−02
	C7	8.21E−01	1.83E−01	**6.97E−02**	3.21E−01
	C8	1.01E−01	8.24E−02	**6.39E−02**	5.18E−01
HE7	C1	1.53E+00	**1.52E+00**	8.00E+01	2.42E+00
	C2	**1.70E+00**	9.62E+00	1.06E+02	3.59E+00
	C3	**1.88E+00**	6.31E+01	1.37E+02	4.82E+00
	C4	**2.58E+00**	7.20E+01	1.59E+02	5.36E+00
	C5	**1.32E+00**	4.12E+00	1.77E+02	2.23E+00
	C6	**1.88E+00**	2.34E+01	1.71E+02	3.26E+00
	C7	**1.68E+00**	4.41E+00	9.48E+01	3.75E+00
	C8	**2.70E+00**	1.01E+02	1.01E+02	5.29E+00
HE9	C1	2.55E+00	3.34E+00	6.95E+00	**2.53E+00**
	C2	**2.16E+00**	3.84E+00	7.51E+00	3.36E+00
	C3	**2.18E+00**	5.33E+00	7.62E+00	5.98E+00
	C4	**2.56E+00**	6.20E+00	7.54E+00	6.54E+00
	C5	**2.05E+00**	3.32E+00	8.19E+00	2.66E+00
	C6	**2.34E+00**	4.24E+00	8.28E+00	4.55E+00
	C7	**2.04E+00**	4.01E+00	6.60E+00	3.54E+00
	C8	**2.44E+00**	6.00E+00	6.53E+00	6.28E+00

Table 4. The CPU running time of each function

	Tr-RMMEDA	Tr-NSGA-II	SPTr-NSGA-II	SPTr-RMMEDA
DMOP3	1.13E+04	2.15E+04	7.13E+03	**1.32E+03**
DIMP2	8.73E+03	3.80E+02	**1.88E+02**	1.21E+03
FDA4	1.70E+04	1.13E+04	**3.69E+03**	6.09E+03
DMOP2	8.46E+03	1.82E+04	5.60E+03	**3.87E+03**
DMOP2$_{iso}$	3.39E+03	3.80E+03	**1.25E+03**	6.58E+03
DMOP2$_{dec}$	5.97E+03	2.26E+04	6.47E+03	**1.73E+03**
FDA5	1.25E+04	1.71E+04	6.53E+03	**4.37E+03**
FDA5$_{iso}$	4.97E+03	4.99E+03	**1.89E+03**	6.74E+03
FDA5$_{dec}$	5.38E+03	2.06E+04	5.94E+03	**4.38E+03**
HE2	1.40E+04	4.72E+04	9.55E+03	**9.09E+03**
HE7	2.24E+04	3.04E+04	5.03E+03	**7.96E+03**
HE9	1.44E+04	1.90E+04	**3.47E+03**	3.88E+03

Table 5. The MIGD of 12 test functions compared with D-NSGA-II-A and D-NSGA-II-B

		Tr-RMMEDA	D-NSGAII-A	D-NSGA-II-B	SPTr-RMMEDA
DMOP3	C1	1.2863E+00	6.3602E+00	5.9420E+00	**3.3944E−01**
	C2	3.5463E−01	1.7099E+00	1.8032E+00	**6.7255E−02**
	C3	3.6507E−02	1.3515E−01	1.7762E−01	**4.3151E−03**
	C4	1.7887E−03	1.3371E−02	1.3662E−02	**1.3911E−03**
	C5	**2.0006E+01**	3.4259E+01(++)	3.4784E+01(++)	2.3269E+01
	C6	**1.3151E+01**	2.5477E+01(++)	2.5007E+01(++)	1.9108E+01
	C7	3.3009E−01	4.6916E−01	5.0449E−01	**5.5857E−02**
	C8	2.2889E−03	6.2000E−03	5.1149E−03	**1.5273E−03**
DIMP2	C1	1.3994E+01(++)	1.0779E+01(++)	**1.0773E+01**	1.1329E+01
	C2	1.1777E+01(++)	6.1396E+00	**6.0993E+00**	8.9868E+00
	C3	6.1650E+00(++)	2.1462E+00	**2.2139E+00**	5.5050E+00
	C4	4.3800E+00(++)	**4.9120E−01**	5.1202E−01	4.4165E+00
	C5	1.1400E+01(++)	**5.8257E+00**	5.9734E+00	9.0067E+00
	C6	4.5364E+00	3.9558E−01	**3.9277E−01**	4.6903E+00
	C7	1.2146E+01(++)	**5.8064E+00**	5.9533E+00	9.2480E+00
	C8	4.5156E+00	**8.0620E−02**	9.1836E−02	4.6632E+00
FDA4	C1	**8.5935E−02**	6.4708E−01(++)	6.5001E−01(++)	9.5432E−02
	C2	**5.9661E−02**	2.5684E−01(++)	2.7022E−01(++)	6.3630E−02
	C3	**3.4118E−02**	3.8644E−02(++)	3.9541E−02(++)	3.8320E−02
	C4	2.8592E−02	1.8032E−02	**1.7852E−02**	3.1015E−02
	C5	**5.9974E−02**	3.4705E+00(++)	3.3771E+00(++)	2.2795E−01
	C6	**2.2941E−02**	3.2088E+00(++)	3.3470E+00(++)	4.3543E−02
	C7	**5.6338E−02**	7.4745E−02(++)	7.4248E−02	6.7860E−02
	C8	2.9195E−02	**1.4137E−02**	1.4436E−02	3.0138E−02

continued

Table 5. continued

		Tr-RMMEDA	D-NSGAII-A	D-NSGA-II-B	SPTr-RMMEDA
DMOP2	C1	3.5707E+00	3.0879E+00	2.7969E+00	**4.5787E−01**
	C2	9.3208E−01	1.7613E−01	1.8053E−01	**1.1683E−01**
	C3	5.6712E−02	9.2279E−02	8.8437E−02	**8.5256E−03**
	C4	2.4209E−03	5.6715E−01	3.3672E−01	**1.8152E−03**
	C5	2.9607E+01	3.7423E+01	3.7335E+01	**2.3767E+01**
	C6	1.9701E+01	3.7423E+01	3.7417E+01	**1.9113E+01**
	C7	9.4594E−01	4.7621E−02	3.8016E−02	**8.1779E−02**
	C8	6.6918E−03	8.9384E−02	2.3242E−03	**1.8401E−03**
$DMOP2_{iso}$	C1	**3.7482E−04**	8.6948E−02(++)	7.6542E−02(++)	3.7893E−04
	C2	3.7430E−04	2.3009E−03	6.6170E−03	**3.7397E−04**
	C3	3.8167E−04	3.8267E−04	3.8241E−04	**3.7660E−04**
	C4	3.7531E−04	3.7810E−04	3.7956E−04	**3.7515E−04**
	C5	**1.2711E−01**	1.3128E−01(++)	1.3841E−01(++)	1.2761E−01
	C6	1.2723E−01	1.2935E−01	1.2926E−01	**1.2703E−01**
	C7	3.7736E−04	1.1516E−03	7.8538E−03	**3.7687E−04**
	C8	3.7787E−04	3.7893E−04	3.7913E−04	**3.7492E−04**
$DMOP2_{dec}$	C1	2.9048E+00	6.9577E+00	7.3607E+00	**9.8003E−01**
	C2	1.6965E+00	1.4666E+01	1.6916E+01	**4.4401E−01**
	C3	2.6503E−01	7.8361E+02	4.7659E+02	**1.2488E−01**
	C4	5.7880E−02	3.5065E+04	2.9362E+04	**5.7634E−02**
	C5	6.3960E+00	5.3260E+05	6.2053E+05	**2.6376E+00**
	C6	2.5014E−01	9.3934E+06	9.1486E+06	**2.2967E−01**
	C7	1.8043E+00	2.8521E+01	2.3125E+01	**3.1359E−01**
	C8	7.1776E−02	2.5679E+03	2.3793E+03	**3.9884E−02**
FDA5	C1	9.4148E−01	1.1460E+00	1.1053E+00	**7.2009E−01**
	C2	8.0605E−01	7.8857E−01	7.9699E−01	**6.5414E−01**
	C3	6.0611E−01(++)	**5.3095E−01**	5.3235E−01	6.0141E−01
	C4	6.3127E−01(++)	5.4535E−01	**5.4509E−01**	6.3126E−01
	C5	**9.8630E−01**	2.5469E+00(++)	2.5525E+00(++)	1.2127E+00
	C6	**4.8942E−01**	3.1909E+00(++)	3.0430E+00(++)	5.1395E−01
	C7	7.8530E−01	6.9525E−01	6.9121E−01	**6.2135E−01**
	C8	6.0839E−01(‡)	**5.2080E−01**	5.2050E−01	6.0710E−01
$FDA5_{iso}$	C1	9.3516E−01	7.8567E−01	7.8876E−01	**6.0538E−01**
	C2	8.2748E−01	8.0192E−01	8.0476E−01	**6.2419E−01**
	C3	6.4934E−01	8.1081E−01	8.1131E−01	**5.9193E−01**
	C4	6.2490E−01	8.0519E−01	8.0919E−01	**6.2485E−01**
	C5	6.8010E−01	6.1900E−01	**6.1835E−01**	7.2335E−01
	C6	4.7161E−01	7.2469E−01	6.6449E−01	**4.7091E−01**
	C7	8.0447E−01	7.8404E−01	7.8613E−01	**6.0168E−01**
	C8	**6.0205E−01**	7.7407E−01(++)	7.6881E−01(++)	6.0213E−01
$FDA5_{dec}$	C1	1.8814E+00(++)	9.4974E−01	**9.4423E−01**	1.3032E+00
	C2	1.6341E+00	**8.5581E−01**	8.6249E−01	1.1515E+00
	C3	1.3257E+00(++)	**7.7800E−01**	7.7964E−01	8.2091E−01
	C4	9.9504E−01	7.7569E−01	7.7636E−01	**7.6108E−01**
	C5	9.6459E+00(++)	1.4692E+00	**1.2889E+00**	1.7701E+00
	C6	**5.5152E−01**	9.1917E−01(++)	8.8310E−01(++)	6.0937E−01
	C7	1.6697E+00(++)	**8.1168E−01**	8.2214E−01	1.0383E+00
	C8	1.2665E+00(++)	7.5331E−01	**7.5236E−01**	8.2960E−01

continued

Table 5. continued

		Tr-RMMEDA	D-NSGAII-A	D-NSGA-II-B	SPTr-RMMEDA
HE2	C1	1.235E+00(++)	8.3571E−02	**8.2342E−02**	5.1780E−01
	C2	8.193E−01(++)	4.1103E−02	**3.9143E−02**	4.0498E−01
	C3	3.602E−01(++)	**3.1264E−02**	3.1374E−02	1.4621E−01
	C4	9.412E−02(++)	3.1389E−02	**3.1218E−02**	4.4132E−02
	C5	7.196E−01(++)	**4.4381E−02**	4.5035E−02	2.2784E−01
	C6	6.250E−02(++)	**3.6443E−02**	3.6448E−02	4.9931E−02
	C7	8.208E−01(++)	**3.8963E−02**	3.9923E−02	3.2136E−01
	C8	1.0090E−01	3.1500E−02	**3.1460E−02**	5.1780E−01
HE7	C1	**1.5310E+00**	4.6152E+00(++)	4.7678E+00(++)	2.4230E+00
	C2	**1.6950E+00**	5.2015E+00(++)	5.0774E+00(++)	3.5941E+00
	C3	**1.8810E+00**	5.2507E+00(++)	5.6113E+00(++)	4.8162E+00
	C4	**2.5830E+00**	6.0205E+00(++)	5.7599E+00(++)	5.3619E+00
	C5	**1.3150E+00**	3.1843E+00(++)	3.0794E+00(++)	2.2293E+00
	C6	**1.8770E+00**	3.8375E+00(++)	3.6621E+00(++)	3.2638E+00
	C7	**1.6790E+00**	5.5184E+00(++)	5.2221E+00(++)	3.7530E+00
	C8	**2.6990E+00**	5.4638E+00(++)	5.8543E+00(++)	5.2944E+00
HE9	C1	2.5510E+00	1.3575E+03	1.3701E+03	**2.5347E+00**
	C2	**2.1620E+00**	1.3701E+03(++)	1.3556E+03(++)	3.3586E+00
	C3	**2.1800E+00**	1.3664E+03(++)	1.3641E+03(++)	5.9792E+00
	C4	**2.5600E+00**	1.3581E+03(++)	1.3534E+03(++)	6.5377E+00
	C5	**2.0450E+00**	1.3584E+03(++)	1.3597E+03(++)	2.6562E+00
	C6	**2.3390E+00**	1.3545E+03(++)	1.3595E+03(++)	4.5507E+00
	C7	**2.0410E+00**	1.3624E+03(++)	1.3608E+03(++)	3.5408E+00
	C8	**2.4350E+00**	1.3608E+03(++)	1.3533E+03(++)	6.2792E+00

5 Conclusion

In this paper, the algorithm (SPTr-RMMEDA) based on special points and transfer component analysis prediction strategy was proposed for DMOPs. In order to improve the convergence, select special points then carry out transfer learning prediction after partly sampling of them and their neighborhoods. Meanwhile, the diversity is introduced adaptively to weight the convergence and diversity. The convergent speed of the population is accelerated, and the convergence and diversity are improved. The statistical results show that the proposed SPTr-RMMEDA is very competitive in DMOPs compared with the other four representative algorithms, especially in the type I and II problems. And comparative experiments have shown that RMMEDA and adaptive prediction contribute equally to the proposed algorithm on different parts. The work presented in this paper still needs further study in the future. In terms of the selection of transfer learning methods and the diversity of introduction strategies, we can conduct a deeper research, and the approach to the type III of DMOPs is also the research direction of our future work.

Acknowledgements. This work was supported by the National Natural Science Foundation of China (Nos. 61876141, 61373111, 61672405, 61871306); and the Provincial Natural Science Foundation of Shaanxi of China (No. 2019JZ-26).

References

1. Azzouz, R., Bechikh, S., Said, L.B.: A dynamic multi-objective evolutionary algorithm using a change severity-based adaptive population management strategy. Soft Comput. **21**(4), 885–906 (2015). https://doi.org/10.1007/s00500-015-1820-4
2. Branke, J.: Evolutionary Optimization in Dynamic Environments. Genetic Algorithms and Evolutionary Computation, vol. 3. Springer, New York (2012). https://doi.org/10.1007/978-1-4615-0911-0, https://books.google.com.hk/books?id=XOLiBwAAQBAJ
3. Branke, J., Kaußler, T., Smidt, C., Schmeck, H.: A multi-population approach to dynamic optimization problems. In: In: Parmee, I.C. (eds.) Evolutionary Design and Manufacture, pp. 299–307. Springer, London (2000)
4. Chen, Z., Bai, X., Yang, Q., Huang, X., Li, G.: Strategy of constraint, dominance and screening solutions with same sequence in decision space for interval multi-objective optimization. Acta Automatica Sinica **41**(12), 2115–2124 (2015). https://doi.org/10.16383/j.aas.2017.c150811
5. Deb, K., Rao N., U.B., Karthik, S.: Dynamic multi-objective optimization and decision-making using modified NSGA-II: a case study on hydro-thermal power scheduling. In: Obayashi, S., Deb, K., Poloni, C., Hiroyasu, T., Murata, T. (eds.) EMO 2007. LNCS, vol. 4403, pp. 803–817. Springer, Heidelberg (2007). https://doi.org/10.1007/978-3-540-70928-2_60
6. Goh, C.K., Tan, K.C.: A competitive-cooperative coevolutionary paradigm for dynamic multiobjective optimization. IEEE Trans. Evol. Comput. **13**(1), 103–127 (2008). https://doi.org/10.1109/TEVC.2008.920671
7. Hatzakis, I., Wallace, D.: Dynamic multi-objective optimization with evolutionary algorithms: a forward-looking approach. In: Proceedings of the 8th Annual Conference on Genetic and Evolutionary Computation, pp. 1201–1208 (2006). https://doi.org/10.1145/1143997.1144187
8. Helbig, M., Engelbrecht, A.: Benchmark functions for CEC 2015 special session and competition on dynamic multi-objective optimization. Department Computer Science, University of Pretoria, Pretoria, Republic of South Africa (2015)
9. Jiang, M., Huang, W., Huang, Z., Yen, G.G.: Integration of global and local metrics for domain adaptation learning via dimensionality reduction. IEEE Trans. Cybern. **47**(1), 38–51 (2017). https://doi.org/10.1109/TCYB.2015.2502483
10. Jiang, M., Huang, Z., Qiu, L., Huang, W., Yen, G.G.: Transfer learning-based dynamic multiobjective optimization algorithms. IEEE Trans. Evol. Comput. **22**(4), 501–514 (2017). https://doi.org/10.1109/TEVC.2017.2771451
11. Nguyen, T.T., Yang, S., Branke, J.: Evolutionary dynamic optimization: a survey of the state of the art. Swarm Evol. Comput. **6**, 1–24 (2012). https://doi.org/10.1016/j.swevo.2012.05.001
12. Schölkopf, B., Platt, J., Hofmann, T.: A kernel method for the two-sample-problem, pp. 513–520 (2007). https://ieeexplore.ieee.org/document/6287330
13. Zhang, Q., Zhou, A., Jin, Y.: RM-MEDA: a regularity model-based multiobjective estimation of distribution algorithm. IEEE Trans. Evol. Comput. **12**(1), 41–63 (2008). https://doi.org/10.1109/TEVC.2007.894202
14. Zhang, X., Tian, Y., Jin, Y.: A knee point-driven evolutionary algorithm for many-objective optimization. IEEE Trans. Evol. Comput. **19**(6), 761–776 (2014). https://doi.org/10.1109/TEVC.2014.2378512
15. Zhou, A., Jin, Y., Zhang, Q.: A population prediction strategy for evolutionary dynamic multiobjective optimization. IEEE Trans. Cybern. **44**(1), 40–53 (2013). https://doi.org/10.1109/TCYB.2013.2245892

Constrained Multi-objective Optimization

An Improved Two-Archive Evolutionary Algorithm for Constrained Multi-objective Optimization

Xinyu Shan[1] and Ke Li[2(✉)]

[1] College of Computer Science and Engineering, University of Electronic Science and Technology of China, Chengdu, Sichuan Province, China
[2] Department of Computer Science, University of Exeter, Exeter EX4 4QF, UK
k.li@exeter.ac.uk

Abstract. Constrained multi-objective optimization problems (CMOPs) are ubiquitous in real-world engineering optimization scenarios. A key issue in constrained multi-objective optimization is to strike a balance among convergence, diversity and feasibility. A recently proposed two-archive evolutionary algorithm for constrained multi-objective optimization (C-TAEA) has be shown as a latest algorithm. However, due to its simple implementation of the collaboration mechanism between its two co-evolving archives, C-TAEA is struggling when solving problems whose *pseudo* Pareto-optimal front, which does not take constraints into consideration, dominates the *feasible* Pareto-optimal front. In this paper, we propose an improved version C-TAEA, dubbed C-TAEA-II, featuring an improved update mechanism of two co-evolving archives and an adaptive mating selection mechanism to promote a better collaboration between co-evolving archives. Empirical results demonstrate the competitiveness of the proposed C-TAEA-II in comparison with five representative constrained evolutionary multi-objective optimization algorithms.

Keywords: Constrained multi-objective optimization · Two-archive evolutionary computation · MOEA/D

1 Introduction

The constrained multi-objective optimization problem (CMOP) considered in this paper is formulated as:

$$
\begin{aligned}
\text{minimize} \quad & \mathbf{F}(\mathbf{x}) = (f_1(\mathbf{x}), \cdots, f_m(\mathbf{x}))^T \\
\text{subject to} \quad & g_j(\mathbf{x}) \geq 0, \quad j = 1, \cdots, q \\
& h_k(\mathbf{x}) = 0, \quad k = 1, \cdots, \ell
\end{aligned}
\tag{1}
$$

This work was supported by UKRI Future Leaders Fellowship (Grant No. MR/S017062/1).

H. Ishibuchi et al. (Eds.): EMO 2021, LNCS 12654, pp. 235–247, 2021.
https://doi.org/10.1007/978-3-030-72062-9_19

where $\mathbf{x} = (x_1, \cdots, x_n)^T \in \Omega$ is a candidate solution. $\Omega = [a_i, b_i]^n \subseteq \mathbb{R}^n$ as the decision space. $\mathrm{F} : \Omega \to \mathbb{R}^m$ consists of m objective functions and $\mathbb{R}^m$ is called the objective space. For a CMOP, the constraint violation of a solution $\mathbf{x}$ can be calculated as:

$$CV(\mathbf{x}) = \sum_{j=1}^{q} \max\{g_j(\mathbf{x}), 0\} + \sum_{k=1}^{\ell} \max\{|h_k(\mathbf{x})| - \delta_k, 0\}, \tag{2}$$

where δ_k is the tolerance value for the k-th equality constraint. $\mathbf{x}$ is regarded as a feasible solution in case $CV(\mathbf{x} = 0)$; otherwise it is an infeasible solution. Given two feasible solutions $\mathbf{x}^1$ and $\mathbf{x}^2 \in \Omega$, $\mathbf{x}^1$ is said to Pareto dominate $\mathbf{x}^2$ (denoted as $\mathbf{x}^1 \preceq \mathbf{x}^2$) if $\mathbf{F}(\mathbf{x}^1)$ is not worse than $\mathbf{F}(\mathbf{x}^2)$ at all objectives and $f_i(\mathbf{x}^1) < f_i(\mathbf{x}^2)$ for at least one objective $i \in \{1, \cdots, m\}$. A solution $\mathbf{x}^* \in \Omega$ is a Pareto-optimal solution of (1) if and only if there does not exist another solution $\mathbf{x} \in \Omega$ that dominates $\mathbf{x}^*$. The set of all Pareto-optimal solutions is called the Pareto-optimal set (PS) while their images in the objective space, i.e., $PF = \{\mathbf{F}(\mathbf{x}) \in \mathbb{R}^m | \mathbf{x} \in PS\}$, is called the Pareto-optimal front (PF).

Due to the population-based property, evolutionary algorithms (EAs) have been widely recognized as an effective approach for multi-objective optimization. We have witnessed a significant amount of efforts devoted to the development of evolutionary multi-objective optimization (EMO) algorithms in the past three decades, e.g., fast and elitist multi-objective genetic algorithm (NSGA-II) [4], indicated-based EA (IBEA) [20] and multi-objective EA based on decomposition (MOEA/D) [19]. As stated in our recent paper[10], it is kind of surprising to note that most EMO algorithms are designed for unconstrained MOPs, which rarely exist in real-world engineering optimization scenarios. The lack of adequate research on constraint handling and algorithm design for CMOPs can hinder the wider uptake of EMO in industry.

It is well known that the balance between convergence and diversity is the cornerstone in EMO algorithm design for unconstrained MOPs. With the consideration of constraints in CMOPs, the feasibility of evolving population becomes an additional issue to handle. As stated in [10], many existing constraint handling techniques in the literature overly emphasize the importance of feasibility thus leading to a loss of diversity and ineffectiveness for problems with many disconnected feasible regions. To strike the balance among convergence, diversity and feasibility simultaneously, Li et al. [10] proposed a two-archive EA (dubbed C-TAEA). Its basic idea is to co-evolving two complementary archives. In particular, one (dubbed as CA) plays as the driving force to push the population towards the PF while the other one (dubbed as DA) mainly tends to maintain the population diversity. Although C-TAEA has shown strong performance on many constrained multi-objective benchmark problems (especially those with many local feasible regions) and a real-world engineering optimization problem. However, partially due to the over emphasis of diversity and the ineffective collaboration between CA and DA, C-TAEA have troubles on problems whose *feasible* PF is dominated by the *pseudo* PF without considering constraints. In this paper, we propose an improved version of C-TAEA (dubbed C-TAEA-II) which maintains the building blocks of C-TAEA but has improved collaboration mechanisms between CA and DA. More specifically, instead of co-evolving both CA and DA separately, the DA in C-TAEA-II is able to progressively guide the CA to escape from local optimal region. In addition, a dynamic mating selection mechanism is proposed to adaptively choose the mating parents from CA and DA according to their evolutionary status.

The remainder of this paper is organized as follows. Section 2 provides a pragmatic overview of the current development of EMO for CMOPs along with an analysis of their limitations. The technical details of C-TAEA-II is delineated in Sect. 3 and its performance is compared with other five state-of-the-art algorithms in Sect. 4. Section 5 concludes this paper and threads some lights of future research.

2 Background

In this section, we start with a pragmatic overview of the current important developments of constraint handling techniques in EMO. Thereafter, we discuss some challenges or drawbacks of C-TAEA in two selected CMOPs which constitute the motivation of this paper.

2.1 Related Works

In this paper, the existing constraint handling techniques are summarized into five categorizes to facilitate our literature review.

- The first group of works derive from the constrained single-objective optimization rigor which mainly rely on penalty functions, e.g., [3,13,17]. Although this idea is straightforward, its performance is sensitive to the setting of the penalty factor(s) associated with the corresponding penalty function. A too large penalty factor can lead to the indifference of infeasible solutions whereas a too small penalty factor can make infeasible solutions be sanctioned. There have been some attempts to set such penalty factor(s) in an adaptive manner, e.g., [7,18], thus to improve the robustness.
- The second category of works mainly rely on feasibility information. Early in [2], Coello Coello and Christiansen proposed a constraint handling method that simply ignores the infeasible solutions. However, this method has no selection pressure when all members of a population are all infeasible ones. Later, Deb et al. proposed a constraint domination principle (CDP) [4] which is exactly the same as the classic Pareto dominance when comparing a pair of feasible solutions while infeasible solutions are compared upon their constraint violations. The CDP becomes one of the most popular constraint handling techniques in EMO afterwards due to its simplicity and effectiveness. For example, Oyama et al. [12] proposed another version of CDP where the dominance relation is compared upon the number of violated constraints. Fan et al. [6] proposed an angle-based CDP where infeasible solutions within a given angle are compared by using the classic Pareto dominance.

- Different from the second category, the third type of approach aims to take advantage of useful infeasible solutions to increase the selection pressure. For example, Peng et al. [14] proposed evolutionary algorithm based on two sets of weight vectors for CMOPs. In particular, it maintains a set of "infeasible" weight vectors used to select the infeasible solutions with smaller constraint violations. This mechanism helps to preserve a population of well distributed infeasible solutions to facilitate the constrained multi-objective optimization.
- The fourth category is based on ϵ functions which use a ϵ as the tolerance threshold for embracing the constraint violations of infeasible solutions (e.g., [15,16]). Note that the setting of ϵ can largely influence the convergence where a large ϵ value enforces a large selection pressure towards the convergence while a small ϵ value implements a fine-tune over a local region.
- The last category uses the idea of the interaction between two co-evolving archives to deal with CMOPs. C-TAEA, the baseline algorithm of this paper is a representative example of this category. It maintains two co-evolving archives. One, dubbed CA, works as a regular EMO algorithm while the other one, dubbed DA mainly aims to provide complementary diversity to the CA. The interaction between these two archives is implemented as a restricted mating selection mechanism.

2.2 Drawbacks of C-TAEA

Although C-TAEA has been reported to have shown a strong performance on a range of constrained benchmark problems, we argue that the interaction or collaboration mechanism between CA and DA is too simple that it only consider the Pareto dominance relation between mating candidates chosen from CA and DA yet ignores their feasibility information.

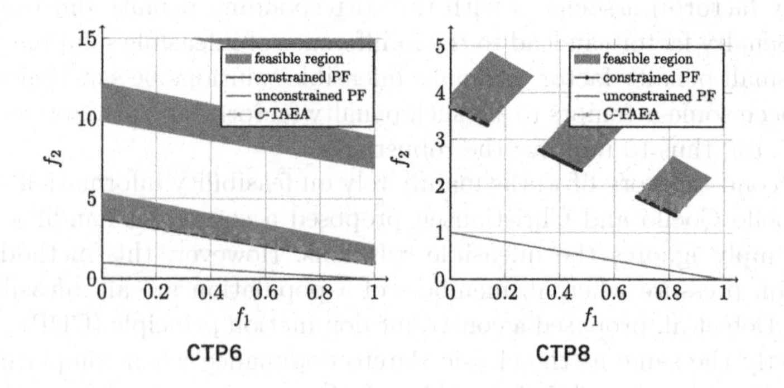

Fig. 1. Comparative results on the CTP6 and CTP8.

Let us used two examples shown in Fig. 1 to further elaborate our above assertion. These are the final populations of solutions obtained by C-TAEA on CTP6 and CTP8 whose *feasible* PF (i.e., the true PF) is dominated by the *pseudo* PF (i.e., the PF without considering constraints) due to the existence of constraints. It is clear that C-TAEA is struggling to converge to the *feasible* PF on both CTP6 and CTP8. In particular, the PF and feasible region of CTP8 is three pieces of disconnected sections while C-TAEA cannot approximate all three sections simultaneously.

From the above observations, we conclude that if the *feasible* PF is dominated by the *pseudo* PF, mating parents coming from the DA, especially when all members of the DA are infeasible, provide negative guidance on offspring reproduction, thus leading to an ineffective convergence of the CA.

3 Proposed Algorithm

The general flow chart of C-TAEA-II, as shown in Fig. 2 follows that of the original C-TAEA. C-TAEA-II maintains two co-evolving archives, dubbed CA and DA, each of which has the same and fixed size N. In particular, CA is the driving force that pushes the population to evolve towards the feasible region thus to approximate the *feasible* PF; while DA shows a complementary effect against CA thus to help CA to jump out of the local optima. To leverage the complementary behavior between CA and DA, an adaptive mating selection mechanism is proposed to adaptively select the most suitable mating parents from them according to their evolution status. In the following paragraphs, we describe the update mechanisms of CA and DA as well as the adaptive mating selection mechanism step by step.

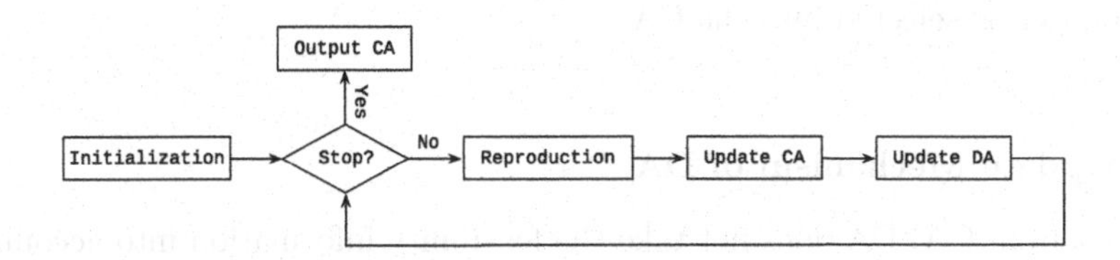

Fig. 2. Flow chart of C-TAEA-II.

3.1 Update Mechanism of CA

CA mainly works at driving the population to evolve towards the *feasible* PF. The pseudo-code of the update mechanism of CA is given in Algorithm 1. Specifically, we first obtain a hybrid population of CA and the offspring population Q, dubbed M_c (line 1 of Algorithm 1). In particular, the feasible solutions in M_c is stored in a temporary archive S_{fea} while the infeasible ones are stored in another temporary archive S_{infea} (lines 2 to 6 of Algorithm 1). Afterwards, there are three different ways to update CA according to the cardinality of S_{fea}.

- If $|S_{\text{fea}}|$ is lower than N, we choose the best $N - |S_{\text{fea}}|$ infeasible solutions from $|S_{\text{infea}}|$ to fill the gap according their constraint violations (lines 7 to 10 of Algorithm 1).
- If $|S_{\text{fea}}|$ equals N, $|S_{\text{fea}}|$ is directly used as the next CA (lines 11 and 12 of Algorithm 1).
- Otherwise, we need to trim S_{fea} until its size equals N. Similar to C-TAEA, we first initialize a set of evenly distributed weight vectors $W = \{\mathbf{w}^1, \cdots, \mathbf{w}^N\}$, each of which specifies a specific subregion in the objective space Δ^i, $i \in \{1, \cdots, N\}$. To have a well balance between convergence and diversity, each solution in S_{fea} is associated with its corresponding subregion as in C-TAEA. Thereafter, the best solution at each subregion having feasible solutions, denoted as $\hat{\mathbf{x}}^i$, $i \in \{1, \cdots, \hat{N}\}$, is chosen to fill the next CA. Specifically,

$$\hat{\mathbf{x}}^i = \underset{\mathbf{x} \in \Delta^i}{\operatorname{argmin}} \{g^{\text{tch}}(\mathbf{x}|\mathbf{w}^i, \mathbf{z}^*)\}, \tag{3}$$

where

$$g^{\text{tch}}(\mathbf{x}|\mathbf{w}^i, \mathbf{z}^*) = \max_{1 \leq j \leq m} \{|f_j(\mathbf{x}) - z_j^*|/w_j^i\}, \tag{4}$$

and $\mathbf{z}^*$ is the approximated ideal point so far. Note that the number of subregions having feasible solutions $\hat{N}$ equals N, these aforementioned selected solutions just fill the new CA. Otherwise, we calculate the absolute difference between the fitness value of the current best solution $\hat{\mathbf{x}}^i$, $i \in \{1, \cdots, \hat{N}\}$, with respect to the remaining feasible solutions in subregion Δ^i. Afterwards, we iteratively choose the solution having the smallest difference value and use it to fill the new CA. Note that the difference values need to be updated once a solution is selected into the CA.

3.2 Update Mechanism of DA

Different from CA, DA does not take the feasibility information into account in its update but mainly aims to improve the diversity of the population as much as possible. The update mechanism of DA is given in Algorithm 2. Like CA, a hybrid population M_d is obtained by combining DA and the offspring population Q (line 1 of Algorithm 2). As in CA, each solution in M_d is associated with its corresponding subregion. Thereafter, the worst solution $\mathbf{x}^w$ from the most crowded subregion Δ^c is removed from M_d until the size of M_d equals N where

$$\mathbf{x}^w = \underset{\mathbf{x} \in \Delta^c}{\operatorname{argmax}} \{g^{\text{tch}}(\mathbf{x}|\mathbf{w}^c, \mathbf{z}^*)\}. \tag{5}$$

3.3 Adaptive Mating Selection Mechanism

As in C-TAEA, the interaction and collaboration between CA and DA is mainly implemented by the mating selection between them. As discussed in the previous sections, C-TAEA can have trouble with problems whose *pseudo* PF dominates

Algorithm 1: Update Mechanism of CA

```
   Input: CA, offspring population Q, weight vector set W
   Output: Updated CA
 1 S ← ∅, S_fea ← ∅, S_infea ← ∅;
 2 M_c ← CA ∪ Q;
 3 foreach x ∈ M_c do
 4     if CV(x) == 0 then
 5         S_fea ← S_fea ∪ {x};
 6     else
 7         S_infea ← S_infea ∪ {x};
 8 if |S_fea| < N then
 9     S ← S_fea;
10     while |S| < N do
11         x^c ← argmin_{x∈S_infea} {CV(x)};
12         S ← S ∪ {x^c};
13 else if |S_fea| == N then
14     S ← S_fea;
15 else
16     Associate each solution in S_fea to the corresponding subregion
       {Δ^1, ··· , Δ^N̂} according to the acute angles to predefined weight
       vectors in W;
17     for i ← 1 to N̂ do
18         x̂^i ← argmin_{x∈Δ^i} {g^tch(x|w^i, z*)};
19         S ← S ∪ {x̂^i};
20         Δ^i ← Δ^i \ {x̂^i};
21     while |S| < N do
22         foreach Δ^i ≠ ∅, i ∈ {1, ··· , N̂} do
23             for j ← 1 to |Δ^i| do
24                 δ^i_j ← |g^tch(x̂^i|w^i, z*) − g^tch(x̂^j|w^i, z*)|;
25             δ^i ← min{δ^i_j}_{j=1}^{|Δ^i|};
26         k ← argmin{δ^i}_{i=1}^{N̂};
27         x̄ ← argmin{δ^k_j}_{j=1}^{|Δ^k|};
28         S ← S ∪ {x̄};
29         Δ^k ← Δ^k \ {x̄};
30 CA ← S;
31 return CA
```

Algorithm 2: Update Mechanism of DA

Input: DA, offspring population Q, weight vector set W
Output: Updated DA

1 $M_d \leftarrow \mathrm{DA} \bigcup Q$;
2 Associate each solution in M_d to a subregion forming $\{\Delta^1, \cdots, \Delta^N\}$;
3 **while** $|M_d| > N$ **do**
4   Find the most crowded subregion Δ^c;
5     $\mathbf{x}^w \leftarrow \underset{\mathbf{x} \in \Delta^c}{\mathrm{argmax}}\{g^{\mathsf{tch}}(\mathbf{x}|\mathbf{w}^c, \mathbf{z}^*)\}$;
6     $\Delta^c \leftarrow \Delta^c \setminus \{\mathbf{x}^w\}$;
7     $M_d \leftarrow M_d \setminus \{\mathbf{x}^w\}$;
8 $\mathrm{DA} \leftarrow M_d$;
9 **return** DA

Algorithm 3: Adaptive Mating Selection

Input: CA, DA, S^t, *choice*
Output: mating parents p_1, p_2

1 $H_m \leftarrow \mathrm{CA} \bigcup S^t$;
2 $\rho_c \leftarrow$ proportion of non-dominated solution of CA in H_m;
3 $\rho_{lc} \leftarrow$ proportion of non-dominated solution of S^t in H_m;
4 **if** $\rho_{lc} > \rho_c$ **then**
5   *utility* = 0;
6   **if** *choice* = 1 **then**
7   *choice* = 2;
8         $\{p_1, p_2\} \leftarrow$ `TournamentSelection` (CA);
9   **else**
10   *choice* = 1;
11         $\{p_1, p_2\} \leftarrow$ `TournamentSelection` (DA);
12 **else**
13   *utility* = 1;
14 **return** p_1, p_2;

the *feasible* PF. This is mainly attributed to the mating selection mechanism in C-TAEA which is merely built upon the Pareto dominance relationship between solutions chosen from CA and DA. In practice, if the solutions in DA dominate those in CA, solutions coming from DA will be chosen as the mating parents although they may bring negative effect to the evolution process. Even worse, those solutions can potentially lie in the infeasible region thus being useless as mating parents.

In order to address the above problem, we propose an adaptive mating selection mechanism that takes the evolutionary status of CA and DA into consideration. The pseudo-code of this adaptive mating selection mechanism is given in Algorithm 3. In particular, we maintain a temporary archive S^t to keep a record of the CA from the last generation to track the evolutionary status of CA. More specifically, we compare the Pareto dominance relationship between S^t and the current CA. If the proportion of non-dominated solutions in the current CA (denoted as ρ_c) is lower than that of S^t (denoted as ρ_t), we set a marker *utility* as 0; otherwise this marker is set as 1. In the meanwhile, we maintain another marker *choice*, which is initially set as 1. In particular, DA will be used as the mating pool in case *choice* equals 2; otherwise, i.e., *choice* equals 1, CA will be chosen as the mating pool. In practice, if *utility* equals 0, it means that CA may be trapped in a local feasible region. Thus, it is recommended to use DA as the mating pool for offspring reproduction. On the other hand, if DA was used as the mating pool last time and the current *utility* is still 0, it suggests that DA may go beyond the *feasible* PF thus it is hard to help CA to move any further. In this case, the next mating pool should be chosen as the CA. This alternation between CA and DA as mating pool can help our C-TAEA-II to adaptively choose the most suitable mating parents for offspring reproduction.

4 Empirical Results

To validate the effectiveness of C-TAEA-II, we compare with five representative constrained EMO algorithms, i.e., C-MOEA/D [9], C-TAEA [10], I-DBEA [1], C-NSGA-III [8], C-MOEA/DD [11]. In our experiments, CTP [5] and DC-DTLZ [10] are chosen to form the benchmark problem suite. In particular, DC-DTLZ problems are scalable to any number of objectives and are with local feasible regions. Here we use the widely used inverted generational distance (IGD) [10]

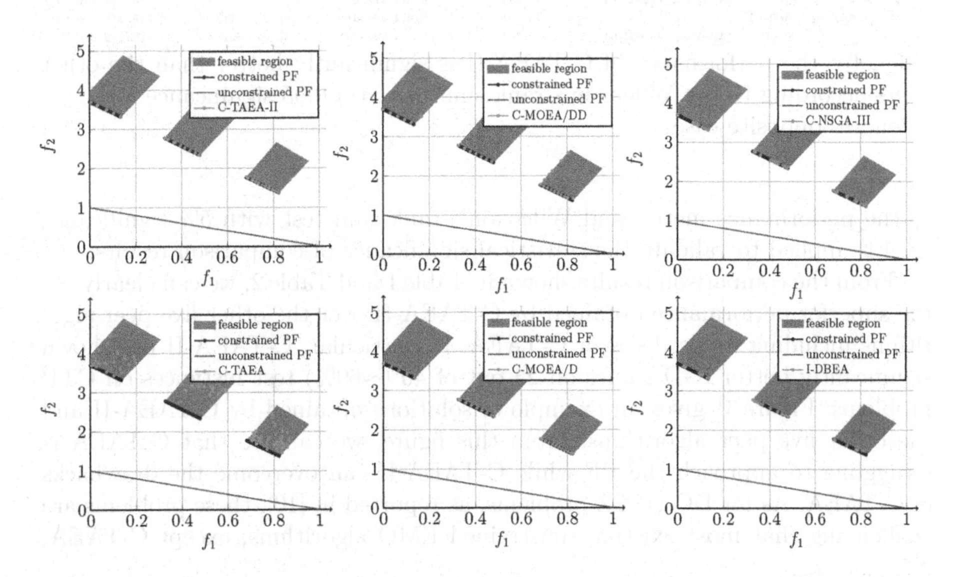

Fig. 3. Solutions obtained by C-TAEA-II and the other five peer algorithms with the best IGD metric value on CTP8.

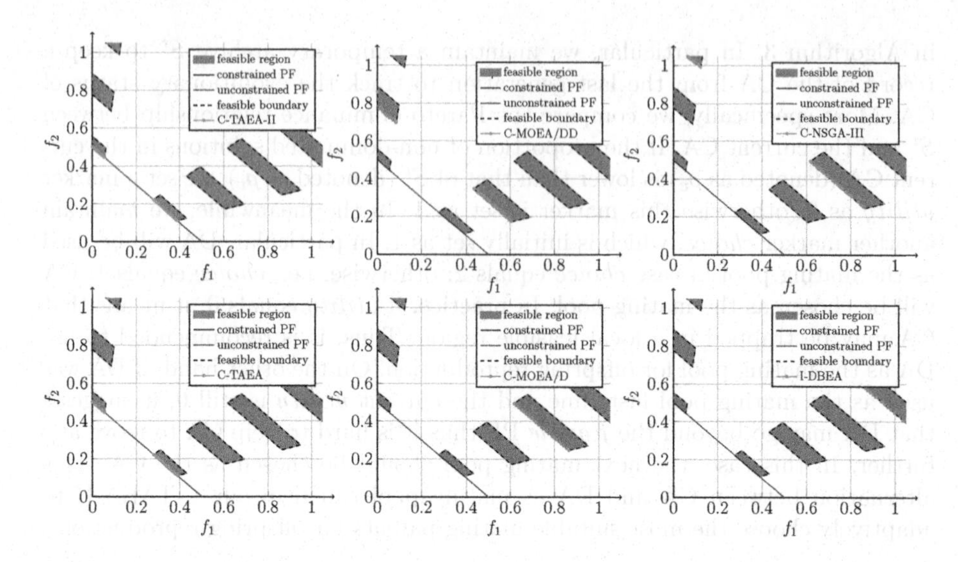

Fig. 4. Solutions obtained by C-TAEA-II and the other five peer algorithms with the best IGD metric value on DC3-DTLZ1.

Table 1. Comparison results on IGD (median and standard deviation) obtained by C-TAEA-II and the other five peer algorithms on CTP problems.

	C-TAEA	C-TAEA-II	C-MOEA/DD	C-MOEA/D	C-NSGA-III	I-DBEA
CTP1	6.39E-3(3.54E-6)†	**4.58E-3(1.52E-8)**	9.95E-3(3.07E-5)†	4.62E-3(1.45E-9)	4.06E-2(1.42E-3)†	1.81E-1(3.14E-3)†
CTP2	1.23E-2(1.49E-5)	3.69E-3(2.27E-7)	6.35E-3(6.79E-7)	**3.59E-3(2.53E-8)**‡	6.95E-3(9.16E-6)	2.19E-1(1.02E-3)
CTP3	3.39E-2(4.25E-5)†	**7.79E-3(5.43E-5)**	1.96E-2(8.28E-6)†	1.47E-2(2.01E-10)†	1.14E-2(1.38E-5)†	4.14E-1(1.77E-2)†
CTP4	1.87E-1(1.34E-3)†	**5.37E-2(7.94E-5)**	8.48E-2(1.12E-4)†	5.53E-2(3.37E-6)†	9.23E-2(2.65E-5)†	3.27E-1(2.55E-2)†
CTP5	5.84E-3(1.08E-6)	1.76E-2(8.36E-5)	2.49E-2(5.52E-6)	4.72E-3(1.07E-6)	**4.29E-3(6.49E-7)**‡	2.03E-2(7.18E-6)
CTP6	5.47E-2(4.29E-4)†	**1.64E-2(7.21E-8)**	1.67E-2(7.41E-8)	1.66E-2(1.41E-9)	2.99E-2(1.73E-6)†	1.27E-1(1.23E-3)†
CTP7	**1.53E-3(2.39E-9)**‡	3.32E-3(4.82E-7)	5.39E-3(1.24E-5)	3.34E-3(1.18E-10)	5.68E-3(2.17E-6)	4.75E-3(6.71E-6)
CTP8	7.03E-2(1.18E-3)†	**1.54E-2(3.84E-6)**	1.82E-2(1.46E-7)†	1.71E-2(1.59E-7)†	2.66E-2(5.82E-6)†	1.31E-1(2.11E-2)†

† denotes the performance of C-TAEA-II is significantly better than the other peers according to the Wilcoxon's rank sum test at a 0.05 significance level;
‡ denotes opposite case.

as the performance metric and Wilcoxon's rank sum test with 5% significance level is applied to validate the statistical significance of comparison results.

From the comparison results shown in Table 1 and Table 2, we can clearly see the superior performance obtained by C-TAEA-II over the other five peer algorithms including its predecessor C-TAEA. In particular, C-TAEA-II has shown significantly better IGD values on 24 out of 40 (~60%) test instances on CTP problems. Figure 3[1] gives an example of solutions obtained by C-TAEA-II and the other five peer algorithms. From this figure, we can see that C-TAEA is struggling to approach the PF while C-TAEA-II can overcome the drawbacks of C-TAEA. As for DC-DTLZ problems, as reported in [10], these problems are challenging that most existing constrained EMO algorithms, except C-TAEA,

[1] Complete results are given in the supplementary document of this paper https://tinyurl.com/yy8jw9bo.

Table 2. Comparison results on IGD (median and standard deviation) obtained by C-TAEA-II and the other five peer algorithms on DC-DTLZ problems.

	m	C-TAEA-II	C-TAEA	C-MOEA/DD	C-MOEA/D	C-NSGA-III	I-DBEA
DC1-DTLZ1	3	**1.51E-2(3.97E-7)**	5.64E-2(1.96E-6)†	6.09E-2(1.81E-4)†	1.54E-1(2.51E-3)†	6.09E-2(7.41E-6)†	5.97E-2(1.99E-4)†
	5	**3.72E-2(5.39E-5)**	6.93E-2(1.85E-4)†	6.46E-2(5.23E-4)†	1.06E-1(1.88E-3)†	8.62E-2(8.36E-4)†	9.97E-2(1.89E-2)†
	8	**7.45E-2(4.93E-5)**	8.46E-2(2.32E-5)†	8.53E-2(3.27E-5)†	8.63E-2(5.37E-3)†	1.07E-1(6.27E-4)†	1.31E-1(5.37E-2)†
	10	**7.56E-2(4.43E-5)**	8.51E-2(5.28E-5)†	8.53E-2(3.71E-5)†	8.62E-2(4.18E-3)†	1.45E-1(4.72E-4)†	1.67E-1(8.42E-2)†
	15	**1.13E-1(3.77E-4)**	1.22E-1(4.92E-4)†	1.18E-1(4.97E-4)†	1.19E-1(6.23E-3)†	1.62E-1(9.24E-3)†	2.32E-1(4.89E-2)†
DC1-DTLZ3	3	4.63E-2(4.97E-5)	**4.58E-2(1.44E-5)**	4.97E-2(1.14E-7)	5.73E-2(4.27E-4)	5.01E-2(1.98E-5)	6.82E-2(9.93E-3)
	5	**1.63E-1(8.29E-5)**	1.65E-1(9.54E-5)†	1.69E-1(6.57E-5)†	6.43E-1(5.63E-4)†	7.26E-1(8.32E-4)†	\
	8	**1.78E-1(1.79E-4)**	1.81E-1(1.24E-4)	1.84E-1(3.36E-8)†	6.02E-1(7.83E-3)†	6.99E-1(5.75E-2)†	\
	10	1.92E-1(5.92E-4)	**1.89E-1(4.72E-4)**	1.92E-1(7.36E-7)	6.21E-1(2.92E-5)	5.85E-1(6.67E-2)	\
	15	**6.45E-1(2.27E-4)**	6.51E-1(1.45E-4)†	6.49E-1(5.33E-5)†	8.45E-1(2.51E-3)†	9.69E-1(1.01E-3)†	\
DC2-DTLZ1	3	**1.86E-2(4.65E-7)**	1.97E-2(3.56E-7)†	\	\	\	\
	5	**5.42E-2(3.43E-7)**	5.44E-2(2.41E-7)	\	\	\	\
	8	8.94E-2(3.79E-5)	**8.47E-2(1.23E-5)‡**	\	\	\	\
	10	**9.06E-2(3.39E-5)**	1.04E-1(5.75E-4)†	\	\	\	\
	15	**1.42E-1(3.42E-4)**	2.18E-1(–)†	\	\	\	\
DC2-DTLZ3	3	**6.05E-2(6.47E-6)**	6.16E-2(8.93E-6)†	\	\	\	\
	5	**1.57E-1(5.93E-5)**	1.65E-1(3.96E-5)†	\	\	\	\
	8	**3.21E-1(3.25E-5)**	1.18E+1(–)†	\	\	\	\
	10	**3.41E-1(3.92E-5)**	3.87E-1(–)†	\	\	\	\
	15	**7.29E-1(2.39E-3)**	7.93E-1(–)†	\	\	\	\
DC3-DTLZ1	3	**2.27E-2(3.73E-6)**	4.47E-2(1.23E-5)†	3.35E-1(3.89E-2)†	2.75E+0(5.09E+0)†	5.01E-1(1.26E-1)†	6.19E-1(2.65E-1)†
	5	**6.31E-2(7.75E-5)**	7.89E-2(2.58E-3)†	3.91E-1(2.01E-2)†	3.18E-1(5.71E-2)†	2.22E-1(8.39E-2)†	4.92E-1(3.97E-1)†
	8	**8.63E-2(4.86E-5)**	1.81E-1(3.49E-3)†	5.89E-1(3.97E-2)†	5.73E-1(6.83E-1)†	5.62E-1(9.43E-2)†	6.82E-1(6.87E-1)†
	10	**1.65E-1(5.32E-5)**	2.07E-1(3.74E-4)†	7.78E-1(4.93E-2)†	7.93E-1(9.87E-1)†	9.03E-1(6.48E-1)†	9.74E-1(8.36E-1)†
	15	**1.81E-1(5.38E-5)**	5.23E-1(4.83E-4)†	1.25E+0(7.49E-2)†	1.08E+0(8.43E-1)†	1.37E+0(9.34E-1)†	1.27E+0(6.38E+0)†
DC3-DTLZ3	3	**7.23E-2(4.68E-5)**	1.12E-1(1.31E-4)†	1.19E+0(2.03E+0)†	1.05E+1(8.49E+1)†	1.45E+0(1.05E+0)†	1.15E+0(9.57E-1)†
	5	**9.17E-2(3.95E-5)**	1.77E-1(2.36E-4)†	2.83E-1(1.14E-2)†	2.66E+0(4.57E+0)†	1.34E+0(1.62E+0)†	\
	8	2.51E-1(4.79E-5)	**2.24E-1(6.54E-5)‡**	4.32E-1(2.31E-2)	2.65E+0(1.18E+1)	2.09E+0(1.85E+0)	\
	10	**2.57E-1(5.47E-5)**	2.75E-1(5.86E-5)†	2.76E-1(1.36E-2)†	7.82E-1(6.23E-2)†	2.06E+0(2.11E+0)†	\
	15	**6.76E-1(7.49E-5)**	6.85E-1(7.28E-4)†	1.16E+0(7.09E-2)†	1.14E+0(4.49E-2)†	2.11E+0(4.21E+0)†	\

† denotes the performance of C-TAEA-II is significantly better than the other peers according to the Wilcoxon's rank sum test at a 0.05 significance level;
‡ denotes the corresponding algorithm significantly outperforms C-TAEA-II.
\ denotes the corresponding fails to find meaningful solutions, while – denotes the standard deviation is not available.

are not able to approximate the PF. C-TAEA-II is the best algorithm where it has shown significantly better IGD values on 137 out of 150 (∼91.3%) test instances on DC-DTLZ problems. Figure 4 gives an example of solutions obtained by C-TAEA-II and the other five peer algorithms. From this figure, we can see that only C-TAEA-II and C-TAEA can find meaningful solutions.

5 Conclusions and Future Works

In this paper, we proposed an improve version of C-TAEA, a state-of-the-art constrained EMO algorithm. Our proposed C-TAEA-II derives from the basic flow of C-TAEA but is featured with two distinctive modifications. The update mechanisms of CA and DA are improved in order to amplify the complementary behavior of both archives. In addition, an adaptive mating selection mechanism is proposed to avoid choosing infeasible solutions beyond the *feasible* PF. From our preliminary experiments on CTP and DC-DTLZ problems, we observed encouraging results that C-TAEA-II has shown statistically better IGD values on 85% comparisons. As for the next step, we plan to further improve the collaboration mechanism between two co-evolving archives and we aim to conduct comprehensive validation on a wider range of benchmark problems and real-world engineering optimization problems.

References

1. Asafuddoula, M., Ray, T., Sarker, R.A.: A decomposition-based evolutionary algorithm for many objective optimization. IEEE Trans. Evol. Comput. **19**(3), 445–460 (2015)
2. Coello, C.A.C., Christiansen, A.D.: MOSES: a multiobjective optimization tool for engineering design. Eng. Optim. **31**(3), 337–368 (1999)
3. Coit, D.W., Smith, A.E., Tate, D.M.: Adaptive penalty methods for genetic optimization of constrained combinatorial problems. Informs J. Comput. **8**(2), 173–182 (1996)
4. Deb, K., Agrawal, S., Pratap, A., Meyarivan, T.: A fast and elitist multiobjective genetic algorithm: NSGA-II. IEEE Trans. Evol. Comput. **6**(2), 182–197 (2002)
5. Deb, K., Pratap, A., Meyarivan, T.: Constrained test problems for multi-objective evolutionary optimization. In: Zitzler, E., Thiele, L., Deb, K., Coello Coello, C.A., Corne, D. (eds.) EMO 2001. LNCS, vol. 1993, pp. 284–298. Springer, Heidelberg (2001). https://doi.org/10.1007/3-540-44719-9_20
6. Fan, Z., Li, W., Cai, X., Hu, K., Lin, H., Li, H.: Angle-based constrained dominance principle in MOEA/D for constrained multi-objective optimization problems. In: Proceedings of the 2016 IEEE Congress on Evolutionary Computation, CEC 2016, pp. 460–467 (2016)
7. Hoffmeister, F., Sprave, J.: Problem-independent handling of constraints by use of metric penalty functions. In: Fogel, L.J., Angeline, P.J., Bäck, T. (eds.) Proceedings of the 5th Annual Conference on Evolutionary Programming, EP 1996, pp. 289–294. MIT Press (1996)
8. Jain, H., Deb, K.: An evolutionary many-objective optimization algorithm using reference-point based nondominated sorting approach, part II: handling constraints and extending to an adaptive approach. IEEE Trans. Evol. Comput. **18**(4), 602–622 (2014)
9. Jan, M.A., Zhang, Q.: MOEA/D for constrained multiobjective optimization: some preliminary experimental results. In: Proceedings of the 2010 UK Workshop on Computational Intelligence, UKCI 2010, pp. 1–6 (2010)
10. Li, K., Chen, R., Fu, G., Yao, X.: Two-archive evolutionary algorithm for constrained multiobjective optimization. IEEE Trans. Evol. Comput. **23**(2), 303–315 (2019)
11. Li, K., Deb, K., Zhang, Q., Kwong, S.: An evolutionary many-objective optimization algorithm based on dominance and decomposition. IEEE Trans. Evol. Comput. **19**(5), 694–716 (2015)
12. Oyama, A., Shimoyama, K., Fujii, K.: New constraint-handling method for multi-objective and multi-constraint evolutionary optimization. Jpn Soc. Aeronaut. Space Sci. Trans. **50**, 56–62 (2007)
13. Panda, A., Pani, S.: A symbiotic organisms search algorithm with adaptive penalty function to solve multi-objective constrained optimization problems. Appl. Soft Comput. **46**, 344–360 (2016)
14. Peng, C., Liu, H., Gu, F.: An evolutionary algorithm with directed weights for constrained multi-objective optimization. Appl. Soft Comput. **60**, 613–622 (2017)
15. Takahama, T., Sakai, S.: Constrained optimization by the ϵ constrained differential evolution with gradient-based mutation and feasible elites. In: Proceedings of 2006 IEEE World Congress on Computational Intelligence, WCCI 2006, pp. 1–8. IEEE (2006)

16. Takahama, T., Sakai, S.: Efficient constrained optimization by the ϵ constrained rank-based differential evolution. In: Proceedings of the IEEE Congress on Evolutionary Computation, CEC 2012, pp. 1–8 (2012)
17. Tessema, B.G., Yen, G.G.: A self adaptive penalty function based algorithm for constrained optimization. In: Proceedings of 2006 IEEE World Congress on Computational Intelligence, WCCI 2006, pp. 246–253. IEEE (2006)
18. Woldesenbet, Y.G., Yen, G.G., Tessema, B.G.: Constraint handling in multiobjective evolutionary optimization. IEEE Trans. Evol. Comput. **13**(3), 514–525 (2009)
19. Zhang, Q., Li, H.: MOEA/D: a multiobjective evolutionary algorithm based on decomposition. IEEE Trans. Evol. Comput. **11**(6), 712–731 (2007)
20. Zitzler, E., Künzli, S.: Indicator-based selection in multiobjective search. In: Yao, X., Burke, E.K., Lozano, J.A., Smith, J., Merelo-Guervós, J.J., Bullinaria, J.A., Rowe, J.E., Tiňo, P., Kabán, A., Schwefel, H.-P. (eds.) PPSN 2004. LNCS, vol. 3242, pp. 832–842. Springer, Heidelberg (2004). https://doi.org/10.1007/978-3-540-30217-9_84

An Improved Epsilon Method with M2M for Solving Imbalanced CMOPs with Simultaneous Convergence-Hard and Diversity-Hard Constraints

Zhun Fan[1,2,3,4](✉), Zhi Yang[1], Yajuan Tang[1], Wenji Li[1], Biao Xu[1], Zhaojun Wang[1], Fuzan Sun[1], Zhoubin Long[1], and Guijie Zhu[1]

[1] Department of Electronic and Information Engineering, Shantou University, Shantou 515063, Guangdong, China
zfan@stu.edu.cn

[2] Key Lab of Digital Signal and Image Processing of Guangdong Province, Shantou University, Shantou 515063, Guangdong, China

[3] Key Laboratory of Intelligent Manufacturing Technology (Shantou University), Ministry of Education, Shantou 515063, Guangdong, China

[4] State Key Lab of Digital Manufacturing Equipment and Technology, Huazhong University of Science and Technology, Wuhan 430000, China

Abstract. When tackling imbalanced constrained multi-objective optimization problems (CMOPs) with simultaneous convergence-hard and diversity-hard constraints, a critical issue is to balance the diversity and convergence of populations. To address this issue, this paper proposes a hybrid algorithm which combines an improved epsilon constraint-handling method (IEpsilon) with a multi-objective to multi-objective (M2M) decomposition approach, namely M2M-IEpsilon. The M2M decomposition mechanism in M2M-IEpislon has the capability to deal with imbalanced objectives. The IEpsilon constraint-handling method can prevent populations falling into large infeasible regions, thus improves the convergence performance of the proposed algorithm. To verify the performance of the proposed M2M-IEpsilon, a series of imbalanced CMOPs with simultaneous convergence-hard and diversity-hard constraints, namely ICD-CMOPs, is designed by using the DAS-CMOPs framework. Six state-of-the-art constrained multi-objective evolutionary algorithms (CMOEAs), including CM2M, CM2M2, NSGA-II-CDP, MOEA/D-CDP, MOEA/D-IEpsilon and PPS-MOEA/D, are employed to compare with M2M-IEpsilon on the ICD-CMOPs. Through the analysis of experimental results, the proposed M2M-IEpsilon is superior to the other six algorithms in solving ICD-CMOPs, which illustrates the superiority of the proposed M2M-IEpsilon in dealing with ICD-CMOPs with simultaneous convergence-hard and diversity-hard constraints.

Keywords: CMOEAs · M2M decomposition · IEpsilon · Imbalanced CMOPs

H. Ishibuchi et al. (Eds.): EMO 2021, LNCS 12654, pp. 248–256, 2021.
https://doi.org/10.1007/978-3-030-72062-9_20

1 Introduction

Many optimization problems usually have conflicting objectives with a set of constraints [6,13,14]. Generally, a constrained multi-objective optimization problem (CMOP) is described as follows [8]:

$$\begin{cases} \text{minimize} & \mathbf{F}(\mathbf{x}) = (f_1(\mathbf{x}), \ldots, f_m(\mathbf{x}))^T \\ \text{subject to} & g_i(\mathbf{x}) \geq 0, i = 1, \ldots, q \\ & h_j(\mathbf{x}) = 0, j = 1, \ldots, p \\ & \mathbf{x} \in \mathbf{R}^n \end{cases} \tag{1}$$

where $\mathbf{F}(\mathbf{x}) = (f_1(\mathbf{x}), \ldots, f_m(\mathbf{x}))^T$ represents an m-dimensional objective function. $\mathbf{x} \in \mathbf{R}^n$ denotes an n-dimensional decision variable. $g_i(\mathbf{x}) \geq 0$ and $h_j(\mathbf{x}) = 0$ represent inequality and equality constraints, respectively. q and p are the number of inequality and equality constraints, respectively.

For any two feasible solutions x_p and x_q, we say that x_p dominates x_q ($x_p \preceq x_q$) when the following two conditions are met:

1. $\forall\, i \in \{1,\ldots,m\}$, $f_i(x_p) \leq f_i(x_q)$
2. $\exists\, j \in \{1,\ldots,m\}$, $f_j(x_p) < f_j(x_q)$

Some real-world optimization problems may have imbalanced objectives and a series of constraints. Therefore, CMOPs with imbalanced objective functions deserve to be solved. To handle with the imbalanced CMOPs, we propose a hybrid CMOEAs, named M2M-IEpsilon, which has adopted two mechanisms. Firstly, to enhance the diversity of the population, an M2M decomposition method [10] is used to handle with imbalanced objectives. Secondly, an improved epsilon constraint-handling method (IEpsilon) [3] is employed to improve the convergence performance of the method. In M2M-IEpsilon, these two mechanisms are combined to solve imbalanced CMOPs [9] with simultaneous diversity-hard and convergence-hard constraints. To verify the effectiveness of the M2M-IEpsilon, a series of imbalanced CMOPs with diversity-hard and convergence-hard constraints is designed by using DAS-CMOPs framework [5], which is named as ICD-CMOPs. The contributions of this paper are listed below:

1. The proposed M2M-IEpsilon decomposes a population into many sub-populations, with each sub-population corresponding to a CMOP, which can assist to solve imbalanced CMOPs.
2. M2M-IEpsilon can dynamically adjust the relaxation of constraint violation according to the proportion of feasible solutions in the current population, which maintains a good balance of searching between the feasible and infeasible regions, and has the ability to get across large infeasible regions.
3. A set of CMOPs, namely ICD-CMOPs, is constructed to assess the diversity and convergence performance of CMOEAs.

The remainder of this paper is organized as follows. Section 2 mainly introduce the M2M decomposition approach and the IEpsilon method. Section 3 describes the proposed M2M-IEpsilon. Section 4 shows the experimental results of M2M-IEpsilon and six other CMOEAs, including CM2M [11], CM2M2 [12], NSGA-II-CDP [2], MOEA/D-CDP [7], MOEA/D-IEpsilon [3] and PPS-MOEA/D [4], on ICD-CMOPs. Section 5 make a summary of this paper.

2 Background

2.1 The Framework of M2M

In M2M [10], the decomposition is carried out in the objective space. Specifically, J unit vectors $v^1, \ldots, v^j$ in $\mathbf{R}_+^m$ are selected. The $\mathbf{R}_+^m$ is split into J sub-regions $P_1, \ldots, P_J$. A sub-region P_j is defined as follows:

$$P_j = \{u \in \mathbf{R}_+^m | \langle u, v^j \rangle \leq \langle u, v^i \rangle \; for \; any \; i = 1, ..., J\} \tag{2}$$

where $\langle u, v^i \rangle$ is the acute angle between the objective vector u and the ith direction vector v^i. To search for optimal solutions in each sub-region, the population of M2M is separated into J sub-populations. The Eq. (3) shows the definition of a subproblem j. In addition, each sub-population adopts two rules in the Ref. [10] to select solutions.

$$\begin{cases} \text{minimize} & \mathbf{F}(x) = (f_1(x), ..., f_m(x)) \\ \text{subject to} & x \in \prod_{i=1}^{n} [a_i, b_i] \\ & F(x) \in P_j \end{cases} \tag{3}$$

Figure 1 shows a decomposition example by using M2M. The objective space $\mathbf{R}_+^m$ is separated into five sub-regions $P_1, \ldots, P_5$. Each sub-region has S individuals, and $v_1, \ldots, v_5$ are five evenly distributed direction vectors.

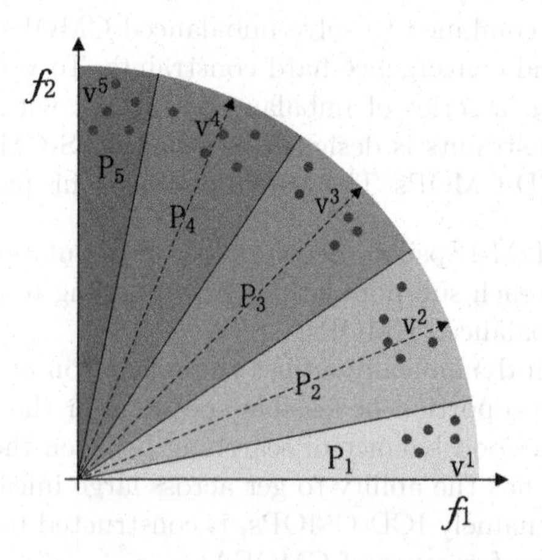

Fig. 1. The population decomposition approach by using M2M.

2.2 The Improved Epsilon Constraint-Handling Method

In paper [3], an improved ε constraint-handling method is proposed, namely IEpsilon, which can dynamically adjust the relaxation of constraint violations based on the ratio of feasible solutions in each generation of the population. The detail description of the setting of $\varepsilon(k)$ can be found in Ref. [3].

$$\varepsilon(k) = \begin{cases} rule\ 1\text{: } \phi(x^{\theta}), & \text{if } k = 0 \\ rule\ 2\text{: } (1-\tau)\varepsilon(k-1), & \text{if } r_k < \alpha \ and\ k < T_c \\ rule\ 3\text{: } (1+\tau)\phi_{max}, & \text{if } r_k \geq \alpha \ and\ k < T_c \\ rule\ 4\text{: } 0, & \text{if } k \geq 0 \end{cases} \quad (4)$$

3 Embedding IEpsilon in M2M

In the M2M framework, the IEpsilon is embedded to solve constraints. The description of M2M-IEpsilon is presented in Algorithm 1. In line 2, the working population is decomposed into J sub-populations which are associated with evenly distributed direction vectors $v^1, \ldots, v^j$, and each sub-population contains S individuals. Line 3 introduces several parameters in M2M-IEpsilon, including $\varepsilon(0)$, r_k, and ϕ_{max}. Lines 5–14 show the process of generate new offspring by using genetic operators. In lines 15–17, $\varepsilon(k)$ is set according to Eq. (4). Lines 18–31 describe the updating process of each sup-population. Specially, if the number of solutions in sup-population P_j is fewer than S, then randomly select $|S - P_j|$ solutions add to P_j. In lines 24–30, if the number of solutions in sup-population P_j is more than S, we apply two strategies to select S solutions. If the number of solutions whose constraint violations are smaller than $\varepsilon(k)$, denoted as $P_{j_{\text{fea}}}$, is equal to or greater than S, we select S solutions from $P_{j_{\text{fea}}}$ by using NSGA-II [2]. Otherwise, we first select all the solutions in $P_{j_{\text{fea}}}$ to the next generation. Then, we sort all solutions in $P_{j_{\text{inf}}}$ according to their constraint violations in an ascending way, and select the first $S - |P_{j_{\text{fea}}}|$ solutions to the next generation.

4 The Analysis of Experimental Study

4.1 Parameter Settings of Seven CMOEAs

The six state-of-the-art CMOEAs, including CM2M [11], CM2M2 [12], NSGA-II-CDP [2], MOEA/D-CDP [7], MOEA/D-IEpsilon [3], PPS-MOEA/D [4], and the proposed M2M-IEpsilon, are carried out on ICD-CMOP1-7. The detailed parameter settings are presented as follows:

1. Population size: $N = 300$.
2. Mutation and crossover probability: $P_m = 1/n$, $CR = 1.0$.
3. Termination condition: all algorithm independently runs for 30 times on ICD-CMOP1-7, then stop when reach 300,000 function evaluations.

Algorithm 1: The Framework of the Proposed Algorithm (M2M-IEpsilon)

Input:
J: the number of the subproblems;
J unit direction vectors: $v^1,...,v^J$;
S: the number of sub-population;
Q: a group of individual solutions;
T_{max}: the maximum generation;
T_c: the control generation for $\varepsilon(k)$;
α: the maximum overall constraint violation;
parameters τ and θ in Eq.(4)

Output: a set of feasible non-dominated solutions.

```
1  Initialization:
2  Decompose the working population into J sub-populations (P_1, ..., P_J), each of
     them consists of S individuals by using Eq.(3)
3  Initialize ε(0), r_k and φ_max according to Eq.(4)
4  while gen ≤ T_max do
5      for j ← 1 to J do
6          foreach x ∈ P_j do
7              y be selected from P_j;
8              A new solutions z be generated by applying genetic operators on x
                 and y;
9              Compute F(z);
10             R := R ∪ {z};
11         end
12         Q := R ∪ (∪_{j=1}^{J} P_j);
13         Use Q to set P_1, ..., P_J according to Eq. (2);
14     end
15     if k > 0 then
16         ε(k) = UpdateEpsilon(τ, r_k, α, φ_max, T_c, k)
17     end
18     for j ← 1 to J do
19         P_{j_fea} = {y ∈ P_j | φ(y) < ε(k)};
20         P_{j_inf} = {y ∈ P_j | φ(y) ≥ ε(k)};
21         if |P_j| ≤ S then
22             randomly select |S − P_j| solutions add to P_j;
23         end
24         if |P_j| > S then
25             if |P_{j_fea}| ≥ S then
26                 select S solutions from P_{j_fea} by using NSGA-II [2].
27             else if |P_{j_fea}| < S then
28                 sort each solution in P_{j_inf} according to their constraint
                     violations in a ascend way, and select the first S − |P_{j_fea}|
                     solutions and all the solutions in P_{j_fea} to the next generation.
29             end
30         end
31     end
32     gen = gen + 1;
33 end
```

4. Parameter setting in CM2M2: the number of infeasible weights $N_1 = 90$, the number of feasible weights $N_2 = 210$.
5. For the bi-objective CMOPs, J is set as 10 in M2M-IEpsilon, CM2M, and CM2M2. For the tri-objective CMOPs, J is set as 15 in M2M-IEpsilon, CM2M, and CM2M2. Where J is the number of sub-problems.
6. Parameter settings of IEpsilon method in M2M-IEpsilon and MOEA/D-IEpsilon: $\theta = 0.05N$, $\alpha = 0.95$, $T_c = 800$, and $\tau = 0.1$.
7. Parameter settings of PPS-MOEA/D could be found in Ref. [4].
8. Parameter settings of MOEA/D-CDP: T is set to 30, n_r is set to 2.

Two commonly performance metrics, including the inverted generation distance metric (IGD) [1] and the hypervolume metric (HV) [15], are employed. A set of imbalanced CMOPs with both diversity-hard and convergence-hard constraints are designed by using the DAS-CMOPs framework [5], which are named as ICD-CMOPs. The constraint functions are designed with convergence-hard and diversity-hard properties, which are generated from DAS-CMOPs [5]. The detail definition of the ICD-CMOPs can be found in the following link: https://github.com/lwjhhxx/Imbalanced-CMOPs.

Table 1. The IGD results of the seven CMOEAs on ICD-CMOP1–7.

Test instance		M2M-IEpsilon	CM2M	CM2M2	NSGA-II	MOEA/D	PPS	IEpsilon
ICD-CMOP1	mean	**1.86E−02**	8.50E−02†	1.49E−01†	3.23E−01†	3.12E−01†	2.97E−01†	2.62E−01†
	std	3.43E−03	3.06E−02	3.27E−02	1.29E−02	4.87E−02	6.52E−02	7.20E−02
ICD-CMOP2	mean	**1.21E−01**	2.26E−01†	1.47E−01	3.09E−01†	2.73E−01†	2.65E−01†	2.66E−01 †
	std	8.59E−02	6.09E−02	6.21E−02	2.56E−02	7.50E−03	5.66E−03	1.60E−02
ICD-CMOP3	mean	**2.04E−01**	3.22E−01†	3.98E−01†	8.54E−01†	4.85E−01†	6.37E−01†	4.44E−01†
	std	8.40E−02	8.05E−02	6.47E−02	8.12E−02	9.17E−02	1.85E−01	9.82E−02
ICD-CMOP4	mean	**2.16E−02**	2.14E−01†	8.50E−02†	3.33E−01†	2.86E−01†	2.86E−01†	2.86E−01†
	std	5.18E−02	7.35E−02	6.05E−02	2.39E−02	1.50E−02	1.22E−02	4.31E−02
ICD-CMOP5	mean	**4.99E−02**	1.27E−01†	1.71E−01†	2.93E−01†	3.24E−01†	3.09E−01†	2.84E−01 †
	std	2.05E−02	2.97E−02	2.31E−02	4.75E−02	4.25E−03	1.44E−02	4.16E−02
ICD-CMOP6	mean	3.92E−01	4.11E−01	**3.15E−01**†	8.05E−01†	7.18E−01†	3.52E−01	7.26E−01†
	std	1.19E−01	1.18E−01	1.02E−02	9.64E−03	1.55E−02	4.74E−02	2.65E−02
ICD-CMOP7	mean	4.92E−01	5.44E−01†	5.35E−01†	7.47E−01†	7.19E−01†	**4.00E−01**†	7.24E−01 †
	std	2.78E−02	4.89E−02	2.52E−02	6.90E−03	9.53E−02	5.56E−02	6.62E−03
Wilcoxon-Test(S-D-I)		–	6−1−0	6−1−0	7−0−0	7−0−0	6−1−0	7−0−0

The IGD results of ICD-CMOP1-7 obtained by M2M-IEpsilon, CM2M, CM2M2, NSGA-II-CDP, MOEA/D-CDP, MOEA/D-IEpsilon and PPS-MOEA/D are presented in Table 1. The Wilcoxon-Test indicates that the proposed M2M-IEsiplon is significantly better than NSGA-II-CDP, MOEA/D-IEpsilon, and MOEA/D-CDP on ICD-CMOP1-7 test instances. For ICD-CMOP6, M2M-IEpsilon has similar performance with CM2M and PPS-MOEA/D. However, for ICD-CMOP1-5 and ICD-CMOP7, M2M-IEpsilon outperforms CM2M and PPS-MOEA/D significantly. M2M-IEpsilon is significantly better than CM2M2 on ICD-CMOPs except ICD-CMOP2.

Table 2. The HV results of the seven CMOEAs on ICD-CMOP1-7.

Test instance		M2M-IEpsilon	CM2M	CM2M2	NSGA-II	MOEA/D	PPS	IEpsilon
ICD-CMOP1	mean	**9.92E−01**	9.14E−01†	8.56E−01†	5.21E−01†	5.51E−01†	5.64E−01†	6.21E−01†
	std	4.34E−03	3.88E−02	1.97E−02	2.46E−02	8.77E−02	1.15E−01	1.33E−01
ICD-CMOP2	mean	**4.98E−01**	4.37E−01†	4.31E−01†	2.60E−01†	3.77E−01†	3.84E−01†	3.66E−01†
	std	5.06E−02	3.03E−02	3.35E−02	5.46E−02	1.93E−02	1.24E−02	4.24E−02
ICD-CMOP3	mean	**3.66E−01**	2.84E−01†	2.55E−01†	2.39E−01†	2.46E−01†	2.54E−01†	2.39E−01†
	std	7.72E−02	6.31E−02	2.83E−02	8.47E−17	2.59E−02	4.05E−02	8.47E−17
ICD-CMOP4	mean	**8.15E−01**	5.84E−01†	7.24E−01†	4.13E−01†	5.11E−01†	5.04E−01†	4.78E−01†
	std	7.00E−02	6.77E−02	8.05E−02	3.13E−02	1.68E−02	1.04E−02	5.83E-02
ICD-CMOP5	mean	**9.47E−01**	8.22E−01†	8.06E−01†	6.14E−01†	5.80E−01†	5.88E−01†	6.06E−01†
	std	2.35E−02	4.12E−02	1.97E−02	5.99E−02	0.00E+00	7.50E−03	5.75E−02
ICD-CMOP6	mean	1.90E−01	1.73E−01†	2.06E−01	7.09E−03†	8.99E−03†	**4.55E−01**	1.12E-02†
	std	6.17E−02	5.81E−02	9.26E−03	7.50E−04	3.96E−03	3.63E−02	1.12E−02
ICD-CMOP7	mean	3.22E−01	2.59E−01†	2.39E−01†	1.41E−01†	1.84E−01†	**5.61E−01**	1.75E−01†
	std	3.33E−02	3.57E−02	2.84E−02	7.00E−03	8.91E−03	3.24E−02	7.55E−03
Wilcoxon-Test(S-D-I)		–	7−0−0	6−1−0	7−0−0	7−0−0	5−2−0	7−0−0

Table 2 shows the HV results of ICD-CMOP1-7 obtained by the seven CMOEAs. For ICD-CMOP1-5, M2M-IEpsilon is significantly better than the rest of CMOEAs. For ICD-CMOP6, M2M-IEsiplon significantly outperforms CM2M, NSGA-II-CDP, MOEA/D-IEpsilon and MOEA/D-CDP, and has no significant difference with CM2M2 and PPS-MOEA/D. For ICD-CMOP7, M2M-IEsiplon significantly outperforms CM2M, CM2M2, NSGA-II-CDP, MOEA/D-IEpsilon and MOEA/D-CDP. The above analysis and observations reveal that the proposed algorithm M2M-IEpsilon significantly better than the rest of CMOEAs on most of the ICD-CMOPs.

To further discuss the advantages of the proposed M2M-IEsiplon in solving ICD-CMOPs, non-dominated solutions with the median HV values achieved by seven algorithms on ICD-CMOP2 plotted in Fig. 2, the proposed M2M-IEpsilon is able to find the whole true PFs while other six CMOEAs can only converge to a part of the true PFs. There are two possible reasons for this. The first one is that the objective functions of ICD-CMOP2 are imbalanced, and only the M2M decomposition method can effectively deal with CMOPs with the imbalanced objective functions. Another reason is that ICD-CMOP2 has diversity-hard and convergence-hard constraints, which makes CM2M and CM2M2 difficult to converge to the whole true PFs, because CM2M and CM2M2 have no specific mechanisms for dealing with constraints to solve CMOPs with simultaneous diversity-hard and convergence-hard constraints. However, the proposed M2M-IEpsilon converge to disconnected PFs, which enables it to have the best performance on the ICD-CMOPs.

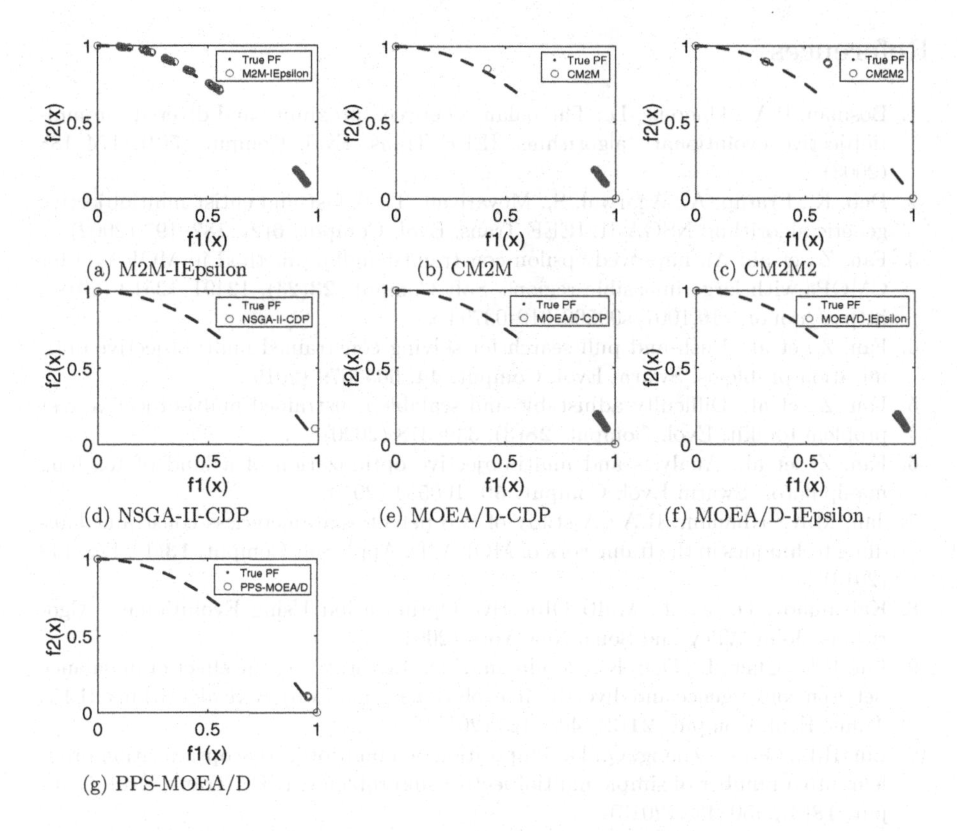

(a) M2M-IEpsilon (b) CM2M (c) CM2M2

(d) NSGA-II-CDP (e) MOEA/D-CDP (f) MOEA/D-IEpsilon

(g) PPS-MOEA/D

Fig. 2. The non-dominated solutions got by seven CMOEAs on ICD-CMOP2.

5 Conclusion

A hybrid CMOEA is proposed in the paper, namely M2M-IEpsilon, which combines the improved epsilon constraint-handling approach and M2M decomposition method to solve imbalanced CMOPs with simultaneous convergence-hard and diversity-hard constraints. A series of problems, namely ICD-CMOPs, is suggested to evaluate the performance of M2M-IEpsilon by using the DAS-CMOPs framework. The ICD-CMOPs consist of imbalanced CMOPs with diversity-hard and convergence-hard constraints. Since M2M-IEpsilon adopts the M2M decomposition method, it is able to solve CMOPs with imbalanced objective functions. The IEpsilon constraint-handling mechanism embedded in M2M-IEpsilon can help the population of M2M-IEpsilon to get across infeasible regions and to improve the diversity performance. To verify this, seven CMOEAs are tested on the ICD-CMOPs. Through the analysis of experimental results, the superiority of the proposed M2M-IEpsilon in dealing with ICD-CMOPs with simultaneous convergence-hard and diversity-hard constraints. In the future, a scheduled work is to combine machine learning techniques with M2M-IEpsilon to solve CMOPs with expensive objective and constraint functions.

References

1. Bosman, P.A., Thierens, D.: The balance between proximity and diversity in multiobjective evolutionary algorithms. IEEE Trans. Evol. Comput. **7**(2), 174–188 (2003)
2. Deb, K., Pratap, A., Agarwal, S., Meyarivan, T.: A fast and elitist multiobjective genetic algorithm: NSGA-II. IEEE Trans. Evol. Comput. **6**(2), 182–197 (2002)
3. Fan, Z., et al.: An improved epsilon constraint-handling method in MOEA/D for CMOPs with large infeasible regions. Soft Comput. **23**(23), 12491–12510 (2019). https://doi.org/10.1007/s00500-019-03794-x
4. Fan, Z., et al.: Push and pull search for solving constrained multi-objective optimization problems. Swarm Evol. Comput. **44**, 665–679 (2019)
5. Fan, Z., et al.: Difficulty adjustable and scalable constrained multi-objective test problem toolkit. Evol. Comput. **28**(3), 339–378 (2020)
6. Fan, Z., et al.: Analysis and multi-objective optimization of a kind of teaching manipulator. Swarm Evol. Comput. **50**, 100554 (2019)
7. Jan, M.A., Khanum, R.A.: A study of two penalty-parameterless constraint handling techniques in the framework of MOEA/D. Appl. Soft Comput. **13**(1), 128–148 (2013)
8. Kalyanmoy, D., et al.: Multi Objective Optimization Using Evolutionary Algorithms. John Wiley and Sons, New York (2001)
9. Liu, H.L., Chen, L., Deb, K., Goodman, E.D.: Investigating the effect of imbalance between convergence and diversity in evolutionary multiobjective algorithms. IEEE Trans. Evol. Comput. **21**(3), 408–425 (2016)
10. Liu, H.L., Gu, F., Zhang, Q.: Decomposition of a multiobjective optimization problem into a number of simple multiobjective subproblems. IEEE Trans. Evol. Comput. **18**(3), 450–455 (2013)
11. Liu, H.L., Peng, C., Gu, F., Wen, J.: A constrained multi-objective evolutionary algorithm based on boundary search and archive. Int. J. Pattern Recogn. Artif. Intell. **30**(01), 1659002 (2016)
12. Peng, C., Liu, H.L., Gu, F.: An evolutionary algorithm with directed weights for constrained multi-objective optimization. Appl. Soft Comput. **60**, 613–622 (2017)
13. Wang, J., Zhou, Y., Wang, Y., Zhang, J., Chen, C.P., Zheng, Z.: Multiobjective vehicle routing problems with simultaneous delivery and pickup and time windows: formulation, instances, and algorithms. IEEE Trans. Cybern. **46**(3), 582–594 (2015)
14. Zhang, M., Li, H., Liu, L., Buyya, R.: An adaptive multi-objective evolutionary algorithm for constrained workflow scheduling in clouds. Distrib. Parallel Databases **36**(2), 339–368 (2018)
15. Zitzler, E., Thiele, L.: Multiobjective evolutionary algorithms: a comparative case study and the strength pareto approach. IEEE Trans. Evol. Comput. **3**(4), 257–271 (1999)

Constrained Bi-objective Surrogate-Assisted Optimization of Problems with Heterogeneous Evaluation Times: Expensive Objectives and Inexpensive Constraints

Julian Blank(✉) and Kalyanmoy Deb

Computational Optimization and Innovation (COIN) Laboratory,
Michigan State University, East Lansing, MI, USA
{blankjul,kdeb}@msu.edu
https://www.coin-lab.org

Abstract. In the past years, a significant amount of research has been done in optimizing computationally expensive and time-consuming objective functions using various surrogate modeling approaches. Constraints have often been neglected or assumed to be a by-product of the expensive objective computation and thereby being available after executing the expensive evaluation routines. However, many optimization problems in practice have separately evaluable computationally inexpensive geometrical or physical constraint functions, while the objectives may still be time-consuming. This scenario probably makes the simplest case of handling heterogeneous and multi-scale surrogate modeling in the presence of constraints. In this paper, we propose a method which makes use of the inexpensiveness of constraints to ensure all time-consuming objective evaluations are only executed for feasible solutions. Results on test and real-world problems indicate that the proposed approach finds a widely distributed set of near-Pareto-optimal solutions with a small budget of expensive evaluations.

Keywords: Multi-objective optimization · Surrogate-assisted optimization · Inexpensive constraints · NSGA-II · Evolutionary algorithm

1 Introduction

Optimization of simulation models is essential and especially important in many interdisciplinary research areas to find optimized solutions for real-world applications. For instance, many engineering design optimization problems [22] require a time-consuming simulation to evaluate the design's performance. A common technique to deal with long-running simulation is to use so-called *surrogate* models or metamodels [1,18,23] to represent the original objective and/or constraint

H. Ishibuchi et al. (Eds.): EMO 2021, LNCS 12654, pp. 257–269, 2021.
https://doi.org/10.1007/978-3-030-72062-9_21

functions. This then allows a computationally cheaper optimization procedure to find potentially good solutions without executing expensive evaluations. However, in real-world problems, a design's feasibility must be checked during the optimization process, as an infeasible solution, no matter how good its objective values are, cannot be accepted. For instance, consider the diffuser inlet design problem [13,24], shown in Fig. 1a. In [20], the diffuser design is modeled by two variables, the length (L) and the angle (θ), to maximize the total pressure ratio and the Mach number at the outlet. The objectives are evaluated using an expensive mesh generation procedure bounding the diffuser's height between 0.1 to 0.8 meters. In turn, these bounds limit the minimum and maximum value of θ for each L. This dependency and some additional constraints result in non-linear constraints that make the shaded region in Fig. 1b feasible. The feasible search space obeys geometric and physical laws, which are relatively easy to satisfy in this problem but can not be expressed simply by box constraints on variables. However, the objective of the problem is to maximize the entropy generation rate by the diffuser, which must start from solving Navier-Stokes questions involving turbulent flow and solving the equations numerically with an adaptive step-size procedure. This process is many times more expensive than finding if a solution is feasible. Thus, a generic surrogate-assisted optimization method that treats all objective and constraint functions as equally expensive may not be most efficient for solving such real-world problems.

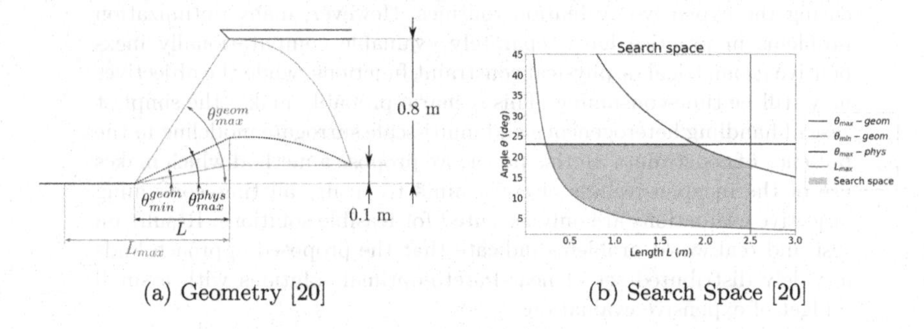

(a) Geometry [20] (b) Search Space [20]

Fig. 1. Geometric constraints of a diffuser inlet design.

In [25], the optimization of geometrical design parameters in IPM motor topology has been discussed. Apart from box constraints, the ratio between variables is bounded. Also, the authors mention that the simulation software allows running a computationally quick feasibility check to identify overlapping edges, negative lengths, or non-conventional geometric relations that will inevitably produce errors after the solver's start. The availability of such quick feasibility checks – without running the time-consuming simulation – results in a relatively computationally inexpensive constraint evaluation. However, to compute the objective function given by the magnetic flux requires an expensive finite element methodology, thereby making the IPM motor design task as another case of heterogeneous surrogate-based optimization.

Even though computationally expensive objective functions have been studied extensively [9,18,26], computationally inexpensive constraints have been paid a little attention in the literature. It is also noteworthy that there can be other combinations of complexities in evaluating them. For instance, the constraint may be expensive to evaluate in other real-world problems, while the objectives may be relatively quick to compute. In this paper, we restrict ourselves to the heterogeneous problems involving inexpensive constraint evaluation and expensive objective evaluation procedures and belabor the consideration of other scenarios in later studies.

In the remainder of this paper, we first present the related work, which considers mostly the more general case of heterogeneous objective function evaluation. In Sect. 3, we propose our method for handling inexpensive constraint functions efficiently on an algorithmic level. The performance of the proposed method is evaluated on a variety of constrained bi-objective test problems in Sect. 4. Conclusions and future works are discussed in Sect. 5.

2 Related Work

In general, optimization problems with computationally expensive objectives and inexpensive constraints belong to the category of heterogeneously expensive target functions. The question being addressed is how an algorithm can use the heterogeneity to improve its convergence properties. Mostly, unconstrained bi-objective problems have been studied so far, which inevitably results in one objective being computationally inexpensive (cheap) and one being expensive or time-consuming. Since only two target values have mostly been considered, authors also refer to the difference as a *delay* in the evaluation between two objective functions, as a problem parameter.

In 2013, Allmendinger et al. [3] proposed three different ways of dealing with missing objective values caused by such a delay: (i) filling the missing objectives with random pseudo values within the boundaries of the objective space, (ii) adding Gaussian noise to a solution randomly drawn from the population, and (iii) replacing by the nearest neighbor's objective values in the design space being evaluated on all objectives. Authors also proposed different evaluation selection strategies, for instance, based on the creation time of an offspring or a priority score obtained by partial or full non-dominated rank. This study was one of the first addressing delayed objective functions and has laid the foundation for further investigations. In 2015, this study was extended by Allmendinger et al. [2] by proposing new ways of dealing with the heterogeneity based on using the cheap objectives to look at one or multiple generations ahead. The experimental study has revealed that the performance is affected by the amount of delay of the objectives and, thus, that some approaches are more suitable than others depending on the amount of delay.

In 2018, Chugh et al. have proposed HK-RVEA [7] as an extension of K-RVEA [6], which handles objective functions with different latencies. Compared to the original K-RVEA, significant changes are related to the surrogate's training and update mechanism, driven by a single-objective evolutionary algorithm.

A comparison regarding bi-objective test problems with previously proposed approaches [2,3] showed that this surrogate-assisted algorithm works especially well in cases with low latencies.

Moreover, a trust region-based algorithm that aims to find a single Pareto-optimal solution with a limited number of function evaluations was proposed by Thomann et al. [27]. The search direction for each step is given by the vector between the current best solution and the minimum of the locally fitted quadratic model. The trust region serves as step size and limits the surrogate's underlying error. Results on bi-objective optimization problems indicate the algorithm's capability to convergence to a Pareto critical point.

Wang et al. [28] proposed T-SAEA that uses a transfer learning approach to exploit knowledge gained about the inexpensive objective before evaluating the more expensive one. A filter-based feature selection finds essential variables of the computationally inexpensive objective. Then, these features serve as a carrier for knowledge transfer to the computationally expensive function and help to improve the metamodel's accuracy.

In general, most studies related to computationally expensive optimization problems do not consider constraints or assume them to be expensive as well. Also, existing literature about heterogeneous expensive target functions focuses on a delay of objective functions but *not* constraints. Because computationally inexpensive constraints are a practical issue that needs to be addressed efficiently, it shall be the focus of this paper.

3 Proposed Methodology

In the following, we propose IC-SA-NSGA-II, an inexpensive constraint handling method using a surrogate-assisted version of NSGA-II [10]. First, we describe three different methods to generate a feasible design of experiments, and, second, the outline of the algorithm. Our proposed method makes explicit use of the computationally inexpensive constraint functions and guarantees a solution's feasibility before running the time-consuming simulation of objective functions.

3.1 Design of Experiments (DOE) to Build Initial Model

In surrogate-assisted optimization [18], the so-called Design of Experiments (DOE) needs to be generated to build the initial model(s). The number of initial points $N^{\texttt{DOE}}$ depends on different factors, such as the number of variables or the complexity of the fitness landscape. A standard method frequently used to generate a well-spaced set of points is Latin Hypercube Sampling (LHS) [19]. However, with the availability of an efficient feasibility check, more sophisticated approaches shall be preferred. Therefore, three different methods returning a well-spaced set of *feasible* solutions using inexpensive constraint functions are described.

Rejection Based Sampling (RBS): A relatively simple approach is modifying LHS, which already provides a well-spaced set of solutions, to consider feasibility.

Such a feasible and well-spaced set of points can be obtained by using LHS to produce a point set P and rejecting all infeasible solutions to obtain a set of *feasible* solutions $P^{(\texttt{feas})}$. Because infeasible solutions have been discarded, the resulting point set does not have the desired number of points ($|P^{(\texttt{feas})}| < N^{\texttt{DOE}}$) and, thus, the process shall be repeated until enough feasible solutions have been found. If the size of the obtained point set exceeds $N^{\texttt{DOE}}$, a subset is selected randomly.

Niching Genetic Algorithm (NGA): The sampling process itself can be seen as an optimization problem where the objective is given by the constraint violation $f(\mathbf{x}) = \text{cv}(\mathbf{x})$, which takes a value zero for a feasible solution $\mathbf{x}$, and a positive value proportional to the sum of normalized constraint violation of all constraints if $\mathbf{x}$ is infeasible. Such an objective function results in a multi-modal optimization problem where a *diverse* solution set with objective values of zeros shall be found. One type of algorithm used for multi-modal problems is niching-based genetic algorithms (NGA) where the diversity is ensured by an ϵ-clearing based environmental survival [21]. For single-objective optimization problems, the ϵ-clearing occurs in the design space and guarantees a distance (usually Euclidean distance is used) from one solution to another. The survival always selects the best performing not already selected or cleared solution and then clears its neighborhood with less than ϵ distance. Thus, the spread of solutions is accomplished by disfavoring solutions in each other's vicinity. Having set the objective to be the constraint violation, a suitable ϵ has to be found. The suitability of a given ϵ depends on the size of the feasible region(s) of the corresponding optimization problem and is not known beforehand. On the one hand, if ϵ is too large, the number of optimal solutions found by the algorithm will not exceed $N^{\texttt{DOE}}$. On the other hand, if ϵ is too small, the solution set's spread has room for improvement. For tuning the hyper-parameter ϵ, we start with $\epsilon = \epsilon_0$ (a number close to 1.0) and execute NGA. If the size of the obtained solution set is less than $N^{\texttt{DOE}}$, we set $\epsilon = 0.9 \cdot \epsilon$ and repeat this procedure until a solution set of at least of size $N^{\texttt{DOE}}$ is found.

Riesz s-Energy Optimization (Energy): The Riesz s-Energy [16] is a generalization of potential energy concept and is defined for the point set $\mathbf{z}$ as

$$U(\mathbf{z}) = \frac{1}{2} \sum_{i=1}^{|\mathbf{z}|} \sum_{\substack{j=1 \\ j \neq i}}^{|\mathbf{z}|} \frac{1}{\left\|\mathbf{z}^{(i)} - \mathbf{z}^{(j)}\right\|^{s}}, \quad \mathbf{z} \in \mathbb{R}^{n \times M}. \tag{1}$$

where the inverse norm to the power s of each pair of points ($\mathbf{z}^{(i)}$ and $\mathbf{z}^{(j)}$) is summed up. It has already been shown in [5] that Riesz s-Energy can be used to achieve a well-spaced point set by executing a gradient-based algorithm. The restriction made there that all points have to lie on the unit simplex has been replaced by the feasibility check provided by the computationally inexpensive constraint. Thus, a point is only replaced by its successor obtained by the gradient update if the successor is *feasible*. The algorithm's initial point set, which is necessary to be provided, is first tried to be obtained by RBS, and if a sufficient number of feasible solutions could not be found by NGA.

Algorithm 1: IC-SA-NSGA-II: **I**nexpensive **C**onstrained **S**urrogate-**A**ssisted NSGA-II.

Input: Number of Variables n, Expensive Objective Function $f(\mathbf{x})$, Inexpensive Constraint Function $g(\mathbf{x})$, Maximum Number of Solution Evaluations $\text{SE}^{\max}$, Number of Design of Experiments N^{DOE}, Exploration Points $N^{(\text{explr})}$, Exploitation Points $N^{(\text{exploit})}$, Number of generations for exploitation k, Multiplier of offsprings for exploration s

```
   /* initialize feas. solutions using the inexpensive function g          */
1  X ← constrained_sampling('energy', N^DOE, g)
2  F ← f(X)
3  while |X| < SE^max do
       /* exploitation using the surrogate                                  */
4      f̂ ← fit_surrogate(X, F)
5      (X^(cand), F^(cand)) ← optimize('nsga2', f̂, g, X, F, k)
6      (X^(cand), F^(cand)) ← eliminate_duplicates(X, X^(cand), F^(cand))
7      C ← cluster('k_means', N^(exploit), F^(cand))
8      X^(exploit) ← ranking_selection(X^(cand), C, crowding(F^(cand)))
       /* exploration using mating and least crowded selection              */
9      X', F' ← survival(X, F)
10     X^(mat) ← mating(X', F', s · N^(explr))
11     X^(explr) ← feas_and_max_distance_selection(X^(mat), X^(cand), X, g)
       /* evaluate and merge to the archive                                 */
12     F^(explr) ← f(X^(explr)); F^(exploit) ← f(X^(exploit));
13     X ← X ∪ X^(explr) ∪ X^(exploit)
14     F ← F ∪ F^(explr) ∪ F^(exploit)
15 end
```

3.2 IC-SA-NSGA-II Algorithm

The outline of IC-SA-NSGA-II is shown in Algorithm 1. The initial design of experiments of size N^{DOE} are obtained by the proposed initialization method based on Riesz s-Energy and then evaluated by executing the expensive simulation $f(\mathbf{x})$ (Lines 1 and 2). Afterwards, while the number of solution evaluations $\text{SE}^{\max}$ is not exceeded (Line 3) the algorithm continues to generate $N^{(\text{exploit})}$ solutions derived from the surrogate for exploitation and $N^{(\text{explr})}$ solutions obtained by mating and a distance-based selection for exploration. The exploitation starts with fitting surrogate model(s) which results in the approximation function $\hat{f}$ (Line 4). Using the surrogates $\hat{f}$ and the computationally inexpensive function g the optimization is continued or in other words simulated assuming $f = \hat{f}$ for k more generations (Line 5). From the last simulated generation, the candidates $\mathbf{X}^{(\text{cand})}$ and $\mathbf{F}^{(\text{cand})}$ are extracted from the optimum, and duplicates with respect to $\mathbf{F}$ are eliminated (Line 6). This ensures $\mathbf{F}^{(\text{cand})}$ to consist of only non-dominated solutions with respect to $\mathbf{F}$. From $\mathbf{X}^{(\text{cand})}$ only

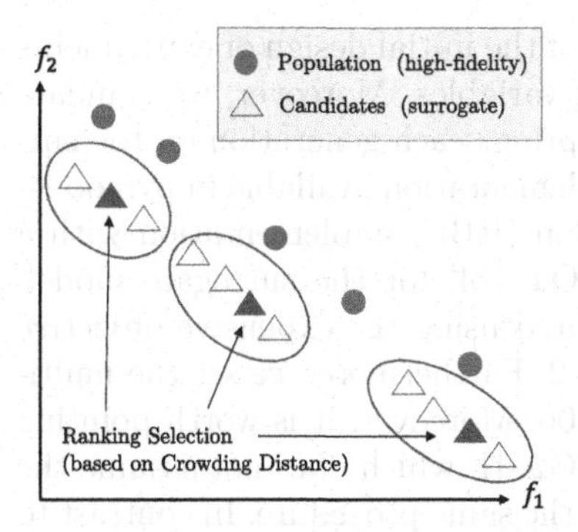

(a) Exploitation: Select solutions from the candidates set obtained by optimizing on the surrogate.

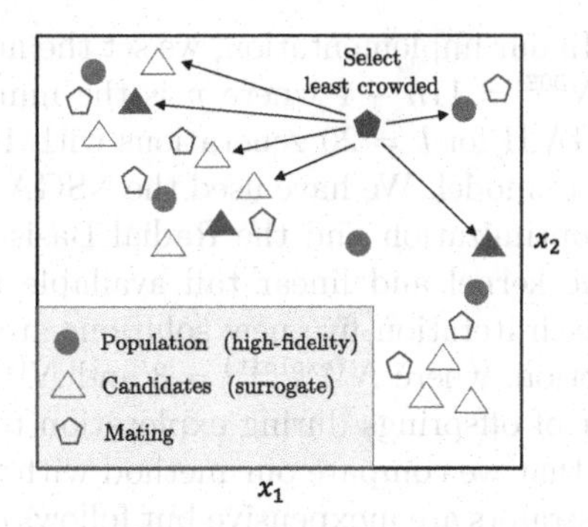

(b) Exploration: Select from a solution set obtained through evolutionary operators by maximizing the distance to existing solutions and candidates.

Fig. 2. The two steps in each iteration: exploitation and exploration.

$N^{(\texttt{exploit})}$ solutions are chosen for expensive evaluation by executing ranking selection [15] in each cluster where the ranking is based on the crowding distance in $\mathbf{F}^{(\texttt{cand})}$. Figure 2a illustrates the exploitation procedure of a population with five solutions (circles). The algorithm found ten candidate solutions (triangles) by optimizing the surrogate and the inexpensive constraint functions. In this example, $N^{(\texttt{exploit})} = 3$ and, thus, the K-means algorithm is instantiated to find three clusters. From each cluster, the solutions obtained by ranking selection based on the crowding distance are assigned to $\mathbf{X}^{(\texttt{exploit})}$.

Besides the exploitation, some exploration is essential to be incorporated into a surrogate-assisted algorithm. The exploration is based on the evolutionary recombination of NSGA-II with post-filtering based on the distance in the design space (Line 9 to 11). First, the environmental survival is executed because the mating should not be based on the archive $\mathbf{X}$ but instead on a subset of more promising solutions $\mathbf{X}'$. Second, mating takes place to produce $s \cdot N^{(\texttt{explr})}$ solutions $\mathbf{X}^{(\texttt{mat})}$ and, third, the set of *feasible* solutions from $\mathbf{X}^{(\texttt{mat})}$ being maximally away from $\mathbf{X}$ and $\mathbf{X}^{(\texttt{cand})}$ are assigned to $\mathbf{X}^{(\texttt{explr})}$. Figure 2b demonstrates this explorative step more in detail. All infeasible solutions generated through mating have already been eliminated. The solution with the maximum distance to others is selected, which represents the least crowded solution with respect to $\mathbf{X}$ and $\mathbf{X}^{(\texttt{cand})}$. The exploration step purposefully chooses solutions not suggested by the surrogate and helps to escape from local optima if necessary. After selecting the first solution with the maximum distance to others, the solution is marked as selected and now considered in the distance calculations for the second iteration in the selection procedure. Finally, the infill solutions $\mathbf{X}^{(\texttt{exploit})}$ and $\mathbf{X}^{(\texttt{explr})}$ are evaluated on the expensive objective functions f and merged with $\mathbf{X}$ and $\mathbf{F}$ (Line 12 to 14).

In our implementation, we set the number of the initial design of experiments to $N^{\texttt{DOE}} = 11n - 1$ where n is the number of variables. Moreover, we simulate NSGA-II for $k = 20$ generations with 100 offsprings each generation on the surrogate model. We have used the NSGA-II implementation available in pymoo [4] for optimization and the Radial Basis Function (RBF) implementation with a cubic kernel and linear tail available in pySOT [14] for the surrogate model. In each iteration five new solutions are evaluated using the expensive objective function, where $N^{(\texttt{exploit})} = 3$ and $N^{(\texttt{explr})} = 2$. Furthermore, we set the multiplier of offsprings during exploration to $s = 100$. Moreover, it is worth pointing out that we compare our method with SA-NSGA-II, which does not assume the constraints are inexpensive but follows overall the same procedure. In contrast to IC-SA-NSGAII, it uses regular Latin Hypercube Sampling for the initial design of experiments and fits a surrogate of the constrained function(s) to evaluate feasibility.

4 Results

The proposed method uses the inexpensiveness of the constraint function(s), and, thus, the performance on *constrained* multi-objective optimization problems shall be evaluated. In this paper, we focus on bi-objective problems with up to ten constraint functions. In contrast to the constraint function, we treat all objectives to be computationally expensive.

To evaluate the algorithm's performance, we use the CTP test problems suite [12], which has been designed to address constraints of varying difficulty. Moreover, the performance on other bi-objective constrained optimization problems frequently used in the literature such as OSY [11], TNK [11], SRN [11], C2DTLZ2 [17], C3DTLZ4 [17], and Car Side Impact (CAR) [17] shall be evaluated. The number of solution evaluations $\texttt{SE}^{\max}$ is kept relatively small to mimic the evaluation budget of time-consuming simulations. In the following, first, the performance of methods proposed to generate a feasible solution set for the design of experiments are *visually* analyzed, and, second, the algorithm's performance on test problems is discussed.

In Fig. 3, the results of Rejection Based Sampling (RBS), Niching GA (NGA), and Riesz s-Energy (Energy) are shown. Compared to RBS and NGA, Energy obtains a very uniform and well-spaced point set in the inside of the feasible region across all problems. Also, it is worth noting that points on the constraint boundary are found, which can be very valuable to start with because, in practice, optima frequently lie on constraint boundaries. For the purpose of visualization, the CTP8 problem with two-variables (and nine feasible disconnected regions) has been investigated. All methods were able to obtain more than one feasible solution in all regions.

Table 1 lists the median values (obtained from 11 runs) of the Inverted Generational Distance (IGD) [8] indicator of 14 constrained bi-objective optimization problems. The obtained results have been normalized with respect to the ideal and nadir point of each problem's True front. The best performing method and

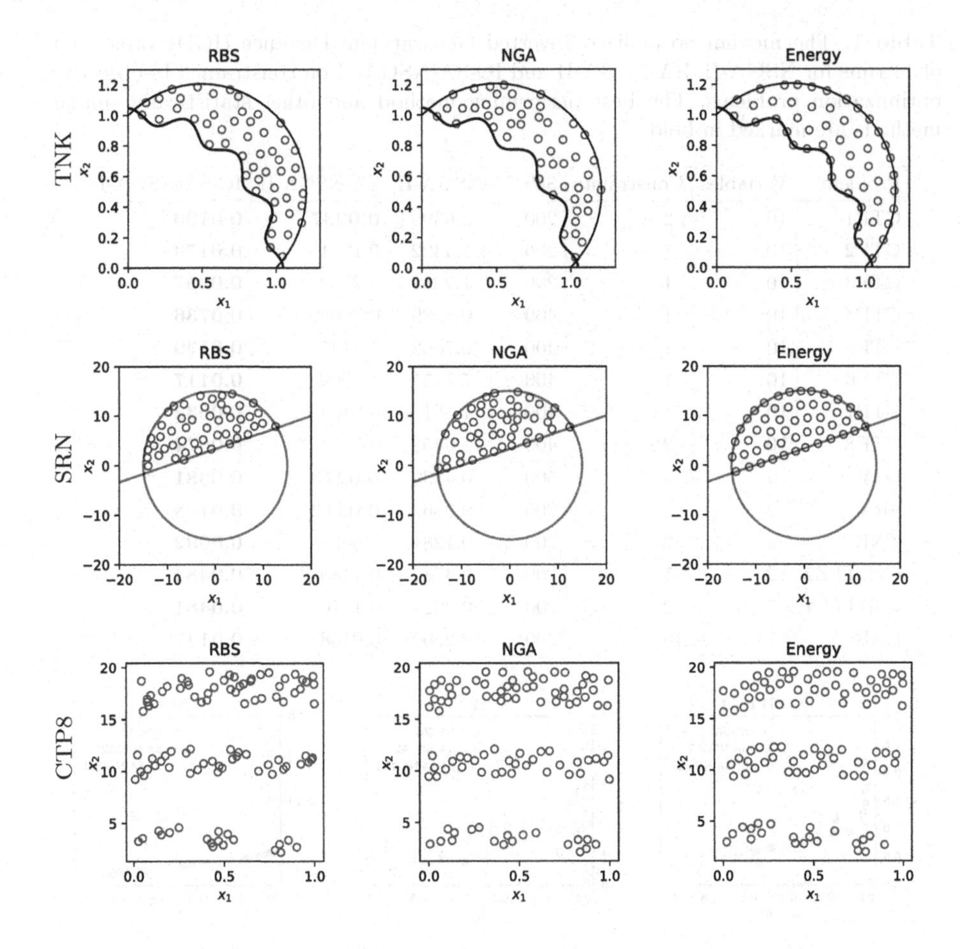

Fig. 3. Sampling the design of experiments only in the feasible space using Rejection-Based Sampling (RBS), Niching Genetic Algorithm (NGA) and Energy Method.

other statistically similar methods (Wilcoxon rank test, $p = 0.05$) are marked in bold. Besides IC-SA-NSGA-II, we ran a more steady-state version of NSGA-II with five offsprings in each generation and an initial population of size $11n - 1$ sampled by LHS. Moreover, SA-NSGA-II, which is our proposed method without the modifications made to exploit the availability of inexpensive constraints. Clearly, IC-SA-NSGA-II outperforms the other approaches for most of the problems. For 11 out of 14 optimization problems, our proposed method shows the best performance significantly; for two problems (CTP1 and SRN), SA-NSGA-II performs statistically similar; and for one problem (OSY), SA-NSGA-II shows slightly better results. NSGA-II is not able to find a near-optimal set of solutions with the limited $\mathrm{SE}^{\max}$ for any of the selected problems.

Table 1. The median normalized Inverted Generational Distance (IGD) values out of 11 runs for NSGA-II, SA-NSGA-II and IC-SA-NSGA-II on constrained bi-objective optimization problems. The best performing method and other statistically similar methods are marked in bold.

Problem	Variables	Constraints	$\text{SE}^{\max}$	NSGA-II	SA-NSGA-II	IC-SA-NSGA-II
CTP1	10	2	200	3.6399	**0.0237**	**0.0196**
CTP2	10	1	200	1.4422	0.1721	**0.0173**
CTP3	10	1	200	1.2282	0.2752	**0.0357**
CTP4	10	1	400	0.8489	0.3969	**0.0736**
CTP5	10	1	400	0.7662	0.1145	**0.0139**
CTP6	10	1	400	7.7155	0.1909	**0.0117**
CTP7	10	1	400	1.5517	0.0164	**0.0032**
CTP8	10	2	400	11.6452	0.5963	**0.0074**
OSY	6	6	500	0.4539	**0.0273**	**0.0381**
SRN	2	2	200	0.0263	**0.0112**	**0.0108**
TNK	2	2	200	0.1281	0.0200	**0.0092**
C2DTLZ2	12	1	200	0.3787	0.1185	**0.0484**
C3DTLZ4	7	2	200	0.2622	0.1210	**0.0481**
CAR	7	10	200	0.2362	0.0168	**0.0147**

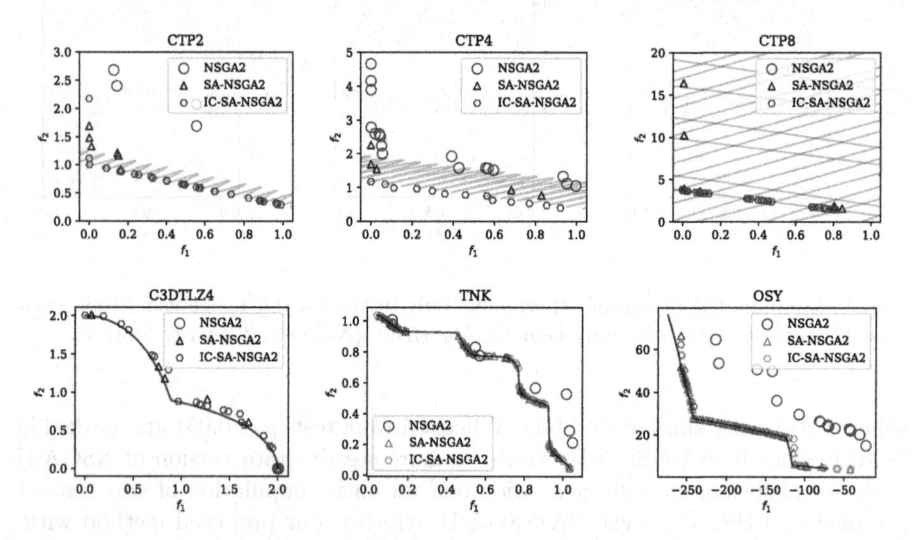

Fig. 4. Solutions in the objective space of representative runs for CTP2, CTP4, CTP8, C3DTLZ4, TNK, and OSY.

Figure 4 shows the obtained solution set for each method of representative runs for CTP2, CTP4, CTP8, C3DTLZ4, TNK, and OSY, for which a well-converged and well-diversified non-dominated solutions are found with 200 to 500 solution evaluations. For the difficult problem CTP2, our proposed method converges near the true optima, which lies on the constraint boundary.

Similarly, for CTP8, where nine feasible islands for which three contain the Pareto-optimal set exist and, thus, a good exploration of the search space is needed. For C3DTLZ4, IC-SA-NSGA-II has obtained a better diversity in the solution set than SA-NSGA-II.

5 Conclusions

The optimization of computationally expensive optimization problems has become more important in practice. Often such problems have physical or geometrical constraints that are relatively computationally inexpensive and can be formulated in equations without running the simulation. To solve these kinds of problems, we have proposed IC-SA-NSGA-II, a surrogate-assisted NSGA-II, which efficiently handles inexpensive constraint functions in the initial design of experiments as well as in each iteration. We have tested our proposed method on 14 constrained bi-objective optimization problems, and our results indicate that efficiently handling the inexpensive constraints helps to converge faster.

This paper has focused on solving constrained optimization problems with inexpensive constraint and expensive objective functions with very limited function calls. However, some other heterogeneous problems with different time scales of objective and constraint evaluations must be considered next. For instance, the constraints can be even more time-consuming than the objective function, or the objectives and constraints can have different time scales within themselves. After addressing such cases, there is a need for developing a unified algorithm for generic heterogeneous optimization problems. Moreover, studies about heterogeneously expensive optimization problems have so far been limit to bi-objective optimization problems. This paper has started to add some more complexity by adding constraint functions. However, the effect of heterogeneity for many-objective optimization problems shall provide more insides into exploiting the discrepancy of evaluation times and asynchronicity.

References

1. Ahrari, A., Blank, J., Deb, K., Li, X.: A proximity-based surrogate-assisted method for simulation-based design optimization of a cylinder head water jacket. Eng. Optim. 1–19 (2020)
2. Allmendinger, R., Handl, J., Knowles, J.: Multiobjective optimization: when objectives exhibit non-uniform latencies. Eur. J. Oper. Res. **243**(2), 497–513 (2015)
3. Allmendinger, R., Knowles, J.: 'Hang on a minute': investigations on the effects of delayed objective functions in multiobjective optimization. In: Purshouse, R.C., Fleming, P.J., Fonseca, C.M., Greco, S., Shaw, J. (eds.) EMO 2013. LNCS, vol. 7811, pp. 6–20. Springer, Heidelberg (2013). https://doi.org/10.1007/978-3-642-37140-0_5
4. Blank, J., Deb, K.: Pymoo: multi-objective optimization in python. IEEE Access **8**, 89497–89509 (2020)
5. Blank, J., Deb, K., Dhebar, Y., Bandaru, S., Seada, H.: Generating well-spaced points on a unit simplex for evolutionary many-objective optimization. IEEE Trans. Evol. Comput

6. Chugh, T., Jin, Y., Miettinen, K., Hakanen, J., Sindhya, K.: A surrogate-assisted reference vector guided evolutionary algorithm for computationally expensive many-objective optimization. IEEE Trans. Evol. Comput. **22**(1), 129–142 (2018)
7. Chugh, T., Allmendinger, R., Ojalehto, V., Miettinen, K.: Surrogate-assisted evolutionary biobjective optimization for objectives with non-uniform latencies. In: Proceedings of the Genetic and Evolutionary Computation Conference. GECCO 2018, New York, NY, USA, pp. 609–616. Association for Computing Machinery (2018)
8. Coello Coello, C.A., Reyes Sierra, M.: A study of the parallelization of a coevolutionary multi-objective evolutionary algorithm. In: Monroy, R., Arroyo-Figueroa, G., Sucar, L.E., Sossa, H. (eds.) MICAI 2004. LNCS (LNAI), vol. 2972, pp. 688–697. Springer, Heidelberg (2004). https://doi.org/10.1007/978-3-540-24694-7_71
9. Deb, K., Hussein, R., Roy, P.C., Toscano-Pulido, G.: A taxonomy for metamodeling frameworks for evolutionary multiobjective optimization. IEEE Trans. Evol. Comput. **23**(1), 104–116 (2019)
10. Deb, K., Pratap, A., Agarwal, S., Meyarivan, T.: A fast and elitist multiobjective genetic algorithm: NSGA-II. Trans. Evol. Comput. **6**(2), 182–197 (2002)
11. Deb, K., Kalyanmoy, D.: Multi-objective Optimization Using Evolutionary Algorithms. John Wiley & Sons Inc., New York (2001)
12. Deb, K., Pratap, A., Meyarivan, T.: Constrained test problems for multi-objective evolutionary optimization. In: Zitzler, E., Thiele, L., Deb, K., Coello Coello, C.A., Corne, D. (eds.) EMO 2001. LNCS, vol. 1993, pp. 284–298. Springer, Heidelberg (2001). https://doi.org/10.1007/3-540-44719-9_20
13. Djebedjian, B.: Two-dimensional Diffuser Shape Optimization (2004)
14. Eriksson, D., Bindel, D., Shoemaker, C.A.: pySOT and POAP: an event-driven asynchronous framework for surrogate optimization. arXiv:1908.00420 (2019)
15. Goldberg, D.E.: Genetic Algorithms for Search, Optimization, and Machine Learning. Addison-Wesley, Reading, Boston (1989)
16. Hardin, D., Saff, E.: Minimal Riesz energy point configurations for rectifiable d-dimensional manifolds. Adv. Math. **193**(1), 174–204 (2005)
17. Jain, H., Deb, K.: An evolutionary many-objective optimization algorithm using reference-point based non dominated sorting approach, part II: handling constraints and extending to an adaptive approach. IEEE Trans. Evol. Comput. **18**(4), 602–622 (2014)
18. Jin, Y.: Surrogate-assisted evolutionary computation: recent advances and future challenges. Swarm Evol. Comput. **1**(2), 61–70 (2011)
19. McKay, M.D., Beckman, R.J., Conover, W.J.: Comparison of three methods for selecting values of input variables in the analysis of output from a computer code. Technometrics **21**(2), 239–245 (1979)
20. Perez, J.L.: Genetic algorithms applied in Computer Fluid Dynamics for multiobjective optimization. Bachelor senior thesis, University of Vermont (2018)
21. Pétrowski, A.: A clearing procedure as a niching method for genetic algorithms. In: IEEE 3rd ICEC 1996, pp. 798–803 (1996)
22. Rao, S.S.: Engineering optimization: Theory and practice. Wiley (2019)
23. Roy, P.C., Hussein, R., Blank, J., Deb, K.: Trust-region based multi-objective optimization for low budget scenarios. In: Deb, K., et al. (eds.) EMO 2019. LNCS, vol. 11411, pp. 373–385. Springer, Cham (2019). https://doi.org/10.1007/978-3-030-12598-1_30
24. Schmandt, B., Herwig, H.: Diffuser and nozzle design optimization by entropy generation minimization. Entropy **13**, 1380–1402 (2011)

25. Stipetic, S., Miebach, W., Zarko, D.: Optimization in design of electric machines: methodology and workflow. In: 2015 International Aegean Conference on Electrical Machines Power Electronics (ACEMP), pp. 441–448 (2015)
26. Stork, J., et al.: Open issues in surrogate-assisted optimization. In: Bartz-Beielstein, T., Filipič, B., Korošec, P., Talbi, E.-G. (eds.) High-Performance Simulation-Based Optimization. SCI, vol. 833, pp. 225–244. Springer, Cham (2020). https://doi.org/10.1007/978-3-030-18764-4_10
27. Thomann, J., Eichfelder, G.: A trust-region algorithm for heterogeneous multiobjective optimization. SIAM J. Optim. **29**(2), 1017–1047 (2019)
28. Wang, X., Jin, Y., Schmitt, S., Olhofer, M.: Transfer learning for Gaussian process assisted evolutionary bi-objective optimization for objectives with different evaluation times. In: Proceedings of the 2020 Genetic and Evolutionary Computation Conference. GECCO 2020, New York, NY, USA, pp. 587–594. ACM (2020)

SAMO-COBRA: A Fast Surrogate Assisted Constrained Multi-objective Optimization Algorithm

Roy de Winter[1,2](✉), Bas van Stein[1], and Thomas Bäck[1]

[1] Leiden Institute of Advanced Computer Science, Leiden University, Leiden, The Netherlands
{r.de.winter,b.van.stein,t.h.w.baeck}@liacs.leidenuniv.nl
[2] C-Job Naval Architects, Hoofddorp, The Netherlands
r.dewinter@c-job.com

Abstract. This paper proposes a novel Self-Adaptive algorithm for Multi-Objective Constrained Optimization by using Radial Basis Function Approximations, SAMO-COBRA. The algorithm automatically determines the best Radial Basis Function-fit as surrogates for the objectives as well as the constraints, to find new feasible Pareto-optimal solutions. The algorithm also uses hyper-parameter tuning on the fly to improve its local search strategy. In every iteration one solution is added and evaluated, resulting in a strategy requiring only a small number of function evaluations for finding a set of feasible solutions on the Pareto frontier. The proposed algorithm is compared to a wide set of other state-of-the-art algorithms (NSGA-II, NSGA-III, CEGO, SMES-RBF) on 18 constrained multi-objective problems. In the experiments we show that our algorithm outperforms the other algorithms in terms of achieved Hypervolume after given a fixed small evaluation budget. These results suggest that SAMO-COBRA is a good choice for optimizing constrained multi-objective optimization problems with expensive function evaluations.

Keywords: Constrained optimization · Multi-objective optimization · Optimization under limited budgets

1 Introduction

According to a survey about adaptive sampling for global meta-modeling, the interest in efficient global meta-models and adaptive sampling techniques has increased over the past years [17]. Additionally, preliminary results of a recently executed questionnaire [5] shows that real world problems often involve continuous optimization variables, constraints, and one or more objectives. The questionnaire responses also indicate that it is often the case that the topological characteristics of the objective space are unknown and long evaluation times make the problems hard to solve. However, to the author's best knowledge, there

H. Ishibuchi et al. (Eds.): EMO 2021, LNCS 12654, pp. 270–282, 2021.
https://doi.org/10.1007/978-3-030-72062-9_22

is no fast multi-objective constrained optimization algorithm capable of approximating the Pareto frontier (PF) with a small number of function evaluations.

In this paper, the authors propose a novel *Self-Adaptive algorithm for Multi-Objective Constrained Optimization by using Radial Basis Function Approximations* (SAMO-COBRA). The algorithm is designed to efficiently determine the Pareto frontier of constrained multi-objective problems. The algorithm models the constraints and objectives with independent Radial Basis Functions (RBFs) using different kernels, automatically chooses the best RBF-fit, automatically tunes hyper-parameters, and aims to find feasible Pareto-optimal solutions. Here a *constrained multi-objective problem* is defined as follows:

$$\begin{aligned} \text{minimize: } & f : \Omega \to \mathbb{R}^k \ , \ f(\mathbf{x}) = (f_1(\mathbf{x}), \dots, f_k(\mathbf{x}))^\top \\ \text{subject to: } & g_i(\mathbf{x}) \leq 0 \ \forall i \in \{1, \dots, m\} \\ & \mathbf{x} \in \Omega \subset \mathbb{R}^d. \end{aligned}$$

In this formulation k is the number of objectives, m is the number of constraints and d is the number of parameters. Without loss of generality, maximisation problems are transformed to minimization problems. A *feasible Pareto-optimal solution* is defined in Definition 1.

Definition 1. ***(Feasible Pareto-optimal solution):*** $\mathbf{x} \in \Omega$ *is called* feasible Pareto-optimal *with respect to* Ω *and* $g_i(\mathbf{x}) \leq 0 \ \forall i \in \{1, \dots, m\}$*, if and only if there is no solution* $\mathbf{x}'$ *for which* $\mathbf{v} = f(\mathbf{x}') = (f_1(\mathbf{x}'), \dots, f_k(\mathbf{x}'))^\top$ *dominates* $\mathbf{u} = f(\mathbf{x}) = (f_1(\mathbf{x}), \dots, f_k(\mathbf{x}))^\top$ *where* $g_i(\mathbf{x}) \leq 0$ *and* $g_i(\mathbf{x}') \leq 0 \ \forall i \in \{1, \dots, m\}$.

The algorithm performance is compared to the performance of NSGA-II [10], NSGA-III [15], CEGO [29], and SMES-RBF [8]. In the results we show that SAMO-COBRA only requires a small number of function evaluations to find feasible Pareto-optimal solutions. This makes SAMO-COBRA a good choice when optimizing problems with expensive or time consuming function evaluations.

The remainder of this paper is organized as follows: In Sect. 2 related work is discussed. In Sect. 3 the newly proposed SAMO-COBRA algorithm is described. In Sect. 4 the experiments conducted to compare SAMO-COBRA to state-of-the-art algorithms are described. In Sect. 5 the results are reported and discussed, and in Sect. 6 the conclusions are drawn.

2 Related Work

Existing work on surrogate assisted optimization is typically limited to a subset of three relevant requirements (multi-objective, constrained, and speed). For example, methods exist for solving constrained single objective problems fast (e.g. SACOBRA [3]), for multi-objective optimization without efficient constraint handling techniques (e.g. SMS-EGO [21] and PAREGO [16]), or for constrained multi-objective optimization without using meta-models, leading to a lot of required function evaluations (e.g. NSGA-II [10], NSGA-III [15], SPEA2 [30], and SMS-EMOA [4]). Some recently proposed algorithms address all three requirements, however with computational efforts that grows cubically in iterations and exponentially for each additional decision parameter due to the use of Kriging surrogates (e.g. CEGO [29], ECMO [25]).

Only very occasionally a surrogate based algorithm is published that deals with both constraints and multiple objectives in an effective manner without using a Kriging surrogate (e.g., Datta's and Regis' SMES-RBF [8]). SMES-RBF is a surrogate-assisted evolutionary strategy that uses cubic Radial Basis Functions as a surrogate for the objectives and constraints to estimate the actual function values. The most promising solution(s) according to a non-dominated sorting procedure are then evaluated on the real objective and constraint function.

3 Constrained Multi-objective Optimization Algorithm

In this section, the new SAMO-COBRA algorithm is introduced. It is designed for dealing with continuous decision variables, multiple objectives, and multiple complex constraints in an efficient manner. The idea behind the algorithm is that in every iteration, for each objective and for each constraint independently, the best transformation and the best RBF kernel is sought. In each iteration the best fit is used to search for a new unseen feasible Pareto efficient point that contributes the most to the HyperVolume (HV) which is computed between a reference point and the Pareto-front (PF). The pseudocode of SAMO-COBRA can be found in Algorithm 1, a Python implementation can be found on GitHub [28]. In the subsections below, the algorithm is explained in more detail.

3.1 Initial Design

Bossek et al. showed empirically that when dealing with sequential model based optimization in most cases it is best to use the Halton sampling strategy with an as small as possible initial sample [6]. Several experiments confirmed the hypothesis that the smallest possible initial sample size also leads to the best results when applied on most constrained multi-objective problems from Sect. 4.

An RBF model that models the relationship between the input space and the output space can already be trained with $d+1$ initial samples. It is therefore advised when using SAMO-COBRA to create an initial Halton sample of size $d+1$ before the sequential optimization procedure starts. Every sample in the initial design is then evaluated (line 2–4 of Algorithm 1) so that all samples have their corresponding constraint and objective scores.

3.2 Radial Basis Function Fitting and Interpolation

RBF interpolation approximates a function by fitting a linear weighted combination of RBFs [3]. The challenge is to find correct weights ($\boldsymbol{\theta}$) and a good RBF kernel $\varphi(\|\mathbf{x}-\mathbf{c}\|)$. An RBF is only dependent on the distance between the input point $\mathbf{x}$ to the center $\mathbf{c}$. The RBFs used in this work take each evaluated point

Algorithm 1: SAMO-COBRA. **Input:** Objective functions $f(\mathbf{x})$, constraint function(s) $g(\mathbf{x})$, decision parameters' lower and upper bounds $[\mathbf{lb}, \mathbf{ub}] \subset \mathbb{R}^d$, reference point $\mathbf{ref} \in \mathbb{R}^k$, number of initial samples N, maximum evaluation budget N_{max}, $RBF_{kernels}$ $(\varphi) = \{cubic,\ gaussian,\ multiquadric,\ invquadric,\ invmultiquadric,\ thinplatespline\}$ **Output:** Evaluated feasible Pareto efficient solutions.

1 **Function** SAMO-COBRA(f, g, $[\mathbf{lb}, \mathbf{ub}]$, $\mathbf{ref}$, N, N_{max}, $RBF_{kernels}$):
2 $\quad \mathbf{X} \leftarrow \{\mathbf{x}_1, \cdots, \mathbf{x}_N\}$ ▷ Generate initial design, $\mathbf{X} \in \mathbb{R}^{d \times N}$
3 $\quad \mathbf{F} \leftarrow f(\mathbf{X})$ ▷ Obtain objective scores, $\mathbf{F} \in \mathbb{R}^{k \times N}$
4 $\quad \mathbf{G} \leftarrow g(\mathbf{X})$ ▷ Obtain constraint scores, $\mathbf{G} \in \mathbb{R}^{m \times N}$
5 $\quad$ **for** $i \leftarrow 1$ **to** $k + m$ **do**
6 $\quad\quad RBF_i^* \leftarrow (\varphi{=}cubic, \text{PLOG}{=}0)$ ▷ Initialize RBF configuration
7 $\quad$ **end**
8 $\quad$ **while** $N < N_{max}$ **do**
9 $\quad\quad \hat{\mathbf{X}} \leftarrow \text{SCALE}(\mathbf{X}, [-1, 1]^d)$ ▷ Scale input space to $[-1, 1]^d$
10 $\quad\quad \tilde{\mathbf{F}} \leftarrow \text{PLOG}(\mathbf{F})$ ▷ See function plog in Eq. (5)
11 $\quad\quad \tilde{\mathbf{G}} \leftarrow \text{PLOG}(\mathbf{G})$ ▷ See function plog in Eq. (5)
12 $\quad\quad \hat{\mathbf{F}} \leftarrow \text{STANDARDIZE}(\mathbf{F})$ ▷ Standardize objective space
13 $\quad\quad \hat{\mathbf{G}} \leftarrow \text{SCALE CONSTRAINT}(\mathbf{G})$ ▷ 0 remains feasibility boundary
14 $\quad\quad$ **for** $\varphi \in RBF_{kernels}$ **do** ▷ For each kernel
15 $\quad\quad\quad$ **for** $i \leftarrow 1$ **to** k **do** ▷ for each objective
16 $\quad\quad\quad\quad \hat{S}_i^{\varphi} \leftarrow \text{FITRBF}(\hat{\mathbf{X}}, \hat{\mathbf{F}}_{(i,\cdot)}, \varphi)$ ▷ Fit RBF without PLOG
17 $\quad\quad\quad\quad \tilde{S}_i^{\varphi} \leftarrow \text{FITRBF}(\hat{\mathbf{X}}, \tilde{\mathbf{F}}_{(i,\cdot)}, \varphi)$ ▷ Fit RBF with PLOG
18 $\quad\quad\quad$ **end**
19 $\quad\quad\quad$ **for** $j \leftarrow 1$ **to** m **do** ▷ for each constraint
20 $\quad\quad\quad\quad \hat{S}_{k+j}^{\varphi} \leftarrow \text{FITRBF}(\hat{\mathbf{X}}, \hat{\mathbf{G}}_{(j,\cdot)}, \varphi)$ ▷ Fit RBF without PLOG
21 $\quad\quad\quad\quad \tilde{S}_{k+j}^{\varphi} \leftarrow \text{FITRBF}(\hat{\mathbf{X}}, \tilde{\mathbf{G}}_{(j,\cdot)}, \varphi)$ ▷ Fit RBF with PLOG
22 $\quad\quad\quad$ **end**
23 $\quad\quad$ **end**
24 $\quad\quad S^* \leftarrow \left\{ S_i^{(RBF_i^*)} \mid \forall i = 1, \ldots, (k+m) \right\}$ ▷ Apply best configuration
25 $\quad\quad \mathbf{PF} \leftarrow \text{PARETO}(\mathbf{X}, \mathbf{F}, \mathbf{G})$ ▷ PF indicator see Definition 1, $\mathbf{PF} \in \{0, 1\}^N$
26 $\quad\quad \mathbf{x}^* \leftarrow \text{MAXIMIZE}(\text{HV}, \mathbf{PF}, \mathbf{ref}, S^*)$ ▷ Find best new solution
27 $\quad\quad \mathbf{x}_{new} \leftarrow \text{SCALE}(\mathbf{x}^*, [\mathbf{lb}, \mathbf{ub}])$ ▷ Scale to original scale
28 $\quad\quad N \leftarrow N + 1$ ▷ Increase iteration counter to new matrix sizes
29 $\quad\quad \mathbf{X} \leftarrow [\mathbf{X}\ \mathbf{x}_{new}]$ ▷ Add new solution, $\mathbf{X} \in \mathbb{R}^{d \times N}$
30 $\quad\quad \mathbf{F} \leftarrow [\mathbf{F}\ f(\mathbf{x}_{new})]$ ▷ Add evaluated objectives, $\mathbf{F} \in \mathbb{R}^{k \times N}$
31 $\quad\quad \mathbf{G} \leftarrow [\mathbf{G}\ g(\mathbf{x}_{new})]$ ▷ Add evaluated constraints, $\mathbf{G} \in \mathbb{R}^{m \times N}$
$\quad\quad$ ▷ Get best RBF configuration and Squared Errors, Section 3.6
32 $\quad\quad RBF^*, \mathbf{SE} \leftarrow \text{SELECTBESTRBF}(\mathbf{SE}, \hat{S}, \tilde{S}, \mathbf{x}^*, \mathbf{F}, \mathbf{G}, \mathbf{PF}, N)$
33 $\quad$ **end**
34 **return** $(\mathbf{F}_{(\cdot,\mathbf{PF})}, \mathbf{G}_{(\cdot,\mathbf{PF})}, \mathbf{X}_{(\cdot,\mathbf{PF})})$

as the centroid of the function, and the weighted linear combination of RBFs always produces a perfect fit through the training points. Besides the perfect fit on the training points, the linear combination of the RBFs can also give a reasonable approximation of the unknown area.

Any function which is only dependent on the distance from a specific point to another point belongs to the group of RBFs. The RBF kernels (φ) considered in this work are the *cubic* with $\varphi(r) = r^3$, *Gaussian* with $\varphi(r) = \exp(-(\epsilon \cdot r)^2)$, *multiquadric* with $\varphi(r) = \sqrt{1 + (\epsilon \cdot r)^2}$, *inverse quadratic* with $\varphi(r) = (1 + (\epsilon \cdot r)^2)^{-1}$, *inverse multiquadric* with $\varphi(r) = (\sqrt{1 + (\epsilon \cdot r)^2})^{-1}$, and *thin plate spline* with $\varphi(r) = r^2 \log r$. Note that we keep the shape/width parameter ϵ for every individual RBF constant such as proposed by Urquhart et al. [27]. Moreover, all shape parameters are fixed to $\epsilon = 1$.

Finding suitable linear weighted combinations $\boldsymbol{\theta}$ of the RBFs can be done by inverting $\boldsymbol{\Phi} \in \mathbb{R}^{n \times n}$ where $\boldsymbol{\Phi}_{i,j} = \varphi(\|\mathbf{x}_i - \mathbf{x}_j\|)$:

$$\boldsymbol{\theta} = \boldsymbol{\Phi}^{-1} \cdot \mathbf{f} \tag{1}$$

Here $\mathbf{f}$ is a vector of length n with the function values belonging to one of the objectives or constraints. Because $\boldsymbol{\Phi}$ is not always invertible, Micchelli introduced RBFs with a polynomial tail, better known as augmented RBFs [18]. In this work we use augmented RBFs with a second order polynomial tail. The polynomial tail is created by extending the original matrix $\boldsymbol{\Phi}$ with $\mathbf{P} = (1, x_{i1}, \ldots, x_{id}, x_{i,1}^2, \ldots, x_{id}^2)$, in its ith row, where x_{ij} is the jth component of vector $\mathbf{x}_i$, for $i = 1, \ldots, n$ and $j = 1, \ldots, d$, $\mathbf{P}^\top$, and zeros $\mathbf{0}_{(2d+1)\times(2d+1)}$, leading to $1 + 2d$ more weights $\boldsymbol{\mu}$ to learn.

$$\begin{bmatrix} \boldsymbol{\Phi} & \mathbf{P} \\ \mathbf{P}^\top & \mathbf{0}_{(2d+1)\times(2d+1)} \end{bmatrix} \begin{bmatrix} \boldsymbol{\theta} \\ \boldsymbol{\mu} \end{bmatrix} = \begin{bmatrix} \mathbf{f} \\ \mathbf{0}_{2d+1} \end{bmatrix} \tag{2}$$

Now that we can compute the weights $\boldsymbol{\theta}$ and $\boldsymbol{\mu}$ with Eq. 1 (Lines 16, 17, 20, 21 of Algorithm 1), we can for each unseen input $\mathbf{x}'$ interpolate/predict the function value (f') by using Eq. (3).

$$f' = \boldsymbol{\Phi}' \cdot \begin{bmatrix} \boldsymbol{\theta} \\ \boldsymbol{\mu} \end{bmatrix} = \sum_{i=1}^{n} \theta_i \varphi(\|\mathbf{x}' - \mathbf{x}_i\|) + \mu_0 + \sum_{l=1}^{d} \mu_l \mathbf{x}_l' + \sum_{l=1}^{d} \mu_l \mathbf{x}_l'^2, \qquad \mathbf{x} \in \mathbb{R}^d \tag{3}$$

3.3 Scaling

In SAMO-COBRA, various scaling and transformation functions are used in lines 9–13 of the algorithm. This is done to improve the predictive accuracy of the RBF surrogate models. The four functions SCALE, PLOG, STANDARDIZE and the SCALE CONSTRAINT are described below.

Scale: The input space/decision variables are scaled into the range $[-1, 1]$ with $x = 2 \cdot (x - x_{lb})/(x_{ub} - x_{lb}) - 1$. By scaling large values in the input space, we can prevent computationally singular (ill-conditioned) coefficient matrices in

Eq. (1). In case we keep large values in the input space, the linear equation solver will terminate with an error, or it will result in a large root mean square error [3]. Additionally, when fitting the RBFs, a small change in one of the variables, is relatively the same small change in all the other variables, making each variable in the basis equally important and equally sensitive.

Standardize: The relationship between the input space and the objective function values is modelled with RBF surrogates. Besides this relationship, Bagheri et al. also exploited similarities between RBF and Kriging surrogates to come up with an uncertainty quantification method [2]. The formula for this uncertainty quantification method is given in Eq. (4).

$$\widehat{U}_{RBF} = \varphi(\|\mathbf{x}' - \mathbf{x}'\|) - \mathbf{\Phi}'^{\top}\mathbf{\Phi}^{-1}\mathbf{\Phi}' \quad (4)$$

The uncertainty ($\widehat{U}_{RBF}$) of solutions far away from earlier evaluated solutions is higher compared to solutions close to earlier evaluated solutions. This uncertainty quantification method can therefore help in exploration and prevent the algorithm from getting stuck in a local optimal solution. However, as can be derived from Eq. (4), the uncertainty quantification method is only dependent on the input space and not on the scale of the objective and/or weights of the RBF models. We therefore standardize the objective values as $y' = (y - \bar{y})/\sigma$ so that the uncertainty scale and the objective scale match. Here σ is the standard deviation of y, and $\bar{y}$ the mean of y.

Scale Constraint: The constraint evaluation function should return a continuous value, namely the amount by which the constraint is violated. Since it is possible to have multiple constraints, and each constraint is equally important, we chose to scale every constraint output with as $c' = c/(\max(c) - \min(c))$, where $\max(c)$ is the maximum constraint violation encountered so far, and $\min(c)$ is the smallest constraint value seen so far. After scaling, the difference between $\min(c)$ and $\max(c)$ becomes 1, for all constraints, making every constraint equally important while 0 remains the feasibility boundary.

Plog: In cases where there are very steep slopes, a logarithmic transformation of the objective and/or constraint scores can be beneficial for the predictive accuracy [23]. Therefore, we transform the scores with the PLOG transformation function. The extension to a matrix argument $\mathbf{Y}$ is defined componentwise, i.e., each matrix element y_{ij} is subject to PLOG.

$$\textsc{Plog}(y) = \begin{cases} +\ln(1+y), & \text{if} \quad y \geq 0 \\ -\ln(1-y), & \text{if} \quad y < 0 \end{cases} \quad (5)$$

3.4 Maximize Hypervolume Contribution

After modelling the relationship between the input space and the response variables with the RBFs, the RBFs are used as cheap surrogates. For each constraint and objective the best RBF configuration is chosen as described in Sect. 3.6. By using Eq. (3) for each unseen input $\mathbf{x}'$ every corresponding constraint and objective prediction can be calculated. The COBYLA (Constrained Optimization BY

Linear Approximations) algorithm [22] is then used to search for a new feasible Pareto-optimal solution $\mathbf{x}' \in [\mathbf{lb},\mathbf{ub}]$. This is achieved by maximizing the hypervolume (HV) a solution $\mathbf{x}'$ adds to the HV between a predefined reference point and the already evaluated solutions on the PF (Line 26 of Algorithm 1). The HV contribution of a solution to the PF is computed with two infill criteria:

1. Compute all objective scores for a given solution $\mathbf{x}'$ with Eq. (3). Then with the interpolated objective scores compute the additional Predicted HV (PHV) score this solution adds to the PF.
2. Compute all objective scores for a given solution $\mathbf{x}'$ with Eq. (3) and subtract the uncertainty of each objective given $\mathbf{x}'$ and Eq. (4), which is similar to the Kriging $\mathcal{S}$-metric Selection (SMS) criterion from Emmerich et al. [4].

Besides the highest possible SMS or PHV score, the potential solution should not violate any of the constraints. This can easily be checked by using the RBF surrogates of the constraints and Eq. (3). COBYLA then searches for a solution which does not violate any of the constraints and has the highest SMS or PHV score. If no feasible solution can be found, the solution with the smallest constraint violation is evaluated. The best candidate is then evaluated on the real objective and constraint functions (Lines 27–31 of Algorithm 1).

3.5 Surrogate Exploration and RBF Adaptation Rules

The surrogate search budget and the number of starting points are updated and optimized every iteration. This is done to limit the time spend on exploring the surrogates and to further increase the chance of finding a solution that adds the most HV to the PF. The problem characteristics (number of variables, constraints, and objectives) influences the optimization problem complexity. Therefore, in the first iteration of SAMO-COBRA, the surrogate evaluation budget and number of starting points are empirically chosen and set at $50 \cdot (d + m + k)$ and $2 \cdot (d + m + k)$ respectively. In every iteration of SAMO-COBRA the convergence of COBYLA is checked. If COBYLA converges every time to a feasible solution, the number of randomly generated points is increased by 10% and the surrogate search budget is decreased by 10%. The opposite update is done if COBYLA did not converge from one of the starting points.

Because in the first iterations, the RBFs do not model the constraints very well yet, an allowed error (ϵ) of 1% for each constraint is built in. If the solution, evaluated on the real constraint function, is feasible, the error margin of this constraint approximation is reduced by 10%. If a solution is infeasible the RBFs surrogate approximation is clearly still wrong, so the error margin of the corresponding constraint is increased by 10%.

3.6 Select Best RBF

In every iteration, the best RBF kernel and transformation strategy is chosen (Line 32 of Algorithm 1). This is done by computing the difference between the

RBF interpolated solution and the solution computed with the real constraint and objective functions. This difference is computed every iteration, resulting in a list of historical RBF approximation errors for each constraint and objective function, for each kernel, with and without the PLOG transformation.

Based on the RBF approximation errors, the best RBF kernel and transformation are chosen. Bagheri et al. show empirically that if only the last approximation error is considered, in the single objective case, the algorithm converged to the best solution faster [1]. This is the case because when closer to the optimum, the vicinity of the last solution is the most important. In the multi-objective case, the vicinities of all the feasible Pareto-optimal solutions are important. We therefore experimented with this and confirmed that the approximation errors of the feasible Pareto-optimal solutions and the last four solutions should be considered. The approximation error of the last four solutions make sure that the algorithm does not get stuck on one RBF configuration and the error of the Pareto efficient solutions make sure that all the vicinity of the optimal solutions are considered. The Mean Squared Error measure is used to quantify which RBF kernel and which transformation function in the previous iterations resulted in the smallest approximation error.

4 Experimental Setup

The SMS and PHV infill criteria in combination with the SAMO-COBRA algorithm are tested on 18 constrained multi-objective test functions. The two variants of the algorithm are compared to CEGO [29], NSGA-II [10], and NSGA-III [15]. Each algorithm is allowed to do $40 \cdot d$ function evaluations after which the HV between the PF and a fixed reference point is computed. The 18 test functions with their sources, their dimensions, feasibility rate, and the average algorithm iteration times are presented in Table 1. The iteration times of NSGA-II and NSGA-III are so low (∼0.001 s) that they are not reported. The experiments are executed on a laptop with i7 processor, 4 cores and a clock speed of 2.5 GHz.

In the experiment we ran the algorithms 10, 10, 10, 100, 100 times with different seeds for the SAMO-COBRA (PHV), SAMO-COBRA (SMS), CEGO, NSGA-II, and NSGA-III algorithm respectively. This number of runs is chosen because the iteration time of SAMO-COBRA and CEGO are higher, and the HV after $40 \cdot d$ function evaluations vary a lot for NSGA-II and NSGA-III.

The implementation of the SMES-RBF algorithm is not provided. We therefore show in one additional experiment on GitHub [28], that SAMO-COBRA requires fewer function evaluations to achieve the same HV as SMES-RBF [8].

4.1 Algorithm Parameter Settings

As already mentioned in Sect. 3.1, a small initial Design of Experiments (DoE) leads to higher final HV. The SAMO-COBRA algorithm is therefore set up with

Table 1. Test functions with citation, the reference points, the number of objectives k, number of parameters d, number of constraints m, feasibility percentage $P(\%)$ based on 1 million random samples, mean running time per iteration in seconds.

Function	Reference point	k	d	m	P(%)	SAMO-COBRA	CEGO
BNH [7]	(140, 50)	2	2	2	96.92	3.5	10.5
CEXP [9]	(1, 9)	2	2	2	57.14	2.2	10.1
SRN [11]	(301, 72)	2	2	2	16.18	4.9	20.6
TNK [11]	(2, 2)	2	2	2	5.05	0.8	8.2
CTP1 [9]	(1, 2)	2	2	2	92.67	0.8	7.3
C3DTLZ4 [26]	(3, 3)	2	6	2	22.22	18.1	338
OSY [7,11]	(0, 386)	2	6	6	2.78	9.8	424
TBTD [13]	(0.1, 50000)	2	3	2	19.46	5.5	151
NBP [12]	(11150, 12500)	2	2	5	41.34	2.8	9.7
DBD [13]	(5, 50)	2	4	5	28.55	34.9	159
SPD [20]	(16, 19000, −260000)	3	6	9	3.27	34.6	405
CSI [15]	(42, 4.5, 13)	3	7	10	18.17	45.3	1h+
SRD [19]	(7000, 1700)	2	7	10	96.92	63.2	172
WB [13]	(350, 0.1)	2	4	5	35.28	19.1	258
BICOP1 [8]	(9, 9)	2	10	1	100	24.2	1h+
BICOP2 [8]	(70, 70)	2	10	2	10.55	12.1	1h+
TRIPCOP [8]	(34, −4, 90)	3	2	3	15.85	3.8	9.6
WP [15]	(83000, 1350, 2.85, 15989825, 25000)	5	3	7	92.06	16.1	225

$d+1$ initial Halton samples. The Kriging surrogates of CEGO require a minimum DoE of size $2{\cdot}d{+}1$. CEGO therefore starts after a DoE of $2{\cdot}d{+}1$ Latin Hypercube Samples. Other hyper-parameters of CEGO are set as originally described [29].

The implementation of Platypus [14] is used for NSGA-II and NSGA-III. The algorithms have several sensitive parameters which influence the final results. We therefore did a grid search for different population sizes and generations for NSGA-II, and a grid search for the number of division that controls the spacing of the reference points for NSGA-III. For the sake of brevity, only the results which gave the highest HV after $40 \cdot d$ function evaluations are reported.

For more detailed parameter settings, the raw results, and the statistical comparisons, please refer to the SAMO-COBRA GitHub page [28].

5 Results

Table 2 shows the mean HV after $40 \cdot d$ function evaluations. The Kruskal-Wallis test confirmed that there is at least one result statistically different from the results from the other algorithms, a post-hoc comparison is done with a

Table 2. Mean HV after $40 \cdot d$ function evaluations for each algorithm on each test function. PHV and SMS represents the SAMO-COBRA variants. The highest mean HV per test function are presented in **bold**. The Wilcoxon rank-sum test (with Bonferroni correction) significance is represented with a grayscale. Background colors represent: $p \leq 0.001$, $p \leq 0.01$, $p \leq 0.05$. Red shows that the algorithm took more than 24 h.

Function	PHV	SMS	CEGO	NSGA-II	NSGA-III
BNH	**5256.0**	5251.0	5218.0	5089.7	4848.8
CEXP	**3.7968**	3.7967	3.7658	3.1544	2.9561
SRN	**62385**	62377	62307	54233	52585
TNK	8.0430	**8.0474**	8.0309	6.4282	6.2948
CTP1	**1.3026**	1.3023	1.2972	1.1929	1.1864
C3DTLZ4	6.3016	**6.4697**	6.2664	5.2489	5.2500
OSY	**100577**	100458	100181	46631	42204
TBTD	4029.3	4027.5	**4055.0**	3421.5	3446.0
NBP	$\mathbf{1.0785 \cdot 10^8}$	$1.0785 \cdot 10^8$	$1.0784 \cdot 10^8$	$9.8573 \cdot 10^7$	$9.5816 \cdot 10^7$
DBD	**228.77**	228.34	227.85	218.47	218.33
SPD	$\mathbf{3.8859 \cdot 10^{10}}$	$3.7602 \cdot 10^{10}$	$3.4057 \cdot 10^{10}$	$2.6069 \cdot 10^{10}$	$2.5689 \cdot 10^{10}$
CSI	**27.800**	25.167	*terminated*	17.296	16.993
SRD	**4205165**	4164992	4148333	2766367	2673599
WB	34.395	34.392	**34.522**	34.0270	33.995
BICOP1	**80.664**	78.160	*terminated*	67.6506	70.911
BICOP2	**4834.33**	4822.13	*terminated*	4816.12	4816.72
TRIPCOP	20610	**20611**	20609	20101	19982
WP	$\mathbf{1.5991 \cdot 10^{19}}$	$1.5698 \cdot 10^{19}$	$1.5350 \cdot 10^{19}$	$1.1623 \cdot 10^{19}$	$1.1797 \cdot 10^{19}$

Wilcoxon rank-sum test (with Bonferroni correction) to find out if there is a significant difference between the best and the lesser results.

Note that the exploiting strategy of the PHV infill criteria leads in most cases to the highest hv after $40 \cdot d$ function evaluations. It is no surprise that this exploiting strategy works well in a constrained multi-objective setting since a similar effect was already shown by Rehbach et al. [24]. He showed that in the single objective case, it is only useful to include an expected improvement infill criterion, if the dimensionality of the problem is low, if it is multimodal, and if the algorithm can get stuck in a local optimum. With the results as presented in Table 2 we can give an advice based on empirical results, as follows: When searching for a set of Pareto-optimal solutions, an uncertainty quantification method should not be used. This is due to the fact that, when searching for a trade-off between objectives, the algorithm is forced to explore more of the objective space in the different objectives, making it less likely to get stuck in a local optimum, and making the uncertainty quantification method redundant.

6 Conclusion and Future Work

In this paper the authors propose SAMO-COBRA, a fast optimization algorithm capable of handling multiple objectives, multiple constraints, and continuous decision variables. Two variants of the algorithm are evaluated and compared to CEGO, NSGA-II, NSGA-III, and SMES-RBF. The experiments show that SAMO-COBRA outperforms the other algorithms in terms of achieved HV after a fixed number of function evaluations. These characteristics make SAMO-COBRA a good choice when optimizing a real-world optimization problem with multiple objectives, multiple constraints, continuous decision variables, and expensive (in terms of time and/or money) function evaluations.

Besides these results, the authors also give the recommendation that an uncertainty quantification method in multi-objective optimization is often not required. This is the case, because the algorithm already automatically explores more of the objective space compared to when optimizing single objective problems, which makes the algorithm less likely to get stuck in a local optimum.

Further research efforts will be put into parallelizing the function evaluations and parallelizing the surrogate exploration so that the wall clock time can be further decreased. Besides parallelization, effort will be put into dealing with mixed integer decision parameters and multi fidelity simulation software.

References

1. Bagheri, S., Konen, W., Bäck, T.: Online selection of surrogate models for constrained black-box optimization. In: 2016 IEEE Symposium Series on Computational Intelligence (SSCI), pp. 1–8. IEEE (2016)
2. Bagheri, S., Konen, W., Bäck, T.: Comparing kriging and radial basis function surrogates. In: Proceedings 27 Workshop Computational Intelligence, pp. 243–259 (2017)
3. Bagheri, S., Konen, W., Emmerich, M., Bäck, T.: Self-adjusting parameter control for surrogate-assisted constrained optimization under limited budgets. Appl. Soft Comput. **61**, 377–393 (2017). https://doi.org/10.1016/j.asoc.2017.07.060
4. Beume, N., Naujoks, B., Emmerich, M.: SMS-EMOA: multiobjective selection based on dominated hypervolume. Eur. J. Oper. Res. **181**(3), 1653–1669 (2007). https://doi.org/10.1016/j.ejor.2006.08.008
5. van der Blom, K., et al.: Towards realistic optimization benchmarks: a questionnaire on the properties of real-world problems. In: Proceedings of the 2020 Genetic and Evolutionary Computation Conference Companion. GECCO 2020, New York, NY, USA, pp. 293–294. Association for Computing Machinery (2020)
6. Bossek, J., Doerr, C., Kerschke, P.: Initial design strategies and their effects on sequential model-based optimization. arXiv preprint arXiv:2003.13826 (2020)
7. Coello, C.A.C., Lamont, G.B., Van Veldhuizen, D.A., et al.: Evolutionary Algorithms for Solving Multi-objective Problems. Genetic and Evolutionary Computation Series. GEVO, vol. 5. Springer, Boston (2007). https://doi.org/10.1007/978-0-387-36797-2
8. Datta, R., Regis, R.G.: A surrogate-assisted evolution strategy for constrained multi-objective optimization. Expert Syst. Appl. **57**, 270–284 (2016). https://doi.org/10.1016/j.eswa.2016.03.044

9. Deb, K.: Multi-objective Optimization Using Evolutionary Algorithms, vol. 16. John Wiley & Sons, New York (2001)
10. Deb, K., Pratap, A., Agarwal, S., Meyarivan, T.: A fast and elitist multiobjective genetic algorithm: NSGA-II. IEEE Trans. Evol. Comput. **6**(2), 182–197 (2002). https://doi.org/10.1109/4235.996017
11. Deb, K., Pratap, A., Meyarivan, T.: Constrained test problems for multi-objective evolutionary optimization. In: Zitzler, E., Thiele, L., Deb, K., Coello Coello, C.A., Corne, D. (eds.) EMO 2001. LNCS, vol. 1993, pp. 284–298. Springer, Heidelberg (2001). https://doi.org/10.1007/3-540-44719-9_20
12. Forrester, A., Sobester, A., Keane, A.: Engineering Design via Surrogate Modelling: A Practical Guide. John Wiley & Sons, New York (2008). https://doi.org/10.2514/4.479557
13. Gong, W., Cai, Z., Zhu, L.: An efficient multiobjective differential evolution algorithm for engineering design. Struct. Multi. Optim. **38**(2), 137–157 (2009). https://doi.org/10.1007/s00158-008-0269-9
14. Hadka, D.B.: Platypus: multiobjective optimization in python (2020). https://platypus.readthedocs.io/
15. Jain, H., Deb, K.: An evolutionary many-objective optimization algorithm using reference-point based non dominated sorting approach, part II: handling constraints and extending to an adaptive approach. IEEE Trans. Evol. Comput. **18**(4), 602–622 (2014). https://doi.org/10.1109/tevc.2013.2281534
16. Knowles, J.: ParEGO: a hybrid algorithm with on-line landscape approximation for expensive multiobjective optimization problems. IEEE Trans. Evol. Comput. **10**(1), 50–66 (2006). https://doi.org/10.1109/tevc.2005.851274
17. Liu, H., Ong, Y.-S., Cai, J.: A survey of adaptive sampling for global metamodeling in support of simulation-based complex engineering design. Struct. Multi. Optim. **57**(1), 393–416 (2017). https://doi.org/10.1007/s00158-017-1739-8
18. Micchelli, C.A.: Interpolation of scattered data: distance matrices and conditionally positive definite functions. Constr. Approximation **2**(1), 11–22 (1986)
19. Mirjalili, S., Jangir, P., Saremi, S.: Multi-objective ant lion optimizer: a multi-objective optimization algorithm for solving engineering problems. Appl. Intell. **46**(1), 79–95 (2016). https://doi.org/10.1007/s10489-016-0825-8
20. Parsons, M.G., Scott, R.L.: Formulation of multicriterion design optimization problems for solution with scalar numerical optimization methods. J. Ship Res. **48**(1), 61–76 (2004). https://doi.org/10.1007/s10489-016-0825-8
21. Ponweiser, W., Wagner, T., Biermann, D., Vincze, M.: Multiobjective optimization on a limited budget of evaluations using model-assisted $\mathcal{S}$-metric selection. In: Rudolph, G., Jansen, T., Beume, N., Lucas, S., Poloni, C. (eds.) PPSN 2008. LNCS, vol. 5199, pp. 784–794. Springer, Heidelberg (2008). https://doi.org/10.1007/978-3-540-87700-4_78
22. Powell, M.J.D.: A direct search optimization method that models the objective and constraint functions by linear interpolation. In: Gomez, S., Hennart, J.P. (eds.) Advances in Optimization and Numerical Analysis. MAIA, vol. 275, pp. 51–67. Springer, Netherlands (1994). https://doi.org/10.1007/978-94-015-8330-5_4
23. Regis, R.G., Shoemaker, C.A.: A quasi-multistart framework for global optimization of expensive functions using response surface models. J. Global Optim. **56**(4), 1719–1753 (2013). https://doi.org/10.1007/s10898-012-9940-1
24. Rehbach, F., Zaefferer, M., Naujoks, B., Bartz-Beielstein, T.: Expected improvement versus predicted value in surrogate-based optimization. arXiv preprint arXiv:2001.02957 (2020). https://doi.org/10.1145/3377930.3389816

25. Singh, P., Couckuyt, I., Ferranti, F., Dhaene, T.: A constrained multi-objective surrogate-based optimization algorithm. In: 2014 IEEE Congress on Evolutionary Computation (CEC). IEEE (2014). https://doi.org/10.1109/cec.2014.6900581
26. Tanabe, R., Oyama, A.: A note on constrained multi-objective optimization benchmark problems. In: 2017 IEEE Congress on Evolutionary Computation (CEC), pp. 1127–1134. IEEE (2017). https://doi.org/10.1109/cec.2017.7969433
27. Urquhart, M., Ljungskog, E., Sebben, S.: Surrogate-based optimisation using adaptively scaled radial basis functions. Appl. Soft Comput. **88**, 106050 (2020)
28. de Winter, R.: SAMO-COBRA: self-adaptive algorithm for multi-objective constrained optimization by using radial basis function approximations (2020). https://doi.org/10.5281/zenodo.4281140
29. de Winter, R., van Stein, B., Dijkman, M., Bäck, T.: Designing ships using constrained multi-objective efficient global optimization. In: Nicosia, G., Pardalos, P. (eds.) Machine Learning, Optimization, and Data Science. LNCS, vol. 11331. Springer, Cham (2018). https://doi.org/10.1007/978-3-030-13709-0_16
30. Zitzler, E., Laumanns, M., Thiele, L.: SPEA2: improving the strength pareto evolutionary algorithm. TIK-report, vol. 103 (2001)

A Fast Converging Evolutionary Algorithm for Constrained Multiobjective Portfolio Optimization

Yi Chen[1(✉)], Hemant Kumar Singh[2], Aimin Zhou[1], and Tapabrata Ray[2]

[1] East China Normal University, Shanghai 200062, China
52164500006@stu.ecnu.edu.cn, amzhou@cs.ecnu.edu.cn
[2] The University of New South Wales, Canberra, ACT 2600, Australia
{h.singh,t.ray}@adfa.edu.au

Abstract. Portfolio optimization is a well-known problem in the domain of finance with reports dating as far back as 1952. It aims to find a trade-off between risk and expected return for the investors, who want to invest finite capital in a set of available assets. Furthermore, *constrained* portfolio optimization problems are of particular interest in real-world scenarios where practical aspects such as cardinality (among others) are considered. Both mathematical programming and meta-heuristic approaches have been employed for handling this problem. Evolutionary Algorithms (EAs) are often preferred for constrained portfolio optimization problems involving non-convex models. In this paper, we propose an EA with a tailored variable representation and initialization scheme to solve the problem. The proposed approach uses a short variable vector, regardless of the size of the assets available to choose from, making it more scalable. The solutions generated do not need to be repaired and satisfy some of the constraints implicitly rather than requiring a dedicated technique. Empirical experiments on 20 instances with the numbers of assets, ranging from 31 to 2235, indicate that the proposed components can significantly expedite the convergence of the algorithm towards the Pareto front.

Keywords: Representation for evolutionary algorithm · Multi-objective portfolio optimization · Constrained portfolio optimization

1 Introduction and Background

Portfolio optimization is a prominent application problem in the field of finance. A number of studies have been reported on this subject since 1952, as reviewed in [9]. The core task is to find the optimal allocation of limited capital among certain given assets so that two conflicting goals, minimizing risk and maximizing returns, can be achieved. Mean-Variance (MV) model is an original portfolio optimization model, which estimates the expected return by mean of the profit

H. Ishibuchi et al. (Eds.): EMO 2021, LNCS 12654, pp. 283–295, 2021.
https://doi.org/10.1007/978-3-030-72062-9_23

and measures the risk by variance of the profit [11]. This model is a quadratic problem since it assumes that the risk of an investment should be assessed by a covariance matrix based on the variance and the correlation among the assets. As a general quadratic problem, it is not difficult to solve using contemporary optimization techniques [14]. However, practical portfolio problems involve a range of constraints, such as cardinality, minimum lot size, pre-selected assets, etc. Incorporating these constraints result in mixed integer/combinatorial models. There are also some exact algorithms, which have been designed and utilized to tackle the constrained portfolio optimization problems [1]. Nonetheless, there is no polynomial-time algorithm for the case when a cardinality constraint is involved [13].

Therefore, meta-heuristic algorithms such as Evolutionary Algorithms (EAs) have gained attention to solve complex instances of portfolio optimization problems. Moreover, population-based metaheuristics are also inherently suited for solving multiobjective optimization problems, since the optimum of such problems consists of a trade-off solution set (Pareto optimal front). They have been used to solve different constrained portfolio optimization problems based on the mean-variance (MV) model, with different numbers of objectives and various different types of constraints. For single-objective problems, Chang et al. [2] used three different meta-heuristics, namely genetic algorithms, tabu search, and simulated annealing, to handle the problem with a fixed amount of invested assets. Gaspero et al. [8] proposed a hybrid algorithm with a local search and a quadratic programming solver. The numerical experiments on a portfolio optimization problem with three constraints (cardinality, quantity, and pre-assignment) showed that it performed better than some of the widely used softwares such as CPLEX and MOSEK. As for the constrained multi-objective problems, Pouya et al. [15] transformed the problem into a single-objective form with a fuzzy normalization. Subsequently, they used a metaheurstic method to solve the transformed problem and obtained improved results over Particle Swarm Optimization [15]. Deb et al. [6] solved the problem using a customized hybrid algorithm with specialized evolutionary operators, repair, local search and clustering. Lwin et al. [10] proposed a learning-guided multi-objective evolutionary algorithm (MODEwAwL), utilizing the information about selection of assets from previous generation to guide the search. It was based on a multiobjective Differential Evolution (DE) algorithm and the study compared results obtained with four well known Multi-Objective Evolutionary Algorithms (MOEAs). The simulation results on a MV model based problem, including four constraints such as cardinality, quantity, pre-assignment, and round lot constraints, indicated that the proposed algorithm improved the performance over the others. Yi et al. [3,4] introduced a tailored representation, Compressed Coding Scheme (CCS) for the same problem as that in [10]. This representation used one gene fragment to represent both the selection and allocation of one asset. The comparison between CCS and the direct representation illustrated that this tailored representation was superior for this problem with regard to the obtained efficient front since it could reduce the interaction between the selection and allocation elements of the problem.

In the methods available to solve constrained multiobjective optimization problems, including those discussed above, the representation of a solution vector is typically done using a binary vector that indicates whether a particular asset is selected or not. Consequently, the length of the solution vector increases significantly as the number of assets increases; resulting in a corresponding (exponential) increase in the solution search space. Therefore, in this paper, we propose and show initial results on a new representation scheme that remains of a fixed size (equal to the cardinality of the portfolio) regardless of the number of assets available to choose from. Moreover, we also propose a customized initialization which guarantees one of the global extremities of the true Pareto front (PF). An evolutionary algorithm constructed with the proposed components shows a commendable speed-up in convergence, delivering a good PF approximation in relatively small number (≈20%) of evaluations compared to the peer algorithms. The mathematical model for the problem is detailed next in Sect. 2, followed by the proposed method in Sect. 3. Numerical experiments and conclusions are presented in Sects. 4 and 5, respectively.

2 Mathematical Model

The constrained portfolio optimization problem studied here has two conflicting objectives (minimum risk, maximum return) and is subject to four constraints: cardinality, quantity, pre-assignment, and round lot constraints [10]. The following notations are used in the formulation:

N	the number of available assets
K	the number of assets in a portfolio, i.e., the cardinality
L	the number of assets in the pre-assignment set
w_i	the proportion of capital invested in the ith asset
ρ_{ij}	the correlation coefficient of the returns of ith and j^{th} assets
σ_i	the standard deviation of ith asset
σ_{ij}	the covariance of ith and j^{th} assets
μ_i	the expected return of the ith asset
v_i	the minimum trading lot of the ith asset
ϵ_i	the lower limit on the investment of the ith asset
δ_i	the upper limit on the investment of the ith asset
y_i	the multiple of the minimum trading lot in the ith asset

$$\sigma_{ij} = \rho_{ij}\sigma_i\sigma_j$$

$$s_i = \begin{cases} 1, & \textbf{if the } i^{th}\ (i = 1, \ldots, N) \text{ asset is chosen} \\ 0, & \text{otherwise} \end{cases}$$

$$z_i = \begin{cases} 1, & \textbf{if the } i^{th} \text{ asset is in the pre-assigned set} \\ 0, & \text{otherwise} \end{cases}$$

With above notations, the problem can be formulated as:

$$\text{Minimize} \quad f_1 = \sum_{i=1}^{N}\sum_{j=1}^{N} w_i w_j \sigma_{ij}, \tag{1}$$

$$\text{Maximize} \quad f_2 = \sum_{i=1}^{N} w_i \mu_i, \tag{2}$$

$$\text{subject to} \quad \sum_{i=1}^{N} w_i = 1, \qquad 0 \leq w_i \leq 1, \tag{3}$$

$$\sum_{i=1}^{N} s_i = K, \tag{4}$$

$$\epsilon_i s_i \leq w_i \leq \delta_i s_i, \qquad i = 1, ..., N, \tag{5}$$

$$s_i \geq z_i, \qquad i = 1, ..., N, \tag{6}$$

$$w_i = y_i v_i, \qquad i = 1, ..., N, \quad y_i \in \mathbb{Z}_+, \tag{7}$$

$$s_i, z_i \in \{0, 1\}, \qquad i = 1, ..., N, \tag{8}$$

The Eqs. (1) and (2) represent the two conflicting objectives, i.e., risk minimization and return maximization. Equation (3) ensures that all available capital should be invested. Equation (4) is the *cardinality constraint* enforcing that exactly K assets are selected. Equation (5) is the *floor and ceiling constraint*, it defines that investment in the ith asset should lie between ϵ_i to δ_i. Further, Eq. (6) designates the *pre-assignment constraint* that certain asset(s) *must be* selected ($z_i = 1$) in the portfolio, eg, based on investors' preferences. Equation (7) defines the *round lot constraint*, specifying a minimum number of lots to be traded for a selected asset. Finally, Eq. (8), *discrete constraint*, limits both s_i and z_i to be binary. In this study, these constraints are set in consistency with the past study [3] as follows:

- Cardinality $K = 10$, floor $\epsilon_i = 0.01$, ceiling $\delta_i = 1.0$, pre-assignment $z_{30} = 1$ and round lot $v_i = 0.008$.

The goal in solving the above problem is to seek a set of portfolios that lie close to/on the Pareto front, with a good diversity to cater to the investors with different risk-return appetite.

3 Proposed Method

The proposed approach follows a canonical evolutionary framework as shown in Algorithm 1. The algorithm uses customized representation and operators (including initialization, offspring generation) to suit the constrained multi-objective portfolio optimization problem considered. The intent behind the customization is to improve the convergence rate, and they assume no more prior knowledge about the problem than already given in the existing works (e.g. [3]). The key components of the algorithm are discussed in the following subsections.

Algorithm 1. Proposed method for multiobjective portfolio optimization

Input: Problem data (See Sect. 2), Population size (N_p), Total evaluations (FE_{max})
Output: PF approximation
1: Set #function evaluation $FE = 0$, archive of evaluated solutions $\mathcal{A} = \emptyset$, generation count $t = 0$
2: Initialize the population P^0 (of size N_p) with the tailored method in Sect. 3.2, and evaluate all solutions. Update $\mathcal{A}$, FE
3: **while** ($FE < FE_{max}$) **do**
4: Generate offspring solutions C^t from current population P^t in three different ways with probabilities, pr_1, pr_2 and pr_3 respectively.
5: Evaluate offspring solutions C^t Update $\mathcal{A}$, FE
6: Rank $T^t = P^t \cup C^t$
7: Select top N_p individuals in T^t as the next generation population P^{t+1} Update $pr1$, $pr2$, $pr3$
8: Set $t = t + 1$
9: **end while**
10: Output the non-dominated solutions $\mathcal{N} \in \mathcal{A}$ as the PF approximation

3.1 Solution Representation

From Sect. 2, one can observe that the solution vector under the standard notations includes $2N$ decision variables. The first N variables $s_i, i = 1, \ldots, N$, are binary, indicating whether i^{th} is selected or not; whereas the next N variables $w_i, i = 1, \ldots, N$, are discrete, indicating the proportion of the capital invested in i^{th} asset. Given that the given number of assets (N) may often comprise a large set, this representation gives rise to a large number of variables.

In this work, we propose and investigate a compact representation, which consists of merely $2K$ variables. The first K variables $a_i, i = 1, \ldots, K, 1 \leq a_i \leq N$, are *categorical*, indicating the labels (ids) of the chosen assets. The next K variables $q_i, i = 1, \ldots, K$, are discrete (positive integers), representing the *lot size* acquired of these K assets.

Note that the above representation helps in implicitly satisfying some of constraints and avoiding repair operators. Since the round lot $v_i = 0.008$, the total number of trading lots can be calculated as $p = \sum_i^N q_i = 1/0.008 = 125$. Now, since, floor $\epsilon_i = 0.01$ and ceiling $\delta_i = 1.0$, therefore, $2 \leq q_i \leq 125$. Thus, having q_i as integers in the given range automatically satisfy the round lot and floor-ceiling constraints. Moreover, the problem converts to a form akin to distributing the available capital (given number of lots ($p = 125$)) lots to a selected number (K) of assets. Such types of problems commonly occur in probability theory, and there is an opportunity to leverage some of the existing techniques to solve them (as detailed in further subsections).

3.2 Initialization

The initialization of the solutions occurs in two parts. First half is a customized initialization to obtain solutions close to the solution with the best return. Of these the one with the maximum return is *guaranteed globally* for the problem, while the others are created by perturbing it. The second half is generated randomly.

Initializing the Best Return Solution(s): In order to maximize the overall return, one would need to buy the maximum quantity of the asset with highest return, while satisfying the prescribed constraints discussed in Sect. 2. We do so in these steps:

- Sort the assets in the decreasing order of their expected return
- Identify the top K assets in the sorted list, and assign the minimum trading lot ($q_i = 2$) to each. This list is referred to as S_K
- If the pre-assigned asset (z_{30}) already exists in the above list, move to next step, otherwise replace the last asset in S_K by the pre-assigned asset.
- Thereafter, assign $q_i = p - 2 \times (K - 1)$ lots to the first asset in S_K (the one with the largest return) and assign $q_i = 2$ lots to the remaining $K - 1$ assets. This assignment satisfies all constraints, sums up to p lots traded, and assigns maximum possible lots to the highest return asset. Thus, it constitutes the solution with maximum overall return, denoted by x_{maxR}
- For generating the rest of the $N_p/2 - 1$ solutions of the first half, the quantity of the first asset in S_k is decreased by 1, while that of the second asset is increased by 1 to construct each subsequent solution.

Initializing the Remaining Solutions: The remaining $N_p/2$ solutions are initialized randomly. For generating each solution, $K - 1$ assets are selected randomly, and the pre-assigned asset z_{30} is appended to the list to complete the selection. As for their lot sizes, the total p round lots are divided among K assets using the *stars and bars* (SaB) approach in probability theory [7]. This particular instance is equivalent to dividing n indistinguishable objects into m partitions. SaB approach solves this problem using a graphical aid by representing n objects as indistinguishable symbols ('stars') in a single row, and placing $m - 1$ separators ('bars') between them to specify the number assigned to each bin. For example, for dividing 5 objects in 3 bins, 2 bars can be placed between the them in different positions. Assuming each bin needs to have at least 1 object, the bars cannot be placed at either of the ends of the row. Thus, a total of $\binom{n-1}{m-1}$ possibilities exist for such an assignment. Correspondingly, for the given portfolio problem, there are $\binom{p-2K-1}{K-1}$ distributions possible where each chosen asset has the minimum round lot $q_i = 2$ allocated. One of such combinations is chosen randomly, that satisfies the minimum round lot constraint.

3.3 Creating Offspring Solutions

The method proposed to create of the offspring solutions attempts to generate offspring solutions targeting different parts of the PF approximation, i.e., some toward low risk solutions, some towards high return solutions, and some across whole PF. The probability of generating the solution in either of these parts is controlled via a set of self-adapting parameters $\{pr_1, pr_2, pr_3\}$, such that $\sum pr_i = 1$. The values are initialized to be equal at the beginning of the run, i.e., $\{1/3, 1/3, 1/3\}$.

The offspring solutions are generated through pairwise recombination among the current population. For two parent solutions px_1 and px_2, the recombination proceeds as follows:

- Let us denote the assets in px_1 and px_2 as apx_1 and apx_2, respectively. The common assets between the two parents are identified as $ap_c = apx_1 \cap apx_2$ Correspondingly, the assets that are not common between the two parents can be identified as $ap_{u1} = apx_1 \backslash ap_c$ and $ap_{u2} = apx_2 \backslash ap_c$.
- The assets of the offspring solutions are then assigned as follows. The common assets (ap_c) are kept in both offspring solutions. The remaining $K - |ap_c|$ assets are chosen randomly from $ap_{u1} \cup ap_{u2}$, while ensuring no asset is assigned to both offspring solutions.
- The lot size for each asset for the offspring solutions are assigned as follows. The lot sizes of the common assets are directly inherited from the parent solutions. For the remaining (not common) assets, first of all, the lot sizes vector $q_i, i = 1 \ldots |K - |ap_c||$, are generated using the same approach as used in initialization (i.e., SaB approach). Then,
  - With probability pr_1, the lot sizes are assigned in decreasing order to the assets with highest to lowest return. This is to encourage generation of solutions towards the PF extremity with high return.
  - With probability pr_3, the lot sizes are assigned in decreasing order by picking the assets with lowest expected risk based on the risk covariance matrix σ. This is to encourage generation of solutions towards the extremity of the PF with lowest risk.
  - With probability pr_2, the lot sizes are assigned randomly, to encourage the solutions across the PF with no bias.

Adjustment of Probabilities $\{pr_1, pr_2, pr_3\}$: After each generation, the probabilities $\{pr_1, pr_2, pr_3\}$ are adjusted by the algorithm to allow focus on unexplored regions as per the above conditions. The adjustment is done as follows:

- The curve length C_l of the current non-dominated set (in the objective space) is calculated assuming it to be piece-wise linear among the non-dominated points.
- The number of solutions in the left (min. risk), middle, and right (max return) are identified as n_1, n_2, n_3 are counted as those within length $l \leq C_l/3$, $C_l/3 \leq l \leq 2C_l/3$ and $2C_l/3 \leq l \leq C_l$ from the one extremity (say min. risk) along the curve. These numbers are then normalized as $\{N_1, N_2, N_3\} = exp(\{n_1, n_2, n_3\}/(n_1 + n_2 + n_3))$. In the final step, the probabilities are calculated as $\{pr_1, pr_2, pr_3\} = \frac{\{1/N_1, 1/N_2, 1/N_3\}}{1/N_1 + 1/N_2 + 1/N_3}$.

The rationale behind the above adjustments is to simply map the sections of the non-dominated set with fewer solutions to high probability of exploration and vice-versa. Other mathematical forms that achieve similar mapping can also be used.

3.4 Ranking and Reduction

For ranking the population, the widely used non-dominated sorting and crowding distance are used [5]. The former targets convergence towards the PF and the latter helps promote diversity among the non-dominated solutions during the search process.

4 Empirical Study

In this section, we discuss the experimental set-up and the results of the benchmarking.

4.1 Experiment Configuration

Two sets of test problems are considered here. One is established recently based on the data from Yahoo Finance website (D1–15) with the number of available assets as high as 2235. The other one is a classic but small scale (D16–20) benchmark suite[1]. Their details are given in Table 1.

Table 1. Twenty benchmark instances considered in this study

Instance	Origin	Name	#Assets	Instance	Origin	Name	#Assets
D1	Korea	KOSPI Composite	562	*D11*	USA	NASDAQ Industrial	808
D2	USA	AMEX Composite	1893	*D12*	USA	NASDAQ Telecom	139
D3	USA	NASDAQ	2235	*D13*	USA	NYSE US100	94
D4	Australia	All ordinaries	264	*D14*	USA	NYSE World	170
D5	Italy	MIBTEL	167	*D15*	USA	S&P 500	469
D6	UK	FTSE ACT250	128	*D16*	Hong Kong	Hang Seng	31
D7	USA	NASDAQ Bank	380	*D17*	Germany	DAX100	85
D8	USA	NASDAQ Biotech	130	*D18*	UK	FTSE 100	89
D9	USA	NASDQ Computer	417	*D19*	US	S&P 500	98
D10	USA	NASDQ Financial00	91	*D20*	Japan	Nikkei	225

For benchmarking with peer algorithms, the popularly used multiobjective performance metric, Hypervolume (HV) [16], is used in a normalized objective space. HV measures the volume of dominated space by a non-dominated set Q relative to a reference point r. The reference point needs to be carefully chosen for computing HV, to ensure that the contributions of the extreme points are counted but do not overwhelmingly contribute more than those of the other points in the set. Thus it is recommended to set r slightly dominated by the nadir point of the true PF (where known). In this work, we have set $r = (1.2, 1.2)$ for this reason in the normalized space. Note that while calculating HV, the second objective (maximize return) is also converted to minimization by negating the values. The second key aspect is to select the normalization bounds.

[1] Datasets can be accessed at https://github.com/CYLOL2019/Portfolio-Instances.

For the datasets from OR-Library (D16–D20), the true PFs can be obtained with a quadratic approach using CPLEX and a mathematical multiobjective method [12]. Therefore, true ideal and nadir points of the true PF are used as normalization bounds for these cases. However, the true PFs are unavailable for some problems in NGINX dataset (D1–D15) due to large number of assets. For these cases, the best known unconstrained Pareto fronts (UCPFs) are considered with the following modification. They are truncated by the best return solution (known to be feasible and globally optimal from Sect. 3). This truncated version of the UCPF is referred to as TUCPF. The PFs, UCPFs and TUCPFs for selected problems are depicted in Fig. 1. It can be seen that PFs and TUCPFs are significantly different from the UCPFs for both instances. Many solutions of UCPFs are infeasible. For instance, on D6, the return for UCPF ranges from 0 to 4, however, the solutions with return higher than 3 are all unavailable. On the other hand, although PFs are very close to TUCPFs on D6, the differences are distinct on D17. These results demonstrate the necessity of utilizing the PFs (where available) or TUCPFs (where PFs are not available) instead of the UCPFs contrary to some past studies that have used UCPFs. Since true PFs are unknown for some problems, we utilize the bounds of TUCPFs for normalizing the obtained PF approximations in those cases.

The proposed method is compared with two recent peer algorithms, namely MODEwAwL [10] and CCS [3]. The number of objective function evaluations are limited to 5000, and each algorithm is run 31 times with a population size of 100 for gathering the statistics of performance (HV).

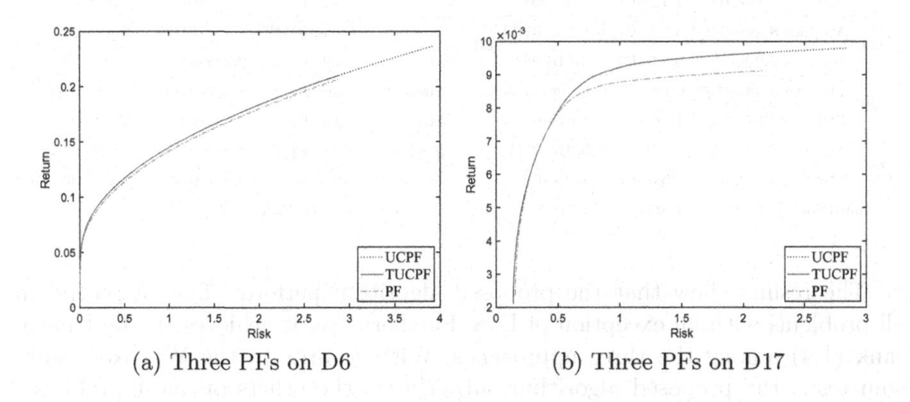

(a) Three PFs on D6 (b) Three PFs on D17

Fig. 1. The three different PFs.

4.2 Results and Comparisons

The statistics of HV obtained using the three algorithms are shown in Table 2. The rank and mean rank of each algorithm on all instances is listed in parenthesis [*] and last row, respectively; where lower value indicates better performance. The best result of each instance is highlighted in gray. The symbol "+/−/="

indicates better, worse, and equal performance relationship between paired algorithms in terms of Wilcoxon rank-sum test (5% significant level). Lastly, if a PF is unavailable (for HV normalization bounds calculation) after running the mathematical approach for 7 days, the datasets are shown with a star(*). In such cases, TUCPFs are used for HV normalization bounds during benchmarking. We note that the problems are unsolvable with a 7 days time limit for the mathematical approaches when the available number of assets exceed 225.

Table 2. HV obtained using the three algorithms. Note: for brevity, we denote MODEwAwL, CCS and proposed method with 'A', 'B' 'C' respectively here.

Data\ Alg.		A	B	C	Data\ Alg.		A	B	C
D1*	Median	7.64e-01[3]	7.84e-01[2]	8.07e-01[1]	D11*	Median	8.57e-01[2]	7.77e-01[3]	8.83e-01[1]
	Std	4.04e-02	9.49e-02	1.44e-04		Std	2.68e-02	1.08e-01	1.60e-03
D2*	Median	7.80e-01[2]	7.70e-01[3]	7.80e-01[1]	D12	Median	8.38e-01[2]	8.11e-01[3]	8.56e-01[1]
	Std	1.10e-02	1.25e-01	1.90e-04		Std	3.71e-02	1.14e-01	3.01e-04
D3*	Median	8.67e-01[2]	6.47e-01[3]	8.85e-01[1]	D13	Median	8.45e-01[2]	8.11e-01[3]	8.45e-01[1]
	Std	2.74e-02	1.03e-01	3.68e-04		Std	1.48e-02	4.62e-02	4.00e-03
D4*	Median	8.68e-01[2]	7.40e-01[3]	8.96e-01[1]	D14	Median	8.42e-01[1]	7.54e-01[3]	8.39e-01[2]
	Std	3.72e-02	1.12e-01	2.63e-04		Std	3.42e-02	1.22e-01	2.26e-03
D5	Median	8.91e-01[2]	8.39e-01[3]	8.92e-01[1]	D15*	Median	9.10e-01[2]	8.04e-01[3]	9.15e-01[1]
	Std	6.56e-03	3.35e-02	5.68e-03		Std	9.08e-03	9.43e-02	3.88e-03
D6	Median	8.07e-01[1]	8.01e-01[3]	8.06e-01[2]	D16	Median	8.13e-01[3]	8.32e-01[1]	8.24e-01[2]
	Std	1.32e-02	7.01e-03	5.20e-04		Std	3.52e-02	2.34e-03	1.11e-03
D7*	Median	8.65e-01[1]	8.33e-01[3]	8.63e-01[2]	D17	Median	8.97e-01[2]	8.84e-01[3]	9.01e-01[1]
	Std	1.85e-02	1.12e-01	2.53e-04		Std	3.11e-02	1.86e-02	4.32e-03
D8	Median	8.40e-01[1]	8.28e-01[3]	8.29e-01[2]	D18	Median	8.33e-01[1]	8.19e-01[2]	8.19e-01[3]
	Std	3.03e-02	6.94e-03	3.11e-03		Std	1.70e-02	3.26e-02	4.22e-03
D9*	Median	8.44e-01[1]	6.99e-01[3]	8.39e-01[2]	D19	Median	8.60e-01[2]	8.45e-01[3]	8.78e-01[1]
	Std	2.97e-02	1.20e-01	1.67e-03		Std	1.12e-02	2.72e-02	2.62e-03
D10	Median	8.99e-01[2]	8.77e-01[3]	9.07e-01[1]	D20	Median	8.82e-01[1]	7.84e-01[3]	8.47e-01[2]
	Std	6.03e-03	1.92e-02	1.58e-03		Std	8.29e-03	2.29e-02	1.60e-02
MeanRank		1.8(A)	2.8(B)	1.4(C)	+/-/=		9/4/7(C&A)	16/1/3(C&B)	-

The results show that the proposed algorithm performs first or second in all problems with an exception of D18. Furthermore, it achieves the best mean rank (1.4) among the three approaches. With regards to the Wilcoxon rank-sum tests, the proposed algorithm outperforms the others on most problems. For example, it shows better performance over MODEwAwL and CCS for large number of instances (9 and 16 times respectively), while it is worse on small number of instances (4 and 1 respectively). Furthermore, the differences are also clear from Fig. 2 in terms of the PF approximations obtained corresponding to the median HV. Due to space limitations, we only show the results for two instances, D3 and D19; but others show a similar trend. The figures show the improved PF approximation obtained by the proposed algorithm. It is able to find solutions covering most regions of the PF, especially the high return end. For example, in Fig. 2(a), the PF approximations of MODEwAwL and CCS only focus on the low risk area, whereas the proposed algorithm achieves much

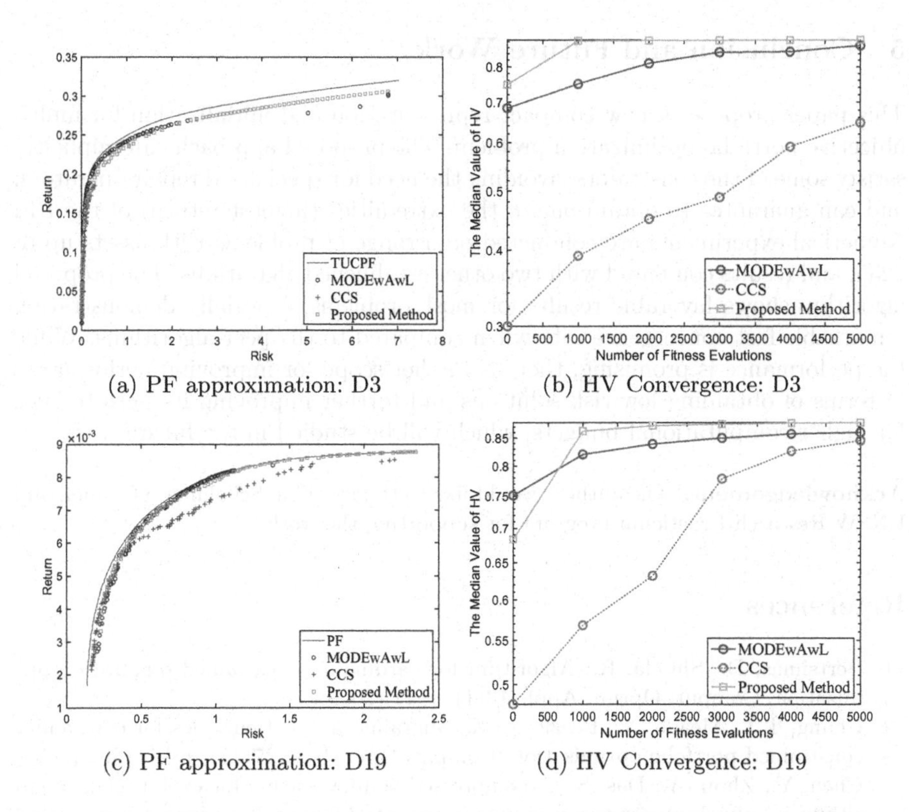

Fig. 2. PF approximations and the corresponding HV convergence curves.

better coverage. Furthermore, the HV convergence plots indicate the significant speedup using the proposed method. It converges to the near-final PF within 1000 fitness evaluations, while the other two algorithms require 5000 fitness evaluations to get to equivalent stage.

On the other end of the spectrum, one relative shortcoming of the proposed algorithm also seems apparent, which is in obtaining the solutions on the low-risk end. For instance, in Fig. 2(c), the PF approximation of CCS can achieve much better low-risk solutions (below 2.5e−4). Moreover, we also observed in preliminary experiments that when large number of evaluations are made available (say 40,000–100,000), the relative advantage of the proposed algorithm is not as significant as its performance for lower evaluation budgets.

To summarize, initial experiments establish that the proposed algorithm is competitive to the state-of-the-art forms and shows significant promise in achieving fast convergence. However, a more in-depth investigation, extensive experimentation and parametric studies need to be further conducted to understand the impact of each component individually, and to improve their performance.

5 Conclusion and Future Work

This paper proposes a new compact representation and initialization for multi-objective portfolio optimization problem. The proposed approach can implicitly satisfy some of the constraints, avoiding the need for specialized repair operators, and can guarantee to obtain one of the extremities (highest return) of the PF. Numerical experiments are conducted on a range of problems with assets up to 2235, and results compared with two other established algorithms. The proposed algorithm shows favorable results for most problems, especially demonstrating remarkable fast convergence rate when compared to the peer algorithms. While the performance is promising, there is further scope for improving performance in terms of obtaining low-risk solutions and further improving its performance for higher computational budgets, which will be studied in the future work.

Acknowledgement. The authors would like to thank China Scholarship Council and UNSW Research Practicum program for supporting the work.

References

1. Bertsimas, D., Shioda, R.: Algorithm for cardinality-constrained quadratic optimization. Comput. Optim. Appl. **43**(1), 1–22 (2009)
2. Chang, T.J., Meade, N., Beasley, J.E., Sharaiha, Y.M.: Heuristics for cardinality constrained portfolio optimisation. Comput. Oper. Res. **27**(13), 1271–1302 (2000)
3. Chen, Y., Zhou, A., Das, S.: A compressed coding scheme for evolutionary algorithms in mixed-integer programming: a case study on multi-objective constrained portfolio optimization (2019)
4. Chen, Y., Zhou, A., Dou, L.: An evolutionary algorithm with a new operator and an adaptive strategy for large-scale portfolio problems. In: Proceedings of the Genetic and Evolutionary Computation Conference Companion, pp. 247–248. ACM (2018)
5. Deb, K., Pratap, A., Agarwal, S., Meyarivan, T.: A fast and elitist multiobjective genetic algorithm: NSGA-II. IEEE Trans. Evol. Comput. **6**(2), 182–197 (2002)
6. Deb, K., Steuer, R.E., Tewari, R., Tewari, R.: Bi-objective portfolio optimization using a customized hybrid NSGA-II procedure. In: Takahashi, R.H.C., Deb, K., Wanner, E.F., Greco, S. (eds.) EMO 2011. LNCS, vol. 6576, pp. 358–373. Springer, Heidelberg (2011). https://doi.org/10.1007/978-3-642-19893-9_25
7. Feller, W.: An introduction to probability theory and its applications (1957)
8. Gaspero, L.D., Tollo, G.D., Roli, A., Schaerf, A.: Hybrid metaheuristics for constrained portfolio selection problems. Quant. Finan. **11**(10), 1473–1487 (2011)
9. Kolm, P.N., Tütüncü, R., Fabozzi, F.J.: 60 years of portfolio optimization: practical challenges and current trends. Eur. J. Oper. Res. **234**(2), 356–371 (2014)
10. Lwin, K., Qu, R., Kendall, G.: A learning-guided multi-objective evolutionary algorithm for constrained portfolio optimization. Appl. Soft Comput. **24**, 757–772 (2014)
11. Markowitz, H.: Portfolio selection. J. Finan. **7**(1), 77–91 (1952)
12. Mavrotas, G., Florios, K.: An improved version of the augmented ε-constraint method (augmecon2) for finding the exact pareto set in multi-objective integer programming problems. Appl. Math. Comput. **219**(18), 9652–9669 (2013)

13. Newman, A.M., Weiss, M.: A survey of linear and mixed-integer optimization tutorials. INFORMS Trans. Educ. **14**(1), 26–38 (2013)
14. Noceda, J., Wright, S.: Numerical optimization. Springer Series in Operations Research and Financial Engineering, ORFE. Springer, Cham (2006). https://doi.org/10.1007/978-0-387-40065-5
15. Pouya, A.R., Solimanpur, M., Rezaee, M.J.: Solving multi-objective portfolio optimization problem using invasive weed optimization. Swarm Evol. Comput. **28**, 42–57 (2016)
16. While, L., Hingston, P., Barone, L., Huband, S.: A faster algorithm for calculating hypervolume. IEEE Trans. Evol. Comput. **10**(1), 29–38 (2006)

Manifold Learning Inspired Mating Restriction for Evolutionary Constrained Multiobjective Optimization

Lianghao Li[1], Cheng He[2](✉), Ran Cheng[2], and Linqiang Pan[1]

[1] Key Laboratory of Image Information Processing and Intelligent Control of Education Ministry of China, School of Artificial Intelligence and Automation, Huazhong University of Science and Technology, Wuhan 430074, China

[2] Guangdong Provincial Key Laboratory of Brain-Inspired Intelligent Computation, Department of Computer Science and Engineering, Southern University of Science and Technology, Shenzhen 518055, China

Abstract. Mating restriction strategies are capable of restricting the distribution of parent solutions for effective offspring generation in evolutionary algorithms (EAs). Studies have shown the importance of these strategies in improving the performance of EAs for multiobjective optimization. Our previous study proposed a specific manifold learning inspired mating restriction (MLMR) strategy. It has shown promising capability of solving multiobjective optimization problems (MOPs) with complicated Pareto set shapes. However, the effect of mating restriction strategies in solving constrained MOPs is yet to be well studied. Here, we investigate the effectiveness of MLMR for solving constrained MOPs. The MLMR strategy is embedded into some representative multiobjective EAs and tested on various benchmark constrained MOPs. Experimental results indicate the encouraging performance of MLMR in constrained multiobjective optimization.

Keywords: Multiobjective optimization · Mating selection · Mating restriction · Constraint handling

1 Introduction

Constrained multiobjective optimization problems (CMOPs) refer to optimization problems with multiple objectives and constraint(s) [17]. They can be mathematically formulated as

$$\begin{aligned} \text{minimize } & f_i(\mathbf{x}) & i = 1, \ldots, M \\ \text{subject to } & g_j(\mathbf{x}) < 0 & j = 1, \ldots, J \\ & h_k(\mathbf{x}) = 0 & k = 1, \ldots, K \\ & \mathbf{x} \in X, & \end{aligned} \tag{1}$$

where $\mathbf{x}$ denotes the decision vector, X denotes the decision space, M, J, K are the numbers of objectives, inequality constraint(s), and equality constraint(s),

H. Ishibuchi et al. (Eds.): EMO 2021, LNCS 12654, pp. 296–307, 2021.
https://doi.org/10.1007/978-3-030-72062-9_24

respectively [52]. If there is no constraint in a CMOP, the problem is called an unconstrained multiobjective optimization problem (MOP). Naturally, the objectives in MOPs/CMOPs are conflicting with each other, leading to a set of optima that trade off among different objectives [15,41]. The collection of those trade-off optimal solutions is called the Pareto optimal set (PS), and the projection of the PS in the objective space is the Pareto optimal front (PF) [37].

In the past two decades, a variety of multiobjective evolutionary algorithms (MOEAs) have been proposed to solve MOPs [2,3,16,31,35,36,38,56], e.g., elitist non-dominated sorting genetic algorithm (NSGA-II) [6], indicator-based EA (IBEA) [55], MOEA based on decomposition (MOEA/D) [48], etc. Generally, an MOEA mainly consists of three components, i.e., mating selection, offspring generation, and environmental selection [11,18]. In the community of multiobjective optimization, most attention is paid to the design of efficient/effective environmental selection strategies [1,8,53]. Since the environmental selection aims to select some candidate solutions from the combination of parent and offspring solutions for survival, its effectiveness may degenerate once the generated offspring population is quite similar to the parent population. There are also some researchers dedicated to the enhancement of reproduction operators for generating better offspring solutions, e.g., polynomial mutation [5], simulated binary crossover (SBX) [4], differential evolution (DE) [39]. Those studies have been reported to perform well on some classical benchmark problems, such as ZDT problems [54] and DTLZ problems [9]. Nevertheless, they may fail to solve MOPs with complicated PSs [31].

By addressing the aforementioned two issues, some studies tried to enhance the qualities of offspring solutions through effective mating selection and mating restriction strategies [31,47]. Once the mating selection distinguishes some high-quality individuals to generate promising offspring solutions, the effectiveness of EAs can be enhanced significantly [10,31]. Specifically, the mating selection aims at selecting appropriate parent solutions for generating a set of offspring solutions with good convergence and distribution. Researchers propose the mating restriction to form subspaces where candidate solutions can be selected as parents during the mating selection [13,20,21,24,27,43]. For instance, Li *et al.* proposed a manifold learning inspired mating restriction (MLMR) strategy, in which the distribution information in both objective space and decision space is considered simultaneously [35]. MLMR adopted the concept of manifold and used the distribution information of solutions in the objective space to guide the restriction of the offspring generation in the decision space.

In spite of the promising performance of existing MOEAs and mating restriction strategies in solving unconstrained MOPs, they may face difficulties in dealing with the constraints and merely find the constrained PF of CMOPs [30]. To solve a CMOP, the constrained MOEAs (CMOEAs) are expected to approximate the PF while considering the feasibility of constraint functions. Many researchers are devoted to the design of CMOEAs or adding constraint handling techniques (CHTs) into unconstrained MOEAs [22,23,26,32,42,45,46]. For example, the constraint dominance principle (CDP) is one of the most widely used CHTs [6].

In CDP, three different dominance relationships are defined to distinguish the qualities of solutions. (1) The feasible solution dominates the infeasible one; (2) the infeasible solution with smaller overall constraint violations dominates the infeasible one with larger overall constraint violation; (3) the Pareto dominance relationship is adopted once two candidate solutions are feasible.

Although the mating restriction strategies are proved to be effective in solving the unconstrained MOPs [29,35], their effectiveness in solving CMOPs is yet to be examined. Motivated by this, this study aims to investigate the effect of MLMR and other mating restriction strategies in solving CMOPs, which is expected to provide some distinct ideas for constraint handling in comparison with conventional CHTs.

The rest of this paper is organized as follows. In Sect. 2, we briefly introduce some background about mating restriction. Then we embed MLMR in some representative MOEAs to empirically investigate its performance on constrained multiobjective optimization as presented in Sect. 3. Finally, conclusions are drawn in Sect. 4.

2 Background

We first introduce some existing mating restriction strategies and then give the details of the adopted MLMR strategy.

2.1 Mating Restriction

In MOEAs, the reproduction among similar parents will produce offspring solutions near the parent solutions to promote exploitation. Meanwhile, they tend to choose different candidate solutions as parents to improve the exploration [31]. Thus, the mating restriction strategies determine the choices of parents based on the algorithm requirements. In recent years, a number of mating restriction strategies have been proposed to enhance the quality of the offspring population [14,34,35,47], which can be roughly classified into two categories according to the metric for measuring the similarity degree between solutions.

The first category involves the mating restriction strategies based on distance, which are most widely used in MOEAs since the distance between individuals is the most intuitive metric to measure the similarity degree. For instances, Le *et al.* proposed an adaptive assortative mating scheme, in which two solutions were considered as parents only if their distance in the decision space was above a threshold in 2007 [28]. Later in 2015, the archive-based steady-state micro genetic algorithm (ASMiGA) was proposed to select one mate solution with the largest distance from neighbors or a random candidate solution from the archive [34]. Very recently, the manifold learning-inspired mating strategy proposed in [35] also adopted the distance based approach. However, different from the other distance based mating restriction strategies, in MLMR, the distribution information in both decision space and objective space is used to determine the mating relationship of the parent population.

The second category consists of the mating restriction strategies based on multiple populations, where multiple populations/archives were used to classify the individuals according to their similarity or other characteristics. For example, in MOEA/D the parents are chosen from the neighbor population constructed by the distance between weight vectors in the objective space [48]. However, choosing parents only from the neighbor population would cause a loss of diversity. In contrary, randomly mating among the whole population could enhance the diversity of the offspring population. Thus, a predefined probability was used to determine the choice of parents from either the neighbor populations or the whole population in the differential evolution based MOEA/D (MOEA/D-DE) [29]. The same predefined probability was used in the self-organizing MOEA (SMEA), where multiple populations were constructed by clustering [47].

2.2 MLMR Restriction

Algorithm 1. The Framework of MLMR Strategy

Input: P (parent population), N (population size).
Output: P (offspring population).
1: $N \leftarrow \lceil N/2 \rceil * 2$
2: $(\mathrm{PF}_1, \mathrm{PF}_2, \ldots, \mathrm{PF}_s) \leftarrow$ Non-dominated-sort(P)
3: **for** $i \leftarrow 1$ *to* s **do**
4: $P_c \leftarrow \mathrm{PF}_i$
5: $\mathbf{MD} \leftarrow$ Manifold distances of P_c
6: $R \leftarrow \frac{\max \mathbf{MD}}{|P_c|-1}$ /*Range of the neighborhood*/
7: **while** $(P_c \neq \emptyset)$ & $(|\mathrm{Pair}| < N/2)$ **do**
8: $j \leftarrow$ The number of current Pair
9: $\mathrm{Pair}^1_{j+1} \leftarrow$ Randomly select a solution in P_c and delete it form P_c
10: $P_n \leftarrow$ Choose the neighbors of the selected solution and delete them form P_c
11: $\mathrm{Pair}^2_{j+1} \leftarrow$ Select the nearest solution in the neighborhood based on the manifold distance
12: **end**
13: **end**
14: **while** $|\mathrm{Pair}| < N/2$ **do**
15: **for** $i \leftarrow 1$ *to* s **do**
16: Pair $\leftarrow$ *Second_Selection*$(\mathrm{Pair}, \mathrm{PF}_i)$
17: **end**
18: **end**
19: $P \leftarrow$ *Reproduction*(Pair, N)

In this study, the MLMR strategy is embedded in some representative CMOEAs for solving CMOPs. The detailed procedures of MLMR are presented in Algorithm 1. Generally, MRML first ranks the candidate solutions in the population using the non-dominated sorting method [50]. Then the manifold distance based mating selection is conducted in the order of rank. If the number of parents is not enough after the mating selection, a secondary mating selection is conducted. After the selection, the selected pairwise parents will generate offspring solutions by balancing both exploration and exploitation.

3 Experimental Results and Analysis

In this section, we embed the MLMR strategy into three representative MOEAs, i.e., NSGA-II [6], SPEA/R [25], and MOMBI-II [19], to validate its effectiveness. The modified MOEAs are named as NSGA-II-MM, SPEA/R-MM, and MOMBI-II-MM, respectively. These modified algorithms are compared with the original algorithms on three widely used CMOP test suites, namely CF test suite, LIR-CMOP test suite, and MW test suite [12,33,49].

3.1 Experimental Settings

As recommended in [49], the dimension of decision vectors is set to 30 for all the test problems. All algorithms are tested on each test instance for 20 independent runs. We also adopt the Wilcoxon rank-sum test at a significance level of 0.05 to compare the statistic results of different algorithms.

In this work, the simulated binary crossover [4] and the polynomial mutation [5] are adopted in the compared algorithms for offspring generation. In the simulated binary crossover, the crossover probability p_c is set to 1.0, and the distribution index of crossover n_c is set to 20. The mutation probability p_m is set to $1/d$, where d is the number of decision variables, and the distribution index of mutation n_m to 20 as recommended in [7].

The inverted generational distance (IGD) [51] and hypervolume (HV) [44] are adopted to evaluate the performance of the compared algorithms. In this work, a set size closest to 10000 of reference points is used for calculating IGD value for all the test problems.

For all the test problems, the population size is set to 100. The maximum number of generation is used as the termination criterion, which is set to 1500 for all the other test instances. All the involved algorithms are implemented on PlatEMO [40], and the recommended parameter values in the literature are adopted for fair comparisons. As for constraint handling, the CDP (as illustrated in Sect. 1) is adopted in all the compared algorithms to tackle CMOPs.

3.2 Results on CF Test Problems

The statistical results of the IGD values obtained by the six compared algorithms on 10 test problems are presented in Table 1. We can see that on most test problems, algorithms with MLMR embedded have achieved smaller IGD values in comparison with their original versions. All the algorithms with MLMR embedded show significant advantages on almost all the test instances, indicating the promising ability of MLMR on tackling CMOPs. Notice that the original SPEA/R shows a competitive performance on some test instances, which could be attributed to its own distance-based mating selection strategy.

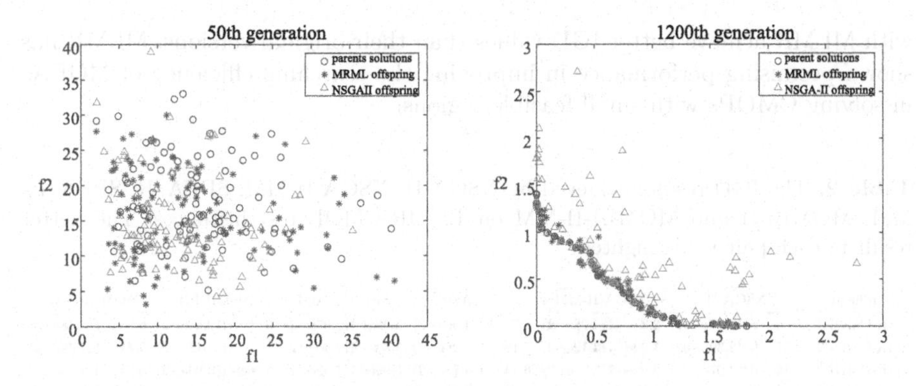

Fig. 1. The offsprings generated by MLMR and the original reproduction strategy in NSGA-II from the same parent population.

Figure 1 displays the offsprings generated from the same parent population by using the MLMR and the original mating selection in NSGA-II on CF1 with complicate PS shape. It can be observed that in the early stage of the evolution, the offsprings generated by MLMR have better convergence; by contrast, the original mating strategy of NSGA-II generates a lot of candidate solutions far away from the PF, leading to the degeneration of the efficiency. In the late stage of evolution, MLMR has generated a set of well-distributed offspring solutions with better convergence in comparison with the mating strategy in NSGA-II. All the results demonstrate that MLMR can improve the performance of MOEAs even on CMOPs with complicate PS shape.

Table 1. The IGD results achieved by NSGA-II, NSGA-II-MM, SPEA/R, SPEA/R-MM, MOMBI-II and MOMBI-II-MM on 10 CF test instances. The better result in each pair is highlighted.

Problem	NSGA-II	NSGA-II-MM	SPEA/R	SPEA/R-MM	MOMBI-II	MOMBI-II-MM
CF1	1.51e−02(2.82e−03)	6.70e−3(9.29e−4)	3.60e−3(4.68e−4)	5.44e−3(1.31e−3)	1.85e−2(9.35e−3)	9.14e−3(5.65e−3)
CF2	4.75e−02(3.22e−02)	2.90e−2(8.08e−3)	3.64e−2(1.11e−2)	1.92e−2(3.95e−3)	3.66e−2(1.49e−2)	1.63e−2(3.47e−3)
CF3	2.65e−01(1.06e−01)	2.21e−1(7.69e−2)	2.38e−1(1.32e−1)	1.93e−1(5.60e−2)	2.55e−1(1.50e−1)	2.04e−1(7.93e−2)
CF4	1.14e−01(5.14e−02)	6.57e−2(1.17e−2)	8.04e−2(4.02e−2)	4.09e−2(8.90e−3)	8.74e−2(3.55e−2)	5.29e−2(1.17e−2)
CF5	3.08e−01(1.11e−01)	2.28e−1(1.01e−1)	2.39e−1(1.30e−1)	2.30e−1(8.38e−2)	3.11e−1(1.31e−1)	2.24e−1(8.39e−2)
CF6	7.65e−02(2.28e−02)	7.13e−2(3.36e−2)	3.82e−2(2.89e−2)	3.43e−2(1.62e−2)	4.60e−2(1.76e−2)	4.21e−2(2.36e−2)
CF7	2.98e−01(1.30e−01)	1.78e−1(7.73e−2)	2.47e−1(1.27e−1)	1.36e−1(5.98e−2)	2.03e−1(7.77e−2)	1.67e−1(1.10e−1)
CF8	2.17e+01(2.05e+01)	1.96e+1(2.30e+1)	1.48e−1(5.84e−2)	1.91e−1(5.21e−2)	2.34e−1(1.16e−1)	1.90e−1(2.94e−2)
CF9	1.50e−01(2.83e−02)	1.45e−1(1.91e−2)	9.33e−2(7.87e−3)	1.21e−1(4.52e−2)	1.70e−1(5.41e−2)	1.30e−1(1.03e−2)
CF10	6.00e+01(2.41e+01)	3.24e+1(1.64e+1)	2.92e−1(4.94e−2)	2.96e−1(1.18e−1)	3.14e−1(1.22e−1)	3.03e−1(1.02e−1)

3.3 Results on LIRCMOP Test Problems

The statistical results of IGD values obtained by the six compared algorithms on 12 LIR-CMOP test problems are presented in Table 2. Note that LIR-CMOPs are a set of CMOPs with large infeasible regions [12]. For instance, the feasible regions of LIR-CMOP1-4 are tiny, and we can observe from Table 2 that MOEAs with MLMR achieve better IGD values than their original versions. MLMR has shown promising performance in improving the searching efficiency of MOEAs in solving CMOPs with small feasible regions.

Table 2. The IGD results achieved by NSGA-II, NSGA-II-MM, SPEA/R, SPEA/R-MM, MOMBI-II and MOMBI-II-MM on 12 LIR-CMOP test instances. The better result in each pair is highlighted.

Problem	NSGA-II	NSGA-II-MM	SPEA/R	SPEA/R-MM	MOMBI-II	MOMBI-II-MM
LIR-CMOP1	3.00e−01(3.17e−02)	2.33e−1(4.08e−2)	6.77e−1(4.99e−3)	5.92e−1(7.97e−2)	6.90e−1(2.54e−2)	6.78e−1(3.38e−3)
LIR-CMOP2	2.57e−01(2.30e−02)	1.93e−1(4.12e−2)	7.13e−1(2.26e−2)	5.33e−1(7.45e−2)	7.76e−1(3.17e−2)	7.11e−1(2.36e−2)
LIR-CMOP3	3.14e−01(3.07e−02)	2.62e−1(5.30e−2)	6.44e−1(4.78e−2)	5.99e−1(7.59e−2)	6.66e−1(5.98e−2)	6.73e−1(4.15e−3)
LIR-CMOP4	2.88e−01(3.21e−02)	2.45e−1(4.24e−2)	6.97e−1(4.31e−2)	5.31e−1(8.79e−2)	7.60e−1(3.94e−2)	6.86e−1(3.46e−2)
LIR-CMOP5	1.22e+00(6.74e−03)	1.23e+0(4.27e−2)	2.79e−1(3.95e−2)	2.43e−1(7.99e−2)	3.40e−1(6.51e−2)	2.46e−1(7.59e−2)
LIR-CMOP6	1.31e+00(1.74e−01)	1.35e+0(2.37e−4)	2.91e−1(6.42e−2)	2.72e−1(9.48e−2)	3.69e−1(8.40e−2)	2.27e−1(8.01e−2)
LIR-CMOP7	4.75e−01(6.33e−01)	1.68e+0(3.93e−4)	8.75e−1(6.95e−2)	5.80e−1(1.30e−1)	9.53e−1(4.09e−2)	6.49e−1(1.18e−1)
LIR-CMOP8	8.88e−01(7.38e−01)	1.60e+0(3.66e−1)	6.75e−1(4.73e−2)	4.03e−1(1.16e−1)	7.58e−1(9.35e−2)	4.93e−1(1.19e−1)
LIR-CMOP9	8.98e−01(1.08e−01)	5.30e−1(1.34e−1)	3.09e−2(9.86e−3)	7.78e−2(1.93e−2)	2.18e−1(6.12e−2)	1.73e−1(4.51e−2)
LIR-CMOP10	8.26e−01(4.60e−02)	6.13e−1(2.01e−1)	1.90e−2(4.35e−3)	3.24e−2(9.54e−3)	7.34e−2(2.60e−2)	5.70e−2(1.64e−2)
LIR-CMOP11	7.04e−01(1.04e−01)	6.38e−1(1.10e−1)	1.68e−1(6.60e−3)	1.62e−1(1.86e−2)	1.60e−1(2.24e−2)	1.84e−1(1.18e−2)
LIR-CMOP12	7.88e−01(1.38e−01)	3.65e−1(1.20e−1)	2.74e−1(9.92e−3)	2.86e−1(2.53e−2)	3.93e−1(3.44e−2)	3.71e−1(5.91e−2)

To approximate the global optima of LIR-CMOP5 to LIR-CMOP8, MOEAs should be capable of searching across the infeasible regions. The results show that MLMR can improve the performance of SPEA/R and MOMBI-II on these problems. As for NSGA-II, the original mating selection has shown better performance. It is worth noting that the original mating selection strategies in SPEA/R and MOMBI-II use the distance information in the objective space, and MLMR uses distance information in both objective and decision space. On the contrary, NSGA-II does not use any distance information during the mating selection. Since LIR-CMOP5 to LIR-CMOP8 call for the capability of MOEAs in passing through the infeasible regions, MOEAs need to generate diverse solutions during the offspring generation. Thus, the distance information in the objective space may reduce the diversity of offsprings and cause degeneration of performance in solving this kind of problems.

LIR-CMOP9 to LIR-CMOP12 have disconnected feasible PFs with large infeasible regions in the objective space. The statistical results in Table 2 demonstrate that MLMR brings significant performance improvement to NSGA-II. Nevertheless, MLMR does not seem to be competitive in comparison with the original mating strategies in SPEA/R and MOMBI-II.

3.4 Results on MW Test Problems

The statistical results of IGD values obtained by the six compared algorithms on six bi-objective MW test problems are presented in Table 3.

The statistical results indicate that MLMR is ineffective in solving MW1, MW2, MW5, and MW6. Specifically, MW3 and MW7 have connected feasible

Table 3. The IGD results achieved by NSGA-II, NSGA-II-MM, SPEA/R, SPEA/R-MM, MOMBI-II and MOMBI-II-MM on 6 MW test instances. The best result in each pair is highlighted.

Problem	NSGA-II	NSGA-II-MM	SPEA/R	SPEA/R-MM	MOMBI-II	MOMBI-II-MM
MW1	**8.76e−02(3.83e−01)**	1.10e+0(1.10e+0)	**4.03e−3(4.73e−4)**	2.48e−1(3.64e−1)	**1.14e−1(1.07e−1)**	2.15e−1(3.42e−1)
MW2	**3.56e−02(1.89e−02)**	4.49e−2(2.06e−2)	**1.29e−2(7.64e−3)**	2.95e−2(9.79e−3)	9.47e−2(1.31e−1)	**5.29e−2(2.59e−2)**
MW3	5.88e−03(2.43e−04)	**5.75e−3(3.80e−4)**	**3.24e−2(1.79e−4)**	3.24e−2(3.39e−4)	8.95e−2(1.04e−1)	**3.78e−2(9.23e−3)**
MW5	**1.61e−01(2.99e−01)**	3.87e−1(3.59e−1)	**6.04e−3(8.87e−4)**	6.36e−3(1.39e−3)	**4.85e−3(1.22e−5)**	2.04e−1(4.07e−1)
MW6	**2.86e−02(1.72e−02)**	6.82e−2(2.98e−2)	**1.12e−2(6.28e−3)**	2.22e−2(9.83e−3)	**2.57e−2(1.99e−2)**	4.41e−2(1.31e−2)
MW7	7.20e−02(1.64e−01)	**7.53e−3(4.66e−3)**	1.07e−1(2.20e−4)	**1.06e−1(5.15e−4)**	1.06e−1(3.74e−4)	**1.05e−1(6.43e−4)**

regions, while other MW problems have disconnected feasible regions or disconnected constrained PFs. We can conclude from these results that MLMR may not be suitable for solving CMOPs with disconnected feasible regions or disconnected constrained PFs.

4 Conclusion

In this work, we investigate the effect of MLMR on conventional MOEAs in solving CMOPs. By embedding the MLMR strategy into three representative CMOEAs, its effect in solving CMOPs is examined over three test suites. Experimental results have indicated the effectiveness of MLMR on CMOPs with complicate PS shapes. To be more specific, MLMR shows its capability in improving the performance of MOEAs on CMOPs with tiny feasible regions and CMOPs connected feasible region. Nevertheless, MLMR may fail to solve CMOPs with disconnected feasible regions or disconnected constrained PF.

All in all, the mating restriction strategies could affect the performance of MOEAs in solving CMOPs. It is desirable to adopt different mating restriction strategies for different types of CMOPs. Future works include investigating the effectiveness of other mating restriction strategies on constrained multiobjective optimization and designing adaptive mating selection/restriction for solving CMOPs.

Acknowledgment. Paper written during a six months stay of L. Li in the Department of Computer Science and Engineering, Southern University of Science and Technology, China, in the middle of 2020. This work is supported in part by the National Natural Science Foundation of China under Grant 61903178 and Grant 61906081; in part by the Program for Guangdong Introducing Innovative and Entrepreneurial Teams under Grant 2017ZT07X386; in part by the Shenzhen Peacock Plan under Grant KQTD2016112514355531; and in part by the Program for University Key Laboratory of Guangdong Province under Grant 2017KSYS008.

References

1. Bader, J., Zitzler, E.: HypE: an algorithm for fast hypervolume-based many-objective optimization. Evol. Comput. **19**(1), 45–76 (2011)

2. Beume, N., Naujoks, B., Emmerich, M.: SMS-EMOA: multiobjective selection based on dominated hypervolume. Eur. J. Oper. Res. **181**(3), 1653–1669 (2007)
3. Corne, D.W., Jerram, N.R., Knowles, J.D., Oates, M.J.: PESA-II: region-based selection in evolutionary multi-objective optimization. In: Proceedings of the Genetic and Evolutionary Computation Conference, pp. 283–290. Citeseer (2001)
4. Deb, K., Agrawal, R.B.: Simulated binary crossover for continuous search space. Complex Syst. **9**(4), 115–148 (1995)
5. Deb, K., Goyal, M.: A combined genetic adaptive search (GeneAS) for engineering design. Comput. Sci. Inform. **26**(4), 30–45 (1996)
6. Deb, K., Pratap, A., Agarwal, S., Meyarivan, T.: A fast and elitist multi-objective genetic algorithm: NSGA-II. IEEE Trans. Evol. Comput. **6**(2), 182–197 (2002)
7. Deb, K.: Multi-Objective Optimization Using Evolutionary Algorithms. John Wiley & Sons, Hoboken (2001)
8. Deb, K., Jain, H.: An evolutionary many-objective optimization algorithm using reference-point-based nondominated sorting approach, part I: solving problems with box constraints. IEEE Trans. Evol. Comput. **18**(4), 577–601 (2014)
9. Deb, K., Thiele, L., Laumanns, M., Zitzler, E.: Scalable Test Problems for Evolutionary Multiobjective Optimization. Springer, London (2005)
10. Eiben, A.E., Smith, J.E.: What is an evolutionary algorithm? Introduction to Evolutionary Computing, pp. 25–48. Springer, Berlin (2015)
11. Eiben, A.E., Smith, J.: From evolutionary computation to the evolution of things. Nature **521**(7553), 476 (2015)
12. Fan, Z., Li, W., Cai, X., Han, H., Goodman, E.: An improved epsilon constraint-handling method in MOEA/D for CMOPs with large infeasible regions. Soft. Comput. **23**(23), 12491–12510 (2019)
13. Gao, J., Fang, L., Wang, J.: A weight-based multiobjective immune algorithm: WBMOIA. Eng. Optim. **42**(8), 719–745 (2010)
14. He, C., Cheng, R., Yazdani, D.: Adaptive offspring generation for evolutionary large-scale multiobjective optimization. IEEE Trans. Syst. Man Cybern. Syst. 1–13 (2020). https://doi.org/10.1109/TSMC.2020.3003926
15. He, C., et al.: Accelerating large-scale multiobjective optimization via problem reformulation. IEEE Trans. Evol. Comput. **23**(6), 949–961 (2019)
16. He, C., Tian, Y., Jin, Y., Zhang, X., Pan, L.: A radial space division based many-objective optimization evolutionary algorithm. Appl. Soft Comput. **61**, 603–621 (2017)
17. He, C., Cheng, R., Zhang, C., Tian, Y., Chen, Q., Yao, X.: Evolutionary large-scale multiobjective optimization for ratio error estimation of voltage transformers. IEEE Trans. Evol. Comput. **24**(5), 868–881 (2020)
18. He, C., Huang, S., Cheng, R., Tan, K.C., Jin, Y.: Evolutionary multiobjective optimization driven by generative adversarial networks (GANs). IEEE Trans. Cybern. (2020)
19. Hernández Gómez, R., Coello Coello, C.A.: Improved metaheuristic based on the R2 indicator for many-objective optimization. In: Proceedings of the 2015 Annual Conference on Genetic and Evolutionary Computation, pp. 679–686. ACM (2015)
20. Ishibuchi, H., Akedo, N., Nojima, Y.: Recombination of similar parents in SMS-EMOA on many-objective 0/1 knapsack problems. In: Coello, C.A.C., Cutello, V., Deb, K., Forrest, S., Nicosia, G., Pavone, M. (eds.) PPSN 2012. LNCS, vol. 7492, pp. 132–142. Springer, Heidelberg (2012). https://doi.org/10.1007/978-3-642-32964-7_14

21. Ishibuchi, H., Narukawa, K., Tsukamoto, N., Nojima, Y.: An empirical study on similarity-based mating for evolutionary multiobjective combinatorial optimization. Eur. J. Oper. Res. **188**(1), 57–75 (2008)
22. Jain, H., Deb, K.: An evolutionary many-objective optimization algorithm using reference-point based nondominated sorting approach, part II: handling constraints and extending to an adaptive approach. IEEE Trans. Evol. Comput. **18**(4), 602–622 (2013)
23. Jan, M.A., Zhang, Q.: MOEA/D for constrained multiobjective optimization: some preliminary experimental results. In: Proceedings of the 2010 UK Workshop on Computational Intelligence (UKCI), pp. 1–6. IEEE (2010)
24. Jaszkiewicz, A.: Genetic local search for multi-objective combinatorial optimization. Eur. J. Oper. Res. **137**(1), 50–71 (2002)
25. Jiang, S., Yang, S.: A strength Pareto evolutionary algorithm based on reference direction for multi-objective and many-objective optimization. IEEE Trans. Evol. Comput. **21**(3), 329–346 (2017)
26. Jiao, R., Zeng, S., Li, C., Pedrycz, W.: Evolutionary constrained multi-objective optimization using NSGA-II with dynamic constraint handling. In: 2019 IEEE Congress on Evolutionary Computation (CEC), pp. 1634–1641. IEEE (2019)
27. Kim, M., Hiroyasu, T., Miki, M., Watanabe, S.: SPEA2+: improving the performance of the strength Pareto evolutionary algorithm 2. In: Yao, X., et al. (eds.) PPSN 2004. LNCS, vol. 3242, pp. 742–751. Springer, Heidelberg (2004). https://doi.org/10.1007/978-3-540-30217-9_75
28. Le, K., Landa-Silva, D.: Adaptive and assortative mating scheme for evolutionary multi-objective algorithms. In: Monmarché, N., Talbi, E.-G., Collet, P., Schoenauer, M., Lutton, E. (eds.) EA 2007. LNCS, vol. 4926, pp. 172–183. Springer, Heidelberg (2008). https://doi.org/10.1007/978-3-540-79305-2_15
29. Li, H., Zhang, Q.: Multiobjective optimization problems with complicated Pareto sets, MOEA/D and NSGA-II. IEEE Trans. Evol. Comput. **13**(2), 284–302 (2009)
30. Li, K., Chen, R., Fu, G., Yao, X.: Two-archive evolutionary algorithm for constrained multiobjective optimization. IEEE Trans. Evol. Comput. **23**(2), 303–315 (2018)
31. Li, X., Zhang, H., Song, S.: A self-adaptive mating restriction strategy based on survival length for evolutionary multiobjective optimization. Swarm Evol. Comput. **43**, 31–49 (2018)
32. Liu, Z., Wang, Y.: Handling constrained multiobjective optimization problems with constraints in both the decision and objective spaces. IEEE Trans. Evol. Comput. **23**(5), 870–884 (2019)
33. Ma, Z., Wang, Y.: Evolutionary constrained multiobjective optimization: test suite construction and performance comparisons. IEEE Trans. Evol. Comput. **23**(6), 972–986 (2019)
34. Nag, K., Pal, T., Pal, N.R.: ASMiGA: an archive-based steady-state micro genetic algorithm. IEEE Trans. Evol. Cybern. **45**(1), 40–52 (2015)
35. Pan, L., Li, L., Cheng, R., He, C., Tan, K.C.: Manifold learning-inspired mating restriction for evolutionary multiobjective optimization with complicated Pareto sets. IEEE Trans. Evol. Cybern. (2019). https://doi.org/10.1109/TCYB.2019.2952881
36. Pan, L., He, C., Tian, Y., Su, Y., Zhang, X.: A region division based diversity maintaining approach for many-objective optimization. Integr. Comput. Aid. Eng. **24**(3), 279–296 (2017)

37. Pan, L., He, C., Tian, Y., Wang, H., Zhang, X., Jin, Y.: A classification-based surrogate-assisted evolutionary algorithm for expensive many-objective optimization. IEEE Trans. Evol. Comput. **23**(1), 74–88 (2018)
38. Pan, L., Li, L., He, C., Tan, K.C.: A subregion division-based evolutionary algorithm with effective mating selection for many-objective optimization. IEEE Trans. Cybern. **50**(8), 3477–3490 (2020)
39. Storn, R., Price, K.: Differential evolution-a simple and efficient heuristic for global optimization over continuous spaces. J. Global Optim. **11**(4), 341–359 (1997)
40. Tian, Y., Cheng, R., Zhang, X., Jin, Y.: PlatEMO: a MATLAB platform for evolutionary multi-objective optimization [educational forum]. IEEE Comput. Intell. Mag. **12**(4), 73–87 (2017)
41. Wang, H., Jin, Y., Yao, X.: Diversity assessment in many-objective optimization. IEEE Trans. Cybern. **47**(6), 1510–1522 (2017)
42. Wang, Y., Li, J.P., Xue, X., Wang, B.C.: Utilizing the correlation between constraints and objective function for constrained evolutionary optimization. IEEE Trans. Evol. Comput. **24**(1), 29–43 (2019)
43. Watanabe, S., Hiroyasu, T., Miki, M.: Neighborhood cultivation genetic algorithm for multi-objective optimization problems. In: Proceedings of the 4th Asia-Pacific Conference on Simulated Evolution and Learning (SEAL-2002), pp. 198–202 (2002)
44. While, L., Hingston, P., Barone, L., Huband, S.: A faster algorithm for calculating hypervolume. IEEE Trans. Evol. Comput. **10**(1), 29–38 (2006)
45. Woldesenbet, Y.G., Yen, G.G., Tessema, B.G.: Constraint handling in multiobjective evolutionary optimization. IEEE Trans. Evol. Comput. **13**(3), 514–525 (2009)
46. Yang, Y., Liu, J., Tan, S.: A constrained multi-objective evolutionary algorithm based on decomposition and dynamic constraint-handling mechanism. Appl. Soft Comput. **89**, 106104 (2020)
47. Zhang, H., Zhou, A., Song, S., Zhang, Q., Zhang, J.: A self-organizing multiobjective evolutionary algorithm. IEEE Trans. Evol. Comput. **20**(5), 1–1 (2016)
48. Zhang, Q., Li, H.: MOEA/D: a multiobjective evolutionary algorithm based on decomposition. IEEE Trans. Evol. Comput. **11**(6), 712–731 (2007)
49. Zhang, Q., Zhou, A., Zhao, S., Suganthan, P.N., Liu, W., Tiwari, S.: Multiobjective optimization test instances for the CEC 2009 special session and competition. University of Essex, Colchester, UK and Nanyang technological University, Singapore, special session on performance assessment of multi-objective optimization algorithms, technical report, pp. 1–30 (2008)
50. Zhang, X., Tian, Y., Cheng, R., Jin, Y.: An efficient approach to non-dominated sorting for evolutionary multi-objective optimization. IEEE Trans. Evol. Comput. **19**(2), 201–213 (2015)
51. Zhou, A., Jin, Y., Zhang, Q., Sendhoff, B., Tsang, E.: Combining model-based and genetics-based offspring generation for multi-objective optimization using a convergence criterion. In: Proceedings of the 2006 IEEE Congress on Evolutionary Computation (CEC), pp. 892–899 (2006)
52. Zhou, A., Qu, B., Li, H., Zhao, S., Suganthan, P.N., Zhang, Q.: Multiobjective evolutionary algorithms: a survey of the state of the art. Swarm Evol. Comput **1**(1), 32–49 (2011)
53. Zhou, A., Zhang, Q., Jin, Y.: Approximating the set of Pareto-optimal solutions in both the decision and objective spaces by an estimation of distribution algorithm. IEEE Trans. Evol. Comput. **13**(5), 1167–1189 (2009)
54. Zitzler, E., Deb, K., Thiele, L.: Comparison of multiobjective evolutionary algorithms: empirical results. Evol. Comput. **8**(2), 173–195 (2000)

55. Zitzler, E., Künzli, S.: Indicator-based selection in multiobjective search. In: Yao, X., et al. (eds.) PPSN 2004. LNCS, vol. 3242, pp. 832–842. Springer, Heidelberg (2004). https://doi.org/10.1007/978-3-540-30217-9_84
56. Ziztler, E., Laumanns, M., Thiele, L.: SPEA2: improving the strength Pareto evolutionary algorithm for multiobjective optimization. In: Evolutionary Methods for Design, Optimization, and Control, pp. 95–100 (2002)

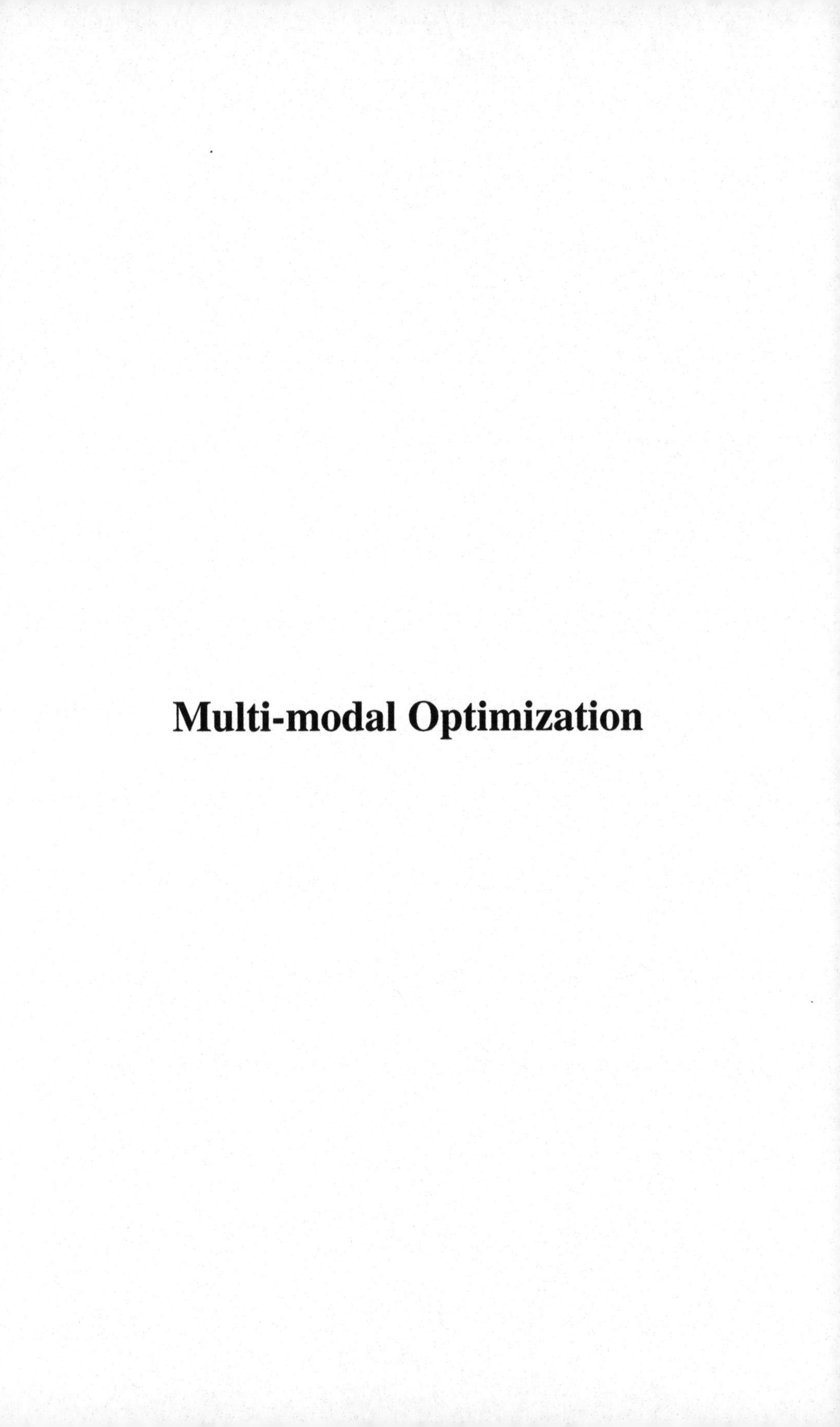

Multi-modal Optimization

Multi^3: Optimizing Multimodal Single-Objective Continuous Problems in the Multi-objective Space by Means of Multiobjectivization

Pelin Aspar[1,2](✉), Pascal Kerschke[1](✉), Vera Steinhoff[1](✉), Heike Trautmann[1](✉), and Christian Grimme[1](✉)

[1] Statistics and Optimization Group, University of Münster, Münster, Germany
{asparp,kerschke,v.steinhoff,trautmann,christian.grimme}@uni-muenster.de
[2] LIACS, University of Leiden, Leiden, The Netherlands

Abstract. In this work we examine the inner mechanisms of the recently developed sophisticated local search procedure SOMOGSA. This method solves multimodal *single-objective* continuous optimization problems by first expanding the problem with an additional objective (e.g., a sphere function) to the *bi-objective* space, and subsequently exploiting local structures and ridges of the resulting landscapes. Our study particularly focusses on the sensitivity of this multiobjectivization approach w.r.t. (i) the parametrization of the artificial second objective, as well as (ii) the position of the initial starting points in the search space.

As SOMOGSA is a modular framework for encapsulating local search, we integrate Gradient and Nelder-Mead local search (as optimizers in the respective module) and compare the performance of the resulting hybrid local search to their original single-objective counterparts. We show that the SOMOGSA framework can significantly boost local search by multiobjectivization. Combined with more sophisticated local search and metaheuristics this may help in solving highly multimodal optimization problems in future.

Keywords: Multiobjectivization · Multimodal optimization · Local search

1 Introduction

The basic idea of multiobjectivization for single-objective (SO) optimization problems is simple [13,19]: instead of optimizing the desired objective alone, introduce a second objective and solve the resulting multi-objective (MO) – more precisely bi-objective – optimization problem using MO solvers (e.g., EMOAs). There are multiple reasons for doing this: First and foremost, local optima are obstacles in search space, which may lead to premature convergence of SO local search methods. And even for global methods, multimodality can be a curse [17].

H. Ishibuchi et al. (Eds.): EMO 2021, LNCS 12654, pp. 311–322, 2021.
https://doi.org/10.1007/978-3-030-72062-9_25

A first promising report on the benefits of transforming SO problems into the MO domain was provided by Knowles et al. [13]. The authors empirically showed that multiobjectivization can reduce the amount of local optima in search space. Follow-up studies showed that helper-objectives can be beneficial [8], while others theoretically demonstrate that an additional objective can improve the search behavior of evolutionary algorithms [16].

A second reason for multiobjectivization is to exploit the power of EMOAs, which are capable of solving many difficult problems [21]. However, the result of multiobjectivization may be ambivalent, and thus have positive and negative effects on an algorithm's performance [1,5]. Still, a central property of multiobjectivization remains: if we introduce an additional objective, there is often more information available that may help in guiding the search. Some authors even report on plateau networks that may level out local optima and make it easier to avoid them [2]. Using new visual representations of MO landscapes [4,10], we were recently able to show that MO landscapes in fact comprise structures that can be exploited to escape local optima [20]. In that work, we not only described the observations in the MO landscapes, but also integrated this behavior in a local search method and demonstrated the general working principle.

Given a SO problem $f_1(x)$ that is to be optimized, with $x \in \mathbb{R}^n$, we additionally introduce a sphere function $f_2(x) = (x_1 - y_1^*)^2 + \cdots + (x_n - y_n^*)^2$ as second objective, with $y^* \in \mathbb{R}^n$ denoting the position of the sphere's center and thus only optimum of f_2. Then, the resulting bi-objective problem $F(x) = (f_1(x), f_2(x))^T$ is solved by a multi-objective local search approach called *multi-objective gradient sliding algorithm* (MOGSA) [3], which essentially moves from a starting point in decision space towards the nearest local efficient set (in the MO sense), explores that local efficient set, and moves on to the next (and dominating) locally efficient set of $F(x)$. Therefore, it exploits the special structure of MO landscapes (for details, see Sect. 2). Tracing the search path from multiple starting points, we were able to show in [20], that we pass better local optima for f_1 in that process than standard local search mechanisms (like Nelder-Mead [15]) could possibly reach for many starting points. Depending on the structure of f_1, those trajectories oftentimes even crossed the global optimum.

In this work, we extend the aforementioned approach (presented in detail in [20]) towards an effective SO local search mechanism. By combining multiobjectivization and MOGSA, our modularized heuristic enables boosting of simple SO local search mechanisms like Gradient Search (GS) and Nelder-Mead (NM).

To assess the performance gain caused by our approach, we evaluate the method for two highly multimodal BBOB problems [6], as well as for the classical Rastrigin function [7]. All three functions posses multiple local optima, which certainly are traps for GS and NM. Using specifically developed measures to systematically examine the behaviour of our approach, we show that the SO variant of MOGSA (SOMOGSA) significantly outperforms both local search mechanisms. Our systematic study of starting points in search space, as well as the examinations of the approach's sensitivity to different positions of the helper objective f_2, suggests that further, possibly more advanced, SO local search mechanisms can easily be incorporated in SOMOGSA. Alternatively, our algorithm could be used within global procedures to support global convergence.

The remainder of this work is structured as follows: Sect. 2 details the idea of SOMOGSA and briefly explains the characteristics of MO landscapes that are exploited by our algorithm. Section 3 describes the conducted experiments and the corresponding results (including performance measures), before Sect. 4 summarizes our findings and highlights perspectives for future work.

2 Single-Objective Optimization via Multiobjectivization

Within this section, we first summarize the fundamental concepts of multiobjectivization. Afterwards, the core components of our hybrid and modularized local search algorithm SOMOGSA will be outlined.

Concept of Multiobjectivization: In the remainder of this work, we aim for the optimization of box-constrained continuous single-objective optimization problems of the form $\min_{x \in [l,u]} f(x)$ with $f : \mathbb{R}^n \rightarrow \mathbb{R}$, and $l, u \in \mathbb{R}^n$ being the problem's (lower and upper) box constraints.

Instead of optimizing the single-objective problem directly – w.l.o.g. we denote this objective f_1 – we herein transform the problem into a bi-objective problem $F = (f_1, f_2)$ by introducing a second objective f_2, which w.l.o.g. shall be minimized as well. This concept is known as *multiobjectivization* [8,13,16,20] and its motivation is that the resulting bi-objective problem landscape possesses structural properties which could potentially be exploited.

One of these characteristics are so-called *locally efficient sets*, i.e., the multi-objective pendants of local optima. Per definition, each of these sets is a connected set of locally efficient points. A *locally efficient point* in turn is an observation, which is not dominated[1] by any other point within the ε-neighborhood $B_\varepsilon(x) \subseteq \mathbb{R}^n$ of that point [11,12]. Interestingly, the existence of locally efficient sets within a MO problem can be very beneficial for optimization algorithms as – in contrast to their SO counterparts – these optima oftentimes do not pose traps (to the algorithms that are optimizing the problem), but instead guide the way to more promising regions of the search space [4].

By making use of the Fritz John [9] necessary conditions, locally efficient points can be identified easily: let $x \in \mathbb{R}^n$ be a locally efficient point and all objective functions of F, i.e., f_1 and f_2, continuously differentiable in $\mathbb{R}^n$. Then, there is a weight vector $v \in [0,1]^m$ with $\sum_{i=1}^m v_i = 1$, such that

$$\sum_{i=1}^{m} v_i \nabla f_i(x) = 0. \tag{1}$$

That is, in case of locally efficient points, the (single-objective) gradients cancel each other out given a suitable weighting vector v. This property provides

[1] For two points $a, b \in \mathbb{R}^m$ we state that a dominates b, if $a_i \leq b_i$ for all $i \in \{1, \ldots, m\}$ and $a_j < b_j$ for at least one $j \in \{1, \ldots, m\}$. As a reminder, within this work, we only consider bi-objective problems, i.e., $m = 2$.

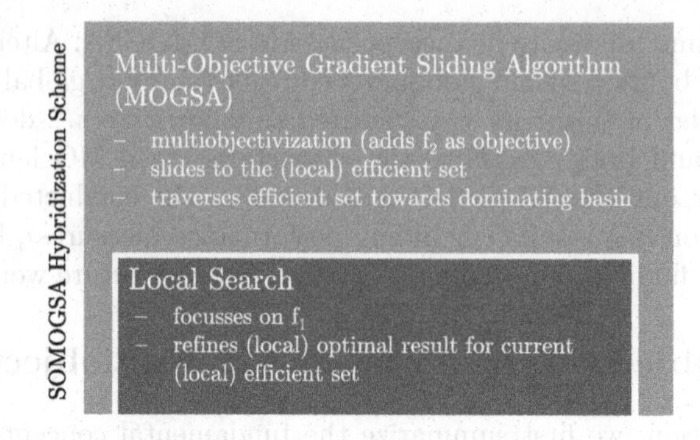

Fig. 1. Hybridization concept of SOMOGSA: While the *multi-objective* local search algorithm MOGSA realizes a descent towards dominating MO basins by approaching and traversing local efficient sets, the encapsulated *single-objective* local search focuses on the refinement of the f_1 values in the visited MO basins.

useful fundamentals for understanding and optimizing multi-objective optimization problems and (as will be shown) for single-objective problems as well. For instance, it can be used for visualizing MO landscapes [10], and is essential for the recently proposed *multi-objective gradient sliding algorithm (MOGSA)* [3,4].

As recently shown in [20], multiobjectivization enables the aforementioned multi-objective algorithm to even optimize single-objective problems. In the following subsection, we will describe this single-objective variant of MOGSA – dubbed SOMOGSA – in more detail.

The Modularized Search Heuristic SOMOGSA: Contrary to previous research (for an extensive review, we refer to Segura et al. [18,19]), we will use a deterministic MO local search algorithm. It efficiently exploits properties of MO landscapes, which we previously identified using a recently developed visualization method based on *gradient field heatmaps* [4]. In our setting the multi-objective problem (MOP) degenerates to a bi-objective problem $F(x) = (f_1(x), f_2(x))^T \in \mathbb{R}^2$. To ensure a simple usage and maximal comprehensibility, we used a simple sphere function $f_2(x) = \sum_{i=1}^{n}(x_i - y_i^*)^2$ with $x, y^* \in \mathbb{R}^n$ as second objective. It comes with the benefits that it is sufficient to guide a MO algorithm, simple enough to avoid unwanted distractions, and allows for simple, analytical determination of the corresponding gradient (which is utilized by SOMOGSA).

The general idea of our proposed hybrid and modularized search heuristic SOMOGSA is given schematically in Fig. 1. While SOMOGSA enables the exploitation of the multiobjective landscape structures, it can encapsulate an arbitrary local search for refinement of f_1 results. In Sect. 3, we will show this for two local searches. The SOMOGSA framework is described in more detail in Algorithm 1. Apart from adding a sphere as second objective (line 1), SOMOGSA essentially (repeatedly) performs the following steps:

1. Perform a MO gradient descent (i.e., 'slide down') into the vicinity of the attracting locally efficient set (lines 5 and 6).

Algorithm 1. SOMOGSA

Require: **a)** start point $x_s \in \mathbb{R}^n$, **b)** function f_1 to be optimized **c)** termination angle $t_\angle \in [0, 180]$ for switching to local search w.r.t. f_1, **d)** step size $\sigma_{MO} \in \mathbb{R}$ for MO gradient descent, **e)** step size $\sigma_{SO} \in \mathbb{R}$ for SO gradient descent w.r.t. f_2, **f)** $y^* \in \mathbb{R}^n$ optimum of f_2

1: $f_2(x) = (x_1 - y_1^*)^2 + \cdots + (x_n - y_n^*)^2$ ▷ *use sphere function for multiobjectivization*
2: $x = x_s$
3: **while** optimum of f_2 not yet reached **do**
4: $x' = x$
5: **while** $|\nabla f_1(x)| > 0$ and $\angle(\nabla f_1(x), \nabla f_2(x)) \leq t_\angle$ **do**
6: $x = x - \sigma_{MO} \cdot \left(\frac{\nabla f_1(x)}{|\nabla f_1(x)|} + \frac{\nabla f_2(x)}{|\nabla f_2(x)|} \right)$ ▷ *MO gradient descent*
7: $x^{t-1} = x = LocalSearch(x, f_1)$ ▷ *local search w.r.t.* f_1
8: **if** $f_1(x) < f_1(x')$ **then** ▷ *check for decreasing* f_1 *value*
9: **if** $|\nabla f_2(x)| > 0$ **then**
10: **while** $\angle(\nabla f_1(x), \nabla f_2(x)) \geq 90°$ and $\angle(\nabla f_2(x^{t-1}), \nabla f_2(x)) \leq 90°$ **do**
11: $x^{t-1} = x$
12: $x = x - \sigma_{SO} \cdot \frac{\nabla f_2(x)}{|\nabla f_2(x)|}$ ▷ *gradient descent towards* f_2
13: **else**
14: $x' = x$
15: **else**
16: leave outer while-loop ▷ f_1 *value deteriorated: break*
17: **return** x', $f_1(x')$

2. Once an (almost) locally efficient point has been reached, SOMOGSA will perform a single-objective local search towards the corresponding (single-objective) local optimum of f_1 (line 7). Here, the user can choose his/her favorite local search strategy.
3. As the local optimum defines one end of the locally efficient set, SOMOGSA now can traverse that set towards the neighboring (better) attraction basin (lines 10–12) by performing a single-objective gradient descent in the direction of the second objective (f_2).

All these steps are repeated until either (a) SOMOGSA has reached the sphere's optimum (line 3), or (b) the objective value of f_1 got worse between two subsequently found local optima of f_1 (line 8). In the latter case, we interrupt the optimization, as SOMOGSA wouldn't improve any longer w.r.t. f_1 but instead waste the remaining budget to eventually reach the optimum of f_2.

As described above, SOMOGSA basically provides a modularized framework, which allows to make use of the strengths of MOGSA. However, the performance of SOMOGSA obviously depends on its configuration (e.g., which problem is used as objective f_2, where the optimum of f_2 is located, which local search strategy is used, etc.). Therefore, in the following, we will investigate the sensitivity of our method w.r.t. different parameters and module choices. In addition, we experimentally demonstrate that SOMOGSA can significantly boost the potential of simple single-objective local search algorithms.

3 Experiments and Results

Setup: Following the general setup in previous work [20], we conduct our experiments on the highly multimodal BBOB functions 21 and 22 [6] (also called Gallagher's 21 and 101 Peak functions) as well as on the classical, regular, but also highly multimodal Rastrigin function [7] in 2D decision space. This Rastrigin version was also used in [20] and provides a simple enough structure for visual interpretation of the algorithms' behavior. For SOMOGSA these functions are fixed as objective f_1 in the multiobjectivized problem. As second objective, we consider $f_2(x) = \sum_{i=1}^{n}(x_i - y_i^*)^2$ with $x, y^* \in \mathbb{R}^n$ (here: $n = 2$). Besides investigating the principal benefits of applying SOMOGSA together with some classical SO local search, we specifically focus on the sensitivity of the approach w.r.t the starting position of the search, as well as the location of the f_2 optimum. We thus discretize the decision space using a regular grid X of $N = d \times d$ starting points with $d = 50$, as larger values of d did not provide additional insights and lower values did not reveal all desired structures. The methods to be assessed are executed for all grid centers of X. Ten positions y^* of the f_2 optimum were generated[2] by using Latin Hypercube Sampling [14] and subsequently mapping those points to their nearest neighbors in X: $(3.5, -1.5)$, $(-1.5, 0.5)$, $(-0.5, 2.5)$, $(2.5, -2.5)$, $(-4.5, -0.5)$, $(-2.5, -3.5)$, $(1.5, 3.5)$, $(4.5, -4.5)$, $(-3.5, 4.5)$, $(0.5, 1.5)$.

As local search mechanisms inside the SOMOGSA framework Gradient Search (GS) and Nelder-Mead (NM) [15] were used, resulting in two variants of SOMOGSA: SOMOGSA+GS and SOMOGSA+NM. As baseline, we also evaluate GS and NM as stand-alone methods on each SO problem. All variants (including stand-alone GS and NM) are allowed a maximum of 4 000 function evaluations, as we observed that especially gradient search may expose slow convergence. As all considered algorithms remain deterministic local searches, a second termination criterion is activated if the method stagnates (often near a local or global SO optimum). In order to exclude this criterion as decisive influence, we evaluated all experiments for precision values $p = 0.01$ and $p = 0.001$. Fortunately, the results qualitatively exhibited the same level of performance due to the simultaneous adjustment of the tolerance of the evaluation metric r_N (see Sect. 3), such that we only report the results for $p = 0.01$.

Methodology: We investigate the sensitivity of SOMOGSA regarding (a) the parametrization of the second objective, (b) the position of the starting point, and (c) the local search strategy. We define well-suited aggregate metrics capturing the performances of each variant in all considered settings (see Fig. 2).

We define the *gain* ($g_{LS} \rightarrow$ max) and the *global gap* ($G_{LS} \rightarrow$ min) as

$$g_{LS}(x_s) = \frac{|f_1(x_b) - f_1(x_s)|}{|f_1(x^*) - f_1(x_s)|}, \qquad G_{LS}(x_s) = \frac{|f_1(x^*) - f_1(x_b)|}{|f_1(x^*) - \max_{x \in X} f_1(x)|},$$

where x_s is a starting point, x_b the best local search result and x^* the known global optimum w.r.t. f_1. In a nutshell, regarding x_b, g_{LS} provides the proportion

[2] LHS implementation of the pyDOE package: https://pythonhosted.org/pyDOE/.

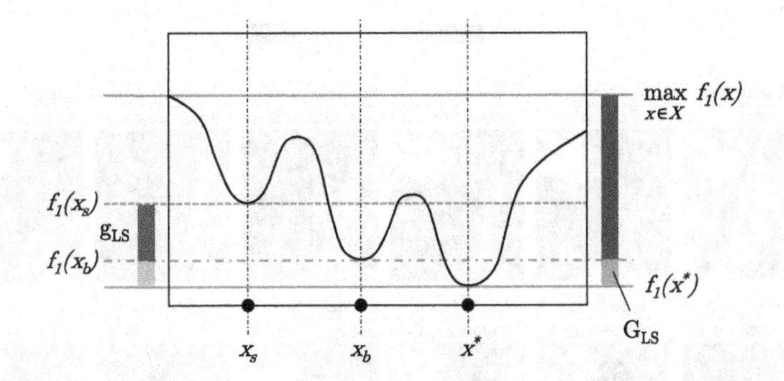

Fig. 2. Illustration of g_{LS} and G_{LS} in a one-dimensional setting.

of the gap between the function value $f(x^*)$ of the optimal solution and the function value $f(x_s)$ of the starting point. G_{LS} expresses the distance of x_b to x^* w.r.t the global maximum distance of any starting point in X to the optimum.

Both measures are evaluated for all combinations of N starting points and problem. The overall aggregated performance of each method can be statistically analyzed and visualized in the decision space, see, e.g., Fig. 3.

Furthermore, we assess the algorithm's capability to reach the optimal solution (of f_1) from any of the starting points of the (discretized) search space X by determining the relative frequency of success, denoted as (success) *ratio*, where $H_N(LS)$ is the absolute frequency of the N algorithm runs that successfully converged to the optimum of f_1 w.r.t. the precision $p = 0.01$:

$$r_N(LS) = \frac{H_N(LS)}{N}$$

Results: Figure 3 shows heatmaps of the g_{LS} measure for the considered search space of Gallagher's 101 Peak function (BBOB function 22). For each SOMOGSA variant and stand-alone LS, we color all starting points x_s – and thus pixels of the corresponding heatmap – according to its $g_{LS}(x_s)$ value, and thereby indicate the gain level – ranging from no gain (blue) to maximum gain (red).

While the first two rows of Fig. 3 show the results related to the application of GS and the corresponding hybridized method SOMOGSA+GS, the two bottom rows depict the corresponding heatmap landscapes regarding NM and the hybridized method SOMOGSA+NM. Note that, for both cases, the respective top left heatmap merely shows the single-objective problem landscape itself, while the second heatmap depicts the g_{LS} values for any starting point, when applying the stand-alone versions of GS or NM, respectively. The remaining heatmaps show the results for each of the considered ten position of f_2.

The heatmaps of the original local search techniques show that local optima are clear traps for these approaches. This is different for SOMOGSA+GS and SOMOGSA+NM. Trapping areas (denoted as blue or green-yellowish structures for GS and NM) vanish from the heatmaps and become red. This indicates that

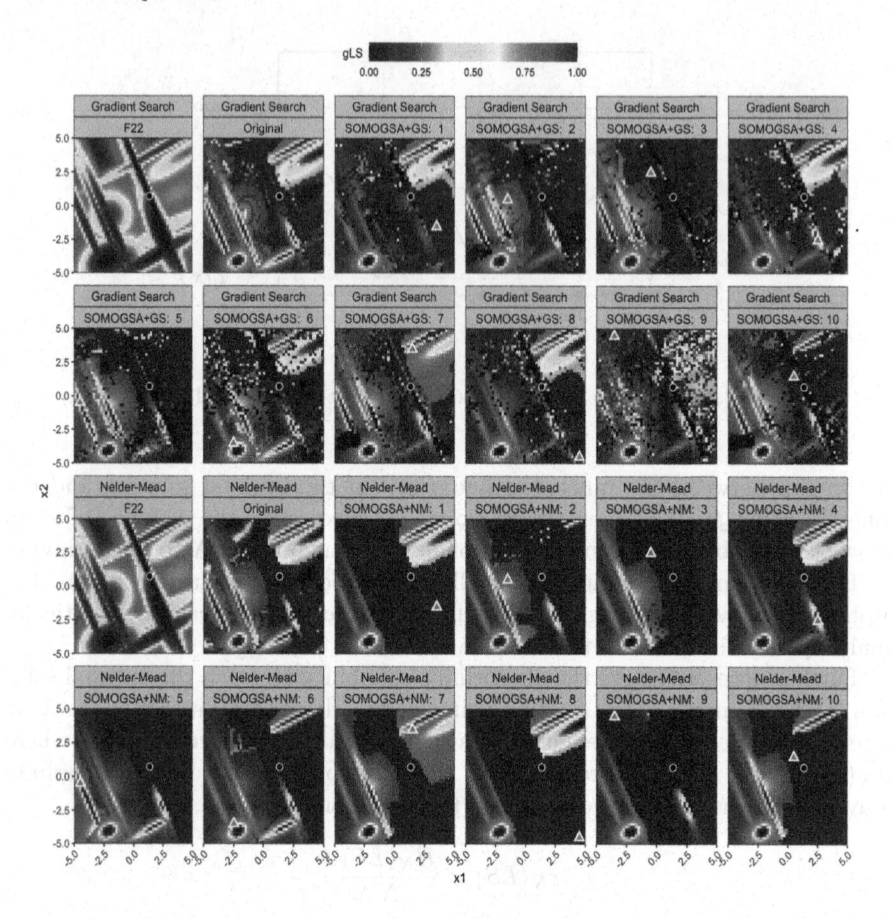

Fig. 3. Heatmaps of the g_{LS} measure for BBOB function 22. First two rows: The top left image shows the landscape of the function itself (the black dot indicates the function's optimum), followed by a heatmap with the g_{LS} performance of the original version of GS on f_1 (F22). The subsequent ten panels display the g_{LS} values of SOMOGSA+GS applied to all ten bi-objective problems. The pink triangles indicate the position of the respective optimum of f_2. Last two rows: Same heatmaps as above, but for NM and its SOMOGSA variants. (Color figure online)

for those starting points a gain is realized by using the SOMOGSA framework. As we observe the complete decision space, we can state that multiobjectivization and using SOMOGSA has the effect of removing traps for the single-objective optimizer. Although we see different resulting structures for varying locations of f_2, the general behaviour is the same for all multiobjectivized problems. Nevertheless, we can also observe an influence of the embedded local search: SOMOGSA+NM seems to produce clearer structures (and thus more robust results) than the simple gradient search mechanism. For the latter we observe many (not red colored) artefacts, which reflect premature convergence.

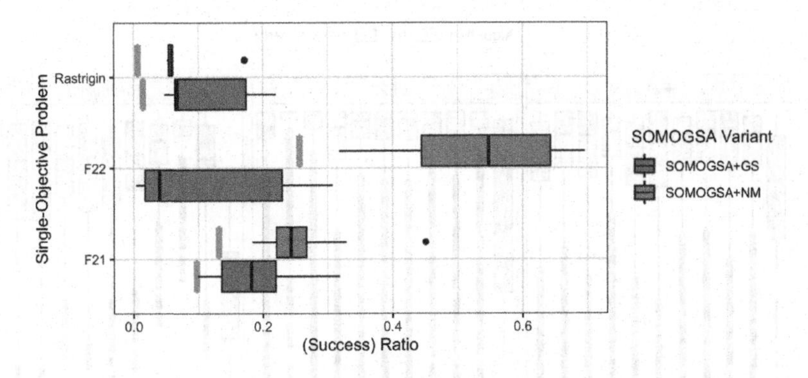

Fig. 4. Boxplots of the ratio values for all considered algorithms w.r.t. BBOB functions 21 and 22, and the Rastrigin function. The gray lines indicate the (success) ratios of the corresponding local search method (GS/NM).

This is also underlined by Fig. 4, which depicts boxplots of the (success) ratio values for all considered cases. More precisely, the red boxplots show the ratios for the three considered functions, solved by SOMOGSA+GS for each of the ten multiobjectivized (bi-objective) problem instances. Accordingly, the blue boxplots show the ratio values w.r.t. SOMOGSA+NM. For each boxplot, the ratio of the respective (original) stand-alone method, GS and NM (applied only to f_1), is shown as gray line. Figure 4 also shows that, for all considered positions of f_2, the multiobjectivization approach reaches the global optimum more often than the pure local search. An exception is SOMOGSA+GS for BBOB problem 22. Here the boxplot reflects what we already observed in Fig. 3: GS seems to get stuck with minimal gain. Furthermore, according to Fig. 4, SOMOGSA+NM proves to be better than or comparable to SOMOGSA+GS regarding all three functions. Therefore, in the following, we focus on this optimizer and take a closer look at its performance in Fig. 5, which displays both g_{LS} and G_{LS} values as boxplots. For each function, the original local search (indicated by red boxes) is compared to the hybridized method (indicated by the blue boxes). For all three functions and all considered sphere positions, this figure also reveals a significant improvement of the performance by adopting multiobjectivization. We tested this by means of a (pairwise) paired Wilcoxon rank sum test (significance level $\alpha = 5\%$) and indicated the significantly better performing variant by a blue *.

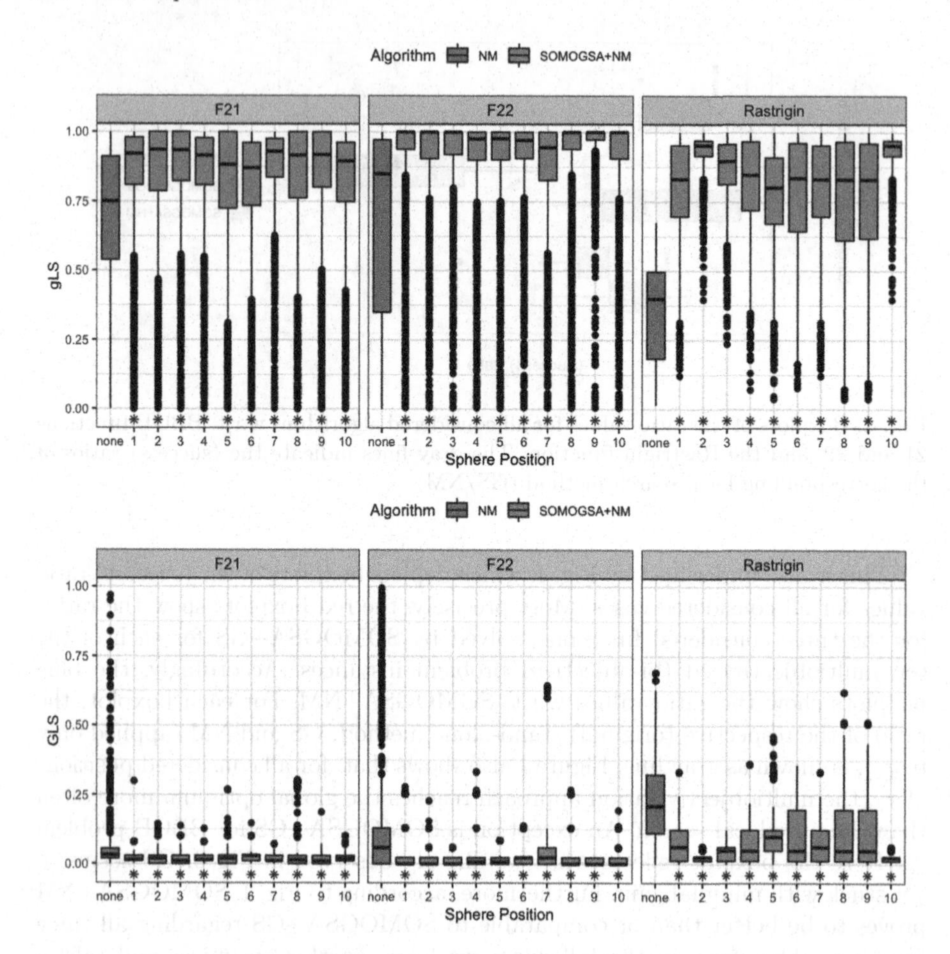

Fig. 5. Boxplots of the g_{LS} (top) and G_{LS} (bottom) measure for BBOB functions 21 and 22, as well as for the Rastrigin function. For each function, the boxplots display the values of NM (red boxplot) and SOMOGSA+NM (blue), respectively. The ten different positions of the helper function are indicated by the numbers 1 to 10. The blue * indicates, which method performs significantly better. (Color figure online)

4 Conclusion

This paper supports previous findings on the benefits of transforming multimodal single-objective problems into multi-objective problems. On selected multimodal BBOB problems we empirically show that SOMOGSA, a hybridization of the previously introduced MOGSA solver, which exploits local structures inside the multi-objective landscape, with simple single-objective local search algorithms like Gradient Search and Nelder Mead, substantially outperforms the single-objective local search optimizers themselves, while the Nelder Mead variant SOMOGSA+NM turned out to be superior to Gradient Search and the respective SOMOGSA variant. For this purpose, we specifically developed suitable performance indicators to assess solver performances on an aggregated level.

Moreover, the experiments revealed that although the performance of SOMOGSA+NM varies w.r.t. the location of the optimum of the artificial second objective – i.e., the sphere function – the extent of variation is not tremendous. This is good news as thus the multi-objective optimization problem is not largely sensitive w.r.t. the location of the sphere optimum. However, an individual configuration of this parameter based on landscape characteristics (ELA features) of the original single-objective problem is a very promising perspective to reach optimal SOMOGSA performance.

Of course, SOMOGSA itself is still a local search mechanism even after hybridization. The proximity to the optimum of individual runs depends on the starting position within the search space. Next steps will include the integration of SOMOGSA into a meta-heuristic which will be able to determine suitable starting points in a sophisticated manner including systematic restarts. Also, the underlying set of multimodal single-objective functions will be largely extended.

Acknowledgement. The authors acknowledge support by the European Research Center for Information Systems (ERCIS).

References

1. Brockhoff, D., Friedrich, T., Hebbinghaus, N., Klein, C., Neumann, F., Zitzler, E.: Do additional objectives make a problem harder? In: Proceedings of the 9th Annual Conference on Genetic and Evolutionary Computation (GECCO), pp. 765–772 (2007)
2. Garza-Fabre, M., Toscano-Pulido, G., Rodriguez-Tello, E.: Multi-objectivization, fitness landscape transformation and search performance: a case of study on the HP model for protein structure prediction. Eur. J. Oper. Res. (EJOR) **243**(2), 405–422 (2015)
3. Grimme, C., Kerschke, P., Emmerich, M.T.M., Preuss, M., Deutz, A.H., Trautmann, H.: Sliding to the global optimum: how to benefit from non-global optima in multimodal multi-objective optimization. In: AIP Conference Proceedings, pp. 020052-1–020052-4. AIP Publishing (2019)
4. Grimme, C., Kerschke, P., Trautmann, H.: Multimodality in multi-objective optimization – more boon than bane? In: Deb, K., et al. (eds.) EMO 2019. LNCS, vol. 11411, pp. 126–138. Springer, Cham (2019). https://doi.org/10.1007/978-3-030-12598-1_11
5. Handl, J., Lovell, S.C., Knowles, J.: Multiobjectivization by decomposition of scalar cost functions. In: Rudolph, G., Jansen, T., Beume, N., Lucas, S., Poloni, C. (eds.) PPSN 2008. LNCS, vol. 5199, pp. 31–40. Springer, Heidelberg (2008). https://doi.org/10.1007/978-3-540-87700-4_4
6. Hansen, N., Finck, S., Ros, R., Auger, A.: Real-parameter black-box optimization benchmarking 2009: noiseless functions definitions. Research Report RR-6829, INRIA (2009). https://hal.inria.fr/inria-00362633

7. Hoffmeister, F., Bäck, T.: Genetic algorithms and evolution strategies: similarities and differences. In: Schwefel, H.-P., Männer, R. (eds.) PPSN 1990. LNCS, vol. 496, pp. 455–469. Springer, Heidelberg (1991). https://doi.org/10.1007/BFb0029787
8. Jensen, M.T.: Helper-objectives: using multi-objective evolutionary algorithms for single-objective optimisation. J. Math. Model. Algorithms **3**(4), 323–347 (2004)
9. John, F.: Extremum Problems with Inequalities as Subsidiary Conditions, Studies and Essays Presented to R. Courant on his 60th Birthday, 8 January 1948 (1948)
10. Kerschke, P., Grimme, C.: An expedition to multimodal multi-objective optimization landscapes. In: Trautmann, H., et al. (eds.) EMO 2017. LNCS, vol. 10173, pp. 329–343. Springer, Cham (2017). https://doi.org/10.1007/978-3-319-54157-0_23
11. Kerschke, P., et al.: Towards analyzing multimodality of continuous multiobjective landscapes. In: Handl, J., et al. (eds.) PPSN 2016. LNCS, vol. 9921, pp. 962–972. Springer, Cham (2016). https://doi.org/10.1007/978-3-319-45823-6_90
12. Kerschke, P., et al.: Search dynamics on multimodal multi-objective problems. Evol. Comput. (ECJ) **27**, 577–609 (2019)
13. Knowles, J.D., Watson, R.A., Corne, D.W.: Reducing local optima in single-objective problems by multi-objectivization. In: Zitzler, E., Thiele, L., Deb, K., Coello Coello, C.A., Corne, D. (eds.) EMO 2001. LNCS, vol. 1993, pp. 269–283. Springer, Heidelberg (2001). https://doi.org/10.1007/3-540-44719-9_19
14. Mckay, M.D., Beckman, R.J., Conover, W.J.: A comparison of three methods for selecting values of input variables in the analysis of output from a computer code. Technometrics **21**(2), 239–245 (1979)
15. Nelder, J.A., Mead, R.: A simplex method for function minimization. Comput. J. **7**(4), 308–313 (1965)
16. Neumann, F., Wegener, I.: Can single-objective optimization profit from multiobjective optimization? In: Knowles, J., Corne, D., Deb, K., Chair, D.R. (eds.) Multiobjective Problem Solving from Nature, pp. 115–130. Springer, Heidelberg (2008). https://doi.org/10.1007/978-3-540-72964-8_6
17. Preuss, M.: Summary and final remarks. Multimodal Optimization by Means of Evolutionary Algorithms. NCS, pp. 171–175. Springer, Cham (2015). https://doi.org/10.1007/978-3-319-07407-8_7
18. Segura, C., Coello Coello, C.A., Miranda, G., León, C.: Using multi-objective evolutionary algorithms for single-objective optimization. 4OR **11**(3), 201–228 (2013)
19. Segura, C., Coello Coello, C.A., Miranda, G., León, C.: Using multi-objective evolutionary algorithms for single-objective constrained and unconstrained optimization. Ann. Oper. Res. **240**(1), 217–250 (2016)
20. Steinhoff, V., Kerschke, P., Aspar, P., Trautmann, H., Grimme, C.: Multiobjectivization of local search: single-objective optimization benefits from multi-objective gradient descent. In: Proceedings of the IEEE Symposium Series on Computational Intelligence (SSCI) (2020). (accepted for publication, preprint available on arXiv: https://arxiv.org/abs/2010.01004)
21. Tran, T.D., Brockhoff, D., Derbel, B.: Multiobjectivization with NSGA-II on the noiseless BBOB testbed. In: Proceedings of the 15th Annual Conference on Genetic and Evolutionary Computation (GECCO) Companion, pp. 1217–1224. ACM (2013)

Niching Diversity Estimation for Multi-modal Multi-objective Optimization

Yiming Peng and Hisao Ishibuchi(✉)

Guangdong Provincial Key Laboratory of Brain-Inspired Intelligent Computation, Department of Computer Science and Engineering, Southern University of Science and Technology, Shenzhen 518055, China
11510035@mail.sustech.edu.cn, hisao@sustech.edu.cn

Abstract. Niching is an important and widely used technique in evolutionary multi-objective optimization. Its applications mainly focus on maintaining diversity and avoiding early convergence to local optimum. Recently, a special class of multi-objective optimization problems, namely, multi-modal multi-objective optimization problems (MMOPs), started to receive increasing attention. In MMOPs, a solution in the objective space may have multiple inverse images in the decision space, which are termed as equivalent solutions. Since equivalent solutions are overlapping (i.e., occupying the same position) in the objective space, standard diversity estimators such as crowding distance are likely to select one of them and discard the others, which may cause diversity loss in the decision space. In this study, a general niching mechanism is proposed to make standard diversity estimators more efficient when handling MMOPs. In our experiments, we integrate our proposed niching diversity estimation method into SPEA2 and NSGA-II and evaluate their performance on several MMOPs. Experimental results show that the proposed niching mechanism notably enhances the performance of SPEA2 and NSGA-II on various MMOPs.

Keywords: Niching · Diversity estimation · Multi-modal multi-objective optimization

1 Introduction

Multi-objective optimization problems (MOPs), which require an optimizer to optimize multiple conflicting objective functions simultaneously, have been actively studied in the past few decades. For consistency, in this paper, all objective functions are assumed to be converted to minimization problems. Due to the trade-off between conflicting objective functions, most MOPs have a set of Pareto optimal solutions (i.e., Pareto set), which cannot be dominated by any solutions. The projection of the Pareto set in the objective space is called the Pareto front. Generally, multi-objective optimization algorithms (MOEAs) try to find a solution set with good convergence (i.e., close to the Pareto front) and

H. Ishibuchi et al. (Eds.): EMO 2021, LNCS 12654, pp. 323–334, 2021.
https://doi.org/10.1007/978-3-030-72062-9_26

good diversity (i.e., well-distributed over the Pareto front). Therefore, diversity maintenance is a critical research topic across the field of evolutionary multi-objective optimization.

Most MOEAs are equipped with some diversity maintenance mechanisms. Naturally, diversity estimation, which is a procedure of assigning a numerical value to each solution reflecting its diversity, is a prerequisite for diversity maintenance. In the past few decades, various of diversity estimation methods have been developed along with the development of MOEAs. For instance, the well-known Pareto-based MOEA called NSGA-II [2] selects solutions based on their Pareto ranks (as a primary criterion) obtained from the non-dominated sorting procedure and their crowding distance values (as a secondary criterion). In NSGA-II, crowding distance is a diversity estimator to estimate the diversity of each solution for environmental selection. Another Pareto-based algorithm called SPEA2 [19] uses the Euclidean distance from each solution to its k-th nearest neighbor in the objective space to estimate the diversity of that solution. In the grid-based MOEA named PESA-II [1], the diversity of each solution is given by the number of the solutions located in the same hyperbox in the objective space. As pointed out in [8], although the approaches used in different diversity estimators vary, the main idea is to estimate the diversity for a solution by measuring the similarity degree between that solution and other solutions in the population. In this paper, the term "diversity" always refers to the diversity in the objective space if not specified. Notice that larger diversity values are more preferable than smaller values.

Recently, multi-modal multi-objective optimization has become an active research topic. In MMOPs, the mapping from the decision space to the objective space is a many-to-one mapping instead of one-to-one mappings in standard MOPs. That is, multiple solutions with different decision values can have the same objective values. Such kind of solutions are termed as equivalent solutions. Figure 1 gives an example of MMOP where solutions marked with the same number have the same objective values. Since it is unlikely for a real-world multi-objective optimization problem to have multiple solutions with exactly the same objective values, the definition of equivalent solutions can be relaxed as follows [16]:

Definition 1 (Equivalent Solutions). *Solutions* $\boldsymbol{x}_1$ *and* $\boldsymbol{x}_2$ *are equivalent solutions iff* $d(\boldsymbol{F}(\boldsymbol{x}_1), \boldsymbol{F}(\boldsymbol{x}_2)) \leq \delta$,

where $\boldsymbol{F}$ is a vector containing all objective functions, d denotes the Euclidean distance function, and δ is a positive threshold parameter specified by the end user.

Solving MMOPs is meaningful since they are common in many real-world applications, e.g., rocket engine design problems [3] and the space mission design problems [13] both can be formulated as MMOPs. In some engineering problems, obtained solutions can become infeasible or difficult to implement due to dynamically changing environments and constraints [9]. In this regard, equivalent solutions will be able to provide alternative implementations for the decision

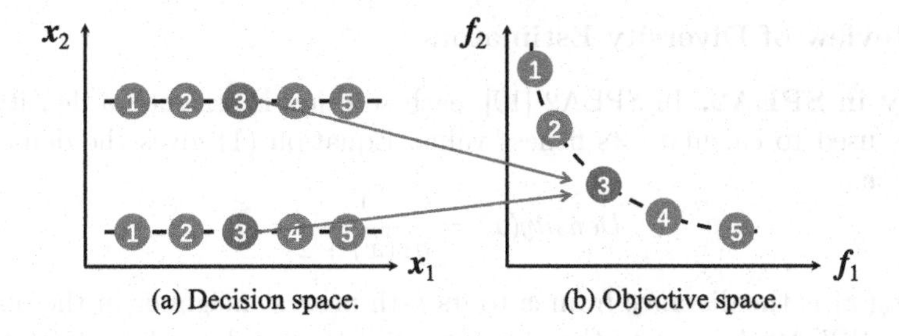

Fig. 1. Illustration of MMOP. The left and right figures show the decision and objective spaces, respectively, and the dash lines in (a) and (b) denote the Pareto set and Pareto front, respectively. Circles marked with the same number are solutions that have the same objective values.

maker. For this reason, when solving MMOPs, it is a good strategy to try to search for as many equivalent solutions as possible if no user preference is given.

As pointed out in the literature [15], standard MOEAs are usually unable to preserve multiple equivalent solutions. Since equivalent solutions are located in the same (or almost the same) position(s) in the objective space, diversity estimators tend to assign high density values to all of them. As a result, equivalent solutions are usually not preferable and likely to be removed in the environmental selection procedure, which leads to the failure of solving MMOPs. To tackle this problem, we propose the use of a simple niching strategy to make standard diversity estimators more efficient when handling MMOPs. Our approach is a simple, efficient, and parameterless mechanism which can be integrated into general diversity estimators in existing MOEAs.

The rest of the paper is organized as follows. Section 2 revisits some representative diversity estimators in MOEAs and discusses their difficulties in the handling of MMOPs. In addition, Sect. 2 also introduces some existing approaches for multi-modal multi-objective optimization. Next, Sect. 3 outlines our proposed niching diversity estimation method. Section 4 reports experimental results. Lastly, concluding remarks and suggested future research directions are presented in Sect. 5.

2 Related Work

In this section, we introduce two representative diversity estimators and discuss the difficulties they meet when handling MMOPs. Subsequently, some existing multi-modal multi-objective optimization algorithms are reviewed.

2.1 Review of Diversity Estimators

Density in SPEA2. In SPEA2 [19], each solution is assigned a density value which is used to calculate its fitness value. Equation (1) gives the density of a solution $\boldsymbol{x}$.

$$Density(\boldsymbol{x}) = \frac{1}{\sigma_k(\boldsymbol{x}) + 2}, \tag{1}$$

where $\sigma_k(\boldsymbol{x})$ is the distance from $\boldsymbol{x}$ to its k-th nearest neighbor in the objective space. In SPEA2, k is set to the square root of the total number of solutions in the current population as a general parameter setting.

Notice that in SPEA2, higher density means worse diversity in the objective space.

Crowding Distance. Crowding distance is proposed along with the NSGA-II algorithm [2] to preserve the diversity of the population in the objective space. The crowding distance of a solution $\boldsymbol{x}$ is given by the average side length of the hypercube constructed by its left and right neighbors in each objective. More precisely, for each objective, the left and right neighbors of $\boldsymbol{x}$ are the solutions at the left and right positions of $\boldsymbol{x}$ for that objective (i.e., in the list obtained by sorting the population in an increasing order of the objective values of that objective). The crowding distance of all boundary solutions (i.e., best solutions in any objectives) are set to ∞ to ensure that they are always selected. In NSGA-II, larger crowding distance values indicate better diversity. Formally, Eq. (2) calculates the crowding distance for a solution $\boldsymbol{x}$.

$$Crowding\text{-}Distance(\boldsymbol{x}) = \begin{cases} \infty & , \boldsymbol{x} \text{ is a boundary solution} \\ \frac{1}{M}\sum_{m=1}^{M}[f_m(\boldsymbol{x}_{rm}) - f_m(\boldsymbol{x}_{lm})] & , \text{otherwise} \end{cases}, \tag{2}$$

where M refers to the number of objectives, and $\boldsymbol{x}_{lm}$ and $\boldsymbol{x}_{rm}$ are the left and right neighbors of solution $\boldsymbol{x}$ regarding the m-th objective, respectively.

2.2 Difficulties When Handling MMOPs

In most diversity estimators in MOEAs, the solution distribution in the decision space is out of consideration, which makes them inefficient on MMOPs. As we have discussed in Sect. 1, in MMOPs, equivalent solutions have the same or almost the same objective values. Consequently, they are usually not preferable in terms of diversity (in the objective space). For this reason, diversity estimators used in MOEAs are often responsible for the loss of equivalent solutions when tackling MMOPs. Figure 2 gives an example when a diversity estimator such as crowding distance produces undesirable effects. In Fig. 2, A and B are two Pareto optimal solutions on different (but equivalent) Pareto subsets (i.e., the upper and lower dash lines in (a)). Although A and B have similar objective values, the decision maker may want to keep both of them since they represent different implementations (i.e., they are different in the decision space). However, a diversity estimator tends to assign bad diversity values to them due to the small difference between their objective values. As a result, some of them are likely to be removed. From this example, we can see that solutions in different regions in the decision space should be considered separately when estimating solution diversity for MMOPs. Following this idea, we propose a niching diversity estimation method in Sect. 3.

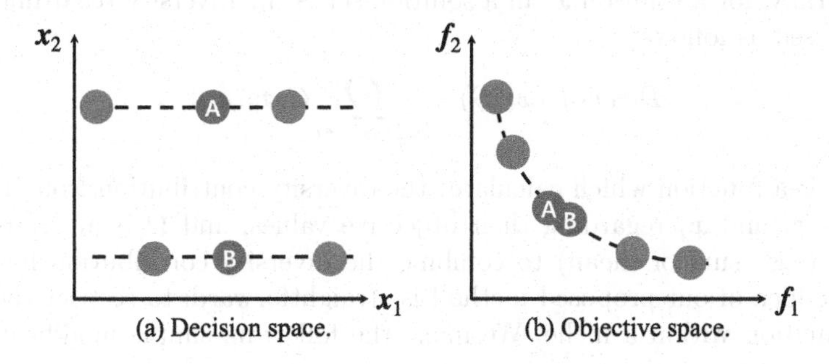

Fig. 2. Explanation of the diversity loss in the decision space caused by diversity estimators when handling an MMOP. The dash lines in (a) and (b) denote the Pareto set and Pareto front, respectively.

2.3 Multi-modal Multi-objective Optimization Algorithms

In most state-of-the-art multi-modal multi-objective evolutionary algorithms (MMEAs), the diversity in the decision space is maintained by niching strategies. Some MMEAs extend existing niching strategies in MOEAs to enable them to maintain the diversity in the objective space as well as in the decision space. For example, in [6], Deb and Tiwari proposed one of the first MMEA called Omni-optimizer which modifies the crowding distance to measure the diversity in the decision space and the objective space simultaneously. Yue et al. proposed a particle swarm optimizer named MO_Ring_PSO_SCD [17] which adopts a similar modified crowding distance and a ring topology to create a niche structure. The DNEA algorithm [10] applies the fitness sharing [4] to both decision and objective spaces and combines them into a single sharing function. Some MMEAs are proposed with dedicated niching strategies in the decision space. Tanabe et al. proposed a decomposition-based MMEA called MOEA/D-AD [15] where multiple solutions can be assigned to a weight vector, and a newly generated solution only competes with other solutions which are assigned to the same weight vector and neighboring to that solution in the decision space. In our previous study [11], we proposed another decomposition-based MMEA which utilizes a clearing strategy in the decision space. Some MMEAs such as the algorithms proposed in [7] and [9] use clustering approaches to maintain the niching structure in the decision space.

3 Proposed Method

3.1 Niching Diversity Estimation

In this section, we outline our proposed niching diversity estimation method for multi-modal multi-objective optimization. Here we first introduce a general representation for most diversity estimators used in MOEAs before diving into the details of the proposed method.

Generally, for a solution $\boldsymbol{x}_i$ in a solution set $\boldsymbol{S}$, its diversity regarding $\boldsymbol{S}$ can be expressed as follows:

$$Diversity(\boldsymbol{x}_i, \boldsymbol{S}) = \underset{\boldsymbol{x}_j \in \boldsymbol{S}, j \neq i}{\Omega} C(\boldsymbol{x}_i, \boldsymbol{x}_j), \tag{3}$$

where C is a function which calculates the diversity contribution from a pair of solutions $\boldsymbol{x}_i$ and $\boldsymbol{x}_j$ regarding their objective values, and Ω is an aggregation function (e.g., sum or mean) to combine the diversity contribution from each pair. The idea of our proposed method is straightforward: to restrict the diversity estimation within a niche. We make the following simple modifications to Eq. (3):

$$Niching\text{-}Diversity(\boldsymbol{x}_i, \boldsymbol{S}) = Diversity(\boldsymbol{x}_i, \boldsymbol{S}'), \tag{4}$$

where $\boldsymbol{S}'$ contains all solutions in $\boldsymbol{S}$ which are in the same niche as $\boldsymbol{x}_i$. In our paper, the closest k solutions in $\boldsymbol{S}$ to $\boldsymbol{x}_i$ in the **decision space** are considered as a niche.

From Eq. (4), we can see that the diversity estimation for each solution is limited to its neighbors in the decision space. With the niching strategy, solution distribution in the decision space is taken into consideration. Take Fig. 2 as an example, if $k = 2$ and crowding distance is used, the two nearest neighbors are selected for each solution (e.g., solution A) in the decision space, and the crowding distance is calculated using the selected neighbors in the objective space. Solution B is unlikely to be chosen as a neighbor of solution A (i.e., B is not likely to be used for the crowding distance calculation of A). In this manner, the diversity of A and B can be estimated in a desirable manner for maintaining the decision space diversity. From this example, we can see that the proposed niching strategy can help diversity estimators in MOEAs to handle MMOPs properly with an appropriate value of k.

Compared to existing approaches we have discussed in Sect. 2.3, our proposed method does not rely on the actual implementations of diversity estimators. It is a general niching strategy that can be conveniently integrated into most diversity estimators in MOEAs.

3.2 SPEA2 and NSGA-II with Niching Diversity Estimation

In this section, we select two classical MOEAs: SPEA2 and NSGA-II to demonstrate the procedure of our proposed niching diversity estimation method into MOEAs. The resulting algorithms are termed Niching-SPEA2 and Niching-NSGA-II, respectively.

In SPEA2 and NSGA-II, diversity estimation is only involved in the environmental selection procedure although the estimated diversity values may be used in other procedures. Therefore, we only describe the modified versions of environmental selection.

In the environmental selection procedure of Niching-SPEA2, the niching strategy is applied to both fitness calculation and archive truncation as outlined in lines 4 and 13 in Algorithm 1. In these two procedures, distance calculation is restricted by the niching strategy. For Niching-NSGA-II, in each generation, non-dominated sorting is employed to rank the whole population into several fronts. Afterward, the crowding distance is computed in each front with the proposed niching strategy.

Algorithm 1: Environmental Selection Procedure of Niching-SPEA2.

```
input  : P: input population;
         N: the number of survivors;
output : Q: population for the next generation;
/* Fitness assignment                                              */
1  S ← the strength value for each solution in P;
2  R ← the raw fitness value for each solution in P;
3  D ← the niching density value for each solution in P;
4  for i = 1, 2, ..., |P| do
5  |  F(i) = R(i) + D(i); // fitness value
6  end
/* Archive truncation                                              */
7  Q ← non-dominated solutions in P;
8  if |Q| < N then
9  |  Q ← best N solutions in P regarding their fitness values.
10 else
11 |  while |Q| > N do
12 |  |  Repeatedly remove the solution with shortest distance to other solutions
   |  |  in the same niche from Q.
13 |  end
14 end
```

4 Numerical Experiments

4.1 Experimental Settings

In our experiments, the performance of Niching-SPEA2 and Niching-NSGA-II as well as the corresponding original algorithms is evaluated on ten MMOPs. Specifically, we use the SYM-PART [12], the Omni-test [6], and the MMF1–8 [17] test problems for benchmarking. For the Omni-test problem, we set the number of decision variables to 3. For the rest of test problems, the default parameter settings in the corresponding papers [12,17] are used. Each algorithm is evaluated on each test problem 31 times independently with population size 100 and 50,000 function evaluations. The niching parameter k in Eq. (4) is set to $\lfloor\sqrt{N}\rfloor$, where N is the size of $\boldsymbol{S}$. This setting of k is based on the suggestions in [14] for statistics and data analysis.

We choose two widely used indicators: IGD$^+$ [5] and IGDX [18] to evaluate the performance of an algorithm in the objective space and the decision space, respectively. Smaller IGD$^+$ and IGDX values indicate better proximity of the obtained solution set to the Pareto front and the Pareto set, respectively.

4.2 Experimental Results

To demonstrate the efficacy of our proposed niching strategy, first we visually examine the distribution of the solution sets found by Niching-NSGA-II and its original version on the SYM-PART test problem. Non-dominated solutions obtained from a single run of each algorithm are shown in Fig. 3 and Fig. 4. In each figure, we select the run with the median IGD$^+$ value among all 31 independent runs as a representative for visual examination.

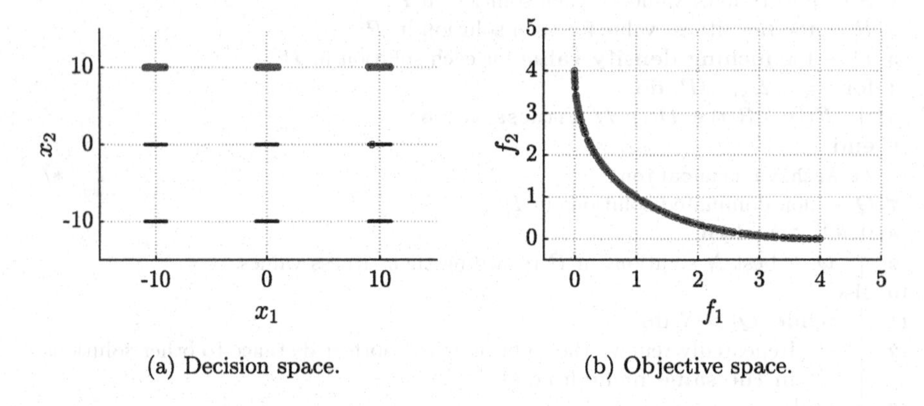

(a) Decision space. (b) Objective space.

Fig. 3. Non-dominated solutions obtained by NSGA-II on SYM-PART. The black lines show Pareto optimal solutions, and red circles show the obtained solutions. (Color figure online)

Figure 3(a) clearly shows that NSGA-II is poorly performed on the SYM-PART test problem. Most obtained solutions are distributed in the upper three Pareto subsets, while almost no solution lies on the other six Pareto subsets. This is because NSGA-II is a standard MOEA without diversity maintenance mechanisms in the decision space. In comparison, Niching-NSGA-II clearly outperforms NSGA-II as shown in Fig. 4(a), where all nine Pareto subsets are covered. This experimental result verifies that our proposed niching strategy can efficiently prevent the loss of equivalent solutions and preserve the diversity in the decision space. Regarding to the distribution in the objective space, in Figs. 3(b) and 4(b), Niching-NSGA-II slightly underperforms NSGA-II on the SYM-PART test

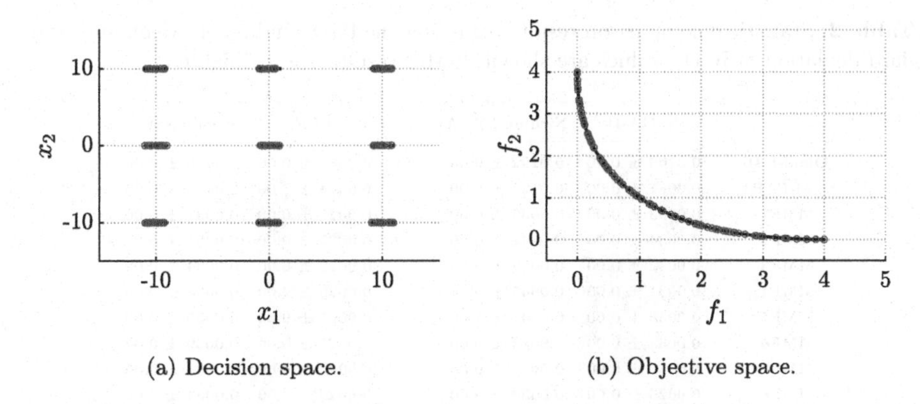

(a) Decision space.

(b) Objective space.

Fig. 4. Obtained non-dominated solutions by Niching-NSGA-II on SYM-PART. The black lines show Pareto optimal solutions, and red circles show the obtained solutions. (Color figure online)

Table 1. Statistical comparison results regarding the IGDX indicator. Mean and standard deviation of IGDX values are shown. Better results are highlighted.

	NSGA-II	Niching-NSGA-II	SPEA2	Niching-SPEA2
Omni-test	1.2610 ± 0.35	0.3647 ± 0.15 +	1.3127 ± 0.36	1.2508 ± 0.23 ≈
SYM-PART	7.1278 ± 2.36	0.0647 ± 0.00 +	6.6313 ± 2.79	2.0840 ± 2.32 +
MMF1	0.1048 ± 0.03	0.0683 ± 0.00 +	0.1015 ± 0.02	0.0639 ± 0.00 +
MMF2	0.0565 ± 0.04	0.0205 ± 0.01 +	0.0734 ± 0.04	0.0422 ± 0.03 +
MMF3	0.0413 ± 0.02	0.0181 ± 0.01 +	0.0386 ± 0.02	0.0292 ± 0.03 +
MMF4	0.1659 ± 0.07	0.0432 ± 0.00 +	0.1312 ± 0.04	0.0426 ± 0.00 +
MMF5	0.2003 ± 0.04	0.1095 ± 0.00 +	0.1843 ± 0.03	0.1054 ± 0.00 +
MMF6	0.2468 ± 0.06	0.0972 ± 0.00 +	0.1995 ± 0.04	0.1516 ± 0.02 +
MMF7	0.0680 ± 0.03	0.0463 ± 0.00 +	0.0751 ± 0.03	0.0320 ± 0.00 +
MMF8	1.6918 ± 0.65	0.0999 ± 0.02 +	1.3505 ± 0.67	0.3508 ± 0.13 +
+/−/≈	baseline	10/0/0	baseline	9/0/1

problem. As reported in [16], this is because equivalent solutions have small (or even zero) contribution to the diversity in the objective space. These observations suggest that there is a clear trade-off between the diversity on the Pareto set in the decision space and the diversity on the Pareto front in the objective space when solving MMOPs.

Table 1 and Table 2 present the statistical comparison results regarding the IGDX and IGD$^+$ indicators, respectively. In each table, the Wilcoxon rank-sum test is performed with $p = 0.05$ to compare the performance of Niching-SPEA2 and Niching-NSGA-II with their original algorithms. The symbols "+", "−", and "≈" in each table indicated that the corresponding algorithm is outperform, underperform, and tied with the baseline in the statistical comparison.

Table 1 clearly shows the superiority of Niching-SPEA2 and Niching-NSGA-II in comparison to the original versions regarding the IGDX indicator. That is, the two modified algorithms have significantly smaller IGDX values than the

Table 2. Statistical comparison results regarding the IGD$^+$ indicator. Mean and standard deviation of IGD$^+$ values are shown. Better results are highlighted.

	NSGA-II	Niching-NSGA-II	SPEA2	Niching-SPEA2
Omni-test	**0.0100 ± 0.00**	0.0192 ± 0.00 −	**0.0081 ± 0.00**	0.0138 ± 0.00 −
SYM-PART	**0.0081 ± 0.00**	0.0124 ± 0.00 −	**0.0074 ± 0.00**	0.0138 ± 0.00 −
MMF1	**0.0033 ± 0.00**	0.0039 ± 0.00 −	**0.0027 ± 0.00**	0.0031 ± 0.00 −
MMF2	**0.0034 ± 0.00**	0.0041 ± 0.00 −	**0.0031 ± 0.00**	0.0051 ± 0.00 −
MMF3	**0.0032 ± 0.00**	0.0039 ± 0.00 −	**0.0029 ± 0.00**	0.0034 ± 0.00 −
MMF4	**0.0031 ± 0.00**	0.0045 ± 0.00 −	**0.0026 ± 0.00**	0.0038 ± 0.00 −
MMF5	**0.0033 ± 0.00**	0.0044 ± 0.00 −	**0.0027 ± 0.00**	0.0039 ± 0.00 −
MMF6	**0.0033 ± 0.00**	0.0043 ± 0.00 −	**0.0027 ± 0.00**	0.0039 ± 0.00 −
MMF7	**0.0034 ± 0.00**	0.0071 ± 0.00 −	**0.0029 ± 0.00**	0.0034 ± 0.00 −
MMF8	**0.0026 ± 0.00**	0.0034 ± 0.00 −	**0.0023 ± 0.00**	0.0034 ± 0.00 −
+/−/≈	baseline	0/10/0	baseline	0/10/0

corresponding original algorithms for almost all test problems. The statistical comparison results further verify that the proposed niching approach can significantly improve the performance of SPEA2 and NSGA-II on various MMOPs. In Table 2, we can see that the modified algorithms with the niching strategy have worse IGD$^+$ values on all test problems than the original ones. This is consistent with the our previous observations on Fig. 3 and Fig. 4 (i.e., there exists a trade-off between the diversity in the decision and the objective spaces). However, from careful examinations of Table 1 and Table 2, we can see for many test problems that large improvement of the IGDX values in Table 1 is obtained at the cost of small deterioration of the IGD$^+$ values in Table 2. The observation above demonstrates that the proposed niching strategy is a promising approach to multi-modal multi-objective optimization problem, whereas the handling of the trade-off remains an open question.

5 Concluding Remarks

In this paper, we proposed a niching diversity estimation method for multi-modal multi-objective optimization. First, we pointed out that standard diversity estimators in MOEAs meet some challenges when handling MMOPs. To address this issue, we proposed a general niching strategy which is applicable to existing MOEAs to enhance their performance on MMOPs. In our proposed niching strategy, only neighboring solutions in the decision space are involved in diversity estimation. In this manner, the proposed niching strategy is able to prevent the loss of equivalent Pareto optimal solutions. In our experimental studies, we incorporated the proposed niching strategy into two classical MOEAs: SPEA2 and NSGA-II. Experimental results on ten MMOPs clearly showed that the performance of the modified algorithms is notably improved compared to the original algorithms. Currently, we employed a simple niching strategy based on the k-th nearest neighbor. The value of k was also simply specified by the square root of the sample size without considering the dimensionality of the decision space. The major contribution of this paper was to clearly illustrate that the incorporation of such a simple niching strategy significantly improved the performance of existing MOEAs on MMOPs. A future research direction can be examining the effect of the value of k and to propose a more effective specification method. More experiments over various test problems using a wide variety of MOEAs can be conducted to examine the effects of our proposed niching mechanism on MOEAs. Moreover, another promising future research issue is developing more sophisticated and efficient niching strategies. In this research direction, the point may be how to handle the trade-off between the decision space performance and the objective space performance.

Acknowledgements. This work was supported by National Natural Science Foundation of China (Grant No. 61876075), Guangdong Provincial Key Laboratory (Grant No. 2020B121201001), the Program for Guangdong Introducing Innovative and Enterpreneurial Teams (Grant No. 2017ZT07X386), Shenzhen Science and Technology Program (Grant No. KQTD2016112514355531), the Program for University Key Laboratory of Guangdong Province (Grant No. 2017KSYS008).

References

1. Corne, D.W., Jerram, N.R., Knowles, J.D., Oates, M.J.: PESA-II: region-based selection in evolutionary multiobjective optimization. In: Proceedings of the 3rd Annual Conference on Genetic and Evolutionary Computation, San Francisco, USA, pp. 283–290, 7–11 July 2001
2. Deb, K., Pratap, A., Agarwal, S., Meyarivan, T.: A fast and elitist multiobjective genetic algorithm: NSGA-II. IEEE Trans. Evol. Comput. **6**(2), 182–197 (2002)
3. Fumiya, K., Tomohiro, Y., Takeshi, F.: A study on analysis of design variables in pareto solutions for conceptual design optimization problem of hybrid rocket engine. In: Proceedings of 2011 IEEE Congress on Evolutionary Computation, New Orleans, LA, pp. 2558–2562, 5–8 June 2011
4. Goldberg, D.E., Richardson, J.: Genetic algorithms with sharing for multimodal function optimization. In: Proceedings of the Second International Conference on Genetic Algorithms, Cambridge, MA, USA, pp. 41–49, 28–31 July 1987
5. Ishibuchi, H., Masuda, H., Tanigaki, Y., Nojima, Y.: Modified distance calculation in generational distance and inverted generational distance. In: Proceedings of Evolutionary Multi-Criterion Optimization, Guimarães, Portugal, pp. 110–125, 29 March–1 April 2015
6. Deb, K., Tiwari, S.: Omni-optimizer: a generic evolutionary algorithm for single and multi-objective optimization. Eur. J. Oper. Res. **185**(3), 1062–1087 (2008)
7. Kramer, O., Danielsiek, H.: DBSCAN-based multi-objective niching to approximate equivalent pareto-subsets. In: Proceedings of the 12th Annual Conference on Genetic and Evolutionary Computation, Portland, Oregon, USA, pp. 503–510, 7–11 July 2010
8. Li, M., Yang, S., Liu, X.: Shift-based density estimation for Pareto-based algorithms in many-objective optimization. IEEE Trans. Evol. Comput. **18**(3), 348–365 (2014)

9. Lin, Q., Lin, W., Zhu, Z., Gong, M., Li, J., Coello, C.A.C.: Multimodal multi-objective evolutionary optimization with dual clustering in decision and objective spaces. IEEE Trans. Evol. Comput. **25**(1), 130–144 (2021)
10. Liu, Y., Ishibuchi, H., Nojima, Y., Masuyama, N., Shang, K.: A double-niched evolutionary algorithm and its behavior on polygon-based problems. In: Proceedings of Parallel Problem Solving from Nature - PPSN XV, Coimbra, Portugal, pp. 262–273, 8–12 September 2018
11. Peng, Y., Ishibuchi, H.: A decomposition-based multi-modal multi-objective optimization algorithm. In: Proceedings of the 2020 IEEE Congress on Evolutionary Computation, Glasgow, UK, pp. 1–8, 19–24 July 2020
12. Rudolph, G., Naujoks, B., Preuss, M.: Capabilities of EMOA to detect and preserve equivalent Pareto subsets. In: Proceedings of Evolutionary Multi-Criterion Optimization, Matsushima, Japan, pp. 36–50, 5–8 March 2007
13. Schütze, O., Vasile, M., Coello Coello, C.A.: Computing the set of epsilon-efficient solutions in multiobjective space mission design. J. Aerosp. Comput. Inf. Commun. **8**(3), 53–70 (2011)
14. Silverman, B.W.: Density Estimation for Statistics and Data Analysis. CRC Press, London (1986)
15. Tanabe, R., Ishibuchi, H.: A decomposition-based evolutionary algorithm for multi-modal multi-objective optimization. In: Proceedings of Parallel Problem Solving from Nature - PPSN XV, Coimbra, Portugal, pp. 249–261, 8–12 September 2018
16. Tanabe, R., Ishibuchi, H.: A review of evolutionary multimodal multiobjective optimization. IEEE Trans. Evol. Comput. **24**(1), 193–200 (2020)
17. Yue, C., Qu, B., Liang, J.: A multiobjective particle swarm optimizer using ring topology for solving multimodal multiobjective problems. IEEE Trans. Evol. Comput. **22**(5), 805–817 (2018)
18. Zhou, A., Zhang, Q., Jin, Y.: Approximating the set of Pareto-optimal solutions in both the decision and objective spaces by an estimation of distribution algorithm. IEEE Trans. Evol. Comput. **13**(5), 1167–1189 (2009)
19. Zitzler, E., Laumanns, M., Thiele, L.: SPEA2: improving the strength Pareto evolutionary algorithm. TIK-report 103 (2001)

Using Neighborhood-Based Density Measures for Multimodal Multi-objective Optimization

Mahrokh Javadi[(✉)] and Sanaz Mostaghim

Faculty of Computer Science, Otto von Guericke University Magdeburg, Magdeburg, Germany
{mahrokh1.javadi,sanaz.mostaghim}@ovgu.de

Abstract. This paper is about a new approach for dealing with Multimodal Multi-objective Optimization Problems (MMOPs). The major challenge in such problems is to discover several equivalent solutions in the decision space which map to the same values in the objective space. Therefore, it is very important to additionally consider the diversity of the solutions in the decision space which is the goal of this paper. We introduce a new algorithm called NxEMMO which is based on a neighborhood-based density measure in the decision space. We have evaluated our proposed approach on a 14 test problems with 2 to 6 decision variables and 2 and 3 objective functions. The experimental analysis confirms the improvement of the results.

Keywords: Multimodal multi-objective optimization · Neighborhood-based density estimation · Non-dominated sorting · Crowding distance

1 Introduction

In many real world applications, conflicting objectives need to be satisfied concurrently. These types of optimization problems are called Multi-objective Optimization Problems (MOPs) and can be formulated as follows:

$$\begin{aligned} &\text{minimize} \vec{f}(\vec{x}) = (f_1(\vec{x}), f_2(\vec{x}), ..., f_m(\vec{x})) \\ &\text{subject to} \qquad \vec{x} \in S \end{aligned}$$

The n-dimensional decision space S is defined over real-valued variables. The domination relation is employed to compare two solution vectors $\vec{x} \in S$ and $\vec{y} \in S$. $\vec{x}$ is said to be dominated by $\vec{y}$ (denoted by $\vec{y} \prec \vec{x}$) if and only if $\forall j \in \{1, \cdots, m\}$, $f_j(\vec{y}) \leq f_j(\vec{x})$, and $\exists k \in \{1, \cdots, m\}$, $f_k(\vec{y}) < f_k(\vec{x})$. A solution is called non-dominated if it is not dominated by any other solution. A set of optimal non-dominated solutions in decision space is called Pareto-Set (PS), and the corresponding objective values are called Pareto-Front (PF).

H. Ishibuchi et al. (Eds.): EMO 2021, LNCS 12654, pp. 335–345, 2021.
https://doi.org/10.1007/978-3-030-72062-9_27

In this paper, the emphasis is on a particular type of MOPs, usually referred as Multimodal Multi-objective Optimization Problems (MMOPs) [14]. In these problems, multiple global or local Pareto-optimal solutions in the decision space map to the same global or local Pareto-optimal solution in the objective space. Seeking these distinct solutions is of interest to decision makers in order to decrease uncertainty by having the opportunity to switch between alternative optimal solutions with the same quality when required. Preservation of these multiple distinct solutions in the decision space is a very challenging task for which we need to explicitly design algorithms for exploration purposes.

Over the past several years, various Multi-objective Optimization Evolutionary Algorithms (MOEAs) have been introduced to address various types of MOPs. These algorithms can be classified into three major categories: Pareto-based MOEAs such as the well-known NSGA-II algorithm [2], Decomposition-based including MOEA/D algorithm [23], and Indicator-based algorithms such as HypE [1]. The main aim of these algorithms is to provide a good coverage of PF by introducing methods to boost the search process towards uncovered areas in the objective space. Nevertheless, a good approximation of PF does not lead to find and preserve all the solutions in the PS, since it could be two distinct solutions in PS mapping to the the same objective values. For this reason, it is important to develop tailored Multimodal Multi-objective Optimization Evolutionary Algorithms (MMOEAs). In this study, to bridge the above-mentioned gap, a neighbourhood density measurement is integrated into the optimization algorithm. In our proposed approach called NxEMMO, we use the Harmonic Average Distance (HAD) [5] which is applied to the k-nearest neighbors of each solution. We additionally propose a new environmental selection process with the capability to preserve the solutions located in disparate areas.

This article is structured as follows: The related studies on available algorithms to handle MMOPs are presented in Sect. 2. In Sect. 3 we describe the proposed algorithm and the novelty of the work. In Sect. 4, the experiments are explained and the results are analysed. We conclude the paper in Sect. 5.

2 Related Work

The major challenge in dealing with MMOPs is to enforce the diversity of the solutions in the decision space. Over the last few years, an increasing amount of MMOEA have been designed to address this issue. One of the first approaches is the Omni-optimizer algorithm [4]. In this work, the authors modified the secondary selection criteria for NSGA-II algorithm by imposing the Crowding Distance (CD) approach in both decision and objective spaces. In the work presented by Liang et al. [14], they propose a Decision Space-based Niching NSGA-II (DN-NSGA-II) algorithm. They use the CD method for the mating selection process in the decision space. Another extension of the NSGA-II is named Double-Niched Evolutionary Algorithm (DNEA) including two sharing functions in both decision and objective spaces [16]. Another mechanism is found in the Mo-Ring-PSO-SCD algorithm [22], using index-based ring topology as a niching technique and a special CD method similar to the Omni-optimizer algorithm to preserve more diverse solutions in the decision space. On the basis of a previous work, a Self-organizing MOPSO (SMPSO-MM) algorithm is proposed using spacial CD and self-organising map network to preserve distinct solutions in decision space with the equivalent objective function values [13]. In [9], we proposed an approach to provide better distribution of solutions in both decision and objective spaces. We proposed a Weighted Sum CD approach (WSCD) along with a Neighbourhood-Based Polynomial Mutation (NBM) approach. In work [8], we studied a grid-based CD method in which the density value of each solution is obtained based on the grid difference computed using the Manhattan distance to its neighbours in decision space. In order to consider the density of solutions among adjacent solutions, the obtained value is additionally multiplied to its CD value in the decision space. One of the recent works dealing with MMOPs is developed by Peng and Ishibuchi [18] called MOEA/D-MM. In this method, the environmental selection in the MOEA/D algorithm is reformulated using the greedy removal and clearing strategy. This algorithm is designed to specifically address MMOPs with large-scale (higher dimensional) decision spaces. Other recent approaches are studied by [7,10,13,15,17,20,21].

3 The Proposed NxEMMO Approach

When dealing with MMOPs, the multi-objective evolutionary algorithms need to preserve the diversity of the solutions in the decision space. One way to achieve this is to increase the selection pressure towards the solutions which are far from each other in the decision space while being identical or close to each other in the objective space. Environmental selection and reproduction procedures are the two important building blocks of MOEAs assisting the diversity of solutions and preventing trapping in the local optima [19]. The main attempt of the current work is to propose a new environmental selection procedure to achieve the above-mentioned goals.

The majority of existing MMOEAs such as [4,9,14] use the CD approach in the decision space to maintain diverse solutions. However, the main difficulty of CD calculation in the decision space comparing to the objective space is that it requires more neighboring solutions. For instance, considering a problem with two objectives and two decision variables, we need two neighboring solutions for each non-dominated solution in order to calculate the CD along a non-dominated front in the objective space. However, we need up to four (i. e., 2^n) neighbors for the same non-dominated solution in the decision space. An example for CD measurement in the decision space is shown in Fig. 1. We require four solutions A, B, D and F to calculate the CD of C, while D, F and G are used for E. We note that C has a larger CD value than E, even if C is located very close to B. This is due to the fact that the solution A has a large impact on the calculation of the CD for C. In order to reduce the influence of such solutions, particularly in large decision spaces, we propose to compute the Harmonic Average Distance (HAD) which is originally introduced by [5] in the objective space. Here, we compute HAD for the solutions in the decision space as the distance between a solution i and its k-nearest neighbours as follows:

$$HAD(i) = \frac{k}{\sum_{j=1}^{k} \frac{1}{d_{ij}}} \tag{1}$$

d_{ij} is considered as the euclidean distance between the solution i and its j-th nearest neighbor. Considering solutions B and F are two nearest neighbors for the solution C, and the two neighbours for solution E are G and D. Let's assume $d_{BC} = 1$, $d_{CF} = 3$, $d_{EG} = 2$ and $d_{ED} = 4$. The harmonic average distance for the solutions C and E are $\frac{6}{4}$ and $\frac{8}{3}$ respectively, i.e., HAD(C) is smaller than HAD(E). However, the CD values have opposite relation.

The proposed NxEMMO algorithm is based on the NSGA-II algorithm with several modifications. The general framework of NxEMMO algorithm is outlined in Algorithm 1. We initialize the population $P(t)$ at generation t with N random individuals (lines 1–2). After evaluating the solutions (line 3), the parents are selected using a mating selection operator (line 5). The offspring Q are generated using the Simulated Binary Crossover (SBX) operator and mutated by the Neighborhood-based Polynomial Mutation (NBM) operator (lines 6–7) [11] and [9]. Before performing the modified environmental selection mechanism (line 10), we normalize the solutions. The proposed environmental selection mechanism (Algorithm 2) is performed on the combination of P and Q. The algorithm starts with the non-dominated sorting as in NSGA-II and the solutions are sorted into several fronts denoted by Front$_i$ for the ith front (line 1). Similar to NSGA-II the solutions from Front$_1$ to Front$_{i-1}$ are passed to the new population ($P(t+1)$) (lines 3 to 6). In case that the solutions in Front$_i$ do not fully fit into the new population, they need to be truncated.

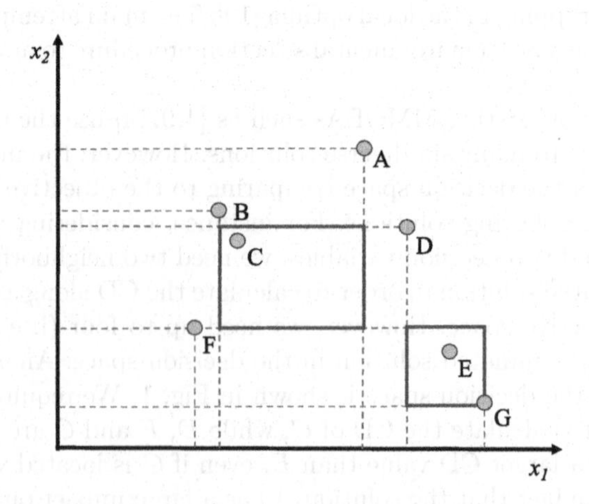

Fig. 1. Example of the CD measurement for two solutions in the decision space

Algorithm 1: NxEMMO Algorithm

```
   Input: Optimization Problem, Search Space S, Population Size N
   Output: Final population P
 1 t = 0 ;
 2 P(t) ← InitPop(P(t));
 3 Evaluate(P(t));
 4 repeat
 5   | P_mate ← Select(P(t));
 6   | P' ← Recombine(P_mate);
 7   | Q ← Mutate(P');
 8   | Evaluate(Q);
 9   | t = t + 1 ;
10   | P(t) ← Modified Environmental Selection(P', Q) ;     // Algorithm 2
11 until Termination criteria is fulfilled;
12 return P
```

Algorithm 2: Modified Environmental Selection: ModifiedEnvSelection

```
   Input: Population: P(t), Offspring Population: Q
   Output: Population P(t + 1)
 1 Front ← fast-non-dominated-sort(P(t) ∪ Q) ; // Front = (Front_1, Front_2, ...)
 2 i = 1 ;
 3 while |P| + |Front_i| ≤ N do
 4   | P(t + 1) ← P(t) ∪ |Front_i|;
 5   | i = i + 1;
 6 end
 7 if i ≥ 2 then
 8   | P(t + 1) ← Addition(Front_i, P(t + 1))
 9 else
10   | P(t + 1) ← Omission(Front_i)
11 end
12 return P(t + 1)
```

The main difference between the proposed NxEMMO algorithm and NSGA-II is the truncation method. In NxEMMO, we propose to replace the CD from NSGA-II with the following. We differentiate between the following two cases. In case that the selected front for truncation is itself $Front_1$, we perform the Nearest Neighbour Distance (NND) mechanism for truncation. NND diversity estimated measurement is originally proposed by Zitzler et al. [25] which is designed to keep the size of a set of solutions to a predefined value. This operation is called Omission (line 10). Otherwise, we perform the HAD for truncation but different from the crowding distance in NSGA-II, we compute HAD between every single solution in Front_i and all other solutions in Front_1 to Front_{i-1}. This means that we set the value for k in Eq. 1 as the total number of solutions in Front_1 to Front_{i-1}. This mechanism is called Addition (line 8). Using this operator, the solution with the largest obtained HAD value is transferred to the new population ($P(t+1)$). After updating the HAD values for the remaining solutions in Front_i, the next individuals are selected iteratively until $R(t+1)$ is filled.

In order to visualise the influence of the Addition function on the diversity of solutions in the decision space, we illustrate an example on the MMF1 test problem [14] in Fig. 2 (left). The solutions in Front_1 to Front_{i-1} are shown as $*$ and the solutions in Front_i are indicated with blue $\diamond$. The solutions in red circle are the selected solutions by the Addition function. Furthermore, the selected solutions based on CD approach are shown with purple $\square$. As can be observed, only those solutions are selected by Addition function which are not located in crowded areas considering all solutions not only in Front_i but also all others in Front_1 to Front_{i-1}. In comparison, the selected solutions by the CD method, are not well distributed over the decision space.

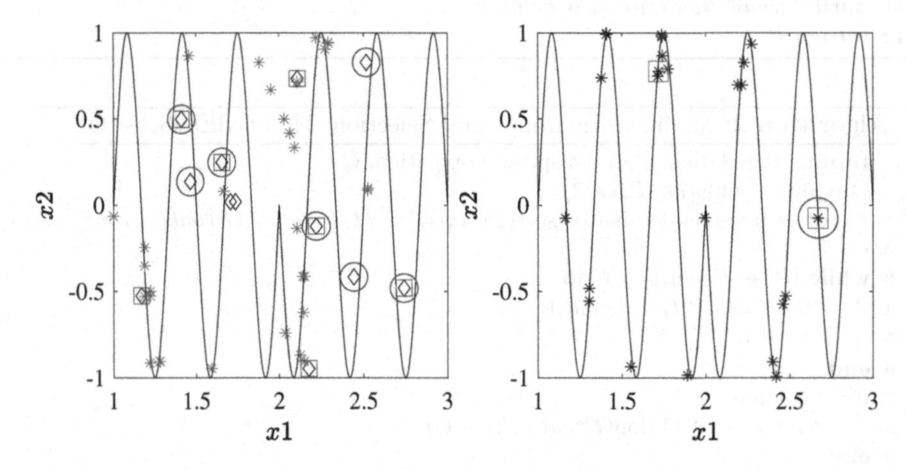

Fig. 2. Example of the Addition (left) and Omission (right) functions on the test problem MMF1

Figure 2 (right) shows an example for the Omission function which is based on the NND mechanism. The Omission function is activated when we only have one front, i. e., the truncation is taking place in $Front_1$. In this example, we aim to omit two solutions. The non-dominated solutions are shown with blue $*$, and the result of the omission process using NND is shown with $\square$. In this figure, there are two solutions (duplicates) located at the exact same position (shown by the red circle). NND removes one of the duplicates and another solution in the crowded area. For comparison purposes, we also illustrate the result of the Omission function, if we take HAD instead of NND. The result of HAD is shown with red circles. As can be observed the two solutions selected by HAD for omission, are the above mentioned duplicates. The HAD mechanism leads to omission of both duplicates which ends up with an empty position in that part of the decision space, while NND keeps one of the duplicates at the same position (the red circle) and eliminates one other solution in a crowded area. This is of great importance, since this concerns the non-dominated front.

4 Experiments

In order to evaluate the performance of the proposed approach, we perform experiments on 14 multimodal multi-objective test problems with different levels of complexity (in terms of the geometry of their PS and PF and the number of optimal solutions). The following test problems have 2 objectives and 2 decision variables: MMF1z, MMF1 to MMF9, Omni-test, SYM-part (simple and rotated) [3,14]. In addition, we take the MMF14 test problem with 2 to 6 decision variables and 2 and 3 objectives [12]. We compare the obtained solutions of NxEMMO to the solutions of the recently proposed MMOEAs such as NSGA-II-GrCD$_{\text{dec}}$ [8], NSGA-II-CD$_{\text{ws}}$ [9], and MOEA/D-MM as a decomposition-based MMOEA algorithm [18]. In addition, we compare with the original NSGA-II approach [2] as a baseline algorithm. We use the standard SBX crossover and the Neighborhood-Based polynomial Mutation (NBM) in NxEMMO, NSGA-II-GrCD$_{\text{dec}}$, and NSGA-II-CD$_{\text{ws}}$ algorithms. For both NSGA-II and MOEA-D/MM algorithms, we use the polynomial mutation as in the original settings. For all the algorithms, we set the probabilities for crossover P_c and mutation P_m to 0.1 and $1/n$ respectively and the distribution index is set to 20. The the neighborhood size for NBM operator is set to 20. The values for the weights in weighted sum CD approach in NSGA-II-CD$_{\text{ws}}$ algorithm are set to be equal for decision and objective spaces. The tuned gird size in NSGA-II-GrCD$_{\text{dec}}$ is set to 20. For MOEA/D-MM algorithm, the Tchebycheff scalarizing function is used, and the remaining parameters are taken from the original paper [18]. The population size for all the algorithms is set to 100 over 10000 function evaluation. The obtained results are conducted over 31 independent run.

To evaluate the performance of the algorithms in the decision space, the Inverted Generational Distance (IGDx) [24] and Pareto-Set Proximity (PSP) [22] are used. The IGDx value shows both the diversity and convergence of the optimal solutions comparing to the PS. The PSP value is computed by the fraction of the cover-rate to its obtained IGDx values, which additionally demonstrate the overlapping ratio between the bounding of obtained results and the PS. In order to measure the distribution of optimal solutions in the objective space, we additionally compute the obtained modified Inverted Generational Distance (IGD$^+$) [6]. Low values of IGDx and IGD$^+$ and large PSP values indicate a good quality of a solution set.

4.1 Results and Discussions

Tables 1 and 2 show the results on the test problems. We report the obtained median and Interquartile Range (IQR) values. To test the statistical significance difference between the best-performed and the rest of the algorithms, the Mann-Whitney-U statistical test is performed on each test problem with the rejection of the null hypothesis of equal medians at 5% significance level. The best performed algorithms are marked in blue shaded. An asterisk (*) represents statistical significance compared to the respective best performing algorithm.

Table 1. The median and IQR values for the MMOPs test problems with 2 objectives and 2 decision variables

Problems		NxEMMO	NSGA-II-GrCD$_{dec}$	NSGA-II-CD$_{ws}$	MOEA/D-MM	NSGA-II
MMF1	IGDx	0.057333 (0.003885)	0.061529 * (0.003658)	0.063365 * (0.002349)	0.096135 *(0.010598)	0.11478* (0.023295)
	PSP	17.4357 (1.2716)	16.1894 * (0.96983)	15.6552 * (0.55573)	10.1255 * (1.2259)	8.4532 * (1.7456)
	IGD+	0.004985 * (0.000491)	0.00461 * (0.000336)	0.004138 * (0.000487)	0.012369 * (0.002251)	0.0039 (0.000136)
MMF1z	IGDx	0.042153 (0.002158)	0.04548 * (0.002751)	0.046743 * (0.002183)	0.085181 * (0.012858)	0.11967 * (0.023685)
	PSP	23.6363 (1.0596)	21.6831 * (1.4293)	21.2745 * (1.1267)	11.5899 * (1.9291)	8.1724 * (1.5981)
	IGD+	0.004527 * (0.000368)	0.004325 * (0.000343)	0.004031 * (0.00025)	0.012124 * (0.002163)	0.003841 (0.000304)
MMF2	IGDx	0.021271 (0.010997)	0.021037 (0.005985)	0.022524 (0.009888)	0.051907 * (0.007774)	0.085467 * (0.064877)
	PSP	47.0041 (22.8022)	47.5307 (11.2784)	44.3962 (16.6322)	18.8317 * (2.8958)	9.9459 * (7.8377)
	IGD+	0.011595 (0.002283)	0.010796 (0.002632)	0.011449 (0.002148)	0.04209 * (0.015426)	0.013448 * (0.015575)
MMF3	IGDx	0.016841 (0.003006)	0.018123 * (0.005107)	0.017635 (0.004109)	0.043114 * (0.009751)	0.068035 * (0.046307)
	PSP	59.2117 (11.1151)	54.6617 * (15.2135)	56.4411 (14.6227)	22.3019 * (5.515)	12.4462 * (9.9079)
	IGD+	0.009627 (0.002095)	0.009779 (0.001319)	0.009541 (0.001992)	0.03908 * (0.016024)	0.009451 (0.007619)
MMF4	IGDx	0.033758 (0.001246)	0.038063 * (0.003046)	0.043562 * (0.003407)	0.058553 * (0.005996)	0.10556 * (0.035845)
	PSP	29.62 (1.1725)	26.1019 * (1.9905)	22.7149 * (1.7544)	16.8302 * (1.9185)	9.1544 * (3.1222)
	IGD+	0.005007 * (0.000336)	0.004708 * (0.000581)	0.00395 * (0.000558)	0.009006 * (0.001183)	0.00347 (0.000179)
MMF5	IGDx	0.56591 (0.00509)	0.59113 * (0.018144)	0.59296 * (0.021784)	0.61116 * (0.027865)	0.61874 * (0.015645)
	PSP	1.2489 (0.010362)	1.1899 * (0.033921)	1.179 * (0.038762)	1.1396 * (0.057833)	1.1292 * (0.06761)
	IGD+	0.009887 * (0.001574)	0.009493 * (0.00139)	0.008523 (0.000716)	0.016301 * (0.003965)	0.008859 * (0.00101)
MMF6	IGDx	0.083267 (0.003702)	0.092411 * (0.005499)	0.10156 * (0.006635)	0.12683 * (0.013485)	0.19387 * (0.057345)
	PSP	12.0077 (0.52738)	10.7622 * (0.67001)	9.7742 * (0.61704)	7.6677 * (1.0447)	4.9954 * (1.6075)
	IGD+	0.005173 * (0.000375)	0.004471 * (0.000341)	0.004105 * (0.00036)	0.010749 * (0.002326)	0.003975 (0.00029)
MMF7	IGDx	0.033066 (0.00205)	0.038008 * (0.003744)	0.036232 * (0.00354)	0.067248 * (0.008353)	0.069254 * (0.017539)
	PSP	30.1618 (1.8358)	25.884 * (2.598)	27.251 * (2.7578)	14.7081 * (1.7308)	13.775 * (4.1412)
	IGD+	0.005234 * (0.000493)	0.005233 * (0.000709)	0.003891 (0.000326)	0.010424 * (0.001453)	0.003858 (0.000161)
MMF8	IGDx	.069569 (0.010388)	0.078789 * (0.005428)	0.090855 * (0.008837)	0.15803 * (0.024995)	0.7981 * (0.33379)
	PSP	14.2175 (1.9557)	12.5812 * (0.87884)	10.8493 * (1.0033)	6.2097 * (0.93669)	1.0041 * (0.39043)
	IGD+	0.071599 * (0.000175)	0.071446 * (0.000206)	0.071129 * (8.6e-05)	0.075571 * (0.001446)	0.070967 (0.000126)
MMF9	IGDx	0.009395 (0.000454)	0.010535 * (0.00052)	0.013456 * (0.001685)	0.024058 * (0.004229)	0.093028 * (0.2199)
	PSP	106.4386 (5.069)	94.5761 * (5.0592)	74.3149 * (9.0069)	41.5646 * (7.5144)	10.7495 * (34.3256)
	IGD+	0.007226 * (0.000346)	0.006367 (0.00049)	0.006848 * (0.000576)	0.024759 * (0.004272)	0.006528 (0.000437)
SYM-PART simple	IGDx	.15533 (0.015901)	0.15829 (0.026981)	0.15666 (0.018678)	0.23649 * (0.029102)	4.8896 * (3.2094)
	PSP	6.4271 (0.62707)	6.3173 (1.0669)	6.3538 (0.73157)	4.2285 * (0.52935)	0.15392 * (0.14778)
	IGD+	0.023712 * (0.002836)	0.024066 * (0.003214)	0.024744 * (0.003663)	0.039777 * (0.010752)	0.009031 (0.000653)
SYM-PART rotated	IGDx	0.16716 (0.020006)	0.1742 (0.022506)	0.16927 (0.029576)	0.33232 * (0.055955)	3.439 * (3.2682)
	PSP	5.935 (0.72076)	5.7166 (0.71325)	5.8827 (0.99934)	2.9901 * (0.50108)	0.18946 * (0.20547)
	IGD+	0.025246 * (0.003639)	0.026858 * (0.005452)	0.026179 * (0.004853)	0.061262 * (0.017728)	0.012021 (0.000855)
Omni-test	IGDx	0.035867 (0.003688)	0.038492 * (0.004029)	0.04554 * (0.007082)	0.078662 * (0.008542)	0.28523 * (0.41252)
	PSP	27.5507 (2.7987)	25.7595 * (2.6943)	21.924 * (3.2973)	12.7127 * (1.4015)	3.4256 * (8.4479)
	IGD+	1.0015 * (0.000313)	1.0012 * (0.00031)	1.001 (0.000163)	1.0057 * (0.001808)	1.0009 (0.000309)

Table 2. The median and IQR values for the MMOPs test problems with m objectives and n decision variables denoted by (m, n)

Problems		NxEMMO	NSGA-II-GrCD$_{dec}$	NSGA-II-CD$_{ws}$	MOEA/D-MM	NSGA-II
MMF14 (2,2)	IGDx	0.009755 (0.000554)	0.011 * (0.000673)	0.011112 * (0.001555)	0.015809 * (0.001303)	0.055163 * (0.23483)
	PSP	102.5104 (5.6823)	90.8343 * (5.4055)	89.9947 * (12.0264)	63.2475 * (5.1912)	18.1279 * (71.2646)
	IGD+	0.00634 * (0.000364)	0.006013 * (0.000332)	0.005009 * (0.000262)	0.013459 * (0.00148)	0.004587 (0.000202)
MMF14 (2,3)	IGDx	0.061717 (0.000547)	0.069971 * (0.004163)	0.27686 * (0.078943)	0.08232 * (0.00316)	0.28135 * (0.14985)
	PSP	16.1963(0.14046)	14.1529 * (0.84656)	0.93739 * (4.1433)	12.126 * (0.42656)	1.1456 * (4.9268)
	IGD+	0.00703 * (0.000573)	0.006198 * (0.000481)	0.004585 (0.000352)	0.013122 * (0.001046)	0.004614 (0.000412)
MMF14 (2,4)	IGDx	0.15991 (0.001459)	0.17021 * (0.004378)	0.33209 * (0.015294)	0.18129 * (0.008009)	0.34708 * (0.08742)
	PSP	6.2536 (0.061754)	5.7898 * (0.18468)	1.1092 * (1.9031)	5.4735 * (0.28719)	1.0033 * (2.5676)
	IGD+	0.00874 * (0.001692)	0.00743 * (0.000919)	0.00445 (0.000389)	0.013832 * (0.001824)	0.004598 (0.000342)
MMF14 (2,5)	IGDx	0.28045 (0.003772)	0.2825 * (0.00889)	0.405 * (0.030056)	0.29064 * (0.006532)	0.43266 * (0.05619)
	PSP	3.561 (0.051713)	3.3742 * (0.16434)	1.0931 * (1.3485)	3.4016 * (0.098865)	0.95643 * (0.98797)
	IGD+	0.010686 * (0.002723)	0.007714 * (0.001634)	0.004613 (0.000347)	0.014384 * (0.00221)	0.004504 (0.000227)
MMF14(2,6)	IGDx	0.39688 * (0.006163)	0.39027 (0.032571)	0.48319 * (0.025679)	0.39108 (0.012884)	0.51255 * (0.059402)
	PSP	2.5191 (0.038893)	2.3889 * (0.20472)	1.2412 * (0.89386)	2.5108 (0.08259)	0.9378 * (1.0087)
	IGD+	0.009607 * (0.001888)	0.007496 * (0.001398)	0.004685 (0.000325)	0.014347 * (0.001797)	0.00456 (0.000276)
MMF14 (3,2)	IGDx	0.028863 (0.002116)	0.030418 * (0.002499)	0.038612 * (0.009857)	0.027346 (0.003925)	0.057272 * (0.016227)
	PSP	34.647 (2.4754)	32.7541 * (2.8163)	25.8985 * (5.8287)	36.5661 (5.0714)	17.4605 * (5.1105)
	IGD+	0.21643 (0.002338)	0.22606 * (0.006805)	0.22765 * (0.005684)	0.22646 * (0.007508)	0.24914 * (0.044195)
MMF14 (3,3)	IGDx	0.065963 (0.001522)	0.077603 * (0.002436)	0.093566 * (0.031933)	0.096688 * (0.00537)	0.098473 * (0.051046)
	PSP	15.1601 (0.35352)	12.8649 * (0.47324)	10.6876 * (3.0811)	10.3053 * (0.56327)	10.1551 * (3.7266)
	IGD+	0.062038 (0.00419)	0.088771 * (0.008063)	0.064592 * (0.003777)	0.099077 * (0.013499)	0.065071 * (0.003574)
MMF14 (3,4)	IGDx	0.1626 (0.00265)	0.17205 * (0.004775)	0.22549 * (0.060824)	0.20149 * (0.010056)	0.24427 * (0.080399)
	PSP	6.15 (0.09987)	5.7297 * (0.18343)	4.162 * (1.1203)	4.9516 * (0.31695)	3.9392 * (1.4713)
	IGD+	0.062882 (0.004794)	0.080258 * (0.012709)	0.0648 (0.004202)	0.10319 * (0.01112)	0.064711 * (0.003161)
MMF14 (3,5)	IGDx	0.28321 (0.003774)	0.28194 (0.010309)	0.36965 * (0.071462)	0.31298 * (0.015284)	0.35853 * (0.080906)
	PSP	3.5309 (0.046894)	3.4452 (0.22479)	2.5605 * (0.58305)	3.1835 * (0.14811)	2.7441 * (0.69071)
	IGD+	0.071212 * (0.005722)	0.084442 * (0.016264)	0.063152 (0.004551)	0.10547 * (0.008923)	0.063997 (0.004941)
MMF14 (3,6)	IGDx	0.39764 * (0.006007)	0.38038 * (0.010943)	0.40883 * (0.012133)	0.41926 (0.015967)	0.4681 * (0.067918)
	PSP	2.5146 * (0.037991)	2.5037 * (0.1042)	1.9384 * (0.43332)	2.3586 (0.101)	1.9291 * (0.39474)
	IGD+	0.067624 * (0.00349)	0.084592 * (0.015221)	0.064076 * (0.003906)	0.10617 * (0.015753)	0.064996 (0.004369)

We observe that the proposed NxEMMO algorithm performs better on 12 out of 13 test instances regarding the obtained median IGDx and PSP values. Even though in MMF2 test case, NSGA-II-Gr-CD_{dec} achieves better IGDx and PSP values, there is no significance statistical difference between the two algorithms.

These results are inline with our expectations. It seems that the modification of the environmental selection and replacing the crowding distance with the approach proposed in this paper leads to better results. The NxEMMO method is able to better identify the solutions located in crowded area in the decision space compared to the crowding distance approach which is used in the NSGA-II-CD_{WS} and NSGA-II-$GrCD_{dec}$ algorithms. We also expect that performing the addition and omission as denoted in Algorithm 2 has led to enhance the approximation of PS. In order to analyse this aspect, we take a closer look at the results on the SYM-PART test case with 9 PSs in Table 1. We note that NxEMMO covers larger number of PSs in comparison with the other approaches. While NSGA-II-$GrCD_{dec}$ gains better IGDx value for MMF2 test problem, no statistically significant difference is observed compared to NxEMMO algorithm. Furthermore, we observe that the NSGA-II algorithm demonstrates its superiority among all the algorithms, when we look at the IGD^+ values in the objective space. The obtained result is in compliance with our expectations, since NSGA-II contributing on distribution of solutions in objective space, while suffering from the preservation of diverse solutions in decision space. We examine the results of NxEMMO on the scalable MMF14 test case with 2 to 6 decision variables and for 2 to 3 objective functions. The IGDx values in Table 2 indicate that even in higher dimensional decision spaces, in 6 out of 10 cases the NxEMMO algorithms performs the best and for the other 2 cases no significance difference can be observed. From the comparison of PSP results, we additionally observe that NxEMMO performs the best, which further shows, that this approach is enhancing the coverage of the PS.

Overall, the above results indicate that the proposed selection mechanism preserves the distribution of solutions in the decision space. Nevertheless, for the test cases with 6 decision variables and 2 objective space, the NSGA-II-Gr-CD_{dec} obtains better performance than NxEMMO algorithm. An interesting observation is that for the 3-objective test problems, the proposed algorithm shows the superiority for the approximation of PF for 2, 3, and 4 decision variables.

5 Conclusion and Future Work

This paper exploited the potential of HAD and NND density-based approaches to handle multimodal multi-objective optimization problems. In the proposed NxEMMO the a density-based environmental selection operator is introduced to increase the distribution of solutions in decision space. We examine the efficiency of the proposed algorithms on various MMOPs and compare it with the recent state-of-the-art MMOEAs and NSGA-II as a baseline. In most of the cases, the results indicate that the proposed approach can enhance the spread of obtained solutions in the decision space. Various research is still remained as future works such as developing new algorithms to deal with multiple local PSs and PFs and designing new performance indicators to properly measure the distribution of solutions in the decision space.

References

1. Bader, J., Zitzler, E.: HypE: an algorithm for fast hypervolume-based many-objective optimization. Evol. Comput. **19**(1), 45–76 (2011)
2. Deb, K., Pratap, A., Agarwal, S., Meyarivan, T.: A fast and elitist multiobjective genetic algorithm: NSGA-II. IEEE Trans. Evol. Comput. **6**(2), 182–197 (2002)
3. Deb, K., Tiwari, S.: Omni-optimizer: a procedure for single and multi-objective optimization. In: Coello Coello, C.A., Hernández Aguirre, A., Zitzler, E. (eds.) EMO 2005. LNCS, vol. 3410, pp. 47–61. Springer, Heidelberg (2005). https://doi.org/10.1007/978-3-540-31880-4_4
4. Deb, K., Tiwari, S.: Omni-optimizer: a generic evolutionary algorithm for single and multi-objective optimization. Eur. J. Oper. Res. **185**(3), 1062–1087 (2008)
5. Huang, V., Suganthan, P., Qin, A., Baskar, S.: Multiobjective differential evolution with external archive and harmonic distance-based diversity measure, pp. 1–25 (2005)
6. Ishibuchi, H., Masuda, H., Tanigaki, Y., Nojima, Y.: Modified distance calculation in generational distance and inverted generational distance. In: Gaspar-Cunha, A., Henggeler Antunes, C., Coello, C.C. (eds.) EMO 2015. LNCS, vol. 9019, pp. 110–125. Springer, Cham (2015). https://doi.org/10.1007/978-3-319-15892-1_8
7. Javadi, M., Ramirez-Atencia, C., Mostaghim, S.: Combining Manhattan and crowding distances in decision space for multimodal multi-objective optimization problems. In: Gaspar-Cunha, A., Periaux, J., Giannakoglou, K.C., Gauger, N.R., Quagliarella, D., Greiner, D. (eds.) Advances in Evolutionary and Deterministic Methods for Design, Optimization and Control in Engineering and Sciences. CMAS, vol. 55, pp. 131–145. Springer, Cham (2021). https://doi.org/10.1007/978-3-030-57422-2_9
8. Javadi, M., Ramirez-Atencia, C., Mostaghim, S.: A novel grid-based crowding distance for multimodal multi-objective optimization. In: 2020 IEEE Congress on Evolutionary Computation (CEC), pp. 1–8. IEEE (2020)
9. Javadi, M., Zille, H., Mostaghim, S.: The effects of crowding distance and mutation in multimodal and multi-objective optimization problems. In: Gaspar-Cunha, A., Periaux, J., Giannakoglou, K.C., Gauger, N.R., Quagliarella, D., Greiner, D. (eds.) Advances in Evolutionary and Deterministic Methods for Design, Optimization and Control in Engineering and Sciences. CMAS, vol. 55, pp. 115–130. Springer, Cham (2021). https://doi.org/10.1007/978-3-030-57422-2_8
10. Kerschke, P., Preuss, M.: Exploratory landscape analysis. In: GECCO 2019 Companion - Proceedings of the 2019 Genetic and Evolutionary Computation Conference Companion, pp. 1137–1155 (2019)
11. Kumar, K., Deb, K.: Real-coded genetic algorithms with simulated binary crossover: Studies on multimodal and multiobjective problems. Complex Syst. **9**, 431–454 (1995)
12. Liang, D., Qu, B., Gong, D., Yue, C.: Problem definitions and evaluation criteria for the CEC 2019 special session on multimodal multiobjective optimization. In: Computational Intelligence Laboratory, Zhengzhou University (2019)

13. Liang, J., Guo, Q., Yue, C., Qu, B., Yu, K.: A self-organizing multi-objective particle swarm optimization algorithm for multimodal multi-objective problems. In: Tan, Y., Shi, Y., Tang, Q. (eds.) ICSI 2018. LNCS, vol. 10941, pp. 550–560. Springer, Cham (2018). https://doi.org/10.1007/978-3-319-93815-8_52
14. Liang, J., Yue, C., Qu, B.Y.: Multimodal multi-objective optimization: a preliminary study. In: 2016 IEEE Congress on Evolutionary Computation (CEC), pp. 2454–2461. IEEE (2016)
15. Lin, Q., Lin, W., Zhu, Z., Gong, M., Li, J., Coello, C.: Multimodal multi-objective evolutionary optimization with dual clustering in decision and objective spaces. IEEE Trans. Evol. Comput. 1–1 (2020)
16. Liu, Y., Ishibuchi, H., Nojima, Y., Masuyama, N., Shang, K.: A double-niched evolutionary algorithm and its behavior on polygon-based problems. In: Auger, A., Fonseca, C.M., Lourenço, N., Machado, P., Paquete, L., Whitley, D. (eds.) PPSN 2018. LNCS, vol. 11101, pp. 262–273. Springer, Cham (2018). https://doi.org/10.1007/978-3-319-99253-2_21
17. Optimization, M.m.M.o., Pal, M., Bandyopadhyay, S.: Decomposition in decision and objective space for multi-modal multi-objective optimization. arXiv preprint arXiv:2006.02628 (2020)
18. Peng, Y., Ishibuchi, H.: A decomposition-based large-scale multi-modal multi-objective optimization algorithm. arXiv preprint arXiv:2004.09838 (2020)
19. Peng, Y., Ishibuchi, H., Shang, K.: Multi-modal multi-objective optimization: problem analysis and case studies. In: 2019 IEEE Symposium Series on Computational Intelligence (SSCI), pp. 1865–1872. IEEE (2019)
20. Sengupta, R., Pal, M., Saha, S., Bandyopadhyay, S.: NAEMO: neighborhood-sensitive archived evolutionary many-objective optimization algorithm. Swarm Evol. Comput. **46**, 201–218 (2019)
21. Tanabe, R., Ishibuchi, H.: A niching indicator-based multi-modal many-objective optimizer. Swarm Evol. Comput. **49**, 134–146 (2019)
22. Yue, C., Qu, B., Liang, J.: A multiobjective particle swarm optimizer using ring topology for solving multimodal multiobjective problems. IEEE Trans. Evol. Comput. **22**(5), 805–817 (2018)
23. Zhang, Q., Li, H.: MOEA/D: a multiobjective evolutionary algorithm based on decomposition. IEEE Trans. Evol. Comput. **11**(6), 712–731 (2007)
24. Zhou, A., Zhang, Q., Jin, Y.: Approximating the set of pareto-optimal solutions in both the decision and objective spaces by an estimation of distribution algorithm. IEEE Trans. Evol. Comput. **13**(5), 1167–1189 (2009)
25. Zitzler, E., Laumanns, M., Thiele, L.: SPEA2: improving the strength pareto evolutionary algorithm. In: Evolutionary Methods for Design Optimization and Control with Applications to Industrial Problems, pp. 95–100 (2001)

Many-objective Optimization

The (M-1)+1 Framework of Relaxed Pareto Dominance for Evolutionary Many-Objective Optimization

Shuwei Zhu[1(✉)], Lihong Xu[1], Erik Goodman[2], Kalyanmoy Deb[2], and Zhichao Lu[3]

[1] Tongji University, Shanghai 201804, China
zswjiang@163.com, xulhk@163.com

[2] Michigan State University, East Lansing, MI 48824, USA
{goodman,kdeb}@egr.msu.edu

[3] Southern University of Science and Technology, Shenzhen 518055, China
mikelzc1990@gmail.com

Abstract. In the past several years, it has become apparent that the effectiveness of Pareto dominance-based multiobjective evolutionary algorithms degrades dramatically when solving many-objective optimization problems (MaOPs). Instead, research efforts have been driven toward developing evolutionary algorithms (EAs) that do not rely on Pareto dominance (e.g., decomposition-based techniques) to solve MaOPs. However, it is still a non-trivial issue for many existing non-Pareto-dominance-based EAs to deal with unknown irregular Pareto front shapes. In this paper, we develop the novel "(M-1)+1" framework of relaxed Pareto dominance to address MaOPs, which can simultaneously promote both convergence and diversity. To be specific, we apply M symmetrical cases of relaxed Pareto dominance during the environmental selection step, where each enhances the selection pressure of M-1 objectives by expanding the dominance area of solutions, while remaining unchanged for the one objective left out of that process. Experiments demonstrate that the proposed method is very competitive with or outperforms state-of-the-art methods on a variety of scalable test problems.

Keywords: Evolutionary algorithm · Many-objective optimization · Pareto dominance · Expanding the dominance area

1 Introduction

Many-objective optimization problems (MaOPs), as refer to the task of optimizing more than three conflicting objectives, have posed great challenges to existing multi-objective evolutionary algorithms that were designed for solving two- or three-objective problems. As the number of objectives M increases, most of Pareto-optimal individuals are mutually non-dominated, resulting in their incomparability. In other words, the Pareto dominance relation cannot effectively distinguish the quality of solutions for MaOPs, mainly due to the loss of selection pressure towards the true Pareto optimal set [4,10].

H. Ishibuchi et al. (Eds.): EMO 2021, LNCS 12654, pp. 349–361, 2021.
https://doi.org/10.1007/978-3-030-72062-9_28

To tackle the scalability problem of Pareto-based evolutionary algorithms (EAs), two main groups of many-objective evolutionary algorithms (MaOEAs) have been developed. The first group is the Relaxed Pareto Dominance (RPD) relation that modifies the Pareto dominance concept, like CDAS [12], S-CDAS [11], and the generalization of Pareto optimality (GPO) [16], to better discriminate among solutions and select elite ones with enhanced selection pressure. The other group modifies the diversity maintenance operations—e.g., the use of associated reference points in NSGA-III [1]. However, most of them, especially the former, have difficulty maintaining the delicate balance between convergence and diversity. Usually, excessive selection pressure tends to deteriorate diversity maintenance—e.g., it may lead the population to converge into a subregion or several small subregions of the Pareto front (PF).

In recent years, a large number of decomposition-based algorithms have been proposed for MaOPs, as they can generate a strong selection pressure towards the PF. However, it has been shown that the performance of decomposition-based algorithms strongly depends on the PF shapes [3], and the predefined reference weights/directions that should be distributed uniformly along the PF play an important role in the performance. Hence, the decomposition-based methods are more likely to fail on irregular MaOPs. More recently, Ishibuchi et al. [5] demonstrated that MaOPs are not always difficult for Pareto dominance-based EAs, and even NSGA-II outperforms recent decomposition-based algorithms on a variety of test MaOPs, apart from the widely-used DTLZ test problems. It was also shown that a slight change of the DTLZ feasible region can clearly improve the convergence ability of NSGA-II for higher numbers of objectives [4].

Therefore, to fully investigate the potential of Pareto dominance in MaOPs, we propose the "(M-1)+1" framework of RPD, (M-1)-RPD for short, which is free of setting reference points/vectors. To solve the problem of losing population diversity in RPD methods, M symmetrical cases of (M-1)-RPD are used simultaneously to balance convergence and diversity, which leads to the MultiRPO (i.e., Multiple (M-1)-RPD-based Optimization) algorithm. Specifically, in each case of (M-1)-RPD, M-1 objectives enhance the selection pressure by expanding dominance area (via RPD), and the one objective left out remains unchanged.

The remainder of this paper is organized as follows: Sect. 2 presents some existing RPD methods. The proposed "(M-1)+1" framework of RPD is described in Sect. 3. The experimental study on test problems is presented in Sect. 4. Finally, Sect. 5 concludes the paper.

2 Existing Relaxed Pareto Dominance Methods

RPD methods use a straightforward way to enhance selection pressure by modifying the objective to expand the dominated area for each solution. There exist many RPD methods in the literature, such as α-dominance [6], CDAS [12], S-CDAS [11], and GPO [16], etc.

First, the early proposed α-dominance [6] directly modifies the objective as: $f_i'(\mathbf{x}) = f_i(\mathbf{x}) + \sum_{j \neq i}^{M} \alpha f_j(\mathbf{x})$, where α is a parameter to control the modification degree. The CDAS approach [12] is a very representative RPD, which has gained much attention recently. It modifies the objective value as: $f_i'(\mathbf{x}) = \frac{\|f(\mathbf{x})\|\sin(\omega_i + S\cdot\pi)}{\sin(S\cdot\pi)}$, where ω_i denotes the declination angle between $\mathbf{x}$ and the i-th axis and $S \in [0.25, 0.5]$ is a key parameter to control the expanding degree. In order to eliminate the parameter S of CDAS, the authors proposed an adaptive version of CDAS, called self-CDAS (S-CDAS) [11]. S-CDAS adaptively determines the expanding degree of a candidate solution $\mathbf{x}$ according to the extreme solutions in the population, thereby modifying the objective value by $f_i'(\mathbf{x}) = \frac{\|f(\mathbf{x})\|\sin(\omega_i + \phi_i)}{\sin(\phi_i)}$, where $\phi_i = \arcsin\frac{\|f(\mathbf{x})\|\cdot\sin(\omega_i)}{\|f(\mathbf{x}) - p_i\|}$ and p_i is the extreme solution in the population with respect to the i-th axis.

As pointed out and analyzed in [16], the existing RPD methods cannot guarantee identity of expanding dominance area for each solution to improve selection pressure if $M > 3$. Nevertheless, a GPO criterion is desired that retains the common features of conventional Pareto optimality, to ensure the identity of dominance envelopes (i.e., the boundaries of the dominance areas) for all solutions [16]. It modifies the objective value as $f_i'(\mathbf{x}) = f_i(\mathbf{x}) + \sum_{k \neq i} \delta_i f_k(\mathbf{x})$, where $\delta_i = \frac{\tan\varphi_i}{\sqrt{(M-1)}}$ is determined by the expanding angle φ. It is worth noting that, in these RPD methods, extra parameters are usually required (except in S-CDAS [11]) that should be well tuned to control the dominance area.

Although the aforementioned RPD methods have shown their capacity to solve MaOPs by enhancing selection pressure, which instead, at the cost of introducing some difficulty in diversity maintenance. To solve this problem, we are inspired by developing a "(M-1)+1" framework of RPD, which can balance convergence and diversity more effectively in addressing MaOPs.

3 Proposed Method

In this section, we first present the overall framework of the proposed MultiRPO method. Thereafter, the key component of MultiRPO—i.e., the (M-1)-RPD-based environmental selection—is described in detail.

3.1 The Overall Framework of MultiRPO

The proposed MultiRPO algorithm starts with the initialization step by randomly generating a population P and evaluating it. Then the evolutionary operators, including mating selection and reproduction operations (typically crossover and mutation), are executed to generate offspring solutions. Here we employ binary tournament selection to select promising parents for recombination. As for the environmental selection step, the (M-1)-RPD-based environmental selection (see Algorithm 2) is used to choose the solutions to survive to the next generation. The evolution procedure is repeated until the termination criterion is met. The overall framework of MultiRPO is presented in Algorithm 1, where the input includes population size N and the parameter for a certain RPD technique. Note that there also exists parameter α in α-dominance, S in CDAS, and φ in GPO. For simplicity, here we just present φ (for GPO) in the input.

Algorithm 1: General framework of MultiRPO

Input: population size N, the parameter φ (for GPO);
Output: final population P.
1 P=Population_initialization();
2 Evaluate population P;
3 **while** *termination criterion is not fulfilled* **do**
4 $P' \leftarrow$ Mating_Selection (P);
5 $Q \leftarrow$ Reproduction (P');
6 $R \leftarrow P \cup Q$;
7 P = (M-1)-RPD-based Environmental_Selection (R, N, φ);
8 **end**

In each generation, the computational complexity is dominated by the environmental selection procedure: each of the M different (M-1)-RPD cases takes $O(MN^2)$ time, so the total complexity is $O(M^2N^2)$, but the M cases of (M-1)-RPD can be executed in parallel with a total complexity reducing to $O(MN^2)$.

3.2 (M-1)-RPD-Based Environmental Selection

In our (M-1)-RPD-based environmental selection, we implement M symmetrical (M-1)-RPD processes, wherein for each process the dominance area expansion is performed in M-1 dimensions, and the remaining dimension remains unexpanded. Here, we take a 3-D problem as an example, and the (M-1)-GPO scheme (GPO is a specific RPD method) is used for a more convenient explanation. Thus, three expanding vectors $\Phi^1 = [0, \varphi, \varphi]$, $\Phi^2 = [\varphi, 0, \varphi]$, and $\Phi^3 = [\varphi, \varphi, 0]$, are used for the three processors with the same expanding angle φ. With this subdivision, the selection pressure of each sub-task is sufficient using a relatively small φ, which remains fixed during the entire optimization process. For an M-objective optimization problem, there are M different, parallel cases of (M-1)-GPO, and each one is responsible for converging to a relatively large subregion.

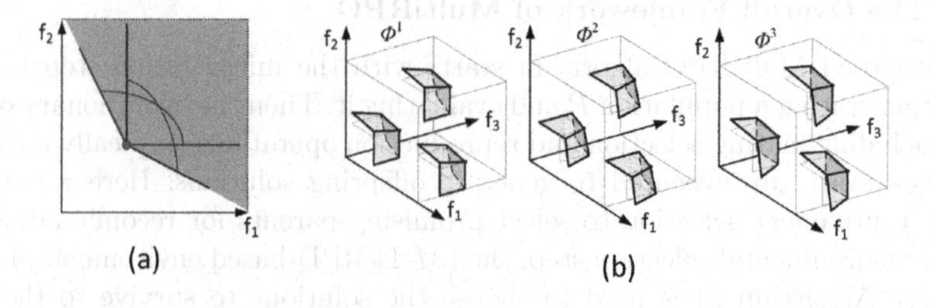

Fig. 1. An illustration of MultiRPO (GPO is used here). (a) the obtained dominating area in bi-objective space; and (b) the dominance envelopes in three 3-objective spaces.

Algorithm 2: (M-1)-RPD-based Environmental_Selection

Input: Population size N, combined population R and expanding angle φ;
Output: The survival solution set P.
1 Initialize $P = \varnothing$ and conduct fast non-dominated sorting on R;
2 Create an M-length vector $T_s = [t, \ldots, t]$, where $t = \lfloor \frac{N}{M} \rfloor$;
3 Let N_t=$t \times M$, and if $N_t < N$, then N-N_t random values of T_s are set as t+1;
4 **for** k=1:M **do**
5 Let expanding vector $\Phi^k = [\varphi, \ldots, \varphi]$ with M copies of the same φ, then set $\Phi^k(k) = 0$ that is used for the (M-1)-RPD$_{f_k}$ case;
6 $Front_k^{(\mathrm{RPD})}$ = (M-1)-RPD-based-Sort(R, Φ^k);
7 **end**
8 Randomly select a solution $\mathbf{x}$ in the first front determined by a certain $Front_k^{(\mathrm{RPD})}$ to append to P, and $R \leftarrow R\backslash\mathbf{x}$, $T_s(k)$=$T_s(k)$-1;
9 Generate a value set $\Gamma = \{1, 2, \ldots, M\}$; %% for randomly selecting each case
10 **for** j=1:M **do**
11 Let R_{nd} be the set of all non-dominated solutions in R;
12 Randomly select a value τ from set Γ, then $\Gamma \leftarrow \Gamma\backslash\tau$;
13 $PF^* = Front_\tau^{(\mathrm{RPD})}$;
14 **for** each solution $\mathbf{x} \in R_{\mathrm{nd}}$ **do**
15 $d_{\min}(\mathbf{x}) = \min_{\mathbf{y}\in P} dist(\mathbf{x}, \mathbf{y})$; %% angular distance is used here
16 **end**
17 **for** $i = 1 : T_s(j)$ **do**
18 Sort solutions of R_{nd} in descending order of $d_{\min}$ values;
19 Let R_T be the subset of the top ranked N_t individuals in R_{nd};
20 $S = \{\mathbf{x} | \mathbf{x} \in R_T, PF^*(\mathbf{x}) = \min(PF^*(R_T))\}$;
21 $\mathbf{s}_i$=arg max($d_{\min}(S)$); %% if $|S| = 1$, then $\mathbf{s}_i$=S
22 $P \leftarrow P \bigcup \mathbf{s}_i$, $R \leftarrow R\backslash\mathbf{s}_i$;
23 **end**
24 **end**

To better explain the proposed framework, Fig. 1 illustrates the dominating area in bi-objective space and the dominance envelopes in three-objective spaces, which is obtained by the MultiGPO (i.e., Multiple (M-1)-GPO-based Optimization) method. In Fig. 1(a) the dominating areas inside the red lines and blue lines, respectively, are achieved by the two complementary (M-1)-GPO cases; while in Fig. 1(b), (M-1)-GPO$_{f_1}$ with Φ^1, (M-1)-GPO$_{f_2}$ with Φ^2, and (M-1)-GPO$_{f_3}$ with Φ^3, are used for the three processors using the same φ value. Specifically, the dominance envelope of (M-1)-GPO$_{f_1}$ is obtained by expanding the dominated area of objectives f_2 and f_3 (f_1 is not expanded), and it is similar for the other two cases. Thanks to the capacity of GPO, the expanding degree is the same of all solutions in each case.

For the environmental selection, fast non-dominated sorting is first performed on the original objectives, for rough selection. The whole process of the (M-1)-RPD-based selection method is described in Algorithm 2, which roughly contains two steps: 1) lines 4–8: simultaneously conducting M symmetrical (M-1)-RPD-based-Sort operations to obtain an $M \times 2N$ matrix $Front^{(\mathrm{RPD})}$ of sorted values, where each row denotes the sorting result of a specific case; and 2) lines 9–24: selecting solutions one-by-one according to each of M rows of $Front^{(\mathrm{RPD})}$. Note that, each row of $Front^{(\mathrm{RPD})}$ is selected in a random order and resampled each generation (see lines 9,12), rather than sequentially adopted from the first row to the M-th one. Moreover the parameter N_t in line 19 is set as $\min(0.5|R|, |R_{\mathrm{nd}}|)$, in order to balance selection pressure enhancement (lines 13, 20) and the max-min subset-selection-based diversity preservation (lines 18–22).

In fact, Algorithm 2 is exactly used for MultiGPO. For other MultiRPO versions, the minor changes are needed in line 5 of Algorithm 2, resulting in the modified objective vector $[f_1'(\mathbf{x}), \dots, f_i(\mathbf{x}), \dots, f_M'(\mathbf{x})]$ for $(M\text{-}1)\text{-RPD}_{f_i}$. For example, $\Phi^k = [S, \dots, S]$ is used in line 5, and set $\Phi^k(k) = 0.5$ (Pareto dominance is achieved by setting $S = 0.5$ [12]) for the MultiCDAS algorithm (via CDAS).

4 Experiments

4.1 Experimental Setting

Firstly, four representative RPD methods—namely α-dominance, CDAS, S-CDAS and GPO—are introduced into our framework, to verify the effectiveness of the proposed (M-1)+1 framework of RPD. Then, for the purpose of performance comparison, we consider some representative decomposition-based MaOEAs, like NSGA-III [1], MOEA/DD [7], and θ-DEA [15], which were also considered in [5]. Also, the NSGA-II [2] considered in [5] and the recent NSGA-II/SDR which based on the strengthened dominance relation (SDR) [14] are all compared here. All methods were executed on the PlatEMO platform [13].

- Test problems: Here, we use 18 newly generated types of test problems (Type1-Type18 for short) from [5] and the MaF2-MaF9 test problems, which have complicated PFs [13] (MaF1 is similar to Type1, so is not used), to compare the performance of the aforementioned methods. These test MaOPs differ in their properties, such as linearity, convexity, concavity, degeneracy, and disconnectedness. Note that the distance function of DTLZ1 and the position function of DTLZ2 were used to generate the Type problems. The number of objectives M is set by $M \in \{5, 8, 10, 15\}$. For the 18 Type test problems, the number of distance variables is specified as $k = 5$, thus the total number of decision variables n is $n = M + 4$; while for the MaF test problems, $n = M + 9$ for MaF2-MaF6, $n = M + 19$ for MaF7, and $n = 2$ for MaF8-MaF9.
- Parameter setting: In all algorithms, simulated binary crossover and polynomial mutation operators [1] are used to generate offspring, with the distribution indices both set as 20. Although the population size N can be arbitrarily assigned for MultiRPO, we set $N = \{210, 240, 275, 240\}$ for the cases $M = \{5, 8, 10, 15\}$, respectively, to make a fair comparison. The maximum generation ($G_{\max}$) is used to determine the termination of all algorithms, and here $G_{\max}$ is set as 200 for the 18 Type problems and 500 for the MaF problems.

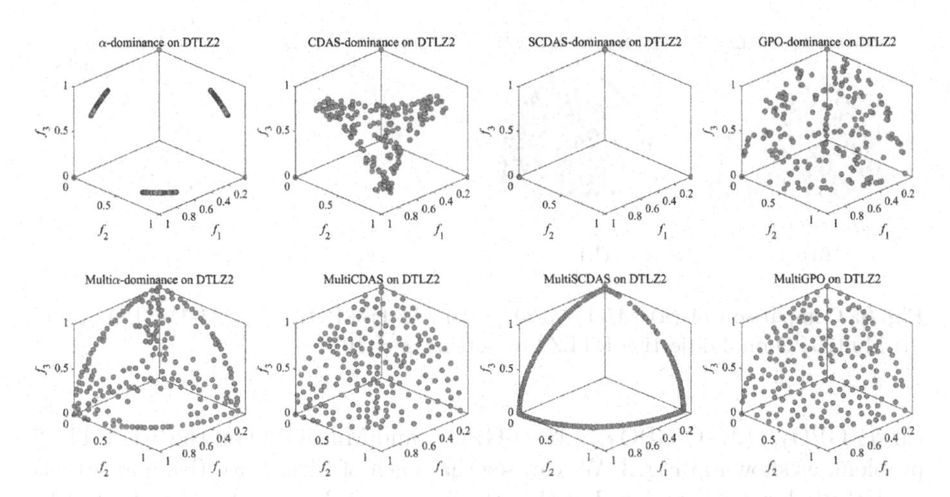

Fig. 2. Comparison of four RPD-based MaOEAs and the corresponding MultiRPO algorithms on 3-objective DTLZ2.

For the parameters that are used to modify the objectives (see Sect. 2), we set the parameter $\alpha = 1/3$ for α-dominance, $S = 0.39$ for CDAS and $\varphi = 3M$ for GPO. Moreover, in MOEA/DD, the neighborhood selection probability is $\delta = 0.9$; while in θ-DEA, the PBI function is used with $\theta = 5$.

- Performance metrics: Due to page limitation, we only present the performance comparison in terms of hypervolume (HV) [1] metric, wherein a larger HV indicates a better result. Before calculating the HV values, all objective values are normalized by the ideal and 1.1 times the values of nadir points. Then the reference point is set to (1.3,...,1.3) for Type problems which partly follows the original paper of Type problems [5], and (1,...,1) for MaF problems. We independently conduct all the experiments for 30 runs, and record the mean value and the standard deviation of each problem instance. Moreover, the Wilcoxon rank-sum test at a significance level of 0.05 is used.

4.2 Effect of the Proposed Framework

To illustrate the effectiveness of our proposed (M-1)+1 framework in balancing convergence and diversity, we compare four representative RPD methods (α-dominance, CDAS, S-CDAS and GPO) and their MultiRPO versions on 3-D DTLZ2 problems, as shown in Fig. 2. Note that the original RPD methods are embedded into NSGA-II to replace the Pareto-dominance-based solution sorting; while the MultiRPO method is as described in Algorithm 1. From Fig. 2, it is obvious that, compared to the four RPD-based methods, the related MultiRPO methods can improve the population diversity significantly.

Taking MultiGPO (the GPO-based MultiRPO version) as an example, we further present its effectiveness in improving population diversity by performing

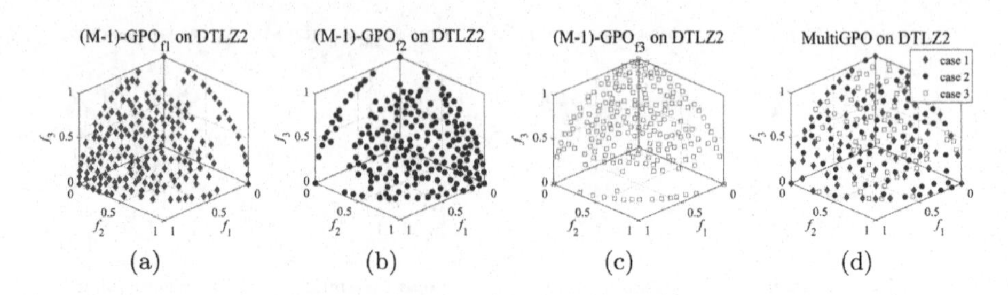

Fig. 3. Comparison of (a) (M-1)-GPO$_{f_1}$, (b) (M-1)-GPO$_{f_2}$, (c) (M-1)-GPO$_{f_3}$, and (d) MultiGPO on 3-objective DTLZ2 by setting $\varphi = 10$.

(M-1)-GPO$_{f_1}$, (M-1)-GPO$_{f_2}$, (M-1)-GPO$_{f_3}$ and MultiGPO on the 3-D DTLZ2 problem, as shown in Fig. 3. We can see that each of Figs. 3 (a), (b), and (c) has a relatively large gap region, but they display a complementary property. In (d), the solution set using these three cases simultaneously can cover the whole PF; also, in each case, the solutions generated can cover the gap region.

Table 1 presents the average HV values of NSGA-II and four MultiRPO methods (i.e., Multi-I–Multi-IV) on some test problems with different properties, including DTLZ2, DTLZ7 (similar to MaF7), MaF6, and Types 1, 7, 13. We see that the Multi-IV (i.e., MultiGPO) outperforms the others mostly. Thus, the MultiGPO method should be compared with other state-of-the-art MaOEAs, as discussed in the next subsection. In addition, we compare GPO and MultiGPO by setting different expanding angles (i.e., φ = 10, 20, 30) on 3-D DTLZ2 and DTLZ7, as shown in Figs. 4 and 5. It is evident that MultiGPO performs much better than GPO in all cases. The diversity of solutions obtained by MultiGPO is acceptable even if φ is fixed at 30, whereas GPO degrades dramatically when φ is increased to provide higher selection pressure. That is to say, the performance of GPO is very sensitive to the setting of φ, but MultiGPO alleviates this problem effectively, which is also achieved by other MultiRPO methods.

Table 1. Average HV values of NSGA-II, and Multi-I–Multi-IV on some test suites. The best for each instance is shown with dark gray background.

	M	NSGA-II	Multi-I	Multi-II	Multi-III	Multi-IV		NSGA-II	Multi-I	Multi-II	Multi-III	Multi-IV
DTLZ2	3	1.75e+0−	1.76e+0−	1.76e+0−	1.74e+0−	1.77e+0	Type1	2.02e+0−	2.04e+0≈	2.04e+0≈	2.03e+0≈	2.04e+0
	5	3.36e+0−	3.52e+0+	3.50e+0−	3.52e+0+	3.51e+0		1.89e+0−	3.68e+0≈	3.68e+0≈	3.60e+0−	3.68e+0
	10	2.41e−1−	1.38e+1+	1.37e+1≈	1.37e+1≈	1.37e+1		0.00e+0−	1.38e+1≈	1.38e+1≈	1.37e+1−	1.38e+1
	15	8.62e−1−	5.12e+1≈	5.10e+1−	5.11e+1−	5.12e+1		0.00e+0−	5.12e+1+	5.08e+1≈	3.45e+1−	5.11e+1
DTLZ7	3	1.10e+0≈	1.11e+0≈	1.11e+0≈	1.11e+0≈	1.10e+0	Type7	1.74e+0−	1.76e+0−	1.76e+0−	1.74e+0−	1.77e+0
	5	1.62e+0−	1.71e+0−	1.30e+0−	1.83e+0−	1.88e+0		2.92e+0−	3.52e+0+	3.50e+0≈	3.36e+0−	3.51e+0
	10	1.12e−3−	4.15e−1−	2.44e−1−	5.36e+0−	5.60e+0		0.00e+0−	1.38e+1+	1.36e+1−	1.35e+1−	1.37e+1
	15	6.58e−5−	1.66e−2−	8.46e−3−	1.67e+1+	7.41e+0		0.00e+0−	5.12e+1+	5.08e+1−	6.44e+0−	5.11e+1
MaF6	3	1.01e+0≈	1.00e+0−	1.00e+0−	8.96e−1−	1.01e+0	Type13	2.17e+0≈	2.17e+0−	2.16e+0−	2.17e+0≈	2.18e+0
	5	1.33e+0+	1.31e+0−	1.31e+0−	1.27e+0−	1.32e+0		2.34e+0−	3.69e+0−	3.69e+0−	3.69e+0−	3.70e+0
	10	0.00e+0−	2.15e+0≈	2.23e+0≈	2.13e+0≈	2.69e+0		0.00e+0−	1.37e+1−	1.37e+1≈	1.38e+1≈	1.38e+1
	15	-	-	-	-	-		0.00e+0−	5.11e+1−	5.12e+1+	5.12e+1+	5.11e+1
+/ − / ≈ (All of the six problems)								1/19/4	6/11/7	1/13/10	3/13/8	

[1] Multi-I∼Multi-IV denotes Multiα-dominance, MultiCDAS, MultiSCDAS and MultiGPO, respectively.
[2]+, −, and ≈ indicate that the compared MOEA performs significantly better than, worse than, or similarly to Multi−IV, respectively, on that problem instance.

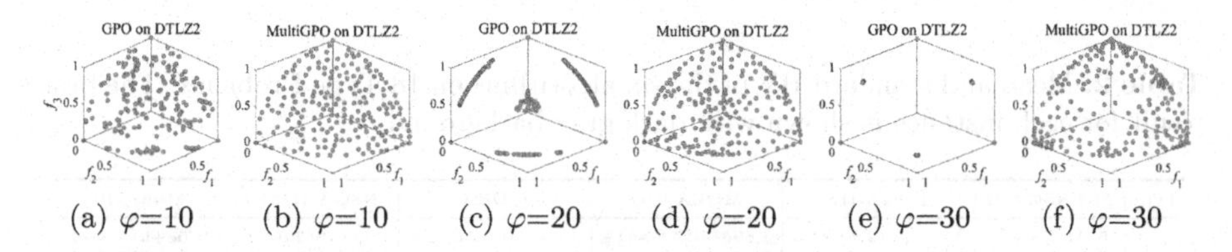

Fig. 4. The comparison of GPO (a, c, e) and MultiGPO (b, d, f) using different expanding angles (i.e., φ = 10, 20, 30) on 3-objective DTLZ2.

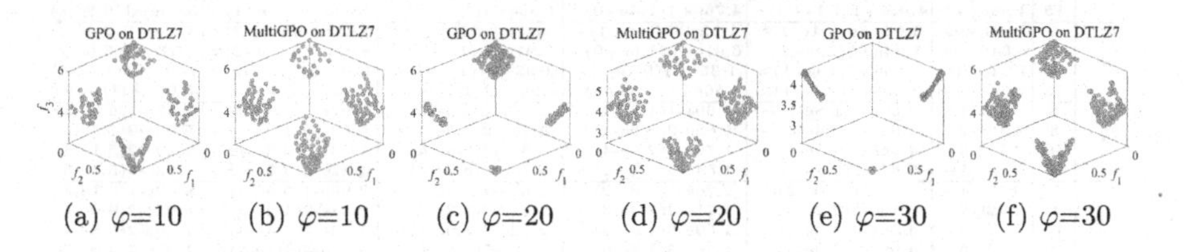

Fig. 5. The comparison of GPO (a, c, e) and MultiGPO (b, d, f) using different expanding angles (i.e., φ = 10, 20, 30) on 3-objective DTLZ7.

4.3 Overall Comparison on the Type and MaF Test Problems

The mean and standard HV values of all methods on Type 1–18 test problems are shown in Table 2. Note that the standard HV result of NSGA-II is not shown here due to space limitation. It is evident that MultiGPO achieves overall better HV performance than other MaOEAs on most instances, which shows the competitive potential of the proposed selection criterion (Algorithm 2) on these problems. To be specific, NSGA-II, NSGA-III, MOEA/DD, θ-DEA, and NSGA-II/SDR obtains better result than MultiGPO in merely 8, 3, 2, 6, and 13 out of 72 cases, respectively. Among the peer methods, NSGA-II/SDR performs a little better, as it can usually obtain the best HV results on Type 7–9 and Type 16–18. This may be owing to the fact that the convergence indicator of SDR, which is measured in terms of the sum of objectives, suits the PF shapes of these kinds of problems—namely, Concave triangular and Convex inverted PFs. As reported in [5], NSGA-II can beat the other examined decomposition-based MaOEAs on half of the 18 Type test problems with 8 and 10 objectives. This can also be witnessed in Table 2, where NSGA-II is beaten by NSGA-III, MOEA/DD, and θ-DEA in just 24, 14, and 23 out of the total 72 comparisons, respectively; however, it is outperformed by both NSGA-II/SDR and MultiGPO.

Table 3 reports the mean and standard HV values of all MaOEAs on the MaF2–MaF9 test problems, which tend to have more complicated PFs. It is also evident that MultiGPO can achieve overall better performance than other MaOEAs on most instances, in terms of HV. To be specific, our MultiGPO method beats NSGA-III, MOEA/DD, θ-DEA, and NSGA-II/SDR in 16, 28, 19, and 23 out of 32 cases, respectively. In addition, MultiGPO can also outperform the competitors on most Type and MaF problems in terms the IGD metric, however, the IGD results are not provided here due to page limitation.

Table 2. Mean and standard HV of the six algorithms on 18 Type Problems. The best result for each instance is shown with dark gray background.

Type	M	NSGA-II	NSGA-III	MOEA/DD	θ-DEA	NSGA-II/SDR	MultiGPO
1	5	1.85e+0−	3.61e+0(2.1e-1)≈	3.69e+0(7.8e-4)+	3.68e+0(8.3e-3)≈	3.63e+0(2.9e-2)−	3.6e+0(3.6e-3)
	8	0.00e+0−	6.53e+0(3.0e+0)−	8.12e+0(2.1e-2)−	8.13e+0(2.8e-2)≈	8.05e+0(5.9e-2)−	8.1e+0(7.6e-4)
	10	0.00e+0−	1.01e+1(5.5e+0)−	1.37e+1(1.1e-2)−	1.37e+1(1.0e-1)≈	1.35e+1(1.1e-1)−	1.3e+1(1.0e-3)
	15	0.00e+0−	4.64e+1(1.2e+1)≈	5.01e+1(4.3e-1)−	5.11e+1(2.4e-2)+	5.02e+1(7.4e-1)−	5.1e+1(3.7e-2)
2	5	3.66e+0≈	3.66e+0(3.4e-2)≈	3.65e+0(2.1e-2)≈	3.66e+0(1.0e-2)≈	3.63e+0(1.5e-2)−	3.66e+0(2.4e-2)
	8	8.08e+0≈	7.95e+0(1.0e-1)−	7.79e+0(2.3e-1)−	8.06e+0(6.3e-2)≈	8.03e+0(6.1e-2)−	8.07e+0(1.4e-1)
	10	1.36e+1≈	1.36e+1(7.7e-2)≈	1.36e+1(1.1e-1)≈	1.36e+1(1.0e-1)≈	1.35e+1(1.0e-1)≈	1.36e+1(1.5e-1)
	15	4.98e+1≈	4.93e+1(1.1e+0)−	4.76e+1(1.4e+0)−	4.96e+1(1.2e+0)≈	5.02e+1(6.4e-1)≈	5.04e+1(6.8e-1)
3	5	3.64e+0≈	3.65e+0(2.1e-2)≈	1.52e+0(1.5e-1)−	3.67e+0(7.8e-3)≈	3.63e+0(1.3e-2)≈	3.65e+0(2.6e-2)
	8	8.04e+0≈	7.94e+0(7.8e-2)−	6.61e+0(1.6e+0)−	7.97e+0(1.1e-1)≈	8.05e+0(4.8e-2)≈	8.07e+0(5.6e-2)
	10	1.34e+1≈	1.36e+1(1.6e-1)≈	1.35e+1(9.7e-2)−	1.37e+1(1.1e-1)≈	1.36e+1(8.3e-2)≈	1.36e+1(1.5e-1)
	15	4.99e+1≈	5.02e+1(7.5e-1)≈	4.56e+1(5.5e+0)−	4.96e+1(1.1e+0)−	5.0e+1(1.3e+0)≈	5.05e+1(1e+0)
4	5	3.76e-1−	3.30e-1(1.8e-2)−	2.60e-1(1.2e-2)−	2.56e-1(3.3e-2)−	3.97e-1(6.0e-3)−	4.18e-1(6.2e-3)
	8	4.13e-2−	3.32e-2(2.4e-3)−	1.62e-2(8.8e-4)−	2.25e-2(3.5e-3)−	5.52e-2(1.0e-3)−	5.93e-2(1.5e-3)
	10	7.76e-3−	4.68e-3(4.1e-4)−	1.70e-3(2.5e-4)−	3.32e-3(5.5e-4)−	1.23e-2(3.2e-4)−	1.29e-2(5.1e-4)
	15	7.84e-5−	5.61e-5(6.8e-6)−	1.54e-5(1.4e-6)−	3.79e-5(8.2e-6)−	1.62e-4(9.5e-6)≈	1.62e-4(5.1e-5)
5	5	3.77e-1−	3.36e-1(1.0e-2)−	2.26e-1(5.6e-2)−	2.61e-1(9.2e-3)−	3.96e-1(5.4e-3)−	4.20e-1(6.1e-3)
	8	4.42e-2−	3.13e-2(2.2e-3)−	1.71e-2(1.3e-3)−	2.34e-2(2.2e-3)−	5.44e-2(1.2e-3)−	6.09e-2(9.0e-4)
	10	9.05e-3−	4.80e-3(3.8e-4)−	1.69e-3(1.6e-4)−	2.99e-3(3.0e-4)−	1.23e-2(3.4e-4)−	1.30e-2(8.5e-4)
	15	9.56e-5≈	4.66e-5(6.0e-6)−	1.56e-5(1.4e-6)−	3.56e-5(6.8e-6)−	1.72e-4(1.7e-5)≈	1.56e-4(4.0e-5)
6	5	3.62e-1−	3.19e-1(7.5e-3)−	1.12e-1(1.2e-2)−	2.79e-1(3.0e-2)−	3.94e-1(6.1e-3)−	4.14e-1(6.4e-3)
	8	4.52e-2−	3.06e-2(6.8e-4)−	1.72e-2(5.6e-3)−	2.25e-2(4.0e-3)−	5.52e-2(9.6e-4)−	5.97e-2(1.2e-3)
	10	9.52e-3−	4.31e-3(2.4e-4)−	1.60e-3(1.2e-4)−	2.34e-3(5.1e-4)−	1.22e-2(2.9e-4)−	1.30e-2(5.3e-4)
	15	1.23e-4−	3.01e-5(1.7e-6)−	1.39e-5(9.7e-7)−	1.86e-5(1.2e-6)−	1.63e-4(1.6e-5)≈	1.95e-4(8.9e-5)
7	5	2.33e+0−	3.15e+0(1.1e+0)≈	3.50e+0(6.7e-3)≈	3.51e+0(8.5e-3)≈	3.50e+0(3.1e-3)≈	3.50e+0(1.3e-2)
	8	0.00e+0−	4.60e+0(3.9e+0)−	8.03e+0(2.0e-2)−	7.93e+0(1.7e-1)−	8.08e+0(2.1e-3)+	8.05e+0(1.6e-2)
	10	0.00e+0−	8.02e+0(6.9e+0)≈	1.37e+1(7.2e-3)+	1.37e+1(1.4e-2)+	1.37e+1(2.6e-3)+	1.36e+1(1.6e-2)
	15	0.00e+0−	4.09e+1(1.8e+1)−	4.93e+1(4.9e-1)−	5.11e+1(4.5e-2)≈	5.11e+1(3.5e-2)≈	5.11e+1(2.6e-2)
8	5	3.39e+0−	3.45e+0(5.9e-2)−	3.45e+0(5.5e-2)≈	3.50e+0(2.2e-2)≈	3.50e+0(1.3e-2)≈	3.49e+0(1.6e-2)
	8	7.71e+0−	7.83e+0(2.3e-1)≈	6.83e+0(1.0e+0)−	7.92e+0(8.0e-2)≈	8.03e+0(8.2e-2)≈	7.97e+0(1.3e-1)
	10	1.30e+1−	1.34e+1(1.8e-1)−	1.35e+1(9.7e-2)≈	1.35e+1(1.8e-1)≈	1.35e+1(2.1e-1)≈	1.36e+1(8.5e-2)
	15	4.77e+1−	4.95e+1(9.1e-1)−	4.48e+1(4.7e+0)−	4.96e+1(9.1e-1)≈	5.09e+1(2.0e-1)≈	5.07e+1(2.7e-1)
9	5	3.35e+0−	3.45e+0(2.3e-2)−	1.03e+0(2.1e-1)−	3.48e+0(1.3e-2)−	3.49e+0(1.7e-2)≈	3.50e+0(2.1e-2)
	8	7.61e+0−	7.88e+0(9.1e-2)−	4.87e+0(2.3e+0)−	7.80e+0(6.4e-2)−	8.08e+0(2.8e-3)+	8.03e+0(5.5e-2)
	10	1.29e+1−	1.35e+1(8.5e-2)−	1.31e+1(3.9e-1)−	1.34e+1(2.1e-1)−	1.37e+1(4.7e-2)+	1.36e+1(8.9e-2)
	15	4.65e+1−	4.98e+1(1.1e+0)−	3.72e+1(1.3e+1)−	4.80e+1(1.7e+0)−	5.09e+1(4.7e-1)≈	5.06e+1(6.5e-1)
10	5	1.73e-1−	1.34e-1(1.3e-2)−	1.31e-1(4.4e-3)−	1.11e-1(7.5e-3)−	1.82e-1(1.2e-3)−	1.84e-1(1.4e-3)
	8	1.60e-2−	7.26e-3(8.0e-4)−	8.29e-3(1.6e-3)−	5.70e-3(1.6e-3)−	1.96e-2(1.6e-3)≈	1.99e-2(1.3e-3)
	10	3.11e-3−	8.01e-4(5.7e-5)−	8.25e-4(1.4e-4)−	9.07e-4(3.6e-4)−	3.80e-3(5.8e-4)≈	3.72e-3(7.0e-4)
	15	3.55e-5≈	6.81e-6(1.2e-6)−	1.01e-5(1.2e-6)−	6.34e-6(1.5e-6)−	4.64e-5(3.1e-5)≈	3.75e-5(8.6e-6)
11	5	1.68e-1−	1.36e-1(8.5e-3)−	1.02e-1(2.8e-2)−	1.20e-1(1.7e-2)−	1.82e-1(1.3e-3)−	1.84e-1(1.9e-3)
	8	1.64e-2−	7.78e-3(8.2e-4)−	8.25e-3(1.4e-3)−	8.90e-3(1.7e-3)−	1.89e-2(1.8e-3)≈	1.91e-2(1.7e-3)
	10	3.18e-3−	8.35e-4(1.2e-4)−	7.83e-4(8.4e-5)−	8.57e-4(2.1e-4)−	3.66e-3(9.5e-4)≈	3.70e-3(5.3e-4)
	15	2.64e-5≈	7.16e-6(1.36e-6)−	8.32e-6(2.7e-6)−	6.72e-6(1.8e-6)−	3.52e-5(1.2e-5)≈	2.90e-5(7.2e-6)
12	5	1.61e-1−	1.32e-1(4.2e-3)−	4.90e-2(9.2e-3)−	1.20e-1(2.1e-2)−	1.81e-1(1.0e-3)−	1.82e-1(1.3e-3)
	8	1.50e-2−	1.16e-2(1.4e-3)−	5.45e-3(3.4e-4)−	7.16e-3(1.0e-3)−	1.99e-2(2.8e-4)≈	2.00e-2(4.0e-4)
	10	2.91e-3−	1.72e-3(3.6e-4)−	6.61e-4(5.7e-5)−	7.81e-4(1.2e-4)−	4.03e-3(1.9e-4)≈	4.15e-3(2.0e-4)
	15	4.46e-5≈	1.32e-5(2.7e-6)≈	5.85e-6(2.1e-7)≈	7.77e-6(1.7e-6)≈	7.64e-5(5.1e-5)≈	4.90e-5(4.7e-5)
13	5	1.98e+0−	3.71e+0(2.2e-3)+	3.68e+0(6.3e-3)−	3.70e+0(2.6e-3)≈	3.68e+0(9.2e-3−	3.70e+0(2.9e-3)
	8	0.00e+0−	7.65e+0(1.2e+0)≈	8.14e+0(5.5e-3)−	8.15e+0(1.0e-3)+	8.11e+0(2.5e-2)−	8.15e+0(9.0e-4)
	10	0.00e+0−	1.34e+1(7.7e-1)−	1.37e+1(1.9e-2)−	1.37e+1(3.5e-3)+	1.37e+1(1.2e-2)−	1.37e+1(1.4e-3)
	15	0.00e+0−	5.11e+1(2.3e-1)≈	5.10e+1(3.6e-2)−	5.11e+1(2.1e-2)+	5.10e+1(6.2e-2)−	5.11e+1(7.6e-3)
14	5	3.70e+0+	3.68e+0(3.6e-2)≈	3.65e+0(2.5e-2)−	3.68e+0(1.2e-2)≈	3.67e+0(7.3e-3)−	3.68e+0(1.0e-2)
	8	8.14e+0+	8.05e+0(8.5e-2)≈	8.05e+0(6.5e-2)≈	8.12e+0(2.6e-2)≈	8.09e+0(2.3e-2)≈	8.10e+0(3.6e-2)
	10	1.37e+1+	1.37e+1(7.7e-2)≈	1.36e+1(7.6e-2)≈	1.37e+1(2.6e-2)≈	1.36e+1(3.4e-2)≈	1.36e+1(8.3e-2)
	15	5.10e+1+	5.08e+1(4.0e-1)+	5.05e+1(6.1e-1)≈	5.09e+1(1.9e-1)+	5.09e+1(2.9e-1)≈	5.04e+1(5.3e-1)
15	5	3.71e+0+	3.70e+0(9.7e-3)+	1.80e+0(2.6e-1)−	3.69e+0(1.3e-2)≈	3.68e+0(1.4e-2)−	3.69e+0(1.7e-2)
	8	8.15e+0+	8.13e+0(3.4e-2)≈	7.71e+0(5.0e-1)−	8.14e+0(5.9e-3)−	8.12e+0(9.8e-3)−	8.14e+0(2.9e-3)
	10	1.37e+1+	1.37e+1(5.6e-3)≈	1.36e+1(1.1e-1)−	1.37e+1(1.4e-2)−	1.37e+1(2.6e-2)−	1.37e+1(3.3e-3)
	15	5.11e+1+	5.10e+1(3.5e-1)≈	4.99e+1(7.2e-1)−	5.10e+1(7.4e-2)−	5.10e+1(9.7e-2)−	5.11e+1(1.8e-2)
16	5	1.01e+0−	9.92e-1(3.9e-2)−	8.80e-1(4.3e-2)−	9.51e-1(4.6e-2)−	1.09e+0(1.4e-2)−	1.12e+0(1.8e-2)
	8	2.15e-1−	2.60e-1(1.8e-2)−	1.60e-1(8.6e-3)−	1.96e-1(2.0e-2)−	3.80e-1(4.1e-3)+	3.39e-1(1.6e-2)
	10	5.99e-2−	9.69e-2(5.7e-3)−	3.80e-2(2.1e-3)−	7.18e-2(3.5e-3)−	1.52e-1(3.2e-3)+	1.18e-1(5.5e-3)
	15	1.13e-3−	3.94e-3(4.4e-4)≈	1.43e-3(2.9e-4)−	2.8e-3(3.2e-4)−	8.41e-3(2.2e-4)+	4.42e-3(4.4e-4)
17	5	1.08e+0−	9.92e-1(3.7e-2)−	7.43e-1(1.5e-1)−	9.62e-1(5.1e-2)−	1.09e+0(9.1e-3)−	1.11e+0(2.4e-2)
	8	2.79e-1−	2.66e-1(1.2e-2)−	1.62e-1(5.0e-3)−	1.96e-1(2.1e-2)−	3.78e-1(5.3e-3)+	3.37e-1(1.8e-2)
	10	9.16e-2−	9.01e-2(4.8e-3)−	4.01e-2(2.3e-3)−	6.74e-2(5.2e-3)−	1.51e-1(2.6e-3)+	1.18e-1(3.5e-3)
	15	2.54e-3−	3.72e-3(3.3e-4)−	1.62e-3(2.6e-4)−	2.67e-3(2.0e-4)−	8.46e-3(3.8e-4)+	4.20e-3(5.8e-4)
18	5	1.04e+0−	9.84e-1(2.0e-2)−	2.87e-1(3.3e-2)−	9.55e-1(2.9e-2)−	1.08e+0(1.2e-2)≈	1.09e+0(2.1e-2)
	8	3.08e-1−	2.33e-1(1.1e-2)−	1.75e-1(5.0e-2)−	2.04e-1(1.8e-2)−	3.71e-1(3.8e-3)+	3.35e-1(7.6e-3)
	10	1.08e-1−	8.60e-2(5.0e-3)−	4.04e-2(4.3e-3)−	6.69e-2(4.8e-3)−	1.50e-1(1.5e-3)+	1.17e-1(3.6e-3)
	15	3.52e-3−	3.50e-3(1.3e-4)−	1.45e-3(1.5e-4)−	2.71e-3(2.4e-4)−	8.64e-3(1.9e-4)+	4.31e-3(3.1e-4)
+/ − / ≈		8/52/12	3/49/20 24/35/13	2/61/9 14/50/8	6/44/22 23/39/10	13/29/30 52/10/10	52/8/12

+, −, and ≈ indicate that the compared MaOEA performs significantly better than, worse than, or similarly to MultiGPO/NSGA-II, respectively, on that problem instance.

Table 3. Mean and standard HV values obtained by six algorithms on MaF2–MaF9 problems. The best result for each instance is shown with dark gray background.

Prob.	M	NSGA-III	MOEA/DD	θ-DEA	NSGA-II/SDR	MultiGPO
MaF2	5	1.879e-1 (1.69e-3)≈	1.575e-1 (3.63e-3)−	1.734e-1 (3.23e-3)−	2.029e-1 (1.40e-3)+	1.879e-1 (2.45e-3)
	8	2.153e-1 (6.84e-3)−	1.588e-1 (3.83e-3)−	1.821e-1 (8.01e-3)−	2.323e-1 (2.25e-3)+	2.236e-1 (2.48e-3)
	10	2.163e-1 (5.59e-3)−	1.825e-1 (6.50e-3)−	1.963e-1 (7.31e-3)−	2.233e-1 (2.79e-3)−	2.263e-1 (2.90e-3)
	15	1.35e9-1 (1.45e-2)−	9.427e-2 (3.47e-3)−	1.243e-1 (1.25e-2)−	2.062e-1 (6.73e-3)−	2.158e-1 (3.36e-3)
MaF3	5	9.989e-1 (4.92e-4)≈	9.916e-1 (2.90e-3)−	9.919e-1 (1.67e-3)−	9.820e-1 (5.36e-3)−	9.996e-1 (2.11e-3)
	8	5.998e-1 (5.07e-1)≈	6.658e-1 (4.30e-1)−	9.971e-1 (1.86e-3)−	9.921e-1 (2.75e-3)−	9.988e-1 (1.29e-3)
	10	9.998e-2 (3.08e-1)≈	5.424e-1 (4.33e-1)≈	9.415e-1 (8.00e-2)+	9.928e-1 (2.33e-3)+	2.769e-1 (4.36e-1)
	15	9.101e-2 (2.58e-1)≈	5.910e-1 (4.20e-1)+	9.162e-1 (7.54e-2)+	9.953e-1 (1.24e-3)+	0.00e+0 (0.0e+0)
MaF4	5	7.578e-2 (5.51e-3)−	5.132e-2 (6.26e-3)−	8.156e-2 (1.15e-2)−	1.314e-1 (1.31e-3)+	1.117e-1 (5.31e-3)
	8	1.965e-3 (1.92e-4)≈	2.453e-5 (1.23e-5)−	1.189e-3 (3.53e-4)−	7.052e-4 (1.70e-4)−	2.027e-3 (1.83e-4)
	10	2.263e-4 (2.38e-5)+	1.315e-7 (2.86e-8)−	2.025e-4 (3.14e-5)+	1.510e-5 (2.83e-6)−	8.466e-5 (1.35e-5)
	15	1.943e-7 (2.67e-8)+	5.53e-13 (1.9e-13)−	1.614e-7 (2.26e-8)+	2.34e-10 (1.9e-10)−	3.988e-9 (1.23e-8)
MaF5	5	8.122e-1 (4.74e-4)+	6.827e-1 (1.17e-2)−	8.124e-1 (3.85e-4)+	1.880e-1 (1.52e-1)−	7.676e-1 (1.06e-2)
	8	9.246e-1 (5.21e-4)+	5.824e-1 (4.83e-2)−	9.260e-1 (2.95e-4)+	1.128e-1 (3.64e-2)−	8.822e-1 (8.76e-3)
	10	9.695e-1 (2.07e-4)+	5.846e-1 (1.90e-2)−	9.709e-1 (2.20e-4)+	1.045e-1 (3.70e-2)−	9.122e-1 (7.18e-3)
	15	9.909e-1 (1.34e-4)+	4.854e-1 (2.63e-2)−	9.913e-1 (9.10e-5)+	1.047e-1 (4.24e-2)−	9.436e-1 (5.63e-3)
MaF6	5	1.237e-1 (1.59e-3)−	1.159e-1 (7.45e-4)−	1.162e-1 (1.42e-3)−	1.244e-1 (2.66e-3)−	1.290e-1 (5.02e-4)
	8	8.808e-2 (3.04e-2)−	9.639e-2 (6.69e-4)−	1.029e-1 (9.38e-4)−	1.038e-1 (1.35e-3)−	1.054e-1 (3.28e-4)
	10	1.899e-2 (3.19e-2)−	9.421e-2 (9.67e-4)≈	8.214e-2 (3.33e-2)≈	9.987e-2 (6.32e-4)+	6.122e-2 (4.33e-2)
	15	9.554e-3 (2.35e-2)≈	9.229e-2 (4.31e-4)+	1.654e-2 (3.48e-2)≈	9.440e-2 (1.26e-3)+	1.121e-2 (2.20e-2)
MaF7	5	2.578e-1 (3.40e-3)−	9.601e-2 (1.57e-2)−	2.201e-1 (1.07e-2)−	2.582e-1 (2.15e-3)−	2.599e-1 (1.16e-2)
	8	2.068e-1 (2.50e-3)+	3.116e-2 (2.69e-2)−	1.849e-1 (2.20e-2)+	2.021e-1 (4.55e-3)+	1.366e-1 (2.83e-2)
	10	1.719e-1 (5.75e-3)+	1.755e-3 (7.52e-3)−	1.868e-1 (8.03e-3)+	1.664e-1 (4.35e-3)+	9.403e-2 (2.38e-2)
	15	7.123e-2 (2.09e-2)≈	3.796e-7 (4.20e-8)−	9.399e-2 (2.37e-2)+	2.545e-3 (1.00e-2)−	6.172e-2 (2.79e-2)
MaF8	5	1.081e-1 (2.35e-3)−	8.205e-2 (4.59e-3)−	8.512e-2 (5.57e-3)−	1.234e-1 (5.55e-4)−	1.265e-1 (4.21e-4)
	8	2.645e-2 (7.80e-4)−	1.887e-2 (1.71e-3)−	1.738e-2 (2.79e-3)−	3.141e-2 (1.74e-4)−	3.209e-2 (9.81e-5)
	10	9.307e-3 (2.65e-4)−	6.005e-3 (2.06e-4)−	5.953e-3 (6.86e-4)−	1.094e-2 (9.54e-5)−	1.121e-2 (5.81e-5)
	15	5.140e-4 (2.85e-5)−	3.134e-4 (1.32e-5)−	3.091e-4 (5.26e-5)−	6.215e-4 (9.36e-6)−	6.602e-4 (1.27e-5)
MaF9	5	1.934e-1 (5.11e-2)−	2.517e-1 (2.01e-3)−	1.253e-1 (3.97e-2)−	2.903e-1 (3.82e-3)−	3.265e-1 (9.57e-4)
	8	2.233e-2 (1.33e-2)−	2.281e-2 (3.45e-3)−	2.323e-2 (1.16e-2)−	4.526e-2 (6.21e-4)−	5.302e-2 (1.71e-4)
	10	9.362e-3 (1.68e-3)−	7.888e-3 (8.58e-5)−	6.161e-3 (1.14e-3)−	1.602e-2 (4.17e-4)−	1.883e-2 (1.41e-4)
	15	8.105e-4 (2.13e-4)−	5.664e-4 (1.11e-5)−	4.172e-4 (1.94e-4)−	1.129e-3 (3.58e-5)−	1.278e-3 (3.15e-5)
+/ − / ≈		8/16/8	2/28/2	11/19/2	9/23/0	

+, −, and ≈ indicate that the compared MaOEA performs significantly better than, worse than, or similarly to MultiGPO, respectively, on that problem instance.

5 Conclusion

In this paper, a new (M-1)+1 framework of relaxed Pareto dominance—i.e., the (M-1)-RPD relation—is proposed for solution ranking in MaOPs, and leads to the MultiRPO methods which adopt different RPD schemes. The MultiRPO method uses M symmetrical (M-1)-RPD relations for solution ranking, which can improve the performance of the original RPD methods to maintain a good balance of convergence and diversity. Among the four MultiRPO methods compared, which used four dominance types—α-dominance, CDAS, S-CDAS and GPO techniques—the MultiGPO version based on GPO [16] outperforms the others. Moreover, MultiGPO is shown to perform better than or comparably to the state-of-the-art on a variety of scalable benchmark problems. Since no reference vectors are employed, the performance of Pareto dominance-related MultiGPO algorithm (or other MultiRPO method) is less dependent on the PF shapes than methods using reference vectors, which deserves more attention, especially in solving problems having irregular PFs. In the future, we will attempt to apply the proposed method on more complicated problems, like neural architecture search [8,9] which is a very useful deep learning task.

Acknowledgments. This work was supported by the Natural Science Foundation of China under Grant 61973337, the U.S. National Science Foundation's BEACON Center, funded under Grant DBI-0939454.

References

1. Deb, K., Jain, H.: An evolutionary many-objective optimization algorithm using reference-point-based nondominated sorting approach, part I: solving problems with box constraints. IEEE Trans. Evol. Comput. **18**(4), 577–601 (2014)
2. Deb, K., Pratap, A., Agarwal, S., Meyarivan, T.: A fast and elitist multiobjective genetic algorithm: NSGA-II. IEEE Trans. Evol. Comput. **6**(2), 182–197 (2002)
3. Ishibuchi, H., Setoguchi, Y., Masuda, H., Nojima, Y.: Performance of decomposition-based many-objective algorithms strongly depends on pareto front shapes. IEEE Trans. Evol. Comput. **21**(2), 169–190 (2017)
4. Ishibuchi, H., Matsumoto, T., Masuyama, N., Nojima, Y.: Effects of dominance resistant solutions on the performance of evolutionary multi-objective and many-objective algorithms. In: Proceedings of Annual Conference on Genetic and Evolutionary Computation (GECCO), pp. 507–515 (2020)
5. Ishibuchi, H., Matsumoto, T., Masuyama, N., Nojima, Y.: Many-objective problems are not always difficult for pareto dominance-based evolutionary algorithms. In: Proceedings of 24th European Conference on Artificial Intelligence (ECAI) (2020)
6. K. Ikeda, H. Kita, S.K.: Failure of Pareto-based MOEAs: does non-dominated really mean near to optimal? In: Proceedings of the 2001 Congress on Evolutionary Computation, vol. 2, pp. 957–962 (2001)
7. Li, K., Deb, K., Zhang, Q., Kwong, S.: An evolutionary many-objective optimization algorithm based on dominance and decomposition. IEEE Trans. Evol. Comput. **19**(5), 694–716 (2015)
8. Lu, Z., Deb, K., Goodman, E., Banzhaf, W., Boddeti, V.N.: NSGANetV2: evolutionary multi-objective surrogate-assisted neural architecture search. In: Vedaldi, A., Bischof, H., Brox, T., Frahm, J.-M. (eds.) ECCV 2020. LNCS, vol. 12346, pp. 35–51. Springer, Cham (2020). https://doi.org/10.1007/978-3-030-58452-8_3
9. Lu, Z., Whalen, I., Dhebar, Y., Deb, K., Goodman, E., Banzhaf, W., Boddeti, V.N.: Multi-objective evolutionary design of deep convolutional neural networks for image classification. IEEE Trans. Evol. Comput. (2020). https://doi.org/10.1109/TEVC.2020.3024708
10. Santos, T., Takahashi, R.H.: On the performance degradation of dominance-based evolutionary algorithms in many-objective optimization. IEEE Trans. Evol. Comput. **22**(1), 19–31 (2018)
11. Sato, H., Aguirre, H.E., Tanaka, K.: Self-controlling dominance area of solutions in evolutionary many-objective optimization. In: Deb, K., et al. (eds.) SEAL 2010. LNCS, vol. 6457, pp. 455–465. Springer, Heidelberg (2010). https://doi.org/10.1007/978-3-642-17298-4_49
12. Sato, H., Aguirre, H.E., Tanaka, K.: Controlling dominance area of solutions and its impact on the performance of MOEAs. In: Proceedings of International Conference on Evolutionary Multi-Criterion Optimization (EMO 2007), pp. 5–20. ACM (2007)
13. Tian, Y., Cheng, R., Zhang, X., Jin, Y.: PlatEMO: a MATLAB platform for evolutionary multi-objective optimization [educational forum]. IEEE Comput. Intell. Mag. **12**(4), 73–87 (2017)

14. Tian, Y., Cheng, R., Zhang, X., Su, Y., Jin, Y.: A strengthened dominance relation considering convergence and diversity for evolutionary many-objective optimization. IEEE Trans. Evol. Comput. **23**(2), 331–345 (2019)
15. Yuan, Y., Xu, H., Wang, B., Yao, X.: A new dominance relation-based evolutionary algorithm for many-objective optimization. IEEE Trans. Evol. Comput. **20**(1), 16–37 (2016)
16. Zhu, C., Xu, L., Goodman, E.D.: Generalization of Pareto-optimality for many-objective evolutionary optimization. IEEE Trans. Evol. Comput. **20**(2), 299–315 (2016)

Handling Priority Levels in Mixed Pareto-Lexicographic Many-Objective Optimization Problems

Leonardo Lai[1], Lorenzo Fiaschi[1], Marco Cococcioni[1](✉), and Kalyanmoy Deb[2]

[1] Pisa University, Largo Lucio Lazzarino 1, 56122 Pisa, Italy
lorenzo.fiaschi@phd.unipi.it, marco.cococcioni@unipi.it
[2] Michigan State University, East Lansing, MI 48824, USA
kdeb@egr.msu.edu

Abstract. This paper studies a class of mixed Pareto-Lexicographic multi-objective optimization problems where the preference among the objectives is available in different *priority levels* (PLs) before the start of the optimization process – akin to many practical problems involving domain experts. Each priority level (PL) is a group of objectives having an identical importance in terms of optimization, so that they must be optimized in the standard Pareto sense. However, between two PLs, a lexicographic preference structure exists. Clearly, finding the entire set of Pareto optimal solutions first and then choosing the lexicographic solutions using the given PL structure is not computationally efficient. A new efficient algorithm is presented here using a recent mathematical breakthrough in handling infinite and infinitesimal quantities: the *Grossone* methodology. The proposal has been implemented within a popular multi-objective optimization algorithm (NSGA-II), thereby obtaining its generalized version named PL-NSGA-II, although other EMO or EMaO algorithms could have also been used instead. A quantitative comparison of PL-NSGA-II performance against existing algorithms is made. Results clearly show the advantage of the proposed Grossone-based methodology in solving such priority-level many-objective problems.

Keywords: Multi-objective optimization · Lexicographic optimization · Evolutionary computation · Grossone methodology

1 Introduction

Multi- and many-objective optimization represents a well-studied and active research topic to solve problems that appear in several areas of engineering and

---

Work partially supported by the Italian MIUR – CrossLab project (Departments of Excellence) and partially by the University of Pisa funded project PRA_2018_81 "Wearable sensor systems: personalized analysis and data security in healthcare".
COIN Report Number 2020025.

H. Ishibuchi et al. (Eds.): EMO 2021, LNCS 12654, pp. 362–374, 2021.
https://doi.org/10.1007/978-3-030-72062-9_29

economics. As the problem dimensionality grows, computational issues arise: ineffectiveness of the standard Pareto dominance and conventional recombination operators, necessity to maintain a significantly larger population, non-linear increase of the required computational power, difficulty to visualize the solutions. Optimization problems that originate from real scenarios often manifest some kind of priority relation among their objectives, since they may not be all of equal preference or interest in practice. In these cases, the objectives can be possibly reorganized based on their preferences, still including all of them in the problem formulation.

The present work introduces a novel way to deal with multi- and many-objective real-world problems exploiting the priority information in a structured way. Specifically, it focuses on *Mixed Pareto-Lexicographic Multi-objective Optimization Problems* (abbreviated as MPL-MOPs), a broad family of problems that assume the existence of priority structure among objectives. A first subclass, characterized by objectives aggregated by priority as "chains", has already been investigated by the authors [14]. This article, instead, considers a category of problems whose objectives can be arranged in priority levels (PL-MPL-MOPs). The common element between this and the previous study is the use of the so-called *Grossone Methodology* (GM) [20], a mathematical framework enabling one to deal with infinite and infinitesimal quantities, in this case the weights by means of which levels are assigned priority.

Section 2 outlines the mathematical description of the problems of our interest, Sect. 3 covers the essential elements of the GM, providing the basic knowledge to deal with PL-MPL-MOPs. Section 4 describes how GM helps solving them, dealing with the mathematical reformulation of the model and the definition of a new and more general dominance relation. Section 5 introduces a Grossone-based generalization of NSGA-II [10], called by the authors PL-NSGA-II, which makes the algorithm able to cope with PL-MPL-MOPs by leveraging the original ideas of this work. Finally, Sect. 6 presents a quantitative comparison of PL-NSGA-II and four other EMO algorithms on PL-variants of two popular benchmarks and an engineering problem. The results confirm the benefits achieved by including the priority information in the optimization procedure. Conclusions are finally drawn in Sect. 7.

2 Mixed Pareto-Lexicographic Problems (MPL-MOPs)

Purely Pareto or purely lexicographic multi-objective optimization problems (MOPs) are commonly found in the literature. The former are analytically formulated like in Eq. (1), where m is the number of objective functions, Ω is the feasible decision space (or search space), and the optimization is performed in the standard Pareto-sense. On the other hand, lexicographic optimization complies with the model in Eq. (2), where the optimization involves scalarizing by means of a lexicographic ranking of the functions, that is, f_1 is infinitely more important than f_2 ($f_1 \gg f_2$), which in turn has priority over f_3 ($f_2 \gg f_3$), and so on.

$$\min\ \{f_1(\mathbf{x}), \ldots, f_m(\mathbf{x}),\} \quad \text{s.t. } \mathbf{x} \in \Omega. \tag{1}$$

$$\text{lexmin}\ \{f_1(\mathbf{x}), \ldots, f_m(\mathbf{x})\}, \quad \text{s.t. } \mathbf{x} \in \Omega. \tag{2}$$

Interestingly, both models represent edge cases of a more general family of problems, which has received little attention in the literature: Mixed-Pareto-lexicographic MOPs (MPL-MOPs). The mathematical description is:

$$\begin{aligned} &\min\ \{f_1(\mathbf{x}), \ldots, f_m(\mathbf{x})\}, \\ &\text{s.t. } \mathbf{x} \in \Omega, \\ &\wp(f_1(\mathbf{x}), \ldots, f_m(\mathbf{x})) \text{ supplied,} \end{aligned} \tag{3}$$

where $\wp(\cdot)$ is a generic distribution of priorities among the objectives. The problem of dealing with arbitrary priority policies is crucial for further improvements of EMO algorithms, since routines able to exploit priority information would definitely help in finding better solutions, as suggested in [12]. In the literature, some studies concerning ad-hoc proposal to cope with priority relations exist, such as ε-constrained methods [11] or scalarization [16]. In spite of their superior performance, they are not specifically designed to deal with MPL-MOPs, resulting in relevant drawbacks such as the inability to find multiple solutions per run, or the need for an appropriate choice of scalarization weights beforehand.

The class of MPL-MOPs is too heterogeneous to design a one-fits-all approach; instead, it is much more reasonable to narrow the focus on specific instances where the shape of the $\wp(\cdot)$ function has peculiar properties or is particularly significant to model real-world scenarios. In this work, $\wp(\cdot)$ groups the objective in priority levels (PLs) and the ordering relation is defined on the levels rather than the functions, as illustrated in Fig. 1. The key idea is: i) each group clusters objectives by importance; ii) a group contains only objectives having the same priority.

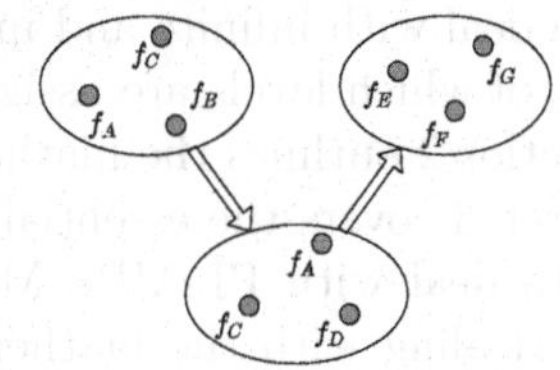

Fig. 1. PL-MPL-MOP structure.

The mathematical formulation of a generic description of a PL-MPL-MOP is:

$$\text{lexmin}\left[\min\begin{pmatrix} f_1^{(1)}(\mathbf{x}) \\ \vdots \\ f_{m_1}^{(1)}(\mathbf{x})\end{pmatrix}, \min\begin{pmatrix} f_1^{(2)}(\mathbf{x}) \\ \vdots \\ f_{m_2}^{(2)}(\mathbf{x})\end{pmatrix}, \ldots, \min\begin{pmatrix} f_1^{(p)}(\mathbf{x}) \\ \vdots \\ f_{m_p}^{(p)}(\mathbf{x})\end{pmatrix}\right], \tag{4}$$

where p is the number of PLs and $f_i^{(j)}$ is the i-th objective in the j-th PL. Note that an objective can repeat in multiple PL. In such problems, the Pareto-optimal solutions of the objectives in the first PL form the decision space for the Pareto-optimization of those in the second PL, and so on. Notice that a purely Pareto optimization problem is a special case of PL-MPL-MOP, where only one PL is considered, while a lexicographic problem is one with multiple PLs having at least one objective in each. In the next section, a possible numerical way to deal with this kind of problems on a computer is presented, as there do not exist tools to concurrently perform Pareto and lexicographic optimization yet.

3 The Grossone Methodology

Grossone Methodology (GM) [20] is a numerical framework that makes it possible to work with infinite and infinitesimal numbers. It has already been successfully applied to many different optimization problems [3–8,14]. The GM fundamental element is the infinite unit *Grossone*, denoted by ①, which allows one to build numerical values composed by finite, infinite and infinitesimal components, known as *gross-scalar* (G-scalar in brief). The latter are indicated as:

$$c = c_{p_m} ①^{p_m} + \ldots + c_{p_0} ①^{p_0} + \ldots + c_{p_{-k}} ①^{p_{-k}},$$

where $m, k \in \mathbb{N}$, the exponents p_i and the digits $c_{p_i} \neq 0$ are called *gross*-powers (G-powers) and *gross*-digits (G-digits), respectively.

The four basic operations between two G-scalars are reasonably intuitive, easy to implement and inherit all the standard properties like associativity, commutativity and existence of the inverse. For instance, consider the following few lines to get a basic understanding of their arithmetical behavior, which is similar to the one of polynomials:

$$① \cdot (① + 2) = ①^2 + 2①, \qquad 0 < \frac{1}{①} = ①^{-1} < ①^0 = 1 < ①^1 = ①,$$
$$\frac{-10.0①^3 + 16.0 + 42.0①^{-3}}{5.0①^3 + 7.0} = -2.0 + 6.0①^{-3}.$$

In addition, the operator $<$ induces a total ordering among the set of G-scalars, and the comparison is made considering the G-powers in descending order: if they differ or they are equal but the corresponding G-digits are different, then the biggest corresponds to the largest G-scalar; otherwise, the next pair of G-powers and G-digits is checked, and so on.

For the rest of the paper, we only consider G-scalars having a finite number of components, each with real-valued G-power and G-digit. In such a scenario, finite numbers are represented by G-scalars with the highest G-power equal to zero, infinitesimal ones with negative highest G-power, and infinite ones with a positive highest G-power. For instance, $-3.4 = -3.4①^0$ and $2 + 0.5①^{-2}$ are both finite numbers, $\pi①^{-1} - ①^{-\sqrt{2}}$ is infinitesimal, while $3①^{\pi} - 70①^{-e}$ is infinite.

4 Use of Grossone in PL-MPL-MOPs

This section shows how ① helps in the design of higher priority-aware routines to deal with PLs better. First, PL-MPL-MOPs are reformulated according to Grossone-based representation where the value of the overall objective function, comprehensive of the Paretian and lexicographic components altogether, can be numerically evaluated. The second and richer part, instead, investigates a novel definition of dominance that possibly fits better the nature of PL-MPL-MOPs. Needless to say, the latter leverages ① as well.

4.1 Problem Reformulation

Similarly to the case of MPL-MOPs with priority chains [14], the lexicographic relation between PLs can be represented both mathematically and in a computer by means of ①. Here, on the other hand, a lexicographic priority between two levels indicates that one group of objectives is infinitely more important than another one. The key idea is to use decreasing powers of ① to represent such ordering, and split up the objectives in well-defined PLs. In mathematical terms, the problem in (4) is reformulated with Grossone as:

$$\min \left[\begin{pmatrix} f_1^{(1)}(\mathbf{x}) \\ \vdots \\ f_{m1}^{(1)}(\mathbf{x}) \end{pmatrix} + ①^{-1} \begin{pmatrix} f_1^{(2)}(\mathbf{x}) \\ \vdots \\ f_{m2}^{(2)}(\mathbf{x}) \end{pmatrix} + \ldots + ①^{1-p} \begin{pmatrix} f_1^{(p)}(\mathbf{x}) \\ \vdots \\ f_{mp}^{(p)}(\mathbf{x}) \end{pmatrix} \right]. \tag{5}$$

In practice, the lexicographic optimization of the priority levels has been transformed into a minimization problem over the weighted average of the PLs contributions. The weights are assigned to the levels by importance: the one with the highest priority is weighted by $①^0$ (the digit, 1, is omitted for brevity), the second one by $①^{-1}$, and so on, until no more PLs remain.

4.2 A Novel Definition of Dominance for PL-MPL-MOPs

When dealing with a PL-MPL-MOP, the Pareto dominance is not suitable because it is oblivious of the priority information. Defining an appropriate *PL-dominance* is not trivial, as demonstrated by the following example, that would be a straightforward but actually a broken definition of PL-Dominance:

Definition 1 (PL-Dominance (INCORRECT)). *Let x and y be two solutions of a PL-MPL-MOP with n PLs: $(\wp_1, \ldots, \wp_n)$ arranged in the order of reducing priority. Let $\wp_i^z$ indicate the i-th PL of solution z, and $x \prec y$ indicates that x Pareto-dominates y. Then, x is said to "PL-dominate" y $(x \prec_{PL} y) \iff \exists i$ such that $\wp_i^x \prec \wp_i^y$ and $\nexists j < i$ such that $\wp_j^y \prec \wp_j^x$.*

This definition surprisingly lacks the transitivity property, i.e. there may exist three elements x, y, z such that x PL-dominates y and y PL-dominates z, but z PL-dominates x. The following counterexample is provided for clarity. Consider the following three solutions x, y, z of a PL-MPL-MOP with two PLs, each with two objectives to be minimized:

$$x = \begin{pmatrix} 1 \\ 6 \end{pmatrix} + ①^{-1} \begin{pmatrix} 1 \\ 1 \end{pmatrix}, \qquad y = \begin{pmatrix} 2 \\ 1 \end{pmatrix} + ①^{-1} \begin{pmatrix} 2 \\ 4 \end{pmatrix}, \qquad z = \begin{pmatrix} 0 \\ 5 \end{pmatrix} + ①^{-1} \begin{pmatrix} 5 \\ 7 \end{pmatrix}$$

With reference to Definition 1, there is a dominance-loop since: i) $x \prec_{\text{PL}} y$ because, although they are non-dominated for the first PL, x performs better than y on the second one; ii) $y \prec_{\text{PL}} z$ for the same reason; iii) $z \prec_{\text{PL}} x$ because it does better on the first PL.

The definition of dominance proposed in this paper leverages the concept of *non-dominated sub-fronts*, a generalization of the notion of fronts which consists in nested fronts generated by the priority. The procedure to partition the population into sub-fronts starts considering only the objectives with the highest priority, and generating a first set of non-dominated fronts as usual, according to Pareto dominance. Then, each front is considered individually, and further split on the basis of the objectives within the second PL, determining new nested fronts (namely, sub-fronts). The same procedure is repeated recursively on the newly-obtained sub-fronts, until no more PLs remain; the result is a population partitioned in a tree-like hierarchy of sub-fronts. Exploiting the natural ordering of the sub-fronts, one can adopt the GM to uniquely identify each leaf-front, i.e. assign an index to each sub-front containing solutions. Such index, which is a G-scalar, is computed as the sum of n components of the form $d_i ①^{p_i}$, where n is the number of PLs, p_i is the G-power associated to the i-th PL (see Equation (5)), and d_i is the position of the parent front at the $1 - i$ level of the hierarchy. For instance, F_i where $i = 3 + 2①^{-1} + 5①^{-2}$ denotes all the solutions in the front 3 for the primary objectives, front 2 for the secondary objectives (within the individuals in the front 3 for the primaries) and front 5 for the tertiaries (among those with primary front 3 and secondary front 2).

Given such calculation of sub-fronts, the new definition of PL-dominance is:

Definition 2 (PL-Dominance (CORRECT)). *Let x and y be two solutions such that $x \in F_i$ and $y \in F_j$. Then, $x \prec_{PL} y \iff i < j$.*

Notice that $\prec_{\text{PL}}$ is not just a function of two arguments (solutions), but has a global dependency on all the other individuals, since it previously requires the assignment of the fitness rank to the whole population.

Going back to the previous counterexample, assuming for simplicity that there are no more elements in the population, it turns out that x, y and z belong to sub-fronts indexed by $2 + 1①^{-1}$, $1 + 1①^{-1}$ and $1 + 2①^{-1}$, respectively. Therefore, it holds true that $y \prec_{\text{PL}} z$ and $z \prec_{\text{PL}} x$, but it becomes false that x dominates y ($x \not\prec_{\text{PL}} y$). The proof of transitivity of the PL-dominance comes straightforwardly from the definition, and it is omitted here for brevity.

5 Proposed PL-NSGA-II Procedure

In this section, the well-known NSGA-II [10] algorithm is enhanced with PL-dominance and GM in order to implement a generalized version able to effectively reckon with PL-MPL-MOPs too. In practice, one needs to modify NSGA-II four core operations, which are: i) solutions non-dominance rank assignment; ii) population sorting by rank; iii) crowding distance assignment to each individual; iv) selection of the new population from the fronts in best-to-worst order, using crowding distance to break ties. The improved version is called PL-NSGA-II. It is only right to say that the choice to improve NSGA-II rather than MOEA/D or NSGA-III is totally arbitrary, and driven by the simplicity to do it.

The first step improvement is implemented by the routine described in Sect. 4.2, which assigns a sub-front to each solution. The sub-front index plays exactly the role of a non-dominance rank for the solution. The second step comes directly by the use of GM, which induces a total ordering among G-scalars. Since the non-dominance rank (i.e., the sub-front index) is a G-scalar, one can resort to the order relation mentioned in Sect. 3 to do so. Both steps are expected to generate a larger number of smaller-sized fronts, a benign effect since one of the main weaknesses of many-objective algorithms is indeed represented by large sets of non-dominated solutions.

The novel way to compute the crowding distance proposed here follows an approach somewhat similar to the one leading to Eq. (5), and it is designed to prioritize solutions that are distant in more important PLs rather than in the lesser ones. Actually, it states that the crowding distance of a solution belonging to a front F_i is the weighted sum of the crowding distances computed at every PL, where the weights are exactly the ones assigned to each of the them. Of course, only the solutions in F_i must be considered for this procedure. Its pseudocode is reported in Algorithm 1.

Since the crowding distance is now a G-scalar too, a total ordering exists and the fourth step is straightforward. The next population is filled with the solutions from the best sub-front, then with those from the second best sub-front, and so on until enough elements have been picked. If a subfront is too large to be included in its entirety, then the solutions with the best crowding distance value are preferred to build the next population.

Altogether, these four improved pieces give birth to PL-NSGA-II, an algorithm which is specifically designed to deal with PL-MPL-MOPs, as well as with standard Pareto and lexicographic problems, being them nothing but corner cases. The next section is devoted to quantitatively assess the performances achieved by the new algorithm PL-NSGA-II.

Algorithm 1. Priority Levels crowding distance assignment.

1: /* *F is the current front in analysis, n is the number of PLs* */
2: **procedure** CROWDING_DIST_ASSIGNMENT(F, n)
3: $p = |F|$
4: **for all** $i \in F$ **do**
5: $CD_i = 0$
6: **for** $q = 1 \dots n$ **do** ▷ For each PL
7: **for** $j = 1 \dots m_q$ **do** ▷ m_q means #objectives in q-th PL
8: $F = sort\left(F, f_j^{(q)}\right)$ ▷ Sort the solutions by the j-th objective
9: $CD_{F[1]}$ += +Inf $①^{1-q}$ ▷ +Inf means "full-scale" (IEEE 754 std.)
10: $CD_{F[p]}$ += +Inf $①^{1-q}$
11: **for** $i = 2 \dots (p-1)$ **do**
12: $CD_{F[i]}$ += $①^{1-q} \frac{f_j^{(q)}(F[i+1]) - f_j^{(q)}(F[i-1])}{f_j^{(q)max} - f_j^{(q)min}}$ ▷ Use of ①
13: **return** CD

6 Results with PL-MPL-MOP

Since there do not exist benchmark problems featuring priority levels yet, the experiments of this section are performed on modified versions of existing problems. We consider two test problems and one engineering design problem here. The performance of PL-NSGA-II is compared with four EMO algorithms: i) NSGA-II, ii) NSGA-III [19], iii) MOEA/D [21], and iv) GRAPH [17] which can handle supplied preference policies. In order to make the comparison fair, solutions from the non-prioritized algorithms (NSGA-II, NSGA-III, and MOEA/D) are filtered after the optimization on the basis of the supplied PL-dominance information. All algorithms have used the SBX crossover operator and polynomial mutation with a population of 200 individuals and run for 600 generations. For each algorithm, the experiment is repeated 50 times to ensure a statistically stable performance comparison through the metric $\Delta(\cdot) = \max\{IGD(\cdot), GD(\cdot)\}$, where IGD is the *inverted generational distance* and GD indicates the *generational distance* [18] (their computation comes after the normalization of the objective space to the unitary hyper-cube). The use of GD and IGD metrics rather than the hypervolume is due to the fact that a clear way to fairly compute the latter in the case of PL-MPL-MOPs is still missing and under investigation. Concerning MOEA/D, NSGA-II and NSGA-III the implementations from pymoo library [1] have been used in this work.

6.1 PL-MaF7 Problem

MaF7 [2] is a widely recognized benchmark problem for multi- and many-objective evolutionary algorithm research. We introduce two priority levels with three objective functions in each level, as follows:

$$\min \left[\begin{pmatrix} \mathbf{x}_1 \\ \mathbf{x}_2 \\ h(\mathbf{x}) \end{pmatrix} + \text{①}^{-1} \begin{pmatrix} f_1^{(2)}(\mathbf{x}) \\ f_2^{(2)}(\mathbf{x}) \\ h(\mathbf{x}) \end{pmatrix} \right],$$
$$\text{s.t. } \mathbf{x} \in [0,\, 1]^{23},$$

where $f_1^{(2)}(\mathbf{x}) = -e^{-0.5\, p(\mathbf{x},\, 0.6)}$, $f_2^{(2)}(\mathbf{x}) = |p(\mathbf{x},\, 0.8) - 0.04|$, $p(\mathbf{x},\, c) = (x_1 - c)^2 + (x_2 - c)^2$, $h(\mathbf{x}) = 6 + \frac{27}{20}\sum_{i=3}^{23} x_i - \sum_{i=1}^{2} x_i\,(1 + \sin(3\pi x_i))$.

The first PL has a Pareto front made by four disjoint regions due to periodic function in $h(\mathbf{x})$, as illustrated in Fig. 2. The second PL has the effect of isolating just one of them, the one closest to the line $x_1 = x_2 = 1$ (Fig. 4).

The function $f_1^{(2)}$ is a bivariate normal distribution centered in (0.6, 0.6), which assigns higher importance to solutions with $(x_1,\, x_2)$ closer to the mean point; $f_2^{(2)}$ favors the candidate solutions whose pair $(x_1,\, x_2)$ is closer to the circumference with center in (0.8, 0.8) and radius 0.2. Finally, the last function selects a single disjoint region: $f_1^{(2)}$ helps keep the optimal part of the front near the origin on the x_1-x_2 plane, whereas h pushes in the opposite direction, i.e., it privileges those points further from the origin.

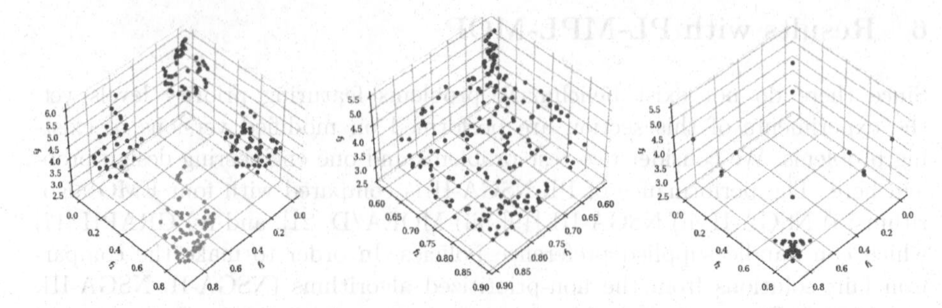

Fig. 2. NSGA-II and true PL-MPL (orange) points for MaF7. (Color figure online)

Fig. 3. Obtained PL-NSGA-II points for PL-MaF7.

Fig. 4. Obtained MOEA/D (2nd-best) points for PL-MaF7.

For each algorithm, the mean and the standard deviation of $\Delta(\cdot)$ are reported in Table 1, ranked from the best to the worst; the best values are boldfaced. Figure 3 reports the primary Pareto front obtained by a single run of PL-NSGA-II, demonstrating the ability of the priority-aware algorithm to accurately generate the right part of the Pareto front, whereas the second-best method (non-priority MOEA/D) cannot find a well-diverse set.

Table 1. Performance ($\Delta(\cdot) = \max\{IGD(\cdot), GD(\cdot)\}$) comparison on MaF7.

Algorithm	Mean	Std. deviation	Avg. # Pareto Sols.
PL-NSGA-II	$\mathbf{1.025e^{-4}} + 3.451e^{-5}①^{-1}$	$\mathbf{4.201e^{-5}} + 1.788e^{-5}①^{-1}$	**200.00**
GRAPH	$0.332 + 0.558①^{-1}$	$0.038 + 0.079①^{-1}$	**200.00**
MOEA/D	$0.003 + 0.001①^{-1}$	$0.001 + 0.001①^{-1}$	60.00
NSGA-III	$0.357 + 0.543①^{-1}$	$0.021 + 0.034①^{-1}$	18.32
NSGA-II	$0.010 + 0.005①^{-1}$	$0.004 + 0.002①^{-1}$	9.44

6.2 PL-MaF11 Problem

MaF11 [2] is a benchmark problem from the Walking Fish Group (WFG) test suite, named there WFG2. We consider a three-objective version of the problem. The original objective functions, now come together in the first PL, are

$$
\begin{aligned}
f_1^{(1)}(\mathbf{x}) &= y_3 + 2\left(1 - \cos\left(y_1 \frac{\pi}{2}\right)\right)\left(1 - \cos\left(y_2 \frac{\pi}{2}\right)\right), \\
f_2^{(1)}(\mathbf{x}) &= y_3 + 4\left(1 - \cos\left(y_1 \frac{\pi}{2}\right)\right)\left(1 - \sin\left(y_2 \frac{\pi}{2}\right)\right), \\
f_3^{(1)}(\mathbf{x}) &= y_3 + 6\left(1 - y_1 \cos^2\left(5 y_1 \pi\right)\right),
\end{aligned}
$$

where

$$y_i = \begin{cases} (t_i^{(3)} - 0.5)\max(1,\, t_3^{(3)}) + 0.5, & i = 1,2,\\ t_3^{(3)}, & i = 3, \end{cases} \qquad t_i^{(3)} = \begin{cases} t_i^{(2)}, & i = 1,2,\\ \frac{1}{5}\sum_{j=3}^{7} t_j^{(2)}, & i = 3, \end{cases}$$

$$t_i^{(2)} = \begin{cases} t_i^{(1)}, & i = 1,2,\\ t_{2i-3}^{(1)} + t_{2i-2}^{(1)} + 2|t_{2i-3}^{(1)} - t_{2i-2}^{(1)}|, & i = 3:7, \end{cases} \qquad t_1^{(1)} = \begin{cases} z_i, & i = 1,2,\\ \frac{|z_i - 0.35|}{|\lfloor 0.35 - z_i \rfloor| + 0.35}, & i = 3:12, \end{cases}$$

with $z_i = x_i/2$ and $x_i \in [0, 2i]$ for $i = 1 : 12$. The second PL, instead, is constructed by the following two objective functions:

$$f_i^{(2)} = -\left(f_1^{(1)} - 2f_2^{(1)} + (-1)^i \frac{11}{4}\right)^2, \quad i = 1,2.$$

It consists in two parabolic cylinders which select only three pieces of the original Pareto front: the middle region (regardless the height) and the two farthest points from the origin on the $f_1^{(1)} - f_2^{(1)}$ plane. It was considered to be a challenging problem, so we use 2,500 generations on a population of 200 individuals to show a satisfying level of convergence [13]. Figure 5 shows the final non-dominated solutions of the original problem found by NSGA-III. Figure 6 and 7 report the optimal front considering both the PLs when approximated by GRAPH and PL-NSGA-II, respectively. Notice that PL-NSGA-II algorithm, being aware of the priority, requires only 600 iterations to achieve such result, in contrast with the 2,500 iterations needed by NSGA-II. Table 2 presents the detailed performance comparison, indicating superior performance of PL-NSGA-II.

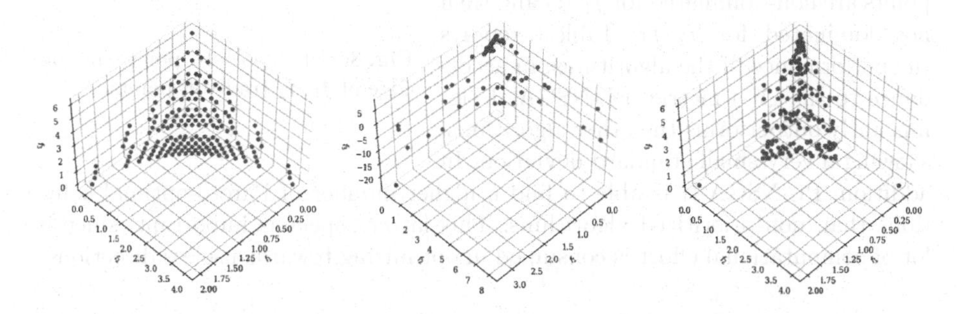

Fig. 5. NSGA-III front of MaF11.

Fig. 6. GRAPH front of MaF11.

Fig. 7. PL-NSGA-II solutions on PL-MaF11.

Table 2. Performance ($\Delta(\cdot) = \max\{IGD(\cdot), GD(\cdot)\}$) comparison for MaF11.

Algorithm	Mean	Std. deviation	Avg. # Pareto Sols.
PL-NSGA-II	**0.642** + 0.240①$^{-1}$	**0.373** + 0.091①$^{-1}$	**200.00**
GRAPH	0.723 + 0.248①$^{-1}$	0.404 + 0.098①$^{-1}$	**200.00**
MOEA/D	0.661 + 0.360①$^{-1}$	0.370 + 0.074①$^{-1}$	3.08
NSGA-II	0.828 + 0.348①$^{-1}$	0.428 + 0.121①$^{-1}$	3.08
NSGA-III	0.852 + 0.363①$^{-1}$	0.441 + 0.123①$^{-1}$	1.52

6.3 Crashworthiness Problem

Finally, we consider the crashworthiness problem [15], which was solved using NSGA-III [9]. Two PLs are formed: the first PL involves minimizing the mass (f_1) and acceleration (f_2), while the second PL minimizes acceleration (f_2) and the toe-board intrusion compliance due to a crash (f_3). This choice reflects the hierarchical interests of a company which seeks to design high performing cars caring also about the driver's safety. Figure 8 shows the obtained PL-NSGA-II points. It clearly shows how PL-NSGA-II points are non-dominated for f_1–f_2 and then non-dominated for f_2–f_3. Table 3 reports the performance of the algorithms, in accordance to the $\Delta(\cdot)$ metric. PL-NSGA-II significantly outperforms the other algorithms, showing a very small standard deviation. In addition, PL-NSGA-II is able to find a higher number of Pareto-optimal solutions than non-prioritized algorithms. This latter aspect is important, since a lot of computational effort is consumed in optimizing towards useless directions.

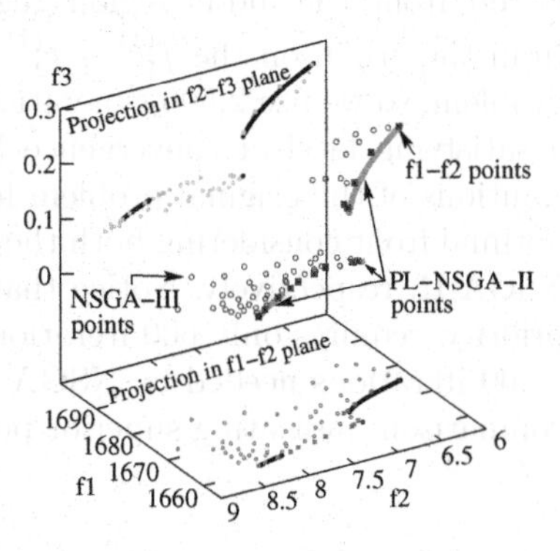

Fig. 8. PL-NSGA-II points lie on edge of f_1–f_2 trade-off points.

Table 3. Performance ($\Delta(\cdot) = \max\{IGD(\cdot), GD(\cdot)\}$) on Crashworthiness.

Algorithm	Mean	Std. deviation	Avg. # Pareto Sols.
PL-NSGA-II	**2.766e^{-7}** + 3.556e^{-7}①$^{-1}$	**9.865e^{-8}** + 1.235e^{-7}①$^{-1}$	**200.00**
GRAPH	0.307 + 0.286①$^{-1}$	0.13 + 0.13①$^{-1}$	**200.00**
MOEA/D	0.158 + 0.293①$^{-1}$	0.022 + 0.001①$^{-1}$	39.00
NSGA-II	0.001 + 0.003①$^{-1}$	0.001 + 0.001①$^{-1}$	16.82
NSGA-III	0.004 + 0.004①$^{-1}$	0.002 + 0.003①$^{-1}$	7.48

Overall, results shows that including the information about the priority structure directly into the algorithms greatly helps finding more and better solutions, without affecting their generality at all.

7 Conclusions

The goal of this paper has been to introduce PL-MPL-MOPs, a sub-category of MPL-MOPs where the objectives are grouped in known levels of priority. Furthermore, the first attempt to improve an already existing algorithm by means of a tool to deal with level-like structured priorities has been presented, implemented and analyzed. It exploits a novel mathematical framework called Grossone methodology – a numerical approach to handle infinite and infinitesimal quantities oriented to scientific calculations. The Grossone and a custom non-dominance definition – the PL-dominance – have enabled a significantly improved performance. This has been demonstrated by illustrative experiments carried out on variations of two standard benchmark problems and an engineering design problem, where a second PL is to turn the problem into a PL-MPL-MOP. The numerical results have demonstrated how the PL-extension of NSGA-II is consistently able to outperform other EMO algorithms of various kinds, some aware of the priority and some not, some designed for many-objective problems and some for multiple objectives only. As a future work, the authors are considering the study of PL-based crossover and mutation operators, as well as the design of more challenging benchmarks for PL-MPL-MOPs. Finally, a deeper analysis of the effectiveness of this technique on real problems is also ongoing.

References

1. Blank, J., Deb, K.: Pymoo: Multi-objective optimization in Python. IEEE Access **8**, 89497–89509 (2020)
2. Cheng, R., Li, M., Tian, Y., et al.: A benchmark test suite for evolutionary many-objective optimization. Complex Intell. Syst. **3**(1), 67–81 (2017)
3. Cococcioni, M., Pappalardo, M., Sergeyev, Y.: Lexicographic multi-objective linear programming using grossone methodology: theory and algorithm. Appl. Math. Comput. **318**, 298–311 (2018)
4. Cococcioni, M., Fiaschi, L.: The Big-M method with the numerical infinite M. Optim. Lett. (2020). https://doi.org/10.1007/s11590-020-01644-6
5. Cococcioni, M., et al.: Solving the lexicographic multi-objective mixed-integer linear programming problem using branch-and-bound and grossone methodology. Commun. Nonlin. Sci. Numer. Simul. **84**, 105177 (2020)
6. De Cosmis, S., De Leone, R.: The use of grossone in mathematical programming and operations research. Appl. Math. Comput. **218**(16), 8029–8038 (2012)
7. De Leone, R.: Nonlinear programming and grossone: quadratic programing and the role of constraint qualifications. Appl. Math. Comput. **318**, 290–297 (2018)
8. De Leone, R., Fasano, G., Roma, M., Sergeyev, Y.D.: Iterative grossone-based computation of negative curvature directions in large-scale optimization. J. Optim. Theory Appl. **186**(2), 554–589 (2020)

9. Deb, K., Jain, H.: An evolutionary many-objective optimization algorithm using reference-point-based nondominated sorting approach, part i: solving problems with box constraints. IEEE Trans. Evol. Comput. **18**(4), 577–601 (2014)
10. Deb, K., Pratap, A., Agarwal, S., Meyarivan, T.: A fast and elitist multiobjective genetic algorithm: NSGA-II. IEEE Trans. Evol. Comput. **6**(2), 182–197 (2002)
11. Deb, K.: Multi-Objective Optimization Using Evolutionary Algorithms. Wiley, Hoboken (2001)
12. Gaur, A., Khaled Talukder, A., Deb, K., Tiwari, S., Xu, S., Jones, D.: Unconventional optimization for achieving well-informed design solutions for the automobile industry. Eng. Optim. **52**, 1542–1560 (2019)
13. Huband, S., et al.: A review of multiobjective test problems and a scalable test problem toolkit. IEEE Trans. Evol. Comput. **10**(5), 477–506 (2006)
14. Lai, L., Fiaschi, L., Cococcioni, M.: Solving mixed pareto-lexicographic multi-objective optimization problems: the case of priority chains. Swarm Evol. Comput. **55**, 100687 (2020)
15. Liao, X., Li, Q., Yang, X., Zhang, W., Li, W.: Multiobjective optimization for crash safety design of vehicles using stepwise regression model. Struct. Multi. Optim. **35**(6), 561–569 (2008)
16. Marler, R.T., Arora, J.S.: Survey of multi-objective optimization methods for engineering. Struct. Multi. Optim. **26**(6), 369–395 (2004)
17. Schmiedle, F., Drechsler, N., Große, D., Drechsler, R.: Priorities in multi-objective optimization for genetic programming. In: Proceedings of the 3rd Annual Conference on Genetic and Evolutionary Computation, pp. 129–136. Morgan Kaufmann (2001)
18. Schutze, O., Esquivel, X., Lara, A., Coello, C.A.C.: Using the averaged Hausdorff distance as a performance measure in evolutionary multiobjective optimization. IEEE Trans. Evol. Comput. **16**(4), 504–522 (2012)
19. Seada, H., Deb, K.: U-NSGA-III: a unified evolutionary optimization procedure for single, multiple, and many objectives: proof-of-principle Results. In: Gaspar-Cunha, A., Henggeler Antunes, C., Coello, C.C. (eds.) EMO 2015. LNCS, vol. 9019, pp. 34–49. Springer, Cham (2015). https://doi.org/10.1007/978-3-319-15892-1_3
20. Sergeyev, Y.D.: Numerical infinities and infinitesimals: methodology, applications, and repercussions on two Hilbert problems. EMS Surv. Math. Sci **4**(2), 219–320 (2017)
21. Zhang, Q., Li, H.: MOEA/D: a multiobjective evolutionary algorithm based on decomposition. IEEE Trans. Evol. Comput. **11**(6), 712–731 (2007)

Many-Objective Pathfinding Based on Fréchet Similarity Metric

Jens Weise([✉]) and Sanaz Mostaghim

Faculty of Computer Science, Otto von Guericke University, Magdeburg, Germany
{jens.weise,sanaz.mostaghim}@ovgu.de
https://ci.ovgu.de

Abstract. Route planning, also known as pathfinding is one of the essential elements in logistics, mobile robotics and other applications, where engineers face many conflicting objectives. One major challenge in dealing with multi-objective pathfinding problems is that several different paths can map to the same objective values. Since typical multi-objective optimization algorithms focus on the selection mechanism in the objective space, such multi-modal solutions cannot be easily found. In this paper, we propose a new methodology for preserving a good diversity of solutions in the decision space, which is tailored for pathfinding problems. We measure the similarity between the solutions in the decision space using the discrete Fréchet distance measurement, which is meant as a replacement for the well-known crowding distance measurement. We examine the proposed approach in three different variations on a variety of benchmark instances.

Keywords: Pathfinding · Multi-objective optimization · Evolutionary algorithm · Path similarities

1 Introduction

Optimal route planning (or pathfinding) is among the most challenging tasks for industrial and logistical applications [26]. Any improvement in the results can have a significant impact on several factors, such as fuel consumption and the environment. Current state-of-the-art route planning algorithms typically consider the travel time, the distance in the optimization. In a recent work, we have proposed a scalable benchmark suite for pathfinding problems [30] and define five objective functions.

There is an extensive amount of literature in the field of route planning and pathfinding in general and especially for vehicle route planning which uses evolutionary algorithms [1]. The most important feature concerns the solution and problem representation, which can define the size of the search space and

This work is funded by the German Federal Ministry of Education and Research through the MOSAIK project (grant no. 01IS18070B).

H. Ishibuchi et al. (Eds.): EMO 2021, LNCS 12654, pp. 375–386, 2021.
https://doi.org/10.1007/978-3-030-72062-9_30

influence the efficiency of the algorithms. However, there is a limited amount of literature using evolutionary algorithms for many-objective pathfinding. Tozer et al. [27] provide an overview of existing approaches and use reinforcement learning to address the problem with six objective functions. Pulido et al. [22] introduce a dimensionality reduction technique in order to minimize dominance checks during the optimization and tested their algorithm on a map from a real-world application. They extended the NAMOA* algorithm, which was first introduced by Mandow [19] and is a multi-objective extension to the well-known A* algorithm [13]. In our previous work, we have proposed the *ASLETISMAC* benchmark suite [30] which includes several maps with various configurations for elevation profiles, neighbourhood metrics and backtracking options. In one our recent studies [29], We have found out that many of the obtained optimal paths have a considerable overlap with each other and it is very challenging to diversify the search in the decision space while solving the many-objective pathfinding problem.

The goal of this paper is to address this problem by introducing a specialized similarity measurement of paths into the multi-objective optimization algorithms. We propose to use the discreet Fréchet distance for the measurement of similarities between two paths and incorporate this measurement in the algorithms three different ways, as a replacement for crowding distance, in the form of an archive, and a combination of both.

The remainder of the paper is structured as follows: In Sect. 2 we introduce the problem. Section 4 presents the proposed similarity measurement. Section 4 is dedicated to the proposed multi-objective algorithm and Sect. 5 describes the experiments and the results. The paper is concluded in Sect. 6.

2 The Pathfinding Problem

Typically, there are two main ways of representing paths: The first is a variable-length chromosome representation of a solution which is often used in combination with the graph-based problem representation [14]. This approach represents a solution as a list of nodes, which can be of different length when computing a path. The second approach is a fixed-length chromosome, representing the directions of travel, together with a list of nodes in a graph or a list of grid cells [4,5]. Grid-based representations for pathfinding problems are shown to be very practical for evolutionary algorithms [1]. Such grid representations can be refined depending on the required resolution of the problem. Moreover, they are often used for benchmarking purposes [25]. Also, they can represent real-world problems by discretizing the problem representation [3]. Grids typically consist of units with adjustable sizes [23]. A solution encoding can consist of a linked-list of units [31], the directions [1], or the coordinates of several waypoints [28]. Considering grids, we can introduce several constraints for movements on specific paths by defining an upper limit for movements (such as speed) on each unit. In this way, we can easily incorporate various linear and non-linear constraints on each unit which can represent speed, ascent, obstacles, and others.

It is comparatively easy to convert a grid into a graph (also called lattice-graphs) by considering units as nodes and their contact-edges as the graph's

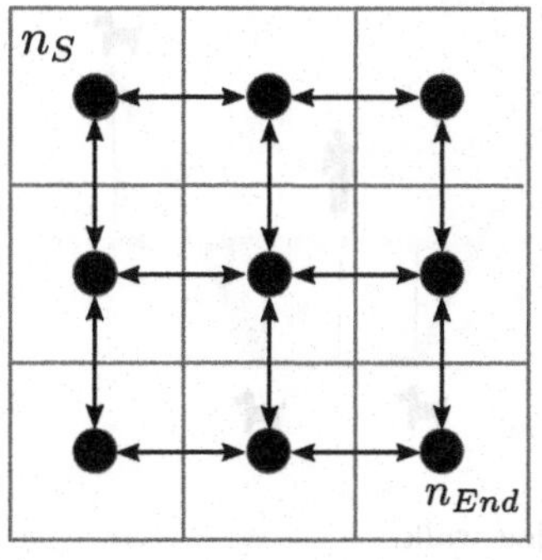

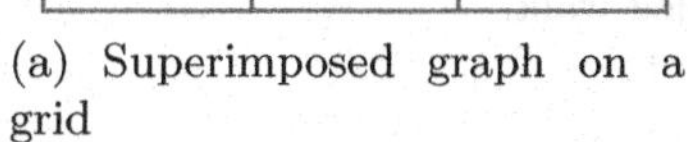
(a) Superimposed graph on a grid

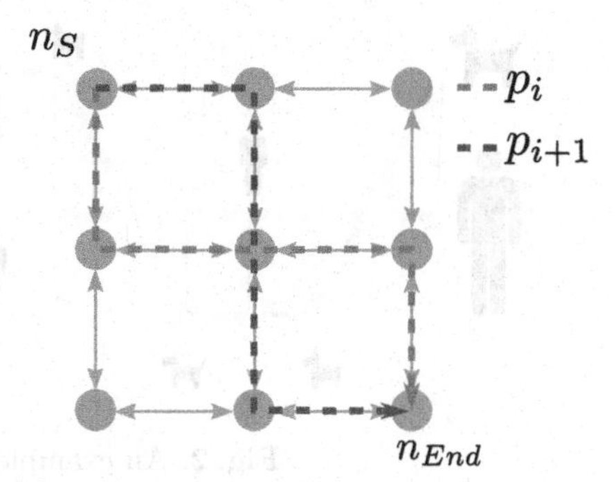

(b) Two paths p_i and p_{i+1} which go along nodes and edges.

Fig. 1. Different discrete Fréchet distance values

edges. This conversion is performed in several applications, e.g. the game industry. The commonly used A* algorithm is an example of pathfinding on a grid which is transferred to a graph [32].

In general, graph-based representations allow higher flexibility in representing real-world problems, which can be considered heterogeneous, compared to grids which are usually homogeneous. In this paper, represent the pathfinding problem as a grid transferred to a graph.

The multi-objective pathfinding problem can be defined as follows. The goal is to find a set of optimal paths $P* = \{p_1, \cdots, p_K\}$ in a graph $G(V, E)$ from a starting node $n_S \in V$ to a pre-defined end node $n_{End} \in V$, i.e., $p_i = (n_S, \cdots, n_{End})$ for path N_i. The paths are evaluated in terms of five objective functions such as the length of a path, travel time, delay, smoothness (or curvature) of a path, and ascent (or elevation) of a path. Figure 1 depicts a graph which is superimposed on a grid (lattice-graph) in the left figure and two paths p_i and p_{i+1} in the right figure.

Given this definition, we can use a variable-length chromosome representation in order to apply the following proposed Fréchet distance metric.

3 Paths Similarities

Path, or curve, similarity measurement is found in several fields. For instance, in handwriting recognition, curves are compared in order to match to letters or words [24]. Other fields include morphing [9] and protein structure alignment [17].

There are different methods of path similarity measurements. Two of them are the Hausdorff distance and the Fréchet distance. With Hausdorff distance, the distance of two subsets (curves) in a metric space can be measured [20], while Fréchet distance also takes the flow of curves into account. Fan et al. used the Fréchet distance on road networks for measuring the resemblance of road

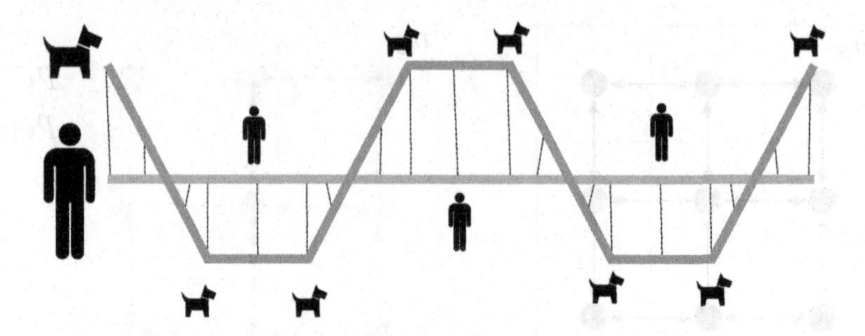

Fig. 2. An example of a dog walk.

tracks [11]. Assuming two subsets in the same metric space, they can have a short Hausdorff distance but a rather large Fréchet distance.

The Fréchet distance is a measurement of similarity for curves in a metric space. Eiter and Mannila described it by using a dog walk analogy [10]. A dog and its owner walk on two different paths, and both can vary their speed but cannot backtrack. Since there is a leash attached to both, the Fréchet distance can be defined as the shortest length of a leash which is required for both to follow their paths as shown in Fig. 2. The dashed lines indicate the leash.

Fréchet distance [10] is mathematically defined as follows:

$$\delta_F(A,B) = \inf_{\alpha,\beta} \max_{t\in[0,1]} \left\{ d\Big(A(\alpha(t)),\, B(\beta(t))\Big) \right\} \tag{1}$$

Here, A and B are curves as a continuous mapping in a metric space S, defined as $A : [0,1] \to S$ and $B : [0,1] \to S$. The reparameterizations α and β are non-decreasing functions from $[0,1]$ to $[0,1]$. In other words, $A(\alpha(t))$ and $B(\beta(t))$ are the positions of the owner and the dog in time-step t and they are non-decreasing as the two entities cannot move backwards. Function d is the distance metric in S, e.g. Euclidean distance, hence it describes the length of the leash between the dog and the owner. Eventually, the Fréchet distance is obtained by computing the infimum (greatest lowest bound) of all paremeterizations α and β of $[0,1]$ of the maximum over all t of the distance d in S between $A(\alpha(t))$ and $B(\beta(t))$. Alt and Godau studied the computational properties of the measurement [2], initially introduced by Frechet in 1906 [12]. They specified an algorithm to compute the distance δ_F in time $\mathcal{O}(ab\log ab)$, where a and b are the number of segments of two curves [10]. Eiter and Mannila proposed a variation of the distance metric, i.e. the coupling distance, or discrete Fréchet distance δ_{dF}, which provides a good approximation of δ_F in time $\mathcal{O}(ab)$. They also show, that δ_{dF} is an upper bound of δ_F. They approximate the two curves A and B by polygonal curves, which is, intuitively, a list of supporting points. Those points specify the endpoints of line segments. The sequence of segment endpoints of a polygonal curve P is denoted as $\sigma(A) = (u_1, \cdots, u_a)$. A *coupling* L between A and B is defined as a sequence $L = (u_{c_1}, v_{d_1}), (u_{c_2}, v_{d_2}), \cdots, (u_{c_m}, v_{d_m})$ of distinct pairs of $\sigma(A) \times \sigma(B)$, with respect to $c_1 = 1, d_1 = 1, c_m = a, d_m = b$. The coupling respects the order of points. Eventually, the length of the coupling is defined as $||L|| = \max_{i=1,\cdots,m} d(u_{c_i}, v_{d_i})$; hence the longest connection in L and the discreet Fréchet distance as shown in Eq. (2) [10].

$$\delta_{dF}(A, B) = \min\{||L|| \mid L \text{ is a coupling between } A \text{ and } B\} \tag{2}$$

Compared to the Hausdorff distance, the Fréchet distance accounts for the flow of the curves while Hausdorff distance measures the distance from one point in one curve to the closest on the second curve. For the remainder of the paper, we refer to the discrete Fréchet distance. As we define a path in this work as $p_i = (n_s, \cdots, n_{End})$, we can represent it analogue to the definition of δ_{dF}; hence $p_i = \sigma(A)$ and $(n_s, \cdots, n_{End}) = (u_1, \cdots, u_a)$, where $n_s = u_1$, $n_{s+1} = u_2$, $n_{End} = u_a$. In other words, A represents a continuously defined curve and $\sigma(A)$ are the segment endpoints. Since a path in a graph is defined by it's nodes, every node $n_i \in p_i$ is in fact a segment endpoint.

4 Fréchet Distance-Based NSGA-II

In order to preserve the diversity of solutions (paths) in the decision space, we use the Fréchet distance path similarity measurement in the well-known NSGA-II algorithm [7]. Although NSGA-II has some drawbacks on many-objective problems, we saw in our previous study [30] that it outperformed NSGA-III is the majority of our problem instances. The problem's Pareto-fronts are usually irregular and degenerate. Due to the use of crowding distance, NSGA-II is more robust to these types of problems, as evenly distributed reference vectors which are used in NSGA-III can lead to the same solution [6]. We use the proposed distance metric in different three ways and aim to understand its impact on the results of the algorithm.

In the first approach, we replace the crowding distance used by the NSGA-II algorithm with the proposed Fréchet distance. In other words, when the next parent population is filled, and the last front cannot be added completely, we compare the solutions by using the Fréchet distance. Also, we use this distance metric in the selection process. For this purpose, we implement a Fréchet distance sorting method. We compute a Fréchet distance matrix for N paths in a population and assign the lowest distance of all $N - 1$ values to each path as described in Algorithm 1.

In this way, paths with less similarity have a higher distance value. Figure 3 illustrates an example with several possible paths on a benchmark instance. Figure 3a shows two of the paths and their respective discrete Fréchet distance of 1, while Fig. 3b shows two rather distinct paths with large distance values of 6.3246. Here, the most south-west path has the highest discrete Fréchet distance. After computing every pair, the paths are sorted by their distance value in descending order. This algorithm is called *NSGA-II-FDWOA* (Fréchet distance without archive).

Algorithm 1: Fréchet distance Density

```
Input: List of Paths P = {p_1, ··· , p_K}
// Map is a key-value store. A path p is the key with the minimum
   Fréchet distance as its value.
Result: Map(Path,minimum Fréchet distance)
// Every Path p_i ∈ P gets ∞ assigned
Map results = Map(Paths,Infinity);
// Here, the Fréchet distance from every path to every other is
   computed
// {(p_i, p_j)|p_i, p_j ∈ P, p_i ≠ p_j}
for i=0 to length(P) do
    for j=i+1 to length(P) do
        frechetOld = Map(p_i);
        // FrechetDistance(p_i,p_j) is implemented according to [10]
           (discreet), input to that function is σ(P) and σ(Q)
        frechetDist=FrechetDistance(p_i,p_j);
        Map(p_i)=min(frechetOld,frechetCurrent);
    end
end
```

In the second approach, we use a limited archive for storing non-dominated solutions and prune solutions by using the proposed discreet Fréchet distance. In this way, we are applying the pruning strategy in decision space rather than in objective space. A variant of this approach is used in the context of multi-modal multi-objective optimization e.g., [8,16]. In this case, the discreet Fréchet distance metric represents a niching method. We call this variant as *NSGA-II-FDAO* (Fréchet distance archive only). We have selected the archive-based approach as it is shown that using an archive for storing the non-dominated solutions can significantly improve the performance [21]. Patil proposes several archive architectures such as a limited archive which contains a maximum number of solutions. After each reproduction phase, all solutions of the population are added to the archive. A new solution S_{new} is checked for domination with the solutions already in the archive. If S_{new} is dominated by all solutions $S_{i,archive}$, it is not added. If it dominates any solution in the archive, they are removed from it. For the case, that S_{new} is non-dominated, and the maximum number of archive members is reached, a solution in a dense area is removed which is done based on clustering, nearest-neighbour or crowding distance.

The third approach combines the two other approaches resulting in *NSGA-II-FD*. In this variation, we replaced the crowding distance with the proposed Fréchet distance (as in the first approach), and the algorithm used an archive (as in the second approach).

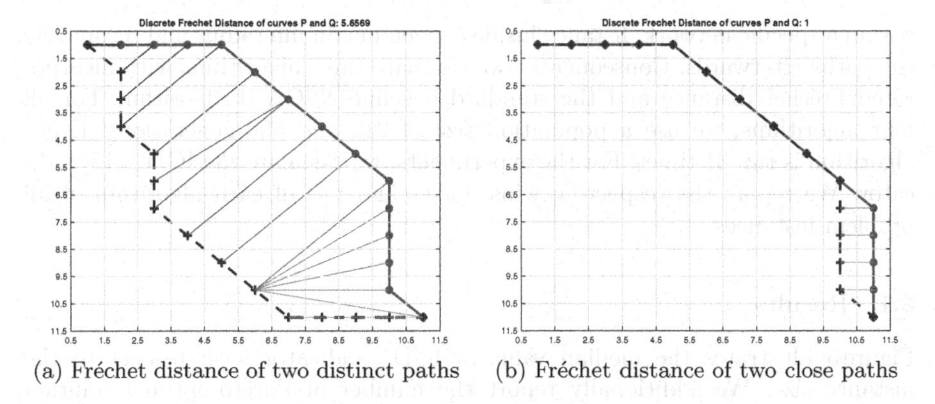

(a) Fréchet distance of two distinct paths (b) Fréchet distance of two close paths

Fig. 3. Different discrete Fréchet distance values

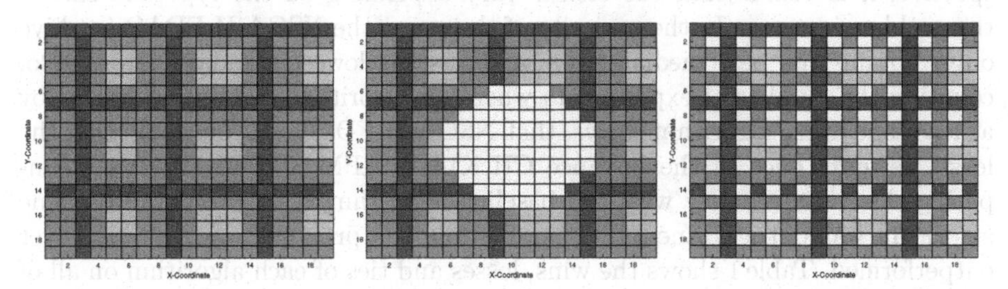

Fig. 4. Examples for various maps with different properties for units representing NO, LA and CH (from left to right)

5 Experiments

In the experiments, we examine the proposed approaches on the benchmark test suite ASLETISMAC [30]. We consider a two-dimensional space with three obstacle types (NO, LA, CH), with K2 and K3 neighbourhood, no backtracking, and grid sizes of 4 to 14. NO indicates no obstacles, while LA and CH introduce bulk and checkerboard obstacles, as shown in Fig. 4. In this way, CH constraints the decision space to a few feasible paths. K2 neighbourhood (only four neighbours for each unit) restricts the number of paths more than the K3 neighbourhood with eight neighbours. All these combinations result in 256 test instances. The instances with the LA obstacle are computed from size five since there are no paths for smaller instances. Given a solution represented by a path N of length K as $N = (n_i, n_{i+1}, \cdots, n_k)$, we evaluate it by five objectives to be minimised: (1) Euclidean length, (2) Delays, (3) Elevation, (4) Travelling time and (5) Smoothness (Curvature). The details can be found in [30].

We use the same operators for pathfinding as in [30] i.e., a one-point crossover, which creates new offspring chromosome by crossing two parent paths at one common point. For the mutation operator, we also use the proposed perimeter mutation operator, which mutates the middle point of two arbitrary points within a specific network distance inside a given maximum radius and reconnects the paths afterwards. Consequently, we compare three algorithms with incorporated Fréchet distance and the standard baseline NSGA-II algorithm. For all four algorithms, we use a population size of 212 and 500 generations. Every algorithm is run 31 times. For the experiments, we calculate the IGD$^+$ [15] indicator. We report the respecting wins, losses and ties of each algorithm of all problem instances.

5.1 Results

Figure 5 illustrates the median value of IGD$^+$ indicator with respect to the instance size. We additionally report the number of Pareto-optimal solution (indicated by # of POS) for each problem instance. From a graphically perspective, it is visible that the results vary depending on the type and size of the problem instance. In the majority of the cases, the NSGA-II-FDAO (archive only) obtains the best median value and has the lowest standard error. Also, counting the number of experiments where an algorithm was outperformed by at least one other algorithm shows, that NSGA-II-FDAO was outperformed the least, i.e. only once in the instance CH_X14_Y14_PM_K3_BF. However, comparing this variant solely with the baseline algorithm, shows that the baseline algorithm was outperformed 127 times, while the proposed variant was never outperformed. Table 1 shows the wins, losses and ties of each algorithm on all of the 256 problems.

Table 1. Wins, Losses and Ties of each algorithm with statistical significance at $p < 0.01$, IGD$^+$ indicator, number of problems in total: 256.

	NSGAII	NSGAIIFD	NSGAIIFDAO	NSGAIIFDWOA
Wins	3	28	70	6
Losses	132	18	1	68
Ties	121	210	185	182

The variant with Fréchet distance only (NSGA-II-FDWOA, no archive), also performed well when compared to the baseline algorithm only. The proposed variation of the algorithm was outperformed twice (concerning the value of the IGD$^+$ indicator), when on the 256 chosen problems, while the baseline algorithm was outperformed 69 times. In the remaining experiments, there was no statistically significant difference.

The mixed variant NSGA-II-FD was never outperformed when only compared to the baseline algorithm, which was outperformed 108 times with respect to the IGD$^+$ indicator.

When analyzing the results, it becomes clear that the high amount of ties indicates room for improvement. This can be due to the fact, that in the Fréchet

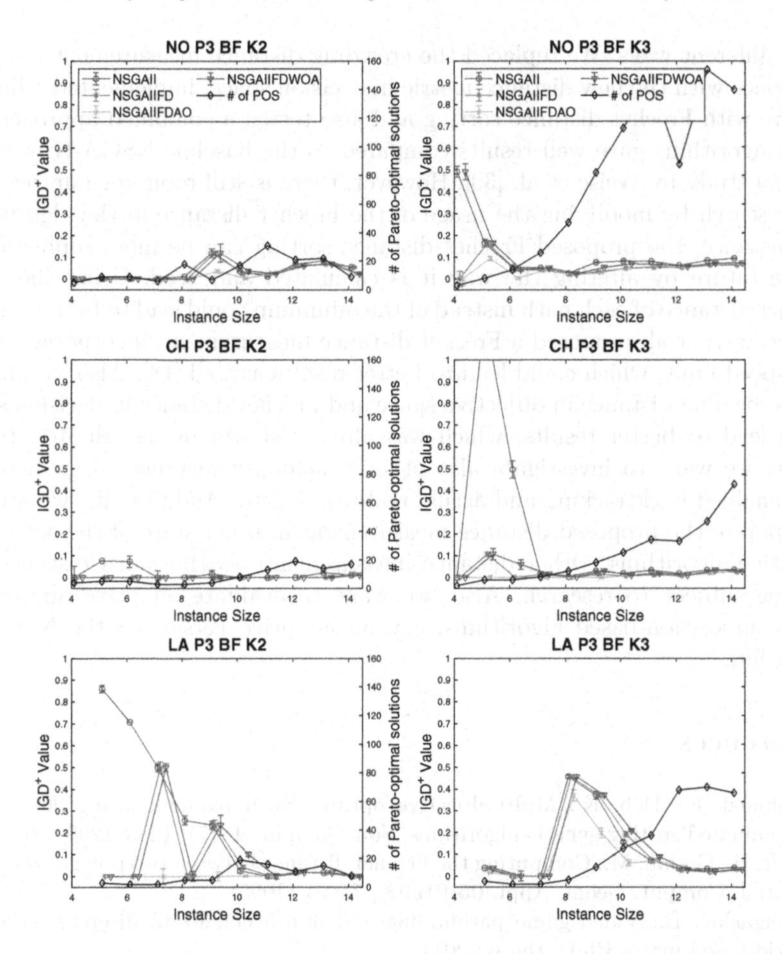

Fig. 5. IGD$^+$ over instance sizes. Top, middle and bottom rows illustrate NO, CH and LA obstacles. Right and left columns show the K2 and K3 neigbourhoods.

distance sorting, given two subsets of distinct paths with relatively short distance values inside the subset, will result in a short distance value for each path, though the two sets can be apart from each other. However, the results show an improvement compared to the baseline algorithm, since the proposed variants won in more problems and lost in less. The archive-only algorithm obtained the most promising results, complying with Patil's statement that only the usage of an archive can have a significant impact on the results.

6 Conclusion

In this paper, we have shown that, for pathfinding problems, a different distance measurement can positively improve the results. We implemented the discreet Fréchet distance calculation, incorporated it into the NSGA-II algorithm in three different ways. We replaced the crowding distance measurement in objective space with the new distance metric in decision space, implemented a limited archive with Fréchet distance sorting and also tested a combined approach. All three algorithms gave well results compared to the baseline NSGA-II from the original study by Weise et al. [30]. However, there is still room for improvement and research by modifying the usage of the Fréchet distance in the algorithms. For instance, the proposed Fréchet distance sorting can be more sophisticated in the future by altering the way it is calculated and used. Using the mean Fréchet distance of each path instead of the minimum could lead to better results. Maheshwari et al. proposed a Fréchet distance measurement, incorporating certain speed limit, which could lead to better results as well [18]. Also, combining both crowding distance in objective space and Fréchet distance in decision space could lead to better results, which was already shown by Javadi [16]. In the future, we want to investigate also other benchmark instances, e.g. instances with enabled backtracking and actual real-world data. Additionally, we want to incorporate the proposed distance measurement in other state-of-the-art many-objective algorithms. Other distance measurements as Hausdorff distance will also be subject to research. Also, we want to evaluate adaptive approaches of decomposition-based algorithms, e.g. an adaptive version of the NSGA-III algorithm.

References

1. Ahmed, F., Deb, K.: Multi-objective optimal path planning using elitist non-dominated sorting genetic algorithms. Soft Comput. **17**(7), 1283–1299 (2013)
2. Alt, H., Godau, M.: Computing the Fréchet distance between two polygonal curves. Int. J. Comput. Geom. Appl. **05**(01n02), 75–91 (1995)
3. Anguelov, B.: Video game pathfinding and improvements to discrete search on grid-based maps. Ph.D. thesis (2011)
4. Beke, L., Weiszer, M., Chen, J.: A comparison of genetic representations for multi-objective shortest path problems on multigraphs. In: Paquete, L., Zarges, C. (eds.) EvoCOP 2020. LNCS, vol. 12102, pp. 35–50. Springer, Cham (2020). https://doi.org/10.1007/978-3-030-43680-3_3
5. Besada-Portas, E., de la Torre, L., Moreno, A., Risco-Martín, J.L.: On the performance comparison of multi-objective evolutionary UAV path planners. Inf. Sci. **238**, 111–125 (2013)
6. Cai, X., Sun, H., Fan, Z.: A diversity indicator based on reference vectors for many-objective optimization. Inf. Sci. **430–431**, 467–486 (2018)
7. Deb, K., Pratap, A., Agarwal, S., Meyarivan, T.: A fast and elitist multiobjective genetic algorithm: NSGA-II. IEEE Trans. Evol. Comput. **6**(2), 182–197 (2002)
8. Deb, K., Tiwari, S.: Omni-optimizer: a procedure for single and multi-objective optimization. In: Coello Coello, C.A., Hernández Aguirre, A., Zitzler, E. (eds.) EMO 2005. LNCS, vol. 3410, pp. 47–61. Springer, Heidelberg (2005). https://doi.org/10.1007/978-3-540-31880-4_4
9. Efrat, A., Guibas, L.J., Har-Peled, S., Mitchell, J.S., Murali, T.M.: New similarity measures between polylines with applications to morphing and polygon sweeping. Discrete Comput. Geom. **28**(4), 535–569 (2002)

10. Eiter, T., Mannila, H.: Computing Discrete Fréchet Distance (1994)
11. Fan, C., Luo, J., Zhu, B.: Fréchet-distance on road networks. In: Akiyama, J., Bo, J., Kano, M., Tan, X. (eds.) CGGA 2010. LNCS, vol. 7033, pp. 61–72. Springer, Heidelberg (2011). https://doi.org/10.1007/978-3-642-24983-9_7
12. Fréchet, M.M.: Sur quelques points du calcul fonctionnel. Rendiconti del Circolo Matematico di Palermo (1884–1940) **22**(1), 1–72 (1906)
13. Hart, P.E., Nilsson, N.J., Raphael, B.: A formal basis for the heuristic determination of minimum cost paths. IEEE Trans. Syst. Sci. Cybern. **4**(2), 100–107 (1968)
14. Hu, J., Zhu, Q., Jun, H., Qingbao, Z.: Multi-objective mobile robot path planning based on improved genetic algorithm. In: 2010 International Conference on Intelligent Computation Technology and Automation, vol. 2, pp. 752–756. IEEE (2010)
15. Ishibuchi, H., Masuda, H., Nojima, Y.: A study on performance evaluation ability of a modified inverted generational distance indicator. In: Proceedings of the 2015 on Genetic and Evolutionary Computation Conference - GECCO 2015. pp. 695–702. ACM Press, New York (2015)
16. Javadi, M., Zille, H., Mostaghim, S.: Modified crowding distance and mutation for multimodal multi-objective optimization. In: Proceedings of the Genetic and Evolutionary Computation Conference Companion, pp. 211–212. ACM, New York (2019)
17. Jiang, M., Xu, Y., Zhu, B.: Protein structure-structure alignment with discrete Fréchet distance. J. Bioinform. Comput. Biol. **06**(01), 51–64 (2008)
18. Maheshwari, A., Sack, J.R., Shahbaz, K., Zarrabi-Zadeh, H.: Fréchet distance with speed limits. Comput. Geom. **44**(2), 110–120 (2011)
19. Mandow, L., De La Cruz, J.L.P.: Multiobjective A* search with consistent heuristics. J. ACM **57**(5), 1–25 (2010)
20. Munkres, J.R.: Topology. Featured Titles for Topology. Prentice Hall, Incorporated (2000)
21. Patil, M.B.: Improved performance in multi-objective optimization using external archive. Sādhanā **45**(1), 70 (2020)
22. Pulido, F.J.J., Mandow, L., Pérez-De-La-Cruz, J.L.L.: Dimensionality reduction in multiobjective shortest path search. Comput. Oper. Res. **64**, 60–70 (2015)
23. Anbuselvi, R.: Path finding solutions for grid based graph. Adv. Comput. Int. J. **4**(2), 51–60 (2013)
24. Sriraghavendra, E., Karthik, K., Bhattacharyya, C.: Fréchet distance based approach for searching online handwritten documents. In: Ninth International Conference on Document Analysis and Recognition (ICDAR 2007), vol. 1, pp. 461–465. IEEE, September 2007
25. Sturtevant, N.R.: Benchmarks for grid-based pathfinding. IEEE Trans. Comput. Intell. AI Games **4**(2), 144–148 (2012)
26. Toth, P., Vigo, D.: Vehicle Routing: Problems, Methods, and Applications. SIAM (2014)
27. Tozer, B., Mazzuchi, T., Sarkani, S.: Many-objective stochastic path finding using reinforcement learning. Expert Syst. Appl. **72**, 371–382 (2017)
28. Weise, J., Mai, S., Zille, H., Mostaghim, S.: On the scalable multi-objective multi-agent pathfinding problem. In: 2020 IEEE Congress on Evolutionary Computation (CEC), pp. 1–8. IEEE, July 2020
29. Weise, J., Mostaghim, S.: A many-objective route planning benchmark problem for navigation. In: Proceedings of the 2020 Genetic and Evolutionary Computation Conference Companion, GECCO 2020, pp. 183–184. ACM, New York, July 2020

30. Weise, J., Mostaghim, S.: Scalable many-objective pathfinding benchmark suite. [Manuscript submitted for publication] (2020). https://ci.ovgu.de/Publications/TEVC_WM_2020-p-910.html
31. Xiao, J., Michalewicz, Z.: An evolutionary computation approach to robot planning and navigation. Soft Comput. Mechatron. **32**, 117–141 (1999)
32. Yap, P.: Grid-based path-finding. In: Cohen, R., Spencer, B. (eds.) AI 2002. LNCS (LNAI), vol. 2338, pp. 44–55. Springer, Heidelberg (2002). https://doi.org/10.1007/3-540-47922-8_4

The Influence of Swarm Topologies in Many-Objective Optimization Problems

Diana Cristina Valencia-Rodríguez(✉) and Carlos A. Coello Coello

CINVESTAV-IPN (Evolutionary Computation Group), Avenue IPN 2508, 07360 San Pedro Zacatenco, Ciudad de México, Mexico
dvalencia@computacion.cs.cinvestav.mx, ccoello@cs.cinvestav.mx

Abstract. Particle Swarm Optimization (PSO) is a bio-inspired metaheuristic that has been successfully adopted for single- and multi-objective optimization. Several studies show that the way in which particles are connected with each other (the swarm topology) influences PSO's behavior. A few of these studies have focused on analyzing the influence of swarm topologies on the performance of Multi-objective Particle Swarm Optimizers (MOPSOs) using problems with two or three objectives. However, to the authors' best knowledge such studies have not been done so far for many-objective optimization problems. This paper provides an analysis of the influence of the ring, star, lattice, wheel, and tree topologies on the performance of SMPSO (a well-known Pareto-based MOPSO) using many-objective problems. Based on these results, we also propose two MOPSOs that use a combination of topologies: SMPSO-SW and SMPSO-WS. Our experimental results show that SMPSO-SW is able to outperform SMPSO in most of the test problems adopted.

Keywords: Swarm topology · Particle Swarm Optimization · Multi-objective Particle Swarm Optimization · Multi-objective optimization · Many-objective optimization

1 Introduction

Particle Swarm Optimization (PSO) is a metaheuristic that simulates the movements of a flock of birds or a school of fish seeking food [8]. PSO adopts a swarm of particles that communicate with each other intending to find the optimal solution. It has been experimentally shown that the communication networks connecting the particles (the *swarm topology*) influence PSO's performance [7,11]. Therefore, selecting the right swarm topology is crucial for obtaining good solutions.

There are numerous extensions of PSO for solving multi-objective optimization problems (see for example [4,5,9,13,18]). However, there are few studies on the influence of swarm topologies on the performance of Multi-Objective Particle Swarm Optimizers (MOPSOs) and such studies only analyze problems with

H. Ishibuchi et al. (Eds.): EMO 2021, LNCS 12654, pp. 387–398, 2021.
https://doi.org/10.1007/978-3-030-72062-9_31

three or fewer objectives [14,15,17]. Consequently, the influence of the swarm topology in the solution of many-objective problems remains as an unexplored topic.

In this paper, we analyze the influence of the ring, star, lattice, wheel, and tree topologies on the performance of the Speed-constrained Multiobjective Particle Swarm Optimizer (SMPSO) [12] using many-objective problems. In particular, we adopt the DTLZ1, DTLZ2, DTLZ3, DTLZ4, and DTLZ7 problems from the Deb-Thiele-Laumanns-Zitzler (DTLZ) [3] test suite with 3, 5, 8, and 10 objectives.[1] As will be seen later on, the results obtained show that the wheel and star topologies provide the best performance concerning the hypervolume and s-energy indicators. Based on these results, we propose here two new MOPSOs that use a combination of these two topologies: the SMPSO-SW and the SMPSO-WS. Our preliminary experimental results show that SMPSO-SW is very competitive with respect to the original SMPSO.

The remainder of this paper is organized in the following way. In Sect. 2, we provide a brief explanation of background concepts. Then, in Sect. 3, we explain what a swarm topology is, and we give some examples. After that, in Sect. 4, we discuss the functionality of SMPSO as well as the way in which we modified it for handling different swarm topologies. In Sect. 5, we present the results of our initial experiments. Based on these results, we propose two new MOPSOs described in Sect. 6. Finally, in Sect. 7, we provide our conclusions and some possible paths for future work.

2 Background

2.1 Multi-objective Optimization

In multi-objective optimization, the aim is to solve problems of the type[2]:

$$\text{minimize } \boldsymbol{f}(\boldsymbol{x}) := [f_1(\boldsymbol{x}), f_2(\boldsymbol{x}), \ldots, f_m(\boldsymbol{x})] \tag{1}$$

subject to:

$$g_i(\boldsymbol{x}) \leq 0 \quad i = 1, 2, \ldots, p \tag{2}$$

$$h_i(\boldsymbol{x}) = 0 \quad i = 1, 2, \ldots, q \tag{3}$$

where $\boldsymbol{x} = [x_1, x_2, \ldots, x_n]^T$ is the vector of decision variables, $f_i : \mathbb{R}^n \to \mathbb{R}$, $i = 1, ..., m$ are the objective functions and $g_i, h_j : \mathbb{R}^n \to \mathbb{R}$, $i = 1, ..., p$, $j = 1, ..., q$ are the constraint functions of the problem.

A few additional definitions are required to introduce the notion of optimality used in multi-objective optimization:

[1] Although many-objective problems are those having more than 3 objectives, our experiments include test problems with 3 objectives to allow a more clear visualization of the effect of dimensionality increase in objective function space.

[2] Without loss of generality, we will assume only minimization problems.

Definition 1. Given two vectors $\boldsymbol{x}, \boldsymbol{y} \in \mathbb{R}^m$, we say that $\boldsymbol{x} \leq \boldsymbol{y}$ if $x_i \leq y_i$ for $i = 1, ..., m$, and that $\boldsymbol{x}$ **dominates** $\boldsymbol{y}$ (denoted by $\boldsymbol{x} \prec \boldsymbol{y}$) if $\boldsymbol{x} \leq \boldsymbol{y}$ and $\boldsymbol{x} \neq \boldsymbol{y}$.

Definition 2. We say that a vector of decision variables $\boldsymbol{x} \in \mathcal{X} \subset \mathbb{R}^n$ is **nondominated** with respect to $\mathcal{X}$, if there does not exist another $\boldsymbol{x}' \in \mathcal{X}$ such that $\boldsymbol{f}(\boldsymbol{x}') \prec \boldsymbol{f}(\boldsymbol{x})$.

Definition 3. We say that a vector of decision variables $\boldsymbol{x}^* \in \mathcal{F} \subset \mathbb{R}^n$ ($\mathcal{F}$ is the feasible region) is **Pareto-optimal** if it is nondominated with respect to $\mathcal{F}$.

Definition 4. The **Pareto Optimal Set** $\mathcal{P}^*$ is defined by:

$$\mathcal{P}^* = \{\boldsymbol{x} \in \mathcal{F} | \boldsymbol{x} \text{ is Pareto-optimal}\}$$

Definition 5. The **Pareto Front** $\mathcal{PF}^*$ is defined by:

$$\mathcal{PF}^* = \{\boldsymbol{f}(\boldsymbol{x}) \in \mathbb{R}^m | \boldsymbol{x} \in \mathcal{P}^*\}$$

Therefore, our aim is to obtain the Pareto optimal set from the set $\mathcal{F}$ of all the decision variable vectors that satisfy (2) and (3).

In this paper, we address problems with more than three objectives (the so-called *many-objective optimization problems*).

2.2 Particle Swarm Optimization

PSO is a bio-inspired metaheuristic proposed by James Kennedy and Russel Eberhart in 1995 [8]. PSO operates with a *swarm* that is a set composed of potential solutions called *particles.* Each particle moves towards promising regions influenced by its previous best position and the best position found so far by the particles in the neighborhood.

Let $\mathbf{x_i}(t)$ be the position of a particle in generation t. Then, PSO updates its position using the following expression:

$$\mathbf{x_i}(t+1) = \mathbf{x_i}(t) + \mathbf{v_i}(t+1). \tag{4}$$

$\mathbf{v_i}(t+1)$ is known as the *velocity vector* and is defined by:

$$\mathbf{v_i}(t+1) = w\mathbf{v_i}(t) + C_1 r_1(\mathbf{x_{p_i}} - \mathbf{x_i}(t)) + C_2 r_2(\mathbf{x_{l_i}} - \mathbf{x_i}(t)) \tag{5}$$

where $r_1, r_2 \in U(0,1)$; C_1 and C_2 are positive constants called *cognitive* and *social* factors, respectively; and w is a positive parameter called *inertia weight.* The position $\mathbf{x_{p_i}}$ is the best solution found by the particle, and $\mathbf{x_{l_i}}$ is the best solution found by the particle's neighbors (known as *leader*). Each particle's neighborhood is determined by the *swarm topology,* which defines the connections of influence among particles.

3 Swarm Topologies

A swarm topology is a social network represented by a graph where each vertex is a particle, and there is an edge between two particles if they influence each other [11]. A topology can change along with the generations (*dynamic topology*) or remain static during the execution (*static topology*). Its formal definition is the following [10]:

Definition 1. *A* ***swarm topology*** *at generation i is a graph* $T_i = (P_i, E_i)$ *where the vertex set* $P_i = \{p_0, p_1, ..., p_{n-1}\}$ *is a set of particles.*

Many swarm topologies have been proposed over the years, each of which having different characteristics. For this study, we selected five static topologies which are representative of the state-of-the-art in the area [7,11]:

- **Lattice**. The graph of this topology represents a two-dimensional lattice. Each particle influences the neighbors above, below, and two on each side. See Fig. 1a.
- **Star** (*gbest*). Each particle in this topology influences the remaining particles in the swarm. See Fig. 1b.
- **Tree**. The particles in this topology are arranged hierarchically, resembling a tree. Each particle influences its father and its children, and viceversa. See Fig. 1c.
- **Wheel**. In this topology, a central particle influences the others in the swarm, and they influence it as well. See Fig. 1d.
- **Ring** (lbest). In this topology, each particle influences its two nearest neighbors. See Fig. 1e.

4 Multi-objective Particle Swarm Optimizer with a Topology Handling Scheme

A single-objective PSO updates the position $\mathbf{x_{p_i}}$ with the position $\mathbf{x_i}(t)$ at each generation if $f(\mathbf{x_i}(t))$ is better than $f(\mathbf{x_{p_i}})$. Then, PSO examines the personal best positions in the particle's neighborhood and selects its leader from it. In contrast, in multi-objective optimization, we often find incomparable solutions. Therefore, most of the MOPSOs store the best positions found so far in a separate set (called *external archive*) and take the leaders from it without specifying the particle's neighborhood.

We selected for the experimental analysis a standard Pareto-based MOPSO that works in this way, the SMPSO [12]. Its core idea is to control the particles' velocity employing a constriction factor χ that prevents high-velocity values and guarantees convergence under certain conditions [1]. This coefficient is defined by:

$$\chi = 2/(2 - \varphi - \sqrt{\varphi^2 - 4\varphi}) \tag{6}$$

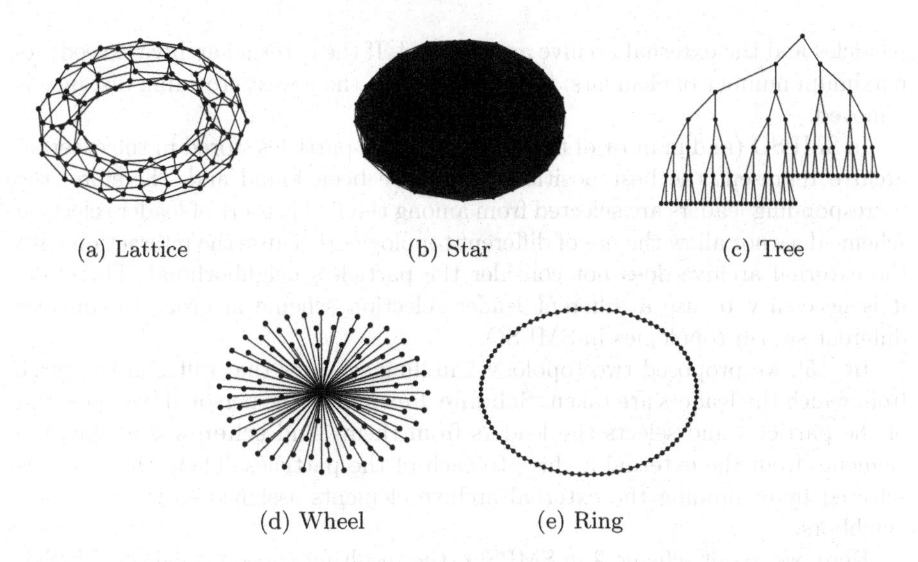

Fig. 1. Swarm topologies of state of the art.

where

$$\varphi = \begin{cases} C_1 + C_2 & \text{if } C_1 + C_2 > 4 \\ 1 & \text{if } C_1 + C_2 \leq 4 \end{cases} \tag{7}$$

Moreover, SMPSO implements a mechanism to constrain the accumulated velocity of the particle i in the dimension j using the following expression:

$$v_{i,j}(t) = \begin{cases} \delta_j & \text{if } v_{i,j}(t) > \delta_j \\ -\delta_j & \text{if } v_{i,j}(t) \leq -\delta_j \\ v_{i,j}(t) & \text{otherwise} \end{cases} \tag{8}$$

where $\delta_j = (upper\_limit_j - lower\_limit_j)/2$, and the jth decision variable is in the range $[lower\_limit_j, upper\_limit_j]$.

Therefore, the particles' velocity is computed using Eq. (5) and then the result is multiplied by the constriction factor defined in Eq. (6), and it is bounded using the rule defined in Eq. (8).

In general, SMPSO works in the following way. First, the swarm is initialized with random values, and the external archive is created with the non-dominated solutions of the swarm. Then, until a maximum number of generations is reached, each particle leader is selected by randomly taking two solutions from the external archive and selecting the one with the largest crowding distance, which measures how isolated is a solution from the others. Next, with the selected leaders, the velocity and position of the particles are computed. Moreover, polynomial-based mutation [2] is applied to the new particles using a probability p_m, and the resulting particles are evaluated. Finally, the personal best positions of the particles and the external archive are updated. If the external archive exceeds its maximum number of elements, the element with the lowest crowding distance is removed.

In SMPSO (and in most of the MOPSOs), the particles stored in the external archive represent the best positions that have been found and, therefore, the corresponding leaders are selected from among them. This sort of leader selection scheme does not allow the use of different topologies, because the interaction with the external archive does not consider the particle's neighborhood. Therefore, it is necessary to use a different leader selection scheme in order to consider different swarm topologies in SMPSO.

In [15], we proposed two topology handling schemes that differ in the place from which the leaders are taken. **Scheme 1** examines the personal best position of the particles and selects the leaders from them. And **scheme 2** assigns the elements from the external archive to each of the particles. Then, the leader is selected by examining the external archive elements assigned to the particle's neighbors.

Here, we adopt scheme 2 in SMPSO (the resulting version is called SMPSO-E2) because this scheme had the best performance in the study reported in [15]. The difference between SMPSO and SMPSO-E2 is that at each generation, SMPSO-E2 assigns the external archive elements to each particle. The archive elements are assigned again if the archive size is smaller than the swarm size. Then, each particle's leader is selected by randomly taking two of its neighbors and picking the one that has the assigned element with the largest crowding distance. After that, the following steps are the same as in the original SMPSO.

5 Experimental Analysis

To carry out our experimental analysis, we compared SMPSO and SMPSO-E2 using the tree, lattice, star, ring, and wheel topologies. For that sake, we performed 30 independent runs of each MOP using the parameters shown in Table 1.

Table 1. Parameters of the MOPSOs adopted in this study

Parameter	Value
Archive size	300
Swarm size	300
Mutation probability (p_m)	$1/n$
Inertia weight (w)	0.1
Max number of generations	2500

We adopted the DTLZ1-DTLZ4 and DTLZ7 problems from the scalable Deb-Thiele-Laumanns-Zitzler (DTLZ) test suite [3]. We tested these problems with

3, 5, 8, and 10 objectives where the decision variables (n) are defined by $n = m + k - 1$ and m is the number of objectives. We used a value of $k = 5$ for DTLZ1, $k = 10$ for DTLZ2-4, and $k = 20$ for DTLZ7.

For assessing performance, we used the hypervolume [19] and s-energy [6] indicators. The hypervolume measures the size of the objective space covered by the approximated set, given a reference point. The larger the space covered, the better the approximation. Therefore, the aim is to maximize the hypervolume indicator. We used as reference points the worst objective function values obtained in the approximated sets for each problem, multiplied by 1.1. On the other hand, the s-energy indicator measures the uniform distribution of a set in a d-dimensional manifold. A lower s-energy value implies an approximation set with higher diversity. Moreover, to evaluate the results' statistical confidence, we applied the Wilcoxon signed-rank test with a significance level of 5%.

Tables 2 and 3 present the mean and the standard deviation of the hypervolume and s-energy indicators. The best values have a gray background, and the symbol "*" means that the difference of the corresponding algorithm with respect to the others is statistically significant.

Table 2. Mean and standard deviation of hypervolume indicator for SMPSO and SMPSO-E2. The best values have a gray background, and the symbol "*" indicates that the result is statistically significant.

	m	SMPSO	SMPSO-E2				
			Lattice	Star	Tree	Wheel	Ring
DTLZ1	3	1.3992e-1 (2.3e-4)	1.4008e-1 (2.5e-4)	1.3999e-1 (2.3e-4)	1.4007e-1 (2.8e-4)	*1.4077e-1 (4.7e-4)	1.4017e-1 (2.9e-4)
	5	7.349e-2 (6.1e-4)	7.4196e-2 (5.1e-4)	7.3538e-2 (4.5e-4)	7.4221e-2 (4.9e-4)	*7.5724e-2 (4.2e-4)	7.4519e-2 (6.1e-4)
	8	2.4274e+12 (7.0e+8)	2.4276e+12 (6.5e+5)	2.4276e+12 (1.0e+7)	2.4276e+12 (1.5e+6)	*2.4276e+12 (1.8e+5)	2.4275e+12 (3.1e+7)
	10	3.6423e+16 (0.e+0)	3.6423e+16 (1.3e+11)	3.6423e+16 (3.6e+9)	3.6423e+16 (1.3e+11)	3.6423e+16 (1.1e+10)	3.6409e+16 (7.5e+13)
DTLZ2	3	9.8288e-1 (2.4e-3)	9.8372e-1 (2.4e-3)	9.8330e-1 (3.e-3)	9.8388e-1 (2.e-3)	*9.8831e-1 (2.9e-3)	9.8334e-1 (2.3e-3)
	5	8.3629e+1 (8.0e-2)	8.3699e+1 (7.8e-2)	8.3619e+1 (8.7e-2)	8.3701e+1 (1.e-1)	*8.4043e+1 (3.4e-2)	8.3682e+1 (6.6e-2)
	8	7.2833e+3 (8.6e+0)	7.2976e+3 (5.4e+0)	7.2836e+3 (1.1e+1)	7.2930e+3 (6.9e+0)	*7.3311e+3 (1.3e+0)	7.2945e+3 (6.5e+0)
	10	1.7587e+5 (4.3e+2)	1.7623e+5 (2.0e+2)	1.7585e+5 (4.4e+2)	1.7618e+5 (2.1e+2)	*1.7688e+5 (5.6e+1)	1.7632e+5 (1.5e+2)
DTLZ3	3	1.0220e+0 (2.e-3)	1.0230e+0 (2.6e-3)	1.0219e+0 (2.2e-3)	1.0230e+0 (3.1e-3)	*1.0287e+0 (3.3e-3)	1.0236e+0 (2.5e-3)
	5	1.5944e+10 (0.e+0)	1.5944e+10 (0.e+0)	1.5944e+10 (0.e+0)	1.5944e+10 (0.e+0)	1.5944e+10 (0.e+0)	1.5944e+10 (0.e+0)
	8	2.9538e+24 (7.e+17)	2.9538e+24 (8.5e+17)	2.9538e+24 (6.6e+17)	2.9538e+24 (8.1e+17)	*2.9538e+24 (0.e+0)	2.9538e+24 (9.7e+17)
	10	9.5943e+30 (7.7e+23)	9.5943e+30 (1.e+24)	9.5943e+30 (8.5e+23)	9.5943e+30 (1.1e+24)	*9.5943e+30 (2.3e+15)	9.5943e+30 (1.4e+24)
DTLZ4	3	1.1089e+0 (1.9e-3)	1.1082e+0 (2.1e-3)	1.1092e+0 (2.2e-3)	1.1082e+0 (2.2e-3)	*1.1159e+0 (2.9e-3)	1.108e+0 (2.5e-3)
	5	7.7119e+1 (2.3e-2)	7.7087e+1 (1.9e-2)	7.7113e+1 (2.1e-2)	7.7098e+1 (2.2e-2)	*7.7139e+1 (2.7e-2)	7.7083e+1 (2.2e-2)
	8	1.4538e+4 (7.2e-1)	1.4537e+4 (1.5e+0)	1.4538e+4 (6.9e-1)	1.4537e+4 (1.2e+0)	1.4537e+4 (2.3e+0)	1.4536e+4 (1.5e+0)
	10	3.4448e+5 (1.3e+1)	3.4438e+5 (7.7e+1)	3.4448e+5 (1.3e+1)	3.444e+5 (5.6e+1)	3.4415e+5 (5.4e+2)	3.443e+5 (1.7e+2)
DTLZ7	3	2.595e+0 (9.4e-3)	2.5939e+0 (1.1e-2)	2.5937e+0 (5.6e-3)	2.593e+0 (1.2e-2)	2.5981e+0 (1.3e-2)	2.5952e+0 (7.9e-3)
	5	4.0609e+0 (4.5e-2)	4.0549e+0 (5.7e-2)	4.0451e+0 (8.6e-2)	4.0871e+0 (5.9e-2)	*4.2474e+0 (5.8e-2)	4.0565e+0 (6.6e-2)
	8	6.3453e+0 (1.4e-1)	6.4385e+0 (1.2e-1)	6.2954e+0 (1.7e-1)	6.4785e+0 (1.1e-1)	*7.0286e+0 (8.2e-2)	6.4752e+0 (1.1e-1)
	10	3.7925e+1 (3.8e+0)	3.8882e+1 (7.9e-1)	3.8786e+1 (2.2e-1)	3.9125e+1 (1.8e-1)	*3.9788e+1 (2.7e-1)	3.9112e+1 (1.9e-1)

Table 2 shows that, on average, the wheel topology has the best performance for the hypervolume indicator in all of the objectives. In comparison, the star topology performs worst. Regarding the other topologies, the tree topology performs better than the lattice, and the lattice topology performs better than the ring. We validated in [16] the topologies' obtained arrangement using the Wilcoxon signed-rank test with a significance level of 5%. Therefore, it seems that the lower the degree of connectivity the topology has, the better the hyper-

Table 3. Mean and standard deviation of s-energy indicator for SMPSO and SMPSO-E2. The best values have a gray background, and the symbol "*" indicates that the result is statistically significant.

	m	SMPSO	SMPSO-E2				
			Lattice	Star	Tree	Wheel	Ring
DTLZ1	3	6.1494e+8 (2.7e+8)	7.1795e+8 (2.3e+8)	*6.0124e+8 (5.4e+8)	6.2656e+8 (1.4e+8)	1.2194e+9 (3.7e+8)	6.3859e+8 (1.3e+8)
	5	3.7068e+10 (2.3e+10)	5.4841e+10 (2.3e+10)	2.8204e+10 (1.8e+10)	6.4885e+10 (2.8e+10)	2.5939e+11 (8.2e+10)	8.9301e+10 (3.1e+10)
	8	2.1012e+11 (6.5e+11)	2.3178e+11 (2.3e+11)	1.3365e+11 (3.1e+11)	2.1332e+11 (2.5e+11)	1.4822e+12 (9.5e+11)	3.1467e+11 (3.6e+11)
	10	7.3677e+10 (2.4e+11)	1.8238e+11 (2.8e+11)	9.4263e+11 (3.2e+12)	2.8388e+11 (5.9e+11)	3.4625e+12 (5.6e+12)	3.0778e+11 (5.7e+11)
DTLZ2	3	7.6337e+9 (8.5e+9)	2.148e+10 (5.e+9)	9.1892e+9 (9.1e+9)	2.0046e+10 (6.3e+8)	2.1769e+10 (7.3e+9)	2.0061e+10 (2.5e+7)
	5	3.6556e+10 (2.3e+10)	9.1259e+10 (4.3e+10)	2.9936e+10 (2.1e+10)	1.2281e+11 (3.2e+10)	1.3777e+11 (1.3e+10)	1.4109e+11 (2.3e+10)
	8	9.6620e+10 (4.8e+10)	1.6801e+11 (6.6e+10)	7.7822e+10 (3.3e+10)	2.1695e+11 (9.2e+10)	6.0208e+11 (8.6e+10)	3.2113e+11 (1.1e+11)
	10	1.7591e+11 (5.2e+10)	2.8241e+11 (8.3e+10)	1.9251e+11 (5.7e+10)	2.9815e+11 (8.1e+10)	9.5665e+11 (1.4e+11)	4.5666e+11 (1.6e+11)
DTLZ3	3	8.3273e+9 (8.3e+9)	1.8324e+10 (4.8e+9)	8.6146e+9 (8.6e+9)	1.9597e+10 (2.7e+9)	2.1457e+10 (7.2e+9)	2.0907e+10 (3.7e+9)
	5	7.7074e+9 (1.4e+10)	3.1648e+10 (2.8e+10)	2.8588e+9 (6.1e+9)	5.1345e+10 (3.7e+10)	1.4193e+11 (2.2e+10)	9.0929e+10 (5.e+10)
	8	8.7195e+9 (1.7e+10)	1.0675e+10 (1.2e+10)	5.3661e+9 (1.0e+10)	2.2667e+10 (3.8e+10)	6.3975e+10 (1.6e+11)	2.4668e+10 (3.3e+10)
	10	1.6612e+10 (2.2e+10)	2.6718e+10 (3.3e+10)	1.0395e+10 (1.6e+10)	2.2187e+10 (2.4e+10)	1.5392e+10 (1.5e+10)	2.7334e+10 (4.1e+10)
DTLZ4	3	1.2133e+10 (1.3e+10)	2.0457e+10 (1.9e+9)	1.519e+10 (8.3e+9)	2.0117e+10 (2.e+8)	2.0134e+10 (2.4e+8)	2.01e+10 (1.6e+8)
	5	8.4434e+10 (3.5e+10)	1.6135e+11 (3.1e+10)	8.3891e+10 (3.e+10)	1.5889e+11 (2.7e+10)	1.6925e+11 (3.1e+10)	1.7089e+11 (2.6e+10)
	8	2.8836e+11 (7.0e+10)	2.8382e+11 (7.2e+10)	2.6838e+11 (7.1e+10)	3.0473e+11 (8.e+10)	5.8014e+11 (3.6e+11)	3.6321e+11 (1.1e+11)
	10	4.6999e+11 (8.2e+10)	4.4684e+11 (8.6e+10)	4.5576e+11 (1.1e+11)	4.4513e+11 (9.1e+10)	4.3781e+11 (1.8e+11)	4.9120e+11 (1.4e+11)
DTLZ7	3	4.3088e+7 (1.1e+7)	7.2996e+8 (3.6e+9)	4.9655e+7 (3.7e+7)	6.3745e+7 (7.4e+7)	1.3143e+9 (4.3e+9)	7.1740e+8 (3.6e+9)
	5	7.8658e+8 (2.1e+9)	7.4775e+8 (1.2e+9)	1.9315e+9 (5.e+9)	4.8294e+9 (7.7e+9)	1.4197e+10 (1.2e+10)	3.5381e+9 (5.6e+9)
	8	3.1266e+10 (2.6e+10)	7.0838e+10 (3.4e+10)	2.5010e+10 (2.3e+10)	9.037e+10 (5.2e+10)	4.0819e+11 (1.2e+11)	9.6846e+10 (4.6e+10)
	10	5.7154e+10 (4.5e+10)	1.4790e+11 (7.0e+10)	5.9874e+10 (3.2e+10)	1.8519e+11 (8.9e+10)	1.1722e+12 (3.4e+11)	1.8915e+11 (7.5e+10)

volume value. Compared with SMPSO, the SMPSO-E2 with wheel topology performed better in almost all problems.

Conversely, Table 3 shows that the star topology had the best performance on average in all of the problems regarding the s-energy indicator; and the wheel topology had the worst performance. If we compare the topologies with the original SMPSO, the SMPSO-E2 with the star topology performed better than the original SMPSO in more problems. However, the Wilcoxon test did not show a statistical significance of the results and in [16] we could not define a statistically confident arrangement between all the topologies.

In summary, we can see that the wheel topology had the best performance regarding the hypervolume indicator and the star topology had the best performance concerning the s-energy indicator. The results also show that there could be a relationship between the degree of connectivity and the hypervolume values.

6 Combining Topologies

The previous experiments showed that the wheel topology had the best performance with respect to the hypervolume indicator. In contrast, the star topology performed best with respect to the s-energy indicator. Therefore, it can be easily inferred that the wheel topology promotes the MOPSO's convergence, and that the star topology improves the solution's distribution. Thus, if we use both topologies, we could balance the exploration and exploitation of the SMPSO-E2.

In order to validate this hypothesis, we modified the SMPSO-E2 to use a dynamic topology that adopts one topology the first half of the generations and changes to another one for the second half of the generations. The version of

SMPSO-E2 that uses the wheel topology the first half of the generations and the star topology during the second half is called SMPSO-WS. Furthermore, the SMPSO-E2 that uses star topology during the first half of the generations and the wheel topology during the second half is called SMPSO-SW.

We performed 30 independent runs of SMPSO, SMPSO-WS, and SMPSO-SW with the parameters and problems that we used in the previous experimental study. For assessing performance, we used again the hypervolume (with the same reference points) and s-energy indicators.

Tables 4 and 5 show the mean and standard deviation of the hypervolume and s-energy indicator. The "*" symbol means that the difference between the corresponding algorithm and the others is statistically significant.

Table 4. Mean and standard deviation of hypervolume indicator for SMPSO, SMPSO-WS, and SMPSO-SW. The best values have a gray background, and the symbol "*" indicates that the result is statistically significant.

	m	SMPSO	SMPSO-WS	SMPSO-SW
DTLZ1	3	1.3992e-1 (2.3e-4)	1.4004e-1 (2.1e-4)	*1.4056e-1 (4.6e-4)
	5	7.349e-2 (6.1e-4)	7.3924e-2 (4.6e-4)	*7.5737e-2 (2.5e-4)
	8	2.4274e+12 (7.0e+8)	2.4276e+12 (5.6e+5)	2.4276e+12 (4.1e+6)
	10	3.6423e+16 (0.e+0)	3.6423e+16 (0.e+0)	3.6423e+16 (1.8e+9)
DTLZ2	3	9.8288e-1 (2.4e-3)	9.8370e-1 (2.7e-3)	*9.8981e-1 (2.7e-3)
	5	8.3629e+1 (8.0e-2)	8.3745e+1 (8.1e-2)	*8.4007e+1 (6.6e-2)
	8	7.2833e+3 (8.6e+0)	7.3003e+3 (5.e+0)	*7.3293e+3 (3.1e+0)
	10	1.7587e+5 (4.3e+2)	1.7637e+5 (1.3e+2)	*1.7681e+5 (7.7e+1)
DTLZ3	3	1.0220e+0 (2.e-3)	1.0228e+0 (2.0e-3)	*1.0281e+0 (3.7e-3)
	5	1.5944e+10 (0.e+0)	1.5944e+10 (0.e+0)	1.5944e+10 (0.e+0)
	8	2.9538e+24 (7.e+17)	2.9538e+24 (3.4e+17)	*2.9538e+24 (0.e+0)
	10	9.5943e+30 (7.7e+23)	9.5943e+30 (4.9e+23)	*9.5943e+30 (1.8e+23)
DTLZ4	3	1.1089e+0 (1.9e-3)	1.1089e+0 (2.1e-3)	*1.1140e+0 (2.9e-3)
	5	7.7119e+1 (2.3e-2)	7.7119e+1 (2.e-2)	*7.7142e+1 (2.4e-2)
	8	*1.4538e+4 (7.2e-1)	1.4538e+4 (7.2e-1)	1.4537e+4 (1.6e+0)
	10	*3.4448e+5 (1.3e+1)	3.4445e+5 (2.9e+1)	3.4431e+5 (1.5e+2)
DTLZ7	3	2.595e+0 (9.4e-3)	2.5919e+0 (8.7e-3)	2.5958e+0 (1.1e-2)
	5	4.0609e+0 (4.5e-2)	4.0212e+0 (6.1e-2)	*4.2420e+0 (6.3e-2)
	8	6.3453e+0 (1.4e-1)	6.3408e+0 (1.3e-1)	*7.0338e+0 (1.1e-1)
	10	3.7925e+1 (3.8e+0)	3.8854e+1 (2.6e-1)	*3.9817e+1 (1.6e-1)

Regarding the hypervolume indicator, it is clear that SMPSO-SW performs best in almost all of the test problems. We assume that this happened because the algorithm examines new regions in the search space at the beginning of the search with the star topology; consequently, the algorithm seems to be able to find promising regions of the search space, which makes it a very successful approach. Then, it exploits these promising regions using the wheel topology, obtaining a refined solution. In contrast, the SMPSO-WS tries to exploit a non-promising region using the wheel topology and then moves to many regions using the star topology thus preventing the algorithm from converging.

In the case of the s-energy indicator, the SMPSO performs best in almost all of the problems. Therefore, neither the SMPSO-WS nor SMPSO-SW had enough impact on this indicator.

Table 5. Mean and standard deviation of s-energy indicator for SMPSO, SMPSO-WS and SMPSO-SW. The best values have a gray background, and the symbol "*" indicates that the result is statistically significant.

	m	SMPSO	SMPSO-WS	SMPSO-SW
DTLZ1	3	6.1494e+8 (2.7e+8)	1.2030e+9 (3.6e+9)	1.1281e+9 (2.3e+8)
	5	3.7068e+10 (2.3e+10)	3.3739e+10 (1.7e+10)	2.1303e+11 (8.6e+10)
	8	*2.1012e+11 (6.5e+11)	2.4204e+11 (1.9e+11)	1.3907e+12 (6.6e+11)
	10	*7.3677e+10 (2.4e+11)	3.3888e+11 (6.8e+11)	4.9779e+12 (4.7e+12)
DTLZ2	3	*7.6337e+9 (8.5e+9)	2.09e+10 (3.5e+9)	2.0074e+10 (4.7e+7)
	5	*3.6556e+10 (2.3e+10)	1.3925e+11 (1.8e+10)	1.3772e+11 (1.8e+10)
	8	*9.6620e+10 (4.8e+10)	4.5010e+11 (3.0e+10)	5.6628e+11 (8.1e+10)
	10	*1.7591e+11 (5.2e+10)	7.123e+11 (9.0e+10)	8.4480e+11 (8.7e+10)
DTLZ3	3	*8.3273e+9 (8.3e+9)	1.9799e+10 (1.6e+9)	2.3675e+10 (1.1e+10)
	5	*7.7074e+9 (1.4e+10)	6.7153e+10 (4.4e+10)	9.3200e+10 (4.6e+10)
	8	8.7195e+9 (1.7e+10)	1.224e+10 (2.6e+10)	6.5998e+10 (2.7e+11)
	10	1.6612e+10 (2.2e+10)	1.3346e+10 (2.4e+10)	1.4024e+10 (1.6e+10)
DTLZ4	3	*1.2133e+10 (1.3e+10)	2.0995e+10 (3.8e+9)	2.1590e+10 (7.2e+9)
	5	*8.4434e+10 (3.5e+10)	1.2783e+11 (4.4e+10)	1.5963e+11 (3.1e+10)
	8	2.8836e+11 (7.0e+10)	2.6689e+11 (5.9e+10)	4.7546e+11 (1.3e+11)
	10	4.6999e+11 (8.2e+10)	4.6444e+11 (9.7e+10)	5.4920e+11 (2.7e+11)
DTLZ7	3	4.3088e+7 (1.1e+7)	4.8593e+7 (2.3e+7)	1.6039e+9 (4.3e+9)
	5	7.8658e+8 (2.1e+9)	5.3764e+8 (7.4e+8)	1.1789e+10 (1.1e+10)
	8	3.1266e+10 (2.6e+10)	3.0083e+10 (1.9e+10)	4.8259e+11 (1.6e+11)
	10	5.7154e+10 (4.5e+10)	6.5817e+10 (3.7e+10)	1.1618e+12 (2.9e+11)

7 Conclusions and Future Work

In this work, we analyzed the influence of the star, ring, wheel, lattice, and tree topologies on the performance of MOPSOs. The experimental results showed a relationship between the connectivity degree of topologies and the hypervolume values. If the connectivity degree decreases, the hypervolume value will increase.

The experimental results also showed that the wheel topology causes better values of the hypervolume indicator. In contrast, the star topology produces better values of the s-energy indicator. Hence, the wheel topology promotes the MOPSO's convergence, and the star topology improves the solution's distribution.

Therefore, in order to balance the exploitation and exploration of SMPSO-E2, we proposed two MOPSOs that use a combination of the wheel and star topologies: the SMPSO-WS and the SMPSO-SW. The SMPSO-SW outperformed the original SMPSO in most of the test problems adopted regarding the hypervolume indicator.

As part of our future work, we plan to test an adaptative topology that changes its connectivity degree depending on the behavior of the MOPSO.

Acknowledgements. The first author acknowledges support from CONACyT and CINVESTAV-IPN to pursue graduate studies in Computer Science. The second author acknowledges support from CONACyT grant no. 1920 and from a SEP-Cinvestav grant (application no. 4).

References

1. Clerc, M., Kennedy, J.: The particle swarm - explosion, stability, and convergence in a multidimensional complex space. IEEE Trans. Evol. Comput. **6**(1), 58–73 (2002). (CEC 2002)
2. Deb, K., Agrawal, R.B.: Simulated binary crossover for continuous search space. Complex Syst. **9**, 115–148 (1995)
3. Deb, K., Thiele, L., Laumanns, M., Zitzler, E.: Scalable test problems for evolutionary multiobjective optimization. In: Abraham, A., Jain, L., Goldberg, R. (eds.) Evolutionary Multiobjective Optimization: Theoretical Advances and Applications, pp. 105–145. Springer, London (2005). https://doi.org/10.1007/1-84628-137-7_6
4. Figueiredo, E.M.N., Ludermir, T.B., Bastos-Filho, C.J.A.: Many objective particle swarm optimization. Inf. Sci. **374**, 115–134 (2016)
5. Han, H., Lu, W., Zhang, L., Qiao, J.: Adaptive gradient multiobjective particle swarm optimization. IEEE Trans. Cybern. **48**(11), 3067–3079 (2018)
6. Hardin, D., Saff, E.: Discretizing manifolds via minimum energy points. Not. Am. Math. Soc. **51**(10), 1186–1194 (2004)
7. Kennedy, J.: Small worlds and mega-minds: effects of neighborhood topology on particle swarm performance. In: Proceedings of the 1999 Congress on Evolutionary Computation (CEC 1999), vol. 3, pp. 1931–1938, July 1999
8. Kennedy, J., Eberhart, R.: Particle swarm optimization. In: Proceedings of the 1995 IEEE International Conference on Neural Networks (ICNN 1995), vol. 4, pp. 1942–1948 (1995)
9. Lin, Q., et al.: Particle swarm optimization with a balanceable fitness estimation for many-objective optimization problems. IEEE Trans. Evol. Comput. **22**(1), 32–46 (2018)
10. McNabb, A., Gardner, M., Seppi, K.: An Exploration of topologies and communication in large particle swarms. In: 2009 IEEE Congress on Evolutionary Computation (CEC 2009), pp. 712–719, May 2009
11. Mendes, R.: Population topologies and their influence in particle swarm performance. Ph.D. thesis, Departamento de Informática, Escola de Engenharia, Universidade do Minho, April 2004
12. Nebro, A.J., Durillo, J.J., García-Nieto, J., Coello Coello, C.A., Luna, F., Alba, E.: SMPSO: a new PSO-based metaheuristic for multi-objective optimization. In: 2009 IEEE Symposium on Computational Intelligence in Multi-Criteria Decision-Making (MCDM 2009), pp. 66–73, March 2009

13. Pan, A., Wang, L., Guo, W., Wu, Q.: A diversity enhanced multiobjective particle swarm optimization. Inf. Sci. **436**, 441–465 (2018)
14. Taormina, R., Chau, K.: Neural network river forecasting with multi-objective fully informed particle swarm optimization. J. Hydroinf. **17**(1), 99–113 (2014)
15. Valencia-Rodríguez, D.C., Coello Coello, C.A.: A study of swarm topologies and their influence on the performance of multi-objective particle swarm optimizers. In: Bäck, T., et al. (eds.) PPSN 2020. LNCS, vol. 12270, pp. 285–298. Springer, Cham (2020). https://doi.org/10.1007/978-3-030-58115-2_20
16. Valencia-Rodríguez, D.C.: Estudio de topologías cumulares y su impacto en el desempeño de un optimizador mediante cúmulos de partículas para problemas multiobjetivo. Master's thesis, Departamento de Computación, CINVESTAV-IPN, México, October 2019, http://delta.cs.cinvestav.mx/~ccoello/tesis/tesis-valencia.pdf.gz
17. Yamamoto, M., Uchitane, T., Hatanaka, T.: An experimental study for multi-objective optimization by particle swarm with graph based archive. In: Proceedings of SICE Annual Conference (SICE 2012), pp. 89–94, August 2012
18. Zhu, Q., et al.: An external archive-guided multiobjective particle swarm optimization algorithm. IEEE Trans. Cybern. **49**(9), 2794–2808 (2017)
19. Zitzler, E.: Evolutionary algorithms for multiobjective optimization: methods and applications. Ph.D. thesis, Swiss Federal Institute of Technology (ETH), Zurich, Suiza, November 1999

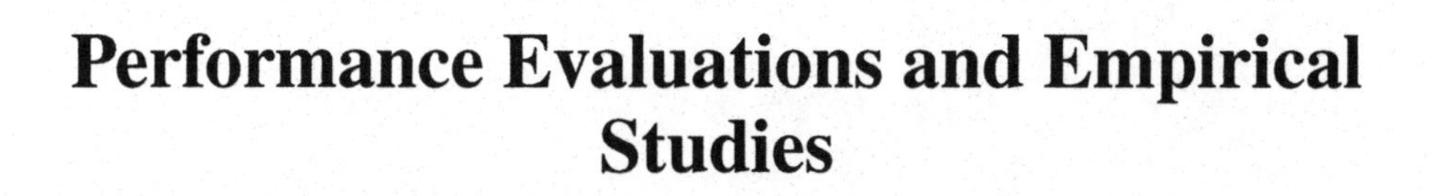

Performance Evaluations and Empirical Studies

An Overview of Pair-Potential Functions for Multi-objective Optimization

Jesús Guillermo Falcón-Cardona[1(✉)], Edgar Covantes Osuna[2], and Carlos A. Coello Coello[1]

[1] CINVESTAV-IPN, Computer Science Department, 07300 Mexico City, Mexico
jfalcon@computacion.cs.cinvestav.mx, ccoello@cs.cinvestav.mx
[2] Tecnológico de Monterrey, School of Engineering and Sciences, 64849 Monterrey, Nuevo León, Mexico
edgar.covantes@tec.mx

Abstract. Recently, an increasing number of state-of-the-art Multi-objective Evolutionary Algorithms (MOEAs) have incorporated the so-called pair-potential functions (commonly used to discretize a manifold) to improve the diversity within their population. A remarkable example is the Riesz s-energy function that has been recently used to improve the diversity of solutions either as part of a selection mechanism as well as to generate reference sets. In this paper, we perform an extensive empirical study with respect to the usage of the Riesz s-energy function and other 6 pair-potential functions adopted as a backbone of a selection mechanism used to update an external archive which is integrated into MOEA/D. Our results show that these pair-potential-based archives are able to store solutions with high diversity discarded by the MOEA/D's main population. Our experimental results indicate that the utilization of the pair-potential-based archives helps to circumvent the known MOEA/D's performance dependence on the Pareto front shapes without meddling with the original definition of the algorithm.

Keywords: Pair-potential functions · Diversity · Manifold discretization · Selection mechanism

1 Introduction

The area of Evolutionary Multi-objective Optimization (EMO) involves the design of population-based Multi-objective Evolutionary Algorithms (MOEAs). Given that evolutionary algorithms use a population which is a set of solutions to a given problem, MOEAs are suited in a natural way for solving Multi-objective Optimization Problems (MOPs) since they can generate multiple trade-offs among the (conflicting) objective functions (i. e., solutions in which an objective

The first author acknowledges support from CINVESTAV-IPN and CONACyT to pursue graduate studies. The third author acknowledges support from CONACyT grant no. 1920 and from SEP-Cinvestav grant (application no. 4).

The original version of this chapter was revised: two equations in section 2.1 have been corrected. The correction to this chapter is available at https://doi.org/10.1007/978-3-030-72062-9_62

H. Ishibuchi et al. (Eds.): EMO 2021, LNCS 12654, pp. 401–412, 2021.
https://doi.org/10.1007/978-3-030-72062-9_32

cannot be improved without worsening another) in a single run [1]. MOPs with four or more objective functions are commonly named Many-objective Optimization Problems (MaOPs) [1].

Well established MOEAs have two basic principles driven by selection. First of all, the goal is to push the current population close to the "true" *Pareto front.* The second goal is to "spread" the population along the front such that it is well covered. The first goal is usually achieved by using selection mechanisms based on the Pareto dominance relation[1], decomposition techniques, or indicator-based selection [1]. The second goal involves the use of diversity mechanisms to promote the spread of the different solutions along the Pareto front.

In general, providing an adequate balance between selection pressure and diversity is one of the main challenges when designing state-of-the-art MOEAs. These efforts have provided many methodologies, algorithms and techniques whose main goal is to improve the performance of MOEAs on MOPs [1]. On the one hand, the selection pressure of Pareto-dominant MOEAs dilutes when tackling MaOPs. This is due to the increasing number of non-dominated solutions found during the evolutionary process that yields to an inefficient random selection process [1]. On the other hand, due to the inherent methods embedded in MOEAs (mutation, crossover, selection, diversity mechanisms, etc.), the search may be biased into choosing a population with a specific distribution, which may favor a certain geometrical shape on the Pareto front.

In this paper, we focus our attention into the latter problem by introducing the usage of the so-called *pair-potential functions* as density estimators in an external population (archive). These functions are commonly used to distribute points on a manifold (or discretizing a manifold) [2]. Given a d-dimensional manifold $\mathcal{A}$ in the search space $\mathbb{R}^m$ $(d \leq m)$ with a given distribution X and described by some geometric property or by some parametrization, the goal is to generate a large number of N points in $\mathcal{A}$ such that they are well-separated and have (nearly) distribution X. In other words, the aim is to generate the smallest population possible that describes the distribution of the full set of elements in $\mathcal{A}$.

One example of such pair-potential functions used in EMO is the Riesz s-energy function, which has been recently used to improve the diversity of solutions [3–5]. We contribute to this line of work by performing an extensive empirical study on the usage of not just the Riesz s-energy function, but also 6 other pair-potential functions on an external population as an updating rule.

Our goal is twofold. Firstly, we want to show that even if a MOEA by itself is not able to maintain a well-distributed set of points during its execution (e. g., due to constraints of its selection mechanism or population size constraints, among others), the pair-potential functions will maintain solutions in the archive with high diversity. This implies that the total energy value of the system (the Pareto front approximation) is minimized. Consequently, the diversity will be improved for a longer period of time compared to the regular execution of the

[1] Given $\boldsymbol{x}, \boldsymbol{y} \in \mathbb{R}^m$, $\boldsymbol{x}$ is said to Pareto dominate $\boldsymbol{y}$ (denoted as $\boldsymbol{x} \prec \boldsymbol{y}$) if and only if $x_i \leq y_i$ for all $i = 1, \ldots, m$ and $\exists j \in \{1, \ldots, m\}$ such that $x_i < y_i$.

MOEA, and without meddling with its original definition. Second, we would like to explore the usage of several pair-potential functions in isolation as an alternative to improve the distribution of the solutions provided by MOEAs. We would like to observe the difference of the diversity achieved between a MOEA and the Pareto front approximation stored in the pair-potential-based archive. This study should provide insights into the performance dependence of a MOEA and the pair-potential functions with respect to the shape of the Pareto front on different MOPs.

The remainder of this paper is organized as follows. In Sect. 2, we introduce the problems of our interest and the definitions of the *pair-potential functions* analyzed in this paper. Sections 3 and 4 establish the algorithmic framework used throughout our experimental analysis and the results obtained from the such experiments, respectively. Finally, we finish with some discussion and concluding remarks in Sect. 5.

2 Preliminaries

In this paper, we consider, without loss of generality, the minimization of functions $\boldsymbol{F}(\boldsymbol{x}) = (f_1(\boldsymbol{x}), f_2(\boldsymbol{x}), \ldots, f_m(\boldsymbol{x}))$, where $\boldsymbol{x}$ is the vector of decision variables that belongs to $\Omega \subseteq \mathbb{R}^n$, which is the decision space. $\boldsymbol{F}(\boldsymbol{x})$ is the vector of $m \geq 2$ objective functions such that $f_i : \Omega \rightarrow \mathbb{R}$ for $i \in \{1, 2, \ldots, m\}$. As there is no single point that minimizes all functions simultaneously, the goal is to find a set of so-called Pareto-optimal solutions that represent the best possible trade-offs among the objective functions. A decision vector $x \in \Omega$ is Pareto optimal if there is no other $y \in \Omega$ such that $\boldsymbol{F}(\boldsymbol{y}) \prec \boldsymbol{F}(\boldsymbol{x})$. The set of all Pareto-optimal decision vectors P^* is called Pareto set and its image in the objective space, given by $PF^* = \boldsymbol{F}(P^*)$, is known as the Pareto front.

2.1 Pair-Potential Functions

Used to distribute points on a manifold (or discretizing a manifold), the *pair-potential functions* [2] are of the form $\mathcal{K} : \mathbb{R}^m \times \mathbb{R}^m \rightarrow \mathbb{R}$ that model the interaction between two given particles. The total energy U of a system with N particles is given as follows:

$$U(\mathcal{A}) = \sum_{i=1}^{N} \sum_{\substack{j=1 \\ j \neq i}}^{N} \mathcal{K}(\boldsymbol{a}_i, \boldsymbol{a}_j), \qquad (1)$$

where $\mathcal{A} = \{\boldsymbol{a}_1, \ldots, \boldsymbol{a}_N\}$ is an approximation set, and $\boldsymbol{a}_j \in \mathbb{R}^m, j = 1, \ldots, N$.

In the following, we describe the pair-potential functions which are of interest in this paper. For all cases, $\boldsymbol{u} = \boldsymbol{F}(\boldsymbol{x}), \boldsymbol{v} = \boldsymbol{F}(\boldsymbol{y}) \in \mathbb{R}^m$ and $\|\cdot\|$ represents the Euclidean distance.

Riesz *s*-energy [2]: Given a parameter $s > 0$, it is defined as follows:

$$\mathcal{K}^{\mathrm{RSE}}(\boldsymbol{u},\boldsymbol{v}) = \frac{1}{\|\boldsymbol{u}-\boldsymbol{v}\|^s}. \tag{2}$$

Gaussian α-energy [2]: Given a parameter $\alpha > 0$, this pair-potential is defined in the following:

$$\mathcal{K}^{\mathrm{GAE}}(\boldsymbol{u},\boldsymbol{v}) = e^{-\alpha\|\boldsymbol{u}-\boldsymbol{v}\|^2}. \tag{3}$$

Coulomb's law [6]: it is given by:

$$\mathcal{K}^{\mathrm{COU}}(\boldsymbol{u},\boldsymbol{v}) = \frac{q_1 q_2}{4\pi\epsilon_0}\cdot\frac{1}{\|\boldsymbol{u}-\boldsymbol{v}\|^2}, \tag{4}$$

where we set $q_1 = \|\boldsymbol{u}\|$ and $q_2 = \|\boldsymbol{v}\|$, and $\frac{1}{4\pi\epsilon_0}$ is the Coulomb's constant.

Pöschl-Teller Potential [7,8]: Given $V_1, V_2, \alpha > 0$, it is given by:

$$\mathcal{K}^{\mathrm{PT}}(\boldsymbol{u},\boldsymbol{v}) = \frac{V_1}{\sin^2(\alpha\,\|\boldsymbol{u}-\boldsymbol{v}\|)} + \frac{V_2}{\cos^2(\alpha\,\|\boldsymbol{u}-\boldsymbol{v}\|)}. \tag{5}$$

Modified Pöschl-Teller Potential [7,8]: Given $D, \alpha > 0$, this function is as follows:

$$\mathcal{K}^{\mathrm{MPT}}(\boldsymbol{u},\boldsymbol{v}) = -\frac{D}{\cosh^2(\alpha\,\|\boldsymbol{u}-\boldsymbol{v}\|)}. \tag{6}$$

General form of Pöschl-Teller Potentials [7,8]: Given $A, B > 0$, it is given by:

$$\mathcal{K}^{\mathrm{GPT}}(\boldsymbol{u},\boldsymbol{v}) = \frac{A}{1+\cos(\|\boldsymbol{u}-\boldsymbol{v}\|)} + \frac{B}{1-\cos(\|\boldsymbol{u}-\boldsymbol{v}\|)} \tag{7}$$

Kratzer Potential [8,9]: Given $V_1, V_2, \alpha > 0$, the Kratzer potential is given as follows:

$$\mathcal{K}^{\mathrm{KRA}}(\boldsymbol{u},\boldsymbol{v}) = V_1\left(\frac{\|\boldsymbol{u}-\boldsymbol{v}\| - 1/\alpha}{\|\boldsymbol{u}-\boldsymbol{v}\|}\right)^2 + V_2. \tag{8}$$

3 Pair-Potential-Based Archives

Following the approaches described in [3,4], we use an iterative selection mechanism based on the calculation of individual contributions to the total energy of a system of N particles (see Eq. (1)). Hence, given an approximation set $\mathcal{A} = \{\boldsymbol{a}_1, \ldots, \boldsymbol{a}_N, \boldsymbol{a}_{N+1}\}$, the individual contribution C of any $\boldsymbol{a} \in \mathcal{A}$ is calculated as follows:

$$C(\boldsymbol{a},\mathcal{A}) = \frac{1}{2}[U(\mathcal{A}) - U(\mathcal{A}\setminus\{\boldsymbol{a}\})], \tag{9}$$

where the term $1/2$ is used since for the adopted pair-potential functions in Eqs. (2) to (8), $\mathcal{K}(\boldsymbol{u},\boldsymbol{v}) = \mathcal{K}(\boldsymbol{v},\boldsymbol{u})$. Since U is to be minimized, the solution to be deleted from $\mathcal{A}$ such that $|\mathcal{A}| = N$ is the one with the maximum contribution value, i. e., $\boldsymbol{a}_{\text{worst}} = \arg\max_{\boldsymbol{a} \in \mathcal{A}} C(\boldsymbol{a}, \mathcal{A})$. Using this selection mechanism, we can iteratively reduce the cardinality of a given approximation set $\mathcal{A}$.

In this paper, we aim to show that the incorporation of an external population (or archive) based on a pair-potential function to a given MOEA, improves its diversity of solutions. Consequently, we selected the MOEA based on decomposition (MOEA/D) [10] as our baseline algorithm since it is known that its performance depends on the Pareto front shape [11]. The underlying idea is to employ the original MOEA/D algorithm (see Algorithm 1) and every time a new solution $\boldsymbol{y}$ is generated, it should be inserted in the archive $\mathcal{A}$ (which is initialized in Line 4) as shown in Lines 9 and 10, respectively. It is worth noting that the archive is only utilized to store solutions, which implies a unidirectional communication from MOEA/D to it. The use of the archive allows MOEA/D to keep solutions that would be possibly deleted due to its design principles. Once the algorithm reaches its stopping criterion, the main population and the archive are returned as two Pareto front approximations.

Algorithm 1. MOEA/D with external archive

Require: N: population size; T: neighborhood size; $\mathcal{K}$: pair-potential function; $A_{\max}$: maximum archive size
Ensure: Main population and archive as Pareto front approximations
1: Initialize N weight vectors $\lambda^1, \ldots, \lambda^N$
2: Determine the T nearest neighbors of each λ^j
3: Initialize the main Population $P = \{\boldsymbol{x}^1, \ldots, \boldsymbol{x}^N\}$
4: Initialize archive $\mathcal{A}$ equals to P.
5: Initialize the reference point z^*
6: **while** Stopping criterion is not satisfied **do**
7:    **for** $j = 1$ to N **do**
8:       Select mating parents from the neighborhood of $\boldsymbol{x}^j$
9:       Generate a new solution $\boldsymbol{y}$ by using variation operators.
10:       Insert($\mathcal{A}, \boldsymbol{y}, \mathcal{K}, A_{\max}$).
11:       Update the reference point z^*
12:       Evaluate new solutions via scalarizing function
13:       Update main population P
14: **return** $\{P, \mathcal{A}\}$

Algorithm 2 describes the process used to insert a new solution $\boldsymbol{y}$ in $\mathcal{A}$. First, $\boldsymbol{y}$ is compared with all the elements in the archive, using the Pareto dominance relation. If $\boldsymbol{y}$ Pareto dominates any $\boldsymbol{a} \in \mathcal{A}$, the latter is deleted from $\mathcal{A}$. However, if any element in the archive weakly Pareto dominates $\boldsymbol{y}$, then $\boldsymbol{y}$ is not inserted and $\mathcal{A}$ is returned. In case that $\boldsymbol{y}$ survives the Pareto-based criterion, it is inserted in the archive. If the maximum archive size $A_{\max}$ is exceeded, we execute the steady-state selection (prior normalization of the archive) where the solution with the worst contribution to the total energy U, using $\mathcal{K}$ as pair-potential function, is iteratively deleted until the desired $A_{\max}$ size is reached.

Regarding the computational complexity of Algorithm 2, it is $\Theta(m|A|^2)$ due to the usage of the fast computation of individual contributions proposed in [4]. Smaller terms consist of the for-loop of Lines 1 to 5 and the normalization procedure in Line 7 with complexity $\Theta(m|A|)$.

Algorithm 2. Insert

Require: $\mathcal{A}$: archive; $\boldsymbol{y}$: solution to be inserted; $\mathcal{K}$: pair-potential function; $A_{\max}$: maximum archive size
Ensure: Updated archive $\mathcal{A}$
1: **for all** $\boldsymbol{a} \in \mathcal{A}$ **do**
2: **if** $\boldsymbol{y} \prec \boldsymbol{a}$ **then**
3: $\mathcal{A} = \mathcal{A} \setminus \{\boldsymbol{a}\}$
4: **else if** $\boldsymbol{a} \preceq \boldsymbol{y}$ **then**
5: **return** $\mathcal{A}$
6: $\mathcal{A} = \mathcal{A} \cup \{\boldsymbol{y}\}$
7: Normalize $\mathcal{A}$
8: **while** $|\mathcal{A}| > A_{\max}$ **do**
9: $\boldsymbol{a}_{\text{worst}} = \arg\max_{\boldsymbol{a} \in \mathcal{A}} C_{\mathcal{K}}(\boldsymbol{a}, \mathcal{A})$
10: $\mathcal{A} = \mathcal{A} \setminus \{\boldsymbol{a}_{\text{worst}}\}$
11: **return** $\mathcal{A}$

4 Experimental Results

This section is devoted to comparing the convergence and diversity properties of the main population returned by MOEA/D and the content of the archive based on the seven selected pair-potential functions[2] which gives rise to MOEA/D$_{\text{RSE}}$, MOEA/D$_{\text{GAE}}$, MOEA/D$_{\text{COU}}$, MOEA/D$_{\text{PT}}$, MOEA/D$_{\text{MPT}}$, MOEA/D$_{\text{GPT}}$, and MOEA/D$_{\text{KRA}}$. MOEA/D with each external archive is tested on several MOPs from the DTLZ, WFG, DTLZ^{-1}, and WFG^{-1} test suites with two and three objective functions. Regarding the DTLZ and DTLZ^{-1} test problems, the number of variables was set to $n = m + K - 1$, where m is the number of objective functions, and $K = 5$ for DTLZ1, $K = 10$ for DTLZ2 and DTLZ5, and $K = 20$ for DTLZ7. Their inverted counterparts have the same value of K. Concerning the WFG and WFG^{-1} problems, n was set to 24 and 26, with the position-related parameters equal to 2 and 4 for two and three objective functions, respectively. For a fair comparison, the population size and the maximum archive size are set to $N = C_{m-1}^{m+H-1} = 120$, where $H = 119$ and 14 for MOPs with two and three objective functions, respectively. N corresponds to the cardinality of the set of weight vectors (constructed by the Das and Dennis method employed by the MOEA/D [10]. For all cases, the stopping criterion was set to 49,920 function evaluations which corresponds to 416 generations. We set the neighborhood size T of MOEA/D equal to 20 and we employed the achievement scalarizing

[2] The source code is available at http://computacion.cs.cinvestav.mx/~jfalcon/PairPotentials/.

Table 1. Average ranking values obtained by all the Pair-Potential-based selection mechanisms for the IGD and PD indicators.

QI	MOP	Dim.	$\mathcal{K}^{RSE}$	$\mathcal{K}^{GAE}$	$\mathcal{K}^{COU}$	$\mathcal{K}^{PT}$	$\mathcal{K}^{MPT}$	$\mathcal{K}^{GPT}$	$\mathcal{K}^{KRA}$
IGD	DTLZ1	2	4.286^4	5.429^6	5.000^5	1.429^1	7.000^7	2.143^2	2.714^3
		3	3.571^3	2.143^2	2.000^1	4.571^4	5.571^7	5.000^5	5.143^6
	DTLZ2	2	3.857^4	6.000^6	1.286^1	2.286^2	7.000^7	3.286^3	4.286^5
		3	3.286^2	1.571^1	4.286^4	5.286^7	5.000^6	4.429^5	4.143^3
	DTLZ5	2	4.000^4	5.857^6	1.286^1	2.286^2	7.000^7	3.286^3	4.286^5
		3	1.571^1	5.286^5	5.286^6	2.571^2	6.857^7	3.571^4	2.857^3
	DTLZ7	2	4.143^4	6.000^6	4.571^5	1.429^1	7.000^7	2.429^2	2.429^3
		3	2.714^2	4.000^5	5.429^6	3.714^4	6.286^7	3.286^3	2.571^1
	WFG1	2	5.286^5	5.286^6	2.429^2	1.857^1	6.429^7	2.857^3	3.857^4
		3	3.571^3	3.143^2	1.000^1	4.571^4	6.143^7	5.000^6	4.571^5
	WFG2	2	3.429^3	5.571^6	4.286^5	1.571^1	7.000^7	2.571^2	3.571^4
		3	3.857^4	3.429^2	1.286^1	3.714^3	6.429^7	4.571^5	4.714^6
	WFG3	2	4.286^4	5.571^6	5.000^5	2.143^2	7.000^7	2.143^3	1.857^1
		3	2.286^2	5.714^6	5.143^5	2.143^1	7.000^7	2.286^3	3.429^4
PD	DTLZ1	2	3.857^4	5.143^6	3.571^2	4.286^5	5.714^7	3.571^3	1.857^1
		3	3.429^2	4.429^6	2.286^1	3.857^4	6.286^7	4.286^5	3.429^3
	DTLZ2	2	4.286^5	2.857^1	3.714^3	3.714^4	3.000^2	4.714^6	5.714^7
		3	4.429^5	1.000^1	5.429^7	4.000^3	5.143^6	3.857^2	4.143^4
	DTLZ5	2	4.286^5	2.571^2	4.000^3	4.000^4	2.143^1	5.000^6	6.000^7
		3	3.571^3	3.857^4	2.000^1	4.571^5	3.000^2	5.571^7	5.429^6
	DTLZ7	2	3.571^3	5.714^6	3.571^4	2.571^1	6.000^7	3.571^5	3.000^2
		3	2.143^1	4.429^5	4.857^6	3.143^3	6.143^7	3.000^2	4.286^4
	WFG1	2	2.857^2	3.857^5	3.143^3	2.714^1	7.000^7	3.714^4	4.714^6
		3	2.429^2	4.571^4	2.143^1	3.429^3	6.143^7	4.714^6	4.571^5
	WFG2	2	3.286^3	6.714^7	4.000^5	1.857^1	5.429^6	2.857^2	3.857^4
		3	3.429^2	3.429^3	2.571^1	4.286^5	6.429^7	4.429^6	3.429^4
	WFG3	2	3.286^3	5.857^6	3.286^4	3.857^5	6.143^7	2.571^1	3.000^2
		3	3.714^3	3.714^4	3.286^2	4.857^6	4.857^7	4.714^5	2.857^1

function for all the test instances. For each test instance, we performed 30 independent executions and the MOEA/D variants used the same random seeds. We selected the Inverted Generational Distance (IGD) [12] and the Pure Diversity (PD) [13] indicators to assess convergence-diversity and diversity, respectively. In order to have statistical confidence of our results, we adopted the one-tailed Wilcoxon rank-sum test, with a significance level of $\alpha = 0.05$ with null hypothesis as follows: "*is MOEA$_A$ statistically different than MOEA$_B$?*". Where MOEA$_A$ is the one having the best indicator value and MOEA$_B$ is each one of the remaining algorithms. This selection was used for all the tables in this paper with the exception of Table 1 because the latter is a summary of results. In Tables 3 and 4, a symbol # is placed when the best algorithm performs significantly better than the others.

4.1 Pair-Potential Functions Parameters Setting

The pair-pontential functions introduced in this paper require the definition of some parameter values. Hence, prior to use the functions in the external archive of MOEA/D, it is mandatory to determine such values. To this aim, we have performed a trial-and-error experimentation to determine good parameter values. Based on the true Pareto fronts of problems DTLZ1, DTLZ2, DTLZ5, DTLZ7, WFG1, WFG2, and WFG3 with two and three objectives, we performed a cardinality reduction, using the seven pair-potential functions, to obtain subsets of size 20, 50, 100, 200, 300, 400, and 500, following the approach described in [4].

The generated subsets were then assessed using IGD and PD. Table 1 shows the statistical average ranking values obtained by the seven selection mechanisms using the parameter values that we found, where a value closer to one is better. Based on the several comparisons performed, we found out that the parameter values in Table 2 allow to produce approximation sets having good IGD and PD values. In contrast to the quality of Riesz s-energy-based reference sets in [4], the IGD and PD results of Table 1 show that the utilization of other pair-potential functions led to better diversity results.

Table 2. Parameter values found by trial-and-error.

Pair-potential function	Parameter values
$\mathcal{K}^{\text{RSE}}$	$s = m - 1$
$\mathcal{K}^{\text{GAE}}$	$\alpha = 512$
$\mathcal{K}^{\text{PT}}$	$V_1 = 5.0,\ V_2 = 3.0,\ \alpha = 0.02$
$\mathcal{K}^{\text{MPT}}$	$D = 1.0,\ \alpha = 10.0$
$\mathcal{K}^{\text{GPT}}$	$A = 1.0,\ B = 1.0$
$\mathcal{K}^{\text{KRA}}$	$V_1 = 5.0,\ V_2 = 3.0,\ \alpha = 10.0$

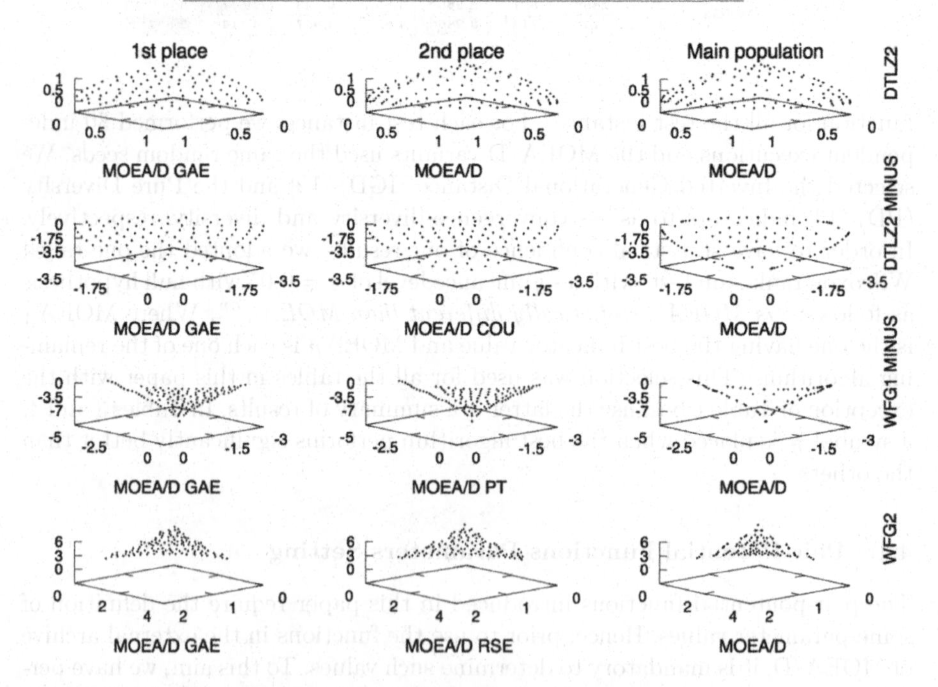

Fig. 1. Comparison of three-objective approximation sets between the 1st and 2nd places in the PD comparison with the resulting main population. The approximation sets correspond to the PD median.

4.2 Improving Diversity Results

According to Ishibuchi *et al.* [11], the performance of MOEA/D strongly depends on the Pareto front shape. This is because the weight vectors that MOEA/D uses, cannot completely intersect irregular Pareto front geometries. To overcome this issue, we propose to aggregate a pair-potential-based archive to store solutions found throughout the evolutionary process with a high diversity degree.

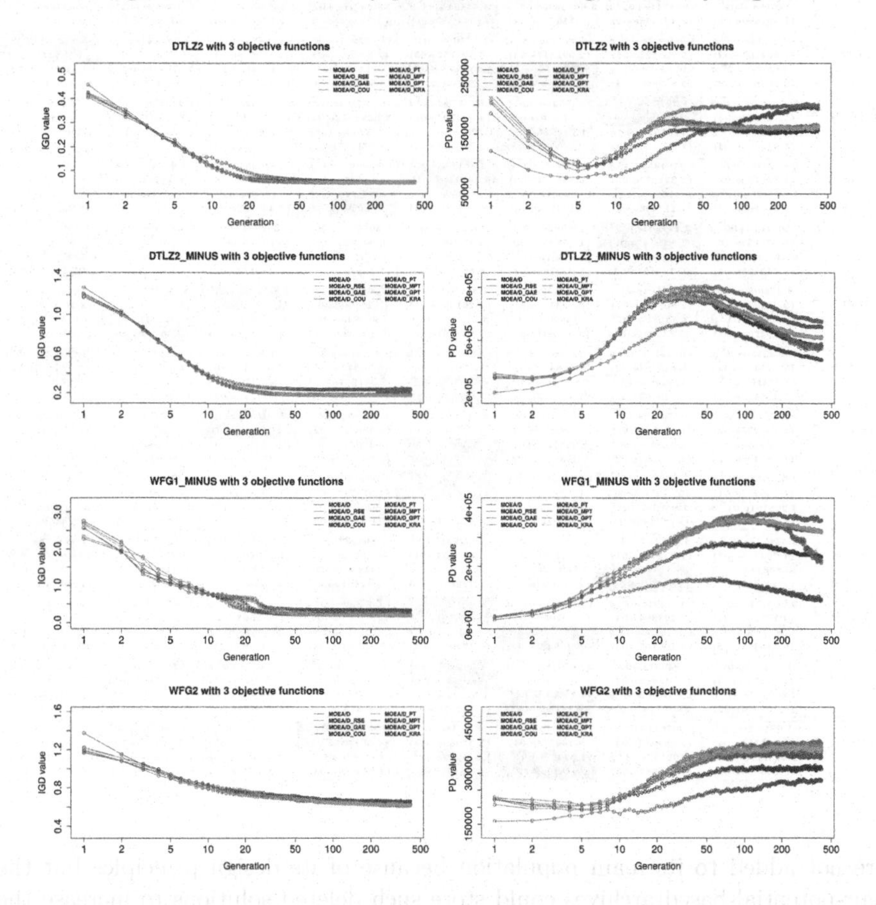

Fig. 2. IGD and PD values of the main population and the Pair Potential-based archives throughout the evolutionary process.

Tables 3 and 4 show IGD and PD comparisons, respectively, between the MOEA/D's main population and the final content of the archives. Regarding IGD, it is clear that the main population had the worst IGD values for almost all the test instances. In contrast, MOEA/D_{KRA} and MOEA/D_{COU} obtained the best IGD values in 7 problems each. These results are followed by MOEA/D_{GAE}, MOEA/D_{RSE}, MOEA/D_{PT}, MOEA/D_{MPT}, and MOEA/D_{GPT}. Hence, these results show that it is possible to tackle the MOEA/D's performance dependence on the Pareto front shape by using a pair-potential-based archive. During the evolutionary process, some solutions with high diversity created by MOEA/D

Table 3. Mean and, in parentheses, standard deviation of the IGD indicator. The two best values are shown in gray scale, where the darker tone corresponds to the best value.

MOP	Dim.	MOEA/D	MOEA/D$_{RSE}$	MOEA/D$_{GAE}$	MOEA/D$_{COU}$	MOEA/D$_{PT}$	MOEA/D$_{MPT}$	MOEA/D$_{GPT}$	MOEA/D$_{KRA}$
DTLZ1	2	4.186999e-02^{7} # (2.019772e-02)	1.838323e-03^{4} # (1.803998e-04)	3.427025e-03^{5} # (2.446102e-04)	1.806034e-03^{2} # (1.068795e-04)	1.811718e-03^{3} # (1.022404e-04)	7.119547e-03^{6} # (5.464629e-04)	4.411732e-02^{8} # (2.421894e-02)	1.775810e-03^{1} (7.145302e-05)
	3	7.997173e-02^{7} # (4.846696e-02)	2.406357e-02^{3} (8.422861e-03)	2.374639e-02^{2} (1.129116e-02)	2.444074e-02^{4} (8.995911e-03)	2.458800e-02^{5} # (8.711923e-03)	3.200721e-02^{6} # (2.119001e-02)	2.490695e-01^{8} # (7.984856e-02)	2.150021e-02^{1} (4.484577e-03)
DTLZ1^{-1}	2	4.247163e+02^{1} (1.882771e-04)	4.247174e+02^{2} # (5.132721e-04)	4.247217e+02^{6} # (8.742813e-04)	4.247179e+02^{5} # (9.144372e-04)	4.247176e+02^{3} # (7.484851e-04)	4.247396e+02^{7} # (3.931103e-03)	4.251433e+02^{8} # (9.760655e-01)	4.247176e+02^{4} # (7.470171e-04)
	3	3.686450e+02^{8} # (1.962899e-01)	3.664744e+02^{3} # (2.303456e-02)	3.665353e+02^{5} # (4.116766e-02)	3.664645e+02^{1} (1.869585e-02)	3.664716e+02^{2} (2.278402e-02)	3.665501e+02^{6} # (2.876059e-02)	3.667952e+02^{7} # (5.769143e-01)	3.664745e+02^{4} # (2.200036e-02)
DTLZ2	2	6.455514e-03^{6} # (1.804195e-03)	3.960807e-03^{5} # (5.226221e-05)	6.730226e-03^{7} # (2.991417e-04)	3.874658e-03^{3} (6.399753e-05)	3.864961e-03^{2} (7.340550e-05)	1.256876e-02^{8} # (5.880958e-04)	3.883124e-03^{4} (6.726117e-05)	3.849461e-03^{1} (7.525377e-05)
	3	5.332670e-02^{8} # (9.806807e-03)	5.071447e-02^{3} # (5.839869e-04)	4.936519e-02^{1} (5.739893e-04)	5.030090e-02^{2} # (9.189512e-04)	5.071447e-02^{4} # (5.839869e-04)	5.245941e-02^{7} # (5.824351e-04)	5.076878e-02^{5} # (4.894479e-04)	5.093289e-02^{6} # (6.223211e-04)
DTLZ2^{-1}	2	5.402829e-02^{7} # (5.182328e-04)	1.367666e-02^{1} (3.730898e-04)	2.348546e-02^{5} # (1.375046e-03)	1.639473e-02^{4} # (4.920727e-04)	1.374464e-02^{2} (2.698812e-04)	4.518382e-02^{6} # (2.905080e-03)	1.453378e-01^{8} # (7.632037e-02)	1.374481e-02^{3} (2.697314e-04)
	3	2.350967e-01^{8} # (5.952031e-04)	1.753651e-01^{3} # (1.822728e-03)	1.739527e-01^{1} (1.839368e-03)	1.778182e-01^{5} # (2.010126e-03)	1.753703e-01^{4} # (1.831097e-03)	1.790863e-01^{6} # (1.910148e-03)	2.334478e-01^{7} # (2.950725e-02)	1.748156e-01^{2} # (1.762841e-03)
DTLZ5	2	6.452047e-03^{6} # (1.831733e-03)	3.916495e-03^{5} # (6.321257e-05)	6.685997e-03^{7} # (2.849174e-04)	3.865443e-03^{1} (7.191131e-05)	3.900809e-03^{3} # (6.945329e-05)	1.289695e-02^{8} # (5.815969e-04)	3.888459e-03^{2} (6.241348e-05)	3.904549e-03^{4} # (8.187004e-05)
	3	1.822996e-02^{8} # (2.699169e-04)	4.416289e-03^{3} (1.117496e-04)	6.693661e-03^{5} # (3.234754e-04)	4.681299e-03^{4} # (1.379174e-04)	4.383775e-03^{1} (1.055094e-04)	1.119680e-02^{6} # (7.429022e-04)	1.351939e-02^{7} # (1.690910e-03)	4.399935e-03^{2} (9.744893e-05)
DTLZ5^{-1}	2	7.500461e+00^{7} # (1.071153e-05)	7.499904e+00^{1} (5.876507e-06)	7.499982e+00^{5} # (1.388855e-05)	7.499921e+00^{4} # (8.378681e-06)	7.499905e+00^{3} (6.337627e-06)	7.500288e+00^{6} # (5.105165e-05)	7.505455e+00^{8} # (7.036907e-03)	7.499905e+00^{2} (6.348228e-06)
	3	8.114210e-01^{7} # (8.743495e-04)	7.988213e-01^{2} (7.192821e-04)	7.993381e-01^{5} # (1.022766e-03)	7.990976e-01^{4} # (6.887294e-04)	7.988213e-01^{3} (7.192821e-04)	8.013794e-01^{6} # (1.040173e-03)	8.454652e-01^{8} # (3.533821e-02)	7.987934e-01^{1} (7.816305e-04)
DTLZ7	2	8.474668e-02^{6} (2.242291e-01)	1.284701e-01^{7} (2.808858e-01)	1.573122e-01^{8} (2.993187e-01)	7.898999e-02^{4} (2.261259e-01)	7.892329e-02^{2} (2.261302e-01)	7.002983e-02^{1} (1.836990e-01)	8.231452e-02^{5} (2.250786e-01)	7.894461e-02^{3} (2.261645e-01)
	3	4.508531e-01^{8} # (3.193705e-01)	3.481312e-01^{2} (3.550939e-01)	4.169200e-01^{6} (3.722431e-01)	3.511664e-01^{3} (3.536501e-01)	3.683843e-01^{4} (3.830220e-01)	4.194404e-01^{7} (3.674776e-01)	3.052841e-01^{1} (3.367699e-01)	4.115675e-01^{5} (3.405298e-01)
DTLZ7^{-1}	2	5.654288e-02^{8} # (2.472440e-02)	1.530213e-02^{6} # (6.816794e-02)	2.847986e-02^{7} # (1.268466e-01)	2.843677e-03^{3} # (1.179169e-04)	2.813414e-03^{2} # (8.125426e-05)	8.698850e-03^{5} # (9.486018e-04)	8.241541e-03^{4} # (1.007504e-02)	2.774438e-03^{1} (8.787205e-05)
	3	2.077316e-01^{8} # (2.168927e-01)	1.232987e-01^{6} (2.746663e-01)	9.900515e-02^{4} (1.965103e-01)	8.226984e-02^{1} (1.790202e-01)	8.341971e-02^{2} (1.787632e-01)	1.190623e-01^{5} # (2.051520e-01)	1.601192e-01^{7} # (1.954057e-01)	9.827485e-02^{3} (1.968313e-01)
WFG1	2	9.401449e-01^{8} (2.191233e-02)	9.376123e-01^{5} (1.565996e-02)	9.386727e-01^{6} # (1.580819e-02)	9.353662e-01^{3} (1.943095e-02)	9.369655e-01^{4} (1.486310e-02)	9.307542e-01^{1} (1.794458e-02)	9.397481e-01^{7} (1.536942e-02)	9.345871e-01^{2} (1.592324e-02)
	3	1.027225e+00^{3} (1.682408e-02)	1.032064e+00^{4} (2.049699e-02)	1.025523e+00^{2} (1.986368e-02)	1.023816e+00^{1} (2.376911e-02)	1.032064e+00^{5} (2.049699e-02)	1.032592e+00^{7} (2.375301e-02)	1.032260e+00^{6} (2.619221e-02)	1.032960e+00^{8} # (1.777136e-02)
WFG1^{-1}	2	2.038353e-01^{8} # (1.666237e-01)	9.865750e-02^{2} (5.922766e-03)	1.026224e-01^{3} # (1.396817e-02)	9.741371e-02^{1} (6.332477e-03)	1.081140e-01^{5} (4.019717e-02)	1.092692e-01^{6} # (7.891856e-03)	1.307802e-01^{7} # (4.813635e-02)	1.081132e-01^{4} (4.019772e-02)
	3	3.268161e-01^{8} # (6.953720e-03)	1.942082e-01^{2} (9.043183e-03)	1.946992e-01^{5} (8.637045e-03)	1.942754e-01^{4} (8.420551e-03)	1.942082e-01^{3} (9.043183e-03)	2.109872e-01^{6} # (7.488072e-03)	2.373561e-01^{7} # (3.868627e-02)	1.927049e-01^{1} (7.845998e-03)
WFG2	2	3.551643e-01^{8} # (1.704628e-01)	3.135232e-01^{4} (1.222467e-01)	3.217534e-01^{5} (1.342997e-01)	2.932231e-01^{3} (1.400955e-01)	2.884977e-01^{2} (1.166928e-01)	3.274172e-01^{6} # (1.215664e-01)	3.286668e-01^{7} (1.753714e-01)	2.884977e-01^{1} (1.166928e-01)
	3	6.086625e-01^{8} # (2.409889e-01)	5.419541e-01^{1} (2.424612e-01)	5.850965e-01^{6} (2.961699e-01)	5.782103e-01^{5} (1.792989e-01)	5.419541e-01^{2} (2.424612e-01)	5.950589e-01^{7} # (2.050419e-01)	5.644249e-01^{4} (1.513871e-01)	5.513425e-01^{3} (2.186715e-01)
WFG2^{-1}	2	3.952074e-02^{8} # (3.178202e-02)	1.113291e-02^{5} # (2.228394e-04)	1.582865e-02^{6} # (5.977065e-04)	1.030592e-02^{1} (1.987543e-04)	1.060899e-02^{2} # (2.035732e-04)	2.099910e-02^{7} # (1.352693e-03)	1.067472e-02^{4} # (1.999304e-04)	1.062263e-02^{3} # (2.010447e-04)
	3	3.086424e-01^{8} # (4.515396e-03)	2.272615e-01^{3} # (3.108107e-03)	2.218605e-01^{1} (3.165435e-03)	2.308400e-01^{6} # (3.572903e-03)	2.272615e-01^{4} # (3.108107e-03)	2.345887e-01^{7} # (2.906680e-03)	2.285305e-01^{5} # (8.686979e-03)	2.265846e-01^{2} # (2.896747e-03)
WFG3	2	1.882592e-02^{5} (1.222206e-02)	1.949911e-02^{6} (1.017810e-02)	2.067928e-02^{7} # (3.835263e-03)	1.565673e-02^{1} (5.160110e-03)	1.877287e-02^{4} (1.230404e-02)	3.407575e-02^{8} # (4.363392e-03)	1.594662e-02^{2} (8.794198e-03)	1.877046e-02^{3} (1.230560e-02)
	3	2.807902e-01^{8} # (7.107544e-02)	1.350266e-01^{2} # (3.408787e-02)	1.198084e-01^{1} (2.105833e-02)	1.525024e-01^{5} (7.142128e-02)	1.350266e-01^{3} # (3.408787e-02)	1.652181e-01^{7} # (4.028171e-02)	1.432406e-01^{4} (6.519973e-02)	1.535460e-01^{6} # (8.419368e-02)
WFG3^{-1}	2	1.091209e-02^{4} (1.181551e-04)	1.097571e-02^{6} # (2.084019e-04)	1.851377e-02^{7} # (8.190213e-04)	1.096048e-02^{5} # (2.379506e-04)	1.085211e-02^{1} (1.491887e-04)	4.007145e-02^{8} # (2.636346e-03)	1.085248e-02^{2} (1.483458e-04)	1.085248e-02^{3} (1.483458e-04)
	3	2.475717e-01^{8} # (7.343098e-04)	1.592187e-01^{4} # (4.859861e-03)	1.544698e-01^{1} (2.368049e-03)	1.574239e-01^{2} # (3.281979e-03)	1.592187e-01^{5} # (4.859861e-03)	1.771357e-01^{7} # (5.028710e-03)	1.634088e-01^{6} # (1.606381e-02)	1.586122e-01^{3} # (3.844772e-03)

are not added to its main population because of its design principles but the pair-potential-based archives could store such deleted solutions to increase the diversity quality. In this light, the PD values in Table 4 indicate that the approximation sets stored in the archive are better than those of the main population in terms of diversity. Concerning PD, MOEA/D$_{GAE}$ presents a more dominant performance since it generated the best diversified approximation sets in 15 out of 28 test instances. The second best results are generated by MOEA/D$_{RSE}$, creating 9 of the best approximation sets. In this case, the main population of MOEA/D was ranked as the worst one. In Fig. 1, we compared the main population and the approximation sets having the best and second best PD value. From the figure, it is clear the lack of diversity of MOEA/D's main population when dealing with irregular Pareto front geometries (three-objective DTLZ2^{-1},

Table 4. Mean and, in parentheses, standard deviation of the PD indicator. The two best values are shown in gray scale, where the darker tone corresponds to the best value.

MOP	Dim.	MOEA/D	MOEA/D_{RSE}	MOEA/D_{GAE}	MOEA/D_{COU}	MOEA/D_{PT}	MOEA/D_{MPT}	MOEA/D_{GPT}	MOEA/D_{KRA}
DTLZ1	2	5.952998e+02^{8}# (1.567214e+02)	9.819786e+02^{1} (1.646415e+02)	9.752486e+02^{2} (1.884763e+02)	8.179121e+02^{7}# (5.079579e+01)	9.415797e+02^{3} (1.273846e+02)	8.941377e+02^{4}# (1.503196e+02)	8.893690e+02^{6}# (5.122440e+02)	8.920221e+02^{5}# (1.001645e+02)
	3	3.894113e+04^{7}# (1.687247e+04)	7.949559e+04^{2}# (1.074164e+04)	8.090461e+04^{1} (1.020890e+04)	7.308939e+04^{5}# (1.163954e+04)	7.941942e+04^{3}# (7.317877e+03)	5.570146e+04^{6}# (1.358686e+04)	2.772268e+04^{8}# (6.117830e+04)	7.898577e+04^{4}# (6.570989e+03)
DTLZ1^{-1}	2	9.829245e+05^{7} (2.304527e+05)	1.095819e+06^{1} (1.261983e+05)	1.062928e+06^{2} (2.245031e+05)	8.127116e+05^{8}# (5.435064e+04)	9.842686e+05^{6}# (1.158998e+05)	1.019312e+06^{3}# (1.732922e+05)	9.994363e+05^{4}# (1.722732e+05)	9.844474e+05^{5}# (1.158242e+05)
	3	1.042068e+07^{8}# (1.625627e+06)	6.120604e+07^{3}# (3.951464e+06)	6.838680e+07^{1} (4.135194e+06)	6.221416e+07^{2}# (5.160296e+06)	6.081194e+07^{4}# (3.932779e+06)	4.635094e+07^{7}# (4.612712e+06)	5.666907e+07^{6}# (7.222818e+06)	5.883329e+07^{5}# (3.875243e+06)
DTLZ2	2	1.291604e+03^{8}# (9.071074e+01)	1.733441e+03^{1} (2.035760e+02)	1.682569e+03^{3} (3.354642e+02)	1.666750e+03^{4} (1.813600e+02)	1.654943e+03^{5} (2.123469e+02)	1.648889e+03^{6} (3.204995e+02)	1.616164e+03^{7}# (1.935316e+02)	1.683851e+03^{2} (1.588133e+02)
	3	1.949957e+05^{2} (1.643671e+04)	1.650157e+05^{3}# (9.689496e+03)	1.997165e+05^{1} (1.076623e+04)	1.575865e+05^{7}# (7.379285e+03)	1.650157e+05^{4}# (9.689496e+03)	1.565059e+05^{8}# (9.949892e+03)	1.608530e+05^{6}# (8.977004e+03)	1.628639e+05^{5}# (6.897749e+03)
DTLZ2^{-1}	2	3.460118e+03^{8}# (1.447837e+02)	5.677262e+03^{4}# (5.377094e+02)	6.295488e+03^{2} (1.501710e+03)	6.623736e+03^{1} (1.405020e+03)	5.668957e+03^{6}# (5.186942e+02)	5.682854e+03^{3}# (1.350085e+03)	4.242075e+03^{7}# (4.710251e+02)	5.669005e+03^{5}# (5.187033e+02)
	3	3.860403e+05^{8}# (2.221565e+04)	5.177377e+05^{3}# (3.165070e+04)	6.040356e+05^{1} (3.459380e+04)	5.720838e+05^{2}# (3.754927e+04)	5.177362e+05^{4}# (3.164902e+04)	4.650628e+05^{7}# (2.956968e+04)	4.706842e+05^{6}# (4.209004e+04)	5.139558e+05^{5}# (1.992277e+04)
DTLZ5	2	1.291604e+03^{8}# (9.071074e+01)	1.733441e+03^{1} (2.035760e+02)	1.682569e+03^{3} (3.354642e+02)	1.666750e+03^{4} (1.813600e+02)	1.654943e+03^{5} (2.123469e+02)	1.648889e+03^{6} (3.204995e+02)	1.616164e+03^{7}# (1.935316e+02)	1.683851e+03^{2} (1.588133e+02)
	3	5.881850e+04^{8}# (5.702902e+03)	7.711397e+04^{2}# (1.114531e+04)	7.251771e+04^{6}# (9.428644e+03)	8.303132e+04^{1} (1.131798e+04)	7.661495e+04^{3}# (1.050952e+04)	7.322066e+04^{5}# (1.224087e+04)	6.869597e+04^{7}# (9.700606e+03)	7.501115e+04^{4}# (1.152780e+04)
DTLZ5^{-1}	2	3.460118e+03^{8}# (1.447837e+02)	5.677262e+03^{4}# (5.377094e+02)	6.295488e+03^{2} (1.501710e+03)	6.623736e+03^{1} (1.405020e+03)	5.668957e+03^{6}# (5.186942e+02)	5.682854e+03^{3}# (1.350085e+03)	4.242075e+03^{7}# (4.710251e+02)	5.669005e+03^{5}# (5.187033e+02)
	3	5.074794e+05^{8}# (1.814572e+04)	6.458456e+05^{4}# (3.138224e+04)	7.234121e+05^{1} (2.971116e+04)	7.106961e+05^{2}# (3.629466e+04)	6.457661e+05^{5}# (3.150492e+04)	6.075672e+05^{6}# (2.984732e+04)	5.094598e+05^{7}# (4.981328e+04)	6.536891e+05^{3}# (3.806468e+04)
DTLZ7	2	1.658939e+03^{3} (4.982475e+02)	1.703469e+03^{1} (6.169178e+02)	1.571216e+03^{6} (5.838278e+02)	1.572009e+03^{5} (4.203846e+02)	1.595144e+03^{4} (4.414085e+02)	1.536629e+03^{7}# (3.978827e+02)	1.512220e+03^{8}# (4.334773e+02)	1.659590e+03^{2} (4.641080e+02)
	3	6.086815e+04^{8}# (1.228435e+04)	1.105173e+05^{2} (3.395800e+04)	1.184883e+05^{1} (3.705969e+04)	1.096011e+05^{3} (3.272136e+04)	1.095923e+05^{4} (3.571369e+04)	7.822746e+04^{6}# (1.720284e+04)	7.605433e+04^{7}# (2.245271e+04)	9.947854e+04^{5}# (3.143508e+04)
DTLZ7^{-1}	2	6.601662e+02^{8}# (4.527766e+01)	9.241120e+02^{1} (1.776790e+02)	8.281332e+02^{6}# (2.052206e+02)	8.697297e+02^{3}# (1.165725e+02)	8.704665e+02^{2}# (1.097159e+02)	7.470244e+02^{7}# (1.274952e+02)	8.500112e+02^{5}# (1.333475e+02)	8.554335e+02^{4}# (1.130821e+02)
	3	1.477186e+04^{8}# (5.663068e+03)	5.307180e+04^{5}# (1.817877e+04)	5.944595e+04^{1} (2.091486e+04)	5.713357e+04^{2}# (1.617265e+04)	5.475668e+04^{3}# (1.521980e+04)	3.871680e+04^{6}# (1.558153e+04)	2.227547e+04^{7}# (1.409303e+04)	5.462697e+04^{4}# (1.877514e+04)
WFG1	2	3.156553e+03^{8}# (6.227043e+02)	4.591890e+03^{1} (6.341935e+02)	4.355820e+03^{3} (6.088311e+02)	4.196410e+03^{7}# (5.728016e+02)	4.286319e+03^{5}# (4.124380e+02)	4.349743e+03^{4} (7.376629e+02)	4.204975e+03^{6}# (4.414066e+02)	4.356841e+03^{2} (4.564431e+02)
	3	4.599881e+05^{7}# (3.509304e+04)	5.184674e+05^{3}# (1.929072e+04)	5.297481e+05^{1} (2.155747e+04)	5.203680e+05^{2}# (2.863031e+04)	5.184674e+05^{4}# (1.929072e+04)	4.461624e+05^{8}# (3.114678e+04)	5.156971e+05^{5}# (1.528391e+04)	5.153043e+05^{6}# (2.276938e+04)
WFG1^{-1}	2	4.634857e+03^{2} (1.046707e+03)	4.235804e+03^{4}# (3.806378e+02)	4.744780e+03^{1} (8.040138e+02)	4.150133e+03^{8}# (4.060752e+02)	4.214520e+03^{5}# (4.438223e+02)	4.324686e+03^{3}# (7.944825e+02)	4.158365e+03^{7}# (4.841233e+02)	4.199205e+03^{6}# (4.117834e+02)
	3	7.715876e+04^{8}# (8.352894e+03)	3.141305e+05^{3}# (2.194826e+04)	3.461392e+05^{1} (1.599937e+04)	3.073141e+05^{5}# (1.906281e+04)	3.141413e+05^{2}# (2.194225e+04)	2.270112e+05^{6}# (1.736704e+04)	2.111147e+05^{7}# (5.681661e+04)	3.102097e+05^{4}# (1.969159e+04)
WFG2	2	2.476658e+03^{8}# (4.968231e+02)	2.790792e+03^{5} (7.035403e+02)	2.843289e+03^{1} (4.917881e+02)	2.701249e+03^{6} (4.563455e+02)	2.828557e+03^{3} (4.748070e+02)	2.600459e+03^{7}# (6.143328e+02)	2.828184e+03^{4} (5.581615e+02)	2.831176e+03^{2} (4.774635e+02)
	3	2.800111e+05^{8}# (5.723262e+04)	3.777302e+05^{2} (6.475809e+04)	3.888116e+05^{1} (7.671066e+04)	3.485334e+05^{6}# (4.423614e+04)	3.777302e+05^{3} (6.475809e+04)	3.127121e+05^{7}# (5.093723e+04)	3.676892e+05^{5}# (5.223327e+04)	3.689925e+05^{4}# (6.743403e+04)
WFG2^{-1}	2	1.736248e+03^{8}# (2.638803e+02)	2.333515e+03^{3} (3.177426e+02)	2.337436e+03^{2} (3.515560e+02)	2.353181e+03^{1} (4.519447e+02)	2.254460e+03^{7} (3.640165e+02)	2.306952e+03^{5} (4.353679e+02)	2.331899e+03^{4} (4.145174e+02)	2.266598e+03^{6} (4.148807e+02)
	3	3.810451e+05^{8}# (1.557674e+04)	5.127600e+05^{5}# (1.979390e+04)	5.715054e+05^{1} (2.972391e+04)	5.192733e+05^{2}# (3.908777e+04)	5.127600e+05^{6}# (1.979390e+04)	4.945208e+05^{7}# (2.548218e+04)	5.163662e+05^{3}# (2.452925e+04)	5.131372e+05^{4}# (2.160505e+04)
WFG3	2	4.737892e+03^{7}# (1.212466e+03)	5.688289e+03^{1} (9.195993e+02)	5.283769e+03^{6}# (9.062012e+02)	4.598677e+03^{8}# (4.967458e+02)	5.323202e+03^{3}# (9.883039e+02)	5.300394e+03^{5}# (8.054953e+02)	5.418351e+03^{2} (8.187545e+02)	5.318340e+03^{4}# (9.899779e+02)
	3	3.530812e+05^{8}# (4.627007e+04)	4.992648e+05^{3} (4.028765e+04)	5.140610e+05^{1} (3.330771e+04)	4.771445e+05^{6}# (3.341461e+04)	4.992648e+05^{4} (4.028765e+04)	4.505951e+05^{7}# (5.602672e+04)	4.994122e+05^{2} (4.829807e+04)	4.943398e+05^{5}# (3.517098e+04)
WFG3^{-1}	2	5.148218e+03^{3} (9.435951e+02)	5.439409e+03^{1} (6.812039e+02)	5.390651e+03^{2} (8.899972e+02)	4.196157e+03^{8}# (2.813812e+02)	5.065853e+03^{6}# (6.770455e+02)	4.860643e+03^{7}# (9.733448e+02)	5.077295e+03^{4}# (6.643410e+02)	5.077295e+03^{5}# (6.643410e+02)
	3	2.293842e+05^{8}# (2.498524e+04)	6.037343e+05^{3}# (2.325666e+04)	6.442096e+05^{1} (4.495150e+04)	5.977539e+05^{5}# (2.421289e+04)	6.037343e+05^{4}# (2.325666e+04)	5.228685e+05^{7}# (2.540989e+04)	5.912077e+05^{6}# (2.997981e+04)	6.046485e+05^{2}# (2.531948e+04)

WFG1^{-1}, and WFG2). In contrast, the use of pair-potential-based archives helps to circumvent such an issue. Additionally, Fig. 2[3] indicate that, in general, the use of the pair-potential-based archives help to increase the diversity throughout the evolutionary process. In consequence, this supports the claim that these archives store solutions that were possibly discarded by the main population even though they have a high diversity.

5 Conclusions and Future Work

In physics, the total potential energy of a system of N particles plays an important role in macroscopic and in molecular fields. To measure the potential energy

[3] All the IGD and PD graphs are available at http://computacion.cs.cinvestav.mx/~jfalcon/PairPotentials/.

due to the interactions of pairs of particles, pair-potential functions are employed. In this paper, we used several pair-potential functions (Riesz s-energy, Gaussian α-energy, Coulomb's law, Pöschl-Teller and Kratzer potential) to increase the diversity of MOEAs. To this aim, we adopted the pair-potential functions as the backbone of a selection mechanism to update an external archive which is coupled to MOEA/D. Our experimental results based on the IGD and PD indicators showed that the pair-potential-based archives store solutions with high diversity that otherwise would be discarded by MOEA/D's main population. Hence, the utilization of the pair-potential-based archives helps to circumvent the known MOEA/D's performance dependence on the Pareto front shapes. As part of our future work, we aim to mathematically analyze the selected pair-potential functions in order to design better selection mechanisms and test them on MaOPs.

References

1. Li, B., Li, J., Tang, K., Yao, X.: Many-objective evolutionary algorithms: a survey. ACM Comput. Surv. **48**(1), 1–35 (2015)
2. Borodachov, S.V., Hardin, D.P., Saff, E.B.: Discrete Energy on Rectifiable Sets. Springer Monographs in Mathematics, 1st edn. Springer, New York (2019). https://doi.org/10.1007/978-0-387-84808-2
3. Falcón-Cardona, J.G., Emmerich, M.T.M., Coello Coello, C.A.: On the cooperation of multiple indicator-based multi-objective evolutionary algorithms. In: Proceedings of CEC 2019, pp. 2050–2057 (2019)
4. Falcón-Cardona, J.G., Ishibuchi, H., Coello Coello, C.A.: Riesz s-energy-based reference sets for multi-objective optimization. In: Proceedings of CEC 2020, pp. 1–8 (2020)
5. Blank, J., Deb, K., Dhebar, Y., Bandaru, S., Seada, H.: Generating well-spaced points on a unit simplex for evolutionary many-objective optimization. IEEE Trans. Evol. Comput. **25**, 1 (2020)
6. Jackson, J.D.: Classical Electrodynamics, 3rd edn. Wiley, Hoboken (1999)
7. Dong, S.-H.: Pöschl-Teller Potential, pp. 95–110. Springer, Dordrecht (2007). https://doi.org/10.1007/978-1-4020-5796-0_7
8. Hamzavi, M., Ikhdair, S.M.: Approximate l-state solution of the trigonometric pöschl-teller potential. Mol. Phys. **110**(24), 3031–3039 (2012)
9. Simons, G., Parr, R.G., Finlan, J.M.: New alternative to the Dunham potential for diatomic molecules. J. Chem. Phys. **59**(6), 3229–3234 (1973)
10. Zhang, Q., Li, H.: MOEA/D: a multiobjective evolutionary algorithm based on decomposition. IEEE Trans. Evol. Comput. **11**(6), 712–731 (2007)
11. Hisao Ishibuchi, Yu., Setoguchi, H.M., Nojima, Y.: Performance of decomposition-based many-objective algorithms strongly depends on pareto front shapes. IEEE Trans. Evol. Comput. **21**(2), 169–190 (2017)
12. Coello Coello, C.A., Cortés, N.C.: Solving multiobjective optimization problems using an artificial immune system. Genet. Program. Evol. Mach. **6**(2), 163–190 (2005). https://doi.org/10.1007/s10710-005-6164-x
13. Wang, H., Jin, Y., Yao, X.: Diversity assessment in many-objective optimization. IEEE Trans. Cybern. **47**(6), 1510–1522 (2017)

On the Parameter Setting of the Penalty-Based Boundary Intersection Method in MOEA/D

Zhenkun Wang[1,2], Jingda Deng[3](✉), Qingfu Zhang[4,5], and Qite Yang[1]

[1] School of System Design and Intelligent Manufacturing, Southern University of Science and Technology, Shenzhen, China
wangzk3@sustech.edu.cn, yangqt@mail.sustech.edu.cn

[2] Department of Computer Science and Engineering, Southern University of Science and Technology, Shenzhen, China

[3] School of Mathematics and Statistics, Xi'an Jiaotong University, Xi'an, China
jddeng@xjtu.edu.cn

[4] Department of Computer Science, City University of Hong Kong, Hong Kong, China
qingfu.zhang@cityu.edu.hk

[5] The City University of Hong Kong Shenzhen Research Institute, Shenzhen, China

Abstract. Multiobjective optimization evolutionary algorithm based on decomposition (MOEA/D) decomposes an multiobjective optimization problem into a number of single-objective subproblems and solves them in a cooperative manner. The subproblems can be designed by various scalarization methods, *e.g.*, the weighted sum (WS) method, the Tchebycheff (TCH) method, and the penalty-based boundary intersection (PBI) method. In this paper, we investigate the PBI method with different parameter settings, and propose a way to set the parameter appropriately. Experimental results suggest that the PBI method with our proposed parameter setting works very well.

Keywords: MOEA/D · Penalty-based boundary intersection method · Scalarization method

1 Introduction

A multiobjective optimization problem (MOP) can be written as:

$$\begin{aligned} &\min F(x) = (f_1(x), f_2(x), \ldots, f_m(x))^T \in \mathbf{R}^m \\ &s.t. \quad x \in \Omega \subset \mathbf{R}^n \end{aligned} \tag{1}$$

This work was supported by the National Natural Science Foundation of China (Grant No. 11991023, 62076197, 61876163 and 62072364) and the Key Project of Science and Technology Innovation 2030 sponsored by the Ministry of Science and Technology of China (Grant No. 2018AAA0101301).

H. Ishibuchi et al. (Eds.): EMO 2021, LNCS 12654, pp. 413–423, 2021.
https://doi.org/10.1007/978-3-030-72062-9_33

where x is the decision variable, $f_i(x)$ is the i-th objective function, and Ω is the feasible region.

Evolutionary algorithms are efficient for solving MOPs [4,16]. Currently, there are three popular algorithm frameworks in the field of evolutionary multiobjective optimization (EMO), including Pareto dominance based, decomposition based, and indicator based evolutionary multiobjective algorithms (EMOAs) [1,5,20,21]. Among them, multiobjective optimization evolutionary algorithm based on decomposition (MOEA/D) [20] has attracted growing research effort in the EMO community.

MOEA/D decomposes an MOP into a number of single-objective subproblems and solves them in a cooperative manner. p-norm scalarization methods and the penalty-based boundary intersection (PBI) method can be used to aggregate objective functions into a single objective function. The performances of PBI method have been investigated in literature [7–9,11,17–19]. It is well known that the penalty parameter θ in the PBI method has a significant effect on the algorithmic performances. However, how to appropriately set the penalty parameter in PBI method is still unclear.

In this paper, we study how to set the penalty parameter θ in PBI method so that the optimal solution of each subproblem in MOEA/D with PBI method would not be far away from the search direction defined by PBI method. We propose a principle and it is quite simple. The setting of θ should depend on the weight vector in PBI method. Numerical experiments suggest that the PBI method with this parameter setting principle can make MOEA/D achieve better performances.

The remainder of this paper is organized as follows. Section 2 introduces and analyzes several scalarization methods used in MOEA/D framework, including p-norm scalarization methods and PBI method. Section 3 discusses the principle to set penalty parameter θ in the PBI method and conducts several numerical experiments. Section 4 concludes the paper.

2 Scalarization Methods in MOEA/D

In MOEA/D, there are many methods for constructing objective functions for subproblems, such as the weighted p-norm ($1 \leq p \leq \infty$) scalarization method, the augmented scalarization method and the PBI method.

2.1 The p-Norm Scalarization Method

The p-norm scalarization method is a traditional approach to transform an MOP to single-objective optimization problem [12]. the objective function for a subproblem in the p-norm scalarization method is:

$$\min_{x \in \Omega} g^{pNorm}(x|\lambda, z^*) = \left[\sum_{i=1}^{m} \lambda_i |f_i(x) - z^*|^p\right]^{\frac{1}{p}} \tag{2}$$

where $z^* \in \mathbf{R}^m$ is a reference vector, and $\lambda \in \mathbf{R}^m$ is a weight vector such that $\lambda > 0$ and $||\lambda||_1 = 1$. In this paper, we assume that z^* is an ideal point, i.e., $z_i^* = \min f_i(x), \forall 1 \leq i \leq m$, so that it is safe to drop the absolute sign.

When $p = 1$, Eq. (2) becomes:

$$\min_{x \in \Omega} g^{WS}(x|\lambda) = \sum_{i=1}^{m} \lambda_i f_i(x), \tag{3}$$

and when $p = \infty$, it is:

$$\min_{x \in \Omega} g^{TCH}(x|\lambda, z^*) = \max_{1 \leq i \leq m} \lambda_i (f_i(x) - z^*). \tag{4}$$

These are respectively the weighted sum (WS) method, and Tchebycheff (TCH) method.

We introduce several basic concepts:

Definition 1 (Dominance). *Given any $u, v \in \mathbf{R}^m$, u is said to dominates v, denoted by $u \prec v$, if and only if $u_i \leq v_i, \forall 1 \leq i \leq m$, and there exists $1 \leq j \leq m$ such that $u_j < v_j$.*

Definition 2 (Strict dominance). *Given any $u, v \in \mathbf{R}^m$, u is said to strictly dominates v, denoted by $u \prec\prec v$, if and only if $u_i < v_i, \forall 1 \leq i \leq m$.*

Definition 3 (Pareto optimality). *A solution $x \in \Omega$ is Pareto optimal if and only if there is no other $x' \in \Omega$ such that $F(x') \prec F(x)$. x is called a Pareto optimal solution and $F(x)$ is called a Pareto optimal vector. All the Pareto optimal solutions make up the Pareto set (PS) and their objective vectors are called the Pareto front (PF).*

Definition 4 (Weak Pareto optimality). *A solution $x \in \Omega$ is weakly Pareto optimal if and only if there is no other $x' \in \Omega$ such that $F(x') \prec\prec F(x)$. x is called a weakly Pareto optimal solution and $F(x)$ is called a weakly Pareto optimal vector.*

Some theoretical results of the p-norm scalarization method are summarized as follows [6,12]:

- Given any weight vector $\lambda > 0$, the p-norm $(1 \leq p < \infty)$ method always finds a Pareto optimal solution, while the TCH method always finds a weakly Pareto optimal solution, and there exists a Pareto optimal solution in the optimal set of a TCH subproblem.
- Given any Pareto optimal solution x^*, there exists a weight vector λ^* such that x^* is an optimal solution of TCH subproblem.
- Given any Pareto optimal solution x^*, if the MOP is convex, there exists a weight vector λ^* such that x^* is an optimal solution of a WS subproblem.
- Given any Pareto optimal solution x^*, if the curvature of the PF of the MOP is less than p^*, there exists a weight vector λ^* such that x^* is an optimal solution of a p-norm $(\max\{1, p^*\} \leq p < \infty)$ scalarization subproblem.

Therefore, when $1 \leq p < \infty$, the p-norm scalarization method cannot find some Pareto optimal solutions for an arbitrary MOP by different settings of λ. On the contrary, the TCH method is able to solve any MOP. Moreover, the TCH method specifies a search direction $d = (\frac{1}{\lambda_1}, \ldots, \frac{1}{\lambda_m})^T$. If there exists an Pareto optimal vector along this direction, the optimal solution of TCH problem would be the corresponding solution of this Pareto optimal vector. On the contrary, other p-norm scalarization methods do not specify search directions, and their optima may be located in any position on the PF, depending on the shape of the PF.

2.2 The PBI Method

In [20], the PBI method was proposed as an alternative of scalarization methods. It is inspired from the boundary intersection method [3]. A subproblem defined by the PBI method is as follows:

$$\min_{x \in \Omega} g^{PBI}(x|\lambda, \theta) = d_1 + \theta d_2, \tag{5}$$

where $d_1 = \frac{|(F(x)-z^*)^T\lambda|}{||\lambda||_2}$, $d_2 = ||F(x) - (z^* + d_1\frac{\lambda}{||\lambda||_2})||_2$, λ is a weight vector, and θ is a penalty parameter. A commonly-used setting of θ is fixed as 5. Note that d_1 and d_2 are functions of $F(x)$, and we can compute g^{PBI}, d_1, and d_2 values given any vector $z \in \mathbf{R}^m$ in the objective space even if z is not in the feasible objective region. For simplification, we also use $g^{PBI}(z), d_1(z), d_2(z)$ for z if there is no confusion.

Much research effort has focused on the influence of the selection of penalty parameter. It is found in [8] that the PBI method performs well in many-objective problems by using a large θ. In [17], extensive numerical experiments are conducted to investigate the practical performances of the PBI method with different settings of θ on multiobjective test instances. In [14], an inverted variant of PBI method is proposed and shows better performances for some special problems. Other works focus on how to adapt the settings of θ in a typical MOP [2,7,11,13,18]. [19] proposes to gradually increase the value of θ to balance the exploration and exploitation in different search stages. Numerical experiments in these papers also show that an inappropriate setting of θ results in failure to fully cover the whole PF. In [15], the relationship between PBI method and augmented Tchebycheff method is analyzed.

2.3 Numerical Experiments

In this subsection, we take a special MOP as an example to show the performances of WS, TCH, and PBI methods. The MOP is a bi-objective problem with box constraints:

$$\min \begin{cases} f_1(x) = (0.5 - x_1 + x_2)^{2.5} \\ f_2(x) = (0.5 + x_1 - x_2)^{1.5} \end{cases} \tag{6}$$
$$s.t. \quad x \in [0, 1]^2$$

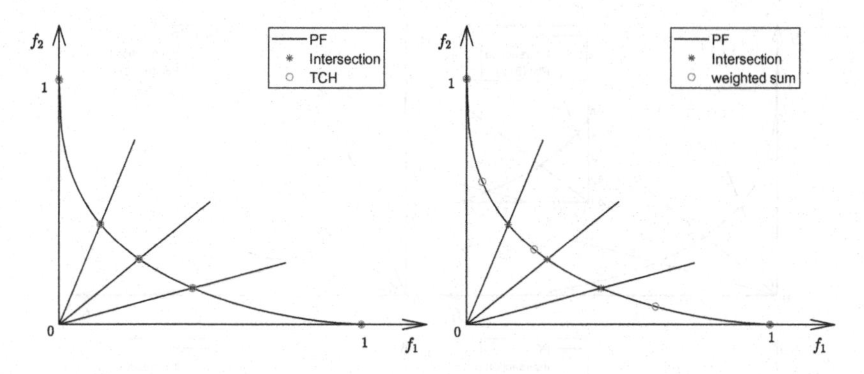

Fig. 1. Optimal solutions found by TCH and WS methods on five subproblems of the MOP (6).

In the experiments, we construct 5 subproblems for all methods and set penalty parameter θ to be $2, 10, 50$ and 200, respectively. MOEA/D-DE [10] is used to optimize these problems and the results are shown in Fig. 1 and Fig. 2. In the figures, points marked by star are the intersections between search directions and the PFs. For the WS method and the PBI method with smaller θ, we find that the optimum significantly deviate their search directions, even if there exist Pareto optimal vectors along the search directions. For the PBI method, as the value of θ increases, the optimal objective vector will move towards to its search direction. For the TCH method, its optimal solution lies on its search direction.

3 Settings of θ in the PBI Method

In this section, we discuss a setting of θ in PBI method. We find that an appropriate setting of θ should be determined by weight vector λ.

3.1 A Principle for Setting θ Based on λ

According to the experiments conducted in Sect. 2.3, it is possible that even if there exists a Pareto optimal solution at the search direction of PBI method, the optima of PBI problem may deviate such solution unintentionally under an inappropriate setting of θ. In this subsection, we propose a principle for setting θ to avoid this situation. It is based on the analysis of the contour shape of a PBI problem and it is easy to understand.

First of all, the contour of a PBI problem is a circular cone. This has been explicitly or implicitly stated in many research papers such as [2,11] from the view of geometry. It is also well-known that $\theta = \cot\alpha$ where α is the angle between the generatrix and the axis of the contour. Here we clarify that a contour $d_1 + \theta d_2 = C$ ($C \in \mathbf{R}$ such that the contour is nonempty) is a circular cone whose vertex is z^0 where $z^0 = z^* + \frac{C}{\|\lambda\|_2}\lambda$, axis l^0 is defined by z^* and z^0, and angle between the generatrix and the axis is $\alpha = \cot^{-1}\theta$, which is a constant for all C.

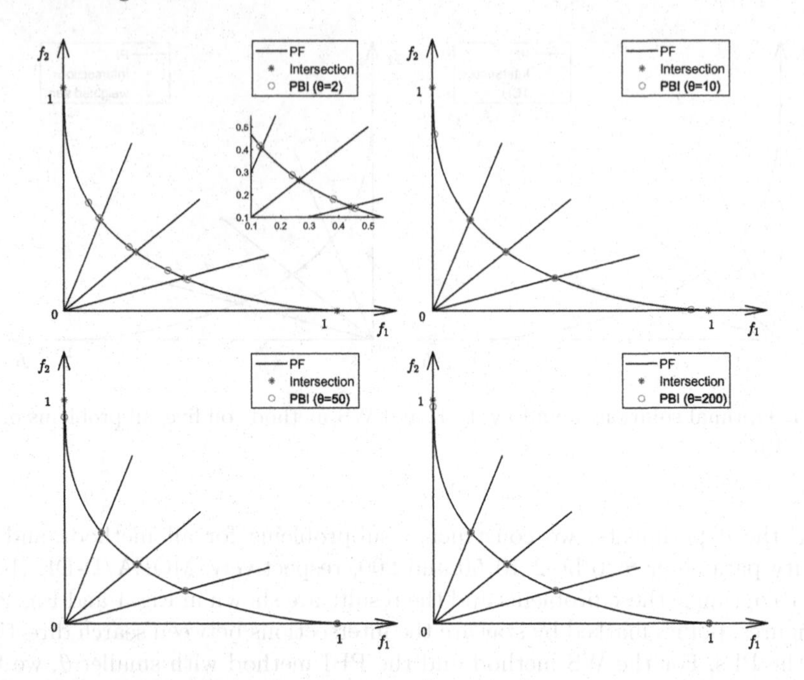

Fig. 2. Optimal solutions found by PBI methods with $\theta = 2, 10, 50, 200$ on five subproblems of the MOP (6).

Then, suppose that there exists a Pareto optimal solution x^* along the search direction of the PBI problem. The contour defined by $C^* = g^{PBI}(x^*|\lambda, \theta)$ is also a circular cone. We claim that in order to ensure that the optima of the PBI problem does not deviate its search direction, all points on the contour except $F(x^*)$ must dominate $F(x^*)$. We prove this by contradiction as follows:

Proof. Suppose there is a point $F \in \mathbf{R}^m$ on the contour such that it does not dominate $F(x^*)$. We can move F slightly as $F' = F - \epsilon\lambda$ ($\epsilon > 0$) such that F' still does not dominate $F(x^*)$. In this case[1], g^{PBI} value of F' is smaller than that of F, since its d_2 value is equal to F and d_1 value is smaller. As a result, g^{PBI} value of F' is smaller than that of $F(x^*)$ because F and $F(x^*)$ are both on the contour. Therefore, x^* is not the optima of the PBI problem, that is, the optima of the PBI problem deviates its search direction.

When all points except $F(x^*)$ on the contour defined by C^* dominate $F(x^*)$, they must be located in the region $F(x^*) - \mathbf{R}_+^m$. Moreover, since the contour shape does not change given a fixed θ for all $C \in \mathbf{R}$, this property should also hold for all C values. Consider any contour of θ with the vertex z^0, axis l^0 and angle α. Then there should be an upper bound for α such that the contour

[1] F' may not be in the feasible objective region. But we are seeking for a common principle, so we have to take the situation into consideration where there is an MOP such that F' is in the feasible objective region.

(circular cone) is just tangent to $z^0 - \mathbf{R}_+^m$. Since $z^0 - \mathbf{R}_+^m$ is a pyramid, the circular cone must be tangent to it at one surface of it, that is, the angle between the surface and the axis l^0 is equal to α as shown in Fig. 3.

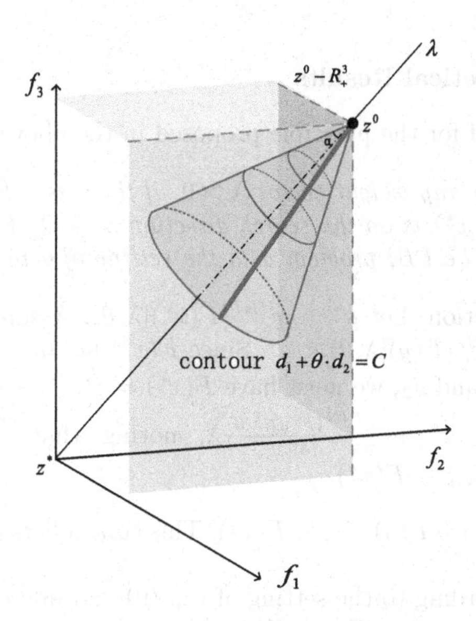

Fig. 3. Illustration of the relationship between the contour of PBI method and $z^0 - \mathbf{R}_+^m$.

Note that the vertex of this pyramid is z^0 and its edges are on m negative axis directions starting from z^0, so surfaces of the pyramid are

$$\tilde{z}^i - \mathbf{R}_+^{m-1}, \forall 1 \le i \le m$$

where $\tilde{z}^i$ is z^0 excluding the i-th element, i.e., $\tilde{z}^i = (z_1^0, \ldots, z_{i-1}^0, z_{i+1}^0, \ldots, z_m^0)^T$, $\forall 1 \le i \le m$. The normal vector of surface $\tilde{z}^i - \mathbf{R}_+^{m-1}$ is the i-th unit coordinate vector, i.e., all of its elements are zero but the i-th element is 1, so it is easy to compute the angle between surface $\tilde{z}^i - \mathbf{R}_+^{m-1}$ and the axis l^0 by:

$$\frac{\pi}{2} - \arccos \frac{\lambda_i}{||\lambda||_2}, \forall 1 \le i \le m. \tag{7}$$

Therefore, we have

$$\alpha = \cot^{-1} \theta \le \frac{\pi}{2} - \max_{1 \le i \le m} \arccos \frac{\lambda_i}{||\lambda||_2}, \tag{8}$$

and finally obtain the following principle for setting θ:

$$\theta \ge \tan \max_{1 \le i \le m} \arccos \frac{\lambda_i}{||\lambda||_2}. \tag{9}$$

Equation (9) is consistent with the calculation in Algorithm 2 in [18]. However, our proposed principle is independently obtained from a different aspect. We also provide some theoretical analysis for our proposed principle in the next subsection.

3.2 More Theoretical Results

Here we give a proof for the principle proposed in the above subsection.

Theorem 1. *Given any weight vector $\lambda > 0$, if there is a Pareto optimal solution x^* such that $F(x^*)$ is on the search direction $z^* + t\lambda$ $(t \geq 0)$, then x^* is an optimal solution of the PBI problem with the setting of θ in Eq. (9).*

Proof. By contradiction. Let $g^* = g^{PBI}(F(x^*)|\lambda, \theta)$. Assume that there exists $y \in \Omega$ such that $g^{PBI}(F(y)|\lambda, \theta) < g^*$. Since $F(x^*)$ lies on $z^* + t\lambda$ $(t \geq 0)$, from the definition of d_1 and d_2, we must have $F(x^*) = z^* + \frac{g^*}{||\lambda||_2}\lambda$.

Consider $\bar{z} = z^* + \frac{g^{PBI}(F(y)|\lambda,\theta)}{||\lambda||_2}\lambda$, noting that $g^{PBI}(F(y)|\lambda, \theta) = g^{PBI}(\bar{z}|\lambda, \theta) < g^*$, so $\bar{z} \prec F(x^*)$.

1. If $F(y) = \bar{z}$, we have $F(y) = \bar{z} \prec F(x^*)$. This contradicts the Pareto optimality of x^*.
2. If $F(y) \neq \bar{z}$, according to the setting of Eq. (9) and inferences in last subsection, we have $F(y) \prec \bar{z} \prec F(x^*)$. This also contradicts the Pareto optimality of x^*.

Therefore, x^* is an optimal solution of the PBI problem with the setting of θ in Eq. (9).

Moreover, we also have the following result:

Corollary 1. *Given any Pareto optimal solution x^* such that $z^* \prec\prec F(x^*)$, there exists a weight vector λ^* such that x^* is an optimal solution of the PBI problem with the setting of θ in Eq. (9).*

Proof. Let $\lambda^* = \frac{F(x^*)-z^*}{||F(x^*)-z^*||_1} > 0$. For the PBI problem $\min_{x\in\Omega} g^{PBI}(x|\lambda^*, \theta)$ where θ satisfies Eq. (9), there is a Pareto optimal solution x^* such that $F(x^*) = z^* + ||F(x^*) - z^*||_1\lambda^*$ is on its search direction. According to Theorem 1, x^* is an optimal solution of this PBI problem.

We have the following comments on the PBI method with our proposed principle for setting θ:

- For an arbitrary MOP, it can find all Pareto optimal solutions whose objective vectors are strictly dominated by z^*. This is an advantage over p-norm scalarization methods for $1 \leq p < \infty$.
- It may find non-Pareto optimal solutions, specially when the PF of the MOP is disconnected.

- It has a search direction, which is similar to the TCH method. However, the reception region of it is smaller than TCH method because its contour is enclosed in the contour of PBI method. This property is beneficial to MOEA/D when a large selection pressure is required, e.g., for many-objective optimization problems.
- For subproblems close to any single objective, i.e., there is an element in λ close to 1, the setting of θ goes up to ∞, while for other subproblems, the requirement of θ is much looser.

3.3 Numerical Experiments

In this subsection, we verify our proposed principle on the MOP in Sect. 2.3. θ values are selected such that the equality holds in Eq. (9) based on different weight vectors. Firstly, we test our principle under the same experimental settings as in Sect. 2.3. The results are shown in Fig. 4. It is found that the optimum of all subproblems are very close to their own search directions.

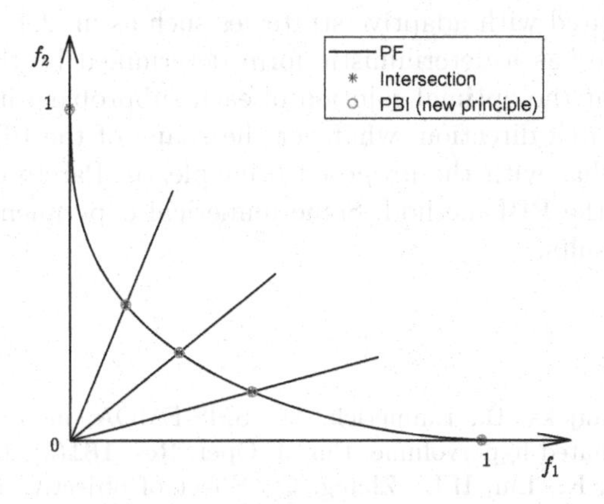

Fig. 4. Optimal solutions found by PBI methods with our proposed principle on five subproblems of the MOP (6)

Then, the proposed principle is compared with fixed settings of θ on a larger number of subproblems. In this experiment, the number of subproblems in MOEA/D-DE is set to be 100. The IGD values of population during the evolution process based on different settings of θ are shown in Fig. 5. From this figure, MOEA/D with our proposed principle has the fastest convergence speed and the smallest final IGD value compared with MOEA/D with fixed settings of θ.

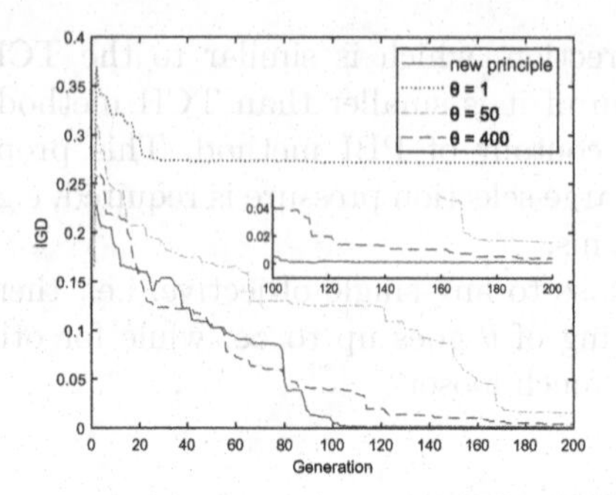

Fig. 5. IGD values during the evolution process for optimizing 100 subproblems based on our proposed principle and fixed values of θ.

4 Conclusion

In this paper, we investigated the setting of the penalty parameter θ in PBI-based MOEA/D. Compared with adaptive strategies such as in [2,7,11,13,18,19], our proposed principle has a deterministic form determined by the weight vector, and it ensures that the optimal solution of each subproblem in MOEA/D does not deviate its search direction, whatever the shape of the PF of the MOP is. We have proved that with the proposed principle, all Pareto optimal solutions can be found by the PBI method. Some numerical experiments are conducted to validate our results.

References

1. Beume, N., Naujoks, B., Emmerich, M.: SMS-EMOA: multiobjective selection based on dominated hypervolume. Eur. J. Oper. Res. **181**(3), 1653–1669 (2007)
2. Chen, L., Deb, K., Liu, H.L., Zhang, Q.: Effect of objective normalization and penalty parameter on penalty boundary intersection decomposition based evolutionary many-objective optimization algorithms. Evol. Comput. **29**, 1–30 (2020)
3. Das, I., Dennis, J.E.: Normal-boundary intersection: a new method for generating the Pareto surface in nonlinear multicriteria optimization problems. SIAM J. Optim. **8**(3), 631–657 (1998)
4. Deb, K.: Multi-objective Optimization Using Evolutionary Algorithms, vol. 16. Wiley, Hoboken (2001)
5. Deb, K., Pratap, A., Agarwal, S., Meyarivan, T.: A fast and elitist multiobjective genetic algorithm: NSGA-II. IEEE Trans. Evol. Comput. **6**(2), 182–197 (2002)
6. Goh, C.J., Yang, X.Q.: Convexification of a noninferior frontier. J. Optim. Theory Appl. **97**, 759–768 (1998). https://doi.org/10.1023/A:1022654528902
7. Guo, J., Yang, S., Jiang, S.: An adaptive penalty-based boundary intersection approach for multiobjective evolutionary algorithm based on decomposition. In: 2016 IEEE Congress on Evolutionary Computation (CEC), pp. 2145–2152 (2016)
8. Ishibuchi, H., Doi, K., Nojima, Y.: Characteristics of many-objective test problems and penalty parameter specification in MOEA/D. In: 2016 IEEE Congress on Evolutionary Computation (CEC), pp. 1115–1122 (2016)

9. Ishibuchi, H., Sakane, Y., Tsukamoto, N., Nojima, Y.: Adaptation of scalarizing functions in MOEA/D: an adaptive scalarizing function-based multiobjective evolutionary algorithm. In: Ehrgott, M., Fonseca, C.M., Gandibleux, X., Hao, J.-K., Sevaux, M. (eds.) EMO 2009. LNCS, vol. 5467, pp. 438–452. Springer, Heidelberg (2009). https://doi.org/10.1007/978-3-642-01020-0_35
10. Li, H., Zhang, Q.: Multiobjective optimization problems with complicated pareto sets, MOEA/D and NSGA-II. IEEE Trans. Evol. Comput. **13**(2), 284–302 (2009)
11. Liu, Y., Ishibuchi, H., Masuyama, N., Nojima, Y.: Adapting reference vectors and scalarizing functions by growing neural gas to handle irregular Pareto fronts. IEEE Trans. Evol. Comput. **24**(3), 439–453 (2020)
12. Miettinen, K.: Nonlinear multiobjective optimization. In: International Series in Operations Research & Management Science. Springer, New York (1999). https://doi.org/10.1007/978-1-4615-5563-6
13. Ming, M., Wang, R., Zha, Y., Zhang, T.: Pareto adaptive penalty-based boundary intersection method for multi-objective optimization. Inf. Sci. **414**, 158–174 (2017)
14. Sato, H.: Analysis of inverted PBI and comparison with other scalarizing functions in decomposition based MOEAs. J. Heuristics **21**(6), 819–849 (2015). https://doi.org/10.1007/s10732-015-9301-6
15. Singh, H.K., Deb, K.: Investigating the equivalence between PBI and AASF scalarization for multi-objective optimization. Swarm Evol. Comput. **53**, 100630 (2020)
16. Tan, K.C., Khor, E.F., Lee, T.H.: Multiobjective Evolutionary Algorithms and Applications. Springer, London (2006). https://doi.org/10.1007/1-84628-132-6
17. Tanabe, R., Ishibuchi, H.: An analysis of control parameters of MOEA/D under two different optimization scenarios. Appl. Soft Comput. **70**, 22–40 (2018)
18. Yang, C., Hu, C., Zou, Y.: PBI function based evolutionary algorithm with precise penalty parameter for unconstrained many-objective optimization. Swarm Evol. Comput. **50**, 100568 (2019)
19. Yang, S., Jiang, S., Jiang, Y.: Improving the multiobjective evolutionary algorithm based on decomposition with new penalty schemes. Soft Comput. **21**(16), 4677–4691 (2016). https://doi.org/10.1007/s00500-016-2076-3
20. Zhang, Q., Li, H.: MOEA/D: a multiobjective evolutionary algorithm based on decomposition. IEEE Trans. Evol. Comput. **11**(6), 712–731 (2007)
21. Zitzler, E., Künzli, S.: Indicator-based selection in multiobjective search. In: Yao, X., et al. (eds.) PPSN 2004. LNCS, vol. 3242, pp. 832–842. Springer, Heidelberg (2004). https://doi.org/10.1007/978-3-540-30217-9_84

A Comparison Study of Evolutionary Algorithms on Large-Scale Sparse Multi-objective Optimization Problems

Yajie Zhang[1], Ye Tian[2(✉)], and Xingyi Zhang[1(✉)]

[1] School of Computer Science and Technology, Anhui University, Hefei, China
[2] Institutes of Physical Science and Information Technology, Anhui University, Hefei, China

Abstract. The performance of most multi-objective evolutionary algorithms (MOEAs) usually degenerates when they are adopted to tackle large-scale multi-objective optimization problems (LSMOPs) involving a large number of decision variables. While LSMOPs attract increasing attention in the evolutionary computation community in recent years, a number of delicate approaches have been proposed to improve the performance of MOEAs on LSMOPs. While many real-world LSMOPs include sparse optimal solutions (i.e., most decision variables are zero) that should be found within a limited budget of function evaluations, few MOEAs have been tailored for these sparse LSMOPs and the performance of existing MOEAs on them has not been well studied. In this paper, we first have a brief review of existing MOEAs for LSMOPs, and then elaborate two customized MOEAs for solving sparse LSMOPs. Next, we select six state-of-the-art MOEAs with different search strategies to conduct an experimental study on eight sparse benchmark problems and four real-world applications. Finally, we outline some future research directions of large-scale optimization involving sparsity.

Keywords: Large-scale multi-objective optimization · Sparse Pareto optimal solutions · Evolutionary algorithm · Real-world applications

1 Introduction

Multi-objective optimization problems (MOPs) consist of multiple conflicting objectives that need to be optimized simultaneously [1–3], which can be mathematically formulated as follows:

$$\begin{array}{ll} \text{Minimize} & \mathbf{F}(\mathbf{x}) = (f_1(\mathbf{x}), f_2(\mathbf{x}), \cdots, f_m(\mathbf{x})) \\ \text{subject to} & \mathbf{x} \in \Omega \end{array}, \quad (1)$$

where Ω is the search space and $\mathbf{x} = (x_1, x_2, \cdots, x_D)$ denotes the decision vector. When the number of decision variables is larger than one hundred, we usually call such kind of problems as large-scale multi-objective optimization problems

H. Ishibuchi et al. (Eds.): EMO 2021, LNCS 12654, pp. 424–437, 2021.
https://doi.org/10.1007/978-3-030-72062-9_34

(LSMOPs) [4]. Multi-objective evolutionary algorithms (MOEAs) are capable of obtaining a set of trade-off solutions in a single run, and a variety of MOEAs performing well on different kinds of MOPs have been proposed during the past two decades [5–7,26]. However, the performance of them often degenerates when they are adopted to tackle LSMOPs due to the huge decision space. Thus, to address the curse of dimensionality [8], various approaches have been tailored for solving LSMOPs during the past ten years, which can be roughly categorized into four types. They are based on the cooperative coevolution framework [27], decision variable analysis [11,12], problem reformulation [9,10], or special offspring generation strategies [13,23,28,29].

The main idea of the algorithms based on cooperative coevolution framework is to divide decision variables into different groups that can be optimized through independent subpopulations in a divide-and-conquer manner. The performance of such kind of MOEAs can be greatly affected by the adopted grouping technique. For example, random grouping [30] are employed in the third-generation cooperative coevolutionary differential evolution algorithm (CCGDE3) [27] considering that dividing variables into random groups provides better results than applying a deterministic division scheme when dealing with nonseparable functions. Besides, other grouping techniques (ordered grouping [31], linear grouping [14] and differential grouping [15]) have also shown effectiveness in solving specific LSMOPs.

MOEA/DVA [11] and LMEA [12] are two well-known algorithms belonging to the category based on decision variable analysis. The key component of MOEA/DVA consists of control property analysis and variable linkage analysis, where the former one divides the decision variables into position variables, distance variables and mixed variables, while the latter one divides distance variables into smaller subgroups of interacting variables. Afterwards, variables in each group are optimized independently through a differential evolutional based optimizer. In contrast to MOEA/DVA treating mixed variables as diversity-related variables, LMEA clusters a decision variable as either convergence-related variable or diversity-related variable. Subsequently, convergence optimization strategy and diversity optimization strategy are employed to optimize the corresponding variables alternately.

For the methods based on problem reformulation, two representative algorithms are WOF [9] and LSMOF [10]. WOF divides the decision variables into different groups and assigns a weight variable to each group, and thus, dimensionality of the problem is greatly reduced by altering variables in the same group at the same time. On the other hand, LSMOF defines a set of reference directions in the decision space and associates them with a number of weight variables to reformulate the problem into a low-dimensional single-objective problem in its first stage. After obtaining enough quash-optimal solutions near the Pareto set, LSMOF spreads such solutions over the approximated Pareto set evenly via an embedded differential evolution algorithm in its second stage.

The last category employs special offspring generation strategies. To improve the search efficiency, LMOCSO [23] uses a two-stage strategy to update particle position, it first pre-updates the position of each particle according to its previous velocity, it updates the position of each pre-updated particle by learning from a leader, and a competitive mechanism is adopted to determine the particles to be updated. Instead of optimizing decision variables directly, LCSA [28] evolves a population of coefficient vectors, by taking advantage of the inherent knowledge of the population, offsprings can thus be obtained via linear combinations of existing individuals. GLMO [29] attempts to embed variable grouping into mutation operators to improve the quality of generated offsprings, and three new mutation operators are presented, they are Linked Polynomial Mutation, Grouped Polynomial Mutation and Grouped and Linked Polynomial Mutation. Recently, He et al. [13] proposed an algorithm to generate promising solutions via constructing direction vectors in the decision space, specifically, in each iteration, two kinds of direction vectors related to convergence and diversity are constructed adaptively, offsprings are then obtained along each direction vector through sampling the built Gaussian distribution.

Despite that these delicate approaches as introduced above can improve the performance of MOEAs on LSMOPs, they still have some drawbacks. To be specific, for the algorithms based on cooperative coevolution framework, their performance can be highly affected by the adopted grouping techniques. For the methods based on decision variable analysis, the process of decision variable analysis can be quite time-consuming, and when the landscape functions are complicated, their analysis results will be inaccurate. For the methods based on problem reformulation, they can speed up convergence via solving small-scale reformulated problems, but the diversity is hard to be maintained. These existing MOEAs have shown their ineffectiveness in solving sparse LSMOPs which widely exist in many applications[16–18,24,25,33], one reason is that few of them consider the sparse nature of the Pareto optimal solutions when evolving the population, thus converging slowly on sparse LSMOPs. To fit this gap, two MOEAs customized for sparse LSMOPs were recently proposed, which make fully use of the sparsity of the problems to speed up the convergence to the Pareto optimal sets.

The remainder of this article is organized as follows. In Sect. 2, the two recently proposed MOEAs for sparse LSMOPs are elaborated. We conduct experimental studies on eight sparse benchmark problems in Sect. 3, and four real-world applications in Sect. 4. In Sect. 5, we outline some future research directions of large-scale sparse multiobjective optimization.

2 Two Algorithms Tailored for Sparse LSMOPs

2.1 SparseEA

SparseEA [19] is the first specially tailored evolutionary algorithm for large-scale sparse multi-objective optimization problems, which has a similar framework to NSGA-II, while SparseEA adopts new strategies to generate the initial population and offsprings to ensure the sparsity of produced individuals, which are also the core components we prepare to introduce in this paper.

A hybrid encoding scheme as $(x_1, x_2, ...) = (dec_1 \times mask_1, dec_2 \times mask_2, ...)$ is adopted in SparseEA, where *dec* denotes the decision variables and *mask* denotes the mask. The population initialization strategy of SparseEA consists of two steps: calculating the scores of variables and generating the initial population based on the obtained scores, which is summarized as follows:

① Set *Dec* to a $D \times D$ random matrix if the decision variables are real numbers, otherwise, a $D \times D$ matrix of ones, and D denotes the number of decision variables, set $Mask$ to a $D \times D$ identity matrix.
② Population Q with D solutions is obtained based on *Dec* and $Mask$, then perform non-dominated sorting on Q to calculate the scores for each variable.
③ Uniformly randomly generate the decision variables *Dec* of N solutions if the decision variables are real numbers, otherwise, set *Dec* to a $N \times D$ matrix of ones. $Mask$ is firstly set to a $N \times D$ matrix of zeros, and then for each row, $rand() \times D$ elements are set to 1 by binary tournament selection according to the scores of decision variables, where *rand()* denotes obtaining a random number between [0, 1].
④ The initial population can be obtained based on *Dec* and $Mask$ in ③.

SparseEA adopts conventional genetic operators for real variables, while it uses new genetic operators for binary variables to flip zero variables or nonzero variables with the same probability. The element to be flipped is decided based on the scores of decision variables, in this way, SparseEA can avoid generating offsprings with the same number of 0 and 1 to ensure the sparsity of the population.

Except for proposing a specially tailored MOEA for sparse LSMOPs, in [19], Tian et al. have also designed a benchmark test suite containing eight sparse MOPs to evaluate the performance of SparseEA at the same time.

2.2 MOEA/PSL

Even though SparseEA is exclusively customized for sparse LSMOPs, it is still somewhat time-consuming, especially, its population initialization strategy requires a large number of function evaluations. To address such issue, MOEA/PSL [20] adopts two unsupervised neural networks (the restricted Boltzmann machine (RBM) and the denoising autoencoder (DAE)) to detect the Pareto optimal subspace during the evolutionary process, in this way, the search space can be greatly reduced and the difficulty mentioned above can be highly alleviated.

In each generation, RBM is responsible for learning a sparse distribution of the decision variables, while DAE is adopted to learn a compact representation. The combination of the learnt sparse distribution and compact representation is regarded as an approximation of the Pareto optimal subspace. Afterwards, conventional genetic operators are executed in the learnt subspace to produce promising offsprings. MOEA/PSL has a similar framework with NSGA-II while some offsprings are generated in the learnt subspace. It is necessary to note that MOEA/PSL adopts the same hybrid encoding scheme as SparseEA. In the following, we elaborate how offsprings are produced in each generation.

ⓐ Input current population P, mating pool P', parameter ρ which controls the ratio of offsprings generated in the learnt subspace and parameter K which denotes the size of hidden layers.
ⓑ Save all the non-dominated solutions in P into *NP*, the binary vectors and real vectors of solutions in *NP* are archived into *NPB* and *NPR*, respectively.
ⓒ Train a RBM with K hidden neurons based on *NPB*, and if the decision variables are real numbers, a DAE should also been trained based on *NPR*.
ⓓ Select two vectors **xb** and **yb** from *NPB* randomly, and if the decision variables are real numbers, select two vectors **xr** and **yr** from *NPR* that have the same locations as **xb** and **yb** in *NPB*.
ⓔ if $\rho >$ *rand()*, reduce **xb** and **yb**, perform single-point crossover and bitwise mutation in the learnt subspace, afterwards, recover the obtained vectors to the original space. If the decision variables are real numbers, conduct similar operations on **xr** and **yr** using simulated binary crossover and polynomial mutation. If $\rho \leq$ *rand()*, perform genetic operators in the original space.
ⓕ repeat step ⓓ - step ⓔ until *NPB* is empty, output obtained offspring set O.

In order to make MOEA/PSL generic to different sparse LSMOPs, ρ and K are designed to be adaptive. Parameter ρ is dynamically adjusted based on the number of successful offsprings generated at the previous generations, while parameter K is changed according to the sparsity of the non-dominated solutions.

3 Experimental Studies on Benchmark Problems

So far in the literature, many large-scale MOEAs have been evaluated on different LSMOPs and shown good performance. However, the comprehensive performance comparisons on sparse LSMOPs remain scarce. In this section, we select six stat-of-the-art large-scale MOEAs from different categories to have an experimental comparison on SMOP1–SMOP8 [19], which have various landscape functions to provide various difficulties to existing MOEAs in obtaining sparse Pareto optimal solutions.

3.1 Experimental Settings

Algorithms: Six algorithms are empirically investigated in this paper, they are WOF-SMPSO [9], LMEA [12], CCGDE3 [27], GLMO [29], SparseEA [19] and MOEA/PSL [20], respectively. For WOF-SMPSO, the number of function evaluations for each optimization of original problem is set to 1000, while the number of evaluations for each optimization of transformed problem is set to 500, the number of chosen solutions is set to 3, the number of groups is set to 4, the ratio of evaluations for optimization of both original and transformed problems

Table 1. IGD values obtained by WOF-SMPSO, LMEA, CCGDE3, GLMO, SparseEA, and MOEA/PSL on SMOP1–SMOP8, where the best result in each row is highlighted. '+', '−' and '≈' indicate that the result is significantly better, significantly worse, and statistically similar to that obtained by MOEA/PSL.

Problem	*Dec*	WOF-SMPSO	LMEA	CCGDE3	GLMO	SparseEA	MOEA/PSL
SMOP1	1000	6.6985e-2 (1.63e-2) −	1.5866e+0 (1.81e-2) −	8.8126e-1 (8.88e-2) −	7.6410e-2 (1.22e-3) −	2.5340e-2 (2.55e-3) −	**1.4568e-2 (3.09e-3)**
	2000	5.5703e-2 (1.05e-2) −	1.6167e+0 (1.42e-2) −	9.6182e-1 (1.03e-1) −	7.8585e-2 (5.28e-4) −	3.2900e-2 (1.99e-3) −	**1.6337e-2 (2.42e-3)**
	5000	4.5108e-2 (5.22e-3) −	1.6335e+0 (6.47e-3)−	9.5750e-1 (1.07e-1) −	7.9465e-2 (2.63e-4) −	3.8251e-2 (1.37e-3) −	**1.7868e-2 (3.82e-3)**
SMOP2	1000	4.5582e-1 (1.51e-1) −	2.2268e+0 (7.54e-3) −	1.8044e+0 (1.44e-1) −	6.9075e-1 (4.75e-2) −	6.5692e-2 (4.44e-3) −	**3.6419e-2 (6.13e-3)**
	2000	3.4336e-1 (1.36e-1) −	2.2359e+0 (6.60e-3) −	1.8346e+0 (1.27e-1) −	7.4145e-1 (7.58e-2) −	8.5930e-2 (3.34e-3) −	**4.4748e-2 (6.05e-3)**
	5000	1.9892e-1 (1.01e-1) −	2.2484e+0 (3.71e-3)−	1.8289e+0 (1.53e-1) −	7.6820e-1 (6.61e-2) −	1.0162e-1 (2.87e-3) −	**5.1891e-2 (8.05e-3)**
SMOP3	1000	6.6731e-1 (1.74e-3) −	2.5887e+0 (1.15e-2) −	2.0253e+0 (7.87e-2) −	7.0279e-1 (9.48e-4) −	**2.8654e-2 (2.72e-3) +**	1.0167e-1 (3.26e-1)
	2000	6.6337e-1 (1.09e-3) −	2.5990e+0 (5.75e-3) −	2.0124e+0 (6.76e-2) −	7.0224e-1 (3.27e-4) −	3.6493e-2 (2.39e-3) −	**1.3851e-2 (1.96e-3)**
	5000	6.6047e-1 (6.62e-4) −	2.6095e+0 (3.47e-3)−	2.0765e+0 (7.27e-2) −	7.1528e-1 (7.56e-2) −	4.4340e-2 (1.40e-3) −	**2.8029e-2 (2.89e-2)**
SMOP4	1000	1.1345e-1 (6.81e-2) −	1.1039e+0 (5.93e-3) −	9.0041e-1 (9.57e-2) −	2.2769e-1 (5.64e-2) −	**4.6729e-3 (2.13e-4) +**	5.0270e-3 (4.41e-4)
	2000	8.7456e-2 (7.46e-2) −	1.1070e+0 (3.77e-3) −	9.7158e-1 (9.98e-2) −	2.8920e-1 (7.20e-2) −	**4.8226e-3 (2.66e-4) ≈**	4.9715e-3 (3.94e-4)
	5000	3.2693e-2 (5.25e-2) ≈	1.1117e+0 (2.50e-3)−	9.7264e-1 (1.25e-1) −	3.0110e-1 (6.00e-2) −	**4.8123e-3 (3.45e-4) ≈**	4.8218e-3 (2.30e-4)
SMOP5	1000	3.4907e-1 (1.41e-3) −	1.0755e+0 (9.10e-3) −	6.3908e-1 (5.33e-2) −	3.4949e-1 (3.63e-4) −	**5.8762e-3 (2.84e-4) +**	7.0848e-3 (4.88e-4)
	2000	3.4819e-1 (9.51e-4) −	1.0913e+0 (1.09e-2) −	6.4593e-1 (4.84e-2) −	3.4945e-1 (3.63e-4) −	**6.0142e-3 (3.14e-4) +**	7.2627e-3 (4.75e-4)
	5000	3.4799e-1 (7.04e-4) −	1.1026e+0 (5.16e-3)−	6.5622e-1 (4.44e-2) −	3.4954e-1 (3.36e-4) −	**6.0042e-3 (2.75e-4) +**	7.2858e-3 (3.85e-4)
SMOP6	1000	2.2532e-2 (4.50e-3) −	4.9404e-1 (3.64e-3) −	2.9715e-1 (2.77e-2) −	1.4484e-2 (6.15e-4) −	**7.2663e-3 (3.98e-4) ≈**	7.5967e-3 (6.98e-4)
	2000	1.7571e-2 (2.66e-3) −	4.9955e-1 (3.82e-3) −	3.0501e-1 (3.27e-2) −	1.4767e-2 (7.71e-4) −	**7.3921e-3 (3.17e-4) +**	8.2052e-3 (6.46e-4)
	5000	1.3859e-2 (2.88e-3) −	5.0365e-1 (1.97e-3)−	3.3183e-1 (3.59e-2) −	1.4685e-2 (8.57e-4) −	7.9026e-3 (1.94e-4) ≈	**7.8546e-3 (4.24e-4)**
SMOP7	1000	1.5267e-1 (5.43e-3) −	2.9411e+0 (2.82e-2) −	1.5156e+0 (5.06e-2) −	1.5722e-1 (3.27e-4) −	8.5253e-2 (8.40e-3) −	**8.4173e-2 (7.98e-2)**
	2000	1.4379e-1 (1.79e-2) −	3.0169e+0 (2.07e-2) −	1.6013e+0 (5.98e-2) −	1.5733e-1 (3.10e-4) −	1.0506e-1 (5.59e-3) −	**1.0238e-1 (8.59e-2)**
	5000	1.3188e-1 (2.78e-2) −	3.0650e+0 (1.67e-2)−	1.6656e+0 (8.92e-2) −	1.5751e-1 (2.21e-4) −	1.2721e-1 (4.04e-3) −	**9.4797e-2 (7.58e-2)**
SMOP8	1000	6.0281e-1 (1.58e-1) −	3.7067e+0 (7.75e-3) −	3.4557e+0 (6.64e-2) −	7.2638e-1 (1.58e-1) −	2.4367e-1 (1.55e-2) ≈	**2.3849e-1 (4.80e-2)**
	2000	5.4849e-1 (4.38e-2) −	3.7247e+0 (3.83e-3) −	3.3090e+0 (5.68e-2) −	6.4276e-1 (6.03e-2) −	2.8719e-1 (1.19e-2) −	**2.5695e-1 (4.96e-2)**
	5000	5.2674e-1 (4.27e-2) −	3.7371e+0 (2.95e-3)−	3.3183e+0 (4.93e-2) −	5.8223e-1 (1.91e-2) −	3.2861e-1 (8.48e-3) −	**2.5661e-1 (4.70e-2)**
+/−/≈		0/23/1	0/24/0	0/24/0	0/24/0	6/13/5	

is set to 0.5, and ordered grouping is adopted. For LMEA, the number of selected solutions for decision variables clustering is set to 2, the number of perturbations on each solution is set to 4, and the number of selected solutions for decision variable interaction analysis is set to 5. For CCGDE3, the number of species is set to 2. For GLMO, the number of groups is set to 4, and NSGA-II is employed as the basic optimizer. In LMEA, GLMO, SparseEA, and MOEA/PSL, the simulated binary crossover and polynomial mutation are employed to produce offsprings, the probability of crossover and mutation are set to 1 and $1/D$, where D is the number of decision variables, and the distribution index of both crossover and mutation is set to 20. In WOF-SMPSO, the PSO operator and polynomial mutation are employed. In CCGDE3, DE operator and polynomial mutation are used for offspring generation, where the control parameters are set to $CR=1$, $F=0.5$, $pm=1/D$, and $\eta=20$.

Problems: For SMOP1–SMOP8, the number of objectives is set to 2, the number of decision variables is set to 1000, 2000, and 5000, and the sparsity of Pareto optimal solutions is set to 0.1, the value of sparsity denotes the ratio of nonzero elements in the decision variables.

Stopping Criteria and Population Size: The maximum number of function evaluations is adopted as the stopping criteria, which is set to $100 \times D$ for each MOEA. The population size is set to 100.

Performance Metrics: The inverted generational distance (IGD) is adopted to measure each obtained solutions set, and roughly 10000 reference points on each Pareto front are sampled to calculate the IGD value. We perform 30 independent runs for each MOEA on each problem, and the Wilcoxon rank sum test with a significance level of 0.05 is adopted to perform statistical analysis. All the experiments are conducted on the PlatEMO[1] [21].

3.2 Experimental Results

Table 1 presents the IGD values obtained by the six compared algorithms on SMOP1–SMOP8 with 1000, 2000 and 5000 decision variables. Obviously, the two specially tailored MOEAs for sparse LSMOPs present the overall superiority compared with the other four state-of-the-art algorithms. MOEA/PSL performs the best on 15 out of 24 problems, while SparseEA performs the best on 9 out of 24 problem instances, which validates the effectiveness and superiority of SparseEA and MOEA/PSL on solving sparse LSMOPs.

Figure 1 shows the Pareto optimal fronts with median IGD values obtained by the six compared algorithms on SMOP3, SMOP5, and SMOP8 with 5000 decision variables among 30 runs. We see that MOEA/PSL and SparseEA perform similarly on SMOP3 and SMOP5, and MOEA/PSL obtains slightly better result than SparseEA on SMOP8. While most obviously, we see that both MOEA/PSL and SparseEA significantly outperform WOF-SMPSO, LMEA, CCGDE3, and GLMO on those three problems.

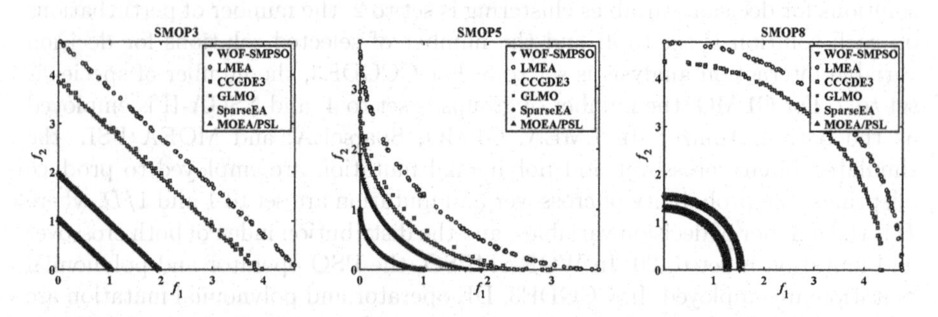

Fig. 1. Pareto optimal fronts with median IGD values obtained by WOF-SMPSO, LMEA, CCGDE3, GLMO, SparseEA, and MOEA/PSL on SMOP3, SMOP5, and SMOP8 with 5000 decision variables.

Due to the page limitation, in Fig. 2, we show only the Pareto optimal sets with median IGD values obtained by the six algorithms on SMOP3 with 5000 decision variables, with the purpose of explaining the superiority of SparseEA and MOEA/PSL over the other four algorithms. We observe that most decision

[1] The source code of algorithms and problems used in this paper can be found from the attached url https://github.com/BIMK/PlatEMO.

variables of the solutions obtained by SparseEA and MOEA/PSL are zero, which can to extend verify the effectiveness of the initialization strategy and genetic operators of SparseEA, and that MOEA/PSL can truly approximate the Pareto optimal subspace. In this way, both algorithms can put more efforts to search in the effective dimensions. However, the other four algorithms without considering the sparse nature of Pareto optimal solutions can not achieve this, especially, the decision variables of the solutions obtained by LMEA and CCGDE3 cover the whole search space, as a result, LMEA and CCGDE3 inevitably waste a lot of computation on low effective dimensions, this also accounts why LMEA and CCGDE3 achieve the worst performance as shown in Table 1 and Fig. 1.

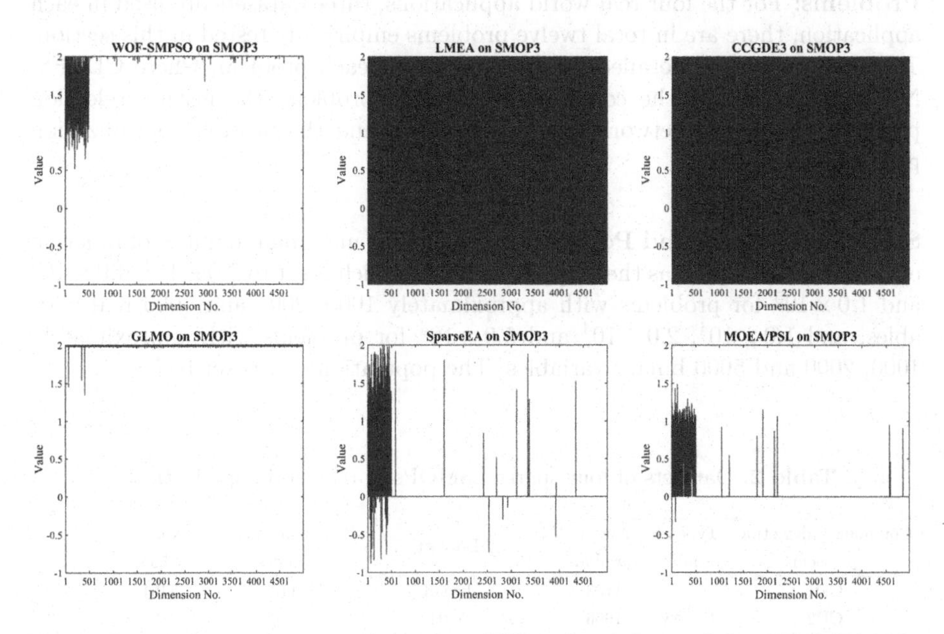

Fig. 2. Pareto optimal sets with median IGD values obtained by WOF-SMPSO, LMEA, CCGDE3, GLMO, SparseEA, and MOEA/PSL on SMOP3 with 5000 decision variables.

4 Experimental Studies on Real-World Applications

To further compare the differences of existing representative large-scale MOEAs on LSMOPs with sparse Pareto optimal solutions, four real-world applications, namely, the community detection problem [24], the feature selection problem [33], the neural network training problem [25], and the portfolio optimization problem [18] are selected to conduct a deeper experimental study.

4.1 Experimental Settings

Algorithms: The selected algorithms and the parameter setting of each algorithm are kept the same as Section. 3 except for that the single-point crossover and bitwise mutation are employed in SparseEA and MOEA/PSL for solving problems with binary variables. For the algorithms that can not be employed to combinatorial MOPs directly, when optimizing binary variables, each algorithm optimizes the same number of real variables within [0,1] and rounds the variables before calculating objective values.

Problems: For the four real-world applications, three datasets are used in each application, there are in total twelve problems empirically tested in this section. Table 2 presents the detailed information about each problem, where CD, FS, NN, and PO denote the community detection problem, the feature selection problem, the neural network training problem, and the portfolio optimization problem, respectively.

Stopping Criteria and Population Size: The maximum number of function evaluations is adopted as the stopping criteria, which is set to 2.0×10^4, 4.0×10^4 and 1.0×10^5 for problems with approximately 1000, 2000 and 5000 real variables, and 1.0×10^4, 2.0×10^4 and 5.0×10^4 for problems with approximately 1000, 2000 and 5000 binary variables. The population size is set to 50.

Table 2. Datasets of four sparse LSMOPs in real-world applications.

Community detection problem	Type of variables	No. of variables	Dataset	No. of nodes	No. of edges	
CD1		1133	Email	1133	5451	
CD2	Binary	1966	Y2H	1966	/	
CD3		4039	Facebook	4039	88234	
Feature selection problem	**Type of variables**	**No. of variables**	**Dataset**	**No. of samples**	**No. of features**	**No. of classes**
FS1		800	Gse72526	61	800	4
FS2	Binary	2000	Coloncancer	62	2000	2
FS3		4434	Gimdata	50	4434	4
Neural network training problem	**Type of variables**	**No. of variables**	**Dataset**	**No. of samples**	**No. of features**	**No. of classes**
NN1		1241	Connectionist bench sonar	208	60	2
NN2	Real	2041	Hill valley	606	100	2
NN3		6241	LSVT vocie rehabilitation	126	310	2
Portfolio optimization problem	**Type of variables**	**No. of variables**	**Dataset**	**No. of instruments**	**Length of each instrument**	
PO1		500	EURCHF	500	50	
PO2	Real	1000	EURCHF	1000	50	
PO3		5000	EURCHF	5000	50	

Performance Metrics: The HV indicator with a reference point (1, 1) is used to measure the results on real-world applications, and the Wilcoxon rank sum test with a significance level of 0.05 is also adopted to perform statistical analysis.

Table 3. HV values obtained by WOF-SMPSO, LMEA, CCGDE3, GLMO, SparseEA and MOEA/PSL on four real-world applications, where the best result in each row is highlighted. '+', '−' and '≈' indicate that the result is significantly better, significantly worse, and statistically similar to that obtained by MOEA/PSL, respectively.

Problem	*Dec*	WOF-SMPSO	LMEA	CCGDE3	GLMO	SparseEA	MOEA/PSL
	1133	**7.1098e-1 (1.30e-3) +**	3.3671e-1 (5.29e-3)−	4.0625e-1 (5.67e-3) −	5.9311e-1 (3.18e-2) −	5.0102e-1 (4.50e-3) −	7.0587e-1 (1.89e-3)
CD	1966	6.5295e-1 (3.96e-3) −	3.3201e-1 (5.45e-3)−	4.0928e-1 (5.74e-3) −	6.4662e-1 (2.75e-2) ≈	5.5418e-1 (4.68e-3) −	**6.5870e-1 (8.70e-3)**
	4039	**7.0511e-1 (1.58e-3) +**	3.2905e-1 (6.01e-3)−	3.8372e-1 (6.55e-3) −	6.2954e-1 (1.90e-2) −	5.4583e-1 (3.35e-3) −	6.9912e-1 (1.19e-3)
	800	9.9640e-1 (1.38e-2) ≈	2.4833e-1 (2.68 e-3)−	2.5212e-1 (3.27e-3) −	9.8617e-1 (2.86e-2) −	**9.9914e-1 (1.13e-6) +**	9.9903e-1 (7.39e-5)
FS	2000	9.8960e-1 (2.61e-2) ≈	4.3046e-1 (3.10e-2)−	4.9063e-1 (3.02e-2) −	9.7700e-1 (3.53e-2) −	**9.9983e-1 (4.52e-6) +**	9.9824e-1 (6.31e-3)
	4434	9.6333e-1 (5.12e-2) ≈	2.7421e-1 (1.40e-3)−	2.8083e-1 (6.27e-3) −	9.1485e-1 (7.13e-2) −	**9.9975e-1 (8.69e-6) +**	9.9936e-1 (1.25e-3)
	1241	3.1473e-1 (1.30e-2) −	3.1123e-1 (1.13e-2)−	8.0550e-2 (9.72e-4) −	3.0896e-1 (1.30e-2) −	8.6351e-1 (1.24e-2) ≈	**8.6562e-1 (8.99e-3)**
NN	2041	2.5261e-1 (1.12e-2) −	2.4372e-1 (7.96e-3)−	6.4116e-2 (3.73e-3) −	2.5511e-1 (1.48e-2) −	7.4686e-1 (1.13e-1) −	**8.5969e-1 (7.77e-2)**
	6241	3.3086e-1 (1.88e-2) −	3.2516e-1 (1.30e-2)−	8.9300e-2 (7.59e-4) −	3.3320e-1 (1.63e-2) −	9.2454e-1 (1.56e-2) ≈	**9.2663e-1 (1.69e-2)**
	500	9.2891e-2 (3.71e-4) −	9.1397e-2 (4.86e-5)−	9.2134e-2 (9.64e-5) −	9.4560e-2 (3.45e-4) −	**1.2375e-1 (1.25e-3) ≈**	1.2364e-1 (2.68e-4)
PO	1000	9.2459e-2 (5.97e-4) −	9.1188e-2 (9.79e-5)−	9.1614e-2 (6.47e-5) −	9.3750e-2 (5.04e-4) −	**1.2370e-1 (1.85e-3) ≈**	1.2297e-1 (2.85e-4)
	5000	9.1914e-2 (2.44e-4) −	9.0962e-2 (1.16e-5)−	9.1134e-2 (1.80e-5) −	9.1826e-2 (8.95e-5) −	1.2385e-1 (1.54e-3) ≈	**1.2423e-1 (3.11e-4)**
+/−/≈		2/7/3	0/12/0	0/12/0	0/11/1	3/4/5	

4.2 Experimental Results

Table 3 lists the HV values obtained by the six compared algorithms on the four real-world applications. We see that SpearseEA and MOEA/PSL have shown their superiority over the other four algorithms. MOEA/PSL and SparseEA perform the best on 5 out of 12 problems, respectively, while WOF-SMPSO performs the best on 2 out 12 problems.

Figure 3 presents the Pareto optimal fronts with median HV values obtained by the six MOEAs on FS3, NN3, and PO3 with approximately 5000 decision variables. We see that for the feature selection problem, the results obtained by MOEA/PSL and SparseEA are quite similar and obviously better than the other four algorithms, even though WOF-SMPSO and GLMO converge well, their diversity can be further improved, LMEA and CCGDE3 perform pretty bad on this problem. For the neural network training problem, the solutions obtained by MOEA/PSL and SparseEA converge well with good diversity, WOF-SMPSO, LMEA and GLMO can not converge, while CCGDE3 can only find a single solution. For the portfolio optimization problem, MOEA/PSL performs slightly better than SparseEA, while the other four algorithms can only converge to a crowded corner.

Figure 4 shows the runtimes consumed by the six algorithms on the four real-world applications with approximately 1000 to 5000 decision variables. We can find that for the feature selection problem, the neural network training problem, and the portfolio optimization problem, the runtimes consumed by SparseEA and MOEA/PSL is not significantly higher than the other four algorithms. For the community detection problem, even though the runtimes of SparseEA and MOEA/PSL is a bit higher than WOF-SMPSO, CCGDE3, and GLMO, it is still acceptable. In short, SparseEA and MOEA/PSL perform well with competitive efficiency.

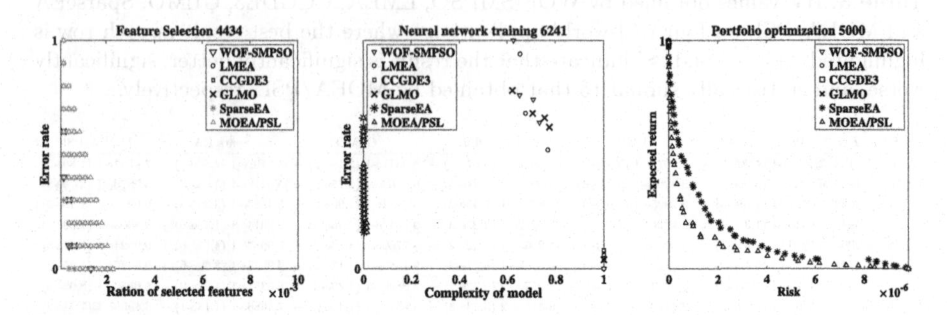

Fig. 3. Pareto optimal fronts with median HV values obtained by WOF-SMPSO, LMEA, CCGDE3, GLMO, SparseEA, and MOEA/PSL on FS3, NN3, and PO3 with approximately 5000 decision variables.

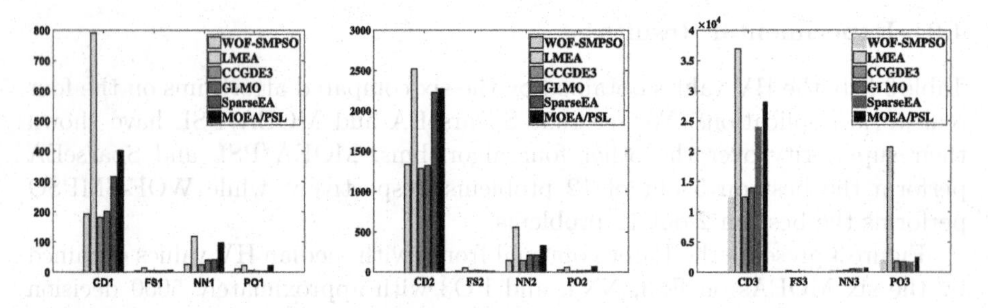

Fig. 4. Runtimes (in second) of WOF-SMPSO, LMEA, CCGDE3, GLMO, SparseEA, and MOEA/PSL on CD, FS, NN, and PO with approximately 1000, 2000, and 5000 decision variables.

5 Conclusions and Future Work

In this paper, we have selected four state-of-the-art large-scale MOEAs, namely, WOF-SMPSO, LMEA, CCGDE3, and GLMO, which belongs to different categories to have a comprehensive comparison with two recently proposed MOEAs, called SparseEA and MOEA/PSL. Experimental studies have been executed on eight benchmark problems and four real-world applications with sparse Pareto optimal solutions, each one has 1000 to 5000 decision variables. Experimental results on those in total 36 problems have demonstrated the effectiveness and superiority of SparseEA and MOEA/PSL on solving large-scale sparse MOPs.

The study on large-scale sparse MOEAs is still in its infancy, and it deserves more attention in the community of evolutionary computation. On the one hand, specific environmental selection strategies can be adopted for solving large-scale sparse MOPs with many objectives. On the other hand, the work can be extended to large-scale sparse constrained MOPs (i.e., time-varying ratio error estimation [22]) and large-scale sparse multimodal MOPs (i.e., multimodal multiobjective optimization in feature selection[32]).

References

1. Ishibuchi, H., Akedo, N., Nojima, Y.: Behavior of multiobjective evolutionary algorithms on many-objective knapsack problems. IEEE Trans. Evol. Comput. **19**(2), 264–283 (2015)
2. Cheng, R., Jin, Y., Olhofer, M., Sendhoff, B.: A reference vector guided evolutionary algorithm for many-objective optimization. IEEE Trans. Evol. Comput. **20**(5), 773–791 (2016)
3. Ponsich, A., Jaimes, A.L., Coello Coello, C.A.: A survey on multiobjective evolutionary algorithms for the solution of the portfolio optimization problem and other finance and economics applications. IEEE Trans. Evol. Comput. **17**(3), 321–344 (2013)
4. Cheng, R., Jin, Y., Olhofer, M., Sendhoff, B.: Test problems for large-scale multiobjective and many-objective optimization. IEEE Trans. Cybern. **47**(12), 4108–4121 (2017)
5. Deb, K., Pratap, A., Agarwal, S., Meyarivan, T.: A fast and elitist multiobjective genetic algorithm: NSGA-II. IEEE Trans. Evol. Comput. **6**(2), 182–197 (2002)
6. Zhang, Q., Li, H.: MOEA/D: a multiobjective evolutionary algorithm based on decomposition. IEEE Trans. Evol. Comput. **11**(6), 712–731 (2007)
7. Tian, Y., Cheng, R., Zhang, X., Jin, Y.: An indicator-based multiobjective evolutionary algorithm with reference point adaptation for better versatility. IEEE Trans. Evol. Comput. **22**(4), 609–622 (2018)
8. Wang, H., Jiao, L., Shang, R., He, S., Liu, F.: A memetic optimization strategy based on dimension reduction in decision space. Evol. Comput. **23**(1), 69–100 (2015)
9. Zille, H., Ishibuchi, H., Mostaghim, S., Nojima, Y.: A framework for large-scale multiobjective optimization based on problem transformation. IEEE Trans. Evol. Comput. **22**(2), 260–275 (2018)
10. He, C., et al.: Accelerating large-scale multiobjective optimization via problem reformulation. IEEE Trans. Evol. Comput. **23**(6), 949–961 (2019)
11. Ma, X., Liu, F., Qi, Y., et al.: A multiobjective evolutionary algorithm based on decision variable analyses for multiobjective optimization problems with large-scale variables. IEEE Trans. Evol. Comput. **20**(2), 275–298 (2016)
12. Zhang, X., Tian, Y., Cheng, R., Jin, Y.: A decision variable clustering-based evolutionary algorithm for large-scale many-objective optimization. IEEE Trans. Evol. Comput. **22**(99), 97–112 (2018)
13. He, C., Cheng, R., Yazdani, D.: Adaptive offspring generation for evolutionary large-scale multiobjective optimization. IEEE Trans. Syst. Man, Cybern. Syst. **PP**, 1–13 (2020)

14. Aelst, S., Wang, X., Zamar, R., Zhu, R.: Linear grouping using orthogonal regression. Comput. Stat. Data Anal. **50**(5), 1287–1312 (2006)
15. Omidvar, M.N., Yang, M., Mei, Y., Yao, X.: DG2: a faster and more accurate differential grouping for large-scale black-box optimization. IEEE Trans. Evol. Comput. **21**(6), 929–942 (2017)
16. Xue, B., Zhang, M., Browne, W.N.: Particle swarm optimization for feature selection in classification: a multi-objective approach. IEEE Trans. Cybern. **43**(6), 1656–1671 (2013)
17. Li, H., Zhang, Q., Deng, J., Xu, Z.: A preferenced-based multiobjective evolutionary approach for sparse optimization. IEEE Trans. Neural Netw. Learn. Syst. **29**(5), 1716–1731 (2018)
18. Lwin, K., Qu, R., Kendall, G.: A learning-guided multi-objective evolutionary algorithm for constrained portfolio optimization. Appl. Soft Comput. **24**, 757–772 (2014)
19. Tian, Y., Zhang, X., Wang, C., Jin, Y.: An evolutionary algorithm for large-scale sparse multi-objective optimization problems. IEEE Trans. Evol. Comput. **24**(2), 380–393 (2020)
20. Tian, Y., Lu, C., Zhang, X., et al.: Solving large-scale multi-objective optimization problems with sparse optimal solutions via unsupervised neural networks. IEEE Trans. Cybern. PP, 1–14 (2020)
21. Tian, Y., Cheng, R., Zhang, X., Jin, Y.: PlatEMO: a MATLAB platform for evolutionary multi-objective optimization. IEEE Comput. Intell. Mag. **12**(4), 73–87 (2017)
22. He, C., Cheng, R., Zhang, C., Tian, Y., Chen, Q., Yao, X.: Evolutionary large-scale multiobjective optimization for ratio error estimation of voltage transformers. IEEE Trans. Evol. Comput. **24**, 868–881 (2020)
23. Tian, Y., Zheng, X., Zhang, X., Jin, Y.: Efficient large-scale multi-objective optimization based on a competitive swarm optimizer. IEEE Trans. Cybern. **50**, 3696–3708 (2019)
24. Zhang, L., Pan, H., Su, Y., Zhang, X., Niu, Y.: A mixed representation-based multiobjective evolutionary algorithm for overlapping community detection. IEEE Trans. Evol. Comput. **47**(9), 2703–2716 (2017)
25. Jin, Y., Sendhoff, B.: Pareto-based multiobjective machine learning: an overview and case studies. IEEE Trans. Syst. Man Cybern. Part C (Appl. Rev.) **38**(3), 397–415 (2008)
26. Emmerich, Michael., Beume, Nicola, Naujoks, Boris: An EMO algorithm using the hypervolume measure as selection criterion. In: Coello Coello, Carlos A., Hernández Aguirre, Arturo, Zitzler, Eckart (eds.) EMO 2005. LNCS, vol. 3410, pp. 62–76. Springer, Heidelberg (2005). https://doi.org/10.1007/978-3-540-31880-4_5
27. Antonio, L.M., Coello Coello, C.A.: Use of cooperative coevolution for solving large scale multiobjective optimization problems. In: IEEE Congress on Evolutionary Computation (CEC), pp. 2758–2765 (2013)
28. Zille, Heiner, Mostaghim, Sanaz: Linear search mechanism for multi- and many-objective optimisation. In: Deb, K., et al. (eds.) EMO 2019. LNCS, vol. 11411, pp. 399–410. Springer, Cham (2019). https://doi.org/10.1007/978-3-030-12598-1_32
29. Zille, H., Ishibuchi, H., Mostaghim, S., Nojima, Y.: Mutation operators based on variable grouping for multi-objective large-scale optimization. In: IEEE Symposium Series on Computational Intelligence (SSCI), Greece (2016)
30. Omidvar, M.N., Li, X., Yang, Z., Yao, X.: Cooperative co-evolution for large scale optimization through more frequent random grouping. In: IEEE Congress on Evolutionary Computation (CEC), pp. 1–8 (2010)

31. Chen, Wenxiang., Weise, Thomas., Yang, Zhenyu, Tang, Ke: Large-scale global optimization using cooperative coevolution with variable interaction learning. In: Schaefer, Robert, Cotta, Carlos, Kołodziej, Joanna, Rudolph, Günter (eds.) PPSN 2010. LNCS, vol. 6239, pp. 300–309. Springer, Heidelberg (2010). https://doi.org/10.1007/978-3-642-15871-1_31
32. Yue, C., Liang, J., Qu, B., Yu, K., Song, H.: Multimodal multiobjective optimization in feature selection. In: IEEE Congress on Evolutionary Computation (CEC), pp. 302–309 (2019)
33. Hamdani, Tarek M., Won, Jin-Myung., Alimi, Adel M., Karray, Fakhri: Multiobjective feature selection with NSGA II. In: Beliczynski, Bartlomiej, Dzielinski, Andrzej, Iwanowski, Marcin, Ribeiro, Bernardete (eds.) ICANNGA 2007. LNCS, vol. 4431, pp. 240–247. Springer, Heidelberg (2007). https://doi.org/10.1007/978-3-540-71618-1_27

EMO and Machine Learning

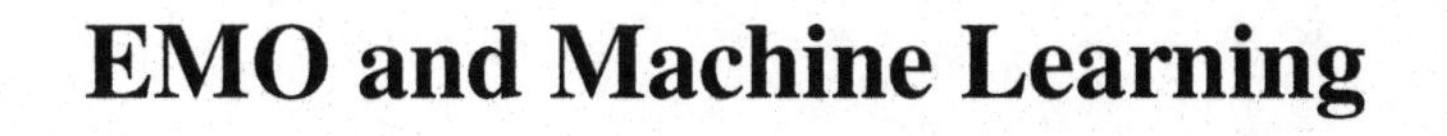

Discounted Sampling Policy Gradient for Robot Multi-objective Visual Control

Meng Xu[1(✉)], Qingfu Zhang[1,2], and Jianping Wang[1,2]

[1] Department of Computer Science, City University of Hong Kong, Hong Kong, China
mxu247-c@my.cityu.edu.hk, {qingfu.zhang,jianwang}@cityu.edu.hk
[2] Shenzhen Research Institute, City University of Hong Kong, Shenzhen, China

Abstract. Robot visual control often involves multiple objectives such as achieving high efficiency, maintaining stability, and avoiding failure. This paper proposes a novel Vision-Based Control method (VBC) with the Discounted Sampling Policy Gradient (DSPG) and Cosine Annealing (CA) to achieve excellent multi-objective control performance. In our proposed visual control framework, a DSPG learning agent is employed to learn a policy estimating continuous kinematics for VBC. The deep policy maps the visual observation to a specific action in an end-to-end manner. The DSPG agent finally can update the policy to obtain the optimal or near-optimal solution using shaped rewards from the environment. The proposed VBC-DSPG model is optimized using a heuristic method. Experimental results demonstrate that the proposed method performs very well compared with some classical competitors in the multi-objective visual control scenario.

Keywords: Multi-objective visual control · Kinematics · Discounted sampling policy gradient · Cosine annealing

1 Introduction

Vision-based control is a mainstream approach to drive the mobile robot to achieve multi-objective control by visual feedback, such as navigation and path planning [1,2]. The error of the feature points on the image plane is used as the feedback to drive the mobile robot towards the targets. Feature error reflects the difference between the current image features and the desired image features. One practical advantage of the VBC method is that there is no need for a

This work was supported by National Key Research and Development Project, Ministry of Science and Technology, China (Grant No. 2018AAA0101301), National Natural Science Foundation of China (Grant No. 61876163), in part by Science and Technology Innovation Committee Foundation of Shenzhen (Grant No. JCYJ20200109143223052) and Hong Kong Research Grant Council (GRF 11200220).

H. Ishibuchi et al. (Eds.): EMO 2021, LNCS 12654, pp. 441–452, 2021.
https://doi.org/10.1007/978-3-030-72062-9_35

geometric model of the target. A key issue is how to achieve an efficient mapping from the 2D feature errors to the velocities in the 3D Cartesian coordinate system. The kinematics bridges two [3]. The multi-objective visual control requires the motion process to consider different objectives such as stability, rapidity, and keeping the target. To achieve the trade-off between different objectives, a feasible way is to use modular controllers [4]. Another way is to regard performance efficiency as the primary objective and others as constraints [5]. The latter can simplify the multi-objective control task into a single objective task with constraints. Previous methods heavily rely on dynamic modeling, and it is difficult to make these dynamic models scalable and generalizable.

The Jacobian matrix transforms the multi-objective visual control problem into a matrix estimation and optimization problem [6]. Three conventional methods are used to approximate the Jacobin matrix for multi-objective visual control. The first one is to employ the current matrix; one is the desired matrix and the other is the average matrix [7]. The universal Jacobian matrix involves a linear combination of the current matrix and the desired matrix. A control parameter ranging from 0 to 1 is used to balance the current matrix and the desired one. It is appropriate to assign a smaller value for the parameter to ensure stable convergence when a mobile robot is close to the target. Instead, a larger value achieves faster convergence [8]. Previous work used the proportional-integral-differential (PID) method to estimate the Jacobian matrix but the performance is limited due to the unsuitable parameters [9]. Until now, the estimation of the matrix still needs human experts' experience. Reinforcement Learning (RL) does not require any prior knowledge about the environment and it drives agents to explore the environment to obtain the best solution via trial and error.

RL methods have been demonstrated to have excellent performance in developing more advanced robots, including multi-objective visual control systems [10]. [11] developed a novel RL method to stabilize a biped robot on a rotating platform. The proposed method addresses the overfitting problem with guaranteed model complexity. However, these conventional methods with tabular RL have a bottleneck due to its discrete action space. It is also helpless for a high-dimensional or continuous control task. Advances in deep learning models have made it possible to extract high-level features from sensory data, and it provides an opportunity of scaling to problems with infinite state spaces. Deep Reinforcement Learning (DRL) applies the computational power of deep learning to relaxing the curse of dimensionality for complex tasks [12]. It can learn a direct mapping from the state space to the policy space in an end-to-end way. Carlos Sampedro et al. developed a deep reinforcement learning controller to achieve the mapping from state space to linear velocity commands for multirotor aerial robots [13]. The learning performance of the DRL based controller is significantly impacted by hyperparameters. A heuristic strategy is an approach to Hyperparameters [14].

This paper proposes a reinforcement learning method with a heuristic strategy to obtain the transient matrix for multi-objective VBC systems. DSPG algorithm employs discounted return sampling which provides an opportunity for the control system with continuous spaces. The proposed method reduces the computation of the performance gradient to an expectation that is estimated by the discounted sampling return. It requires only a fixed computing interval instead of a predetermined environmental model. The deep policy network produces control actions via an on-policy learning method. An appropriate action hence is chosen to improve the performance of the conventional multi-objective visual control, in terms of the convergence rate and stability. The hyperparameter for the proposed RL based controller is optimized by a heuristic strategy, which can give better performance to the learning model of mobile robots.

Major contributions of this work are stated as follows:

1) A VBC method with the DSPG algorithm (VBC-DSPG) is proposed for the multi-objective control. The DSPG algorithm is used to select the time-varying Jacobian matrix online.
2) A parameter tuning method with the Cosine Annealing for the VBC-DSPG method is proposed. The learning rate of the RL model is tuned to improve the learning performance of the proposed method.
3) Related experiments are conducted to investigate the potential of the suggested scheme in a renowned robot platform.

This paper is organized as follows. Section 2 gives a brief introduction to the Multi-objective visual control model. Section 3 gives details of the proposed VBC-DSPG method with a Cosine Annealing. Section 4 demonstrates the experiment setup and experimental results. In the last section, the conclusions are presented.

2 Vision-Based Multi-objective Control

2.1 Multi-objective Visual Control Model

In a visual control process, a robot uses an RGB-D vision to extract features from the current image, which is $\mathbf{S}_c = (\mathbf{s}_1^c, \mathbf{s}_2^c, \ldots, \mathbf{s}_N^c)^{\mathrm{T}}$. To use visual control methods with the closed-loop feedback mechanism, we denote the desired features as $\mathbf{S}_* = (\mathbf{s}_1^*, \mathbf{s}_2^*, \ldots, \mathbf{s}_N^*)^{\mathrm{T}}$. Then, the errors $\mathbf{e}(t) = [\mathbf{s}_1^c - \mathbf{s}_1^*, \mathbf{s}_2^c - \mathbf{s}_2^*, \ldots, \mathbf{s}_N^c - \mathbf{s}_N^*]^{\mathrm{T}} \in \Re^{2N\times 1}$ between the current features and the desired features can be calculated. The feature errors $\mathbf{e}(t)$ is a time-varying vector and its change depends on the motion of the robot. The velocities for the robot are derived from the time derivative for the vector $\mathbf{e}(t)$ and it is as follows

$$\dot{\mathrm{e}}(t) = d\mathbf{e}(t)/dt = \left(d\left[\mathbf{s}_1^c - \mathbf{s}_1^*\right]/dt, d\left[\mathbf{s}_2^c - \mathbf{s}_2^*\right]/dt, \ldots, d\left[\mathbf{s}_N^c - \mathbf{s}_N^*\right]/dt\right). \quad (1)$$

Multi-objective visual control requires the movement of a robot to simultaneously satisfy three different objectives, which are less fluctuation, short time, and no loss of target. The Jacobian matrix J transforms the change rate for feature errors into velocities of the robot. The conversion process is as follows

$$\dot{\mathbf{e}}(t) = \mathbf{J}\left[\omega_t, v_t^x, v_t^y\right]^{\mathrm{T}} = \mathbf{J}[\boldsymbol{\omega}, \mathbf{v}]^{\mathrm{T}}, \mathbf{J} \in \Re^{2N*3}, \quad (2)$$

where N is the number of the features. The previous study [4] used $\dot{\mathbf{e}}(t) = -\varphi\mathbf{e}(t)$ to ensure an exponential decoupled decrease for the error vector $\mathbf{e}(t)$. Equation (3) gives the visual control law using the universal Jacobian matrix.

$$\begin{matrix} [\omega_t^r, v_t^x, v_t^y]^{\mathrm{T}} = -\varphi\hat{\mathbf{J}}^{+}\mathbf{e}(t), \hat{\mathbf{J}}^{+} \in \Re^{3*2N} \\ \mathbf{J} = x\mathbf{J}_1 + (1-x)\mathbf{J}_2 \end{matrix}, \tag{3}$$

where $\mathbf{J}^{+}$ is the pseudo-inverse for the Jacobian matrix.$\varphi$is a constant, which is the servo gain [4]. Based on the Jacobian matrix, a multi-objective visual control model is given by,

$$\begin{matrix} \text{minimize } F(x) = [f_1(x), f_2(x), \ldots, f_m(x)], x \in [0,1] \\ \text{subject to} \begin{cases} \varphi(\mathbf{e}(t), \mathbf{J}(x)) = [\boldsymbol{\omega}, \mathbf{v}]^{\mathrm{T}} \\ \mathbf{J}(x) = x\mathbf{J}_1 + (1-x)\mathbf{J}_2 \end{cases} \end{matrix}, \tag{4}$$

where x is the control parameter and its value ranges from 0 to 1. $f_1(x)$, $f_2(x), \ldots, f_m(x)$ represents the objectives for a perfect control performance, which are defined in a specific task. $\varphi(\bullet)$ is a conversion function mapping the feature errors to velocities of the robot. In this work, these objectives are less fluctuation, short time, and the low probability of target loss. An appropriate value for the control parameter x can satisfy the high-efficiency multi-objective control.

2.2 Computation of the Jacobian Matrix

Figure 1 describes a coordinate transformation model for visual features. In the transformation model, the coordinate for the origin $O_1\,(u_o, v_o)$ on the image plane is mapped to the coordinate for the origin O_2 on the camera plane.

If the coordinate in pixel for the point $\mathbf{a}$ on the camera plane is $\mathbf{a}(u, v)$, the 3D point concerning the point $\mathbf{a}$ on the Cartesian coordinate system is $\mathbf{A}$. Take the point $\mathbf{a}$ and the point $\mathbf{A}$, for example, Eq. (5) gives the transformation from a point on the camera plane to a 3D point in the Cartesian coordinate system. The value of the pixel for the point $\mathbf{a}$ is $\mathbf{d}\,(u', v')$ on the image plane.

$$\begin{cases} X_a = Z_a d_x\,(u - u_o)\,/f^2 \\ Y_a = Z_a d_y\,(v - v_o)\,/f^2 \end{cases}, \tag{5}$$

where the focal length of the RGB-D camera is f. The scaling constants from the image plane to the camera plane in the x-axis and y-axis directions are represented by ϑ_x and ϑ_y. For the $j-th$ feature point (u_j, v_j), the expression for the Jacobian matrix is given by Eq. (6).

$$\mathbf{J}_j = \begin{bmatrix} -\left(f/Z_a\vartheta_x\right) & 0 & v_j\left(\vartheta_y/\vartheta_x\right) \\ 0 & -\left(f/Z_a\vartheta_x\right) & u_j\left(\vartheta_x/\vartheta_y\right) \end{bmatrix}, \tag{6}$$

For the N features, the whole expression for the Jacobian matrix is $\mathbf{J} = [\mathbf{J}_1, \mathbf{J}_2, \ldots, \mathbf{J}_N]^{\mathrm{T}} \in \Re^{2N*3}$.

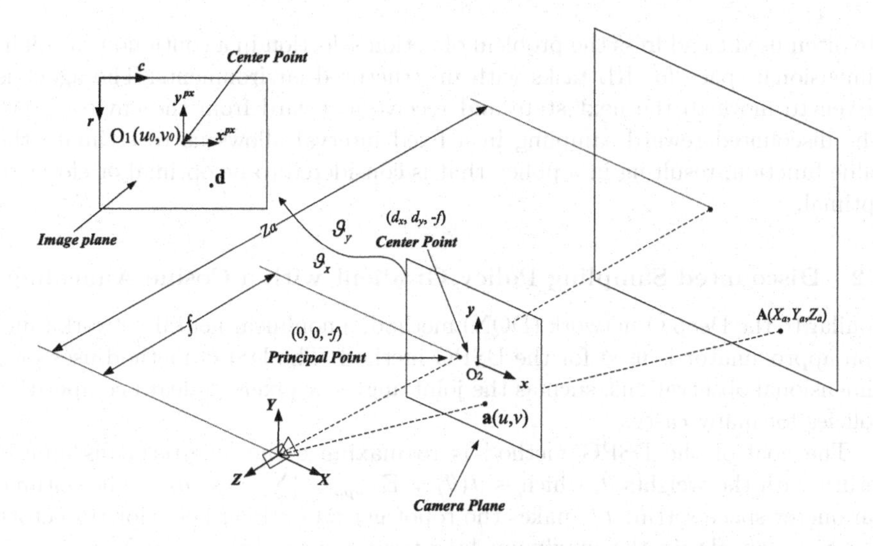

Fig. 1. Coordinate transformation model for visual features.

3 A Discounted Sampling Policy Gradient with a Heuristic Strategy for Multi-objective Visual Control

3.1 Reinforcement Learning

An RL agent selects an action a_t using a policy $\pi(s, a)$ when receives state s_t. Then, the agent gets a reward r_t from the environment and moves to the next state s_{t+1}. This process is repeated until the RL agent reaches the target position. The action policy π is constantly updated with the exploration of the agent. The RL method maximizes the expectation of the cumulative rewards $R(s_t) = r_t + \gamma R(s_{t+1}) = \sum_{i=0}^{\infty} \gamma^i r_{t+i}$ for each state. γ is the discount factor. The value function is used to represent the expectation for the cumulative reward. The value function for the state s_t is given by,

$$\begin{aligned} V(s_t) &= \mathrm{E}\left[r_{t+1} + \gamma r_{t+2} + \gamma^2 r_{t+3} + \ldots \mid s = s_t\right] \\ &= \sum_a \pi(s, a) \sum_{s_{t+1}} P^a_{s_{t+1}} \left(r_t + \gamma V(s_{t+1})\right) \end{aligned}. \tag{7}$$

The Q-learning method uses the action-value function to represent the expectation for the cumulative reward. The action-value function for Q-learning in one-step prediction to the RL problems can be represented by,

$$Q(s, a) = \mathrm{E}\left[\sum_{k=0}^{\infty} \gamma^k r_{t+k+1} \mid s_t = s, a_t = a\right]. \tag{8}$$

For the RL problems, the policy gradient optimizes the policy with the gradient in the policy space, instead of the value function. This method is practical when encountering a task with a stochastic action space [15]. DSPG algorithms are often used to address the problem of action selection in a continuous or high-dimensional space for RL tasks with unstructured environments. The agent is driven to move to the next state and receive a reward from the environment. The discounted reward sampling in a fixed interval allows us to estimate the value function, resulting in a policy that is considered to be optimal or closer to optimal.

3.2 Discounted Sampling Policy Gradient with a Cosine Annealing

Similar to the Deep Q-network (DQN) method, a nonlinear neural network function approximator is used for the DSPG method. The DSPG method uses low-dimensional observations, such as the joint angles or pixels, to learn competitive policies for many cases.

The goal of the DSPG method is to maximize the long-term discounted return with the weights θ, which is $J(\theta) = \mathrm{E}_{\tau \sim p_\theta(\tau)} \left[\sum_t r\left(s_t, a_t\right)\right]$. The optimal parameter space setting θ^* makes the robot get an optimal behavior trajectory τ, which can obtain the maximum long-term return $r(\tau) = \max \sum_t r\left(s_t, a_t\right)$. The better the trajectory means the robot can make a wise decision to select a time-varying matrix $\mathbf{J}$. s_0 is the initial state. The probability of a trajectory with a differentiable distribution function $p_\theta(\tau)$ is shown in Eq. (9).

$$p_\theta(\tau) = p\left(s_0\right) \prod_{t=0}^{T-1} p\left(s_{t+1} \mid s_t, a_t\right) \pi_\theta\left(a_t \mid s_t\right), \tag{9}$$

where $\pi_\theta\left(a_t \mid s_t\right)$ is the parameterized policy and $p\left(s_{t+1} \mid s_t, a_t\right)$ is the state transition probability. The gradient for the objective function is given by,

$$\begin{aligned} \nabla_\theta J(\theta) &= \int p_\theta(\tau) r(\tau) d\tau \\ &= \mathrm{E}_{\tau \sim p_\theta(\tau)}\left[\left(\sum_t \nabla_\theta \log\left(\pi_\theta\left(a_t \mid s_t\right)\right)\right)\left(\sum_t r\left(s_t, a_t\right)\right)\right] \\ &\approx \frac{1}{N} \sum_{j=1}^N \sum_t \left[\nabla_\theta \log\left(\pi_\theta\left(a_{t,j} \mid s_{t,j}\right)\right)\left(\sum_{t'=t} r\left(s_{t',j}, a_{t',j}\right)\right)\right], \end{aligned} \tag{10}$$

where the objective function samples N trajectories. The sampled returns are used to approximate the value function. In an episode, the effect of current reward on current policy decreases with the increase of time step. So, a discounted return sampling method for the expectation is proposed. The objective function with a discounted return sampling is given by,

$$\begin{aligned} \nabla_\theta J(\theta) &= \int p_\theta(\tau) r(\tau) d\tau \\ &\approx \frac{1}{N} \sum_{j=1}^N \sum_t \left[\nabla_\theta \log\left(\pi_\theta\left(a_{t,j} \mid s_{t,j}\right)\right)\left(\sum_{t'=t} \gamma^{t-t'} r\left(s_{t',j}, a_{t',j}\right)\right)\right], \end{aligned} \tag{11}$$

where γ is the discount factor. η_t is the learning rate. The updating law for the policy network is given by,

$$\theta_{t+1} \leftarrow \theta_t + \eta_t \nabla_{\theta_t} J\left(\theta_t\right). \tag{12}$$

Hyperparameter strategies evaluate the performance of each configuration on the learning tasks. This method can complete the training process of automatically parameter testing for a machine learning model. However, the time cost of the evaluation for a configuration is expensive. In the process of Hyperparameter tuning, re-evaluation is necessary, which will be very inefficient for complex models. Some heuristic methods, such as evolutionary algorithms, are difficult to apply in the tuning of Hyperparameter. Previous works often leverage the Simulated Annealing (SA) to optimize the Hyperparameters [16]. As a heuristic method, SA simulates the annealing process in the thermodynamics energetics. A rule of the agent learning process should be satisfied: more current experience should be reserved in the early stage of learning. Conversely, the more prior experience should be preserved. A Cosine Annealing scheme is employed to tune the learning rate for the RL controller. The updating law with the Cosine Annealing is given by,

$$\eta_t \leftarrow \eta_{\min} + \frac{1}{2}\left(1 + \cos\left(\frac{T_t}{T_{\max}}\pi\right)\right)(\eta_{\max} - \eta_{\min}), \tag{13}$$

where $\eta_{\max}$ and $\eta_{\min}$ are the maximum and minimum of learning rate respectively, which are constant. T_t represents the number of current epochs. $T_{\max}$ is the maximum epochs. An exploration noise N sampled from a noise process is added to the policy network to develop an exploration action policy. The action policy with an exploration noise is given by,

$$a_t = \pi_\theta(s_t) + \mathrm{N}. \tag{14}$$

3.3 An Adaptive Multi-objective Visual Control Model with DSPG (VBC-DSPG)

DSPG method employs the policy network to select a time-varying matrix and uses a sample discounted return to estimate the value of the action function. An RGB-D camera receives image features by the contour recognition methods. Then, the current observation s_t is calculated. The observation information consists of a 4-dimensional vector defined by Eq. (15).

$$\mathbf{S}_t = (\mathbf{e}_x, d\mathbf{e}_x/dt, \mathbf{e}_y, d\mathbf{e}_y/dt), \tag{15}$$

where $\mathbf{e}_x, \mathbf{e}_y$ represents the vector for the feature errors of the current positions concerning the desired positions defined in the image plane respectively. The current policy network receives the vector of the current observation as the input and outputs a distribution $\pi_\theta(a_t \mid s_t)$ for the action space. This work defines a mapping function to ensure that the output action is limited between 0 and 1 because the control parameter x ranges from 0 to 1. Equation (16) gives the expression for the mapping function.

$$x = f(a_t) = \frac{1}{\pi}\arctan(a_t) + 0.5. \tag{16}$$

As illustrated in Fig. 2, the DSPG agent consists of a feed-forward neural network with an input layer, an output layer, and hidden layers of 20 and 40

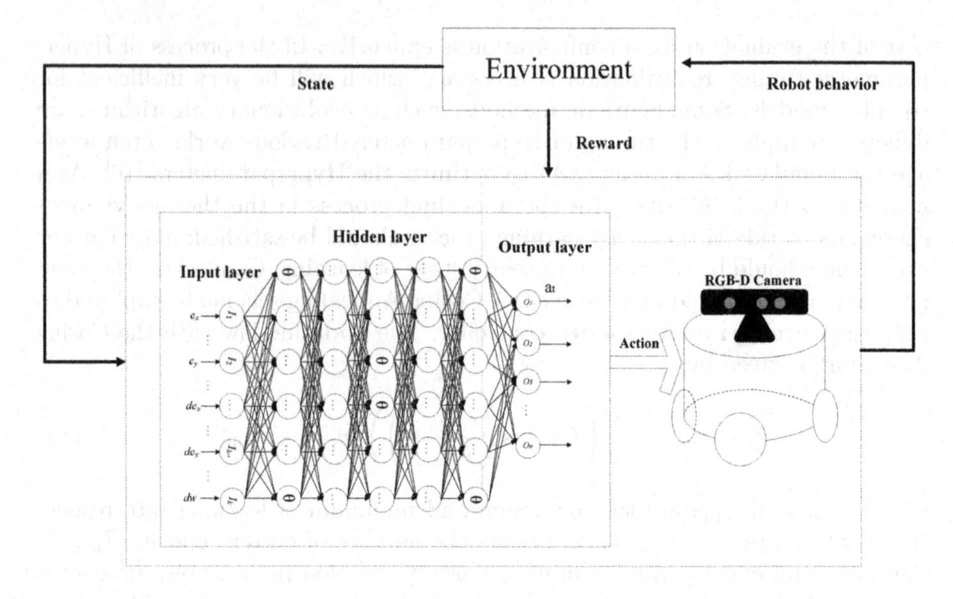

Fig. 2. Structure of the proposed DSPG model.

units. In hidden layers, the Sigmoid function is utilized as the activation function mapping an input to an output.

In the VBC-DSPG method, the shaping reinforcement signal r_t in Eq. (17) is used to update the policy.

$$r_t = -\frac{R}{2}\left(\sum_{i=1}^{N}\sqrt{(e_x^i)^2 + (e_y^i)^2} + \sqrt{(de_x^i/dt)^2 + (de_y^i/dt)^2}\right) / \left(N\sqrt{row^2 + col^2}\right), \quad (17)$$

where the weight and length of the image plane are represented by row and col respectively. The environment gives the reward R to the RL agent depending on the actual situation. The number of feature points is N. The robot receives a higher reward if it is closer to the target.

The main motivation for employing the reinforcement signal is to drive the robot to achieve the target as soon as possible. In Eq. (17), the reward term penalizes the learning agent when the feature errors are big. The weights of the policy network are updated in the way that the robot adaptively selects a time-varying matrix for Multi-objective visual control.

4 Experiments

4.1 Experimental Environment

As shown in Fig. 3, we construct a simulation model with the same dynamics as the real device using commercial robotic software, Webots7.0.3. In the simulation, the robot learns policy using the DSPG algorithm and when the policy converges, this policy is optimized using the Cosine Annealing. Noise and disturbance are added in the simulation environment to create a complex task environment close to the real-world. Then, the experiments on a mobile robot are performed to appraise the practicality of the proposed method. Finally, the proposed method and the conventional pseudo inverse kinematics methods are used to select the matrix. Experimental results are recorded to test whether the proposed method has better performance.

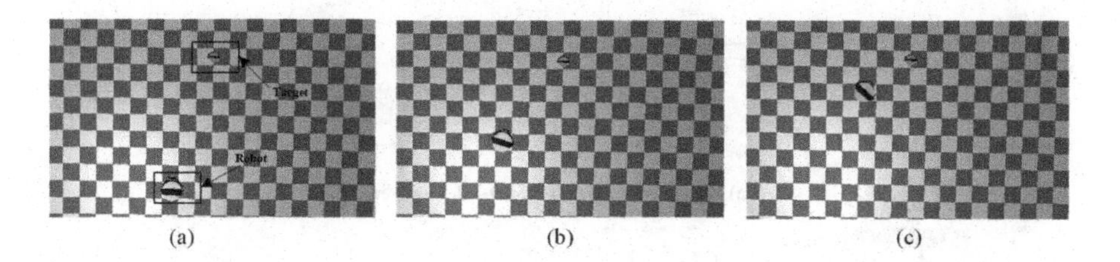

Fig. 3. A visual control task in the simulation environment.

4.2 Training for RL

An episodic RL setting is used to train the RL agent. This control task requires the mobile robot to move from an initial position to a target position, as shown in Fig. 3. The visual control process of a robot should satisfy multiple objectives: less fluctuation, short time, and no loss of target. In each episode, the initial position for the RL agent is randomly selected, and the desired position is (0.463 m, 0.593 m, 37.8°). In each time step, the DSPG agent acts with an added exploration noise. The gain φ° is set to 0.35. The hyperparameters for the DSPG agent are shown below. The initial value for the learning rate η is 0.5. The discount γ is 0.95. The reward R is 10. The maximum epochs $T_{\max}$ is 1000.

The rules for the RL agent are as follows. 1). This episode will be terminated if the robot loses some features. 2). This episode will be terminated if the feature errors remain a certain pixel for a certain time. 3). This episode will be terminated if the robot does not arrive at the desired position after the maximum number of steps. 4). After a new action is selected by the RL agent, the matrix is updated using Eq.(3). 5). The robot returns to the starting position if a new episode starts.

When this episode is terminated negatively, the reward −10 is given to the RL agent. In other episodes, the RL agent is rewarded with the reward function. With this configuration, the agent learns a stable motion behavior after considerable training episodes. Since the RL agent is given a negative reward before reaching the desired position, the reward is a negative value per episode. After training, agents will learn a policy from state space to behavior space.

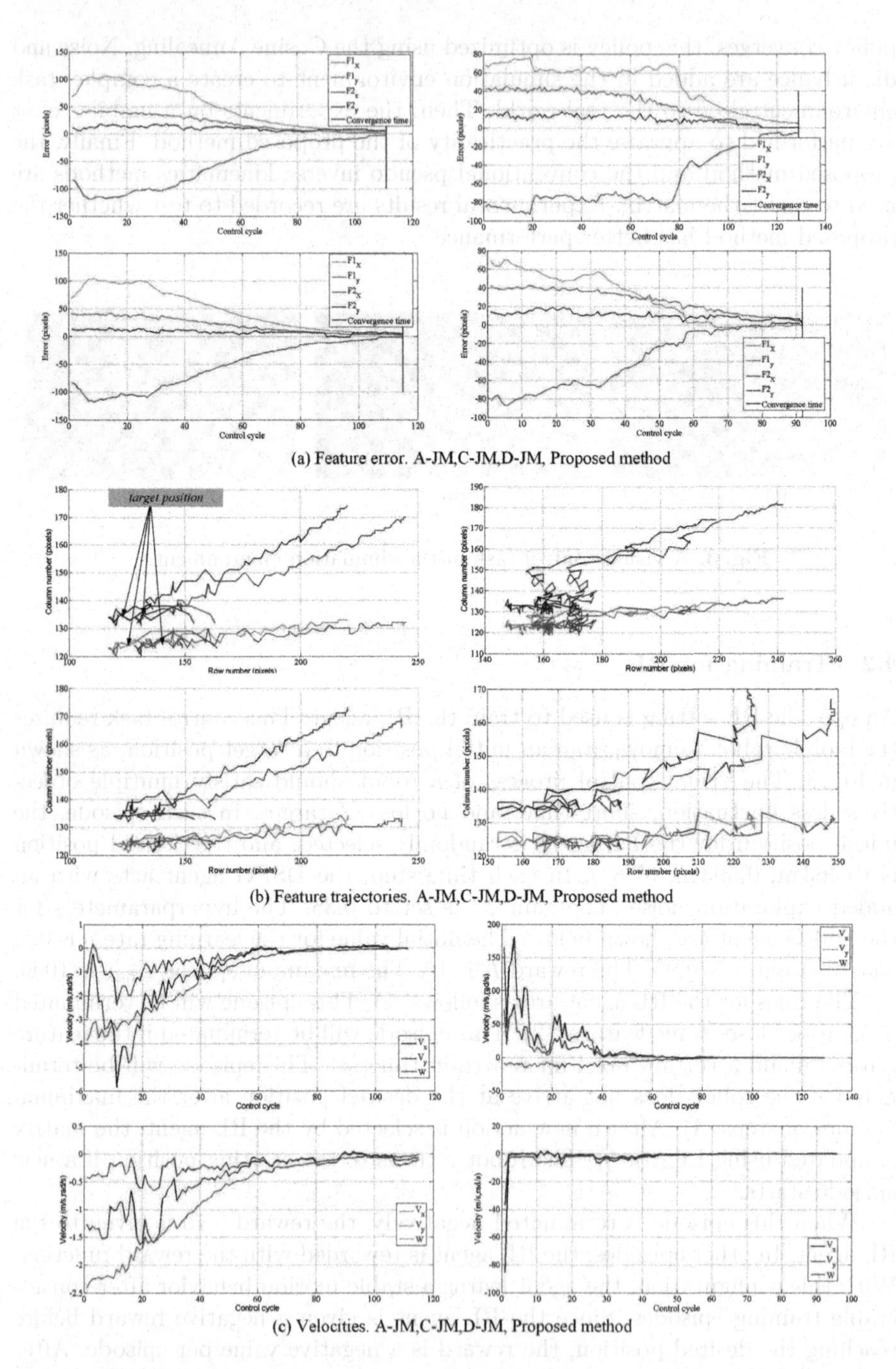

(a) Feature error. A-JM,C-JM,D-JM, Proposed method

(b) Feature trajectories. A-JM,C-JM,D-JM, Proposed method

(c) Velocities. A-JM,C-JM,D-JM, Proposed method

Fig. 4. Experimental results, in terms of the Feature error, Feature trajectories, and Velocities. For each item, from left to right, from top to bottom are A-JM, C-JM, D-JM, and Proposed method.

4.3 Simulation

In the simulation, to verify the effectiveness of the proposed method, the proposed VBC-DSPG method and three competitors were run on the platform. The initial position is (0.0289 m, 0.062 m, 0.07°) and the desired position is (0.463 m, 0.593 m, 37.8°). Three competitors include the Jacobian matrix method using the current one (C-JM) [1], the matrix using the desired one (D-JM) [1], and the matrix using the average one (A-JM) [4]. The four methods were tested 50 times and the average values for these 50 times are shown in Fig. 4.

Experimental results show that the C-JM method converges in around 129 control cycles. The D-JM method converges in around 115 control cycles. The A-JM method converges in around 111 control cycles. The convergence time for the proposed method is around 92 control cycles. Although the three conventional competitors reach the desired position, the convergence rates for the three competitors are slower than the proposed method. Different methods select different Jacobian matrix, which results in different behavior. In Fig. 4, at the beginning of the control process, the proposed method chooses a larger velocity, which drives the robot to reach the target position faster.

However, in the beginning, excessive-velocity may lead to instability, difficulty in control, and even loss of target, resulting in danger, such as C-JM methods. Figure 4 shows the feature trajectories for the four methods. Conventional methods cause one dimension to reach the desired position, while others do not. So, there are many fluctuations in the feature trajectory. The proposed method enables the robot to learn a policy to achieve better multi-objective control performance.

5 Conclusion

This paper developed a new multi-objective visual control method that integrates discounted sampling policy gradient and a heuristic strategy. A DSPG based scheme is proposed to handle the multi-objective visual control challenges of mobile robots by selecting an appropriate Jacobian matrix adaptively. The policy network is obtained by a training process. Then, the network can automatically calculate a time-varying matrix to achieve a robust control performance. The proposed method has been extensively compared with three conventional methods.

References

1. Santamarianavarro, A., Andradecetto, J. : Uncalibrated image-based visual servoing. In: 2013 IEEE International Conference on Robotics and Automation, pp. 5247–5252 (2016)
2. Fraichard, Thierry, Levesy, Valentin: From crowd simulation to robot navigation in crowds. IEEE Robot. Autom. Lett. **5**(2), 729–735 (2020)
3. Chaumette, F., Hutchinson, S.: Visual servo control. I. Basic approaches. IEEE Robot. Autom. Mag. **13**(4), 82–90 (2006)

4. Wang, B., et al.: Parallel structure of six wheel-legged robot model predictive tracking control based on dynamic model. In: 2019 Chinese Automation Congress, pp. 5143–5148 (2019)
5. Sun, W., et al.: Multi-objective control for uncertain nonlinear active suspension systems. Mechatronics **24**(4), 318–327 (2014)
6. Malis, E: Improving vision-based control using efficient second-order minimization techniques. In: 2004 International Conference on Robotics and Automation, pp. 1843–1848 (2004)
7. Marey, M., Chaumette, F.: Analysis of classical and new visual servoing control laws. In: 2008 International Conference on Robotics and Automation, pp. 3244–3249 (2008)
8. Watanabe, K., et al.: Image-based visual PID control of a micro helicopter using a stationary camera. Adv. Robot. **22**(2–3), 381–393 (2008)
9. Kober, J., Bagnell, J.A., Peters, J.: Reinforcement learning in robotics: a survey. Int. J. Robot. Res. **32**(11), 1238–1274 (2013)
10. Zhang, J., et al.: VR-goggles for robots: real-to-sim domain adaptation for visual control. IEEE Robot. Autom. Lett. **4**, 1148–1155 (2019)
11. Xi, A., et al.: Balance control of a biped robot on a rotating platform based on efficient reinforcement learning. IEEE/CAA J. Automatica Sinica **6**(4), 938–951 (2019)
12. Agostinelli, F., et al.: Solving the Rubik's cube with deep reinforcement learning and search. Nat. Mach. Intell. **1**(8), 356–363 (2019)
13. Sampedro, C., et al.: Image-based visual servoing controller for multirotor aerial robots using deep reinforcement learning. In: 2018 International Conference on Intelligent Robots and Systems, pp. 979–986 (2018)
14. Sehgal, A., et al.: Deep reinforcement learning using genetic algorithm for parameter optimization. In: 2019 Third IEEE International Conference on Robotic Computing, pp. 596–601 (2019)
15. Wang, Lixin., Wang, Maolin, Yue, Ting: A fuzzy deterministic policy gradient algorithm for pursuit-evasion differential games. Neurocomputing **362**, 106–117 (2019)
16. Samma, H., et al.: Q-learning-based simulated annealing algorithm for constrained engineering design problems. Neural Comput. Appl. **32**(9), 1–15 (2019). https://doi.org/10.1007/s00521-019-04008-z

Lexicographic Constrained Multicriteria Ordered Clustering

Jean Rosenfeld, Dimitri Van Assche, and Yves De Smet(✉)

Université libre de Bruxelles, Ecole polytechnique de Bruxelles,
50 Avenue Franklin Rooseveld, 1050 Bruxelles, Belgium
yves.de.smet@ulb.ac.be

Abstract. Researchers in the multicriteria analysis community have recently started to study the application of clustering techniques. Different methods have been proposed leading to complete or partial multicriteria partitions. Recently, a method based on the exploitation of the ordinal properties of a valued preference relation has been developed in order to obtain a totally ordered clustering. The underlying idea is to find a partition that minimizes the inconsistencies between the obtained ordered clusters and the input preference matrix. The exact algorithm gives the guarantee to achieve this goal based on a lexicographic order. In this contribution, we propose to further extend their approach to the case where minimum and maximum capacities are imposed on the cluster sizes. This leads to add a constraint satisfaction test (that is based on the levels and the ranks of the graph built during the procedure). The algorithm is evaluated regarding its execution time and illustrated on the academic ranking of world universities.

Keywords: Multiple criteria decision aid · Multicriteria clustering · Capacity constraint

1 Introduction

Over the last decades, multicriteria decision aid has been extensively used to manage real world problems. In this context, three main problem settings are usually considered: choice, ranking and sorting [18]. Let A denote a set of alternatives that are evaluated according to a set of criteria $F = \{f_1, f_2, \ldots, f_q\}$. The *choice* problem consists in the selection of a subset $A' \subseteq A$. This subset should be as small as possible and be composed of the alternatives considered as the best ones. In a procurement context, the selection of the best offers (according to the price, time of delivery, quality of the supplier, etc.) is a typical example. The *ranking* problem consists in determining a complete or partial pre-order over the set of alternatives [16]. The ranking of universities (such as the Academic Ranking of World Universities - also commonly called *Shanghai* ranking according to academic performances), countries (regarding the human development or environmental performances) are common examples of this kind of problems.

H. Ishibuchi et al. (Eds.): EMO 2021, LNCS 12654, pp. 453–464, 2021.
https://doi.org/10.1007/978-3-030-72062-9_36

Sorting consists in assigning each alternative to a predefined ordered category. The set of alternatives is then divided in several groups. For example, patients suffering from cancer can be sorted in four categories (defined by the World Health Organization (WHO)) depending on the disease severity. In this case, categories are supposed to be known a priori. Recently, a new kind of problem has been investigated by the multicriteria decision aid community; multicriteria clustering. Here the aim consists in detecting groups of alternatives as well as potential relations between these groups by explicitly taking into account the multicriteria nature of the problem.

Clustering is a well-known topic in data mining. Many methods have been developed (k-means, hierarchical approaches, etc.) [1,10]. These generally rely on a distance measure in order to form groups as homogeneous as possible and that are as heterogeneous as possible between themselves. By definition the distance used in this context is a symmetric measure. Here relies the main difference with multicriteria decision aid where the comparison between alternatives leads most of the time to non symmetric (preference) relations. This observation was at the origin of the works on multicriteria clustering. This distinctive feature opens new perspectives such as, for instance, the existence of preference relations between the clusters. Finally, let us note that the main difference with sorting methods lies in the fact that categories are not supposed to be known beforehand. On the contrary, the objective is here to identify these groups (therefore, multicriteria clustering could be viewed as a pre-process to sorting methods).

To the best of our knowledge, De Smet and Guzmàn [4] were the first who investigated the multicriteria clustering problem based on binary preferences. Their approach relies on the definition of a particular distance which includes the multicriteria preferential information of the alternatives comparison. The model was limited to the detection of nominal clusters. De Smet and Eppe [3] developed an extension of this first work to build relations between the clusters. De Smet and al. [5] also developed an exact algorithm to detect a totally ordered clustering based on the lexicographic optimization [6] of a penalty vector. More precisely they start by searching totally ordered partitions that minimize the worst penalty. Among these, they refine the search to those which minimize the second highest penalty and so on until an unique partition is obtained. Another method of multicriteria clustering has been proposed by Rocha et al. [15]. It distinguishes first the clustering approach and then a multicriteria application to find relations between the clusters. Finally, Meyer and Olteanu [12] recently summarized and formalized this new topic.

In this contribution, we will focus on totally ordered clustering based on ordinal preferences existing between each pair of alternatives. We will consider the method developed by De Smet et al. [5] and propose an extension in order to manage capacity constraints on the clusters sizes. Let us point out that capacity constraints have already received some attention in sorting contexts [11].

Imposing constraint capacities on clusters sizes in a multicriteria context might have a number of applications. Let us consider the following illustrative examples:

- **Junior football player**: A famous international football team is proceeding to the recruitment of young players. After passing a series of standard physical tests, 100 children between 5 and 7 years old have to be grouped in 5 classes based on their performances. The first class is assumed to represent the best players and so on. Each class is managed by two coaches. Therefore, its size should be less than or equal to 25 children (in order to properly manage the players) and should be at least equal to 11 (in order to be able to form a team in each group);
- **Disaster management:** After a major earthquake, 150 victims have to be urgently treated. Depending on the severity of their injuries, they will be sent to an hospital (with a capacity constraint of 60 people maximum), to mobile medical units (with a total capacity of 80 people). The victims who will not be treated directly will have to wait under the supervision of nurses. Within this last group, a ranking will have to be established in order to provide a priority of treatments when spaces become available.

Finally, let us note that the problem of limiting the sizes of the groups in traditional clustering has been developed since 1984 [13]. The problem is known as the capacitated clustering problem and one of its applications is the sales forces area distribution where the clients are distributed among clusters that should be similar but with a limit on the size in order to have a balance between the sales forces [8]. Often those problems can be solved by integer mathematical programming but at the cost of important computation times. This is the reason why heuristics have been commonly proposed to work on these problems. This is the case, for instance, in [14] where they use tabu search and simulated annealing, or in [19]. A modified version of the k-means is proposed in [7] where a constraint is added in the assignment step (in order to stop the assignment when the cluster size limit is reached).

In this contribution, we propose to develop an exact algorithm in the context of multicriteria clustering. The paper is organized as follow. First, we formalize the problem in Sect. 2. Then, a dedicated algorithm is presented in Sect. 3. This approach is illustrated in Sect. 4. Finally, we conclude and give directions for future research.

2 The Problem

Let $A = \{a_1, a_2, ..., a_n\}$ be the set of alternatives and let $F = \{f_1, f_2, \dots, f_q\}$ be the set of criteria. In the following model, the number of clusters K and the preference matrix π are considered as inputs. The value π_{ij} represents the strict preference value of alternative a_i over alternative a_j. When there is no preference between a_i and a_j, we assume that $\pi_{ij} = 0$. In the following we assume that all preference values are different. If it is not the case, identical values can be differentiated by adding, for instance, an arbitrary small value ϵ.

An ordered partition of A into K clusters, noted $P_K(A) = [C_1, C_2, ..., C_K]$ is defined as follows:

- $A = \bigcup_{i=1,\ldots,K} C_i$,
- $\forall i \neq j : C_i \cap C_j = \emptyset$,
- $C_1 \succ C_2 \succ \ldots \succ C_K$,

where $C_i \succ C_{i+1}$ denotes that cluster C_i has a lower rank than cluster C_{i+1} (in other words, C_i is considered to be better than C_{i+1}).

The output of the method represents the cluster which each alternative belongs to. Two main assumptions can be made about this partition:

- Two alternatives belonging to the same cluster should be considered as being indifferent or similar. Therefore, their mutual preference values should be as low as possible;
- If an alternative a_i is assigned to a cluster which is better than the cluster of another alternative a_j, a_i is considered to be better than a_j. Therefore π_{ji} should be as low as possible.

In addition, we assume that clusters are subject to capacity constraints. More formally, we assume that:

$$x_h \leq |C_h| \leq X_h \tag{1}$$

where X_h and x_h represent, respectively, the maximum number and the minimum number of alternatives in the cluster C_h (to be consistent, we have $x_h \leq X_h$ and $\sum_h x_h \leq n \leq \sum_h X_h$).

The problem consists in determining the best partition among a number of possibilities (which increases very fast with the number of alternatives). To do so, one will try to minimize the inconsistencies with respect to the two previous conditions. Two types of inconsistencies have been distinguished:

- $a_i \in C_l$ and $a_j \in C_m$ with $l < m$ while $\pi_{ji} > 0$. Indeed, in an ideal case, we should have $\pi_{ij} > 0$ and $\pi_{ji} = 0$;
- $a_i, a_j \in C_l$ while $\pi_{ij} > 0$ or $\pi_{ji} > 0$. Indeed, in an ideal case, we should have $\pi_{ij} = \pi_{ji} = 0$.

Of course the severity of these inconsistencies will depend on the values of π_{ij}. There are different ways to minimize these inconsistencies. For instance, one could consider to minimize the sum of all inconsistencies. In this contribution, we follow the model proposed by De Smet et al. [5]. The idea is to base the optimization on the ordinal properties of the preference matrix. We will first consider all the K ordered partitions that minimize the worst inconsistency value. Among them, we will then identify those that minimize the second worst inconsistency. We repeat this step until a unique K ordered partition remains.

3 A Dedicated Algorithm

In this section, we present an extension of the algorithm proposed by De Smet et al. [5] in order to address capacity constraints. However, we decided to use different notations in order to simplify the presentation.

Let $I = \{1, 2, \ldots, n\}$ be a set of indices and $H : \pi \rightarrow I \times I : \pi_{ij} \rightarrow (i, j)$. The H function allows to determine the ordered pair of elements corresponding to a given preference value. Let ρ be the $n \cdot (n-1)$ dimensional vector of the elements of π sorted in decreasing order (let us recall that all these elements are assumed to be different).

The proposed method consists of successively considering the values of ρ and verify if the corresponding elements can be put in clusters such that the first elements is in a better cluster than the second one (in other words to have an ordered partition that is as compatible as possible with the π matrix). Therefore, we will consider a binary matrix M. $M_{ij} = 1$ means that alternative a_i is put in a better cluster than alternative a_j. M is initially a null matrix. Obviously, we have $M_{H(\rho(1))} = 1$. Indeed, at the beginning, the elements corresponding to the highest preference value of the preference matrix should be placed in such a way that the induced K ordered partition is compatible with this value. The elements of M give the relative positions of elements. It will be built step by step. Of course it should respect constraints:

- M should be acyclic since we want to obtain a total order on the set of clusters;
- the longest path within the M matrix should not exceed length $(K-1)$ since the number of clusters is limited to K;
- capacity constraints have to be respected.

In other words, one will successively consider the elements of the ρ vector in a descending order. At iteration k, we will investigate if the pair of elements (a_i, a_j) given by $H(\rho(k)) = (i, j)$ are such that a_i can be put in a better cluster than a_j. This will be authorized if the three previous conditions are simultaneously satisfied. Let M^k be the binary matrix at iteration k. We have:

1. $M^0 = 0_{n \times n}$;
2. $M^1 = M^0$ and $M^1_{H(\rho(1))} = 1$;

   ...

3. $M^{k+1} = M^k$ and $M^{k+1}(H(\rho(k+1))) = 1$.

If M^{k+1} simultaneously satisfies the three conditions, the considered preference $\rho(k+1)$ is considered to be compatible with the partition that is currently built. If it is not the case, the preference value is added to the penalty vector and $M^{k+1}(H(\rho(k+1))) = 0$.

In order to manage these constraints, let us define the following notions. At first, since we consider a given iteration, we will omit the index k in the M matrix in order to simplify the notations. For a given $a_i \in A$, let us define $W(a_i) = \{a_j \in A | M_{ij} = 1\}$, the set of alternatives that are currently put in a worse cluster than a_i. Similarly, let us define $B(a_i) = \{a_j \in A | M_{ji} = 1\}$, the set of alternatives that are currently put in a better cluster than a_i. Let $w(0) = \{a_i \in A | W(a_i) = \emptyset\}$ and $w(k) = \{a_i \in A | W(a_i) \subset w(0) \cup w(1) \cup \ldots \cup w(k-1)\}$. In other words, $w(0)$ denotes the subset of alternatives that have no *successors* in the M matrix. These alternatives can potentially be put in the worst cluster. It is easy to see that $a_i \in w(k)$ is equivalent to say that the longest path starting from a_i is of length k. For instance, let us consider a problem with $K = 5$. If $a_i \in w(3)$ it means there is a path of length 3 starting from a_i in the M matrix. As a consequence a_i could be placed in C_1 or C_2. Otherwise the maximum number of clusters would not be satisfied. It is easy to see that when w can be determined for all the elements of A, the M matrix is acyclic. In a similar way, let us define $b(0) = \{a_i \in A | B(a_i) = \emptyset\}$ and $b(k) = \{a_i \in A | B(a_i) \subset b(0) \cup b(1) \cup \ldots \cup b(k-1)\}$. The meaning of $b(k)$ is similar to $w(k)$ except that we now consider *predecessors* (instead of successors). For instance, if $a_i \in b(2)$ it could not be placed in C_1 or C_2 (since there exists a path of length 2 with a_i as the last node). Let us note that b and w are the so-called *ranks* and *levels* of the graph induced by the M adjacency matrix.

As already said, the M matrix will be built progressively. To summarize, when $a_i \in b(k)$ and $a_i \in w(l)$, this alternative could not be placed in clusters that are characterized by ranks greater than $K - l$ or lower than $k + 1$. All the clusters from C_{k+1} to C_{K-l} are possible assignments that are compatible with the desired number of clusters K. Finally, when $k + 1 = K - l$, the assignment of a_i is precisely determined. During the execution of the algorithm the possible assignments related to a_i will be progressively refined.

These notions help us verifying at each step that the M matrix is acyclic and respect the desired number of clusters. The partition of A with respect to b and w allows us to build a K ordered partition. Now we still have to take into account the constraints on capacities.

Let n_{kl} be the number of alternatives that could be placed in $C_k \cup \cdots \cup C_l$ (with $k < l$) during the execution of the algorithm. For the sake of simplicity, let us consider the case $K = 3$. The conditions for maximum capacity constraints can be elicited as follows:

$$\begin{cases} n_1 \leq X_1 \\ n_2 \leq X_2 \\ n_3 \leq X_3 \\ n_1 + n_2 + n_{12} \leq X_1 + X_2 \\ n_2 + n_3 + n_{23} \leq X_2 + X_3 \\ n_1 + n_2 + n_3 + n_{12} + n_{23} + n_{13} \leq X_1 + X_2 + X_3. \end{cases} \tag{2}$$

It is obvious that the simultaneous satisfaction of all these constraints leads to an incomplete partition that is compatible with the maximum capacities. Of course, some of the alternatives are not yet assigned to a precise cluster (especially during the first iterations) but it leaves room for a future assignment that will respect the maximum capacities.

The general condition can be written as follows: $\forall k \in I, \forall l \in I, k \leq l$:

$$\sum_{J \subseteq \{k,k+1\ldots,l\}, |J| \leq 2} n_J \leq \sum_{k \leq m \leq l} X_m. \tag{3}$$

In a similar way the conditions for minimum capacity constraints can be formulated as follows:

$$\begin{cases} n_1 + n_{12} + n_{13} \geq x_1 \\ n_2 + n_{12} + n_{23} + n_{13} \geq x_2 \\ n_3 + n_{23} + n_{13} \geq x_3 \end{cases} \tag{4}$$

More generally, we have: $\forall k \in I$

$$\sum_{J \subseteq I, |J| \leq 2, J \cap \{k\} \neq \emptyset} n_J \geq x_k. \tag{5}$$

At each step of the algorithm, these conditions will be verified in order to be sure that the capacity constraints could be satisfied in the future steps.

Algorithm 1. Determine K constrained ordered partition

Require: n, π, K and $(x_h, X_h)\, \forall h = 1, \ldots, K$
  Create ρ {a decreasing vector constituted by the elements of π}
  Create $M^0 : 0_{n \times n}$
  $M^1 = M^0$ and $M^1_{H(\rho(1))} = 1$
  p $\leftarrow \emptyset$ {a penalty vector}
  $k = 1$
  while $k \leq n \cdot (n-1)/2$ **and** $\rho(k+1) > 0$ **do**
    $k \leftarrow k + 1$
    $M^k = M^{k-1}$ and $M^k_{H(\rho(k))} = 1$
    Based on M^k, compute b and w
    if M^k has a cycle **or** M^k has a path bigger than $K - 1$ **or** M^k does not respect constraints 3 and 5 **then**
      $M^k_{H(\rho(k))} = 0$
      p(end+1) $\leftarrow \rho(k)$
    end if
  end while

Algorithm 1 allows to obtain a K ordered partition that satisfies the capacity constraints. In addition, there is no other K ordered partition that lexicographically minimizes the penalty vector. More formally, let p be the penalty vector obtained at the end of Algorithm 1 and $\tilde{p}$ be the penalty vector associated to another K ordered partition. Let $T = \min\{|p|, |\tilde{p}|\}$, we have:

$$p \succ_{\mathcal{L}} \tilde{p} \Leftrightarrow p_i < \tilde{p}_i, i = \min[k \in \{1, 2, \ldots, T)\}|p_k \neq \tilde{p}_k] \tag{6}$$

Without loss of generality, we may assume the k first components of p and $\tilde{p}$ are the same (k might be equal to 0). By assumption, we have $p(k+1) > \tilde{p}(k+1)$. Following our approach, the value $p(k+1)$ was put in the penalty vector due to (at least) one of these conditions: M has a cycle, M has a path longer than $K-1$ or M does not respect the capacity constraints.

Since the value $p(k+1)$ does not appear in the penalty vector of the new partition, we have $M_{H(\tilde{p}(k+1))} = 1$ and the induced graph is not a K ordered partition satisfying the capacity constraints.

Let us point out that sometimes Algorithm 1 can lead to imprecise assignments. This happens, for instance, when there are many zero values in the π matrix (leading to a situation where the alternatives cannot be properly discriminated). Finally, it is important to note that this approach is based on the assumption that preference values are different. As already stressed, this can easily be satisfied if identical values are discriminated by adding an arbitrary small value to these preferences. This step should not impact the penalty vector in a significant way (and therefore the partition structure). Nevertheless, we suggest to repeat different trials in order to assess the solution robustness.

4 Validation

Obviously, a minimal requirement is being able to replicate the results of the algorithm developed by De Smet et al. [5]. Let us consider the optimal solution for a given problem without constraint with $K = 3$. This solution assigns n_1 (respectively n_2, n_3) for the clusters 1 (respectively 2, 3). If we propose capacity constraints that are compatible with these values, the results given by the new approach should be the same. Many tests have been done with several datasets (and different parameters). Fortunately, results have always been the same.

In what follows, we will investigate computational time issues in Subsect. 4.1 and study qualitative results based on a real dataset in Subsect. 4.2.

4.1 Computation Time

The proposed algorithm was tested on different artificial datasets for different numbers of alternatives and clusters. In these cases, all the π_{ij} values have been drawn from a $[0, 1]$ uniform random distribution (theses value being assumed to be stochastically independent). The performances are shown in Table 1.

Table 1. Execution time in seconds on artificial datasets.

# alternatives	3 clusters	4 clusters	5 clusters	6 clusters
100	2	2	2	2
300	71	70	87	62
500	381	382	327	311
700	900	952	1104	879

In addition, tests have also been conducted on real datasets:

- The Human Development Index (HDI) [9];
- The Academic Ranking of World Universities (ARWU) [17].

Table 2. Execution time in seconds on HDI dataset.

# alternatives	3 clusters	4 clusters	5 clusters	6 clusters
10	0.01	0.01	0.01	0.02
50	0.6	1.14	1.35	1.80
100	6.9	14	19	22.4
150	31	62	87	103

In Table 2, execution time in seconds is shown on randomly sampled subsets from the HDI dataset (those obtained on the ARWU dataset being similar).

For both tests in Table 1 and 2, the capacity constraints have been set as the same minimum and maximum capacities:

$$min\_capacity = \frac{number\_of\_alternatives}{(number\_of\_clusters) * 1.5} \tag{7}$$

$$max\_capacity = \frac{number\_of\_alternatives * 1.5}{(number\_of\_clusters)} \tag{8}$$

4.2 Qualitative Results

In this section, we show the results obtained while applying the algorithm on a real dataset: the Academic Ranking of World Universities (ARWU). In order to compute the input preference matrix we decided, without loss of generality, to use the linear PROMETHEE preference function [2]. For the sake of simplicity, the indifference and the preference thresholds correspond respectively to the second and the eighth decile of the differences of evaluations for each criterion. Here we consider the 100 first institutions of the Shanghai ranking. Weight values are the same as in the initial ranking. When the model is tested without constraints for $K = 4$, we get the results represented on Fig. 1. It is represented in the PCA two-dimensional space. This ordered partition puts Harvard, the first university in the ranking, alone in a single cluster. This shows how Harvard is better than the other alternatives (which is coherent with the very high evaluations of this university on all the criteria). This partition assigns 1 alternative to C_1, 9 alternatives to C_2, 25 to C_3 and 65 to C_4.

The obtained partition is clearly unbalanced (a single alternative in the first cluster and 65 in the worst one). In order to better balance this output, one would like to investigate a partition that respects the following conditions; at least 5 universities in the best cluster and at most 60 in the last cluster. The results are presented on Fig. 2. Its distribution is as follows: 5 alternatives in C_1, 7 alternatives in C_2, 28 in C_3 and 60 in C_4.

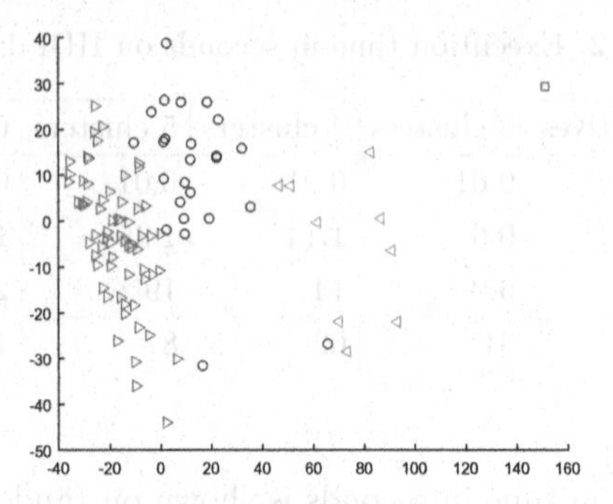

Fig. 1. Model with no size constraints applied to the 100 first institutions of the Shanghai Ranking of World Universities, K = 4 - PCA representation. We have $C_1(\square), C_2(\triangleleft), C_3(\circ)$ and $C_4(\triangleright)$.

The confusion matrix F is defined as follows:

$$F = \begin{pmatrix} 1 & 0 & 0 & 0 \\ 4 & 5 & 0 & 0 \\ 0 & 2 & 23 & 0 \\ 0 & 0 & 5 & 60 \end{pmatrix}$$

where $F(i,j)$ denotes the number of elements belonging in C_i in the unconstrained version of the method that are placed in C_j when constraints are taken into account. As expected, these new constraints imply a reduction of the quality of the clustering in terms of lexicographic optimization. Figure 3 shows the evolution of the 100 first terms of the inconsistency matrix. One can see the higher lexicographic values of the model without constraints. In fact, the first term of the inconsistency matrix is equal to 0.57 (respectively 0.83) for the unconstrained (respectively constrained) model.

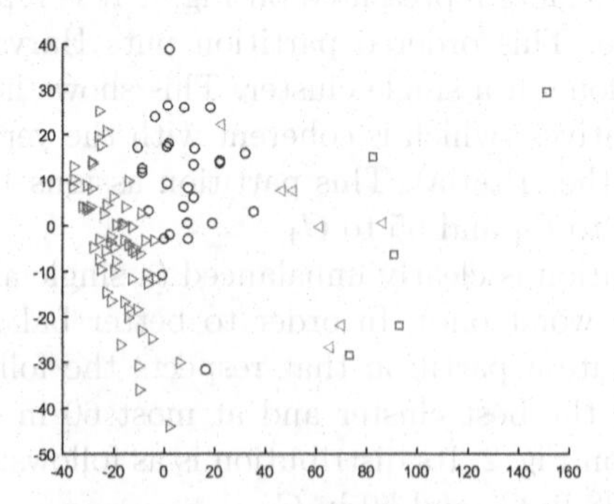

Fig. 2. Model $x_1 = 5$ and $X_4 = 60$ applied to the 100 first institutions of the Shanghai Ranking of World Universities, K=4 - PCA representation. We have $C_1(\square), C_2(\triangleleft), C_3(\circ)$ and $C_4(\triangleright)$.

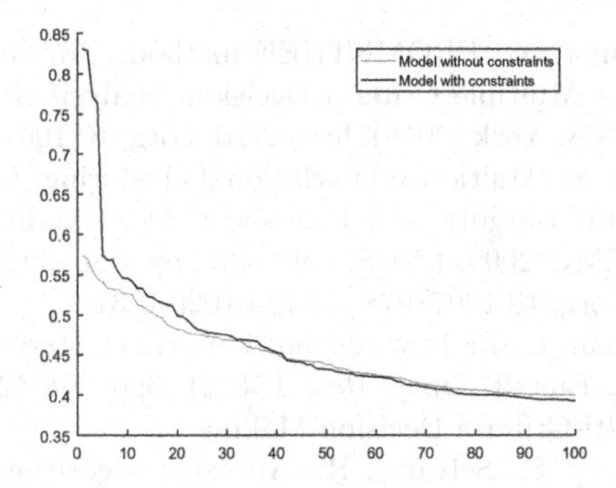

Fig. 3. Comparison of the inconsistency values for the models with and without constraints - Shanghai ranking

5 Conclusion

In this contribution, we have proposed an exact method to address capacity constraints in the context of multicriteria ordered clustering. The distinctive feature of this approach is to minimize a vector of inconsistencies based on a lexicographic order following the work of De Smet et al. [5]. The execution time issue of the proposed method has been studied and results have been illustrated on a real data set. Multicriteria clustering is a recent research topic. If some works have tried to formalize it, there are still a number of questions to be addressed. To the best of our knowledge this work is the first to investigate capacity constraints in this specific context (while similar questions have already been raised in multicriteria sorting methods or in *traditional* clustering). A common question in clustering is to properly determine the number of clusters to create. Some indicators might help the analyst answering this question. Here the question is even more complex since the input information is much richer; minimum and maximum constraints may be provided. To our point of view, these two kinds of parameters are related and need further investigations. The proposed solution heavily relies on the fact that we only use the ordinal properties of the preference matrix (by using a lexicographic order). Of course, other objectives can be used (such as minimizing the sum of inconsistent preferences, etc.). In these cases, managing capacity constraints would become more difficult (and the solution based on the computation of the levels and the ranks of a graph would not be appropriate anymore).

References

1. Bradley, P.S., Fayyad, U.M.: Refining initial points for K-means clustering. In: Proceedings of the Fifteenth International Conference on Machine Learning, ICML 1998, San Francisco, CA, USA, pp. 91–99. Morgan Kaufmann Publishers Inc. (1998)

2. Brans, J.-P., De Smet, Y.: PROMETHEE methods. In: Greco, S., Ehrgott, M., Figueira, J.R. (eds.) Multiple Criteria Decision Analysis. ISORMS, vol. 233, pp. 187–219. Springer, New York (2016). https://doi.org/10.1007/978-1-4939-3094-4_6
3. De Smet, Y., Eppe, S.: Multicriteria relational clustering: the case of binary outranking matrices. In: Ehrgott, M., Fonseca, C.M., Gandibleux, X., Hao, J.-K., Sevaux, M. (eds.) EMO 2009. LNCS, vol. 5467, pp. 380–392. Springer, Heidelberg (2009). https://doi.org/10.1007/978-3-642-01020-0_31
4. De Smet, Y., Guzmàn, L.M.: Towards multicriteria clustering: an extension of the k-means algorithm. Eur. J. Oper. Res. **158**(2), 390–398 (2004). Methodological Foundations of Multi-Criteria Decision Making
5. De Smet, Y., Nemery, P., Selvaraj, R.: An exact algorithm for the multicriteria ordered clustering problem. Omega **40**(6), 861–869 (2012). Special Issue on Forecasting in Management Science
6. Ehrgott, M.: Multicriteria Optimization, 2nd edn., p. 323. Springer, Heidelberg (2005). https://doi.org/10.1007/3-540-27659-9
7. Ganganath, N., Cheng, C., Tse, C.K.: Data clustering with cluster size constraints using a modified K-means algorithm. In: 2014 International Conference on Cyber-Enabled Distributed Computing and Knowledge Discovery, pp. 158–161 (2014)
8. Höppner, F., Klawonn., F.: Clustering with size constraints. In: Jain, L.C., Sato-Ilic, M., Virvou, M., Tsihrintzis, G.A., Balas, V.E., Abeynayake, C. (eds.) Computational Intelligence Paradigms. Studies in Computational Intelligence, vol. 137, pp. 167–180. Springer, Heidelberg. https://doi.org/10.1007/978-3-540-79474-5_8
9. Human Development Index (2015). http://hdr.undp.org/en/content/human-development-index-hdi
10. Kanungo, T., et al.: An efficient k-means clustering algorithm: analysis and implementation. IEEE Trans. Pattern Anal. Mach. Intell. **24**, 881–892 (2002)
11. Köksalan, M., Mousseau, V., Özpeynirci, S.: Multi-criteria sorting with category size restrictions. Int. J. Inf. Technol. Decis. Making **16**(01), 5–23 (2017)
12. Meyer, P., Olteanu, A.-L.: Formalizing and solving the problem of clustering in MCDA. Eur. J. Oper. Res. **227**(3), 494–502 (2013)
13. Mulvey, J.M., Beck, M.P.: Solving capacitated clustering problems. Eur. J. Oper. Res. **18**(3), 339–348 (1984)
14. Osman, I.H., Christofides, N.: Capacitated clustering problems by hybrid simulated annealing and tabu search. Int. Trans. Oper. Res. **1**(3), 317–336 (1994)
15. Rocha, C., Dias, L.C., Dimas, I.: Multicriteria classification with unknown categories: a clustering-sorting approach and an application to conflict management. J. Multi-Criteria Decis. Anal. **20**(1–2), 13–27 (2013)
16. Roy, B.: The outranking approach and the foundations of electre methods. In: Bana e Costa, C.A. (eds.) Readings in Multiple Criteria Decision Aid, pp. 155–183. Springer, Heidelberg. https://doi.org/10.1007/978-3-642-75935-2_8
17. ShanghaiRanking Consultancy (2015). http://www.shanghairanking.com
18. Vincke, P.: Multicriteria Decision-Aid. Wiley, New York (1992)
19. Zhu, S., Wang, D., Li, T.: Data clustering with size constraints. Knowl. Based Syst. **23**(8), 883–889 (2010)

Local Search is a Remarkably Strong Baseline for Neural Architecture Search

Tom Den Ottelander[1(✉)], Arkadiy Dushatskiy[1], Marco Virgolin[1], and Peter A. N. Bosman[1,2]

[1] Centrum Wiskunde and Informatica, Amsterdam, The Netherlands
{tdo,arkadiy.dushatskiy,marco.virgolin,peter.bosman}@cwi.nl
[2] Delft University of Technology, Delft, The Netherlands

Abstract. Neural Architecture Search (NAS), i.e., the automation of neural network design, has gained much popularity in recent years with increasingly complex search algorithms being proposed. Yet, solid comparisons with simple baselines are often missing. At the same time, recent retrospective studies have found many new algorithms to be no better than random search (RS). In this work we consider the use of a simple Local Search (LS) algorithm for NAS. We particularly consider a multi-objective NAS formulation, with network accuracy and network complexity as two objectives, as understanding the trade-off between these two objectives is arguably among the most interesting aspects of NAS. The proposed LS algorithm is compared with RS and two evolutionary algorithms (EAs), as these are often heralded as being ideal for multi-objective optimization. To promote reproducibility, we create and release two benchmark datasets, named MacroNAS-C10 and -C100, containing 200K saved network evaluations for two established image classification tasks, CIFAR-10 and CIFAR-100. Our benchmarks are designed to be complementary to existing benchmarks, especially in that they are better suited for multi-objective search. We additionally consider a version of the problem with a much larger architecture space. While we find and show that the considered algorithms explore the search space in fundamentally different ways, we also find that LS substantially outperforms RS and even performs nearly as good as state-of-the-art EAs. We believe that this provides strong evidence that LS is truly a competitive baseline for NAS against which new NAS algorithms should be benchmarked.

Keywords: Neural Architecture Search · Local Search · Evolutionary algorithm · Random search · Multi-objective NAS · NAS baseline

1 Introduction

Deep learning has achieved excellent results for machine learning tasks in heterogeneous fields, including image recognition [19], natural language processing [17], games [21,24], and genomics [20]. While the availability of large amounts of data [11] and hardware advancements [9] have been key enabling factors for deep

H. Ishibuchi et al. (Eds.): EMO 2021, LNCS 12654, pp. 465–479, 2021.
https://doi.org/10.1007/978-3-030-72062-9_37

learning, the success of the field is also due to the ingenious design of competent network architectures (consider, e.g., the invention of residual connections [8]).

Proper and innovative architecture design requires experience, intuition, and expensive trial-and-error. This has sparked research on techniques to automate this task, i.e., the field of Neural Architecture Search (NAS) has emerged [7, 26]. NAS research is quickly gaining popularity: 2019 alone counts almost 250 publications[1]. Most NAS proposals present new, typically increasingly complex, NAS algorithms. However, recent work questions whether the need for many of these algorithms is actually justified: [30] and [28] showed that several NAS algorithms are no better than Random Search (RS) when validated properly.

While RS can be considered perhaps the simplest baseline, it is also by far the worst form of heuristic search in many cases, with the worst scalability. For this reason, it is often not even considered in many modern heuristic design studies. Yet, from RS to more complex search heuristics such as EAs, is still quite a leap. In the gap, e.g., lie almost equally simple and classical, yet far less random in nature, Local Search (LS) techniques. Still, (classic) LS (in the space of encodings) has, to the best of our knowledge, not been applied to NAS before, except for in [25], a work that happened concurrently to ours (see Sect. 2).

In this paper, we consider LS for NAS and in doing so, we make three contributions: (1) We propose a simple and parameter-less LS algorithm for multi-objective (accuracy vs. architecture complexity) NAS. (2) To enable quick and easy benchmarking, we release[2] two datasets containing cached performances for over 200,000 Convolutional Neural Network (CNN) architectures for CIFAR-10 and CIFAR-100. Our datasets, named MacroNAS-C10 and MacroNAS-C100, are designed to be complementary to similar existing proposals [5,29]. We remark that we designed these datasets to be able to compare NAS approaches, and not necessarily to obtain State-of-the-Art (SotA) neural networks. (3) We provide evidence on the validity of LS being a superior baseline for NAS by conducting several experiments. In particular, we consider NAS in different search spaces; We compare LS with RS as well as with a classic and a SotA Evolutionary Algorithm (EA); We include an experiment on CIFAR-100 where the possible architectures are less constrained, leading to a large search space containing over 104 Billion possible neural architectures.

In this work, we consider NAS in a multi-objective setting. Single-objective NAS, i.e., searching only for good accuracy-networks, has already been amply investigated in [28,30], where it was found that very different search algorithms, including RS, actually perform very similar. Nevertheless, additional experiments that show the performance of LS for single-objective NAS can be found in the Appendix of our online preprint [15] and provide similar insights to those found in [25]. Multi-objective NAS on the other hand has been studied less (see Sect. 2), and is arguably a more interesting setting, where sophisticated search

[1] https://www.automl.org/automl/literature-on-neural-architecture-search/.

[2] Benchmark datasets: https://github.com/ArkadiyD/MacroNASBenchmark. Source code for reproducibility: https://github.com/tdenottelander/MacroNAS.

algorithms may potentially be needed. Alongside RS, we consider EAs, since they are a natural fit for multi-objective optimization, and are often used for NAS.

2 Related Work

NAS can be broadly categorized by the level at which the search is performed, i.e., macro- and micro-level [7] (different nomenclatures exist [26]). Macro-level search aims at composing (wiring) atomic *cells* of different type. A cell is a computational graph (a sub-net) composed of different operations/layers (convolutions, activation functions, etc.). In micro-level search, the position of cell types within the architecture is pre-decided (often a same cell type is repeated in multiple places), while it is the cells' internal wiring that is optimized. Both levels can also be searched at the same time [12,14], potentially along with hyper-parameter settings [14,27]. In the remainder of the paper, we will refer to "architecture" and "network" (for brevity, "net") interchangeably.

Search algorithms considered for NAS vary wildly. Reinforcement Learning (RL), Bayesian optimization approaches, and EAs are among the most popular [7]; the latter being typically employed in multi-objective scenarios. In this paper, we mostly focus on performing multi-objective NAS at macro-level. Therefore, in the following we report on recent works in this direction, and focus on whether comparisons against a baseline (i.e. RS) were made.

The work in both [3,10] builds upon NSGA-II [4] to obtain a front of nets that trade off accuracy and compactness. Although nets with SotA performance are ultimately obtained, the search algorithm is not compared with any baseline. In [6], a multi-objective EA is proposed that includes Lamarckian mechanisms: instead of training nets from scratch, child nets inherit the weights of their parents and are only fine-tuned. Although the method is shown to outperform RS, it is unclear whether this is due to a difference with the way the EA searches, or solely because of weight inheritance. In [12], NSGA-Net is presented, which evolves fixed-length architectures at both macro- and micro-level. The authors show, through experimental analysis, that NSGA-Net is superior to RS. Summarizing, most of the literature work considers only RS as a baseline (sometimes in not entirely equal conditions), or no baseline at all.

Importantly, the only work we are aware of that investigated the use of LS similarly to us is [25], which appeared on arXiv shortly after the first version of our paper appeared on the same platform. The work shares strong similarities with ours, as it investigates the use of a simple LS algorithm for NAS, and reaches the same conclusion, i.e., that LS is a strong baseline for NAS. Despite these similarities, our works can be considered complementary to each other: the authors of [25] focus on single-objective NAS benchmarks and provide interesting theoretical underpinnings; we instead focus on multi-objective NAS, and, among other contributions, propose new NAS benchmark datasets.

Since we release benchmark datasets of saved net evaluations for macro-level NAS, we also refer to important related works of this nature. NAS-Bench-101 [29] contains cached performance indicators for 423,000 nets on CIFAR-10, while NAS-Bench-201 [5] does the same also for CIFAR-100 and ImageNet, but for far less architectures (15,625). Both datasets are built for micro-level NAS. We did not consider NAS-Bench-201 because the limited number of possible architectures can limit the analysis of the search behavior for long run-times. We built new benchmark datasets for three reasons. First, we intended to provide the community with a macro-level search benchmark dataset for better reproducible NAS: we are not aware of other works proposing this. Second, NAS-Bench-101 includes infeasible solutions (the MacroNAS datasets do not), and choosing how to handle them substantially changes the search behavior of an algorithm. Lastly, for NAS-Bench-101 it has already been shown that RS is a competitive baseline in a single-objective setting [29]. We will show however that, in fact, in the search space of NAS-Bench-101 it is hard to do any better than RS, even when considering a multi-objective setting (see Sect. 5.1).

3 Objectives, Encodings, Search Algorithms

We now present the objectives to optimize, the encoding schemes we use, and the algorithms we consider in the experimental analysis. For a detailed description of the MacroNAS datasets we introduce, we refer the reader to the our preprint [15].

3.1 Objectives

The first objective f_1 is to maximize accuracy (fraction of correct net predictions). Particularly, we set f_1 to the validation accuracy (acc_{val}), i.e., the accuracy of the net for a set of data that is not used for training the net, to assess whether the net generalizes. Because the validation set is still used in the NAS optimization loop, we consider the ability of nets with good acc_{val} to generalize to a second set of completely held-out data, i.e., the test set, in Sect. 5.4.

The second objective f_2 is to maximize a net's efficiency, i.e., to minimize the Mega Multiply–ACcumulate operations (MMACs, 1 MMAC $\simeq$ 2 FLOPs), which is the number of GPU computations a net requires to make a prediction. MMACs is a popular complexity metric [22] and is deterministic (in contrast to actual time taken). Because large differences in scale of the objectives can influence search mechanisms (e.g., the crowding distance [4]), we set MMACs to be in the same range as acc_{val}, i.e., we normalize MMACs to lie in $[0, 1]$. The min/max MMAC values for the normalization are trivial to determine as it suffices to consider the smallest/largest achievable nets. For the sake of a consistent optimization direction in both objectives, the secondary objective f_2 becomes maximizing $1 - normalize(\text{MMACs})$. Since no MMACs were reported for NAS-Bench-101, for the experiment in Sect. 5.1 we use $1 - normalize(\#parameters)$.

3.2 Encodings

We use a direct encoding that represents feed-forward CNNs. The encoding is an array of discrete variables $[x_1, \ldots, x_\ell]$, where each x_i takes values (cell types) in Ω_i, a position-dependent alphabet.

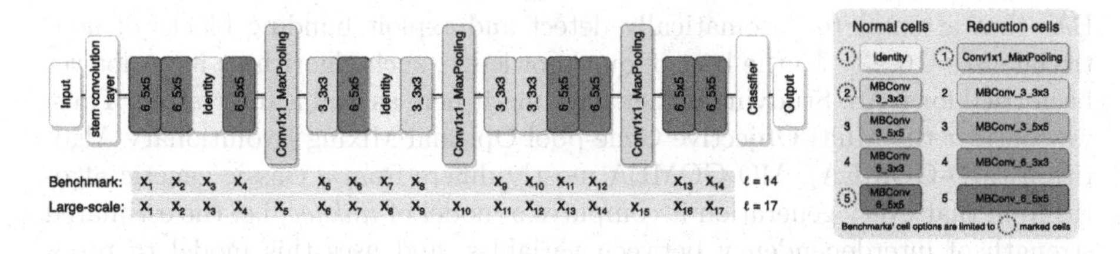

Fig. 1. Best acc_{val} net on CIFAR-10 (left) and cell types (right). For MacroNAS, 3 types are possible for normal cells (types 1, 2 and 5) and reduction cells (positions 5, 10, and 15) are fixed to type 1. For the large-scale experiment, 5 types are available for all cells. For details on cell types see the Appendix of our preprint [15].

In the MacroNAS datasets, the number of searchable cells (length of the encoding) is $\ell = 14$. In particular, only *normal* cells, i.e., cells that maintain the height and width of the input they receive [31], can be optimized. Conversely, cells that reduce spatial dimensions (*reduction* cells) are fixed. The alphabet Ω_i is identical for each position i and contains 3 options. One option is the *identity*, i.e., a placeholder cell that forwards information as is. The other two options are commonly used convolutional blocks from literature (described in the Appendix of our preprint [15]). Figure 1 shows an example of an encoding. Even though the encoding has a fixed length, it can represent variable-length architectures by means of identity cells. Note that the mapping from encoding to (computationally unique) architecture is redundant. The size of the search space (total number of encodings) is $3^{14} = 4{,}782{,}969$, which maps to 208,537 architectures. The latter number is sufficiently small to allow us to evaluate all architectures and save their performances to create the benchmark datasets.

For the large-scale experiment, we increase the number of variables ℓ to 17 by including the optimization of reduction cells, and also increasing the alphabet size $|\Omega_i|$ to 5, $\forall i$. Here, the position i influences what Ω_i is used, as options for normal cells are different from options for reduction cells (see Fig. 1). The total search space of $5^{17} = 762{,}939{,}453{,}125$ maps to 104,086,030,125 architectures.

The encoding for NAS-Bench-101 [29] (micro-level) is different. Here, both the type of operations in a (repeated) cell, and their connections, can be optimized. The encoding is a binary string of length 21 that represents connections, concatenated with a ternary string of length 5 that represents operations.

3.3 Search Algorithms

The first algorithm we consider is NSGA-II [4]. It is by far the most popular multi-objective EA in general. We use most commonly adopted parameter settings of NSGA-II: 2-point crossover, single-variable mutation with probability $p_m = 1/\ell$, tournament size 2 and population size $n_{\text{pop}} = 100$.

Fundamentally better algorithmic design is foundational to EA research. This means building better problem-specific EAs or ones capable of doing better at large classes of interesting problems. In the latter direction lie linkage learning

EAs that attempt to automatically detect and exploit building blocks of non-trivial sizes [16,23]. To the best of our knowledge, such algorithms have not yet been tried for NAS. Studying if they have merit here as well is interesting. Thus, we consider the Multi-Objective Gene-pool Optimal Mixing Evolutionary Algorithm (MO-GOMEA). MO-GOMEA mostly differs from a classic genetic algorithm in that every generation it computes a model of *linkage*, i.e. the estimated strength of interdependency between variables, and uses this model to propagate potential building blocks during variation [23]. If the NAS search space exhibits any linkage, this algorithm can likely exploit it. Also, MO-GOMEA uses clusters to partition the population in objective space, so it is generally able to improve upon specific parts of the front by recombining within niches. We use one of the latest implementations for discrete optimization [13], which includes the Interleaved Multi-start Scheme (IMS). The IMS allows to avoid the manual setting of population size parameter by evolving populations of increasing sizes in interleaved fashion. For the initial population size, we use $n_{\text{pop}} = 8$ (default).

We further consider RS as a standard baseline. RS generates new nets repeatedly by sampling each cell uniformly at random from the alphabets Ω_i.

Finally, we consider a simple random restart LS algorithm:

1. Initialize a random architecture (as in RS);
2. Sample a scalarization coefficient $\alpha \sim \mathcal{U}(0, 1)$;
3. Consider all variables in random order, as follows:
   3.1 For each variable x_i, evaluate the net obtained by setting x_i to each option in Ω_i. Keep the best according to $\alpha \times f_1 + (1 - \alpha) \times f_2$;
4. Repeat step 1 until the computation budget (evaluations/time) is exhausted.

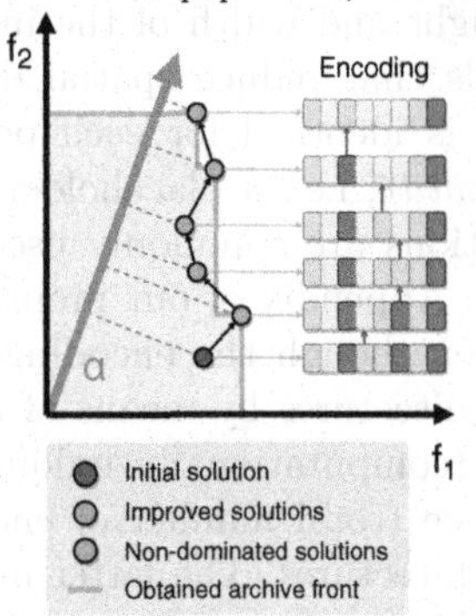

Fig. 2. Example iteration on one random scalarization coefficient α in LS. Only solutions that improve the scalarized objective are shown.

In other words, we perform one round of first-improvement LS at the level of variables, but best-improvement LS at the level of options for one variable. Figure 2 illustrates one iteration of our LS algorithm (from now on, we simply refer to it as LS).

4 Experimental Setup

All search algorithms store an archive $\mathcal{A}$ of best-found nets according to (strict) Pareto domination. Formally, it is: $\boldsymbol{x} \succ \boldsymbol{y}$ ("$\boldsymbol{x}$ dominates $\boldsymbol{y}$") iff $f_i(\boldsymbol{x}) \geq f_i(\boldsymbol{y}) \wedge \boldsymbol{f}(\boldsymbol{x}) \neq \boldsymbol{f}(\boldsymbol{y})$, for each objective f_i. The archive thus satisfies $\nexists \boldsymbol{x}, \boldsymbol{y} \in \mathcal{A} : \boldsymbol{x} \succ \boldsymbol{y}$. We use no size limitation for the archive, and update the archive each time a new net is discovered (the time cost is insignificant compared to evaluating a net's performance, see below).

To compare algorithms, we consider the improvement of the hypervolume enclosed by $\mathcal{A}$, using the origin as reference point. For the experiments of Sects. 5.1 and 5.2 (where the Pareto front is known), we also investigated the inverse generational distance [2], and obtained very similar results (omitted for brevity). We only count the evaluations of unique architectures as evaluating a net is typically costly (minutes to hours). Caching is used to not re-evaluate a same architecture twice.

We handle the net with *identity* cells for all variables differently from the others. This trivial net can be considered fundamentally uninteresting to search for, but it still influences multi-objective metrics. We therefore set NSGA-II and MO-GOMEA to include one instance of it in the population at initialization. Similarly, we include it in the archive of RS and LS from the beginning.

For experiments concerning the benchmark datasets, i.e., where performances for different architectures are pre-stored, we perform 30 runs for each algorithm. For the large-scale search space, where pre-evaluating all nets is simply unfeasible and a single evaluation takes on average 6.5 min on Nvidia RTX 2080 Ti GPU, we perform 6 runs for each algorithm. To assess whether differences between algorithms are significant, we adopt pairwise non-parametric Mann-Whitney-U tests with Bonferroni correction.

5 Results

First, we compare the algorithms on NAS-Bench-101 and on MacroNAS-C10/C100. Next, we analyze the performance on MacroNAS-C100, and on the large-scale version for this task, in more detail. Lastly, we discuss what the observed differences mean when considering the nets' ability to generalize.

5.1 Preliminary Experiments

We begin by analyzing the results for NAS-Bench-101 (Fig. 3 left). For this benchmark, differences among the algorithms are negligible. This can also be

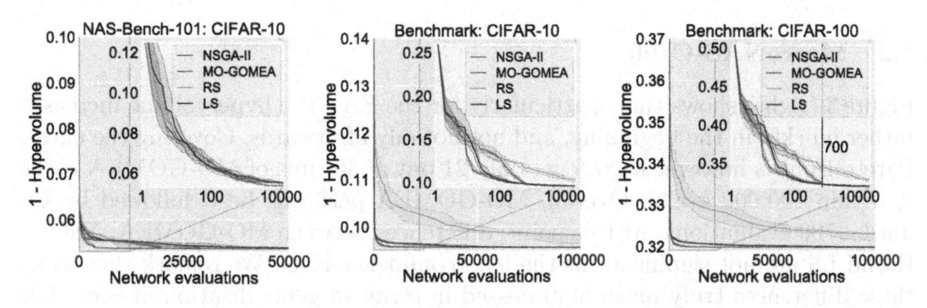

Fig. 3. Convergence graphs on NAS-Bench-101 (left), MacroNAS-C10 (middle) and MacroNAS-C100 (right). Medians (solid lines) and 25/75th percentiles (bands) of 30 runs are shown. Note that only the horizontal axes are the same, and are in logarithmic scale in the zoomed-in views (insets).

seen in Fig. 4, where fronts obtained by the algorithms are similar at different points in time. These results are in line with the ones reported for the single-objective experiments conducted in [29]: differences among fundamentally different algorithms are relatively small.

Regarding the macro-level search spaces, i.e., on MacroNAS-C10 and MacroNAS-C100 (Fig. 3 middle and right, respectively), a notable difference is found between RS and the other algorithms. There are however also small differences between the EAs and LS. In the zoomed-in views it can be seen that, for the first 100 evaluations, NSGA-II performs as good as RS: this is because the initial population ($n_{pop} = 100$) is evaluated. Meanwhile, MO-GOMEA already performed optimization on the small initial populations instantiated by its IMS.

The hypervolume obtained for CIFAR-100 is not as good as the one obtained for CIFAR-10. This is not surprising because CIFAR-100 is considered a harder classification task. On the MacroNAS benchmarks, the best acc_{val} for CIFAR-10 is 0.925, the one for CIFAR-100 is 0.705. Furthermore, note that the fact that the hypervolumes for NAS-Bench-101 reach better values, is mainly because f_2 is different here (based on number of net parameters instead of MMACs). Still, NAS-Bench-101 contains nets with slightly larger acc_{val} (yet easily found by all algorithms), up to 0.951. Since CIFAR-100 is the harder task, we focus on this task in the next sections.

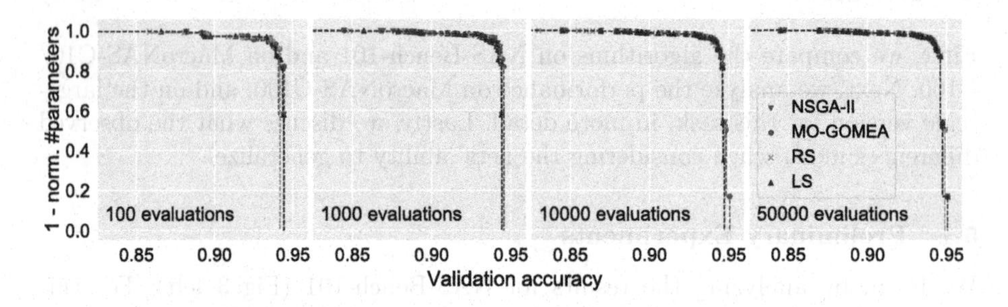

Fig. 4. Evolving archives for one example run on NAS-Bench-101.

5.2 MacroNAS-C100

Figure 3 (right) shows that, particularly for the EAs, the hypervolume increases rather quickly in the beginning, and not notably afterwards. Covering the entire Pareto front is however hard (e.g., only 21 out of 30 runs of MO-GOMEA cover it within 100,000 evals). Overall, MO-GOMEA performs best, followed by LS until ~700 evaluations. At this point, differences between MO-GOMEA, NSGA-II and LS are not significant at the 99% confidence level. We remark that what these differences truly mean is discussed in terms of generalization in Sect. 5.4. RS is clearly the inferior approach.

To obtain a better understanding of the search the algorithms perform, we present the fronts discovered for CIFAR-100 at different time steps in Fig. 5 and Fig. 11 (left) (we found very similar results for MacroNAS-C10). The plots for 10 and 100 evaluations confirm the fact that MO-GOMEA and LS are the quickest

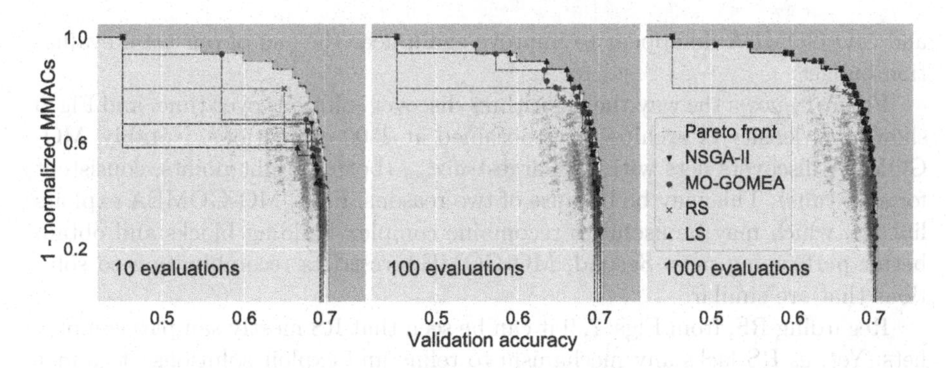

Fig. 5. Evolving archives for one example run on the MacroNAS-C100 benchmark dataset. Gray points represent all possible architectures.

at approaching the front. Note that the archive at 10 evaluations (Fig. 5 left) of LS is already obtained during the very first iteration of LS.

Compared to MO-GOMEA, LS is especially quick at finding the less-densely populated areas in the objective space (low MMACs). This is likely because making networks with less MMACs is straightforward, as more layers can be set to identity, which is effectively searched by the LS (with the proper objective weighting) that tries every option in each cell. Moreover, the EAs need to rely on the encodings present in the initial population, where an identity cell is sampled in position i only with chance $1/|\Omega_i|$. As shown in Fig. 6, evaluating small architectures is necessary to cover the part of the Pareto front containing efficient nets, as the number of nets there is smaller.

It is clear that RS is inferior to the other algorithms, as it struggles to find efficient nets that occur only sparsely in the search space. This is due to the fact that RS does not build up from efficient nets and thus solely relies on the way it samples: sampling a net with many identity cell is rare. For example, the probability of sampling the full-identity net[3] is $1/|\Omega_i|^\ell$.

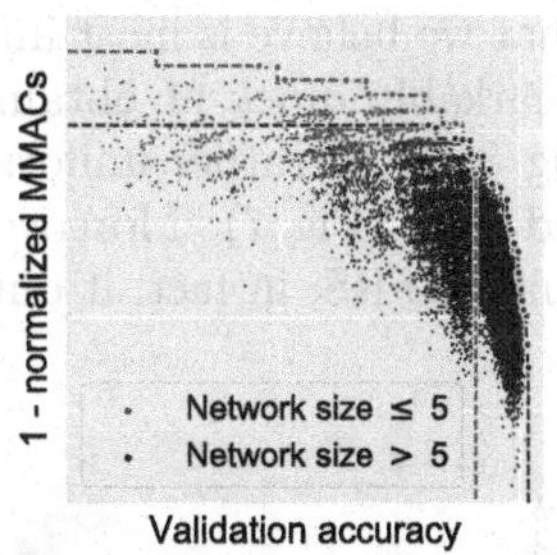

Fig. 6. Distribution of two classes of nets and their Pareto fronts.

5.3 Large-Scale NAS: CIFAR-100

Figure 10 (right; consider only acc_{val}, i.e., solid lines) shows how the hypervolume improves over time for the large-scale experiment on CIFAR-100. MO-GOMEA and LS obtain a good hypervolume similarly quickly, as observed for MacroNAS-C100. NSGA-II converges more slowly than MO-GOMEA and LS, but it is capable of outperforming them in terms of hypervolume towards the end. Nonetheless, both LS

[3] Recall that this net is pre-inserted in the archive because uninteresting.

and MO-GOMEA do appear to improve again near the end of our total evaluation budget.

Figure 8 shows the way the algorithms discover solutions over time, and Fig. 9 shows the density of architectures obtained at 2500 evaluations. Notably, MO-GOMEA discovers nets with the largest acc_{val} (bottom-right points, consistent for all 6 runs). This may be because of two reasons. First, MO-GOMEA exploits linkage, which may be useful to recombine complex building blocks and obtain better performing nets. Second, MO-GOMEA restricts recombination to solutions that are similar.

Regarding RS, from Figs. 7, 9 it can be seen that RS mostly samples complex nets. Yet, as RS lacks any mechanism to refine and exploit solutions, it cannot match MO-GOMEA's capability of discovering the most accurate nets. The main issue of RS is, as aforementioned, that it samples efficient nets too rarely.

NSGA-II behaves interestingly: it searches progressively from more complex to simpler architectures (Figs. 7, 8). This happens because in the beginning it is more likely to generate complex nets (as RS does), while later on NSGA-II's crowding distance operator steers the search towards the areas where points are most scattered, i.e., where efficient nets are located. Moreover, it is likely that its non-linkage informed operators are less likely to find the best performing nets.

LS is found to be slightly inferior to the EAs in terms of final hypervolume (Fig. 10 right, solid lines), and it does not ultimately obtain the best front (Figs. 8, and 11 middle-right) . Having only 6 repetitions, statistical testing results in no significant differences for the hypervolume at the end of the runs (i.e., at 2500 evaluations). What we believe is most important is that LS is remarkably quick at obtaining fairly good nets (Fig. 8), with rather uniform spread in terms of trade-offs (Fig. 7). Moreover, LS has no issues similar to RS, in fact, it outperforms it markedly.

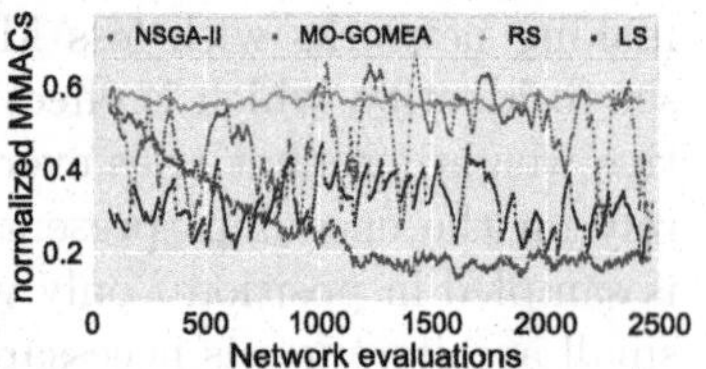

Fig. 7. MMACs of evaluated nets throughout one example run, smoothed for readability (moving avg. filter of size 75).

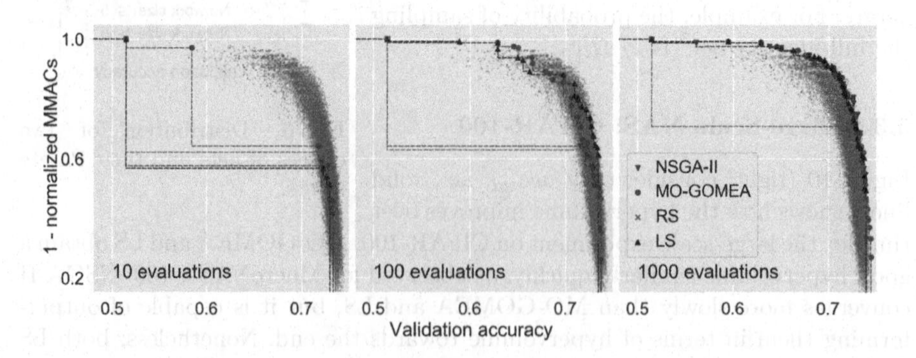

Fig. 8. Evolving archives for one example run of the large-scale NAS for CIFAR-100. Gray points represent all architectures ever discovered.

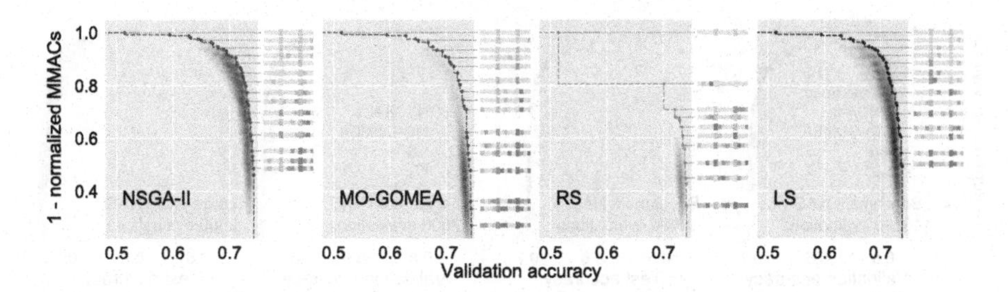

Fig. 9. Fronts obtained at 2500 evaluations by the search algorithms in one example run on the large-scale experiment on CIFAR-100, with a visual description of the nets. The colored cells correspond to the ones displayed in Fig. 1. Heatmaps display the distribution of nets evaluated by the algorithms.

5.4 Generalization

Because the validation is used during the search, a risk exists to implicitly start overfitting to it [1]. Hence, a second held-out set is used, i.e., given (a front of) best-acc_{val} nets, their acc_{test} is measured.

As can be seen from Figs. 10 (left) and 11 (left-most two), on MacroNAS-C100 RS clearly remains the worst performing algorithm when considering generalization to the test set. However, MO-GOMEA and NSGA-II lose the lead to LS, between 100 and 18,000 evaluations. This is a consequence of the aforementioned phenomenon: despite the fact that nets are not trained on the set of data on which acc_{val} is measured, acc_{val} is used to select what nets are best, hence the EAs' better search translates to more overfitting to this set. LS is slightly

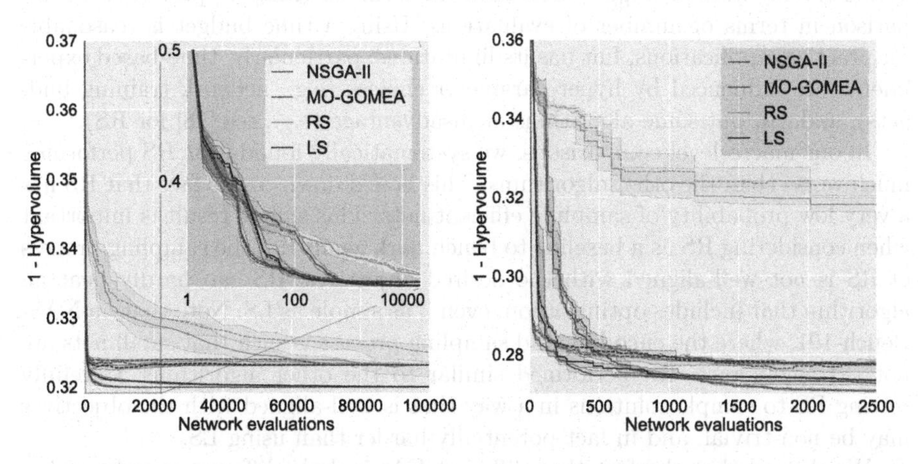

Fig. 10. Convergence graphs regarding acc_{val} (solid lines) and acc_{test} (dashed lines) on MacroNAS-C100 (left) and on CIFAR-100 on the enlarged search space (right). Medians (lines) and 25/75th percentiles (bands) of respectively 30 and 6 runs are shown for all algorithms. Note the different ranges for the vertical axes and the logarithmic scale for the horizontal axis in the zoomed-in view.

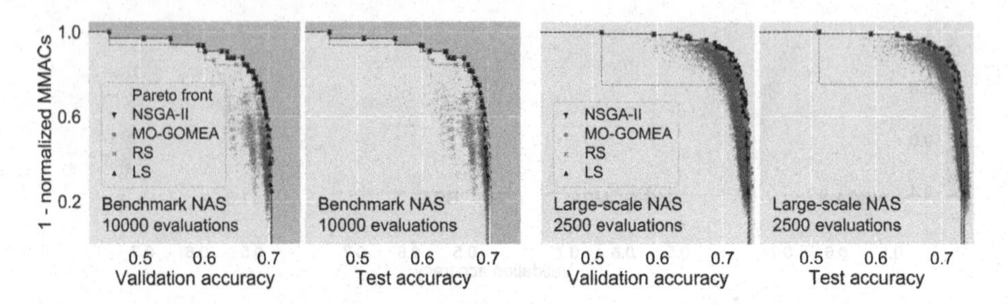

Fig. 11. Validation and test archives for 10,000 evaluations on MacroNAS-C100 (resp. first and second from the left) and 2500 evaluations on large-scale NAS (resp. second and first from the right). Archives are from one example run.

less powerful at searching (worse hypervolume in terms of acc_{val}), yet the nets generalize well to the test set. Crucially, there are no significant differences (at 99% confidence) among LS and the EAs at 100,000 evaluations.

The findings for the enlarged search space reflect the ones found for MacroNAS-C100: the difference in performance between LS and the EAs becomes smaller when considering acc_{test} (see Fig. 10 right). Figure 11 (right-most two) displays this especially well, as the most accurate nets discovered by MO-GOMEA turn out to be less accurate when tested again. Again, no statistical differences are observed at the end of the runs between LS and the EAs.

6 Discussion and Conclusion

While several works compare algorithms in terms of time, we preferred a comparison in terms of number of evaluations. Using a time budget is reasonable for practical applications, but has its limitations: particularly, time-based experiments are influenced by hyper-parameter choices (e.g., network training budgets), and can put some algorithms at disadvantage (e.g., see [18] for RS).

In our macro-level comparisons, we systematically found that RS performed much worse than the other algorithms, This boiled down to the fact that RS has a very low probability of sampling efficient nets. This simple result is important when considering RS as a baseline to benchmark against: if the sampling process of RS is not well-aligned with the desired objectives, RS can hardly beat an algorithm that includes optimization, even if as simple as LS. Note that on NAS-Bench-101, where the encoding and sampling process is such that small nets are not extremely rare, RS performed similar to the other algorithms. Carefully setting RS to sample solutions in a way that is well-aligned with the objectives may be non-trivial, and in fact potentially harder than using LS.

We showed that the fact that different EAs include different search mechanisms, such as linkage learning and cluster-restricted mating for MO-GOMEA, and non-dominated sorting and crowding distance for NSGA-II, does lead to notable differences in what nets are found at what point in time. This is important to consider in practical applications where the evaluation budget is limited.

LS was found to be competitive in this respect, as it quickly obtained a rather uniform scattering of nets with different trade-offs. Moreover, LS proved capable of ultimately competing very well when generalization is accounted for: the capability of SotA EAs to optimize at a very refined level with respect to acc_{val} does not always translate to better acc_{test} due to implicit overfitting.

In conclusion, through several experiments, we have shown that a simple, parameter-less LS algorithm can be a competitive baseline for NAS. We found that the proposed LS algorithm is competitive with state-of-the-art EAs, even up to thousands of evaluated architectures, in a multi-objective setting. By a greedy single-variable search mechanism and a random scalarization of the objectives, LS discovers nets with rather uniform trade-offs between accuracy and complexity in few evaluations. Performance gaps with EAs evident at search time are often lost when generalization is assessed. Importantly, we found that LS outperforms RS consistently and markedly. We therefore recommend to adopt LS as a baseline for NAS.

Acknowledgements. This work is part of the research programme Commit2Data with (project #628.011.012), which is financed by the Dutch Research Council (NWO), and part of a project (#15198) that is included in the research program Technology for Oncology, which is financed by the Netherlands Organization for Scientific Research (NWO), the Dutch Cancer Society (KWF), the TKI Life Sciences & Health, Asolutions, Brocacef, and Cancer Health Coach.

References

1. Bishop, C.M.: Pattern Recognition and Machine Learning. In: Information Science and Statistics. Springer, New York (2006). https://www.springer.com/gp/book/9780387310732, ISBN: 978-0-387-31073-2
2. Bosman, P.A.N.: The anticipated mean shift and cluster registration in mixture-based EDAs for multi-objective optimization. In: Proceedings of the Genetic and Evolutionary Computation Conference, GECCO 2010, pp. 351–358. Association for Computing Machinery, New York (2010). https://doi.org/10.1145/1830483.1830549
3. Chu, X., Zhang, B., Xu, R., Ma, H.: Multi-objective reinforced evolution in mobile neural architecture search. arXiv preprint arXiv:1901.01074 (2019)
4. Deb, K., Pratap, A., Agarwal, S., Meyarivan, T.: A fast and elitist multiobjective genetic algorithm: NSGA-II. IEEE Trans. Evol. Comput. **6**(2), 182–197 (2002)
5. Dong, X., Yang, Y.: NAS-Bench-201: extending the scope of reproducible neural architecture search. In: International Conference on Learning Representations (2020)
6. Elsken, T., Metzen, J.H., Hutter, F.: Efficient multi-objective neural architecture search via Lamarckian evolution. In: International Conference on Learning Representations (2019)
7. Elsken, T., Metzen, J.H., Hutter, F.: Neural architecture search: a survey. J. Mach. Learn. Res. **20**(55), 1–21 (2019)

8. He, K., Zhang, X., Ren, S., Sun, J.: Deep residual learning for image recognition. In: Proceedings of the IEEE Conference on Computer Vision and Pattern Recognition, pp. 770–778 (2016)
9. Jouppi, N.P., et al.: In-datacenter performance analysis of a tensor processing unit. In: Proceedings of the 44th Annual International Symposium on Computer Architecture, pp. 1–12 (2017)
10. Kim, Y.H., Reddy, B., Yun, S., Seo, C.: NEMO: neuro-evolution with multiobjective optimization of deep neural network for speed and accuracy. In: ICML 2017 AutoML Workshop (2017)
11. Kitchin, R.: The Data Revolution: Big Data, Open Data, Data Infrastructures and Their Consequences. Sage, Thousand Oaks (2014)
12. Lu, Z., et al.: NSGA-Net: neural architecture search using multi-objective genetic algorithm. In: Proceedings of the Genetic and Evolutionary Computation Conference, GECCO 2019, pp. 419–427. Association for Computing Machinery, New York (2019). https://doi.org/10.1145/3321707.3321729
13. Luong, N.H., Poutré, H.L., Bosman, P.A.N.: Multi-objective Gene-pool optimal mixing evolutionary algorithm with the interleaved multi-start scheme. Swarm Evol. Comput. **40**, 238–254 (2018)
14. Miikkulainen, R., et al.: Evolving deep neural networks. In: Artificial Intelligence in the Age of Neural Networks and Brain Computing, pp. 293–312. Elsevier (2019)
15. Ottelander, T.D., Dushatskiy, A., Virgolin, M., Bosman, P.A.N.: Local search is a remarkably strong baseline for neural architecture search. arXiv preprint arXiv:2004.08996 (2020)
16. Pelikan, M., Sastry, K., Cantú-Paz, E.: Scalable Optimization via Probabilistic Modeling: From Algorithms to Applications (Studies in Computational Intelligence). Springer, Heidelberg (2006). https://doi.org/10.1007/978-3-540-34954-9
17. Radford, A., Wu, J., Child, R., Luan, D., Amodei, D., Sutskever, I.: Language models are unsupervised multitask learners. OpenAI Blog (2019)
18. Real, E., et al.: Large-scale evolution of image classifiers. In: Precup, D., Teh, Y.W. (eds.) Proceedings of the 34th International Conference on Machine Learning, vol. 70, pp. 2902–2911. PMLR, International Convention Centre, Sydney, Australia (2017)
19. Russakovsky, O., et al.: ImageNet large scale visual recognition challenge. Int. J. Comput. Vis. **115**(3), 211–252 (2015). https://doi.org/10.1007/s11263-015-0816-y
20. Senior, A.W., et al.: Improved protein structure prediction using potentials from deep learning. Nature **577**, 706–710 (2020)
21. Silver, D., et al.: Mastering the game of Go without human knowledge. Nature **550**(7676), 354–359 (2017)
22. Sze, V., Chen, Y.H., Yang, T.J., Emer, J.S.: Efficient processing of deep neural networks: a tutorial and survey. Proc. IEEE **105**(12), 2295–2329 (2017)
23. Thierens, D., Bosman, P.A.N.: Optimal mixing evolutionary algorithms. In: Proceedings of the 13th Annual Conference on Genetic and Evolutionary Computation, GECCO 2011, pp. 617–624. Association for Computing Machinery, New York (2011). https://doi.org/10.1145/2001576.2001661
24. Vinyals, O., et al.: AlphaStar: Mastering the Real-Time Strategy Game StarCraft II (2019). https://deepmind.com/blog/alphastar-mastering-real-time-strategy-game-starcraft-ii/
25. White, C., Nolen, S., Savani, Y.: Local search is state of the art for NAS benchmarks. arXiv preprint arXiv:2005.02960 (2020)
26. Wistuba, M., Rawat, A., Pedapati, T.: A survey on neural architecture search. arXiv preprint arXiv:1905.01392 (2019)

27. Wong, C., Houlsby, N., Lu, Y., Gesmundo, A.: Transfer learning with neural AutoML. In: Bengio, S., Wallach, H., Larochelle, H., Grauman, K., Cesa-Bianchi, N., Garnett, R. (eds.) Advances in Neural Information Processing Systems, vol. 31, pp. 8356–8365. Curran Associates, Inc. (2018)
28. Yang, A., Esperança, P.M., Carlucci, F.M.: NAS evaluation is frustratingly hard. In: International Conference on Learning Representations (2020)
29. Ying, C., Klein, A., Christiansen, E., Real, E., Murphy, K., Hutter, F.: NAS-Bench-101: towards reproducible neural architecture search. In: Chaudhuri, K., Salakhutdinov, R. (eds.) Proceedings of the 36th International Conference on Machine Learning, vol. 97, pp. 7105–7114. PMLR, Long Beach (2019)
30. Yu, K., Sciuto, C., Jaggi, M., Musat, C., Salzmann, M.: Evaluating the search phase of neural architecture search. In: International Conference on Learning Representations (2020)
31. Zoph, B., Vasudevan, V., Shlens, J., Le, Q.V.: Learning transferable architectures for scalable image recognition. In: Proceedings of the IEEE Conference on Computer Vision and Pattern Recognition, pp. 8697–8710 (2018)

A Study on Realtime Task Selection Based on Credit Information Updating in Evolutionary Multitasking

Yumeng Cao[1], Yaqing Hou[1(✉)], Liang Feng[2], Hongwei Ge[1], Qiang Zhang[1(✉)], and Xiaopeng Wei[1]

[1] School of Computer Science and Technology, Dalian University of Technology, Dalian 116024, China
cymaabb123@mail.dlut.edu.cn

[2] College of Computer Science, Chongqing University, Chongqing 400000, China

Abstract. Recently, evolutionary multi-tasking (EMT) has been proposed as a new search paradigm for optimizing multiple problems simultaneously. Since the beneficial knowledge can be transferred among tasks to speed up the optimization process, EMT shows better performance in many problems compared with single-task evolutionary search algorithms. Notably, existing works on EMT have been devoted to the evolutionary search with two tasks. However, when multiple tasks (i.e., number of tasks >2) are involved, the existing methods might fail as it is necessary to decide which is the most suitable task to be selected for performing the knowledge transfer. To address this issue, we propose an online credit information based task selection algorithm to enhance the performance of EMT. Specifically, a credit matrix is introduced to express the transfer qualities among tasks. Then, we design two kinds of credit information and propose an updating mechanism to adjust the credit matrix online. After that, a greedy task selection mechanism is proposed for balancing the exploration and exploitation of the task selection process. Besides, we also propose an adaptive transfer rate to enhance positive transfer while reduce the impact of negative transfer. In the experiment, we compare our method with the existing works. The results clearly demonstrate the efficacy of the proposed method.

Keywords: Evolutionary multi-tasking · Task selection · Multi-population

---

This work was supported in part by the National Key Research and Development Program of China under Grant 2018YFC0910500, in part by the Liaoning Key R&D Program under Grant 2019JH210100030, in part by the Liaoning United Foundation under Grant U1908214, in part by the National Natural Science Foundation of China under Grant 61906032, and in part by the Fundamental Research Funds for the Central Universities under Grant DUT18RC(3)069.

H. Ishibuchi et al. (Eds.): EMO 2021, LNCS 12654, pp. 480–491, 2021.
https://doi.org/10.1007/978-3-030-72062-9_38

1 Introduction

Recently, evolutionary multi-tasking (EMT) has been proposed as a novel search paradigm for optimizing multiple different problems currently [1]. The idea is mainly inspired by the remarkable ability of human brain that can handle multiple tasks together and the fact that the real world problems seldom exist in isolation. The correlation among tasks has made EMT a great success in solving distinct optimization problems, including continuous, combinatorial, multi, many-objective optimization tasks, etc. [2–4].

In recent years, a lot of effort has been put into the study of EMT [4–12]. In particular, Gupta et al. [2] proposed multifactorial evolutionary algorithm (MFEA), which is one of the most popular instantiations of the paradigm. MFEA uses a common population to optimize all tasks simultaneously by encoding the solutions in a unified search space and sharing beneficial knowledge via implicit genetic transfer. Further, Feng et al. [13] proposed another EMT framework, namely explicit evolutionary multitasking algorithm (EEMTA). Instead of using a uniform coding scheme, EEMTA employs problem-specific encoding with multiple populations in different optimization problems respectively. In their work, each population employs a specific evolutionary solver for task optimization, which brings additional flexibility while the optimization performance could also be enhanced.

In the study of EMT problem, most existing works have been devoted to the simultaneous evolutionary optimization with two tasks, wherein useful knowledge could be transferred from one other directly. However, when it comes to the many-task optimization problems (i.e., number of tasks >2), it is necessary to decide which is the most suitable task to be selected and when to perform the knowledge transfer. Note that different optimization tasks possess distinct domain-specific knowledge along with its evolution process. This trait indicates that the useful knowledge is time-sensitive. The knowledge of one task can be useful for instructing the search process of others at a certain generation while fails when the evolution process progresses. The out-of-date knowledge will not only fail to accelerate the search speed, but also suffer from the negative transfer and hence inhibit the evolutionary search progress [14]. Therefore, it is of great significance to select an appropriate task for conducting knowledge transfer during multitasking evolution.

To the best of our knowledge, previous studies on task selection in EMT have introduced a credit matrix to guide task selection based on the framework of EEMTO [15]. In this case, knowledge transfer is carried out explicitly through the autoencoder transfer [16] across multi-populations. In order to make proper task selections, the credit matrix is designed to estimate the correlations among various tasks and used to instruct the task selection process. The method performs knowledge transfer among all tasks and takes the number of excellent solutions transferred from other tasks to initialize the credit matrix. The credit matrix is then updated by adding new values acquired in the same way every certain intervals. Each task selects another task for knowledge transfer simply based on the value of the credit matrix. This algorithm requires a certain amount of additional evaluation cost, and uses a relatively fixed feedback mechanism and selection mechanism.

Keeping this in mind, towards reducing the additional evaluation cost and further enhancing search performance, we propose an improved task selection algorithm. We design new feedback information to update the credit matrix in real time, which is available without additional evaluation. Also, we propose a greedy selection mechanism to enhance the exploration of evolutionary algorithm. Moreover, an adaptive transfer rate is proposed to strengthen the effect of positive knowledge transfer while suppress the impact of negative knowledge transfer. To evaluate the performance of the proposed method, we conduct empirical studies on 7 commonly used benchmarks in EMT optimization problems.

The rest of this paper is organized as follows. Section 2 first gives a brief introduction of the researches about EMT and the existing work of task selection. Then, the details of the proposed algorithm for task selection based on EEMTA are presented in Sect. 3. Section 4 provides the experimental study conducted to evaluate the effectiveness and efficiency of the proposed approach. Lastly, we give a conclusion about our work in Sect. 5.

2 Background

MFEA [2] is the most popular framework in solving evolutionary multitasking optimization problems. Based on this framework, a lot of works have been carried out in various directions. For example, Gupta et al. [4] extended the EMT paradigm in the multi-objective optimization problem, which is called multi-objective multifactorial evolutionary algorithm (Mo-MFEA). Gong et al. [8] designed an algorithm to allocate the computational resources dynamically according to the computational complexities of tasks. Moreover, Zheng et al. [9] improved MFEA so that it can automatically adapt the intensity of cross-task knowledge transfer, which improves the performance of the algorithm (SRMTO). More related studies about MFEA are available in [10–12].

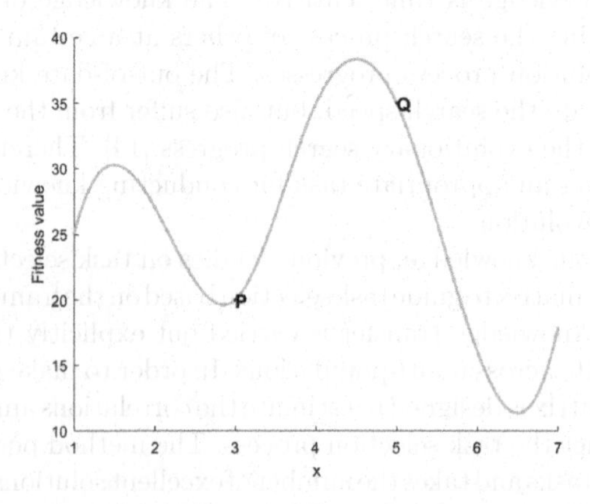

Fig. 1. An illustrative example of certain task in one dimension. Y-axis: fitness value; X-axis: generation.

As mentioned, task selection is important to improve the performance of EMT. However, few effort has been put into it to improve the efficiency of EMT in previous works. Recently, Shang et al. [15] introduced a task selection method based on the credit matrix. Nevertheless, as reported, the every update of the credit matrix requires an additional evaluation cost, thus reduce the number of updates, which damages the efficiency of knowledge transfer. If the spacing algebra between updates is too large, the efficiency of knowledge transfer will be greatly reduced. Otherwise, it will lead to an expansive evaluation cost. We introduce new feedback information to update the credit matrix without additional evaluation costs. In addition, the updating mechanism of the credit matrix in their work is one-way increasing, causing the feedback information in the later stage easily affected by earlier knowledge. We design a bidirectional reward-punishment mechanism to update the credit matrix. Besides, the existing task selection method always selects a best task. However, a solution with a better fitness value does not always reflect a better search direction. As shown in Fig. 1, the solution P has a better fitness value, but it is easy to fall into local optimization. The current fitness value of the solution Q is not as good as P, while it is easier to search in a better direction. Therefore, we employs a greedy task selection mechanism, which gives more opportunity for exploration. Moreover, the existing task selection method adopts a fixed transfer rate which is common in most transfer modes in evolutionary multitasking [2,4,8]. To the best of our knowledge, SREMT [9] uses an adaptive transfer rate by employing an ability vector in implicit knowledge transfer. However, there is no proper adaptive transfer rate in explicit evolutionary multitasking optimization. In contrast, We adopt an adaptive transfer rate between tasks during the search to for higher flexibility.

3 Proposed Method for Task Selection

In order to enhance the performance of explicit evolutionary many-tasking optimization (EEMTO), we propose a new task selection method. The whole workflow of our proposed method is shown in Fig. 2. Firstly, in Sect. 3.1, we introduce a new credit information, which can be collected without additional evaluations. Secondly, a reward-punishment mechanism is described to update the credit matrix using the collected information in Sect. 3.2. Thirdly, a greedy task selection mechanism is described in Sect. 3.3 for instructing task selection. Lastly, we employ a flexible transfer rate to improve the performance of knowledge transfer in Sect. 3.4. Specifically, the credit matrix is recomputed after a large interval G for accuracy. The details of the task selection algorithm are depicted in Algorithm 1.

3.1 Credit Information in Evolutionary Process

We hope to utilize information that is readily available during evolution for instructing the task selection process. To this end, we divide each population

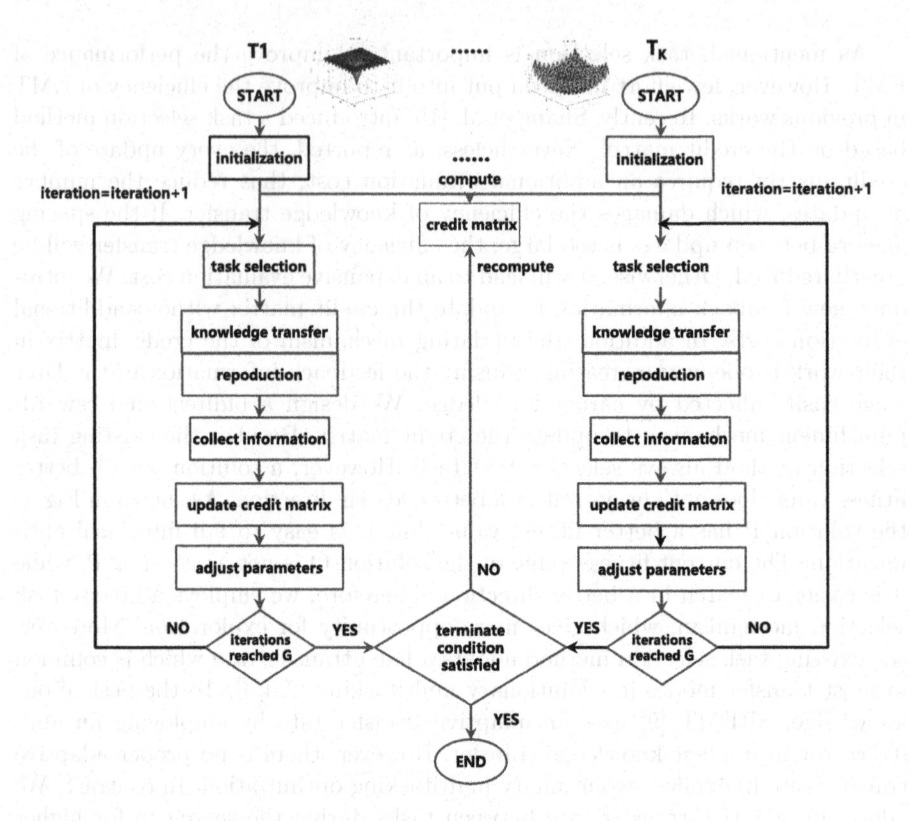

Fig. 2. Workflow of the proposed method (G denotes the interval of computing credit matrix).

into three categories, including the set of original parent individuals, the set of individuals obtained through crossover and variation, and the set of individuals acquired from another task by knowledge transfer. Among which, we design two kinds of credit information, including *weights* and *quality*. That is each set has its own *weights* and *quality* per generation, which can be defined by Eq. (1) and Eq. (2):

$$weights = \sum_{i=1}^{ct} NP - rank(i) \tag{1}$$

$$quality = \frac{weights}{ct} \tag{2}$$

where NP is the population size, $rank(i)$ denotes the fitness ranking of individual i in its population, and ct represents the number of the individuals in one set. *weights* is used to evaluate the overall performance of the set and *quality* aims to evaluate the average optimal performance of individuals in the set. Typically, a higher value of the *quality* indicates better average optimization performance of individuals in the set.

Algorithm 1. Pseudo code of the proposed method for task selection

Require: K: number of optimization tasks. K benchmark functions.optimization solvers GA or DE
Ensure: the result of solutions found of K tasks
1: Assign a solver to each task
2: initialize K populations of K tasks randomly, compute fitness of each individual of each task
3: compute credit matrix W according to auto-encoder transfer
4: **while** stopping conditions are not satisfied **do**
5: **for** every task T_i **do**
6: select a task T_k for knowledge transfer according to equation (6)
7: evolution with knowledge transfer from the selected task T_k
8: collect feedback information *weithts* and *quality* according to equation (1) and (2)
9: compute the *reward* value in this evolution according to equation (4)
10: update W_{ii} and W_{ik} in the credit matrix using *reward* according to equation (5)
11: update the parameters used in the evolution
12: **end for**
13: **end while**

3.2 Realtime Updating Mechanism of Credit Matrix

In this section, we propose an updating mechanism to adjust the credit matrix per generation, thus enabling knowledge transfer in real time. The credit matrix, denoted by W, is a $K \times K$ matrix which is computed based on the baseline method [15]. The proposed realtime updating mechanism allows each task to select itself when making selections. This is due to the fact that the transferred individuals obtained by other tasks do not always perform better than the individuals generated during their own optimization. We update elements of the credit matrix according to the information collected, specifically, the updating mechanism is defined by Eq. (3)–(5):

$$effect = \frac{quality}{quality^{old}} \quad (3)$$

$$reward = log(effect)W_{ii} \quad (4)$$

$$W_{ij} = W_{ij} + reward \quad (5)$$

where, W_{ii} indicates the realtime performance of optimization on Task i itself, which will be updated after each generation. And W_{ij} denotes the realtime performance of transferred solutions from Task j to Task i, which will be updated if Task i acquired transferred solutions from Task j during this generation. The two values are updated in a same way through a reward mechanism. Moreover, $quality^{old}$ is the value of the set which consists of original parent individuals in the new generation, and $quality$ is the value of other two sets. $effect$ is defined to evaluate the performance of the two sets. Specially, we employ a log function to make the reward positive when $effect > 1$, and negative otherwise.

3.3 Greedy Task Selection Mechanism

In our method, to make suitable task selections, we adopt a greedy task selection mechanism which is commonly used in reinforcement learning [17]. Specifically, we assign a probability to each task first, and then make selections according to the probability. Instead of assigning the best task in the credit matrix with a probability of 1, we assign the highest probability to it. The rest probability is reserved for other tasks considering more exploration. The probability allocation for each task being selected could be defined by Eq. (6):

$$P_{ik} = \begin{cases} \dfrac{W_{ik}}{\sum_{j=1}^{K} W_{ij} - W_{iq}} \epsilon, k \neq q, \\ 1 - \epsilon, k = q, \end{cases} \tag{6}$$

where, P_{ik} is the probability that Task i selects Task k for knowledge transfer. Task q denotes the best task to be selected by Task i in the credit matrix. ϵ represents the probability of exploring knowledge in other tasks, which is assigned based on the proportion of other tasks in the credit matrix. In addition to using a greedy task selection mechanism, we also employ a variable exploration probability. If one task is continuously selected by a certain task in the selection history, the exploration probability of the task will be improved. On the contrary, if one task make selections too randomly, the exploration probability of the task will be reduced.

3.4 Adaptive Transfer Rate in Task Selection

We adaptively change the transfer rate *rmp* to enhance the positive transfer effect and reduce the negative transfer, where *rmp* is adjusted by the definition Eq. (7) and Eq. (8):

$$eff\_rate = \frac{ct}{rmp \times NP} \tag{7}$$

$$rmp = \begin{cases} Amp \times rmp, eff\_rate = 1, \\ rmp \times eff\_rate, else \end{cases} \tag{8}$$

where *ct* is the number of transferred individuals preserved in the new generation. eff_rate represents the proportion of effective transferred individuals. *Amp* is the amplification coefficient and eff_rate is the magnificine to adjust the transfer rate *rmp*. *rmp* is then changed accordingly, greater effective transfer ratio leads to a greater transfer rate.

4 Empirical Study

In this section, empirical study is conducted to evaluate the efficacy of the proposed task selection method. Specially, we introduce the experimental setup in Sect. 4.1. Then we compare the proposed method with other works and discuss the results in Sect. 4.2.

Table 1. Summary of the shift operation on each optimization function.

Function	Griewank (T1)	Rastrigin (T2)	Ackley (T3)	Schwefel (T4)	Sphere (T5)	Rosenbrock (T6)	Weierstrass (T7)
Interval for shift	[−1,1]	[−5,5]	[−3,3]	[−10,10]	[−5,5]	[−10,10]	[−0.1,0.1]

4.1 Experimental Setup

First of all, we employ the common-used evolutionary multi-tasking optimization benchmarks proposed in [18], 7 various classical single-objective continuous optimization functions, i.e., "Griewank", "Rastrigin", "Ackley", "Schwefel", "Sphere", "Rosenbrock" and "Weierstrass", which are labeled as Task 1 (T1) to Task 7 (T7), respectively, to develop the evolutionary multi-task optimization problem in our study. Since most of these functions share a same global optimum, we perform random rotation and shift on these functions separately and independently to complicate these problems, so that we could conduct better investigation of the task selection algorithm in EMT. In particular, the interval of random shift for each function is summarized in Table 1.

Next, to evaluate the performance of the proposed method for task selection, we compare the proposed method with baseline method (i.e. [15]), random task selection method and the traditional single-task method. We conduct the experiment using two optimization solvers: genetic algorithm (GA) [19] and differential evolutionary algorithm (DE) [20]. Note that for the same optimization task, the same solver is used in all compared methods for fairness. Other important parameters are presented:

- Population Size: NP = 100 for all solvers.
- Number of sampled solutions for learning the task mappings: NoS = 100.
- Maximum Function Evaluations: MaxFEs = 50000 for all solvers.
- Independent number of runs: runs = 20 for all methods.
- Evolutionary operators in GA:
  SBX crossover: pc = 1, $\eta_c = 2$.
  Polynomial mutation: pm = 1/d, $\eta_m = 5$.
- Evolutionary operators in DE:
  F = 0.5, CR = 0.6.
- Interval of large scale solution transfer across tasks to update credit matrix:
  G = 50 in the proposed method.
  G = 10 in the baseline method.
- Number of transferred solutions to compute credit matrix: Q = 10.
- Transfer rate:
  Rmp = $[0.1, 0.3]$ in the proposed method, and initial value is set to 0.2.
  Rmp = 0.2 in all the compared methods.
- Exploration probability:
  ϵ in the proposed method is set to $[0.1, 0.3]$, and initial value is set to 0.1.

Table 2. Averaged results of four task selection algorithms with two solvers randomly selected, across 20 independent runs, respectively. The superior performance is highlighted in bold.

Tasks	Proposed selection method	Baseline selection method	Random selection method	Single-task method
T1	$\mathbf{1.34e-05 \pm 9.64e-06}$	$1.70e-04 \pm 9.58e-05$	$3.72e-05 \pm 1.59e-05$	$2.15e-04 \pm 7.84e-05$
T2	$\mathbf{459.70 \pm 21.39}$	479.37 ± 22.47	467.75 ± 16.71	480.56 ± 17.85
T3	$\mathbf{1.38e-04 \pm 4.25e-05}$	$4.89e-04 \pm 8.41e-05$	$6.33e-04 \pm 2.85e-04$	3.17 ± 7.72
T4	$1.374e+04 \pm 419.47$	$1.385e{+}04{\pm}508.81$	$1.372e+04 \pm 525.71$	$\mathbf{1.371e+04 \pm 575.05}$
T5	$\mathbf{7.11e-08 \pm 3.96e-08}$	$1.10e-06 \pm 6.58e-07$	$1.20e-06 \pm 4.66e-07$	$2.56e-06 \pm 9.19e-07$
T6	$\mathbf{58.15 \pm 31.86}$	173.06 ± 302.17	114.53 ± 123.88	$775.95 \pm 1.65e+03$
T7	$\mathbf{13.59 \pm 2.37}$	13.90 ± 2.98	14.83 ± 2.15	67.17 ± 2.27

4.2 Results and Discussions

1) Solution quality: The averaged fitness value obtained by the proposed method against with the baseline method, random task selection method and single-task method, across 20 independent runs, on all the 7 optimization tasks are summarized in Table 2. In the table, superior performance is highlighted in bold.

In Table 2, it can be observed that our proposed task selection method obtained superior averaged fitness values on 6 tasks over the other one. In particular, on T1, T3, T5, and T6, the proposed task selection algorithm obtains much better solution qualities than the other methods. For instance, on T5, the proposed method achieves averaged fitness value approaching to $e-8$ class, while the other two only converges to averaged fitness values around $e-6$ class. On T2 and T7, our task selection method obtains a little bit better results. On T2, the fitness value achieved by the baseline method is almost the same as that of single-task method while our proposed method gets a better result. It indicates that appropriate knowledge transfer is effective on Task 2, however, too many negative transfer in the later process will slow the search. On T4, four methods obtain similar results, which means knowledge transfer from other 6 tasks may not work very well on T4. In this situation, the baseline method even gets worst result due to negative transfer while our proposed method reduces the negative transfer during exploring. It can also be observed that random selection method does better than baseline method in some functions such as T1, T2 and T6, and performs worse in other functions. It illustrates that knowledge transfer with high frequency could be useful and effective, but it need more accurate task selection to avoid negative transfer. These competitive results demonstrate the efficiency of our proposed method.

2) Convergence speed: The averaged convergence trend of our proposed task selection method, the baseline method, random task selection method and single-task method, across 20 independent runs, on all the 7 optimization tasks are depicted in Fig. 3 and Fig. 4, respectively. Concretely, Y-axis represents the aver-

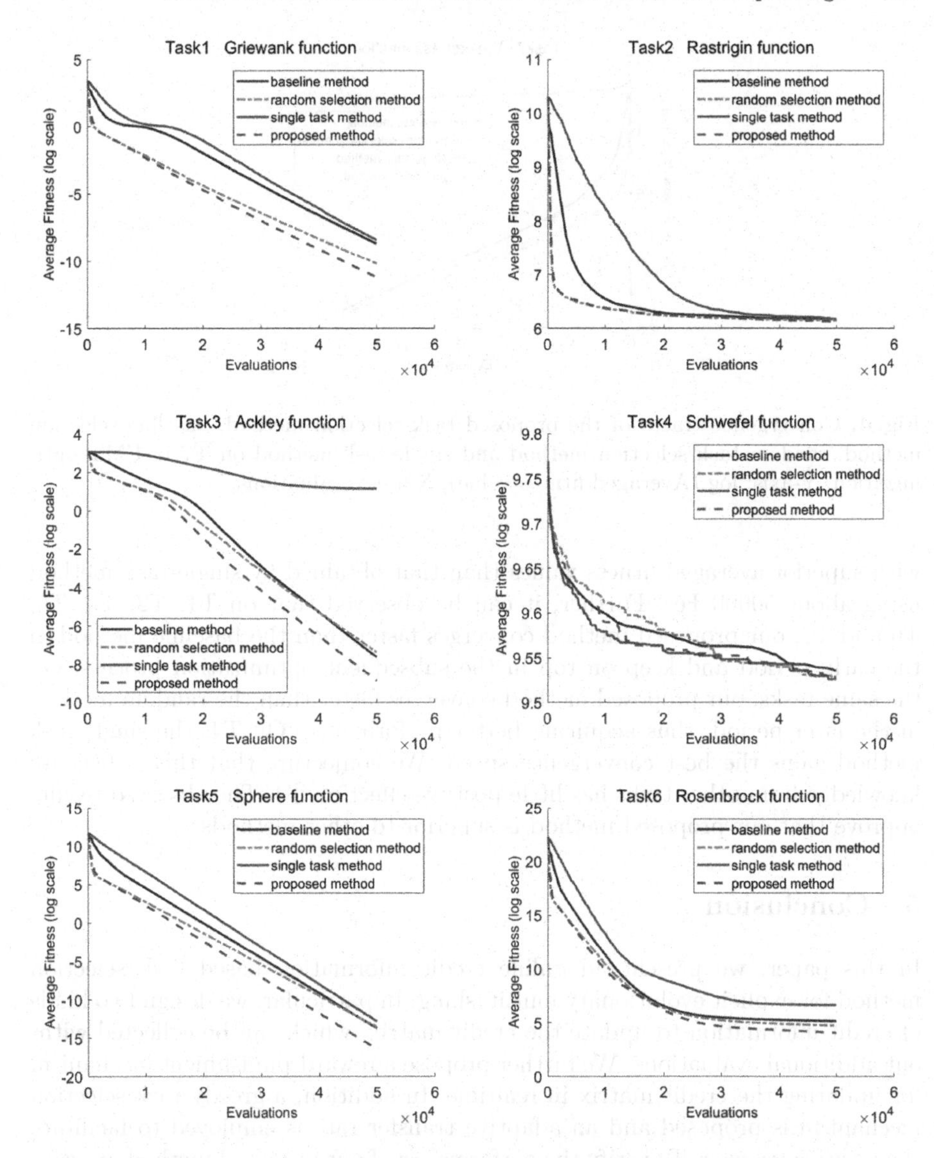

Fig. 3. Convergence traces of the proposed task selection method, baseline selection method, random task selection method and single-task method on T1–T6 in EMT optimization. Y-axis: log (Averaged fitness value); X-axis: evaluations.

aged objective values in log scale for clarity and intuition, while the X-axis gives the number of function evaluations.

From these above figures, we can observe that on most of tasks, the three methods with transfer learning converge much faster than the traditional single-task method. Especially in T3 and T7, the three methods with knowledge transfer take only around 15000 Fes and less than 5000 Fes to arrive at the solutions

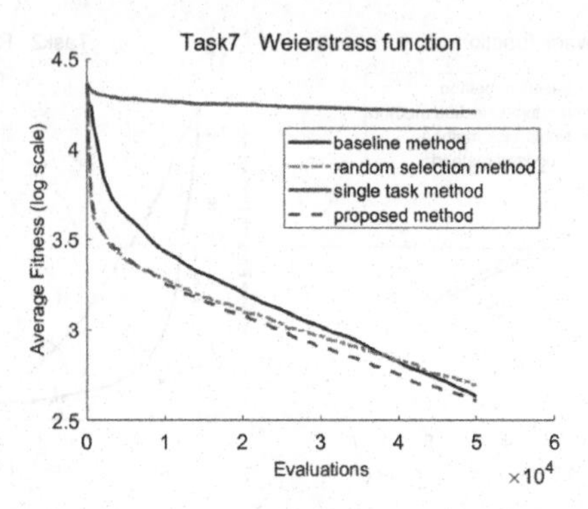

Fig. 4. Convergence traces of the proposed task selection method, baseline selection method, random task selection method and single-task method on T7 in EMT optimization. Y-axis: log (Averaged fitness value); X-axis: evaluations.

with superior averaged fitness values than that obtained by single-task method using about 50000 Fes. Further, it can be observed that on T1, T2, T3, T5, T6 and T7, our proposed method converges faster than the baseline method in the early period and keep on top in the subsequent optimization process. On the same tasks, our proposed method converges faster than the random method in the later period, thus acquiring better performance. On T4, the single-task method gains the best convergence speed. We conjecture that this is because knowledge from other tasks has little positive effect on T4. The observed results approve that our proposed method is superior to other methods.

5 Conclusion

In this paper, we present an online credit information based task selection method for explicit evolutionary multitasking. In particular, we design two kinds of credit information to update the credit matrix, which can be collected without additional evaluations. We further propose a reward-punishment mechanism for updating the credit matrix in realtime. In addition, a greedy task selection mechanism is proposed and an adaptive transfer rate is employed to facilitate the positive transfer. To verify the performance of our proposed method, we conduct experiments on the EMT optimization problem in our proposed method compared with the baseline task selection method, the random task selection method, and the traditional single-task method. The final results confirmed the superiority of the proposed method.

References

1. Ong, Y.-S., Gupta, A.: Evolutionary multitasking: a computer science view of cognitive multitasking. Cogn. Comput. **8**, 125–142 (2016). https://doi.org/10.1007/s12559-016-9395-7

2. Gupta, A., Ong, Y.S., Feng, L.: Multifactorial evolution: toward evolutionary multitasking. IEEE Trans. Evol. Comput. **20**(3), 343–357 (2016)
3. Liu, C.H., Ting, C.K.: Computational intelligence in music composition: a survey. IEEE Trans. Emerg. Top. Comput. Intell. **1**(1), 2–15 (2017)
4. Gupta, A., Ong, Y., Feng, L., Tan, K.C.: Multiobjective multifactorial optimization in evolutionary multitasking. IEEE Trans. Cybern. **47**(7), 1652–1665 (2017)
5. Bali, K.K., Gupta, A., Feng, L., Ong, Y.S., Siew, T.P.: Linearized domain adaptation in evolutionary multitasking. In: 2017 IEEE Congress on Evolutionary Computation (CEC), pp. 1295–1302 (2017)
6. Zhong, J., Feng, L., Cai, W., Ong, Y.: Multifactorial genetic programming for symbolic regression problems. IEEE Trans. Syst. Man Cybern. Syst. **50**, 1–14 (2018)
7. Ding, J., Yang, C., Jin, Y., Chai, T.: Generalized multitasking for evolutionary optimization of expensive problems. IEEE Trans. Evol. Comput. **23**(1), 44–58 (2019)
8. Gong, M., Tang, Z., Li, H., Zhang, J.: Evolutionary multitasking with dynamic resource allocating strategy. IEEE Trans. Evol. Comput. **23**(5), 858–869 (2019)
9. Zheng, X., Qin, A.K., Gong, M., Zhou, D.: Self-regulated evolutionary multitask optimization. IEEE Trans. Evol. Comput. **24**(1), 16–28 (2020)
10. Wen, Y., Ting, C.: Parting ways and reallocating resources in evolutionary multitasking. In: 2017 IEEE Congress on Evolutionary Computation (CEC), pp. 2404–2411 (2017)
11. Liaw, R., Ting, C.: Evolutionary many-tasking based on biocoenosis through symbiosis: a framework and benchmark problems. In: 2017 IEEE Congress on Evolutionary Computation (CEC), pp. 2266–2273 (2017)
12. Tang, Z., Gong, M., Zhang, M.: Evolutionary multi-task learning for modular extremal learning machine. In: 2017 IEEE Congress on Evolutionary Computation (CEC), pp. 474–479 (2017)
13. Feng, L., et al.: Evolutionary multitasking via explicit autoencoding. IEEE Trans. Cybern. **49**(9), 3457–3470 (2019)
14. Pan, S.J., Yang, Q.: A survey on transfer learning. IEEE Trans. Knowl. Data Eng. **22**(10), 1345–1359 (2010)
15. Shang, Q., Zhang, L., Feng, L., Hou, Y., Liu, H.L.: A preliminary study of adaptive task selection in explicit evolutionary many-tasking. In: 2019 IEEE Congress on Evolutionary Computation (CEC) (2019)
16. Feng, L., Ong, Y., Jiang, S., Gupta, A.: Autoencoding evolutionary search with learning across heterogeneous problems. IEEE Trans. Evol. Comput. **21**(5), 760–772 (2017)
17. Tokic, M.: Adaptive ε-greedy exploration in reinforcement learning based on value differences. In: Dillmann, R., Beyerer, J., Hanebeck, U.D., Schultz, T. (eds.) KI 2010. LNCS (LNAI), vol. 6359, pp. 203–210. Springer, Heidelberg (2010). https://doi.org/10.1007/978-3-642-16111-7_23
18. Da, B., et al.: Evolutionary multitasking for single-objective continuous optimization: benchmark problems, performance metric, and baseline results (2017)
19. Schatten, A.: Genetic algorithm tutorial. Stats. Comput. **4**(2), 65–85 (2002)
20. Storn, R., Price, K.: Differential evolution: a simple and efficient heuristic for global optimization over continuous spaces. J. Global Optim. **11**, 341–359 (1997). https://doi.org/10.1023/A:1008202821328

Multi-objective Neural Architecture Search with Almost No Training

Shengran Hu, Ran Cheng(✉), Cheng He, and Zhichao Lu

Guangdong Provincial Key Laboratory of Brain-inspired Intelligent Computation, Department of Computer Science and Engineering, Southern University of Science and Technology, Shenzhen 518055, China

Abstract. In the recent past, neural architecture search (NAS) has attracted increasing attention from both academia and industries. Despite the steady stream of impressive empirical results, most existing NAS algorithms are computationally prohibitive to execute due to the costly iterations of stochastic gradient descent (SGD) training. In this work, we propose an effective alternative, dubbed *Random-Weight Evaluation* (RWE), to rapidly estimate the performance of network architectures. By just training the last linear classification layer, RWE reduces the computational cost of evaluating an architecture from hours to seconds. When integrated within an evolutionary multi-objective algorithm, RWE obtains a set of efficient architectures with state-of-the-art performance on CIFAR-10 with less than two hours' searching on a single GPU card. Ablation studies on rank-order correlations and transfer learning experiments to ImageNet have further validated the effectiveness of RWE.

Keywords: Neural architecture search · Performance estimation · Multi-objective optimization · Evolutionary algorithms

1 Introduction

Deep Convolutional Neural Networks (CNNs) have achieved remarkable success in various computer vision tasks. One of the key reasons behind this success is the innovation on CNN architectures [9,15,20]. Despite the steady stream of promising improvements over state-of-the-art across a wide range of tasks and datasets, these early architectural advancements take years of efforts from researchers and practitioners with vast computational resources. Neural Architecture Search (NAS), on the other hand, aims to alleviate this painstaking process by automating the design of CNN architectures. However, existing NAS algorithms typically require enormous computation overheads to search. For example, early works [23,37] require thousands of GPU days to complete one search, making them impractical for real-world deployments.

H. Ishibuchi et al. (Eds.): EMO 2021, LNCS 12654, pp. 492–503, 2021.
https://doi.org/10.1007/978-3-030-72062-9_39

The main computation bottleneck of a NAS algorithm resides in the step of evaluating architectures' performance. A thorough training of an architecture can take days, even weeks on a single GPU card depending on the complexity of architectures and the scale of the datasets. Hereby, advocating for substitute measurements becomes a common theme in existing NAS algorithms. For instances, there are works [19,23,37] using the performance measured on architectures with reduced sizes and training epochs; and there are works [17,21] using performance measured on shared (instead of trained) weights. Conceptually, both of these two methods are restricted by the generality of search spaces, where the former requires search spaces to be modular and the later requires search spaces to be sequential [21,25] (as oppose to multi-branched [17,19,37]).

To address the aforementioned issues, we propose the Random-Weight Evaluation (RWE), a flexible and effective method to accelerate the performance evaluation of architectures. By leveraging the expressive power of randomly initialized convolution filters [13,26], RWE freezes the backbone[1] part of CNN architectures, and only trains the last classification layer. The subsequent performance becomes the indicator to select architectures. In short, our key contributions are summarized below:

- In this work, we propose RWE to expedite the performance estimation of CNN architectures. RWE is conceptually flexible as it is independent of search spaces, and empirically effective as we show later in the paper that it significantly reduces the evaluation wall-clock time from days to less than a minute. At the same time, RWE reliably estimates the performance of CNN architectures, measured in rank-order correlations.
- We further design a multi-objective evolutionary NAS algorithm to effectively utilize the proposed evaluation method. On CIFAR-10, the proposed algorithm achieves a set of efficient architectures with state-of-the-art performance, yielding 2.98% Top-1 error and 1.5M parameters with less than two hours on a single GPU card.

2 Related Works

In this section, we provide a brief overview on the topics that are closely related to technicalities of our approach.

Performance Estimation: A commonly used technique, in early NAS works [23,37], to expedite the performance evaluations involves down-scaling the architecture sizes, by reducing the number of layers, channels, and training epochs. The main limitation of this method is that it is only applicable to modular search spaces, where a CNN architecture is constructed by repeatedly stacking modular blocks. Extending this method to the search space that allows non-modular

[1] All layers prior to the task-specific heads, e.g., the last linear layer in case of object classification.

architectures is not trivial. Concurrently, there is a study [34] showing that extensively reducing the number of training epochs leads to a poor rank-order correlation between predicted and true performance.

Instead of training every architecture from scratch, the *weight sharing* method attempts to speed up the peformance evaluations by sharing the weights among architectures sampled during search. NAS algorithms in this category [17,21] typically construct a supernet (prior to the search), such that all searchable architectures become subsets (of the supernet; i.e. subnets) and weights are directly inherited instead of randomly initialized. Then the evaluation of an architecture becomes an inference on the validation set, which is much cheaper than training. Despite the efficiency gained during search, the supernet, required by the weight sharing method, can be more time consuming to train than a complete search [1], and may not be feasible for all search spaces, e.g. [16].

Expressive Power of Randomly Initialized Convolution Filters: Convolution filters, even with randomly initialized weights, are surprisingly powerful in extracting meaningful feature representations from visual inputs [12]. A number of existing works have shown that randomly initialized convolution filters can achieve comparable performance to CNNs with fully trained convolution filters on both vision and control tasks [6,13]. Few trials have been attempted to estimate the performance of a neural network from randomly initialized weights in the literature. More specifically, Saxe *et al.* use randomly initialized weights to predict the ranking of shallow CNNs' performance [26]; Rosenfeld and Tsotsos show that the performance ranking of widely used CNN architectures can be predicted by training a tiny fractions of the weights in the CNNs [24]. In this work, we train the last classification layer while freezing all other weights at initial values, and use this performance as the indicator to compare architectures. We demonstrate that our approach scales to modern CNN search spaces containing deep and complex CNNs.

Multi-objective NAS: Early NAS algorithms [17,23,37] are primarily driven by a single objective of maximizing predictive accuracy. However, real-world applications oftentimes require the CNN architectures to balance other completing objectives, such as power consumption, inference latency, memory footprint, to name a few. A portfolio of recently emerged NAS works scalarizes multiple objectives into a composite measurement that simultaneously promotes predictive performance and penalizes architecture complexity [2,30]. In this work, we opt for evolutionary multi-objective optimization to approximate the entire efficient frontier in one run [3].

3 Proposed Approach

The problem of designing optimal architectures under multiple objectives for a target dataset $\mathcal{D} = \{\mathcal{D}_{trn}, \mathcal{D}_{vld}, \mathcal{D}_{tst}\}$ can be formulated as the following bilevel optimization problem [19],

$$\begin{aligned} \underset{\boldsymbol{\alpha}}{\text{minimize}} \quad & f_1(\boldsymbol{\alpha}; \boldsymbol{w}^*(\boldsymbol{\alpha})), f_2(\boldsymbol{\alpha}), ..., f_m(\boldsymbol{\alpha}) \\ \text{subject to} \quad & \boldsymbol{w}^*(\boldsymbol{\alpha}) \in \underset{\boldsymbol{w}}{\operatorname{argmin}} \, \mathcal{L}(\boldsymbol{w}; \boldsymbol{\alpha}), \\ & \boldsymbol{\alpha} \in \Omega_{\alpha}, \quad \boldsymbol{w} \in \Omega_{w}, \end{aligned}$$

where the upper lever variable $\boldsymbol{\alpha}$ defines a candidate architecture, and the lower level variable $\boldsymbol{w}(\boldsymbol{\alpha})$ represents the weights with respect to it. $\mathcal{L}(\boldsymbol{w}; \boldsymbol{\alpha})$ is the cross-entropy loss on the $\mathcal{D}_{trn}$ for a candidate architecture $\boldsymbol{\alpha}$ with weights $\boldsymbol{w}$. The first objective f_1 represents the classification error on $\mathcal{D}_{vld}$, which depends on both architectures and weights. Other objectives $f_2, ..., f_m$ only depend on architectures, such as the number of parameters, floating-point operations (FLOPs), latency, etc. In our approach, we use predictive performance to approximate the ground-truth one and take the predictive performance and FLOPs as two objectives to be optimized, which aims to balance between the performance and the complexity of architectures.

3.1 Random-Weight Evaluation

The core of our approach, Random-Weight Evaluation (RWE), is shown in Algorithm 1. First, the architecture $\boldsymbol{\alpha}$ to be evaluated would get decoded into a CNN *net* without the last classification layer, which is a linear classifier. Notice that, the number of channels and layers of the decoded architecture should be indicated in this step. Then, the weights of *net* would be randomly initialized and then kept fixed throughout the whole algorithm. We use a modified version of the *Kaiming initialization* [8] to initialize the *net* (default setting in PyTorch). After the initialization of *net*, a list of linear classifiers *list_clsfr* would be initialized. Those classifiers get trained on the *features* that are extracted by the *net*, and each classifier could only be exposed to a part of the *features*. Then all of the classifiers are combined as the last classification layer in the inference phrase via neural network ensemble technique [7]. More specifically, the length of the *list_clsfr* is set to five, and each classifier gets trained on 4/5 *features*. Compared with the case that only one linear classifier gets trained, a list of linear classifiers can help stabilizing RWE. Finally, the architecture is validated by taking the labels that get approved by most classifiers as a result. After calculating the classification error rate and FLOPs of the architecture, these two measurements of performance and complexity of architectures will be returned as two objectives to be optimized.

3.2 Search Space and Encoding

Our proposed RWE is conceptually flexible and can be applied to various search spaces. In this work, we adopt the NASNet [37] search space, which is widely used in various NAS algorithms [17,22,23]. A pictorial overview of this search space and encoding is shown in Fig. 1.

Algorithm 1: Random-Weight Evaluation

Input : An architecture $\boldsymbol{\alpha}$, a training and validation dataset $\mathcal{D}_{trn}$, $\mathcal{D}_{vld}$, the number of linear classifiers L.

Output: The predictive classification error rate *Error* on $\mathcal{D}_{vld}$ and the floating-point operations *FLOPs* of $\boldsymbol{\alpha}$.

1 *net* ← Decode the architecture $\boldsymbol{\alpha}$ into CNN;
2 Randomly initialize the *net*;
3 Freeze the weights throughout the whole algorithm;
4 *list_clsfr* ← Randomly initialize a list of linear classifiers with length L;
5 $\mathcal{S}$ ← Split the $\mathcal{D}_{trn}$ uniformly into L subsets;
6 **for** $i \leftarrow 1$ **to** L **do**
7   **if** $L \neq 1$ **then**
8     $\mathcal{D}_{trn,i} \leftarrow \bigcup \mathcal{S}[j]$ for $j = 1, ..., L$ and $j \neq i$;
9   **else**
10     $\mathcal{D}_{trn,i} \leftarrow \mathcal{D}_{trn}$;
11   **end**
12   *Features* ← Infer the *net* on $\mathcal{D}_{trn,i}$;
13   Train the linear classifier *list_clsfr*[i] with the *Features* as input.
14 **end**
15 **foreach** *image, target* $\in \mathcal{D}_{vld}$ **do**
16   **foreach** *linear classifier* $\in$ *list_clsfr* **do**
17   Infer the model for *image* with *linear classifier*;
18   **end**
19   *ensemble_inference* ← the label approved by most classifiers.
20   Compare the *ensemble_inference* with *target* and record the result;
21 **end**
22 *Error* ← Calculate the classification error rate;
23 *FLOPs* ← Calculate the FLOPs of *net*;
24 **return** *Error, FLOPs*.

Network: The macro network architecture is prefixed to be a stack of several cells. We search for two cells: a normal cell and a reduction cell. All normal cells in an architecture share the same structure (but different weights), which is the same case for the reduction cell. The normal cell keeps the resolution and the number of channels for input tensors, while the reduction cell downsamples the resolution and double the number of channels for input tensors (Fig. 1a RIGHT).

Cell and Node: The structures of both normal and reduction cells are defined by five nodes (Node 2 to Node 6), each of which chooses two inputs from previous nodes and two operations applied on two inputs respectively. For a node i, we search for two inputs from previous nodes $N_1^{cell,i}$ and $N_2^{cell,i}$ and two operations $O_1^{cell,i}$ and $O_2^{cell,i}$ applied on them, where the *cell* could be n or r, representing normal cells and reduction cells. For those nodes that are not chosen to become the inputs of another node, their outputs would be concatenated to the output node (Node 7) (Fig. 1a MIDDLE and LEFT).

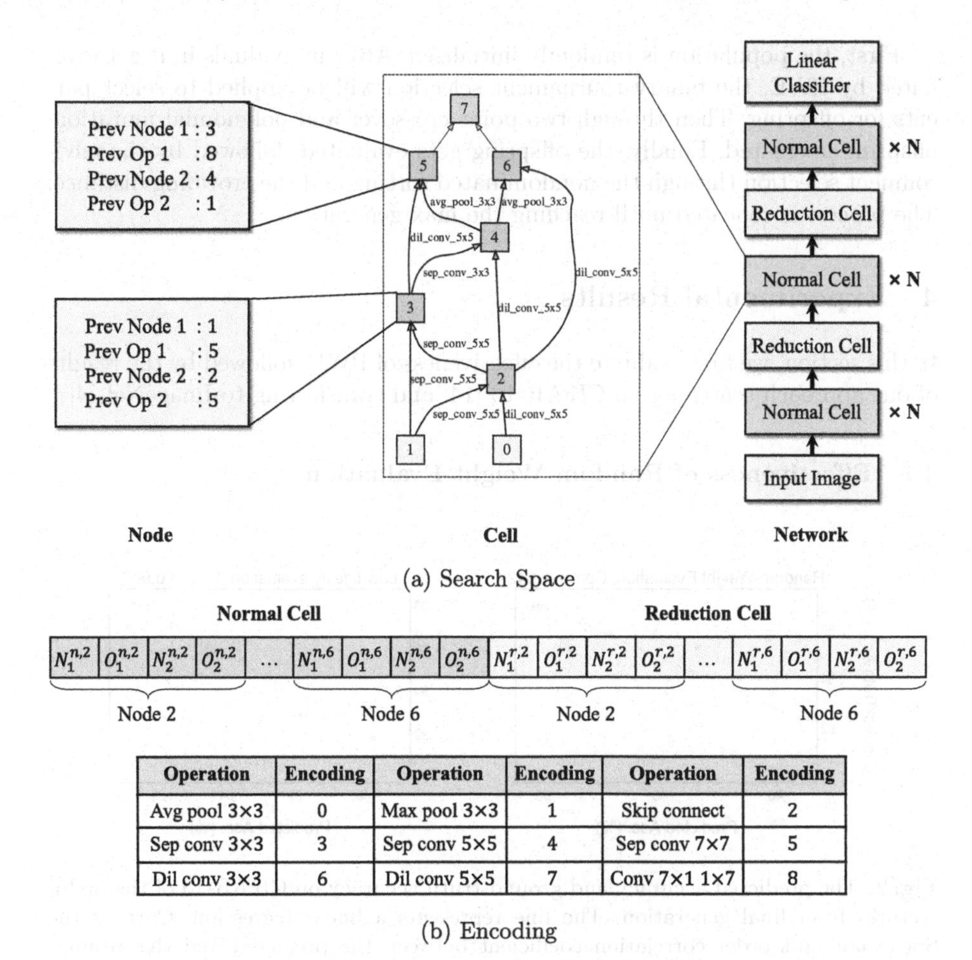

Operation	Encoding	Operation	Encoding	Operation	Encoding
Avg pool 3×3	0	Max pool 3×3	1	Skip connect	2
Sep conv 3×3	3	Sep conv 5×5	4	Sep conv 7×7	5
Dil conv 3×3	6	Dil conv 5×5	7	Conv 7×1 1×7	8

(b) Encoding

Fig. 1. The search space [37] adopted in our approach. (a) RIGHT: the macro network architecture of CNN. MIDDLE: an example structure of normal cells and reduction cells. LEFT: the searching elements for a node. (b) TOP: a 40-integer vector defining an architecture. BOTTOM: candidate operations and their encoding.

Encoding: We use an integer vector to encode an architecture, as shown in Fig. 1b. Each input and operation to be applied is encoded into an integer denoting a choice. A cell consists of five nodes, and each node is represented using four integers. The integer vectors for a normal and a reduction cell are concatenated into a 40-integer vector.

3.3 Search Strategy

In our approach, we adopt the classic multi-objective evolutionary algorithm NSGA-II [3] as the searching framework. The search process is briefly summarized below.

First, the population is randomly initialized. After individuals in it get evaluated by RWE, the binary tournament selection will be applied to select parents for offspring. Then through two-point crossover and polynomial mutation, offspring is created. Finally, the offspring gets evaluated, followed by the environment selection through the nondominated sorting and the crowding distance. The process is repeated until reaching the max generation.

4 Experimental Results

In this section, we first evaluate the effectiveness of RWE, followed by the results of our approach searching on CIFAR-10 [14] and transferring to ImageNet [4].

4.1 Effectiveness of Random-Weight Evaluation

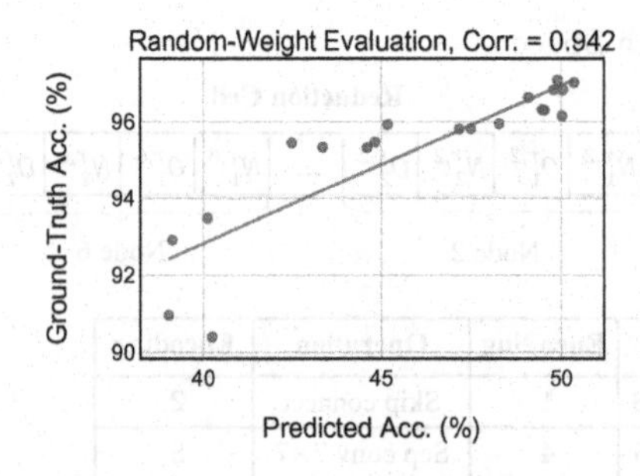

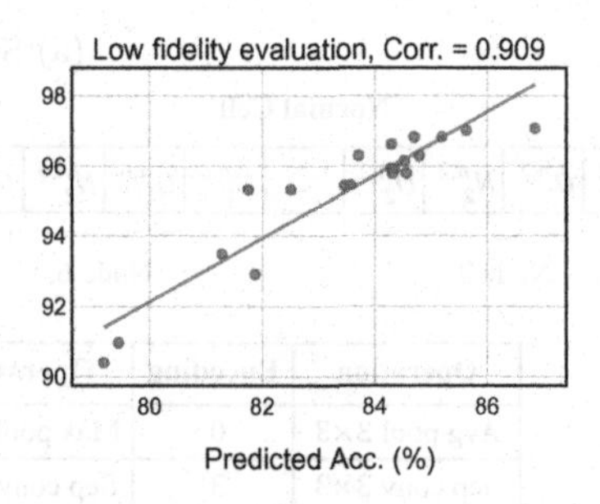

Fig. 2. The predicted accuracy and ground-truth accuracy on CIFAR-10 of the architectures from final generation. The line represents a linear regression. Corr. is the Spearman rank-order correlation coefficient between the predicted and the ground-truth accuracy.

To demonstrate the effectiveness of RWE, we calculate the Spearman rank-order correlation coefficient between the predicted and the ground-truth accuracy on CIFAR-10, which is the larger the better, equal to 1.0 if the predicted rank is the same as the ground-truth one. The result is shown in Fig. 2. The ground-truth accuracy is obtained through the setting described in Sect. 4.2. Figure 2 LEFT shows the predictions by RWE, as described in Algorithm 1. Figure 2 RIGHT shows the predictions by low fidelity evaluation commonly adopted [22,23,37]. These predictions are obtained by a twenty-epoch training on the architecture with 8 layers and 16 initial channels. The correlation coefficients for these two methods are 0.942 and 0.909, respectively. The result shows that, in this circumstance, RWE predicts the accuracy slightly better than the low fidelity evaluation.

4.2 Searching on CIFAR-10

In this work, we search on CIFAR-10, a dataset that is widely used for benchmarking image classification. CIFAR-10 is a 10-category dataset, consisting of 60K images with resolution of 32×32. The dataset is split into a training set and a testing set which consists of 50K and 10K images, respectively. Furthermore, we split the training set (80%–20%) to create the training and validation set for the searching stage.

In our searching stage, the population size N is set to 20 and we search for 30 generations. For RWE, every architecture is decoded into a CNN with 10 channels and 5 layers, where the second and fourth layers are reduction cells, and others are normal cells. The length of the linear classifiers list $L = 5$. The preprocessing for input images only contains the normalization, without the standard data augment techniques that introduce the randomness. The training for the linear classifiers uses the SGD optimizer with a batch size set of 512 and the momentum of 0.9. Also, the learning rate is set to 0.25 initially and then decay to zero by the cosine annealing schedule [18]. We train each linear classifier for 30 epochs, and the search takes one hour and fifteen minutes with a single Nvidia 2080Ti. Figure 3 TOP shows the bi-objective Pareto fronts for different generations in the searching stage. Figure 3 BOTTOM shows that our approach saves orders of magnitude search cost compared with other NAS algorithms.

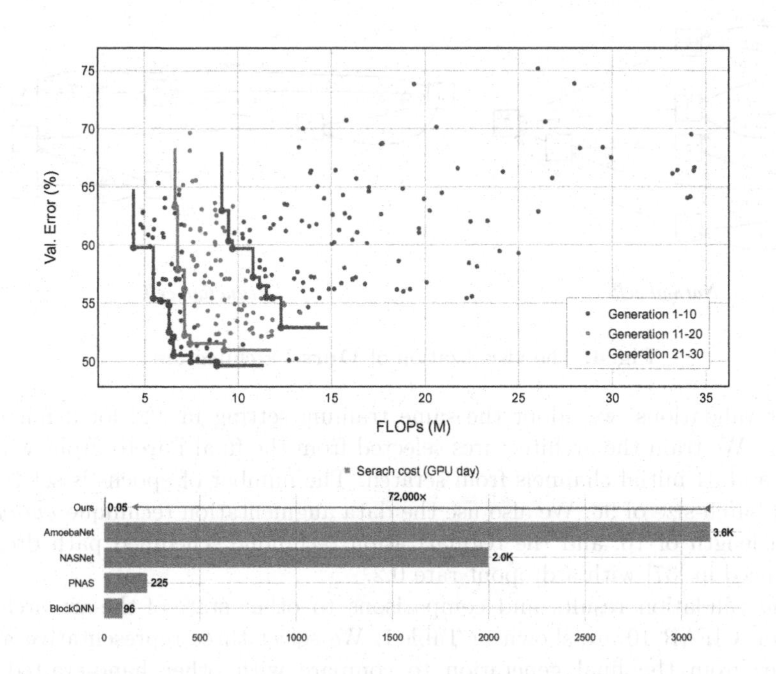

Fig. 3. TOP: progression of pareto fronts for different generations. BOTTOM: search cost comparison.

Table 1. Comparisons with other state-of-the-art architectures on CIFAR-10. ⇕denotes the work that shares the same setting with ours and the performance are reported in [22]. †denotes the work that adopts the *cutout* [5] technique.

Architecture	Test error (%)	Params (M)	FLOPs (M)	Search cost (GPU days)	Search method
Wide ResNet [33]	4.17	36.5	–	–	Manual
DenseNet-BC [11]	3.47	25.6	–	–	Manual
BlockQNN† [36]	3.54	39.8	–	96	RL
SNAS† [32]	3.10	2.3	–	1.5	Gradient
NASNet-A†⇕ [37]	2.91	3.2	532	2,000	RL
DARTS†⇕ [17]	2.76	3.3	547	4	Gradient
NSGA-Net⇕ + macro space [22]	3.85	3.3	1290	8	Evolution
Ours† + macro space	**4.27**	**2.79**	**1074**	**0.14**	Evolution
AE-CNN + E2EPP [27]	5.30	4.3	–	7	Evolution
Hier. Evolution [16]	3.75	15.7	–	300	Evolution
AmoebaNet-A†⇕ [23]	2.77	3.3	533	3,150	Evolution
NSGA-Net†⇕ [22]	2.75	3.3	535	4	Evolution
Ours-s†	**4.05**	**0.9**	**203**	**0.05**	Evolution
Ours-m†	**3.37**	**1.2**	**249**	**0.05**	Evolution
Ours-l†	**2.98**	**1.5**	**340**	**0.05**	Evolution

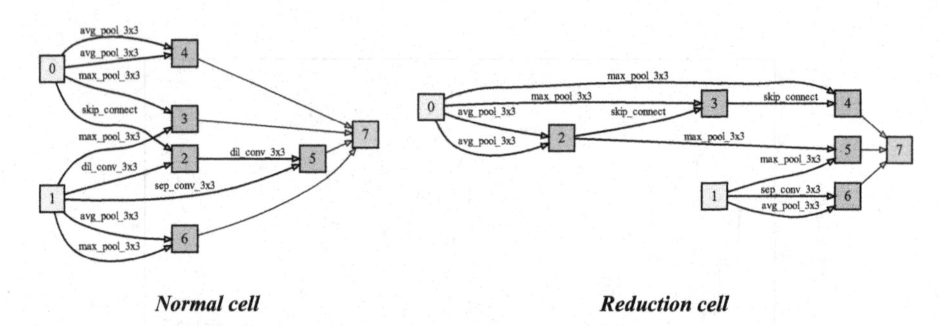

Fig. 4. The visualization of **Ours-l** architecture.

For validations, we adopt the same training setting in [22] for a fair comparison. We train the architectures selected from the final Pareto Front with 20 layers and 34 initial channels from scratch. The number of epochs is set to 600 with a batch-size of 96. We also use the data augmentation technique *cutout* [5] with a length of 16, and the regularization technique scheduled path dropout introduced in [37] with a dropout rate 0.2.

The validation results and comparisons to other state-of-the-art architectures on CIFAR-10 are shown in Table 1. We select three representative architectures from the final generation to compare with other hand-crafted and search-generated architectures. The chosen architecture with the lowest error rate (**Ours-l**, as shown in Fig. 4) results in a 2.98% classification error and 340M FLOPs, which is competitive to other works in error rate but has fewer

FLOPs. Also, compared with other NAS algorithms, our approach has much less computational cost measured in GPU days.

To further validate the effectiveness of RWE, we also apply our NAS algorithm on the macro search space [31] adopted in [22]. The results in Table 1 show that our chosen architecture has competitive performance but fewer FLOPs, similar to the case in micro search space.

4.3 Transferring to ImageNet

It is a common approach that architectures get searched on CIFAR-10 and then get transferred to other datasets or tasks [32,37]. To test the transferability of our algorithm, we transfer our architecture with lowest error rate to ImageNet [4] dataset, which is one of the most challenging datasets for image classification. It is consisted of 1.28M images for the training set and 50K images for the validation set. The images are of various resolution and unevenly distributed in 1000 categories. We adopt some common data augmentation techniques, including the random resize and crop, the random horizontal flip, and the color jitter. We adjust our network architecture for ImageNet based on [17]. We train the model on 4 Nvidia Tesla V100 with the SGD optimizer for 250 epochs, with the batch size 1024 and resolution 224×224. The learning rate is set to 0.5, with the momentum 0.9 and the weight decay 3×10^{-5}. The linear learning rate scheduler is used and as a result, the learning rate decays from 0.5 to 1×10^{-5} linearly during the training. Also, the warm-up strategy is adopted to increase the learning rate from 0 to 0.5 over the first five epochs. The label smooth [29] technique is also used with a smooth ratio of 0.1. The results of comparisons of our approach to other state-of-the-art architectures on ImageNet is shown in Table 2.

Table 2. Comparisons with other state-of-the-art architectures on ImageNet. $^{\Uparrow}$denotes the architectures that are searched on CIFAR-10 and transferred to ImageNet.

Architecture	Test error (%)		Params (M)	FLOPs (M)
	Top-1	Top-5		
MobileNetV1 [10]	31.6	–	2.6	325
InceptionV1 [28]	30.2	10.1	6.6	1448
ShuffleNetV1 [35]	28.5	–	3.4	292
MobileNetV2 [25]	28.0	9.0	3.4	300
NASNet-C$^{\Uparrow}$ [37]	27.5	9.0	4.9	558
SNAS$^{\Uparrow}$ [32]	27.3	9.2	4.3	533
Ours-I$^{\Uparrow}$	**27.6**	**9.4**	**3.7**	**363**

5 Conclusion

In this paper, we proposed a novel performance estimation strategy, Random-Weight Evaluation (RWE), to reduce the search cost of a NAS algorithm. RWE fixes majority of the weights at randomly initialized values, and only trains the last linear classifier layer. We integrated RWE in a multi-objective NAS algorithm, achieving state-of-the-art performance while reducing the computational cost drastically. In particular, RWE obtained a novel architecture on CIFAR-10, yielding 2.98% top-1 classification error and 1.5M parameters with less than two hours searching on a single GPU card. When transferred to ImageNet, the obtained architecture achieved 27.6% top-1 classification error.

Acknowledgement. This work was supported by the National Natural Science Foundation of China (Grant No. 61903178 and 61906081), the Program for Guangdong Introducing Innovative and Enterpreneurial Teams (Grant No. 2017ZT07X386), the Shenzhen Peacock Plan (Grant No. KQTD2016112514355531), and the Program for University Key Laboratory of Guangdong Province (Grant No. 2017KSYS008).

References

1. Cai, H., Gan, C., Wang, T., Zhang, Z., Han, S.: Once for all: train one network and specialize it for efficient deployment. In: ICLR (2020)
2. Cai, H., Zhu, L., Han, S.: ProxylessNAS: direct neural architecture search on target task and hardware. In: ICLR (2019)
3. Deb, K., Pratap, A., Agarwal, S., Meyarivan, T.: A fast and elitist multiobjective genetic algorithm: NSGA-II. IEEE Trans. Evol. Comput. **6**(2), 182–197 (2002)
4. Deng, J., Dong, W., Socher, R., Li, L.J., Li, K., Fei-Fei, L.: ImageNet: a large-scale hierarchical image database. In: CVPR (2009)
5. DeVries, T., Taylor, G.W.: Improved regularization of convolutional neural networks with cutout. arXiv preprint arXiv:1708.04552 (2017)
6. Gaier, A., Ha, D.: Weight agnostic neural networks. In: NeurIPS (2019)
7. Hansen, L.K., Salamon, P.: Neural network ensembles. IEEE Trans. Pattern Anal. Mach. Intell. **12**(10), 993–1001 (1990)
8. He, K., Zhang, X., Ren, S., Sun, J.: Delving deep into rectifiers: surpassing human-level performance on imagenet classification. In: ICCV (2015)
9. He, K., Zhang, X., Ren, S., Sun, J.: Deep residual learning for image recognition. In: CVPR (2016)
10. Howard, A.G., et al.: MobileNets: efficient convolutional neural networks for mobile vision applications. arXiv preprint arXiv:1704.04861 (2017)
11. Huang, G., Liu, Z., Van Der Maaten, L., Weinberger, K.Q.: Densely connected convolutional networks. In: CVPR (2017)
12. Jarrett, K., Kavukcuoglu, K., Ranzato, M., LeCun, Y.: What is the best multi-stage architecture for object recognition? In: ICCV (2009)
13. Juefei-Xu, F., Naresh Boddeti, V., Savvides, M.: Local binary convolutional neural networks. In: CVPR (2017)
14. Krizhevsky, A., Hinton, G., et al.: Learning multiple layers of features from tiny images. Technical report, Citeseer (2009)
15. Krizhevsky, A., Sutskever, I., Hinton, G.E.: ImageNet classification with deep convolutional neural networks. In: Advances in Neural Information Processing Systems (2012)
16. Liu, H., Simonyan, K., Vinyals, O., Fernando, C., Kavukcuoglu, K.: Hierarchical representations for efficient architecture search. In: ICLR (2018)
17. Liu, H., Simonyan, K., Yang, Y.: DARTS: differentiable architecture search. In: ICLR (2019)
18. Loshchilov, I., Hutter, F.: SGDR: stochastic gradient descent with warm restarts. arXiv preprint arXiv:1608.03983 (2016)
19. Lu, Z., et al.: Multi-objective evolutionary design of deep convolutional neural networks for image classification. IEEE Trans. Evol. Comput. 1 (2020)

20. Lu, Z., Deb, K., Boddeti, V.N.: MUXConv: information multiplexing in convolutional neural networks. In: CVPR (2020)
21. Lu, Z., Deb, K., Goodman, E., Banzhaf, W., Boddeti, V.N.: NSGANetV2: evolutionary multi-objective surrogate-assisted neural architecture search. In: Vedaldi, A., Bischof, H., Brox, T., Frahm, J.-M. (eds.) ECCV 2020. LNCS, vol. 12346, pp. 35–51. Springer, Cham (2020). https://doi.org/10.1007/978-3-030-58452-8_3
22. Lu, Z., et al.: NSGA-Net: neural architecture search using multi-objective genetic algorithm. In: GECCO (2019)
23. Real, E., Aggarwal, A., Huang, Y., Le, Q.V.: Regularized evolution for image classifier architecture search. In: AAAI (2019)
24. Rosenfeld, A., Tsotsos, J.K.: Intriguing properties of randomly weighted networks: generalizing while learning next to nothing. In: CRV (2019)
25. Sandler, M., Howard, A., Zhu, M., Zhmoginov, A., Chen, L.C.: MobileNetV2: inverted residuals and linear bottlenecks. In: CVPR (2018)
26. Saxe, A.M., Koh, P.W., Chen, Z., Bhand, M., Suresh, B., Ng, A.Y.: On random weights and unsupervised feature learning. In: ICML (2011)
27. Sun, Y., Wang, H., Xue, B., Jin, Y., Yen, G.G., Zhang, M.: Surrogate-assisted evolutionary deep learning using an end-to-end random forest-based performance predictor. IEEE Trans. Evol. Comput. **24**(2), 350–364 (2020)
28. Szegedy, C., et al.: Going deeper with convolutions. In: CVPR (2015)
29. Szegedy, C., Vanhoucke, V., Ioffe, S., Shlens, J., Wojna, Z.: Rethinking the inception architecture for computer vision. In: CVPR (2016)
30. Tan, M., et al.: MnasNet: platform-aware neural architecture search for mobile. In: CVPR (2019)
31. Xie, L., Yuille, A.: Genetic CNN. In: ICCV (2017)
32. Xie, S., Zheng, H., Liu, C., Lin, L.: SNAS: stochastic neural architecture search. In: ICLR (2019)
33. Zagoruyko, S., Komodakis, N.: Wide residual networks. In: BMVC (2016)
34. Zela, A., Klein, A., Falkner, S., Hutter, F.: Towards automated deep learning: efficient joint neural architecture and hyperparameter search. arXiv preprint arXiv:1807.06906 (2018)
35. Zhang, X., Zhou, X., Lin, M., Sun, J.: ShuffleNet: an extremely efficient convolutional neural network for mobile devices. In: CVPR (2018)
36. Zhong, Z., et al.: BlockQNN: efficient block-wise neural network architecture generation. IEEE Trans. Pattern Anal. Mach. Intell. 1 (2020)
37. Zoph, B., Vasudevan, V., Shlens, J., Le, Q.V.: Learning transferable architectures for scalable image recognition. In: CVPR (2018)

On the Interaction Between Distance Functions and Clustering Criteria in Multi-objective Clustering

Adán José-García[1](✉) and Julia Handl[2]

[1] Department of Computer Science, University of Exeter, Exeter, UK
a.jose-garcia@exeter.ac.uk

[2] Alliance Manchester Business School, University of Manchester, Manchester, UK
julia.handl@manchester.ac.uk

Abstract. Multi-criterion algorithms for clustering have gained some traction due to their ability to cater to a diverse range of cluster properties. Here, we investigate the interaction between the clustering criteria employed in a multi-objective algorithm and the distance functions on which these criteria operate. We do so by contrasting the multi-criterion evolutionary algorithm Δ-MOCK with a bi-objective version of the evolutionary multi-view clustering approach MVMC, which uses a single clustering criterion but can incorporate multiple dissimilarity matrices. Using a benchmark suite representing a diverse range of cluster properties, we illustrate that comparable results to Δ-MOCK can be achieved using MVMC with two complementary distance functions. We then establish the mathematical equivalence of Δ-MOCK's connectivity objective to a compactness criterion operating on a redefined distance function. We conclude by discussing the implications of our findings for future work on the representation of clusters in multi-objective evolutionary clustering.

Keywords: Data clustering · Multi-objective optimisation · Multi-view clustering

1 Introduction

There is an array of work that indicates that cluster structure is in the eyes of the beholder, i.e., that different individuals may visually perceive different structures in the same dataset [18], and that efforts to mathematically capture this intangible and somewhat individual human intuition explain the existence of an array of different cluster quality measures [8]. Yet, in analysing the capabilities of a given cluster quality measure, it is easy to forget that its performance ultimately depends on two aspects: its mathematical definition and the dissimilarity function it operates on. Whilst there is wide-spread empirical understanding that the choice of distance function can make a significant difference to practical clustering performance, this is often motivated in terms of data types (e.g. the use of the Cosine distance for text data). The specific interaction between clustering criteria and distance functions is rarely considered.

H. Ishibuchi et al. (Eds.): EMO 2021, LNCS 12654, pp. 504–515, 2021.
https://doi.org/10.1007/978-3-030-72062-9_40

Here, we explore this interaction in the context of the problem of multi-objective data clustering. The motivation for this investigation is two-fold: (i) to help contrast and position recent work on multi-criterion and multi-view evolutionary clustering (both of which rely on the use of state-of-the-art multi-objective optimisers); (ii) to identify opportunities for the use of simpler representations for multi-criterion evolutionary clustering, such as cluster prototypes, whilst retaining an algorithm's ability to identify complex, elongated clusters. Specifically, we make use of the recent development of a multi-objective algorithm for multi-view clustering (MVMC [11]), an algorithm developed for the partitioning of data in the presence of multiple feature sets or dissimilarity matrices. We demonstrate empirically, that, through the simultaneous use of two suitable dissimilarity measures alone, the algorithm can achieve competitive performance to the multi-criterion clustering algorithm Δ-MOCK [6]. Finally, we illustrate that Δ-MOCK itself can be reformulated as a bi-objective multi-view optimisation problem, involving the minimisation of two compactness-based criteria in distinct dissimilarity spaces.

2 Background

Following on from our introductory comments on cluster quality, it is clear that human perception of dissimilarity (or distance) between data points is ambiguous in its own right, and that assumptions about dissimilarity will further influence the detectability of certain types of structures, by humans or algorithms. The impact of dissimilarity on the structure perceivable in a dataset is best illustrated in the context of simple two-dimensional datasets. Figure 1 shows four datasets that exhibit a range of different shapes, defined in two dimensions. This original visualisation reflects the actual coordinates of all data points and does not yet introduce any assumption regarding the definition of distance in this space. We can then apply different dissimilarity functions to these data, in this example, the Euclidean distance and the maximum edge distance [2], a graph-based approach.

The Euclidean distance is well known and its definition will not be reiterated here. In contrast, the maximum edge distance is less commonly known. It is defined over the Minimum Spanning Tree (MST) of the graph representing the original pair-wise dissimilarities in a given dataset (so it implicitly operates on a second distance function).

$$d_{\mathrm{MED}}(i,j) = \max\{e \in E_{P_{ij}}\} \tag{1}$$

where P_{ij} is the component of the MST that represents the path between data points i and j, $E_{P_{ij}}$ is the set of edges contained in MST component P_{ij}, and e are the elements of that set, represented by their lengths, i.e., the original dissimilarities between the edges' endpoints. Intuitively, this measure then redefines the dissimilarity as the longest edge encountered when traversing the MST between items i and j.

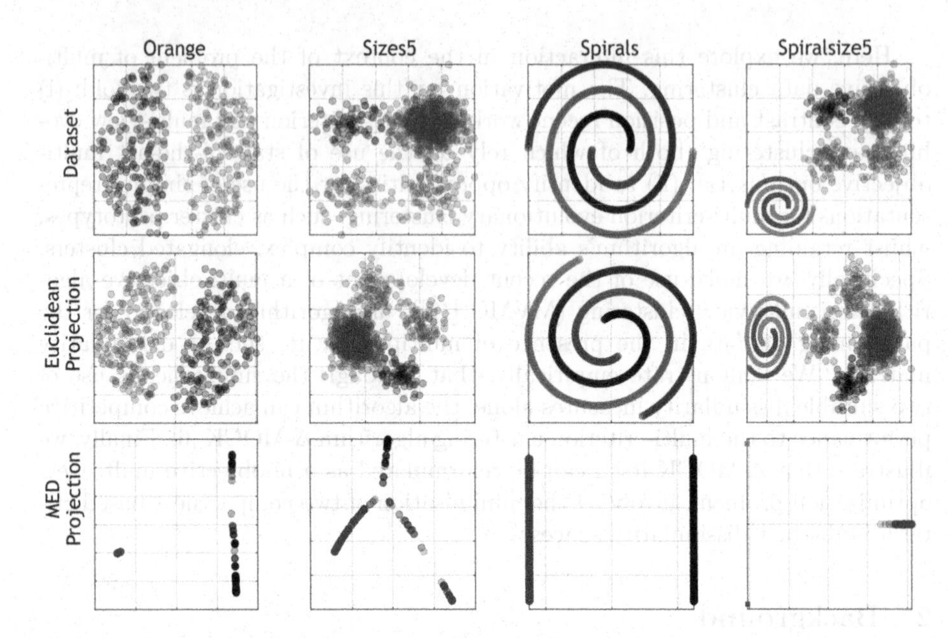

Fig. 1. The effect of different distance functions. (top) Original data; (centre) Embedding of the associated Euclidean distance; (bottom) Embedding of the MED distances.

Once a dissimilarity matrix for a dataset has been obtained, we can use the approach of multi-dimensional scaling to re-embed the data into two dimensions. It is evident from the results in Fig. 1 that the choice of distance functions makes a fundamental difference to the detectability of particular types of clusters. Specifically, the MED distance favours the clear spatial distinction of elongated cluster structures (e.g., on the `Spiral` data) but is less capable than the Euclidean distance in separating cluster with overlap (e.g., on the `Sizes5` data). Given the definition of the two distance measures, this observation is perhaps unsurprising, but it raises the question of whether improved clustering performance may be achieved from a combination of these measures.

3 Multi-criterion Versus Multi-view Clustering

Previous consideration of cluster analysis as a multi-criterion optimisation problem often stems from the informal definition of a data partition as a grouping that minimises intra-class homogeneity and maximises between-class separation. Delattre and Hansen [5] proposed a first distinct mathematical formulation that reflected this and proposed an exact solution approach for their particular pair of clustering criteria. More recently, the development of evolutionary approaches for the problem has permitted the consideration of a broader range of possible cluster quality measures [14]. On the whole, empirical comparison of these algorithms to single-criterion approaches has demonstrated their improved robustness to varying cluster properties. However, the model selection stage, i.e., the identification of a final clustering solution, poses some additional challenges, compared to a single-criterion approach. In particular, a Pareto optimisation approach to the problem implies that multiple optimal trade-off solutions may exist, even for a fixed (known) number of clusters.

Different to the traditional work on multi-criterion clustering, the emphasis in multi-view clustering lies on the integration of multiple features or dissimilarity spaces, rather than the integration of multiple clustering objectives [4,11]. Problems with multiple views arise in a range of application domains, and new algorithms for unsupervised and supervised multi-view learning have been steadily gaining traction in the machine learning literature [19]. The fact that individual views may arise as distinct features or dissimilarity spaces is unique to the problem of unsupervised multi-view learning.[1] Given an effective multi-view clustering algorithm, this opens up an alternative use: given a dataset defined in terms of a *single feature space*, multiple distance functions can be considered and integrated when attempting to improve the robustness of the partition. In the examples in Fig. 1 above, this would be similar to identifying those partitions that are supported by the Euclidean and MED projection, simultaneously.

It is evident from the above that both the problem statement of multi-criterion and that of multi-view clustering give rise to a multi-objective formulation of the data clustering problem. However, existing work typically treats these problems as distinct. This is due to the separate motivations for the problems (set out above) and, partially, because certain aspects of the algorithm design (such as the choice of cluster representation) do not allow a straightforward transfer of algorithms. In the following, and keeping with the preceding section of this paper, we will aim to cleanly separate between the two problem statements through use of the following terminology: We will refer to *multi-criterion clustering* when describing the optimisation of several, complementary cluster quality criteria. We will refer to *multi-view clustering* when describing the optimisation of a single cluster-quality measure with respect to multiple dissimilarity spaces. We will refer to *multi-objective clustering* as the super-class that encompasses both problem statements. Ultimately, we aim to illustrate that the boundary between the two problems is fuzzy, due to the predication of all internal clustering criteria on a distance function. In the following, we briefly review MVMC and Δ-MOCK, the multi-view and multi-criterion algorithms that will underpin the associated experiments and analysis.

MVMC. This multi-objective evolutionary approach to multi-view clustering was developed to identify all optimal trade-offs between available data views, achieving scalability to a significant number of views, through the use of a many-objective optimiser [11]. Specifically, MVMC uses a decomposition-based optimiser, MOEA/D [16], as the underlying search engine for its clustering approach. Furthermore, it employs a medoid-based representation, a representation that is more general than centroids, as it can be used both for problems defined in terms of feature spaces or dissimilarity matrices. In its current implementa-

[1] In supervised multi-view learning, a feature representation of each view is required, as the end-goal is typically a feature-to-output mapping.

tion, MVMC uses a fixed number of medoids, so requires the desired number of clusters as input.

MVMC focuses on a single cluster-quality measure, but aims to optimise this with respect to every view, thus giving rise to a multi-objective optimisation problem in its own right. Let $\mathbf{C}^r$ and $\mathbf{w}^r$ be the partition and weight vector, respectively, corresponding to the r-th subproblem. Also, let $\{D_1, \ldots, D_M\}$ denote the M dissimilarity matrices, which represent the different data views and are each considered by a separate objective. MVMC then uses the within-cluster scatter as the optimisation criterion, which, for the m-th objective of the r-th subproblem, is computed as:

$$f_m(\mathbf{C}^r) = \sum_{c_k \in \mathbf{C}^r} \sum_{i,j \in c_k} d_m(i,j) \ , \tag{2}$$

where $d_m(i,j)$ is the dissimilarity between the points i and j as defined in D_m.

Δ-MOCK. The multi-objective evolutionary algorithm Δ-MOCK uses a flexible representation (the locus-based adjacency scheme [15]), which avoids the emphasis on spherical clusters typically associated with a prototype-based encoding, and makes no prior assumptions on cluster number. It does so in an efficient manner, by exploiting information from the minimum spanning tree of the data to narrow down the search space.

In its standard setup, Δ-MOCK formulates multi-objective clustering as a bi-objective problem. The criterion of within-cluster sum of squares assesses the compactness of a partition as follows:

$$\text{Var}(\mathbf{C}) = \frac{1}{N} \sum_{k=1}^{K} \sum_{i \in c_k} d(i, \mu_k)^2 \tag{3}$$

where N represents the number of data points, K represents the number of clusters, c_k and μ_k represent the partition and the centroid of cluster k, respectively, and $d(\cdot, \cdot)$ is a distance function. Typically, this is the Euclidean distance, but other choices are possible.

In contrast, the criterion of connectivity is more complex and caters towards density preservation in the dataset:

$$\text{Conn}(\mathbf{C}) = \sum_{i=1}^{N} \sum_{l=1}^{L} \rho(i,l) \tag{4}$$

where L is a user-defined parameter specifying the size of the relevant neighbourhood, and $\rho(i,l) = \frac{1}{l}$ is a penalty term incurred for $l \leqslant L$ only, when data point i and its l-th nearest neighbour are not co-assigned in a given clustering solution. The two objectives are known to have opposing trends with regards to the number of clusters, allowing Δ-MOCK to efficiently explore a range of possible cluster numbers within the same optimisation run.

Δ-MOCK's reliance on the minimum spanning tree reduces its suitability for multi-view clustering settings, i.e., settings in which multiple feature spaces may need to be considered during cluster analysis. This is because the minimum spanning tree is constructed with respect to a single dissimilarity matrix, potentially biasing the search towards one of the views available. On the other hand, Δ-MOCK's focus on complementary clustering criteria has been demonstrated to result in robustness towards a broad array of cluster properties. In the following, we consider whether equivalent levels of robustness could be obtained from the alternative application of a multi-view method, using a single clustering criterion in conjunction with multiple dissimilarity measures, on a single feature space. Fundamentally, this approach simply "shifts" the challenge of adequate (flexible and robust) cluster recognition from the clustering criterion to the definition of dissimilarity.

4 Experimental Setup

This section describes the datasets, the performance assessment measure, and the settings adopted for this investigation.

Datasets. A total of 20 datasets are considered for the experiments of this study. The description of these datasets in terms of the number of patterns (N), number of dimensions (D) and number of clusters (K) is summarised in Table 1. Figure 2 illustrates the diversity of cluster structures covered by our benchmark data, which was compiled from datasets introduced in a range of multi-objective data clustering [3,7,10] and supervised learning studies [12], and presents varying sizes, degrees of overlap, and types of cluster shapes, including complex combinations within the same dataset. This collection of test datasets is available at: https://mvc-repository.github.io.

Performance Assessment. The Adjusted Rand Index (ARI) is used to assess the clustering performance [9]. ARI is defined in the range [~0, 1]; the larger the value for ARI, the better the correspondence between the partition obtained and the true partition of the data (which is known for our test datasets). Each run of MVMC and Δ-MOCK produces a set of candidate partitions. From this set, the best solution in terms of the ARI measure is selected and considered in the results reported in Sect. 5.

Reference Methods. We compare MVMC and Δ-MOCK against two well-known and conceptually different single-objective clustering algorithms: k-means [13] and Single-Link (SL) [20]. Two MVMC configurations are considered: MVMC_{WGS} and MVMC_{SIL}. MVMC_{WGS} minimises the within-cluster scatter and is representative of the original design of the algorithm [11]. Here, we include a second version of the algorithm that directly optimises the Silhouette Width [17], often considered a more effective measure of cluster validity (as it considers both within and between-cluster variation of a partition).

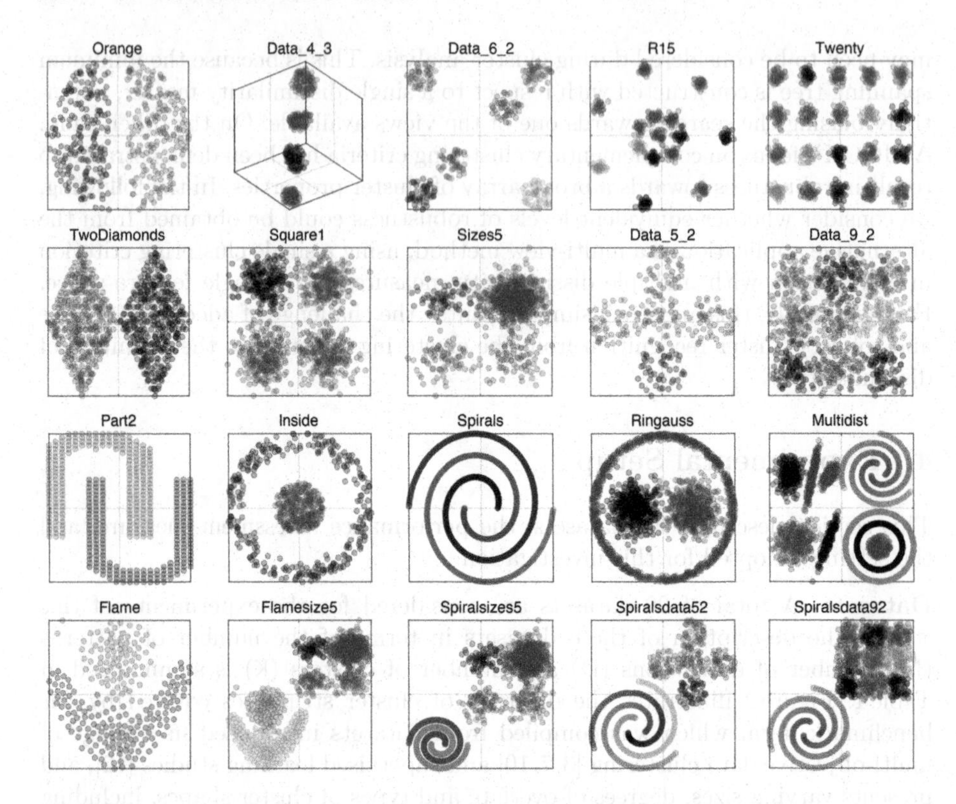

Fig. 2. This figure illustrates the diversity of properties covered by our collection of test datasets. The figures in the first row contain well-separated clusters. The figures in the second row correspond to overlapping clusters. The figures in the third row consist of nonlinearly separable clusters. Finally, the figures in the last row are datasets having mixtures of different data distributions.

Parameter Settings. The settings for MVMC adopted in our experiments are as follows. The population size is $\mathsf{NP} = 100$, the number of generations is $\mathsf{G}_{\max} = 100$, the recombination probability is $\mathsf{Pr} = 0.5$, the mutation probability is $\mathsf{Pm} = 0.03$, and the neighbourhood size is $\mathsf{T} = 20$. Regarding the Δ-MOCK algorithm, the settings considered are as follows: $\mathsf{NP} = 100$, $\mathsf{G}_{\max} = 100$, crossover and mutation of 1.0, and the maximum number of clusters is $kmax = 2\mathsf{K}$, where K is the true number of clusters.

For all algorithms, except SL (due to its deterministic nature), a total of 31 independent executions were performed for each dataset. Finally, the Kruskal–Wallis test, considering a significance level of $\alpha = 0.05$ and Bonferroni correction, was applied to compare the ARI values obtained.

5 Findings

We now proceed to analyse the empirical results obtained on our benchmark of single-view problems, focusing on the relative performance of MVMC and

Table 1. ARI values obtained by the different clustering algorithms k-means, SL, δ-MOCK, and MVMC. The best ARI value scored for each dataset has been shaded and highlighted in bold and, additionally, the statistically best ($\alpha = 0.05$) results are highlighted in boldface.

Dataset	N	D	K	k-means$^{\blacktriangle}$	k-means$^{\blacktriangledown}$	SL$^{\blacktriangle}$	SL$^{\blacktriangledown}$	Δ-MOCK	MVMC$^{\blacktriangle\blacktriangledown}_{SIL}$	MVMC$^{\blacktriangle\blacktriangledown}_{WGS}$
Orange	400	2	2	**1.000±0.00**	**1.000±0.00**	**1.000**	**1.000**	**1.000±0.00**	**1.000±0.00**	**1.000±0.00**
Data_4_3	400	3	4	**1.000±0.00**	**1.000±0.00**	**1.000**	**1.000**	**1.000±0.00**	**1.000±0.00**	**1.000±0.00**
Data_6_2	300	2	6	**1.000±0.00**	**1.000±0.00**	**1.000**	**1.000**	**1.000±0.00**	**1.000±0.00**	**1.000±0.00**
R15	600	2	15	**0.966±0.05**	0.975±0.00	0.548	0.751	**0.992±0.00**	**0.992±0.00**	**0.993±0.00**
Twenty	1000	2	20	0.966±0.04	**1.000±0.00**	**1.000**	**1.000**	**1.000±0.00**	**1.000±0.00**	**1.000±0.00**
TwoDiamonds	800	2	2	**1.000±0.00**	0.895±0.01	0.000	0.000	**0.999±0.00**	**0.999±0.00**	**1.000±0.00**
Square1	1000	2	4	**0.973±0.00**	0.906±0.01	0.000	0.000	**0.946±0.18**	**0.978±0.00**	**0.979±0.00**
Size5	1000	2	4	**0.920±0.00**	0.739±0.01	0.025	0.015	**0.961±0.01**	**0.970±0.00**	**0.962±0.00**
Data_5_2	250	2	5	0.870±0.01	0.750±0.02	0.189	0.394	**0.916±0.02**	**0.946±0.04**	**0.930±0.02**
Data_9_2	900	2	9	**0.831±0.00**	0.506±0.03	0.000	0.000	0.748±0.01	0.826±0.01	**0.838±0.01**
Part2	417	2	2	0.265±0.00	**1.000±0.00**	**1.000**	**1.000**	**1.000±0.00**	**1.000±0.00**	**1.000±0.00**
Inside	600	2	2	0.008±0.01	**1.000±0.00**	**1.000**	**1.000**	**1.000±0.00**	**1.000±0.00**	**1.000±0.00**
Spirals	1000	2	2	0.074±0.00	**1.000±0.00**	**1.000**	**1.000**	**1.000±0.00**	**1.000±0.00**	**1.000±0.00**
Ringauss	2000	2	3	0.252±0.00	**0.963±0.02**	0.001	0.001	0.473±0.02	**0.971±0.00**	**0.972±0.00**
Multidist	3012	2	11	0.532±0.03	**0.991±0.00**	0.805	0.805	0.764±0.02	0.967±0.06	**0.984±0.04**
Flame	240	2	2	0.462±0.02	**0.910±0.05**	0.013	0.013	**0.963±0.01**	**0.935±0.03**	**0.967±0.00**
Flamesize5	240	2	6	0.926±0.01	0.824±0.00	0.489	0.657	**0.971±0.01**	**0.948±0.01**	**0.976±0.00**
Spiralsizes5	2000	2	6	0.659±0.00	0.833±0.02	0.555	0.782	**0.987±0.00**	**0.974±0.04**	**0.980±0.02**
Spiralsdata52	562	2	8	0.342±0.00	0.934±0.01	0.772	0.808	0.735±0.03	0.948±0.00	**0.960±0.01**
Spiralsdata92	1212	2	12	0.610±0.02	0.623±0.02	0.128	0.130	0.677±0.02	0.750±0.12	**0.878±0.01**

Δ-MOCK, in particular. We then highlight how these results can be explained in terms of a reformulation of Δ-MOCK's objectives.

5.1 Empirical Results

The empirical results of our experiments are summarised in Table 1. For many of the datasets, Δ-MOCK and MVMC are broadly similar in terms of the solutions obtained. It is evident that both Δ-MOCK and the two versions of MVMC can perform robustly across a range of different cluster shapes, outperforming k-means and Single-link in that regard. As may be expected, k-means with the Euclidean distance (k-means$^{\blacktriangle}$) performs poorly on those datasets containing elongated clusters (e.g., `Inside`, `Spiral`, `Multidist`). Its performance for these datasets can be improved through a switch to the MED-distance (k-means$^{\blacktriangledown}$), but it then starts to struggle on datasets with significant overlap (e.g., `Data_5_2`, `Data_9_2`). Single-link's performance is problematic for a wider range of datasets (those containing overlap and/or outliers), and a switch to the MED-distance is of limited help in those instances.

MVMC's competitive performance with respect to Δ-MOCK is perhaps surprising given that, at the face of it, MVMC optimises a compactness-based clustering criterion. It is clear from these results that the simultaneous use of the Euclidean and MED-distance helps leverage the structural information in the data to the same extent (or better) than the use of multiple clustering criteria.

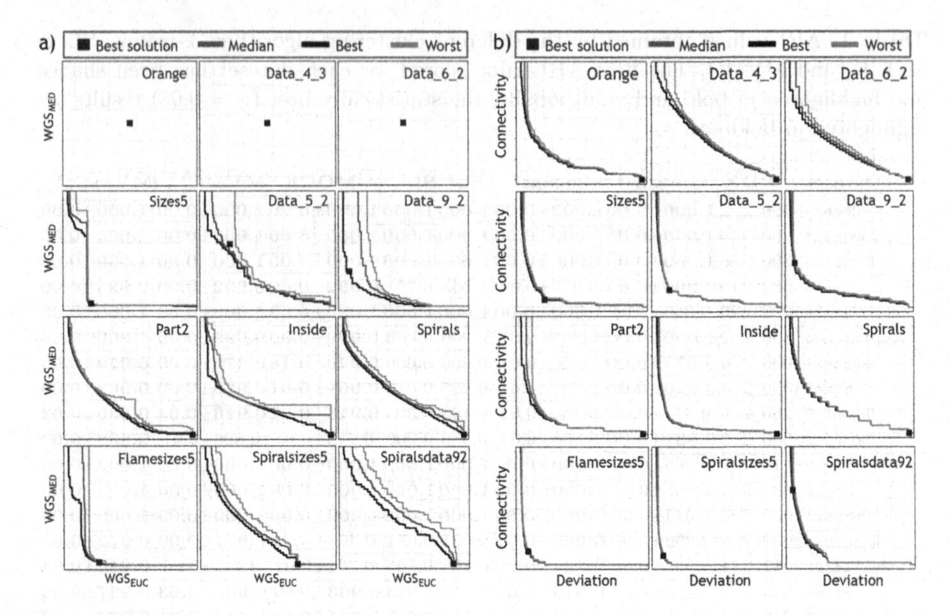

Fig. 3. Comparison of the EAFs computed from the PFAs obtained from all independent execution of MVMC and Δ-MOCK approaches: (a) MVMC's results when minimising the within-group scatter (WGS) using the Euclidean distance (x-axis) and MED distance (y-axis), and (b) Δ-MOCK's results when minimising the overall deviation (x-axis) and the connectivity (y-axis) criteria. For each subfigure, the best, average, and the worst PFAs are plotted in black, blue, and grey lines, respectively. Additionally, each subfigure includes a square to illustrate the best solution (maximum ARI value). (Color figure online)

The results reported in Table 1 focus on best ARI solutions only. Solution-selection in multi-objective clustering requires the identification of a single solution from the Pareto front approximations (PFAs), which, as discussed earlier in this paper, remains a challenging problem. On the other hand, the shape and span of the PFAs can provide additional insight regarding the strength of the signal and the type of data structures in a given dataset. A detailed discussion of this aspect is beyond the scope of this paper, but a selection of attainment fronts for MVMC and Δ-MOCK is included for completeness (see Fig. 3).

We would like to include some additional observations regarding our experimental setup and how this may influence the results observed. In terms of the experimental design, Δ-MOCK is singled out here (compared to all other algorithms), as it attempts to optimise across a range of possible number of clusters. This can play to its advantage in some comparisons (see discussion regarding Single-link in the second next paragraph), but it also presents a disadvantage, as solutions for particular numbers of clusters may become dominated by solutions at a higher level in the cluster hierarchy. These dominant solutions are typically those that identify fewer, but more pronounced clusters, representing super-sets of the correct groups. An example of this would be the identification of just two clusters in the `Spiralsdata92` dataset, merging the three Spirals and the nine Gaussians, respectively. In our results, we can see that Δ-MOCK is indeed running into difficulties on five of the twenty datasets: the results obtained on `Data_9_2`, `Ringauss`, `Multidist`, `Spiralsdata52` and `Spiralsdata92` suggest that the best ARI solution underestimates the correct number of clusters, indicating that these solutions might dominate the solutions with the correct number of clusters. All five of these datasets contain a hierarchical structure that is readily evident to a human observer.

Just like k-means and Single-link, MVMC optimises for a fixed number of clusters k, and this goes some way to explain differences in the attainment fronts shown in Fig. 3. These show a much wider reach of Δ-MOCK's fronts for some datasets (`Orange`, `Data_4_3` and `Data_6_2`, in particular), which reflect the exploration of different values of k. In order to produce partitions for different values of k, MVMC would need to be re-run multiple times (with different values of k), as is standard practice for algorithms such as k-means. Alternatively, its current implementation would require further changes to representation and variation operators, to allow for variation of k during the search. Finally, keeping k flexible during the optimisation would require a change to clustering criterion suited to the comparison of partitions across different values of k. One example of such a criterion is the Silhouette Width, so the results for $\text{MVMC}_{\text{SIL}}^{\blacktriangle\blacktriangledown}$ take a first promising step in that direction.

Finally, it should be noted that, for datasets with pronounced outliers, Single-link's performance may be expected to improve somewhat if solutions with a larger number of clusters were considered. This is because Single-link is well known to overestimate the number of clusters by assigning outliers to singleton clusters. However, the relative performance of Δ-MOCK, k-means and Single-link across a range of values of k has been extensively studied elsewhere [6], and is not included here for compactness sake.

5.2 Mathematical Analysis

In the following, we analyse Δ-MOCK's formulation, in order to shed light on the similarities and differences between Δ-MOCK and MVMC.

The similarity between the algorithms' first objective is straightforward to establish, with both algorithms optimising a compactness-based criterion. While Δ-MOCK minimises the total squared distance between all data points and their cluster centre, MVMC minimises the total distance between all pairs of data points in the same clusters. The key differences here are (i) the avoidance of direct use of the cluster prototypes in MVMC's objective, and (ii) the use of squared distances in MOCK, which has the effect of heavily penalising outliers. Largely, the two objectives are, however, highly aligned and will support the recognition of similar partitions.

The differences in the second objective are more interesting. Considering Δ-MOCK's connectivity objective:

$$\min\ \text{Conn}(\mathbf{C}) = \min\ \sum_{i=1}^{N} \sum_{l=1}^{L} \rho(i,l) \tag{5}$$

this is equivalent to a maximisation problem over a similarity measure $s(i, j)$:

$$\max \text{ Conn}^s(\mathbf{C}) = \max \sum_{k=1}^{K} \sum_{i \in c_k} \sum_{j \in c_k \setminus i} s(i, j) \quad (6)$$

were, $s(i, j)$ returns $\frac{1}{l}$ if j is the l-th nearest neighbour of i and $l \leqslant L$, and 0 otherwise. Note that $s(i, j)$ is not a metric, as it is not symmetric.

This can be reformulated as a minimisation problem over a non-metric dissimilarity measure $d(i, j) = 1 - s(i, j)$ as follows:

$$\min \text{ Conn}^d(\mathbf{C}) = \min \sum_{k=1}^{K} \sum_{i \in c_k} \sum_{j \in c_k \setminus i} d(i, j) - \sum_{k=1}^{K} |c_k|(|c_k| - 1) \quad (7)$$

Considering this last definition, it is apparent that Δ-MOCK's second objective can be rewritten as a compactness-based criterion (essentially within-cluster scatter plus an additional bias term that counterbalances the measure's natural bias to small, equally-sized clusters), formulated over a more complex (non-metric) distance function $d(i, j)$. It is thus not dissimilar to MVMC's second objective, which calculates within-cluster scatter with respect to a second view.

6 Discussion and Conclusions

So why does this matter? Our work helps demonstrate how, in the context of an evolutionary clustering approach, the suitable choice of a distance measure can overcome the normal limitations of prototype-based encodings, allowing for the design of an algorithm that can cater for a wide range of cluster structures, whilst exploiting the advantages of a simple representation scheme. Effectively, the distance function transforms the input data into a simpler space, facilitating the effective use of a compactness-based criterion—this is not unlike the approach taken by spectral [21] and recent deep clustering approaches [1].

Furthermore, our findings suggest that the separation between multi-criterion and multi-view data clustering is somewhat fuzzy, and that there may be valuable scope for cross-fertilisation between the two tasks. In particular, we have demonstrated, empirically, that results competitive to multi-criterion clustering can be obtained when a multi-view algorithm is run with an appropriately powerful pair of distance measures, on a single feature space. Finally, we have shown that this can be explained through shared commonalities between Δ-MOCK and the multi-view algorithm MVMC: this was done through a mathematical reformulation of the relative contributions of Δ-MOCK's clustering criterion and distance function, resulting in a transfer of complexity from the criterion to the distance level. Mathematically, the underlying optimisation problem remains equivalent, but the reformulation explicitly highlights the parallels to a multi-view approach.

References

1. Aljalbout, E., Golkov, V., Siddiqui, Y., Strobel, M., Cremers, D.: Clustering with deep learning: taxonomy and new methods. arXiv:1801.07648 (2018)
2. Bayá, A.E., Granitto, P.M.: How many clusters: a validation index for arbitrary-shaped clusters. IEEE/ACM Trans. Comput. Biol. Bioinform. **10**(2), 401–14 (2013)
3. de Carvalho, F.A., Lechevallier, Y., de Melo, F.M.: Partitioning hard clustering algorithms based on multiple dissimilarity matrices. Pattern Recogn. **45**(1), 447–464 (2012)
4. de Carvalho, F.A.T., Lechevallier, Y., Despeyroux, T., de Melo, F.M.: Multi-view clustering on relational data. In: Guillet, F., Pinaud, B., Venturini, G., Zighed, D. (eds.) Advances in Knowledge Discovery and Management. SCI, vol. 527, pp. 37–51. Springer, Cham (2014). https://doi.org/10.1007/978-3-319-02999-3_3
5. Delattre, M., Hansen, P.: Bicriterion cluster analysis. IEEE Trans. Pattern Anal. Mach. Intell. **2**(4), 277–291 (1980)
6. Garza-Fabre, M., Handl, J., Knowles, J.: An improved and more scalable evolutionary approach to multiobjective clustering. IEEE Trans. Evol. Comput. **22**(4), 515–535 (2018)
7. Handl, J., Knowles, J.: An evolutionary approach to multiobjective clustering. IEEE Trans. Evol. Comput. **11**(1), 56–76 (2007)
8. Hennig, C.: What are the true clusters? Pattern Recogn. Lett. **64**, 53–62 (2015)
9. Hubert, L., Arabie, P.: Comparing partitions. J. Classif. **2**(1), 193–218 (1985). https://doi.org/10.1007/BF01908075
10. José-García, A., Gómez-Flores, W.: Automatic clustering using nature-inspired metaheuristics: a survey. Appl. Soft Comput. **41**, 192–213 (2016)
11. Jose-Garcia, A., Handl, J., Gomez-Flores, W., Garza-Fabre, M.: An evolutionary many-objective approach to multiview clustering using feature and relational data. Under review to Applied Soft Computing (2020)
12. Kanaan-Izquierdo, S., Ziyatdinov, A., Perera-Lluna, A.: Multiview and multifeature spectral clustering using common eigenvectors. Pattern Recogn. Lett. **102**, 30–36 (2018)
13. MacQueen, J.: Some methods for classification and analysis of multivariate observations. In: Proceedings of the Fifth Berkeley Symposium on Mathematical Statistics and Probability, pp. 281–297. University of California Press (1967)
14. Mukhopadhyay, A., Maulik, U., Bandyopadhyay, S.: A survey of multiobjective evolutionary clustering. ACM Comput. Surv. (CSUR) **47**(4), 1–46 (2015)
15. Park, Y., Song, M.: A genetic algorithm for clustering problems. In: Proceedings of the Third Annual Conference on Genetic Programming, vol. 1998, pp. 568–575 (1998)
16. Zhang, Q., Li, H.: MOEA/D: a multiobjective evolutionary algorithm based on decomposition. IEEE Trans. Evol. Comput. **11**(6), 712–731 (2007)
17. Rousseeuw, P.J.: Silhouettes: a graphical aid to the interpretation and validation of cluster analysis. J. Comput. Appl. Math. **20**, 53–65 (1987)
18. Santos, J.M., de Sá, J.M.: Human clustering on bi-dimensional data: an assessment. Technical report, INEB - Instituto de Engenharia Biomedica (2005)
19. Sun, S.: A survey of multi-view machine learning. Neural Comput. Appl. **23**, 2031–2038 (2013). https://doi.org/10.1007/s00521-013-1362-6
20. Theodoridis, S., Koutrumbas, K.: Pattern Recognition, 4th edn. Elsevier Inc., Amsterdam (2009)
21. Von Luxburg, U.: A tutorial on spectral clustering. Stat. Comput. **17**(4), 395–416 (2007). https://doi.org/10.1007/s11222-007-9033-z

Surrogate Modeling and Expensive Optimization

Investigating Normalization Bounds for Hypervolume-Based Infill Criterion for Expensive Multiobjective Optimization

Bing Wang[✉], Hemant Kumar Singh, and Tapabrata Ray

School of Engineering and Information Technology,
The University of New South Wales, Canberra, ACT 2600, Australia
{bing.wang,h.singh,t.ray}@adfa.edu.au
http://www.mdolab.net/

Abstract. While solving *expensive* multi-objective optimization problems, there may be stringent limits on the number of allowed function evaluations. Surrogate models are commonly used for such problems where calls to surrogates are made in lieu of calls to the true objective functions. The surrogates can also be used to identify *infill* points for evaluation, i.e., solutions that maximize certain performance criteria. One such infill criteria is the maximization of predicted hypervolume, which is the focus of this study. In particular, we are interested in investigating if better estimate of the normalization bounds could help in improving the performance of the surrogate assisted optimization algorithm. Towards this end, we propose a strategy to identify a better ideal point than the one that exists in the current archive. Numerical experiments are conducted on a range of problems to test the efficacy of the proposed method. The approach outperforms conventional forms of normalization in some cases, while providing comparable results for others. We provide critical insights on the search behavior and relate them with the underlying properties of the test problems.

1 Introduction

Multiobjective optimization problem (MOP) refers to optimization problems with more than one conflicting objectives [15]. Such problems are commonly encountered in several domains including engineering, finance and operations research, among others. The optimum solution of such problems consists of not one but multiple trade-off solutions, referred to as Pareto set (PS) in the design (variable space) and Pareto front (PF) in the objective space.

In certain real-world problems, evaluating each candidate design may involve significant expense [20]. While typically the *expense* refers to computational effort, such as time-consuming simulations, it can also relate to other scenarios such as a physical experiments or financial cost (e.g. of a third-party analysis). In order to deal with such problems, surrogate models are often incorporated within metaheuristic search methods to significantly cut down the number of

H. Ishibuchi et al. (Eds.): EMO 2021, LNCS 12654, pp. 519–530, 2021.
https://doi.org/10.1007/978-3-030-72062-9_41

evaluations required to obtain satisfactory solutions [4,6]. Surrogate assisted optimization (SAO) operates by building approximation models of the response functions, and using them in lieu of true evaluations strategically during the course of search. A number of surrogate models exist, such as polynomial regression, radial basis functions, Kriging, support vector machines, etc., and often multiple models are used to improve the approximations [4,6].

The solutions that are identified for true evaluation based on the surrogate models are often referred to as *infill* solutions. An infill solution is typically obtained by extremizing/improving a certain criteria based on the predicted values. For example, one way to identify infill solution(s) could be to simply optimize the predicted objective value(s) from the approximate model [4]. Another widely used method in the context of single-objective optimization was proposed by Jones et al. [10], referred to as efficient global optimization (EGO). In this method, an expected improvement (EI) is calculated taking into account both predicted mean and variance of a Kriging model, built based on existing truly evaluated designs. EI is then maximized to obtain the infill solution, followed by updating the model with a newly evaluated point. On similar lines, several infill criteria have since been proposed to deal with expensive multi-objective optimization. These include scalarized objective functions [12], hypervolume (HV) improvement, expected HV improvement, Euclidean distance improvement, maxmin distance improvement, etc. [22]. Of these, this study focuses on studying HV improvement based infill criteria.

HV is a particularly interesting metric in the field of multi-objective optimization since it is *Pareto-compliant*, implying that a solution (set) that dominates another solution (set) will have a better (higher) HV. Consequently, the PF has the maximum HV in the objective space. Originally proposed for performance comparison [24], HV has since also been frequently used as a driving metric for environmental selection in indicator-based evolutionary algorithms [2,3,21]. One of the key user inputs in calculating the HV is the reference point, which can have an impact in the assessed relative quality of the compared PF approximations [9,16]. It is generally recommended to set the reference point to be slightly dominated by the nadir point of the non-dominated set. This allows for the HV contributions of the extreme points to be included while not being significantly higher than the rest of the non-dominated set. The second aspect that needs to be taken into account is the normalization of the objective vectors. This is because the range of objective values can be quite different for different objectives, leading to the HV being biased towards certain objectives. Combining both these aspects, a common way to drive a population based on HV would be to normalize the current non-dominated solutions linearly between 0 and 1 using *current estimates of ideal and nadir points*, and then use reference point $R = (1.1, 1.1, \ldots 1.1)$, as done in, e.g. [21]. It has been shown in some studies that the choice (of using 1.1) may not be ideal especially for large number of objectives [9,16], but for low objective dimensions it generally works well.

Normalization is a very important aspect in contemporary evolutionary multi-objective optimization algorithms that depend on *decomposition* of objective space. A few variations of normalization have been studied in various frameworks such as NSGA-III [5], MOEA/D [7], and DBEA [18]. The conventional method of normalization based on the bounds of the current non-dominated set helps in removing bias between objectives, which is critical for the spatial diversity that these algorithms aim to achieve. However, such normalization could also result in the algorithm losing track of the state of population relative to global objective space, which could slow down the resulting convergence and/or diversity of the population (discussed in next section). While the effects of certain normalization methods have been discussed in the above studies for respective algorithms, these algorithms do not deal with expensive MOPs. In the context of SAO algorithms, this aspect has not been highlighted in the literature. In this paper, we take a step towards addressing the above research gap in the context of HV based infill criteria. In particular, we investigate whether providing a better estimate of ideal point could improve the performance of the algorithms.

Following this introduction, the basic idea and algorithmic framework is outlined in Sect. 2, followed by numerical experiments and discussion in Sect. 3. Concluding remarks and future work are discussed in Sect. 4.

2 Proposed Approach

2.1 Normalization Bounds

As mentioned in the previous section, the standard way to normalize the solutions is based on the current non-dominated (ND) solutions. The potential drawback in doing so is that if the current non-dominated solutions are concentrated in a small region near the PF, the infill point obtained by maximizing the predicted HV will only have an opportunity to lie within a small portion of the PF. Consequently, the improvement in the quality of the obtained PF approximation will be relatively small. The situation is illustrated in Fig. 1a. It can be seen that the current situation of non-dominated points is localized towards a region of the PF. If the normalization limits are set based on them, and the reference point set as shown (slightly dominated by Z^N), then the infill point could potentially only be generated close to the part of the PF that lies within the dotted black lines; restricting the diversity that could be achieved by HV maximization.

As evident from the Fig. 1a, the main reason the infill criteria may not be able to improve diversity to cover the PF in this scenario is because the current estimates of the ideal and nadir points (Z^I and Z^N respectively) are not accurate w.r.t. the *true* values, say Z^{*I} and Z^{*N}, of the PF. The main proposal of this study is to improve that by *searching for* the normalization bounds close to Z^{*I} and Z^{*N} in order to specify the reference point. We believe that a surrogate assisted framework is in-fact a very suitable paradigm to do so, since the search for the extreme points can be conducted on the surrogate models instead of true function. Only once the individual surrogate objectives have been optimized, the resulting solutions can be truly evaluated; incurring one evaluation for each objective's minimum. This is in stark contrast for following the same process in a non-surrogate optimization frameworks, where obtaining the extreme points

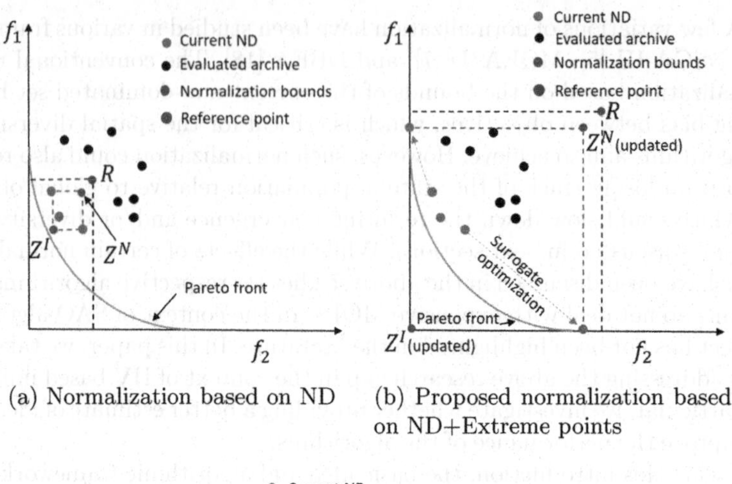

(a) Normalization based on ND (b) Proposed normalization based on ND+Extreme points

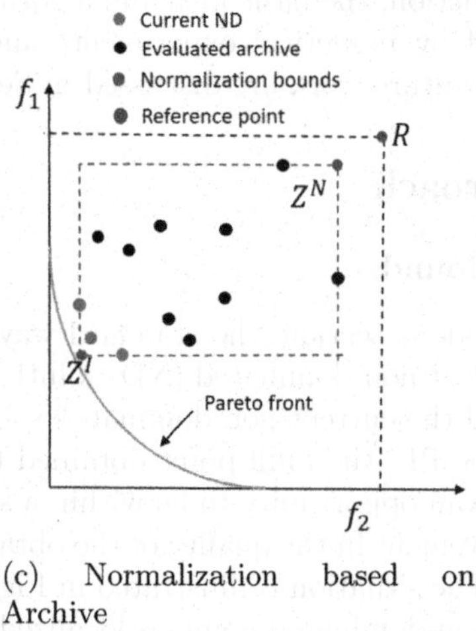

(c) Normalization based on Archive

Fig. 1. Illustration of normalization schemes

themselves could take up substantial cost [18]. The intended result is as shown in Fig. 1b. Starting from the currently known best values for each objective, an individual search is conducted on the surrogate model built based on existing archive of truly evaluated points. This yields the two extreme points, which are then evaluated, yielding the updated normalization bounds (Z^I(updated) and Z^N(updated) in Fig. 1b) and the corresponding reference point. It allows for the infill search to span and identify good non-dominated solutions in an expanded region to potentially uncover more diverse solutions along the PF.

Note that occasionally in the literature there is also another type of normalization followed, which is based on the whole archive of the solutions evaluated so far [13,22]. While this may also help in expanding the search region, it also has a potential drawback. If the evaluated population happens to extend well beyond the true PF range (e.g., as shown in Fig. 1c), the infill solutions (based on HV) will have a tendency to distribute over a large non-dominated region weekly dominated by the extremities, rather than focusing the distribution on the PF itself.

Thus, it can be inferred that the proposed method tries to overcome potential limitations of setting the reference point and normalization range based on the current non-dominated set and the archive. Of course, two challenges are anticipated outright in executing the above hypothesis to work as intended.

- Based on the existing models, the points obtained by minimizing the individual objective do not improve as far as their true optimum. This could be due to high non-linearity of the objective function not reflected in the surrogate model, or the limitations of the search algorithm itself. Therefore, the search is conducted again to update these extreme solutions *whenever* new infill solutions that go beyond the current bounds are discovered. A realistic expected scenario is thus that the extremities will get updated a few times during the search, progressively getting closer to the true ideal/nadir points.
- The second is that the individual objective function(s) are *multimodal*. In such a case, while minimizing an individual objective (say f_2), the resulting extreme point has the minimum value of f_1 but it can occur for multiple values of $\mathbf{x}$, which could yield different values of f_1, some being far from the true extreme point of the PF. For example, in Fig. 1b, one may find an extreme point on the f_1 axis that is far away (towards the right) from the point shown. In the parlance of evolutionary multi-objective optimization, such points are often referred to as *dominance resistance* solutions (DRS) [1]. In order to reduce the possibility of selecting such points, the solutions explored during the surrogate optimization (of, say f_2) that have nearly the same (within a small threshold ϵ) value as the f_2 are assumed equivalent, and one among them with least f_1 is chosen as the extreme solution before evaluating it.

2.2 Implementation

The pseudo-code of the overall framework is presented in Algorithm 1. The algorithm starts from generation of an initial set of designs using the Latin Hypercube Sampling (LHS), a widely used method for design of experiments. The designs are then truly evaluated and surrogate models are built for each objective function. An archive of all truly evaluated solutions is maintained and updated throughout the search. The extreme points of the search are identified by minimizing the predicted values of surrogate model of each objective. The obtained solutions are evaluated, and maximum and minimum values among the existing non-dominated solutions are used as the normalization bounds. Using reference point $R = (1.1, \dots 1.1)$, an HV maximization is conducted to obtain the infill solution, which is then evaluated and added to the archive. The surrogate models, archive and number of evaluations are updated, and the process repeats until the evaluation budget is exhausted.

Algorithm 1. Proposed HV infill-based algorithm

Input: Max. number of function evaluations FE_{max}, Initial sample size N, Normalization bounds search related parameters, Infill search related parameters
Output: PF approximation

1: Initialize archive of evaluation solutions, $\mathcal{A} = \emptyset$, No. of evaluations $FE = 0$.
2: Sample N initial designs & evaluate them. Update $\mathcal{A}$, FE.
3: **while** ($FE < FE_{max}$) **do**
4: Build surrogate models $\mathcal{K} = \{\mathcal{K}_1, \mathcal{K}_2 \ldots \mathcal{K}_M\}$ using the dataset $\{\mathbf{x}, \mathbf{f}\}$ in $\mathcal{A}$
5: **if** first interation or infill point is better than current ideal on any objective **then**
6: search for extreme solutions $\mathbf{E}_1, \mathbf{E}_2 \ldots \mathbf{E}_M$ by optimizing $f_1 \ldots f_M$ individually using $\mathcal{K}$
7: **end if**
8: Evaluate the extreme solutions $\mathbf{E}$. Update $\mathcal{A}, FE, \mathcal{K}$
9: Set Z^I, Z^N based on non-dominated solutions $\mathcal{N} \in \mathcal{A}$
10: Normalize the solutions in $\mathcal{A}$ using Z^I, Z^N as the bounds to get $\overline{\mathcal{A}}$ and $\overline{\mathcal{N}}$
11: Identify infill solution $\mathbf{x}_{HV}$ by maximizing HV with reference point $R = (1.1, \ldots 1.1)$
12: Evaluate $\mathbf{x}_{HV}$. Update $\mathcal{A}, FE$
13: **end while**
14: Output the non-dominated solutions $\mathcal{N} \in \mathcal{A}$ as the PF approximation

In the above process, we have used Kriging [14] as the surrogate model for approximating objective functions in Lines 4, 8. For minimization of surrogate objectives in Line 6 as well as HV maximization for infill identification in Line 11, we have used differential evolution (DE) algorithm [19]. For removing the DRS solutions, for each objective, the tolerance $\epsilon = 1e-5$ is used.

3 Numerical Experiments

To demonstrate the performance of the proposed approach, numerical experiments are conducted on a range of test problems widely used in evolutionary multi-objective optimization domain. These include problems from the ZDT [23], DTLZ [11] and WFG [8] series, to cover a diversity of problem characteristics such as nonlinearity, convex/concave PF, difficulty in achieving convergence, difficulty in achieving diversity, discontinuous PF, etc. The settings used for the experiments are listed in Table 1. The parameter setting for ZDT, DTLZ and infill search follows commonly used initialization configuration, while for WFG, as the problems are more difficult, initial sample size is increased to improve approximation accuracy. The adjustments in ruggedness and bias parameter is to mitigate approximation challenge on surrogate, as the focus of this study is on normalization rather than accuracy of surrogate. Kriging model is built from implementation of DACE toolbox [14] in Python. The regression function is set to be zero order polynomial, while the correlation model is set to be Gaussian. HV is used as the metric for performance comparison of the final solutions, in the objective space normalized by the true ideal and nadir points, and reference point as (1.1, 1.1). A total of 29 independent runs are conducted to observe the statistical behavior. The performance is compared among three normalization methods and reference point set using:

- $NormR_{ND}$, based on current non-dominated solutions,
- $NormR_A$, based on full archive and
- $NormR_{NDE}$, the proposed method based on the non-dominated solutions augmented with extreme point search.

Table 1. Settings for numerical experiments

Problem-specific settings				
	Max evals.	Initial samples	# Variables (D)	Others
ZDT, DTLZ	100	$11D-1$	6	Ruggedness parameter $= 2\pi$ and scaling factor for $g(\mathbf{x}) = 1$ for DTLZ1, DTLZ3
WFG	250	200	6	Bias parameter for $b_{poly} = 0.5$ for WFG1
Differential Evolution (DE) settings				
	Pop. size	Generations	Crossover rate	Scaling factor
Extreme point search	100	100	0.8	0.8
Infill search	100	100	0.8	0.8

3.1 Results and Discussion

The median HV values obtained by the three algorithms for all problems considered is shown in Table 2. The best (highest) median value is highlighted in bold for each problem. Wilcoxon Ranksum test with significance level 0.05 is used for establishing statistical significance between the obtained results. The symbols $\downarrow, \uparrow, \approx$ are used to denote cases where the compared algorithm is statistically worse, better or equivalent, respectively, to the proposed approach. Individually, $NormR_{NDE}$ is better than $NormR_{ND}$ for 8 instances, worse for 2 instances and equivalent for 7. When compared to $NormR_A$, the numbers are 7, 0 and 10, respectively.

Thus, generally speaking, the proposed method $NormR_{NDE}$ performs equivalent or better for majority of the problems compared to both, $NormR_A$ and $NormR_{ND}$. Notably, for certain problems (e.g. ZDT1, ZDT3) the performance of $NormR_{ND}$ is much worse than the other two methods, and for certain other problems (e.g. DTLZ1, DTLZ3) $NormR_A$ is much worse than the other two methods. In contrast, the proposed $NormR_{NDE}$ is typically only marginally inferior to the other methods in certain problems, reflecting on its consistent performance.

The relative performance and their correlation with the problem characteristics is further investigated in the following subsections. We divide the discussion based individually on the three test suites, i.e., ZDT, DTLZ, and WFG.

ZDT Problems: For ZDT problems, typically all the randomly sampled initial points lie far from the PF in the objective space. Also, the range of the PF is covered well in terms of f_1 (as it is simply equal to x_1), but not f_2. The range based on both, the archive and the initial non-dominated solutions are therefore far from the true ideal and nadir points. The extreme point search in this case is particularly helpful as it yields close to the true ideal point, with the lowest value of f_2 covered. For ZDT1, this is illustrated in Fig. 2a–b and is the main reason

Table 2. Median HV obtained for all problems across 29 independent runs. Wilcoxon rank sum test is used to test significance, with significance level set as 0.05 ($\downarrow, \uparrow, \approx$) are used to denote whether the compared algorithm is statistically worse, better or equivalent, respectively, to the proposed approach.

Test problem	$NormR_A$ (archive)	$NormR_{ND}$ (current ND)	$NormR_{NDE}$ (current ND+E)
ZDT1	0.8536 ↓	0.4292 ↓	**0.8582**
ZDT2	0.5228 ↓	0.5233 ↓	**0.5295**
ZDT3	**0.5911** ≈	0.2370 ↓	0.4080
ZDT6	0.3973 ≈	0.3518 ≈	**0.4129**
WFG1	**0.6690** ≈	0.6607 ↓	0.6647
WFG2	0.6696 ↓	0.6431 ↓	**0.6845**
WFG3	0.6543 ≈	0.6399 ↓	**0.6565**
WFG4	**0.3409** ≈	0.3384 ↓	0.3400
WFG5	**0.3578** ≈	0.3577 ≈	0.3576
WFG6	**0.3423** ≈	0.3404 ≈	0.3402
WFG7	0.2926 ≈	0.2954 ≈	**0.2976**
WFG8	0.2119 ≈	**0.2145** ≈	0.2128
WFG9	0.3523 ≈	**0.3566** ≈	0.3548
DTLZ1	0.2466 ↓	**0.6291** ≈	0.6256
DTLZ2	0.3739 ↓	**0.4073** ↑	0.4058
DTLZ3	0.0407 ↓	**0.1793** ↑	0.1065
DTLZ4	0.1071 ≈	0.1039 ≈	**0.1083**
DTLZ7	0.5309 ↓	0.3510 ↓	**0.5390**

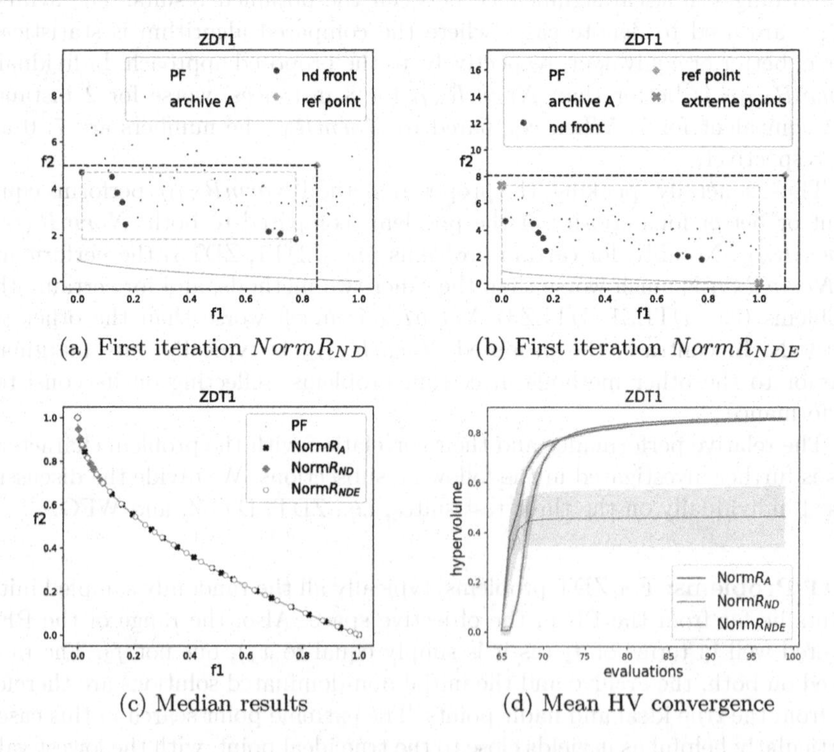

(a) First iteration $NormR_{ND}$ (b) First iteration $NormR_{NDE}$

(c) Median results (d) Mean HV convergence

Fig. 2. Typical scenario observed for ZDT1 problem

why the performance and convergence of $NormR_{NDE}$ is better than $NormR_{ND}$, as is clear from the distribution obtained in the median run in Fig. 2c and HV convergence plot in Fig. 2d. Due to a slightly larger coverage of objective space, $NormR_A$ is also able to perform better than $NormR_{ND}$.

WFG Problems: For most WFG problems, the initial population lies much closer to the PF compared to the ZDT problems, but the coverage of the PF is limited. In such cases, if the proposed algorithm could expand the range, the performance should significantly improve. However, the search for extremities yielded only slightly better points than the current non-dominated set in this case, as illustrated for WFG4 in Fig. 3a–b. This may be attributed to relatively higher complexity of WFG problem formulations, which the approximation models may not be able to capture accurately using the dataset with the given number of points. The small improvement in the extremities correspondingly result in marginal improvements in the performance, as illustrated in Fig. 3c–3d. Overall, the three methods show similar performance in this case.

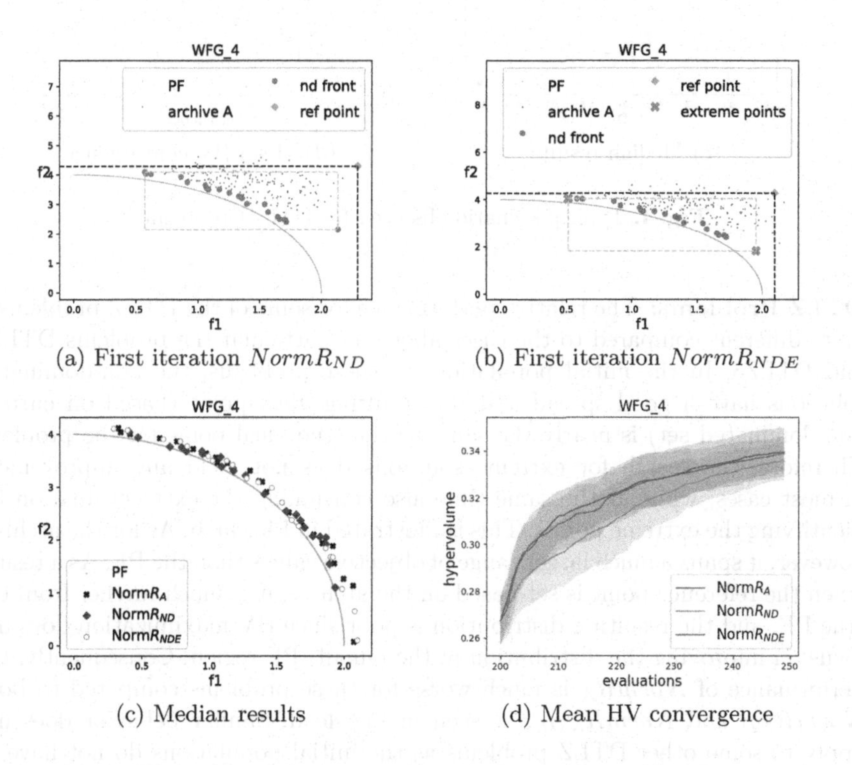

(a) First iteration $NormR_{ND}$ (b) First iteration $NormR_{NDE}$

(c) Median results (d) Mean HV convergence

Fig. 3. Typical scenario observed for WFG4 problem

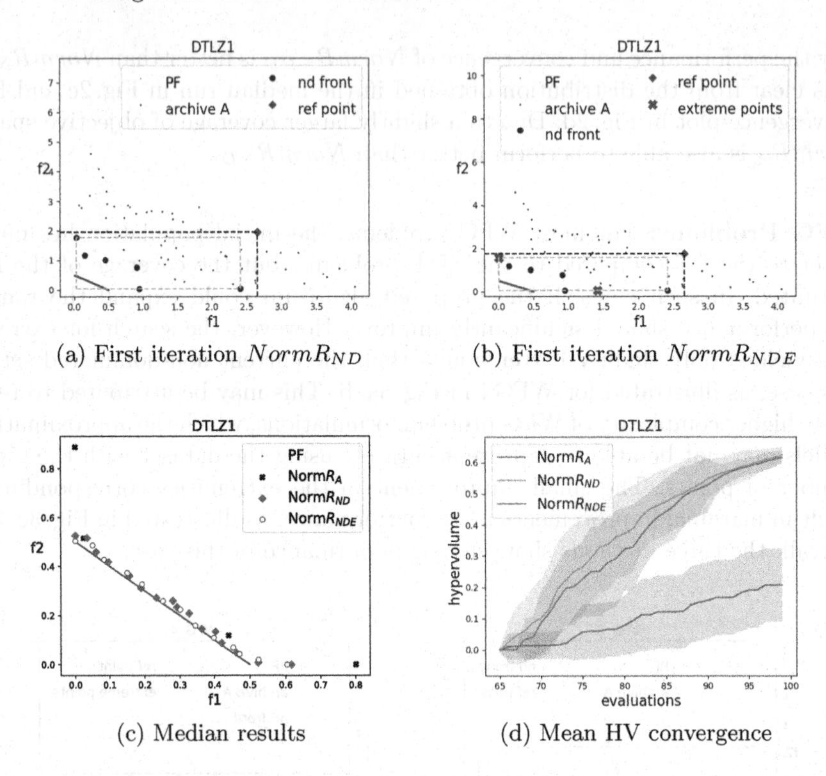

(a) First iteration $NormR_{ND}$ (b) First iteration $NormR_{NDE}$

(c) Median results (d) Mean HV convergence

Fig. 4. Typical scenario observed for DTLZ1 problem

DTLZ Problems: The relative performance for some of the DTLZ problems is quite different compared to the cases above, in particular the problems DTLZ1 and DTLZ3. In the initial population for these problems, the non-dominated solutions have a good spread and the resulting ideal point (based on current non-dominated set) is nearly the same as the true ideal point for the problem. Therefore, the search for extreme solutions does not yield any improvement in most cases; while at the same time also consuming the extra evaluation for identifying the extreme points. This is illustrated in Fig. 4a–b. As for the archive, however, it spans a much larger range of objective values than the PF. As a result, when the reference point is set based on the archive, it is much further from the true PF, and the resulting distribution is poor since HV maximization does not focus on improving the distribution in the (small) PF region. Consequently, the performance of $NormR_A$ is much worse for these problems compared to both $NormR_{ND}$ and $NormR_{NDE}$, as seen in Fig. 4c–d. Similar behavior does not apply to some other DTLZ problems as the initial populations do not have as large a range of objectives as DTLZ1/3.

4 Conclusions and Future Work

In this paper, we have presented a method to improve the normalization bounds and reference point for an algorithm using HV based infill criterion to handle computationally expensive multiobjective optimization problems. The numerical experiments demonstrate the consistency of the proposed approach as compared to two other common approaches. The advantages gained were particularly notable for problems where the initial sampling yields solutions far away from the PF in the objective space, and where the extreme point search is able to find a better estimate of the true extremities of the PF.

A number of future directions could be pursued to extend the work. First would be to improve the scalability of the method to deal with higher number of objectives. For $\geq$3 objective problems, the extremities may not be obtained by simply optimizing each objective individually [17], hence more sophisticated extreme point identification methods could be incorporated. Moreover, other infill criteria, such as those discussed in [22], could be analyzed and improved. Source code of the proposed algorithm can be found in http://www.mdolab.net/.

Acknowledgements. The authors would like to acknowledge Discovery Project DP190102591 from the Australian Research Council.

References

1. Asafuddoula, M., Singh, H.K., Ray, T.: An enhanced decomposition-based evolutionary algorithm with adaptive reference vectors. IEEE Trans. Cybern. **48**(8), 2321–2334 (2017)
2. Bader, J., Zitzler, E.: HypE: an algorithm for fast hypervolume-based many-objective optimization. Evol. Comput. **19**(1), 4 (2011)
3. Beume, N., Naujoks, B., Emmerich, M.: SMS-EMOA: multiobjective selection based on dominated hypervolume. Eur. J. Oper. Res. **181**, 1653–1669 (2007)
4. Bhattacharjee, K.S., Singh, H.K., Ray, T.: Multiple surrogate-assisted many-objective optimization for computationally expensive engineering design. J. Mech. Des. **140**(5), 1–10 (2018)
5. Blank, J., Deb, K., Roy, P.C.: Investigating the normalization procedure of NSGA-III. In: Deb, K., et al. (eds.) EMO 2019. LNCS, vol. 11411, pp. 229–240. Springer, Cham (2019). https://doi.org/10.1007/978-3-030-12598-1_19
6. Habib, A., Singh, H.K., Chugh, T., Ray, T., Miettinen, K.: A multiple surrogate assisted decomposition-based evolutionary algorithm for expensive multi/many-objective optimization. IEEE Trans. Evol. Comput. **23**(6), 1000–1014 (2019)
7. He, L., Ishibuchi, H., Trivedi, A., Srinivasan, D.: Dynamic normalization in MOEA/D for multiobjective optimization. In: IEEE Congres on Evolutionary Computation (CEC) (2020)
8. Huband, S., Hingston, P., Barone, L., While, L.: A review of multiobjective test problems and a scalable test problem toolkit. IEEE Trans. Evol. Comput. **10**(5), 477–506 (2006)

9. Ishibuchi, H., Imada, R., Setoguchi, Y., Nojma, Y.: How to specify a reference point in hypervolume calculation for fair performance comparison. Evol. Comput. **26**(3), 411–440 (2018)
10. Jones, D.R., Schonlau, M., Welch, W.J.: Efficient global optimization of expensive black-box functions. J. Global Optim. **13**, 455–492 (1998). https://doi.org/10.1023/A:1008306431147
11. Deb, K., Thiele, L., Laumanns, M., Zitzler, E.: Scalable test problems for evolutionary multiobjective optimization. In: Abraham, A., Jain, L., Goldberg, R. (eds.) Evolutionary Multiobjective Optimization. AI&KP, pp. 105–145. Springer, London (2005). https://doi.org/10.1007/1-84628-137-7_6
12. Knowles, J.: ParEGO: a hybrid algorithm with on-line landscape approximation for expensive multiobjective optimization problems. IEEE Trans. Evol. Comput. **10**(1), 50–66 (2006)
13. Liu, Y., Ishibuchi, H., Yen, G.G., Nojima, Y., Masuyama, N., Han, Y.: On the normalization in evolutionary multi-modal multi-objective optimization. In: IEEE Congres on Evolutionary Computation (CEC) (2020)
14. Lophaven, S.N., Nielsen, H.B., Sondergaard, J.: A Matlab kriging toolbox. Technical report, Technical University of Denmark (2002)
15. Miettinen, K.: Nonlinear Multiobjective Optimization, vol. 12. Springer, Heidelberg (2012)
16. Singh, H.K.: Understanding hypervolume behavior theoretically for benchmarking in evolutionary multi/many-objective optimization. IEEE Trans. Evol. Comput. **24**(3), 603–610 (2020)
17. Singh, H.K., Isaacs, A., Ray, T.: A Pareto corner search evolutionary algorithm and dimensionality reduction in many-objective optimization problems. IEEE Trans. Evol. Comput. **15**(4), 539–556 (2011)
18. Singh, H.K., Yao, X.: Improvement of reference points for decomposition based multi-objective evolutionary algorithms. In: Shi, Y., et al. (eds.) SEAL 2017. LNCS, vol. 10593, pp. 284–296. Springer, Cham (2017). https://doi.org/10.1007/978-3-319-68759-9_24
19. Storn, R., Price, K.: Differential evolution-a simple and efficient heuristic for global optimization over continuous spaces. J. Glob. Optim. **11**(4), 341–359 (1997). https://doi.org/10.1023/A:1008202821328
20. Wang, G.G., Shan, S.: Review of metamodeling techniques in support of engineering design optimization. J. Mech. Des. **129**(4), 370–380 (2007)
21. Zapotecas-Martínez, S., López-Jaimes, A., García-Nájera, A.: LIBEA: a Lebesgue indicator-based evolutionary algorithm for multi-objective optimization. Swarm Evol. Comput. **44**, 404–419 (2019)
22. Zhan, D., Cheng, Y., Liu, J.: Expected improvement matrix-based infill criteria for expensive multiobjective optimization. IEEE Trans. Evol. Comput. **21**(6), 956–975 (2017)
23. Zitzler, E., Deb, K., Thiele, L.: Comparison of multiobjective evolutionary algorithms: empirical results. IEEE Trans. Evol. Comput. **8**(2), 173–195 (2000)
24. Zitzler, E., Thiele, L.: Multiobjective optimization using evolutionary algorithms—a comparative case study. In: Eiben, A.E., Bäck, T., Schoenauer, M., Schwefel, H.-P. (eds.) PPSN 1998. LNCS, vol. 1498, pp. 292–301. Springer, Heidelberg (1998). https://doi.org/10.1007/BFb0056872

Pareto-Based Bi-indicator Infill Sampling Criterion for Expensive Multiobjective Optimization

Zhenshou Song[1], Handing Wang[1](✉), and Hongbin Xu[2]

[1] School of Artificial Intelligence, Xidian University, Xi'an 710071, China
zssong@stu.xidian.edu.cn, hdwang@xidian.edu.cn

[2] School of Software Engineering, South China University of Technology, Guangzhou 510006, China
sehongbin_xu@mail.scut.edu.cn

Abstract. Infill sampling criteria play a crucial role in saving expensive evaluations for surrogate-assisted multiobjective evolutionary algorithms. Promoting convergence and maintaining diversity in the population are the two main goals of designing a new infilling sampling criterion, which is naturally a bi-objective optimization problem. In this paper, a Pareto-based bi-indicator infill sampling criterion is proposed to select candidate solutions which will be re-evaluated using expensive objective functions. The proposed criterion is embedded into a Gaussian process assisted evolutionary algorithm to solve expensive multiobjective optimization problems. In the proposed criterion, we introduce two indicators measuring the convergence and diversity as two optimization objectives. Empirical studies on the UF and DTLZ problems demonstrate that the proposed algorithm is more effective than other state-of-the-art surrogate-assisted evolutionary algorithms using the same number of expensive function evaluations.

Keywords: Gaussian process · Expensive multiobjective optimization · Infill sampling criterion

1 Introduction

In last two decades, a large number of multiobjective evolutionary algorithms (MOEAs) have been successfully applied to a wide range of multiobjective optimization problems (MOPs) [1]. An MOP can be defined as follow:

$$\begin{aligned} &\text{minimize} \quad \mathbf{F}(\mathbf{x}) = (f_1(\mathbf{x}), f_2(\mathbf{x}), ..., f_m(\mathbf{x}))^T, \\ &\text{subject to} \quad x_{L_i} \le x_i \le x_{U_i}, i \in \{1, ..., d\}, \end{aligned} \tag{1}$$

where $\mathbf{x} = (x_1, ..., x_d)^T$, d is the number of decision variables, x_{L_i} and x_{U_i} are the lower and upper bounds of x_i, and $f_1, f_2, ..., f_m$ are m objective functions.

Supported by the National Natural Science Foundation of China (No. 61976165).

H. Ishibuchi et al. (Eds.): EMO 2021, LNCS 12654, pp. 531–542, 2021.
https://doi.org/10.1007/978-3-030-72062-9_42

Due to the conflicting nature of objectives, i.e. the improvement of one objective may degenerate another objective, multiple objectives cannot reach a single optimum simultaneously. Thus, without considering any user preferences, MOEAs usually output a set of Pareto optimal solutions which cannot further optimize any objective unless degenerating other objectives. The collection of all Pareto trade-off solutions is called the Pareto set (PS), and the mapping of the PS in objective space is called the Pareto front (PF). The concept of the dominance relationship is critical for obtaining a Pareto solution [2]. To compare the performance of two given solutions $\mathbf{x}^1$ and $\mathbf{x}^2$, $\mathbf{x}^1$ dominates $\mathbf{x}^2$ if and only if $\forall i \in \{1, 2, ..., m\}, f_i(\mathbf{x}^1) \leq f_i(\mathbf{x}^2)$ and $\exists j \in \{1, 2, ..., m\}, f_i(\mathbf{x}^1) < f_j(\mathbf{x}^2)$.

Based on the dominance relationship, many Pareto-based MOEAs have been developed, e.g. the elitist non-dominated sorting genetic algorithm (NSGA-II) [3]. Apart from the Pareto-based MOEAs, some MOEAs using different selection mechanisms have also been proposed for solving MOPs. Indicator-based MOEAs incorporate a performance indicator, such as $I_\epsilon+$ [4,5], hypervolume (HV) [4], and R2 [6], as their fitness functions in the optimization process. Aggregation-based MOEAs decompose an MOP into several single objective optimization problems by conventional aggregation approaches [7]. Furthermore, some hybrid algorithms inspired by the above approaches have been proposed to solve MOPs. For example, in addition to using non-dominated sorting, NSGA-III [8], an improved NSGA-II, employs reference vectors (RVs, also known as reference points) to select candidate solutions during the environmental selection for providing good convergence and well-distribution of the solution set. Note that most existing research on the above MOEAs that require a large number of function evaluations, which is based on an implicit assumption that the function evaluation of objectives of candidate solutions is easily achieved and cheap [9].

However, such assumption is impractical in many real-world optimization problems. Instead, evaluations of the objectives in real-word MOPs can be economically or computationally expensive. For example, in an optimization problem involving computational fluid dynamic (CFD) simulations, a single CFD simulation could take an hour [10]. It is impractical that allowing thousands of simulations to evaluate objectives during the optimization process. In recent years, a new approach using surrogate models as the replacement of expensive evaluations has been successfully proven effective in solving expensive MOPs. Surrogate models can be embedded into MOEAs to yield surrogate-assisted evolutionary algorithms (SAEAs) [10]. Since only a few of function evaluations are available, one essential challenge for SAEAs to solve is which candidate solution should be chosen to be re-evaluated using the expensive function evaluation. In other words, how to select solutions with benefits on convergence and diversity in the population, so that the number of expensive function evaluations can be reduced as much as possible [9].

There exist many infilling sampling criteria for dealing with this challenge, in which RVs have played an important role in maintaining diversity. In the Pareto based efficient global optimization (ParEGO) [11], one RV is randomly selected to aggregate the MOP into a single-objective problem, then the expected improvement (EI) [12] is applied for selecting the candidate solution to re-evaluate. As the improvement, the efficient global optimization based MOEA/D (MOEA/D-EGO) [13] simultaneous optimizes a set of single objective problems aggregated by the pre-defined RVs. EI is also employed in MOEA/D-EGO for selecting the best solution of single sub-problem. Apart from EI, angle penalized distance (APD) [14] is also applied in the Kriging assisted RV guided evolutionary algorithm (KRVEA) [15] for selecting the best candidate solution on one RV. Different from MOEA/D-EGO, KRVEA divided RVs into several clusters before the candidate solutions selection. The RV as the center of each cluster is selected for calculating the APD value of each assisted candidate solutions. In the ordinal regression surrogate-assisted evolutionary algorithm (OREA) [16], the candidate solution with the best predicted rank will be selected for re-evaluation, where the diversity is also maintained by RVs. The classification based surrogate-assisted evolutionary algorithm (CSEA) [17] selects candidate solutions in the 'good' class to re-evaluate, and it relies on reference points that compare with candidate solutions to maintain diversity. Furthermore, there are some other criteria without using the reference information exists. Taking selection-based efficient global optimization (SMS-EGO) [18] as an example, it uses an HV-based metric (S-metric) to select the candidate solution for expensive evaluation.

Although implementations of these sampling criteria are different, these above sampling criteria are ultimately designed for two objectives, which are promoting convergence and maintaining diversity in the population. Therefore, designing a new infilling sampling criterion can be considered as a bi-objective optimization problem. On the basis of this thought, we propose a Pareto-based bi-indicator infill sampling criterion (PBISC) for expensive MOPs optimization, where the infilling sampling process of the algorithm is considered as a bi-objective optimization. The major features of PBISC are as follows:

1. A convergence indicator (CI) and a diversity indicator (DI) are introduced in PBISC to measure the convergence improvement and the diversity maintenance. Different from other diversity indicators, such as the pure diversity (PD) [19], DI can also be used for measuring convergence in some special cases, which can accelerate the population converge to the true PF.
2. Different from existing sampling criteria, PBISC manages to transform the selection into a multiobjective optimization problem by considering two indicators simultaneously. Furthermore, PBISC does not need to pre-define the number of re-evaluated solutions in each iteration.

Due to the excellent performance of NSGA-III in MOPs optimization [8], we embed PBISC into the NSGA-III framework yield PB-NSGAIII for verifying the effectiveness of PBISC, where the Gaussian process (GP) regression [20] is used as the surrogate model to replace most of the expensive function evaluations.

The remainder of this paper is organized as follows. Section 2 introduces the preliminaries including the NSGA-III and GP. In Sect. 3, a detailed description of the proposed algorithm is given. The parameter settings, the experimental results and analysis are given in Sect. 4. Finally, the conclusion and future work are provided in Sect. 5.

2 Preliminaries

2.1 Non-dominated Sorting Genetic Algorithm III

NSGA-III [8] follows the framework of NSGA-II, but its mechanism for selecting new individuals is essentially different from that in NSGA-II. It combines the non-dominated sorting approach from NSGA-II with the aggregation approach from MOEA/D, which avoids the degradation caused by the increasing objective dimension and ensures the diversity of the population. That is the reason why we select NSGA-III as the optimization framework. The main components of NSGA-III are presented in Algorithm 1.

Algorithm 1. Pseudocode of NSGA-III

Input: The maximum number of generations t_{max}; Population size N; Reference points Z.
Output: Non-dominated solutions from population $P_{t_{max}}$.
1: Generate an initial population P_1 and set counter $t = 1$.
2: **while** $t < t_{max}$ **do**
3: Generate offspring Q_t by crossover and mutation operations.
4: Combine parents and offspring populations, $C_t = P_t \cup Q_t$.
5: $(P_{t+1}) = \textit{Environmental selection}(C_t, N, Z)$.
6: Select parents P_{t+1} for next generation.
7: **end while**

At the t-th environmental selection, non-dominated sorting is performed on C_t which is divided into multiple fronts. Solutions in the first front have the highest selection priority to be kept as parent for the next generation (P_{t+1}) while those solutions in the following front have the lower selection priority [21]. Solutions in the lowest front (F_l) are progressively selected according to the RV-based approach until the size of the new population is equal to N. In this phase, each member of P_{t+1} and F_l is associated with a specific RV using the shortest perpendicular distance of each solution from the RV. Note that after normalizing the objective values, the origin is the ideal point (formed by best objective function values available). The rest k solutions are chosen from the RV with the least associated individuals, where $k = N - |P_{t+1}|$.

2.2 Gaussian Process

The GP regression model is also known as the Kriging model [22]. GP is a stochastic process that subject to the joint normal distribution of random variables. Since it can output variances as the uncertainty information, GP has become a very attractive surrogate model utilized in many evolutionary algorithms for finding optimal solutions. It is based on the assumption that the prior distribution of the objective function $y(x)$ is subject to joint normal distribution with a constant mean μ and a constant variance σ^2. For any solution $\mathbf{x}$, the GP approximation function can be given as follows:

$$y(\mathbf{x}) = \mu + \epsilon(\mathbf{x}), \epsilon(\mathbf{x}) \sim N(0, \sigma^2), \tag{2}$$

For any two input data $\mathbf{z}^1$, $\mathbf{z}^2$ with their corresponding errors $\epsilon(\mathbf{z}^1)$, $\epsilon(\mathbf{z}^2)$, their correlation $R(\epsilon(\mathbf{z}^1), \epsilon(\mathbf{z}^2))$ can be measured by a covariance function. In this work, we use the following squared exponential function as the covariance function:

$$R(\epsilon(\mathbf{z}^1), \epsilon(\mathbf{z}^2)) = \exp(-\sum_{i=1}^{d} \theta_i |\mathbf{z}^1, \mathbf{z}^2|^{l_i}), \tag{3}$$

where θ is the scale parameter and l_i is the length-scale parameter, and d is the dimension of $\mathbf{z}^1$.

As recommended in [20], the hyperparameters in Eq. (3) are learned by maximizing the log marginal likelihood function:

$$\frac{1}{(2\pi)^{n/2}(\sigma^2)^{n/2}|\mathbf{Co}|^{1/2}} \exp(-\frac{(\mathbf{f} - \mathbf{1}\mu)^T \mathbf{Co}^{-1}(\mathbf{f} - \mathbf{1}\mu)}{2\sigma^2}), \tag{4}$$

where $|\mathbf{Co}|$ is the correlation matrix, and $\mathbf{f}$ is the achieved objective vector. More details could be found in [20].

3 Proposed Algorithm

In this section, we will give the implementation of the PB-NSGAIII and discuss the PBISC in details. The pseudo-code of the PB-NSGAIII is presented in Algorithm 2, which can be divided into the following phases:

- **Phase I: Initialization** (Line 1). Initial training data *Arch* are sampled from decision space using the Latin hypercube sampling [23], where individuals are evaluated by expensive objective functions. The initial data are employed as the initial population P_1 for evolution. Initialing RVs according to [8].
- **Phase II: Building Surrogate Models** (Line 2). Using *Arch* to build a GP model for approximating each objective function.
- **Phase III: Using Surrogate Models** (Lines 5 to 10). In this phase, GP models are treated as the approximated objective functions in NSGA-III to evaluate the offspring for several iterations, i.e. repeat steps 2–7 of Algorithm 1 with the GP models instead of the expensive objective functions. After a set of offspring with the potential benefits for population $P_{t_{max}}$ are obtained.
- **Phase IV: Infill Sampling** (Line 11). We introduce two indicators to measure the potential improvement of convergence and diversity of each individual in $P_{t_{max}}$. Convergence and diversity are regarded as two objectives to be optimized. The non-dominated sorting approach is applied to screen the solution in the first front for re-evaluation.

- **Phase V: Updating** (Lines 12 to 15). Add new samples into the training data and update the current population P_{FE} after re-evaluating the new samples by expensive objective functions. GP models are rebuilt using all training data.

Algorithm 2. Pseudocode of PB/NSGA-III

Input: The maximum number of expensive function evaluations: FE_{max}; Population size: N; Reference vectors: Z; The predefined number of iterations before updating GP models: t_{max}.
Output: Non-dominated solutions from population P_{FE}.
/*Initialization*/
1: Generate an initial training data set *Arch* using Latin hypercube sampling, initialize the number of function evaluations: $FE = N$.
/*Building GP models*/
2: Train a GP model for each expensive objective function by *Arch*.
3: **while** $FE \leq FE_{max}$ **do**
4: $P_t = P_{FE}$; $t = 1$.
/*Using Surrogate Models*/
5: **while** $t < t_{max}$ **do**
6: Generate offspring Q_t by crossover and mutation operations with GP models instead of the expensive objective functions.
7: Combine parents and offspring populations, $C_t = P_t \cup Q_t$.
8: $(P_{t+1}) = Environmental\ selection(C_t, N, Z)$.
9: Select parents P_{t+1} for next iteration.
10: **end while**
/*Infill Sampling*/
11: Select individuals using the Pareto-based bi-indicators infill sampling criterion from $P_{t_{max}}$, recorded as $Offspring$.
/*Updating*/
12: $Arch = Arch \cup Offspring$; use *Arch* to update GP models.
13: $C_{p+q} = P_{FE} \cup Offspring$.
14: $FE = FE + |offspring|$.
15: $P_{FE} = Environmental\ selection(C_{p+q}, N, Z)$.
16: **end while**

The main idea of proposed infill sampling criterion is to transform the selection into an MOP. As mentioned in Sect. 1, promoting convergence and maintaining diversity in the population are two goals of SAEAs in the phase of infill sampling. Simultaneously optimizing those two objectives is a positive way to enhance the performance of SAEAs. Here, we first introduce two indicators (termed convergence indicator (CI) and diversity indicator (DI)) to measure the convergence and the diversity improvement of each candidate solution, respectively. In PBISC, the first objective is promoting the population converge to the true PF. However, directly calculating the distance between the candidate solution and PF is impossible. Thus, we employ the distance between the candidate

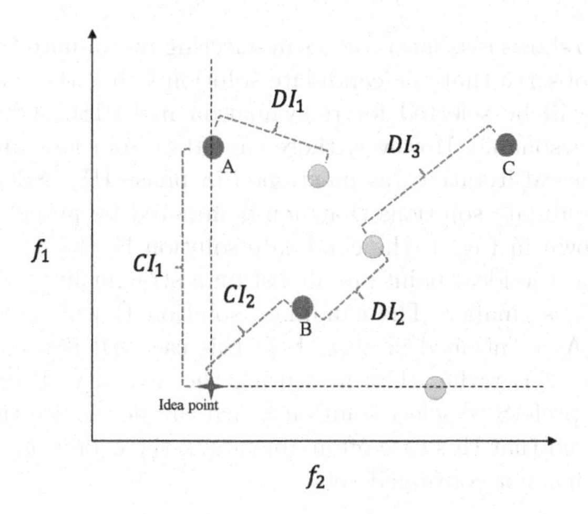

Fig. 1. A schematic diagram of DI and CI, where the green circles are the offspring generated by gray circles, and the four-pointed star is the ideal point (Color figure online)

solution and the ideal point as CI to measure the convergence improvement. The ideal point is formed by the best objective function values available among the candidate solution and the current population P_{FE}. The smaller the CI, the closer the solution is to the true PF. The second objective is maintaining the diversity in the population. Inspired by Two_Arch2 [5], we use the distance between the candidate solution and the closest solution in the parent population as DI to measure the increase in maintaining diversity. The larger DI is, the distribution of the population is better. The mathematical description of DI can be formulated as follow:

$$DI(\mathbf{x}^*) = \min_{p=1:N}\{|\mathbf{x}^*, \mathbf{x}_p|^2\}, \quad where \;\; \mathbf{x}_p \in P_{FE}. \tag{5}$$

where $\mathbf{x}^*$ is candidate solution, $\mathbf{x}_p$ is the p-th individual in the current population P_{FE}. The introduced CI and DI are regarded as two objectives of an MOP, which can be formulated as follow:

$$\begin{aligned} \min_{\mathbf{x}^*} \quad & \mathbf{G}(\mathbf{x}^*) = (g_1, g_2), \\ & g_1 = CI(\mathbf{x}^*), \\ & g_2 = -DI(\mathbf{x}^*). \end{aligned} \tag{6}$$

Note that, since DI is expected to be maximized, we use a negative sign in the Eq. (6) to convert it into a minimization indicator. The schematic diagram of DI and CI is given in Fig. 1. Once the non-dominated candidate solutions in the present generation are generated in the **Phase III**, the non-dominated sorting approach [3] is employed to sort solutions into different fronts according to their corresponding **G** according to Eq. (6). The candidate solutions on the first front will be re-evaluated using the expensive objective functions.

Figure 1 also shows two special cases in selecting re-evaluated solutions. From Fig. 1, we can observe that the candidate solution C has the largest DI, which means that C will be selected for re-evaluation in PBISC. Obviously, such a selection is unreasonable. However, those candidate solutions are generated by NSGA-III in several iteration (as mentioned in phase III). NSGA-III can provide a set of candidate solutions that non-dominated by parent population. In addition, as shown in Fig. 1, the candidate solution B, the corresponding parent solution, and the ideal point are almost on a straight line, where the bigger DI_2 is, the CI_2 is smaller. The candidate solution B will be selected due to its DI and CI. As mentioned in Sect. 1, in this case, DI is used for measuring the convergence of B rather than measuring the diversity. Therefore, the proposed criterion preferS to select solution which can accelerate the convergence. It should be noted that this case often appears in the evolutionary process that the population has not converged yet.

4 Empirical Studies

In order to validate the effectiveness of PB-NSGAIII, numerical experiments are conducted. UF1-7 [24] are chosen as the 2-objective test problems while UF8-10 and DTLZ1-7 [25] are used as the 3-objective test problems in this section. The performance of the proposed algorithm is compared with that of other state-of-the-art SAEAs including the classification based pre-selection multiobjective evolutionary algorithm (CPS-MOEA) [26], CSEA [17], KRVEA [15], MOEA/D-EGO [13]. NSGA-III [8] is also included in the comparison to provide an intuitive appreciation of how differently an MOEA performs with and without surrogate models on expensive MOPs.

In order to evaluate the performance of the compared algorithms on these test problems, Inverted Generational Distance (IGD) [27] is chosen in this paper. IGD evaluates both the convergence and the diversity of the solution set, which can be formulated as follow:

$$IGD(PF_{true}, PF^*) = \frac{\sum_{\mathbf{v} \in PF_{true}} d(\mathbf{v}, PF^*)}{|PF_{true}|}, \tag{7}$$

where PF_{true} is a set of uniformly distributed solutions on the true PF, PF^* is the obtained PF, $\mathbf{v}$ is an individual of PF_{true} and $d(\mathbf{v}, PF^*)$ is the shortest distance from $\mathbf{v}$ to PF^*. The smaller metric IGD is, the closer to the obtained PF approximates to the true PF. The Wilcoxon rank-sum (WRS) test is executed to compare the results achieved by the proposed algorithm and other five compare algorithms under comparison at a significance level of 0.05. Symbols "↑" and "↓" denote that the proposed algorithm is significantly superior to and inferior to the compared algorithms, respectively. The symbol "≈" indicates that there is no statistically significant difference between the proposed algorithm and the compared algorithms.

4.1 Parameter Settings

The parameter setting of the proposed algorithms are listed below.

- The number of variables for all test problems are set as 10.
- The initial population size, $N = 100$.
- The maximum number of function evaluations, $FE_{max} = 300$.
- The predefined number of iterations before updating GP models is set as suggested in [14], $t_{max} = 10$.
- The range of hyperparameter of the design and analysis of computer experiments (DACE) [28] for building GP models is set to $[10^{-5}, 10^{5}]$.

Furthermore, we adopt the recommended setting in the original literature for specific parameters of the compared algorithms.

4.2 Comparison Experiments

The results of achieved IGD values under comparison over 30 independent runs are reported in Table 1, where the best results are highlighted. Among that, we perform the WRS test for the comparison of PB-NSGAIII with other five compared algorithms. All the compared algorithms have been ranked and the average ranking is listed in Table 1. We also list the obtained ranking of PB-NSAG-III in each problem. It can be observed that PB-NSGAIII obtains the best performance with 10 best results on the 17 test problems, followed by KRVEA with 6 best results. PB-NSGAIII also achieves the best overall performance with the highest average ranking.

Among the UF test problems, it is can be seen that PB-NSGAIII performs poorly on a few instances, mainly on UF6-8. As described in [24], UF6 is a disconnected problem which has an irregular PF. UF7 replaces the disconnected PF of UF6 with a continuous one. UF8 has a discontinuous Pareto set on decision space. Both PF shapes of UF6 and UF7 are a straight line, which means that the CI value is different, even though these solutions are located on true PF. PB-NSGAIII tends to select the middle solution. However, KRVEA has achieved the best performance of these problems, we attribute the success to reference vectors which can maintain the diversity in different shapes of PF.

Both DTLZ1 and DTLZ3 are difficult for MOEAs to converge due to their multimodal landscapes. It can be observed that PB-NSGAIII performs best on the two problems, followed by CSEA. This might because that DI provides more convergence pressure over the multimodal problems. DTLZ2 is easy for MOEAs to converge but need to maintain the diversity. PB-NSGAIII also achieves the best performance on DTLZ2, followed by KRVEA. It can be concluded that the proposed PB-NSGAIII is effective in accelerating convergence and maintaining diversity on expensive many objective optimization. DTLZ4 is a sensitive problem, which is hard to obtain a well-distributed initial training data. It can be seen that KRVEA achieves the best performance on DTLZ4, followed by PB-NSGAIII. This might be attributed to the use of uncertainty information in

Table 1. Statistic results of the five compared algorithms and PB-NSGAIII on 17 test problems. The best results for each problems are highlighted. Symbols "↑", "≈", and "↓" indicate that PB-NSGAIII is statistically significantly superior to, equivalent to and inferior to the particular compared algorithm, respectively.

Problems	NSGAIII	CPS-MOEA	CSEA	KRVEA	MOEA/D-EGO	PB-NSGAIII
UF1	7.05E+00↑	7.97E+00↑	4.04E+00↑	2.99E+00↑	3.33E+00↑	**2.70E+00**(1)
UF2	5.83E−01↑	5.85E−01↑	2.38E−01↑	**1.20E−01**≈	2.62E−01↑	1.20E−01(2)
UF3	3.29E−01↑	3.45E−01↑	1.64E−01↑	7.64E−02↑	1.86E−01↑	**6.11E−02**(1)
UF4	1.39E+00↑	1.41E+00↑	1.11E+00↑	8.26E−01↑	9.74E−01↑	**4.68E−01**(1)
UF5	1.52E−01↑	1.37E−01↑	1.28E−01↑	1.00E−01↑	1.11E−01↑	**8.49E−02**(1)
UF6	3.28E+00↑	3.26E+00↑	2.41E+00↑	**1.12E+00**↓	2.52E+00↑	1.43E+00(2)
UF7	3.47E+00↑	3.65E+00↑	1.97E+00↑	**1.14E+00**↓	1.76E+00↑	1.50E+00(2)
UF8	6.21E−01↑	6.74E−01↑	3.61E−01↑	**1.91E−01**↓	3.32E−01↑	2.73E−01(2)
UF9	1.26E+00↑	1.44E+00↑	6.15E−01↑	2.96E−01↑	4.66E−01↑	**2.23E−01**(1)
UF10	1.30E+00↑	1.51E+00↑	7.19E−01↑	2.33E−01≈	4.63E−01↑	**2.20E−01**(1)
DTLZ1	9.34E+01↑	8.56E+01↑	5.64E+01≈	9.12E+01↑	8.23E+01↑	**4.96E+01**(1)
DTLZ2	3.07E−01↑	3.37E−01↑	2.12E−01↑	1.34E−01↑	3.34E−01↑	**7.43E−02**(1)
DTLZ3	2.62E+02↑	2.29E+02↑	1.64E+02↑	2.53E+02↑	2.02E+02↑	**1.30E+02**(1)
DTLZ4	6.68E−01↑	5.89E−01↑	4.57E−01↑	**3.22E−01**≈	6.40E−01↑	3.58E−01(2)
DTLZ5	2.26E−01↑	2.54E−01↑	1.16E−01↑	1.15E−01↑	2.66E−01↑	**1.55E−02**(1)
DTLZ6	6.14E+00↑	3.96E+00↑	4.99E+00↑	3.16E+00↑	**2.01E+00**↓	2.79E+00(2)
DTLZ7	5.17E+00↑	4.86E+00↑	1.76E+00↑	**1.42E−01**≈	2.47E−01↑	1.78E−01(2)
Average ranking	5.29	5.29	3.47	2.06	3.47	1.41
↑/≈/↓	17/0/0	17/0/0	16/1/0	10/4/3	16/0/1	

KRVEA and the usage of DI indicator in PB-NSGAIII for choosing the farthest candidate solution to expensive re-evaluate. DTLZ5, DTLZ6, and DTLZ7 have an irregular surface PF, where the PF of DTLZ5 and DTLZ6 are degenerate curves and the PF of DTLZ7 is discontinuous. It can be observed that PB-NSGAIII achieves the best or second performance on these three problems, which has been shown competitive compared with the other five algorithms.

In the comparison between PB-NSGAIII and KRVEA, the WRS test results show that PB-NSGAIII achieves 10 "better", 3 "worse", and 4 "similar" results on test problems. Compared with other four algorithms, PB-NSGAIII achieves almost all the best results, expect for DTLZ6.

5 Conclusions

This paper proposes a Pareto-based bi-indicator infill sampling criterion (PBISC) for expensive multiobjective optimization. PBISC transforms the infill sampling of SAEAs into a multiobjective optimization problem. In PBISC, we first introduce DI and CI to measure the diversity and the convergence improvement of the candidate solution, respectively. DI also represents the convergence improvement in some special cases, which can accelerate the convergence. These two indicators are regarded as two objectives, and the non-dominated sorting is employed to sort the candidate solutions. Solutions in the first front will be picked out for expensive evaluations without the predefined number of selected solutions. Furthermore, we have embed PBISC into GP-assisted NSGA-III algorithm for solving expensive MOPs.

The overall performance on UF and DTLZ test problems proves that PBISC is promising for maintaining diversity and promoting convergence of SAEAs. However, PBISC does not distinguish similar candidate solutions, which would lead to a waste of expensive evaluations. Thus, it deserves further efforts to distinguish similar candidate solutions. Further, the proposed criterion can be embedded into other MOEAs. Finally, it deserves further extend PBISC to solve expensive many-objective optimization problems.

References

1. Zhou, A., Qu, B.Y., Li, H., Zhao, S.Z., Zhang, Q.: Multiobjective evolutionary algorithms: a survey of the state of the art. Swarm Evol. Comput. **1**(1), 32–49 (2011)
2. Yu, P.L.: Cone convexity, cone extreme points, and nondominated solutions in decision problems with multiobjectives. J. Optim. Theory Appl. **14**(3), 319–377 (1974)
3. Deb, K., Pratap, A., Agarwal, S., Meyarivan, T.: A fast and elitist multiobjective genetic algorithm: NSGA-II. IEEE Trans. Evol. Comput. **6**(2), 182–197 (2002)
4. Zitzler, E., Künzli, S.: Indicator-based selection in multiobjective search. In: Yao, X., et al. (eds.) PPSN 2004. LNCS, vol. 3242, pp. 832–842. Springer, Heidelberg (2004). https://doi.org/10.1007/978-3-540-30217-9_84
5. Wang, H., Jiao, L., Yao, X.: Two_Arch2: an improved two-archive algorithm for many-objective optimization. IEEE Trans. Evol. Comput. **19**(4), 524–541 (2014)
6. Brockhoff, D., Wagner, T., Trautmann, H.: On the properties of the R2 indicator. In: Proceedings of the 14th Annual Conference on Genetic and Evolutionary Computation, pp. 465–472 (2012)
7. Zhang, Q., Li, H.: MOEA/D: a multiobjective evolutionary algorithm based on decomposition. IEEE Trans. Evol. Comput. **11**(6), 712–731 (2007)
8. Deb, K., Jain, H.: An evolutionary many-objective optimization algorithm using reference-point-based nondominated sorting approach, part i: solving problems with box constraints. IEEE Trans. Evol. Comput. **18**(4), 577–601 (2013)
9. Jin, Y.: Surrogate-assisted evolutionary computation: recent advances and future challenges. Swarm Evol. Comput. **1**(2), 61–70 (2011)
10. Jin, Y., Sendhoff, B.: A systems approach to evolutionary multiobjective structural optimization and beyond. IEEE Comput. Intell. Mag. **4**(3), 62–76 (2009)
11. Knowles, J.: ParEGO: a hybrid algorithm with on-line landscape approximation for expensive multiobjective optimization problems. IEEE Trans. Evol. Comput. **10**(1), 50–66 (2006)
12. Mockus, J., Tiesis, V., Zilinskas, A.: The application of Bayesian methods for seeking the extremum. Towards Global Optim. **2**(117–129), 2 (1978)
13. Zhang, Q., Liu, W., Tsang, E., Virginas, B.: Expensive multiobjective optimization by MOEA/D with gaussian process model. IEEE Trans. Evol. Comput. **14**(3), 456–474 (2009)
14. Cheng, R., Jin, Y., Olhofer, M., Sendhoff, B.: A reference vector guided evolutionary algorithm for many-objective optimization. IEEE Trans. Evol. Comput. **20**(5), 773–791 (2016)

15. Chugh, T., Jin, Y., Miettinen, K., Hakanen, J., Sindhya, K.: A surrogate-assisted reference vector guided evolutionary algorithm for computationally expensive many-objective optimization. IEEE Trans. Evol. Comput. **22**(1), 129–142 (2016)
16. Yu, X., Yao, X., Wang, Y., Zhu, L., Filev, D.: Domination-based ordinal regression for expensive multi-objective optimization. In: 2019 IEEE Symposium Series on Computational Intelligence (SSCI), pp. 2058–2065. IEEE (2019)
17. Pan, L., He, C., Tian, Y., Wang, H., Zhang, X., Jin, Y.: A classification-based surrogate-assisted evolutionary algorithm for expensive many-objective optimization. IEEE Trans. Evol. Comput. **23**(1), 74–88 (2018)
18. Ponweiser, W., Wagner, T., Biermann, D., Vincze, M.: Multiobjective optimization on a limited budget of evaluations using model-assisted $\mathcal{S}$-metric selection. In: Rudolph, G., Jansen, T., Beume, N., Lucas, S., Poloni, C. (eds.) PPSN 2008. LNCS, vol. 5199, pp. 784–794. Springer, Heidelberg (2008). https://doi.org/10.1007/978-3-540-87700-4_78
19. Wang, H., Jin, Y., Yao, X.: Diversity assessment in many-objective optimization. IEEE Trans. Cybern. **47**(6), 1510–1522 (2016)
20. Williams, C.K., Rasmussen, C.E.: Gaussian Processes for Machine Learning, vol. 2. MIT Press Cambridge, MA (2006)
21. Wang, H., Yao, X.: Corner sort for Pareto-based many-objective optimization. IEEE Trans. Cybern. **44**(1), 92–102 (2014)
22. Krige, D.G.: A statistical approach to some mine valuation and allied problems on the Witwatersrand: By DG Krige. Ph.D. thesis, University of the Witwatersrand (1951)
23. McKay, M.D., Beckman, R.J., Conover, W.J.: A comparison of three methods for selecting values of input variables in the analysis of output from a computer code. Technometrics **42**(1), 55–61 (2000)
24. Zhang, Q., Zhou, A., Zhao, S., Suganthan, P.N., Liu, W., Tiwari, S.: Multiobjective optimization test instances for the CEC 2009 special session and competition (2008)
25. Deb, K., Thiele, L., Laumanns, M., Zitzler, E.: Scalable test problems for evolutionary multiobjective optimization. In: Evolutionary Multiobjective Optimization, pp. 105–145. Springer (2005)
26. Zhang, J., Zhou, A., Zhang, G.: A classification and Pareto domination based multiobjective evolutionary algorithm. In: 2015 IEEE Congress on Evolutionary Computation (CEC), pp. 2883–2890. IEEE (2015)
27. Bosman, P.A., Thierens, D.: The balance between proximity and diversity in multiobjective evolutionary algorithms. IEEE Trans. Evol. Comput. **7**(2), 174–188 (2003)
28. Lophaven, S.N., Nielsen, H.B., Søndergaard, J.: DACE: a Matlab kriging toolbox, vol. 2. Citeseer (2002)

MOEA/D with Gradient-Enhanced Kriging for Expensive Multiobjective Optimization

Fei Liu[1(✉)], Qingfu Zhang[1,2], and Zhonghua Han[3]

[1] Department of Computer Science, City University of Hong Kong, Hong Kong, China
fliu36-c@my.cityu.edu.hk, qingfu.zhang@cityu.edu.hk

[2] The City University of Hong Kong Shenzhen Research Institute, Shenzhen, China

[3] Institute of Aerodynamic and Multidisciplinary Design Optimization, National Key Laboratory of Science and Technology on Aerodynamic Design and Research, School of Aeronautics, Northwestern Polytechnical University, Xi'an, China
hanzh@nwpu.edu.cn

Abstract. Expensive multiobjective optimization problem poses a big challenge. In many real-world engineering design problems, the time-consumed function evaluation is done by solving partial differential equations. The partial derivatives of a candidate solution can be calculated as a byproduct. Naturally, these problems can be solved more efficiently if gradient information is used. This paper proposes such a method, called MOEA/D-GEK, which combines MOEA/D and gradient-enhanced Kriging to solve expensive multiobjective problem. The gradient information is used for the construction of the Kriging model. Experimental studies on a set of test instances and a real-world aerodynamic design problem show high efficiency and effectiveness of our proposed method.

Keywords: Multiobjective optimization · Expensive optimization · Surrogate model · Gradient-enhanced Kriging · Pareto optimality

1 Introduction

Multiobjective evolutionary algorithms (MOEAs) [29] have been attracting growing attention from many application areas. Surrogate-assisted MOEAs [6,14], which are able to obtain reasonably good solutions with a limited number of evaluations, are well-suited for expensive engineering design optimization. In many engineering design optimization problems, the function evaluation

This work was supported by the National Natural Science Foundation of China (Grant No: 61876163) and ANR/RGC Joint Research Scheme sponsored by the Research Grants Council of the Hong Kong Special Administrative Region, China and France National Research Agency (Project No: A-CityU101/16).

H. Ishibuchi et al. (Eds.): EMO 2021, LNCS 12654, pp. 543–554, 2021.
https://doi.org/10.1007/978-3-030-72062-9_43

requires to solve partial differential equations [25], where partial derivatives can be simultaneously calculated by the adjoint method. Naturally, we can utilize the gradient information to further accelerate the optimization process. This paper is to develop an efficient multiobjective evolutionary algorithm with gradient-enhanced Kriging (GEK).

Much effort has been made to use surrogate models for solving expensive multiobjective optimization problems. Among other surrogate models [13,18,24,26], Kriging (i.e., Gaussian stochastic process model) [17,19,22,28] has gained popularity since it can not only estimate the objective function value, but also quantify the estimation error. Kriging models have been used in several multiobjective optimization algorithms such as (e.g. [7,17,28]). Gradient information has not been used in these algorithms.

Gradient-based multiobjective optimization algorithms work very well [5,8,9] when derivatives are available. These algorithms are theoretically sound and often converge to the Pareto front (PF) fast. However, these algorithms can be easily trapped in a local Pareto optimal point.

Gradient-enhanced Kriging (GEK) provides a principled approach to incorporate gradient information for expensive optimization. Two major approaches along this line are direct GEK [10,11], and indirect GEK [2]. Direct GEK treats gradient values as random and includes them as responses in the Kriging equation system. Indirect GEK uses the first-order Taylor's expansion to revert gradient values to function values in a neighborhood of the evaluated point, and these estimated function values are then used in Kriging modeling. Direct GEK is more accurate in theory and no additional parameters are involved. This paper proposes MOEA/D-GEK, which combines MOEA/D and direct GEK for expensive multiobjective optimization.

The remaining of this paper is organized as follows. Section 2 gives a description of the targeted problem, Sect. 3 details the proposed MOEA/D-GEK, Sect. 4 presents the experimental studies, and the last section concludes the paper.

2 Problem Definition

We consider the following multiobjective expensive optimization problem:

$$\begin{aligned} &min.\ \mathrm{F}(\mathbf{x}) = (f_1(\mathbf{x}), \cdots, f_m(\mathbf{x}))^T \\ &s.t.\ \mathbf{x} = (x_1, \cdots, x_d)^T \in \prod_{j=1}^{d} [a_j, b_j] \end{aligned}, \tag{1}$$

where $\mathrm{F}(\mathbf{x}) = (f_1(\mathbf{x}), \cdots, f_m(\mathbf{x}))^T$ represents m independent objectives and $\mathbf{x} = (x_1, \cdots, x_d)^T$ represents d-dimensional variables with the boundaries $-\infty < a_j < b_j < +\infty$, $j = 1, \cdots, d$. In addition, all the objective functions $f_i(\mathbf{x})$ are differentiable. We assume that (i) the evaluation of $f_i(\mathbf{x})$ value will require a time-consuming computer simulation, thus a very large number of function evaluations are not affordable; (ii) the gradient value of each $f_i(\mathbf{x})$ can be obtained at a cost of about one $f_i(\mathbf{x})$-value evaluation; (iii) the objective functions are multimodal, therefore, there are many locally Pareto optimal solutions; and (iv) decision makers will want to have an as-good-as-possible approximation to the PF in the objective space, for understanding the trade-off relationship among different objectives, and making their final decision. This kind of optimization problem can be found in many engineering design fields [12,20,25] when adjoint methods [1,2], or other methods can be used for computing gradients.

3 Proposed Algorithm

The main idea of MOEA/D-GEK is to integrate GEK into the framework of MOEA/D to replace the expensive evaluation and enable cooperation between subproblems so that the optimization process can be accelerated. The surrogates are built only for original objective functions, and subproblems are constructed using these models. After the construction of subproblems, MOEA/D is performed to generate new samples and a batch of best candidates are selected. The best candidates are evaluated and used to update surrogate models. Thanks to the use of gradients, the number of required samples during the optimization can be significantly reduced compared with other surrogate-assisted MOEAs. The framework of MOEA/D-GEK is given in Algorithm 1 and the main steps are introduced in the following subsections.

3.1 Gradient-Enhanced Kriging (GEK)

Gradient-enhanced Kriging [11] uses the gradient information for constructing the model. The prediction of GEK is defined as a weighted sum of both function value responses and partial derivatives. We assume that the sampled dataset $(\mathbf{S}, \mathbf{y_S})$ for a d-dimensional problem with n samples is of the form

$$\begin{aligned} &\mathbf{S} = [\mathbf{x}^{(1)}, ..., \mathbf{x}^{(n)}, \mathbf{x_g}^{(1)}, ..., \mathbf{x_g}^{(n')}]^{\mathrm{T}} \in \mathbb{R}^{(n+n') \times d}, \mathbf{x} = (x_1, .., x_d) \in \mathbb{R}^d \\ &\mathbf{y_S} = [y^{(1)}, ..., y^{(n)}, \partial y^{(1)}, ..., \partial y^{(n')}]^{\mathrm{T}} \in \mathbb{R}^{n+n'} \end{aligned}, \tag{2}$$

where $\mathbf{x_g}^{(j)}$, $j \in \{ 1, \cdots, n' \}$ represents the locations at which the j-th partial derivatives are used and $\partial y^{(j)}, j \in \{ 1, \cdots, n' \}$ represent the corresponding j-th partial derivatives. Note that the method itself is proposed in a general form and does not assume that the gradient information is available at every tested sample site. For simplicity, full partial derivatives are utilized in this paper, i.e., $n' = n \times d$.

In a stationary random process, the statistical prediction of the unknown function at an untested position $\mathbf{x}$ is given by a linear combination of the function values and their partial derivatives as follows:

$$\hat{y}(\mathbf{x}) = \sum_{i=1}^{n} w^{(i)} y^{(i)} + \sum_{j=1}^{n \times d} \lambda^{(j)} \partial y^{(j)}, \tag{3}$$

where $\hat{y}(\mathbf{x})$ is the predicted value at an untried site $\mathbf{x}$, $w^{(i)}$ and $\lambda^{(j)}$ are the weight coefficients for function responses and partial derivatives, respectively. It assumes a random process corresponding to unknown function $y(\mathbf{x})$:

Algorithm 1. MOEA/D-GEK

Input:

N_I: the number of initial samples;
N_G: the number of new samples generated in each generation;
N_E: the number of candidates selected in each generation;
$\lambda_1, \cdots, \lambda_{N_G}$: uniformly distributed weighting coefficient vectors.

Initialization:

generate N_I samples $\mathbf{x}^{(1)}, \cdots, \mathbf{x}^{(N_I)}$ from the decision space by using a design of experiments (DOE) method (e.g. Latin hypercube sampling, LHS) and evaluate the function responses and partial derivatives of all the N_I samples.

While stopping conditions are not satisfied:

Step 1 Modeling: build m gradient-enhanced Kriging models for objective functions through the sampled data. (Subsect. 3.1)

Step 2 Updating reference point: the reference point is updated adaptively by choosing from the minimum value in the existed data set and minimum predicted value by GEK. (Subsect.3.2)

Step 3 Suggesting new samples: the objective function of an acquisition problem for i-th subproblem is defined as $\xi^i(\mathbf{x})$. MOEA/D is used to find the optimal solution of maximizing $\xi^i(\mathbf{x})$. In turn, N_G new samples are obtained. (Subsect. 3.3)

Step 4 Selecting candidates and evaluations: N_E points are selected from N_G new samples. They are evaluated by expensive simulations and the resulting data is augmented to update surrogate models. (Subsect. 3.4)

Output: the best solutions so far.

$$Y(\mathbf{x}) = \beta_0 + Z(\mathbf{x}), \tag{4}$$

where β_0 is an unknown constant and $Z(\mathbf{x})$ is a stationary random process having zero mean and a covariance of:

$$\begin{aligned} Cov[Z(\mathbf{x}^{(i)}), Z(\mathbf{x}^{(j)})] &= \sigma^2 R(\mathbf{x}^{(i)}, \mathbf{x}^{(j)}) \\ Cov[Z(\mathbf{x}^{(i)}), \frac{\partial Z(\mathbf{x}^{(j)})}{\partial x_k^{(j)}}] &= \sigma^2 \frac{\partial R(\mathbf{x}^{(i)}, \mathbf{x}^{(j)})}{\partial x_k^{(j)}} \\ Cov[\frac{\partial Z(\mathbf{x}^{(i)})}{\partial x_l^{(i)}}, Z(\mathbf{x}^{(j)})] &= \sigma^2 \frac{\partial R(\mathbf{x}^{(i)}, \mathbf{x}^{(j)})}{\partial x_l^{(i)}} \\ Cov[\frac{\partial Z(\mathbf{x}^{(i)})}{\partial x_l^{(i)}}, \frac{\partial Z(\mathbf{x}^{(j)})}{\partial x_k^{(j)}}] &= \sigma^2 \frac{\partial^2 R(\mathbf{x}^{(i)}, \mathbf{x}^{(j)})}{\partial x_l^{(i)} \partial x_k^{(j)}} \end{aligned}, \tag{5}$$

where $\frac{\partial R(\mathbf{x}^{(i)},\mathbf{x}^{(j)})}{\partial x_l^{(i)}}$ and $\frac{\partial R(\mathbf{x}^{(i)},\mathbf{x}^{(j)})}{\partial x_k^{(j)}}$ are the partial derivatives w.r.t. the l-th component of $\mathbf{x}^{(i)}$ and k-th component of $\mathbf{x}^{(j)}$, respectively. σ^2 is the variance of $Z(\cdot)$ and R is the spatial correlation function.

We can also treat $\hat{y}(\mathbf{x})$ as random, and minimize its MSE

$$MSE[\hat{y}(\mathbf{x})] = E[(\sum_{i=1}^{n} w^{(i)} y^{(i)} + \sum_{j=1}^{n\times d} \lambda^{(j)} \partial y^{(j)} - \mathbf{Y}(\mathbf{x}))^2]. \tag{6}$$

Subject to unbiased estimation, the predictor of $\hat{y}(\mathbf{x})$ is:

$$\begin{aligned} &\hat{y}(\mathbf{x}) = \hat{\beta} + \mathbf{r^T}(\mathbf{x}) \underbrace{\mathbf{R}^{-1}(\mathbf{y_s} - \beta_0 \mathbf{F})}_{=:\, \mathbf{V_{krig}}} , \\ &\text{where } \hat{\beta} = (\mathbf{F^T R^{-1} F})^{-1} \mathbf{F^T R^{-1} y_s} , \\ &\mathbf{F} = [1, \cdots, 1, 0, \cdots, 0]^T \in \mathbf{R}^{n+n\times d}, \\ &\mathbf{R} = (R(\mathbf{x}^{(i)}, \mathbf{x}^{(j)}))_{i,j} \in \mathbf{R}^{n\times(d+1)\times n\times(d+1)}, \\ &\mathbf{r} = (R(\mathbf{x}^{(i)}, \mathbf{x}))_i \in \mathbf{R}^{n\times(d+1)}. \end{aligned} \tag{7}$$

The MSE of the GEK prediction is

$$MSE[\hat{y}(\mathbf{x})] \equiv \hat{s}^2(\mathbf{x}) = \sigma^2\{1.0 - \mathbf{r^T R^{-1} r} + (\mathbf{r^T R^{-1} F} - 1)^2/(\mathbf{F^T R^{-1} F})\}. \tag{8}$$

3.2 Updating Reference Point

Reference point $\mathbf{z}_j{}^* = (z_1{}^*, \cdots z_m{}^*)$ is vital for decomposition-based multiobjective optimization algorithms. In the literature, reference point is set as the minimum objective value of tested samples $z_j{}^* = f_{j,\min}$, $j = \{1, \cdots, m\}$. When a small number of samples are used, the reference point can be far away from the optimum. In this paper, an adaptive method is proposed to update the reference point. It is defined as follows:

$$z_j{}^* = \begin{cases} \hat{f}_{j,min} \ , \ \text{if } |f_{j,\min} - f_{j,last\,\min}| > \mathrm{K} \cdot \left|\hat{f}_{j,min} - f_{j,last\,\min}\right| \ \ or\ ns = N_I \\ f_{j,\min} \ , \ \text{if } |f_{j,\min} - f_{j,last\,\min}| < \mathrm{K} \cdot \left|\hat{f}_{j,min} - f_{j,last\,\min}\right| \ \ and\ ns{>}N_I \end{cases},$$

$$j = \{1, \cdots, m\} \tag{9}$$

where $z_j{}^*$ is the value of reference point on j-th objective, $f_{j,last\,\min}$ and $f_{j,\min}$ represent the minimum value of the responses of all tested points on j-th objective in last generation and present generation, respectively. $\hat{f}_{j,min}$ represents the minimum value predicted by GEK model on j-th objective in present generation. $K \in [0, 1]$ is the trust coefficient of GEK. In addition, ns is the number of tested samples and $z_j{}^* = \hat{f}_{j,min}$ at the beginning of optimization by default. In this way, reference point is set as the minimum predicted value of GEK models at the beginning of optimization and is updated to the minimum objective value of tested samples when little improvement is observed. Because GEK model is believed to be accurate enough, the reference point is likely to be optimized.

3.3 Expected Improvement of Subproblems

In each generation, MOEA/D is used for maximizing the expected improvement metric values of all N_G subproblems. The subproblems are built using the prediction of GEK models. Tchebycheff aggregation is used to decompose a multiobjective problem into a number of single-objective optimization subproblems.

$$f^{te}(\mathbf{x}|\lambda) = \max_{1\leqslant j\leqslant m} \{\lambda_j(f_j(\mathbf{x}) - z_j{}^*)\}. \tag{10}$$

For simplicity, we assume that all these individual functions are independent to each other statistically. Thus, the predictions of different objective functions are:

$$f_j(\mathbf{x}) \sim N(\hat{f}_j(\mathbf{x}), \hat{s}_j^2(\mathbf{x})), j = 1, \cdots, m. \tag{11}$$

Therefore, using Tchebycheff aggregation, we have

$$\lambda_j \left[f_j(\mathbf{x}) - z_j{}^*\right] \sim N\left\{\lambda_j(\hat{f}_j(\mathbf{x}) - z_j{}^*),\ [\lambda_j \hat{s}_j(\mathbf{x})]^2\right\}. \tag{12}$$

In order to obtain a good approximation of PF, a proper strategy to define the objective function $\xi^i(\mathbf{x})$ of subproblem is of significant importance. A couple of infill-sampling criteria have been developed for gaussian process. Among them, expected improvement (EI) [15] enables good balance between exploration and exploitation. The formula of EI on each subproblem is as follows:

$$E^i[I(\mathbf{x})] = \begin{cases} [y^i{}_{\min} - \hat{y}^i(\mathbf{x})]\Phi(\frac{y^i{}_{\min} - \hat{y}^i(\mathbf{x})}{\hat{s}^i(\mathbf{x})}) + \hat{s}^i(\mathbf{x})\phi(\frac{y^i{}_{\min} - \hat{y}^i(\mathbf{x})}{\hat{s}^i(\mathbf{x})})\ ,\ \hat{s}^i(\mathbf{x}) > 0 \\ 0\ ,\ \hat{s}^i(\mathbf{x}) = 0 \end{cases}, \tag{13}$$

where $\Phi(\cdot)$ and $\phi(\cdot)$ are cumulative distribution and probability density function of a standard normal distribution, respectively. $y^i_{\min}$ is the minimum of Tchebycheff aggregation of tested samples and $y^i(\mathbf{x})$ is the Tchebycheff aggregation built on the prediction of GEK models. The details of computing $\hat{s}^i(\mathbf{x})$ please refer to [28].

3.4 Selecting Best Candidates

After N_G new samples are generated, the best candidates are selected to evaluate for updating surrogate models. We use the K-means clustering to cluster all the weight vectors as well as the solutions into N_E clusters. Then N_E solutions are selected from N_E clusters. The selected solutions should satisfy the following two conditions: (i) have the highest ξ-value in their clusters; (ii) the Euclidean distances to the tested samples are larger than 10^{-4}.

4 Experimental Studies

4.1 Test Instances

Experimental studies are first carried out on some MOP test instances. All these instances are minimization problems. The basic information are listed in Table 1 and the detailed description can be found in reference papers.

Table 1. A list of MOP test instances.

Problem	No. of variables	No. of objectives	Feature
ZDT1 [3]	8	2	Convex
ZDT2 [3]	8	2	Nonconvex
ZDT3 [3]	8	2	Discontinues
ZTLZ2 [4]	6	3	Nonconvex
F1 [21]	10	2	Convex/complicated pareto set
F2 [21]	10	2	Convex/complicated pareto set

4.2 Experimental Settings

It should be noticed that the number of function evaluations used in MOEA/D-EGO equals to the sum of function evaluations and gradient calculations used in MOEA/D-GEK. It comes out that the maximum number of the test samples used in MOEA/D-GEK is half of that in MOEA/D-EGO. It is based on the assumption that the cost of gradient calculation is about the cost of one expensive function evaluation as introduced in the problem definition. We suppose the assumption is the same on the test instances so that we can verify the proposed method on these benchmarks. The experimental settings are as follows:

a) The number of initial test samples N_I: N_I is set to be $5d$ and $10d$ for MOEA/D-GEK and MOEA/D-EGO, respectively, where d is the number of design variables;
b) The maximum number of test samples: Maximum is 100 and 200 for MOEA/D-GEK and MOEA/D-EGO, respectively;
c) The number of independent runs: 10 independent runs on each test instance;
d) The number of weight vectors and subproblems: N_G is set to be 200 for two-objective instances and 300 for three-objective instances;
e) The number of selected candidates N_E: 10 candidates are selected, except for F1 and F2, where 5 candidates are selected by MOEA/D-GEK;
f) The trust coefficient K: It is set as 0 for F1 and F2 and 1 for other instances.

4.3 Experimental Results

Table 2 shows the IGD [27] value of two methods and Fig. 1 is the corresponding convergence process with respect to the iteration number. On all the instances, MOEA/D-GEK has larger initial IGD values because of less initial points are used. It outperforms MOEA/D-EGO on the final approximations after few iterations. Distinct improvements on optimization efficiency are observed, especially on ZDT problems. Results also indicate that the gradient-enhanced models are accurate enough for optimization even with a small number of initial samples.

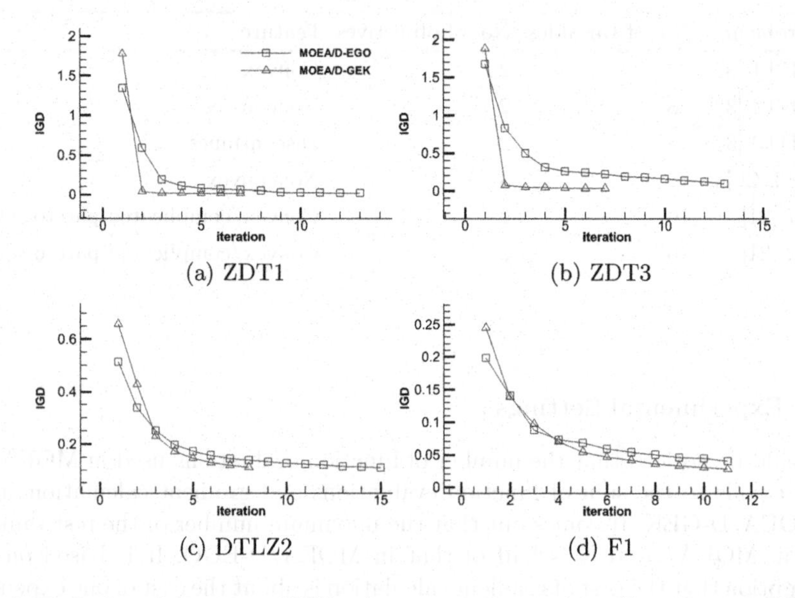

Fig. 1. Evolution of the mean of IGD values versus the number of iterations on different test instances.

Table 2. IGD values of final approximations obtained by MOEA/D-EGO and MOEA/D-GEK (ten independent runs).

	MOEA/D-EGO				MOEA/D-GEK			
	Evaluations	Average	Best	Worst	Evaluations	Average	Best	Worst
ZDT1	200f	0.0182	0.0127	0.0271	100f + 100g	0.0123*	0.0107	0.0154
ZDT2	200f	0.0408	0.0157	0.0762	100f + 100g	0.0135*	0.0129	0.0144
ZDT3	200f	0.0986	0.0291	0.3316	100f + 100g	0.0322*	0.0252	0.0442
DTLZ2	200f	0.1102*	0.0937	0.1235	100f + 100g	0.1116	0.1064	0.1180
F1	200f	0.0410	0.0281	0.0526	100f + 100g	0.0283*	0.0205	0.0390
F2	200f	0.1882	0.1399	0.2239	100f + 100g	0.1680*	0.1249	0.2112

4.4 Airfoil Aerodynamic Design Optimization

The proposed method is further validated on an airfoil aerodynamic design optimization problem. The objective is to improve the lift and reduce the drag of airfoil NACA0012 at the design point of $Ma = 0.3, \alpha = 2.0, \mathrm{Re} = 6.37 \times 10^6$, where Ma represents Mach number, Re represents Reynolds number and α is the angle of attack. The problem is defined as:

$$\begin{aligned} min.\ & \mathrm{F}(\mathbf{x}) = (f_1(\mathbf{x}), f_2(\mathbf{x}))^T \\ & f_1(\mathbf{x}) = 50 \cdot C_d,\ f_2(\mathbf{x}) = -C_l\ , \\ s.t.\ & x_i \in \{-0.25, 0.25\}, i = 1, \cdots, d \end{aligned} \tag{14}$$

where C_d and C_l are the drag and lift coefficients, $x_i, i = 1, \cdots, d$ are the design variables, where $d = 14$.

In this paper, Free Form Deformation (FFD) method is employed to perturb the shape of the airfoil. The mesh is generated around a geometric configuration using a hyperbolic mesh generator. Then, a RANS flow solver with an adjoint solver are used to obtain drag and lift coefficients and corresponding partial derivatives with respect to different design variables (FFD parameters). The geometric parameterization and mesh generator together with flow simulation are performed on open-source software packages in the framework of MACH-Aero [16,23]. Both MOEA/D-GEK and MOEA/D-EGO are used in the optimization and each performs five times with different initial samples. For each optimization, 10 new solutions are selected in each generation and the total number of evaluated points are 100 and 200 for MOEA/D-GEK and MOEA/D-EGO, respectively.

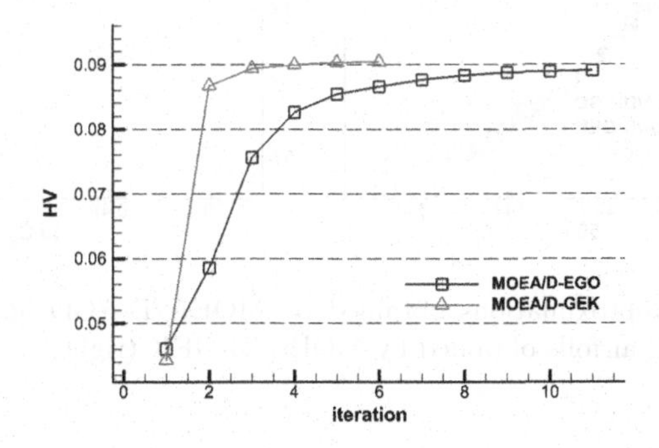

Fig. 2. Evolution of the mean of HV values versus the number of iterations on airfoil aerodynamic design problem.

Figure 2 compares the evolution of the mean of HV [30] values. Table 3 is a comparison of average time cost (on 3.0 GHz CPU) and average HV values of final approximations. We adopt HV as the performance metric, where reference point is defined by the maximum objective values of non-dominated

Table 3. Time cost on evaluations in one optimization and the mean of HV value of final approximations.

	No. of CFD	One CFD cost	No. of adjoint	One adjoint cost	Total expense of evaluations	HV
MOEA/D-EGO	200	2.41 min	/	/	8 h 2 min	0.0891
MOEA/D-GEK	100	2.41 min	100	1.75 min	6 h 56 min	0.0904*

solutions obtained by all independent runs. Results show that, the final approximation of MOEA/D-GEK outperforms that of MOEA/D-EGO on HV value with less computational cost. Figure 3 shows the approximated PFs obtained by two methods. The approximations overlap in the middle part while MOEA/D-GEK dominates MOEA/D-EGO in the top and bottom areas. Three optimized airfoils are selected from the final approximation obtained by MOEA/D-GEK on the right side of Fig. 3. Compared with the baseline, No.1 airfoil has an 8.6% performance improvement on C_d and No.3 airfoil improves C_l by a remarkable 155.5%. It is validated that MOEA/D-GEK is efficient and effective on this multiobjective airfoil design problem. The proposed method generates better solutions with a less total computational budget than MOEA/D-EGO does.

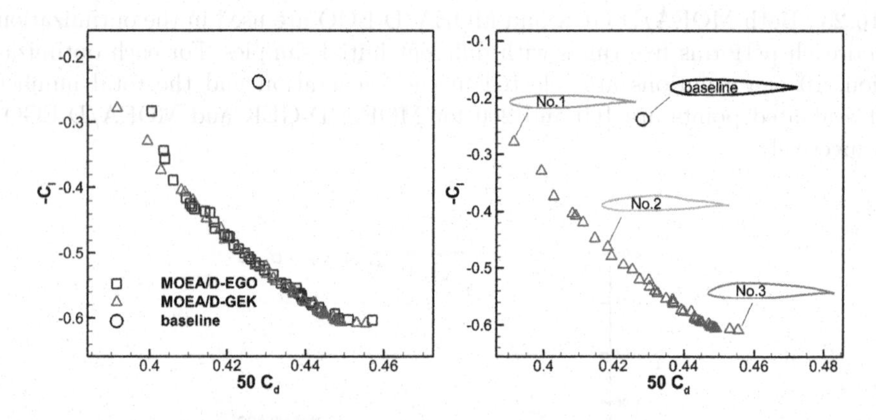

Fig. 3. The final approximations obtained by MOEA/D-EGO and MOEA/D-GEK (left) and optimized airfoils obtained by MOEA/D-GEK (right).

5 Conclusion

In many real-world multiobjective optimization problems, the function evaluation is expensive, while the partial derivatives can be simultaneously obtained. We proposed MOEA/D-GEK, which combines MOEA/D with gradient-enhanced Kriging to use the affordable gradient information for expensive multiobjective optimization. Experimental studies are carried out on a set of test instances and a real-world airfoil aerodynamic design optimization problem. Results show the high efficiency of MOEA/D-GEK. MOEA/D-GEK generates better solutions with limited computational overhead than MOEA/D-EGO. In our future work, we will investigate other advanced gradient-enhanced approaches.

References

1. Anderson, W.K., Venkatakrishnan, V.: Aerodynamic design optimization on unstructured grids with a continuous adjoint formulation. Comput. Fluids **28**(4–5), 443–480 (1999)
2. Dalbey, K.R.: Efficient and Robust Gradient Enhanced Kriging Emulators. Sandia National Laboratories, Albuquerque, NM, USA (2013)
3. Deb, K.: Multi-objective genetic algorithms: problem difficulties and construction of test problems. Evol. Comput. **7**(3), 205–230 (1999)
4. Deb, K., Thiele, L., Laumanns, M., Zitzler, E.: Scalable multi-objective optimization test problems. In: Proceedings of the 2002 Congress on Evolutionary Computation. CEC2002 (Cat. No. 02TH8600), vol. 1, pp. 825–830. IEEE (2002)
5. Désidéri, J.A.: Multiple-gradient descent algorithm (mgda) for multiobjective optimization. C.R. Math. **350**(5–6), 313–318 (2012)
6. Díaz-Manríquez, A., Toscano, G., Barron-Zambrano, J.H., Tello-Leal, E.: A review of surrogate assisted multiobjective evolutionary algorithms. Comput. Intell. Neurosci. **2016** (2016)
7. Emmerich, M.T., Giannakoglou, K.C., Naujoks, B.: Single-and multiobjective evolutionary optimization assisted by gaussian random field metamodels. IEEE Trans. Evol. Comput. **10**(4), 421–439 (2006)
8. Fliege, J., Drummond, L.G., Svaiter, B.F.: Newton's method for multiobjective optimization. SIAM J. Optim. **20**(2), 602–626 (2009)
9. Fliege, J., Svaiter, B.F.: Steepest descent methods for multicriteria optimization. Math. Methods Oper. Res. **51**(3), 479–494 (2000)
10. Forrester, A.I., Keane, A.J.: Recent advances in surrogate-based optimization. Progress Aerosp. Sci. **45**(1–3), 50–79 (2009)
11. Han, Z.H., Görtz, S., Zimmermann, R.: Improving variable-fidelity surrogate modeling via gradient-enhanced kriging and a generalized hybrid bridge function. Aerosp. Sci. Technol. **25**(1), 177–189 (2013)
12. Han, Z.H., Zhang, Y., Song, C.X., Zhang, K.S.: Weighted gradient-enhanced kriging for high-dimensional surrogate modeling and design optimization. AIAA J. **55**(12), 4330–4346 (2017)
13. Herrera, M., Guglielmetti, A., Xiao, M., Filomeno Coelho, R.: Metamodel-assisted optimization based on multiple kernel regression for mixed variables. Struct. Multidisciplinary Optim. **49**(6), 979–991 (2014). https://doi.org/10.1007/s00158-013-1029-z
14. Jin, Y., Wang, H., Chugh, T., Guo, D., Miettinen, K.: Data-driven evolutionary optimization: an overview and case studies. IEEE Trans. Evol. Comput. **23**(3), 442–458 (2018)
15. Jones, D.R., Schonlau, M., Welch, W.J.: Efficient global optimization of expensive black-box functions. J. Global Optim. **13**(4), 455–492 (1998)
16. Kenway, G., Kennedy, G., Martins, J.: A cad-free approach to high-fidelity aerostructural optimization. In: 13th AIAA/ISSMO Multidisciplinary Analysis Optimization Conference, p. 9231 (2010)

17. Knowles, J.: Parego: a hybrid algorithm with on-line landscape approximation for expensive multiobjective optimization problems. IEEE Trans. Evol. Comput. **10**(1), 50–66 (2006)
18. Kourakos, G., Mantoglou, A.: Development of a multi-objective optimization algorithm using surrogate models for coastal aquifer management. J. Hydrol. **479**, 13–23 (2013)
19. Krige, D.G.: A statistical approach to some basic mine valuation problems on the Witwatersrand. J. South. Afr. Inst. Min. Metallurgy **52**(6), 119–139 (1951)
20. Lalau-Keraly, C.M., Bhargava, S., Miller, O.D., Yablonovitch, E.: Adjoint shape optimization applied to electromagnetic design. Opt. Express **21**(18), 21693–21701 (2013)
21. Li, H., Zhang, Q.: Multiobjective optimization problems with complicated Pareto sets, MOEA/D and NSGA-II. IEEE Trans. Evol. Comput. **13**(2), 284–302 (2008)
22. Lin, X., Zhang, Q., Kwong, S.: An efficient batch expensive multi-objective evolutionary algorithm based on decomposition. In: 2017 IEEE Congress on Evolutionary Computation (CEC), pp. 1343–1349. IEEE (2017)
23. Mader, C.A., Kenway, G.K., Yildirim, A., Martins, J.R.: ADflow: an open-source computational fluid dynamics solver for aerodynamic and multidisciplinary optimization. J. Aerosp. Inf. Syst. 1–20 (2020)
24. Regis, R.G.: Multi-objective constrained black-box optimization using radial basis function surrogates. J. Comput. Sci. **16**, 140–155 (2016)
25. De los Rayes, J.C.: Numerical PDE-Constrained Optimization. Springer, Berlin (2015)
26. Tabatabaei, M., Hakanen, J., Hartikainen, M., Miettinen, K., Sindhya, K.: A survey on handling computationally expensive multiobjective optimization problems using surrogates: non-nature inspired methods. Struct. Multidisciplinary Optim. **52**(1), 1–25 (2015). https://doi.org/10.1007/s00158-015-1226-z
27. Zhang, Q., Li, H.: MOEA/D: a multiobjective evolutionary algorithm based on decomposition. IEEE Trans. Evol. Comput. **11**(6), 712–731 (2007)
28. Zhang, Q., Liu, W., Tsang, E., Virginas, B.: Expensive multiobjective optimization by MOEA/D with gaussian process model. IEEE Trans. Evol. Comput. **14**(3), 456–474 (2009)
29. Zhou, A., Qu, B.Y., Li, H., Zhao, S.Z., Suganthan, P.N., Zhang, Q.: Multiobjective evolutionary algorithms: a survey of the state of the art. Swarm Evol. Comput **1**(1), 32–49 (2011)
30. Zitzler, E., Thiele, L., Laumanns, M., Fonseca, C.M., Da Fonseca, V.G.: Performance assessment of multiobjective optimizers: an analysis and review. IEEE Trans. Evol. Comput. **7**(2), 117–132 (2003)

Exploring Constraint Handling Techniques in Real-World Problems on MOEA/D with Limited Budget of Evaluations

Felipe Vaz[1], Yuri Lavinas[2](✉), Claus Aranha[2], and Marcelo Ladeira[1]

[1] University of Brasilia, Brasilia, Brazil
mladeira@unb.br
[2] University of Tsukuba, Tsukuba, Japan
lavinas.yuri.xp@alumni.tsukuba.ac.jp, caranha@cs.tsukuba.ac.jp

Abstract. Finding good solutions for Multi-objective Problems (MOPs) is considered a hard problem, especially when considering MOPs with constraints. Thus, most of the works in the context of MOPs do not explore in-depth how different constraints affect the performance of MOP solvers. Here, we focus on exploring the effects of different Constraint Handling Techniques (CHTs) on MOEA/D, a commonly used MOP solver when solving complex real-world MOPs. Moreover, we introduce a simple and effective CHT focusing on the exploration of the decision space, the Three Stage Penalty. We explore each of these CHTs in MOEA/D on two simulated MOPs and six analytic MOPs (eight in total). The results of this work indicate that while the best CHT is problem-dependent, our new proposed Three Stage Penalty achieves competitive results and remarkable performance in terms of hypervolume values in the hard simulated car design MOP.

Keywords: Multi objective optimization · Real world problems · Constraint handling techniques · MOEA/D

1 Introduction

Multi-objective Optimization Problems (MOPs) appear in many application contexts in which conflicting objective functions need to be simultaneously optimized. Finding good sets of solutions for continuous MOPs is generally considered a hard problem, particularly in real-world MOPs. These real-world MOPs usually have two main characteristics: (1) unknown and irregular shape of the Pareto front and (2) a highly unfeasible (constrained) objective space. These constraints invalidate some solutions, which makes finding a set of feasible solutions a challenging task.

There are several promising multi-objective evolutionary algorithms (MOEAs) that can be used to solve real-world MOPs [13,19,22]. They can modify their behavior when searching for solutions according to the MOP in question

H. Ishibuchi et al. (Eds.): EMO 2021, LNCS 12654, pp. 555–566, 2021.
https://doi.org/10.1007/978-3-030-72062-9_44

taking into account the constraints of each MOP. However, few of these algorithms have their performance evaluated in constrained MOPs or the constraints in the MOPs studied are simple to address [18].

In this work, we focus on solving real-world MOPs with complicated constraints, using the Multi-Objective Evolutionary Algorithm Based on Decomposition, MOEA/D [21], one of these popular MOEA algorithms. The main features of this algorithm are to decompose the MOP into multiple subproblems and solve them in parallel. Decomposing the MOP in various single-objective subproblems makes the algorithm very flexible for dealing with constraints MOP, because adding a penalty value is straightforward. Given this nature of the single-objective subproblems, MOEA/D can handle well different constraint handling techniques (CHTs) while finding a diverse set of optimal solutions.

This work aims to investigate and explore the effects and behavior of constraint handling techniques in MOEA/D when solving real-world MOPs with a limited budget of evaluations. For that, we compare real-world analytic MOPs, compiled together by Tanabe et al. [18] and two simulated MOPs: (1) the problem of selecting landing sites for a lunar exploration robot [17]; and (2) the problem of optimization of car designs [9]. To further enhance the performance of MOEA/D, we propose an efficient CHT that works well with problems that require an exploration of the unfeasible search space.

The remainder of this paper is organized as follows: Sect. 2, presents the different types of CHTs that we study in this work as well as our new CHT. Section 3 shows the experimental settings, such as which parameters are used in MOEA/D, the evaluation criteria, and the results of the experiments 3. Finally, Sect. 4 presents our conclusion remarks.

2 Constraint Handling Techniques

Our goal is to study constraint handling techniques in MOEA/D when solving real-world MOPs yet to be explored in the context of MOEA/D.

Although other types of CHTs exist [15], this study focus on penalty-based techniques since they are easy to understand, simple to implement and extend to multi-objective (in case they were proposed for single-objective optimization). All the constraint handling techniques (CHTs) presented here adapt a constrained MOP into an unconstrained problem, by altering the objectives values according to the CHT under use. They follow the next pattern:

$$f^{agg}_{penalty}(x) = f^{agg} + penalty * v(x), \tag{1}$$

where f^{agg} is the aggregation function, a principal component of MOEA/D. This aggregation function is used to measure the quality of solutions for each subproblem [3], generated based on weight vectors using a decomposition method. Since the f^{agg} does not consider the constraint violation, MOEA/D can substitute this aggregation function by the $f^{agg}_{penalty}$, making this EA algorithm able to consider the constraints violation for each solution. The other variables in the equation above is the total violation of the solution, $v(x)$, and one or more penalty factors, named *penalty*.

Static CHT. This penalty is one of the simplest CHTs. This class of penalty is generally characterized by maintaining its penalty value constant during all the evolutionary process. For more information on this penalty refer to the MOEADr package [2]. In this paper, the term "static penalty" is used to reference penalties defined by the following equation:

$$f_{penalty}^{agg}(x) = f^{agg} + \beta * v(x) \tag{2}$$

where x is one solution, β is an user defined value penalization constant, and v is the violation.

Multi-staged CHT. This CHT has different static penalty values for multiple levels of violation (refer to Homaifar et al. [5]). This way, more unfeasible solutions have higher penalties, while almost feasible solutions can be kept in the population for longer, making it possible to achieve a higher exploration. Its main downside is the high number of parameters that depend highly on the problem. Its penalty is defined by:

$$f_{penalty}^{agg}(x) = f^{agg} + \sum_{i=1}^{m}(R_{k,i} * max[g_i(x), 0]^2), \tag{3}$$

where $R_{k,i}$ are the penalty coefficients, m is the number of constraints, $g_i(x)$ is the penalty of constraint i, and $k = 1, 2, ..., l$ where l is the number of levels of violations defined by the user.

Dynamic CHT. This CHT considers the current generation's number to determine the penalty value. In general, the penalty value is small at the beginning of the search and increases across the generations. The idea is to focus on the diversity of feasible solutions and then later focus on the convergence of those solutions. There are several variations of this penalty, and Joines and Houck [8] proposed the one used throughout this study. It is defined by:

$$f_{penalty}^{agg}(x) = f^{agg} + (C * t)^{\alpha} * v(x) \tag{4}$$

where C and α are constants defined by the user, and t is the number of the current generation.

Self-adaptive CHT. The main characteristics of this CHT are the absence of any user-defined parameters and the focus on balancing the number of both feasible and unfeasible solutions in the population, favoring unfeasible solutions with good objective functions (for more information, see Tessema and Yen [20]). This CHT uses the ratio of feasible solutions in the current population to determine the penalty value in a two-penalty based approach. This way, the objective function is modified, having two components: a distance measure and an adaptive penalty. The distance measure changes the aggregated objective function by also considering the effect of the individual's violation:

$$d(x) = \begin{cases} v(x), & \text{if } r_f = 0 \\ \sqrt{f^{agg"}(x)^2 + v(x)^2}, & \text{otherwise.} \end{cases} \quad (5)$$

The rate of feasible solutions n the incumbent population $r_f = n_f/n_u$, where n_f is the number of feasible, n_u is the number of unfeasible and $f^{agg"}$ is the normalized aggregated objective value. The penalty of this CHT is defined by:

$$p(x) = (1 - r_f)M(x) + (r_f)N(x), \quad (6)$$

where

$$M(x) \begin{cases} 0, & \text{if } r_f = 0, \\ v(x), & \text{otherwise} \end{cases} \quad (7)$$

$$N(x) \begin{cases} 0, & \text{if x is a feasible individual,} \\ f^{agg"}(x), & \text{if x is an unfeasible individual.} \end{cases} \quad (8)$$

Three Step CHT. In this work, we propose a new penalty-based CHT, the Three Step Penalty. The core idea of this penalty is to get a high exploration of the unfeasible space. This idea comes from the fact that there is a need to search through an unfeasible area to get to better feasible solutions. This penalty has three stages: (1) a non-penalty stage, (2) a low penalty stage, and (3) a high penalty stage. The first stage is responsible for exploring the unfeasible space. The second stage has a low penalty to let the population find the nearby feasible region. At last, the third stage increases the penalty to guarantee the convergence to a feasible region. Its penalty is defined by, where P_i is the penalty value of step i, and t is the number of the current generation:

$$f^{agg}_{penalty}(x) = f^{agg} + \begin{cases} v(x) * P_1, & \text{if } 25 > t \\ v(x) * P_2, & \text{if } 50 > t \geq 25 \\ v(x) * P_3, & \text{if } t \geq 50 \end{cases} \quad (9)$$

3 Comparison Study

To illustrate the difference in the performance of the different CHT, we compare MOEA/D with the following distinct CHT: (1) Low value and (2) High value Static, (3) Multi-staged, (4) Slow and (5) Fast growth Dynamic, (6) Self-adaptive, and (7) Three Stage Penalty against MOEA/D with no CHT ("No Penalty"). The goal is to verify if there is any improvement, and by how much, in MOEA/D performance when using CHT.

3.1 Problems

The real-world analytical multi-objective optimization problems used were selected from the new test suite introduced by Tanabe et al. [18]: (1) bar truss design (CRE21), the objectives are to minimize the structural weight and displacement resulting from the joint; (2) design of welded beams (CRE22), the objectives are to minimize the cost and the final deflection of the welded beams; (3) disc brake design (CRE23), the objectives are to minimize brake mass and decrease downtime; (4) side impact design of the car (CRE31), the objectives are to minimize the average weight of the car, the average speed of the column V responsible for supporting the impact load and the force experienced by a passenger; (5) conceptual submarine project (CRE32), the objectives are to minimize the cost of transportation, the weight of the light vessel and the annual cargo transport capacity; (6) water resource planning (CRE51), the objectives are to minimize the cost of the drainage network, the cost of installing the storage center, the cost of installing the treatment center, the expected cost of flood damage and the expected economic loss due to the flood.

The real-world simulated MOPs used are the three recently proposed by the Japanese evolutionary computing society: (1) Mazda benchmark problem (MAZDA): this is a discrete optimizing problem with the goal of designing MAZDA cars, where the objectives are to maximize the number of parts common to three different cars and minimize the total weight of the car [9]; (2) lunar landing site selection (MOON): the goal is to select landing sites coordinates (x,y) for a lunar exploration robot to land with the objectives of minimizing the number of continuous shaded days, minimizing the inverse of the total communication time[1], and minimizing the tilt angle [17].

3.2 Experimental Parameters

We use the chosen components of MOEA/D as introduced by Li and Zhang [21]. The parameters of the crossover and mutation operators and the probability of using the neighborhood with these operators were fine-tuned using irace [1,14] using the DTLZ unconstrained problems [4]. We make these parameter choices to isolate the contribution of each CHT. Table 1 summarizes the experimental parameters.

Details of these parameters can be found in the documentation of package MOEADr and the original MOEA/D reference [2,3,21]. All objectives and violations were linearly scaled at every iteration to the interval $[0, 1]$. The scaling of the violations uses the maximum and minimum constraint violations of the new and the incumbent populations as reference for the scaling. Finally, the Weighted Tchebycheff scalarization function [16] was used.

The parameters of the CHTs can be found in Table 2. There are two instances of dynamic penalty to test the effects of slow growth and high growth of the penalty value. We name these two instances as Slow growth and Fast growth

[1] i.e. maximizing the total communication time.

Table 1. Experimental parameter settings.

MOEA/D parameters	Value
Polynomial mutation params	$\eta_m = 56$ (2 obj)
	$\eta_m = 26$ (3 obj)
	$\eta_m = 31$ (5 obj)
	$p_m = 0.55$ (2 obj)
	$p_m = 0.15$ (3 obj)
	$p_m = 0.05$ (5 obj)
Simulated Binary Crossover (SBX)	$\eta_m = 24$ (2 obj)
	$\eta_m = 74$ (3 obj)
	$\eta_m = 29$ (5 obj)
	$p_c = 0.7$ (2 obj)
	$p_c = 0.65$ (3 obj)
	$p_c = 0.7$ (5 obj)
Neighborhood size	$T = 20\%$ of the population size
Neighborhood probability	$\delta = 0.981$ (2 obj)
	$\delta = 0.471$ (3 obj)
	$\delta = 0.971$ (5 obj)
SLD decomposition param	$h = 299$ (2 obj)
	$h = 14$ (3 obj)
	$h = 8$ (5 obj)
Population size	$N = 300$ (2 and 3 obj)
	$N = 210$ (5 obj)
Experiment Parameters	Value
Repeated runs	21
Computational budget	30000 evaluations

Dynamic Penalty, respectively. There are also two static penalty cases. The first one, called Low value Static, has a low penalty value ($\beta = 1$), aiming for a more balanced search and the second one, called High value Static, has a high penalty value ($\beta = 1000$), that greatly penalizes unfeasible solutions. We recall the reader that since the self-adaptive penalty requires no parameters, this CHT is not included in Table 2.

3.3 Experimental Evaluation

We compare the different strategies using the Hypervolume (HV, higher is better) indicator. For reproducibility purposes, all the code and experimental scripts are available online[2].

[2] https://github.com/LordeFelipe/MultiObjectiveProjects.

Table 2. Experimental CHT parameters.

CHT parameters	Value
Low value Static	$\beta = 1$
High value Static	$\beta = 1000$
Dynamic	$C = 0.05$
	$\alpha = 2$
	$C = 0.02$
	$\alpha = 2$
Multi-staged	$R_1 = 1$ if $v(x) < 0.25$
	$R_2 = 2$ if $0.5 > v(x) \geq 0.25$
	$R_3 = 4$ if $0.75 > v(x) \geq 0.5$
	$R_4 = 8$ if $v(x) \geq 0.75$
Three Step	$P_1 = 0$ if $25 > t$
	$P_2 = 1$ if $50 > t \geq 25$
	$P_3 = 1000$ if $t \geq 50$

For the calculation of HV for the real analytical problems, the objective function was scaled to the $(0, 1)$ interval, with reference points set to $(1.1, 1.1)$, for two-objective problems; and $(1.1, 1.1, 1.1)$, for three-objective ones. For the real-world simulated MOPs, the reference point was the recommended in the competitions of the Japanese Society for Evolutionary Computation $(1.1, 0)$ for the two-objective car design problem [6] and $(1, 0, 1)$ for the three-objective moon landing problem [7].

3.4 Experimental Results

Here, we present and discuss how is the performance in the real-world constrained problems described above of the following CHT: Low-value Static, High value Static, Slow growth Dynamic, Fast growth Dynamic, Multi-staged, Self-adaptive and Three Stage penalties, and for control, we evaluate the performance of MOEA/D with no CHT ("No penalty").

3.5 Results of the CHTs on the Analytical Problems

Table 3 shows the mean hypervolume (HV, higher is better) of the final population of each CHT on the analytical Benchmark Problems. The results show that the performance of MOEA/D can be improved by any CHT, with no clear difference in HV values between CHTs in these experiments. Because of that, it seems that any focus on the exploration of the feasible space is sufficient for drastically improving the performance of MOEA/D in these problems.

Table 3. Hypervolume mean results of the analytical problems. Best HV values are highlighted in bold. It is clear that for this set of problems, using any CHT is the best choice over MOEA/D with no CHT penalty ("No penalty"). The newly proposed Three Stage CHT performs the best in the CR21, CR22 and CRE31 with competitive performance on all of the other problems.

CHT	CRE21	CRE22	CRE23	CRE31	CRE32	CRE51
No penalty	0.00	0.05	0.84	0.75	0.00	0.88
Low value static	1.09	**1.20**	1.16	**0.81**	0.99	**1.01**
High value static	1.09	**1.20**	**1.18**	**0.81**	**1.01**	1.00
Multi-staged	1.09	**1.20**	1.17	**0.81**	0.99	**1.01**
Slow growth dynamic	1.09	**1.20**	0.98	**0.81**	**1.01**	1.00
Fast growth dynamic	1.07	**1.20**	1.12	**0.81**	1.00	1.00
Self-adaptive	1.10	**1.20**	1.16	**0.81**	1.00	1.00
Three stage	**1.11**	**1.20**	1.11	**0.81**	0.82	0.99

3.6 Results of the CHTs on the Simulated Problems

Now, we describe our results on the MAZDA problem and MOON benchmark problems in Table 4. In contrary to the results of the CHTs on the analytical problems, here, it is noticeable that the right choice of the CHTs has a deep impact on the performance of MOEA/D. This fact suggests that these real-world simulated problems are harder to solve when compared to the analytical benchmark problems and that finding the right technique to explore the feasible search space has high relevance. Given the performance of MOEA/D in terms of hypervolume, the best CHT for the MAZDA benchmark problem is the Three-Stage Penalty, our newly proposed technique, and the best for the moon landing problem is the Fast growth Dynamic Penalty.

3.7 Hypervolume Anytime Performance of MOEA/D with CHT

Here we show the anytime performance of the different CHT in terms of HV values. Figures 1 and 2 illustrate the hypervolume anytime (higher is better) of the eight different MOEA/D variants (seven with different CHT and one without CHT) for the MOON and MAZDA problems, respectively. The hypervolume shown in both figures is the mean HV in the 21 executions for each problem.

These results indicate that, at the start of the search (in the first few generations), the static, self-adapting, and multi-staged CHTs have a better performance. Our interpretation of such results comes from the observation that these CHTs put more pressure on the population to get feasible solutions right from the beginning.

On the other side, dynamic and three-stage penalties initially have a poor performance but get better quickly when the penalty value rises, thus increasing the pressure towards a population with more feasible solutions. We understand

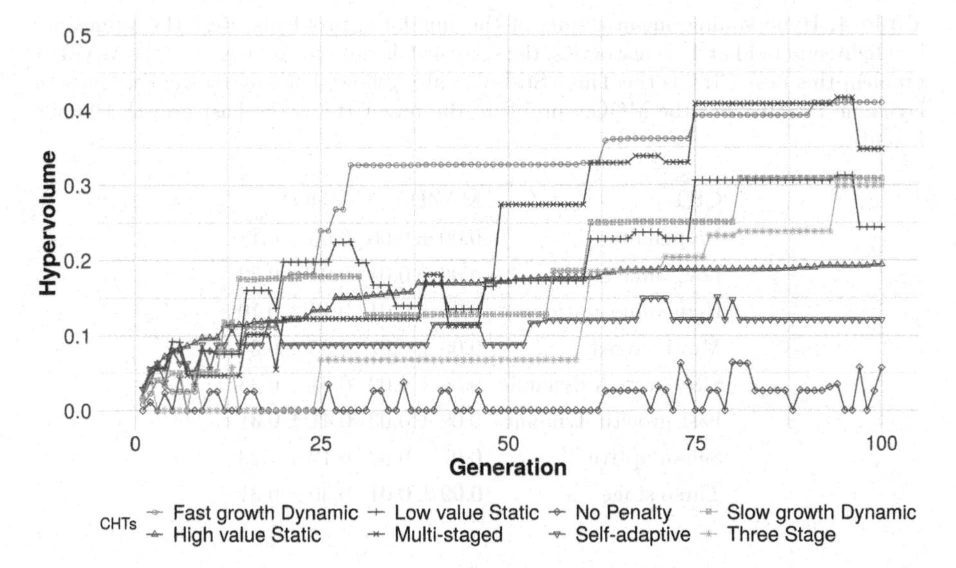

Fig. 1. Anytime mean HV (higher is better) performance of the different CHT in MOEA/D in the moon landing problem. The Three Stage CHT has the competitive performance with most of the other CHT, but is overcomed by the Fast growth Dynamic CHT.

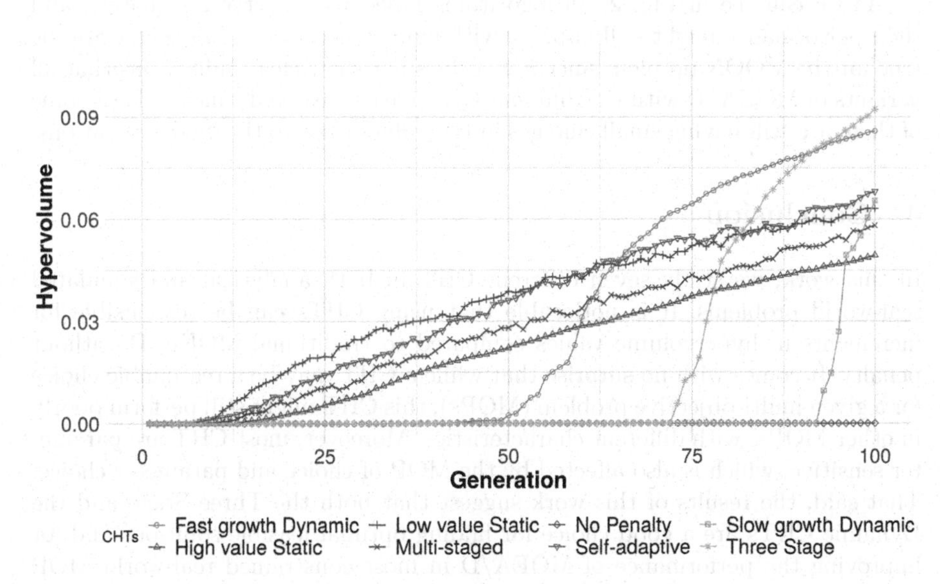

Fig. 2. Anytime mean hypervolume (HV, higher is better) performance of the different CHT in MOEA/D in the MAZDA car problem. The Three Stage CHT achieves the best values at the end of the search execution.

Table 4. Hypervolume mean results of the simulated problems. Best HV values are highlighted in bold and in case of ties, the standard deviation was used. For the MAZDA problem this best CHT is the Three State Penalty followed closely by the Fast growth Dynamic Penalty. For the MOON problem, the best CHT is the Fast growth Dyamic Penalty.

CHT	MAZDA	MOON
No penalty	0.00 ± 0.00	0.06 ± 0.18
Low value static	0.06 ± 0.01	0.24 ± 0.32
High value static	0.05 ± 0.01	0.19 ± 0.10
Multi-staged	0.06 ± 0.02	0.35 ± 0.34
Slow growth dynamic	0.06 ± 0.02	0.31 ± 0.33
Fast growth dynamic	0.09 ± 0.03	$\mathbf{0.41 \pm 0.31}$
Self-adaptive	0.07 ± 0.02	0.12 ± 0.23
Three stage	$\mathbf{0.09 \pm 0.01}$	0.30 ± 0.31

that these Figs. (1 and 2) suggest that increasing the pressure at the right generation may have a deep impact on MOEA/D's hypervolume performance in the simulated problems studied here.

As we can see in Fig. 2, all algorithms failed to converge adequately, and their performance might still improve with more generations. This might also be true for the MOON problem since Fig. 1 does not show clear indications that all variants of MOEA/D with the different CHT have converged, since at least some of them are still having small changes in HV values close to the final generations.

4 Conclusion

In this work, we study seven different CHT in both analytical and simulated real-world problems. It is noticeable that using CHTs can be responsible for increments in hypervolume values compared to traditional MOEA/D without penalty. It comes with no surprise that while a CHT may be a reasonable choice for a given multi-objective problem (MOPs), this CHT might still perform poorly in other MOPs, with different characteristics. Moreover, most CHT are parameter sensitive, which is also affected by the MOP of choice and parameter choice. That said, the results of this work suggest that both the Three Stage and the Dynamic CHTs are a good choice for finding optimal feasible solutions and for improving the performance of MOEA/D in most constrained real-world MOP study in this work. We highlight the increase of performance showed by our newly proposed penalty, particularly in the real-world simulated car design.

The Three Stage CHT has few parameters and we recall the reader that the idea behind this technique is simple yet efficient: first, the Three State CHT lets MOEA/D have a free exploration of the search space, with no penalty to solutions that violate constraints, them this CHT increases the penalty of solutions that violate constraints to an intermediate value, forcing MOEA/D to prefer solutions that are feasible while keeping good performing unfeasible solutions to only at the final stage start to heavily penalize unfeasible solutions and given high preference to feasible solutions. This way MOEA/D can further explore the search space while being able to select optimal feasible solutions at the end of the run.

To further improve the reliability of the results, it would be interesting to test the CHTs with different MOEA/D parameters and also expand the number of tested problems. Future works also include finding ways to improve the performance of MOEA/D on simulated real-world problems using MOEA/D variants that use resource allocation (RA) [10–12] since they show promising results in MOP without constraints. We also plan to study applying CHT as the method for RA as it can control the allocation of resources to explore the unfeasible space more efficiently.

References

1. Bezerra, L.C., López-Ibánez, M., Stützle, T.: Automatic component-wise design of multiobjective evolutionary algorithms. IEEE Trans. Evol. Comput. **20**(3), 403–417 (2016)
2. Campelo, F., Aranha, C.: MOEADr: Component-wise MOEA/D implementation (2018). https://cran.R-project.org/package=MOEADr, r package version 1.2.0
3. Campelo, F., Batista, L., Aranha, C.: The MOEADr package: a component-based framework for multiobjective evolutionary algorithms based on decomposition. J. Stat. Softw. (2020). https://arxiv.org/abs/1807.06731
4. Deb, K., Thiele, L., Laumanns, M., Zitzler, E.: Scalable test problems for evolutionary multiobjective optimization. Evolutionary Multiobjective Optimization, pp. 105–145. Springer, London (2005). https://doi.org/10.1007/1-84628-137-7_6
5. Homaifar, A., Qi, C.X., Lai, S.H.: Constrained optimization via genetic algorithms. Simulation **62**(4), 242–253 (1994)
6. Japanese Society for Evolutionary Competition: Evolutionary competition 2017 (2017). http://is-csse-muroran.sakura.ne.jp/ec2017/EC2017compe.html
7. Japanese Society for Evolutionary Competition: Evolutionary competition 2018 (2018). http://www.jpnsec.org/files/competition2018/EC-Symposium-2018-Competition-English.html
8. Joines, J.A., Houck, C.R.: On the use of non-stationary penalty functions to solve nonlinear constrained optimization problems with GA's. In: Proceedings of the First IEEE Conference on Evolutionary Computation. IEEE World Congress on Computational Intelligence, pp. 579–584. IEEE (1994)
9. Kohira, T., Kemmotsu, H., Akira, O., Tatsukawa, T.: Proposal of benchmark problem based on real-world car structure design optimization. In: Proceedings of the Genetic and Evolutionary Computation Conference Companion, pp. 183–184. ACM (2018)
10. Lavinas, Y., Aranha, C., Ladeira, M., Campelo, F.: MOEA/D with random partial update strategy. In: 2020 IEEE Congress on Evolutionary Computation (CEC), pp. 1–8 (2020)

11. Lavinas, Y., Aranha, C., Ladeira, M.: Improving resource allocation in MOEA/D with decision-space diversity metrics. In: Martín-Vide, C., Pond, G., Vega-Rodríguez, M.A. (eds.) TPNC 2019. LNCS, vol. 11934, pp. 134–146. Springer, Cham (2019). https://doi.org/10.1007/978-3-030-34500-6_9
12. Lavinas, Y., Aranha, C., Sakurai, T.: Using diversity as a priority function for resource allocation on MOEA/D. In: Proceedings of the Genetic and Evolutionary Computation Conference Companion, pp. 215–216. ACM (2019)
13. Li, H., Zhang, Q.: Multiobjective optimization problems with complicated Pareto sets, MOEA/D and NSGA-II. IEEE Trans. Evol. Comput. **13**(2), 284–302 (2009)
14. López-Ibáñez, M., Dubois-Lacoste, J., Cáceres, L.P., Birattari, M., Stützle, T.: The irace package: iterated racing for automatic algorithm configuration. Oper. Res. Perspect. **3**, 43–58 (2016)
15. Mezura-Montes, E., Coello Coello, C.A.: Constraint-handling in nature-inspired numerical optimization: past, present and future. Swarm Evol. Comput. **1**(4), 173–194 (2011). https://doi.org/10.1016/j.swevo.2011.10.001. http://www.sciencedirect.com/science/article/pii/S2210650211000538
16. Miettinen, K.: Introduction to multiobjective optimization: noninteractive approaches. In: Branke, J., Deb, K., Miettinen, K., Słowiński, R. (eds.) Multiobjective Optimization. LNCS, vol. 5252, pp. 1–26. Springer, Heidelberg (2008). https://doi.org/10.1007/978-3-540-88908-3_1
17. Nishiyama, M., et al.: Selection of landing sites for lunar lander with "KAGUYA" data using multi-objective optimization (in Japanese with English abstract). In: Space Science Informatics Symposium FY2014 (2015)
18. Tanabe, R., Ishibuchi, H.: An easy-to-use real-world multi-objective optimization problem suite. Appl. Soft Comput. **89**, 106078 (2020)
19. Trivedi, A., Srinivasan, D., Sanyal, K., Ghosh, A.: A survey of multiobjective evolutionary algorithms based on decomposition. IEEE Trans. Evol. Comput. **21**(3), 440–462 (2017)
20. Woldesenbet, Y.G., Tessema, B.G., Yen, G.G.: Constraint handling in multi-objective evolutionary optimization. In: 2007 IEEE Congress on Evolutionary Computation, pp. 3077–3084 (2007)
21. Zhang, Q., Li, H.: MOEA/D: a multiobjective evolutionary algorithm based on decomposition. IEEE Trans. Evol. Comput. **11**(6), 712–731 (2007)
22. Zitzler, E., Künzli, S.: Indicator-based selection in multiobjective search. In: Yao, X., et al. (eds.) PPSN 2004. LNCS, vol. 3242, pp. 832–842. Springer, Heidelberg (2004). https://doi.org/10.1007/978-3-540-30217-9_84

Dimension Dropout for Evolutionary High-Dimensional Expensive Multiobjective Optimization

Jianqing Lin, Cheng He, and Ran Cheng(✉)

Shenzhen Key Laboratory of Computational Intelligence, University Key Laboratory of Evolving Intelligent Systems of Guangdong Province, Department of Computer Science and Engineering, Southern University of Science and Technology, Shenzhen 518055, China

Abstract. In the past decades, a number of surrogate-assisted evolutionary algorithms (SAEAs) have been developed to solve expensive multiobjective optimization problems (EMOPs). However, most existing SAEAs focus on low-dimensional optimization problems, since a large number of training samples are required (which is unrealistic for EMOPs) to build an accurate surrogate model for high-dimensional problems. In this paper, an SAEA with Dimension Dropout is proposed to solve high-dimensional EMOPs. At each iteration of the proposed algorithm, it randomly selects a part of the decision variables by Dimension Dropout, and then optimizes the selected decision variables with the assistance of surrogate models. To balance the convergence and diversity, those candidate solutions with good diversity are modified by replacing the selected decision variables with those optimized ones (i.e., decision variables from some better-converged candidate solutions). Eventually, the new candidate solutions are evaluated using expensive functions to update the archive. Empirical studies on ten benchmark problems with up to 200 decision variables demonstrate the competitiveness of the proposed algorithm.

Keywords: Surrogate-assisted optimization · High-dimensional · Multiobjective optimization · Dimension dropout

1 Introduction

In the science and engineering areas, most practical optimization problems have multiple conflicting objectives [10,12], and the evaluation of an objective function could be computationally expensive. Consequently, these problems usually require extremely high computational cost to be optimized, which are also known as expensive multiobjective optimization problems (EMOPs) [9]. For example, in the car cab design problem, a structural simulation takes tens of minutes or even hours [2].

H. Ishibuchi et al. (Eds.): EMO 2021, LNCS 12654, pp. 567–579, 2021.
https://doi.org/10.1007/978-3-030-72062-9_45

Recently, surrogate-assisted evolutionary algorithms (SAEAs), which combine traditional evolutionary algorithms with surrogate model techniques, are the mainstream methods to solve EMOPs. They use a surrogate model instead of a part of computationally expensive function evaluation. The most commonly used surrogate models include polynomial regression (PR) [34], Gaussian process (GP) [19,34], radial basis function (RBF) [23,31], Neural Networks (NN) [22], etc. Since the surrogate model is computationally cheap, it can significantly reduce the required computational cost [11]. As an essential part of SAEAs, model management mainly involves how to update the current surrogate model, including the selection of offspring individuals to be re-evaluated by expensive function. Then, the re-evaluated individuals will be added to training data for updating the surrogate model. When selecting individuals to be re-evaluated, it is crucial to balance convergence and diversity.

In recent decades, a number of SAEAs have been proposed to solve EMOPs effectively. The most representative SAEAs include ParEGO [17], MOEA/D-EGO [33], K-RVEA [4,5], CSEA [22], etc. However, the aforementioned SAEAs only focus on low-dimensional optimization with less than 30 decision variables [5,14,17,33]. It is mainly due to the lack of training samples for obtaining accurate surrogate models and the unbearable time cost for training the models with high-dimensional input. To address these issues, a surrogate-assisted differential evolution with Dimension Dropout, called D2SADE, is proposed. We focus on the selection of training data to manage convergence and diversity, given a large number of decision variables. The main contributions of this paper can be summarized as follows:

* **Dimension Dropout:** Inspired by the Dropout strategy in deep learning, we expand Dropout to high-dimensional EMOPs by selecting d from D decision variables to be optimized randomly at each iteration.
* **Infill criterion:** A novel infill criterion is used to select individuals from the optimized d-dimension individuals that are well-converged via surrogate-based search.
* **Replacement mechanism:** This mechanism chooses diversity-related solutions from existing non-dominated solutions. Then the corresponding dimensions of these solutions will be replaced by those individuals optimized by the infill criterion.

In the next section, some preliminaries are presented. The details of the proposed algorithm are illustrated in Sect. 3. The experimental studies and result analysis are given in Sect. 4. Finally, Sect. 5 concludes this paper.

2 Preliminaries

2.1 Multiobjective Optimization

Multiobjective optimization problems (MOPs) are commonly seen in the science and engineering areas [8,21], e.g., structure optimization [27], industrial dispatch

[1], and robotics [16]. An MOP aims to simultaneously optimize more than two often conflicting objectives, which can be mathematically formulated as follows.

$$\begin{aligned} \text{minimize} \quad & F(\mathbf{x}) = (f_1(\mathbf{x}), f_2(\mathbf{x}), \ldots, f_m(\mathbf{x})) \\ \text{subject to} \quad & \mathbf{x} \in X, \end{aligned} \tag{1}$$

where X is the search space of decision variables, $m (\geq 2)$ is the number of objectives, and $\mathbf{x} = (x_1, \ldots, x_D)$ is the decision vector with D decision variables.

The goal of multiobjective optimization is to obtain a set of representative Pareto-optimal solutions [6]. Specifically, a Pareto-optimal solution is the one not Pareto dominated by other solutions in X. The Pareto dominance of a solution $\mathbf{x}_1$ and another solution $\mathbf{x}_2$ is defined as $(\mathbf{x}_1 \preceq \mathbf{x}_2)$ if and only if $f_i(\mathbf{x}_1) \leq f_i(\mathbf{x}_2)$ for all $1 \leq i \leq m$, and $f_j(\mathbf{x}_1) < f_j(\mathbf{x}_2)$ for at least one $1 \leq j \leq m$. Specifically, if the evaluation of any objective function in an MOP is computationally expensive, the problem is called an expensive MOP, i.e., EMOP.

2.2 Dropout

Dropout originally refers to temporarily discarding neural network units from the network according to a certain probability during the training process of the deep neural network [13,18]. It can effectively alleviate the occurrence of over-fitting, and an example of the Dropout in a neural network is shown in Fig. 1. After dropout, the complexity of the neural net is significantly reduced.

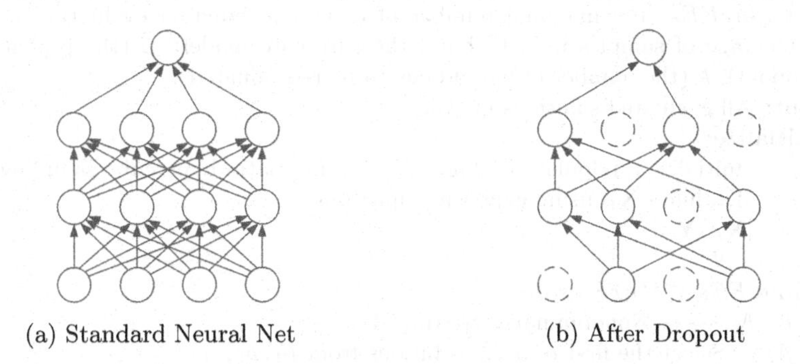

Fig. 1. An example of the dropout in neural network.

2.3 Radial Basis Function (RBF)

The RBF is a simple and widely used surrogate model. On the basis of the training samples $\{(\mathbf{x}_i, y_i) | i = 1, \ldots, N\}$, it approximates a continuous function as follows:

$$\hat{f}(\mathbf{x}) = \sum_{i=1}^{N} w_i \phi(\mathbf{x}_i, \mathbf{x}), \tag{2}$$

where w_i is the weight coefficient and $\phi(\cdot,\cdot)$ represents the kernel function. Commonly used kernel functions include linear kernel function, polynomial kernel function, Gaussian kernel function, etc. The weight vector $\mathbf{w} = (w_1, ..., w_N)$ is calculated as follows:

$$\mathbf{w} = (\Phi^T\Phi)^{-1}\Phi^T\mathbf{y}, \tag{3}$$

where $\mathbf{y} = (y_1, \ldots, y_N)$ is the output vector of the training data and Φ is the matrix computed as follows:

$$\Phi = \begin{bmatrix} \phi(\mathbf{x}_1, \mathbf{x}_1) & \cdots & \phi(\mathbf{x}_1, \mathbf{x}_N) \\ \vdots & \ddots & \vdots \\ \phi(\mathbf{x}_N, \mathbf{x}_1) & \cdots & \phi(\mathbf{x}_N, \mathbf{x}_N) \end{bmatrix}. \tag{4}$$

3 Proposed Method

This work proposes D2SADE to solve high-dimensional EMOPs, which introduces the Dimension Dropout and some related strategies to achieve a good balance between exploration and exploitation. Generally, D2SADE mainly consists of the Dimension Dropout, the infill criterion, and the replacement mechanism. In what follows, we first demonstrate the general framework of the proposed algorithm, followed by the details of the three components. The proposed D2SADE is

Algorithm 1. The pseudocode of D2SADE

Input: $MaxFEs$ (the maximal number of expensive function evaluations), α (the percentage of samples used to build the surrogate model), β (the probability of dropout), k (the number of individuals to be re-evaluated);
Output: All evaluated solutions in A_1

1: **Initialize:**
2: $\mathbf{X}_P \leftarrow$ Initialize a population of size N by using Latin hypercube sampling
3: $\mathbf{Y}_P \leftarrow$ Evaluate $\mathbf{X}_P$ using expensive functions
4: $A_1 \leftarrow [\mathbf{X}_P, \mathbf{Y}_P]$
5: $FEs \leftarrow N$
6: **while** $FEs < MaxFEs$ **do**
7: $F_1, F_2, \cdots \leftarrow$ Nondominated_Sorting(A_1)
8: $A_2 \leftarrow$ Select the first $\alpha \cdot FEs$ solutions from $F_1, F_2, \ldots$
9: /*Dimension Dropout*/
10: $S \leftarrow$ Randomly select d indices from $[1, 2, \ldots, D]$, where $d = \beta \cdot D$
11: $A_{2d} \leftarrow [\mathbf{X}_{A_2} \backslash_S, \mathbf{Y}_{A_2}]$
12: /*Infill criterion*/
13: $X_d \leftarrow k$ d-dimensional candidate solutions optimized by *RBF-based Search*
14: /*Replacement mechanism*/
15: $X_D \leftarrow k$ non-dominated solutions with the biggest crowding distances in A_2
16: $X_{new} \leftarrow$ Replace the corresponding decision dimensions of X_D with X_d
17: $Z_{new} \leftarrow$ Evaluate X_{new} by using expensive functions
18: $A_1 \leftarrow A_1 \cup Z_{new}$
19: $FEs \leftarrow FEs + k$
20: **end while**

presented in Algorithm 1. First, N solutions are sampled using Latin hypercube sampling [24] and evaluated by the expensive functions. Then these evaluated candidate solutions are merged into an empty archive A_1. Afterwards, the non-dominated sorting is used to sort members in A_1 and the top α percent solutions in A_1 are merged into an empty archive A_2. During the evolution, three phases are implemented, i.e., Dimension Dropout (line 9–11), infill criterion (line 12–13), and replacement mechanism (line 14–16). The Dimension Dropout discards the decision dimensions from D dimensions to d randomly for the solutions in A_2. Next, on the basis of these d-dimensional individuals, d-dimensional RBF models are built and the RBF-based search is used to optimize the d-dimensional variables. During the optimization, the infill criterion is used to select k individuals (denoted as X_d) which are well converged. In what follows, the replacement mechanism chooses k solutions with diversity in A_1 (denoted as X_D). The corresponding dimensions of X_D will be replaced by X_d to form new individuals X_{new}, and X_{new} will be evaluated using expensive functions for updating the archives A_1. Finally, the obtained A_1 will be output as the final solutions.

3.1 Dimension Dropout

The Dimension Dropout discards the decision dimensions from D dimensions to d randomly for solutions in A_2. Then a d-dimensional surrogate model is constructed for each objective function with d-dimensional input. Eventually, the RBF-based search is used to optimize the selected d dimensions. An example of the Dimension Dropout is shown in Fig. 2, where dimensions $\{2, 7, \ldots, D-1\}$ are selected.

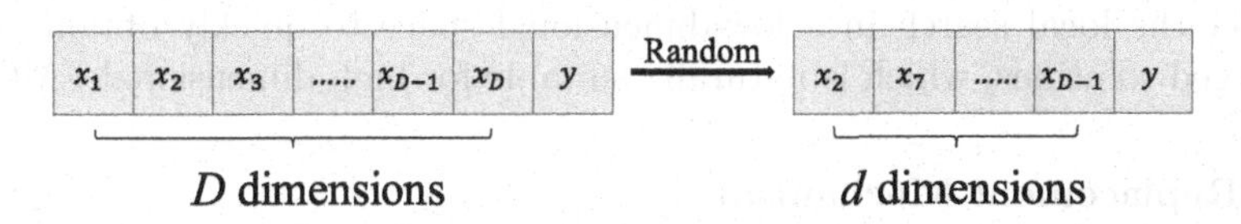

Fig. 2. An example of the Dimension Dropout, which reduces the dimension from D to d ($d < D$).

For high-dimensional EMOPs, a large number of training samples are required to construct highly accurate surrogate models. Dimension Dropout reduces the dimension of decision variables from D to d, which helps build relatively accurate d-dimensional surrogate models with the same number of training samples. Compared with the surrogate models built with D-dimensional training samples, although not all decision variables can be optimized at every iteration, the surrogate model built with d-dimensional training samples can better help optimize the d-dimensional variables and accelerate convergence. Different from the principal component analysis (PCA) [20] method, which maps data from high-dimensional space to low-dimensional space, Dimension Dropout retains the original information of the selected d dimensions.

3.2 Infill Criterion

During the optimization, the infill criterion is used to select k d-dimensional well-converged individuals according to the RBF-based Search in Algorithm 2.

Algorithm 2. RBF-based Search

Input: λ_{max} (maximum number of iterations), A_{2d} (the data to construct RBF), k (the number of individuals to be re-evaluated)
Output: k d-dimension better-converged candidate individuals selected from $P_{\lambda_{max}}$
1: $RBF \leftarrow$ Train RBF for each objective function by using individuals in A_{2d}
2: $P_0 \leftarrow$ Consider A_{2d} as the parent population
3: $\lambda \leftarrow 0$
4: **while** $\lambda < \lambda_{max}$ **do**
5: $Q_\lambda \leftarrow$ Generate offspring according to P_λ and evaluate with RBF
6: $R_\lambda \leftarrow P_\lambda \cup Q_\lambda$
7: $P_{\lambda+1} \leftarrow$ Select by non-dominated sorting from R_λ
8: $\lambda \leftarrow \lambda + 1$
9: **end while**

For the surrogate model, RBF is used due to its insensitivity to the increment in the dimension of the function to be approximated [7,15]. Compared with Kriging model [26], RBF is more suitable for high-dimensional problems, since the computation time for training Kriging models will become unbearable when the number of training samples is large.

In RBF-based Search, we use the Differential Evolution(DE) algorithm [25] as the optimizer. Compared with genetic algorithm, DE is more suitable for solving high-dimensional optimization problems [29,30].

The combination of the used infill criterion and Dimension Dropout aims to conduct the local search in a low-dimensional space to quickly obtain better-converged solutions, which is naturally suitable for high-dimensional EMOPs.

3.3 Replacement Mechanism

To maintain the diversity of the candidate solutions, we choose k solutions with good diversity in A_1 by replacement mechanism, donated as X_D.

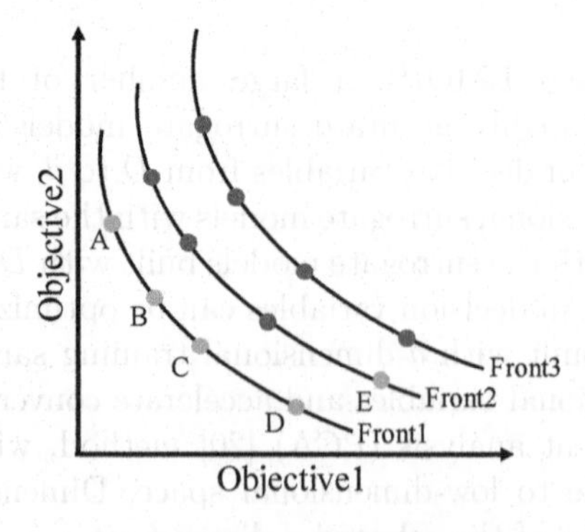

Fig. 3. Illustration of which solutions are selected in replacement mechanism.

To be more specific, the non-dominated sorting is first used to obtain those non-dominated solutions, and then their crowding distance are calculated. If k is less than or equal to the number of non-dominated solutions, we select k non-dominated solutions with the largest crowding distances. If k is greater than the number of non-dominated solutions, we select all the non-dominated solutions and the solutions with the largest crowding distances in the next front just as shown in Fig. 3. Then the corresponding dimensions of X_D will be replaced by X_d to form new individuals X_{new}. Finally, we re-evaluate X_{new} for updating the archives, and an example of the replacement mechanism is shown in Fig. 4.

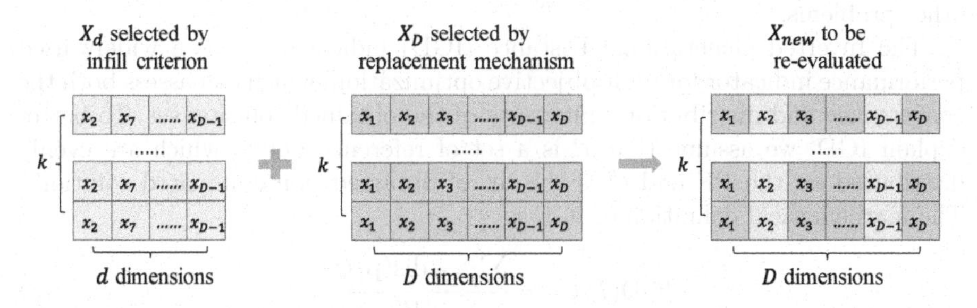

Fig. 4. An example of the proposed replacement mechanism, where the dropout dimensions of k selected solutions will be replaced by the optimized decision variable values accordingly.

4 Experimental Studies

4.1 Experimental Settings

For fair comparisons, the two compared algorithms and our proposed D2SADE are all implemented in PlatEMO [28] using MATLAB, and all algorithms are run on a computer with Intel Core i7 processor and 8 GB of RAM.

1) Parameters Settings: The percentage of samples used to build the surrogate model is set to $\alpha = 0.7$. The probability of Dimension Dropout is set to $\beta = 0.5$, and the number of individuals to be re-evaluated is $k = 5$. The maximum number of generations in the RBF-based search is set to $\lambda_{max} = 100$. For the compared algorithms, the recommended parameter settings in the literature are used.

2) Reproduction Operators: In D2SADE, the crossover rate CR is set to 1, the scaling factor is set to $F = 0.5$, p_m is set to $1/D$, and η is set to 20.
3) Termination Condition: For all the test problems, the size of the initial data is set to 500, and the maximum function evaluations using the expensive function (denoted as $MaxFEs$) is set to 1000.

4.2 Test Problems and Performance Indicator

In this paper, we use ten IMF problems selected from [3]. Among these test problems, the number of objectives is three in IMF4 and IMF8, and it is two in other problems.

The Inverted Generational Distance (IGD) indicator [32] is a widely used performance indicator for multiobjective optimization, which can assess both the convergence and distribution uniformity of the obtained solution set. To better explain IGD, we assume that P is a set of reference points which are evenly distributed on the PF and Q is the set of obtained non-dominated solutions. The mathematical definition of IGD is

$$\mathrm{IGD}(P, Q) = \frac{\sum_{\mathbf{p} \in P} \mathrm{dis}(\mathbf{p}, Q)}{|P|}, \tag{5}$$

where $\mathrm{dis}(\mathbf{p}, Q)$ is the minimum Euclidean distance between $\mathbf{p}$ and points in Q, and $|P|$ is the number of reference points in P. A smaller IGD value indicates better performance of the algorithm.

4.3 Effectiveness of the Dropout

To verify the validity of Dimension Dropout strategy, this section we design a comparative algorithm which is the same as D2SADE except for the Dimension Dropout, named D2SADE∗. Table 1 summarizes the statistical results of the IGD values obtained by the two comparison algorithms on IMF1-IMF10. Our proposed Dropout Strategy has performed the better on these 20 test problems.

As shown in Table 1, compared with D2SADE∗, D2SADE obtains better IGD values in most of the test problems. It indicates that the Dimension Dropout helps build a d-dimensional relatively accurate RBF model, and the optimization of d-dimensional variables at each generation can accelerate expensive multiobjective optimization.

4.4 General Performance

In this subsection, D2SADE is compared with two representative SAEAs, i.e., MOEA/D-EGO [33] and KRVEA [5]. K-RVEA is an SAEA that relies on a set of adaptive reference vectors for expensive many-objective optimization. MOEA/D-EGO is MOEA/D-based SAEA with the Gaussian stochastic process

Table 1. IGD Results Obtained by D2SADE∗ and D2SADE on 20 IMF Test Instances. The Best Result in Each Row is Highlighted.

Problem	Dim	D2SADE∗	D2SADE
IMF1	100	3.10E+00(3.49E−01)	2.28E+00(4.59E−01)
	200	8.75E+00(1.81E+00)	5.76E+00(6.06E−01)
IMF2	100	5.33E+00(4.01E−01)	2.91E+00(5.36E−01)
	200	1.12E+01(2.15E+00)	7.37E+00(9.35E−01)
IMF3	100	7.51E+00(1.23E+00)	4.64E+00(1.22E+00)
	200	1.64E+01(6.86E−01)	8.53E+00(8.16E−01)
IMF4	100	6.92E+01(8.53E+00)	5.52E+01(8.40E+00)
	200	3.13E+02(3.50E+01)	2.03E+02(2.53E+01)
IMF5	100	2.87E−01(8.19E−03)	2.88E−01(6.62E−03)
	200	3.08E−01(2.94E−03)	3.06E−01(1.12E−02)
IMF6	100	4.87E−01(1.06E−02)	4.56E−01(2.37E−02)
	200	5.44E−01(8.19E−03)	4.68E−01(1.13E−02)
IMF7	100	4.35E−01(1.06E−02)	4.33E−01(1.82E−02)
	200	4.72E−01(1.00E−02)	4.92E−01(3.24E−02)
IMF8	100	4.34E+00(1.41E−01)	4.05E+00(2.18E−01)
	200	8.80E+00(3.17E−01)	8.81E+00(4.74E−01)
IMF9	100	5.51E−01(4.76E−03)	5.41E−01(1.83E−02)
	200	5.92E−01(2.69E−03)	5.88E−01(1.94E−03)
IMF10	100	5.31E+02(3.95E+01)	4.69E+02(3.92E+01)
	200	1.43E+03(1.04E+02)	1.07E+03(1.18E+02)

model. The experimental results of IGD value provided by K-RVEA, MOEA/D-EGO, and D2SADE are recorded in Table 2. It can be observed that D2SADE obtains better IGD values in most of the 20 test problems compared with MOEA/D-EGO and K-RVEA. In particular, D2SADE performs much better in IMF1-IMF3, but MOEA/D-EGO has a better performance on IMF7.

Table 2. IGD Results Obtained by K-RVEA, MOEA/D-EGO and D2SADE on 20 IMF Test Instances. The Best Result in Each Row is Highlighted.

Problem	Dim	MOEA/D-EGO	K-RVEA	D2SADE
IMF1	100	1.12E+01(9.94E−01)	1.17E+01(6.11E−01)	**2.28E+00(4.59E−01)**
	200	1.31E+01(3.94E−01)	1.32E+01(6.52E−01)	**5.76E+00(6.06E−01)**
IMF2	100	1.51E+01(1.13E+00)	1.52E+01(1.62E+00)	**2.91E+00(5.36E−01)**
	200	1.63E+01(4.46E−01)	1.72E+01(5.17E−01)	**7.37E+00(9.35E−01)**
IMF3	100	1.29E+01(1.54E+00)	1.39E+01(1.79E+00)	**4.64E+00(1.22E+00)**
	200	1.66E+01(8.90E−01)	1.65E+01(7.61E−01)	**8.53E+00(8.16E−01)**
IMF4	100	1.24E+02(4.14E+01)	**4.74E+01(1.09E+01)**	5.52E+01(8.40E+00)
	200	3.60E+02(7.35E+00)	3.57E+02(2.71E+01)	**2.03E+02(2.53E+01)**
IMF5	100	3.03E−01(1.46E−02)	3.10E−01(1.99E−02)	**2.88E−01(6.62E−03)**
	200	3.04E−01(9.99E−03)	**3.03E−01(7.40E−03)**	3.06E−01(1.12E−02)
IMF6	100	5.01E−01(2.83E−02)	4.73E−01(2.56E−02)	**4.56E−01(2.37E−02)**
	200	5.42E−01(1.49E−02)	5.39E−01(8.24E−03)	**4.68E−01(1.13E−02)**
IMF7	100	**3.90E−01(2.98E−02)**	4.22E−01(1.02E−02)	4.33E−01(1.82E−02)
	200	**4.49E−01(2.72E−02)**	4.78E−01(6.56E−03)	4.92E−01(3.24E−02)
IMF8	100	4.29E+00(2.11E−01)	4.12E+00(1.60E−01)	**4.05E+00(2.18E−01)**
	200	9.41E+00(3.77E−01)	9.16E+00(5.06E−01)	**8.81E+00(4.74E−01)**
IMF9	100	5.58E−01(4.43E−04)	5.56E−01(2.54E−03)	**5.41E−01(1.83E−02)**
	200	5.93E−01(1.92E−03)	5.95E−01(1.22E−03)	**5.88E−01(1.94E−03)**
IMF10	100	9.49E+02(5.91E+01)	9.84E+02(2.92E+01)	**4.69E+02(3.92E+01)**
	200	2.07E+03(1.52E+02)	2.10E+03(8.08E+01)	**1.07E+03(1.18E+02)**

5 Conclusion

In this paper, we have proposed an SAEA with Dimension Dropout, called D2SADE, for solving high-dimensional EMOPs with up to 200 decision variables. Due to the effect of the Dimension Dropout, D2SADE has shown encouraging performance in solving high-dimensional EMOPs.

Generally, the Dimension Dropout in D2SADE is adopted for selecting d-dimension decision variables from D-dimension original decision variables. It can help build d-dimensional relatively high-precision RBF models, and the algorithm only needs to optimize the d-dimensional variables at each iteration.

An infill criterion and a replacement mechanism are adopted to balance the convergence and diversity of the population. The infill criterion selects k d-dimensional well-converged individuals assisted by RBF models. In replacement mechanism, k D-dimensional solutions with the biggest crowding distances are selected from the set of non-dominated solutions. The corresponding dimensions in k D-dimensional solutions are replaced by the k d-dimensional individuals selected by infill criterion to form k new D-dimensional individuals.

Some systematic comparisons have been conducted on a set of EMOPs with up to 200 decision variables. The proposed D2SADE is compared with two representative algorithms, i.e., K-RVEA and MOEA/D-EGO. The experimental results indicate the superiority of D2SADE in solving EMOPs with high-dimension decision variables.

This paper indicates that D2SADE is promising in solving high-dimension EMOPs. Furthermore, it is promising to extend D2SADE to solving expensive many-objective optimization problems, and it is also desirable to apply it to real-world applications.

Acknowledgment. This work was supported in part by the National Natural Science Foundation of China under Grant 61903178 and Grant 61906081; in part by the Program for Guangdong Introducing Innovative and Entrepreneurial Teams under Grant 2017ZT07X386; in part by the Shenzhen Peacock Plan under Grant KQTD2016112514355531; and in part by the Program for University Key Laboratory of Guangdong Province under Grant 2017KSYS008.

References

1. Abido, M.A.: Multiobjective evolutionary algorithms for electric power dispatch problem. IEEE Trans. Evol. Comput. **10**(3), 315–329 (2006)
2. Akhtar, T., Shoemaker, C.A.: Multi objective optimization of computationally expensive multi-modal functions with RBF surrogates and multi-rule selection. J. Glob. Optim. **64**(1), 17–32 (2016)
3. Cheng, R., Jin, Y., Narukawa, K., Sendhoff, B.: A multiobjective evolutionary algorithm using gaussian process-based inverse modeling. IEEE Trans. Evol. Comput. **19**(6), 838–856 (2015)
4. Cheng, R., Jin, Y., Olhofer, M., Sendhoff, B.: A reference vector guided evolutionary algorithm for many-objective optimization. IEEE Trans. Evol. Comput. **20**(5), 773–791 (2016)
5. Chugh, T., Jin, Y., Miettinen, K., Hakanen, J., Sindhya, K.: A surrogate-assisted reference vector guided evolutionary algorithm for computationally expensive many-objective optimization. IEEE Trans. Evol. Comput. **22**(1), 129–142 (2018)
6. Deb, K., Pratap, A., Agarwal, S., Meyarivan, T.: A fast and elitist multiobjective genetic algorithm: NSGA-II. IEEE Trans. Evol. Comput. **6**(2), 182–197 (2002)
7. Er, M., Wu, S., Lu, J., Toh, H.: Face recognition with radial basis function (RBF) neural networks. IEEE Trans. Neural Netw. **13**(3), 697–710 (2002)
8. He, C., Cheng, R., Danial, Y.: Adaptive offspring generation for evolutionary large-scale multiobjective optimization. IEEE Trans. Syst. Man Cybern. Syst. (2020). https://doi.org/10.1109/TSMC.2020.3003926
9. He, C., Cheng, R., Jin, Y., Yao, X.: Surrogate-assisted expensive many-objective optimization by model fusion. In: 2019 IEEE Congress on Evolutionary Computation (CEC), pp. 1672–1679. IEEE (2019)
10. He, C., Cheng, R., Zhang, C., Tian, Y., Chen, Q., Yao, X.: Evolutionary large-scale multiobjective optimization for ratio error estimation of voltage transformers. IEEE Trans. Evol. Comput. **24**(5), 868–881 (2020). https://doi.org/10.1109/TEVC.2020.2967501

11. He, C., Huang, S., Cheng, R., Tan, K.C., Jin, Y.: Evolutionary multiobjective optimization driven by generative adversarial networks (GANs). IEEE Trans. Cybern. (2020). https://doi.org/10.1109/TCYB.2020.2985081
12. He, C., Tian, Y., Jin, Y., Zhang, X., Pan, L.: A radial space division based many-objective optimization evolutionary algorithm. Appl. Soft Comput. **61**, 603–621 (2017). https://doi.org/10.1016/j.asoc.2017.08.024
13. Hinton, G.E., Srivastava, N., Krizhevsky, A., Sutskever, I., Salakhutdinov, R.R.: Improving neural networks by preventing co-adaptation of feature detectors. Comput. Sci. **3**(4), 212–223 (2012)
14. Hussein, R., Deb, K.: A generative kriging surrogate model for constrained and unconstrained multi-objective optimization. In: Proceedings of the Genetic and Evolutionary Computation Conference 2016, pp. 573–580 (2016)
15. Kattan, A., Galvan, E.: Evolving radial basis function networks via GP for estimating fitness values using surrogate models. In: 2012 IEEE Congress on Evolutionary Computation, pp. 1–7 (2012)
16. Kim, J., Kim, Y., Choi, S., Park, I.: Evolutionary multi-objective optimization in robot soccer system for education. IEEE Comput. Intell. Mag. **4**(1), 31–41 (2009)
17. Knowles, J.: ParEGO: a hybrid algorithm with on-line landscape approximation for expensive multiobjective optimization problems. IEEE Trans. Evol. Comput. **10**(1), 50–66 (2006)
18. Li, C., Gupta, S., Rana, S., Nguyen, V., Venkatesh, S., Shilton, A.: High dimensional Bayesian optimization using dropout. In: Proceedings of the Twenty-Sixth International Joint Conference on Artificial Intelligence, pp. 2096–2102 (2017)
19. Liu, B., Zhang, Q., Gielen, G.G.E.: A Gaussian process surrogate model assisted evolutionary algorithm for medium scale expensive optimization problems. IEEE Trans. Evol. Comput. **18**(2), 180–192 (2014)
20. Ma, M., Li, H., Huang, J.: A multi-objective evolutionary algorithm based on principal component analysis and grid division. In: 2018 14th International Conference on Computational Intelligence and Security (CIS), pp. 201–204 (2018)
21. Pan, L., He, C., Tian, Y., Su, Y., Zhang, X.: A region division based diversity maintaining approach for many-objective optimization. Integr. Comput. Aided Eng. **24**(3), 279–296 (2017)
22. Pan, L., He, C., Tian, Y., Wang, H., Zhang, X., Jin, Y.: A classification-based surrogate-assisted evolutionary algorithm for expensive many-objective optimization. IEEE Trans. Evol. Comput. **23**(1), 74–88 (2019)
23. Regis, R.G.: Evolutionary programming for high-dimensional constrained expensive black-box optimization using radial basis functions. IEEE Trans. Evol. Comput. **18**(3), 326–347 (2014)
24. Stein, M.: Large sample properties of simulations using Latin hypercube sampling. Technometrics **29**(2), 143–151 (1987)
25. Storn, R., Price, K.: Differential evolution-a simple and efficient heuristic for global optimization over continuous spaces. J. Glob. Optim. **11**(4), 341–359 (1997)
26. Sun, C., Jin, Y., Cheng, R., Ding, J., Zeng, J.: Surrogate-assisted cooperative swarm optimization of high-dimensional expensive problems. IEEE Trans. Evol. Comput. **21**(4), 644–660 (2017)
27. Sun, G., Pang, T., Fang, J., Li, G., Li, Q.: Parameterization of criss-cross configurations for multiobjective crashworthiness optimization. Int. J. Mech. Sci. **124**, 145–157 (2017)
28. Tian, Y., Cheng, R., Zhang, X., Jin, Y.: PlatEMO: a matlab platform for evolutionary multi-objective optimization [educational forum]. IEEE Comput. Intell. Mag. **12**(4), 73–87 (2017)

29. Wang, H., Rahnamayan, S., Wu, Z.: Parallel differential evolution with self-adapting control parameters and generalized opposition-based learning for solving high-dimensional optimization problems. J. Parallel Distrib. Comput. **73**(1), 62–73 (2013)
30. Wang, H., Wu, Z., Rahnamayan, S.: Enhanced opposition-based differential evolution for solving high-dimensional continuous optimization problems. Soft Comput. **15**(11), 2127–2140 (2011)
31. Wang, Y., Yin, D., Yang, S., Sun, G.: Global and local surrogate-assisted differential evolution for expensive constrained optimization problems with inequality constraints. IEEE Trans. Cybern. **49**(5), 1642–1656 (2019)
32. Yen, G.G., He, Z.: Performance metric ensemble for multiobjective evolutionary algorithms. IEEE Trans. Evol. Comput. **18**(1), 131–144 (2014)
33. Zhang, Q., Liu, W., Tsang, E., Virginas, B.: Expensive multiobjective optimization by MOEA/D with Gaussian process model. IEEE Trans. Evol. Comput. **14**(3), 456–474 (2010)
34. Zhou, Z., Ong, Y., Nguyen, M., Lim, D.: A study on polynomial regression and gaussian process global surrogate model in hierarchical surrogate-assisted evolutionary algorithm. In: 2005 IEEE Congress on Evolutionary Computation, vol. 3, pp. 2832–2839 (2005)

Multiobjective Optimization with Fuzzy Classification-Assisted Environmental Selection

Jinyuan Zhang and Hisao Ishibuchi(✉)

Guangdong Provincial Key Laboratory of Brain-Inspired Intelligent Computation, Department of Computer Science and Engineering, Southern University of Science and Technology, Shenzhen 518055, China
{zhangjy,hisao}@sustech.edu.cn

Abstract. Most environmental selection strategies in multiobjective evolutionary algorithms (MOEAs) select solutions based on their objective function values. However, the objective evaluations of many real-world problems are very time-consuming. The use of a large number of objective evaluations will inevitably reduce the efficiency of MOEAs. This paper proposes a fuzzy classification-assisted environmental selection (FAES) scheme to reduce the number of objective evaluations of MOEAs. The proposed method uses a fuzzy classifier to choose promising solutions in environmental selection. In the proposed method, first, solutions in the previous generations are classified into two classes using the Pareto dominance relation. The non-dominated solutions are positive class, and the dominated solutions are negative class. Next, the classified solutions are used to build a fuzzy classifier. Then, the built classifier is used to predict the membership degree of each of the current and offspring solutions. Only the offspring solutions, whose membership degrees to the positive class are larger than their parents', are evaluated. The offspring solutions with smaller membership degrees are discarded with no objective evaluations. Therefore, the number of objective evaluations can be reduced. Finally, the evaluated offspring solutions are used in the environmental selection together with the current solutions. The proposed FAES strategy is integrated into an MOEA in computational experiments. Experimental results show the efficiency of the proposed FAES on reducing the number of objective evaluations.

Keywords: Multiobjective evolutionary algorithms · Fuzzy classification · Environmental selection · Surrogate models

1 Introduction

Evolutionary algorithms (EAs) have been widely used to solve multiobjective optimization problems (MOPs), due to their ability to search for a set of Pareto optimal solutions of MOPs [20]. Many multiobjective evolutionary algorithms (MOEAs) have been proposed in the last decades, i.e., the dominance-based

H. Ishibuchi et al. (Eds.): EMO 2021, LNCS 12654, pp. 580–592, 2021.
https://doi.org/10.1007/978-3-030-72062-9_46

algorithms [5,22], the indicator-based algorithms [2,21], and the decomposition-based algorithms [13,19]. These MOEAs mainly contain three components: (1) initialization strategies to initialize a set of solutions, (2) reproduction operators to generate a set of new solutions, and (3) environmental selection operators to select promising solutions for the new population.

Environmental selection operators in MOEAs choose promising solutions based on their objective function values. Since many evaluated solutions are discarded for their worse objective function values or worse fitness values, these objective value-based methods waste the computation resources. Especially, many objective evaluations are time-consuming for those expensive problems.

In terms of the above concern, many methods have been proposed for reducing the number of objective evaluations. Among them, surrogate model-based approach [8,15] is the representative one. In these surrogate-based MOEAs, surrogate models are built to approximate the objective functions or approximate the fitness value of each solution. Since the surrogates are with cheap computation resources, the time required for objective evaluations is significantly reduced. As a special case of the surrogate-based approach, classification-based MOEAs have been attracted much attention recently. The motivation of using classifiers to assist MOEAs is that the environmental selection of MOEAs can be treated as a classification problem. The selected solutions by environmental selection can be regarded as positive solutions and the discarded solutions are negative ones. These classification-based MOEAs mainly focus on using classifiers to learn the relation between solutions. Loshchilov et al. [12] proposed to use the combination of a one-class classifier and a regression support vector machine (SVR) to predict the dominance relation between the solutions and the current Pareto set. Bandaru et al. [1] proposed to use a multi-class classification model to learn the Pareto dominance relation of solutions. Bhattacharjee et al. [3] proposed to use the support vector machine (SVM) to assist NSGA-II. Zhang et al. [16–18] and Lin et al. [11] proposed to use classifiers to preselect candidate offspring solutions of MOEAs. Pan et al. [14] proposed a classifier and surrogate combined algorithm for expensive many-objective optimization problems. These classification-based MOEAs have shown their efficiency in improving the performance of MOEAs. Current classification-based methods mostly focus on deciding the quality of solutions only by the predicted labels of solutions. Since the accuracy of the built classifiers cannot be ensured, only predicting the quality of solutions by the labels may lead to misclassification. Based on this concern, we propose to use the fuzzy classifier to assist MOEAs to reduce the number of objective evaluations. Fuzzy classifiers not only predict the label of a solution, but also provide the membership degree of a solution. Therefore, we can usually obtain a kind of certainty about the classification result when we use a fuzzy classifier.

In this paper, we propose a general fuzzy classification-assisted environmental selection (FAES) strategy for MOEAs to reduce the number of objective evaluations. Our idea is to use solutions in the previous generations[1] as training dataset. Since the non-dominated solutions and the dominated solutions in MOEAs can generally be regarded as two classes, we use them as the two patterns of the training dataset to build a fuzzy classifier. Next, we use the model to predict the membership degree of each of the current and offspring solutions. Then, the membership degrees to the positive class of each offspring and one of its parents are compared. Only the offspring solutions with larger membership degrees are evaluated and combined with the current population to generate the new population. The offspring solutions with smaller membership degrees are discarded directly. In this manner, only the better offspring solutions are evaluated. Therefore, the number of objective evaluations is reduced.

The rest of this paper is summarized as follows. First, Sect. 2 presents the framework of the proposed FAES-based MOEA. The details of the training data definition and the employed fuzzy classifier are also presented. Then, Sect. 3 dedicates to a experimental study of the FAES strategy. Finally, Sect. 4 concludes this paper with some potential work in the future.

2 The Proposed FAES-Based MOEA

In this section, we propose a fuzzy classification-assisted environmental selection (FAES) scheme for MOEAs. First, Sect. 2.1 presents the general algorithm framework of FAES-based MOEAs. Next, Sect. 2.2 explains the training dataset definition strategy of the proposed FAES. Then, the employed fuzzy classifier is explained in Sect. 2.3. Finally, Sect. 2.4 proposes the FAES strategy.

2.1 Algorithm Framework

The FAES-based MOEA mainly has the following procedures. First, solutions in the previous generations are used as training patterns. Second, a fuzzy classifier is built by using the training dataset. Third, the offspring solutions are generated in the reproduction procedure. Fourth, the membership degree of each of current and offspring solutions is predicted by the model. Finally, all solutions are integrated into the FAES strategy to generate the new population. The algorithm framework of FAES-MOEA is presented in Algorithm 1. m_{x+} and m_{x-} represent the membership degrees to the positive class and the negative class of solution x, respectively. Likewise, m_{y+} and m_{y-} are the membership degrees to the positive class and the negative class of solution y.

2.2 Training Dataset

Current solutions are usually the best optimal solutions that MOEAs found so far. To build a classifier that can efficiently predict the quality of solutions,

[1] In this paper, if the current generation is the t-th generation, the previous generations can be defined as the last g generations (i.e., the $(t-g)$th generation, the $(t-g+1)$th generation,$\cdots$, the $(t-1)$th generation).

Algorithm 1: Framework of FAES-MOEA

1 Initialize the population $P = \{x^1, x^2, \cdots, x^N\}$, evaluate solutions in P;
2 Set $P_+ = NDS(P)$ and $P_- = P \backslash P_+$;
3 **while** *termination condition is not satisfied* **do**
4 Train a fuzzy classifier FC based on the P_+ and P_-;
5 Generate an offspring population $Q = \{y^1, \cdots, y^N\}$;
6 Predict the fuzzy membership degrees of each solution in P by $[m_{x+}, m_{x-}] = FC(x)$;
7 Predict the fuzzy membership degrees of each solution in Q by $[m_{y+}, m_{y-}] = FC(y)$;
8 Set $\{P, P_+, P_-\} = FAES(P, Q, N)$;
9 **end**
10 **return** P.

solutions in the previous generations (besides current solutions) are the best ones to be chosen as the training dataset. Therefore, we use solutions in the previous generations to build the classifier. The non-dominated solutions and dominated solutions in the previous generations are defined as two classes of the training dataset.

Let $P = NDS(T)$ be the non-dominated sorting scheme. It is used to distinguish non-dominated solutions from T. T is the set of solutions in the previous generations. P_+ is the positive class of the training dataset, and P_- is the negative class of the training dataset. The two classes are updated as,

$$P_+ = NDS(T),$$

and

$$P_- = T \backslash P_+.$$

It should be noted that P_+ contains non-dominated solutions in the previous generations, P_- contains dominated solutions in the previous generations.

2.3 Fuzzy Classifier

Fuzzy classifiers are a kind of methods that classify fuzzy data by precise methods [10]. Instead of restricting each sample to an exact class, a fuzzy classifier predicts the certainty of each sample to each class. This certainty is defined as a membership degree, which is obtained by the membership function. The membership function is learned from the training data automatically.

In this paper, we employ the Fuzzy-KNN (FKNN) classifier [9] that uses fuzzy similarity in the classification inference process. First, Fuzzy-KNN computes the fuzzy similarity between the testing sample and each training sample. Then, the K nearest neighbors of the testing sample are selected. Finally, the membership degree of the testing sample to each class is computed.

2.4 Fuzzy Classification-Assisted Environmental Selection

Let Q^E be the solution set that contains each offspring solution with better membership degree (with respect to the positive class) than one of its parent solutions. Let $P = ENS(Q, N)$ denote the environmental selection, N is the size of the population.

The proposed fuzzy classification-assisted environmental selection (FAES) strategy is presented in Algorithm 2. First, the membership degrees to the positive class of each offspring and one of its parents are compared. Second, the offspring solutions with better membership degrees are evaluated and stored in Q^E. Third, if Q^E is empty, all the offspring solutions are evaluated and stored in Q^E. Fourth, solutions in $Q^E \cup P$ are applied to the environmental selection to generate the new population. Finally, the training dataset P_+ is updated by the non-dominated solutions in $Q^E \cup P$, and P_- is updated by the dominated solutions.

Algorithm 2: $\{P, P_+, P_-\} = FAES(P, Q, N)$

```
1  for i = 1 : N do
2  |  if m^i_{y+} ≥ m^i_{x+} then
3  |  |  Evaluate y^i;
4  |  |  Q^E = Q^E ∪ y^i;
5  |  end
6  end
7  if Q = ∅ then
8  |  Evaluate all of y;
9  |  Q^E = Q;
10 end
11 P = ENS(Q^E ∪ P, N);
12 Set P+ = NDS(Q^E ∪ P) and P- = Q^E ∪ P\P+;
13 return {P, P+, P-}.
```

3 Experimental Study

This section studies the efficiency of the FAES strategy. In the following sections, Sect. 3.1 shows the experimental settings. Section 3.2 compares the performance of FAES with different training datasets. Section 3.3 studies the performance of FAES with different settings of K. Section 3.4 compares the performance of FAES-based MOEA and the original MOEA.

3.1 Experimental Settings

We use NSGA-II [5] as a basic MOEA for experiments. The DTLZ1–DTLZ7 and WFG1–WFG9 test problems [6,7] with two objectives are used as the benchmark problems. For each test problem, the number of decision variables is $n = 10$; each algorithm runs 30 times independently. The population size is $N = 100$. The maximum number of objective evaluations is 60,000. The other algorithm parameters are the same as in [5]. The inverted generational distance (IGD) [4] and hypervolume (HV) [23] indicators are used to evaluate the performance of each algorithm in the experiments.

3.2 Effects of Specifications of Training Datasets

Table 1. Means$_{std}$ of IGD and HV values obtained by FAES-NSGA-II with four training datasets on DTLZ1–DTLZ7, WFG1–WFG9.

		1G	3G	10G	50G
DTLZ1	IGD	2.13e-01$_{2.43e-01}$[4]	1.67e-01$_{2.46e-01}$[1]	1.67e-01$_{1.98e-01}$[2]	1.97e-01$_{2.67e-01}$[3]
	HV	1.44e+00$_{6.95e-06}$[3]	1.44e+00$_{2.92e-06}$[1]	1.44e+00$_{7.80e-06}$[2]	1.44e+00$_{1.09e-04}$[4]
DTLZ2	IGD	5.27e-03$_{3.14e-04}$[4]	5.20e-03$_{2.82e-04}$[3]	5.11e-03$_{1.58e-04}$[2]	5.05e-03$_{2.40e-04}$[1]
	HV	1.27e+00$_{1.59e-02}$[3]	1.27e+00$_{1.28e-02}$[2]	1.27e+00$_{1.09e-02}$[4]	1.28e+00$_{1.22e-02}$[1]
DTLZ3	IGD	6.76e-01$_{7.31e-01}$[1]	1.05e+00$_{9.47e-01}$[2]	1.12e+00$_{1.01e+00}$[3]	1.32e+00$_{1.05e+00}$[4]
	HV	1.44e+00$_{7.21e-06}$[1]	1.44e+00$_{4.39e-04}$[4]	1.44e+00$_{4.44e-04}$[3]	1.44e+00$_{1.96e-04}$[2]
DTLZ4	IGD	8.84e-02$_{1.39e-01}$[2]	1.35e-01$_{1.55e-01}$[4]	1.08e-01$_{1.37e-01}$[3]	6.84e-02$_{1.12e-01}$[1]
	HV	1.35e+00$_{2.92e-02}$[2]	1.35e+00$_{5.43e-02}$[4]	1.35e+00$_{3.79e-02}$[3]	1.35e+00$_{1.87e-02}$[1]
DTLZ5	IGD	5.17e-03$_{2.67e-04}$[4]	5.13e-03$_{2.03e-04}$[3]	5.06e-03$_{2.24e-04}$[1]	5.11e-03$_{2.56e-04}$[2]
	HV	1.27e+00$_{1.38e-02}$[1]	1.27e+00$_{1.52e-02}$[4]	1.27e+00$_{1.61e-02}$[3]	1.27e+00$_{1.10e-02}$[2]
DTLZ6	IGD	2.66e-02$_{1.78e-02}$[1]	2.86e-02$_{1.15e-02}$[3]	3.45e-02$_{1.70e-02}$[4]	2.82e-02$_{1.16e-02}$[2]
	HV	1.43e+00$_{2.20e-03}$[3]	1.43e+00$_{1.68e-03}$[2]	1.43e+00$_{1.40e-03}$[1]	1.43e+00$_{4.08e-03}$[4]
DTLZ7	IGD	3.45e-02$_{1.11e-01}$[3]	3.45e-02$_{1.11e-01}$[2]	1.98e-02$_{7.98e-02}$[1]	4.90e-02$_{1.33e-01}$[4]
	HV	1.38e+00$_{1.65e-02}$[3]	1.38e+00$_{1.68e-02}$[2]	1.38e+00$_{1.20e-02}$[4]	1.38e+00$_{1.95e-02}$[1]
WFG1	IGD	1.04e+00$_{8.31e-02}$[1]	1.07e+00$_{7.05e-02}$[3]	1.07e+00$_{8.02e-02}$[2]	1.09e+00$_{6.79e-02}$[4]
	HV	1.09e+00$_{2.62e-02}$[1]	1.09e+00$_{3.08e-02}$[3]	1.08e+00$_{3.76e-02}$[4]	1.09e+00$_{3.28e-02}$[2]
WFG2	IGD	1.29e-02$_{1.53e-03}$[1]	1.35e-02$_{1.36e-03}$[4]	1.31e-02$_{9.40e-04}$[3]	1.30e-02$_{1.30e-03}$[2]
	HV	1.16e+00$_{7.20e-03}$[2]	1.16e+00$_{7.85e-03}$[3]	1.16e+00$_{1.63e-02}$[1]	1.15e+00$_{7.15e-03}$[4]
WFG3	IGD	2.04e-01$_{3.37e-03}$[2]	2.03e-01$_{3.04e-03}$[1]	2.07e-01$_{6.67e-03}$[4]	2.04e-01$_{4.47e-03}$[3]
	HV	1.11e+00$_{1.01e-02}$[1]	1.10e+00$_{7.43e-03}$[3]	1.10e+00$_{1.01e-02}$[4]	1.11e+00$_{8.66e-03}$[2]
WFG4	IGD	1.40e-02$_{8.94e-04}$[4]	1.40e-02$_{7.17e-04}$[3]	1.37e-02$_{6.99e-04}$[2]	1.36e-02$_{7.75e-04}$[1]
	HV	8.66e-01$_{1.46e-02}$[2]	8.66e-01$_{1.68e-02}$[3]	8.62e-01$_{1.40e-02}$[4]	8.69e-01$_{1.40e-02}$[1]
WFG5	IGD	6.80e-02$_{2.84e-04}$[3]	6.80e-02$_{3.72e-04}$[4]	6.80e-02$_{4.33e-04}$[2]	6.78e-02$_{3.11e-04}$[1]
	HV	9.11e-01$_{1.70e-02}$[3]	9.11e-01$_{1.46e-02}$[4]	9.16e-01$_{1.89e-02}$[2]	9.18e-01$_{1.40e-02}$[1]
WFG6	IGD	8.24e-02$_{2.06e-02}$[3]	8.01e-02$_{2.80e-02}$[1]	8.29e-02$_{2.79e-02}$[4]	8.23e-02$_{2.51e-02}$[2]
	HV	9.40e-01$_{1.03e-02}$[4]	9.43e-01$_{1.34e-02}$[2]	9.43e-01$_{1.19e-02}$[3]	9.44e-01$_{1.24e-02}$[1]
WFG7	IGD	1.74e-02$_{1.03e-03}$[4]	1.73e-02$_{1.07e-03}$[2]	1.74e-02$_{1.07e-03}$[3]	1.71e-02$_{1.43e-03}$[1]
	HV	9.07e-01$_{1.59e-02}$[1]	9.05e-01$_{1.94e-02}$[3]	9.03e-01$_{2.19e-02}$[4]	9.07e-01$_{1.69e-02}$[2]
WFG8	IGD	3.03e-01$_{7.54e-03}$[3]	3.01e-01$_{1.10e-02}$[2]	2.98e-01$_{9.93e-03}$[1]	3.03e-01$_{1.98e-02}$[4]
	HV	9.38e-01$_{1.49e-02}$[3]	9.33e-01$_{2.15e-02}$[4]	9.41e-01$_{1.03e-02}$[2]	9.41e-01$_{1.25e-02}$[1]
WFG9	IGD	2.51e-01$_{3.83e-02}$[1]	2.58e-01$_{8.78e-05}$[2]	2.58e-01$_{1.05e-04}$[4]	2.58e-01$_{1.03e-04}$[3]
	HV	9.28e-01$_{2.76e-02}$[4]	9.32e-01$_{1.69e-02}$[3]	9.38e-01$_{1.26e-02}$[1]	9.37e-01$_{1.55e-02}$[2]
mean	IGD	2.56	2.50	2.56	2.38
rank	HV	2.31	2.94	2.81	1.94

This section investigates the performance of FAES with four settings of the training dataset size. The solutions in the previous generation, the three pervious generations, the ten previous generations, and the fifty previous generations (denoted as 1G, 3G, 10G, 50G) are chosen as the training patterns. The Fuzzy-KNN with $K = 7$ is chosen as a classifier. The other parts are the same as in Algorithm 1. The four training datasets are applied to FAES-NSGA-II and studied on the DTLZ and WFG test problems. The statistical IGD and HV values achieved by FCAS-NSGA-II with four settings of the training dataset size are shown in Table 1.

Table 2. Means$_{std}$ of IGD and HV values obtained by FAES-NSGA-II with four classifiers (Fuzzy-KNN with 4 K values) on DTLZ1–DTLZ7, WFG1–WFG9.

		K=1	K=3	K=5	K=7
DTLZ1	IGD	3.92e-01$_{3.32e-01}$[4]	2.87e-01$_{3.87e-01}$[3]	1.62e-01$_{1.71e-01}$[1]	1.97e-01$_{2.67e-01}$[2]
	HV	1.44e+00$_{1.75e-05}$[2]	1.44e+00$_{1.25e-04}$[4]	1.44e+00$_{2.01e-05}$[1]	1.44e+00$_{1.09e-04}$[3]
DTLZ2	IGD	5.19e-03$_{2.53e-04}$[4]	5.04e-03$_{1.94e-04}$[1]	5.16e-03$_{2.91e-04}$[3]	5.05e-03$_{2.40e-04}$[2]
	HV	1.27e+00$_{1.45e-02}$[4]	1.27e+00$_{1.24e-02}$[3]	1.27e+00$_{1.04e-02}$[2]	1.28e+00$_{1.22e-02}$[1]
DTLZ3	IGD	1.97e+00$_{1.48e+00}$[3]	2.17e+00$_{1.33e+00}$[4]	1.53e+00$_{1.61e+00}$[2]	1.32e+00$_{1.05e+00}$[1]
	HV	1.44e+00$_{2.04e-04}$[1]	1.44e+00$_{9.38e-04}$[4]	1.44e+00$_{3.60e-04}$[3]	1.44e+00$_{1.96e-04}$[2]
DTLZ4	IGD	7.81e-02$_{1.16e-01}$[3]	9.74e-02$_{1.39e-01}$[4]	6.50e-02$_{1.10e-01}$[1]	6.84e-02$_{1.12e-01}$[2]
	HV	1.34e+00$_{5.79e-02}$[4]	1.36e+00$_{3.23e-02}$[2]	1.37e+00$_{3.31e-02}$[1]	1.35e+00$_{1.87e-02}$[3]
DTLZ5	IGD	5.25e-03$_{3.02e-04}$[4]	5.05e-03$_{1.65e-04}$[1]	5.13e-03$_{2.51e-04}$[3]	5.11e-03$_{2.56e-04}$[2]
	HV	1.27e+00$_{1.06e-02}$[1]	1.27e+00$_{1.32e-02}$[3]	1.27e+00$_{1.34e-02}$[4]	1.27e+00$_{1.10e-02}$[2]
DTLZ6	IGD	1.32e-01$_{9.41e-02}$[4]	3.67e-02$_{3.21e-02}$[3]	3.37e-02$_{3.37e-02}$[2]	2.82e-02$_{1.16e-02}$[1]
	HV	1.43e+00$_{1.75e-03}$[3]	1.43e+00$_{2.39e-03}$[1]	1.43e+00$_{1.99e-03}$[2]	1.43e+00$_{4.08e-03}$[4]
DTLZ7	IGD	6.36e-02$_{1.51e-01}$[4]	4.89e-02$_{1.33e-01}$[1]	4.90e-02$_{1.33e-01}$[2]	4.90e-02$_{1.33e-01}$[3]
	HV	1.38e+00$_{2.21e-02}$[1]	1.38e+00$_{1.98e-02}$[4]	1.38e+00$_{1.95e-02}$[2]	1.38e+00$_{1.95e-02}$[3]
WFG1	IGD	1.19e+00$_{1.16e-01}$[4]	1.08e+00$_{7.71e-02}$[2]	1.07e+00$_{7.26e-02}$[1]	1.09e+00$_{6.79e-02}$[3]
	HV	1.10e+00$_{4.47e-02}$[1]	1.09e+00$_{3.26e-02}$[2]	1.09e+00$_{4.52e-02}$[4]	1.09e+00$_{3.28e-02}$[3]
WFG2	IGD	1.41e-02$_{1.31e-03}$[4]	1.33e-02$_{1.21e-03}$[3]	1.27e-02$_{1.06e-03}$[1]	1.30e-02$_{1.30e-03}$[2]
	HV	1.15e+00$_{7.03e-03}$[2]	1.15e+00$_{8.74e-03}$[3]	1.16e+00$_{8.60e-03}$[1]	1.15e+00$_{7.15e-03}$[4]
WFG3	IGD	2.06e-01$_{7.70e-03}$[4]	2.04e-01$_{4.02e-03}$[2]	2.03e-01$_{3.56e-03}$[1]	2.04e-01$_{4.47e-03}$[3]
	HV	1.10e+00$_{1.00e-02}$[4]	1.10e+00$_{7.31e-03}$[3]	1.10e+00$_{1.09e-02}$[2]	1.11e+00$_{8.66e-03}$[1]
WFG4	IGD	1.42e-02$_{8.48e-04}$[4]	1.38e-02$_{9.20e-04}$[3]	1.35e-02$_{6.96e-04}$[1]	1.36e-02$_{7.75e-04}$[2]
	HV	8.71e-01$_{1.52e-02}$[1]	8.63e-01$_{1.92e-02}$[4]	8.63e-01$_{1.33e-02}$[3]	8.69e-01$_{1.40e-02}$[2]
WFG5	IGD	6.81e-02$_{5.77e-04}$[4]	6.78e-02$_{3.08e-04}$[2]	6.80e-02$_{4.27e-04}$[3]	6.78e-02$_{3.11e-04}$[1]
	HV	9.12e-01$_{1.61e-02}$[4]	9.15e-01$_{1.51e-02}$[2]	9.13e-01$_{1.53e-02}$[3]	9.18e-01$_{1.40e-02}$[1]
WFG6	IGD	8.52e-02$_{3.06e-02}$[3]	8.34e-02$_{3.02e-02}$[2]	8.80e-02$_{2.65e-02}$[4]	8.23e-02$_{2.51e-02}$[1]
	HV	9.40e-01$_{1.49e-02}$[4]	9.42e-01$_{1.26e-02}$[2]	9.40e-01$_{1.30e-02}$[3]	9.44e-01$_{1.24e-02}$[1]
WFG7	IGD	1.78e-02$_{1.08e-03}$[4]	1.66e-02$_{1.21e-03}$[1]	1.66e-02$_{1.27e-03}$[2]	1.71e-02$_{1.43e-03}$[3]
	HV	9.06e-01$_{2.04e-02}$[3]	9.08e-01$_{2.09e-02}$[1]	9.06e-01$_{2.31e-02}$[4]	9.07e-01$_{1.69e-02}$[2]
WFG8	IGD	3.09e-01$_{1.14e-02}$[4]	3.05e-01$_{5.81e-03}$[3]	3.00e-01$_{1.11e-02}$[1]	3.03e-01$_{1.98e-02}$[2]
	HV	9.21e-01$_{2.12e-02}$[4]	9.33e-01$_{1.85e-02}$[3]	9.39e-01$_{2.23e-02}$[2]	9.41e-01$_{1.25e-02}$[1]
WFG9	IGD	2.58e-01$_{1.97e-04}$[4]	2.52e-01$_{3.71e-02}$[1]	2.58e-01$_{1.79e-04}$[3]	2.58e-01$_{1.03e-04}$[2]
	HV	9.38e-01$_{2.08e-02}$[2]	9.33e-01$_{2.78e-02}$[4]	9.39e-01$_{1.57e-02}$[1]	9.37e-01$_{1.55e-02}$[3]
mean	IGD	3.81	2.25	1.94	2
rank	HV	2.56	2.81	2.38	2.25

Table 1 shows the mean and std (standard deviation) of IGD and HV values obtained by FAES-NSGA-II with each of the four training datasets on the 16 test problems after 60,000 objective evaluations. The rank values of the four training datasets on each test problem are also presented in Table 1. Table 1 clearly shows that FAES-NSGA-II with each training dataset can get the best IGD or HV values. For IGD values, FAES-NSGA-II with 50G and 1G get 5 best values; 3G and 10G get 3 best results. For HV, 50G obtains 7 best results; 1G gets 5 best results; 10G gets 3 best results, and 3G gets 1 best result. In terms of mean rank values, FAES-NSGA-II with 50G is the best both for IGD values and HV values. The results suggest that 50G (i.e., the use of the solutions in the previous 50 generations) is the best setting of the training dataset for FAES-NSGA-II.

Based on the above results, we employ solutions in the fifty previous generations as the training dataset in the following experiments.

3.3 Effects of the Settings of K

This section investigates the performance of FAES with different values of K in the Fuzzy-KNN. The four settings of K ($K = 1, 3, 5, 7$) are examined. The other parts are the same as in Algorithm 1. The four classifiers are applied to FAES-NSGA-II and studied on the DTLZ and WFG test problems. The statistical IGD and HV values achieved by these algorithms are shown in Table 2.

Table 2 shows the mean and std (standard deviation) of IGD and HV values achieved by FAES-NSGA-II with each of the four values of K on the 16 test instances after 60,000 objective evaluations. The rank of each classifier for each test problem is also shown for each performance indicator in Table 2. The mean rank value of each classifier over all test problems for each performance indicator is shown at the bottom of Table 2. Table 2 clearly shows that there does not exist a classifier that always outperforms the other classifiers. However, with respect to the mean rank values, we can see that the order from the best to the worst for IGD is $K = 5$, $K = 7$, $K = 3$, and $K = 1$. For HV, the order from the best to the worst is $K = 7$, $K = 5$, $K = 1$, and $K = 3$. Thus, FAES-NSGA-II with $K = 5$ and $K = 7$ works better than FAES-NSGA-II with the other two values with respect to the mean rank values over the 16 test problems.

Based on the above analysis, we can conclude that there is no large differences in the performance of FAES-NSGA-II among the four settings of K. Considered the experimental results in Table 2, in the following section, $K = 7$ is chosen as the default classifier in FAES.

3.4 Performance Comparison with the Original MOEA

To investigate the benefits of the proposed FAES strategy over the original MOEA, this section studies the comparison between FAES-NSGA-II and the original NSGA-II.

Table 3 presents the median, mean, and std (standard deviation) of objective evaluations used by the two compared algorithms when achieving the same IGD and HV values on 16 test problems. It should be noted that the target is the value that both of FAES-NSGA-II and NSGA-II can achieve. The Wilcoxon rank-sum test is employed to compare the number of objective evaluations taken by the two compared algorithms. "+, −, ∼" denote the FAES-based algorithm is better than, worse than, or similar to the original algorithm at the 95% significance level. Table 3 shows that FAES-NSGA-II achieves the same IGD value with fewer objective evaluations than NSGA-II on 12 test problems. On DTLZ3, DTLZ4, WFG1, and WFG9, the two algorithms required similar number of objective

Table 3. Median, mean, std of objective evaluations required by FAES-NSGA-II and NSGA-II when achieving the same IGD and HV values on DTLZ1–DTLZ7, WFG1–WFG9.

	metrics	values	median	mean	std	median	mean	std
			FAES-NSGA-II			NSGA-II		
DTLZ1	IGD	2.00e+00	3.20e+04	3.28e+04(+)	6.62e+03	3.80e+04	3.92e+04	5.25e+03
	HV	1.43e+00	2.61e+03	2.71e+03(+)	5.84e+02	2.80e+03	3.79e+03	4.60e+03
DTLZ2	IGD	6.00e-03	8.75e+03	1.48e+04(+)	1.48e+04	1.46e+04	2.29e+04	1.67e+04
	HV	1.25e+00	1.64e+03	3.23e+03(+)	7.83e+03	2.90e+03	7.70e+03	1.43e+04
DTLZ3	IGD	6.00e+00	3.66e+04	3.51e+04(∼)	7.59e+03	3.77e+04	3.69e+04	9.33e+03
	HV	1.43e+00	2.64e+03	2.68e+03(+)	5.48e+02	3.20e+03	3.18e+03	6.44e+02
DTLZ4	IGD	6.00e-01	4.23e+03	4.51e+03(∼)	1.43e+03	4.70e+03	4.91e+03	1.20e+03
	HV	1.20e+00	4.13e+03	4.19e+03(+)	6.95e+02	4.60e+03	6.58e+03	9.23e+03
DTLZ5	IGD	7.00e-02	2.11e+03	2.07e+03(+)	3.16e+02	2.60e+03	2.79e+03	5.01e+02
	HV	1.20e+00	1.00e+03	9.63e+02(+)	2.05e+02	1.10e+03	1.14e+03	2.74e+02
DTLZ6	IGD	2.00e-01	3.58e+04	3.61e+04(+)	1.06e+03	4.52e+04	4.52e+04	1.25e+03
	HV	1.40e+00	1.96e+04	1.97e+04(+)	6.25e+02	2.51e+04	2.57e+04	4.25e+03
DTLZ7	IGD	5.00e-01	3.46e+03	3.76e+03(+)	1.20e+03	3.90e+03	4.40e+03	1.62e+03
	HV	1.35e+00	5.17e+03	5.02e+03(+)	1.09e+03	6.30e+03	6.89e+03	5.55e+03
WFG1	IGD	1.30e+00	2.01e+04	2.22e+04(∼)	8.15e+03	2.35e+04	2.29e+04	5.26e+03
	HV	1.00e+00	3.46e+04	3.49e+04(+)	3.43e+03	3.97e+04	4.02e+04	5.92e+03
WFG2	IGD	2.00e-02	1.52e+04	1.67e+04(+)	4.72e+03	1.87e+04	1.89e+04	3.05e+03
	HV	1.10e+00	4.69e+03	5.31e+03(+)	2.20e+03	5.60e+03	5.86e+03	2.22e+03
WFG3	IGD	2.50e-01	5.63e+03	5.89e+03(+)	1.35e+03	7.30e+03	7.64e+03	1.53e+03
	HV	1.05e+00	4.04e+03	4.24e+03(+)	1.02e+03	5.50e+03	5.97e+03	1.54e+03
WFG4	IGD	2.00e-02	1.39e+04	1.43e+04(+)	2.45e+03	1.89e+04	1.95e+04	3.15e+03
	HV	8.00e-01	2.51e+03	2.49e+03(+)	6.08e+02	3.20e+03	3.33e+03	1.07e+03
WFG5	IGD	7.00e-02	1.19e+04	1.42e+04(+)	7.95e+03	2.48e+04	2.55e+04	6.99e+03
	HV	8.70e-01	3.42e+03	5.46e+03(+)	9.90e+03	4.30e+03	5.71e+03	5.81e+03
WFG6	IGD	2.00e-01	3.42e+03	3.51e+03(+)	5.82e+02	4.60e+03	4.60e+03	7.98e+02
	HV	9.00e-01	5.40e+03	5.75e+03(+)	1.77e+03	6.60e+03	9.57e+03	7.72e+03
WFG7	IGD	3.00e-02	5.80e+03	6.06e+03(+)	7.39e+02	9.20e+03	9.23e+03	9.06e+02
	HV	8.50e-01	3.03e+03	3.14e+03(+)	7.12e+02	3.60e+03	3.71e+03	8.03e+02
WFG8	IGD	4.50e-01	5.58e+03	5.63e+03(+)	2.14e+03	6.00e+03	6.42e+03	1.30e+03
	HV	8.50e-01	5.70e+03	6.45e+03(+)	2.29e+03	7.00e+03	7.57e+03	2.98e+03
WFG9	IGD	3.00e-01	1.63e+03	1.68e+03(∼)	2.98e+02	1.80e+03	1.87e+03	4.35e+02
	HV	8.90e-01	1.59e+03	1.72e+03(∼)	8.53e+02	1.60e+03	1.86e+03	1.33e+03
+/-/∼	IGD			12/0/4				
	HV			15/0/1				

evaluations when achieving the same IGD value. FAES-NSGA-II achieves the same HV value with fewer objective evaluations than NSGA-II on 15 test problems, and they need similar number of objective evaluations on WFG9.

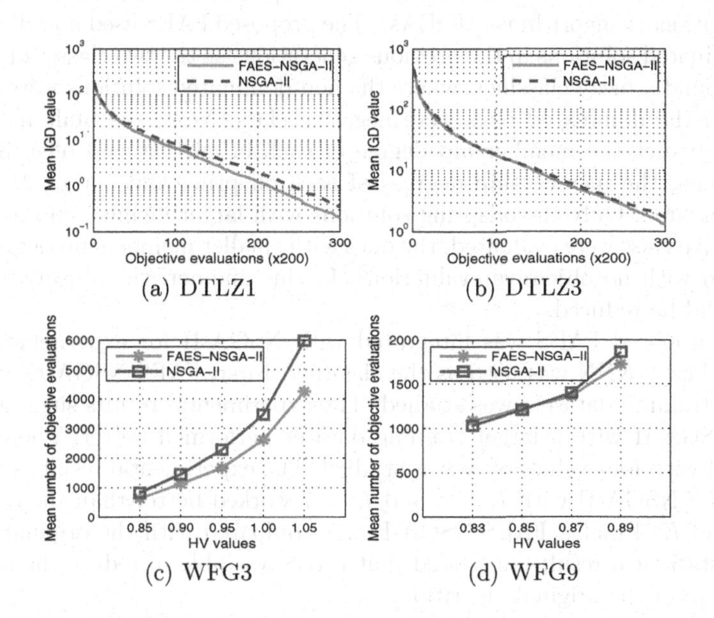

Fig. 1. The IGD value achieved by FAES-NSGA-II and NSGA-II when taking the same number of objective evaluations on DTLZ1 and DTLZ3 are in (a)(b). The mean number of objective evaluations needed by FAES-NSGA-II and NSGA-II when achieving the same HV values on WFG3 and WFG9 are in (c)(d).

Figure 1(a)(b) show the statistical results of mean IGD values versus the number of objective evaluations of FAES-NSGA-II and NSGA-II on DTLZ1 and DTLZ3. Figure 1(c)(d) present the statistical results of the number of objective evaluations versus different HV values of FAES-NSGA-II and NSGA-II on WFG3 and WFG9. Figure 1(a)(b) suggest that FAES-NSGA-II converges faster than NSGA-II on DTLZ1, and the two algorithms perform similarly on DTLZ3. Figure 1(c)(d) show that, for WFG3, FAES-NSGA-II achieves the target with fewer objective evaluations than NSGA-II. For WFG9, NSGA-II takes fewer objective evaluations on the first two targets, while FAES-NSGA-II takes fewer for the last two targets. The results in Table 3 and Fig. 1 are consistent. We can conclude that, for achieving the same IGD and HV values, FAES-NSGA-II usually takes fewer objective evaluations than NSGA-II on most test problems.

Studies in this section show that the FAES strategy is able to assist NSGA-II to reduce the number of objective evaluations. This indicates that the fuzzy classifier can successfully predict the quality of solutions and FAES can help to discard the unpromising solutions as early as possible.

4 Conclusion

This paper proposed a fuzzy classification-assisted environmental selection (FAES) strategy for reducing the number of objective evaluations of multiobjective evolutionary algorithms (MOEAs). The proposed FAES used non-dominated and dominated solutions in the previous generations as the two classes of training data to build a fuzzy classifier, where the non-dominated solutions were positive class and the dominated ones were negative class. Next, the built model was used to predict the membership degree of each of current and offspring solutions. Then, the membership degrees of each offspring and one of its parents were compared. Only the offspring solutions with larger membership degrees to the positive class were evaluated, the ones with smaller membership degrees were discarded with no objective evaluations. In this manner, the objective evaluations could be reduced.

The proposed FAES was integrated with NSGA-II for experimental studies. The Fuzzy-KNN was used as the classifier. First, FAES-NSGA-II with four kinds of training datasets was studied. The experimental results suggested that FAES-NSGA-II with a larger training dataset performed better. Then, FAES-NSGA-II with four values of K was studied. The experimental results suggested that FAES-NSGA-II with $K = 5$ and $K = 7$ worked better than the other two settings of K. Finally, FAES-NSGA-II was compared with the original NSGA-II. The statistical results suggested that FAES was able to reduce the objective evaluations of the original algorithm.

This paper demonstrated the efficiency of the proposed FAES on reducing the number of objective evaluations in MOEAs. The effectiveness of the proposed strategy can be examined for other MOEAs. The use of other classifiers together with other settings of training data should be further addressed. In the future, we will also apply the proposed strategy to solve other kinds of problems, such as combinatorial optimization problems, and real-world application problems.

Acknowledgement. This work was supported by National Natural Science Foundation of China (Grant No. 61876075), Guangdong Provincial Key Laboratory (Grant No. 2020B121201001), the Program for Guangdong Introducing Innovative and Enterpreneurial Teams (Grant No. 2017ZT07X386), Shenzhen Science and Technology Program (Grant No. KQTD2016112514355531), the Program for University Key Laboratory of Guangdong Province (Grant No. 2017KSYS008).

References

1. Bandaru, S., Ng, A.H., Deb, K.: On the performance of classification algorithms for learning pareto-dominance relations. In: Proceedings of 2014 IEEE Congress on Evolutionary Computation, Beijing, China, pp. 1139–1146. IEEE, July 2014
2. Beume, N., Naujoks, B., Emmerich, M.: SMS-EMOA: multiobjective selection based on dominated hypervolume. Eur. J. Oper. Res. **181**(3), 1653–1669 (2007)
3. Bhattacharjee, K.S., Ray, T.: Selective evaluation in multiobjective optimization: a less explored avenue. In: Proceedings of 2015 IEEE Congress on Evolutionary Computation, Sendai, Japan, pp. 1893–1900. IEEE, May 2015

4. Coello Coello, C.A., Reyes Sierra, M.: A study of the parallelization of a coevolutionary multi-objective evolutionary algorithm. In: Monroy, R., Arroyo-Figueroa, G., Sucar, L.E., Sossa, H. (eds.) MICAI 2004. LNCS (LNAI), vol. 2972, pp. 688–697. Springer, Heidelberg (2004). https://doi.org/10.1007/978-3-540-24694-7_71
5. Deb, K., Pratap, A., Agarwal, S., Meyarivan, T.: A fast and elitist multiobjective genetic algorithm: NSGA-II. IEEE Trans. Evol. Comput. **6**(2), 182–197 (2002)
6. Deb, K., Thiele, L., Laumanns, M., Zitzler, E.: Scalable test problems for evolutionary multiobjective optimization. In: Abraham, A., Jain, L., Goldberg, R. (eds.) Evolutionary Multiobjective Optimization, pp. 105–145. Springer, London (2005). https://doi.org/10.1007/1-84628-137-7_6
7. Huband, S., Barone, L., While, L., Hingston, P.: A scalable multi-objective test problem toolkit. In: Coello Coello, C.A., Hernández Aguirre, A., Zitzler, E. (eds.) EMO 2005. LNCS, vol. 3410, pp. 280–295. Springer, Heidelberg (2005). https://doi.org/10.1007/978-3-540-31880-4_20
8. Jin, Y.: A comprehensive survey of fitness approximation in evolutionary computation. Soft Comput. **9**(1), 3–12 (2003)
9. Keller, J.M., Gray, M.R., Givens, J.A.: A fuzzy k-nearest neighbor algorithm. IEEE Trans. Syst. Man Cybern. **4**, 580–585 (1985)
10. Klir, G., Yuan, B.: Fuzzy Sets and Fuzzy Logic. Prentice Hall, Upper Saddle River (1995)
11. Lin, X., Zhang, Q., Kwong, S.: A decomposition based multiobjective evolutionary algorithm with classification. In: Proceedings of 2016 IEEE Congress on Evolutionary Computation, Vancouver, Canada, pp. 3292–3299. IEEE, July 2016
12. Loshchilov, I., Schoenauer, M., Sebag, M.: A mono surrogate for multiobjective optimization. In: Proceedings of Genetic and Evolutionary Computation Conference, Portland, USA, pp. 471–478. ACM, July 2010
13. Murata, T., Ishibuchi, H.: MOGA: multi-objective genetic algorithms. In: Proceedings of 1995 IEEE Congress on Evolutionary Computation, Perth, Australia, pp. 289–294. IEEE, November 1995
14. Pan, L., Cheng, H., Tian, Y., Wang, H., Zhang, X., Jin, Y.: A classification-based surrogate-assisted evolutionary algorithm for expensive many-objective optimization. IEEE Trans. Evol. Comput. **23**(1), 74–88 (2019)
15. Tabatabaei, M., Hakanen, J., Hartikainen, M., Miettinen, K., Sindhya, K.: A survey on handling computationally expensive multiobjective optimization problems using surrogates: non-nature inspired methods. Struct. Multi. Optim. **52**(1), 1–25 (2015)
16. Zhang, J., Zhou, A., Tang, K., Zhang, G.: Preselection via classification: a case study on evolutionary multiobjective optimization. Inf. Sci. **465**, 388–403 (2018)
17. Zhang, J., Zhou, A., Zhang, G.: A classification and Pareto domination based multiobjective evolutionary algorithm. In: Proceedings of 2015 IEEE Congress on Evolutionary Computation, Sendai, Japan, pp. 2883–2890. IEEE, May 2015
18. Zhang, J., Zhou, A., Zhang, G.: A multiobjective evolutionary algorithm based on decomposition and preselection. In: Gong, M., Pan, L., Song, T., Tang, K., Zhang, X. (eds.) BIC-TA 2015. CCIS, vol. 562, pp. 631–642. Springer, Heidelberg (2015). https://doi.org/10.1007/978-3-662-49014-3_56
19. Zhang, Q., Li, H.: MOEA/D: a multiobjective evolutionary algorithm based on decomposition. IEEE Trans. Evol. Comput. **11**(6), 712–731 (2007)
20. Zhou, A., Qu, B.Y., Li, H., Zhao, S.Z., Suganthanb, P.N., Zhang, Q.: Multiobjective evolutionary algorithms: a survey of the state of the art. Swarm Evol. Comput. **1**, 32–49 (2011)

21. Zitzler, E., Künzli, S.: Indicator-based selection in multiobjective search. In: Yao, X., et al. (eds.) PPSN 2004. LNCS, vol. 3242, pp. 832–842. Springer, Heidelberg (2004). https://doi.org/10.1007/978-3-540-30217-9_84
22. Zitzler, E., Laumanns, M., Thiele, L.: SPEA2: improving the strength Pareto evolutionary algorithm. Technical Report 103, Computer Engineering and Networks Laboratory (TIK), Swiss Federal Institute of Technology (ETH) Zurich, Gloriastrasse 35, CH-8092 Zurich, Switzerland (2001)
23. Zitzler, E., Thiele, L.: Multiobjective optimization using evolutionary algorithms — a comparative case study. In: Eiben, A.E., Bäck, T., Schoenauer, M., Schwefel, H.-P. (eds.) PPSN 1998. LNCS, vol. 1498, pp. 292–301. Springer, Heidelberg (1998). https://doi.org/10.1007/BFb0056872

Surrogate-Assisted Multi-objective Particle Swarm Optimization for Building Energy Saving Design

Xiao-ke Liang, Yong Zhang(✉), and Dun-wei Gong

Information and Control Engineering, China University of Mining and Technology, Xuzhou, China
yongzh401@126.com

Abstract. Evolutionary optimization has been successfully used in building energy saving design due to its global search capability. However, existing algorithms generally need high computation cost because they need to evaluate individuals/solutions by a time-consuming energy consumption software. In view of this, this paper proposes a surrogate-assisted multi-objective particle swarm optimization (PSO) method for building energy efficiency design. Firstly, a management strategy of surrogate model is developed to effectively update the surrogate model; then, a multi-objective evolutionary simulation platform for the energy saving design of buildings is established by integrating the proposed multi-objective PSO algorithm with the building energy simulation software, EnergyPlus. The proposed algorithm is applied in an office building at Beijing of China. Experimental results show that it can obtain highly competitive solutions on the basis of significantly reducing the running cost.

Keywords: Particle swarm optimization · Building energy efficiency · Multi-objective · Surrogate model

1 Introduction

Energy is one of the most important resources of society. With the development of industry, energy consumption continues to increase, and the importance of reducing energy consumption has become increasingly prominent [1–3]. Among the total energy consumption, the construction industry spends about 40% of the total energy consumption. And, 57% of the energy consumption comes from heating, ventilation, air conditioning, and lighting. Therefore, improving the energy efficiency of buildings has become an international issue [4].

In essence, the problem of building energy saving design is a typical multi-objective optimization one, which includes two conflicting performance indexes at least, i. e., building energy consumption and users discomfort. Due to the development of multi-objective evolutionary computation (EC) technology, multi-objective building energy saving design has attracted the attention of

H. Ishibuchi et al. (Eds.): EMO 2021, LNCS 12654, pp. 593–604, 2021.
https://doi.org/10.1007/978-3-030-72062-9_47

researchers in recent years. Typical evolutionary optimization technologies such as NSGA-II [5] and multi-objective artificial bee colony [6] have been used in this kind of problems. However, since the need to evaluate individuals by using expensive building energy consumption software, these algorithms still have the disadvantage of high computation cost [7]. This seriously limits the application of EC-based building energy saving design method.

Surrogate-assisted evolutionary algorithms (SAEAs) were proposed in the 1980s [8]. Since the computation cost of constructing surrogate model is much lower than the evaluation cost of real objective function, this kind of algorithm can significantly save the computational cost of EA when solving expensive optimization problems. To solve expensive single-objective optimization problems, many effective SAEA algorithms have been proposed [9–12]. In recent years, surrogate-assisted multi-objective EAs have also been paid more and more attention by scholars. Zhang et al. [13] used the Kriging surrogate model to accelerate the search speed of NSGA-II, and presented a multi-objective evolutionary algorithm for the design of twin-web disk. Rosales-Prez et al. [14] trained support vector regression machine by using all non-dominated solutions, and proposed a multi-objective EA for SVM model selection. However, the computational cost of these algorithms is still relatively high because they need to calculate the real function values of a large number of non-dominated solutions.

In order to reduce the computational cost of EAs in dealing with the problem of building energy saving design, this paper proposes a surrogate-assisted multi-objective PSO method. The innovations of this paper are as follows: 1) Proposing a surrogate-assisted multi-objective PSO method for building energy saving design problems for the first time; 2) Establishing a multi-objective evolutionary simulation platform for energy efficiency design of buildings, By combing the proposed multi-objective PSO algorithm with the software, EnergyPlus.

2 Related Work

2.1 Multi-objective Optimization

Multi-objective optimization problems exist in the real world widely [15]. First, we give some key definitions.

Definition 1: Multi-objective optimization problem (MOP). Taking a minimization problem as an example, a MOP problem can be defined as,

$$\begin{aligned} minF(X) &= (f_1(X), f_2(X), ..., f_M(X)) \\ s.t.h_j(X) &= (\leq)0, j = 1, 2, ..., J \\ X = &(x_1, x_2, ..., x_n), x_i^L \leq x_i \leq x_i^U, i = 1, 2, ..., n. \end{aligned} \tag{1}$$

Where, $F(X)$ is the objective function or performance index, $h_j(X)$ represents the $j-th$ equality/inequality constraint. X is the decision variable, x_i^L and x_i^U are the upper and lower bounds of the $i-th$ variable.

Definition 2: Pareto dominance. A vector $Z = (z_1, z_2, ..., z_M)$ is said to dominate $Y = (y_1, y_2, ..., y_M)$, if

$$\forall i \in 1,2,...,M, z_i \leq y_i, and \exists j \in 1,2,...,M, z_i < y_i \tag{2}$$

and mark it as $Z \prec Y$.

Definition 3: Pareto optimal solution. For the problem $F(X)$ with the feasible region of Ψ, a solution $X^* \in \Psi$ is Pareto optimal, if there is no solution X satisfying:

$$\forall i \in 1,2,...,M, f_i(X) \leq f_i(X^*), and \exists j \in 1,2,...,M, f_i(X) < f_i(X^*) \tag{3}$$

Furthermore, all Pareto optimal solutions of the problem constitute the Pareto optimal solution set (PS); the objective values corresponding to PS constitute the Pareto-optimal front of the problem (PF).

2.2 Particle Swarm Optimization

PSO is inspired by the swarm behaviors of birds looking for food [16,17]. In each iteration, each particle updates its position by learning two empirical knowledge. One is the optimal position found by the particle itself, that is, the personal leader; another is the optimal position currently found by its neighborhoods, that is, the global leader. Supposing that the $i-th$ particle is $X_i = (x_{i,1}, x_{i,2}, ..., x_{i,N})$, the global and personal leaders are $Gb = (gb_1, gb_2, ..., gb_N)$ and $Pb_i = (pb_{i,1}, pb_{i,2}, ..., pb_{i,N})$ respectively, the particle is updated as follows:

$$v_{i,j}(t+1) = \omega v_{i,j}(t) + c_1 r_1 (pb_{i,j}(t) - x_{i,j}(t)) + c_2 r_2 (gb_j(t) - x_{i,j}(t)) \tag{4}$$

$$x_{i,j}(t+1) = x_{i,j}(t) + v_{i,j}(t+1)) \tag{5}$$

Where, t is the number of iterations, ω is the inertia weight, c_1 and c_2 are the two learning factors, r_1 and r_2 are the random numbers within [0, 1].

3 The Proposed Surrogate-Assisted Multi-objective PSO Method

In this section, a surrogate-assisted multi-objective PSO method (Sa-MOPSO) is presented to solve the problem of building energy saving design. First, the problem of multi-objective building energy efficiency design is introduced; next, Sa-MOPSO is developed; finally, the implementation strategy of Sa-MOPSO and corresponding simulation platform are introduced.

3.1 Description of Multi-objective Building Energy Saving Design

Building energy consumption and users discomfort are the two main indicators which must be considered in building energy saving design. Generally, the reduction of building energy consumption will inevitably lead to the increase of users discomfort, so these two indicators are conflicting with each other.

In this section, the problem of building energy saving design is described a multi-objective optimization one based on the two indicators, i.e., the building's annual energy consumption (BEC) and the users discomfort hours (UDH). Generally, it is difficult to guarantee the practicability of the built mathematical model, because the factors affecting energy consumption vary with the building type or climate. In order to simplify the modeling process of engineers and ensure the accuracy and practicability of built models, researchers have successively developed different building design simulation software, among which EnergyPlus is the most representative and widely used one. Similar to the literature [5–7], this paper uses EnergyPlus to generate the multi-objective optimization model of the problem. After setting the values of these decision variables, the software EnergyPlus will output the values of BEC and UDH in one year.

3.2 The Proposed Sa-MOPSO

Figure 1 shows the basic framework of the proposed Sa-MOPSO. The framework mainly consists of two parts: surrogate-based particle update and surrogate model management. Similar to traditional multi-objective PSO algorithms, the main function of the first part is to update the particles and save non-dominated solutions generated during the iterations; the main difference from traditional algorithms is that it uses a surrogate model to estimate the fitness of new particles, replacing the real objective evaluation function. This part also uses an archive set to store non-dominated solutions generated during the iterations, and we use the archive set to update the global leader (Gb) of each particle. In addition, another important function of the archive set is to provide new samples for updating the surrogate model. In the iterative process of swarm, the archive set is updated according to the method in [18].

The main purpose of the surrogate model management is to update the surrogate model. The management strategy is as follows: Select Pareto optimal solutions in the archive set as representative solutions; Use the building simulation software, EnergyPlus, to really evaluate these solutions, and generate new samples by these solutions and their real function values; add these new samples into the sample set, *Data*, and use *Data* to retrain the surrogate model. When the algorithm is terminated, Pareto optimal solutions in *Data* are selected as the final result.

1) Management of surrogate model

Firstly, a specified number of initial samples are generated based on Latin hypercube sampling, and then these samples are used to construct the initial surrogate model. According to the suggestion in [19], the number of samples

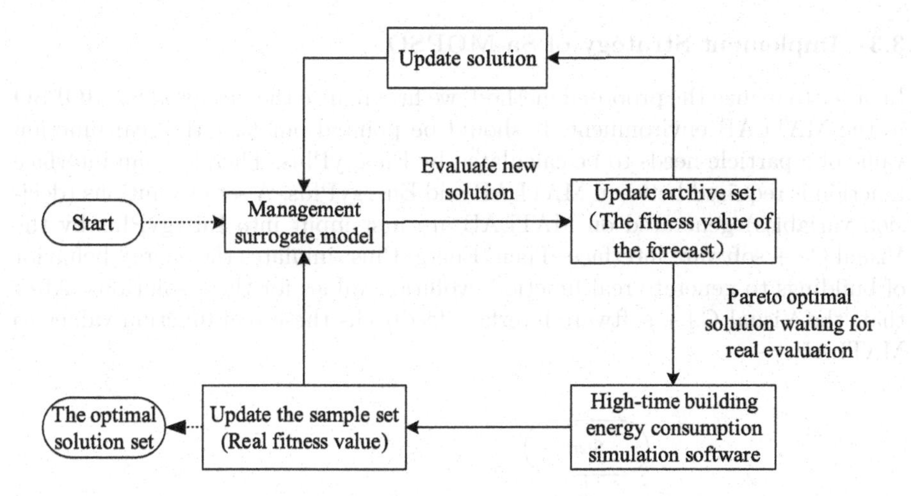

Fig. 1. The framework of Sa-MOPSO.

required for the initial surrogate model is set to be $q = \frac{1}{2}(n+1)(n+2)$, where n is the dimension of decision variable. These initial samples also are saved into the sample set, *Data*.

In order to maintain or improve the accuracy of the surrogate model, it is necessary to continuously update the model in the evolution process. The specific management method of surrogate model is as follows. Every T_0 iterations, all new Pareto solutions are selected from the archive set and evaluated by using the real objective function; after adding these solutions into the sample set, *Data*, the surrogate model is retraining by using Data. Here, the parameter T_0 is used to control the update frequency of the surrogate model.

2) Update of particle position

Traditional PSO methods are not sensitive to the inertial weight and learning factors, because they need to rely on them to control the global exploration and local exploitation of swarm. In view of this, a bare-bones particle updating strategy with less control parameters is proposed in the reference [18]. This strategy uses a Gaussian distribution about the global and local leaders to update particle, without setting the above control parameters. Specifically, the update formula is as follows:

$$x_{i,j}(t+1) = \begin{cases} N\left(\frac{r_3 \times pb_{i,j}(t) + (1-r_3) \times gb_{i,j}(t)}{2}, \left|pb_{i,j}(t) - gb_{i,j}(t)\right|\right), & \text{if} U(0,1) < 0.5 \\ gb_{i,j}(t), & \text{otherwise} \end{cases} \quad (6)$$

Where, $N(a, b)$ is the Gaussian distribution function with the mean a and the variance b; r_3 is a random number within [0,1].

3.3 Implement Strategy of Sa-MOPSO

In order to realize the proposed method, we first realize the proposed Sa-MOPSO in the MATLAB environment. It should be pointed out that the real function value of a particle needs to be calculated by EnergyPlus. Therefore, an interface function is required between MATLAB and EnergyPlus. A set of solutions (decision variables) generated on MATLAB are first input into EnergyPlus by the Visual C++ software interface. Then, EnergyPlus simulates the energy behavior of buildings to generate real function evolution values for these solutions. After that, the Visual C++ software interface feedbacks these real function values to MATLAB.

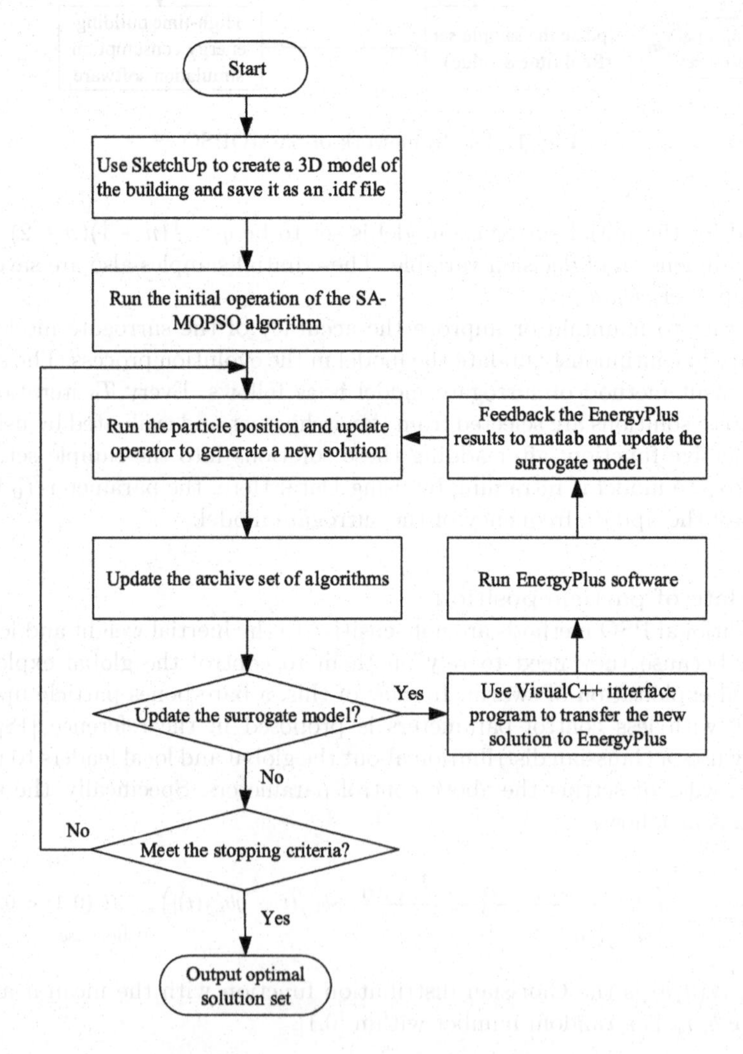

Fig. 2. The implement of Sa-MOPSO.

Figure 2 shows the execution framework of the proposed algorithm. Firstly, we use the software SketchUp to draw the 3D model of optimized building; Secondly, run the initial operation of Sa-MOPSO on MATLAB to generate the initial swarm and the initial sample set, and use these initial samples to construct the initial surrogate model; Next, update the positions of particles, and use the method in [20] to update the archive set; Then, determine whether the surrogate model needs to be updated. If yes, transfer new solutions from the archive set to EnergyPlus, and output the real function value of each new solution, i.e., the building energy consumption values and the discomfort hours. After MATLAB reads these real function values, these new solutions will be used to update the sample set *Data* and retrain the surrogate model. After that, the particle update operator in Sa-MOPSO continues to generate new solutions. Cycle through the above steps until the algorithm stops. Finally, Pareto optimal solutions in *Data* are the final result of the algorithm. Note that, in order to further improve the accuracy of the final output, in the last T_1 iterations the proposed Sa-MOPSO will evaluate all new particles by the real objective function, i.e., EnergyPlus.

4 Application Cases

In this experiment, we take the office building at Beijing as an example. As the most high-rise office building area and densely populated area in China, Beijing's office buildings are typical across the country [20].

The software SketchUp is used to create the architectural shape of optimized office room. The benchmark style of the office model is showed in Fig. 3. The length, width and height are 8.8 m, 3.6 m and 3.9 m, respectively. The initial length and height of the windows are 1.7 m and 1.6 m. The office model uses embedded lighting.

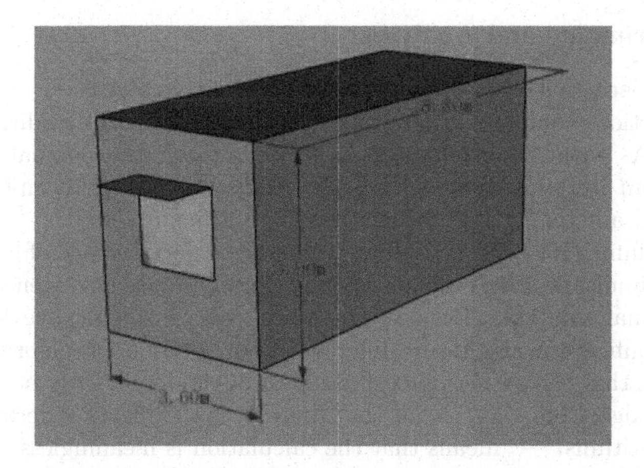

Fig. 3. Outline of the office room at Beijing.

4.1 Decision Variables

In this article, we consider 12 coefficients (or decision variables) including building orientation, window length, window width, glazing u-factor, glazing solar heat gain coefficient, thickness, wall solar absorptance, personnel density, lighting power density, equipment power density, heating setpoint temperature and cooling setpoint temperature. Table 1 gives the ranges of these variables.

Table 1. Decision variables and their ranges.

Decision variables	Unit	Range	Reference value
Building orientation	°	[0,360)	0
Window length	m	(0,3.6)	1.7
Window height	m	(0,3.9)	1.6
Glazing u-factor	$\omega/(m^2 \cdot k)$	(2,6)	4.3
Glazing solar heat gain coefficient	–	(0,0.7)	0.65
Thickness	m	(0,0.1)	0.1
Wall solar absorptance	–	(0.1,1)	0.6
Personnel density	–	(0.1,1)	0.2
Lighting power density	ω/m^2	[6,12]	9
Equipment power density	ω/m^2	[10,18]	14
Heating setpoint temperature	°C	[18,23]	20
Cooling setpoint temperature	°C	[24,28]	26

4.2 Experiments and Analyses

1) Comparison with traditional multi-objective EAs

This section compares the proposed algorithm with traditional multi-objective EAs without surrogate model. For fairness, the population size and the maximum iteration times of all algorithms are set to 20 and 50. Table 2 shows the parameter settings for all these algorithms in detail.

Hypervolume (HV) [21] is used to evaluate the performance of an algorithm, which can simultaneously evaluate the distribution and convergence of a set of Pareto optimal solutions. Besides, the t-test with a significance level of 0.05 is used to confirm the significant difference between different algorithms, where "R+" means that the performance of Sa-MOPSO is significantly better than the comparison algorithm, "=" means that there is no significant difference between the two algorithms, "–" means that the calculation is meaningless.

All algorithms are run 20 times. Table 3 shows their HV values for the office room at Beijing. Table 4 shows the running time of these algorithms. We can see that: (1) Comparing with the other four comparison algorithms, the average HV

Table 2. Parameter settings for all the algorithms.

Algorithms	Parameter setting
Sa-MOPSO	The archive size, $N_a = 50$
NSGA-II	The cross probability = 0.7; The mutation probability = 0.02;
	Tolerance 0.001
MOPSO	The learning factors $c_1 = c_2 = 2$; the inertia weight = 0.5;
	The archive size $N_a = 50$
MOABC	The maximum number of searches per food is 5
BBMOPSO-A	The archive size $N_a = 50$

value of Sa-MOPSO is significantly better than that of NSGA-II [13], MOPSO [4], MOABC [14], and BBMOPSO-A [15], although its results show a larger variance. (2) With the help of the surrogate model, the running time of Sa-MOPSO is greatly reduced. Compared with the fastest algorithm, BBMOPSO-A, among the comparison algorithms, its calculation time is more than 2 times faster.

Table 3. HV values obtained by different algorithms for the office room at Beijing.

Algorithms	HV (best)	HV (worst)	HV (average)	HV (std)	T-test
Sa-MOPSO	**46618.79**	21813.02	**30360.20**	9060.62	–
NSGA-II	16724.62	6376.02	12981.56	5173.07	R+
MOPSO	26024.96	22438.35	24610.73	**1481.52**	R+
MOABC	27958.26	18326.33	25555.79	4170.77	R+
BBMOPSO-A	27978.35	**24414.97**	26381.27	1489.32	=

Table 4. Running time of Sa-MOPSO and the four comparison Algorithms (unit: second).

Algorithms	Time (best)	Time (worst)	Time (average)	Time (std)
Sa-MOPSO	**865.48**	**1202.93**	**1084.44**	**105.62**
NSGA-II	3673.40	4345.03	4008.20	302.55
MOPSO	6116.32	7868.88	6997.85	619.69
MOABC	3239.11	3908.09	3479.39	286.20
BBMOPSO-A	3283.08	4854.85	3804.96	632.07

2) Comparison with multi-objective EAs with surrogate model

This section compares the proposed algorithm with a multi-objective EAs with surrogate model, the hybrid algorithm with on-line landscape approximation for expensive multi-objective optimization problems (ParEGO) [22]. Tables 5 and 6 shows the HV and running time of Sa-MOPSO and ParEGO. It can be seen that ParEGO and Sa-MOPSO have similar performance in the running time, but Sa-MOPSO has obtained better HV value than ParEGO.

Table 5. HV values obtained by two algorithms for the office room at Beijing.

Algorithms	HV (best)	HV (worst)	HV (average)	HV (std)	T-test
Sa-MOPSO	**47607.08**	14597.69	**31008.69**	**11573.34**	–
ParEGO	31236.43	15605.75	24701.32	4560.37	=

Table 6. Running time of Sa-MOPSO and ParEGO (unit: second).

Algorithms	Time (best)	Time (worst)	Time (average)	Time (std)
Sa-MOPSO	914.41	**1285.34**	1130.50	**132.27**
ParEGO	**701.83**	1841.88	**1071.82**	484.26

5 Conclusions

Aiming at the problem of building energy saving design, this paper proposed a surrogate-assisted multi-objective PSO algorithm, Sa-MOPSO. Integrating MATLAB with the software EnergyPlus, the specific implementation method of Sa-MOPSO was also given. Taking the office building at Beijing as an example, comparing with the existing typical EA methods (NSGA-II, MOPSO, MOABC, BBMOPSO-A and ParEGO), experimental results show that the proposed algorithm can obtain a high-quality Pareto optimal solution set in less running time, and it is a highly competitive optimization method to solve the problem of building energy saving design.

Acknowledgment. This work was supported by key research and development program in Xuzhou (No. KC20184).

References

1. Tian, W., Song, J., Li, Z., Wilde, P.D.: Bootstrap techniques for sensitivity analysis and model selection in building thermal performance analysis. Appl. Energy **135**, 320–328 (2014)

2. Garcia Sanchez, D., Lacarriere, B., Musy, M., Bourges, B.: Application of sensitivity analysis in building energy simulations: combining first-and second-order elementary effects methods. Energy Build. **68**, 741–750 (2014)
3. Ferrara, M., Fabrizio, E., Virgone, J., Filippi, M.: A simulation-based optimization method for cost-optimal analysis of nearly Zero Energy Buildings. Energy Build. **84**, 442C–457 (2014)
4. Delgarm, N., Sajadi, B., Kowsary, F., et al.: Multi-objective optimization of the building energy performance: a simulation-based approach by means of particle swarm optimization (PSO). Appl. Energy **170**, 293–303 (2016)
5. Delgarm, N., Sajadi, B., Delgarm, S., et al.: A novel approach for the simulation-based optimization of the buildings energy consumption using NSGA-II: case study in Iran. Energy Build. **127**, 552–560 (2016)
6. Delgarm, N., Sajadi, B., Delgarm, S., et al.: Multi-objective optimization of building energy performance and indoor thermal comfort: a new method using artificial bee colony (ABC). Energy Build. **131**(11), 42–53 (2016)
7. Zhang, Y., Yuan, L., Cheng, S.: Building energy performance optimization: a new multi-objective particle swarm method. In: Tan, Y., Shi, Y., Niu, B. (eds.) ICSI 2019. LNCS, vol. 11655, pp. 139–147. Springer, Cham (2019). https://doi.org/10.1007/978-3-030-26369-0_13
8. Grefenstette, J.J., Fitzpatrick, J.M.: Genetic search with approximate function evaluation. In: International Conference on Genetic Algorithms, pp. 112–120 (1985)
9. Tong, H., Huang, C., Liu, J., et al.: Voronoi-based efficient surrogate-assisted evolutionary algorithm for very expensive problems. In: IEEE Congress on Evolutionary Computation 2019, Wellington, pp. 1996–2003. IEEE (2019)
10. Sun, C., Jin, Y., Cheng, R., et al.: Surrogate-assisted cooperative swarm optimization of high-dimensional expensive problems. IEEE Trans. Evol. Comput. **21**(4), 644–660 (2017)
11. Wang, H., Jin, Y., Doherty, J.: Committee-based active learning for surrogate-assisted particle swarm optimization of expensive problems. IEEE Trans. Cybern. **47**(9), 2664–2677 (2017)
12. Liu, B., Sun, N., Zhang, Q., et al.: A surrogate model assisted evolutionary algorithm for computationally expensive design optimization problems with discrete variables. In: IEEE Congress on Evolutionary Computation 2016, Canada, pp. 1650–1657. IEEE (2019)
13. Zhang, M., Gou, W., Li, L., et al.: Multidisciplinary design and multi-objective optimization on guide fins of twin-web disk using Kriging surrogate model. Struct. Multi. Optim. **51**(1), 361–373 (2017)
14. Rosales-Prez, A., Gonzalez, J.A., Coello Coello, C.A., et al.: Surrogate-assisted multi-objective model selection for support vector machines. Neurocomputing **150**, 163–172 (2015)
15. Hamdy, M., Nguyen, A.T., Hensen, J.L.M.: A performance comparison of multi-objective optimization algorithms for solving nearly-zero-energy-building design problems. Energy Build. **121**, 57–71 (2016)
16. Eberhart, S.Y.: Particle swarm optimization: developments, applications and resources. Congr. Evol. Comput. **1**(1), 81–86 (2001)
17. Kennedy, J., Eberhart, R.: Particle swarm optimization. In: Proceedings. IEEE International Conference on Neural Networks, vol. 4, pp. 1942–1948 (1995)
18. Zhang, Y., Gong, D.W., Ding, Z.H.: A bare-bones multi-objective particle swarm optimization algorithm for environmental/economic dispatch. Inf. Sci. **192**, 212–227 (2012)

19. Giunta, A.A., Wojtkiewicz, S.F., Eldred, M.S.: Overview of modern design of experiments methods for computational simulations. In: The 41st AIAA Aerospace Sciences Meeting and Exhibit, pp. 1–18 (2003)
20. Yang, C.F., Li, H.J., Rezgui, Y., et al.: High throughput computing based distributed genetic algorithm for building energy consumption optimization. Energy Build. **76**, 92–101 (2014)
21. Zitzler, E., Thiele, L.: Multiobjective evolutionary algorithms: a comparative case study and the strength Pareto approach. IEEE Trans. Evol. Comput. **3**(4), 257–271 (1999)
22. Knowles, J.: ParEGO: a hybrid algorithm with on-line landscape approximation for expensive multiobjective optimization problems. IEEE Trans. Evol. Comput. **10**(1), 50–66 (2006)

Solving Large-Scale Multi-Objective Optimization via Probabilistic Prediction Model

Haokai Hong[1], Kai Ye[1], Min Jiang[1](✉), and Kay Chen Tan[2]

[1] School of Informatics, Xiamen University, Fujian 361005, China
minjiang@xmu.edu.cn

[2] Department of Computer Science, City University of Hong Kong, Kowloon, Hong Kong

Abstract. The main feature of large-scale multi-objective optimization problems (LSMOP) is to optimize multiple conflicting objectives while considering thousands of decision variables at the same time. Since the purpose of effective LSMOP algorithm is escaping from local optimum in large search space, the current research is focused on decision variable analysis or grouping, which easily leads to excessive computational complexity due to the large-scale decision variables. In order to maintain the diversity of the population while avoiding the computational complexity caused by large-scale decision variables, we propose a Probabilistic Prediction Model based on trend prediction model (TPM) and Generating-Filtering strategy to tackle LSMOP. Since TPM has an individual-based evolution mechanism, the computational complexity of the proposed algorithm is independent of decision variables, which maintains low complexity of the evolutionary algorithm while ensuring that the algorithm can converge to the Pareto optimal Front(POF). We compared the proposed algorithm with several state-of-the-art algorithms for different benchmark functions. The experimental results and complexity analysis have demonstrated that the proposed algorithm has significant improvement in terms of its performance and computational efficiency in large-scale multi-objective optimization.

Keywords: Evolutionary multi-objective optimization · Large-scale optimization · Probabilistic prediction model · Trend prediction model

1 Introduction

Many optimization problems in the real world involve hundreds or even thousands of decision variables and multiple conflicting goals. These problems are defined as large-scale multi-objective optimization problems (LSMOP) [14]. For example, in the problem of designing protein molecular structure, thousands of

This work was supported by the National Natural Science Foundation of China (No. 61673328).

H. Ishibuchi et al. (Eds.): EMO 2021, LNCS 12654, pp. 605–616, 2021.
https://doi.org/10.1007/978-3-030-72062-9_48

decision variables are considered. Since the size of the search space has an exponential relationship with the number of decision variables, which leads to the problem of dimensionality explosion in LSMOP [18], finding the optimal solution in LSMOP is more challenging than ordinary multi-objective optimization problems(MOP). Therefore, an excellent large-scale multi-objective optimization algorithm (LSMOA) should overcome this problem through searching in the decision space effectively, which is essential for solving LSMOP.

In order to solve LSMOP, various LSMOAs have been presented in recent years. These proposed LSMOAs can be roughly divided into the following three categories. The first category of LSMOAs is benefits from the decision variable analysis. A large-scale evolutionary algorithm based on the decision variable clustering (LMEA) [22] is representative in this category. The second category of LSMOAs is based on the decision variable grouping. This type of LSMOA divides the decision variables into several groups, and then optimizes each group of decision variables. The third category is based on problem conversion. A representative framework of this category is called Large-scale Multi-Objective Optimization Framework (LSMOF) [5].

In the methods of decision variable analysis and decision variable grouping, due to that fact that the algorithm relies on the research on decision variables, the algorithm complexity is relatively high. We believe that solving LSMOP, should not only consider how to escape from local optimum, but also a factor that has to be considered is to reduce the impact of the number of decision variables on the complexity of the algorithm. With the advantages of trend prediction model (TPM), the method not only maintains the diversity of the population [10], but also keeps the computational complexity independent of decision variables.

In this paper, we use a probabilistic prediction model to obtain a novel LSMOA algorithm. This method is a probabilistic prediction model for solving large-scale multi-objective optimization based on trend prediction model (LT-PPM). Our proposed algorithm selects individuals and expands them from the previous generation to achieve the purpose of expanding and generating a new offspring population via TPM, which ensures the diversity of the offspring population. The contributions of this work are summarized as follows.

1. The proposed algorithm framework is an effective tool to solve the problem of large decision variable space and overcome the tendency to fall into a local optimum in LSMOP. Our method uses the trend prediction model to maintain the quality and diversity of the population simultaneously to escape from the local optimum, which provides a novel idea to solve LSMOP.
2. Taking into account the high dimensionality of the decision variables of the LSMOP problem, the complexity of our proposed method is not directly related to the scale of the decision variables, but only related to the number of objectives and populations. While escaping from the local optimum, the influence of large-scale decision variables on the computational complexity is also eliminated in our method.

The rest of this paper is organized as follows. Section 2 introduces the existing LSMOA used for LSMOP. Section 3 introduces the LT-PPM algorithm in detail. Section 4 demonstrates the experimental results of LT-PPM compared with several state-of-the-art LSMOAs on LSMOP. Finally, the conclusion is drawn and the future work is discussed in Sect. 5.

2 Preliminary Studies

In this section, the definition of large-scale multi-objective optimization problems is presented. And then existing methods to solve LSMOPs are introduced.

2.1 Large-Scale Multi-objective Optimization

The mathematical form of LSMOPs is as follows:

$$\begin{aligned} \textbf{Minimize } F(x) = \langle f_1(x), f_2(x), ..., f_m(x) \rangle \\ s.t.\ x \in \Omega \end{aligned} \tag{1}$$

where $x = (x_1, x_2, \ldots, x_n)$ is the n-dimensional decision vector, $F = (f_1(x), f_2(x), ..., f_m(x))$ is the m-dimensional objective vector.

It should be noticed that the dimension of decision vector n is more than 100 in LSMOP [16]. When dealing with multi-objective optimization problems, evolutionary algorithms (EAs) are usually based on individual evolution strategies, genetic operators are used to evolve individuals, so that EAs have good global convergence and versatility. The evolutionary framework we use in this paper is an emerging branch in this field called the Probabilistic Prediction Model (PPM) [7,11], PPM promotes the evolution of the entire population by establishing a probability model on the Pareto-optimal Front.

2.2 Related Works

In this part, we will introduce some state-of-the-art large-scale multi-objective optimization algorithms. Much progress has been made in solving LSMOPs. Most of existing methods can be divided into three categories: based on decision variable analysis, based on decision variable grouping and based on problem transformation.

The first category is based on the analysis of decision variables. Zhang et al. [22] proposed a large-scale evolutionary algorithm (LMEA) based on clustering of decision variables. Another representative of this type of algorithm is MOEA/DVA [13], in which the original LSMOP is decomposed into many simpler sub-MOPs.

The second type of LSMOAs is decision variable grouping-based framework [19]. The method proposed by Antonio et al. [1] maintained several independent subgroups. Each subgroup is a subset of the same-length decision variables obtained by variable grouping. All sub-populations work together to optimize

LSMOP in a divide-and-conquer manner. Zille et al. [24] proposed a weighted optimization framework (WOF) for solving LSMOP. Decision variables are divided into many groups, and each group is assigned a weight variable. Then, in the same group, the weight variable is regarded as a sub-problem of a sub-space from the original decision space.

The third category is based on problem transformation. In order to enhance the pressure of convergence, one intuitive solution is to directly convert the definition of traditional Pareto dominance, such as θ-dominance [20] and fuzzy dominance [17]. Another kind of idea that belongs to this category is to take traditional advantages into consideration. MOEAs that fall into this category include NSGA-II based on substitute distance assignment [12] and Knee point driven Evolutionary Algorithm (KnEA) [23]. Recently, transfer learning [6] is also used in solving multi-objective optimization porblems [8,9], in addition, He et al. [4] designed a GANs driven evolutionary multi-objective optimization (GMOEA), which is also based on problem transformation.

As the number of decision variables increases linearly, the complexity (as well as volume) of the search space will increase exponentially, and the methods of analysis or grouping of decision variables presented above cannot tackle the curse of dimensionality. Therefore, in this work, we focus on how to use TPM to maintain population diversity to avoid falling into local optimum while avoiding the curse of dimensionality caused by large-scale decision variables to better solve LSMOP.

3 A Probabilistic Prediction Model Based on Trend Prediction Model

3.1 Overview

Based on the Generating-Filtering strategy, LT-PPM combines the importance sampling method with the non-parametric density estimation function to propose a different selection method for candidate solution. Specifically, LT-PPM uses a kernel density estimation method, a non-parametric probability density estimation method to assign a sparseness to each individual, and then uses importance sampling to select subsequent solutions. This method of selecting candidate solutions aims to ensure the balance between exploitation and exploration in terms of time and space: on the one hand, the sparseness will change with generation, on the other hand, sparseness itself will also guide the method to achieve the balance in the large-scale search space.

We first give the definition of particles, and then in the remainder of this chapter we will introduce importance sampling based on kernel density estimation, trend prediction model, LT-PPM and the complexity analysis of the algorithm.

Definition 1 (Particle)
For an optimization problem with n optimization goals, the particle is a two-tuple $p = (x, \boldsymbol{d})$*, where* x *represents a solution of the multi-objective optimization*

problem, and $\mathbf{d}$ *is a unit vector in n-dimensional space representing the direction of the particle.*

Note that the direction vector of the first generation population is randomly initialized. In the subsequent offspring population, the direction vector of each individual is the direction of movement when their parent particle produce them.

3.2 Importance Sampling Based on Kernel Density Estimation

The probability density of the population in different regions reflects the degree of density in the search space: where the probability density is high, the population is denser. Therefore, the density of particle p_k can be defined as its kernel density estimate, The **Sparseness** of particle p_k is defined as the reciprocal of **Density**. Based on the **Sparseness** of each particle, Importance Sampling for our method can be defined. Since each particle is randomly selected, particle p_k obeys a uniform distribution $U(p_k) = 1/n$. Then we have importance distribution,

$$IS(p_k) = \frac{\mathbf{sparseness}(p_k)}{\sum_i \mathbf{sparseness}(p_i)} \tag{2}$$

The number of offspring that the particle p_k will produce is proportional to $S(p_k)$. The expectation value of sparseness describes the sampling process, and its calculation formula is: $E_U[S(p_k)]$. However, in MOPs, it is not easy to determine how many offspring the selected particle should produce, so we do not want to directly use this method for sampling. The original sampling process can be rewritten as,

$$\mathbf{E}_U[S(p_k)] = \mathbf{E}_S[S(p_k) \cdot \omega(p_k)] = \mathbf{E}_S[U(p_k)] \tag{3}$$

where $\omega(p_k) = \frac{U(p_k)}{S(p_k)}$.

This means that we can select particles by the importance distribution S, and then produce the same number of offspring. Algorithm 1 summarizes the above sampling process. In each iteration, first, the sparseness of each particle is calculated, and then the importance distribution S is constructed, then an individual is sampled on the distribution S, and the individual and its velocity are returned. This sampling method does not rely on the analysis of decision variables, avoiding the influence of large-scale decision variables on computational complexity.

3.3 Trend Prediction Model

After selecting the particles to be expanded via above selection process, it is necessary to sample on a specific distribution represented by selected particles to generate new individuals. Physically, if an object is not affected by external forces, it will maintain its original state of motion. Therefore, if a particle moves in the direction $\boldsymbol{v}$, the displacement direction $\boldsymbol{s}$ of the particle at the next moment is equal to $\boldsymbol{v}$.

Algorithm 1: ISample(POS)

Input: POS (Pareto-optimal Set), h (bandwidth)
Output: p_i (sampled particle), x_k(solution of p_i), v_k(velocity of p_i)
1 $n =$ size of POS ;
2 **for** p_i *in* POS **do**
3   **Density**$(p_k) = \frac{1}{nh}\sum_{i=1}^{n}\kappa(\frac{p_k-p_i}{h})$;
4   **Sparseness**$(p_k) = \frac{1}{\textbf{Density}(p_k)}$;
5 **for** x_i *in* POS **do**
6   Constructe $IS(x_i)$ according to Equation 2 ;
7 Sampled x_k based on $IS(x_i)$;
8 **return** x_k, v_k

In order to effectively move to the POF, the randomness is added to the movement of the particle, so that the direction of movement only provides information on the trend of the movement instead of directly determining the place at the next moment. Such trend information can well guide the particles to update, thereby approaching the POF faster and more accurately.

Definition 2 (Trend Prediction Model, TPM)
For any given direction vector $v \in \mathbb{R}^n$ and distance measurement function $d(\cdot,\cdot)$. A probability distribution P on $D \subset \mathbb{R}^n$ is called as a trend prediction model (TPM).

For the two vectors t_1 and t_2 in the model, if it satisfies:

$$\forall t_1, t_2 \in D, d(v, t_1) \le d(v, t_2) \rightarrow P(t_1) \ge P(t_2) \tag{4}$$

then it means that the vector t_1 is closer to the direction vector v than t_2, and t_1 should have a higher probability of occurrence.

In the trend prediction model, particle movement has a moving trend, which is indicated by the unit vector v. The method aims to design the **TPM** distribution so that the closer the sampling vector to v, the higher the sampling probability. For this, **TPM**$_h$ is introduced. Given the distance measurement method d is the cos distance, the domain D is the unit hypersphere, and sampled from the Gaussian distribution with the expectation of 0 and the variance of $1/h$, which is the bandwidth. Since the domain is a unit hypersphere, **TPM**$_h$ is actually a probability distribution on a direction vector. In this distribution, directions close to the vector v have a higher probability of being selected (that is, the angle with v is small), otherwise the probability is low.

After sampling an angle θ on the **TPM**$_h$ system, it is necessary to randomly select one among all the vectors whose angle is θ with v. In order to sample all vectors whose angle is θ with v, we can construct a rotation matrix R via Schmidt Orthogonalization to rotate the vector v to the X axis. Then, randomly select

Algorithm 2: randpick($\boldsymbol{v}, \theta$)

Input: $\boldsymbol{v} \in \mathbb{R}^n$ (Unit direction vector)
Output: $\boldsymbol{u}$ (Random direction vector)
1 $n =$ Dimension of $\boldsymbol{v}$;
2 $t =$ The subscript of the first non-zero vector in $\boldsymbol{v}$;
3 **for** $i = 1, 2, \ldots, n-1$ **do**
4   **if** $i = t$ **then**
5         $\beta_i =$ The vector that only the nth element is 1 and other elements are 0;
6   **else**
7         $\beta_i =$ The vector that only the ith element is 1 and other elements are 0;
8 $R =$ Apply **Schmidt orthogonalization** to the matrix $[\boldsymbol{v}, \boldsymbol{\beta}_1, \ldots, \boldsymbol{\beta}_{n-1}]$;
9 $\boldsymbol{u} =$ Sampled from the $n-1$ dimensional unit hyperspherical coordinate system ;
10 $\boldsymbol{u} = R^{-1} \cdot [\cos(\theta)\boldsymbol{u}]^T$;
11 **return** $\boldsymbol{u}$

a vector on the $n-1$ dimensional hypersphere. Finally, the sampled vector is rotated back to the original coordinate space through the inverse transformation $R^{(-1)}$.

Algorithm 2 is the process of generating offspring direction vectors according to the parent individual by importance sampling. The complete trend prediction model is given in Algorithm 3, where ISample (POS) is Algorithm 1. In the specific strategy of generating offspring, after sampling the direction vector by Algorithm 2, it is necessary to continue to determine the particle's movement step length. The actual step length is sampled from a normal distribution with a expectation of 0 and a variance of h.

Algorithm 3: TPM(POS, h, N) (Trend Prediction Model)

Input: POS (Pareto-Optimal Set), h (bandwidth), N (number of samples)
Output: S(Sampled set)
1 $S = \emptyset$;
2 **for** $i = 1, 2, \ldots, N$ **do**
3     $< p, v >=$ **ISample**(POS) ;
4     $\theta =$ Sampled from **Norm**$(0, 1/h)$;
5     $u =$ **randpick**(v, θ) ;
6     $\lambda =$ Sampled from **Norm**$(0, h)$;
7     $\lambda = \lceil \lambda \rceil$;
8     $S = S \cup \{p + \lambda u\}$;
9 **return** S ;

3.4 Large Scale TPM Based Probabilistic Prediction Model (LT-PPM)

Combining the importance sampling and trend prediction model, the complete algorithm implementation of LT-PPM is given in Algorithm 4.

Algorithm 4: Large scale TPM based PPM

Input: N(size of restricted population), P(size of population), e (the maximum evaluations), r(attenuation coefficient)

Output: POS(*the* **POS**)

1 Random initial P population POS ;
2 **while** *Used evaluations* $\leq e$ **do**
3 $\quad S = sample(POS, 1/h, P)$;
4 $\quad$ Update POS and based on S ;
5 $\quad$ Delete the densest individuals repeatedly until $|POS| \leq N$ $h = h \cdot r$;
6 **return** POS ;

The algorithm framework of LT-PPM is briefly illustrated in conjunction with Figs. 1(a) to 1(d). In a certain generation, the LT-PPM algorithm maintains a population and calculates the corresponding POS and POF (the solid line in Fig. 1(a) represents the POF of this generation, and the dashed line represents the real POF). Then use the non-parametric kernel density estimation method to estimate the probability density of the distribution corresponding to the POS, and use the importance sampling method to select a particle. The principle is that particles with higher sparseness will have a higher probability of being selected (Fig. 1(b)), while particles in dense particles have a lower probability of being selected. After that, apply the trend prediction model to the selected particles to obtain a series of new solutions (Fig. 1(c), where gray dots represent the newly generated population). Finally, update POS and POF according to the newly generated individuals (Fig. 1(d)).

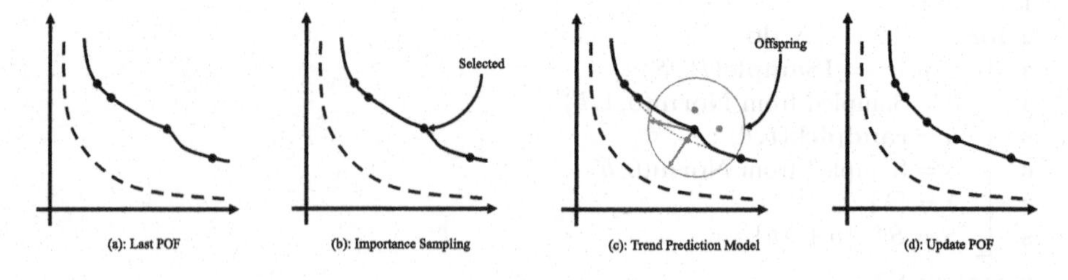

Fig. 1. Algorithm framework of LT-PPM

3.5 Complexity Analysis

Assuming a n objectives LSMOP with m population, then the time to calculate all objectives is $\mathcal{O}(n)$. In our algorithm, m individuals are sampled in each iteration, and each individual requires $\mathcal{O}(m)$ time to sample, therefore the time complexity of the sampling process in each iteration is $\mathcal{O}(m^2)$. Suppose the evaluation is e, the actual interation times is e/m, so the total complexity of the algorithm is $\mathcal{O}(nme)$.

The complexity of the algorithm is proportional to the square of the number of population and the number of objectives, and has no direct relation with the scale of decision variables of the problem. Therefore, our proposed LT-PPM can avoid the influence of large-scale decision variables on the computational complexity in solving large-scale multi-objective optimization problems.

4 Experiments and Analysis

4.1 Algorithms in Comparison and Test Problems

The proposed LT-PPM is compared with several state-of-the-art algorithms, including KnEA [23], LMEA [22], LMOCSO [16], NSGAIII [3] and RMMEDA [21]. According to [15], the parameters of KnEA, LMEA, RMMEDA and LMOCSO algorithms are set to default values.

The remaining parameters are set as follows,

1. Population size: The population size N is set to 300.
2. Termination condition: The maximum evaluation times e of all compared LSMOAs is set to 100000.
3. Parameter of LT-PPM: The bandwidth attenuation coefficient r is set to 0.9.

The large-scale multi-objective test problem LSMOP1-LSMOP9 [2] are used for experiments. For all test problems, the number of optimization objectives is set to 3.

4.2 Performance on LSMOP Problems and Discussion

Two widely used performance indicators, the Schott's Spacing (SP) and the Inverted Generational Distance (IGD) are adopted to evaluate the performance of the compared algorithms.

Table 1 and Table 2 show the statistical results of the average SP and IGD values for 10 runs respectively, where the best result on each test instance is shown in bold.

From Table 1 and Table 2, it obviously that the performance of the proposed LT-PPM is better than the other five comparative convergence algorithms. Table 1 lists the SP values of the six comparison algorithms. LT-PPM performed best in 7 out of 9 test instances, followed by NSGAIII with 2 best results. In Table 2, LT-PPM got 6 of the 9 best results, LMOCSO got 3 best results.

Table 1. Mean values of sp metric obtained by compared algorithms over LSMOP problems with 3-objectives

Problem	Des.	LTPPM	KnEA	LMEA	LMOCSO	NSGAIII	RMMEDA
LSMOP1	1000	**4.15E−01**	4.66E−01	4.72E−01	4.75E−01	4.37E−01	4.36E−01
LSMOP2	1000	3.10E−01	1.10E−01	5.34E−01	6.19E−02	**1.68E−02**	4.80E−01
LSMOP3	1000	**1.76E−01**	7.68E−01	1.07E+00	6.67E−01	8.62E−01	9.27E−01
LSMOP4	1000	2.97E−01	1.63E−01	5.12E−01	1.11E−01	**5.24E−02**	4.81E−01
LSMOP5	1000	**4.39E−01**	6.16E−01	7.25E−01	7.13E−01	6.61E−01	7.61E−01
LSMOP6	1000	**3.08E−01**	1.01E+00	1.15E+00	1.30E+00	9.77E−01	1.02E+00
LSMOP7	1000	**2.86E−01**	7.48E−01	9.46E−01	3.57E−01	8.59E−01	8.73E−01
LSMOP8	1000	**3.39E−01**	7.16E−01	7.23E−01	9.56E−01	5.24E−01	7.39E−01
LSMOP9	1000	**3.66E−01**	7.47E−01	9.46E−01	8.93E−01	8.58E−01	7.54E−01

The above experimental results show that for most test instances, LT-PPM can obtain a set of solutions with good convergence and diversity. It can be seen that the proposed LT-PPM can obtain competitive performance on the LSMOP test problem, which confirms the effectiveness of LT-PPM in solving the challenging large-scale decision variable problem.

Table 2. Mean values of IGD metric obtained by compared algorithms over LSMOP problems with 3-objectives

Problem	Des.	LTPPM	KnEA	LMEA	LMOCSO	NSGAIII	RMMEDA
LSMOP1	1000	**6.50E−01**	4.54E+00	1.07E+01	1.29E+00	5.36E+00	8.77E+00
LSMOP2	1000	3.85E−02	2.91E−02	3.38E−02	**2.57E−02**	3.05E−02	3.38E−02
LSMOP3	1000	**4.89E+00**	1.28E+01	5.52E+01	9.86E+00	1.42E+01	1.64E+01
LSMOP4	1000	1.09E−01	8.60E−02	1.11E−01	**8.33E−02**	1.05E−01	1.11E−01
LSMOP5	1000	**6.42E−01**	1.14E+01	1.89E+01	2.99E+00	9.52E+00	1.09E+01
LSMOP6	1000	**1.91E+00**	1.22E+03	3.08E+04	3.53E+02	3.68E+03	7.15E+03
LSMOP7	1000	2.81E+00	1.09E+00	5.43E+02	**8.24E−01**	1.08E+00	1.08E+00
LSMOP8	1000	**1.98E−01**	5.63E−01	5.92E−01	4.96E−01	5.83E−01	5.75E−01
LSMOP9	1000	**1.35E+00**	9.43E+00	1.40E+02	2.43E+00	2.15E+01	4.07E+01

5 Conclusion and Future Works

This paper proposes an algorithm for solving LSMOP based on trend prediction model, called LT-PPM. Since the TPM has an individual-based evolution mechanism, the method maintains low complexity of the evolutionary algorithm while ensuring that the algorithm can converge to the Pareto optimal Front (POF). Our method is compared with several other state-of-the-art LSMOAs on 9 LSMOPs. The experimental results show that LT-PPM has good performance in convergence and distribution.

This paper presents a preliminary work on the application of trend prediction model to large-scale evolutionary calculations. In our future work a wealth of advanced machine learning techniques can be integrated to solve LSMOP more effectively.

Acknowledgement. This work was supported by the National Natural Science Foundation of China (No. 61673328).

References

1. Antonio, L.M., Coello, C.A.C.: Use of cooperative coevolution for solving large scale multiobjective optimization problems. In: 2013 IEEE Congress on Evolutionary Computation, pp. 2758–2765. https://doi.org/10.1109/CEC.2013.6557903
2. Cheng, R., Jin, Y., Olhofer, M., Sendhoff, B.: Test problems for large-scale multiobjective and many-objective optimization. IEEE Trans. Cybern. **47**(12), 4108–4121 (2017). https://doi.org/10.1109/TCYB.2016.2600577
3. Deb, K., Jain, H.: An evolutionary many-objective optimization algorithm using reference-point-based nondominated sorting approach, part I: solving problems with box constraints. IEEE Trans. Evol. Comput. **18**(4), 577–601 (2014). https://doi.org/10.1109/TEVC.2013.2281535
4. He, C., Huang, S., Cheng, R., Tan, K.C., Jin, Y.: Evolutionary multiobjective optimization driven by generative adversarial networks (GANs). IEEE Trans. Cybern. 1–14 (2020). https://doi.org/10.1109/TCYB.2020.2985081
5. He, C., et al.: Accelerating large-scale multiobjective optimization via problem reformulation. IEEE Trans. Evol. Comput. **23**, 949–961 (2019)
6. Jiang, M., Huang, W., Huang, Z., Yen, G.G.: Integration of global and local metrics for domain adaptation learning via dimensionality reduction. IEEE Trans. Cybern. **47**(1), 38–51 (2017). https://doi.org/10.1109/TCYB.2015.2502483
7. Jiang, M., Huang, Z., Jiang, G., Shi, M., Zeng, X.: Motion generation of multi-legged robot in complex terrains by using estimation of distribution algorithm. In: 2017 IEEE Symposium Series on Computational Intelligence (SSCI), pp. 1–6 (2017). https://doi.org/10.1109/SSCI.2017.8285444
8. Jiang, M., Huang, Z., Qiu, L., Huang, W., Yen, G.G.: Transfer learning-based dynamic multiobjective optimization algorithms. IEEE Trans. Evol. Comput. **22**(4), 501–514 (2018). https://doi.org/10.1109/TEVC.2017.2771451
9. Jiang, M., Wang, Z., Guo, S., Gao, X., Tan, K.C.: Individual-based transfer learning for dynamic multiobjective optimization. IEEE Trans. Cybern. 1–14 (2020). https://doi.org/10.1109/TCYB.2020.3017049
10. Jiang, M., Wang, Z., Hong, H., Yen, G.G.: Knee point based imbalanced transfer learning for dynamic multi-objective optimization. IEEE Trans. Evol. Comput. 1 (2020). https://doi.org/10.1109/TEVC.2020.3004027
11. Jiang, M., Qiu, L., Huang, Z., Yen, G.G.: Dynamic multi-objective estimation of distribution algorithm based on domain adaptation and nonparametric estimation. Inf. Sci. **435**, 203–223 (2018). https://doi.org/10.1016/j.ins.2017.12.058. http://www.sciencedirect.com/science/article/pii/S0020025517311775
12. Köppen, M., Yoshida, K.: Substitute distance assignments in NSGA-II for handling many-objective optimization problems. In: Obayashi, S., Deb, K., Poloni, C., Hiroyasu, T., Murata, T. (eds.) EMO 2007. LNCS, vol. 4403, pp. 727–741. Springer, Heidelberg (2007). https://doi.org/10.1007/978-3-540-70928-2_55

13. Ma, X., et al.: A multiobjective evolutionary algorithm based on decision variable analyses for multiobjective optimization problems with large-scale variables. IEEE Trans. Evol. Comput. **20**(2), 275–298 (2016)
14. Ponsich, A., Jaimes, A.L., Coello, C.A.C.: A survey on multiobjective evolutionary algorithms for the solution of the portfolio optimization problem and other finance and economics applications. IEEE Trans. Evol. Comput. **17**(3), 321–344 (2013)
15. Tian, Y., Cheng, R., Zhang, X., Jin, Y.: PlatEMO: a MATLAB platform for evolutionary multi-objective optimization [educational forum]. IEEE Comput. Intell. Mag. **12**(4), 73–87 (2017). https://doi.org/10.1109/MCI.2017.2742868
16. Tian, Y., Zheng, X., Zhang, X., Jin, Y.: Efficient large-scale multi-objective optimization based on a competitive swarm optimizer. IEEE Trans. Cybern. **50**, 3696–3708 (2019)
17. Wang, G., Jiang, H.: Fuzzy-dominance and its application in evolutionary many objective optimization. In: International Conference on Computational Intelligence & Security Workshops (2007)
18. Wang, H., Jiao, L., Shang, R., He, S., Liu, F.: A memetic optimization strategy based on dimension reduction in decision space. Evol. Comput. **23**(1), 69–100 (2015)
19. Yang, Z., Tang, K., Yao, X.: Large scale evolutionary optimization using cooperative coevolution. Inf. Sci. **178**(15), 2985–2999 (2014)
20. Yuan, Y., Xu, H., Wang, B., Yao, X.: A new dominance relation-based evolutionary algorithm for many-objective optimization. IEEE Trans. Evol. Comput. **20**(1), 16–37 (2016). https://doi.org/10.1109/TEVC.2015.2420112
21. Zhang, Q., Zhou, A., Jin, Y.: RM-MEDA: a regularity model-based multiobjective estimation of distribution algorithm. IEEE Trans. Evol. Comput. **12**(1), 41–63 (2008). https://doi.org/10.1109/TEVC.2007.894202
22. Zhang, X., Tian, Y., Cheng, R., Jin, Y.: A decision variable clustering-based evolutionary algorithm for large-scale many-objective optimization. IEEE Trans. Evol. Comput. **22**(1), 97–112 (2018). https://doi.org/10.1109/TEVC.2016.2600642
23. Zhang, X., Tian, Y., Jin, Y.: A knee point-driven evolutionary algorithm for many-objective optimization. IEEE Trans. Evol. Comput. **19**(6), 761–776 (2015). https://doi.org/10.1109/TEVC.2014.2378512
24. Zille, H., Ishibuchi, H., Mostaghim, S., Nojima, Y.: A framework for large-scale multiobjective optimization based on problem transformation. IEEE Trans. Evol. Comput. **22**(2), 260–275 (2018). https://doi.org/10.1109/TEVC.2017.2704782

MCDM and Interactive EMO

An Artificial Decision Maker for Comparing Reference Point Based Interactive Evolutionary Multiobjective Optimization Methods

Bekir Afsar[1](✉), Kaisa Miettinen[1], and Ana B. Ruiz[2]

[1] University of Jyvaskyla, Faculty of Information Technology, P.O. Box 35 (Agora), FI-40014 University of Jyvaskyla, Finland
{bekir.b.afsar,kaisa.miettinen}@jyu.fi

[2] Department of Applied Economics (Mathematics), Universidad de Málaga, C/Ejido 6, 29071 Málaga, Spain
abruiz@uma.es

Abstract. Comparing interactive evolutionary multiobjective optimization methods is controversial. The main difficulties come from features inherent to interactive solution processes involving real decision makers. The human can be replaced by an artificial decision maker (ADM) to evaluate methods quantitatively. We propose a new ADM to compare reference point based interactive evolutionary methods, where reference points are generated in different ways for the different phases of the solution process. In the learning phase, the ADM explores different parts of the objective space to gain insight about the problem and to identify a region of interest, which is studied more closely in the decision phase. We demonstrate the ADM by comparing interactive versions of RVEA and NSGA-III on benchmark problems with up to 9 objectives. The experiments show that our ADM is efficient and allows repetitive testing to compare interactive evolutionary methods in a meaningful way.

Keywords: Decision making · Aspiration levels · Performance comparison · Many-objective optimization · Interactive methods

1 Introduction

Many real-world applications involve optimizing a set of conflicting objectives over a set of feasible solutions. This type of problems are known as *multiobjective optimization problems* (MOPs). Typically, no solution exists optimizing all the

Electronic supplementary material The online version of this chapter (https://doi.org/10.1007/978-3-030-72062-9_49) contains supplementary material, which is available to authorized users.

H. Ishibuchi et al. (Eds.): EMO 2021, LNCS 12654, pp. 619–631, 2021.
https://doi.org/10.1007/978-3-030-72062-9_49

objectives at the same time, and we look for so-called Pareto optimal solutions, at which an improvement of any objective always implies a sacrifice in, at least, one of the others. Mathematically, Pareto optimal solutions are incomparable, and we need information about the preferences of a *decision maker* (DM), an expert in the problem domain, to identify the *most preferred solution* (MPS).

In interactive methods [15,16], the DM plays an active role in directing a solution process. (S)he sees information about solutions available, and expresses and possibly changes her/his preferences in an iterative process, which ends once the DM is satisfied. Often, the interactive solution process can be divided into two phases [16]. In the learning phase, the DM explores different solutions to find a *region of interest* (ROI), i.e., solutions suiting her/his preferences. Then, in the decision phase, (s)he fine-tunes the search in this ROI to find her/his MPS.

In most interactive evolutionary multiobjective optimization (EMO) methods, preferences are used to approximate a ROI [2,14,21]. Comparing interactive EMO methods is important to reveal the strengths and weaknesses of new proposals against old ones but the quantitative assessment of methods involving DMs is still open [13]. The DM's preferences evolve while (s)he learns about the trade-offs in the problem and the feasibility of one's preferences in the interactive solution process. Therefore, the subjectivity of real DMs, human fatigue, or other limiting factors make it hard to design experiments with human DMs.

Alternatively, the human DM can be replaced somehow. In this regard, interactive methods can be divided into non ad-hoc and ad-hoc methods [19], depending on whether the DM can be replaced by a value function or not, respectively. Here we focus on ad-hoc interactive EMO methods. To compare such methods, we can use an *artificial DM* (ADM) simulating the DM's actions. Comparing methods with ADMs is cheaper and less time consuming than with human DMs. Some literature of ADMs already exists. To generate reference points, the ADMs developed in [1,18] have a pre-defined steady part that does not change during the solution process and current context that evolves. In [10], a cone is used based on a pre-defined MPS, and the learning of a DM is simulated by narrowing the cone's angle in the solution process. These ADMs need a goal point (initial aspiration levels or a MPS, respectively) to be reached by the ADM. However, this may hinder the exploration of the search and, thus, performances depend on that point. Furthermore, algorithms are run individually, i.e., preferences are generated based on the output of each single algorithm.

We propose an ADM for comparing the performance of interactive EMO methods based on reference points. By performing iterations, our ADM adjusts the preferences according to the insight gained during the solution process. At each iteration, a reference point is generated based on the solutions obtained so far by all compared algorithms. To simulate varying search objectives in different phases, the ADM produces reference points in a different way for the learning and the decision phases. In the learning phase, the reference points simulate exploration to examine the whole set of Pareto optimal solutions in search for a potential ROI. Then, the reference points of the decision phase mimic a progressive convergence towards a MPS inside the previously identified ROI. To compare the performances of the algorithms, the obtained solutions are evaluated according to the reference point generated at each iteration.

For comparability, the ADM assigns the same computational resources (i.e. number of evaluations or generations per iteration) to all algorithms compared. Note that the quality of the results depends on the algorithms, not on the ADM. Since our ADM allows assigning a different pre-fixed number of iterations to the learning and decision phases, it can be applied for different needs. E.g., more iterations can be performed in the learning phase to evaluate the search adaptation capacity when the preferences change drastically. This simulates an exploratory behaviour to gain knowledge about the conflicting objectives or feasible solutions. To compare the ability to search within a specific region to fine-tune solutions, one can increase the number of iterations of the decision phase in the ADM.

In the literature, few works analyze practical properties inherent to interactive solution processes. The reason may be the lack of a framework to simulate the iterative process followed by a DM. Our ADM is aimed at comparing interactive reference point-based EMO methods. However, we do not consider human factors related to decision making, such as cognitive biases like anchoring, which deserve further research. The proposed ADM contributes to existing research by generating reference points in a different manner for the learning and decision phases. To our best knowledge, this is the first ADM to differentiate two phases explicitly. It runs algorithms simultaneously and uses their output to generate the reference points, without requiring a pre-defined goal point, as in [1,10,18].

In the following, Sect. 2 introduces the basic concepts and notation needed. The new ADM is described in Sect. 3. Next, Sect. 4 demonstrates the performance of our proposal. Finally, conclusions are drawn in Sect. 5.

2 Background

A *multiobjective optimization problem* that minimizes k (with $k \geq 2$) conflicting objective functions $f_i : S \rightarrow \mathbb{R}$ ($i = 1, \ldots, k$) can be formulated as follows:

$$\begin{aligned} &\min \ \{f_1(\mathbf{x}), \ldots, f_k(\mathbf{x})\} \\ &\text{s.t.} \ \ \mathbf{x} \in S, \end{aligned} \tag{1}$$

where $S \subset \mathbb{R}^n$ is the *feasible set* of *decision vectors* $\mathbf{x} = (x_1, \ldots, x_n)^T$ in the decision space. Corresponding *objective vectors* of the form $\mathbf{f}(\mathbf{x}) = (f_1(\mathbf{x}), \ldots, f_k(\mathbf{x}))^T$ constitute a *feasible objective region* $Z = \mathbf{f}(S) \subset \mathbb{R}^k$ in the objective space.

Because of the conflicting nature of the objectives, a single solution optimizing all of them at the same time does not exist. On the contrary, there is a set of *Pareto optimal* solutions, at which none of the objectives can be improved without deteriorating, at least, one of the others. Given $\mathbf{z}, \mathbf{z}' \in \mathbb{R}^k$, we say that $\mathbf{z}$ *dominates* $\mathbf{z}'$ if $z_i \leq z'_i$ for all $i = 1, \ldots, k$ and $z_j < z'_j$ for at least one index j. If $\mathbf{z}$ and $\mathbf{z}'$ do not dominate each other, we say that they are *(mutually) non-dominated.* Then, $\mathbf{x} \in S$ is *Pareto optimal* if there does not exist any $\mathbf{x}' \in S$ such that $\mathbf{f}(\mathbf{x}')$ dominates $\mathbf{f}(\mathbf{x})$. The corresponding objective vector $\mathbf{f}(\mathbf{x}) \in Z$ is called a *Pareto optimal objective vector*. All Pareto optimal solutions form a *Pareto optimal set* in the decision space, denoted by E. Its image in the objective space is known as a *Pareto optimal front* (PF), denoted by $\mathbf{f}(E)$.

The ranges of the objective functions defined by their worst and best possible values in the PF form so-called nadir and ideal points, respectively. The upper bounds define the *nadir point* $\mathbf{z}^{\text{nad}} = (z_1^{\text{nad}}, \ldots, z_k^{\text{nad}})^T$, i.e., $z_i^{\text{nad}} = \max_{\mathbf{x} \in E} f_i(\mathbf{x})$ $(i = 1, \ldots, k)$, while the lower bounds constitute an *ideal point* $\mathbf{z}^{\star} = (z_1^{\star}, \ldots, z_k^{\star})^T$ as $z_i^{\star} = \min_{\mathbf{x} \in E} f_i(\mathbf{x}) = \min_{\mathbf{x} \in S} f_i(\mathbf{x})$ $(i = 1, \ldots, k)$. In practice, the nadir point is difficult to calculate because the set E is unknown and, thus, $\mathbf{z}^{\text{nad}}$ is commonly estimated (see e.g. [15,20]).

We consider methods that include preference information in the form of a reference point $\mathbf{q} = (q_1, \ldots, q_k)^T$. Here, q_i is a so-called aspiration level, that is, a desirable value for the objective f_i provided by the DM $(i = 1, \ldots, k)$.

3 Artificial Decision Maker

We propose an ADM to compare the performances of reference point based interactive EMO algorithms, which runs all algorithms simultaneously. To make a meaningful comparison, the ADM provides the same computational resources (i.e. number of function evaluations or generations per iteration) and produces reference points in a similar manner for all algorithms.

As mentioned before, we consider the two phases of an interactive solution process: the learning and the decision phases. Accordingly, our ADM has two strategies to produce reference points. In the learning phase, the ADM explores the objective space to see the solutions available by providing, at each iteration, a reference point in the least explored area of the PF. Eventually, a ROI is found at the end of the learning phase. In the decision phase, the ADM aims to refine solutions inside this ROI in search for a MPS, so it provides a reference point in this ROI at each iteration of this phase. At the beginning, a pre-fixed number of iterations is assigned to each phase. Respectively, we denote by L and D the number of iterations performed in the learning and in the decision phases.

The ADM uses solutions obtained so far from all algorithms to be compared. Thus, new reference points are generated according to the responses of the algorithms. For this, the ADM first merges the solutions and then eliminates dominated ones to build a *composite front*, as shown in Fig. 1a. Thus, the composite front includes only the non-dominated solutions obtained by all algorithms from the first iteration until the current iteration. Note that the ADM does not need any true PF information to generate reference points.

Besides preference generation, the proposed ADM evaluates the responsiveness of the algorithms for the given reference point after each iteration by applying performance metrics that take into account the given reference point. Moreover, it calculates cumulative metric values to evaluate the performances of the algorithms separately in the learning and in the decision phases.

In the literature, decomposition-based EMO algorithms have been proposed to handle problems with $k > 3$. They usually decompose the original MOP into

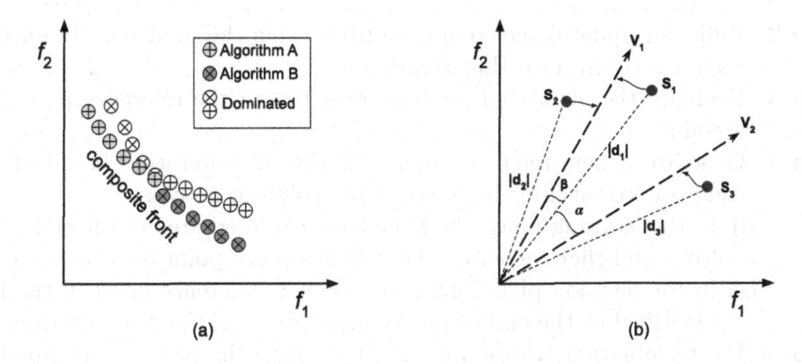

Fig. 1. (a) A composite front. (b) Solution assignment to reference vectors.

a group of sub-problems by dividing the whole PF into a group of subsets to enhance the performance of algorithms. We have adapted this idea in our ADM to find the exploration degree of the different parts of the whole PF (any other metric giving information about sub-areas of the PF can be also applied for this).

The following approach is used for finding areas of the composite front to be explored at each iteration of the learning phase. First, the composite front is divided equally into a group of subsets by using reference vectors which are uniformly distributed on the PF, as in [3,4]. Therefore, the ADM creates a set of uniformly distributed reference vectors using the canonical simplex-lattice design method [5]. In this, the number of reference vectors is adjusted by a lattice resolution (l), as $\binom{l+k-1}{k-1}$, where l is a pre-fixed parameter in the proposed ADM. With these vectors, the ADM assigns solutions in the composite front to reference vectors according to the angle between them. In the example in Fig. 1b, we have two reference vectors (V_1, V_2) and three solutions (S_1, S_2, S_3). Since the angle (β) between S_1 and V_1 is smaller than the angle (α) between S_1 and V_2, solution S_1 is assigned to vector V_1. Similarly, each solution is assigned to the reference vector with the minimum angle (S_2 is assigned to V_1, S_3 is assigned to V_2). The number of assigned solutions to each reference vector gives us information about the exploration degree of different parts of the composite front. In this way, the ADM divides the whole composite front into sub-areas to be able to find the least explored area of the PF. Below, we describe how distances of the solutions to the ideal point are taken into account in generating a reference point. To summarize, the main steps of the proposed ADM are:

Step 0. Initialize all algorithms and provide the first reference point randomly.

Step 1. Run all algorithms with the same computational budget (number of generations or function evaluations) and the previously obtained reference point.

Step 2. Build (or update) the composite front using the solutions obtained by each algorithm until this iteration.
Step 3. Evaluate the algorithm performances taking the reference point into account.
Step 4. Generate a new reference point for the next iteration based on the composite front and the phase of the solution process:
- **a)** In the learning phase, find the least explored area of the composite front and then, generate the next reference point for that area.
- **b)** In the decision phase, generate the new reference point in the ROI identified at the end of the learning phase to fine-tune solutions.

Step 5. If a termination criterion is met, terminate the process and calculate cumulative metric values for each phase. Else, continue with Step 1.

Preference Generation in the Learning Phase: In this phase, our ADM explores potential regions of the PF to examine those that have been poorly covered so far. Thus, the ADM tries to localize unexplored areas of the composite front and find more solutions in the least explored area by providing the next reference point inside it. As mentioned, a set of uniformly distributed reference vectors is generated on the composite front first. Then, all solutions are assigned to reference vectors as previously described, and the least explored area is identified based on the reference vector which has the minimum number of assigned solutions. Out of the four reference vectors shown in Fig. 2a, V_2 is selected.

The location of the reference point on the selected reference vector is determined by using solutions assigned to it. To this end, the ADM calculates the distances of these solutions to the ideal point of the current composite front, and selects the solution with the minimum distance. This distance $|d|$ sets the next reference point on the selected reference vector, as shown in Fig. 2a.

After L iterations, the reference vector that has the maximum number of solutions is identified. We denote it as V_D, and solutions assigned to it constitute the ROI to be further studied in the decision phase.

Preference Generation in the Decision Phase: In the D iterations of the decision phase, the reference vector V_D is employed to generate reference points in order to get progressively closer to the PF by refining solutions in the ROI.

At each iteration, our ADM finds the solution in the ROI with the minimum distance $|d|$ to the ideal point of the composite front, and $|d|$ is used to generate the next reference point as shown in Fig. 2b. All the reference points generated in this phase lie on the reference vector V_D. With iterations, the new reference points get closer to the ideal point of the composite front, given that the solutions generated by the algorithms (used to update the composite front) converge to the PF. With this, the ADM performs a finer search in the ROI to find a MPS.

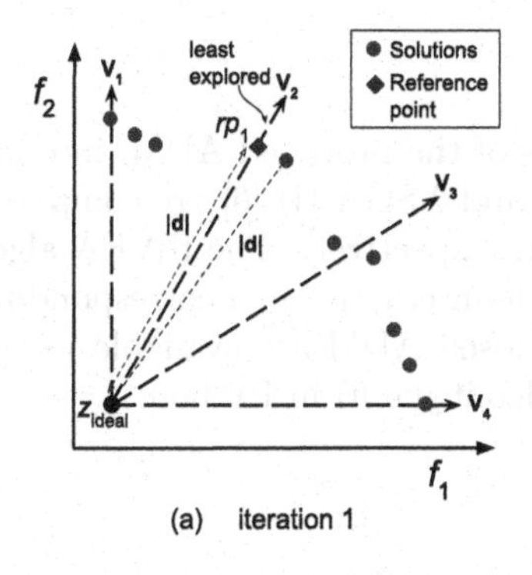

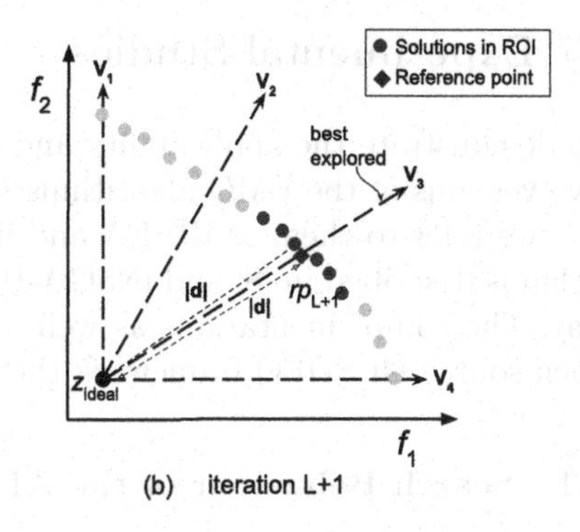

Fig. 2. Preference generation in (a) the learning phase and (b) the decision phase.

Evaluation: If no preference information is taken into account in the solution process, having a sufficient number of solutions, good closeness to the PF (convergence), good spread over the PF, and good uniformity amongst solutions define the quality of the obtained solutions [12]. To evaluate and compare the performances of reference point based EMO methods, one should redefine the term 'quality' of the solutions. Since preference information is used to guide the solution process, the above features should be measured for the preferred ROI, which is defined in our case by a reference point, instead of for the whole PF. However, to evaluate interactive processes, aspects related to the interaction with the DM should be quantified, but to the best of our knowledge, there do not exist quality indicators for assessing the performance of interactive methods. Therefore, the proposed ADM analyses the performance of each interactive method at each iteration using metrics developed for reference point based EMO methods, where preferences are provided a priori, before the solution process.

Once the ADM gets the solution sets produced by the algorithms at each iteration, performance metrics (denoted by m_i) are calculated for each set. At the end of the solution process, the ADM finds cumulative metric values for each phase, to evaluate the performances of the algorithms depending on the needs of each phase. For the learning phase, the metric values are considered until iteration L as $\sum_{i=1}^{L} m_i$, and for the decision phase, from iteration $L+1$ until the termination of the algorithm as $\sum_{i=L+1}^{L+D} m_i$.

In the literature, some performance metrics for a priori EMO methods have been proposed (see e.g. [9,11,17,22]). Any of these metrics could be utilized here. We use the R-metric [11] to measure the responsiveness of each algorithm for the provided reference points. The R-metric applies regular performance metrics (e.g., IGD) for the solutions in the ROI that is defined by the reference point. The size of the ROI is controlled by a parameter Δ. We employ the R-IGD (using IGD) since it is computationally efficient for a high number of objectives and it measures both convergence and diversity of solutions. The lower the R-IGD value, the better is the quality of the solutions of an algorithm in the ROI.

4 Experimental Studies

To demonstrate the applicability and usefulness of the proposed ADM, interactive versions of the EMO algorithms RVEA [3] and NSGA-III [6] are compared and we refer to them as iRVEA and iNSGA-III, respectively. The iRVEA algorithm is described in [8] and iNSGA-III was made interactive in a corresponding way. Their implementations as well as the proposed ADM are available in the open source DESDEO framework (https://desdeo.it.jyu.fi) in Python.

4.1 Search Behaviour of the ADM

To have a better understanding of the proposed ADM, we first illustrate its behaviour in the learning and decision phases, respectively. For this purpose, we conducted two experiments, one for each phase. The ADM iterated iRVEA and iNSGA-III for 20 times, using 50 generations per iteration for both algorithms.

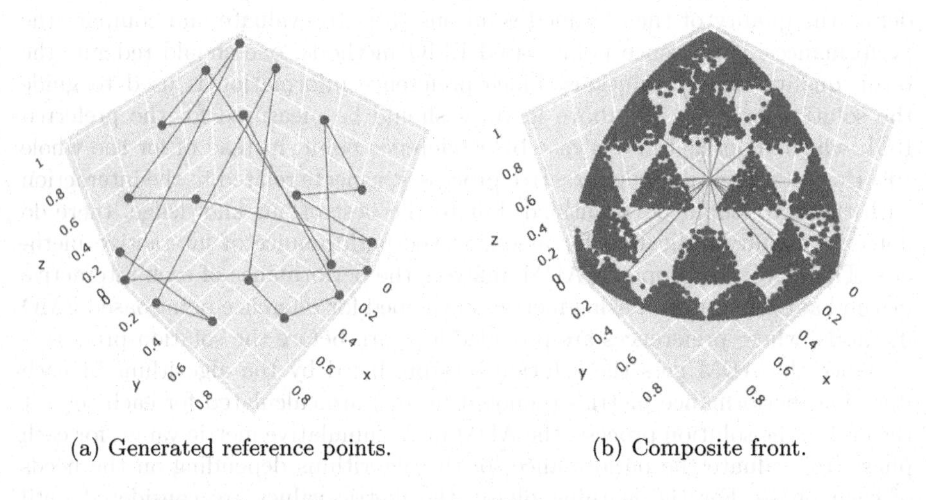

(a) Generated reference points. (b) Composite front.

Fig. 3. Search behaviour of the ADM in the learning phase.

Figure 3a illustrates the generated reference points as dots and Fig. 3b contains the composite front obtained by the algorithms after 20 iterations for the three-objective DTLZ2 problem [7]. As shown, the ADM explores the whole objective space. In Fig. 3a, the continuous line shows the search path taken by the ADM. After each iteration, the ADM found the least explored area, where the next reference point was generated. Figure 3b shows the composite front obtained by the two algorithms.

In the second experiment, we studied the ADM in the decision phase. After obtaining the first (randomly generated) reference point, the ADM found the most explored area and generated next reference points there. Therefore, it converged in one direction to refine the solutions in the same area. Due to page limitations, we cannot visualize this here. See Fig. S1 in the supplementary document (http://www.mit.jyu.fi/optgroup/extramaterial.html).

4.2 Experimental Settings

We conducted experiments on the benchmark problems DTLZ1-4 [7] with the number of objectives (k) ranging from 3 to 9, resulting in 28 different problems. The number of variables was set as $10+k-1$ [7]. For each algorithm, we executed 50 and 100 generations per iteration. All other parameters of iRVEA and iNSGA-III were set as in [3] and [6], respectively. We set the ADM parameters as follows: the number of iterations L for the learning phase 4, the number of iterations D for the decision phase 3, the lattice resolution 5 and the parameter Δ for the R-metric 0.3 for the learning phase and 0.2 for the decision phase. We made 21 independent runs with the proposed ADM by using the same experimental settings. We assumed that the solution process starts with a learning phase and continues with a decision phase. However, this could be easily adapted if one prefers several phases in different orders, instead.

4.3 Numerical Results

In Table 1, we give a concrete example of reference points generated by the ADM for one run with the three-objective DTLZ1 problem. As can be seen, the switch from one phase to another takes place between iterations 4 and 5. Additionally, the R-IGD metric values are listed per iteration to show the responses of algorithms after each iteration. As mentioned before, the cumulative metric values are eventually considered for both phases.

Table 1. Reference points and R-IGD values at each iteration for DTLZ1.

Iteration	Reference point	iRVEA	iNSGA-III
1	[0.9475, 0.9475, 2.8426]	6.00E−01	5.70E−01
2	[0.0000, 0.9417, 0.6278]	7.39E−01	7.38E−01
3	[0.0000, 0.0000, 0.5090]	5.52E−01	5.52E−01
4	[0.0000, 0.5067, 0.0000]	5.52E−01	5.52E−01
5	[0.0000, 0.2843, 0.1895]	7.64E−01	7.63E−01
6	[0.0000, 0.2684, 0.1789]	7.65E−01	7.62E−01
7	[0.0000, 0.2680, 0.1786]	7.64E−01	7.63E−01

Table 2. Cumulative R-IGD values for test problems with 4, 7 and 9 objectives.

Problem	k	Phase	iRVEA		iNSGA-III	
			Mean	Std. dev.	Mean	Std. dev.
DTLZ1	4	Learning	2.74E+00	3.49E−01	**2.62E+00**	1.66E−01
		Decision	2.03E+00	3.76E−01	**1.98E+00**	2.69E−01
	7	Learning	3.30E+00	4.49E−01	**2.73E+00**	2.13E−01
		Decision	**2.27E+00**	3.01E−01	2.38E+00	3.89E−01
	9	Learning	3.07E+00	2.59E−01	**2.80E+00**	2.76E−01
		Decision	**2.16E+00**	2.99E−01	2.46E+00	6.23E−01
DTLZ2	4	Learning	4.71E−01	3.43E−01	**2.21E−01**	1.21E−01
		Decision	5.73E−01	7.67E−01	**2.41E−01**	2.64E−01
	7	Learning	8.28E−01	5.28E−01	**4.15E−01**	2.10E−01
		Decision	1.47E+00	1.57E+00	**6.10E−01**	5.00E−01
	9	Learning	9.68E−01	5.32E−01	**4.93E−01**	1.78E−01
		Decision	1.23E+00	1.53E+00	**5.01E−01**	3.36E−01
DTLZ3	4	Learning	**4.47E−01**	3.09E−01	5.66E−01	2.31E−01
		Decision	1.42E−01	9.69E−02	**9.20E−02**	5.82E−02
	7	Learning	8.91E−01	5.14E−01	**8.26E−01**	2.95E−01
		Decision	**3.43E−01**	1.39E−01	7.50E−01	5.05E−01
	9	Learning	**7.03E−01**	3.34E−01	1.02E+00	3.70E−01
		Decision	**3.75E−01**	2.20E−01	9.87E−01	7.22E−01
DTLZ4	4	Learning	3.60E−01	4.43E−01	**2.19E−01**	1.92E−01
		Decision	3.68E−01	3.91E−01	**3.43E−01**	3.95E−01
	7	Learning	3.01E+00	2.19E+00	**7.29E−01**	3.70E−01
		Decision	1.18E+00	6.74E−01	**6.69E−01**	4.48E−01
	9	Learning	8.25E−01	3.11E−01	**7.60E−01**	2.83E−01
		Decision	8.26E−01	4.48E−01	**6.90E−01**	4.20E−01

Table 2 summarizes the cumulative R-IGD values of iRVEA and iNSGA-III for the problems considered with 4, 7 and 9 objectives and 50 generations per iteration. We report the phase of the solution process (learning, iterations 1–4, or decision, iterations 5–7), and the cumulative R-IGD metric values. The mean and the standard deviation of the metric values of 21 independent runs are presented and the best results are highlighted in bold.

An interesting observation can be made for DTLZ1 in Table 2. The R-IGD values of iNSGA-III are better than those of iRVEA in the learning phases for all numbers of objectives. This means that iNSGA-III responded better to the drastic changes of the provided preference by the ADM. In contrast, iRVEA outperformed iNSGA-III in the decision phases as the number of objectives increased. We conclude that iRVEA exploited better than iNSGA-III on DTLZ1.

For problems DTLZ2 and DTLZ4, iNSGA-III performed better than iRVEA for all numbers of objectives in both phases. A special case was seen for DTLZ3 with 4 objectives when iRVEA responded better in the learning phase, and iNSGA-III refined solutions better in the decision phase. The situation was the opposite for 7 objectives. Moreover, iRVEA results were better for both the learning and the decision phases for the 9-objective DTLZ3 problem. The results for other numbers of objectives can be found in Table S1 in the supplementary document. As mentioned, we also made the experiments by increasing the computational budget to 100 generations per iteration for both algorithms. We provide these results in Table S2 in the above-mentioned supplementary document.

We have shown that the proposed ADM enables us to study the suitability of each algorithm in the learning phase or in the decision phase. This type of information is hard, or even impossible, to get by just using a performance metric, specially if the metric is not formulated for interactive methods. Observe that the ADM has been designed to evaluate the search conducted by the algorithms distinguishing both phases, and it does not change the algorithms themselves. One should note that the findings of the analysis should not be generalized. The objective of this consideration was to demonstrate how the ADM can be applied, not to find any winner among the algorithms compared.

5 Conclusions

In this paper, an ADM was proposed for comparing reference point based interactive EMO methods. The ADM provides reference points in a different manner for the learning and the decision phases to reflect various objectives in the interactive solution process. In the learning phase, the ADM generates reference points to find more solutions in the least explored areas, while in the decision phase, it refines solutions inside the ROI identified. During the solution process, the ADM evaluates the performances of the algorithms by applying performance metrics that take into account the preference information and, at the end of the solution process, it calculates the metric values for each phase separately.

Experiments on benchmark problems were conducted to demonstrate how the ADM can be used to compare interactive EMO methods. To make the comparison meaningful, the same computational budgets were given to the algorithms compared and the R-IGD was used to evaluate their performances.

The proposed ADM has a modular structure and can be adapted for different needs. The comparison is meaningful since all algorithms to be compared apply the same preference information. In our future research, we intend to make it adaptive to decide automatically when to switch from the learning phase to the decision phase. We also plan to consider different types of preference information and develop quality indicators dedicated for interactive methods.

Acknowledgements. This research was partly funded by the Academy of Finland (grants 322221 and 311877) and is related to the thematic research area Decision Analytics utilizing Causal Models and Multiobjective Optimization (DEMO), https://www.jyu.fi/demo, at the University of Jyvaskyla. Ana B. Ruiz would like to thank the support of the Spanish government (project ECO2017-88883-R) and the Andalusian regional government (PAI group SEJ-532 and project UMA18-FEDERJA-024).

References

1. Barba-González, C., Ojalehto, V., García-Nieto, J., Nebro, A.J., Miettinen, K., Aldana-Montes, J.F.: Artificial decision maker driven by PSO: an approach for testing reference point based interactive methods. In: Auger, A., Fonseca, C.M., Lourenço, N., Machado, P., Paquete, L., Whitley, D. (eds.) PPSN 2018. LNCS, vol. 11101, pp. 274–285. Springer, Cham (2018). https://doi.org/10.1007/978-3-319-99253-2_22
2. Branke, J., Deb, K., Miettinen, K., Slowinski, R. (eds.): Multiobjective Optimization. Interactive and Evolutionary Approaches. Springer, Heidelberg (2008). https://doi.org/10.1007/978-3-540-88908-3
3. Cheng, R., Jin, Y., Olhofer, M., Sendhoff, B.: A reference vector guided evolutionary algorithm for many-objective optimization. IEEE Trans. Evol. Comput. **20**(5), 773–791 (2016)
4. Chugh, T., Jin, Y., Miettinen, K., Hakanen, J., Sindhya, K.: A surrogate-assisted reference vector guided evolutionary algorithm for computationally expensive many-objective optimization. IEEE Trans. Evol. Comput. **22**(1), 129–142 (2018)
5. Cornell, J.A.: Experiments with Mixtures: Designs, Models, and the Analysis of Mixture Data. Wiley, Hoboken (2011)
6. Deb, K., Jain, H.: An evolutionary many-objective optimization algorithm using reference-point-based nondominated sorting approach, part I: solving problems with box constraints. IEEE Trans. Evol. Comput. **18**(4), 577–601 (2013)
7. Deb, K., Thiele, L., Laumanns, M., Zitzler, E.: Scalable multi-objective optimization test problems. In: Proceedings of the 2002 Congress on Evolutionary Computation, pp. 825–830 (2002)
8. Hakanen, J., Chugh, T., Sindhya, K., Jin, Y., Miettinen, K.: Connections of reference vectors and different types of preference information in interactive multiobjective evolutionary algorithms. In: Proceedings of the 2016 IEEE Symposium Series on Computational Intelligence, pp. 1–8 IEEE (2016)
9. Hou, Z., Yang, S., Zou, J., Zheng, J., Yu, G., Ruan, G.: A performance indicator for reference-point-based multiobjective evolutionary optimization. In: Proceedings of the 2018 IEEE Symposium Series on Computational Intelligence, pp. 1571–1578. IEEE (2018)
10. Huber, S., Geiger, M.J., Sevaux, M.: Simulation of preference information in an interactive reference point-based method for the bi-objective inventory routing problem. J. Multi-Criteria Decis. Anal. **22**(1–2), 17–35 (2015)
11. Li, K., Deb, K., Yao, X.: R-Metric: evaluating the performance of preference-based evolutionary multiobjective optimization using reference points. IEEE Trans. Evol. Comput. **22**(6), 821–835 (2018)
12. Li, M., Yao, X.: Quality evaluation of solution sets in multiobjective optimisation: a survey. ACM Comput. Surv. **52**(2), 1–38 (2019)

13. López-Ibáñez, M., Knowles, J.: Machine decision makers as a laboratory for interactive EMO. In: Gaspar-Cunha, A., Henggeler Antunes, C., Coello, C.C. (eds.) EMO 2015. LNCS, vol. 9019, pp. 295–309. Springer, Cham (2015). https://doi.org/10.1007/978-3-319-15892-1_20
14. Meignan, D., Knust, S., Frayret, J.M., Pesant, G., Gaud, N.: A review and taxonomy of interactive optimization methods in operations research. ACM Trans. Interact. Intell. Syst. **5**(3), 17:1–17:43 (2015)
15. Miettinen, K.: Nonlinear Multiobjective Optimization. Kluwer Academic Publishers, Boston (1999)
16. Miettinen, K., Ruiz, F., Wierzbicki, A.P.: Introduction to multiobjective optimization: interactive approaches. In: Branke, J., Deb, K., Miettinen, K., Słowiński, R. (eds.) Multiobjective Optimization. LNCS, vol. 5252, pp. 27–57. Springer, Heidelberg (2008). https://doi.org/10.1007/978-3-540-88908-3_2
17. Mohammadi, A., Omidvar, M.N., Li, X.: A new performance metric for user-preference based multi-objective evolutionary algorithms. In: 2013 IEEE Congress on Evolutionary Computation, Proceedings, pp. 2825–2832. IEEE (2013)
18. Ojalehto, V., Podkopaev, D., Miettinen, K.: Towards automatic testing of reference point based interactive methods. In: Handl, J., Hart, E., Lewis, P.R., López-Ibáñez, M., Ochoa, G., Paechter, B. (eds.) PPSN 2016. LNCS, vol. 9921, pp. 483–492. Springer, Cham (2016). https://doi.org/10.1007/978-3-319-45823-6_45
19. Steuer, R.E.: Multiple Criteria Optimization: Theory, Computation and Application. Wiley, Hoboken (1986)
20. Szczepanski, M., Wierzbicki, A.P.: Application of multiple criteria evolutionary algorithm to vector optimization, decision support and reference-point approaches. J. Telecommun. Inf. Technol. **3**(3), 16–33 (2003)
21. Xin, B., Chen, L., Chen, J., Ishibuchi, H., Hirota, K., Liu, B.: Interactive multiobjective optimization: a review of the state-of-the-art. IEEE Access **6**, 41256–41279 (2018)
22. Yu, G., Zheng, J., Li, X.: An improved performance metric for multiobjective evolutionary algorithms with user preferences. In: Proceedings of the 2015 IEEE Congress on Evolutionary Computation, pp. 908–915. IEEE (2015)

To Boldly Show What No One Has Seen Before: A Dashboard for Visualizing Multi-objective Landscapes

Lennart Schäpermeier(✉), Christian Grimme, and Pascal Kerschke

Statistics and Optimization, University of Münster, Münster, Germany
{schaepermeier,christian.grimme,kerschke}@uni-muenster.de

Abstract. Simultaneously visualizing the decision and objective space of continuous multi-objective optimization problems (MOPs) recently provided key contributions in understanding the structure of their landscapes. For the sake of advancing these recent findings, we compiled all state-of-the-art visualization methods in a single R-package (moPLOT). Moreover, we extended these techniques to handle three-dimensional decision spaces and propose two solutions for visualizing the resulting volume of data points. This enables – for the first time – to illustrate the landscape structures of three-dimensional MOPs.

However, creating these visualizations using the aforementioned framework still lays behind a high barrier of entry for many people as it requires basic skills in R. To enable any user to create and explore MOP landscapes using moPLOT, we additionally provide a dashboard that allows to compute the state-of-the-art visualizations for a wide variety of common benchmark functions through an interactive (web-based) user interface.

Keywords: Multi-objective optimization · Continuous optimization · Visualization · Graphical user interface · R-package · Software

1 Introduction

When analyzing and developing optimization methods, scientists and practitioners from Operations Research traditionally use methods for visualizing problems and their characteristics. But also teaching Operations Research without graphical representation of search and objective space (often together in a functional landscape) is unthinkable: no textbook of Operations Research can do without the classical visual presentation of problem characteristics, showing multimodality, minima, maxima, plateaus, etc. [19].

For most of the last two decades, there has been only one approach in (continuous) multi-objective optimization [6], which provided analogous representations to the single-objective domain. Apart from the representation of the Pareto front, and in a few cases the Pareto set [5,17], only few researchers invested significant

H. Ishibuchi et al. (Eds.): EMO 2021, LNCS 12654, pp. 632–644, 2021.
https://doi.org/10.1007/978-3-030-72062-9_50

effort into the advancement of visualization [10,23]. Accordingly, the methodology for visualization in the domain of multi-objective optimization is still in its infancy. Yet, recently a few more methods have been introduced [15,20], which help to study multi-objective problems from a new perspective.

In fact, the recent insights into the decision space of multi-objective optimization problems (MOPs) have contributed greatly to revealing structural characteristics and to the understanding of benchmark problem landscapes [20] and contributed to the development of algorithmic ideas. That lead to an enhanced multi-objective gradient descent, which follows local efficient sets until sliding into dominating basins of attraction, eventually even leading to the Pareto set in many cases [11,12]. Further, these insights helped in the development of advanced single-objective local search, which is capable of escaping local optima by multiobjectivization and exploiting the known structures [1,22].

In order to provide the new visualization techniques for researchers and practitioners dealing with MO problems an `R`-package, `moPLOT`, has been published. However, this package is not accessible for anyone without knowledge of the scripting language `R`. This makes it difficult to analyze even well-known benchmarks. In addition, the current visualization techniques are often limited to bi-objective problems and two-dimensional decision spaces. Although visualization is naturally limited to three dimensions, the current methods do not exploit this potential w.r.t. search space. In contrast to common single-objective landscapes, solution quality (in the MO context expressed by mutual dominance or incomparability of solutions) is not visualized as height but only colored accordingly. In principle, that leaves room for considering visualization of a third dimension in decision space. Of course, this implies that a volume of solutions is shown, with the inner solutions being hidden by solutions further out. Accordingly, a technique for visually accessing the inner solutions has to be provided. Addressing all aforementioned issues, the key contributions of this work are the following:

1. We extend the R-package `moPLOT` by an interactive and user-friendly (web-based) dashboard, which allows users with little programming experience to visually explore the problem landscapes of common continuous MO benchmarks with two and three objectives and in two- and three-dimensional decision space settings.
2. Enabled by the interactive dashboard environment, we (a) extend the existing methods to cope with up to three dimensions, and (b) propose two views on the resulting volume of points, paving the way for visualizing MOPs with up to three variables and/or objectives.

While the first contribution enables many researchers to peek into before unseen (and for many still surprising) structures of MOPs, the latter allows to deepen the understanding of these structures and provides some generalization of the currently 2D-based interpretation of MO landscapes.

The remainder of this paper first provides some background on continuous multi-objective optimization and the state-of-the-art techniques for visualizing MOPs, followed by an introduction to our dashboard application and its core features. In Sect. 4, we then present our extensions of the existing visualizations to three-dimensional decision and objective spaces. Finally, Sect. 5 gives some concluding remarks and an outlook on future research potentials.

2 Background

In the following, we will briefly introduce important concepts and methods that are underlying this work. While Sect. 2.1 introduces the mathematical terminology used within this work, Sect. 2.2 will give an overview of the state-of-the-art methods for illustrating the landscapes of continuous MOPs.

2.1 Continuous Multi-objective Optimization

We consider box-constrained continuous MOPs $f : \mathbb{R}^p \rightarrow \mathbb{R}^k$ with p decision variables and k objectives, which are w.l.o.g. to be minimized. The box constraints limit the decision space to a feasible domain of $\mathcal{X} := [\mathbf{l}, \mathbf{u}] \subset \mathbb{R}^p$:

$$f(\mathbf{x}) = (f_1(\mathbf{x}), \ldots, f_k(\mathbf{x})) \stackrel{!}{=} \min \qquad \text{with} \qquad l_i \leq x_i \leq u_i,\ i = 1, \ldots, p. \quad (1)$$

For the remainder of this work, we consider two- and three-dimensional decision and objective spaces, i.e., $p \in \{2, 3\}$ and $k \in \{2, 3\}$.

In general, the solution of a MOP is the Pareto set $\mathcal{X}^* \subseteq \mathcal{X}$ of non-dominated solutions, i.e., the set of solutions $\mathbf{x}^* \in \mathcal{X}$ for which there exists no solution $\mathbf{x} \in \mathcal{X}$ with $f_i(\mathbf{x}) \leq f_i(\mathbf{x}^*)$, $i = 1, \ldots, k$, and $f_i(\mathbf{x}) < f_i(\mathbf{x}^*)$ for at least one i. The image of the Pareto set under f is known as the Pareto front $f(\mathcal{X}^*)$.

Similar to single-objective continuous optimization, local optimality in the multi-objective case is also defined relative to an ε-ball $B_\varepsilon(\mathbf{x}) \subseteq \mathcal{X}$ [11]. That is, a solution $\mathbf{x}$ is considered locally optimal (i.e., it is a locally efficient point), if it is not dominated by any other solution $\mathbf{y} \in B_\varepsilon(\mathbf{x})$ for some ε.

A useful concept when studying local optimality, and an important tool for some of the visualizations, is the multi-objective gradient (MOG)[1]. It points towards a common descent direction for all objectives, if one exists, and vanishes if a locally efficient point is reached. In the bi-objective case, a multi-objective gradient is found by the sum of the normalized single-objective gradients [15]. This definition, however, does not generalize for $k > 2$.

A more general definition of a MOG is given by the shortest vector in the convex hull of the normalized single-objective gradients [8]. In the bi-objective case, this can be simplified to almost yield the previous definition, up to an additional factor of $\frac{1}{2}$:

$$\nabla f(\mathbf{x}) := (\mathbf{u}_1 + \mathbf{u}_2)/2 \qquad \text{where} \qquad \mathbf{u}_i := \nabla f_i(\mathbf{x}) / \|\nabla f_i(\mathbf{x})\| \text{ with } i = 1, 2$$

if all single-objective gradients are non-zero and $\nabla f(\mathbf{x}) := 0$ otherwise.

[1] Note that we focus here on the case where the box constraints are inactive. Some discussion of how to handle the decision boundary can be found, e.g., in [20].

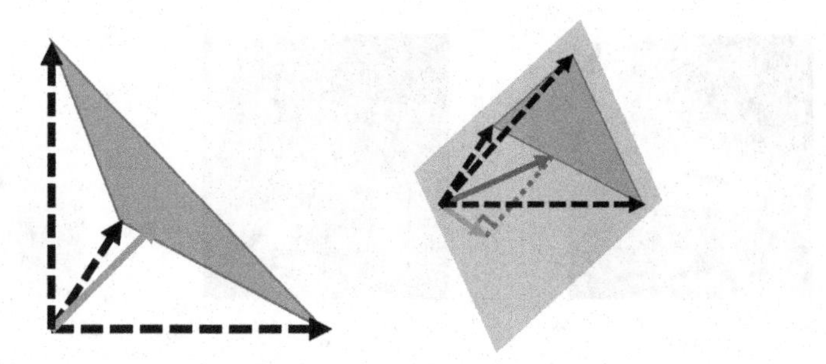

Fig. 1. Main cases for the tri-objective MOG in 3D. The (normalized) single-objective gradients are dashed. The shortest vector to the plane spanned by the three gradients is shown in red. If it lies inside their convex hull, this is the MOG (left image). Otherwise (right image), if it lies outside the convex hull, the MOG is given by the shortest bi-objective gradient (blue). (Color figure online)

For the tri-objective scenario ($k = 3$), we distinguish two cases. (1) For $p = 2$ decision variables: If the convex hull of the single-objective gradients does not contain the zero vector, the MOG is given by the bi-objective MOG of the two single-objective gradients with the largest angle between them, otherwise it is defined as 0. (2) For $p = 3$ decision variables, we can compute the orthogonal projection of the origin onto the plane that is spanned by the three normalized single-objective gradients. If the projection is located within the gradients' convex hull, this gives the MOG. Otherwise, the MOG touches the boundary of the convex hull, and it can be found as the shortest of the *pairwise* bi-objective gradients. A schematic illustration of this case is given in Fig. 1.

2.2 Visualizations of Continuous Multi-objective Landscapes

To our knowledge, there exist three main approaches for visualizing the interaction of decision and objective space of (two-dimensional) continuous MOPs. The cost landscape by Fonseca [6], the gradient field heatmap by Kerschke and Grimme [15], and the PLOT approach by this paper's authors [20]. Figure 2 gives an overview of the different visualizations for an exemplary MOP.

A commonality of all three techniques is that they are based on a discretized version of the MOP's decision space. On the basis of this grid, each technique proceeds differently and focuses on particular properties of the landscape.

Cost Landscape. Being the first technique for illustrating the decision space of a MOP's landscape, Fonseca's *cost landscapes* [6] focus on *global* optimality of points in the decision space – but mainly ignore local relationships. Therefore, the Pareto ranking is computed w.r.t. the evaluated points, and each point is assigned a height value corresponding to the number of points it is dominated by (plus one). These height values are then plotted on a logarithmic color scale, lending more weight to the (near) globally optimal points.

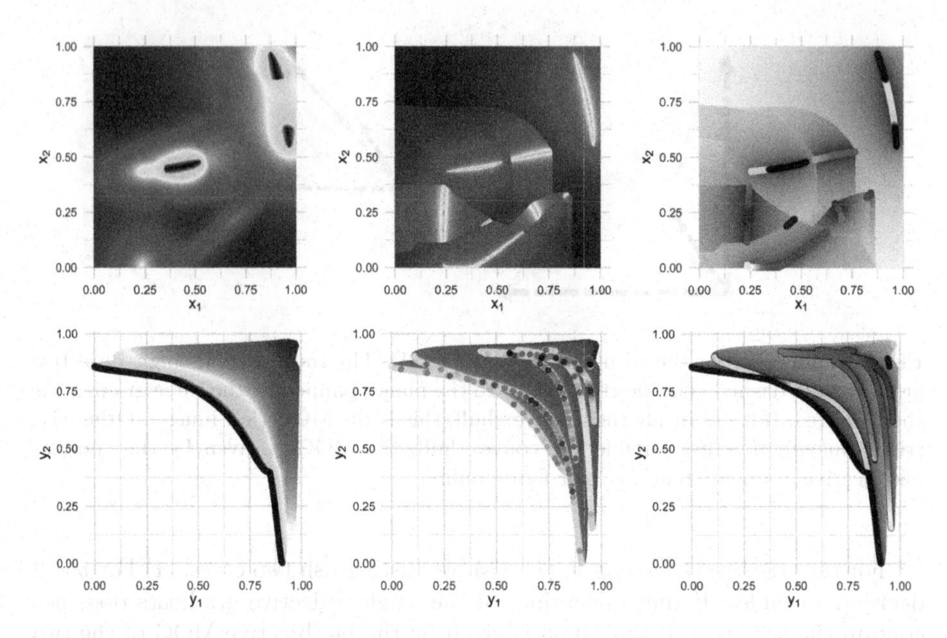

Fig. 2. Comparison of the cost landscape, gradient field heatmap, and PLOT visualizations (left to right), shown in the decision (top) and objective space (bottom). The color scales range from blue/black (better) to red/white (worse), respectively. The test function is a bi-objective MOP created with the MPM2 problem generator [26] (with 2 decision variables, 3 peaks, random topology, and seeds 4 and 8, respectively). The visualizations reveal several locally optimal sets, which are frequently intersected by a superposed basin of attraction, and a disconnected globally optimal set.

Gradient Field Heatmap. In contrast to the previous approach, the *gradient field heatmap* by Kerschke and Grimme [15] gives insight into the (approximate) placements of *locally* efficient sets and their corresponding basins of attraction. This is achieved by computing the MOG (see Sect. 2.1) for each point in a discretized decision space, and subsequently moving to the neighbouring cell closest to this descent direction. This process is repeated until a locally efficient point is reached or an already seen cell is revisited. Along this integration path, the lengths of the MOGs are cumulated and interpreted as the "height" value of the starting point. As with the cost landscape, the heights are then plotted on a logarithmic color scale.

Note that a more recent, but less detailed visualization of local dominance and basins of attraction is provided in [9]. Therein, points from locally dominance-neutral regions or from basins of attraction are represented in binary fashion.

PLOT. The *Plot of Landscapes with Optimal Trade-offs* (PLOT) [20] is the most recent and most sophisticated of the visualization techniques, combining the advantages of the two previous approaches. Based on the single- and multi-objective gradients, locally efficient points are approximated up to the resolution of the grid. Having these points determined explicitly, they can (1) be used to improve the visualization quality of the gradient field heatmap, and (2) be visualized in isolation from the remainder of the decision space. Within PLOT, the decision space is shown as a gray-scaled version of the gradient field heatmap, and all locally efficient points are colored according to their Pareto ranking. This provides complete information on the location(s) of the locally efficient points, their relation to global optima w.r.t. dominance, and their basins of attraction. Thereby it provides a comprehensive view of the structure of local and global optima for multi-objective benchmark problems.

2.3 The Plotting Library moPLOT

Although joint visualizations of the decision and objective space are essential to further our understanding of MOPs, so far, there exist only few works which considered these approaches. While one could argue that the gradient field heatmaps and PLOT visualizations are still rather new tools, the cost landscapes have been proposed two decades ago. Therefore, it is more likely that the limited usage of these approaches rather results from a lack of publicly available tools. We thus combined all the aforementioned visualization techniques within a single R-package called moPLOT (https://github.com/kerschke/moPLOT).

3 The moPLOT Dashboard

As stated in the previous subsection, the R-package moPLOT provides all functionalities to produce the different visualizations. However, for users who do not have experience in using R, it will be very cumbersome to produce these images. To lower this burden and facilitate the usage of moPLOT, we therefore decided to enhance our plotting library with a graphical user interface (GUI).

When designing the interface, we wanted to satisfy a wide variety of requirements: it should (1) be accessible as a standalone and ideally web-based dashboard application that is independent of R, (2) contain all state-of-the-art methods for visualizing MOPs, (3) allow to produce plots for the decision and objective space, (4) support interactive usage (e.g., zooming into regions of interest), (5) offer easy access to a wide variety of common MOPs (from well-established test suites like DTLZ [7] or ZDT [29], to more recent benchmarks like bi-objective BBOB [24] or the CEC 2019 test suite [28]), (6) enable the import of precomputed data (to avoid repetitive time-consuming computations), and (7) allow to export the created visualizations to the user's hard disk.

In the following, we will briefly describe the chosen architecture and technology underlying our application, and afterwards give a short tour through the dashboard.

3.1 Implementation

We developed a user-friendly dashboard that meets all the desired requirements, available at https://schaepermeier.shinyapps.io/moPLOT. At its core,

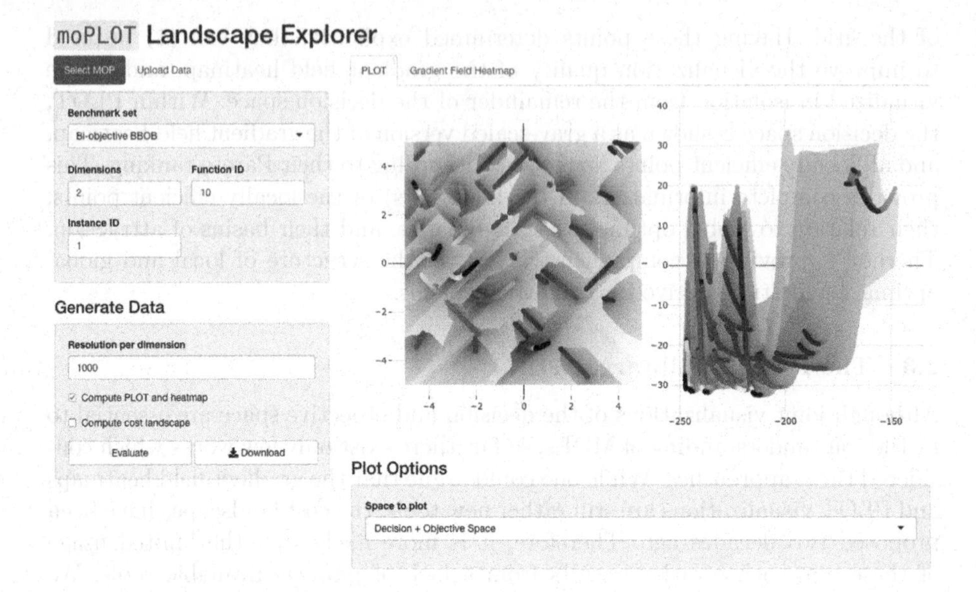

Fig. 3. Overview of the dashboard showing the PLOT visualization for a bi-objective BBOB function (Function ID: 10, Instance ID: 1) in the decision and objective space. In the left column, the user selects the test function, its parameters, and for which visualizations to generate data. Then, on the right, one of the available visualizations is selected, including some options for the chosen technique.

the `moPLOT` dashboard is a web application based on the R-package `shiny` [3]. Plots are produced using the `ggplot2` [27] package and their interactivity is ensured by the graphing library `plotly` [21]. In addition, we use the R-package `smoof` [2] to easily access a plethora of benchmark functions.

3.2 A Guided Tour Through the moPLOT Dashboard

Figure 3 shows a screenshot of the dashboard's starting page. Initially, the user chooses a MOP (that is to be visualized) by first selecting the corresponding benchmark set (if applicable) and afterwards specifying the desired function. If a MOP is parametrizable, its parameters are made available in the interface, with default values already filled in. In the given example, we selected a function from the bi-objective BBOB [24].

After selecting the MOP, controls for the resolution of the visualizations and for generating the required visualization data will be revealed. They contain options on whether to compute the data for the PLOTs and gradient field heatmaps, or for the (computationally much more expensive) cost landscapes. Most of the time, these controls can be left at their corresponding default values. In the example, a grid resolution of $1,000 \times 1,000$ is used and data is generated only for the gradient field heatmap and PLOT approaches.

Subsequent to data generation, the visualizations are shown in the main view of the dashboard. Here, the user can choose between all visualizations for which the data is available. Separate views of the decision and objective space, as well as a joint view of both spaces, as shown in the example, are available.

4 Visualization of Three-Dimensional MOPs

Building on this new and interactive visualization environment, some of the previously introduced concepts can be enhanced to enable the visualization of three-dimensional MOPs. Extending the methods to handle three-dimensional decision spaces is beneficial for studying tri-objective MOPs in particular: While the Pareto sets of the latter are, in general, degenerated in two-dimensional decision spaces, they can be better studied in three-dimensional decision spaces.

In the following, we first discuss our extensions of the existing approaches for 3D decision spaces in Sect. 4.1. Afterwards, we introduce two techniques that enable the interactive exploration of an MOP's characteristics, inspired by MRI scans and isosurface-based "onion layers", respectively (Sect. 4.2).

4.1 Extension of Visualization Techniques

The amount of changes required to adapt each visualization technique to 3D decision and objective spaces varies greatly depending on the specific visualization.

Cost Landscape. For the cost landscape, virtually no changes need to be implemented. The Pareto ranking is not affected by a change in decision space dimensionality and naturally scales with the numbers of objectives. Yet, it may become more computationally expensive. The only issue is the visualization of a 3D volume of data points (called *voxels* in graphics research), which will be addressed in the following subsection.

Gradient Field Heatmap. The gradient field heatmap requires some changes to handle 3D decision and objective spaces, respectively. To cover a 3D decision space, the approach used for following the gradient from one cell to the next one is changed such that decisions are made per decision space dimension. More precisely, for each variable x_i, $i = 1, 2, 3$, we compute the angle between the MOG and the plane spanned by $\{x_1, x_2, x_3\} \setminus \{x_i\}$. If that angle exceeds 22.5° (is below −22.5°), a step will be made in positive (negative) direction of x_i, otherwise one will keep the x_i value of the current cell. This decision is made for all $p = 3$ dimensions and by combining all results, we identify the next cell.

Moving from two to three objectives requires an additional change. Originally, the gradient field heatmap was only defined for bi-objective problems using the sum of normalized single-objective gradients as MOG. As introduced in Sect. 2.1, the MOG based on the convex hull can be used to generalize the MOG for more objectives. With the updated MOG definition, the remainder of the approach can remain unchanged.

PLOT. As PLOT builds on the other two visualizations, some of the main extensions have already been discussed. The only further component requiring changes is the approximation of the locally efficient points in a 3D decision space.

For a 2D PLOT, first-order critical points are determined based on the convex hull of the single-objective gradients in triangular neighborhoods. If the zero vector is contained in its interior, the points in this neighborhood are considered first-order critical. This directly extends to 3D using tetrahedral neighbourhoods.

Then, a stability analysis of the MOG vector field (oriented towards the descent direction) is conducted to ensure second-order optimality. It is considered stable, if the eigenvalues of its Jacobi matrix all have non-positive real parts. In 2D, this condition can be checked robustly using the MOG's divergence [20]. Analogously, for 3D, robust conditions based on matrix invariants are available [4]. Note that the critical points are degenerate (i.e. not singular) w.r.t. the MOG vector field, such that we can expect at least one eigenvalue to be zero.

4.2 Visualizing a 3D Volume of Points

Visualization of 3D volume elements (i.e., the voxels or regular cells, which fill out the considered 3D decision space) comes with the challenge of outer voxels obscuring the view from inner voxels. For overcoming this issue, we propose to use two different approaches: An MRI scan-like slicing of the volume [13] and an onion-like layer-based visualization based on isosurfaces [25].

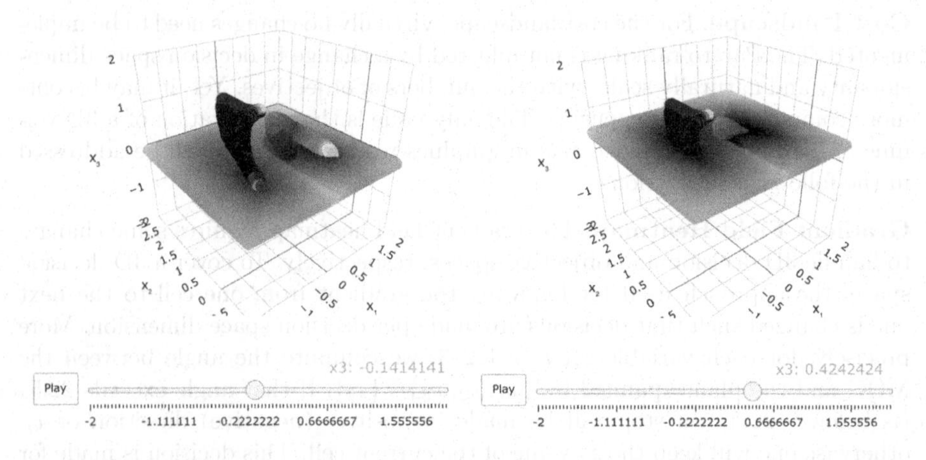

Fig. 4. The MRI scan technique applied to the PLOT visualization for a tri-objective MOP with one locally optimal (red) and one globally optimal (blue) set in the decision space. The user can slide the heatmap plane along the x_3 direction (another dimension may also be chosen), showing the intersection of the attraction basins of the two sets. (Color figure online)

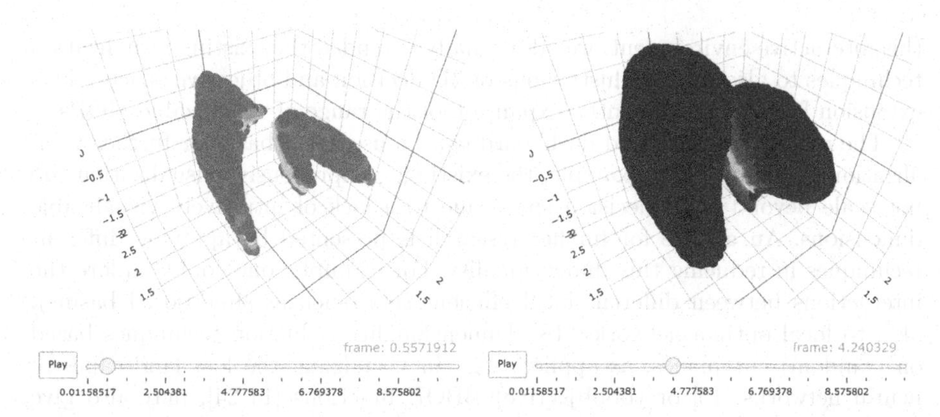

Fig. 5. The onion layers technique applied to the gradient field heatmap visualization for the same tri-objective MOP as in Fig. 4. The value of the isosurface can be adjusted by the user. Again, it can be observed that the basin of the locally optimal set is superposed by the basin of the globally optimal set.

MRI Scan. A general technique to reduce the amount of data shown is by *slicing* it along one of the three coordinate axes. Then, the placement along this axis can be controlled by the user, providing a complete picture of the visualization data. As this resembles an MRI scan, we use this name to refer to this method.

In the case of the PLOT technique, as there are only few locally efficient points, these can be permanently visualized, applying the slicing only to the gradient field heatmap. An example of this is given in Fig. 4.

Onion Layers. Based on the gradient field heatmap, we propose an alternative view on attraction basins that literally encapsulate the (locally) efficient sets in 3D decision space. Therefore, we consider the discretized isosurface[2] w.r.t. the computed voxels at a given height value. Different height thresholds reveal layers of the discretized decision space and allow the analysis of the volume and the attraction basins. As we could interpret the different isosurfaces as layers surrounding the efficient sets, we name this approach onion layer visualization.

In addition to illustrating the approximate location of the locally efficient points, the *onion layers* also reveal the intersection of their corresponding attraction basins. An example of this method is given in Fig. 5.

5 Conclusions

In this paper, we introduced the R-package `moPLOT` and its accompanying web-based, interactive and user-friendly dashboard. Thereby, we lowered the technical barrier to access and interactively explore the state-of-the-art techniques for visualizing continuous multi-objective optimization problems (MOPs). Building on

[2] A level set of a function $f : \mathbb{R}^n \to \mathbb{R}$ w.r.t. value $c \in \mathbb{R}$ is the set of points $\{(x_1, \ldots, x_n) \in \mathbb{R}^n : f(x_1, \ldots, x_n) = c\}$. The isosurface is a 3D level set.

this interactive environment, we additionally extended the existing visualization techniques to also enable illustrations of 3D decision and objective spaces. This extension leads to a substantial expansion of the range of visualizable MOPs.

The `moPLOT` package and dashboard can be used as a basis for further visualization techniques. For example, the existing, comprehensive visualizations do not scale beyond a 3D decision space due to a lack of intuitively visualizable dimensions. An avenue for further research is presented by applying different techniques in reducing this dimensionality. For example, one could explore the interactions between different local efficient sets (such as superposed basins), akin to local optima networks [18]. Dimensionality reduction techniques based on (random) cut planes, as applied, e.g., in visualizing the loss landscapes of neural networks [16] or (bi-objective) BBOB functions [14,24], may also give insights into the decision space of higher-dimensional MOPs.

Beyond gaining insights on the optimization landscapes themselves, another future extension can cover the visualization of optimizer trajectories. This would allow to study the interactions between landscape properties and the performance of different optimization algorithms.

Acknowledgements. The authors acknowledge support by the *European Research Center for Information Systems (ERCIS).*

References

1. Aspar, P., Kerschke, P., Steinhoff, V., Trautmann, H., Grimme, C.: Multi^3: optimizing multimodal single-objective continuous problems in the multi-objective space by means of multiobjectivization. In: Ishibuchi, H., et al. (eds.) Proceedings of the 11th International Conference on Evolutionary Multi-Criterion Optimization (EMO), EMO 2021, LNCS 12654, pp. 311–322. Springer, Cham (2021)
2. Bossek, J.: smoof: single- and multi-objective optimization test functions. R J. **9**, 103 (2017)
3. Chang, W., Cheng, J., Allaire, J., Xie, Y., McPherson, J.: shiny: web Application Framework for R (2020). https://CRAN.R-project.org/package=shiny, R package version 1.5.0
4. Chong, M.S., Perry, A.E., Cantwell, B.J.: A general classification of three-dimensional flow fields. Phys. Fluids A: Fluid Dyn. **2**(5), 765–777 (1990)
5. Coello Coello, C.A., van Veldhuizen, D.A., Lamont, G.B.: Evolutionary Algorithms for Solving Multi-Objective Problems, 2nd edn. Springer, Boston (2007). https://doi.org/10.1007/978-0-387-36797-2
6. da Fonseca, C.M.M.: Multiobjective Genetic Algorithms with Application to Control Engineering Problems. Ph.D. Thesis, Department of Automatic Control and Systems Engineering, University of Sheffield, September 1995
7. Deb, K., Thiele, L., Laumanns, M., Zitzler, E.: Scalable test problems for evolutionary multiobjective optimization. In: Abraham, A., Jain, L., Goldberg, R. (eds.) Evolutionary Multiobjective Optimization. Advanced Information and Knowledge Processing, pp. 105–145. Springer, London (2005). https://doi.org/10.1007/1-84628-137-7_6
8. Désidéri, J.A.: Multiple-Gradient Descent Algorithm (MGDA) for multiobjective optimization. C. R. Math. **350**(5–6), 313–318 (2012)

9. Fieldsend, J.E., Chugh, T., Allmendinger, R., Miettinen, K.: A feature rich distance-based many-objective visualisable test problem generator. In: Proceedings of the 2019 Genetic and Evolutionary Computation Conference (GECCO), pp. 541–549. ACM (2019)
10. Filipič, B., Tušar, T.: Visualization in multiobjective optimization. In: Proceedings of the 2020 Genetic and Evolutionary Computation Conference (GECCO) Companion, pp. 775–800. ACM (2020)
11. Grimme, C., Kerschke, P., Emmerich, M.T.M., Preuss, M., Deutz, A.H., Trautmann, H.: Sliding to the global optimum: how to benefit from non-global optima in multimodal multi-objective optimization. In: AIP Conference Proceedings, pp. 020052-1-020052-4. AIP Publishing (2019)
12. Grimme, C., Kerschke, P., Trautmann, H.: Multimodality in multi-objective optimization – more boon than bane? In: Deb, K., et al. (eds.) EMO 2019. LNCS, vol. 11411, pp. 126–138. Springer, Cham (2019). https://doi.org/10.1007/978-3-030-12598-1_11
13. Hansen, C.D., Johnson, C.R. (eds.): The Visualization Handbook. Elsevier, Amsterdam (2005)
14. Hansen, N., Finck, S., Ros, R., Auger, A.: Real-Parameter Black-Box Optimization Benchmarking 2009: Noiseless Functions Definitions. Technical report. RR-6829, INRIA (2009)
15. Kerschke, P., Grimme, C.: An expedition to multimodal multi-objective optimization landscapes. In: Trautmann, H., et al. (eds.) EMO 2017. LNCS, vol. 10173, pp. 329–343. Springer, Cham (2017). https://doi.org/10.1007/978-3-319-54157-0_23
16. Li, H., Xu, Z., Taylor, G., Studer, C., Goldstein, T.: Visualizing the Loss Landscape of Neural Nets. In: Advances in Neural Information Processing Systems, pp. 6389–6399 (2018)
17. Miettinen, K.: Nonlinear Multiobjective Optimization. International Series in Operations Research & Management Science, vol. 12. Springer, Boston (1998). https://doi.org/10.1007/978-1-4615-5563-6_11
18. Ochoa, G., Tomassini, M., Vérel, S., Darabos, C.: A study of NK landscapes' basins and local optima networks. In: Proceedings of the 10th Annual Conference on Genetic and Evolutionary Computation, pp. 555–562 (2008)
19. Preuss, M.: Multimodal Optimization by Means of Evolutionary Algorithms. Springer, Cham (2015)
20. Schäpermeier, L., Grimme, C., Kerschke, P.: One PLOT to show them all: visualization of efficient sets in multi-objective landscapes. In: Bäck, T., Preuss, M., Deutz, A., Wang, H., Doerr, C., Emmerich, M., Trautmann, H. (eds.) PPSN 2020, Part II. LNCS, vol. 12270, pp. 154–167. Springer, Cham (2020). https://doi.org/10.1007/978-3-030-58115-2_11
21. Sievert, C.: Interactive Web-Based Data Visualization with R, plotly, and shiny. Chapman and Hall/CRC, London (2020). https://plotly-r.com
22. Steinhoff, V., Kerschke, P., Aspar, P., Trautmann, H., Grimme, C.: Multiobjectivization of local search: single-objective optimization benefits from multi-objective gradient descent. In: Proceedings of the IEEE Symposium Series on Computational Intelligence (SSCI), Canberra, Australia (2020)
23. Tušar, T., Filipič, B.: Visualization of Pareto front approximations in evolutionary multiobjective optimization: a critical review and the prosection method. IEEE Trans. Evol. Comput. (TEVC) **19**(2), 225–245 (2015)
24. Tušar, T., Brockhoff, D., Hansen, N., Auger, A.: COCO: The Bi-Objective Black Box Optimization Benchmarking (bbob-biobj) Test Suite. arXiv preprint abs/1604.00359 (2016)

25. Udupa, J., Herman, G.: Display of 3-D digital images: computational foundations and medical applications. IEEE Comput. Graph. Appl. **3**(05), 39–46 (1983)
26. Wessing, S.: The Multiple Peaks Model 2. Technical report TR15-2-001, TU Dortmund University, Germany (2015)
27. Wickham, H.: ggplot2: Elegant Graphics for Data Analysis. Springer, New York (2016). https://doi.org/10.1007/978-0-387-98141-3. https://ggplot2.tidyverse.org
28. Yue, C., Qu, B., Yu, K., Liang, J., Li, X.: A novel scalable test problem suite for multimodal multiobjective optimization. Swarm Evol. Comput. **48**, 62–71 (2019)
29. Zitzler, E., Deb, K., Thiele, L.: Comparison of multiobjective evolutionary algorithms: empirical results. Evol. Comput. **8**(2), 173–195 (2000)

Interpretable Self-Organizing Maps (iSOM) for Visualization of Pareto Front in Multiple Objective Optimization

Deepak Nagar[1(✉)], Palaniappan Ramu[1(✉)], and Kalyanmoy Deb[2(✉)]

[1] ADOPT Laboratory, Department of Engineering Design, Indian Institute of Technology Madras, Chennai, India
{ed19s004,palramu}@smail.iitm.ac.in

[2] COIN Laboratory, Computer and Electrical Engineering, Michigan State University, East Lansing, USA
kdeb@egr.msu.edu

Abstract. Visualization techniques in design space exploration with high dimensional data are helpful in enhancing the decision making in the context of multiple objective optimization. Visualization of Pareto solutions obtained is crucial to understand the trade-off between the objectives as it enables intuitive decision making. However, such a task is not trivial beyond three dimensions. In this work, we propose using interpretable self-organizing map (iSOM), to visualize Pareto solutions for MOO problems involving n objectives ($n > 3$). iSOM enable simplified component plane plots that allow visual inspection of the Pareto fronts and also allow identifying clusters in the Pareto front and the corresponding design variables. Proposed approach is successfully demonstrated on 3 analytical examples.

Keywords: Multi-objective optimization · Pareto solution set · Pareto front · Visualization · Self-organizing maps · iSOM.

1 Introduction

The basic paradigm of design with multiple objectives is tradeoff. Betterment in one objective is at the expense of other objective(s). Often designers are faced with multiple objectives and it is important to understand the tradeoffs to make wise design decisions. Techniques such as Multiobjective Evolutionary Alogrithms(MOEAs) allow understanding these tradeoffs. During the conceptual or early design stages of a design process, designer is interested in exploring the design space that will correspond to good designs, or tradeoffs in the case of multiple objectives. The ability to explore the design space permits the designer to make more informed decisions. Its useful if the exploration has a visualization aspect. Holden et al.[3] discuss about future designer who will be equipped with plenty of data and the need for advanced visualization techniques to understand tradeoffs for making quick design decisions. Hence it is desirable

H. Ishibuchi et al. (Eds.): EMO 2021, LNCS 12654, pp. 645–655, 2021.
https://doi.org/10.1007/978-3-030-72062-9_51

to develop appropriate visualization techniques that are inherently understandable, to explore designs with tradeoffs, in the event of multiple objectives, and high dimensions.

The solution to a multiple objective problem is not a single point but a set of nondominated solutions (Pareto solutions), being chosen as optimal if no objective can be improved without compromising at least one other objective. In order to make a trade-off decision, it is useful for a decision maker to visualize the pareto front and pick solution that matches their preference. There are many visualization techniques used in multiple objective optimization for visualizing Pareto data. Few of those techniques include Scatterplot Matrix, Parallel axis plots, Nested axes plot, Tile plot, RadVis-3D, Self Organizing Maps (SOM), Chernoff faces and Hyperspace Pareto Frontier (HPF) [4,6,12]. While, each of the listed techniques have their own advantages and disadvantages, most of them become cluttered in high dimensions and are not easily interpretable. However, SOM is known to preserve the topology while taking the high-dimensional data as input and producing a low-dimensional, generally two-dimensional, representation of the input data. This property is attractive from visualizing high dimensions and understanding design space including tradeoffs among objectives. Obayashi et al.[8] use conventional SOM (cSOM) to visualize tradeoffs in multiple objectives and further, use the resulting SOM to generate a new SOM that generates clusters of design variables, which indicate roles of the design variables for design improvements and tradeoffs. Parashar et al. [9] use conventional SOM to visualize Pareto front of multiple objectives. Witowski et al. [14] investigate the use of Parallel Coordinate Plot, Hyper Radial Visualization and SOM for visualization of Pareto data in the context of a shape optimization case crash application executed with LS-OPT. Suzuki et al. [11] proposes a technique for visualizing Pareto optimal solution sets using growing hierarchical self-organizing maps (GHSOMs).GHSOM is a variation of conventional SOM which generates multiple maps while input spaces are divided at each layer. Thole et al. [12] develop an interpretable SOM (iSOM) that avoids self intersections and is inherently understandable. They use it for design space exploration and adaptive sampling. We propose the use of iSOM for exploring the Pareto front generated by using nondominated sorting genetic algorithm II (NSGA-II). The contribution of the proposed work will be the simplified visualization and understanding of the SOM plots of the Pareto fronts. It also permits direct mapping of regions of interest in the Pareto front to the corresponding patches in the design variable space. We demonstrate the proposed approach on Examples with 2–4 objectives and 2–10 design variables. Rest of the paper is organized as follows. Discussions on SOM and iSOM are provided in Sect. 2. Section 3 discusses the results followed by conclusion in Sect. 4.

2 Self-Organizing Maps (SOM)

Self-Organizing Maps (SOM) are a type of Artificial Neural Network which is based on unsupervised learning [6,10,13]. It provides a mapping from N-dimensional input to a low dimensional, generally 2D space. Mapping preserves the topology of the data so that similar data points will be mapped to nearby locations on the map so that the relative distances between input data points is conserved. This allows the user to identify the pattern in the input data. SOM implementation consist of nodes (neurons) associated with a weight vector of the same dimension as the input data vectors. The nodes are arranged in a 2D space using a hexagonal or rectangular topology. If there are n design variables in the dataset namely $[x_1, x_2, ..., x_n]$ and response y, then each input vector (corresponding to a DoE point) to SOM will be $v_i = [x_{i1}, x_{i2}, ..., x_{in}, y_i]$ and each SOM node w_j will have weight vectors $[m_{j1}, m_{j2}, ..., m_{jn}, m_{jout}]$. First n weights correspond to design variables and the last weight corresponds to response. SOMs commonly use the unified distance matrix (U-Matrix) to store each node's average distance to its closest neighbors. In the final SOM output, different colors are used to represent each node's distance to adjacent nodes. A large distance (typically denoted by red color) between the neurons correspond to a large gap between the codebook values in the input space.

2.1 Conventional SOM (cSOM) Algorithm

Training a SOM algorithm iteratively consists of choosing a winner neuron called Best Matching Unit (BMU). The BMU is the neuron that has the least distance from the selected data point. Upon identifying the BMU, the weights of the neighboring nodes are modified based on an update rule. The BMU selection and updation are repeated until a prescribed number of iterations or error convergence metrics are met. The output of the iterative algorithms is the trained SOM weights and these weights are used to construct the U-matrix (Unified Distance Matrix) and component plane plots. The U-Matrix representation of the SOM is a single plot that captures the distances between the neurons (in all the dimensions). It shows the local distance structure of a topology preserving projection of a high dimensional data set. This is sufficient to detect cluster boundaries. It is composed of the distances from the map units to it's nearest neighbors which can be visualized in various ways, but most often a color map is used to visualize the distance values. Low denote that the nodes are closer to each other, while high values convey cluster borders where the values are rapidly changing from one value to other.

A component plane plot helps in visualizing weights corresponding to a particular attribute (input or output) for all the map units in the SOM. In the component plane each node has the same position regardless of which attribute is being visualized, making it easy to compare different attributes. For example, if SOM is trained over data with columns $[x, y, f(x, y)]$ to give weights $[m_x, m_y, m_f\ ]$, the x component plane will represent values of m_x and so on. Component planes that look similar suggest that the respective variables are correlated.

2.2 Interpretable SOM (iSOM) Algorithm

cSOM tends to fold and self intersect [7] which affects the visualization and interpretation of design space. iSOM avoids these folds and self intersections while preserving the topography of the function that is approximated [12]. iSOM which is inherently interpretable can further be used explore the design space. iSOM algorithm is similar to the regular SOM but for BMU selection. The BMUs are computed using the input space alone as against using both input and response values in the conventional SOM. Then, the weights of the respective node and the ones in its neighborhood are updated along the response values.This modification on algorithm allows avoiding the folds or intersections. In addition, the component plane plots are also in an ordered fashion permitting simplified understanding of component planes and their comparison. For further details the reader is refered to [12].

2.3 Methodology

We propose using iSOM to visualize Pareto fronts so that the tradeoff can be visualized directly, aiding in design decisions. Upon obtaining the Pareto front, the Pareto solution set is subjected to iSOM training. iSOM trains along the response dimension at a time; however, in MOO problems we have more than one objective. If iSOM for each objective is trained based on separate map grid, then mapping back to the input space becomes impossible. Hence, a weighted sum of all the responses is used to initialize the SOM grid. The weights for each of the response is proportional to the range of the response. The same grid is used to train Pareto solutions of each objective function on an individual basis. The resulting plots will consist of component planes of the input variables and the objective functions. The component plane plots of the objective functions can be used to understand the tradeoffs. During visualization of Pareto fronts using iSOM, U-matrix is constructed using updated weights corresponding to the objective functions.

3 Numerical Examples

In this section, we discuss the examples on which the proposed approach will be illustrated. We consider two examples from [15] and the ZDT1 example [2].

Example-1
The first example we consider here consists of four-objectives and two-input variables and stated as follows:

$$F1 = \sqrt{(x_1 - 50)^2 + (x_2 - 50)^2} \tag{1}$$

$$F2 = \sqrt{(x_1 - 50)^2 + (x_2 - 150)^2} \tag{2}$$

$$F3 = \sqrt{(x_1 - 150)^2 + (x_2 - 50)^2} \tag{3}$$

$$F4 = \sqrt{(x_1 - 150)^2 + (x_2 - 150)^2} \tag{4}$$

Example-2
The second example consists of three objectives with two-input variables, stated as follows:

$$F1 = 0.5 \times (x_1^2 + x_2^2) + sin(x_1^2 + x_2^2) \tag{5}$$

$$F2 = \frac{(3x_1 - 2x_2 + 4)^2}{8} + \frac{(x_1 - x_2 + 1)^2}{27} + 15 \tag{6}$$

$$F3 = \frac{1}{x_1^2 + x_2^2 + 1} - 1.1 \times \exp(-x_1^2 - x_2^2) \tag{7}$$

Example-3
The third example is a biobjective ZDT1 problem with ten input variables.

$$F1 = x_1 \tag{8}$$

$$F2 = g \cdot \left(1 - \sqrt{\frac{F1}{g}}\right) \tag{9}$$

$$\mathrm{g}\,(x_2, \ldots, x_n) = 1 + \frac{9}{n-1} \sum_{i=2}^{n} x_i \tag{10}$$

$$0 \leq x_i \leq 1; n = 10$$

In all the examples stated above, the objectives are to be minimized. NSGA-II, the well known fast sorting and elite multi objective genetic algorithm [1] is used to generate the Pareto data. The Pareto front for examples 1, 2 and 3 are presented in Figs. 1, 2 and 3 respectively.

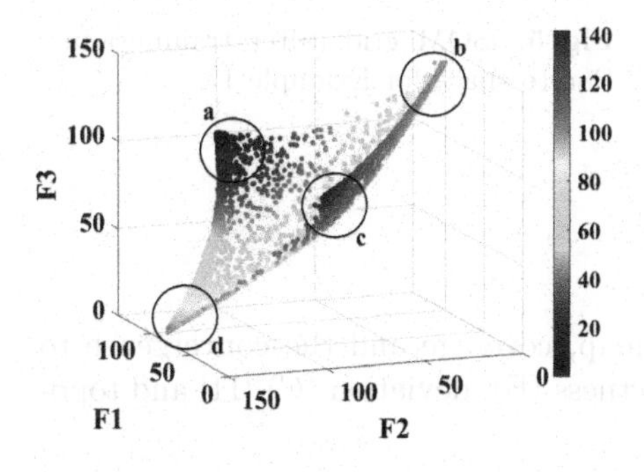

Fig. 1. Pareto front for Example 1

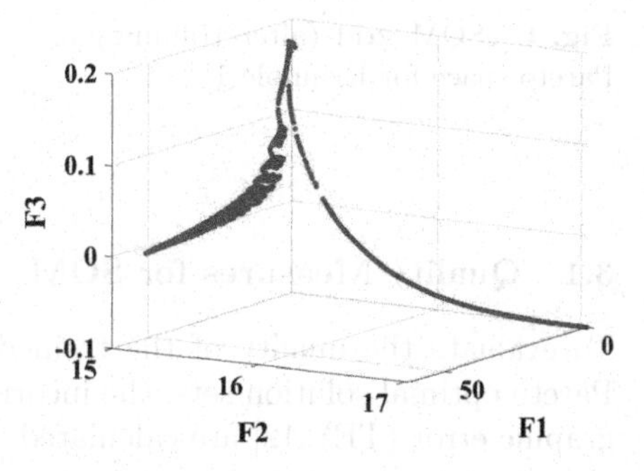

Fig. 2. Pareto front for Example 2

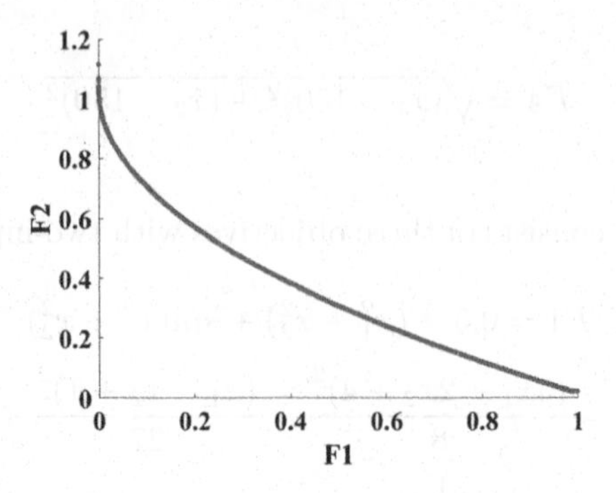

Fig. 3. Pareto front for Example 3

Pareto optimal solution sets serve as the input data to train the SOM. In iSOM, each objective function is trained one at a time in the design variable space and the weights of these objective functions are used to generate the U-matrix. It is to be noted that the map grid for both cSOM and iSOM are initialized with same number of nodes and map grids after training for Examples 1 and 2 are shown in Figs. 4, 5 and Figs. 6, 7 respectively.

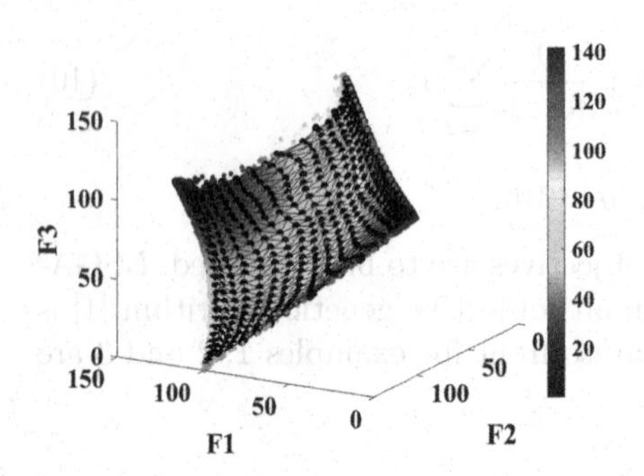

Fig. 4. cSOM grid (after training) in Pareto space for Example 1

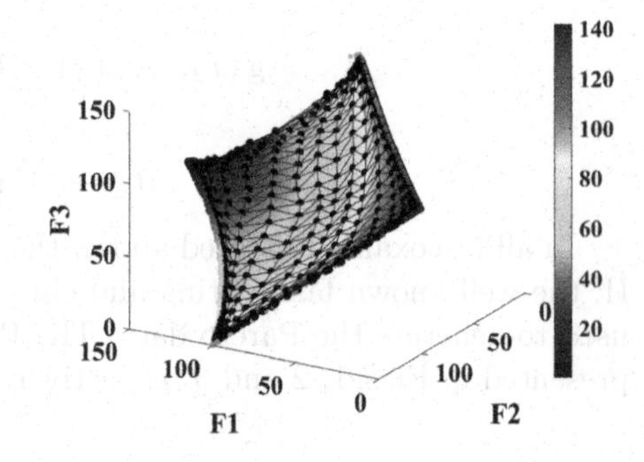

Fig. 5. iSOM grid (after training) in Pareto space for Example 1

3.1 Quality Measures for SOM

To evaluate the quality of the trained map, coverage and the convergence to Pareto optimal solution sets, the incorrectness (E), deviation (C) [11] and topographic error (TE) [12] are calculated.

Incorrectness (E) is the average of errors between node on the map and its nearest Pareto optimal solution [11]. Therefore, it means that the smaller the value of E, more the node on the map converge to the Pareto optimal solution.

Deviation (C) is the average of errors between Pareto optimal solution and its nearest node [11]. Therefore, it means that the smaller the value of C, more uniformly the node on the map covers the input Pareto optimal solution set.

Topographic error (TE) measure the percentage of data vectors for which the first- and second-BMUs are not adjacent units. Therefore, it means that smaller the value of TE, more accurately the structure of the input space is preserved by SOM [12]. It allows capturing local discontinuities [5].

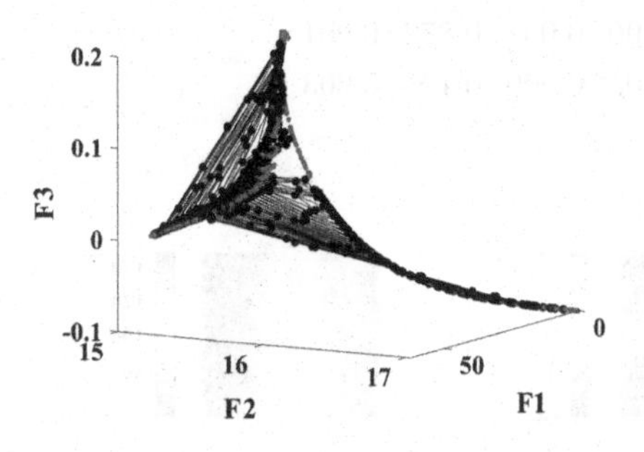

Fig. 6. cSOM grid (after training) in Pareto space for Example 2

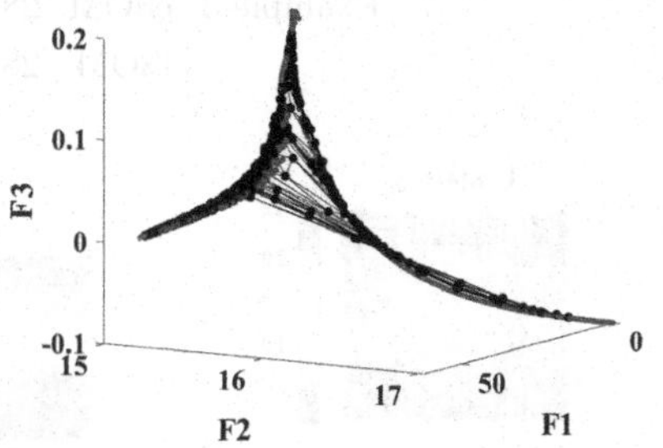

Fig. 7. iSOM grid (after training) in Pareto space for Example 2

4 Results and Discussion

Figures 8, 9, 10, 11, 12 and 13 show visualization maps of cSOM, and iSOM for Pareto optimal solution sets (Pareto front) for Examples 1, 2 and 3. Figures 4, 5, 6 and 7 show the orientation of trained map grid for SOM and iSOM along with the Pareto optimal solutions (input data) on the objective space for Examples 1 and 2. It should be noted that coloured points show Pareto optimal solution and black points are the reference vector values(nodes). iSOM shows better convergence of nodes on the Pareto front, which indicates that iSOM captures more accurate variation of the Pareto front than the cSOM. Incorrectness (E), deviation (C), and Topographic error (TE) for maps obtained for each Example are presented in Table 1.

In Example-1 the Pareto front is not complex due to which both SOM and iSOM captures the correct variation of the Pareto front and generate similar visualization maps as shown in Fig. 8 and 9 but in iSOM more nodes on the map converge to the Pareto optimal solution as shown in Fig. 4 and 5 and this can also be validated by comparing the Incorrectness (E) value. Smaller the value of E, more nodes on the map converge to the Pareto optimal solution. Here U-matrix did not provide much information as there is no cluster or discontinuity on the Pareto front. It is observed that U-matrix presents very low values at the corners which denotes data points are closer to each other in that region whereas in the middle region they are not. The original Pareto front presented in Fig. 1 supports the inference made from observing the U-matrix.

Table 1. Number of data points (K), Number of nodes (N_v), incorrectness (E), Topographic error (TE), and deviation (C) of the obtained maps.

		K	N_v	E	TE	C
Example-1	cSOM	3500	625	1.383	0.094	1.776
	iSOM	3500	625	1.298	0.022	2.253
Example-2	cSOM	3500	625	0.070	0.620	0.072
	iSOM	3500	625	0.038	0.173	0.092
Example-3	cSOM	2813	900	0.001	0.872	0.001
	iSOM	2813	900	0.009	0.128	0.003

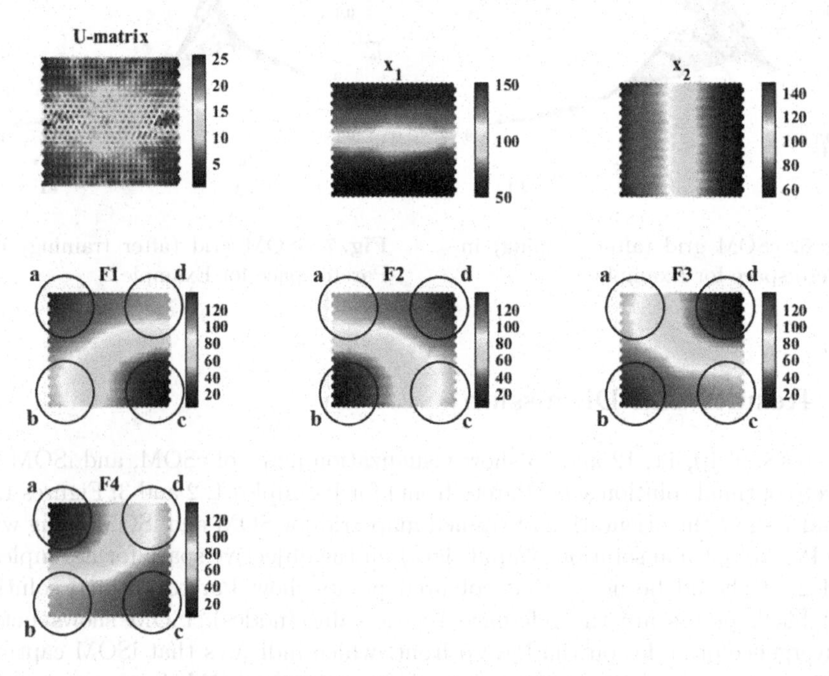

Fig. 8. cSOM for Pareto front of Example 1

To observe the tradeoff between the objectives, four regions (a, b, c and d) are highlighted on the objective component planes as shown in Figs. 8 and 9. In region (d), objective function($F1$) attains its maximum value whereas in region (a, c and b), other objective functions ($F2, F3$ and $F4$) attain their maximum values respectively. The variation in the input variables corresponding to these regions can be observed by highlighting the same regions in the respective component planes. We can use these component planes to generate more information about our Pareto front. For example, by observing component planes of $F1$ and $F4$ in Figs. 8 and 9, it can be inferred that $F1$ and $F4$ show conflicting behavior. $F1$ shows an increasing trend from region b to d whereas $F4$ shows decreasing

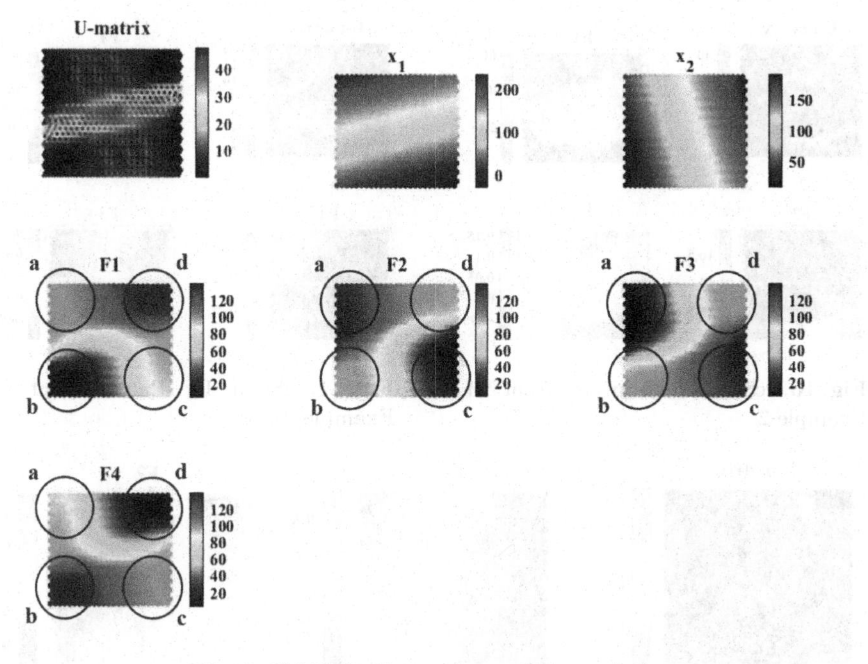

Fig. 9. iSOM for Pareto front of Example 1

trend for the same region. Similar conflicting behavior can be observed for $F2$ and $F3$ from region c to a. Again, the inferences are supported by the actual Pareto front presented in Fig. 1. The objective of our work is to interpret the Pareto function space using SOM component planes. Hence, only response component planes will be presented from here onwards.

In Example-2 the Pareto front is complex due to which there are folds and self intersections in the cSOM grid as shown in Fig. 6 and it does not captures the correct variation of the Pareto front. Whereas in iSOM there are no folds or self intersection in the iSOM grid as shown in Fig. 7 due to which iSOM captures the correct variation of the Pareto front.This is also validated by comparing the Topographic error (TE) value. Smaller the value of TE, better is the topography preservation. This can also be proved by comparing the component planes of the objective functions generated by cSOM and iSOM. Lets consider a region on component plane where F1 is minimum as shown in Fig. 10 and 11. Comparing on how F2 and F3 are varying in this region, one can observe that iSOM captures the fact that F2 and F3 have conflicting behavior in this region, where as cSOM does not provide such information about F2 and F3. This can be validated from the actual Pareto front as shown in Fig. 2.

Example-3 is a high dimension example which has a Pareto front with less variation in the objective space due to which both cSOM and iSOM capture the correct variation of the Pareto front as shown in Fig. 12 and 13. Figures reveal that both the objectives are majorly governed by the design variable x_1 but

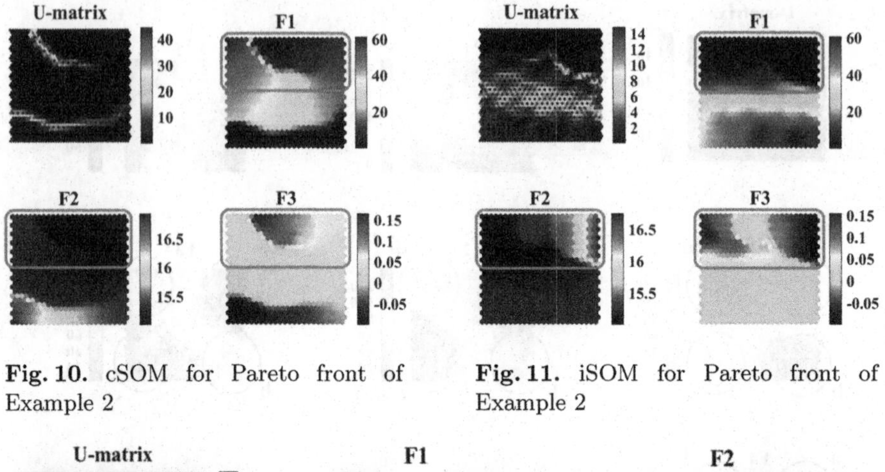

Fig. 10. cSOM for Pareto front of Example 2

Fig. 11. iSOM for Pareto front of Example 2

Fig. 12. cSOM for Pareto front of Example 3

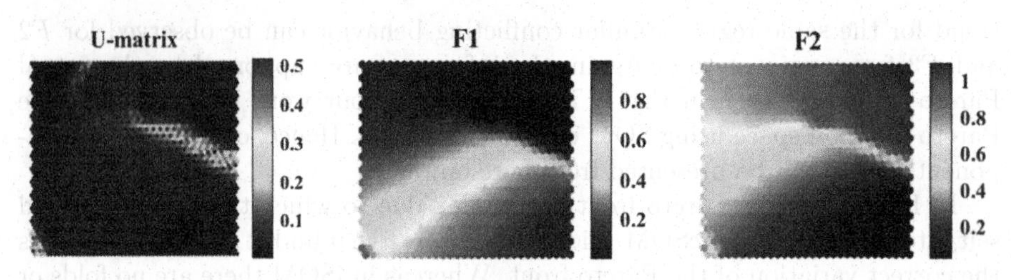

Fig. 13. iSOM for Pareto front of Example 3

iSOM preserves the structure of function space more accurately and this can be validated by comparing the Topographic error (TE) value, smaller the value of TE, the more accurately the structure of the input space is preserved by SOM.

5 Conclusion

iSOM was used to plot the Pareto fronts of multiple objective functions. It was observed that iSOM performs better in terms of Topographic Error, Deviation and Incorrectness. These metrics measure the extent of original function preservation and the extent by which the SOM nodes are close to the points in Pareto front. In addition, iSOM plots also allow understanding clusters in the input space that correspond to regions of interest in the Pareto front.

Acknowledgments. This work is supported in part by American Express Lab for Data Analytics, Risk and Technology, Indian Institute of Technology Madras.

References

1. Deb, K., Pratap, A., Agarwal, S., Meyarivan, T.: A fast and elitist multiobjective genetic algorithm: NSGA-II. IEEE Trans. Evol. Comput. **6**(2), 182–197 (2002). https://doi.org/10.1109/4235.996017. ISSN 1089778X
2. Deb, K., Chaudhuri, S., Miettinen, K.: Towards estimating nadir objective vector using evolutionary approaches. In: GECCO 2006 - Genetic and Evolutionary Computation Conference, vol. 1, pp. 643–650 (2006). https://doi.org/10.1145/1143997.1144113
3. Holden, C., Keane, A.: Visualization methodologies in aircraft design. In: 10th AIAA/ISSMO Multidisciplinary Analysis and Optimization Conference, p. 4449 (2004)
4. Ibrahim, A., Martin, M.V.: 3D-RadVis: Visualization of Pareto Front in Many-Objective Optimization, July 2016
5. Kiviluoto, K.: Topology preservation in self-organizing maps. In: IEEE International Conference on Neural Networks - Conference Proceedings, vol. 1, pp. 294–299 (1996)
6. Kohonen, T.: Exploration of very large databases by self-organizing maps. In: IEEE International Conference on Neural Networks - Conference Proceedings, vol. 1 (1997). https://doi.org/10.1109/ICNN.1997.611622. ISSN 10987576
7. Lopez-Rubio, E.: Improving the quality of self-organizing maps by self-intersection avoidance. IEEE Trans. Neural Netw. Learn. Syst. **24**(8), 1253–1265 (2013). https://doi.org/10.1109/TNNLS.2013.2254. ISSN 2162237X
8. Obayashi, S., Sasaki, D.: Visualization and data mining of Pareto solutions using self-organizing map. In: Fonseca, C.M., Fleming, P.J., Zitzler, E., Thiele, L., Deb, K. (eds.) EMO 2003. LNCS, vol. 2632, pp. 796–809. Springer, Heidelberg (2003). https://doi.org/10.1007/3-540-36970-8_56
9. Parashar, S., Pediroda, V., Poloni, C.: Self Organizing Maps (SOM) for design selection in robust multi-objective design of aerofoil. In: 46th AIAA Aerospace Sciences Meeting and Exhibit, Aerospace Sciences Meetings. American Institute of Aeronautics and Astronautics, January 2008. https://doi.org/10.2514/6.2008-914
10. Song, L., Guo, Z., Li, J., Feng, Z.: Optimization and knowledge discovery of a three-dimensional parameterized vane with nonaxisymmetric endwall. J. Propul. Power **34**(1), 234–246 (2018). https://doi.org/10.2514/1.B36014. ISSN 07484658
11. Suzuki, N., Okamoto, T., Koakutsu, S.: Visualization of Pareto optimal solution sets using the growing hierarchical self-organizing maps. Electron. Commun. Jpn. **100**(1), 3–17 (2017). https://doi.org/10.1002/ecj.11915. ISSN 19429541
12. Thole, S.P., Ramu, P.: Design space exploration and optimization using self-organizing maps. Struct. Multidiscip. Optim. **62**, 1071–1088 (2020). https://doi.org/10.1007/s00158-020-02665-6. ISSN 1615–1488
13. Torkkola, K., Gardner, R., Kaysser-Kranich, T., Ma, C.: Exploratory analysis of gene expression data using self-organizing maps. In: Proceedings of the Joint Conference on Information Sciences, vol. 5, issue 2, pp. 782–785 (2000)
14. Witowski, K., Liebscher, M., Goel, T.: Decision making in multi-objective optimization for industrial applications-data mining and visualization of Pareto data. In: Proceedings of the 7th European LS-DYNA Conference, Salzburg, Austria (2009)
15. Zhen, L., Li, M., Cheng, R., Peng, D., Yao, X.: Multiobjective Test Problems with Degenerate Pareto Fronts, pp. 1–20 (2018). http://arxiv.org/abs/1806.02706

Applications

An Investigation of Decomposition-Based Metaheuristics for Resource-Constrained Multi-objective Feature Selection in Software Product Lines

Yi Xiang[1], Xue Peng[2(✉)], Xiaoyun Xia[3], Xianbing Meng[1], Sizhe Li[1], and Han Huang[1(✉)]

[1] School of Software Engineering, South China University of Technology, Guangzhou 510006, China
hhan@scut.edu.cn
[2] School of Mathematics and Systems Science, Guangdong Polytechnic Normal University, Guangzhou, China
[3] College of Mathematics Physics and Information Engineering, Jiaxing University, Jiaxing 314001, China

Abstract. The multi-objective feature selection from software product lines is a problem that has attracted increasing attention in recent years. However, current studies on this problem suffer from two main limitations: (1) resource constraints are naturally adhered to the feature selection process, but they are completely ignored or inadequately handled, and (2) there is a strong preference to the use of evolutionary algorithms for the feature selection problem, and the suitability of other multi-objective metaheuristics remains to be fully explored. To address the above two limitations, this paper proposes the multi-objective feature selection with multiple linear and non-linear resource constraints, and investigates the performance of decomposition-based metaheuristics on the proposed problem. We construct a number of problem instances using both artificial and real-world software product lines, considering 2, 3 and 4 objectives. Experimental results show that, within the decomposition-based framework, reproduction operators that are based on probabilistic models (PM) perform better than genetic operators. Moreover, we demonstrate that adaptation of weight vectors can further improve the performance. Finally, we show that PMAD (the combination of *PM*-based reproduction operators, adaptation of weight vectors and decomposition-based framework) is better than several state-of-the-art algorithms when handling this problem.

Keywords: Multi-objective feature selection · Decomposition-based framework · Resource constraints · Probabilistic model · Adaptive weight vectors

H. Ishibuchi et al. (Eds.): EMO 2021, LNCS 12654, pp. 659–671, 2021.
https://doi.org/10.1007/978-3-030-72062-9_52

1 Introduction

In the past few years, the multi-objective feature selection (MOFS) [21] from software product lines (SPLs) has attracted growing attention in both evolutionary multi-objective (EMO) and search-based software engineering (SBSE) communities [18]. An SPL defines a family of software products, which differ from each other but share some common functionalities [3]. The commonality and variability of SPLs are often compactly represented by a feature model (FM), in which each feature is used to describe one of the system functionalities [10]. An FM defines constraints among features, and thus specifies which combinations of features are valid. The MOFS aims at choosing a subset of features that conform to all the constraints defined by the FM. At the same time, this feature selection process should be 'optimal' with respect to multiple and often conflicting objectives like maximizing the number of selected features and minimizing product cost [21]. Since a set of features means to a configuration or a product of an SPL, equally we name it a solution, which stands for a configuration or a product in software engineering [10].

In practice, real-world FMs may have hundreds or even thousands of features, and may involve a huge number of constraints. For example, the Linux X86 kernel, which was widely investigated in prior work [9,20,29], contains 6,888 features and 343,944 constraints. It is extremely difficult or even impossible to manually select optimal features in such a large and highly-constrained decision space. This selection is made even harder when multiple (e.g., five or eight [10]) objectives should be optimized simultaneously. Therefore, automatic techniques are needed to reduce the selection efforts. In the past decades, many multi-objective evolutionary algorithms (MOEAs) have been adopted as automatic techniques to handle the MOFS problem [7,9,10,20,21,27,29–31].

However, current studies on this problem suffer from two limitations. First, resource constraints are naturally adhered to the optimal feature selection process [8,23,24], but they are completely ignored or inadequately handled. Resource constraints are related to stakeholders' requirements [8]. A resource can be the CPU time, memory, or cost for a product [2,8,19,23]. The consumed resources must not exceed available values. Indeed, most prior works did not consider resource constraints. In our recent work [26], we have formulated one or two linear resource constraints when defining the MOFS problem. But, nonlinear resource constraints are not introduced even though they indeed exist in practice [19]. Second, Pareto-dominance-based MOEAs (e.g., NSGA-II [5] and VaEA [28]) and indicator-based MOEAs (e.g., IBEA [33]) have been extensively adopted to handle the feature selection problem in previous works, e.g., [9,10,21,29]. However, the use of decomposition-based framework [32] seems to be very limited [6,10], and therefore needs to be fully explored.

This paper systematically investigates the use of decomposition-based framework for the MOFS problem with resource constraints, and proposes two enhancements in both decision and objective spaces. The first one of these is the reproduction operators (in the decision space) that are based on a *probabilistic model* (PM) [22]. The second enhancement is to periodically adapt weight vectors during the search so as to fit the irregular Pareto fronts (PFs) of the problem instances. We use the name PMAD[1] (*Probability Model, Adaptation and Decomposition*) to denote the combination of the PM-based reproduction operators, adaptation of weight vectors, and decomposition-based framework. Main contributions of this paper are summarized as follows.

- The formal definition of MOFS with multiple linear and nonlinear resource constraints. Introducing nonlinear constraints makes this work significantly different from [26], where only linear resource constraints are considered.
- We experimentally show that, within the decomposition-based framework, PM-based reproduction operators outperforms genetic operators, and the mixed use of genetic and PM-based operators [26]. Moreover, it is necessary to use adaptive weight vectors so as to improve the diversity of products. The above findings are meaningful, but have not been revealed previously.
- A comprehensive experimental study to compare the proposed PMAD with several state-of-the-art algorithms, including SATVaEA [29], a tailored algorithm for the MOFS problem.

The rest of the paper is organized as follows. Section 2 gives formal definition of MOFS with resource constraints. Section 3 describes the proposed method in detail. Next, experimental studies are given in Sect. 4. Finally, we conclude the paper and outline possible directions for future studies in Sect. 5.

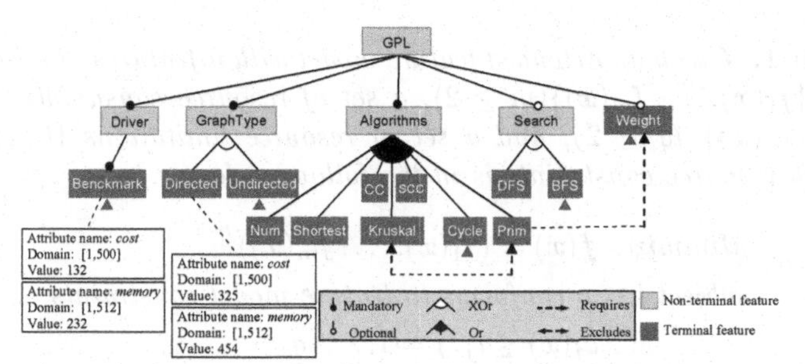

Fig. 1. Extended feature model of Graph Product Line (GPL) [15].

2 Problem Statement

The traditional FM mainly reflects functional aspects of an SPL [1]. To express non-functional properties, FMs can be extended by adding attributes to features [2], and they are referred to as *extended feature models*. The model shown in Fig. 1 is such an FM. Table 1 gives all the attributes used in this paper to construct the

[1] Source codes of PMAD are publicly available at https://doi.org/10.5281/zenodo.4275041.

feature selection problem. For each attribute, we list its name, brief description, domain and unit (for numerical attributes only). It should be mentioned that we select these attributes because they were widely used in previous works to build feature selection problems [9,10,19,21,29].

Table 1. Attributes used to construct MOFS problem with resource constraints.

Name	Description	Domain	Unit
used_before	Indicate whether a feature has been used before	Boolean $\{true, false\}$	/
defects	How many defects detected in a feature	Integer [0, 10]	/
cost	Specify the cost of a feature	Real [1, 500]	Monetary unit
memory	Specify the memory size of a feature	Integer [1, 512]	Kilobytes
time	Specify time-related resources required by a feature [19]	Integer [10, 500]	Milliseconds

Definition 1. *Given an extended feature model with n features, a set of objective functions* $\{f_1(\boldsymbol{x}), \ldots, f_m(\boldsymbol{x})\} (m \geq 2)$, *a set of resource consumption functions* $\{c_1(\boldsymbol{x}), \ldots, c_q(\boldsymbol{x})\}$ $(q \geq 2)$, *and a set of resource limitations* $\{l_1, \ldots, l_q\}$, *the MOFS with resource constraints is in the following form.*

$$\textit{Minimize } \boldsymbol{f}(\boldsymbol{x}) = (f_1(\boldsymbol{x}), \ldots, f_m(\boldsymbol{x}))^T \quad (1)$$

$$\textit{subject to } \boldsymbol{x} \textit{ conforms to feature model constraints} \quad (2)$$

$$c_j(\boldsymbol{x}) \leq l_j, j = 1, \ldots, q \quad (3)$$

where $\boldsymbol{x} = (x_1, \ldots, x_n)^T \in \{0,1\}^n$ *is the decision vector, in which the i-th component* x_i *takes 1 if the i-th feature is selected, and 0 otherwise.*

In (2), feature model constraints, also referred to as functional constraints, are derived from parent-child relations (including *mandatory*, *optimal*, *Or* and *Xor* in Fig. 1) and cross-tree constraints (CTCs, including *requires* and *excludes*) [4]. According to [1,9,17], feature model constraints can be expressed as boolean formula in conjunctive normal form (CNF).

In principle, the number and the formula of objective functions can be flexibly specified by stakeholders based on their own requirements. In this paper, we define the following four objective functions.

$$f_1(\boldsymbol{x}) = n - \sum_{j=1}^{n} x_j \cdot \mathcal{T}(x_j), \qquad f_2(\boldsymbol{x}) = \sum_{j=1}^{n} x_j \cdot \mathcal{T}(x_j) \cdot (1 - used\_before_j),$$

$$f_3(\boldsymbol{x}) = \sum_{j=1}^{n} x_j \cdot \mathcal{T}(x_j) \cdot defects_j, \quad f_4(\boldsymbol{x}) = \sum_{j=1}^{n} x_j \cdot \mathcal{T}(x_j) \cdot cost_j$$

In the above calculations, $\mathcal{T}(x_j)$ returns 1 if the variable x_j represents a terminal feature, and 0 otherwise. A terminal feature (e.g., 'Benckmark' in Fig. 1) is a leaf while a non-terminal feature (e.g., 'GraphType') is an interior node of an FM, and is usually used to group other features. Distinguishing between terminal and non-terminal features makes this work distinct from most of the existing works (e.g., [9,10,21,29]), in which features are not differentiated. In the formula for $f_1(\boldsymbol{x})$, $\sum_{j=1}^{n} x_j \cdot \mathcal{T}(x_j)$ calculates the number of selected terminal features. We seek to maximize this number so as to provide system functionalities as many as possible. Equivalently, $f_1(\boldsymbol{x})$ should be minimized. The second objective $f_2(\boldsymbol{x})$ calculates the number of terminal features that were not used before. The third objective $f_3(\boldsymbol{x})$ calculates the number of total defects presenting in all the currently selected terminal features. Finally, the fourth objective $f_4(\boldsymbol{x})$ indicates the total cost of terminal features selected in the current solution. Clearly, all the objectives need to be minimized.

Likewise, the number and the formula of resource consumption functions can be also specified by users. In this paper, the following three resource consumption functions are defined.

$$c_1(\boldsymbol{x}) = \sum_{j=1}^{n} x_j \cdot \mathcal{T}(x_j) \cdot memory_j, \qquad c_2(\boldsymbol{x}) = \sum_{j=1}^{n} x_j \cdot \mathcal{T}(x_j) \cdot cost_j,$$

$$c_3(\boldsymbol{x}) = \max_{j=1}^{n} \; (x_j \cdot \mathcal{G}(x_j) \cdot time_j).$$

The first consumption function calculates the total memory required to run the current configuration. The second consumption function is actually the fourth objective function $f_4(\boldsymbol{x})$. Quite often, the budget to configure a product is limited. In the calculation of the third consumption function, $\mathcal{G}(x_j)$ returns 1 if the variable x_j represents both a terminal feature and a focused feature, and 0 for otherwise. A focused feature, indicated by the symbol ▲ in Fig. 1, is the feature to which stakeholders pay special attention, or the feature that is very important and its performance should be guaranteed. If the feature *time* represents CPU time, and the execute time for a group of focused terminal features is limited. Then, we need to compute the maximum CPU time among these features, and check whether this CPU time is larger than the limitation or not. It is worth mentioning that the first two resource consumption functions are linear with respect to decision variables $x_1, \ldots, x_n$. Given the fact that non-linear functions are needed to aggregate attribute values for some real SPLs, e.g., the RFW product line [19], we introduce the third function that is non-linear (because *max* is used to aggregate values). To our best knowledge, non-linear resource constraints in this context have not been defined before us. Finally, note that, for each FM, we can construct 2-, 3- and 4-objective instances by choosing the first two, three and four optimization functions, respectively. In all the problem instances, all the three resource constraints are considered.

Algorithm 1. PMAD framework

Input: N (population size), T (neighborhood size), R (weight vectors update period)
Output: The final population P
1: Initialize N weight vectors $\Lambda = \{\boldsymbol{\lambda}_1, \ldots, \boldsymbol{\lambda}_N\}$. For each $i = 1, \ldots, N$, set $B(i) = \{i_1, \ldots, i_T\}$, where $\boldsymbol{\lambda}_{i_1}, \ldots, \boldsymbol{\lambda}_{i_T}$ are T closest weight vectors to $\boldsymbol{\lambda}_i$ (regarding the Euclidean distance)
2: Randomly generate an initial population $P = \{\boldsymbol{x}^1, \ldots, \boldsymbol{x}^N\}$. Repair solutions violating feature model constraints by using two SAT solvers [26]
3: Evaluate $\boldsymbol{x}^1, \ldots, \boldsymbol{x}^N$ by working out objectives and constraint violations
4: **while** the termination condition is not met **do**
5: $\{\boldsymbol{z}^*, \boldsymbol{z}^{max}\} \leftarrow$ *normalization* (P) // Normalize the population with $\widetilde{f}_i(\boldsymbol{x}) = \frac{f_i(\boldsymbol{x}) - z_i^*}{z_i^{max} - z_i^*}$
6: $\boldsymbol{z}^{**} \leftarrow \{-\epsilon, \ldots, -\epsilon\}$, where $\epsilon = 0.1$ [26] // Initialize the reference point
7: **for** $i = 1, \ldots, N$ **do**
8: Generate a new solution $\boldsymbol{y}$ for the i-th subproblem using PM-based reproduction operators
9: Repair, if necessary, $\boldsymbol{y}$ using two SAT solvers [26]
10: Evaluate $\boldsymbol{y}$ and normalize the objective vector $\boldsymbol{f}(\boldsymbol{y})$ to $\widetilde{\boldsymbol{f}}(\boldsymbol{y})$ using $\boldsymbol{z}^*$ and $\boldsymbol{z}^{max}$ obtained in Line 5
11: Update $\boldsymbol{z}^{**}$ based on $\widetilde{\boldsymbol{f}}(\boldsymbol{y})$: For each $j = 1, \ldots, m$, if $\widetilde{f}_j(\boldsymbol{y}) < z_j^{**}$, then set z_j^{**} to $\widetilde{f}_j(\boldsymbol{y})$.
12: $P \leftarrow$ *updateSubproblems* $(i, \boldsymbol{y}, \boldsymbol{z}^{**})$
13: **end for**
14: Adapt weight vectors for every R generations
15: **end while**
16: **return** P

3 The Proposed PMAD Method

The PMAD framework is outlined in Algorithm 1. This is a general framework for MOFS problem with both feature model constraints and resource constraints. In this framework, we adopt two SAT solvers to repair infeasible solutions which violate feature model constraints, following the practice in [26]. According to Liang et al. [14], SAT-based analysis of real-world FMs is easy. To handle resource constraints, we use constraint violations (CV) [5] as the metric to compare solutions. In general, solutions with small CV values are preferred. The definition of CV is given as $CV(\boldsymbol{x}) = \sum_{j=1}^{q} \max\{c_j(\boldsymbol{x}) - l_j, 0\}$. Clearly, CV is equal to 0 for solutions without resource constraint violations. In fact, Algorithm 1 follows the general procedures presented in [26]. Due to space limitations, we do not go into details on most of them. Instead, we focus on the PM-based reproduction and the adaptation of weight vectors.

3.1 PM-based Reproduction Operators (Line 8)

One of our goals is to explore the use of PM-based reproduction operators for the feature selection problem. To this end, we select the scale adaptive reproduction (SAR) operator [22] and the estimation of distribution (EoD) operator [26]. Both of them are developed previously for binary optimization, which is just the case in this paper. In the previous work [26], we have shown that the combination of crossover and EoD operators performs in general better than the traditional way, i.e., *crossover+mutation*. However, it remains unknown if the performance can be further improved by replacing crossover with another PM-based operator. This is the reason why we introduce the SAR operator. For details on SAR and EoD, please refer to [22] and [26], respectively.

Algorithm 2. Weight vector adaptation

Input: Λ (a set of weight vectors), P (the population)
Output: Λ (the set of adapted weight vectors)
1: Initialize a new weight vector pool: $W \leftarrow \emptyset$ // The maximum capacity of W is N
2: $\mathbf{A}_{N\times N} \leftarrow calulateAngleMatrix(P)$
3: For each solution $\boldsymbol{x}_i$ ($i = 1, \ldots, N$), calculate its density according to Eq. (4)
4: Identify indexes of the first T solutions with the minimum density: $I_1^{min}, \ldots, I_T^{min}$, and those of the first $2T$ solutions with the maximum density: $I_1^{max}, \ldots, I_{2T}^{max}$
5: **for** $i \leftarrow 1$ to $2T$ **do**
6: $\quad \overline{\boldsymbol{\lambda}} \leftarrow 0.5 \times (\boldsymbol{\lambda}_{I_i^{max}} + \boldsymbol{\lambda}_{B(I_i^{max}, T)})$
7: $\quad$ Remove the first element if W is full, and add $\overline{\boldsymbol{\lambda}}$ into W
8: **end for**
9: $\Lambda' \leftarrow \Lambda \backslash \{I_1^{min}, \ldots, I_T^{min}\}$
10: $\{\boldsymbol{\pi}, \boldsymbol{d}\} \leftarrow$ *association* (W, Λ') /* *Associate each* $\overline{\boldsymbol{\lambda}} \in W$ *with the closest weight vector in* Λ', $\boldsymbol{\pi} = (\pi_1, \ldots, \pi_{|W|})$ *and* $\mathbf{d} = (d_1, \ldots, d_{|W|})$, *where* π_i *is the index of the weight vector with which* $\overline{\boldsymbol{\lambda}}_i$ *is associated, and* d_i *is the corresponding Euclidean distance.* */
11: **for** $i \leftarrow 1$ to T **do**
12: $\quad k \leftarrow \arg\max_{1 \le j \le |W|} d_j$
13: $\quad$ Update Λ: replace $\boldsymbol{\lambda}_{I_i^{min}} \in \Lambda$ with $\overline{\boldsymbol{\lambda}}_k \in W$
14: $\quad$ Remove $\overline{\boldsymbol{\lambda}}_k$ from W, and insert it at the end of Λ'
15: $\quad$ Update $\boldsymbol{\pi}$ and $\boldsymbol{d}$ accordingly
16: **end for**
17: Redetermine neighbor structure for each weight vector in Λ
18: **return** Λ

3.2 Adaptation of Weight Vectors (Line 14)

The procedure for weight vector adaptation is given in Algorithm 2. The function *calulateAngleMatrix* in Line 2 returns the angle between each pair of solutions in P, and these angles are stored in an $N \times N$ matrix $\mathbf{A} = (a_{ij})$, where a_{ij} denotes the angle between the i-th solution ($\boldsymbol{x}_i \in P$) and the j-th solution ($\boldsymbol{x}_j \in P$). Based on $\mathbf{A}$, the density of the i-th solution $\boldsymbol{x}_i$ is defined as

$$De(\boldsymbol{x}_i) = \frac{\sum_{j=1}^{T} a_{i,\pi(j)}}{T}, \quad (4)$$

where $a_{i,\pi(j)}$ ($j = 1, \ldots, T$) is the j-th smallest angle to $\boldsymbol{x}_i$. That is to say, the density of a solution is defined as the average angle to its T closest solutions.

Line 4 in Algorithm 2 identifies the indexes of the first T solutions with the minimum density, i.e., $I_1^{min}, \ldots, I_T^{min}$. These solutions are located at crowded regions and the corresponding weight vectors are to be replaced subsequently. At the same time, the indexes of the first $2T$ solutions with the maximum density, $I_1^{max}, \ldots, I_{2T}^{max}$, are also worked out. The corresponding weight vectors are utilized to generate new weight vectors (see Lines 5–8). Specifically, for each $\boldsymbol{\lambda}_{I_i^{max}}$ and its T-th neighbor $\boldsymbol{\lambda}_{B(I_i^{max},T)}$ (which is farthest to $\boldsymbol{\lambda}_{I_i^{max}}$ among all its neighbors), we compute the midpoint $\overline{\boldsymbol{\lambda}}$, a new weight vector to be added into the weight vector pool W. Note that we simply remove the first element if W is full. In other words, the set W is maintained by a queue data structure.

In Line 10, the function *association* associates each $\overline{\boldsymbol{\lambda}}_i \in W$ with the closest weight vector in $\Lambda' = \Lambda \backslash \{I_1^{min}, \ldots, I_T^{min}\}$. The index of the closest weight vector is denoted by π_i and the corresponding Euclidean distance is denoted by d_i. Since added weight vectors should not be too close to the existing ones [13], we add each time the new weight vector with the maximum d value. Specifically, we first find out the index of the maximum d value (Line 12), then update Λ by replacing $\boldsymbol{\lambda}_{I_i^{min}} \in \Lambda$ with $\overline{\boldsymbol{\lambda}}_k \in W$ (Line 13). Thereafter, $\overline{\boldsymbol{\lambda}}_k$ is removed from W, and inserted at the end of Λ' (Line 14). Finally, according to Line 15, $\boldsymbol{\pi}$ and $\boldsymbol{d}$ should be updated accordingly: For each $j \in \{1, 2, \ldots, |W|\}$ and $j \neq k$, if *distance*$(\overline{\boldsymbol{\lambda}}_j, \overline{\boldsymbol{\lambda}}_k)$, the distance between the two weight vectors, is smaller than d_j, then set d_j and π_j to *distance*$(\overline{\boldsymbol{\lambda}}_j, \overline{\boldsymbol{\lambda}}_k)$ and $|\Lambda'| + 1$, respectively.

4 Experimental Study

In the empirical study, 22 feature models are used to construct problem instances. For each model, according to Sect. 2, we can construct 2-, 3- and 4-objective problem instances. Therefore, we have $22 \times 3 = 66$ instances in total. In the evaluations, we employ two performance metrics, i.e., Hypervolume (HV) [34] and IGD+ [11]. Due to page issues, details on instance construction, the setting of population size and termination condition, parameters used in the algorithms, and supplementary results (including visualized comparisons) are provided online https://doi.org/10.5281/zenodo.4275041. Here, we only present several critical results and the reached conclusions.

4.1 Effect of PM-based Reproduction Operators

Regarding reproduction in decomposition-based framework, we compare the combination of two PM-based reproduction operators (i.e., *SAR+EoD*) with two strategies, i.e., *Crossover+Mutation* [9,10,20] and *Crossover+EoD* [26], which were widely used before or have been proposed very recently. Note that *Crossover+Mutation* and *Crossover+EoD* are actually MOEA/D [32] and MOEA/D-EoD [26], respectively. We give in Fig. 2 the percentage of the best HV and IGD+ results obtained by each strategy. Concerning both metrics, as shown clearly, *SAR+EoD* is the best-performing strategy, followed by *Crossover+EoD*. Finally, *Crossover+Mutation* is the most ineffective strategy. It is worth mentioning that *SAR+EoD* dramatically improves over the two strategies. Specifically, *SAR+EoD* shows an improvement over *Crossover+EoD* by 24% (i.e., 62%−38%) and 50% (i.e., 71%−21%) with respect to HV and IGD+, respectively. Compared with *Crossover+Mutation*, the best HV and IGD+ results obtained by *SAR+EoD* are increased by 62% (i.e., 62%−0%) and 63% (i.e., 71%−8%), respectively. The ineffectiveness of *Crossover+Mutation* is probably due to that it tends to blindly generate new solutions, without considering the occurrence probability of variables. The above results lead us to the follow conclusion: *Within the decomposition-based framework, PM-based reproduction operators perform better than genetic operators, and the mixed use of genetic and PM-based operators, when handling the MOFS problem.*

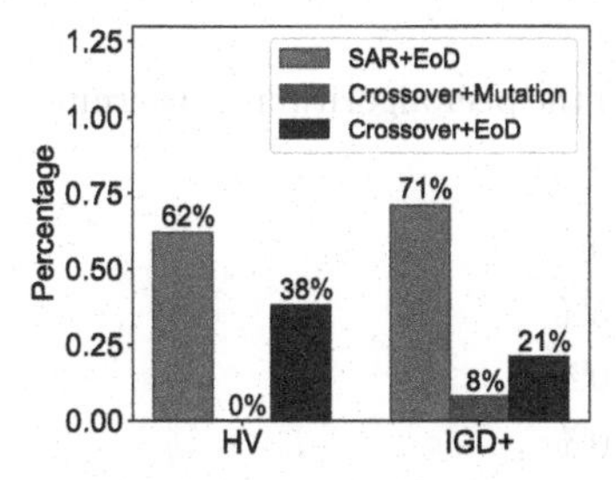

Fig. 2. Percentage of the best HV and IGD+ results obtained by each strategy.

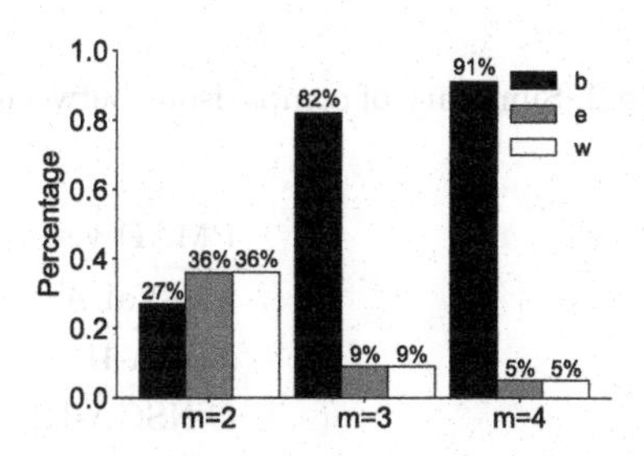

Fig. 3. For each m, the percentage of problem instances on which *With adaptation* performs better than, equally to and worse than *Without adaptation*.

4.2 Adaptive Weight Vectors Make a Difference

In this part, we compare the algorithm with adaptive weight vectors (denoted by *With adaptation*) against the one without adaptive weight vectors (denoted by *Without adaptation*) to investigate whether adaptive weight vectors make a difference or not. We calculate the percentages of problem instances on which *With adaptation* performs better than ('b'), equally to ('e') and worse than ('w') *Without adaptation*. Compared with *Without adaptation*, as shown in Fig. 3, *With adaptation* performs better on 27% of all the 2-objective instances, but worse on 36% instances. It seems that *Without adaptation* is slightly better than *With adaptation* on the 2-objective problem instances. For 3- and 4-objective cases, however, the situation is totally different. The algorithm with adaptive weight vectors outperforms its counterpart on 82% and 91% instances for $m = 3$ and $m = 4$, respectively. Inversely, it is defeated on only 9% and 5% instances, respectively. It is clear that the adaptation of weight vectors makes a great difference when the number of objectives is larger than two. The above results emphasize the following: *In decomposition-based framework, the adaptation of weight vectors indeed makes a difference, particularly for 3- and 4-objective cases.*

4.3 Overall Performance Evaluations

In this section, we compare PMAD with SATVaEA [29], NSGA-II [5] and C-NSGA-III [12]. Both NSGA-II and C-NSGA-III are well-known in the EMO community, while SATVaEA is a tailored algorithm for the MOFS problem. Note that C-NSGA-III also uses adaptive weight vectors, but are adjusted in a different way. The Wilcoxon's rank sum test [25] results of HV obtained by the four algorithms are summarized in Table 2. It is found that PMAD outperforms SATVaEA, NSGA-II and C-NSGA-III on 89%, 97% and 100% problem instances, respectively. Clearly, PMAD obtains the best overall performance, and shows a significant improvement over the algorithms under comparison.

Table 2. Summary of comparisons between PMAD and the peer algorithms concerning HV.

PMAD v.s.	b	e	w
SATVaEA	89%	6%	5%
NSGA-II	97%	3%	0%
C-NSGA-III	100%	0%	0%

5 Conclusions and Future Work

In this paper, we propose a prototype for the multi-objective feature selection with multiple linear and non-linear resource constraints, and investigate the performance of decomposition-based metaheuristics on the proposed problem. We demonstrate that, within the decomposition-based framework, PM-based reproduction operators are better than the traditional genetic operators (i.e., crossover and mutation), and the mixed use of genetic and PM-based operators. Moreover, we show that adaptive weight vectors can significantly improve the performance of the algorithm. Finally, experimental results on a number of problem instances indicate that PMAD is able to obtain better overall performance than several state-of-the-art algorithms.

Since our primary goal in this paper is to explore the possibility and suitability of decomposition-based metaheuristics for the MOFS problem, we currently use existing PM-based reproduction operators. In the future, it is interesting to develop new PM-based operators tailored for the problem. Moreover, the adaptation of weight vectors is designed based on the observation that boundaries are intensively covered if fixed weight vectors are utilized. We intend to explore the use of strategies suggested in the EMO community [16] to adapt weight vectors in the context of multi-objective feature selection.

Acknowledgment. This paper is supported by National Natural Science Foundation of China (61906069, 61703183, 62006081), Guangdong Basic and Applied Basic Research Foundation (2019A1515011411, 2019A1515011700), Project Funded by China Postdoctoral Science Foundation (2019M662912, 2020M672630), Science and Technology Program of Guangzhou (202002030355, 202002030260), Public Welfare Technology Application Research Plan of Zhejiang (LGG19F030010), and Fundamental Research Funds for the Central Universities (2019MS088).

References

1. Batory, D.: Feature models, grammars, and propositional formulas. In: Obbink, H., Pohl, K. (eds.) SPLC 2005. LNCS, vol. 3714, pp. 7–20. Springer, Heidelberg (2005). https://doi.org/10.1007/11554844_3
2. Benavides, D., Segura, S., Ruiz-Corts, A.: Automated analysis of feature models 20 years later: a literature review. Inf. Syst. **35**(6), 615–636 (2010)
3. Clements, P., Northrop, L.: Software Product Lines: Practices and Patterns. Addison-Wesley Longman Publishing Co., Inc., Boston (2001)
4. Czarnecki, K., Eisenecker, U.: Generative Programming: Methods, Tools, and Applications. ACM Press/Addison-Wesley Publishing Co. 1515 Broadway, New York, United States (2000)
5. Deb, K., Pratap, A., Agarwal, S., Meyarivan, T.: A fast and Elitist multiobjective genetic algorithm: NSGA-II. IEEE Trans. Evol. Comput. **6**(2), 182–197 (2002)
6. do Nascimento Ferreira, T., Kuk, J.N., Pozo, A., Vergilio, S.R.: Product selection based on upper confidence bound MOEA/D-Dra for testing software product lines. In: 2016 IEEE Congress on Evolutionary Computation (CEC), pp. 4135–4142 (2016)
7. Guo, J., et al.: SMTIBEA: a hybrid multi-objective optimization algorithm for configuring large constrained software product lines. Softw. Syst. Model. **18**, 1447–1466 (2019). https://doi.org/10.1007/s10270-017-0610-0
8. Guo, J., White, J., Wang, G., Li, J., Wang, Y.: A genetic algorithm for optimized feature selection with resource constraints in software product lines. J. Syst. Softw. **84**(12), 2208–2221 (2011)
9. Henard, C., Papadakis, M., Harman, M., Traon, Y.L.: Combining multi-objective search and constraint solving for configuring large software product lines. In: The 37th International Conference on Software Engineering, vol. 1, pp. 517–528, May 2015
10. Hierons, R.M., Li, M., Liu, X., Segura, S., Zheng, W.: SIP: optimal product selection from feature models using many-objective evolutionary optimization. ACM Trans. Softw. Eng Methodol. **25**(2), 17:1–17:9 (2016)
11. Ishibuchi, H., Masuda, H., Tanigaki, Y., Nojima, Y.: Difficulties in specifying reference points to calculate the inverted generational distance for many-objective optimization problems. In: IEEE Symposium on Computational Intelligence in Multi-Criteria Decision-Making (MCDM), pp. 170–177. IEEE (2014)

12. Jain, H., Deb, K.: An evolutionary many-objective optimization algorithm using reference-point based nondominated sorting approach, part II: handling constraints and extending to an adaptive approach. IEEE Trans. Evol. Comput. **18**(4), 602–622 (2014)
13. Li, H., Landa-Silva, D.: An adaptive evolutionary multi-objective approach based on simulated annealing. Evol. Comput. **19**(4), 561–595 (2011)
14. Liang, J.H., Ganesh, V., Czarnecki, K., Raman, V.: SAT-based analysis of large real-world feature models is easy. In: Proceedings of the 19th International Conference on Software Product Line, SPLC 2015, pp. 91–100. ACM, New York (2015). https://doi.org/10.1145/2791060.2791070
15. Lopez-Herrejon, R.E., Chicano, F., Ferrer, J., Egyed, A., Alba, E.: Multi-objective optimal test suite computation for software product line pairwise testing. In: IEEE International Conference on Software Maintenance, pp. 404–407 (2013)
16. Ma, X., Yu, Y., Li, X., Qi, Y., Zhu, Z.: A survey of weight vector adjustment methods for decomposition based multi-objective evolutionary algorithms. IEEE Trans. Evol. Comput. **24**(4), 634–649 (2020)
17. Mendonca, M., Wasowski, A., Czarnecki, K.: SAT-based analysis of feature models is easy. In: Proceedings of the 13th International Software Product Line Conference, SPLC 2009, pp. 231–240. Carnegie Mellon University, Pittsburgh (2009)
18. Ramírez, A., Romero, J.R., Ventura, S.: A survey of many-objective optimisation in search-based software engineering. J. Syst. Softw. **149**, 382–395 (2019)
19. Roos-Frantz, F., Benavides, D., Ruiz-Cortés, A., Heuer, A., Lauenroth, K.: Quality-aware analysis in product line engineering with the orthogonal variability model. Softw. Qual. J. **20**, 519–565 (2012). https://doi.org/10.1007/s11219-011-9156-5
20. Sayyad, A.S., Ingram, J., Menzies, T., Ammar, H.: Scalable product line configuration: a straw to break the camel's back. In: 2013 28th IEEE/ACM International Conference on Automated Software Engineering (ASE), pp. 465–474, November 2013
21. Sayyad, A.S., Menzies, T., Ammar, H.: On the value of user preferences in search-based software engineering: a case study in software product lines. In: 2013 35th International Conference on Software Engineering (ICSE), pp. 492–501, May 2013
22. Wang, B., Xu, H., Yuan, Y.: Scale adaptive reproduction operator for decomposition based estimation of distribution algorithm. In: 2015 IEEE Congress on Evolutionary Computation (CEC), pp. 2042–2049, May 2015
23. White, J., Dougherty, B., Schmidt, D.C.: Selecting highly optimal architectural feature sets with filtered Cartesian flattening. J. Syst. Softw. **82**(8), 1268–1284 (2009). sI: Architectural Decisions and Rationale
24. White, J., Doughtery, B., Schmidt, D.C.: Filtered cartesian flattening: an approximation technique for optimally selecting features while adhering to resource constraints. In: Software Product Lines, 12th International Conference, SPLC 2008, Limerick, Ireland, 8–12 September 2008, Proceedings. Second Volume (Workshops), pp. 209–216 (2008)
25. Wilcoxon, F.: Individual comparisons by ranking methods. Biom. Bull. **1**(6), 80–83 (1945)
26. Xiang, Y., Yang, X., Zhou, Y., Huang, H.: Enhancing decomposition-based algorithms by estimation of distribution for constrained optimal software product selection. IEEE Trans. Evol. Comput. **24**(2), 245–259 (2020)
27. Xiang, Y., Yang, X., Zhou, Y., Zheng, Z., Li, M., Huang, H.: Going deeper with optimal software products selection using many-objective optimization and satisfiability solvers. Empir. Softw. Eng. **25**, 591–626 (2020)

28. Xiang, Y., Zhou, Y., Li, M., Chen, Z.: A vector angle based evolutionary algorithm for unconstrained many-objective problems. IEEE Trans. Evol. Comput. **21**(1), 131–152 (2017)
29. Xiang, Y., Zhou, Y., Zheng, Z., Li, M.: Configuring software product lines by combining many-objective optimization and SAT solvers. ACM Trans. Softw. Eng. Methodol. **26**(4), 141–1446 (2018)
30. Xue, Y., Li, M., Shepperd, M., Lauria, S., Liu, X.: A novel aggregation-based dominance for Pareto-based evolutionary algorithms to configure software product lines. Neurocomputing **364**, 32–48 (2019)
31. Xue, Y., et al.: IBED: combining IBEA and DE for optimal feature selection in software product line engineering. Appl. Soft Comput. **49**, 1215–1231 (2016)
32. Zhang, Q., Li, H.: MOEA/D: a multiobjective evolutionary algorithm based on decomposition. IEEE Trans. Evol. Comput. **11**(6), 712–731 (2007)
33. Zitzler, E., Künzli, S.: Indicator-based selection in multiobjective search. In: Yao, X., et al. (eds.) PPSN 2004. LNCS, vol. 3242, pp. 832–842. Springer, Heidelberg (2004). https://doi.org/10.1007/978-3-540-30217-9_84
34. Zitzler, E., Thiele, L.: Multiobjective evolutionary algorithms: a comparative case study and the strength Pareto approach. IEEE Trans. Evol. Comput. **3**(4), 257–271 (1999)

Operator-Adapted Evolutionary Large-Scale Multiobjective Optimization for Voltage Transformer Ratio Error Estimation

Changwu Huang[1], Lianghao Li[2], Cheng He[1(✉)], Ran Cheng[1], and Xin Yao[1]

[1] Guangdong Provincial Key Laboratory of Brain-Inspired Intelligent Computation, Department of Computer Science and Engineering, Southern University of Science and Technology, Shenzhen 518055, China
{huangcw3,xiny}@sustech.edu.cn

[2] Key Laboratory of Image Information Processing and Intelligent Control of Education Ministry of China, School of Artificial Intelligence and Automation, Huazhong University of Science and Technology, Wuhan 430074, China

Abstract. Large-scale multiobjective optimization problems (LSMOPs) exist widely in real-world applications, and they are challenging for existing evolutionary algorithms due to their massive volume of search space. Despite that a number of large-scale multiobjective evolutionary algorithms (LSMOEAs) have been proposed in recent years, their effectiveness in solving LSMOPs remains unsatisfactory. One main reason is that most existing LSMOEAs may fail to balance convergence enhancement and diversity maintenance, especially for solving real-world problems. To address this issue, we propose to use a hybridized LSMOEA with adaptive operator selection (AOS) to handle real-world LSMOPs. Specifically, the proposed hybridized LSMOEA with AOS (AOS-LSMOEA) includes multiple different offspring generation and environmental selection strategies extracted from some existing LSMOEAs. Then it uses the AOS to adaptively determine the application rates of different offspring generation and environmental selection operators in an online manner. The proposed approach is capable of taking advantage of existing LSMOEAs, and the AOS enables the algorithm to choose suitable operators for solving different LSMOPs. In this study, the proposed algorithm is expected to solve the voltage transformer ratio error estimation (TREE) problems effectively. Experimental results show that AOS-LSMOEA achieves significant performance improvement due to the hybridization of different operators and the adoption of AOS method.

Keywords: Large-scale optimization · Multiobjective optimization · Adaptive operator selection · Voltage transformer ratio error estimation

H. Ishibuchi et al. (Eds.): EMO 2021, LNCS 12654, pp. 672–683, 2021.
https://doi.org/10.1007/978-3-030-72062-9_53

1 Introduction

Multiobjective optimization problems (MOPs) involve multiple, often conflicting, objectives that need to be optimized simultaneously [5]. There are many MOPs in real-world applications, such as flowshop scheduling [18], portfolio optimization [24], and voltage transformers ratio error estimation (TREE) problems [16]. Different from single-objective optimization problems, there is a set of trade-off optimal solutions instead of a single optimum in an MOP. Specifically, these trade-off optima constitute the Pareto optimal set (PS), and the projection of the PS in objective space forms the Pareto optimal front (PF) [23]. Due to their population based characteristics for obtaining multiple solutions in a single run, evolutionary algorithms (EAs) are naturally suitable and efficient for multiobjective optimization [17]. In recent decades, a variety of multiobjective EAs (MOEAs) have been proposed, including the Pareto-based MOEAs (e.g., NSGA-II [7] and SPEA2 [37]), the indicator-based MOEAs (e.g., IBEA [35] and SMS-EMOA [2]), and the decomposition based MOEAs (e.g., MOEA/D [31] and MOEA/D-M2M [20]). Existing MOEAs have shown to be effective in solving conventional MOPs, but their performance drops drastically when the number of decision variables increases to hundreds or even more.

MOPs with a large number of decision variables, usually more than 100, are also known as large-scale MOPs (LSMOPs) [3,4]. With the linear increment in the number of decision variables, the volume and the complexity of the search space will increase exponentially. How to use a limited size of the population to search an enormous search space with an acceptable computational cost is the main challenge of solving LSMOPs. Recently, with a growing interest in large-scale multiobjective optimization, four main types of approaches are proposed to tackle LSMOPs [22]. The cooperative coevolution (CC) framework based approaches formed the first type, which handles large-scale decision variables in a divide-and-conquer manner [25]. Specifically, the CC method decomposes the decision variables into several groups and cooperatively optimizes all the groups of decision variables, e.g., third-generation CC differential evolution algorithm [1]. The second type of approaches tries to reformulate the original LSMOPs into some simpler problems with low-dimensional decision variables. For instances, the weighted optimization framework used decision variable grouping methods and some transforming functions to reduce the number of decision variables [33]; the large-scale multiobjective optimization framework (LSMOF) [14] use problem reformulation to improve the efficiency of MOEAs on LSMOPs. The third type of approaches dedicate to the decision variable analysis based approaches, e.g., MOEA based on decision variable analysis (MOEA/DVA) [21] and decision variable clustering-based large-scale EA (LMEA) [32]. The last type of approaches try to enhance the effectiveness of offspring generation in conventional MOEAs, i.e., generating evenly distributed offspring solutions with better convergence in comparison with conventional MOEAs, e.g., competitive swarm optimizer based EA (LMCSO) [30] and direction guided adaptive offspring generation based EA (DGEA) [13].

Existing LSMOEAs have shown their advantages in terms of either efficiency or effectiveness due to the use of different decision variable handling strategies, leading to their success in solving some benchmark problems (e.g., ZDT problems [34], DTLZ problems [8], WFG problems [26], and LSMOP problems [3]). However, their performance could be unsatisfactory in solving real-world applications, such as the voltage transformer ratio error estimation (TREE) [16]. It could be time-consuming and error-prone to design a specific LSMOEA for solving TREE problems. An intuitive and easy way is to take advantages of existing LSMOEAs. Motivated by this, we design a hybridized framework of LSMOEA with adaptive operator selection (AOS), which is abbreviated to AOS-LSMOEA hereafter, for solving the TREE problems. Generally, the proposed AOS-LSMOEA approach is capable of using the crucial components or operators in existing LSMOEAs and adaptively selecting suitable one from them to use. It can not only enhance the effectiveness of existing LSMOEAs in solving TREE problems but also be applicable to other black-box LSMOPs.

In the remainder of this work, we first introduce some background about the AOS-LSMOEA, including the iterated problem reformulation based large-scale MOEA (iLSMOA) and the AOS, in Sect. 2. The schema and details of our proposed AOS-LSMOEA approach are given in Sect. 3, and the experimental studies are provided in Sect. 4. Finally, conclusions are drawn in Sect. 5.

2 Background

In this section, we first give a short introduction to the iterated problem reformulation based large-scale MOEA (iLSMOA) [15], which is the basis of the proposed AOS-LSMOEA. Then, the AOS method used in this work is described.

2.1 iLSMOA

In this study, we adopt iLSMOA as the basis to construct the hybridized framework for large-scale multiobjective optimization mainly for two reasons. First, iLSMOA is a flexible framework that can easily involve different components in most existing LSMOEAs. Second, the iterated mechanism in iLSMOA is validated to be more effective than its counterpart that uses a two-step strategy in large-scale multiobjective optimization [15].

The general framework of iLSMOA is given in Algorithm 1, which uses the problem reformulation based single-objective optimization and conventional MOEA in an iterative manner. To begin with, a population of size N is randomly generated from the original LSMOP Z, and Z is reformulated into a low-dimension single-objective optimization problem Z'. Then the single-objective optimization followed by environmental selection is conducted to obtain a set of well-converged solutions $\mathcal{P}$. Afterwards, $\mathcal{P}$ is further evolved using conventional MOEA for diversity maintenance. Finally, the above procedures are repeated in an iterative manner until the termination criterion is fulfilled. To be more specific, the original iLSMOA uses the MOEA/D-DE [19] during the evolution by MOEA to enhance the diversity of the population.

Algorithm 1. The framework iLSMOA [15].

Input: Z (original LSMOP), N (population size), r (number of reference solutions), $g_{\max}$ (maximum iteration).
Output: $\mathcal{P}$ (final population).
1: $\mathcal{P} \leftarrow$ Initialization(N, Z)
2: **while** NotTerminated **do**
3: $Z' \leftarrow$ ProblemReformulation($\mathcal{P}, r, Z$).
4: $A \leftarrow$ SingleObjectiveOptimization($Z', g_{\max}$).
5: $\mathcal{P} \leftarrow$ EnvironmentalSelection(A, N).
6: $\mathcal{P} \leftarrow$ EvolveByMOEA($\mathcal{P}, N, g_{\max}$).
7: **end while**

2.2 Adaptive Operator Selection

AOS is a recent paradigm to tackle the operator selection problem of EAs in an adaptive manner or online fashion. The aim of AOS is to select a suitable operator from the given candidate operators set to use during the search process. Usually, the selection is conducted on the basis of the recent performance or some measure of the quality of candidate operators. AOS mainly involves two components: (1) credit assignment, which defines how to assign credit or reward to an operator based on the impact brought by its recent application on the current search process; and (2) operator selection, which selects the operator to be applied next based on the previously collected rewards. These two components used in this work are described in the following.

Credit Assignment. The goal of credit assignment is to provide a reward to an operator after it has been applied based on its performance regarding the progress of the search. Most credit assignment methods use fitness improvement as the reward [10,11]. Since in this work we apply AOS to an LSMOEA, the fitness improvement cannot be used directly as a reward in credit assignment. We propose to adopt the hypervolume (HV) indicator [36], which is widely used to measure the performance of MOEAs, to measure the impact of operator application, and the improvement of HV is taken as the reward in credit assignment.

At the beginning of generation or iteration g, the maximum value of each objective from the objective vectors of current population $\mathcal{P}^{(g)}$ is taken as the reference point $f_{RP}^{(g)}$ for calculating the HV value $\mathrm{HV}^{(g)}$. Then, the evolutionary operations are performed to evolve the population, and the HV value of the updated population is also updated to $\mathrm{HV}^{(g+1)}$. Afterwards, the reward of evolutionary operator application at generation g is measured by the HV improvement $R^{(g)} = \max\left((\mathrm{HV}^{(g+1)} - \mathrm{HV}^{(g)}), 0\right)$.

Operator Selection. Based on the reward or credit values received from the credit assignment, the operator selection scheme maintains an up-to-date empirical quality estimate for each candidate operator and use it to update the application rates of candidate operators. The two most promising and commonly used operator selection schemes are the probability matching (PM) [12] and adaptive pursuit (AP) [27]. We adopt the AP method for operator selection, since it has been demonstrated that AP achieves superior performance than PM [28].

The AP scheme can be mathematically formalized as follows. Suppose we have a set of K candidate operators $\mathcal{A} = \{a_1, \cdots, a_K\}$, AP maintains a probability vector $\mathbf{P}^{(g)} = [P_1^{(g)}, \cdots, P_K^{(g)}]$ $(0 \le P_i^{(g)} \le 1$ and $\sum_{i=1}^{K} P_i^{(g)} = 1)$ and a quality vector $\mathbf{Q}^{(g)} = [Q_1^{(g)}, \cdots, Q_K^{(g)}]$ at time g (in this work time is equivalent to iteration or generation count of evolutionary process), where $P_i^{(g)}$ and $Q_i^{(g)}$ are the selection probability and empirical quality estimate of the candidate operator $a_i \in \mathcal{A}$, respectively. Initially, the probability and quality vectors are initialized as $P_i^{(0)} = \frac{1}{K}$, $Q_i^{(0)} = 1.0$ for $i = 1, \cdots, K$. At generation g, the j-th operator a_j is selected with probability $P_j^{(g)}$ through a roulette-wheel selection scheme, i.e.,

$$a_j \leftarrow \text{RouletteWheelSelection}(\mathbf{P}^{(g)}). \tag{1}$$

After the application of operator a_j, a reward $R_j^{(g)}$ is obtained by the credit assignment. Then $Q_j^{(g)}$ is updated to account for the currently received reward $R_j^{(g)}$, via the exponential recency-weight average updating mechanism, as,

$$Q_j^{(g+1)} = Q_j^{(g)} + \alpha \left[R_j^{(g)} - Q_j^{(g)} \right], \tag{2}$$

where α $(0 < \alpha \le 1)$ is the learning rate. Subsequently, the current best operator is chosen $i^* = \arg\max_i (Q_i^{(g+1)})$ and its selection probability is increased as

$$P_{i^*}^{(g+1)} = P_{i^*}^{(g)} + \beta \left[P_{\max} - P_{i^*}^{(g)} \right], \tag{3}$$

while the selection probability of other candidate operators are decreased as

$$P_i^{(g+1)} = P_i^{(g)} + \beta \left[P_{\min} - P_i^{(g)} \right] \quad \text{for } i = 1, \cdots, K \text{ and } i \ne i^*, \tag{4}$$

where $\beta \in [0, 1]$ is the learning rate which controls the greediness of the winner-takes-all strategy, $P_{\min}$ is the minimal selection probability value which avoids the selection probability decreases to zero, and $P_{\max} = 1 - (K-1)P_{\min}$ is the maximal selection probability.

3 The Proposed AOS-LSMOEA

In this study, we propose the AOS-LSMOEA for solving the real-world LSMOPs, that is, the TREE problems. To enhance the search ability, multiple specific offspring generation and environment selection operators are included in AOS-LSMOEA. Besides, AOS is used to automatically select suitable operators in each iteration (or generation) of the search process. The overall implementation of AOS-LSMOEA is presented in the form of pseudo-code in Algorithm 2.

Algorithm 2. The framework of AOS-LSMOEA.

Input: N (population size), $g_{\max}$ (maximum iteration), $\mathcal{P}_{\text{OffGen}}$ (pool of candidate offspring generation opeartors), $\mathcal{P}_{\text{EnvSel}}$ (pool of candidate environmental selection opeators), $P_{\min}$ (minimal selection probability), α and β (learning rates).
Output: $\mathcal{P}$ (final population).
1: OffGenSelector $\leftarrow$ AdaptivePursuit($\mathcal{P}_{\text{OffGen}}$, $P_{\min}$, α, β).
2: EnvSelSelector $\leftarrow$ AdaptivePursuit($\mathcal{P}_{\text{EnvSel}}$, $P_{\min}$, α, β).
3: $\mathcal{P} \leftarrow$ Initialization(N).
4: **while** NotTerminated **do**
5: OffGenOperator $\leftarrow$ Select an operator by OffGenSelector according to (1).
6: EnvSelOperator $\leftarrow$ Select an operator by EnvSelSelector according to (1).
7: $A \leftarrow$ OffGenOperator($\mathcal{P}$)
8: $\mathcal{P} \leftarrow$ EnvSelOperator(A, N).
9: $\mathcal{P} \leftarrow$ EvolveByMOEA($\mathcal{P}, N, g_{\max}$).
10: $R \leftarrow$ Calculate the reward according to credit assignment.
11: Update quality vector $\mathbf{Q}$ of OffGenSelector and EnvSelSelector according to (2).
12: Update selection probability vector $\mathbf{P}$ of OffGenSelector and EnvSelSelector according to (3) and (4).
13: **end while**

Given a pool of candidate offspring generation operators $\mathcal{P}_{\text{OffGen}}$ and a pool of candidate environmental selection operators $\mathcal{P}_{\text{EnvSel}}$, two operator selectors (i.e., OffGenSelector and EnvSelSelector) using AP are instantiated respectively for selecting the offspring generation and environmental selection operators. After initializing the population $\mathcal{P}$ of size N, the algorithm enters the main loop. In each iteration, the two selectors first select an offspring generation and environmental selection operator for generating new offspring and updating the population. Next, the reward for the selected operators is calculated according to the credit assignment scheme. Then the quality vector and probability vector of the two selectors are updated based on the received reward following the updating rules of AP method. The algorithm is stopped until the termination criterion is fulfilled.

In the proposed AOS-LSMOEA, the offspring generation strategies used in LSMOF [14], LMOCSO [30], NSGA-II [7], MOEA/D-DE [19], and direction guided offspring generation in DGEA [13] are included as offspring generation operators, i.e., $\mathcal{P}_{\text{OffGen}} = \{\text{LSMOF}, \text{LMOCSO}, \text{NSGA-II}, \text{MOEA/D-DE}, \text{DGEA}\}$. And the environmental selection strategies in NSGA-II [7], IBEA [35], and SPEA2 [37] are taken as the possible environment selection operators as well as without using any selection (WithoutSel), i.e., $\mathcal{P}_{\text{EnvSel}} = \{\text{WithoutSel}, \text{NSGA-II}, \text{IBEA}, \text{SPEA2}\}$. It is worth noting that LSMOF, LMOCSO, and DGEA are representative LSMOEAs while NSGA-II and MOEA/D-DE are popular MOEAs. The hybridization of these five offspring generation strategies is expected to generate offspring solutions with both good convergence and diversity. Moreover, the environmental selection strategies in NSGA-II, IBEA, and SPEA2 are taken as the possible environment selection strategies as well as without using any selection, aiming to maintain diverse candidate solutions. The parameter values recommended in [27] are used in AP method, i.e., $P_{\min} = 0.1$, $\alpha = 0.8$, and $\beta = 0.8$.

4 Experimental Studies

To validate the performance of the proposed AOS-LSMOEA, experimental studies on a set of seven TREE problems are performed. Since this study mainly focuses on unconstrained multiobjective optimization, the constraint functions in those TREE problems are ignored. The iLSMOA using DGEA as offspring generation operator and SPEA2 as environmental selection operator is compared with the proposed AOS-LSMOEA. Additionally, AOS-LSMOEA is compared with its counterpart using a uniformly random operator selection strategy (denoted as UR-LSMOEA) and the same candidate operators to assess the effectiveness of AOS used in our algorithm. Each algorithm is run 25 times independently on each test problems, and the maximal number of fitness function evaluation is set as 20000 in all the runs. The Wilcoxon rank-sum test [9] is used to compare the results obtained by AOS-LSMOEA and the compared algorithms with a significance level of 0.05. The "+", "−" and "≈" signs summarize the Wilcoxon rank-sum test results, i.e., the number of problems on which the performance of the corresponding algorithm is significantly better than, worse than, or almost similar to that of the AOS-LSMOEA, respectively. All the investigated algorithms are implemented in PlatEMO [29] with MATLAB 2019b. The experiments were run on the same computing platform, i.e., Dell Precision T3630 with Intel Core i7-8700 processor and 32 GB RAM.

4.1 Experimental Results

Since we do not know the PFs of real-world test problems, the inverted generational distance (IGD) [6] metric cannot be computed, and thus we use the hypervolume (HV) [36] indicator to evaluate the algorithm's performance. For each algorithm, the average HV (HV_{avg}) and standard deviation of HV (HV_{std}) over 25 independent runs on each test problem are listed in Table 1. The maximal HV_{avg} achieved on each test problem is highlighted in bold. The dimension (D) of each problem is also listed in the table.

From Table 1, the proposed AOS-LSMOEA obtains the maximal HV_{avg} on all the seven test problems. Statistically, AOS-LSMOEA performs significantly better than iLSMOA on all the seven test problems. In the comparison between AOS-LSMOEA and UR-LSMOEA, the former has statistically better performance on two problems, and they have similar performance on the remaining five problems. It can be stated that the proposed AOS-LSMOEA generally performs better than iLSMOA and UR-LSMOEA. Additionally, the box plots of the HV from each algorithm on each problem are given in Fig. 1. It can be found that AOS-LSMOEA has the highest median HV values, and the box plots of AOS-LSMOEA are shorter than that of UR-LSMOEA on all the test problems. Experimental results show that UR-LSMOEA generally achieves better performance than iLSMOA, and that the proposed AOS-LSMOEA performs better

Table 1. The average HV (HV_{avg}) and standard deviation of HV (HV_{std}) over 25 independent runs of each algorithm on each problem.

Problem	D	iLSMOA		UR-LSMOEA		AOS-LSMOEA	
		HV_{avg}	HV_{std}	HV_{avg}	HV_{std}	HV_{avg}	HV_{std}
TREE1	3000	0.5909 (−)	0.0382	0.7769 (≈)	0.0279	**0.7795**	0.0258
TREE2	3000	0.6648 (−)	0.0288	0.8245 (−)	0.0270	**0.8363**	0.0094
TREE3	6000	0.5000 (−)	0.0286	0.8506 (≈)	0.0844	**0.8732**	0.0099
TREE4	6000	0.0877 (−)	0.0009	0.7968 (≈)	0.1706	**0.8470**	0.1241
TREE5	6000	0.4002 (−)	0.0291	0.8770 (≈)	0.0625	**0.9021**	0.0197
TREE_DSJ	30000	0.1114 (−)	0.0301	0.8863 (−)	0.0511	**0.9091**	0.0083
TREE_DZ	30000	0.5289 (−)	0.0216	0.7437 (≈)	0.1247	**0.7744**	0.0977
+/−/≈		0/7/0		0/2/5		–	

than both iLSMOA and UR-LSMOEA. This first indicates that the search ability can be enhanced by applying multiple operators. Considering that the only difference between AOS-LSMOEA and UR-LSMOEA is that AOS is not adopted in UR-LSMOEA, it can be confirmed that the superiority of AOS-LSMOEA over UR-LSMOEA should be credited to the AOS approach. Thus, it can be concluded that the proposed AOS-LSMOEA is effective for LSMOPs.

Furthermore, the average run-time (RT_{avg}) and standard deviation of run-time (RT_{std}) over 25 independent runs on each problem are recorded and listed in Table 2. It is apparent that AOS-LSMOEA does not increase computing time significantly. On each problem, the RT_{avg} of AOS-LSMOEA is only several seconds longer than that of iLSMOA and UR-LSMOEA. In other words, the additional computational cost brought by AOS in AOS-LSMOEA is negligible. It can be stated that the AOS-LSMOEA significantly improves the performance of iLSMOA on TREE problems. At the same time, the computational cost is

Table 2. The average run-time (RT_{avg}) and standard deviation of run-time (RT_{std}) over 25 independent runs of each algorithm on each problem (Unit: Second).

Problem	iLSMOA		UR-LSMOEA		AOS-LSMOEA	
	RT_{avg}	RT_{std}	RT_{avg}	RT_{std}	RT_{avg}	RT_{std}
TREE1	**11.3435**	0.9638	13.5427	1.3334	16.5715	1.3836
TREE2	**16.8035**	1.2226	20.4451	1.3342	23.6534	1.5737
TREE3	**19.6647**	1.2161	22.4597	1.5112	25.3052	1.7684
TREE4	**19.3748**	1.1434	21.7596	1.2912	24.3863	1.6936
TREE5	**21.5849**	1.4275	24.7026	1.6295	27.6274	1.9116
TREE_DSJ	**99.3003**	9.3173	100.1800	8.9203	105.0038	9.6405
TREE_DZ	**98.3424**	9.0984	100.6455	9.1502	104.3981	9.5728

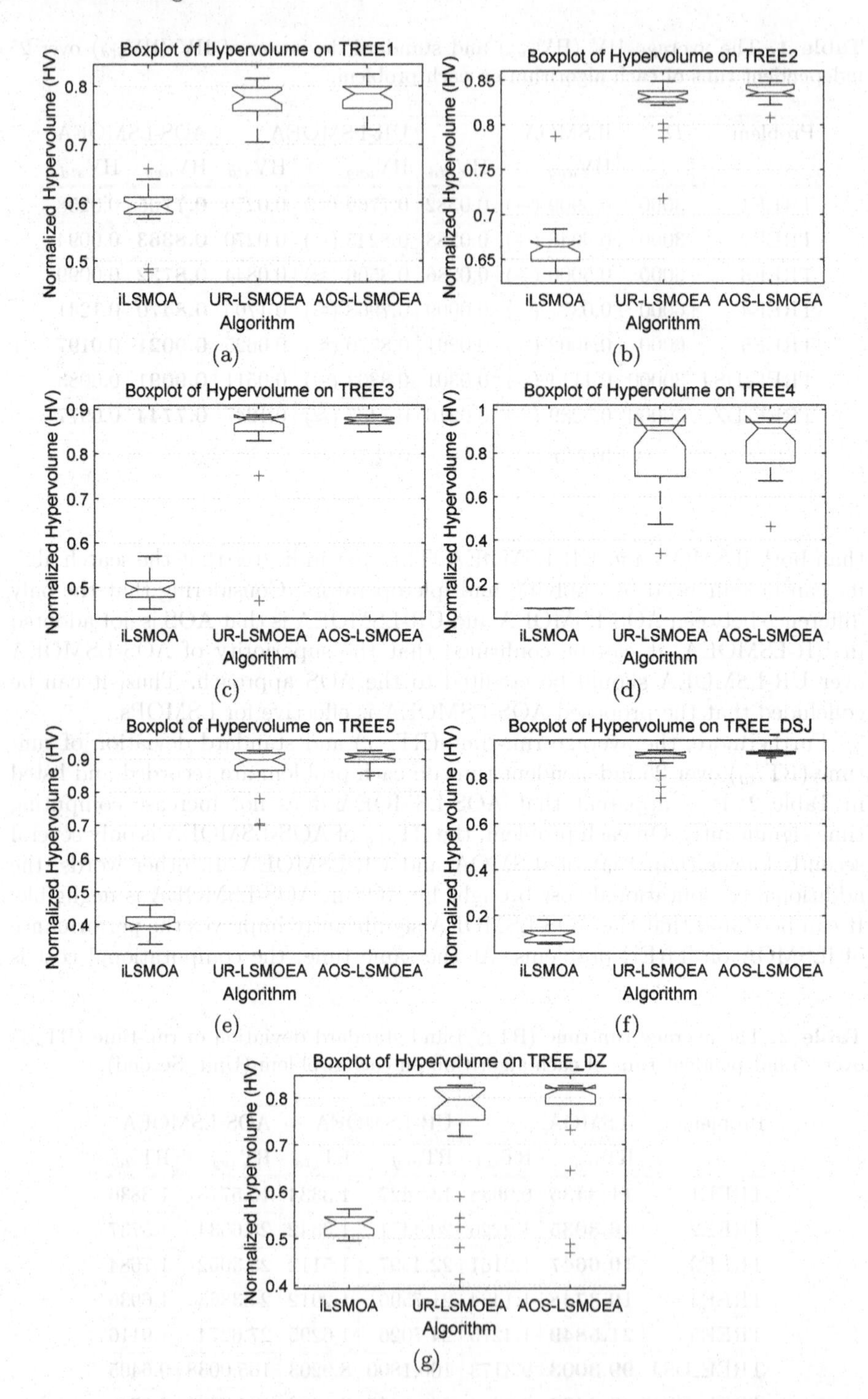

Fig. 1. Box plot of hypervolume values over 25 runs of each algorithm on each problem.

only increased slightly. Therefore, the proposed AOS-LSMOEA is an effective and efficient algorithm for solving LSMOPs.

5 Conclusion

This paper proposed a hybridized LSMOEA framework with adaptive operator selection (AOS), termed AOS-LSMOEA, for solving the real-world LSMOPs, that is, the TREE problems. In the proposed AOS-LSMOEA, multiple offspring generation and environmental selection operators are included, and the AOS method automatically selects suitable operators to use in each iteration. Experimental studies show that AOS-LSMOEA performs better than iLSMOA and UR-LSMOEA. The superiority of AOS-LSMOEA can be attributed to two aspects. On the one hand, by including multiple operators in the algorithm, the search ability can be enhanced since different operators have different strength during the search. On the other hand, by adopting AOS, proper operators can be identified and selected from the set of available operators so that the performance of AOS-LSMOEA is better than UR-LSMOEA, which uniformly random operator selection strategy. This work also shows that it is promising to use automatic algorithm configuration methods, such as adaptive operator selection and automatic parameter tuning, to design new MOEAs or LSMOEAs to solve challenging MOPs.

Acknowledgements. This work was supported by the Guangdong Basic and Applied Basic Research Foundation (Grant No. 2019A1515110575), the Guangdong Provincial Key Laboratory (Grant No. 2020B121201001), the Program for Guangdong Introducing Innovative and Entrepreneurial Teams (Grant No. 2017ZT07X386), the Shenzhen Science and Technology Program (Grant No. KQTD2016112514355531), the Program for University Key Laboratory of Guangdong Province (Grant No. 2017KSYS008), and the National Natural Science Foundation of China (No. 61903178 and 61906081).

References

1. Antonio, L.M., Coello, C.A.C.: Use of cooperative coevolution for solving large scale multiobjective optimization problems. In: 2013 IEEE Congress on Evolutionary Computation, pp. 2758–2765 (2013)
2. Beume, N., Naujoks, B., Emmerich, M.: SMS-EMOA: multiobjective selection based on dominated hypervolume. Eur. J. Oper. Res. **181**(3), 1653–1669 (2007)
3. Cheng, R., Jin, Y., Olhofer, M., Sendhoff, B.: Test problems for large-scale multiobjective and many-objective optimization. IEEE Trans. Cybern. **47**(12), 4108–4121 (2017)
4. Cheng, R.: Nature inspired optimization of large problems. Ph.D. thesis, University of Surrey (2016)
5. Coello, C.A.C., Lamont, G.B., Van Veldhuizen, D.A., et al.: Evolutionary Algorithms for Solving Multi-Objective Problems, vol. 5. Springer, New York (2007). https://doi.org/10.1007/978-0-387-36797-2

6. Coello Coello, C.A., Reyes Sierra, M.: A study of the parallelization of a coevolutionary multi-objective evolutionary algorithm. In: Monroy, R., Arroyo-Figueroa, G., Sucar, L.E., Sossa, H. (eds.) MICAI 2004. LNCS (LNAI), vol. 2972, pp. 688–697. Springer, Heidelberg (2004). https://doi.org/10.1007/978-3-540-24694-7_71
7. Deb, K., Pratap, A., Agarwal, S., Meyarivan, T.: A fast and elitist multi-objective genetic algorithm: NSGA-II. IEEE Trans. Evol. Comput. **6**(2), 182–197 (2002)
8. Deb, K., Thiele, L., Laumanns, M., Zitzler, E.: Scalable test problems for evolutionary multiobjective optimization. In: Advanced Information and Knowledge Processing. Springer, London (2005)
9. Derrac, J., García, S., Molina, D., Herrera, F.: A practical tutorial on the use of nonparametric statistical tests as a methodology for comparing evolutionary and swarm intelligence algorithms. Swarm Evol. Comput. **1**(1), 3–18 (2011)
10. Fialho, Á., Da Costa, L., Schoenauer, M., Sebag, M.: Extreme value based adaptive operator selection. In: Rudolph, G., Jansen, T., Beume, N., Lucas, S., Poloni, C. (eds.) PPSN 2008. LNCS, vol. 5199, pp. 175–184. Springer, Heidelberg (2008). https://doi.org/10.1007/978-3-540-87700-4_18
11. Fialho, Á., Da Costa, L., Schoenauer, M., Sebag, M.: Analyzing bandit-based adaptive operator selection mechanisms. Ann. Math. Artif. Intell. **60**(1–2), 25–64 (2010)
12. Goldberg, D.E.: Probability matching, the magnitude of reinforcement, and classifier system bidding. Mach. Learn. **5**, 407–425 (1990)
13. He, C., Cheng, R., Yazdani, D.: Adaptive offspring generation for evolutionary large-scale multiobjective optimization. IEEE Trans. Syst. Man Cybern. Syst. 1–13 (2020)
14. He, C., Li, L., Tian, Y., Zhang, X., Cheng, R., Jin, Y., Yao, X.: Accelerating large-scale multiobjective optimization via problem reformulation. IEEE Trans. Evol. Comput. **23**(6), 949–961 (2019)
15. He, C., Cheng, R., Tian, Y., Zhang, X.: Iterated problem reformulation for evolutionary large-scale multiobjective optimization. In: 2020 IEEE Congress on Evolutionary Computation (CEC), pp. 1–8. IEEE (2020)
16. He, C., Cheng, R., Zhang, C., Tian, Y., Chen, Q., Yao, X.: Evolutionary large-scale multiobjective optimization for ratio error estimation of voltage transformers. IEEE Trans. Evol. Comput. **24**(5), 868–881 (2020)
17. He, C., Tian, Y., Jin, Y., Zhang, X., Pan, L.: A radial space division based many-objective optimization evolutionary algorithm. Appl. Soft Comput. **61**, 603–621 (2017)
18. Ishibuchi, H., Murata, T.: Multiobjective genetic local search algorithm and its application to flowshop scheduling. IEEE Trans. Syst. Man Cybern. Part C **28**(3), 392–403 (1998)
19. Li, H., Zhang, Q.: Multiobjective optimization problems with complicated Pareto sets, MOEA/D and NSGA-II. IEEE Trans. Evol. Comput. **13**, 284–302 (2009)
20. Liu, H.L., Gu, F., Zhang, Q.: Decomposition of a multiobjective optimization problem into a number of simple multiobjective subproblems. IEEE Trans. Evol. Comput. **18**(3), 450–455 (2014)
21. Ma, X., et al.: A multiobjective evolutionary algorithm based on decision variable analyses for multiobjective optimization problems with large-scale variables. IEEE Trans. Evol. Comput. **20**(2), 275–298 (2015)
22. Mahdavi, S., Shiri, M.E., Rahnamayan, S.: Metaheuristics in large-scale global continues optimization: a survey. Inf. Sci. **295**, 407–428 (2015)
23. Pan, L., He, C., Tian, Y., Wang, H., Zhang, X., Jin, Y.: A classification-based surrogate-assisted evolutionary algorithm for expensive many-objective optimization. IEEE Trans. Evol. Comput. **23**(1), 74–88 (2018)

24. Ponsich, A., Jaimes, A.L., Coello, C.A.C.: A survey on multiobjective evolutionary algorithms for the solution of the portfolio optimization problem and other finance and economics applications. IEEE Trans. Evol. Comput. **17**(3), 321–344 (2013)
25. Potter, M.A., De Jong, K.A.: A cooperative coevolutionary approach to function optimization. In: Davidor, Y., Schwefel, H.-P., Männer, R. (eds.) PPSN 1994. LNCS, vol. 866, pp. 249–257. Springer, Heidelberg (1994). https://doi.org/10.1007/3-540-58484-6_269
26. Huband, S., Hingston, P., Barone, L., While, L.: A review of multiobjective test problems and a scalable test problem toolkit. IEEE Trans. Evol. Comput. **10**(5), 477–506 (2006)
27. Thierens, D.: An adaptive pursuit strategy for allocating operator probabilities. In: Beyer, H., O'Reilly, U., (eds.) Proceedings of Genetic and Evolutionary Computation Conference (GECCO), pp. 1539–1546. ACM (2005)
28. Thierens, D.: Adaptive strategies for operator allocation. In: Lobo, F.G., Lima, C.F., Michalewicz, Z. (eds.) Parameter Setting in Evolutionary Algorithms, Studies in Computational Intelligence. SCI, vol. 54, pp. 77–90. Springer, Heidelberg (2007). https://doi.org/10.1007/978-3-540-69432-8_4
29. Tian, Y., Cheng, R., Zhang, X., Jin, Y.: PlatEMO: a MATLAB platform for evolutionary multi-objective optimization [educational forum]. IEEE Comput. Intell. Mag. **12**(4), 73–87 (2017)
30. Tian, Y., Zheng, X., Zhang, X., Jin, Y.: Efficient large-scale multiobjective optimization based on a competitive swarm optimizer. IEEE Trans. Cybern. **50**(8), 3696–3708 (2020)
31. Zhang, Q., Li, H.: MOEA/D: a multiobjective evolutionary algorithm based on decomposition. IEEE Trans. Evol. Comput. **11**(6), 712–731 (2007)
32. Zhang, X., Tian, Y., Cheng, R., Jin, Y.: A decision variable clustering-based evolutionary algorithm for large-scale many-objective optimization. IEEE Trans. Evol. Comput. **22**(1), 97–112 (2016)
33. Zille, H., Ishibuchi, H., Mostaghim, S., Nojima, Y.: A framework for large-scale multiobjective optimization based on problem transformation. IEEE Trans. Evol. Comput. **22**(2), 260–275 (2018)
34. Zitzler, E., Deb, K., Thiele, L.: Comparison of multiobjective evolutionary algorithms: empirical results. Evol. Comput. **8**(2), 173–195 (2000)
35. Zitzler, E., Künzli, S.: Indicator-based selection in multiobjective search. In: Yao, X., et al. (eds.) PPSN 2004. LNCS, vol. 3242, pp. 832–842. Springer, Heidelberg (2004). https://doi.org/10.1007/978-3-540-30217-9_84
36. Zitzler, E., Thiele, L.: Multiobjective evolutionary algorithms: a comparative case study and the strength Pareto approach. IEEE Trans. Evol. Comput. **3**(4), 257–271 (1999)
37. Ziztler, E., Laumanns, M., Thiele, L.: SPEA2: improving the strength Pareto evolutionary algorithm for multiobjective optimization. In: Evolutionary Methods for Design, Optimization, and Control, pp. 95–100 (2002)

Multi-objective Reinforcement Learning Based Multi-microgrid System Optimisation Problem

Jiangjiao Xu[1], Ke Li[1](✉), and Mohammad Abusara[2]

[1] Department of Computer Science, University of Exeter, Exeter, UK
[2] Department of Engineering, University of Exeter, Exeter, UK
{J.Xu,K.Li,M.Abusara}@exeter.ac.uk

Abstract. Microgrids with energy storage systems and distributed renewable energy sources play a crucial role in reducing the consumption from traditional power sources and the emission of CO_2. Connecting multi microgrid to a distribution power grid can facilitate a more robust and reliable operation to increase the security and privacy of the system. The proposed model consists of three layers, smart grid layer, independent system operator (ISO) layer and power grid layer. Each layer aims to maximise its benefit. To achieve these objectives, an intelligent multi-microgrid energy management method is proposed based on the multi-objective reinforcement learning (MORL) techniques, leading to a Pareto optimal set. A non-dominated solution is selected to implement a fair design in order not to favour any particular participant. The simulation results demonstrate the performance of the MORL and verify the viability of the proposed approach.

Keywords: Multi-microgrid · Multi-objective reinforcement learning · Independent system operator · Market operator · Pareto Front

1 Introduction

Over the past decade, the leading domestic electricity price paid by households is fixed for their energy consumption, predominantly gas and electricity. At the same time, non-hydro renewable energy sources (RES), such as wind, solar, tidal and geothermal power, continues to penetrate the power generation market in meaningful ways. The percentage of these sources has risen from 0.37 % in 1990 to 5.93 % in 2016 [1]. It is well understood that wholesale price variability is an essential characteristic of deregulation in the electricity market. Energy consumers who are sensitive to energy prices may vary their electricity consumption based on the dynamic price signals [2]. It means that dynamic electricity prices can reduce the demand for peak load period, while the use of renewable energy

K. Li was supported by UKRI Future Leaders Fellowship (Grant No. MR/S017062/1).

H. Ishibuchi et al. (Eds.): EMO 2021, LNCS 12654, pp. 684–696, 2021.
https://doi.org/10.1007/978-3-030-72062-9_54

and energy storage systems can significantly reduce the use of fossil fuels, thereby reducing power generation costs and carbon dioxide emissions.

A wide range of demand side management studies in dynamic pricing schemes have been conducted. Reference [3] presented a demand response approach based on the dynamic energy pricing, which accomplishes the optimal load control of the devices by establishing a virtual power transaction process. A decision model for smart grid considering the demand response and market energy pricing is proposed to interact between the market retail price and energy consumers [4]. In [5], to improve the system operation and optimise the power flow, a coordinated operating strategy for the gas and electricity integrated energy system is proposed in a multi-energy system. However, the above studies in demand side management only optimise the energy prices from an operational perspective and do not acknowledge the influence of price variations in the energy market and customer demand on the planning level. Moreover, the majority of existing articles study single utility objective optimisation problem solely, e.g., reduce the overall cost [6], maximise costumers' utility [7] and modelling demand figure [8]. When a multi-microgrid system is designed, there will be a coupling interaction between microgrids, independent system operator (ISO) and the power grid. There are often some conflicts between these participants during planning. The impact of dynamic pricing in a multi-microgrid system for a multi-objective problem has not been investigated comprehensively.

To investigate a comprehensive model and balance all participants, we consider designing a multi-microgrid system, including three microgrids, an independent system operator (ISO) and a main power grid [9]. Microgrids are connected to each other, but renewable energy generation cannot be dispatched between each other. A dynamic pricing scheme will be implemented in this multi-microgrid system to balance the system operation for all participants. In this case, a multi-objective optimal approach needs to be proposed to balance all objectives without favour any single participant.

Multi-Objective Reinforcement learning is an outstanding algorithm to address multi-objective problems for complex strategic interactions. In [10], a reinforcement learning environment is typically formalised by employing a Markov decision process (MDP). A Q-learning algorithm was introduced to approximate optimal Q values in [11] iteratively. In a multi-objective optimisation problem, the objective space will contain two or more dimensions, and regular MDPs will be generalised to multi-objective MDPs. Many approaches of MORL rely on a single-policy algorithm to learn Pareto optimal solutions [12]. The most straightforward idea is to convert the multi-objective problem into a standard single-objective problem by utilising a scalarisation function [13–15].

However, this transformation may not be suitable to solve a non-linear problem that lies in non-convex regions of the Pareto front. In this paper, we develop a L_p metrics based MORL algorithm to create a fair multi-microgrid system design based on the Approximate Pareto Front (APF), which can achieve high-quality solutions to non-linear multi-objective functions. To the best of our knowledge, the proposed MORL is the first time to be used in a multi-objective optimisation smart grid scenario. Then the Pareto Front will be applied to investigate the connection and influence between various objective functions, which can support to produce a fair result for all participants. Our main contributions are:

It combines real-time various energy prices with planning scenarios in practice and considers the impact of real-time changing electricity prices and renewable energy on the design of multi-microgrid systems by implementing a method of a MORL algorithm. The main grid provider sales revenue, the university office utilises the renewable energy to operate the energy storages and the users save energy consumption cost. Three conflict objectives are included in the multi-microgrid planning scenario.

A multi-objective formulation is developed to a future multi-microgrid framework. To solve a multi-objective problem (MOP), we develop a MORL algorithm considering the dynamic electricity prices and the operation of energy storage, e.g., charging/discharging/idle, which can generate an APF to provide a fair and effectiveness operating planning for a multi-microgrid network.

The rest of this report is organised as follows. Section 2 describes the main structure of the multi-microgrid network and discusses the mathematical system models of three participants. The proposed MORL problem formulation is presented in Sect. 3, and Sect. 4 describes the methodology of MORL algorithms in detail. In Sect. 5, the numerical simulation results based on the Pareto are presented. Finally, Sect. 6 is the conclusion and future work.

2 Multi-microgrid Description

A multi-microgrid system based on the Penryn Campus, University of Exeter is shown in Fig. 1. It consists of RESs, Microgrids, independent system operator and Power Grid. Information and communication technology (ICT) systems are implemented to exchange the information among microgrids, i.e., price, power demand and generation. A high-level multi-microgrid optimisation system is considered to be designed in this paper. Detailed mathematical models of the power grid, ISO and microgrid will be presented as follows. Let $\mathcal{N} = 1, 2, ..., N$ and $\mathcal{N}_s = 1, 2, ..., N_s$ denote the set of the microgrids and the set of the microgrids with energy storages, respectively, where $N_s \leq N$.

2.1 Microgrid Model

The mathematical models of a microgrid model will show the power balance among energy storages, microgrids and the power grid. For microgrid n without energy storage, it has

$$p_{d_n}(t) = p_{g_n}(t) + p_{r_n}(t). \tag{1}$$

where $p_{g_n}(t)$ is the power transmission between microgrid n and the power grid at time t. If $p_{g_n}(t)$ is positive, the power grid will transmit the power to the microgrid n, otherwise, the microgrid n will send the power back to the grid. $p_{r_n}(t)$ and $p_{d_n}(t)$ are the power generation from the renewable energy sources and power demand in microgrid n, respectively.

For microgrid n with energy storage system, then it satisfies

$$p_{d_n}(t) = p_{g_n}(t) + p_{r_n}(t) + s_n(t) - s_n(t-1). \quad (2)$$

$$subject\ to \quad \forall\, t \in \mathcal{T}$$

$$0 \leq s_n(t) \leq \overline{s}_n \quad (3a)$$

$$j s_n(t) - s_n(t-1) j \leq \Delta s_n \quad (3b)$$

where (3a) is the maximum storage capacity constraints, $s_n(t)$ is the stored energy of microgrid n at time t. And $\overline{s}_n$ denotes the maximum capacity of the storage. (3b) is the maximum charging/discharging rate constraints, Δs_n is the maximum charging/discharging power from time t to $t+1$.

To consider the shiftable loads, the power demand $p_{d_n}(t)$ can be rewritten as

$$p_{d_n}(t) = f_{d_n}(\lambda(t), l_{b_n}(t)) = (1 + h_n(\lambda(t)))l_{b_n}(t) \quad (4)$$

where $\lambda(t)$ is the electricity price and $l_{b_n}(t)$ is the nominal value of the baseload at time t. Since the baseload has almost no fluctuations in practice and the baseload prediction technology can achieve high-precision prediction results, we will assume that $l_{b_n}(t)$ is a known data in advance.

Different household users may have different responses to the same price. The responses of different household users to various price schemes can be modelled by adopting a utility function from microeconomics [16]. For each user, the utility function denotes the level of user's satisfaction corresponding to the energy consumption. The overall function of multi-microgrid can be expressed as

$$\begin{aligned} \max_{\lambda(t)} : F_w &= f_w(p_{d_1}(t), ..., p_{d_N}(t), \lambda(t)) \\ &= \sum_{n=1}^{N} (f_u(p_{d_n}(t), \omega_n) - f_c(\lambda(t), p_{d_n}(t))) \end{aligned} \quad (5)$$

where

$$f_c(\lambda(t), p_{d_n}(t)) = \lambda(t) * p_{d_n}(t) \quad (6a)$$

$$f_u(p_{d_n}(t), \omega_n) = \begin{cases} \omega_n p_{d_n}(t) - \dfrac{\alpha}{2} p_{d_n}(t)^2, & if \quad 0 \leq p_{d_n}(t) \leq \dfrac{\omega_n}{\alpha} \\ \dfrac{\omega_n}{\alpha}, & if \quad p_{d_n}(t) \geq \dfrac{\omega_n}{\alpha} \end{cases} \quad (6b)$$

where F_m is the overall welfare function. $f_c(\lambda(t), p_{d_n}(t))$ and $f_u(p_{d_n}(t), \omega_n)$ are the cost function imposed by the energy provider and the quadratic utility function of the user corresponding to linear decreasing marginal benefit, respectively. $p_{d_n}(t) = (1 + h(\lambda(t))b_n(t))$ is the consuming power, $n = 1, 2, \ldots, N$ is the number of microgrid. $b_n(t)$ is the base load and $h(\lambda(t))$ is the positive/negative value in percentage at different profiles of price signal. ω_n is the value which may change among users and also at different intervals of the day. α is a predetermined parameter. For each advertised price value $\lambda(t)$, each user tries to modify the energy consumption to maximise its own welfare. It can be accomplished by setting the derivative of F_m equal zero, which means that the marginal benefit of the user would be equal to the advertised price.

2.2 ISO Model

The ISOs in this paper mainly act as an emergency energy provider to afford emergency demand response programs. It will store as much energy as possible to achieve a secure level. To provide the maximum emergency energy and increase the life of the batteries, the utility function can be given as follows:

$$\max_{\lambda(t), p_{g_n}(t)} : F_s = \sum_{n=1}^{N_s} s_n(t). \tag{7}$$

$$\begin{aligned} & \textit{subject to} \quad \forall\, t \in \mathcal{T} \\ & \underline{s}_n \leq s_n(t) \leq \overline{s}_n. \end{aligned} \tag{8}$$

where $\underline{s}_n$ is the minimum stored energy level for emergency operation, $n \in N_s$.

2.3 Power Grid Model

The power grid mainly injects the power to the microgrid when renewable energy in a microgrid is not enough. However, it can also absorb the power from the microgrid when renewable energy in microgrid has surplus energy. The mathematical models of the power grid can be written as

$$p_g(t) = \sum_{n=1}^{N} p_{g_n}(t). \tag{9}$$

where $p_g(t)$ is the total power distribution from power grid and all microgrids.

To obtain the maximum interest of the power grid, the derived from providing power $p_g(t)$ at the main power grid can be written as

$$\begin{aligned} \max_{\lambda(t), p_{g_n}(t)} : F_g &= f_g(\lambda(t), p_g(t)) \\ &= \lambda(t) p_g(t) - C_{p_g}(t) \end{aligned} \tag{10}$$

where

$$C_{p_g}(t) = a_g p_g(t)^2 + b_g p_g(t) + c_g \tag{11}$$

where $C_{p_g}(t)$ is the quadratic cost functions. $a_g > 0$ and $b_g, c_g \geq 0$ are the pre-determined generator parameters.

The main research scenario, including demand and generation data based on the Penryn Campus, University of Exeter, was presented. The university office of general affairs will act as the ISO to purchase electrical energy from the power company and combine the existing RESs and energy storages to generate a new time-varying electricity price. Students living in the university student apartments need to pay their own electricity bills for the use of various electrical appliances, e.g., washing machine, dryers and freezer. If the electricity price changes at different times, students may change their electricity consumption behaviour to reduce their electricity bills. At the same time, the university can use time-varying electricity prices to reduce the load in a high electric period and optimise the operation of energy storage systems to reduce the purchase of electricity from the main grid. A number of smart meters installed are 4.7 million and 4.5 million in British homes in 2018 and 2019, respectively [17]. Therefore, this designed scenario is also very practical in a smart grid environment in the local community.

3 Multi-objective Problem Formulation

This section will describe a multi-objective problem formulation to maximise the benefits of three participants for multi-microgrid system design. A multi-objective reinforcement learning (MORL) technique is then proposed to optimise the problem in a real-time market scenario.

To solve these three objectives F_w, F_s and F_g at the same time, a MOP formulation is given as

$$\min_{\lambda(t)} -F_w = -f_w(p_{d_1}(t), ..., p_{d_N}(t), \lambda(t)) \tag{12a}$$

$$\min_{\lambda(t), p_{g_n}(t)} \quad -F_s = -\sum_{n=1}^{N_s} s_n(t) \tag{12b}$$

$$\min_{\lambda(t), p_{g_n}(t)} \quad -F_g = -f_g(\lambda(t), p_g(t)) \tag{12c}$$

$$subject\ to \quad (1) - (4), (8)\ and\ (9)$$

where $\lambda(t)$ and $p_{g_n}(t)$ are the two variables related to the ISO determined by the current generation of renewable energy and the status of energy storage during the period. To solve the problem considering all the constraints, an additional function is introduced in the following way

$$\begin{aligned} F_a = &\sum_{n=1}^{N_s}[max(|s_n(t) - s_n(t-1) - \Delta s_n, 0) \\ &+ max(s_n(t) - \overline{s}_n, 0) + max(\underline{s}_n - s_n(t), 0)] \end{aligned} \tag{14}$$

where F_a is determined by the stored energy. The multi-microgrid is stable and satisfies the all constraints if and only if $F_a = 0$. According to the equation (14), the resulting MOP (12) can be rewritten by

$$\min_{\lambda(t), p_{g_n}(t)} \quad F_m = [-F_w \ \ -F_s \ \ -F_g \ \ F_a]^T \tag{15}$$

To solve the MOP, the Pareto optimality will be utilised to satisfy all requirements.

4 Proposed Algorithm for Multi-Microgrid Optimisation

To obtain the Pareto optimal set for MOP, a multi-objective Q learning algorithm is proposed in this section. The MORL framework is based on a scalarized single-policy algorithm that employs scalarization functions to reduce the dimensionality of the multi-objective environment to a single and scalar dimension.

A scalarization function can be defined as

$$F = f(\mathbf{x}, \mathbf{w}) \tag{16}$$

where in the case of MORL, $\mathbf{x}$ and $\mathbf{w}$ in the objective functions are the $\mathbf{Q}$ vector and the weight vector, respectively. The scalar Q values can be extended to $\mathbf{Q}$ vector that contains each Q value for each objective. When an action is chosen, the function F will be applied to the $\mathbf{Q}$ vector to achieve a single, scalar $SQ(s, a)$ estimate.

$$SQ(s, a) = \sum_{n=1}^{N} w_n \cdot Q_n(s, a) \tag{17}$$

where $n \in N$ is stand for the index of each objective function. And the weight vector should satisfy the following equation $\sum_{n=1}^{N} w_n = 1$.

However, this linear scalarization function has a fundamental limitation that can only find policies in convex regions of the Pareto optimal set [18]. Then a scalarization function based on the L_p metrics is proposed in this paper [19]. The proposed L_p metrics measure the distance between a point $\mathbf{x}$ in the multi-objective space and a utopian point $\mathbf{z}^*$ which is an adjustable parameter during the learning process. The measured distance between $\mathbf{x}$ and $\mathbf{z}^*$ for each objective can be given as follows

$$L_p(x) = (\sum_{n=1}^{N} w_n |x_n - z_n^*|^p)^{1/p}. \tag{18}$$

where $1 \leq p \leq \infty$. In the case of $p = \infty$, the metric can be known as the weighted L_∞ or the Chebyshev metric

$$L_\infty(x) = \max_{n=1,\ldots,N} w_n |x_n - z_n^*|. \tag{19}$$

where in the case of MORL, x_n can be replaced by $Q_n(s, a)$ to obtain the $SQ(s, a)$ with state s and action a

$$SQ(s, a) = \max_{n=1,\ldots,N} w_n |Q_n(s, a) - z_n^*|. \tag{20}$$

RL elements, including state and action spaces, reward function, learning and exploration rates, and discount factor, are described in detail in the following subsections:

State Space. The state variables are time of day (ToD_j) and State of Charge (SoC_k).

$$s|s_{j,k} = (ToD_j, SoC_k) \tag{21}$$

where the time of day ToD is discretised into 24 h $j = 1, 2, ..., 24$, the State of Charge SoC is divided into 8 levels from 30% to 100%.

Action Space. Action Space is the combination of Price and Charging/Discharging.

$$A = \{a|(PriceSignal, charging/discharging/idle)\} \tag{22}$$

where a total of 24 actions can be selected. The price is divided into 8 values which are set from 1.5 to 5.0.

Algorithm 1: Scalarized ϵ greedy strategy

1: **Initialise** *SQList*
2: **for** each action $a \in A$ **do**
3: $\mathbf{x} \leftarrow \{Q_1(s,a), ..., Q_m(s,a)\}$
4: $SQ(s,a) \leftarrow f(\mathbf{x}, \mathbf{w})$
5: Append $SQ(s,a)$ to *SQList*
6: **End for**
7: **return** ϵ **greedy(*SQList*)**

Algorithm 2: Multi-objective Q-learning algorithm)

1: **Initialise** $Q_n(s,a)$
2: **for** each episode **do**
3: Initialise state s
4: **repeat**
5: Select action a using ϵ greedy strategy
6: Take action and observe new state $s' \in S$
7: Obtain reward vector $\mathbf{r}$ and select new action
8: **for** each objective n **do**
9: $Q_n(s,a) = Q_n(s,a) + \alpha_t(r_n + \gamma Q_n(s', a') - Q_n(s,a))$
10: **end for**
11: $s \rightarrow s'$
12: **until** s is terminal
13: **end for**

Reward. The reward value $r_n(t)$ for each objective is the immediate incentive gained by taking a specific action at state s. The reward function of each objective is designed to minimise the objective function. Then these obtained reward values will be saved to the extended Q table.

According to these $SQ(s,a)$ values, an appropriate action selection strategy, i.e., scalarised ϵ greedy strategy, can be taken to select an action. The detailed scalarised ϵ greedy strategy can be found in Algorithm 1.

The proposed multi-objective Q-learning algorithm is shown in Algorithm 2. First of all, the Q values for each objective are initialised. Then the algorithm starts each episode in state s and select action via Scalarized ϵ greedy strategy. In terms of the selected action, the algorithm will transit to a new state $s^{'}$ and produce the reward vector $\mathbf{r}$. More precisely, these reward values are updated for each objective individually and single objective reward update rule will be extended to a multi-objective environment. Then the best scalarised action for the next state $s^{'}$ will be chosen via ϵ greedy. As long as each action and state is fully sampled, convergence can be guaranteed.

5 Simulation Results and Performance

The numerical simulation results will be performed to assess the performance of the proposed multi-objective reinforcement learning algorithm. It is assumed there are three microgrids $\mathcal{N} = 3$ and two microgrids with energy storage $\mathcal{N}_f = 2$. The two maximum storage capacities are 200 kWh and 250 kWh, respectively. Let Δs_n equals 10% of the maximum capacity. The typical power demand corresponding to price λ can be obtained as discussed in [20]. And the baseload l_{b_n} from the Penryn Campus, Exeter University. An example of baseload and generation on Nov 17, 2019 is shown in Fig. 1. The power demand can increase/decrease in terms of the price value when the price signal varied.

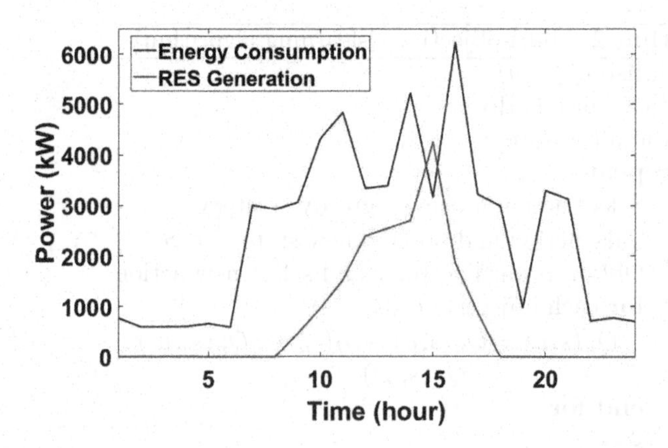

Fig. 1. Demand and RES generation of Microgrid 1 on Nov 17, 2019.

Figure 2 show an example of an extreme value of Approximated Pareto Front (APF) that minimises the first objective function $-F_w$. However, the other two functions are affected by objective function one and cannot converge to the minimum. Obviously, each objective function will affect each other, so it is necessary to provide a fair design for all participants.

One particular solution P^* in Pareto optimal set that can maximise the minimum improvement in all dimensions is discussed. Figure 3 shows an example

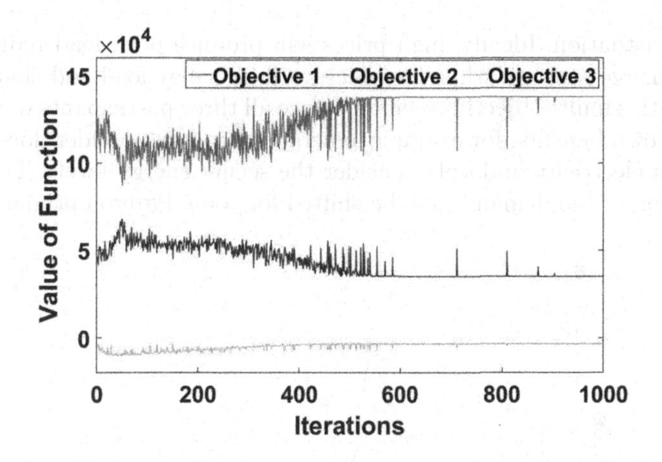

Fig. 2. Sample of Convergence Rate of MORL.

of the APF and reveals the connection between these three objectives. Three distinct solutions p_1^*, p_2^* and p_3^* are the extreme dominated solutions for three objective functions, respectively. It means that each solution will advantage to every single objective function. To ensure fairness for all objective functions, one particular solution P^* based on the APF will be picked to give no advantage to any single objective. As can be seen in Fig. 3, the optimal solution is located in the centre of the APF graphically. It refers that the proposed MORL approach can provide a fair solution to all three participants.

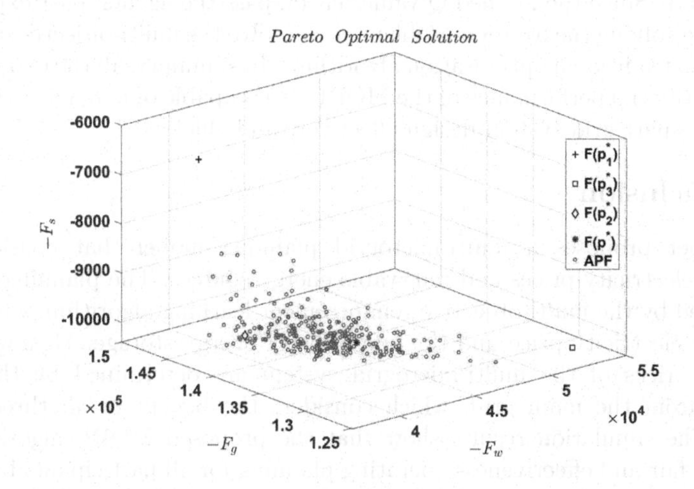

Fig. 3. APFs and non-dominated vectors $F(p^*)$ sampled.

The fluctuation of price signal plays a significant role in smart grid energy management. Figure 4 proves that the proposed method can generate outstanding price fluctuation. Ideally, high prices will produce peak load reduction and discharge energy storage, while low prices will fill valley load and charge energy storage. In this multi-objective scenario, since all three participants want to maximise their own benefits, for example, emergency energy provider does not mind the price of electricity and only consider the secure energy levels. Thus, only a small portion of the demand may be shifted for some Pareto optimal solutions.

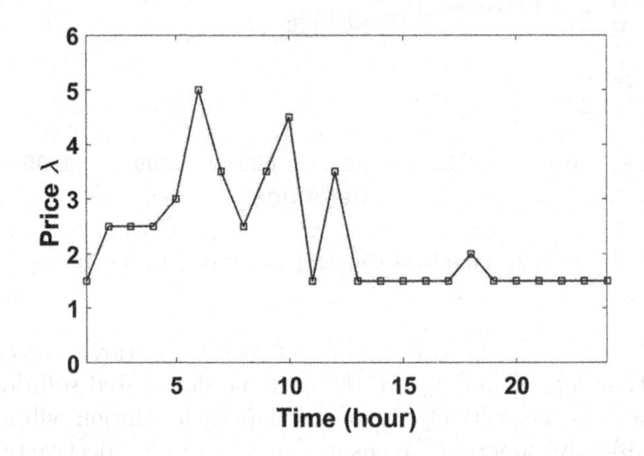

Fig. 4. Dynamic price signal λ by using the proposed MORL approach.

The Pareto optimal set from the experimental results in Fig. 3 shows that the proposed MORL is able to benefit to a singular participant or balance to all participants. Since the learned Q values encompass the agents' past experiences without re-solving the decision problem, it can solve the multi-objective problems faster than traditional optimisation algorithms. In summary, all the experimental results verify the performance of the MORL. It is capable of managing the multi-objective smart grid system design effectively and efficiently.

6 Conclusion

This paper proposes a multi-microgrid planning model that considers the dynamic electricity prices and renewable energy sources. The planning scenario is analysed by the multi-objective reinforcement learning algorithm, which optimises the electricity price and the operation of energy storage. Meanwhile, the dynamic prices of the multi-microgrid system are determined by the power demand from the main grid, which considers the benefit of all three participants. The simulation results show that the proposed MORL algorithm can provide a fair and effectiveness operating planning for all participants by control the energy storage operation and modify the real-time energy tariff. It shows the ability of MORL to learn the optimal control policy. The optimised coordinated operation can help to improve the utilisation of renewable energy, increase the operating life of batteries, reduce the operation cost of multi-microgrid, save bills for customers and maximise the profit for the power grid.

References

1. Agency, I.E.: Electricity Information 2016 (2016)
2. Sinha, A.K., Kumar, N.: Demand response management of smart grids using dynamic pricing. In: 2016 International Conference on Inventive Computation Technologies (ICICT), vol. 1, pp. 1–4 (2016)
3. Yu, M., Hong, S.H.: A real-time demand-response algorithm for smart grids: a Stackelberg game approach. IEEE Trans. Smart Grid **7**(2), 879–888 (2016)
4. Wei, W., Liu, F., Mei, S.: Energy pricing and dispatch for smart grid retailers under demand response and market price uncertainty. IEEE Trans. Smart Grid **6**(3), 1364–1374 (2015)
5. Bai, L., Li, F., Cui, H., Jiang, T., Sun, H., Zhu, J.: Interval optimization based operating strategy for gas-electricity integrated energy systems considering demand response and wind uncertainty. Appl. Energy **167**, 270–279 (2016)
6. Salinas, S., Li, M., Li, P., Fu, Y.: Dynamic energy management for the smart grid with distributed energy resources. IEEE Trans. Smart Grid **4**(4), 2139–2151 (2013)
7. Fahrioglu, M., Alvarado, F.L.: Using utility information to calibrate customer demand management behavior models. In: 2002 IEEE Power Engineering Society Winter Meeting. Conference Proceedings (Cat. No. 02CH37309), vol. 1, p. 26 (2002)
8. Dong, Q., Yu, L., Song, W., Yang, J., Wu, Y., Qi, J.: Fast distributed demand response algorithm in smart grid. IEEE/CAA J. Automatica Sinica **4**(2), 280–296 (2017)
9. Dimeas, A.L., Hatziargyriou, N.D.: Operation of a multiagent system for microgrid control. IEEE Trans. Power Syst. **20**(3), 1447–1455 (2005)
10. Mannor, S., Shimkin, N.: A geometric approach to multi-criterion reinforcement learning. J. Mach. Learn. Res. **5**, 325–360 (2004)
11. Tsitsiklis, J.N.: Asynchronous stochastic approximation and Q learning. In: Proceedings of 32nd IEEE Conference on Decision and Control, vol. 1, pp. 395–400 (1993)
12. Zeng, F., Zong, Q., Sun, Z., Dou, L.: Self-adaptive multi-objective optimization method design based on agent reinforcement learning for elevator group control systems. In: 2010 8th World Congress on Intelligent Control and Automation, pp. 2577–2582 (2010)
13. Li, K., Deb, K., Zhang, Q., Kwong, S.: An evolutionary many-objective optimization algorithm based on dominance and decomposition. IEEE Trans. Evol. Comput. **19**(5), 694–716 (2015)
14. Chen, R., Li, K., Yao, X.: Dynamic multiobjectives optimization with a changing number of objectives. IEEE Trans. Evol. Comput. **22**(1), 157–171 (2018)
15. Li, K., Chen, R., Fu, G., Yao, X.: Two-archive evolutionary algorithm for constrained multiobjective optimization. IEEE Trans. Evol. Comput. **23**(2), 303–315 (2019)
16. Green, J.R., Mas-Colell, A., Whinston, M.: Microeconomic Theory. Oxford University Press, New York (1995)
17. Deane, L.: One million faulty smart meters were installed in British homes. The Daily Mail, March 2020
18. Das, I., Dennis, J.E.: A closer look at drawbacks of minimizing weighted sums of objectives for pareto set generation in multicriteria optimization problems. In: Structural Optimization, vol. 14, pp. 63–69 (1997)

19. Dunford, N., Schwartz, J.T., Bade, W.G., Bartle, R.G.: Linear Operators: General Theory. Part I. Interscience Publishers (1998)
20. Yu, N., Yu, J.: Optimal TOU decision considering demand response model. In: International Conference on Power System Technology 2006, pp. 1–5 (2006)

Pareto Optimization for Influence Maximization in Social Networks

Kai Wu[1], Jing Liu[2(✉)], Chao Wang[2], and Kaixin Yuan[2]

[1] School of Artificial Intelligence, Xidian University, Xi'an 710071, China
kwu@xidian.edu.cn

[2] Guangzhou Institute of Technology, Xidian University, Guangzhou 510555, China
neouma@mail.xidian.edu.cn

Abstract. Influence maximization is the problem of finding a small subset of seed nodes to maximize the spread of influence in a social network and is an NP-hard problem. In this paper, Pareto optimization is employed for influence maximization (POIM) where the task of finding a set of influential nodes is reformulated as a bi-objective problem. It has been demonstrated is Pareto optimization is quite effective in coping with NP-hard problems. We theoretically compare POIM with a greedy algorithm, which is a widely used algorithm for mining influential seed nodes. We prove that POIM can obtain the best-so-far theoretically guaranteed approximation performance. Empirical studies verify the theoretical results and show the effectiveness of POIM.

Keywords: Pareto optimization · Influence maximization · Evolutionary algorithm

1 Introduction

With the development of the Internet, the prosperity of online social network sites and social media sites, such as Facebook and Twitter, is increasing. One important role of a social network is to carry and model the spread of influence and information in a word-of-mouth manner. It is a fundamental problem to find a small subset of influential nodes in a social network. In this case, these influential nodes can influence the maximum number of nodes in a social network. Here, we give the following hypothetical scenario as an example. A company plans to target a small number of influential individuals of the network by giving them free products. The company wishes that these initial users would enjoy the product and start recommending their friends on the social network to use it, and their friends would recommend their friends' friends, and so on, and thus through the word-of-mouth affect a large population in the social network would adopt the application. The problem is how to choose the initial users so that these initial users affect the largest number of people in the network.

We can call the above problem influence maximization. This problem is formulated as a discrete optimization problem [1]. For a parameter k, it aims to find k nodes in one network which can influence the maximum number of nodes; we refer it to as

H. Ishibuchi et al. (Eds.): EMO 2021, LNCS 12654, pp. 697–707, 2021.
https://doi.org/10.1007/978-3-030-72062-9_55

influence spread. This problem is an NP-hard problem, which has been proved by Kempe et al. [1]. Moreover, Kempe et al. also presented a greedy algorithm (GA) to solve the influence maximization problem that appeared in three models, such as the independent cascade (IC) model, the weight cascade model, and the linear threshold model. Kempe et al. [1] proved that the influence spread optimized by GA is within $(1 - 1/e)$ of the optimal influence spread [1]. Since GA runs Monte Carlo simulations to obtain an exact estimate of the influence spread, this algorithm is inefficient. Although several works are proposed to address this task [5–7, 11], they retain or sacrifice the $(1 - 1/e)$ approximation guarantee. Besides, several nature-inspired algorithms are proposed to cope with influence maximization [12, 13]. However, they provide no guarantee.

In this paper, we employ the Pareto optimization (PO) algorithm to solve it by reformulating it as a bi-objective optimization problem. Recently, Pareto optimization is very powerful for several NP-hard problems, such as the subset selection problem [14], the ensemble pruning problem [15], several constrained optimization problems [16, 17], network reconstruction [21, 22], and hyperparameter optimization [23]. In terms of these problems, Pareto optimization is shown to be superior to other methods both theoretically and empirically. This paper proposes the POIM (Pareto Optimization for Influence Maximization). POIM first formulates influence maximization as a bi-objective problem. In this problem, POIM maximizes the influence spread and minimizes the seed set simultaneously. If the size of the seed set obtained by Pareto optimization (we term it as k_p) is smaller than the given parameter k, we need to choose new nodes. In this process, we employ GA to select seed nodes until the size of the seed set is k.

We theoretically compare the efficiency of POIM with GA and prove that POIM achieves the approximation guarantee $(1 - 1/e)$ on the optimal influence spread if $k_p = k$ and the approximation guarantee $(1 - e^{-1-1/k})$ on the optimal influence spread if $k_p < k$. The experimental results prove the theoretical guarantee and demonstrate the superior performance of POIM.

2 Related Work

Domingos and Richardson [5] first consider influence maximization as an algorithmic problem and design a probabilistic solution. This problem is then formulated as a discrete optimization problem [1]. As this optimization problem is NP-hard [1], the most known works attempt to obtain approximate solutions instead of the exact solution. These works broadly fall into two categories: GAs and heuristic algorithms.

In the first category, Kempe et al. [1] first presented a GA on the IC model and the linear threshold model, showing that the influence spread is within $(1 - 1/e)$ of the optimal influence spread and their GA significantly outperforms the centrality-based heuristics and classic degree in terms of influence spread. Based on GA, many algorithms have been proposed to solve their inefficiency and lacked scalability. Due to the submodularity property of influence maximization objective, Leskovec et al. [6] proposed the CELF algorithm to select the seed nodes. This method reduces the number of evaluations on the influence spread of nodes. Besides, this proposal was further extended to CELF++ [7]. Chen et al. [8] presented the NewGreedy algorithm which obtains a smaller network by removing the edges that will not denote propagation from the original network. Then NewGreedy conducted the influence diffusion on the smaller graph.

Moreover, the NewGreedy algorithm has been further developed into the MixedGreedy algorithm [8] by integrating the advantages of both CELF and NewGreedy. Since the submodularity of influence spread is not strictly guaranteed in existing GAs, Static-Greedy [9] was proposed to strictly guarantee the submodularity. Unfortunately, those GAs needs many Monte Carlo simulations for influence spread estimation, which leads to the ineffectiveness of GAs.

In the second category, several heuristic methods have been proposed to improve the scalability of GA. Chen et al. [8] presented a degree discount heuristic (DegreeDiscount) to decrease running time. This method assumes that the influence spread increases with the degree of nodes [8]. To reduce the computational complexity of GA, Wang et al. [10] proposed a CGA algorithm. In CGA, first, the network is divided into communities; then Monte Carlo simulations are conducted within each community. Besides, Chen et al. [11] employed maximum influence paths for influence spread estimation, called PMIA. PMIA is the best heuristic so far. However, although these algorithms guarantee a faster running time, they provide no guarantees or significantly weaker ones. Besides, several works employ evolutionary algorithms and nature-inspired algorithms to cope with influence maximization. Jiang et al. [12] employed a simulated annealing algorithm with several heuristics to speed up influence spread estimation. Gong et al. [13] employed the memetic algorithm with communities (CMA) for the influence maximization problems. Instead of GA in CGA, CMA uses a memetic algorithm to optimize each community. However, these algorithms have not been theoretically guaranteed.

3 Influence Maximization

Based on the definition in [1], the IC model is defined as follows. Given a network $G = (V, E)$ with edge labels pp: $E \rightarrow [0, 1]$. m represents the number of edges in G, and n represents the number of nodes in G. In the network G, $pp(u, v)$ represents the propagation probability of the edge $(u, v) \in E$. v is activated by u through the edge (u, v) in the next step with the probability $pp(u, v)$ after u is activated.

Consider a seed set $S \subseteq V$, the procedure of the IC model performs as follows. $S_t \subseteq V$ is the set of nodes that are activated at the steps $t \geq 0$, with $S_0 = S$. At step $t + 1$, each node $u \in S_t$ may activate its out-neighbors $v \in V \backslash \cup_{0 \leq i \leq t} S_i$ with probability $pp(u, v)$. The process stops until $S_t = \emptyset$. In the IC model, each activated node only has one chance to activate its out-neighbors at the step after itself is activated. Each node keeps as an activated node if it is activated. The influence spread of S (denoted as σ_S) is the number of activated nodes. In order to optimize the influence maximization problem, one needs to estimate the influence spread for any given seed set. Due to the intractability of exact evaluation of the influence spread on a typically sized graph, we run Monte Carlo simulations for R times to estimate the influence spread by averaging over all simulations.

Based on Definition 1, influence maximization originally aims at selecting a set of most influential nodes so that their aggregated influence in the graph is maximized. Then, we introduce the notion of submodular and monotone for a general set function, i.e. influence spread σ. The following proposition states an important property of σ.

Definition 1 (Influence Maximization). *Given an integer k, the influence maximization in one specific model is to select a seed set $S \subseteq V$ such that.*

$$\arg\max_{S \subseteq V} \sigma_S \; s.t. \; |S| = k \quad (1)$$

where $|\cdot|$ represents the size of a set.

Definition 2 (Submodularity). *Let σ be a nonnegative set function. σ is submodular if it satisfies.*

$$\sigma_{S \cup \{v\}} - \sigma_S \geq \sigma_{T \cup \{v\}} - \sigma_T$$

For all elements v and all pairs of sets $S \subseteq T$.

Definition 3 (Monotonicity). *Suppose σ is submodular and takes only non-negative values. σ is monotone if it satisfies $\sigma_{S \cup \{v\}} \geq \sigma_S$ all elements v and set S.*

Proposition 1 (Theorem 2.2 of [1]). *The influence spread σ under independent cascade model is monotone and submodular and $\sigma_\varnothing = 0$.*

Algorithm 1 Greedy Algorithm

Input:

An integer parameter k;

Network G;

Output:

The set of influential nodes: S.

1: $t \leftarrow 0$ and $S = \varnothing$;
2: **while** ($t < k$) **do**
3: For each node u, use repeated sampling to approximate $\sigma_{S \cup \{u\}}$;
4: Add the node with the largest estimate for $\sigma_{S \cup \{u\}}$ to S;
5: **end while**
6: **return** S.

It is shown in Theorem 2.4 of [1] that influence maximization is NP-hard for the IC model. However, since σ is monotone and submodular (Proposition 1), a GA is employed to achieve approximation guarantee $1 - e^{-1}$ for σ [1, 2]. A simple GA is shown in Algorithm 1. GA iteratively finds a new node u that maximizes the incremental change of the influence spread into the seed set S until k nodes are found.

4 POIM

The influence maximization in Eq. (1) can be divided into two parts. First, we transfer the influence maximization problem into the following

$$\arg\max_{s \subseteq V} \sigma_S \quad s.t. \quad |S| \leq k \quad (2)$$

Instead of optimizing this constrained optimization problem, we reformulate the influence maximization problem as a bi-objective task which is defined as follows:

$$\arg\min_{S \subseteq V}\{-\sigma_S, |S|\} \tag{3}$$

In the bi-objective formulation, one optimizes the influence maximization of set S, meanwhile, the other keeps the size small. The two objectives described in Eq. (3) are conflicting, namely, a set with a larger influence spread could have a larger size. Pareto optimization is employed to solve the two objectives simultaneously.

Algorithm 2 Pareto Optimization

Input:

The number of nodes in seed set: k;
Network: G;
The number of iterations: T;

Output:

The set of influential nodes: S.

1: Set $s = \{0\}^n$, $P = \{s\}$, and $t = 0$;
2: **for** ($t = 1$ to T) **do**
3: Find s from P uniformly at random;
4: Obtain s^+ from s by flipping each bit of s with probability $1/n$;
5: Calculate $f_2(s^+)$ and $f_1(s^+)$;
6: **If** $\neg\exists\, x \in P$ such that z *strictly dominates* s^+ **then**
7: $U = \{x \in P \mid s^+$ *weakly dominates* $x\}$
8: $P \leftarrow (P \backslash U) \cup \{s^+\}$;
9: **end if**
10: **end for**
11: **return** $S = \arg\max_{S \in P, |S| \le k} \sigma_S$.

In this paper, we employ the binary vector to represent the subsets membership indication. $s \in \{0, 1\}^n$ stands for a set S of V, and $s_i = 1$ if the i-th node is in S and $s_i = 0$ otherwise. For the first objective $f_1(s) = -\sigma_S$ in Eq. (3), we define that $f_1(s) = 0$ if $s = \{0\}^n$ or $|s| > k + 1$ and $f_1(s) = -\sigma_S$ otherwise.

For this problem, we need to compare solutions in terms of two objective values. For example, for two solution s and s^+, s *weakly dominates* s^+ if s has a smaller or equal value on both the objectives $(f_1(s) \le f_1(s^+) \wedge f_2(s) \le f_2(s^+))$; s is said to *strictly dominates* s^+ when s has a smaller value on one objective, and meanwhile has a smaller or equal value on another objective (s *weakly dominates* $s^+ \wedge f_1(s) < f_1(s^+) \vee f_2(s) < f_2(s^+)$). s^+ and s are incomparable if both s does not dominate s^+ and s is not dominated by s^+.

The procedure of Pareto optimization is shown in Algorithm 2. In line 1, the algorithm starts from an empty set, and the archive P contains only the solution with the empty set. PO then iteratively improves the solutions in P. In each iteration, PO generates a new solution s^+ by randomly flipping bits of s selected from the current P (lines 3 and 4). In line 5, the solution s^+ is evaluated. Then, s^+ is compared with solutions in the current P. If the newly generated s^+ strictly dominates any solution in P, it will be added into P. Besides, if the solutions in U are strictly dominated by the solution s^+, they will be cleaned. After T iteration, the best solution with the largest influence spread and $|S| \le k$ is selected.

After performing Pareto optimization on Eq. (3), we judge whether the solution of Pareto optimization $|S_P| < k$. We then conduct the greedy algorithm step to optimize Eq. (1) with the initialized seed set $S = S_P$ if $|S_P| < k$. Otherwise, we set the solution of POIM S to S_P. In Pareto optimization, even if we find one solution $|S| = k$ ($S \in P$), we don't ensure that this solution has the largest influence spread. Thus, it is necessary to execute this process. The procedure of POIM is shown in Algorithm 3.

Algorithm 3 POIM

Input:

An integer parameter k;

Network G;

Output:

The set of influential nodes: S.

1: Obtain S by optimizing problem (1) using Pareto optimization;
2: **if** ($|S| < k$) **then**
3: $l = ||S| - k|$, i=0;
4: **while** ($i < l$) **do**
5: Let X° be a variable maximizing the influence speed of $S \cup \{X^{\circ}\}$;
6: Let $S = S \cup \{X^{\circ}\}$, and $i \leftarrow i$+1;
7: **end while**
8: **end if**
9: **return** S.

5 Theoretical Analysis

Let OPT be the optimal function value of Eq. (1).

Lemma 1. *Let σ be a non-negative, monotone submodular function. For any $S \subseteq V$, there exists one variable $X^{\circ} \in V \backslash S$ such that*

$$\sigma_{S \cup \{X^{\circ}\}} - \sigma_S \geq \frac{1}{k}(OPT - \sigma_S)$$

Proof. The proof idea is inspired by Lemma 1 in [4]. That lemma has proved that.

$$\sigma_{S \cup \{X^{\circ}\}} - \sigma_S \geq \frac{\gamma_{S,k}}{k}(OPT - \sigma_S)$$

where $\gamma_{S,k}$ is the submodularity ratio of σ concerning a set S and a parameter $k \geq 1$ [3]. We can obtain $\gamma_{S,k} \geq 1$ when σ is submodular. □

Theorem 1. *For problem (2), let σ be a non-negative, monotone submodular function. Pareto optimization with the expected running time $O(nRmk^2)$ selects a set S of variables with $\sigma_S \geq (1 - e^{-1}) \cdot OPT$ and $|S| \leq k$.*

Proof. Let S^* denote an optimal set of nodes of Eq. (2) and $\sigma_{S^*} = OPT$. Let $I_{\max}$ be the maximum value of $i \in [0, k]$ such that there exists a solution S with $|S| \leq i$ and $\sigma_S \geq (1 - 1(1 - k^{-1})^i) \cdot OPT$. That is, $I_{\max} = \max\{i \in [0, k] \mid \exists S \in P, |S| \leq i \wedge \sigma_S \geq (1 - (1 - k^{-1})^i) \cdot OPT\}$. In the following, we will analyze the expected running time until $I_{\max} = k$, which implies that a $(1 - e^{-1})$-approximate solution is found ($(1 - (1 - k^{-1})^k) \geq (1 - e^{-1})$).

Since POIM starts with the solution $\{0\}^n$, $I_{\max} = 0$. Suppose that $I_{\max} = i < k$. Let S denote the corresponding solution with the value i ($|S| \leq i$ and $\sigma_S \geq (1 - (1 - k^{-1})^i) \cdot OPT$). First, we are to analyze that $I_{\max}$ cannot decrease. In POIM, there are two cases for current solution S. (a) S is kept in P. In this case, $I_{\max}$ obviously will not decrease; (b) S is replaced by the newly generated solution S^+ (by lines 7 and 8 of Algorithm 2). In this case, S^+ weakly dominates S, which implies that the newly generated S^+ must have a smaller size and greater σ value ($|S^+| \leq |S| \leq i \wedge \sigma_{S^+} \geq \sigma_S \geq (1 - (1 - k^{-1})^i) \cdot OPT$). Thus, it is easy to see that $I_{\max}$ will not decrease.

Then, we are to analyze that $I_{\max}$ can increase by flipping a specific 0 bit of S. From Lemma 1, we know that flipping one specific 0 bit of S can generate a new solution S^+ and we can obtain

$$\sigma_{S^+} \geq \left(1 - \frac{1}{k}\right)\sigma_S + \frac{1}{k} \cdot OPT \geq \left(1 - \left(1 - \frac{1}{k}\right)^{i+1}\right) \cdot OPT$$

S^+ will be added into P (line 6 to 9 of Algorithm 2) because $|S^+| = |S| + 1 \leq i + 1$; Otherwise, S^+ must be strictly dominated by one solution in the current archive P. This case demonstrates that the value of $I_{\max}$ has already been greater than i, which contradicts with the previous assumption $I_{\max} = i$. Thus, $I_{\max} = i + 1$.

Finally, we are to show the expected running time for selecting S^+. The solutions in P are incomparable and each value of one property can correspond to at most one solution in P. The largest size of P is $P_{\max}$. As we have known, the solutions with $|S| \geq k + 1$ have 0 value. Thus, $|S|$ has $k + 1$ number of distinct values, which implies that $P_{\max} \leq k + 1$. Owing to the uniform selection, the probability of finding S in line 3 of Algorithm 2 is at least $1/P_{\max}$. $I_{\max}$ can increase by flipping a specific 0 bit of S with the probability of at least $\frac{1}{P_{\max}} \cdot \frac{1}{n}\left(1 - \frac{1}{n}\right)^{n-1} \geq \frac{1}{en(k+1)}$, where $\frac{1}{n}\left(1 - \frac{1}{n}\right)^{n-1}$ is the probability of flipping a specific bit of S and keeping other bits unchanged in line 4 of Algorithm 2. Then, it needs at most $2ekn$ running time in order to increase $I_{\max}$. Since k such steps are enough to obtain $I_{\max} = k$, the expected running time is $enk(k + 1)$. Owing to Monte Carlo simulation takes $O(Rm)$ time, the time complexity is $O(nRmk^2)$. □

Theorem 2. *For influence maximization, S_P is the solution of Pareto optimization. POIM selects a set S of variables with $|S| = k$ and $\sigma_S \geq (1 - e^{-1-1/k}) \cdot OPT$ if $|S_P| < k$, and $\sigma_S \geq (1 - e^{-1}) \cdot OPT$ if $|S_P| = k$.*

Proof. This proof is divided into two cases.

(1) $|S| = |S_P| = k$. From Theorem 1, it is easy to see that $\sigma_S \geq (1 - e^{-1}) \cdot OPT$.
(2) $|S_P| < |S| = k$. $l = ||S_P| - k| \geq 1$. We use GA to obtain the remaining seeds (Algorithm 3).

From Lemma 1, when the new seed X° is included in S, we can obtain

$$\sigma_{S_P \cup \{X^{\circ}\}} - \sigma_{S_P} \geq \frac{1}{k}(OPT - \sigma_{S_P})$$

The above inequality holds for each iteration $i = 1, \cdots, l$. By accumulating the above inequality for each iteration of GA, we can obtain

$$\sigma_S \geq \frac{l}{k}OPT + \left(1 - \frac{l}{k}\right)\sigma_{S_P}$$
$$\geq \frac{l}{k}OPT + \left(1 - \frac{l}{k}\right) \cdot \left(1 - \left(1 - \frac{1}{k}\right)^k\right) \cdot OPT$$
$$= \left(1 - \left(1 - \frac{l}{k}\right)\left(1 - \frac{1}{k}\right)^k\right) \cdot OPT \geq \left(1 - \left(1 - \frac{1}{k}\right)^{k+1}\right) \cdot OPT \geq \left(1 - e^{-1-\frac{1}{k}}\right) \cdot OPT$$

The second inequality holds owing to Theorem 1. By computing the second equation, the following equation holds. Since $(1 - l/k) \leq (1 - 1/k)$, the third inequality holds. Then, the fourth inequality holds because of $(1 - x) \leq e^{-x}$. □

6 Empirical Study

6.1 Experimental Setup

Compared with other proposals, we know that the general GA has the best accuracy. Thus, in this paper, we just compare POIM with GA. For POIM, the number of iterations T is set to be $\lfloor enk(k+1) \rfloor$ as suggested by Theorem 1. We use the spread simulation method [1] to evaluate the influence spread of the solution of POIM and GA. In practice, we set R to 10000. Note that all experiments are conducted on the IC model. Three real-world networks are used to show the performance of POIM. Since Monte Carlo simulation is very time-consuming, we just select small-scale networks to validate the performance of our proposal, namely, Dolphin [18], ZK [19], and Football [20] networks. Table 1 gives the properties of the networks. All experiments are conducted on a server with 2.20 GHz Intel (R) Xeon CPU E5-2660 and 64G memory.

Table 1. Statistics of three real-life networks

Dataset	Dolphin	ZK	Football
#Node	62	34	115
#Edge	159	78	616

6.2 On Optimization Performance

We first discuss the effect of *pp* on the performance of POIM. *pp* is increased from 0.01 to 0.09 in steps of 0.02. Here, we set $k = 5$. The experimental results are shown in Fig. 1. In, Fig. 1, We obtain the value of each point by averaging 30 independent runs. We can find that *pe* has an important effect on the performance of POIM and GA. Moreover, POIM outperforms GA on Dolphin and Football networks. The performance of POIM is worse than GA on the ZK network. With the increase of *pp*, the advantage of POIM over GA becomes greater on Dolphin and Football networks.

We then discuss the effect of k on the performance of POIM. We set $pe = 0.01$ and increase k from 3 to 11 in steps of 2. In each case, bold fonts denote that POIM is significantly better than the corresponding method by the *t-test* with a confidence level of 0.05. From Table 2, we can see that POIM is shown to significantly exceed all the compared methods in 11 of 15 cases. On the ZK network, POIM loses four times to GA. However, we still find that the advantage of the two algorithms is small. From the results, we can see that POIM is worse than GA in dealing with the small network (on ZK network), but POIM is better than GA in the relatively large-scale network (on Dolphin and Football networks). Note that all the above experiments for POIM don't take the greedy algorithm step of POIM to make $k = k_p$.

6.3 On Efficiency

Since the $enk(k + 1)$ time is a theoretical upper bound for POIM being as good as GA, in this section, we empirically examine how tight this bound is. We consider the running time as the number of objective function evaluations. The algorithms are evaluated on the IC model with $pp = 0.01$ and $k = 5$. We plot the curve of σ over the running time for POIM on Dolphin and Football networks, as shown in Fig. 2. In the following experiments, we don't execute the GA step of POIM. The x-axis is in kn, the running time of GA. We can observe that POIM takes about only 24.7% (4.02kn) and 13.9% (2.27kn) of the theoretical time to achieve a better performance than GA, respectively

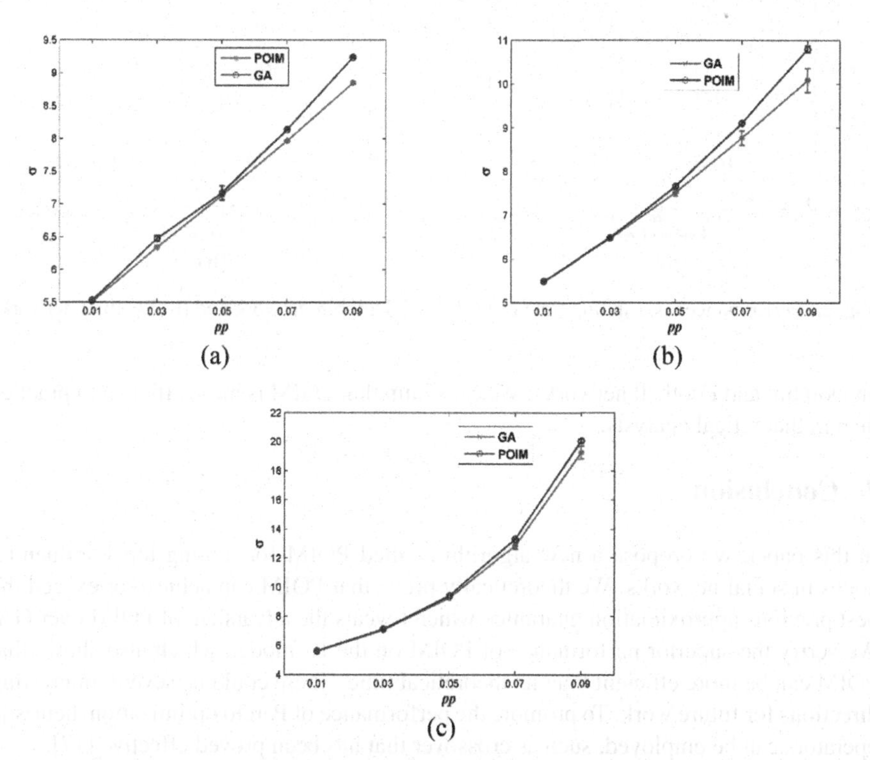

Fig. 1. The influence spreads of different algorithms under the IC model as *pp* is varied. (a) ZK network, (b) Dolphin network, and (c) Football network.

Table 2. The influence spreads of different algorithms under the IC model as k is varied.

k	Algorithms	Dolphin	ZK	Football
3	POIM	**3.324 ± 0.008**	**3.408 ± 0.011**	**3.376 ± 0.004**
	GA	3.321 ± 0.004	3.406 ± 0.007	3.371 ± 0.003
5	POIM	**5.492 ± 0.004**	5.505 ± 0.004	**5.623 ± 0.006**
	GA	5.480 ± 0.004	**5.531 ± 0.007**	5.608 ± 0.008
7	POIM	**7.645 ± 0.007**	7.552 ± 0.005	**7.862 ± 0.008**
	GA	7.622 ± 0.009	**7.570 ± 0.006**	7.842 ± 0.008
9	POIM	**9.765 ± 0.006**	9.569 ± 0.005	**10.092 ± 0.006**
	GA	9.752 ± 0.014	**9.591 ± 0.016**	10.067 ± 0.016
11	POIM	**11.878 ± 0.008**	11.575 ± 0.003	**12.320 ± 0.007**
	GA	11.836 ± 0.021	**11.600 ± 0.008**	12.295 ± 0.018

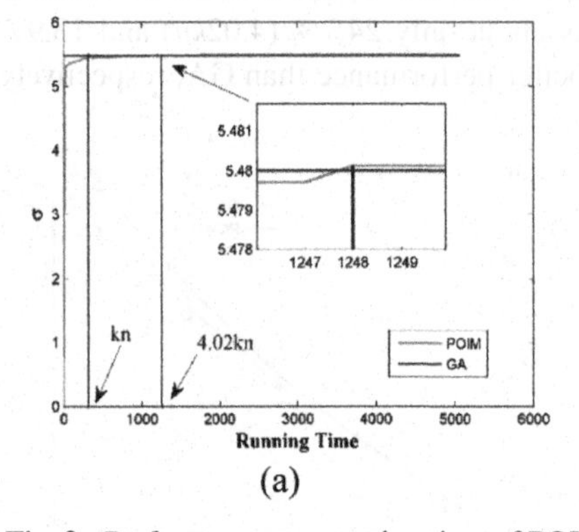

(a)

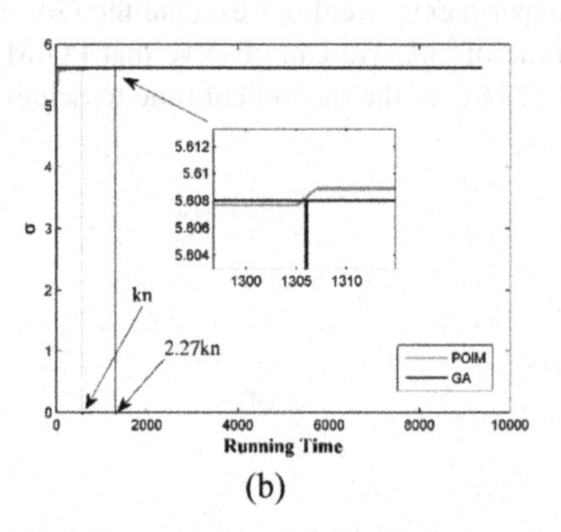

(b)

Fig. 2. Performance v.s. running time of POIM. (a) Dolphin network and (b) Football network.

on Dolphin and Football networks. We can claim that POIM is more efficient in practice than in theoretical analysis.

7 Conclusion

In this paper, we propose a new algorithm called POIM for mining top k influential nodes in social networks. We theoretically prove that POIM can achieve or exceed the best previous approximation guarantee which reveals the advantage of POIM over GA. We verify the superior performance of POIM on the IC model, which also shows that POIM can be more efficient than its theoretical time. There could be several interesting directions for future work. To promote the performance of Pareto optimization, heuristic operators can be employed, such as crossover that has been proved effective [17].

References

1. Kempe, D., Kleinberg, J.M., Tardos, E.: Maximizing the spread of influence through a social network. In: KDD, pp. 137–146 (2003)
2. Nemhauser, G.L., Wolsey, L.A., Fisher, M.L.: An analysis of approximations for maximizing submodular set functions—I. Math. Program. **14**(1), 265–294 (1978). https://doi.org/10.1007/BF01588971
3. Das, A., Kempe, D.: Submodular meets spectral: greedy algorithms for subset selection, sparse approximation and dictionary selection. In: ICML, pp. 1057–1064 (2011)
4. Qian, C., Shi, J.C., Yu, Y., Tang, K., Zhou, Z.H.: Parallel Pareto optimization for subset selection. In: IJCAI, pp. 1939–1945 (2016)
5. Domingos, P., Richardson, M.: Mining the network value of customers. In: KDD, pp. 57–66 (2001)
6. Leskovec, J., Krause, A., Guestrin, C., Faloutsos, C., VanBriesen, J., Glance, N.: Cost-effective outbreak detection in networks. In: KDD, pp. 420–429 (2007)
7. Goyal, A., Lu, W., Lakshmanan, L.V.: CELF++: optimizing the greedy algorithm for influence maximization in social networks. In: WWW, pp. 47–48 (2011)
8. Chen, W., Wang, Y., Yang, S.: Efficient influence maximization in social networks. In: KDD, pp. 199–208 (2009)
9. Cheng, S., Shen, H., Huang, J., Zhang, G., Cheng, X.: Staticgreedy: solving the scalability-accuracy dilemma in influence maximization. In: CIKM, pp. 509–518 (2013)
10. Wang, Y., Cong, G., Song, G., Xie, K.: Community-based greedy algorithm for mining top-k influential nodes in mobile social networks. In: KDD, pp. 1039–1048 (2010)
11. Chen, W., Wang, C., Wang, Y.: Scalable influence maximization for prevalent viral marketing in large-scale social networks. In: KDD, pp. 1029–1038 (2010)
12. Jiang, Q., Song, G., Gao, C., Wang, Y., Si, W., Xie, K.: Simulated annealing based influence maximization in social networks. In: AAAI, pp. 127–132 (2011)
13. Gong, M., Song, C., Duan, C., Ma, L., Shen, B.: An efficient memetic algorithm for influence maximization in social networks. IEEE Comput. Intell. Mag. **11**(3), 22–33 (2016)
14. Qian, C., Yu, Y., Zhou, Z.H.: Subset selection by Pareto optimization. In: NIPS, pp. 1774–1782 (2015)
15. Qian, C., Yu, Y., Zhou, Z.H.: Pareto ensemble pruning. In: AAAI, pp. 2935–2941 (2015)
16. Qian, C., Yu, Y., Zhou, Z.H.: On constrained Boolean Pareto optimization. In: IJCAI, pp. 389–395 (2015)
17. Qian, C., Yu, Y., Zhou, Z.H.: An analysis on recombination in multi-objective evolutionary optimization. Artif. Intell. **204**, 99–119 (2013)
18. Lusseau, D., Schneider, K., Boisseau, O.J., Haase, P., Slooten, E., Dawson, S.M.: The bottlenose dolphin community of doubtful sound features a large proportion of long-lasting associations. Behav. Ecol. Sociobiol. **54**(4), 396–405 (2003). https://doi.org/10.1007/s00265-003-0651-y
19. Zachary, W.W.: An information flow model for conflict and fission in small groups. J. Anthropol. Res. **33**(4), 452–473 (1977)
20. Girvan, M., Newman, M.E.J.: Community structure in social and biological networks. Proc. Natl. Acad. Sci. **99**, 7821–7826 (2002)
21. Wu, K., Liu, J., Hao, X., Liu, P., Shen, F.: An evolutionary multi-objective framework for complex network reconstruction using community structure. IEEE Trans. Evol. Comput. (2020). https://doi.org/10.1109/TEVC.2020.3020423
22. Wu, K., Liu, J., Chen, D.: Network reconstruction based on time series via memetic algorithm. Knowl.-Based Syst. **164**, 404–425 (2019)
23. Wu, K., Liu, J.: Classification-based optimization with multi-fidelity evaluations. In: 2019 IEEE Congress on Evolutionary Computation (CEC), pp. 1126–1131 (2019)

Parallel Algorithms for the Multiobjective Virtual Network Function Placement Problem

Joseph Billingsley[1(✉)], Ke Li[1], Wang Miao[1], Geyong Min[1], and Nektarios Georgalas[2]

[1] Department of Computer Science, University of Exeter, Exeter, UK
{jb931,k.li,wang.miao,g.min}@exeter.ac.uk
[2] Research and Innovation, British Telecom, London, UK
nektarios.georgalas@bt.com

Abstract. Datacenters are critical to the commercial and social activities of modern society but are also major electricity consumers. To minimize their environmental impact, we must make datacentres more efficient whilst keeping the quality of service high. In this work we consider how a key datacenter component, Virtual Network Functions (VNFs), can be placed in the datacenter to optimise the conflicting objectives of minimizing service latency, packet loss and energy consumption. Multiobjective Evolutionary Algorithms (MOEAs) have been proposed to solve the Virtual Network Function Placement Problem (VNFPP), but state of the art algorithms are too slow to solve the large problems found in industry. Parallel Multiobjective Evolutionary Algorithms (PMOEAs) can reduce execution time by distributing the optimization process over many processes. However, this can hamper the search process as it is inefficient to share information between processes. This paper aims to determine whether PMOEAs can efficiently discover good solutions to the VNFPP. We found that PMOEAs can solve the VNFPP 5–10× faster than a sequential MOEA without harming solution quality. Additionally, we found that one parallel algorithm, PPLS/D, found *better* solutions than other MOEAs and PMOEAs in most test instances. These results demonstrate that PMOEAs can solve the VNFPP faster, and in some instances better, than sequential MOEAs.

Keywords: Network function virtualisation · Multi-objective optimisation · Parallel multiobjective evolutionary algorithms

---

This work was supported by EPSRC Industrial CASE and British Telecom (Grant No. 16000177) and UKRI Future Leaders Fellowship (Grant No. MR/S017062/1).

H. Ishibuchi et al. (Eds.): EMO 2021, LNCS 12654, pp. 708–720, 2021.
https://doi.org/10.1007/978-3-030-72062-9_56

1 Introduction

Recent research indicate that datacenters will be responsible for between 3% and 5% of total energy consumption worldwide by 2030 [6]. With the disastrous impact climate change could have, there are environmental as well as business imperatives to improve the efficiency of datacenters. From 2010 onwards, datacenters became more energy efficient by reducing energy spent on 'overhead' [7] i.e. energy consumed by fans, pumps, transformers and other auxillary equipment. Despite these efforts the total energy consumed by datacenters doubled from 2010–2020 [19] and there are diminishing returns to reducing overhead further. Recent work has identified that the computing components of the datacenter, e.g. servers and switches, are now where the greatest efficiency improvements can be made [19].

Network function virtualization is a technology that will allow datacenter components to be used more efficiently. A network function is a datacenter component that performs a specific task such as load balancing or packet inspection. Services, such as phone call handling or video streaming, are formed by directing traffic through network functions in a prescribed order. Traditionally, these functions were provided by 'middleboxes' implemented in purpose-built hardware. However, middleboxes cannot easily be reconfigured, added or removed from the datacenter and hence consume energy even when not required. Virtual network functions (VNFs) provide the same functionality as middleboxes but run on software that is executed on a virtual machine. New instances of VNFs can be created or destroyed in seconds [1] allowing the datacenter to spend energy on services only when necessary.

To use VNFs efficiently, the optimal number of instances of each VNF must be placed in the datacenter. In the VNF Placement Problem (VNFPP) there are multiple conflicting quality of service (QoS) metrics (e.g. expected latency, packet loss) for each service in the network, which must also be balanced against the energy consumption of the datacenter. A VNFPP instance defines a set of services to provide and a datacenter topology. A solution to the problem defines where VNFs for each service should be placed and how packets should traverse the datacenter to form services.

Many means of solving the VNFPP have been explored however no algorithm has had widespread acceptance. Whilst exact [8,9,29], heuristic [23,33,34] and metaheuristic [10,36,39] optimization methods have been considered, existing algorithms are limited by the speed they can evaluate solutions to large problems. Packet level models of a datacenter such as discrete event simulators, accurately measure the datacenter metrics [32] but are slow to converge. Surrogate models are used in many works [4,23,24,33–35,40] as indicative measures of the solution quality. However, it has never been shown that optimizing a surrogate is analogous to optimizing the datacenter metrics. The fastest, accurate models use queueing theory. Queueing theory models can closely approximate the datacenter metrics efficiently by modelling the behavior of queues in the datacenter. Despite this, most queueing theory models consider a simplified representation of the datacenter where queues are allowed to be infinite and hence packets are never dropped [2,21,31] whereas in practice, queues are finite and packet loss is significant. In a previous paper we proposed a fast and accurate queueing model that corrected this issue [12], however on large problems this model is slow to converge.

One means of remedying this issue is to utilise the capabilities of modern CPUs, in particular support for high degrees of multithreading. Evolutionary algorithms are parallelisable and appropriate for the VNFPP as they can efficiently find approximate solutions to challenging optimisation problems. Parallel evolutionary algorithms can achieve sub or near linear [13,20,27,28,38,41] and even super-linear [5,30] reductions in execution time for increasing numbers of threads. However, increased parallelisation can also come at the cost of poorer solutions [13] or increased execution time [20].

In this work we determine whether PMOEAs can efficiently and effectively solve the VNFPP. First, in Sect. 2 we formally specify the VNFPP problem. Next, in Sect. 3 we describe the PMOEAs considered in this work. Section 4 presents the results of our evaluation. Finally, Sect. 5 summarizes the overall contributions.

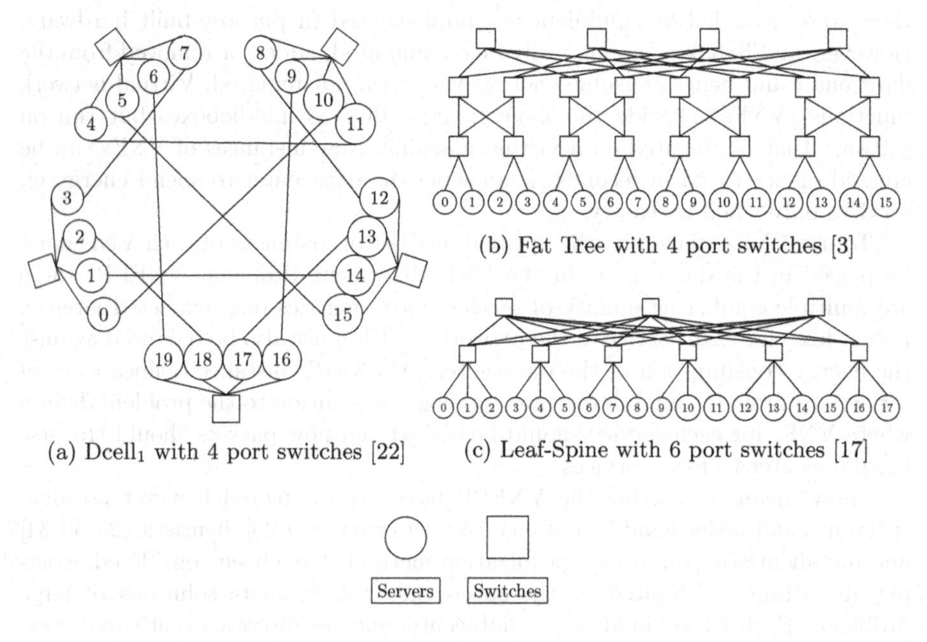

(a) Dcell_1 with 4 port switches [22]

(b) Fat Tree with 4 port switches [3]

(c) Leaf-Spine with 6 port switches [17]

Fig. 1. Common datacenter topologies

2 Problem Formulation

In this section, we describe the VNFPP and its constraints in detail. In the VNFPP, services must be placed on virtual machines in a datacenter so as to maximize multiple QoS objectives whilst minimizing energy consumption. The datacenter is a collection of servers interconnected by a network topology (see Fig. 1). A service is a sequence of VNFs. Packets arrive at the first VNF in the service and must visit each VNF in sequence. A solution to the VNFPP places one or more instances of each VNF in the datacenter and specifies paths between VNFs to form services.

The VNFPP has three objectives: two QoS metrics, latency and packet loss, and a cost metric, energy consumption. As service latency, packet loss and energy consumption conflict [12] we formulate the VNFPP as a multi-objective optimization problem. Further, as services in a datacenter can number in the thousands it is not possible to treat each service quality metric as a separate objective. Instead, we aim to minimize the average service latency and the average service packet loss over all services, as well as the total energy consumption. Formally:

$$\text{minimize } \mathbf{F}(R_S) = (W, P, E)^T \tag{1}$$

where R_S is the set of routes for each service, W is the average service latency, P is the average service packet loss and E is the total electricity consumption of the datacenter. Several models exist to determine W, P and E for the VNFPP. This work uses the queueing theory model we proposed in [10] and extended in [11].

Finally, for a solution to the VNFPP to be feasible, it must satisfy two constraints. First, every VNF consumes some resources on a server. The total resources consumed by VNFs on a server cannot exceed the total resources of the server. Second, the solution must define one or more valid paths for each service. To be valid, a path must visit each VNF of the service in the order it appears in the service.

3 Parallel MOEAs for the VNFPP

In this section we outline the algorithms evaluated in this work and the genetic operators used to solve the VNFPP. First we describe the PMOEAs we evaluated and then we describe the genetic operators we have constructed to solve the VNFPP.

3.1 Parallel MOEA Frameworks

PMOEAs can be grouped into three categories: master/slave algorithms, multiple-deme and isolated-deme algorithms [14]. Each parallel algorithm distributes tasks over N_T threads. In this work we evaluated a PMOEA from each category against a serial MOEA that does not use multithreading. The remainder of this section describes these algorithms in detail.

A serial MOEA is an evolutionary algorithm that only uses a single process. For this category, we selected the well known NSGA-II [18] algorithm, illustrated in Fig. 2a. NSGA-II maintains a population of N parent solutions. A population of N child solutions are generated via mutation and crossover of the parent population. Both populations are combined and then fast nondominated sorting and crowding distance procedures are used to determine which solutions survive to become the parent solutions of the next generation. This process repeats until the stopping condition is met. NSGA-II has been shown to be effective on a range of multiobjective optimization problems previously, including the VNFPP [12]. Additionally, several parallel variants of NSGA-II exist, allowing for a fair comparison between single and multithreaded algorithms.

A master/slave algorithm has a single 'master' thread that allocates tasks to 'slave' threads. Once the slave threads are finished, the master thread aggregates the results. In this work we parallelise NSGA-II using this architecture, hereafter referred to as MS-NSGA-II (illustrated in Fig. 2b). In this implementation each thread performs crossover, mutation and evaluation to produce N/N_T solutions. Afterwards the solutions from each thread are aggregated to form the child population. As in NSGA-II, fast nondominated sorting and crowding distance procedures are used to select the parent population for the next generation. This process repeats until the stopping condition is met.

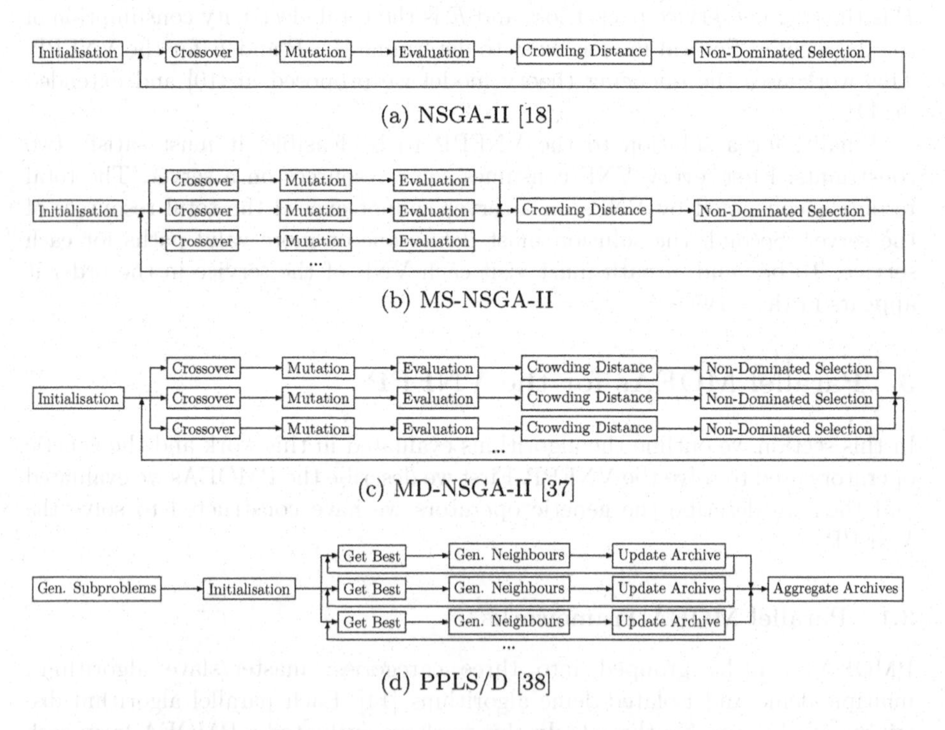

Fig. 2. The four MOEAs we considered in this work. NSGA-II is a serial algorithm whilst the remaining parallelize some or all steps.

A multiple-deme algorithm evolves multiple subpopulations in parallel and periodically exchanges solutions between them. In this work we use the algorithm proposed by Roberge et al. [37], hereafter referred to as MD-NSGA-II, which uses a multiple-deme architecture (illustrated in Fig. 2c). Roberge et al. first produce a set of initial solutions and group them into N_T equal sized subpopulations. The variation, evaluation and selection operators from NSGA-II are performed on each subpopulation in parallel. After a set number of generations, the subpopulations are aggregated into a new population and then randomly redistributed back into N_T subpopulations. This process repeats until the stopping condition is met.

An isolated-deme algorithm also evolves multiple subpopulations but does not exchange information between the subpopulations. In this work we use the isolated-deme algorithm PPLS/D [38] (illustrated in Fig. 2d). PPLS/D first decomposes the objective space into N_T scalar objective optimization subproblems. The algorithm maintains a population of unexplored solutions and an archive of non-dominated solutions for each subproblem. First, a solution is added to the unexplored population of each subproblem. Next, for each subproblem the algorithm selects the best solution from the unexplored population and generates a set of neighboring solutions. A neighboring solution is added to the unexplored population and the non-dominated archive if it is: 1) closer to the subproblem that generated it than any other subproblem, as measured by the angle between the solution and the subproblems and 2) is either a better solution to the subproblem than the current best solution or if it is not dominated by any solutions in the non-dominated set of the subproblem. Once the stopping condition is met, the algorithm aggregates the non-dominated archives from each subproblem and returns the overall non-dominated solutions.

PPLS/D was designed for unconstrained optimization problems, so required modification to fit the VNFPP. In our variant, an infeasible solution is accepted into the unexplored population only if there are no more feasible solutions in the population.

Additionally, to determine how PPLS/D is affected by the number of subproblems used, we generate N subproblems that are placed into N_T equally sized groups. Each subproblem in a group is executed in series and each group is executed in parallel. This technique produces solutions with the same solution quality as if greater numbers of threads were available whilst not unfairly benefitting PPLS/D.

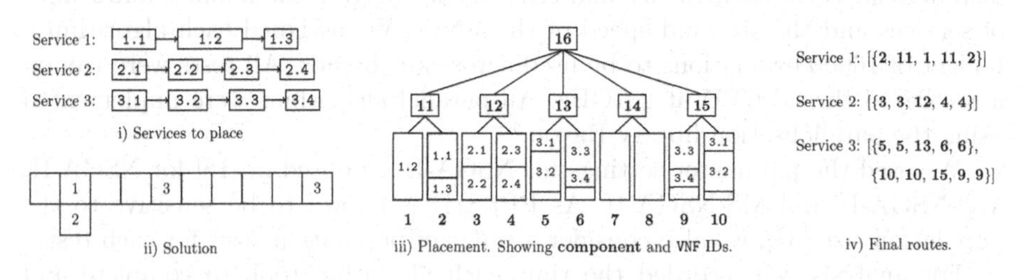

Fig. 3. A partial mapping from genotype to phenotype showing the assignment of services to servers.

3.2 Genetic Operators

As this work focusses on the benefits of parallelisation, we use existing genetic operators that were designed for the VNFPP. In particular we use operators we demonstrated to be effective in earlier work [12]. These operators use a genotype-phenotype representation where genetic operators consider a simplified solution representation (the genotype) which is converted into the 'true' solution representation (the phenotype) to be evaluated. In our implementation, the genotype is an array of vectors where each vector represents a server in the datacenter (see Fig. 3 ii). Each vector contains zero or more service instances. The phenotype is the set of paths for each service (see Fig. 3 iv). The genotype-phenotype mapping has two parts: expansion and routing as shown in Fig. 3. The expansion step iterates over each server in the genotype. When a service instance is encountered, the algorithm places each VNF of the service on the closest server that can accommodate it. The routing step finds the set of shortest paths between each VNF and combines them to form a path.

MD-NSGA-II and MS-NSGA-II require initialization, mutation, crossover and evaluation operators. PPLS/D additionally requires a local search operator. Since all services are considered equally important, the initialization operator always generates genotypes where each service has the same number of service instances. The operator first determines the maximum number of service instances that can be placed in a datacenter whilst meeting this constraint. Then it generates a range of solutions between the maximum and minimum number of service instances that can be placed. The mutation operator has an equal probability of 1) swapping the contents of two servers, 2) adding a random service instance to a random server and 3) removing a random service instance from a server. The same operator is used to generate neighbors in PPLS/D. Crossover is performed using the N-Point crossover.

4 Experimental Results

In this section, we first describe our test methodology and then present the results. Our experiments were configured as follows. We generated 30 VNFPP instances for each problem size and topology by varying the number and length of services and the size and speed of the VNFs. We assigned each algorithm a budget of 16000 evaluations to be used across all threads. All tests were run on a 8 core/16 thread CPU at 2.6 GHz. All parallel algorithms were implemented using the parallelisation library Rayon.[1]

We used the parameter settings for NSGA-II proposed in [18] for NSGA-II, MD-NSGA-II and MS-NSGA-II. As PPLS/D is known to be sensitive to the population size [38], we also consider a range of population sizes for each test.

For analysis, we recorded the time each algorithm took to complete and the hypervolume of the final population. Since the three objectives typically have orders of magnitude different values it is necessary to first normalize the

[1] https://github.com/rayon-rs/rayon.

objectives. The utopian and nadir points were estimated by taking the best and worst objective values respectively from all tests. We used the reference point $1 + \epsilon$ for each objective where ϵ is a small number.

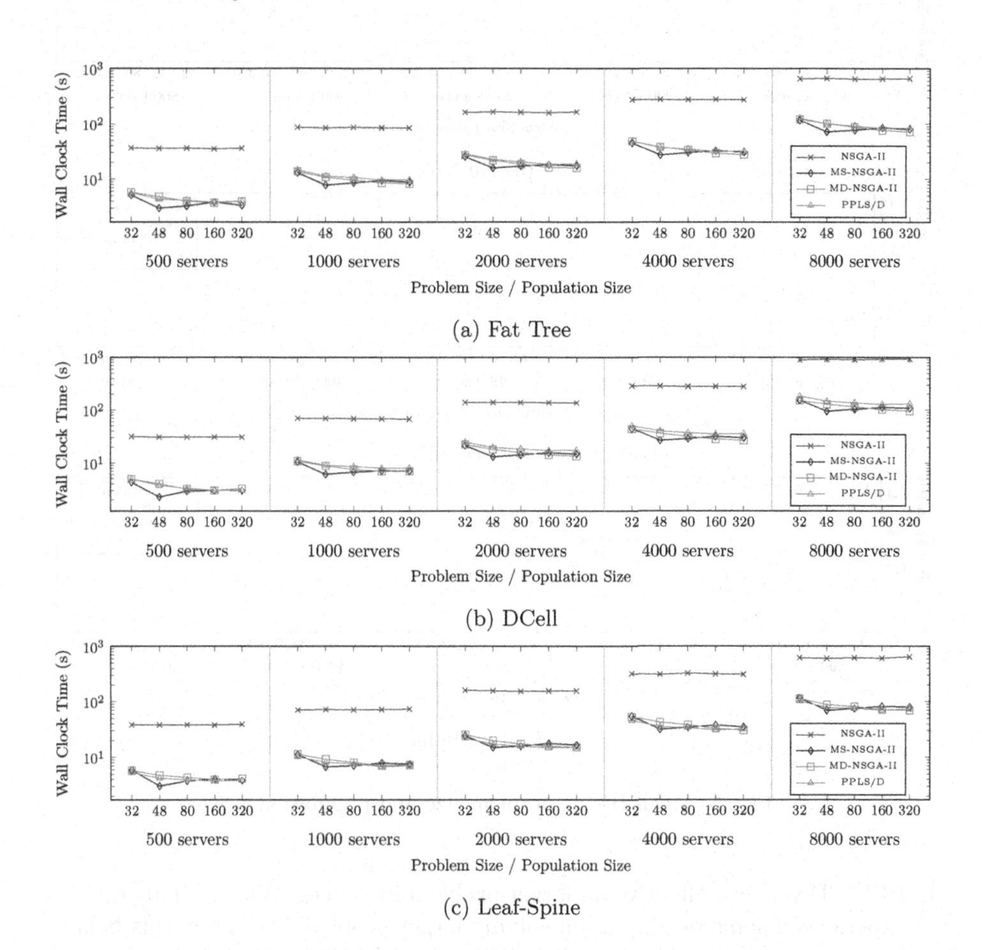

Fig. 4. Mean execution time vs problem size and population size

Figure 4 and Fig. 5 demonstrate our two key findings. First, Fig. 4 shows that PMOEAs are typically 5–10× faster than the sequential NSGA-II. As the problem size increases, the time required to evaluate a solution also increases. However, this step can be parallelized reducing the overall execution time.

Second, Fig. 5 shows that the speed improvements from parallelisation did not come at the cost of solution quality. This demonstrates that PMOEAs solve VNFPPs faster and at no cost to solution quality.

Further, when the best parameter settings are considered for each algorithm, PPLS/D found *better* solutions on average than other algorithms on the majority of problems. These results must be considered with three caveats:

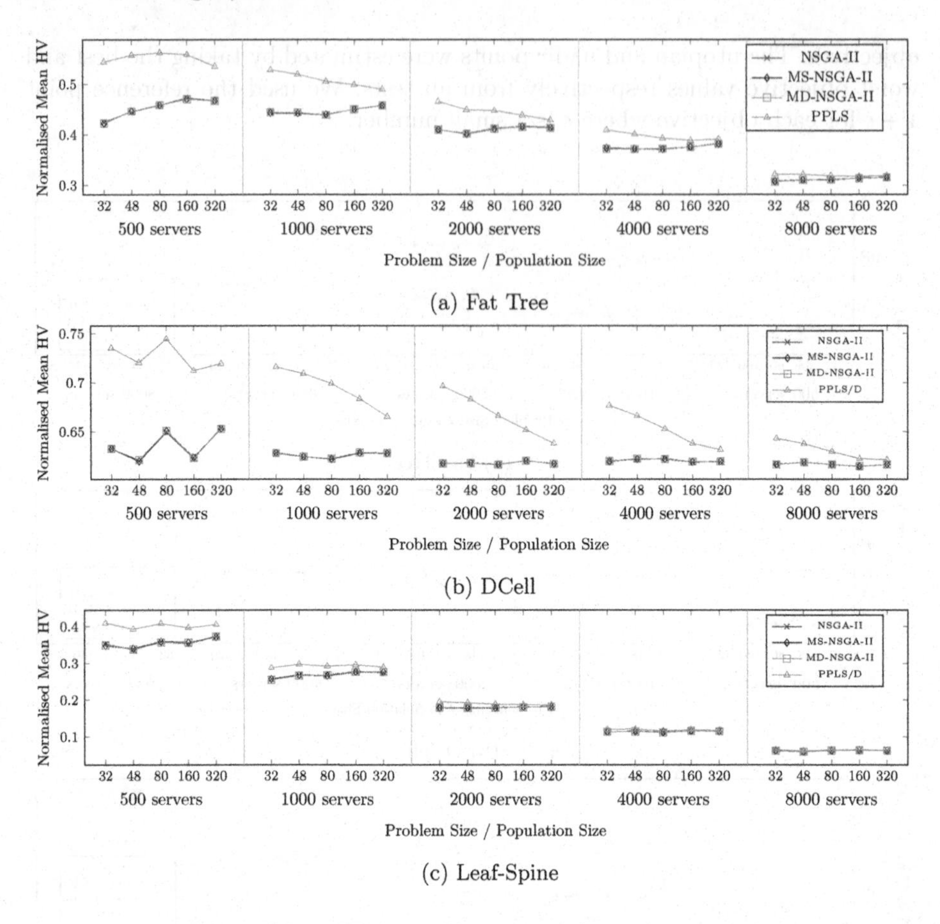

Fig. 5. Mean HV vs problem size and population size

1. PPLS/D was less effective on larger problem instances. Whilst all algorithms experienced some decline when solving larger problem instances, this behavior is particularly apparent with PPLS/D. PPLS/D relies on local search operators to explore the solution space. On large problem instances, the local search area is very large. It is likely that on large problems PPLS/D only explores the neighborhood of the initial population. Alternative local search operators may be able to explore the search space more effectively, however this remains for future research.
2. PPLS/D is sensitive to the population size whereas the other algorithms were not. In the majority of instances, PPLS/D produced better sets of solutions when a small population was used. Shi et al. [38] proposed this was due to large population sizes restricting each subproblem to a small region of the solution space. In PPLS/D, a subproblem can only consider a solution if it is the closest subproblem to the solution (see Sect. 3). With a large population, the solutions generated by the local search are more likely to be near to

a different subproblem, reducing the number of solutions considered, and slowing the search process. This appears particularly important on larger problem instances wherein the setting of the population size was the most important factor in whether PPLS/D outperformed the other algorithms.
3. PPLS/D maintains an archive population and hence stores more solutions on the Pareto front than the other algorithms used. PPLS/D typically produced 1–2 orders of magnitude more non-dominated solutions than the other algorithms. This in part explains the improved hypervolume for PPLS/D as more solutions can better approximate the Pareto front.

Despite these caveats, it is still notable that PPLS/D outperformed other algorithms using only local search operators. Some of the drawbacks of PPLS/D may be able to be resolved with alternative local search operators in future work.

5 Conclusion

In this work we determined whether PMOEAs could be used to solve the VNFPP more efficiently than sequential MOEAs without harming solution quality. We found that not only is this true, but that in certain instances PMOEAs produce better representations of the Pareto front than a comparable sequential algorithm. These results indicate that PMOEAs have potential to solve practical VNFPPs and are worthy of further research. Future work will explore alternative neighbor generation strategies to provide improved local search capabilities. Additional research could extend PMOEAs from single CPU, multithreaded systems to multi-cpu distributed systems in pursuit of larger speedups. In addition, more multi-objective evolutionary algorithms are worthwhile being considered as the baseline optimizer (e.g., [15,16,25,26]).

References

1. Abrita, S.I., Sarker, M., Abrar, F., Adnan, M.A.: Benchmarking VM startup time in the cloud. In: Zheng, C., Zhan, J. (eds.) Bench 2018. LNCS, vol. 11459, pp. 53–64. Springer, Cham (2019). https://doi.org/10.1007/978-3-030-32813-9_6
2. Agarwal, S., Malandrino, F., Chiasserini, C., De, S.: Joint VNF placement and CPU allocation in 5G. In: INFOCOM 2018: IEEE Conference on Computer Communications, pp. 1943–1951. IEEE (2018)
3. Al-Fares, M., Loukissas, A., Vahdat, A.: A scalable, commodity data center network architecture. In: SIGCOMM: 2008 Conference on Applications, Technologies, Architectures, and Protocols for Computer Communications, pp. 63–74 (2008)
4. Alameddine, H.A., Qu, L., Assi, C.: Scheduling service function chains for ultra-low latency network services. In: CNSM 2017: Conference on Network and Service Management, pp. 1–9. IEEE Computer Society (2017)
5. Alba, E.: Parallel evolutionary algorithms can achieve super-linear performance. Inf. Process. Lett. **82**(1), 7–13 (2002)
6. Andrae, A., Edler, T.: On global electricity usage of communication technology: trends to 2030. Challenges **6**(1), 117–157 (2015)

7. Avgerinou, M., Bertoldi, P., Castellazzi, L.: Trends in data centre energy consumption under the european code of conduct for data centre energy efficiency. Energies **10**(1470), 1–18 (2017)
8. Bari, M.F., Chowdhury, S.R., Ahmed, R., Boutaba, R.: On orchestrating virtual network functions. In: CNSM 2011: Proceedings of the 11th International Conference on Network and Service Management, pp. 50–56 (2015)
9. Baumgartner, A., Reddy, V.S., Bauschert, T.: Combined virtual mobile core network function placement and topology optimization with latency bounds. In: EWSDN 2015: Proceedings of the 4th European Workshop on Software Defined Networks, pp. 97–102 (2015)
10. Billingsley, J., Li, K., Miao, W., Min, G., Georgalas, N.: A formal model for multi-objective optimisation of network function virtualisation placement. In: Deb, K., et al. (eds.) EMO 2019. LNCS, vol. 11411, pp. 529–540. Springer, Cham (2019). https://doi.org/10.1007/978-3-030-12598-1_42
11. Billingsley, J., Li, K., Miao, W., Min, G., Georgalas, N.: Multi-objective virtual network function placement: formal model and effective algorithm (2020, unpublished)
12. Billingsley, J., Li, K., Miao, W., Min, G., Georgalas, N.: Routing-led placement of VNFs in arbitrary networks. In: WCCI 2020: World Congress on Computational Intelligence (2020)
13. Branke, J., Schmeck, H., Deb, K., Maheshwar, R.S.: Parallelizing multi-objective evolutionary algorithms: cone separation. In: Proceedings of the IEEE Congress on Evolutionary Computation, CEC 2004, Portland, OR, USA, 19–23 June 2004, pp. 1952–1957. IEEE (2004)
14. Cantú-Paz, E., Goldberg, D.E.: On the scalability of parallel genetic algorithms. Evol. Comput. **7**(4), 429–449 (1999)
15. Cao, J., Kwong, S., Wang, R., Li, K.: AN indicator-based selection multi-objective evolutionary algorithm with preference for multi-class ensemble. In: 2014 International Conference on Machine Learning and Cybernetics, Lanzhou, China, 13–16 July 2014, pp. 147–152. IEEE (2014)
16. Chen, R., Li, K., Yao, X.: Dynamic multiobjectives optimization with a changing number of objectives. IEEE Trans. Evol. Comput. **22**(1), 157–171 (2018)
17. CISCO: spine and leaf architecture: design overview white paper. https://www.cisco.com/c/en/us/products/collateral/switches/nexus-7000-series-switches/white-paper-c11-737022.html. Accessed 18 Sept 2020
18. Deb, K., Agrawal, S., Pratap, A., Meyarivan, T.: A fast elitist non-dominated sorting genetic algorithm for multi-objective optimization: NSGA-II. In: Schoenauer, M., et al. (eds.) PPSN 2000. LNCS, vol. 1917, pp. 849–858. Springer, Heidelberg (2000). https://doi.org/10.1007/3-540-45356-3_83
19. Dodd, N., et al.: Development of the EU green public procurement (GPP) criteria for data centres. Technical report, Server Rooms and Cloud Services, Publications Office of the European Union (2020)
20. El-Alfy, E.M., Alshammari, M.A.: Towards scalable rough set based attribute subset selection for intrusion detection using parallel genetic algorithm in MapReduce. Simul. Model. Pract. Theory **64**, 18–29 (2016)
21. Gouareb, R., Friderikos, V., Aghvami, A.H.: Delay sensitive virtual network function placement and routing. In: ICT 2018: 25th International Conference on Telecommunications, pp. 394–398. IEEE (2018)

22. Guo, C., Wu, H., Tan, K., Shi, L., Zhang, Y., Lu, S.: DCell: a scalable and fault-tolerant network structure for data centers. In: SIGCOMM 2008: Conference on Applications, Technologies, Architectures, and Protocols for Computer Communications, pp. 75–86 (2008)
23. Guo, H., et al.: Cost-aware placement and chaining of service function chain with VNF instance sharing. In: NOMS 2020: IEEE/IFIP Network Operations and Management Symposium, pp. 1–8. IEEE (2020)
24. Kuo, T., Liou, B., Lin, K.C., Tsai, M.: Deploying chains of virtual network functions: on the relation between link and server usage. IEEE/ACM Trans. Netw. **26**(4), 1562–1576 (2018)
25. Li, K., Chen, R., Fu, G., Yao, X.: Two-archive evolutionary algorithm for constrained multiobjective optimization. IEEE Trans. Evol. Comput. **23**(2), 303–315 (2019)
26. Li, K., Deb, K., Zhang, Q., Kwong, S.: An evolutionary many-objective optimization algorithm based on dominance and decomposition. IEEE Trans. Evol. Comput. **19**(5), 694–716 (2015)
27. von Lücken, C., Barán, B., Sotelo, A.: Pump scheduling optimization using asynchronous parallel evolutionary algorithms. CLEI Electron. J. **7**(2) (2004)
28. Luo, J., Fujimura, S., Baz, D.E., Plazolles, B.: GPU based parallel genetic algorithm for solving an energy efficient dynamic flexible flow shop scheduling problem. J. Parallel Distrib. Comput. **133**, 244–257 (2019)
29. Miotto, G., Luizelli, M.C., da Costa Cordeiro, W.L., Gaspary, L.P.: Adaptive placement & chaining of virtual network functions with NFV-PEAR. J. Internet Serv. Appl. **10**(1), 1–19 (2019). https://doi.org/10.1186/s13174-019-0102-2
30. Mühlenbein, H., Schomisch, M., Born, J.: The parallel genetic algorithm as function optimizer. Parallel Comput. **17**(6–7), 619–632 (1991)
31. Oljira, D.B., Grinnemo, K., Taheri, J., Brunström, A.: A model for QoS-aware VNF placement and provisioning. In: NFV-SDN 2017: IEEE Conference on Network Function Virtualization and Software Defined Networks, pp. 1–7. IEEE (2017)
32. Pongor, G.: OMNeT: objective modular network testbed. In: MASCOTS 1993: Proceedings of the International Workshop on Modeling, Analysis, and Simulation on Computer and Telecommunication Systems, pp. 323–326 (1993)
33. Qi, D., Shen, S., Wang, G.: Towards an efficient VNF placement in network function virtualization. Comput. Commun. **138**, 81–89 (2019)
34. Qu, L., Assi, C., Shaban, K.B., Khabbaz, M.J.: A reliability-aware network service chain provisioning with delay guarantees in NFV-enabled enterprise datacenter networks. IEEE Trans. Netw. Serv. Manag. **14**(3), 554–568 (2017)
35. Rankothge, W., Le, F., Russo, A., Lobo, J.: Optimizing resource allocation for virtualized network functions in a cloud center using genetic algorithms. IEEE Trans. Netw. Serv. Manag. **14**(2), 343–356 (2017)
36. Rankothge, W., Ma, J., Le, F., Russo, A., Lobo, J.: Towards making network function virtualization a cloud computing service. In: IM 2015: Proceedings of the 2015 IFIP/IEEE International Symposium on Integrated Network Management, pp. 89–97 (2015)
37. Roberge, V., Tarbouchi, M., Labonté, G.: Comparison of parallel genetic algorithm and particle swarm optimization for real-time UAV path planning. IEEE Trans. Ind. Inform. **9**(1), 132–141 (2013)
38. Shi, J., Zhang, Q., Sun, J.: PPLS/D: parallel pareto local search based on decomposition. IEEE Trans. Cybern. **50**(3), 1060–1071 (2020)

39. Soualah, O., Mechtri, M., Ghribi, C., Zeghlache, D.: Energy efficient algorithm for VNF placement and chaining. In: CCGRID 2017: International Symposium on Cluster, Cloud and Grid Computing, pp. 579–588. IEEE Computer Society/ACM (2017)
40. Vizarreta, P., Condoluci, M., Machuca, C.M., Mahmoodi, T., Kellerer, W.: QoS-driven function placement reducing expenditures in NFV deployments. In: ICC 2017: IEEE International Conference on Communications, pp. 1–7. IEEE (2017)
41. Zhang, Y., Yu, W., Chen, X., Jiang, J.: Parallel genetic algorithm to extend the lifespan of internet of things in 5G networks. IEEE Access **8**, 149630–149642 (2020)

Using Multi-objective Grammar-Based Genetic Programming to Integrate Multiple Social Theories in Agent-Based Modeling

Tuong Manh Vu(✉), Eli Davies, Charlotte Buckley, Alan Brennan, and Robin C. Purshouse

University of Sheffield, Sheffield, UK
{t.vu,c.m.buckley,a.brennan,r.purshouse}@sheffield.ac.uk

Abstract. Different theoretical mechanisms have been proposed for explaining complex social phenomena. For example, explanations for observed trends in population alcohol use have been postulated based on norm theory, role theory, and others. Many mechanism-based models of phenomena attempt to translate a single theory into a simulation model. However, single theories often only represent a partial explanation for the phenomenon. The potential of integrating theories together, computationally, represents a promising way of improving the explanatory capability of generative social science. This paper presents a framework for such integrative model discovery, based on multi-objective grammar-based genetic programming (MOGGP). The framework is demonstrated using two separate theory-driven models of alcohol use dynamics based on norm theory and role theory. The proposed integration considers how the sequence of decisions to consume the next drink in a drinking occasion may be influenced by factors from the different theories. A new grammar is constructed based on this integration. Results of the MOGGP model discovery process find new hybrid models that outperform the existing single-theory models and the baseline hybrid model. Future work should consider and further refine the role of domain experts in defining the meaningfulness of models identified by MOGGP.

Keywords: Inverse generative social science · Agent-based modeling · Multi-objective optimization · Grammar-based genetic programming

1 Introduction

1.1 Background

Agent-based modeling (ABM) is a bottom-up methodology that models a system as a collection of heterogeneous agents and their interactions. Since ABM allows the modeling of a complex system at a fine-grained resolution, it has become an

H. Ishibuchi et al. (Eds.): EMO 2021, LNCS 12654, pp. 721–733, 2021.
https://doi.org/10.1007/978-3-030-72062-9_57

established tool for generative social science: how micro-level agent behaviours and their interactions can generate macro-level social phenomenon. Following the motto of generative social science – "If you didn't grow it, you didn't explain it." [7] – the *forward* approach is where, if an agent-based model with a defined set of mechanisms can produce the target social phenomenon, the model is a candidate explanation for the phenomenon. However, this does not mean that the candidate is unique – there can be other models that can generate the same target phenomenon. In the *inverse* problem, from a target phenomenon, the aim is to find possible explanatory agent-based models. This process is known as *inverse generative social science* [18] or *model discovery* [10]. The literature on model discovery is very limited and was reviewed by Vu and colleagues [17].

A social phenomenon can be explained by different theories. An agent-based model encoded with mechanisms from a single social theory can be a candidate explanation. But one can wonder which theory is better or what can be missing from a theory. There is also a possibility that a single theory cannot explain the phenomenon and there is a need to combine multiple theories. However, the integration of multiples theories is a major endeavour. To address that gap, this paper proposes using genetic programming to explore combinations of social theories with the aim of better explanatory capability.

In genetic programming [14], a population of computer programs or candidates is evolved over many generations using a fitness function. Computer programs with high fitness are selected probabilistically for crossover and mutation operators to generate the next generation. To mitigate the problem of invalid computer programs created by the random nature of the genetic operators, we adopted grammar-based genetic programming (GGP) [12] in which a grammar is employed for genotype-to-phenotype mapping to enforce a particular structure and guide the evolutionary process. In this paper, the GGP was used in the model discovery process to explore the search space of multi-theory models.

1.2 Aim of the Present Study and Organization of the Paper

This paper aims to demonstrate how to use multi-objective GGP (MOGGP) to integrate single-theory models into a hybrid model to provide better explanation of a social phenomenon. As a case study, the paper uses two established agent-based models based on social norm theory and social role theory for the integration and model discovery process. In Sect. 2, the integration and MOGGP model discovery processes are both introduced. Section 3 briefly introduces the existing models that form the basis for integration. Results and discussion of the MOGGP process are provided in Sect. 4. Lastly, Sect. 5 concludes the paper and suggests possible future works.

2 Model Discovery Process

The flow chart depicting the model discovery process is shown in Fig. 1. This is an extension of the recent approach by Vu and colleagues [17]. There are three roles in the model discovery process: *domain expert*, *analyst*, and *modeler*. The domain expert is a person with deep knowledge about the social science underpinning the model and who can assess the theoretical credibility of the model structure. The analyst assesses commonalities and differences between social theories with the domain expert, as well as abstracting and defining the building blocks of entities and mechanisms. The modeler designs, implements, verifies and validates the agent-based models.

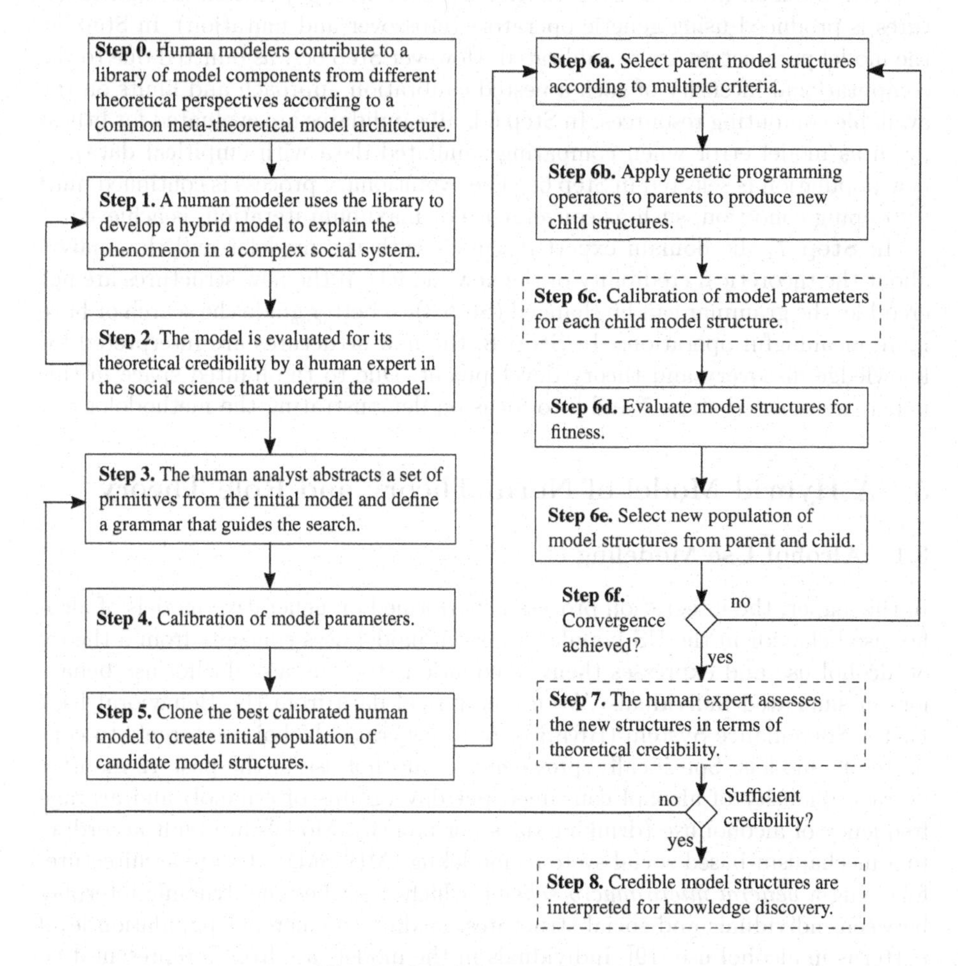

Fig. 1. Model discovery process

Step 0 represents a pre-condition for the model integration process. We assume that a library of theory building blocks, implemented as model components according to a common meta-theoretical software architecture is available [19]. In **Step 1**, a human modeler uses the library to develop an initial hybrid model. In **Step 2**, the theoretical credibility of the model is evaluated by a human expert. The model may need to be adjusted and redeveloped (returning to Step 1). In **Step 3**, the human analyst abstracts a set of primitives from the hybrid model and defines the grammar for the genetic programming. In **Step 4**, model parameters are calibrated. In **Step 5**, the hybrid model built by the human modeler with the best calibrated setting is cloned for the whole initial population.

The evolutionary process is performed in **Step 6**. Parent candidates are selected based on fitness in Step 6a. In Step 6b, a new population of child structures is produced using genetic operators (crossover and mutation). In Step 6c, the model parameters are re-calibrated. However Step 6c was omitted due to the computational intensity of such a nested calibration approach and limits on the available computing resources. In Step 6d, all candidates are evaluated for fitness (such as model error when comparing simulated data with empirical data). A new population is selected in Step 6e. The evolutionary process is continued until a stopping condition, such as convergence or maximum iteration, is achieved.

In **Step 7**, the domain expert discusses with the modeler and the analyst about the theoretical credibility of the new models. If the new structures are not credible, the grammar can be changed (Step 3) to better guide the search or have more meaningful operations. In **Step 8**, the new structures are interpreted for knowledge discovery and theory development. Due to the limited space in this paper, we omitted steps 7 and 8 to focus on demonstrating the methodology.

3 A Hybrid Model of Norm Theory and Role Theory

3.1 Alcohol Use Modeling

In this paper, the integration process is performed on generative models of alcohol use behavior in the US population. Each model uses concepts from a theory of alcohol use and expresses them as equations to generate alcohol use behaviors in simulated individuals. We use empirical data from the Behavioral Risk Factor Surveillance System (BRFSS) [6] to generate alcohol use targets to calibrate our models. Specifically, prevalence of alcohol use in the past 12 months, average quantity of alcohol consumed per day (grams of ethanol) and average frequency of alcohol use (drinking days per month). Models are built according to a mechanism-based social systems modeling (MBSSM) software architecture, following a *general micro-macro scheme* which describes the dynamic interplay between individuals and social structures, leading to emergent population-level patterns in alcohol use [19]. Individuals in the models are from a representative population-level US microsimulation model 1980–2015 that accounts for births, deaths and migration over time [2]. Individuals in the microsimulation have socio-demographic properties and alcohol use variables at baseline that are generated from several US data sources including the BRFSS, US Census [16] and American Community Survey [13].

3.2 Social Norms Model

The social norms model is described in detail in [15]. This model uses concepts from social norm theory to generate a disposition to consume alcohol for each modeled individual. Disposition (*NormDisp*) is determined by three variables: descriptive norm, injunctive norm and autonomy, which all range between 0 and 1. The descriptive norm is an individual's appraisal of drinking behavior in their reference group (age group and sex) in terms of whether individuals are drinking at all (prevalence) and how much they are drinking (quantity). The descriptive norm is determined by calculating the prevalence and quantity of drinking in the reference group and applying a perception bias that adjusts the descriptive norm to be closer to the agent's own drinking. A descriptive norm of 1 would indicate that 100% of individuals in the reference group are perceived to be drinking. The injunctive norm refers to the perceived acceptability of drinking in society for an individual like them (i.e., in their reference group), initialized using data from the US National Alcohol Survey [9]. An injunctive norm of 1 would indicate that it is completely acceptable for an individual in a reference group to drink. Autonomy refers to the individual's desire to ignore the norms; an autonomy of 0.7 would indicate that the agent only pays attention to norms 30% of the time. These concepts are combined in Eq. 1 to determine, for individual i, disposition $NormDisp$ to consume drink k. The injunctive norms I are combined with descriptive norms D and weighted by the inverse of autonomy a. This term is combined with the individual's desire u to consume drink k weighted by autonomy. Over time, the injunctive and descriptive norms are updated in response to individuals' drinking behavior in the model.

$$NormDisp_i[k] = u_i[k]a_i + (1 - a_i)\sqrt{D_i[k]I_i[k]} \tag{1}$$

3.3 Social Roles Model

The social roles model is described in detail in [17] and is based on social role theory [1] that describes how individuals' marital, parental and employment responsibilities affect their desire and ability to drink. Disposition to drink according to social role theory (*RoleDisp*) is generated using Eq. 2 and is determined by an individual's desire to drink, u, for drink k multiplied by their probability of having the opportunity to drink inside (*ProbOpIn*) and outside (*ProbOpOut*) the home. Social role theory suggests that the roles individuals hold affect their ability to participate in drinking situations and therefore regulate the daily opportunities to consume alcohol [11]. Opportunity to drink in and out of the home is determined by role load (the stress that results from needing to perform a role), which is calculated using each individual's role status and their level of involvement in that role. Role strain – "the experience of stress associated with positions or expected role" [1] – is suggested to lead to alcohol use as a means of coping with a set of roles that are too complex (role load) or lacking roles that provide meaning (role deprivation). Role strain (*RoleStrain*) is weighted by β, describing the size of the effect role strain has on drinking behavior. On each day, *RoleDisp* is calculated for each individual for $k = 1$ and compared to a uniform random number between 0 and 1 to determine if they will drink. For each individual that drinks, disposition is calculated for $k = 2$, and continues for progressive values of k until they stop drinking.

$$RoleDisp_i[k] = u_i[k](ProbOpOut_i[k]+ProbOpIn_i[k])(1+\beta RoleStrain_i[k]) \quad (2)$$

3.4 Integration: A Hybrid Model of Norm Theory and Role Theory

We postulated that the social theories have different influence on an individual's k^{th} drink. For example, let us imagine an employed individual, *Agent Alan*, engages in a drinking occasion. Agent Alan's first two drinks are based solely on social norms. Agent Alan's chance of consuming a third drink is lower because he is aware he has to go to work the next day – the balance of probabilities being based on both norms and roles. Agent Alan never drinks four drinks because of his prospective employment commitments, so the disposition to consume a fourth drink is based only on roles. Extending this hypothesis, we decided to perform model discovery on the drinking disposition of individual agents.

Both the social norms and social roles models operate by calculating a disposition (*Disp*) for each agent on each day to determine whether to drink and, if so, how much to drink. In the hybrid model, the social norms and roles dispositions are calculated on each day. The models are combined by using a weighting (w, bounded between 0 and 1) of the two calculated dispositions ($NormDisp$ and *RoleDisp*) as shown in Eq. 3. The hybrid model was implemented using MBSSM [19] and Repast HPC libraries [4].

$$Disp_i[k] = wRoleDisp_i[k] + (1 - w)NormDisp_i[k] \quad (3)$$

3.5 Parameter Calibration

The hybrid model contains 65 parameters that represent unknown effect sizes from social role theory and social norm theory plus the weighting for combining the dispositions. Each parameter is assigned a prior distribution reflecting the range of acceptable values. We sampled 10,000 parameter settings from the joint prior distribution using the lhs R package [3]. Each model setting was run once for the years 1984–2004 and an implausibility metric (Eq. 4) was calculated to compare the model output to alcohol use BRFSS target data for prevalence, quantity and frequency of male and female drinking.

$$z_1 = \frac{1}{KM}\sum_{k=1}^{K}\sum_{m=1}^{M}\frac{|y_m^{\star}[k] - y_m[k]|}{\sqrt{(s_m[k])^2 + (d_m)^2}}, \quad (4)$$

where M is the number of output measures, K is the number of observations, $s_m[k]$ is the observed standard error for output m at time point k, and $(d_m)^2$ is the variance of the model discrepancy[1] for output m, which is taken as 10% of the possible output range for each output.

3.6 Multi-objective Grammar-Based Genetic Programming

In the hybrid model, an agent can have maximum of 30 drinks within a day, so k is ranged from 1 to 30. After parameter calibration, agents used the following calibrated equation to calculate all the drinking dispositions: $Disp_i[k] = 0.99667 RoleDisp_i[k] + (1 - 0.99667) NormDisp_i[k]$. For model discovery, the MOGGP can be set up to work with 30 equations for each of the kth drinks. However, for simplification, we decided to use only six equations by grouping the drinks into the following categories: drink 1, drink 2, drink 3 to 4, drink 5 to 7, drink 8 to 11, drink 12+. For the primitives, terminals consist of norms disposition and role disposition and the allowed functions are weighted sum and geometric mean. Additionally, a list of both arbitrary and calibrated constants are included. The design of the MOGGP is captured in the grammar:

```
<p> ::= mediatedFirstDrink=<e>; mediated2Drinks=<e>;
  mediated3to4Drinks=<e>; mediated5to7Drinks=<e>;
  mediated8to11Drinks=<e>; mediated12MoreDrinks=<e>;
<v> ::= RoleDisp | NormDisp
<e> ::= (<c>*<e>+<c>*<e>) | sqrt(<e>*<e>) | <v>
<c> ::= 0.1 | 0.2 | 0.3 | 0.4 | 0.5 | 0.6 | 0.7 | 0.8 | 0.9 |
  0.99667 | 0.00333
```

For the initial population, we used the same structure designed and calibrated by the modeler. Then a MOGGP was configured to minimize both model error and complexity. The first objective, model error, represents the ability of the simulation model to reproduce the pattern observed in the real world and is captured by Eq. 4. The second objective, complexity, is defined by the number of nodes in the structures and discourages complex structures that overfit the data and are difficult for humans to interpret. In this paper, we used the NSGA-II optimizer [5] to develop an even, sample-based representation of the Pareto front representing the trade-off between model error and complexity.

The MOGGP was implemented using the PonyGE2 toolkit [8], which supports both NSGA-II optimization and the complexity evaluation of a structure via number of nodes in the tree. Model error evaluation of a new model structure required additional scripting: (i) modifying the source code of the drinking disposition in agent decision making; (ii) re-compiling the Repast HPC model; (iii) running the simulation; and (iv) calculating the model error by comparing simulated data against empirical data.

The evolutionary setup was as follows: 500 candidates per population for 50 generations, 75% subtree crossover, 25% subtree mutation, tree depth maximum of 17, and other default settings of PonyGE2. The source code is available at https://bitbucket.org/r01cascade/emo2021_hybrid_norms_roles_gp and is

[1] Model discrepancy is the error in a model output that arises because the model is not a perfect representation of reality.

licensed under the GNU General Public License version 3. It is computationally intensive to do a complete run of MOGGP; the process took 2.5 days on an Intel i9 9980XE processor with 36 cores.

4 Results and Discussion

Both model error and complexity reduced over the generations. The search converged at generation 36, showing no change to the Pareto front. Figure 2 shows the final 14 non-dominated structures found by the MOGGP, the structure of the human hybrid model, and the two single-theory models (before integration into the hybrid model). From this point onward, the structures found by MOGGP will be referred to by their complexity value, i.e. the structure with complexity 19 will be called GP19. On the Pareto front, there are six extremely complex structures but with minimal improvement on the model errors (GP231, GP235, GP247, GP259, GP283, GP287). We decided to exclude these models from the discussion because it is very challenging to interpret them. Table 1 shows the structures of the human hybrid model and the remaining eight structures discovered by MOGGP.

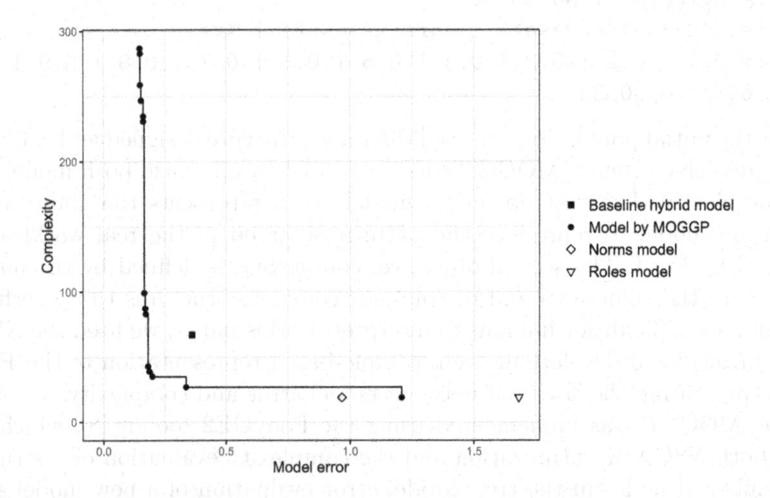

Fig. 2. Pareto front of the structures discovered by the MOGGP process versus the original baseline hybrid structure developed by the modeler

All the dispositions in the simplest model GP19 equate to $NormDisp$ which means that for a single theory, the social norms model performs better than the roles model. This is the same when comparing the two single-theory models before integration in Fig. 2. Another observation is that GP19 has the same higher error even though its mechanisms are the same as the single-theory norms model. This is because GP19 used the calibrated parameters from the hybrid

Table 1. Hybrid model structures in detail, excluding six extremely complex structures. Each row of a structure relates to the ordinal drinking categories: drink 1, drink 2, drink 3 to 4, drink 5 to 7, drink 8 to 11, drink 12+.

ID	Complexity	Model error	Simplified structure of the dispositional equations
Human	67	0.3629	(0.964*RoleDisp + 0.036*NormDisp); (0.964*RoleDisp + 0.036*NormDisp); (0.964*RoleDisp + 0.036*NormDisp); (0.964*RoleDisp + 0.036*NormDisp); (0.964*RoleDisp + 0.036*NormDisp); (0.964*RoleDisp + 0.036*NormDisp);
GP19	19	1.2089	NormDisp; NormDisp; NormDisp; NormDisp; NormDisp; NormDisp;
GP27	27	0.3382	0.964*RoleDisp + 0.036*NormDisp; RoleDisp; RoleDisp; RoleDisp; RoleDisp; RoleDisp;
GP35	35	0.1974	0.036*NormDisp + 1.7005*RoleDisp; RoleDisp; RoleDisp; NormDisp; NormDisp; NormDisp;
GP39	39	0.1843	0.036*NormDisp + 0.9293*RoleDisp + 0.7712*sqrt(NormDisp*RoleDisp); RoleDisp; RoleDisp; NormDisp; RoleDisp; NormDisp;
GP43	43	0.1799	0.036*NormDisp + 1.7005*RoleDisp; RoleDisp; NormDisp; 0.1*NormDisp + 0.964*RoleDisp; RoleDisp; NormDisp;

(*continued*)

Table 1. *continued*

ID	Complexity	Model error	Simplified structure of the dispositional equations
GP83	83	0.1709	0.036*NormDisp + 1.83006*RoleDisp; 0.036*NormDisp + 0.964*RoleDisp; 0.964*NormDisp + 0.036*RoleDisp; 1.1581*NormDisp; 0.7*NormDisp + 0.964*RoleDisp; NormDisp;
GP87	87	0.1684	0.036*NormDisp + 1.71438*RoleDisp; 0.964*RoleDisp + 0.036*sqrt(NormDisp*NormDisp); 0.036*NormDisp + 0.964*RoleDisp; 1.89459*NormDisp; 0.036*NormDisp + 0.036*RoleDisp; RoleDisp;
GP99	99	0.1648	0.036*NormDisp + 1.71438*RoleDisp; 0.036*NormDisp + 0.964*RoleDisp; 0.036*NormDisp + 0.964*RoleDisp; 1.89459*NormDisp; 0.036*NormDisp + 0.9*RoleDisp; 0.7712*NormDisp + 0.8072*RoleDisp;

model. It would perform the same if we recalibrate GP19 in Step 6c. Additionally, Fig. 2 shows that the two single theory models and GP19 performed poorly in term of model error when compared to the other hybrid models. Thus it can be concluded that, in this case study, integration of multiple theories better explains the phenomenon than a single theory.

Interestingly, in GP27 (the next model after GP19 in term of complexity), the first equation is the same as the baseline model, $0.964 * RoleDisp + 0.036 * NormDisp$, and the remaining equations are $RoleDisp$. This means that just by including a fraction of the norm disposition into the first drink disposition, the roles model was improved (into a hybrid model) and outperformed the norms model. The next structure is GP35 where the first three equations are dominated by role disposition and the last three equations are only influenced by norm disposition. This suggests that the first four drinks are driven by roles and the heavy drinking of 5+ drinks is driven by norms. Figure 3 shows the 1984–2012 time series of the four structures discussed (human, GP19, GP27, GP35).

For the remaining structures (GP39, GP43, GP83, GP87, GP99), the roles and norms disposition alternate between different categories. Since the domain expert asserted that there is no theory supporting this behavioral approach, we conclude that these structures are not theoretical plausible. In future work, the next iteration of the grammar should include this restriction.

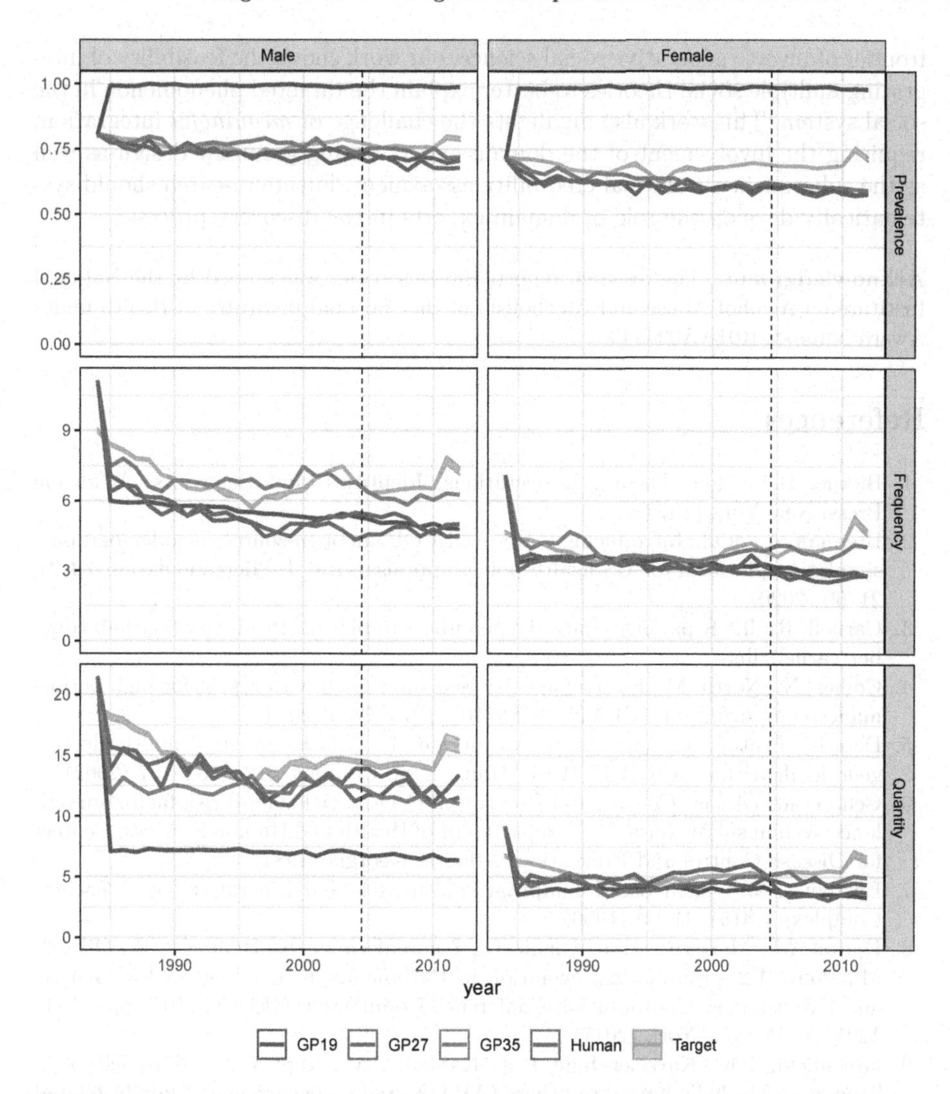

Fig. 3. Time series of the human model, GP19, GP27, and GP35 versus the empirical target data (mean and 95% confidence interval). The vertical dashed line separates the calibration period (1984–2004) and validation period (2005–2012).

5 Conclusion

This paper presents a novel method that utilizes multi-objective grammar-based genetic programming to integrate multiple social theories and discover new model structures. The case study of alcohol use modeling has shown that a multi-theory model can better explain the real world phenomenon than a single-theory model and our model discovery method offers a promising approach to generate novel combinations of multiple mechanisms of these theories. In the research frontier of inverse generative social science, our work shows the feasibility of integrating multiple social theories to better explain the targeted phenomenon in the social system. This work also highlights the challenge of *meaningful* integration, requiring the involvement of the domain expert during grammar design as well as the subsequent theoretical credibility assessment. Future research should systematically develop the role of domain experts in the discovery process.

Acknowledgment. The research reported in this paper was funded by the National Institute on Alcohol Abuse and Alcoholism of the National Institutes of Health under Award Number R01AA024443.

References

1. Biddle, B.J.: Role Theory: Expectations, Identities, and Behaviors. Academic Press, New York (1979)
2. Brennan, A., et al.: Introducing CASCADEPOP: an open-source sociodemographic simulation platform for US health policy appraisal. Int. J. Microsimulation **13**(2), 21–60 (2020)
3. Carnell, R.: lhs R package: Latin Hypercube Samples (2019). https://github.com/bertcarnell/lhs
4. Collier, N., North, M.: Parallel agent-based simulation with repast for high performance computing. SIMULATION **89**(10), 1215–1235 (2013)
5. Deb, K., Pratap, A., Agarwal, S., Meyarivan, T.: A fast and elitist multiobjective genetic algorithm: NSGA-II. IEEE Trans. Evol. Comput. **6**(2), 182–197 (2002)
6. Centers for Disease Control and Prevention (CDC): Behavioral risk factor surveillance system survey data. U.S. Department of Health and Human Services, Centers for Disease Control and Prevention, Atlanta, Georgia (1984)
7. Epstein, J.M.: Agent-based computational models and generative social science. Complexity **4**(5), 41–60 (1999)
8. Fenton, M., McDermott, J., Fagan, D., Forstenlechner, S., Hemberg, E., O'Neill, M.: PonyGE2: grammatical evolution in Python. In: Proceedings of the Genetic and Evolutionary Computation Conference Companion, GECCO 2017, pp. 1194–1201. ACM, New York (2017)
9. Greenfield, T.K., Karriker-Jaffe, K.J., Kaplan, L.M., Kerr, W.C., Wilsnack, S.C.: Trends in Alcohol's harms to others (AHTO) and co-occurrence of family-related AHTO: the four US National Alcohol Surveys, 2000–2015. Subst. Abuse Res. Treat. **9**(Suppl 2), 23–31 (2015)
10. Gunaratne, C., Garibay, I.: Alternate social theory discovery using genetic programming: towards better understanding the artificial anasazi. In: GECCO 2017 - Proceedings of the 2017 Genetic and Evolutionary Computation Conference, pp. 115–122. ACM Press, Berlin (2017)
11. Knibbe, R.A., Drop, M.J., Muytjens, A.: Correlates of stages in the progression from everyday drinking to problem drinking. Soc. Sci. Med. **24**(5), 463–473 (1987)
12. Koza, J.: Genetic Programming: On the Programming of Computers by Means of Natural Selection. MIT Press, Cambridge (1992)
13. Manson, S., Schroeder, J., Van Riper, D., Ruggles, S.: IPUMS National Historical Geographic Information System: Version 14.0 [Database]. IPUMS, Minneapolis (2019)

14. Poli, R., Langdon, W.B., McPhee, N.F., Koza, J.R.: A Field Guide to Genetic Programming. Lulu Press, Morrisville (2008)
15. Probst, C., et al.: The normative underpinnings of population-level alcohol use: an individual-level simulation model. Health Educ. Behav. **47**(2), 224–234 (2020)
16. Ruggles, S., et al: IPUMS USA: Version 9.0 [Dataset]. IPUMS, Minneapolis (2019)
17. Vu, T.M., et al.: Multiobjective genetic programming can improve the explanatory capabilities of mechanism-based models of social systems. Complexity **2020**, 1–20 (2020)
18. Vu, T.M., Probst, C., Epstein, J.M., Brennan, A., Strong, M., Purshouse, R.C.: Toward inverse generative social science using multi-objective genetic programming. In: GECCO 2019 - Proceedings of the 2019 Genetic and Evolutionary Computation Conference, GECCO 2019, pp. 1356–1363. Association for Computing Machinery, Prague, July 2019
19. Vu, T.M., et al.: A software architecture for mechanism-based social systems modelling in agent-based simulation models. J. Artif. Soc. Soc. Simul. **23**(3), 1 (2020)

Change Detection in SAR Images Based on Evolutionary Multiobjective Optimization and Superpixel Segmentation

Tianqi Gao, Wenyuan Qiao, Hao Li, Maoguo Gong(✉), Gengchao Li, and Jie Min

School of Electronic Engineering, Key Laboratory of Intelligent Perception and Image Understanding of Ministry of Education of China, Xidian University, Xi'an 710071, China
gong@ieee.org

Abstract. Synthetic aperture radar (SAR) imagery has been widely used in the field of remote sensing image change detection. However, its disadvantage of strong coherent multiplicative noise reduces the accuracy of change detection results. This paper proposes a novel SAR image change detection method, which is mainly comprised of three steps. Firstly, the difference image (DI) which is generated by log-ratio operator is segmented into superpixels by Simple Linear Iterative Clustering (SLIC) Algorithm. Secondly, superpixels are encoded uniformly in order to be utilized as the training samples, and deep neural network is used to extract deep features of DI. Finally, this paper designs an improved clustering algorithm which is optimized by Non-dominated Sorting Genetic Algorithm (NSGA-II). When the deep features of DI are used to cluster, Bhattacharyya distance between two categories of samples is selected as the similarity measurement. Taking the logarithmic likelihood function of clustering algorithm and the Bhattacharyya distance between the two categories as two optimization objectives, NSGA-II algorithm is used to optimize the model, and a set of pareto optimal solutions are thus generated. Compared with various indexes for accuracy evaluation, the map which has the highest accuracy is the final change detection map. Experimental results on real synthetic aperture radar datasets show that the proposed method is superior to other classical change detection methods, which demonstrates its effectiveness, feasibility, and superiority of the proposed method.

Keywords: Change detection · Non-dominated sorting genetic algorithm · Superpixel segmentation · Deep neural network

This work was supported in part by the Key-Area Research and Development Program of Guangdong Province under Grant 2020B090921001 and in part by the National Natural Science Foundation of China under Grant 62036006 and Grant 61906146, and in part by the Key Research and Development Program of Shaanxi Province under Grant 2018ZDXM-GY-045.

H. Ishibuchi et al. (Eds.): EMO 2021, LNCS 12654, pp. 734–745, 2021.
https://doi.org/10.1007/978-3-030-72062-9_58

1 Introduction

Change detection is a technology to recognize the difference of target state or identify changes of objects by analyzing the information obtained at different dates in the same region [1]. As an important representative of microwave remote sensing image, SAR image is imaged by synthetic aperture radar, which has strong penetration ability and improved range resolution and azimuth resolution [2]. However, the inherent speckle noise in SAR images produces the intensity distortion [3] of signal correlation in the image, which reduces the accuracy of change detection.

Domestic and foreign researchers have proposed a variety of change detection methods for different remote sensing image detection targets. And these methods can be divided into three types: pixel-based change detection, feature-based change detection and object-based change detection [4]. The traditional change detection takes pixel as the basic unit of change, which can locate the changing edge. However, it often produces salt-and-pepper noise, which greatly reduces the detection accuracy [5]. Object-based change detection takes the object (a series of adjacent pixels with similar spectrum) as the basic unit [6]. It makes full use of the spatial information of pixels to extract the spectrum, spatial context, geometry, texture and other information of the object from the complex ground features, which effectively suppresses the occurrence of salt-and-pepper noise and greatly improves the separability of the object in the feature space [7].

In recent years, the research of deep learning has developed rapidly. As a multi-layer neural network model, it uses a large number of training data to obtain useful feature information, so as to improve the accuracy of classification or prediction. Nowadays, there are many neural network models, which have been widely used in remote sensing field and achieved ideal results. M. Castelluccio et al. applied CNN to remote sensing image classification, fully considered the two-dimensional spatial information and obtained good classification effect [8]. Chen et al. successfully applied AE to the classification of hyperspectral remote sensing image [9]. Li et al. used DBN in feature extraction and classification process of hyperspectral image, and verified the advantages of deep learning in remote sensing image classification process by comparing SVM [10]. Zhan et al. used CNN to learn the semantic differences between the changed pixels and the unchanged pixels [7]. Combined with superpixel segmentation, it obtained good detection results.

This paper proposes a novel method for SAR image change detection. The difference image is segmented into superpixels by SLIC Algorithm. Then we encode these super pixel blocks according to the gray-value of each pixel in order to facilitate them as samples into the neural network for training. And we select the neural network of stacked denoising autoencoders to extract deep features. In order to better analyze the difference image of SAR images, this paper selects clustering algorithm to analyse DI, and selects the Bhattacharyya distance between the two categories as the similarity measurement. This paper uses Non-dominated Sorting Genetic Algorithm (NSGA-II), which takes the reciprocal of logarithmic likelihood function of clustering algorithm and the Bhattacharyya distance between the two categories as two optimization objectives. Then we can obtain a set of Pareto optimal solutions, which satisfies the maximization of the logarithmic likelihood function of clustering algorithm and the minimum of the Bhattacharyya distance between two categories at the same time so that can choose compromise solutions between improving the accuracy of parameter estimation and enhancing the separability of samples.

The rest of this paper is organized as follows: Sect. 2 describes some related background including the change detection model of remote sensing image and the procedures of NSGA-II. Section 3 describes the proposed change detection algorithm with NSGA-II. In Sect. 4, we present the experimental settings and show the experimental results and analysis. Finally, concluding remarks are presented in Sect. 5.

2 Related Background

2.1 The Model of Change Detection for Remote Sensing Images

Change detection is to compare the remote sensing image of one time point with that of another time point in the same zone, so as to detect the differences between the images. For two observations of T1 and T2 in the same zone with different times, it can be expressed as a set:

$$I_1 : R^l \rightarrow R^p \tag{1}$$

$$I_2 : R^l \rightarrow R^q \tag{2}$$

where R^p and R^q can be regarded as a mapping of R^l from geographical space to feature space. Then change detection can be understood as a classification problem. From R^p and R^q, we can extract and describe features, and then measure and classify them. Finally, the image can be divided into two categories: the change class and the unchanged class. And the classification result map is generated as $B(\mathrm{x}) : R^l \rightarrow (0, 1)$:

$$B(x) = \begin{cases} 1, \text{changed} \\ 0, \text{unchanged} \end{cases} \tag{3}$$

At present, the process of change detection for remote sensing imagery is mainly divided into four steps: (1) Image preprocessing, (2) Extracting change information (generating DI), (3) Detecting changes (analyze DI), (4) Accuracy evaluation. The four steps are as follows:

1) Preprocessing. Image preprocessing is the registration and correction of two remote sensing images to achieve the requirement of pixel overlap at the same position of two images.
2) Generating DI. In order to find the differences of images, it is necessary to extract the target features and describe them, and then select some appropriate difference operation to generate a difference image with the same size as the original image so that we can extract useful change information.

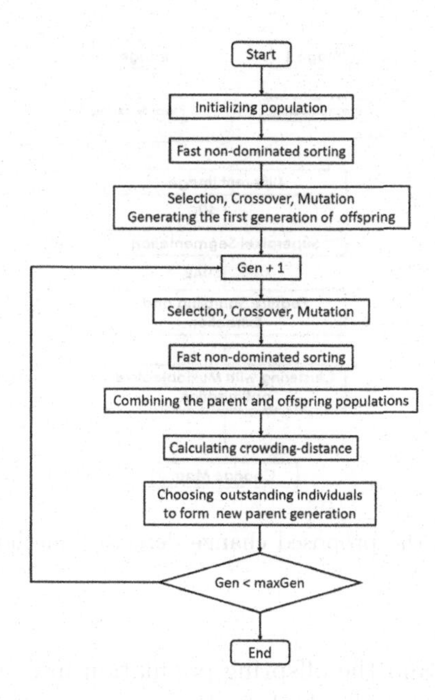

Fig. 1. Flow diagram of Non-dominated Sorting Genetic Algorithm.

3) Analyzing DI. After generating the difference image, supervised or unsupervised algorithm can be used to analyze it. The extracted features are divided into two categories: the changed and the unchanged. Finally, a binary change detection map can be generated.
4) Accuracy evaluation. Commonly used indexes will be detailed in Sect. 4.

2.2 Non-dominated Sorting Genetic Algorithm (NSGA-II)

The basic idea of NSGA-II is as follows: Firstly, an initial population of N is generated randomly. After non-dominated sorting, the first generation of offspring population is obtained through three basic operations of genetic algorithm: selection, crossover and mutation. Secondly, starting from the second generation, the parent population and the offspring population are merged to make a fast and non-dominated sorting. At the same time, the crowding-distance of the individuals in each non-dominated layer is calculated. According to the non-dominated relationship and the crowding-distance of individuals, the outstanding individuals are selected to form a new parent population. Finally, by the basic operation of genetic algorithm, a new generation population is generated. And so on until the termination criteria of the program is satisfied. The flow diagram of NSGA-II is shown in Fig. 1.

NSGA-II proposes a fast non-dominated sorting method, which reduces the complexity of the algorithm. And it proposes the concept of crowding-distance as a comparison standard in the same level, which ensures the diversity of the population effectively. Moreover, this algorithm uses elitism, which combines

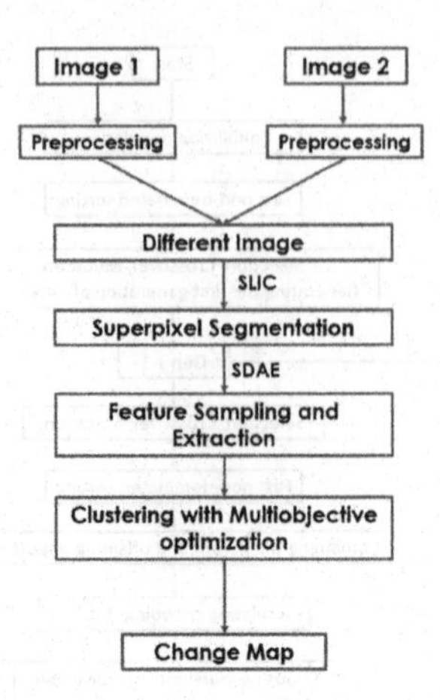

Fig. 2. Flow diagram of the proposed change detection method with multi-objective optimization.

the parent population and the offspring population into the next generation to ensure that the excellent individuals in the parent generation will not be lost. Therefore, NSGA-II has high computational efficiency and strong robustness.

3 Methodology

In this section, we will introduce a novel change detection method for SAR images based on features of superpixels which is optimized by NSGA-II in clustering. The specific flow of this method is shown in Fig. 2.

3.1 SLIC Segmentation Algorithm

Because SLIC algorithm has the lowest complexity and the highest execution efficiency, we use it to segment images into superpixels. In SLIC algorithm, the number of superpixels k or the expected side length S of superpixels should be given in advance. The specific implementation steps are as follows:

1. Initialization of the image segmentation blocks. Firstly, the image is divided into patches with the same size $S \times S$. If the number of pixels is N and the number of pre-generated patches is k, then there is a relationship: $S = \sqrt{\frac{N}{k}}$
2. Cluster centers initialization. In each patch, we need randomly sample a pixel as the initial clustering center. In order to avoid the cluster center located in the noise point or edge position, it is necessary to calculate the gradient of all pixels in the area 3×3 near the sampling point, and set the seed point at the position with the minimum gradient.
3. Calculation of the distance between each pixel and its cluster center. SLIC algorithm only needs to calculate the feature distance in the range of $2S \times 2S$

around each pixel to find the nearest cluster center. SLIC algorithm considers chromaticity distance d_c and spatial distance d_s. The final distance is D', and the formula is as follows:

$$\begin{aligned} d_c &= \sqrt{(l_j - l_i)^2 + (a_j - a_i)^2 + (b_j - b_i)^2} \\ d_s &= \sqrt{(x_j - x_i)^2 + (y_j - y_i)^2} \\ D' &= \sqrt{(\tfrac{d_c}{m})^2 + (\tfrac{d_s}{S})^2} \end{aligned} \tag{4}$$

where D′ calculates the final distance between the pixel and the nearest cluster center, where m is the weight of geometric information and S is the expected side length of superpixels.

4. Iterative optimization. According to the previous step, we can calculate the new cluster center. And continuing to iterate until it converges or reaches the maximum number of iterations.

3.2 Change Feature Sampling and Extraction

In our framework, we use SLIC algorithm to map the change metrics image from pixel feature space to superpixel feature space. Furthermore, these superpixels need to be utilized as the training samples so that they have to be transformed into patches with the same size. So the choice of the coding criterion of superpixels is essential for feature learning. Here we develope a criterion which is suitable for training samples in the neural network. These superpixels were stretched into homogeneous sample vectors in dimension n, where the value of n depends on the maximum number of pixels contained in each superpixel. The shape of the vectors influences the change detection result. Patches that have insufficient dimensions were performed with zero padding operation. From this, a set of homogenous samples containing different edge information were obtained as the training data of the network for feature learning to get the changed information of difference image.

Then we use stacked denoising autoencoders to extract deep features. The simple AE network contains only one hidden layer, and learns features of data only once. SDAE has $n_l - 1$ hidden layers so that it has $n_l - 1$ connection weight matrices $W_1, W_2, W_3, \ldots\ldots W_{n_l-1}$. After getting values of each layer, we use BP algorithm to learn the features of each DAE. In the training process, the front feedback operation is performed to get the output of each layer: $h_l, (l \in \{0, 1, 2, 3, \ldots, n_l\})$, where $h_0 = D_i$, $h_l = f(W_l h_{l-1} + b_l)$. And the output of SDAE is $o_i = h_{n_l} = f(W_{n_l} h_{n_l-1} + b_{n_l})$. Through back-propagation algorithm, we can fine-tune the parameters of each layer of the model to make the model converge further. The formula for minimizing the loss function is as follows:

$$\min_{\theta} \sum_{i=1}^{m} L(o_i, y_i) = \sum_{i=1}^{m} \frac{1}{2} \|o_i - y_i\|^2, \theta = \{W_1, b_1 \dots W_{nl}, b_{nl}\} \quad (5)$$

In BP algorithm, the update rules of weight matrix W_l and matrix of bias term b_l are as follows:

$$W_l = W_l - \eta \nabla W_l \quad (6)$$

$$b_l = b_l - \eta \nabla b_l \quad (7)$$

where η is learning rate, ∇W_l and ∇b_l are respectively expressed as the gradient of the loss function to W_l and b_l. And in the process of BP, we need to calculate the residual of each layer. The residual formula of each hidden layer is: $\delta_i^l = a_i^l(1 - a_i^l) \sum_{j=1}^{s_l+1} W_{ji}^l \delta_i^{l+1}$, where $s_l + 1$ represents the total number of nodes in the layer l, and a_i^l represents the output value of the ith unit in the layer l. By using the residual error and adjustment coefficient, SDAE can adjust the network parameters of each layer.

3.3 Clustering Method Optimized by NSGA-II

Clustering is a process of grouping samples according to certain similarity requirements. When clustering DI, we want to distinguish the changed class and the unchanged class, so the number K of components of classification is determined to be 2. Then the corresponding clustering form is as follows:

$$p(x) = \pi_1 \mathrm{N}(x | \mu_1, \Sigma_1) + \pi_2 \mathrm{N}(x | \mu_2, \Sigma_2) \quad (8)$$

where $\pi_1 + \pi_2 = 1$. In the above formula, there are six unknown parameters: $(\pi_1, \mu_1, \sum_1, \pi_2, \mu_2, \sum_2)$. Only by solving these six parameters can we determine which category each data point belongs to, so we need to estimate the parameter.

The most effective method of determining its parameter value is Maximum Likelihood Estimation. For data set X and parameter Θ, the maximum likelihood estimation P is:

$$P(X | \Pi, \Theta) = \prod_{i=1}^{N} f(x_i | \Pi, \Theta) = \prod_{i=1}^{N} (\sum_{j=1}^{K} \pi_j P(x_i | \Theta_j)) \quad (9)$$

where $\Pi = \{\pi_1, \pi_2, ..., \pi_k\}$. And since logarithm is monotonous, we can take logarithm of the above formula to obtain its logarithmic likelihood function:

$$L(\Pi, \Theta | X) = \sum_{i=1}^{N} \log(\sum_{j=1}^{K} \pi_j P(x_i | \Theta_j)) \quad (10)$$

By Maximum Likelihood Estimation, the maximum value of parameters Θ^* can be obtained so that all parameters used in clustering can be estimated accurately.

However, because this clustering method relies on the probability distribution of two categories, it is inevitable that two distributions could overlap in many cases, and it is difficult to determine which category the data points falling in the overlapping area should be divided into. Therefore, it is necessary to introduce a similarity measure to calculate the overlap between the two components and evaluate the separability between the two categories.

In this paper, Bhattacharyya distance between the probability distribution of two types of samples is selected as the similarity measurement standard. In statistics, Bhattacharyya distance is often used to calculate the separability of probability distribution of two samples. It is closely related to the Bhattacharyya coefficient which measures the overlap between two statistical samples or populations.

For two discrete probability distributions P and Q on the same domain x, the Bhattacharyya distance between them is:

$$D_B(p, q) = -\ln(BC(p, q)) \tag{11}$$

where BC is the Bhattacharyya coefficient, and its definition is $BC(p, q) = \sum_{x \in X} \sqrt{p(x)q(x)}$.

Based on this, on the one hand, we want to maximize the logarithmic likelihood function for estimating parameters, and on the other hand, we want to minimize the distance between the two samples for increasing separability. Therefore, we model this clustering problem as a multiobjective optimization problem. Here we introduce a multi-objective optimization genetic algorithm named NSGA-II. We set up two optimization objectives, one is the reciprocal of logarithmic likelihood function, the other is the Bhattacharyya distance between the two categories, so the problem is transformed into two minimization problems. Pareto optimal solutions are accordingly generated through NSGA-II. Among all the Pareto optimal solutions obtained, each solution is put into clustering to get a detection map. By comparing all the accuracy evaluation indexes, the map which has generally the best performance is the final change detection map we need.

4 Experimental Study

4.1 Experimental Datasets and Evaluation Criteria

The real SAR imagery dataset we use in our experiment is the Ottawa dataset. It represents a section (290 × 350 pixels) of two SAR images over the city of Ottawa. The images are captured in May 1997 and August 1997, respectively, by the RADARSAT SAR sensor. The dataset was provided by Defense Research and Development Canada, Ottawa. The two images and corresponding available ground truth are shown in Fig. 3.

In order to verify the effectiveness of the proposed algorithm, we also experimented two classical algorithms of change detection as comparison. Fuzzy c-means clustering algorithm (FCM) is a classical clustering method based on the

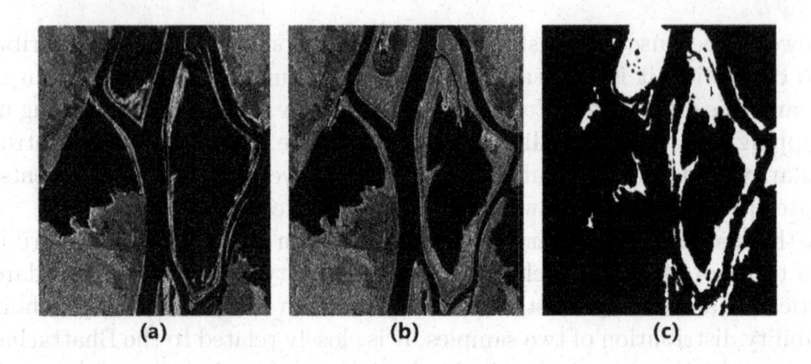

Fig. 3. Ottawa dataset. (a) Image acquired in May 1997. (b) Image acquired in August 1997. (c) Ground truth change map created by on-the-spot investigation and expert knowledge.

optimization of fuzzy objective function, which often achieve accurate clustering results. FCM_S is a modified method of FCM proposed by Ahmed et al. which is standard by introducing the classification of directly adjacent pixels. By adding spatial information to the 8 neighborhood points of pixels, FCM_S improves the robustness of noisy images.

It is necessary to use some criteria to assess the performance of different methods, and the quantitative analysis of change detection results is set as follows. First, false negative (FN) and false positive (FP) are calculated. FN means the number of changed pixels which are undetected and FP means the number of unchanged pixels which are wrongly classified as changed ones. TP and TN are the number of true changed pixels and unchanged pixels. Second, the overall error (OE) can be computed by $OE = (FN + FP)/N$, where N means the total number of pixels. Third, we can calculate the percentage correct classification (PCC). It is given by $PCC = (\text{TP} + \text{TN})/(\text{TP} + \text{FP} + \text{TN} + \text{FN})$. Finally, kappa coefficient (KC) is usually applied to evaluate the effect of classification. The higher the value of KC is, the higher the classification accuracy is. KC is calculated as follows:

$$\text{KC} = \frac{PCC - PRE}{1 - PRE} \tag{12}$$

where $PRE = \frac{(FN+TP)\cdot(FP+TP)}{(TP+FN+TN+FP)^2} + \frac{(FN+TN)\cdot(FP+TN)}{(TP+FN+TN+FP)^2}$.

4.2 Experimental Results

For the real SAR image dataset, we follow the proposed algorithm. First, DI is generated by log-ratio operator, and then superpixel segmentation is performed by SLIC algorithm. After these superpixels are uniformly coded, they are input into sdae as training samples for feature extraction. Finally, a set of Pareto optimal solutions are obtained by using the improved clustering algorithm.

When using SDAE network to extract features, the dimensions of input layer in the network were decided by the size of the feature vectors, which can influence the reconstruction results to a certain extent. Thus, in order to get the suitable

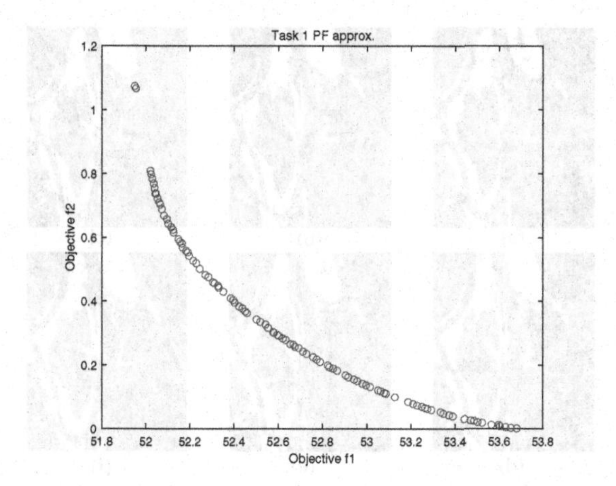

Fig. 4. The pareto front of two objective optimization of the Ottawa dataset.

size, different sizes of samples were tested. Here, we need to manually set the expected side length S and the compactness m of superpixels. Based on our experimental results, Setting the value of S to 5 and the value of m to 10 is the best choice. So, the size of homogeneous neighbor vector is set to 25.

After change features are extracted by SDAE, NSGA-II algorithm is used to optimize the clustering algorithm. Meanwhile, the experiment parameters in optimized clustering algorithm are displayed in Table 1. After applying our optimization algorithm to the Ottawa dataset, we get a uniformly-distributed PF including one hundred Pareto optimal solutions with different effects of detail preserving and noise removing. The Pareto Front of Ottawa dataset is shown in Fig. 4. As shown in Fig. 4, the objective function F1 is the reciprocal of logarithmic likelihood function of clustering algorithm which we have magnified 10^6 times, while the objective function F2 is the Bhattacharyya distance between two categories. From the one hundred Pareto optimal solutions on the PF, we can choose a compromise solution between improving the accuracy of parameter estimation and enhancing the separability of samples.

Table 1. The parameters in optimized clustering algorithm

Parameter	Meaning	Value
gen	The maximun of generations	100
pc	The probability of crossover	0.5
pm	The probability of mutation	0.01
pop	The population size	100

From the PF, we select six points and display their corresponding final change detection maps in Fig. 5. By comparing the six images in Fig. 5, we can find the fact that Fig. 5(a) and (b) contain less noise than the other four maps but fail to

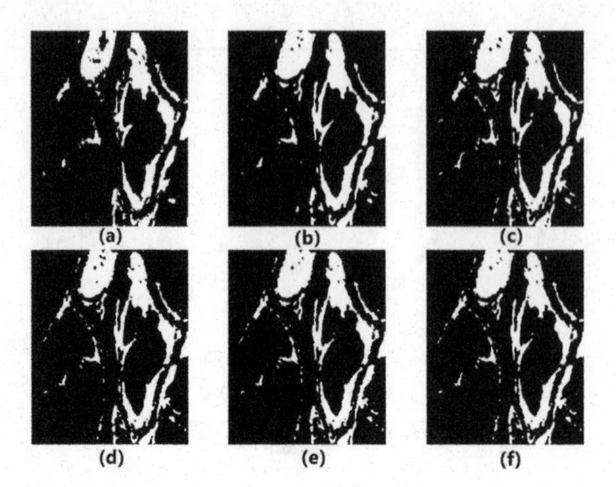

Fig. 5. Six different change detection maps selected from all the 100 results of the Ottawa dataset.

Table 2. The results obtained by three methods

Methods	FN	FP	OE	PCC	KC
FCM	3742	1089	4.76%	95.24%	0.8185
FCM_S	2715	1203	3.86%	96.14%	0.8797
Proposed method	2070	1185	3.21%	96.79%	0.8822

detect some crucial changed regions, which proves that the choice of the scheme of log likelihood function and Bhattacharyya distance has a certain impact on the result of change detection in the process of clustering optimization. Finally, we select the map which has the best result of comprehensive index among the one hundred maps as the final change detection map.

In order to demonstrate the availability of the proposed method in depth, we calculate the average values of the best PCC and KC over 30 times running, and the results are recorded in Table 2. For comparison, the results of FCM and FCM_S on the same datasets are also listed in Table 2. By comparing the values of FN, FP, OE, PCC and KC obtained by the proposed method with those of FCM and FCM_S, we can conclude that our method can obtain better results.

5 Conclusion

In this paper, we have presented a novel change detection method for SAR images, which uses neutral network to extract superpixel-based change features and an improved clustering algorithm which is optimized by Non-dominated Sorting Genetic Algorithm. In the proposed method, SLIC segmentation algorithm is used to segment the difference image into superpixels. The segmentation results are encoded uniformly according to coding criteria to be input into SDAE network for feature extraction, and then the extracted feature images are clustered by the improved clustering algorithm. When using clustering algorithm, we select NSGA-II algorithm to improve the effect of clustering. In order to enhance the separability between the two types of samples, we select the Bhattacharyya distance between two categories as the similarity measurement standard. And the logarithmic likelihood function of clustering algorithm and the Bhattacharyya distance between the two categories are optimized by NSGA-II algorithm at the same time, so as to obtain a set of Pareto optimal solutions which satisfy the maximum of logarithmic likelihood function and the maximum separability between the two categories. Among all the Pareto optimal solutions, each solution is put into clustering to get a detection map. By comparing all the indexes for accuracy evaluation, the map which has the best result of comprehensive index is the final change detection map.

In order to evaluate the new algorithm, real data sets are employed in experiments. And two classical methods are our compared algorithms. The experimental results on real SAR data sets have demonstrated its effectiveness and reliability of the proposed method on change detection for SAR images.

References

1. Lu, D., Mausel, P., Brondizio, E., Moran, E.: Change detection techniques. Int. J. Remote Sens. **25**(12), 2365–2401 (2004)
2. Chen, Y., Zhang, R., Yi, D.: Multi-polarimetric SAR image compression based on sparse representation and super-resolution. In: 2012 International Conference on Audio, Language and Image Processing, pp. 705–709. IEEE (2012)
3. Lee, J.-S., Pottier, E.: A new statistical model for Markovian classification of urban areas in high-resolution SAR images. CRC Press (2017)
4. Hazel, G.G.: Object-level change detection in spectral imagery. IEEE Trans. Geosci. Remote Sens. **39**(3), 553–561 (2001)
5. Hussain, M., Chen, D., Cheng, A., Wei, H., Stanley, D.: Change detection from remotely sensed images: from pixel-based to object-based approaches. ISPRS J. Photogramm. Remote. Sens. **80**, 91–106 (2013)
6. Chen, G., Hay, G.J., Carvalho, L.M., Wulde, M.A.: Object-based change detection. Int. J. Remote Sens. **33**(14), 4434–4457 (2012)
7. Gong, M., Zhan, T., Zhang, P., Miao, Q.: Superpixel-based difference representation learning for change detection in multispectral remote sensing images. IEEE Trans. Geosci. Remote Sens. **55**(5), 2658–2673 (2017)
8. Castelluccio, M., Poggi, G., Sansone, C.: Land use classification in remote sensing images by convolutional neural networks. arXiv preprint arXiv:1508.00092 (2015). IEEE Trans. Geosci. Remote Sens
9. Chen, Y., Lin, Z., Zhao, X., Wang, G., Gu, Y.: Deep learning-based classification of hyperspectral data. IEEE J. Sel. Top. Appl. Earth Obs. Remote Sens. **7**(6), 2094–2107 (2014)
10. Li, T., Zhang, J., Zhang, Y.: Classification of hyperspectral image based on deep belief networks. In: 2014 IEEE International Conference on Image Processing (ICIP), pp. 5132–5136 (2014)

Multi-objective Emergency Resource Dispatch Based on Coevolutionary Multiswarm Particle Swarm Optimization

Si-Chen Liu[1], Chunhua Chen[2], Zhi-Hui Zhan[1](✉), and Jun Zhang[3]

[1] School of Computer Science and Engineering, South China University of Technology, Guangzhou 51006, China
[2] School of Software Engineering, South China University of Technology, Guangzhou 51006, China
[3] Victoria University, Melbourne, VIC 8001, Australia

Abstract. Emergency resource dispatch plays a significant role in the occurrence of emergency events. An efficient schedule solution not only can deliver the required resources in time, but also can reduce the loss in the disaster area. Recently, many scholars are dedicated to dealing with emergency resource dispatch problem (ERDP) by constructing a model with one objective (e.g., the cost to transport resources or the satisfaction degree of people in the disaster area) and solving the model with single objective optimization algorithms. In this paper, we build a multi-objective model that considers both the cost objective and the satisfaction objective, which takes into account multiple retrieval depots and multiple kinds of resources. We propose to solve this multi-objective ERDP optimization model via the recently famous coevolutionary multiswarm particle swarm optimization (CMPSO) algorithm. Based on multiple populations for multiple objectives (MPMO) framework, the CMPSO algorithm uses two populations to optimize the above two objectives respectively, and leads particles to find Pareto optimal solutions by storing information of different populations in a shared archive. We construct ERDP with various scales to validate the feasibility of the applied CMPSO algorithm. Moreover, by setting the satisfaction objective as the constraint, we also compare the results obtained by CMPSO with those obtained by constrained single objective particle swarm optimization (PSO) algorithm. Experimental results show that: 1) the nondominated solutions obtained by CMPSO perform well in both convergence and diversity on two objectives; 2) the results on the cost objective obtained by CMPSO are generally superior to those of PSO under same degree of satisfaction.

Keywords: Emergency resource dispatch · Multi-objective optimization · Particle swarm optimization

1 Introduction

When faced with severe disasters, it is essential to get reasonable schedule solutions to deliver required resources to the disaster area in time [1], for reducing the loss of

H. Ishibuchi et al. (Eds.): EMO 2021, LNCS 12654, pp. 746–758, 2021.
https://doi.org/10.1007/978-3-030-72062-9_59

human lives and properties. Compared with normal resource dispatch problem, emergency resource dispatch places more emphasis on time and efficiency [2–4]. That is, we need to pay more attention on both time and efficiency to get appropriate schedule solutions finally.

Evolutionary algorithms (EAs) are possessed of self-adaptive [5], self-organized [6], and self-learning [7], that can efficiently deal with complex problem while traditional methods cannot solve [8]. Especially, EAs have been successfully applied in lots of scheduling problems. For example, Chen *et al.* [9] put forward an ant colony system approach for cloud computing resources scheduling. Liu *et al.* [10] proposed a genetic algorithm-based simultaneous scheduling technique to solve flexible job shop scheduling problem. Shi *et al.* [11] applied the differential evolution algorithm for task scheduling problem. The ant colony optimization algorithm has also shown its ability in solving cloud workflow schedule problem [12].

Recently, many researchers have addressed emergency resource dispatch problem (ERDP) by using different methods which can be classified into two categories, evolutionary methods and non-evolutionary methods. In the first category, Liu *et al.* [13] built a model to minimize the number of retrieval depots under the premise of earliest response time. Aiming at minimizing travel time, Li *et al.* [14] adopted an improved simulated annealing algorithm for solving. Gao *et al.* [15] constructed a model from the aspects of transport time, cost, and resource demand satisfaction rate, and applied a single objective optimization algorithm with the linear weight combination method. Xiong *et al.* [16] put forward a model with two objectives to minimize the transport time and the number of rescue points. Although [15] and [16] built a multi-objective model, they did not directly apply multi-objective optimization algorithm to solve. Besides, the linear weight combination method adopted in [15] requires extra cost of parameters setting, and need to run many times to get a variety of solutions. While in the second category, Tang and Sun [17] established two optimal models under railway emergency rescue situation and used a voting analytic hierarchy process to solve the models. Wu *et al.* [18] investigated a bi-objective rescue vehicle scheduling problem for multi-point forest fires and constructed the model with optimizing total fire-extinguishing time and the number of used vehicles, then the authors developed an integer program technique, a dynamic programming algorithm, and a fast greedy heuristic method to address the problem.

In this paper, we build a multi-objective model and adopt a coevolutionary multi-swarm particle swarm optimization (CMPSO) [19] algorithm to solve the ERDP. When dispatching the emergency resources, optimizing the total cost for delivering and the satisfaction degree of the people in the disaster area simultaneously are necessarily required from the aspects of economy and efficiency. Thus, our model considers both cost objective and satisfaction objective. Moreover, the multi-objective ERDP model takes into account multiple retrieval depots and multiple kinds of resources. The cost objective measures the total cost from all retrieval depots to disaster area in two terms. The first term is the maximum time cost for delivering resources from different retrieval depots, and the second term takes the consumption of manpower during transportation into account. The satisfaction objective represents the subsidized degree of people in the disaster area which is positively correlated with the number of transported resources.

Apparently, the cost objective is in contraction with the satisfaction objective, and therefore, we want to optimize and balance these two objectives. Since CMPSO is an efficient algorithm in solving multi-objective problem, herein, we apply CMPSO to solve the multi-objective ERDP.

The contributions of this paper are summarized as follows.

1) Different from most existing works that model the ERDP with one or two objectives and solve the model just via a single objective optimization algorithm, in this paper, we comprehensively consider the ERDP and construct a multi-objective model with extended CMPSO algorithm for solving.
2) We design two kinds of experiments to test the capability of CMPSO in addressing multi-objective ERDP. By comparing with two state-of-the-art multi-objective EAs (MOEAs) and constrained particle swarm optimization (PSO) algorithm, the CMPSO algorithm shows its advantage for dealing with multi-objective real-world application.

The rest of the paper is organized as follows. Section 2 describes the multi-objective ERDP model. The CMPSO algorithm is completely introduced in Sect. 3. Section 4 provides the experiments and results analysis. Finally, the conclusion and future work are given in Sect. 5.

2 Model Description

2.1 Notations Specification

The multi-objective ERDP model takes into account multiple retrieval depots, multiple kinds of resources, and one disaster area. Suppose K resources are required in the disaster area denoted by $T_1, T_2,\ldots, T_K$, and there are R retrieval depots around the disaster area denoted by $P_1, P_2,\ldots, P_R$. Given the number of demand resources in the disaster area as $Demand_1, Demand_2,\ldots, Demand_K$, the number of resources in each retrieval depot as $Obtain_{ij}$ ($1 \leq i \leq R$, $1 \leq j \leq K$), and the distances between each retrieval depot and disaster area as $Dist_1, Dist_2,\ldots, Dist_R$. We use an array $X[R][K]$ to represent a schedule solution where $X[i][j]$ is the number of jth resource transported by i^{th} retrieval depot. The array $supply[K]$ is defined as the total number of jth resource provided by all retrieval depots in a schedule solution, and the array $select[R]$ is used to indicate whether i^{th} retrieval depot will transport resource to the disaster area or not, as shown in (1) and (2).

$$supply[j] = \sum_{i=1}^{R} X[i][j] \quad 1 \leq j \leq K \tag{1}$$

$$select[i] = \begin{cases} 1, & \text{if } i^{\text{th}} \text{ retrieval depot transports resources to the disaster area} \\ 0, & \text{otherwise} \end{cases} \quad 1 \leq i \leq R \tag{2}$$

2.2 Assumptions

In order to build the model reasonably, we make the following assumptions.

Assumption 1: assume the speeds of delivering resources from different retrieval depots to disaster area are the same and do not change;
Assumption 2: assume the weight of each resource is in the same numerical level;
Assumption 3: assume the number of required resources doesnot change over time.

2.3 Model Construction

Emergency resource dispatch solution is related to time, cost, and satisfaction degree of people in the disaster area. In this paper, we model ERDP as a multi-objective optimization problem.

The first objective is the cost objective that consists of time cost and fee cost, as shown in (3–5). Time cost is expressed as the longest time required to deliver resources and can be determined by the distance from the retrieval depot to the disaster area according to assumption 1. Fee cost is affected by drain on manpower to transport resources. More transported resources and longer transportation distance will cause greater consumption on manpower. Based on assumption 2, different kinds of resources with same quantity will consume same amount of manpower. Therefore, the cost objective is positively correlated with the number of transported resources and the distances between retrieval depot and disaster area.

$$obj_1 = time\_cost + fee\_cost \tag{3}$$

$$time\_cost = \max_{1 \le i \le R} \{select[i] \times Dist_i\} \tag{4}$$

$$fee\_cost = \sum_{i=1}^{R} \sum_{j=1}^{K} select[i] \times X[i][j] \times Dist_i \tag{5}$$

The second objective is the satisfaction objective which refers to the adequacy of resources in the disaster area. According to assumption 3, the demands in the disaster area are known and fixed, thus the satisfaction degree of people in the disaster area will become higher with the increase of the provided resources. In order to minimize the objective, the satisfaction objective (6) is rewritten as (7).

$$satisfaction = \sum_{j=1}^{K} \frac{supply[j]}{Demand_j} \tag{6}$$

$$obj_2 = \sum_{j=1}^{K} \frac{Demand_j}{supply[j] + 1.0} \tag{7}$$

It is mentioned that in order to avoid *supply*[*j*] equals zero and cannot be divided, the denominator is added by one in (7).

In conclusion, the complete model of ERDP is shown as

$$\text{minimize}\begin{cases} obj_1 \\ obj_2 \end{cases}$$
$$\text{s.t.} \sum_{i=1}^{R} Obtain_{ij} \geq Demand_j \quad 1 \leq j \leq K \tag{8}$$

where the inequality constraint ensures that the total resources in all retrieval depots satisfy all demands in the disaster area.

3 CMPSO Approach

Since we build a multi-objective ERDP model where the cost objective has conflict with the satisfaction objective, we propose to apply a multi-objective optimization algorithm to solve the model. In this paper, we adopt the recently famous CMPSO [19] to address ERDP, which has shown the availability in dealing with multi-objective applications [12, 20] and many-objective optimization problems [21].

3.1 Traditional PSO

PSO is a swarm intelligence [22] algorithm developed by Kennedy and Eberhart [23] which used to solve single objective optimization problem. In PSO, each particle *in* in the population owns a position vector $x_{in} = [x_{in,1}, x_{in,2},\ldots, x_{in,D}]$ and a velocity vector $v_{in} = [v_{in,1}, v_{in,2},\ldots, v_{in,D}]$ with D dimensions. Besides, each particle *in* has an array $pBest_{in} = [pBest_{in,1}, pBest_{in,2},\ldots, pBest_{in,D}]$ to record the individually historically optimal solution. The best solution among all the *pBest* is called population historically optimal solution and is denoted by $gBest = [gBest_1, gBest_2,\ldots, gBest_D]$. In the first generation, the position is randomly initialized within $[x_{l,d}, x_{u,d}]$, where $x_{l,d}$ and $x_{u,d}$ are the lower bound and upper bound of dth dimension in the searching space respectively, and the velocity is randomly initialized in the range of $[-0.2 \times (x_{u,d} - x_{l,d}), 0.2 \times (x_{u,d} - x_{l,d})]$. After initialization, $pBest_{in}$ is assigned as the initial position x_{in} and *gBest* is set as the best solution in the initialized population. During the evolution, each particle updates its velocity and position according to

$$v_{in,d} = \omega \times v_{in,d} + c_1 \times r_{1d} \times (pBest_{in,d} - x_{in,d}) + c_2 \times r_{2d} \times (gBest_d - x_{in,d}) \tag{9}$$

$$x_{in,d} = x_{in,d} + v_{in,d} \tag{10}$$

where ω is an inertia weight which equals 0.9 at first, and linearly decreases to 0.4 in the end of evolution. c_1 and c_2 are acceleration coefficients and are suggested to set as 2.0 [23]. r_{1d} and r_{2d} are two random real numbers within [0, 1]. Note that when $x_{in,d}$ or $v_{in,d}$ exceeds the bound, it will be corrected to the corresponding boundary value.

After updating velocity and position of all particles, these newly generated particles are evaluated. If a new particle x_p gets a better objective value, the *pBest* of x_p will be updated as x_p. In the same way, if the best particle x_g among all updated *pBest* performs better than before, *gBest* will be updated as x_g.

The updating operation in velocity, position, *pBest*, and *gBest* will be iterated in turn. When meeting the termination condition, PSO algorithm finishes and *gBest* is output as the final solution.

3.2 Multi-objective Optimization Problems (MOPs)

Without loss of generality, a minimization MOP can be illustrated as follows

$$\begin{aligned} &\text{minimize } F(x) = (f_1(x), f_2(x), \ldots, f_M(x))^T \\ &\text{s.t. } x \in R^D \end{aligned} \tag{11}$$

where $x = [x_1, x_2, \ldots, x_D]$ is a solution with D dimensions in the searching space, M is the number of objectives, and $F(x) = [f_1(x), f_2(x), \ldots, f_M(x)]$ is the objective vector of a given solution x in the objective space. In MOPs (e.g., $M = 2$), the objectives always conflict with each other, that is, if a solution improves in one objective, it will degrade in another objective. Some basic definitions of minimization MOPs are given in the following.

Definition 1: *Pareto dominance.*
Assume x_a and x_b are two solutions in the searching space, if x_a and x_b satisfy (12) and (13), then we say x_a dominates x_b, or x_b is dominated by x_a.

$$\exists o \in [1, M] f_o(x_a) < f_o(x_b) \tag{12}$$

$$\forall o \in [1, M] f_o(x_a) \le f_o(x_b) \tag{13}$$

Definition 2: *Pareto optimal solution.*
A solution $U = [U_1, U_2, \ldots, U_D]$ is called a Pareto optimal solution if and only if U is not dominated by any solutions in the population.

Definition 3: *Pareto set.*
Pareto set (PS) consists of all Pareto optimal solutions, which is expressed as

$$PS = \left\{ x \in R^D | x \text{ is a Pareto optimal solution} \right\} \tag{14}$$

Definition 4: *Pareto front.*
Pareto front (PF) is the set of objective values of all the solutions in PS, which is defined as

$$PF = \{F(x) | x \in PS\} \tag{15}$$

3.3 CMPSO Algorithm

CMPSO is a variant algorithm of PSO which focuses on solving multi-objective optimization problem. The main idea of CMPSO is using M populations to optimize M objectives separately. Besides, CMPSO develops an archive technique to store extra information to guide the evolution of M populations. In this paper, we propose to apply CMPSO algorithm to search for PS in ERDP.

Solution Encoding. As mentioned in Sect. 2.1, $X[i][j]$ represents the number of jth resource transported by i^{th} retrieval depot. Herein, we use the number of delivered resources to encode the solutions. In particular, the value of each dimension in X is set to real number, which aims at retaining the characteristics of particles during the evolution. When calculating the two objectives, the value of each dimension in X will be rounded.

Solution Updating. Similar with traditional PSO, the position and velocity of each particle *in* in m^{th} population are randomly initialized in the searching space in CMPSO. After that, $pBest_{in}^{m}$ is assigned by the initial position x_{in}^{m} and $gBest^{m}$ is set as the best solution in the mth objective. During the evolution, the velocity and position of each particle *in* are updated as

$$\begin{aligned} v_{in,d}^{m} = {} & \omega \times v_{in,d}^{m} + c_1 \times r_{1d} \times (pBest_{in,d}^{m} - x_{in,d}^{m}) + c_2 \times r_{2d} \times (gBest_{d}^{m} - x_{in,d}^{m}) \\ & + c_3 \times r_{3d} \times (A_{ra,d}^{m} - x_{in,d}^{m}) \end{aligned} \tag{16}$$

$$x_{in,d}^{m} = x_{in,d}^{m} + v_{in,d}^{m} \tag{17}$$

where the update operation in velocity is influenced by three parts: individually historically optimal solution, population historically optimal solution, and a random solution *ra* in the *Archive*. The settings of ω, r_{1d}, and r_{2d} are the same as (9). c_1, c_2, and c_3 are acceleration coefficients and set to 4.0/3. r_{3d} is same as r_{1d} and r_{2d}. Note that the updating rule of *Archive* is described in the next section.

Under the MPMO framework, each population focuses on minimizing one objective, therefore, it is designed to update $pBest_{in}^{m}$ and $gBest^{m}$ when getting a better performance on the optimized objective. It is worth mentioning that the solution update rule in CMPSO is easier than other MOEAs, since the latter aims at optimizing all the objectives in one population, thus they cannot distinguish the solutions based on only one objective.

Initialization and Updating Rule of Archive. CMPSO defines a candidate set to store potential solutions and an *Archive* to record useful solutions. In the first stage, *Archive* and candidate set are empty. After that, all $pBest_{in}^{m}$ and solutions in the *Archive* are added to the candidate set. Moreover, the candidate set also contains all the disturbed solutions [19] in the *Archive*. Then, the dominated solutions in the candidate set will be deleted. If the size of candidate set does not exceed the predefined size *NA*, *Archive* is updated by all the solutions in the candidate set. Otherwise, the best *NA* solutions with higher crowding degree will be selected into *Archive* according to

$$density[in] = \sum_{m=1}^{M} \frac{f_{in+1}^{m} - f_{in-1}^{m}}{f_{max}^{m} - f_{min}^{m}} \tag{18}$$

where each objective is sorted in an ascending order, f_{max}^{m} and f_{min}^{m} represent the maximum value and minimum value in m^{th} objective respectively. For more details of computing crowding degree, please refer to [24].

3.4 Post Processing of CMPSO

Since the final solutions of CMPSO are composed of PS, and only one solution will be implemented in reality, we design a post processing method to reduce the number of PS, and provide decision-makers with post processed schedule solutions as the reference.

The main steps of post processing method are described as following five steps.

Step 1: Normalizing the PF obtained by CMPSO;
Step 2: The *k*-means clustering method [25] is carried out on the normalized objectives to classify the solutions into several clusters, here we define the number of class as five (i.e., $k = 5$);
Step 3: Select five solutions nearest to each cluster center, termed schedule solutions;
Step 4: Adjust each dimension in the schedule solutions to the integer that nearest to the multiple of 5. After adjustment, the schedule solutions are called post processed schedule solutions. For example, if $X[0][2]$ in one schedule solution equals 1048.2, then it will be changed to 1050 since 1050 is the closest to 1048.2.
Step 5: Provide the post processed schedule solutions to decision-makers.

Figure 1 shows the flowchart of CMPSO approach to schedule ERDP.

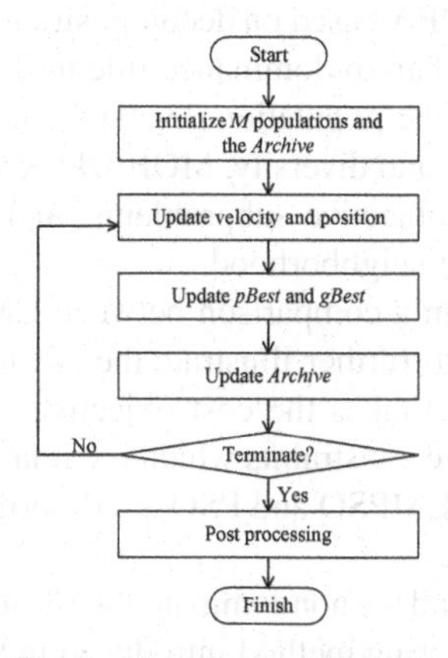

Fig. 1. Flowchart of CMPSO.

4 Experimental Results

4.1 Experimental Design

Data Design. Experiments are conducted on three types of test data with different scales. Case 1: 10 retrieval depots, 4 resources. Case 2: 25 retrieval depots, 5 resources. Case 3: 50 retrieval depots, 10 resources. Test data in [16] is adopted for case 1, while test data in case 2 and case 3 are generated in a random way. Due to the space limitation, only the $Obtain_{ij}$ ($1 \le i \le R$, $1 \le j \le K$) in case 1 is listed, as shown in Table 1. All the test data are available on https://github.com/sc621/TestData.

Table 1. The number of 4 kinds of resources obtained in 10 retrieval depots.

Retrieval depots	Resource T_1	Resource T_2	Resource T_3	Resource T_4
P_1	350	60	3000	700
P_2	360	70	2500	650
P_3	350	75	4000	600
P_4	370	80	4000	550
P_5	300	60	4000	550
P_6	300	70	2000	600
P_7	500	100	6000	1000
P_8	350	60	2000	550
P_9	400	80	3500	500
P_{10}	400	70	3000	600

Competitor Algorithms. In order to discuss the characteristic of CMPSO in solving ERDP, we adopt two state-of-the-art MOEAs called nondominated sorting genetic algorithm (NSGA-II) [24] and MOEA based on decomposition (MOEA/D) [26] for comparison. NSGA-II put forward a Pareto dominance rule to distinguish solutions, a crowding degree formula to ensure the population diversity, and an elite selection technique for balancing the convergence and diversity. MOEA/D decomposed the multi-objective problem into a set of single objective subproblems, and optimized each subproblem through the information of the neighborhood.

In addition, we also design a comparison between CMPSO and a single objective optimization algorithm PSO to further illustrate the advantages of CMPSO in dealing with MOPs. Specifically, PSO takes the cost objective (obj_1) as the single objective and optimizes obj_1 under some constraints which are related to obj_2. The main steps of comparison method between CMPSO and PSO are described as following three steps.

Step 1: Applying CMPSO to find the nondominated solutions, and five schedule solutions are obtained by k-means clustering method introduced in Sect. 3.4;
Step 2: PSO is used to minimize obj_1 and take obj_2 of the schedule solution got by Step 1 as the constraint. Note that only one schedule solution obtained by CMPSO is adopted in each comparative experiment. During the evolution in PSO, the obj_2 of all particles must be smaller than or equal the obj_2 of this adopted schedule solution. The particle with minimum obj_1 is selected as the best particle when PSO finishes; Also note that PSO independently runs 20 times for each comparative experiment, and the average minimum obj_1 is computed as the final result of PSO.
Step 3: Compare obj_1 between the schedule solution obtained by CMPSO and the final result obtained by PSO under the condition that obj_2 is fixed.

Parameters Setting. The parameters setting of CMPSO and PSO are presented in Table 2, while the parameters in NSGA-II and MOEA/D are set following their original papers. Besides, the lower bound $x_l^{i,j}$ and upper bound $x_u^{i,j}$ in all tested algorithms are set as 0 and $Obtain_{ij}$ ($1 \le i \le R$, $1 \le j \le K$) respectively and the algorithms are finished when the running time for evolution is more than 1 s.

Table 2. Parameters setting in CMPSO and PSO.

Algorithms	Population size	ω	c_1	c_2	c_3	r_{1d}	r_{2d}	r_{3d}	*NA*
CMPSO	50 × 2 (100)	0.9–0.4	4.0/3			[0, 1]			100
PSO	100	0.9–0.4	2.0		–	[0, 1]		–	–

4.2 Experimental Analysis

Comparisons Among All the Tested Algorithms. The PFs obtained by all the tested algorithms are plot in Fig. 2. It can be seen from Fig. 2 that NSGA-II can only obtain a subregion of PF, and the solutions obtained by MOEA/D are gathered in the objective space and dominated by the solutions obtained by CMPSO. Although some solutions obtained by CMPSO are dominated by the solutions obtained by NSGA-II (i.e., in case 3), CMPSO shows the capability to efficiently search for well-distributed Pareto optimal solutions and achieves the better balance between convergence and diversity. Under the MPMO framework, each population is able to search widely in its own objective in CMPSO, yielding the broader PF. Moreover, the consistency of experimental results in Fig. 2(a), Fig. 2(b), and Fig. 2(c) shows the stability and robustness of CMPSO.

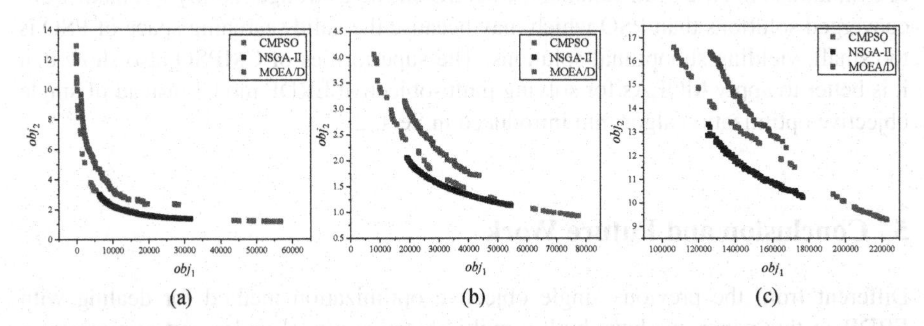

Fig. 2. PFs obtained by all the tested algorithms in (a) case 1 (b) case 2, (c) case 3.

Comparisons Between CMPSO and PSO. In order to analyze the convergence of CMPSO, we compare the schedule solutions in CMPSO with the best particles in PSO. Figure 3 illustrates the clustering results of CMPSO, while the comparisons results between CMPSO and PSO are shown in Fig. 4.

In Fig. 3, the points with same color belong to the same class, and each black point is the solution that is the nearest to clustering center and is defined as a schedule solution. Figure 4 shows that the obj_1 of the schedule solutions got by CMPSO surpass the best

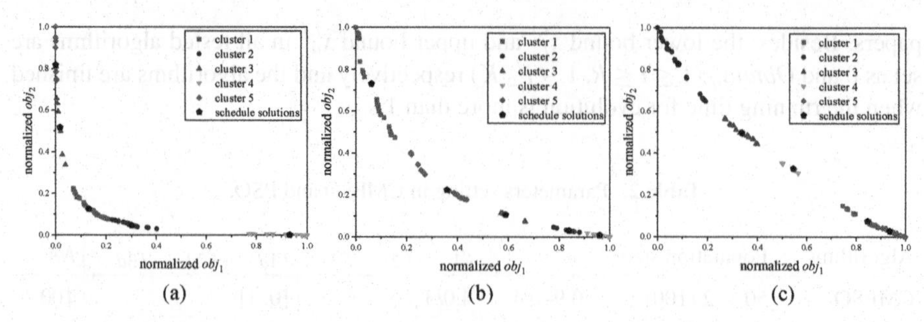

Fig. 3. Clustering results of the normalized PF in (a) case 1, (b) case 2, (c) case 3.

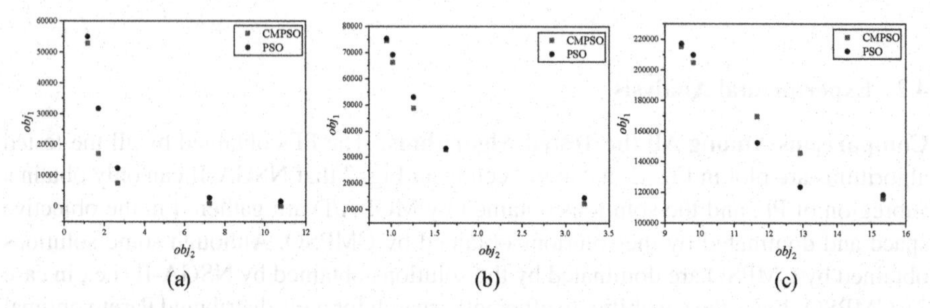

Fig. 4. Comparisons between CMPSO and PSO in (a) case 1, (b) case 2, (c) case 3.

particles obtained by PSO under the condition that obj_2 is fixed. Specifically, CMPSO is better than PSO in all 5 situations in both case 1 and case 2, and beats PSO in 3 out of 5 situations in case 3. In summary, CMPSO shows a stronger ability to obtain well-converged solutions than PSO which may because the valid searching space of PSO is too small, yielding suboptimal solutions. The superiorities of CMPSO also show that it is better to apply MOEAs for solving multi-objective ERDP model instead of single objective optimization algorithm introduced in Sect. 2.

5 Conclusion and Future Work

Different from the previous single objective optimization method for dealing with ERDP, in this paper, we have built a multi-objective model under cost-minimization and satisfaction-maximization framework. This model is solved by an efficient multi-objective optimization algorithm CMPSO. Experimental results on three types of test data with different scales show that CMPSO algorithm is able to obtain well-converged and widely-distributed Pareto optimal solutions to ERDP. When satisfaction objective is fixed, the overall performance of the schedule solutions on the cost objective obtained by CMPSO is also generally better than those obtained by the single objective optimization algorithm PSO. In the future work, ERDP will be extended to more complex model and we will propose a new algorithm under MPMO framework to solve it.

Acknowledgments. This work was supported in part by the National Key Research and Development Program of China under Grant 2019YFB2102102, in part by the Outstanding Youth Science Foundation under Grant 61822602, in part by the National Natural Science Foundations of China (NSFC) under Grant 61772207 and Grant 61873097, in part by the Key-Area Research and Development of Guangdong Province under Grant 2020B010166002, in part by the Guangdong Natural Science Foundation Research Team under Grant 2018B030312003, and in part by the Guangdong-Hong Kong Joint Innovation Platform under Grant 2018B050502006.

References

1. Liu, M., Qian, X., Liu, Z., Guo, C., Li, J.: Emergency resource optimized dispatch model among multiple disaster places under one way maximum transport-capacity constraint condition. In: Proceedings of the RSETE 2011, Nanjing, China, pp. 1170–1174 (2011)
2. Zhong, S., He, T., Li, M., Rui, L., Xia, G., Zhu, Y.: An emergency resource scheduling model based on edge computing. In: Han, S., Ye, L., Meng, W. (eds.) AICON 2019. LNICST, vol. 287, pp. 353–366. Springer, Cham (2019). https://doi.org/10.1007/978-3-030-22971-9_31
3. Hong, W., Wen-tao, Z., Jing, W.: The method of emergency rescue program generation in civil aviation airport of China. In: Wang, M. (ed.) KSEM 2013. LNCS (LNAI), vol. 8041, pp. 431–443. Springer, Heidelberg (2013). https://doi.org/10.1007/978-3-642-39787-5_35
4. Guedes, R., Furtado, V., Pequeno, T.: Pareto sets as a model for dispatching resources in emergency centres. Peer-to-Peer Netw. Appl. **12**(4), 865–880 (2019)
5. Federici, L., Benedikter, B., Zavoli, A.: EOS: a parallel, self-adaptive, multi-population evolutionary algorithm for constrained global optimization. In: CEC 2020, Glasgow, United Kingdom, pp. 1–10 (2020)
6. Nguyen, S., Tran, B., Alahakoon, D.: Dynamic self-organising swarm for unsupervised prototype generation. In: CEC 2020, Glasgow, United Kingdom, pp. 1–8 (2020)
7. Zhong, J., Feng, L., Cai, W., Ong, Y.-S.: Multifactorial genetic programming for symbolic regression problems. IEEE Trans. Syst. Man Cybern. Syst. **50**(11), 4492–4505 (2018)
8. Jian, J., Zhan, Z., Zhang, J.: Large-scale evolutionary optimization: a survey and experimental comparative study. Int. J. Mach. Learn. Cybernet. **11**(3), 729–745 (2020)
9. Chen, Z., et al.: Deadline constrained cloud computing resources scheduling through an ant colony system approach. In: ICCCRI 2015, Singapore, pp. 112–119 (2015)
10. Liu, B., Qiu, S., Li, M.: Simultaneous scheduling strategy: a novel method for flexible job shop scheduling problem. In: CEC 2020, Glasgow, United Kingdom, pp. 1–8 (2020)
11. Shi, X., Zhang, X., Xu, M.: A self-adaptive preferred learning differential evolution algorithm for task scheduling in cloud computing. In: AEECA 2020, Dalian, China, pp. 145–148 (2020)
12. Chen, Z., et al.: Multiobjective cloud workflow scheduling: a multiple populations ant colony system approach. IEEE Trans. Cybern. **49**(8), 2912–2926 (2019)
13. Liu, C., He, J., Shi, J.: The study on optimal model for a kind of emergency material dispatch problem. Chin. J. Manage. Sci. **9**(3), 29–36 (2001)
14. Lei, L., Li, Z., Lai, X., Yu, X.: An optimal model for emergency resource dispatching based on simulated annealing algorithm. In: Proceedings of the 26th Chinese Control and Decision Conference, Changsha, pp. 71–73 (2014)
15. Gao, Z., Yan, A., Yang, Y., Qiu, X.: Multi-objective resource scheduling mechanism for emergency rescue. J. Beijing Univ. Posts Telecommun. **40**, 1–4 (2017)
16. Xiong, G., Yang, J.: Multi-objective dispatch model of emergency management under multi-resource combinations. In: 2011 International Conference on Business Management and Electronic Information, Guangzhou, vol. 5, pp. 216–219 (2011)

17. Tang, Z., Sun, J.: Multi objective optimization of railway emergency rescue resource allocation and decision. Int. J. Syst. Assur. Eng. Manag. **9**, 696–702 (2018)
18. Wu, P., Chu, F., Che, A., Zhou, M.: Bi-objective scheduling of fire engines for fighting forest fires: new optimization approaches. IEEE Trans. Intell. Transp. Syst. **19**(4), 1140–1151 (2018)
19. Zhan, Z., Li, J., Cao, J., Zhang, J., Chung, H.S., Shi, Y.: Multiple populations for multiple objectives: a coevolutionary technique for solving multiobjective optimization problems. IEEE Trans. Cybern. **43**(2), 445–463 (2013)
20. Zhou, S., Zhan, Z., Chen, Z., Kwong, S., Zhang, J.: A multi-objective ant colony system algorithm for airline crew rostering problem with fairness and satisfaction. IEEE Trans. Intell. Transp. Syst. 1–15 (2020). https://doi.org/10.1109/TITS.2020.2994779
21. Liu, X., Zhan, Z., Gao, Y., Zhang, J., Kwong, S., Zhang, J.: Coevolutionary particle swarm optimization with bottleneck objective learning strategy for many-objective optimization. IEEE Trans. Evol. Comput. **23**(4), 587–602 (2019)
22. Kennedy, J., Eberhart, R.C., Shi, Y.: Swarm Intelligence. Morgan Kaufmann, San Mateo (2001)
23. Kennedy, J., Eberhart, R.C.: Particle swarm optimization. In: Proceedings of ICNN 1995, Perth, Australia, vol. 4, pp. 1942–1948 (1995)
24. Deb, K., Pratap, A., Agarwal, S., Meyarivan, T.: A fast and elitist multiobjective genetic algorithm: NSGA-II. IEEE Trans. Evol. Comput. **6**(2), 182–197 (2002)
25. Macqueen, J.: Some methods for classification and analysis of multivariate observations. In: Proceedings of 5th Berkeley Symposium Mathematical Statistics and Probability, vol. 1, pp. 281–297 (1967)
26. Zhang, Q., Li, H.: MOEA/D: a multiobjective evolutionary algorithm based on decomposition. IEEE Trans. Evol. Comput. **11**(6), 712–731 (2007)

Prediction of Blast Furnace Temperature Based on Evolutionary Optimization

Tenghui Hu[1], Xianpeng Wang[2(✉)], Yao Wang[1], Zhiming Dong[1], and Xinyu Zhuang[3]

[1] Liaoning Key Laboratory of Manufacturing System and Logistics, Shenyang 110819, China
1910258@stu.neu.edu.cn, dongzm@stumail.neu.edu.cn

[2] Key Laboratory of Data Analytics and Optimization for Smart Industry, Northeastern University, Shenyang 110819, China
wangxianpeng@ise.neu.edu.cn

[3] Lenovo (Beijing) Ltd., Beijing 100085, China
zhuangxy1@lenovo.com

Abstract. Blast furnace temperature, generally characterized by silicon content, is an important indicator to characterize the stability of furnace conditions in the blast furnace ironmaking process. To achieve its accurate prediction, a prediction model based on multi-objective evolutionary optimization and non-linear ensemble learning is proposed in this paper. First, the input features of each sub-learner are optimized through a modified discrete multi-objective evolutionary algorithm to obtain a set of highly accurate and diverse sub-learners. Subsequently, a nonlinear ensemble method based on extreme learning machine is employed to ensemble the obtained sub-learners to further improve the accuracy and robustness of the prediction model. Furthermore, the effectiveness of the proposed prediction method is verified by experiments based on actual production data.

Keywords: Silicon content prediction · Extreme learning machine · Multi-objective evolutionary optimization · Nonlinear ensemble

1 Introduction

Blast furnace (BF) ironmaking, as the beginning of the steel production process, has a critical impact on the quality of steel. Figure 1 shows a typical blast furnace structure. During the production process, coal and iron ore are charged

This research was supported by the National Key Research and Development Program of China (2018YFB1700404), the Fund for the National Natural Science Foundation of China (62073067), the Major Program of National Natural Science Foundation of China (71790614), the Major International Joint Research Project of the National Natural Science Foundation of China (71520107004), and the 111 Project (B16009).

H. Ishibuchi et al. (Eds.): EMO 2021, LNCS 12654, pp. 759–768, 2021.
https://doi.org/10.1007/978-3-030-72062-9_60

into the blast furnace layer by layer through the charging system, while the preheated air mixed with pulverized coal is blown into the BF by the air supply system. A series of physical and chemical reactions happen between descending materials and ascending gas, which reduce iron oxides to liquid iron and produces two by-products, slag and blast furnace gas. In order to improve the quality of hot metal and reduce energy consumption, it is necessary to maintain BF temperature within a certain range during the ironmaking process. However, as the internal temperature of the BF can reach up to more than 2000 degrees Celsius [7], there is still no direct approach to measuring the temperature, so BF temperature is generally inferred from the silicon content in hot metal. Therefore, accurate prediction of silicon content is of great importance for blast furnace iron production.

Blast furnace ironmaking is a typical black box model due to the complex series of physical and chemical changes that occur during the process. Thus, it is difficult to develop a satisfactory mechanistic model for predicting silicon content [11]. For this reason, many data-driven models have been established. An algorithm based on fuzzy c-means (FCM) and exogenous nonlinear autoregressive model (NARX) is proposed by Diane Otília Lima Fontes et al. [3], in which the FCM was used to identify natural groupings in data and NARX presented a model for the prediction of silicon content. In [9], an enhanced extreme learning machine framework was proposed to predict silicon content after an outlier detection was performed on the real production data.

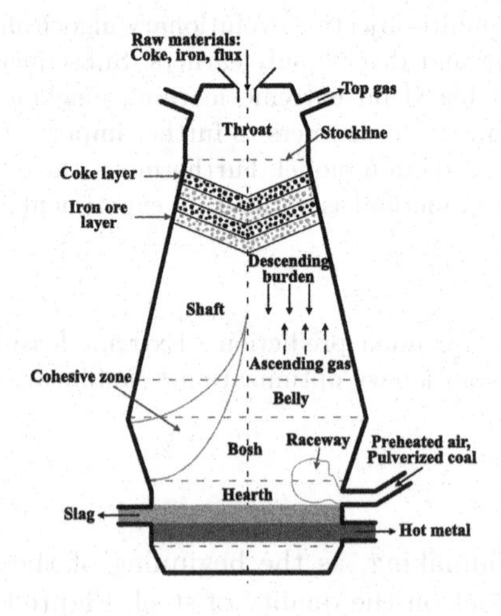

Fig. 1. Schematic diagram of a typical blast furnace structure

Although many different methods have been proposed for this problem, these methods still have the following two problems:

- When constructing prediction model, most of the current research works only selected several features as the input variables through correlation analysis [4,10]. However, these selection results are only based on some experimental

data. When encountering new production conditions of BF, the same feature selection methods may produce totally different results. Thus, these models have poor robustness.
- Compared with a single learner, ensemble models usually have better prediction accuracy and stronger robustness. Nevertheless, the generation of sub-learners is not an easy task, because the successful construction of an ensemble model requires that sub-learners should have both high accuracy and good diversity [5].

To address these issues, this paper proposes a novel model that combines multi-objective optimization and nonlinear ensemble learning (MOO-NEL) for silicon content prediction, of which the sub-learner is Extreme Learning Machine (ELM). The proposed model employs an evolutionary algorithm to optimize the input features of each sub-learner to produce a set of highly accurate and diversity sub-learners. In addition, different from the existing multi-objective evolutionary ensemble learning (MOEEL), in which the sub-learners are usually integrated by linear methods such as simple average and weighted average [6], this paper proposes a nonlinear ensemble method based on ELM to further improve the prediction accuracy by re-learning the outputs of the sub-learners.

2 Multi-objective Optimization and Nonlinear Ensemble Learning Based Prediction Model

Compared with existing silicon content prediction methods and MOEEL, the proposed model has the following main characteristics:

- The input features of each sub-learner are encoded as the individual in the population of the multi-objective evolutionary algorithm (MOEA).
- A discrete MOEA based on NSGA-II [2] and DE is developed. The MOEA follows the main framework of NSGA-II, except that the GA operators of NSGA-II are replaced with the discrete evolution operators of DE.
- To avoid over-fitting and achieve better ensemble performance, the ELM is used to integrate the sub-learners into an ensemble model.

In the following, we will first give the overall framework of the proposed MOO-NEL method, followed by some detailed descriptions.

2.1 Overall Framework of MOO-NEL

The overall framework of the proposed method is shown in Algorithm 1.

The detailed operations of each step in Algorithm 1 are described as follows.

Step 1 (line 1): Input training data, and initialize the parameters.
Step 2 (line 2): Randomly generate N solutions as the initial population *Pop*.
Step 3 (line 3): Create an empty combined population *Combined-Pop* of size $2N$ and add the individuals in the initial *Pop* into *Combined-Pop*.

Step 4 (line 5-line 10): Perform mutation and crossover on each individual in *Pop* to generate another N new individuals, train the corresponding ELMs, and calculate their accuracy measures. Put these new individuals into *Combined-Pop*.

Step 5 (line 11): For each individual in *Combined-Pop*, calculate their diversity measures by Pairwise Failure Crediting (PFC) [1] metric.

Step 6 (line 12-line 14): According to the accuracy and diversity measures, use the fast non-dominated sorting method in NSGA-II to sort the $2N$ individuals and then select the best N individuals as the new population for the next generation.

Step 7 (line 4, line 15): Determine if the termination condition has been met. If so, proceed to Step 8, otherwise, repeat Step 4 to Step 7.

Step 8 (line 16): Train the ELMs corresponding to the individuals in the final *Pop*. Integrate the obtained ELMs by another ELM to construct the final ensemble model.

Algorithm 1. *MOO-NEL*

Input: Training data, parameters of the discrete MOEA, such as population size N and maximum number of iterations *MaxIteration*.
Output: Ensemble model.

```
1: Initialize the population Pop = {X_1, X_2, ···, X_N} and set g = 0;
2: Combined-Pop = Pop;
3: while g < MaxIteration do
4:     for each solution X_i (i := 1 to N) in the Pop do
5:         V_i = MUTATION(X_i);
6:         U_i = CROSSOVER(V_i, X_i);
7:         GET-ACCURACY(U_i);
8:         Combined-Pop = Combined-Pop + U_i;
9:     end for
10:    GET-DIVERSITY(Combined-Pop);
11:    Fast-non-dominated-sorting(Combined-Pop);
12:    Pop = Get-next-generation(Combined-Pop);
13:    g = g + 1;
14: end while
15: Get-ensemble-model(Pop)
```

2.2 Coding and Decoding

Each individual in the population is coded by a binary vector. The 0s and 1s in this binary vector reflect the selection of optional input features by the corresponding sub-learner.

To evaluate an individual, we need to decode it into an ELM model with the corresponding input features and training data. After the ELM is trained, the prediction accuracy and diversity of the obtained model are used to evaluate this individual.

2.3 Accuracy and Diversity Measures

In order to be more suitable for practical applications, MOO-NEL uses hit rate (HR), a commonly used evaluation indicator in practice, as the measure for the prediction accuracy of the sub-learners, i.e., the first objective of the MOEA. The definition of HR is shown in Eq. (1)

$$HR = \frac{1}{l}\left(\sum_{k=1}^{l} H_k\right) \times 100\% \tag{1}$$

where l is the number of samples, and H_k is Heaviside function, defined as

$$H_k = \begin{cases} 1, & |\hat{y_k} - y_k| < 0.1 \\ 0, & others \end{cases} \tag{2}$$

where $\hat{y_k}$ is the predicted value of the forecast model on the k-th sample, and y_k is the true value of the k-th sample.

The PFC method [1], which characterized the diversity of a sub-learner by comparing the output of it with the outputs of the other sub-learners in the population, is used as the second objective to measure the diversity of each sub-learner. The application of PFC allows us to optimize the diversity of sub-learners as an explicit target of MOEAs, but it does not specify how to obtain the failure pattern of regression problems. In order to apply PFC in silicon content prediction problem, combined with the above-mentioned accuracy measure, we specify that a prediction is considered correct when it hits the true value and wrong otherwise.

2.4 Discrete Mutation and Crossover Operators

A new mutation operation consisting of the following three steps is proposed to handle the binary variable problem, where x_p^t , x_q^t and x_r^t are three individuals randomly selected from the population of the t-th generation, and F is the differential weight.

Step 1: Get a binary vector x_{temp} by applying the Exclusive-OR operation on x_q^t and x_r^t, since the two solutions are binary vectors.
Step 2: For each 1 in x_{temp}, generate a random number *rand*. If *rand* is smaller than F, the corresponding 1 will be changed to 0; otherwise, no operation is performed.
Step 3: OR operation is performed between the obtained vector x_{temp} and x_p^t to obtain the perturbed mutation vector.

The crossover operator is the same as that used in traditional DE.

2.5 ELM Based Nonlinear Ensemble Method

Due to the complicated environment of BF smelting, the silicon content is affected by many factors, which makes its dynamics very complicated. So, linear ensemble methods may not make the best use of the outputs of the sub-learners. Therefore, this paper proposes a nonlinear ensemble method based on ELM to integrate the obtained sub-learners, in which the outputs of sub-learners are taken as input vectors, and the raw silicon content is taken as target. The ensemble method can further improve the prediction accuracy of silicon content by re-learning the outputs of the sub-learners in a nonlinear way.

3 Experimental Result

3.1 Experimental Data

The experimental data came from the practical BF ironmaking process of one major iron & steel enterprise in China, and has been denoised and normalized. A set of experiment data was obtained, in which each sample consisted of nineteen optional input features and one output variable. In the following experiments, the population size for all evolutionary algorithms is set to 40 and the number of iterations is set to 500. And all the results are obtained through 10 times 4-fold cross-validation experiments.

3.2 Validation of Feature Selection

To illustrate the effect of feature selection on silicon content prediction, this section gives a comparison of the proposed method with its feature-selection-free version (denoted as EN-ELM), in which the input features of sub-learners are fixed to be the nine variables with the highest correlation with silicon content, according to correlation analysis methods [4,10]. EN-ELM takes the input weights and hidden layer thresholds of the sub-learners as the optimization object. Except the feature selection, the experimental conditions of EN-ELM are the same as the proposed method.

Table 1. Effects of feature selection based on cross-validation

	HR		RMSE	
	Mean	Variance	Mean	Variance
MOO-NEL	89.95%	6.31E−06	0.06295	4.12E−08
EN-ELM	88.04%	6.83E−06	0.06562	6.17E−07

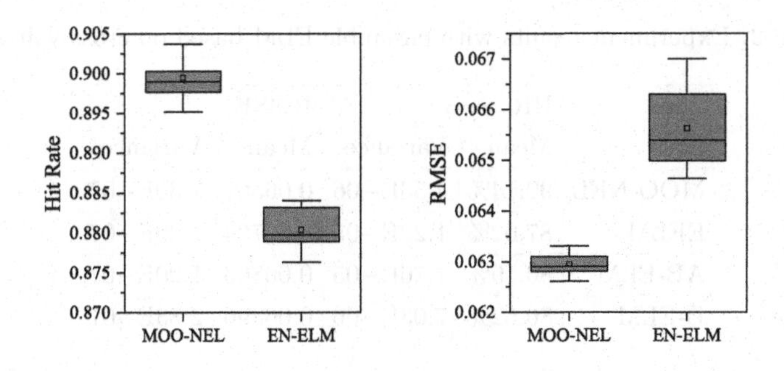

Fig. 2. Effects of feature selection based on cross-validation

The experimental results are shown in Table 1 and Fig. 2. It can be seen that compared with EN-ELM, MOO-NEL not only has higher HR and lower root mean square error (RMSE), but also has a significant improvement in robustness. Another noteworthy thing is that in MOO-NEL, the sub-learners are original ELMs, whose input weights and hidden layer biases are randomly assigned, while in EN-ELM, the parameters of all sub-learners are directly optimized by a MOEA. It is clear that the application of feature selection plays an important role in improving the accuracy and robustness of the prediction of silicon content in hot metal.

3.3 Comparison with Other Silicon Content Prediction Models

Comparison with Ensemble ELM

- Extended ELM [8] (EELM): Run the original ELM for P times and calculate the average values of prediction as final prediction results.
- Adaboost ELM (AB-ELM): Adaboost algorithm with ELM as the sub-learner.
- Bagging ELM (B-ELM): Bagging algorithm with ELM as the sub-learner.

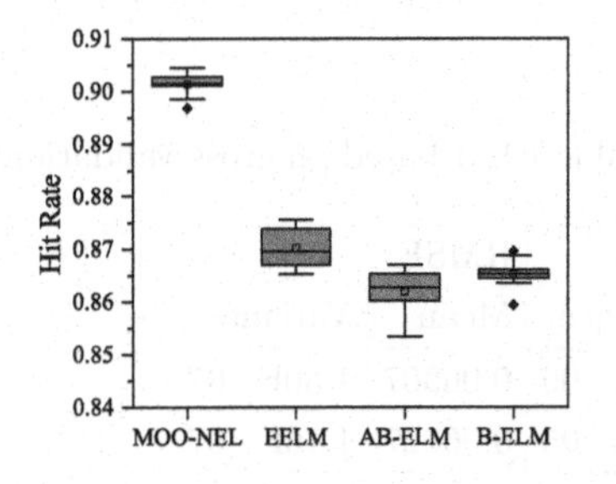

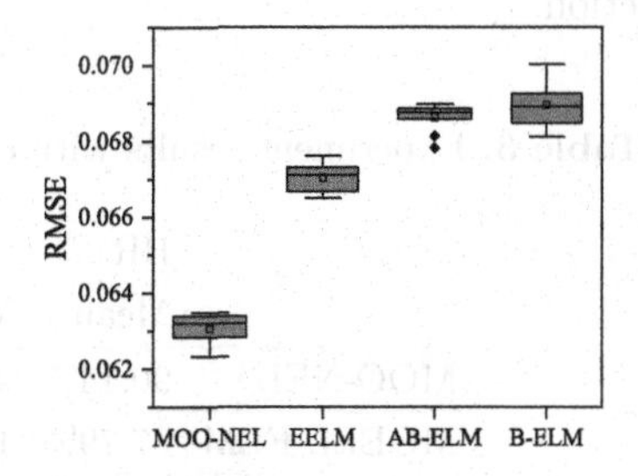

Fig. 3. Experiment results with ensemble ELM based on cross-validation

Table 2. Experiment results with ensemble ELM based on cross-validation

	HR		RMSE	
	Mean	Variance	Mean	Variance
MOO-NEL	90.14%	4.53E−06	0.06307	1.30E−07
EELM	87.02%	1.22E−05	0.06704	1.39E−07
AB-ELM	86.19%	1.76E−05	0.06863	1.20E−07
B-ELM	86.52%	7.03E−06	0.06896	2.83E−07

As shown in Table 2 and Fig. 3, MOO-NEL achieves the best performance among these ensemble models. We know that the diversity among sub-learners is a key issue to construct an ensemble model with good generalization ability. Compared to these classical ensemble modeling methods, the search range of EA-based MOO-NEL is much wider, and the sub-learners used in subsequent ensemble are collaboratively optimized through the evolutionary process of the population. These two features are conducive to diversity among sub-learners and thus make MOO-NEL achieve good generalization ability.

Comparison with Traditional MOEEL-ELM. According to the previous analysis, EA-based ensemble learning models usually have better learning capabilities than those classical models. Therefore, we further carried out a comparison experiment between the proposed method and the traditional MOEEL method.

MOEEL-ELM is a typical representative of the traditional multi-objective evolutionary ensemble learning, in which the input weights and hidden layer thresholds of sub-learners are taken as evolutionary objects. And after obtaining the final sub-learners, they are integrated by the average method.

We can see from Table 3 and Fig. 4 that the performance of MOO-NEL is obviously better than MOEEL-ELM. The results demonstrate the two improvements of MOO-NEL, i.e., feature selection and nonlinear ensemble, compared with MOEEL-ELM, can enhance the accuracy and robustness of silicon content prediction.

Table 3. Experiment results with ensemble ELM based on cross-validation

	HR		RMSE	
	Mean	Variance	Mean	Variance
MOO-NEL	90.14%	4.53E−06	0.06307	1.30E−07
MOEEL-ELM	87.79%	1.00E−05	0.06627	1.76E−07

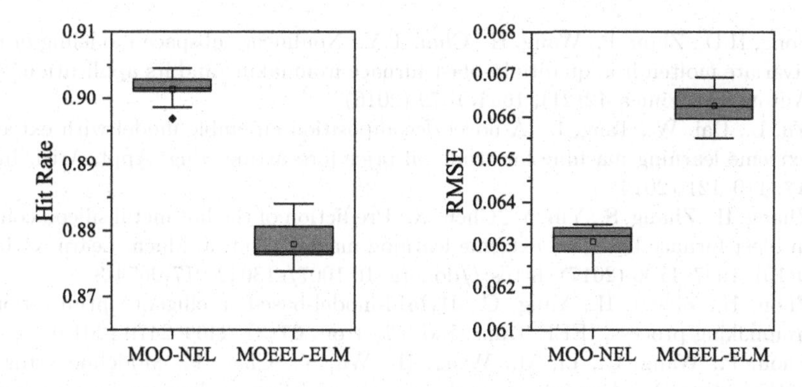

Fig. 4. Experiment results with ensemble ELM based on cross-validation

4 Conclusion

To realize accurate prediction of BF temperature, this paper proposes a prediction model based on multi-objective optimization and nonlinear ensemble learning, i.e., MOO-NEL. The model takes the input feature selection of the sub-learners as individuals in the population, and optimize their accuracy and diversity to obtain a set of sub-learners with high prediction accuracy and strong diversity. Then, these sub-learners are integrated through a nonlinear ensemble method based on ELM, which can further improve the accuracy and robustness of the prediction model. Experiments based on actual production data show that, MOO-NEL performs better than existing models whose input features are a sub-set of the available features, and therefore provides more valuable reference in-formation to on-site operators.

References

1. Chandra, A., Yao, X.: Ensemble learning using multi-objective evolutionary algorithms. J. Math. Model. Algorithms **5**(4), 417–445 (2006)
2. Deb, K., Pratap, A., Agarwal, S., Meyarivan, T.: A fast and elitist multiobjective genetic algorithm: NSGA-II. IEEE Trans. Evol. Comput. **6**(2), 182–197 (2002)
3. Fontes, D.O.L., Vasconcelos, L.G.S., Brito, R.P.: Blast furnace hot metal temperature and silicon content prediction using soft sensor based on fuzzy c-means and exogenous nonlinear autoregressive models. Comput. Chem. Eng. **141**, 107028 (2020)
4. Jiang, Z.H., Dong, M.L., Gui, W.H., Yang, C.H., Xie, Y.F.: Two-dimensional prediction for silicon content of hot metal of blast furnace based on bootstrap. Acta Automatica Sinica **42**(5), 715–723 (2016)
5. Krogh, A., Vedelsby, J.: Neural network ensembles, cross validation, and active learning. In: Advances in Neural Information Processing Systems, vol. 7, pp. 231–238 (1994)
6. Peimankar, A., Weddell, S.J., Jalal, T., Lapthorn, A.C.: Multi-objective ensemble forecasting with an application to power transformers. Appl. Soft Comput. **68**, 233–248 (2018)

7. Song, H.D., Zhou, P., Wang, H., Chai, T.Y.: Nonlinear subspace modeling of multivariate molten iron quality in blast furnace ironmaking and its application. Acta Automatica Sinica **42**(21), 1664–1679 (2016)
8. Yu, L., Dai, W., Tang, L.: A novel decomposition ensemble model with extended extreme learning machine for crude oil price forecasting. Eng. Appl. Artif. Intell. **47**, 110–121 (2016)
9. Zhang, H., Zhang, S., Yin, Y., Chen, X.: Prediction of the hot metal silicon content in blast furnace based on extreme learning machine. Int. J. Mach. Learn. Cybern. **9**(10), 1697–1706 (2017). https://doi.org/10.1007/s13042-017-0674-8
10. Zhou, H., Zhang, H., Yang, C.: Hybrid-model-based intelligent optimization of ironmaking process. IEEE Trans. Ind. Electron. **67**(3), 2469–2479 (2019)
11. Zhou, P., Wang, C., Li, M., Wang, H., Wu, Y., Chai, T.: Modeling error pdf optimization based wavelet neural network modeling of dynamic system and its application in blast furnace ironmaking. Neurocomputing **285**, 167–175 (2018)

Multiobjective Optimization Design of Broadband Dual-Polarized Base Station Antenna

Dawei Ding[1(✉)], Haibo Zhang[1], Xianheng Ding[1], and Xiuyin Zhang[2]

[1] Anhui University, Hefei 230601, China
19119@ahu.edu.cn
[2] South China University of Technology, Guangzhou 510641, China

Abstract. A compact broadband dual-polarized antenna is proposed for 3G/4G/5G base station applications. The main radiator is composed of two crossed dipoles contoured by cubic Hermite spline interpolation method to obtain compact size. In order to shorten design period, a multiobjective optimization technique is adopted through synthesizing multiple performance characteristics simultaneously and thus avoiding extensive structural parameter analysis. For antenna designers, this paper provides a novel design method for broadband dual-polarized base station antennas with stable radiation patterns and small gain variations from the perspective of optimization. The proposed dual-polarized antenna achieves a bandwidth of 66.4% (1.83–3.65 GHz). The measured port isolation is higher than 28 dB. Within the working band, the realized gain is more than 8.1 dBi, and gain variations are smaller than 0.5 dB. The half-power beamwidth (HPBW) is within 65° ± 3°. The cross polarization discrimination (XPD) level is better than 20 dB at boresight and better than 13 dB within a sector of ±60°.

Keywords: Base station antenna · Broadband antenna · The fifth generation (5G) communication system · Multiobjective evolutionary algorithm

1 Introduction

With the rapid development of modern mobile communications, broadband antennas supporting multi-network communication systems have attracted much attention. They could reduce the overall number of antennas, save the installation space, and save the costs of construction and operation of base stations [1]. Dual-polarized antenna is widely used for polarization diversity to reduce multipath fading effect and increase system channel capacity [1]. In order to serve 3G (1880–2170 MHz), 4G (2300–2400 MHz and 2570–2690 MHz), and sub-6 GHz 5G (3300–3600 MHz) communication systems, the dual-polarized antenna covering 1880–3600 MHz is much appreciated [2, 3].

In recent years, many kinds of dual-polarized base station antennas have been proposed, such as path antennas [4, 5], crossed dipole antennas [6–9], slot antennas [10], and magneto-electric (ME) dipole antennas [11]. Both dual-polarized patch and slot antennas have low profile and low cost. Nevertheless, most patch and slot antennas designed

H. Ishibuchi et al. (Eds.): EMO 2021, LNCS 12654, pp. 769–778, 2021.
https://doi.org/10.1007/978-3-030-72062-9_61

for base station applications have complicated configuration and cannot maintain stable radiation patterns within operating band. Various shapes of crossed dipoles have been proposed in [6–9]. Most of them have wide impedance bandwidth, simple structure, and ease of mass production. However, most of them have large aperture size, and their radiation patterns intrinsically vary significantly in the working band, as the patterns are generated mainly due to the radiation of one dipole for each polarization. In order to obtain stable radiation patterns, shaped grounds are employed. However, adding walls increases the size, weight, and cost of the antennas. Two dual-polarized antennas based on ME dipoles invented by using the complementary antenna were proposed in [11] with wide impedance bandwidth, small gain variations, stable radiation patterns, and low cross polarization level. However, they suffer from bulky radiator sizes. Unfortunately, for these dual-polarized antenna designs, there is no efficient method to achieve excellent impedance matching and stable radiation patterns at the same time.

In this paper, a compact broadband dual-polarized antenna is proposed for 3G/4G/5G base station applications. The proposed dual-polarized antenna is contoured by cubic Hermite spline interpolation method, rather than quadratic Bézier spline approximation method employed in [6]. Compared with the designs in [6, 7], this structure has more design freedom and makes better use of the limited design space, which is more beneficial for miniaturization. In addition, multiobjective evolutionary algorithm based on decomposition (MOEA/D) [12–14] is adopted to shorten the whole design period through synthesizing multiple performance characteristics simultaneously and thus avoiding extensive parameter analysis. To the best of our knowledge, it is the first paper to provide efficient design method for broadband dual-polarized antennas with stable radiation patterns and small gain variations for antenna designers from the perspective of optimization.

2 Multiobjective Optimization Technique for Antenna Designs

2.1 Multiobjective Optimization Problems in Antenna Designs

For most modern antenna designs, there are several performance characteristics to be synthesized, such as reflection coefficient, port isolation, antenna gain, and beam- width. It is hard to find a single design to trade off all performance characteristics because a design improving one performance characteristic might destroy another one [12]. It is much important for antenna designers to rapidly find candidate designs through trading off all performance characteristics and avoiding extensive parameter analysis.

The antenna design problems with multiple performance characteristics could be mathematically formulated as a multiobjective optimization problem (MOP).

$$F(x) = (f_1(x), f_2(x), \ldots, f_m(x)), \quad (1)$$

where x represents a structural parameter, m is the number of performance characteristics, and $f_i(x)$ ($i = 1,2,\ldots,m$) means the i-th mathematically-modeled objective function of performance characteristic. The value of $f_i(x)$ could be used to evaluate the corresponding performance characteristic.

2.2 MOEA/D-Based Optimization Technique

Population-based multiobjective metaheuristic algorithm is a popular computational intelligence method for solving MOPs due to its global searching capability regardless of the complexity of design objectives and the ability of obtaining several tradeoff designs rapidly without extensive structural parameter analysis [12]. These methods generate several tradeoff designs through a continuous iterative process of population, where each population consists of N structural parameters.

Due to the fast convergence speed, great population diversity, and excellent algorithm robustness, multiobjective evolutionary algorithm based on decomposition (MOEA/D) has become one of the most popular methods for solving MOPs [12–15].

The flowchart of MOEA/D is shown in Fig. 1. In MOEA/D, a MOP is decomposed into N single-objective optimization subproblems $g^j(x)$ at first, where λ^j is weight vector. Each subproblem is an aggregation of all objective functions and is optimized by using the information from its neighboring subproblems, which is the core idea of MOEA/D and brings the fast convergence speed. Differential evolution (DE) operators [16], including crossover, mutation and selection, are employed to speed up searching. The detailed descriptions on MOEA/D could be found in [13].The detailed descriptions on DE operators could be found in [16].

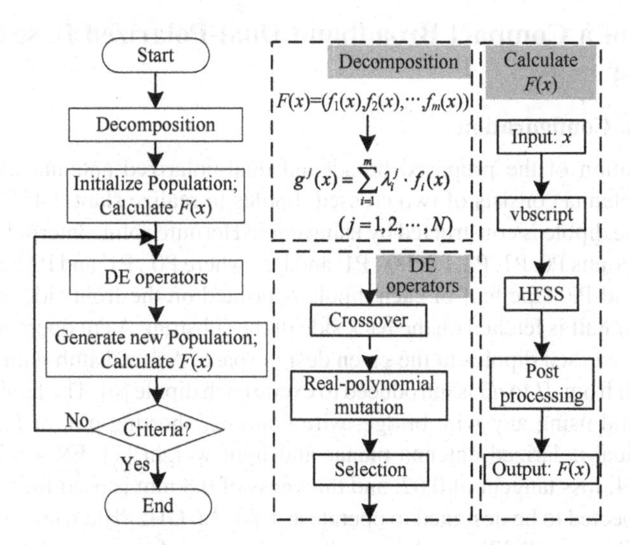

Fig. 1. Flowchart of MOEA/D-based optimization technique for antenna designs.

2.3 Procedures for Antenna Design

In this paper, MOEA/D-based optimization technique is suggested to rapidly find tradeoff designs through optimizing several electrical characteristics simultaneously and avoiding extensive parameter analysis. Multiple tradeoff designs could be obtained in a single run for antenna designers making decision. A complete design procedure for antenna designers is given as follows.

Step 1: Determine design variable, i.e. structural parameter x, and its searching space.
Step 2: Determine design objectives according to performance specifications, and then establish their mathematical models (i.e. objective functions). The number of objective functions is suggested to be less than five because too many design objectives would slow down the searching speed extremely [12–15].
Step 3: Select suitable electric-magnetic (EM) simulator for objection function evaluation, i.e. calculating $F(x)$. Then connect EM simulator with MOEA/D. Both simulation accuracy and simulation time should be balanced for EM solver selection. Figure 1 shows an example of the connection of High Frequency Structure Simulator (HFSS) and MOEA/D.
Step 4: Set algorithm parameters of MOEA/D, run it, and then obtain multiple tradeoff designs.
Step 5: Select a final design from the obtained tradeoff designs according to the application-dependent requirements, such as the robustness of antenna structural parameter and the balance of all performance specifications, or conduct further optimization based on the obtained promising designs.

In the following section, the aforementioned antenna design procedures will be demonstrated step by step.

3 Design of a Compact Broadband Dual-Polarized Base Station Antenna

3.1 Antenna Configuration

The configuration of the proposed broadband dual-polarized antennais illustrated in Fig. 2. The antenna consists of two crossed dipoles to realize slant ±45°polarizations. Each arm ofthe dipole is constructed by using cubic Hermite spline interpolation method with contour points P0, P1, P2, P3, P0′, P1′ and P2′, where P0′, P1′ and P2′ are symmetric with P0, P1 and P2. One half of each dipole is printed on the front side of a dielectric while the other half is fetched on the back side of the substrate.A cut operation is utilized to confine the crossed dipoles to the given design space. A metal stub with the linearly-changed width from *d1* to *d2* is introduced to excite each dipole [6]. The feeding structure is used to avoid using any wire bridge, wire bonding, shorting pin, or feeding probe, making the dual-polarized antenna planar and light weight [7]. FR-4 with dielectric constant of 4.4, loss tangent of 0.02, and thickness of 0.8 mm is used for substrate. The antenna is expected to be designed to operate at 1.88–3.6 GHz. The width of the radiator W is set as 40.8 mm (0.37λ, λ: the wavelength at center frequency) for compactness. The height h is set as 0.25λ. The ground plane width G is set as λ.

The dipole with this kind of irregular structure could provide more design freedom. Each dipole is equivalent to a two-element antenna array, where its feeding structure provides excitation for the dipole. Therefore, through properly choosing the geometric parameters of the dipoles and the feeding structures, it is expected to achieve stable radiation characteristics within the whole working band.

In this paper, because contour points P0 and P0′ are located by *d1*, *d2* and *h1*, eight geometric parameters, (*x*1, *y*1, *x*2, *y*2, *x*3, *d1*, *d2*, *h1*), are optimized to obtain good impedance matching, high isolation, and stable radiation performances.

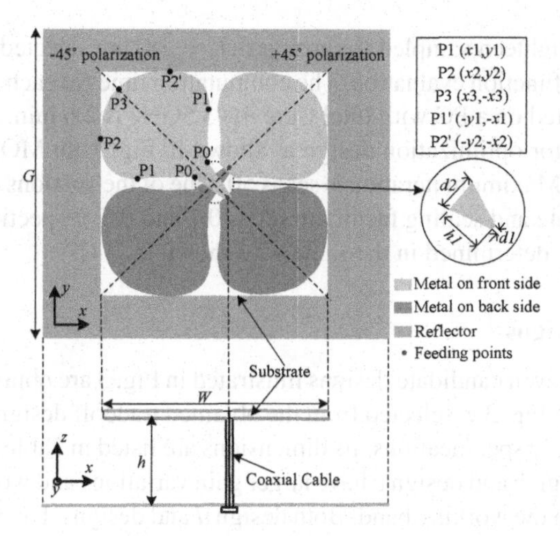

Fig. 2. Antenna configuration.

3.2 Optimization Settings

In order to obtain wideband dual-polarized antenna with stable radiation characteristics within the whole working band 1.88–3.6 GHz, four objective functions referring to return loss, port isolation, antenna gain variation, beamwidth, and cross polarization discrimination level are set as follows.

$$f_1 = \max\left(\frac{10}{\min\limits_{\omega\in[\omega_1,\omega_2]}(|S_{11}(\omega)|)}, \frac{25}{\min\limits_{\omega\in[\omega_1,\omega_2]}(|S_{21}(\omega)|)}\right) \tag{2}$$

$$f_2 = std\left(\left\{Gain(\omega_i)|_{i=1,2,\ldots,K}\right\}\right) \tag{3}$$

$$f_3 = \max_{\omega\in[\omega_1,\omega_2]}\left|\frac{BW(\omega) - 65}{5}\right| \tag{4}$$

$$f_4 = \max\left(\frac{20}{|Cx(\theta = 0°)|}, \frac{10}{\min\limits_{\theta\in[-60°,60°]}|Cx(\theta)|}\right) \tag{5}$$

BW in (4) represents beamwidth. *Cx* in (5) signifies cross polarization level. *Std*(•) calculates the standard deviation of a set of antenna gain values sampled at *K* frequencies. $[\omega_1, \omega_2]$ denotes the operating band 1.88–3.6 GHz. Those constants in (2)-(5) are used for normalization.

Objective functions (2)-(5) give the mathematical models of different electrical characteristics of antenna. It's obviously seen that the smaller the value, the better the performance. For example, when antenna gain variations are small, the standard deviation of the sampled antenna gain is small. When the beamwidths at all sampled frequencies are close to 65°, objective function f_3 would be small.

Therefore, this broadband dual-polarized antenna design could attribute to a quad-objective optimization problem with eight design variables.

In this paper, eighteen sampled frequencies ($K = 18$) are selected, and HFSS is utilized for objective function evaluation. The computation time for each objective function evaluation conducted on a PC with Intel Core 4@3.5GHz is 2.6 min. MOEA/D is combined with HFSS for optimization design as shown in Fig. 1. In MOEA/D, population size N is set as 20. Maximum iteration is set as 50. One of the versions of DE, rand/1/bin, where crossover rate and scaling factor are set as 0.1 and 0.5, respectively, is employed. Weight vectors are determined in the same way shown in [14].

3.3 Antenna Designs

Eight hours later, seven candidate designs illustrated in Fig. 3 are obtained by MOEA/D. Design a located in Fig. 3 is selected from the obtained tradeoff designs according to the balance of all electric specifications. Its dimensions are listed in Table 1. Compared with design *a*, both design *b* and design *c* have larger gain variations and worse consistency of beamwidths within the working band. Both design *d* and design *k* have worse impedance matching performance. In addition, there is no better design than design a in terms of all electrical specifications.

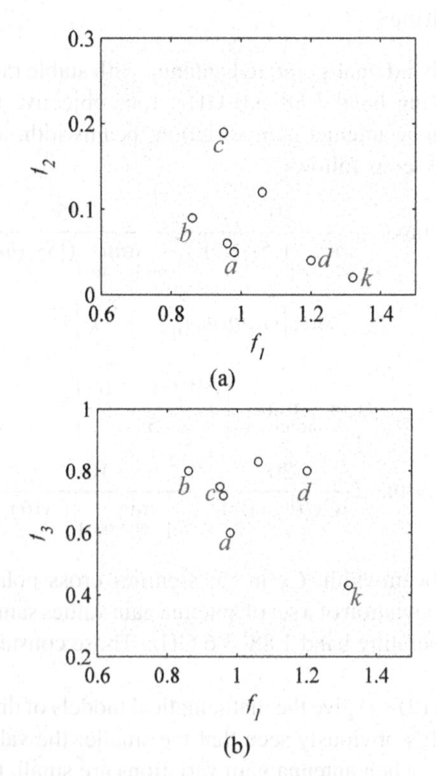

Fig. 3. Obtained tradeoff designs mapped in (a) f_1–f_2 space and (b) f_1–f_3 space.

Table 1. Parameters of the fabricated design.

Parameter	$x1$	$y1$	$x2$	$y2$	$x3$	$d1$	$d2$	$d3$
Value (mm)	−9.5	1.3	−16.7	3.6	−20.4	1.1	1.4	0.72

Fig. 4. Prototype of design *a*.

Design *a* is fabricated in Fig. 4 and is measured by using vector network analyzer, Agilent E8362B, and anechoic chamber. The measured results are illustrated in Fig. 5. Great agreement between simulated results and measured results are observed. In the operating band, the measured 10-dB impedance bandwidth is 1820 MHz (1.83–3.65 GHz), and the measured port isolation is higher than 28 dB. For each port, the realized antenna gain is more than 8.1 dBi, gain ripple is smaller than 0.5 dB, and HPBW is within $65^{\circ} \pm 3^{\circ}$. In addition, for each port, the cross polarization discrimination level is better than 20 dB at boresight and better than 13 dB within a sector of $\pm 60^{\circ}$. Therefore, the proposed design is promising for 3G/4G/5G mobile communication systems.

3.4 Comparison

Table 2 [6] lists the comparison among this design and several state-of-the-art dual-polarized base station antenna designs. It's observed that the proposed design has compact size and wide impedance matching bandwidth, along with smaller gain variations and more stable radiation patterns.

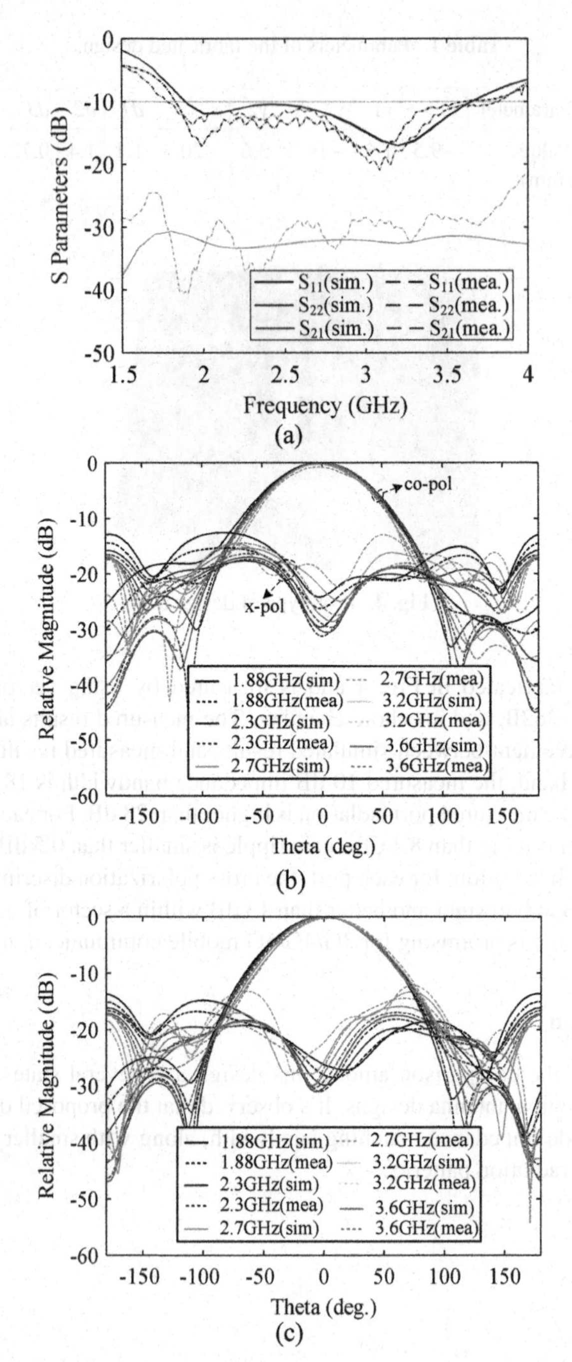

Fig. 5. Measured results. (a) *S* parameters, (b) co-pol and x-pol on *xoz*-plane at port 1, (c) co-pol and x-pol on *xoz*-plane at port 2, and (d) antenna gain and HPBW for two ports.

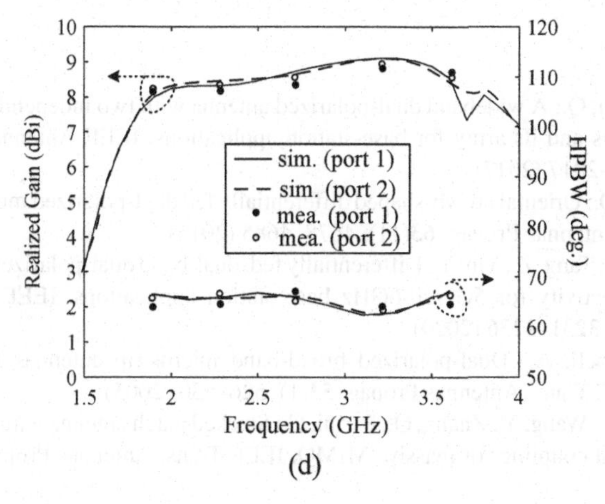

Fig. 5. (*continued*)

Table 2. Comparisons among several dual-polarized antennas.

Design	Working band (GHz)	Radiator size (λ^2)	Gain (dBi)	Gain variation (dB)	HPBW (°)
[6]	1.427–2.9(68%)	0.54 × 0.54	6	4	65 ± 11
[7]	1.7–2.7(45%)	0.42 × 0.42	7.5	2	66 ± 6
[8]	1.62–2.77(52.3%)	0.47 × 0.47	NA	NA	NA
[9]	3.14–5.04(46.5%)	0.37 × 0.37	7.8	0.6	71.8 ± 2.5
This design	1.83–3.65(66.4%)	**0.37 × 0.37**	**8.1**	**0.5**	**65 ± 3**

NA: Not Available.

4 Conclusion

In this paper, a compact broadband dual-polarized antenna covering 1.88–3.6 GHz is proposed by using Cubic Hermite spline interpolation method. Through establishing mathematical models and then using MOEA/D-based optimization technique, stable radiation patterns and small gain variations are rapidly obtained without extensive structural parameter analysis. Measured results show that the proposed antenna has smaller size, wide impedance matching bandwidth, stable radiation patterns and small gain variations. Therefore, the antenna is suitable for 3G/4G/5G base station applications. It is concluded that the introduced spline interpolation method is efficient for antenna miniaturization. What's more, it is concluded that a novel, efficient and rapid design method based on multiobjective evolutionary algorithm for broadband dual-polarized base station antennas with stable radiation performances is promising for antenna designers.

Acknowledgments. This project is partially funded by National Nature Science Foundation of China (NSFC) under Grant 61801194 and China Post-doctoral Science Foundation under grant 2018M643083.

References

1. Zheng, D., Chu, Q.: A wideband dual-polarized antenna with two independently controllable resonant modes and its array for base-station applications. IEEE Antennas Wirel. Propag. Lett. **16**, 2014–2017 (2017)
2. Luo, Y., Chu, Q.: Oriental crown-shaped differentially fed dual-polarized multidipole antenna. IEEE Trans. Antennas Propag. **63**(11), 4678–4685 (2015)
3. Li, Y., Zhao, Z., Tang, Z., Yin, Y.: Differentially fed, dual-band dual-polarized filtering antenna with high selectivity for 5G sub-6GHz base station applications. IEEE Trans. Antennas Propag. **68**(4), 3231–3236 (2020)
4. Gao, S., Sambell, A.: Dual-polarized broad-band microstrip antennas fed by proximity coupling. IEEE Trans. Antennas Propag. **53**(1), 526–530 (2005)
5. Gao, Y., Ma, R., Wang, Y., Zhang, Q., Parini, C.: Stacked patch antenna with dual-polarization and low mutual coupling for massive MIMO. IEEE Trans. Antennas Propag. **64**(10), 4544–4549 (2016)
6. Zhang, Q., Gao, Y.: A compact broadband dual-polarized antenna array for base stations. IEEE Antennas Wirel. Propag. Lett. **17**(6), 1073–1076 (2018)
7. Cui, Y., Li, R., Fu, H.: A broadband dual-polarized planar antenna for 2G/3G/LTE base stations. IEEE Trans. Antennas Propag. **62**(9), 4836–4840 (2014)
8. Liu, Y., Wang, S., Li, N.: A compact dual-band dual-polarized antenna with filtering structures for sub-6 GHz base station applications. IEEE Antennas Wirel. Propag. Lett. **17**(10), 1764–1768 (2018)
9. Ye, L., Cao, Y., Zhang, X., Gao, Y., Xue, Q.: Wideband dual-polarized omnidirectional antenna array for base-station applications. IEEE Trans. Antennas Propag **67**(10), 6419–6429 (2019)
10. Liu, Y., Wang, S., Wang, X., Jia, Y.: A differentially fed dual-polarized slot antenna with high isolation and low profile for base station application. IEEE Antennas Wirel. Propag. Lett. **18**(2), 303–307 (2019)
11. Xue, Q., Liao, S., Xu, J.: A differentially-driven dual-polarized magneto-electric dipole antenna. IEEE Trans. Antennas Propag **61**(1), 425–430 (2013)
12. Marler, R., Arora, J.: Survey of multiobjective optimization methods for engineering. Struct. Multi. Opt. **26**(6), 369–395 (2004)
13. Zhang, Q., Li, H.: MOEA/D: a multiobjective evolutionary algorithm based on decomposition. IEEE Trans. Evol. Comput. **11**(6), 712–731 (2007)
14. Li, H., Zhang, Q.: Multiobjective optimization problems with complicated Pareto sets, MOEA/D and NSGA-II. IEEE Trans. Evol. Comput. **13**(2), 284–302 (2009)
15. Trivedi, A., Srinivasan, D., Sanyal, K., Ghosh, A.: A survey of multiobjective evolutionary algorithms based on decomposition. IEEE Trans. Evol. Comput. **21**(3), 440–462 (2017)
16. Price, K., Storn, R., Lampinen, J.: Differential Evolution: A Practical Approach to Global Optimization (Natural Computing Series). Springer, New York (2005). https://doi.org/10.1007/3-540-31306-0

Correction to: An Overview of Pair-Potential Functions for Multi-objective Optimization

Jesús Guillermo Falcón-Cardona, Edgar Covantes Osuna, and Carlos A. Coello Coello

Correction to: Chapter "An Overview of Pair-Potential Functions for Multi-objective Optimization" in: H. Ishibuchi et al. (Eds.): *Evolutionary Multi-Criterion Optimization*, LNCS 12654, https://doi.org/10.1007/978-3-030-72062-9_32

The original version of this chapter was revised. Two equations in section 2.1 have been corrected.

$$\mathcal{K}^{COU}(\boldsymbol{u}, \boldsymbol{v}) = \frac{q_1 q_2}{4\pi\epsilon_0} \cdot \frac{1}{\|\boldsymbol{u} - \boldsymbol{v}\|^2} \tag{4}$$

$$\mathcal{K}^{\mathrm{PT}}(\boldsymbol{u}, \boldsymbol{v}) = \frac{V_1}{\sin^2(\alpha\|\boldsymbol{u} - \boldsymbol{v}\|)} + \frac{V_2}{\cos^2(\alpha\|\boldsymbol{u} - \boldsymbol{v}\|)} \tag{5}$$

The updated version of this chapter can be found at https://doi.org/10.1007/978-3-030-72062-9_32

H. Ishibuchi et al. (Eds.): EMO 2021, LNCS 12654, p. C1, 2021.
https://doi.org/10.1007/978-3-030-72062-9_62

Author Index

Printed in the United States
by Baker & Taylor Publisher Services